Neurobiology of Attention

Neurobiology of Attention

Edited by

Laurent Itti
University of Southern California, Los Angeles, USA

Geraint Rees
University College London, London, UK

John K. Tsotsos
York University, Toronto, Canada

ELSEVIER
ACADEMIC
PRESS

Amsterdam • Boston • Heidelberg • London
New York • Oxford • Paris • San Diego
San Francisco • Singapore • Sydney • Tokyo

Elsevier Academic Press
30 Corporate Drive, Suite 400, Burlington, MA 01803, USA
525 B Street, Suite 1900, San Diego, California 92101-4495, USA
84 Theobald's Road, London WC1X 8RR, UK

This book is printed on acid-free paper. ∞

Copyright © 2005, Elsevier Inc. All rights reserved.

No part of this publication may be reproduced or transmitted in any form or by any means, electronic or mechanical, including photocopy, recording, or any information storage and retrieval system, without permission in writing from the publisher.

Permissions may be sought directly from Elsevier's Science & Technology Rights Department in Oxford, UK: phone: (+44) 1865 843830, fax: (+44) 1865 853333, e-mail: permissions@elsevier.com.uk. You may also complete your request on-line via the Elsevier homepage (http://elsevier.com) by selecting "Customer Support" and then "Obtaining Permissions."

Library of Congress Cataloging-in-Publication Data
Application submitted

British Library Cataloguing in Publication Data
A catalogue record for this book is available from the British Library

ISBN: 0-12-375731-2

For all information on all Elsevier Academic Press publications visit our Web site at
www.books.elsevier.com

Printed in the United States of America
05 06 07 08 09 9 8 7 6 5 4 3 2 1

Dedications

To my parents and April,

To Rebecca and Amelia,

For my father, Kostas Tsotsos 1920–2004, from whom I learned what *idealism* means

Table of Contents

Contributors xiii

Foreword by *Michael A. Arbib* xix

Preface xxi

A Brief and Selective History of Attention xxiii

A Tour of This Volume xxxiii

I
FOUNDATIONS

1. Computational Foundations for Attentive Processes 3
   JOHN K. TSOTSOS

2. Capacity Limits for Spatial Discrimination 8
   MICHAEL J. MORGAN AND JOSHUA A. SOLOMON

3. Directed Visual Attention and the Dynamic Control of Information Flow 11
   CHARLES H. ANDERSON, DAVID C. VAN ESSEN, AND BRUNO A. OLSHAUSEN

4. Selective Attention as an Optimal Computational Strategy 18
   GREG BILLOCK, CHRISTOF KOCH, AND DEMETRI PSALTIS

5. Surprise: A Shortcut for Attention? 24
   PIERRE BALDI

6. A Heteromodal Large-Scale Network for Spatial Attention 29
   M-MARSEL MESULAM, DANA M. SMALL, RIK VANDENBERGHE, DARREN R. GITELMAN, AND ANNA C. NOBRE

7. Parietal Mechanisms of Attentional Control: Locations, Features, and Objects 35
   JOHN T. SERENCES, TAOSHENG LIU, AND STEVEN YANTIS

8. Visual Cortical Circuits and Spatial Attention 42
   JOHN H. REYNOLDS

9. Psychopharmacology of Human Attention 50
   JENNIFER T. COULL

10. Neuropharmacology of Attention 57
    JEAN A. MILSTEIN, JEFFREY W. DALLEY, AND TREVOR W. ROBBINS

11. Identifying the Neural Systems of Top-Down Attentional Control: A Meta-analytic Approach 63
    BARRY GIESBRECHT AND GEORGE R. MANGUN

12. Attention Capture: The Interplay of Expectations, Attention, and Awareness 69
    MICHAEL S. AMBINDER AND DANIEL J. SIMONS

13. Change Blindness 76
    RONALD A. RENSINK

14. Development of Covert Orienting in Young Infants 82
    JOHN E. RICHARDS

15. Prior Entry 89
    DAVID I. SHORE AND CHARLES SPENCE

16. Inhibition of Return 96
    RAYMOND M. KLEIN AND JASON IVANOFF

17. Guidance of Visual Search by Preattentive Information 101
    JEREMY M. WOLFE

18. The Top in Top-Down Attention 105
    CHRIS FRITH

19. Allocation of Attention in Three-Dimensional Space 109
    PAUL ATCHLEY

20. Covert Attention and Saccadic Eye Movements 114
    JOHN M. FINDLAY

21. Prefrontal Selection and Control of Covert and Overt Orienting 117
    NARCISSE P. BICHOT AND JEFFREY D. SCHALL

22. Dissociation of Selection from Saccade Programming 124
    KIRK G. THOMPSON

23. Space- and Object-Based Attention 130
    MICHAEL C. MOZER, SHAUN P. VECERA

24. Attention and Binding 135
    LYNN C. ROBERTSON

25. Top-Down Facilitation of Visual Object Recognition 140
    MOSHE BAR

26. Spatial Processing of Environmental Representations 146
    JAMES R. BROCKMOLE AND RANXIAO FRANCES WANG

27. Decision and Attention 152
    ANDREI GOREA AND DOV SAGI

28. Visual Attention and Emotional Perception 160
    LUIZ PESSOA AND LESLIE G. UNGERLEIDER

29. The Difference between Visual Attention and Awareness: A Cognitive Neuroscience Perspective 167
    VICTOR A.F. LAMME

30. Reaching Affects Saccade Trajectories 175
    STEVEN P. TIPPER

31. The Premotor Theory of Attention 181
    LAILA CRAIGHERO, GIACOMO RIZZOLATTI

32. Cross-Modal Consequences of Human Spatial Attention 187
    JON DRIVER, MARTIN EIMER AND EMILIANO MACALUSO

33. Attention and Scene Understanding 197
    VIDHYA NAVALPAKKAM, MICHAEL ARBIB AND LAURENT ITTI

II
FUNCTIONS

34. Visual Search and Popout in Infancy 207
    SCOTT A. ADLER

35. Attention in Conditioning 213
    PETER DAYAN

36. Electrophysiology of Reflexive Attention 219
    JOSEPH B. HOPFINGER

37. Natural Scene Statistics and Salient Visual Features 226
    CHRISTOPH ZETZSCHE

38. Salience of Feature Contrast 233
    HANS-CHRISTOPH NOTHDURFT

39. Stimulus-Driven Guidance of Visual Attention in Natural Scenes 240
    DERRICK J. PARKHURST AND ERNST NIEBUR

40. Contextual Guidance of Visual Attention 246
    MARVIN M. CHUN

41. Gist of the Scene 251
    AUDE OLIVA

42. Temporal Orienting of Attention 257
    IVAN C. GRIFFIN AND ANNA C. NOBRE

43. Visual Search: The Role of Memory for Rejected Distractors 264
    TODD S. HOROWITZ

44. The Neuropsychology of Visual Feature Binding 269
    GLYN W. HUMPHREYS AND M. JANE RIDDOCH

45. Visual Saliency and Spike Timing in the Ventral Visual Pathway 272
    RUFIN VANRULLEN

46. Object Recognition in Cortex: Neural Mechanisms, and Possible Roles for Attention   279
    MAXIMILIAN RIESENHUBER

47. Binding Contour Segments into Spatially Extended Objects   288
    PIETER R. ROELFSEMA AND HENK SPEKREIJSE

48. Scanpath Theory, Attention, and Image Processing Algorithms for Predicting Human Eye Fixations   296
    CLAUDIO M. PRIVITERA AND LAWRENCE W. STARK

49. The Feature Similarity Gain Model of Attention: Unifying Multiplicative Effect of Spatial and Feature-based Attention   300
    JULIO C. MARTÍNEZ-TRUJILLO AND STEFAN TREUE

50. Biasing Competition in Human Visual Cortex   305
    SABINE KASTNER AND DIANE M. BECK

51. Nonsensory Signals in Early Visual Cortex   311
    DAVID RESS AND DAVID J. HEEGER

52. Effects of Attention on Auditory Perceptual Organization   317
    ROBERT P. CARLYON AND RHODRI CUSACK

53. Attention in Language   324
    ANDRIY MYACHYKOV AND MICHAEL I. POSNER

54. Attention and Spatial Language   330
    LAURA A. CARLSON AND GORDON D. LOGAN

55. The Sustained Attention to Response Test (SART)   337
    TOM MANLY AND IAN H. ROBERTSON

56. ERP Measures of Multiple Attention Deficits Following Prefrontal Damage   339
    LEON Y. DEOUELL AND ROBERT T. KNIGHT

57. Nonspatially Lateralized Mechanisms in Hemispatial Neglect   345
    MASUD HUSAIN

58. Visual Extinction and Hemispatial Neglect after Brain Damage: Neurophysiological Basis of Residual Processing   351
    PATRIK VUILLEUMIER

59. Attention in Split-Brain Patients   358
    TODD C. HANDY AND MICHAEL S. GAZZANIGA

60. Divided Attention in the Normal and the Split Brain: Chronometry and Imaging   363
    MARCO IACOBONI

III
MECHANISMS

61. Neurophysiological Correlates of the Attentional Spotlight   372
    EDGAR A DEYOE AND JULIE BREFCZYNSKI

62. Spatially-Specific Attentional Modulation Revealed by fMRI   377
    DAVID C. SOMERS AND STEPHANIE A. MCMAINS

63. The Neural Basis of the Attentional Blink   383
    RENÉ MAROIS

64. Neurophysiological Correlates of the Reflexive Orienting of Spatial Attention   389
    JILLIAN H. FECTEAU, ANDREW H. BELL, MICHAEL C. DORRIS, AND DOUGLAS P. MUNOZ

65. Specifying the Components of Attention in a Visual Search Task   395
    GREGORY J. ZELINSKY

66. Neural Evidence for Object-based Attention   401
    KATHLEEN M. O'CRAVEN

67. Location- or Feature-based Targeting of Spatial Attention   407
    RIK VANDENBERGHE

68. Dimension-based Attention in Pop-out Search   412
    JOSEPH KRUMMENACHER AND HERMANN J. MÜLLER

69. Irrelevant Singletons Capture Attention   418
    JAN THEEUWES

70. Attentional Modulation of Apparent Stimulus Contrast   428
    JULIO C. MARTINEZ-TRUJILLO AND STEFAN TREUE

71. Attentional Suppression Early in the Macaque Visual System   429
    ORBAN, G.A., PAUWELS, K., VAN HULLE, M.M., AND VANDUFFEL, W.

72. Attentional Modulation in the Human Lateral Geniculate Nucleus and Pulvinar   435
    SABINE KASTNER, KEITH A. SCHNEIDER, AND DANIEL H. O'CONNOR

73. Transient Covert Transient Attention Increases Contrast Sensitivity and Spatial Resolution: Support for Signal Enhancement   442
    MARISA CARRASCO

74. External Noise Distinguishes Mechanisms of Attention   448
    ZHONG-LIN LU AND BARBARA ANNE DOSHER

75. Attentional Modulation and Changes in Effective Connectivity   454
    CHRISTIAN BÜCHEL

76. Attentional Modulation of Surround Inhibition   460
    BARBARA ZENGER-LANDOLT

77. Attentional Processes in Texture Perception   466
    CHARLES CHUBB

78. Mechanisms of Perceptual Learning   471
    BARBARA ANNE DOSHER AND ZHONG-LIN LU

79. Lateral Interactions between Targets and Flankers Require Attention   477
    ELLIOT FREEMAN

80. Attention and Changes in Neural Selectivity   485
    SCOTT O. MURRAY

81. Attentional Effects on Motion Processing   490
    AMY A. REZEC AND KAREN R. DOBKINS

82. ERP Studies of Selective Attention to Nonspatial Features   496
    ALICE M. PROVERBIO AND ALBERTO ZANI

83. Effects of Attention on Figure-Ground Responses in the Primary Visual Cortex during Working Memory   502
    HANS SUPÈR

84. Electrophysiological and Neuroimaging Approaches to the Study of Visual Attention   507
    ANTÍGONA MARTÍNEZ AND STEVEN A. HILLYARD

85. The Timing of Attentional Modulation of Visual Processing as Indexed by ERPs   514
    ALBERTO ZANI AND ALICE M. PROVERBIO

86. Selective Visual Attention Modulates Oscillatory Neuronal Synchronization   520
    PASCAL FRIES AND ROBERT DESIMONE

87. Putative Role of Oscillations and Synchrony in Cortical Signal Processing and Attention   526
    WOLF SINGER

88. Attention to Tactile Stimuli Increases Neural Synchrony in Somatosensory Cortex   534
    P. N. STEINMETZ, S. S. HSIAO, K. O. JOHNSON AND E. NIEBUR

89. Crossmodal Attention in Event Perception   538
    KATSUMI WATANABE AND SHINSUKE SHIMOJO

IV
SYSTEMS

90. The FeatureGate Model of Visual Selection   547
    KYLE R. CAVE, MIN-SHIK KIM, NARCISSE P. BICHOT AND KENITH V. SOBEL

91. Probabilistic Models of Attention Based on Iconic Representations and Predictive Coding   553
    RAJESH P. N. RAO AND DANA H. BALLARD

92. The Selective Tuning Model for Visual Attention   562
    JOHN K. TSOTSOS

93. The Primary Visual Cortex Creates a Bottom-up Saliency Map   570
    LI ZHAOPING

94. Models of Bottom-up Attention and Saliency   576
    LAURENT ITTI

95. Saliency in Computer Vision   583
    GÉRARD MEDIONI AND PHILIPPOS MORDOHAI

96. Contextual Influences on Saliency   586
    ANTONIO TORRALBA

97. A Neurodynamical Model of Visual Attention   593
    GUSTAVO DECO, EDMUND T. ROLLS AND JOSEF ZIHL

98. How the Detection of Objects in Natural Scenes Constrains Attention in Time   600
    FRED H. HAMKER

99. Memory-Driven Visual Attention: An Emergent Behavior of Map-Seeking Circuits   605
    DAVID W. ARATHORN

100. The Role of Short-Term Memory in Visual Attention   610
    GUSTAVO DECO AND EDMUND T. ROLLS

101. Scene Segmentation through Synchronization   618
    GÜNTHER PALM AND ANDREAS KNOBLAUCH

102. Attentive Wide-Field Sensing for Visual Telepresence and Surveillance   624
    JAMES H. ELDER, FADI DORNAIKA, BOB HOU AND RONEN GOLDSTEIN

103. Neuromorphic Selective Attention Systems   633
    GIACOMO INDIVERI

104. The Role of Visual Attention in the Control of Locomotion   638
    M. ANTHONY LEWIS

105. Attention Architectures for Machine Vision and Mobile Robots   642
    LUCAS PALETTA, ERICH ROME AND HILARY BUXTON

106. Attention for Computer Graphics Rendering   649
    HECTOR YEE AND SUMANTA PATTANAIK

107. Linking Attention to Learning, Expectation, Competition, and Consciousness   652
    STEPHEN GROSSBERG

108. Attention-Guided Recognition Based on "What" and "Where" Representations: A Behavioral Model   663
    ILYA A. RYBAK, VALENTINA I. GUSAKOVA, ALEXANDER V. GOLOVAN, LUBOV N. PODLADCHIKOVA AND NATALIA A. SHEVTSOVA

109. A Model of Attention and Recognition by Information Maximization   671
    KERSTIN SCHILL

Index   677

Contributors

Scott A. Adler, York University, Department of Psychology and Centre for Vision Research, 333 Behavioural Sciences Building, 4700 Keele St., Toronto, Ontario, Canada, M3J 1P3

Michael S. Ambinder, University of Illinois at Urbana-Champaign, Psychology, 1817 Valley Road, Champaign, IL, USA, 61820

Charles H. Anderson, Washington University in St. Louis, Philosophy-Neuroscience-Psychology Program, One Brookings Drive, Campus Box 1073, St. Louis, MO, USA, 63130-4899

David W. Arathorn, Center for Computational Biology (CCB), 1 Lewis Hall, Montana State University, Bozeman, MT, USA, 59717

Michael A. Arbib, University of Southern California, Hedco Neuroscience Building, Room 5, 3641 Watt Way, Los Angeles, CA, USA, 90089-2520

Paul Atchley, University of Kansas, Dept. of Psychology, 1415 Jayhawk Blvd., Lawrence, KS, USA, 66045

Pierre Baldi, University of California, Irvine, School of Information and Computer Science, Irvine, CA, USA, 92697-3425

Dana H. Ballard, University of Rochester, Department of Computer Science, Rochester, NY, USA, 14627-0226

Moshe Bar, NMR Center at MGH, Harvard Medical School, 149 Thirteenth Street, Charlestown, MA, USA, 02129

Diane M. Beck, Princeton University, Department of Psychology, Center for the Study of Brain, Mind and Behavior, Green Hall, Princeton, NJ, USA, 08544

Andrew H. Bell, Queen's University, Centre for Neuroscience Studies, 2nd floor, Botterell Hall, Kingston, Ontario, Canada, K7L 3N6

Narcisse P. Bichot, National Institute of Mental Health, Laboratory of Neuropsychology, Bethesda, MD, USA, 20892

Greg Billock, California Institute of Technology, Psaltis Optics Group, Mail Code 136-93, Pasadena, CA, USA, 91125

Julie A. Brefczynski-Lewis, Medical College of Wisconsin, Cell Biology, Neurobiology and Anatomy, 8701 Watertown Plank Road, Milwaukee, WI, USA, 53226

James R. Brockmole, Michigan State University, Psychology, 129 Psychology Research Building, East Lansing, MI, USA, 48824

Christian Büchel, Hamburg University, NeuroImage Nord, Martinistr. 52, Hamburg, Germany, 20246

Hilary Buxton, University of Sussex, Cognitive and Computing Sciences, Falmer, Brighton, United Kingdom, BN1 9QH

Laura A. Carlson, University of Notre Dame, Dept. of Psychology, 118-D Haggar Hall, Notre Dame, IN, USA, 46556

Robert P. Carlyon, Medical Research Council, Cognition & Brain Sciences Unit, 15 Chaucer Rd, Cambridge, UK, CB2 2EF

Marisa Carrasco, New York University, Psychology & Neural Science, 6 Washington Pl., New York City, NY, USA, 10003

Kyle R. Cave, University of Massachusetts Amherst, Psychology Department, Tobin Hall, Tobin Hall, Amherst, MA, United States, 01003

Charles Chubb, University of California, Irvine, Cognitive Sciences, Social Science Plaza A2101, Irvine, CA, USA, 92697-5100

Marvin M. Chun, Yale University, Dept. of Psychology, 2 Hillhouse Ave, PO Box 208205, New Haven, CT, USA, 06520-8205

Jennifer T. Coull, Centre National de la Recherche Scientifique, Laboratoire de Neurobiologie de la Cognition, 31 Chemin Joseph-Aiguier, Marseille, Bouche du Rhone, France, 13009

Laila Craighero, Dipartimento di Scienza biomediche e Terapie avanzate, Sezione Fisiologia Umana, Università di Ferrara, via Fossato di Mortara 17/19, 44100 Ferrara (Italy)

Rhodri Cusack, MRC Cognition and Brain Sciences Unit, 15 Chaucer Road, Cambridge, UK, CB2 2EF

Jeffrey W. Dalley, University of Cambridge, Department of Experimental Psychology, Downing St, Cambridge, UK, CB2 3EB

Peter Dayan, University College London, Gatsby Computational Neuroscience Unit, 17 Queen Square, London, UK, WC1N 3AR

Gustavo Deco, Institució Catalana de Recercai Estudis Avancats and Universität Pompeu Fabra, Passeig de Circumval lació, 8, Barcelona, Spain, 08003

Robert Desimone, Laboratory of Neuropsychology, NIMH, NIH, Bldg. 49 Rm. 1B80, 9000 Rockville Pike, Bethesda, MD 20892-4415

Leon Y. Deouell, The Hebrew University of Jerusalem, Department of Psychology, Mount Scopus, Jerusalem, Israel, 91905

Edgar A. DeYoe, Medical College of Wisconsin, Department of Radiology, 8701 Watertown Plank Rd, Milwaukee, WI, USA, 53226

Karen R. Dobkins, University of California, at San Diego, Department of Psychology, 9500 Gilman Dr. 0109, La Jolla, CA, USA, 92093-0109

Michael C. Dorris, Queen's University, Centre for Neuroscience Studies, 4th Floor, Botterell Hall, Kingston, Ontario, K7L 3N6

Barbara Anne Dosher, University of California, Irvine, Department of Cognitive Sciences and Institute of Mathematical Behavioral Sciences, Irvine, CA, USA, 92697-5100

Fadi Dornaika, CNRS, University of Technology of Compiegne, BP 20529, 60205 Compiegne Cedex, France

Jon Driver, University College London, Institute of Cognitive Neuroscience, 17 Queen Square, London, UK, WC1N 3AR

Martin Eimer, University of London, School of Psychology, Birbeck College, Malet Street, London WC1E 7HX, UK

James H. Elder, York University, Centre for Vision Research, 4700 Keele Street, Toronto, Ontario, Canada, M3J 1P3

Jillian H. Fecteau, Queen's University, Centre for Neuroscience Studies, Department of Physiology, 4th Floor Botterell Hall, Kingston, Ontario, Canada, K7L 3N6

John M. Findlay, University of Durham, Centre for Vision and Visual Cognition, Department of Psychology, South Road, Durham, UK, DH1 3LE

Elliot D. Freeman, University College London, Institute of Cognitive Neuroscience, 17 Queen Square, London, England, UK, WC1N 3AR

Pascal Fries, F. C. Donders Centre for Cognitive Neuroimaging, University of Nijmegen, Adelbertusplein 1, 6525 EK Nijmegen, The Netherlands and Department of Biophysics, University of Nijmegen, Geert Grooteplein 21, 6525 EZ Nijmegen, The Netherlands

Chris D. Frith, University College London, Wellcome Department of Imaging Neuroscience, Institute of Neurology, 12 Queen Square, London, UK, HA1 3JW

Michael S. Gazzaniga, Dartmouth College, Center for Cognitive Neuroscience, 6162 Moore Hall, Hanover, NH, USA, 03755-3569

Barry Giesbrecht, University of California, Santa Barbara, Dept. of Psychology, Santa Barbara, CA, USA, 93106

Darren R. Gitelman, Northwestern University, Neurology, Radiology, Cognitive Neurology and Alzheimer's Disease Center, 320 East Superior Street, Searle 11-470, Chicago, IL, USA, 60611

Ronen Goldstein, Centre for Vision Research, York University, 4700 Keele Street, Toronto, Ontario, Canada M3J 1P3

Alexander V. Golovan, Rostov State University, AB Kogan Research Institute for Neurocybernetics, 194/1 Stachka Avenue, Rostov-on-Don, Russia, 344090

Andrei Gorea, Centre National de la Recherche and René Descartes University, Laboratoire de Psychologie Expérimentale, 71 Ave. Edouard Vaillant, Boulogne-Billancourt, France, 92774

Ivan C. Griffin, University of Oxford, Dept. of Experimental Psychology, South Parks Road, Oxford OX1 3UD, UK

Stephen Grossberg, Boston University, Dept. of Cognitive and Neural Systems, 677 Beacon Street, Boston, MA, USA, 02215

Valentina I. Gusakova, Rostov State University, AB Kogan Research Institute for Neurocybernetics, 194/1 Stachka Avenue, Rostov-on-Don, Russia, 344090

Fred H. Hamker, Westfälische Wilhelms-Universität Münster, Psychology, Westfälische Wilhelms-Universität Münster, Fliednerstrasse 21, Münster, Germany, 48149

Todd C. Handy, University of British Columbia, Dept. of Psychology, 2136 West Mall, Vancouver, BC, Canada, V6T 1Z4

David J. Heeger, New York University, Department of Psychology and Center for Neural Science, 6 Washington Place, 8th floor, New York, NY, USA, 10003

Steven A. Hillyard, University of California, San Diego, Neurosciences, 9500 Gilman Drive, La Jolla, CA, USA, 92903-0608

Joseph B. Hopfinger, University of North Carolina at Chapel Hill, Dept. of Psychology, Chapel Hill, NC, USA, 27599

Todd S. Horowitz, Brigham and Women's Hospital, Harvard Medical School, Visual Attention Laboratory, 64 Sidney Street, Suite 170, Cambridge, MA, USA, 02139

Bob Hou, Centre for Vision Research, York University, 4700 Keele Street, Toronto, Ontario, Canada M3J 1P3

Steven S. Hsiao, Johns Hopkins University, Neuroscience, 338 Krieger Hall, 3400 N. Charles St., Baltimore, MD, USA, 21218

Glyn W. Humphreys, University of Birmingham, School of Psychology, Edgbaston, Birmingham, UK, B15 2TT

Masud Husain, Imperial College London, Division of Neuroscience & Psychological Medicine, Charing Cross Hospital, Fulham Palace Road, London, UK, W6 8RF

Marco Iacoboni, David Geffen School of Medicine at UCLA, Dept. of Psychiatry and Biobehavioral Science, Ahmanson Lovelace Brain Mapping Center, 660 Charles E. Young Drive South, Los Angeles, CA, USA, 90095

Giacomo Indiveri, University of Zurich—ETH Zurich, Institute of Neuroinformatics, Winterthurerstrasse 190, Zurich, Switzerland, CH-8057

Laurent Itti, University of Southern California, Computer Science, Hedco Neuroscience Building, HNB-30A, 3641 Watt Way, Los Angeles, CA, USA, 90089-2520

Jason Ivanoff, Dalhousie University, Department of Psychology, Halifax, Nova Scotia, Canada, B3H 4J1

Kenneth O. Johnson, Johns Hopkins University, Neuroscience, Zanvyl Krieger Mind/Brain Institute, 3400 N. Charles St. Baltimore, MD, USA, 21218

Sabine Kastner, Princeton University, Dept. of Psychology, Center for the Study of Brain, Mind, and Behavior, Green Hall, Princeton, NJ, USA, 08544

Min-shik Kim, Yonsei University, 134 Sinchon-dong, Seodaemun-gu, Seoul, 120-749, Korea

Raymond M. Klein, Dalhousie University, Department of Psychology, Halifax, Nova Scotia, Canada, B3H 4J1

Robert T. Knight, University of California at Berkeley, Helen Wills Neuroscience Institute and Dept. of Psychology, Berkeley, CA, USA, 94720

Andreas Knoblauch, University of Ulm, Department of Neural Information Processing, Oberer Eselsberg, Ulm, Germany, 89069

Christof Koch, California Institute of Technology, Division of Biology, 139-74, 1200 E. California Blvd, Caltech, Pasadena, CA, USA, 91125

Joseph Krummenacher, Ludwig-Maximilian-University Munich, Department of Psychology, Leopoldstrasse 13, Munich, Germany, 80802

Victor A. F. Lamme, University of Amsterdam, Department of Psychology, Cognitive Neuroscience Group, Room 626, Department of Psychology, Roetersstraat 15, Amsterdam, The Netherlands, 1018 WB

M. Anthony Lewis, Iguana Robotics, Inc., P.O. Box 625, Urbana, IL, USA, 61803

Taosheng Liu, New York University, Dept. of Psychology, Meyer Bldg, 6 Washington Place, Room 970, New York, NY 10003

Gordon D. Logan, Vanderbilt University, Dept. of Psychology, Nashville, Tennessee, USA, 37203

Zhong-Lin Lu, University of Southern California, Depts. of Psychology and Biomedical Engineering, and Neuroscience Graduate Program, Los Angeles, CA, USA, 90089-1061

Emiliano Macalvso, Laboratorio di Neuroimmagini, Fondazione Santa Lucia, Instituto di Ricovero e Cura a Carrattere Scientifico, Via Ardeatina, 306, Roma, Italy, 00179

George R. Mangun, University of California, Davis, Center for Mind and Brain, Depts. of Neurology and Psychology, Davis, CA, USA, 95616

Tom Manly, UK Medical Research Council, Cognition and Brain Sciences Unit, Box 58, Addenbrooke's Hospital, Hills Road, Cambridge, UK, CB2 2QQ

René Marois, Vanderbilt University, Dept. of Psychology, 530 Wilson Hall, 111 21st Avenue, Nashville, TN, USA, 37203

Antígona Martínez, University of California, San Diego, Neurosciences, 9500 Gilman Drive, La Jolla, CA, USA, 92093-0608

Julio C. Martinez-Trujillo, York University, Centre for Vision Research, 4700 Keele Street, York University, CSB, Toronto, Ontario, Canada, M3J 1P3

Stephanie A. McMains, Department of Psychology & Program in Neuroscience, Boston University, 64 Cummington St., Boston, MA, 02215

Gérard Medioni, University of Southern California, Computer Science, SAL 200 mc 0781, Los Angeles, CA, USA, 90035

Marsel Mesulam, Northwestern University, Cognitive Neurology and Alzheimer's Disease Center, 320 East Superior St., Chicago, IL, USA, 60611

Jean A. Milstein, University of Cambridge, Experimental Psychology, Downing St., Cambridge, UK, CB2 3EB

Philippos Mordohai, University of Southern California, Electrical Engineering, 3737 Watt Way, PHE 204, Los Angeles, CA, USA, 90089

Michael J. Morgan, City University, London, UK

Michael C. Mozer, University of Colorado, Computer Science Department, and Institute of Cognitive Science, Boulder, CO, USA, 80309-0430

Hermann J. Müller, Ludwig-Maximilians-University Munich, Department of Psychology, General and Experimental Psychology, Leopoldstr. 13, Munich, Germany, 80802

Douglas P. Munoz, Queen's University, Centre for Neuroscience Studies, Botterell Hall, Kingston, Ontario, Canada, K7L 3N6

Scott Murray, University of Washington, Dept. of Psychology, Seattle, WA 98195-1525

Andriy Myachykov, Department of Psychology, Room 223, 58 Hillhead Street, Glasgow, Scotland, UK G12 8QB

Vidhya Navalpakkam, University of Southern California, Computer Science, Hedco Neuroscience Building, Room 06, 3641 Watt Way, Los Angeles, CA, USA, 90089-2520

Ernst Niebur, Johns Hopkins University, Mind/Brain Institute and Dept. of Neuroscience, 3400 N. Charles St., Baltimore, MD, USA, 21210

Anna Christina Nobre, University of Oxford, Department of Experimental Psychology, South Parks Road, Oxford, UK, OX1 3UD

Hans-Christoph Nothdurft, Visual Perception Lab (VPL), c/o MPI Biophysical Chemistry, Göttingen, Germany, 37070

Daniel H. O'Connor, Princeton University, Psychology, Green Hall, Princeton, NJ, USA, 08544

Kathleen M. O'Craven, Rotman Research Institute, Baycrest Centre, 3560 Bathurst St., Toronto, Ontario, Canada, M4N 2P5

Aude Oliva, Massachusetts Institute of Technology, Brain and Cognitive Sciences, 77 Massachusetts Avenue—NE20-463, Cambridge, MA, USA, 02139

Bruno A. Olshausen, Redwood Neuroscience Institute, 1010 El Camino Real, Suite 380, Menlo Park, CA, USA, 94025

Guy A. Orban, Katholieke Universiteit Leuven Medical School, Laboratorium voor Neuro- en Psychofysiologie, Campus Gasthuisberg, Herestraat 49, Leuven, Belgium, B-3000

Lucas Paletta, Joanneum Research, Institute of Digital Image Processing, Wastiangasse 6, Graz, Austria, A-8010

Günther Palm, University of Ulm, Neural Information Processing, Oberer Eselsberg, Ulm, Germany, 89069

Derrick J. Parkhurst, Johns Hopkins University, The Mind/Brain Institute, 3400 N. Charles St., Baltimore, MD, USA, 21218

Sumanta N. Pattanaik, University of Central Florida, Department of Computer Science, Orlando, FL, USA, 32816

Karl Pauwels, Katholieke Universiteit Leuven Medical School, Laboratorium voor Neuro- en Psychofysiologie, Campus Gasthuisberg, Herestraat 49, Leuven, Belgium, B-3000

Luiz Pessoa, Brown University, Department of Psychology, Department of Psychology, 89 Waterman St., Providence, RI, USA, 02912

Lubov N. Podladchikova, Rostov State University, AB Kogan Research Institute for Neurocybernetics, 194/1 Stachka Avenue, Rostov-on-Don, Russia, 344090

Michael I. Posner, University of Oregon, Department of Psychology, 1227 University of Oregon, Eugene, OR, USA, 97403-1227

Claudio M. Privitera, University of California, Neurology Units—School of Optometry, 485 Minor Hall, Berkeley, CA, USA, 94720-2020

Alice M. Proverbio, University of Milano-Bicocca, Department of Psychology, Piazza dell'Ateneo nuovo 1, Italy, 20154

Demetri Psaltis, California Institute of Technology, Psaltis Optics Group, MS 136-93, Pasadena, CA, USA, 91125

Rajesh P. N. Rao, University of Washington, Department of Computer Science and Engineering, Box 352352, Seattle, WA, USA, 98195-2350

Geraint Rees, University College London, Institute of Cognitive Neuroscience, 17 Queen Square, London, UK, WC1N 3AR

Ronald A. Rensink, University of British Columbia, Depts. of Psychology and Computer Science 2136 West Mall, Vancouver, BC, Canada, V6T 1Z4

David Ress, Stanford University, Depts. of Psychology and Radiology, Bldg. 420/488, Stanford, CA, USA, 94305-2130

John H. Reynolds, Salk Institute for Biological Studies, P.O. Box 85800, La Jolla, CA, USA, 92186-5800

Amy A. Rezec, University of California, at San Diego, Department of Psychology, 9500 Gilman Dr. 0109, La Jolla, CA, USA, 92093-0109

John E. Richards, University of South Carolina, Department of Psychology, Columbia, SC, USA, 29208

M. Jane Riddoch, University of Birmingham, School of Psychology, Behavioural Brain Sciences Centre, Birmingham, West Midlands, UK, B15 2TT

Maximilian Riesenhuber, Georgetown University Medical Center, Dept. of Neuroscience, Research Building, WP-12, 3970 Reservoir Rd. NW, Washington, DC, USA, 20007

Giacomo Rizzolatti, Università di Parma, Dipartimento di Neuroscienza, Sezione Fisiologia, via Volturno 3, Parma, Italy, 43100

Trevor W. Robbins, University of Cambridge, Department of Experimental Psychology, Downing St., Cambridge, Cambridgeshire, UK, CB2 3EB

Ian H. Robertson, Trinity College, Institute of Neuroscience, Dublin, Ireland

Lynn C. Robertson, Veterans Administration Martinez, CA and University of California, Berkeley, Dept. of Psychology, and Helen Wills Neurosciences Institute, Tolman Hall, Berkeley, CA, USA, 94720

Pieter R. Roelfsema, Netherlands Ophthalmic Research Institute, Meibergdreef 47, Amsterdam, The Netherlands, 1105 BA

Edmund T. Rolls, Oxford University, Department of Experimental Psychology, Oxford, UK, OX1 3UD

Erich Rome, Fraunhofer Institute for Autonomous Intelligent Systems, Robot Control Architectures (ARC), Schloss Birlinghoven, 53754, Sankt Augustin, Germany

Ilya A. Rybak, Drexel University, School of Biomedical Engineering, Science and Health Systems, Philadelphia, PA, USA, 19104

Dov Sagi, The Weizmann Institute of Science, Dept. of Neurobiology/Brain Research, Rehovot, Israel, 76100

Jeffrey D. Schall, Vanderbilt University, Psychology Department, Vanderbilt Vision Research Center, Nashville, TN, USA, 37240

Kerstin Schill, Universität Bremen, FB3/Computer Science, Kognitive Neuroinformatik, P.O. Box 330440, Bremen, Germany, 28334

Keith A. Schneider, Princeton University, Dept. of Psychology, Center for the Study of Brain, Mind, and Behavior, Green Hall, Princeton, NJ, USA, 08544

John T. Serences, Johns Hopkins University, Dept. of Psychological and Brain Sciences, 3400 N. Charles St., Baltimore, MD, USA, 21218

Natalia A. Shevtsova, Rostov State University, AB Kogan Research Institute for Neurocybernetics, 194/1 Stachka Avenue, Rostov-on-Don, Russia, 344090

Shinsuke Shimojo, California Institute of Technology, Computation and Neural Systems, Pasadena, CA, USA, 91125

David I. Shore, McMaster University, Department of Psychology, 1280 Main Street West, Hamilton, Ontario, Canada, L8S 4K1

Daniel J. Simons, University of Illinois, Urbana-Champaign, Department of Psychology, 603 E. Daniel Street, Champaign, IL, USA, 61820

Wolf Singer, Max Planck Institute for Brain Research, Deutschordenstr. 46, D-60528 Frankfurt am Main, Germany, D-60528

Dana M. Small, Northwestern University, Cognitive Neurology and Alzheimer's Disease Center, 320 East Superior Street, Chicago, IL, USA, 60611

Kenith V. Sobel, School of Psychology, Victoria University of Wellington, PO Box 600, Wellington, New Zealand

Joshua A. Solomon, City University, Department of Optometry and Visual Science, London, UK, EC1V 0HB

David C. Somers, Boston University, Psychology Dept. & Program in Neuroscience, 64 Cummington St, Boston, MA, USA, 02215

Henk Spekreijse, Academic Medical Centre (University of Amsterdam), P.O. Box 12011, Amsterdam, Netherlands, 1100 AA

Charles Spence, Department of Experimental Psychology, University of Oxford, South Parks Road, Oxford, UK, OX1 3UD

Lawrence W. Stark, University of California, Neurology Units—School of Optometry, 485 Minor Hall, Berkeley, CA, USA, 94720-2020

Peter N. Steinmetz, University of Minnesota, Biomedical Engineering, 7-126 Basic Sciences and Biomedical Engineering, 312 Church St. SE, Minneapolis, MN, USA, 55455

Hans Supèr, Netherlands Ophthalmic Research Institute, Vision & Cognition, Meibergdreef 47, Amsterdam, NH, The Netherlands, 1105 BA

Jan Theeuwes, Vrije Universiteit, Cognitive Psychology, van der Boechorststraat 1, Amsterdam, The Netherlands, 1081BT

Kirk G. Thompson, National Eye Institute, National Institutes of Health, Laboratory of Sensorimotor Research, Bldg. 49, Rm. 2A50, Bethesda, MD, USA, 20892

Steven P. Tipper, University of Wales, Centre for Clinical and Cognitive Neuroscience, School of Psychology, Penrallt Rd., Bangor, Gwynedd, UK, LL57 2AS

Antonio Torralba, Massachusetts Institute of Technology, Computer Science and Artificial Intelligence Laboratory, The Stata Center, Building 32, office D462, 32 Vassar Street, Cambridge, MA, USA, 02139

Stefan Treue, German Primate Center, Cognitive Neuroscience Lab, Kellnerweg 4, Goettingen, Germany, 37077

John K. Tsotsos, York University, Centre for Vision Research, 4700 Keele St., Toronto, Ontario, Canada, M3J 1P3

Leslie G. Ungerleider, National Institute of Mental Health, Lab Brain & Cognition, 10 Center Drive, Bldg. 10, Room 4C104, Bethesda, MD, USA, 20892-1366

David C. Van Essen, Washington University in St. Louis, Anatomy and Neurobiology, Box 8108, 660 S. Euclid, St. Louis, MO, USA, 63110

Marc M. Van Hulle, Katholieke Universiteit Leuven, Laboratorium voor Neuro- en Psychofysiologie, Campus Gasthuisberg, Herestraat 49, Leuven, Belgium, B-3000

Rik Vandenberghe, Neurology Department UZ Gasthuisberg Laboratory of Experimental Neurology, KU Leuven, Cognitive Neurology and Alzheimer's Disease Center, Northwestern University, Chicago, IL

Wim Vanduffel, Katholieke Universiteit Leuven, Laboratorium voor Neuro- en Psychofysiologie, Campus Gasthuisberg, Herestraat 49, Leuven, Belgium, B-3000

Rufin VanRullen, CNRS, Centre de Recherche Cerveau et Cognition, 133 Route de Narbonne, 31062 Toulouse Cendex, France

Shaun P. Vecera, University of Iowa, Department of Psychology, Iowa City, IA, USA, 52242

Patrik O. Vuilleumier, Laboratory for Behavioral Neurology & Imaging of Cognition, Dept. of Neurosciences & Clinical of Neurology, University Medical Center, 1 rve Michel-Servet, 1211 Geneva 4, Switzerland

Ranxiao Frances Wang, University of Illinois, Department of Psychology, 603 E. Daniel St., Champaign, IL, USA, 61820

Katsumi Watanabe, National Institute of Advanced Industrial Science and Technology, Institute of Human Science and Biomedical Engineering, Visual Cognition Group, AIST Tsukuba Central 6, 1-1-1, Higashi, Tsukuba, Ibaraki 305-8566, Japan

Jeremy M. Wolfe, Brigham And Women's Hospital, Harvard Medical School, Visual Attention Lab, 64 Sidney St., Cambridge, MA, USA, 02139

Steven Yantis, Johns Hopkins University, Dept. of Psychological and Brain Sciences, 3400 N. Charles St., Baltimore, MD, USA, 21218

Hector Y. Yee, PDI/DreamWorks, 1800 Seaport Blvd., Redwood City, CA, USA, 94063

Alberto Zani, Institute of Molecular Bioimaging and Physiology, National Council of Research (CNR), c/o LITA Bldg., Via Fratelli Cervi 93, Segrate (Milano), Italy, 20090

Gregory J. Zelinsky, State University of New York at Stony Brook, Dept. of Psychology, Stony Brook, New York, USA, 11794-2500

Barbara Zenger-Landolt, Computation and Neural Science, California Institute of Technology, Pasadena, CA 91125

Christoph Zetzsche, University of Bremen, Cognitive Neuroinformatics, FB 3 Mathematics and Informatics, P.O. Box 330 440, Bremen, Germany, 28334

Li Zhaoping, University College London, Department of Psychology, London, UK

Josef Zihl, Ludwig-Maximilians–Universität München, Department of Psychology, Leopold Str. 13, 80802 Munich, Germany

Foreword: Neurobiology of Attention

How is it that we (humans, primates, mammals) can attend to so much that is relevant to our survival in a complex environment? And why is it that we need to attend rather than simultaneously sample information from our total environment in "complete detail"? How can we refine our scientific understanding of attention through psychological analysis of humans, both normals and patients with various disorders? What can we gain in understanding by human brain imaging, and how can this understanding be enriched by analysis of circuitry and neurophysiology in monkeys and other mammals? How can we, starting either from psychophysics or neurophysiology, develop detailed computational models of the neural circuitry underlying attention, and what can such models contribute to the design of computer vision systems in particular and robots and biomedical technology in general?

The present volume, *Neurobiology of Attention*, answers these intriguing questions and many more, by gathering over 100 short articles by leading authorities, each of which introduces the non-specialist reader to part of the puzzle, and then encourages the reader to build out from this first "island of knowledge" by seeking out related articles that extend the treatment of the topic either in breadth or depth. This seeking out is aided by the 'tour' commentary written by the editors, which identifies themes and inter-relationships running through the book, as well as extensive cross-referencing among the articles. Specialists in any subfield of this broad topic will also find many articles that deserve their attention.

This fascinating handbook presents the articles under five general headings:

1. **Theoretical and Anatomical Foundations**: What is the nature of attention in mammals? The answers are diverse and involve theoretical issues of capacity limitation and attentional bottlenecks; issues of what groups sensory data together to constitute a focus of attention, and the ways in which attention may be deployed over time, whether to build up an ever richer understanding of some scene or to divide attention among several tasks. The discussion ranges from specific analyses of psychophysical and neurological data to deep philosophical questions of intentionality and consciousness.

2. **Characterizing the Functions of Attention**: The theme of this section is complemented by the more detailed mechanism-oriented theme of the next section. These two themes provide two successive passes onto the functions and mechanisms of attention; first in more general terms in this section and looking at focused experimental results in the next. For example, how do lateral inhibitory interactions mediate the competition between different parts of the image to prove "most salient" and thus capture the next shift of attention, and what grouping mechanisms determine which features in the image cooperate to between them determine a salient object or a salient movement?

3. **Mechanisms of Attention**: Building upon the functions exposed in the previous theme, this theme is articulated around a number of pointed experiments aimed at understanding the detailed implementation and characteristics of attentional functions. If we are looking for a pen, how do we find it? If we notice a person acting, how do we find and attend to the goal of that action? As in other places, the power of *Neurobiology of Attention* here is that it approaches such questions from different perspectives—cognitive science,

physiology, anatomy, brain imaging, and modeling, addressing data from animals and from humans with and without specific disorders—to the mutual enrichment of each of them.
4. **Systems Computational Neuroscience and Applications:** Here the emphasis turns from processes to comprehensive algorithms associated with mechanisms serving attention in the mammalian brain and to technical applications, such as using attention to maximize utilization of machine resources, neuromorphic computer vision systems; robotics systems and surveillance systems; and biomedical applications.

Dear reader: A feast is laid before you. You may taste the courses in any order you prefer, and your understanding of attention will be well nourished indeed.

Michael A. Arbib
Los Angeles

Preface

Key to the survival of many biological organisms is their ability to selectively focus neural processing resources onto the most relevant subsets of all available sensory inputs. Study of the brain mechanisms responsible for this selection has flourished over the past century, under the broad heading of attention research. In the last 20 years, the availability of new experimental techniques has created an explosion of new studies and findings. While attention research initially had mostly interested specialists of human psychology, it is now one of the central topics of modern neuroscience research, bringing together computational, philosophical, clinical, anatomical, physiological, and neuropsychological scientists. As a consequence, embracing the breadth of recent knowledge on attention has become an increasingly challenging task for any researcher.

The format of this book is very different from conventional edited volumes. Rather than a few relatively long chapters, there are 109 short chapters that concisely summarize a particular topic or empirical finding. In complement, two additional chapters, "A Brief and Selective History of Attention" and "A Tour of This Volume", propose several different pathways for navigating the text. The inspiration for this format came from Michael A. Arbib's very successful *Handbook of Brain Theory and Neural Networks* (MIT Press, 2nd Ed., 2004). This phenomenal encyclopedic text provides concise yet comprehensive coverage of the main topics in neural network research, through a broad series of short and focused topical research articles. Our goal for the present volume was to attempt to emulate this format, but on the broad topic of recent, state-of-the-art attention research. The individual chapters provide rapidly absorbable and focused summaries of current knowledge, along with pointers to the key references.

In organizing the contributors and structure of the volume, we have attempted to make the present handbook approachable to the widest possible readership. This includes students, who may like the structured organization of the volume and the Tour, introductory and review articles, and the short and focused research articles describing recent findings on their topics of current interest. Specialists in the field of attention research may enjoy the extensive coverage, from anatomy to medical disorders, via behavioral studies and theoretical models. Importantly, researchers in connected disciplines may also appreciate the Tour and encyclopedic organization, review articles, and the possibility of gaining state-of-the-art knowledge on current attention research while being exposed to the "big picture." Finally, readers with only a casual interest in attention may find that the index will facilitate rapid access to specific key topics and findings that may have sparked their interest.

The Tour chapter proposes one possible read through the book, but there are many others. As the reader will quickly realize, many of the contributed chapters attack several aspects of attention, and thus are useful in several contexts. In organizing the volume into sections, we have attempted to provide multiple views of the central topics of attention research, from broader perspective and foundation articles to detailed novel experimental findings and systems-level implementations. Cross-referencing between chapters invites the reader to hop from chapter to chapter across the entire volume, finding their own path through the body of knowledge summarized in this handbook.

We thank Johannes Menzel, Kirsten Funk, Angela Dooley, and the entire Elsevier team for making this volume possible. Without their hard work and guidance, this encyclopedic endeavor would not have been possible. We are indebted to the authors, not only for a superb collection of high-quality articles, but also for their patience, tenacity, encouragements, and valuable organizational suggestions throughout the evolution of the project. We thank Dr. Arbib for providing invaluable knowledge and advice in guiding us throughout the several years it has taken to work towards this goal. Finally, our thanks go to our reviewers, some of whom have also contributed chapters to the volume, including Neil Bruce, Ran Carmi, Nitin Dhavale, Jianwei Lu, Nathan Mundhenk, Vidhya Navalpakkam, Antonio Rodriguez, Albert Rothenstein, Michael Tombu, and Andrei Zaharescu.

The original cover design was contributed by April Tsui and summarizes the contents of the volume by playing with the viewer's attentional system and rapid visual orienting capabilities.

Laurent Itti, Geraint Rees, and John Tsotsos

A Brief and Selective History of Attention

John K. Tsotsos, Laurent Itti, Geraint Rees

1. Roots

Attention: from the Latin attenti, from attentus, the past participle of attendere, meaning "to heed."

Although the word existed in Roman times, there is little reference to any scientific basis for the human capacity for attention until Descartes (1649), who related attention to movements of the pineal body acting on the animal spirit:

Thus when one wishes to arrest one's attention so as to consider one object for a certain length of time, this volition keeps the gland tilted towards one side during that time.

In contrast, Hobbes (1655) believed:

While the sense organs are occupied with one object, they cannot simultaneously be moved by another so that an image of both arises. There cannot therefore be two images of two objects but one put together from the action of both.

The first extended treatment of attention may be due to Malebranche (1674) who claimed that:

Attention is necessary for conserving evidence in our knowledge.

He believed that the seeker of truth must avoid strong sensations and passions that may lead to distraction. Malebranche recommended detailed procedures for cultivating attention, the study of geometry being a prime element. Leibnitz (1765) and Wolff (1734) then introduced the idea of *apperception*, the process that admits perceptions into consciousness:

In order for the mind to become conscious of perceived objects, and therefore for the act of apperception, attention is required.

This made a connection between consciousness and attention for the first time. Hebart (1824) believed, like Leibnitz, that ideas can exist in the mind without being conscious, but he attributed this to inhibition from competing ideas. He developed an elaborate algebraic model of attention using differential calculus:

He is said to be attentive, whose mind is so disposed that it can receive an addition to its ideas: those who do not perceive obvious things are, on the other hand, lacking in attention.

2. Phenomenonology and Early Psychological Accounts

Helmholtz (1860) believed that nervous stimulations are perceived directly, never the objects themselves, and there are mental activities (judgment theory) that enable us to form an idea as to the possible causes of the observed actions on the senses. He wrote:

We are not in the habit of observing our sensations accurately except as they are useful in enabling us to recognize external objects. On the contrary, we are wont to disregard all those parts of the sensations that are of no importance so far as external objects are concerned.

These activities are instantaneous, unconscious, and cannot be corrected by the perceiver by better knowledge. He disagreed with Panum (1858) who believed that attention is an activity entirely subservient to the conscious will of the observer. Attention becomes difficult to hold once interest in an object fades. The greater, the disparities between the intensities of two impressions, the harder it is to keep the attention on the weaker one. Panum studied this in the specific context of binocular rivalry; but more generally, he observed that we are able to "see" only a certain number of objects simultaneously. He therefore concluded that it makes sense that the field of view is first filled with the strongest objects. In studying an object, first attention, and then the eye, is directed to those contours that are seen by indirect vision.

Hamilton (1859) first raised the question of the span of attention:

The doctrine that the mind can attend to, or be conscious of, only a single object at a time would in fact involve the conclusion that all comparison and discrimination are impossible? [...]. Suppose that the mind is not limited to the simultaneous consideration of a single object, a question arises, How many objects can it embrace at once?

Brentano (1874) developed a competing theory known as *act psychology*. In this view, an act is a mental activity that affects percepts and images rather than objects. Examples of acts include attending, picking out, laying stress on something, and similar actions. This was the first discussion of the possibility that a subject's actions play a dominant role in perception. Metzger (1974) summarizes aspects of action that contribute to perception: bringing stimuli to receptors, enlarging the "accessible area," foveation (the act of centering the central, highest-resolution part of the retina onto an object), optimization of the state of receptors, slowing down of fading and local adaptation, exploratory movement, and finally the search for principles of organization within visual stimuli.

A further connection between attention and consciousness was drawn by Wundt (1874), who described attention as an inner activity causing ideas to be present in consciousness to differing degrees. The focus of attention can narrow or widen, reflecting these degrees of consciousness. G. E. Müller's (1904, 1923) *Komplextheorie*, or theory of collective attention, was based explicitly on acts of the subject, obviously following the lead of Brentano. For Titchener (1908), attention was an intensive attribute of a conscious experience equated with "sensible clearness." He compared attention to a wave, but with only one peak (focus). He further argued that the fundamental effect of attention is to increase clarity, while Kulpe (1902) suggested that attention enhanced not clarity but discriminability. Petermann (1929) argued against the subject being a passive battlefield of stimuli. He proposed an attention-direction theory, based on actions (like Brentano), as the mechanism for an active attentive process. Both Petermann and Müller needed to introduce criteria for the action of attention but were subsequently criticized for using attention as a magical principle that can do everything and explain nothing.

William James (1890) is credited for perhaps the best-known plain language description of attention:

Everyone knows what attention is. It is the taking possession by the mind, in clear and vivid form, of one out of what seem several simultaneously possible objects or trains of thought.

James specified two domains in which these objects occur, sensory and intellectual. He listed three physiological processes which he believed played a role in the implementation of attention: the accommodation or adjustment of the sensory organs, the anticipatory preparation from within of the ideational centers concerned with the object to which the attention is paid, and an afflux of blood to the ideational center.

3. Early Behavioral Findings

Early experimentalists played an important role in shaping views on how to address questions related to attention. Jevons (1871), following Hamilton's suggestion, investigated the span of attention by throwing sets of black beans into a round box and examining his ability to estimate the number by a single mental act (subitizing). This was later studied in a controlled manner using the tachistoscope. Cattell (1885) first manipulated the tachistoscope to control temporal span and thus prevent fixation changes. Attention not only had a span in space, as was studied by Hamilton and Jevons, but also in time. Binocular rivalry was thought to be a result of attention fluctuating over time but with a constant stimulus by Breese (1899) and Helmholtz (1860).

Attention was also studied beyond the visual modality. The "complication" experiment (which grew out of the difficulties astronomers faced in recording transit times of stars) was introduced by von Tschisch (1885). Subjects were required to watch a moving pointer and specify its location when a sound or some other non-visual stimulus occurred. Even though the two events differed in modality (the visual pointer and the sound, for example) and were presented simultaneously, one was registered in consciousness before the other. James (1890) believed this may be due to concentration of attention on each in turn. Angell and Pierce (1892) corroborated this experimentally. The effects of attention on behavioral measures of perception and action were further investigated by many others, including Exner (1882) and Cattell (1885).

Around this time, Helmholtz (1896/Mackeben and Nakayama 1989) introduced the idea that attention could be deployed in a covert fashion, independent of eye movements:

The electrical discharge illuminated the printed page for a brief moment during which the image of the sheet became visible and persisted as a positive after-image for a short while. Hence, perception of the image was limited to the duration of the after-image. Eye movements of measurable size could not be performed during the duration of the flash and even those performed during the short persistence of the after-image could not shift its location on the retina. Nonetheless, I found myself able to choose in advance which part of the dark field off to the side of the constantly fixated pinhole I wanted to perceive by indirect vision. Consequently, during the electrical illumination, I in fact perceived several groups of letters in that region of the field [...]. The letters in most of the remaining part of the field, however, had not reached perception, not even those that were close to the point of fixation.

Such a demonstration is simple and compelling and so represents powerful evidence for the existence of covert attention without independent eye movements. The similarities and differences between movements of the eye and those of spatial attention have been a focus of research ever since.

After the exuberant early behavioral experiments at the end of the 19th century, Ribot (1889), Watson (1919), Holt (1915), Dashiell (1928), Mowrer (1938), and others attempted to describe mental phenomena solely in behaviorist terms. To them, attention was a by-product of behavior, and identified with the postural or motor processes that facilitate reception of stimuli. Mowrer (1939) studied the effects of expectancy on reaction time and response amplitude, and connected these effects with what he called the "preparatory set," later called attention. He believed that this mechanism is central rather than peripheral.

The Gestalt school did not believe in attention. Köhler's 1947 book, for example, only barely mentions attention. Gestaltists could have been expected to take attention seriously. However, they believed that the patterns of electrochemical activity in the brain are able to sort things out by themselves, and to achieve an organization that best represents the visual world, reconciling any conflicts along the way. The resulting internal organization includes portions that seem more prominent than others. Thus attention became an emergent property and not a process in its own right. Figure-ground concerns loomed larger for them and the figure would dominate perceptions within a scene, thus emerging as the focus of attention rather than being explicitly computed as such. Berlyne (1974) tells us that Edgar Rubin, well known for his vase/profile illusion of figure-ground perception, actually read a paper at a meeting in Jena in 1926 titled "On the nonexistence of attention." Healthy skepticism continues to this day.

4. Early Neuroanatomy and Neurophysiology

Early findings from neuroanatomy and neurophysiology contributed to knowledge about basic brain function that would form the foundation of contemporary understanding of the neural mechanisms of attention. Helmholtz (1866) demonstrated that the rate of nerve conduction was not infinitely fast, but only a relatively slow 100 meters/second. Thus, every mental operation requires a period of time for its accomplishment. Sechenov (1863/1965) argued that the highest levels of the nervous system exercise inhibitory control over lower levels, and thus one might expect inhibition to be associated with central attention.

Hebb (1949) extended Cajal's (1909) single neuron doctrine by considering assemblies of neurons, functional entities that formed the basis of a neural theory for perception and learning. He emphasized that the perceiver's attention or expectations must form an integral part of the perceptual process. Pavlov (1927) described two basic internal aspects of behavior: facilitation and inhibition. An unexpected stimulus is likely to induce the orientation reflex, and at the same time there is an interruption of response to other stimuli that may be present. This "negative induction" was the tendency for a strong focus of excitation in one region of the cerebral cortex to generate inhibition in neighboring regions and consequently to weaken responses controlled by those regions. Sokolov (1963) studied the orienting reflex and elaborated it into a general view about the alignment of the central nervous system with sources of stimulation. This reflex combined outward signs and inward systems designed to improve processing of selected signals.

5. Information Processing and the First Articulated Models of Attention

Selective hearing is the domain where early research motivated the development of the first information processing models of attention. Cherry (1953) and Poulton (1953) first exposed subjects to two or more verbal messages simultaneously by presentation to different ears. The messages were distinguished by gender of speaker, by content, and by language or acoustical characteristics. Subjects were instructed to attend to one particular characteristic or were given no instructions at all, and then were asked questions about the messages. Some experiments involved "shadowing," repeating verbal stimuli as they were received. Subjects showed surprisingly little awareness for the content or even characteristics of unattended stimuli, suggesting that unattended stimuli were rejected from further processing. In an intellectual landmark, Broadbent summarized this empirical data and presented a theoretical synthesis in 1958. Broadbent proposed that humans can be viewed as limited-capacity information processing systems. He suggested a model of processing that included a short-term store to act to extend the duration of a stimulus; that stimuli could be partitioned into channels (modalities); then a selective filter selects among channels, and a limited capacity channel then processes the selected channel. This is the prototypical *early selection* model of attention and represents the first theoretical account to relate psychological phenomena to information processing concepts from computer science. Broadbent's clear empirical and theoretical synthesis stimulated the development of alternative accounts. Deutsch & Deutsch (1963), Norman (1968), Moray (1969), and MacKay (1973) proposed *late selection* models. In their view, limited awareness of unattended stimuli has little to do with perceptual processing but instead represents a failure of perception to subsequently access memory or control action, a view later developed by Duncan (1980). Late selection theories propose that all information is completely processed and recognized before it receives the attention of a limited capacity processor. Recognition can occur in parallel, and stimulus relevance determines what is attended. These models typically included some form of memory within the process. How attention may interact with perceptual memory was first investigated by Sperling (1960), who observed that cues that follow a visual display by a short fraction of a second can be used to select information from a brief sensory memory system that seems to hold much more than the subject was able to report (the partial report task).

Treisman (1964) was not completely convinced by the stark dichotomy between early and late selection models, and suggested a hybrid view. Her model included a filter that attenuates (in a graded fashion) unattended signals, leaving them incompletely analyzed. Importantly, this filter can operate at different stages of information processing. Treisman thus introduced the notion that the effects of attention may be hierarchically described. But these levels need not be all based on perception, as Neisser (1967) noted. He claimed that we can attend to our own thoughts as well as to external stimuli, and that attention worked by matching stimulus sequences to internal processes. Moray (1969) added that attention, although restricted to one channel at a time, is allowed to switch over to a different channel when the importance of channels changes. This switching required time, and in this way he hoped to explain some of the temporal characteristics of attention.

The influence of computer technology and information processing can be seen perhaps most strongly in the theory of Kahneman (1973), who argued that attention is a general-purpose limited resource. Some activities are more demanding and therefore require more mental effort than others. The total available processing capacity may be increased or decreased by other factors such as arousal. Several activities can be carried out at the same time, provided that their total effort does not exceed the available capacity. Rules or strategies exist that determine allocation of resources to various activities and to various stages of processing. Attentional capacity will therefore reflect the demands made at the perceptual level, the level at which the input is interpreted or committed to memory, and the response selection stage. Kahneman thus believed in the existence of a Central Processor that operates a Central Allocation Policy, constantly evaluating the demands made by each task and adjusting attention accordingly. It is clear that Kahneman's inspiration came from consideration of the operating systems of computers; but beyond the metaphoric use, the connection is a weak one, if not irrelevant.

Milner (1974) considered the role that attention plays in visual recognition. He claimed that the unity of a figure at the neuronal level is defined by synchronized firing activity and that attention acts in two ways: to select relevant figure from among others and to activate the feedback pathways from the cell assembly to the early visual cortex for precise localization. Milner suggested that feedback pathways communicate attentional instructions.

Perhaps the first mathematically articulated theory relevant to attentional processing was Grossberg's Adaptive Resonance Theory (1975, 1976). ART algorithms are clustering algorithms that obey the following governing principles: bottom-up activation can drive a cell if strong enough; top-down priming can modulate a cell; a cell becomes active if it receives large enough top-down and bottom-up activation; top-down activation, even small, can negate bottom-up activation; and, feedback leads to resonance and convergence. He believed that top-down attentional mechanisms should occur in every cortical area where learning can occur and suggested specific circuitry for interactions.

An important aspect of attentional processing missing from these models, and not well understood to this day, is the temporal nature of the process. C. Eriksen and colleagues (Eriksen and Collins 1969, Colegate, Hoffman and Eriksen 1973) performed the first experiments attempting to illuminate the subject. They used spatial cueing in selection tasks where the display was made up of arrays of characters and subjects were asked to respond to the cued characters. They investigated primarily cue-stimulus onset timings (SOA) and the resulting response times (RT) and accuracy. Accuracy improved as the cues were presented earlier than the stimulus up to around 100 msec. with the benefit peaking at about 200 msec.

and plateauing afterwards, showing that it takes time for the cue to take effect.

6. Neurophysiology

Modern techniques in neurophysiology have led to major advances in the understanding of brain mechanisms of attention through experiments in awake behaving animals and in humans. With the invention of human electroencephalography (Berger 1929) many stimuli were found to generally increase "arousal" or attentiveness in subjects. In particular, activation of the reticular formation can produce these effects. Increased arousal was found to accompany an increased capacity to take in and process information. Detailed electrophysiological investigation of the brain mechanisms of attention began with event-related potential (ERP) studies in animals (Hernandez-Peon et al. 1956, Horn 1965) and humans (Spong et al. 1965, Groves and Eason 1969, Harter and Salmon 1972, Hillyard et al. 1973, Naatanen 1975, Simson et al. 1977), followed closely by single unit recordings in monkeys (Goldberg and Wurtz 1972, Goldberg and Bushnell 1981, Bushnell et al. 1981, Mountcastle et al. 1981, Moran and Desimone 1985). These experiments revealed that sensory responses to auditory or visual stimuli were modulated by the animal's attentional state. For example, Hernandez-Peon, Scherrer and Jouvet (1956) measured activity evoked by the ticking of a metronome in neurons in the cochlea nucleus of a cat. When the cat was shown a mouse, the neural activity evoked by the metronome disappeared. They concluded that top-down signals exist that alter sensory processing to prevent distraction. Initially, such top-down effects of attention appeared relatively modest, or restricted to brain areas traditionally associated with spatial attention (e.g., the parietal lobe; Wurtz et al. 1982). However, as the scope of the research proceeded and experimental paradigms became more sophisticated, it became clear that attentional modulation of sensory responses could be found throughout the hierarchy of sensory processing areas and indeed in virtually every cortical area. Such empirical findings challenged early theoretical accounts of attention as a unitary sensory filter and began to stimulate the development of more sophisticated theoretical models.

Non-invasive ways of measuring brain processes in humans began to appear at the end of the 1980s, including magnetoencephalography (MEG), positron emission tomography (PET), and subsequently functional MRI (fMRI). Such techniques measure the activity in relatively large neuronal populations, either directly (MEG/ERP) or indirectly through measuring changes in local cerebral hemodynamics. In the decade leading up to this volume, there has been an explosion in the use of functional imaging to characterize attentional functions in humans that is beyond the scope of this history and indeed is the focus of much of this volume. In general, the principles that characterize attentional modulation of sensory processing measured using functional imaging techniques are remarkably similar to those determined by single unit electrophysiology, though important issues remain in understanding the relationship between these two measures.

7. Integrated Theories of Attention

Broadbent's (1958) synthesis, discussed above, opened up the prospect of integrated psychological theories of attention. Continuing from where late selection models left off were Shiffrin and Schneider (1977), who claimed that automatic processes, such as parallel visual analysis, do not require significant mental capacity, are resistant to suppression, and do not require deliberate intent by the observer to have effect. Their model required a final comparison stage where a target in memory may be checked against the resulting visual representations. Interestingly, they also included the practice effect in their model, that is, how responses to visual search tasks change with training. Their results led them to believe that even the final comparison stage could be made automatic with sufficient practice. In general, the changes in processing induced by practice remain poorly understood.

Feature Integration Theory was first presented in a seminal paper by Treisman & Gelade (1980). Here they introduced for the first time several ideas that have subsequently pervaded the attentional literature, and strongly influence the field to this day:

[. . .] the feature-integration theory suggests that we become aware of unitary objects in two different ways—through focal attention or through top-down processing. The first route to object identification depends on focal attention, directed serially to different locations, to integrate the features registered within the same spatio-temporal "spotlight" into a unitary percept [. . .]. The second way [. . .] is through top-down processing. In a familiar context, likely objects can be predicted. Their presence can then be checked by matching their disjunctive features to those in the display, without also checking how they are spatially conjoined.

They argued that separate feature maps existed, all connected to a master map of locations over which an attentional spotlight would move, illuminating particular locations that were to be attended. The "illumination" metaphor to describe an analog selection of a contiguous portion of an image seems to have its roots in Shulman, Remington and McLean (1979). When more than one visual object is present and the visual system is overloaded, incorrect combinations of features belonging to different objects occur, known as "illusory conjunctions" (Treisman and Schmidt, 1982). These are failures of binding, as pointed out first by Rosenblatt (1961). Rosenblatt's classical view of binding is one in which one kind of visual feature, such as an object's shape or color, must be correctly associated with another feature, such as its location, to provide a unified representation of that object. Such explicit association or binding is important when more than one visual object is present, in order to avoid incorrect combinations of features belonging to different objects.

Although Feature Integration Theory, and in fact most models, concentrates on covert attention alone, it is clear that fixation changes (whether overt or covert) are tightly coupled, and a considerable amount of research has studied such a relationship. Top-down influences on overt attention were perhaps most clearly demonstrated by Yarbus (1967). He showed that eye movements changed depending on the question asked of the subject; that is, the reason for looking at an image in the first place. For a very conventional picture of a family in their living room he asked observers several different questions pertaining to different aspects of the scene (such as the nature of the clothing worn by the family, or where the family were located in the room) and recorded the ensuing eye movements. The eye movement patterns were radically different and consistent across observers, thus providing strong evidence for the role of task in how we acquire information from an image.

Noton and Stark (1971) reported a connection between the eye movement patterns observed during learning of a visual pattern and the subsequent viewing of that pattern. During learning, subjects followed a characteristic scanpath. When later presented with the pattern again, subjects usually followed a very similar scanpath for at least the first few fixations. This suggested that the internal representation of a pattern in memory is a network of features, and thus attention shifts move from feature to feature.

In one theoretical synthesis of research on both overt and and covert attention, Posner (1980) proposed that attention had three major functions. First, it provided the ability to process high priority signals, or *alerting*. Second, it permitted *orienting* and overt foveation of a stimulus. Finally, it allowed *search* to detect targets in cluttered scenes. Orienting improves efficiency of target processing in terms of acuity, permitting events at the foveated location to be reported more rapidly as well as at a lower threshold. Overt foveation is strongly linked to movement of covert attention. Overt orienting, whether of the eyes or the head or both, is termed *exogenous* while covert fixation shifts are called *endogenous*. Exogenous control of gaze direction is controlled reflexively by external stimulation while endogenous gaze is controlled by internally generated signals. Covert fixations are not observable and thus must be inferred from performance of some task.

The psychological "binding problem" has also motivated consideration of possible physiological mechanisms. Von der Malsburg (1981) sought to understand how active cells can express relationships among themselves and developed his Correlation Brain Theory. He suggested that synapses could be modulated, so that they switch between conducting and non-conducting states, and that the modulation is governed by correlations in the temporal structure of signals. In this way, momentarily useless connections are deactivated, and interference between different memory traces is reduced and memory capacity increased; thus synapses were dynamically modulated. He also claimed that the brain does not contain complex feature detector cells; rather, timing correlations signal objects.

When stimuli are available for just a brief period (approximately 100 ms) only limited amounts of spatial information can be extracted by the visual system, according to Bergen and Julesz (1983). Eye movements are not possible and the time during which the after-image of the stimulus is available for inspection is cut by presentation of a masking pattern. Bergen and Julesz showed that under these conditions a small pattern is easily detected against a background made up of many others, only if it differs from the background patterns in certain local features. If so, its detectability is almost independent of the number of background elements, suggesting that a parallel process is responsible for the computation. Detection of patterns that do not differ from the background in such features requires a serial focal attention process.

An early attempt to connect previous ideas and speculate on the neural correlate of attentional control was presented by Crick (1984). He suggested that the spotlight of Treisman and Gelade is controlled by the reticular complex of the thalamus, and that the searchlight is expressed by rapid bursts of firing from subsets of thalamic neurons. Feature conjunctions are mediated by rapidly modifiable synapses (Malsburg synapses) by these bursts. Activation of Malsburg synapses produces transient cell assemblies connecting neurons at different levels.

A bias against returning attention to previously attended locations was first reported by Posner and Cohen (1984). They argued that the mechanism responsible for the effect "evolved to maximize sampling of the visual environment." Posner, Rafal, Choate, and Vaughan (1985) coined the term "inhibition of return" (IOR) to refer to this inhibitory component. This terminology derives from the notion that attention is drawn reflexively to the location of a salient cue. Following the withdrawal of attention back to fixation, attention is then relatively impaired for returning to the previously cued location. Posner and colleagues (1985) thus characterized IOR as an effect of attention on attention.

Koch and Ullman (1985) presented a model of visual attention that has played a major role in the discipline, based largely on Feature Integration Theory, but providing a detailed mathematical foundation. Their model included a saliency map (Treisman and Gelade's master map of locations) that was an encoding of overall feature saliency at each location in an image. Saliency was determined by a weighted sum of all features computed at each point. A winner-take-all competition would choose the strongest and thus assumed most salient point. All features at that point would then be routed to a central representation for further processing. Due to the mathematical characteristics of a winner-take-algorithm, selected points must be inhibited or eliminated from the competition in order for the next most salient item to be selected. This kind of inhibition of return is not really the same as that of Posner and Cohen; the underlying intent of each is different. In their model, the time to move attention was proportional to logarithmic distance between stimuli. Interestingly, Sagi and Julesz (1985) documented the independence of performance in visual search on distance, suggesting fast non-inertial shifts of attention. Koch and Ullman's theory did not include single cell or synaptic modulations.

8. Disorders of Attention

The identification of neuropsychological deficits of attention following brain damage has played an important role in shaping notions of attentional mechanisms in the human brain. Poppelreuter (1917) introduced the word inattention, thus conceiving of the notion of a brain disorder affecting attention. Brain (1941) and Critchley (1966) proposed that unilateral neglect following parietal damage reflected an attentional disorder. However, not until Kinsbourne (1970) and Heilman & Valenstein (1972) was the notion that an attentional impairment underlays unilateral neglect more generally accepted. Unilateral neglect is a common and disabling deficit after unilateral brain damage, particularly following strokes centred on the right inferior parietal lobe. Many of the ideas and methods developed in the study of normal attention have been applied fruitfully to the study of such patients, particularly the demonstration of residual processing for neglected information and the modulation of visual extinction (one of the components of the neglect syndrome) by grouping processes.

9. Conclusions

Like attention, all histories are selective and this chapter is no exception. We have chosen to cover the field until the mid-1980s. Since then, a great deal has happened to attention research that history has yet to judge, and this chapter will make no attempt to second-guess the future. The interested reader can read the comprehensive and detailed treatments by Posner (1978), Allport (1993), Desimone and Duncan (1995), Maunsell and Ferrera (1995), Rafal and Robertson (1995), Wolfe (1996) for a review of visual search; Pashler (1998), Driver & Mattingley (1998) for a review of disorders of attention; Chapter 11 of Palmer (1999), Colby and Goldberg (1999), Kastner and Ungerleider (2000) for a review of brain areas where attentional effects have been observed; Logothetis et al. (2001) and Bandettini and Ungerleider (2001) for a review of brain imaging techniques; and Itti and Koch (2001) for a review of computational models of attention. Additional material on the history of the field can be found in Pillsbury (1908), Treisman (1969), Berlyne (1974), and Metzger (1974). In any case, most if not all of the chapters in this volume provide some level of background to the work presented, thus filling in much of the events since 1985 in a manner not subject to our interpretation. This may be the best way to permit the reader to play a role in the historical judgment of research in the past 20 years.

All attention researchers recognize the diversity of possible definitions of the topic and this has sometimes led to healthy expenditure of a great deal of critical angst (Allport 1993). Indeed, Groos wrote as early as 1896 that:

"To the question, 'What is Attention'? there is not only no generally recognized answer, but the different attempts at a solution even diverge in the most disturbing manner."

Pillsbury (1908) agreed, saying that attention was "in disarray." Spearman (1937) commented on the diversity of meanings associated with the word:

"For the word attention quickly came to be associated ... with a diversity of meanings that have the appearance of being more chaotic even than those of the term 'intelligence' ".

Even as recently as 1998, Sutherland in reviewing recent attention books claimed that:

"Over the past 50 years, the sheer ingenuity displayed by psychologists working on attention rivals if it does not exceed that of cosmologists studying black holes. Indeed, there is a similarity in their results—after many thousands of experiments, we know only marginally more about attention than about the interior of a black hole."

Of course, quite a bit is known about black holes nowadays (Hawking et al. 2004). Nevertheless, the point remains that the very diversity of the field can be seen as an impediment to understanding. We disagree. In preparing this volume, we celebrate the intellectual diversity of opinion and multiplicity of empirical findings that characterize contemporary attention research. Diversity stimulates ingenuity, interdisciplinarity, and a wide diversity of empirical work, and it may well be the case that a full explanation of attention is not possible from a single viewpoint alone. The explosion of interest in cognitive neuroscience and the development of non-invasive neuroimaging techniques in the last decade have led to an ever-increasing body of knowledge about attention and the brain on which this volume is focused. The papers in this volume represent a necessarily selective but comprehensive snapshot of the current enthusiasm, both empirical and theoretical, that characterizes attention research.

Regardless of one's methodology, discipline, and intuitions, there is only one core issue that justifies attentional processes: *information reduction*. Humans are faced with immense amounts of information, through not only their senses but also through the need to manage their memory and knowledge, and to utilize that information at appropriate moments. The search for how this occurs in support of the complex sensory-guided behaviors we humans exhibit is what motivates the research that is represented in this volume.

References

Allport, A. (1993). Attention and control: have we been asking the wrong questions? A critical review of twenty-five years. In "Attention and Performance XIV: Synergies in Experimental Psychology, Artificial Intelligence, and Cognitive Neuroscience." MIT Press, Cambridge, MA.

Angell, J. R., and Pierce, A. H. (1892). Experimental research on the phenomena of attention. *American Journal of Psychology* **4**, 528–541.

Bandettini, P. A., and Ungerleider, L. G. (2001). "From neuron to BOLD: new connections." *Nat Neurosci.* **4**(9), 864–866.

Bergen, J. R., and Julesz, B. (1983). "Parallel versus serial processing in rapid pattern discrimination." *Nature* **303**, 696–698.

Berger, H. (1929). "Über das Elektrenkephalogramm des Menschen." *Archiv fuer Psychiatrie und Nervenkrankheiten* **87**, 527–570.

Berlyne, D. E. (1974). Attention, in "Handbook of Perception Volume 1, Historical and Philosophical Roots of Perception," edited by E. C. Carterette and M. P. Friedman, Academic Press, New York.

Brain, W. R. (1941). Visual disorientation with special reference to lesions of the right cerebral hemisphere. *Brain* **64**, 244–272.

Breese, B. B. (1899). On inhibition. *Psychological Monographs*, **3**(1).

Brentano, F. (1874). "Psychologie vom Empirischen Standpunkt." Meiner, Leipzig.

Broadbent, D. (1958). "Perception and Communication." Pergamon Press, NY.

Bushnell, K. C., Goldberg, M. E., and Robinson, D. L. (1981). Behavioral enhancement of visual responses in monkey cerebral cortex I: Modulation in posterior parietal cortex related to selective visual attention. *J Neurophysiol* **46**, 755–772.

Cajal, S. R. (1909). "Histology of the Nervous System of Man and Vertebrates." Translated by N. Swanson and L. W. Swanson. New York: Oxford University Press, 1995).

Cattell, J. McK. (1885). The inertia of the eye and the brain. *Brain* **8**, 295–312.

Cherry, E. C. (1953). Some experiments on the recognition of speech, with one and with 2 ears. *Journal of the Acoustical Society of America* **25**(5), 975–979.

Colby, C. L., and Goldberg, M. E. (1999). Space and attention in parietal cortex. *Annu. Rev. Neurosci.* **22**, 319–349.

Colegate, R. L., Hoffman, J. E., and Eriksen, C. W. (1973). Selective encoding from multielement visual displays. *Perception and Psychophysics* **14**, 217–224.

Crick, F. (1984). Function of the thalamic reticular complex: The searchlight hypothesis. *Proc. Natl. Acad. Sci. USA* **81**, 4586–4590.

Critchley, M. (1966). The enigma of Gerstmann's syndrome. *Brain* **89**, 183–198.

Dashiell, J. F. (1928). "Fundamentals of General Psychology." Houghton, Boston.

Descartes, R. (1649). "Les Passions de l'âme." Le Gras, Paris.

Desimone, R., and Duncan, J. (1995). Neural mechanisms of selective Attention. *Annual Review of Neuroscience* **18**, 193–222.

Deutsch, J., and Deutsch, D. (1963). Attention: Some theoretical considerations. *Psych. Review* **70**, 80–90.

Driver, J., and Mattingley, J. B. (1998). Parietal neglect and visual awareness. *Nat. Neurosci.* **1**(1), 17–22.

Duncan, J. (1980). The locus of interference in the perception of simultaneous stimuli. *Psychological Review*, **87**, 272–300.

Eriksen, C. W., and Collins, J. F. (1969). Temporal course of selective attention. *J. Experimental Psychology* **80**, 254–261.

Exner, S. (1882). ZurKenntniss von der Wechselwirkung der Erregungen im Centralnervensystem. *Archiv fuer die gesamte Physiologie des Menschen und der Tiere* **28**, 487–506.

Goldberg, M. E., and Bushnell, M. C. (1981). Behavioral enhancement of visual responses in monkey cerebral cortex II: Modulation in frontal eye fields specifically related to saccades. *J. Neurophysiol* **46**, 773–787.

Goldberg, M. E., and Wurtz, R. H. (1972). Activity of superior colliculus in behaving monkeys II: Effect of attention on neuronal responses. *J. Neurophysiol.* **35**, 560–574.

Groos, K. (1986). "Die Spiele der Thiere." Fischer, Jena.

Grossberg, S. (1975). A neural model of attention, reinforcement, and discrimination learning. *International Review of Neurobiology*, **18**, 263–327.

Grossberg, S. (1976). Adaptive pattern classification and universal recoding, II: Feedback, expectation, olfaction, and illusions. *Biological Cybernetics*, **23**, 187–202.

Groves, P. M., and Eason, R. G. (1969). Effects of attention and activation on the visual evoked cortical potential and reaction time. *Psychophysiology* **5**, 394–398.

Hamilton, W. (1859). "Lectures on Metaphysics and Logic, Vol. 1., Metaphysics." Blackwood, Edinburgh & London.

Harter, M. R., and Salmon, L. E. (1972). Intra-modality selective attention and evoked cortical potentials to randomly presented patterns. *Electroenceph Clin Neurophsyiol* **32**, 605–613.

Hawking, S. W., Miller, R., and Sagan, C. (2004). "A Brief History of Time: From the Big Bang to Black Holes." Bantam.

Hebart, J. F. (1824). "Psychologie als Wissenschaft neu Gegründet auf Erfahrung, Metaphsyik und Mathematik." Unzer, Konigsberg.

Heilman, K. M., and Valenstein, E. (1972). Frontal lobe neglect in man. *Neurology* **22**, 660–664.

von Helmholtz, H. (1886/1962). "Physiological Optics, Vol. 3." (3rd editor) Translated by J. P. C. Southall. NewYork: Dover.

von Helmholtz, H. (1896/1989). "Physiological Optics" (1896—2nd German edition, translated by M. Mackeben, from Nakayama and Mackeben, *Vision Research* 29:11, 1631–1647, 1989).

Hebb, D. O. (1949). "The Organization of Behavior." Wiley, New York.

Hernández-Peón, R., Scherrer, H., and Jouvet, M. (1956). Modification of electrical activity in the cochlear nucleus during attention in unanesthetized cat. *Science*, **123**, 331–332.

Hillyard, S. A., Hink, R. F., Schwent, V. L., and Picton, T. W. (1973). Electrical signs of selective attention in the human brain. *Science* **182**, 177–180.

Hobbes, T. (1655). "Elementorum Philosophiae Sectio Prima de Corpore." Crook, London.

Holt, E. B. (1915). "The Freudian Wish and its Place in Ethics." Holt, New York.

Horn, G. (1965). Physiological and psychological aspects of selective perception. In: "Advances in Animal Behavior" (D. S. Lehrmann and R. A. Hinde, Eds.), Vol. 1, 155–215. New York: Academic Press.

Itti, L., and Koch, C. (2001). Computational modelling of visual attention. *Nature Reviews Neuroscience Vol.* **2**, 194–203.

James, W. (1890). "Principles of Psychology." Holt, New York.

Jevons, W. S. (1871). The power of numerical discrimination. *Nature* **3**, 218–282, London.

Kahneman, D. (1973). "Attention and Effort." Prentice-Hall, Engelwood Cliffs, NJ.

Kastner, S., and Ungerleider, L. G. (2000). Mechanisms of visual attention in the human cortex. *Annual Review Neurosci.* **23**, 315–341.

Kinsbourne, M. (1970). The cerebral basis of lateral asymmetries in attention. *Acta Psychologica*, **33**, 193–197.

Koch, C., and Ullman, S. (1985). Shifts in selective visual attention: Towards the underlying neural circuitry. *Human Neurobiology* **4**, 219–227.

Köhler, W. (1947). "Gestalt Psychology: An Introduction to New Concepts in Modern Psychology." Livernight, New York.

Kulpe, O. (1902). Uber die objectivirung und subjectivirung von sinneseindruken. *Philosophische Studien* **19**, 508–536.

Leibnitz, G. W., (1765). Nouveaux Essais sur L'Entendement Humain, in R. E. Raspe (Ed.), "Oeuvres Philosophiques de feu M. Leibnitz." Screuder, Amsterdam & Leipzig.

Logothetis, N. K., Pauls J., Augath M., Trinath T., and Oeltermann A. (2001). Neurophysiological investigation of the basis of the fMRI signal. *Nature.* **412**(6843), 150–157.

MacKay, D. G. (1973). Aspects of the theory of comprehension, memory and attention. *Quarterly Journal of Experimental Psychology* **25**, 22–40.

Malebranche, N. (1674) "De la Recherche de la Vérité." Pralard, Paris.

Maunsell, J., and Ferrera, V. (1995). Attentional Mechanisms in Visual Cortex, in "The Cognitive Neurosciences." ed. by M. Gazzaniga, 451–461, MIT Press.

Metzger, W. (1974). Consciousness, Perception and Action, in "Handbook of Perception Volume 1, Historical and Philosophical Roots of Perception," edited by E. C. Carterette and M. P. Friedman, Academic Press, New York.

Milner, P. (1974). A model for visual shape recognition, *Psych. Rev.* **81**, 521–535.

Moran, J., and Desimone, R. (1985). Selective attention gates visual processing in the extrastriate cortex. *Science* **229**, 782–784.

Moray, N. (1969). "Attention: Selective Processes in Vision and Hearing." Hutchison, London.

Mowrer, O. H. (1938). Preparatory set (expectancy): A determinant in motivation and learning. *Psychological Review* **45**, 62–91.

Mountcastle, V., Andersen, R., and Motter, B. (1981). The influence of attentive fixation upon the excitability of the light-sensitive neurons of the posterior parietal cortex. *J. Neuroscience* **1**, 1218–1225.

Müller, G. E. Die Gesichtspunkte und die Tatsachen der psychophysischen Methodik. In L. Asher & K. Spiro (eds) Ergebnisse der Physiologie, 1903, Jahrgang II, Abtheilung II. 1904, p. 267–516.

Müller, G. E., Komplextheorie und Gestalttheorie. Göttingen: Vandenhoek & Ruprecht, 1923.

Naatanen, R. (1975). Selective attention and evoked potentials in humans—a critical review. *Biol Psychol* **2**, 237–307.

Neisser, U. (1967). "Cognitive Psychology." Appleton, New York.

Norman, D. (1968). Toward a theory of memory and attention. *Psych. Review* **75**, 522–536.

Noton, D., Stark, L. (1971). Scanpaths in eye movements during pattern perception. *Science* **171**(3968), 308–311.

Palmer, S. (1999). "Vision Science." MIT Press, Cambridge, MA.

Panum, P. L. (1858). In: Physiologische Untersuchungen ueber das Sehen mit zwei Augen. Kiel, Germany: Schwers.

Pashler, H. (1998). The Psychology of Attention. MIT Press, Cambridge, MA.

Pavlov, I. P. (1927). "Conditioned Reflexes." Oxford University Press, London & NewYork.

Petermann, B. (1929). "Die Wertheimer-Koffka-Köhlerische Gestalttheorie und das Gestaltproblem." Barth, Liepzig.

Pillsbury, W. B. (1908). "Attention." MacMillan, New York.

Poppelreuter, W. (1917). Die psychischen Schadigungen durch Kopfschuss im Kriege 1914/1916. Voss, Leipzig.

Posner, M. I. (1978). "Chronometric Explorations of Mind." Erlbaum, Hillsdale, NJ.

Poulton, E. C. (1953). Two channel listening. *J. of Experimental Psychology* **46**, 91–96.

Posner, M. I. (1980). "Orienting of Attention". *Quarterly Journal of Experimental Psychology* **32**, 1, 3–25.

Posner, M. I., and Cohen, Y. (1984). In *Attention and Performance*, Vol. X (H. Bouma and D. Bouwhuis, Eds.), 531–556. Erlbaum, Hillsdale, New Jersey.

Posner, M. I., Rafal, R. D., Choate, L. S., and Vaughan, J. (1985). Inhibition of return: Neural basis and function. *Cognitive Neuropsychology*, **2**(3), 211–228.

Rafal, R., Robertson, L. (1995). The Neurology of Visual Attention. *In* The Cognitive Neurosciences, Edited by MS Gazzaniga. MIT Press, Cambridge, MA.

Ribot, T. (1889). "La Psychologie del'Attention." Alcan, Paris.

Rosenblatt, F. (1961). "Principles of Neurodynamics: Perceptions and the Theory of Brain Mechanisms." Washington, DC: Spartan Books.

Sagi, D., and Julesz, B. (1985). Fast noninertial shifts of attention. *Spat Vis.* **1**(2), 141–149.

Sechenov, I. M. (1863/1965). "Reflexes of the Brain," translated by S. Belsky. Cambridge, Mass., MIT Press.

Shiffrin, R. M., and Schneider, W. (1977). Controlled and automatic human information processing II: Perceptual learning, automatic attending, and a general theory. *Psychological Review* **84**, 127–190.

Shulman, G. L., Remington, R., and McLean, J. P. (1979). Moving attention through visual space. *J. Experimental Psychology* **92**, 428–431.

Simson, R., Vaughn, H. G. Jr, and Ritter, W. (1977). The scalp topography of potentials in auditory and visual discrimination tasks. *Electroenceph Clin Neurophysiol* **42**, 528–535.

Sokolov, E. N. (1963). "Perception and the Conditioned Reflex." Pergamon, Oxford.

Spearman, C. E. (1937). "Psychology Down the Ages." MacMillan, New York.

Sperling, G. (1960). The information available in brief visual presentations. *Psychological Monographs: General and Applied* **74**(498), 1–21.

Spong, P., Haider, M., and Lindsley, D. B. (1965). Selective attentiveness and cortical evoked responses to visual and auditory stimuli. *Science* **148**, 395–397.

Sutherland, S. (1998). Book reviews. *Nature* **392**(26), 350.

Titchener, E. B. (1908). "Lectures on the Elementary Psychology of Feeling and Attention." MacMillan, New York.

Treisman, A. (1964). The effect of irrelevant material on the efficiency of selective listening. *American J. Psychology* **77**, 533–546.

Treisman, A. (1969). Strategies and models of selective attention. *Psychological Review* **76**, 282–299.

Treisman, A., and Gelade, G. (1980). A feature integration theory of attention. *Cognitive Psychology* **12**, 97–136.

Treisman, A., and Schmidt, H. (1982). Illusory conjunctions in the perception of objects. *Cognitive Psychology* **14**(1), 107–141.

von der Malsburg, C. (1981). The correlation theory of brain function, Internal Rpt. 81–2, Dept. of Neurobiology, Max-Planck-Institute for Biophysical Chemistry, Gottingen, Germany.

von Tschisch, W. (1885). Uber die Zeitverhältnisse der Apperception einfacher und Zusammengesetzter Vorstellungen, untersucht mit Hülfe der Complications-methode. *Philosophische Studien (Wundt)* **2**, 603–634.

Watson, J. B. (1919). "Psychology from the Standpoint of a Behaviorist." Lippincott, Philadelphia.

Wolfe, J. (1996). Visual search, in "Attention" (ed. Pashler, H.), 13–74, University College London, London.

Wolff, C. (1734). "Psychologia Rationalis." Renger, Frankfort & Leipzig.

Wundt, W. (1874). Grundzüge der Physiologischen Psychologie. Engelmann, Liepzig.

Wurtz, R. H., Goldberg, M. E., and Robinson, D. L. (1982). Brain mechanisms of visual attention. *Scientific American*, **246**(6), 100–107.

Yarbus, A. L. (1967). "Eye Movements and Vision." New York: Plenum.

A Tour of This Volume

Michael Tombu, Neil Bruce, Albert Rothenstein, John K. Tsotsos

INTRODUCTION

This tour of the chapters in this volume is organized along the following lines, representative of the topics our authors have covered:

Conceptual Issues
   Attention
   Saliency
   The Binding Problem
Brain Structures
   Anatomy
   Neurochemistry
Perception and Action
   Attention and Perception
   Inhibition of Return
   Preattentive Features
   Top-Down Control
   Visual Attention and Eye Movements
   Interplays between Perception and Action
   Object and Scene Understanding
   Development and Learning
Brain Mechanisms
   Enhancement and Inhibition
   Cross-Modal Attention
Models of Attention
   Biologically Realistic Models
   Machine Vision

The goal of this chapter is to provide an overview, summarizing very briefly each chapter and suggesting some reading trajectories through the book. But this is not the only way the chapters of this book may be organized. As one reads through the book, it becomes abundantly clear that there are many opinions about what is important and thus how one might organize a broad overview of attention. Moreover, it will be also clear that it is not so easy to pigeon-hole the chapters along any lines, let alone the ones above. It is often the case that one cannot address only one of the above topics in isolation in any research effort. Many of the chapters need to touch on several of the above aspects. We have tried to provide multiple homes for such works within each of the above headings when the additional topic is a major element of the chapter. It is important to remember that our organization strategy is subjective, artbitrary, and reflects our own prejudices!

So, if there are so many differing opinions on attention, why bother with this volume and our attempt to provide a comprehensive view of the field? It may be that what is required for better progress is "the big picture"—but what is the big picture for attention? Perhaps one of the papers in this volume might point the way. Perhaps some insightful graduate student will see the differing points of view represented here and draw a key connection. Regardless of all the negative viewpoints presented over the years and all the conflicting theories, the topic is still one that intrigues many; and it is clear the enthusiasm and energy that we devote is not waning. Such a diversity of opinion is in our opinion very positive for the field. Not only does it stimulate ingenuity, inter-disciplinarity, and a wide diversity of empirical work, but it may well be the case that a full explanation of attention is not possible from a single view alone. This volume is presented with the hope that it will make a small but positive step towards our complete understanding of the phenomenon of attention.

1.0 Conceptual Issues

1.1 Attention

Visual attention is often operationalized as a selection mechanism responsible for filtering out unwanted information in a visual scene. In this book Tsotsos (1) uses computational complexity proofs to quantify the pervasive, but often unsupported, claims that capacity limits exist in vision. These proofs lead to constraints on biologically plausible architectures for visual attention, and demonstrate that the brain is not solving the generic vision problem, but reshapes it to make it solvable by the available processing power. Tsotsos casts vision as a massive search problem, but there are other, complementary, points of view to consider. Anderson et al. (3) present the dynamic routing model of visual attention, with emphasis on the circuitry for translation and scale invariance, and the reference frames used in the process. The major predictions and supporting physiological evidence are briefly reviewed, together with some open questions. Baldi (5) presents the key ideas of his computational theory of surprise, and discusses its potential importance for attention research, arguing that over short time scales, relevance and surprise have significant overlap. The author concludes that surprise can be used as a computational shortcut to relevance, which in turn must await confirmation by additional and slower processes.

Many people equate attention with awareness. Lamme (29) examines the differences between attention and awareness. He proposes that visual attention is the product of visual information on its initial feed-forward pass through the visual system interacting with memory that guides and shapes how this information is processed. For an organism to become aware of a stimulus, it is not enough for a stimulus to activate high-level areas such as frontal cortex; there must be recurrent processing between V1 and extrastriate areas.

Lu and Dosher (74) distinguish three classes of attentional mechanisms, each with a characteristic performance pattern. They partition experiments in the literature according to the mechanisms they act upon, producing a task taxonomy. Mozer and Vecera (23) discuss the nature of object-based attention and its interaction with spatial attention drawing the important distinction between types of attention.

Of course, some argue that the effects observed empirically may have explanations that do not require attention at all. Although selective attention is a computational necessity for the visual system and is generally viewed as being responsible for the facilitation provided when a precue immediately precedes a target, Morgan and Solomon (2) outline recent findings on the effect of precues and mask-onset asynchronies in visual search experiments. They conclude that pre-cueing benefits are not necessarily a consequence of focal attention, and suggest that spatial cues help observers identify the most appropriate local mechanisms for spatial discrimination. Recent ERP findings suggest that it is possible to perform some object recognition tasks in the absence of attention. Riesenhuber (46) outlines a purely feed-forward model that can account for this data. The general framework of the model is described and implications for visual attention are discussed.

1.2 Saliency

Determining which stimulus in a visual display is the most important to an observer requires the integration of top-down and bottom-up factors to discover which stimulus is the most salient. Saliency plays an important role guiding attention in many models of visual attention, either via a saliency map or distributed representation of saliency. Determining exactly how the visual system computes saliency is therefore an important goal in the study of visual attention.

The issue of salience from feature contrast is a ubiquitous phenomenon in the attention literature but remains a contentious issue, especially from a neurobiological perspective. Nothdurft (38) considers the relationship between response variability in the primary visual cortex, and perceptual salience. A number of experiments are presented that support the idea that salience is reflected in response differences of cells in V1, but also that salience is necessarily encoded among other visual areas, indicating a distributed representation of salience. Krummenacker and Müller (68) review recent results on the role of feature- and dimension-specific processes underlying pop-out in visual search. These findings also indicate that the saliency computation is not a simple automatism, but that it is spatially specific and subject to bottom-up learning and top-down modulation. Apparently contradictory findings are presented by Theeuwes (69), indicating that saliency is determined in parallel across the visual field, irrespective of top-down goals.

VanRullen (45) argues that representations in the ventral processing stream are intrinsically biased to reflect more salient stimuli, as opposed to via a saliency map. At each layer of the ventral stream, the most salient stimuli have the highest activity. Because the input to successive stages emphasizes what was salient at the previous stage, representations are biased to favor salient stimuli. In addition, top-down factors can also affect saliency by lowering thresholds to relevant stimulus locations or features.

Zelinksy (65) proposes that the term attention is underspecified and can be subdivided into its component parts. A model for performing visual search is proposed. In the first component of the model the target relevant features are identified and these features are used to guide search by correlating points in the search array with the target, which acts as a saliency map. In the second component a threshold that increases over time is applied to the saliency map and the centroid is constantly calculated. Once the centroid exceeds a minimum distance from the current locus of fixation, an eye movement is executed. This process repeats until the target is fixated. Parkhurst and Niebur (39) explore the nature of bottom-up, stimulus driven mechanisms of attention. Attentional allocation in complex natural scenes is observed by measuring the eye movements made by observers. A comparison is made between stimulus salience, as determined by a biologically motivated computational model of the primate visual system, and fixation locations, indicating that there exists a significant correlation between fixation locations and stimulus salience. Zhaoping (93) considers the notion that a saliency map resides within the primary visual cortex. A model is presented that describes contextual influences in V1 based on layer 2 and 3 pyramidal cells, horizontal intracortical connections, and inhibitory interneurons. The model predicts a variety of experimental phenomena related to pop-out, conjunction search, and search asymmetry. An opposing view comes from Tsotsos (92), who does not subscribe to a single saliency map model, instead arguing that salience is a distributed localized computation.

Finally, Itti (94) reviews a number of computational architectures that focus on the bottom-up deployment of attention. A historical perspective on models that revolve around the notion of saliency is presented, including discussion of how such models have been employed to subserve recognition tasks.

1.3 The Binding Problem

One important question about object identification is how different parts or attributes of an object are bound into a whole. This problem is known as the *binding problem* and several authors propose solutions to this elusive problem in this book.

Given that different object properties are processed in a distributed fashion in the human brain and that damage to different brain areas can selectively disrupt processing of individual object properties while leaving others intact, the binding problem asks how distributed object properties are integrated by the human brain to form a single percept. Robertson (24) investigates links between spatial attention and the binding problem, exploring evidence from patient populations and relating them to theory. Mozer and Vecera (23) suggest that object-based attention can be explained by an attentional mechanism that groups low-level features into objects.

Martínez and Hillyard (84) review a number of ERP studies, comparing the selection of locations and features. They suggest that the integration of features into objects is contingent upon the prior selection of spatial location, which acts as a gain control over perceptual processing. Proverbio and Zani (82) also review a series of ERP studies, and draw conclusions on the interaction between the dorsal and ventral processing streams, as well as on the nature of the binding process needed to perceive multidimensional objects. Focusing on a different brain area, Humphreys and Riddoch (44) examine neuropsychological evidence indicating that the parietal lobe is important in binding information computed in different areas and that there may be different ways in which information can be bound.

Evidence of the action of binding is presented by O'Craven (66), who shows that neural activity related to unattended features of attended objects is enhanced. On each trial of an fMRI experiment, participants were presented with a transparent face and house stimulus superimposed upon one another in which either the house or the face stimulus was in motion. Participants attended to either the house, the face, or the moving stimulus and indicated if the stimulus was the same as on the previous trial. The resulting pattern of brain activity indicates that attention is allocated to objects and spreads to unattended features of attended objects, but not to unattended objects that share spatial position with the attended object.

Several authors propose algorithms and mechanisms that are claimed to solve the binding problem. Roelfsema and Spekreijse (47) propose an algorithm for binding contours into object representations. The initial feed-forward sweep of visual information provides a neural activation map to which the grouping criteria of connectedness and co-linearity are applied to form object representations. Visual attention, in the form of enhanced firing rates, binds the properties of the object to one another. In contrast, Tsotsos (92) puts forward a different suggestion. He proposes that localized, distributed saliency computations are sufficient for feature binding required for complex motion recognition. A recurring theme in these chapters is that synchronization of neural firing may represent a mechanism of attentional selection as well as binding. Singer (87) reviews the evidence that synchronized neural firing is used by the brain to enhance the saliency of neural responses, for perceptual grouping, and for the definition of relations in memory. Steinmetz et al. (88) review the literature on somatosensory attention, with emphasis on the effect of attention on the synchronization of neural firing. Periodic and non-periodic synchronization are discussed. Fries and Desimone (86) study the effect of attention on the synchronization of V4 neurons, and find an enhancement of gamma-frequency, and a reduction of low-frequency synchronization. The potential mechanisms and importance of these effects are discussed.

2.0 Brain Structures

2.1 Anatomy

Determining the centers involved in the control of attention is critical to any complete understanding of attention. There is basic agreement on the existence of a distributed network of brain areas associated with attention. However, the precise functions subtended by each element of that network remain unclear. Serences et al. (7) review the evidence for the involvement of the parietal cortex in the goal-directed control of attention. This review is presented within the framework of the biased competition model. Giesbrecht and Mangun (11) present a meta-analysis of a series of studies, with the goal of disambiguating the attentional orienting response from other cognitive operations evoked by cueing. The study suggests that bilateral subregions in the superior frontal sulcus and intraparietal sulcus play a crucial role in the control of attentional orienting. Mesulam et al. (6) present clinical evidence for a distributed and heterogeneous set of centers involved in the control of attention, and interpret the various symptoms of hemispatial neglect within this framework, labeling it a "network syndrome." Kastner et al. (72) review brain imaging evidence for attentional modulation at the thalamic level, both in the lateral geniculate nucleus and in the pulvinar. These findings demonstrate the usefulness of fMRI in the study of subcortical nuclei, despite their small sizes and deep locations. Büchel (75) reviews the basic concepts of effective connectivity in neuroimaging, and introduces several mathematical methods to assess it. Evidence is presented that modulatory influences from parietal cortex are sufficient to account for a significant component of the attentional modulation of V5/hMT responses to inputs from V1/V2. Vandenberghe (67) describes closely matched location and feature selection experiments, in an attempt to determine which specific brain areas are involved in each case. Results are discussed in the context of previous findings, and explanations provided for discrepancies.

The attentional blink (AB) is a transient impairment in detecting the second of two targets in a rapid serial visual presentation stream when the second target follows the first by 500 ms or less. ERP and fMRI results reviewed by Marois (63) indicate that the capacity limitation associated with the AB is not located in the visual cortex, but instead in the parieto-frontal network. Specifically, second targets that are not consciously reported nonetheless still activate higher areas of visual cortex, such as the parahippocampal place area, but only correctly reported second targets activate the lateral frontal cortex. Humphreys and Riddoch (44) suggest that the parietal lobe plays an important role in binding based on neuropsychological observations.

One approach to the study of visual attention is to examine the consequences of cortical damage to the visual system. Impairments can be studied in infants with developing visual systems, as well as in adults. In adults impairments can be induced or they can be studied in patients with brain damage. In this book all of these approaches to the study of visual attention are described. Manly and Robertson (55) describe an experimental paradigm that characterizes properties of attentional drift. The task requires a key press response following each presentation of a single random digit; and in 11% of trials, a no-go response to a predetermined digit. The task is designed to allow the possibility for subjects to shift into a mode of absentminded, automatic response circumvented only by the effort to sustain attention. It is demonstrated that the prefrontal cortex appears to play a significant role in sustaining attention. Deouell and Knight (56) present data from studies involving event related potentials that elucidate the role of the lateral prefrontal cortex (LPFC) in attentional control. The LPFC is suggested as a key factor in sustaining attention, detection of novelty and change, and inhibition and enhancement of information. Some of the aforementioned functions may reflect interaction between remote brain areas. Studies concerning lateralized deficits following right hemisphere stroke abound in the literature. Husain (57) discusses several recent studies that also observe deficits affecting both ipsilesional and contralesional sides of space. Non-spatially lateralized deficits include a more severe form of attentional blink, difficulty in encoding transiently presented visual stimuli, and difficulties in sustaining attention. Inferior parietal and lateral frontal cortices are implicated as regions responsible for non-spatially lateralized mechanisms involved in neglect. Vuilleumier (58) describes recent behavioral and neuroimaging studies on spatial neglect and extinction. Although right parietal damage may eliminate conscious perception in contralesional space, brain activity related to contralesional stimuli may still be observed in ventral occipito-temporal pathways. Visual awareness appears to correlate with greater activity among a distributed network of visual areas and, in particular, frontoparietal regions. Handy and Gazzaniga (59) consider the effect that severing the corpus callosum, which disrupts inter-hemispheric communication, has on the control of attention. Experiments are reviewed in which split-brain patients perform a variety of tasks involving visual search, visual orienting, endogenous versus exogenous deployment of attention, and attentional allocation during dual-task situations. The form of attention required, or neural processes demanded by the attention task, appear to account for variations in hemispheric control of such processes. When many copies of the same stimulus are presented to a subject in a go-no-go task, response times are faster than those resulting from presentation of a single copy of the stimulus, a result known as the redundant target effect. Iacoboni (60) details experiments that offer insight concerning the paradoxical finding that the redundant target effect is larger in split-brain patients when two copies of the same stimulus are split between hemifields. A number of behavioral and imaging studies are described with results that may account for the observed paradox.

2.2 Neurochemistry

Studying the effects of drugs on attentional processing can also help to elucidate processing associated with attention. Coull (9) investigates the effect of drugs acting on the noradrenergic or cholinergic neuromodulatory neurotransmitter systems on human attentional performance. Despite broad differences, there is insufficient evidence to propose a neurochemical specialization of these two neurotransmitter systems for distinct attentional processes, and future directions of study are proposed. Milstein et al. (10) survey evidence for the neurochemical modulation of attentional systems in animals, with the focus primarily on rat neurochemistry and behavioral pharmacology. The main distinctions between the effects of acetylcholine, noradrenaline, and dopamine are presented.

3.0 Perception and Action

3.1 Attention and Perception

One important role of the visual system is to quickly locate salient objects in our environment. Given that the visual environment usually contains more information than the visual system can handle at any one time, a mechanism, selective attention, is necessary to guide the visual system to objects of interest. This can be accomplished in a number of ways, and both exogenous bottom-up and endogenous top-down factors play a role. Bottom-up factors usually refer to properties of the stimulus, whereas top-down factors refer to the goals of the observer. Several such factors are examined in this book.

Attentional capture, the observation that task-irrelevant stimuli can nonetheless involuntarily impair performance, is examined and critiqued by Ambinder and Simons (12). Implicit attentional capture demonstrates that a salient yet irrelevant distracter can influence task performance, although the irrelevant distracter may be affected by top-down factors. Top-down factors appear to also play a role in explicit attentional capture where the nature of the task performed by subjects influenced the subject's ability to detect infrequent irrelevant objects displayed along with the target stimulus.

Change blindness describes the failure to notice changes in a visual scene that occur in plain view, even when expected, in the absence of focal attention. Interestingly, when searching for a changing item among non-changing items, four items can be attended simultaneously, whereas only one item can be attended when searching for a non-changing item among changing objects. Rensink (13) uses this search asymmetry, as well as other results from the change blindness paradigm, to draw far-reaching conclusions regarding attention, the representation of information, and the processes involved in perception.

Prior entry refers to the finding that attended stimuli are perceived prior to unattended stimuli and usually involves presenting observers with two stimuli to which subjects must indicate which was presented first (or second). Shore and Spence (15) discuss confounds with this paradigm that have led some researchers to argue that prior entry is the result of a response bias, as well as more recent research that controls for potential confounds. The results suggest that while response bias plays a role in prior entry, a residual prior entry effect remains. Possible mechanisms for prior entry are discussed.

Griffin and Nobre (42) review a number of experiments pertaining to temporal cueing, which indicate that temporal expectancies may be flexibly used to allocate attention selectively to a cued temporal window, improving task performance. Studies concerning neural mechanisms of temporal orienting, and implications of temporal phenomena in attention research are discussed. The need for careful consideration of temporal factors in attentional paradigms is raised given the pervasive influence of temporal expectancies.

Gorea and Sagi (27) review signal detection theory and examine the consequences of making multiple simultaneous decisions about perceptual events. When subjects perform two visual detection tasks with different contrast sensitivities, they show that a single response criterion is employed, which results in over reporting of the higher contrast stimulus and under reporting of the lower contrast stimulus. These results are related to selective attention and extinction.

While attention appears to have a beneficial impact on attended stimuli, a consequence of this selection is that unattended stimuli are filtered out of later visual processing and, as a result, unattended stimuli are not able to impact upon later processing. Research has demonstrated this prediction in numerous domains; however an exception is the area of emotional faces, to which researchers have given special significance. Pessoa and Ungerleider (28) review studies that demonstrate that when attention is fully occupied by another task, even the processing of emotional faces is eliminated, thus demonstrating that like other types of visual processing, processing of emotion also requires attention.

3.2 Inhibition of Return

In a simple cueing paradigm subjects fixate a central location and are presented with a cue at one of two peripheral locations, which is followed at some delay by a target in one of the peripheral locations. At short cue-target intervals objects respond to validly cued targets faster than invalidly cued trials (facilitation); whereas, at long cue-target intervals inhibition of return (IOR), slower responses to targets at the cued than at the uncued location are observed. Klein and Ivanoff (16) discuss some of the factors affecting IOR and its possible role as a foraging facilitator. Fecteau et al. (64) show that the temporal dynamics of facilitation and IOR, as well as the spatial frame of reference (environmental vs. retinotopic) depend on the effector (manual vs. saccadic) used to respond to the target. Facilitation and inhibition result from changes in the sensory representation of the target. Observed differences related to the effector used to respond to the task are argued to result from differences in how different networks use this visual information. Recent studies appear to contradict the traditional belief that IOR allows visual search to be conducted in a manner that results in rejected distractors *never* being revisited. Horowitz (43) reviews a number of experiments that challenge the classic view of "sampling without replacement" using multiple target search, attentional reaction time methods, and the method of randomized search. Horowitz discusses the possibility that a compromise between a memoryless search and the standard IOR model may resolve controversies that arise in the context of these experiments.

Richards (14) examines the development of IOR in infants and finds that only older infants show IOR.

3.3 Preattentive Features

Early visual processing is massively parallel. However, due to subsequent convergence onto a single processing stream, stimuli must be selected for additional processing, while others are filtered out. This act of selection is often referred to as visual attention. Wolfe (17) explores the preattentive features that guide this selective process and compiles lists of feature dimensions, dividing them into probable, possible, and improbable preattentive feature dimensions. Atchley (19) reviews research that examines attention in the depth dimension and concludes that attention can be focused not only on a 2-dimensional (2D) location, but also to a point in depth likely in a viewer centered frame of reference. While attention in depth has a greater extent in simple visual displays, it becomes more focused in more complex situations. As is the case in 2D situations, attention spreads along objects that extend in depth. Freeman (79) presents a number of recent studies on the influence of attention in flanker tasks. These results indicate that attention modulates early vision mechanisms sensitive to collinear structures. Rezec and Dobkins (81) present a comparative review of a wide variety of results in the study of the influence of attention on motion processing, including attempts to correlate psychophysical and neurophysiological effects. Zenger-Landolt (76) compares the results of a detection task under various attentional loads. This analysis results in a simple model in which distracters contribute in a non-linear way to the target's gain control. Chubb (77) discusses texture processing, proposing that it is based on a relatively small number of "visual substances," whose linear combinations are used to make perceptual decisions. The role of attention in modulating this combination is investigated.

3.4 Top-Down Control

An important feature of the visual system is that it can be guided by the goals and preferences of the organism in question. Understanding how these top-down effects are implemented can provide insights into how the visual system interacts with other cortical regions.

It is one thing to conceive of top-down information being useful for visual selection, but quite another to delineate where such information comes from in the brain. Bar (25) proposes a top-down mechanism for the facilitation of object recognition. According to his hypothesis, coarse low spatial frequency information from the magnocellular pathway is passed directly from early visual areas to prefrontal cortex. Prefrontal cortex generates a subset of potential objects based on this low spatial frequency information that is then passed to temporal cortex to guide object recognition. A similar strategy is proposed in the next contribution. Chun (40) considers the manner in which top-down information derived from perceptual experience impacts on the guidance of attention, a phenomenon termed contextual cueing. The point is made that scenes contain a number of regularities such as stability in the spatial layout of objects as a function of time, and predictability in movement patterns. Experiments are described in which contextual cueing is incorporated in the form of a variety of regularities in the stimulus presentation, such as the repetition of a particular spatial arrangement of target and distracters. It is shown that such regularities improve search performance, demonstrating that contextual cueing aids in the top-down guidance of attention. Another view of the source of top-down control appears in the chapter by Serences et al. (7) focusing on the locus of top-down control, reviewing evidence for the involvement of the parietal cortex in the goal-directed control of attention.

How does top-down information affect perception? The earlier discussion of attentional capture by Ambinder and Simons (12) mentioned the importance of top-down factors for both implicit and explicit attentional capture. Implicit attentional capture demonstrates that an irrelevant distracter may be affected by top-down factors. In addition, top-down factors appear to also play a role in explicit attention capture where the nature of the task performed by subjects influenced the subject's ability to detect infrequent irrelevant objects displayed along with the target stimulus. Using fMRI, Kastner and Beck (50) examine competition for neural representation among stimuli. They show that stimuli compete, by mutual suppression, for representation in visual cortex. Asking subjects to attend to one of the stimuli can substantially reduce these suppressive interactions, and spatial attention also results in increases in baseline firing rates at the attended location prior to stimulus presentation. These findings demonstrate the top-down influences of spatial attention on processing in visual cortex.

Finally, Frith (18) examines the different meanings of "top-down" and argues that they are not always compatible. At a physiological level "top-down" refers to reentrant processing, whereas at the psychological level it refers to biasing processing in favor of one class of stimuli. The links between working memory and top-down control, as well as what constitutes the top of top-down control, are discussed.

3.5 Visual Attention and Eye Movements

Eye movements are important for shifting overt attention within a visual scene as they refocus the fovea on areas of interest. Research indicates that neural structures responsible for overt eye movements are also implicated in covert attention. Several chapters review important findings in this area.

While many studies have focused on examining covert attention independent of overt attention, Findlay (20) stresses the point that in most real world situations, covert attention is almost always followed by a saccade to the focus of covert attention, resulting in overt attention at the target location. Findlay stresses that the likely primary purpose of covert attention for humans is to provide the visual system with a preview of the location that is the target of a saccadic eye movement. This preview allows the visual system to begin analyzing the target and if it is not a desired target, begin planning its next saccade.

Bichot and Schall (21) examine two classes of neurons in the frontal eye field (FEF). Research is reviewed that concludes that there is a class of visually responsive neurons that code for behavioral relevance, taking both bottom-up and top-down factors into account. These neurons produce a "saliency map" of the visual scene that predicts the latency and the location of a saccade, but they are not involved in saccade production for which movement related neurons appear to be responsible.

Thompson (22) examines the oculomotor and visual system properties of the FEF. Evidence dissociating visual selection from saccade production in the FEF is reviewed. Visually responsive neurons indicate the location of behaviorally relevant stimuli regardless of whether a saccade is produced. Zetzsche (37) considers the strategy underlying saccadic fixations in the context of the statistical properties of fixated regions. Consideration of higher order statistics suggests that non-redundant image features such as contours, edges, and occlusions are an important factor in the selection of fixation points. Zetzsche concludes that the strategy underlying saccadic fixation is in some measure determined by statistical regularities of natural scene statistics.

3.6 Interplays between Perception and Action

Once objects are identified they must also be represented into a knowledge structure so that the observer can navigate and interact with the environment as well as make decisions about objects. The traditional view of environmental knowledge posits a hierarchical network of spatial representations. When spatial relations are updated within one spatial representation, updating spreads to other representations. Brockmole and Wang (26) reviews evidence that suggests that this is not the case. Instead representations appear to be independent and representation switching is required to update other representations.

Another important function of the visual system is to link perception with action. In fact, there is evidence that perception and action processing are influenced by one another in the human brain. Several authors in this book review evidence supporting this claim. Tipper (30) examines the links between saccade production and reaching. In studies of visually guided reaching, the path of a saccade to a target is affected by factors that affect reaching, but that have no effect on the path a saccade takes in the absence of reaching. This result argues that there is interaction between neural systems that controls eye and hand movements. Craighero and Rizzolatti (31) discuss the pre-motor theory of attention, according to which spatial attention results from activation of motor circuits that would be used in the event of a movement to the attended location. Via feedback, systems processing at the attended location in extrastriate and striate areas is also enhanced. Imaging, micro-stimulation, and behavioral evidence supporting the pre-motor theory of attention are discussed.

3.7 Object and Scene Understanding

In order to be able to understand complex scenes the visual system must be able to locate points of interest quickly and direct spatial attention to them. A remarkable talent of humans is the ability to recognize the context of a scene, objects, and other salient visual information at a single glance. Several contributions focus on different aspects of this recognition process. Oliva (41) describes a series of efforts aimed at qualifying the image properties that enable such efficient categorization, or sampling of what is termed the *gist* of the scene. Coarse spatial scale information appears to allow enough structural cues to arrive at a reliable perceptual *gist*. Navalpakkam et al. (33) propose a four phase model for scene understanding that incorporates both bottom-up and top-down guidance of visual attention to find areas of interest in a scene. A fast global process captures the gist of the scene, while visual attention is applied to salient locations for additional processing. Information acquired during each attentional fixation is then used to update the internal representation of the scene. Arguing against such a viewpoint, Riesenhuber (46) proposes a purely feed-forward model that can perform object recognition without attention. Proverbio and Zani (82) review a series of ERP studies, and draw conclusions on the interaction between the dorsal and ventral processing streams, as well as on the nature of the binding process needed to perceive multidimensional objects. Roelfsema and Spekreijse (47) propose an algorithm for binding contours into object representations. The initial feed-forward sweep of visual information provides a neural activation map to which the grouping criteria of connectedness and co-linearity are applied to form object representations. Visual attention in the form of enhanced firing rates bind the properties of the object to one another. Evidence supporting this algorithm is discussed.

According to scanpath theory, a top-down cognitive model of a visual scene is constructed that contains information about the scene, the locations of regions of interest (ROIs), which are likely determined by salience, and a sequential ranking of the ROIs to be examined with eye fixations. Top-down controls execute eye fixations to the ROIs whereas bottom-up information is compared to the hypothesized visual scene from the cognitive model. Privitera and Stark (48) discuss the importance of understanding how attentional shifts operate in humans for the computer vision community. The fMRI studies of Ress and Heeger (51) indicate that cortical activity agrees with behavioral performance, perhaps revealing that activity in visual cortex is representative of a subject's perception of a scene and not merely a representation of the sensory environment. Bar (25) proposes a top-down mechanism for the facilitation of object recognition. According to this hypothesis, coarse low spatial frequency information from the magnocellular pathway is passed directly from early visual areas to prefrontal cortex. Prefrontal cortex generates a subset of potential objects based on this low spatial frequency information that is then passed to temporal cortex to guide object recognition.

Object recognition that employs attention and object-based attention are very different. Mozer and Vecera (23) discuss the nature of object-based attention and its interaction with spatial attention. Views of object-based attention vary from viewer-centered groupings of low-level features, to object-centered representations that decompose complex objects into their component parts. A review of the data suggests that object-based attention effects can be explained by an attentional mechanism that groups low-level features into objects, while retaining a viewer-centered frame of reference. This view is computationally the simplest and therefore the most parsimonious. As previously discussed, O'Craven (66) shows that unattended features of attended objects show neural enhancement associated with attention in neural areas specialized for unattended features.

3.8 Development and Learning

Studying the developing visual system can provide insights into its function. Two such studies are discussed in the book, one of which has been reviewed previously. Richards (14) examines whether infants can covertly shift attention and explores the neural mechanisms that are involved. Infants as young as 3 months of age show cue-related facilitation; however, only older infants show IOR. The superior colliculus, which is relatively mature at birth, may be involved in response facilitation, whereas the frontal eye fields and fusiform gyrus may be involved in IOR. Adler (34) reviews a number of studies that consider the development of attentive mechanisms in infants. Particular focus is placed on the emergence of mechanisms for visual search and pop-out. Adler concludes that pop-out behavior appears to be exhibited as early as 3 months of age whereas serial search is typically not observed prior to 6 months of age.

Billock et al. (4) explore the idea of using selective attention to guide an unsupervised learning algorithm. They choose a robot arm guidance task as a test-bed, and discuss the implications of their results in the context of the biological functions of attention and awareness. The authors conclude with a brief overview of the categories of problems for which this approach is appropriate. Dosher and Lu (78) review the literature on visual perceptual learning, separating two mechanisms: tuning of the task and enhancing the stimulus. A brief overview of the role of attention in perceptual learning is presented. Dayan (35) considers two possible ideal learner models in the context of animal conditioning. Learning is allocated on the basis of relative uncertainty regarding associations, or relative reliability of stimuli. Selective attention emerges as a consequence of statistical optimality, and not any notion of limited capacity.

4.0 Brain Mechanisms

4.1 Enhancement and Inhibition

While models of attention abound in the literature, most hypothesize that attention biases competition for neural representation in the visual system either in favor of attended stimuli, against unattended stimuli, or some combination of the two. Determining the exact form of the attentional modulation in visual cortex can help to restrict possible models of visual attention to those that predict the pattern of modulation observed in visual cortex. Investigators have used several approaches to characterize how attention modulates processing in the visual cortex, including fMRI, event related brain potentials (ERPs), and single unit recording. In addition, these methods can also be used to establish similarities and differences between potentially different forms of attention, such as space- and feature-based attention. In this book, several investigators review results that demonstrate attentional modulation in different areas of visual cortex. Attention can enhance baseline firing rates at an attended location prior to the presentation of a target, enhance the representation of an attended stimulus, and suppress representations of unattended stimuli. Determining the topography of attentional enhancement and suppression is another promising use of fMRI technology that can help to further elucidate how attention is manifested in the visual system.

The interaction of feature-based and spatial attention is considered by Martinez-Trujillo and Treue (49), who describe single cell recordings in macaque area MT. Spatial attention inhibits neural firing at the unattended location and enhances it at the attended location. Similarly, attending to a neuron's preferred direction of motion enhances firing rates, while attending to a neuron's null direction inhibits firing rates—even when attention is directed to a stimulus outside of the receptive field of the neuron under investigation. Interestingly, space- and feature-based attention appear to operate in similar manners and their effects are additive. Reynolds (8) further examines spatial attention, drawing a parallel between the effects of spatial attention and contrast-dependent response modulations. He reviews evidence indicating that spatial attention improves stimulus processing of a lone stimulus in a manner equivalent to a 51% increase in stimulus contrast. In addition, attending to one of two stimuli in a neuron's receptive field has similar effects to increasing the contrast of the attended stimulus. Both phenomena appear to affect processing by increasing the suppressive effect of the attended/higher contrast stimulus on the unattended/lower contrast stimulus.

Suppressive effects of attention are investigated in the next three contributions. Kastner and Beck (50) provide strong evidence from fMRI for mutual suppression when stimuli are in competition for neural representation. Asking subjects to attend to one of a number of stimuli reduces suppressive interactions, and directed spatial attention leads to increased baseline firing rates at the attended location prior to stimulus presentation. Murray (80) investigates the hypothesis that attention sharpens population-level responses, thus restricting activity to those neurons that code for specific features of the stimulus. Implications and possible mechanisms are discussed. Supèr (83) investigates primary visual cortex neural responses in visual memory tasks, and the effect of attention on these responses. He finds differential enhancement and suppression of activity, consistent with a push-pull mechanism of attention. Orban et al. (71) present evidence for suppressive attention-dependent modulation of activity in the early levels of the visual system, and the role of the reticular thalamic nucleus in this effect. A computational model supporting these findings is also presented. Carrasco (73) presents three studies that investigate the effects of attention on early visual processes. The results show that attention increases contrast sensitivity and spatial resolution via signal enhancement, a conclusion supported by earlier single-unit and fMRI studies. Martinez-Trujillo and Treue (70) present evidence that the effect of attentional modulation is similar to that of changes in the stimulus contrast. In addition to the known result that both modulations cause a multiplicative scaling of responses, the non-linearity of this effect is also shown to be common. Hopfinger (36) outlines recent findings concerning the neural dynamics of reflexive attention derived from studies utilizing ERPs. Experiments indicate that reflexive attention manifests in an initial facilitation of sensory processing among early and late stages of sensory processing. Such facilitation was quickly followed by a relative inhibition for cued location stimuli.

It was not so long ago that it was difficult to conceive of attention affecting individual sensory neuronal responses, let alone affecting some specific aspect of early visual processing. Yet that is the current understanding, and the following contributions further elaborate this. Somers and McMains (62) review experiments showing attentional modulation in visual cortex as early as area V1. This attentional modulation takes at least two forms: enhancement of an attended stimulus and inhibition of an unattended stimulus. This pattern of results is inconsistent with spotlight models of attention, which do not predict inhibition of unattended stimuli, but is consistent with the Selective Tuning (92) model, which does. In addition, experiments demonstrating that attention can be simultaneously focused on two loci are reviewed, further restricting possible models of attention. Using fMRI, DeYoe and Brefczynski (61) examine the topography of spatial attention in occipital and parietal cortex. Subjects were presented with a circular stimulus divided into 18 segments of which one segment was covertly attended. Attention-related increases in activity were observed contralateral to the attended segment in virtually all known occipital visual areas as well as bilaterally in parietal cortex. Furthermore, shifts of attention were accompanied by corresponding shifts of attention-related activity in visual cortex, suggesting a neural correlate to a focus of attention. Attention-related increases in activity in visual cortex gradually decreased surrounding the attended segment supporting a gradient model of spatial attention. Performing a stimulus detection task resulted in increases in activity in visual cortex (V1, V2, and V3), even in the absence of visual stimulation, in an fMRI study by Ress and Heeger (51). The amount of activation was spatially selective, dependent on task difficulty, and was related to behavioral performance accuracy—larger responses were observed for both hits and false alarms than for either misses or correct rejections. These results indicate that cortical activity agrees with behavioral performance, indicating that activity in visual cortex is representative of a subject's perception of a scene and not merely a representation of the sensory environment.

4.2 Cross-Modal Attention

Driver and colleagues (32) review the cross-modal consequences of spatial attention. When subjects perform one auditory and one visual task, behavioral performance is better when the auditory and visual stimuli share a common spatial location than when they do not. ERP and fMRI results show that attending to a stimulus (e.g., tactile) that shared a common location with an irrelevant stimulus from another modality (e.g., visual) enhanced sensory processing of the irrelevant stimulus. The implications for the operation of spatial attention in humans are discussed. Myachykov and Posner (53) examine the role of attention in language. Attention is important in the assignment of syntactic roles; for example the attended object is most often assigned to the subject of the sentence. Neuroimaging results are reviewed that examine neural areas involved in attention-demanding aspects of language. Findings from visual search paradigms indicate that performance for detecting a target defined by a spatial relationship between constituent parts depends on the number of items in the search array. This finding, as well as those from a flicker paradigm, indicates that spatial relations can only be decoded with the aid of attention. Carlson and Logan (54) discuss the important role of attention in spatial language and examine at which steps of decoding spatial relationships attention plays a role. Watanabe and Shimojo (89) investigate attention in a cross-modal event perception task using an ambiguous motion display. The results are consistent with an automatic shift of attention from the visual stimulus to the sensory transient used in the experiment. Auditory streaming is the ability to divide auditory information into separate streams, for example, the ability to divide the vocal components of a song from the instrumental components. Carlyon and Cusack (52) examine four approaches to studying the effects of attention on the ability to perform auditory streaming. Results from these approaches indicate that while streaming can take place in the absence of full attention, attention does affect the streaming process, indicating that attention can effect perceptual organization.

5.0 Models of Attention

5.1 Biologically Realistic Models

The models in this group take the neurobiology and psychology of attention seriously in their design, and compare their performance directly to biological systems. Some use simulations, while others use real images as input and embed attention in a larger vision system. They mostly attempt to make testable predictions based on their underlying theories. Deco, Rolls and Zihl (97) detail a neurodynamical model of attention that builds on the concept of biased competition. The model incorporates feature extraction in early visual areas and separate ventral and dorsal pathways. Top-down bias may be initiated at the level of the posterior parietal module in the dorsal stream, or the inferior temporal module of the ventral stream, allowing the emergence of two separate modes of attention. Deco later collaborated with Rolls to add short-term memory to the biased competition model. Deco and Rolls (100) propose a model of the prefrontal cortex that describes mechanisms involved in short-term memory. The proposed architecture combines recurrent excitatory working memory mechanisms with attentional biased competition. Simulations successfully reproduce neurophysiological and fMRI experiments related to the deployment of spatial attention.

A very different sort of model is described by Palm and Knoblauch (101), who consider the issue of binding in distributed neural assemblies. In their model, scene segmentation and binding is achieved through interaction among two reciprocally connected visual areas with simultaneous state switching achieving binding among neuron groups. Simulations demonstrate compatibility with experimentally observed phenomena relating to synchronization, and results are discussed in the context of attention and biased competition.

Broad models (in the sense of addressing many issues within one framework) are represented by contributions from Grossberg, Cave et al., and Anderson et al. Grossberg (107) discusses the role of attention in cortical stability and organization. Central to the discussion is the tradeoff between neural stability and plasticity, and a discussion of Adaptive Resonance Theory, which describes how attention may aid in resolving such a tradeoff. Issues such as learning, competition, expectation, and consciousness are discussed in the context of their relationship to attention. Cave et al. (90) describe an implemented hierarchical neural network model of attention, intended to account for results of a number of experimental efforts. Attention directs the flow of the network such that connections from a number of input units within a local area converge on a single set of output units with features from a single location passing to the next layer. Competition is governed by closing gates at locations with non task-relevant features, and holding gates open in response to higher local feature contrast. Anderson et al. (3) present the dynamic routing model of visual attention, with emphasis on the circuitry for translation and scale invariance, and the reference frames used in the process. The major predictions and supporting physiological evidence are briefly reviewed, together with some open questions.

Memory is clearly an important component of an overall vision system and attention clearly must interact with memory. Still, very few models explicitly include memories and provide a theoretical basis for their inclusion. Arathorn (99) describes a model of memory-driven attention based on map seeking circuits. Objects in the visual field are located on the basis of a contextually valid mapping between a stored memory representation, and attributes of the object under consideration. The architecture seeks a transformation between a visual pattern and stored memory representations facilitating recognition.

Saliency issues dominate the next two contributions, from Zelinksy and Zhaoping. Zelinksy (65) proposes that the term attention is underspecified and can be subdivided into its component parts. A model for performing visual search is proposed. In the first component of the model the target relevant features are identified and these features are used to guide search by correlating points in the search array with the target, which acts as a saliency map. In the second component a threshold that increases over time is applied to the saliency map and the centroid is constantly calculated. Once the centroid exceeds a minimum distance from the current locus of fixation an eye movement is executed. This process repeats until the target is fixated. Zhaoping (93) considers the notion that a saliency map resides within the primary visual cortex. A model is presented that describes contextual influences in V1 based on layer 2 and 3 pyramidal cells, horizontal intracortical connections, and inhibitory interneurons. The model predicts a variety of experimental phenomena related to pop-out, conjunction search, and search asymmetry.

The controversy between bottom-up and top-down processes in attention is alive and well in the modeling community. Itti (94) reviews a number of computational architectures that focus on the bottom-up deployment of attention. A historical perspective on models that revolve around the notion of saliency is presented, including discussion of how such models have been employed to subserve recognition tasks. Applications of attention in artificial systems are discussed with the problem of evaluating the relative quality of magazine covers from a marketing perspective, provided as an example for the wide ranging potential of bottom-up computational attention in different application domains. Parkhurst and Niebur (39) explore the nature of bottom-up, stimulus driven mechanisms of attention. Attentional allocation in complex natural scenes is observed by measuring the eye movements made by observers. A comparison is made between stimulus salience as determined by a biologically motivated computational model of the primate visual system, and fixation locations, indicating that there exists a significant correlation between fixation locations and stimulus salience. In contrast, Tsotsos (92) describes a model that incorporates both top-down and bottom-up processes. Inspired by first principles arguments related to computational complexity and information routing, Tsotsos describes a proposal called Selective Tuning that explains attention at the levels of behavior and computation. The Selective Tuning Model consists of a multi-layer pyramid structure, with feed-forward and feedback connections implementing selection via a hierarchy of winner-take-all processes, influenced by task-relevant bias. The model demonstrates extensive predictive power and exhibits behavior compatible with experimental observations through application to digitized images.

Rao & Ballard, and Hamker, address how eye movements might be integrated into an attentive framework. Rao and Ballard (91) review two attention models that are motivated by probabilistic principles. The first model observes overt attention directed on the basis of iconic representation and saliency maps, with a comparison made to eye movement patterns. A second model based on predictive coding is presented and it is demonstrated that attentive behavior emerges as a result of selective filtering of predictive error signals. Hamker (98) describes a model of attention involving top-down guided object detection combining attention and recognition as co-dependent components of a common network. The model includes early stimulus-driven, salience-based detection causing V4 and IT activation, and feature specific feedback initiated by IT causing gain based on expectation among earlier visual areas. Competition within frontal eye fields and inhibition of return guide fixation.

5.2 Machine Vision

A number of contributions describe the value of attentive processes in a variety of applications, from computer vision to robotics to graphics. Medioni and Mordohai (95) describe a means of detecting perceptually salient structures in data sets of arbitrary dimensionality. Saliency is represented in the context of a second order symmetric tensor, with a tensor-voting framework based on Gestalt principles selecting structures based on smooth continuation and simplicity. The method is demonstrated as being robust to noise, and offers a means of considering computer vision problems within the domain of perceptual organization. Context plays a major role in the next chapter as well. Torralba (96) presents a model that incorporates scene priors to modulate saliency measures associated with regions of an image. Contextual influence is included based on scene priors represented in a Bayesian formulation. Improvement in the selection of salient locations by the proposed model relative to context-free selection and variation in object recognition performance in human experiments demonstrates the importance of considering context.

Robot locomotion that is guided by attentive vision systems seems a very natural application given the huge amount of data that a moving robot can visually acquire. Lewis (104) considers the exact functions of attention in the context of robot locomotion. Potentially destabilizing objects are detected through a measure of novelty based on disparity between the expected and observed visual stimuli. Flagging novel events allows motor action to respond to meaningful visual stimuli while ignoring irrelevant information. Still in the domain of robot behavior, Paletta and colleagues (105) address saliency and feature selection, and attention in object recognition. The authors emphasize the value of attention in various machine vision applications, and highlight the necessity for attentive behavior in emerging technologies.

Attentional windows and how they behave under intelligent control for computer vision is a theme in the next two contributions. Rybak and colleagues (108) propose a model of visual perception guided by the movement of an attentional window. The attentional window selects a retinal image for further processing by an intermediate stage that computes invariant second order features and subsequently a high level subsystem consisting of parallel "what" and "where" structures that store the invariant features and attentional window movements, respectively. The model is tested in the context of tasks of face and object recognition in complex scenes. Elder and colleagues (102) present a two component machine vision system inspired by the drop-off in resolution of the eye with increasing distance from the center of the optical axis. Events are detected within a large, fixed, low-resolution field of view for further interpretation and recognition within a high resolution, moving field of view. The proposed system offers a solution for achieving high-resolution surveillance with a wide field of view, overcoming the tradeoff between these two elements inherent in using a single sensor.

Indiveri (103) describes an attention-based architecture for constructing multi-chip vision systems. Details of a two chip VLSI device are presented, implementing a winner-take-all competition, which directs a pan-tilt address-event representation retina on the basis of contrast transients. The infrastructure affords the ability to produce selective attention behaviors in a hardware context, and demonstrates potential utility in engineering applications. Yee and Pattanaik (106) borrow ideas from visual attention to expedite illumination computation in computer graphics rendering. Global illumination computation is directed on the basis of a saliency map derived from important image cues. The authors report an order of magnitude improvement in computational speed of rendering.

Finally, the once commonly seen link between computational vision and artificial intelligence re-emerges in a new model of scene analysis. Schill (109) proposes a model of saccadic scene analysis based on maximizing information sampled from the scene. A parallel stage composed of linear and nonlinear neural operators selects informative regions of the scene for further analysis by a knowledge-based reasoning system. Based on Dempster-Shafer theory for uncertain reasoning, the knowledge-based component incorporates higher-level cognitive influences on recognition.

Summary

This quick tour is not intended to be a substitute for reading the chapters! There is nothing like the authors' own words and you are encouraged to enjoy them. The tour presented above is necessarily very abbreviated and certainly cannot do justice to each of the successive chapters, especially with respect to those contributions that cross one or more boundaries of the categorization presented here. That is part of the beauty and maturity of this field; it is becoming increasingly common that one must address multiple topics within attention together. Perhaps this is how, slowly, the "big picture" will develop.

SECTION I

FOUNDATIONS

CHAPTER

1

Computational Foundations for Attentive Processes

John K. Tsotsos

ABSTRACT

Notions such as capacity limits pervade the attention literature. This presentation attempts to make these concrete and to discover constraints on plausible solutions to vision. Through the proofs, approximations, and optimizations to find architectures that plausibly do not violate biological constraints, important problems such as information routing and signal interference can be addressed. Perhaps the most important conclusion is that the brain is not solving the generic vision problem. Rather, the generic problem is reshaped through approximations so that it becomes solvable by the amount of processing power available for vision. Selective attention in feature, image, and object space plays a necessary role.

I. THEORETICAL BACKGROUND

A. Introduction

One of the most frustrating things about studying attention is that research is so often accompanied by vague discussions of *capacity limits*, *bottlenecks*, and *resource limits*. How can these notions be made more concrete? The area of computer science known as computational complexity is concerned with the theoretical issues dealing with the cost of achieving solutions to problems in terms of time, memory, and processing power as a function of problem size. It thus provides the necessary theoretical foundation on which to base an answer to the capacity question.

It is reasonable to ask whether computational complexity has relevance for real problems? Stockmeyer and Chandra (1988) present a compelling argument. The most powerful computer that could conceivably be built could not be larger than the known universe, could not consist of hardware smaller than the proton, and could not transmit information faster than the speed of light. Such a computer could consist of at most 10^{126} pieces of hardware. It can be proved that, regardless of the ingenuity of its design and the sophistication of its program, this ideal computer would take at least 20 billion years to solve certain mathematical problems that are known to be solvable in principle (e.g., the well-known traveling salesman problem with a sufficiently large number of destinations). As the universe is probably less than 20 billion years old, it seems safe to say that such problems defy computer analysis. There exist many real problems for which this argument applies (see Garey and Johnson, 1979, for a catalog), and they form the foundation for the theorems presented here.

With respect to neurobiology, many have considered complexity constraints in the past but mostly in a qualitative manner. All roads lead to the same conclusions: the brain cannot fully process all stimuli in parallel in the observed response times. But this is like saying there is a capacity limit: *This does not constrain a solution*. By arguing in this manner we are no closer to knowing what exactly the brain is doing to solve this problem. This chapter takes the position that a more formal analysis of vision at the appropriate level of abstraction will help to reveal quantitative constraints on visual architectures and processing. First, however, it is important to address the applicability of this analysis for the neurobiology of the brain.

B. Can Human Vision Be Modeled Computationally?

This nontrivial issue is important because if it could be proved that human brain processes cannot be modeled computationally (and this is not tied to current computer hardware), then modeling efforts are futile. A proof of *decidability* is sufficient to guarantee that a problem can be modeled computationally (Davis, 1958, 1965). Decidability should thus not be confused with tractability. Tractability refers to the sort of problem Stockmeyer and Chandra (1988) described: a tractable problem is one for which enough resources can be found and enough time can be allocated so that the problem can be solved reasonably. An intractable problem may be decidable; but for an undecidable problem, one cannot determine its tractability. Intractable problems are those that have exponential complexity in space and/or time; that is, the mathematical function that relates processing time/space to the size of the input is exponential in that input size. There are several classes of such problems with differing characteristics and NP-complete is one of those classes. To show that vision is decidable, then it must first be formulated as a decision problem. This means that if it is the case that for some problem we wish to know of each element in a countably infinite set A, whether or not that element belongs to a certain set B which is a proper subset of A, then that problem can be formulated as a decision problem. Such a problem is decidable if there exists a Turing machine that computes "yes" or "no" for each element of A in answer to the decision question. A Turing machine is a hypothetical computing device consisting of a finite state control, a read–write head, and a two-way infinite sequence of labeled tape squares. A program then provides input to the machine, is executed by the finite state control, and computations specified by the program read and write symbols on the squares of the tape.

This formulation for the totality of visual performance does not currently exist, but does exist for several subproblems. One of the relevant decidable perceptual problems is visual search (Tsotsos, 1989).

This, however, is not a proof that human vision can be modeled computationally. If no subproblem of vision could be found to be decidable, then it might be that perception as a whole is undecidable and thus cannot be computationally modeled. But, what if there are other undecidable vision subproblems? Even if some other aspect of vision is determined to be undecidable, this does not mean that all of vision is also undecidable or that other aspects of perception cannot be modeled computationally. Hilbert's 10th problem in mathematics and the halting problem for Turing machines are two examples of famous undecidable problems. The former does not imply that mathematics is not possible, whereas the latter does not mean that computers are impossible. It seems that most domains feature both decidable and undecidable subproblems and these coexist with no insurmountable difficulty.

II. THE COMPUTATIONAL COMPLEXITY OF VISION

What is the generic vision problem? Given a sequence of images for each pixel, determine whether it belongs to some particular object or other spatial construct, localize all objects in space, detect and localize all events in time, determine the identity of all the objects and events in the sequence, and relate all objects and events to the available world knowledge. This section briefly summarizes the steps of an argument that concludes that the generic vision problem as defined here is intractable if its solution is limited to strictly feedforward processes in the brain. On the other hand, if task and domain guidance is permitted, the problem becomes tractable. The reader should follow the references for full details.

A. A Simple Counting Argument

Purely feedforward, unconstrained visual processing seems to have an inherent exponential nature if one considers a blind (without guidance) search process among all possible combinations of pixels in an image. This can be partially tackled by including hierarchical organization, pyramidal abstraction, separate visual maps, and spatiotemporally localized receptive fields as part of the processing machinery that performs the search (Tsotsos, 1987).

B. Finishing off the Counting Argument with Proofs

To show the decidability of visual search, two theoretical abstract problems have been defined and proofs of their complexity were presented. The first is *unbounded visual search*. This was intended to model recognition where no task guidance to optimize search is permitted. It corresponds to recognition with all top-down connections in the visual processing hierarchy removed or disabled. In other words, this is pure data-directed vision, as Marr believed was possible. The second problem is *bounded visual search*. This is recognition with knowledge of a target and task in advance, and that knowledge is used to optimize the process. The basic theorems, proved in (Tsotsos, 1989) and later confirmed by Rensink (1989), are:

THEOREM 1. Unbounded visual search is NP-complete.

THEOREM 2. Bounded visual search has time complexity linear in the number of test image pixel locations.

The results are broad and powerful. The first tells us that the pure data-directed approach to vision (and, in fact, to perception in any sensory modality) is computationally intractable in the general case. Marr's "in principle" solution to vision cannot be put into practice and his implementation level of description is not feasible. The second tells us that visual search takes time linearly dependent on the size of the input, something that has been observed in a huge number of experiments. Even small amounts of task guidance can turn an exponential problem into one with linear time complexity.

C. Dispensing with Objections to this Analysis

Two main objections to this analysis must be addressed.

1. Does the Brain Handle Worst-Case Scenarios?

Perhaps the most obvious objection is that a worst-case analysis as required for the above proofs implies that biological vision handles the worst-case scenarios. This is incorrect. Worst cases do not only occur for the largest possible problem size; rather, the worst-case time complexity function for a problem gives the worst-case number of computations for any problem size. This worst case may result simply because of unfortunate ordering of computations (e.g., a linear search through a list of items would take a worst-case number of comparisons if the item sought is the last one). Thus, worst-case situations in the real world may happen frequently for any given problem size (Tsotsos, 1990).

2. Empirical Analysis Addressing Median-Case Analysis

Some claim that biological vision systems are designed around average or best-case assumptions; it is likely that expected case analysis more correctly reflects biological systems. Parodi et al. (1998) addressed this criticism by developing an algorithm for generating random instances of polyhedral scenes and examining median-case complexity of labeling the scenes. The key results are the following:

1. Blind depth-first search has median-case time complexity that is exponential in the number of junctions.
2. Informed best-first search has median-case time complexity that is linear in the number of junctions.

These results were achieved empirically with the following experimental strategies. In the first experiment, they investigated the median-case complexity of blind depth-first search. Time for the search stage is computed as the number of times that the depth-first-search stack containing all nodes, which have been visited but not explored, is updated. One hundred different random scenes were generated for each size of scene (total number of junctions in the scene), and the median number of algorithmic steps was computed for each set. The result was fit to a straight line in logarithmic space as a function of number of scene junctions, thus demonstrating exponential behavior even in the median case.

In the second experiment, they tested the median-case complexity for informed best-first search. This search exploits the following heuristic rules:

- The node that has the smallest set of legal labeling is chosen, breaking ties arbitrarily.
- The structure of the domain is used to guide search so that the more common labelings are tried first.

As a result of these two experiments (plus more that can be found in the paper), the objection that worst-case analysis is not appropriate is defused. Even in the median case, it has been shown that complexity is exponential if no knowledge is used to tune search, whereas even very modest amounts of general knowledge can convert an exponential process to a linear one.

III. COMPLEXITY CONSTRAINS THE ARCHITECTURE FOR VISUAL PROCESSING

A. How to Deal with an Intractable Problem in Practice?

If a problem is NP-complete, how is it possible to deal with it? NP-Completeness eliminates the possibility of developing a completely optimal and general algorithm. Garey and Johnson (1979) provide a number of guidelines. The relevant guideline for this analysis is: Use natural parameters to guide the search for approximate algorithms. There are a number of ways a problem can be exponential. Consider the natural parameters of a problem and attempt to reduce the exponential effect of the largest valued parameters.

B. Constraints on a Model

In Tsotsos (1988, 1990), a sequence of modifications to the problem of unbounded visual search are given, driven by the size of the perception problem in terms of number of photoreceptors, feature types, size of visual areas, connectivity of visual areas, and so on, to transform the problem into a tractable one. The result is a visual search problem that is tractable in time and tractable in space (requires no more processing machinery than the brain may afford), but is not guaranteed to always find optimal solutions. The solutions found are approximate ones, quite acceptable most of the time, but may lack precision or completeness. The claim is that this is the form of the visual search problem that the brain is actually solving; a conclusion of this sort is the only possible one because it has been proved that optimal solutions for the general problem of visual search lead to intractability.

In the unbounded visual search problem, the worst-case time complexity is $O(N2^{PM})$, that is, it requires a number of operations on the order of $N2^{PM}$, discarding constant factors. The natural parameters of this computation are N (number of objects in the world), P (size of image in pixels), and M (number of features computed at each pixel). N is very large and any reduction leads to linear improvements. Reduction in P can lead to exponential improvement, as does reduction in M.

1. Architecture

The above time complexity function can be reduced to at least $O(P^{1.5}2^M \log_2 N)$ using a number of simple optimizations and approximations:

1. Hierarchical organization takes search of model space from $O(N)$ to $O(\log_2 N)$.
2. Search within a pyramidal abstraction (a layered representation, each layer with decreasing spatial resolution with bidirectional connections between adjacent layers) operates in a top-down fashion, beginning with a smaller more abstract image, and is then refined locally, thus reducing P.
3. Logically (not necessarily physically) separate visual maps permit selection of features of relevance and, thus, reduce M.
4. Spatiotemporally localized receptive fields reduce the number of possible receptive fields from 2^P to $O(P^{1.5})$ (this assumes a hexagonal grid, and hexagonal, contiguous receptive fields of all possible sizes centered at all locations in the image array).

After application of the four constraints, $O(P^{1.5}2^M \log_2 N)$ is the worst-case time complexity. Where does attention come in? Attention can further reduce the $P^{1.5}$ term if one selects the receptive field that is to be processed. This is not a selection of location only, but rather a selection of a local region and its size at a particular location. This represents spatial selectivity. Feature selectivity can further reduce the M term, that is, which subset of all possible features actually is represented in the image or is important for the task at hand. Object selectivity can further reduce the N term, reflecting once again task-specific information.

2. Information Routing

It is important to consider how information moves around in this pyramidal representation. Spatial data are represented at low resolution in the output layer and at high resolution in the input layer. One strategy that minimizes overall connection numbers and lengths is that units at the output layer access high-resolution spatial representations through the available pyramid connections rather than connecting directly to the layer of interest. This is easy to accomplish because the direct connections for any receptive field in any layer provide the pathway. However, there is a problem. The receptive fields higher up in the pyramid have large receptive fields that would typically include a great deal of stimulus content beyond the parts that may be of interest. Because the connections from the whole receptive field converge onto single units, the stimuli interfere with one another. This signal interference degrades the signal; however, this can be eliminated. A "beam" that affects the full hierarchy to control the interference of distracters within a receptive field can accomplish this, as shown in Fig. 1.1.

Can there be more than one of these beams at a one time? It should be clear that distance between beams plays a critical role. If the beams overlap, then the potential for signal interference is large. If the beams are sufficiently separate, then it may be possible to attend to more than one item at a time. It may be that the default policy is for single beam or attentional focus unless task demands dictate otherwise.

3. The Vision Problem Is Reshaped

The brain is **not** solving the general vision problem. Of the mechanisms introduced in Section III.B.1, hierarchical organization and logical map separation do not affect the nature of the vision problem. The other mechanisms have the following effects:

- Pyramidal abstraction affects the problem through the loss of information and signal combination.
- Spatiotemporally localized receptive fields force the system to look at features across a receptive

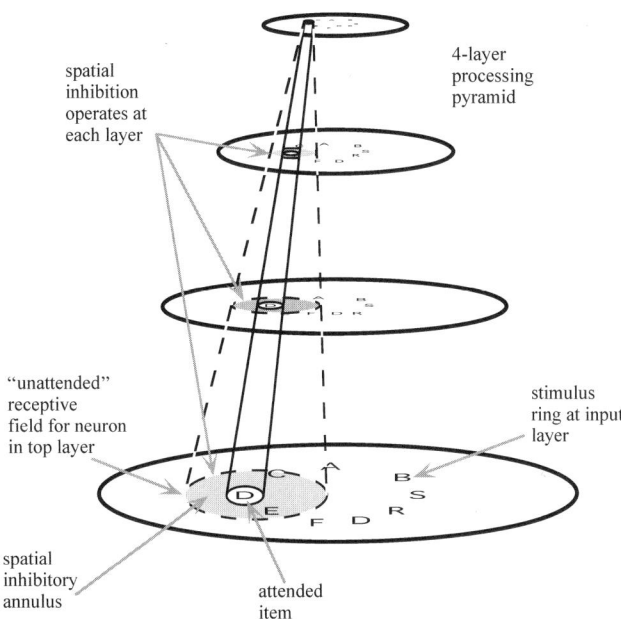

FIGURE 1.1 Effect of spatial attentive selection in a hypothetical four-layer processing hierarchy. Adapted, with permission, from Tsotsos (1990).

field instead of finer-grain combinations, and thus arbitrary combinations, of locations are disallowed.
- Attentional selection further limits what is processed in the space, feature, and object domains. The knowledge that task-related attentive processes use can be as generic as statistical regularities.

As a result of these, the generic problem as defined in Section II has been altered.

IV. CONCLUSIONS

This chapter quantifies the processing requirements for vision and thus attempts to make notions such as capacity limits concrete and to discover constraints on plausible solutions to vision. It is possible through the proofs and architectural changes to find architectures that plausibly do not violate biological constraints on numbers of neurons, connections, and processing time. Perhaps the most important conclusion is that the problem the brain is actually solving is not the generic vision problem. Rather, the problem is reshaped through approximations and attentional selection so that it becomes solvable by the amount of processing power available for vision.

References

Davis, M. (1958). "Computability and Unsolvability," McGraw–Hill, New York.
Davis, M. (1965). "The Undecidable," New York: Hewlett Raven Press.
Garey, M., and Johnson, D. (1979). "Computers and Intractability: A Guide to the Theory of NP-Completeness." Freeman, San Francisco.
Parodi, P., Lancewicki, R., Vijh, A., and Tsotsos, J. K. (1998). Empirically-derived estimates of the complexity of labelling line drawings of polyhedral scenes. *Artif. Intell.* **105**, 47–75.
Rensink, R. (1989). "A New Proof of the NP-Completeness of Visual Match," Technical Rep. 89-22, Dept. of Computer Science, University of British Columbia.
Stockmeyer, L., and Chandra, A. (1988). Intrinsically difficult problems. *In* "Trends in Computing," Vol. 1, pp. 88–97. Scientific American Inc., New York.
Tsotsos, J. K. (1987). A complexity level' analysis of vision. *In* "Proceedings, International Conference on Computer Vision, London, England."
Tsotsos, J. K. (1988). How does human vision beat the computational complexity of visual perception? *In* "Computational Processes in Human Vision: An Interdisciplinary Perspective" (Z. Pylyshyn, Ed.), pp. 286–338. Ablex Press, Norwood, NJ.
Tsotsos, J. K. (1989). The complexity of perceptual search tasks. *In* "Proceedings, International Joint Conference on Artificial Intelligence, Detroit, Michigan," pp. 1571–1577.
Tsotsos, J. K. (1990). A complexity level analysis of vision. *Behav. Brain Sci.* **13**, 423–455.

CHAPTER 2

Capacity Limits for Spatial Discrimination

Michael J. Morgan and Joshua A. Solomon

ABSTRACT

When observers are searching for a single target among multiple distracters, accuracy usually decreases with the number of distracters (the set-size effect). Set-size effects have been measured to seek evidence for a limit on the capacity for spatial discriminations. Experiments with unmasked displays have found no evidence for a sensory bottleneck. On the other hand, postmasked displays produce set-size effects too large to be explained by uncertainty alone. When mask-onset asynohronies are short (e.g., 100 ms), orientation-defined targets frequently seem to be adjacent to their true positions and tilt thresholds increase with the square root of the number of cued items. These results suggest that the mean of an ensemble can be computed when there is too little time to evaluate each item individually. Precues can improve discrimination even in the absence of distracters, but as multiple precues are almost as effective as single ones, most of this enhancement is preattentive.

I. INTRODUCTION

Attention has been likened to a spotlight, but does it really help us see better? Specifically, we address the controversy regarding whether we can better analyze one part of a scene when allowed to ignore others. There are at least three popular paradigms in which the ability to attend is systematically modified. Here we concentrate on two: visual search and spatial cuing. Concurrent tasks, which tap observers' abilities to generate independent responses to multiple visual targets, have also been used to assess attention's effect on spatial discrimination. However, limited memory can produce failures of independence as easily as a sensory bottleneck. For that reason, we reserve the results from these dual-report experiments for another discussion.

II. FEATURE SEGREGATION

In visual search experiments, threshold tilt for an orientation-defined target increases as more non-targets are added to the display. This result does not necessarily imply better acuity with fewer display items. The more distracters there are, the more likely observers will be to mistake one for the target. Even when observers know which item is the target, the presence of distracters may interfere with its analysis. For this reason, Palmer et al. (1993) advocated fixing the number of display items while varying ("relevant") set size with spatial cues. This type of experiment, at the intersection of the spatial cuing and visual search paradigms, is the focus of our discussion.

With long exposures (Morgan et al., 1998; Solomon and Morgan, 2001) and/or no postmask (Palmer, et al., 1993; Palmer, 1994; Baldassi and Verghese, 2002), set-size effects are usually consistent with signal detection theory (SDT, see Appendix), which assumes that the analysis of each display item is independent from the analysis of every other display item. One consequence of this assumption is that, when observers do mistake a distracter for the target, its position in the display should be independent of the target's actual position. As shown in Fig. 2.1, this is true for long, but not short, postmasked exposures. In other words, the less time

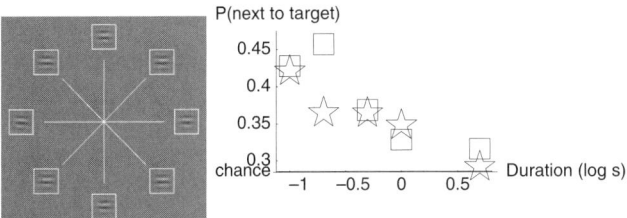

FIGURE 2.1 The proximity effect. Stimulus and results. Left: Display with seven horizontal distracters and one tilted target. Right: Frequencies of mislocations directed to elements adjacent to the target. Stars and boxes indicate different observers. Data from Solomon and Morgan (2001).

available for visual processing, the more likely it is that adjacent items will be analyzed together.

III. FEATURE INTEGRATION

Information from more than one item can be summarized in a statistic. For example, there is evidence for a process that estimates the mean orientation of patterns surrounding each point in the visual field (Keeble, et al., 1995). This textural analysis necessarily lacks the spatial resolution of SDT's independent analyzers, but it may be faster. When observers were asked to identify the direction of tilt (i.e., clockwise vs anticlockwise) in a brief, postmasked display similar to that in Fig. 2.1, threshold for a single cued target was 0.22 times threshold for an uncued target among 15 (untilted) distracters (see Fig. 2.2). This finding led Morgan, et al. (1998) to propose a textural process for orientation identification. In their "averaging model," threshold is determined by the signal-to-noise ratio within a mechanism that computes the mean orientation from independent samples of each potential target. When there is only one real target, threshold should rise with the square root of the number of cued items. Thus, threshold with seven distracters should be $\sqrt{8} \approx 2.8$ times threshold with zero distracters. For comparison, SDT predicts the ratio should be ~2.5. The similarity of these predictions allowed Baldassi and Verghese (2002) to dismiss earlier support for the averaging model (Baldassi and Burr, 2000).

SDT and the averaging model give very different predictions when distracters are replaced by additional targets (holding the total number of stimuli constant). In this case, the averaging model predicts an inverse proportionality between threshold and number of targets. Parkes, et al. (2001) confirmed this prediction using brief, but unmasked nine-item displays. Threshold for a single target among eight distracters was nine times threshold for nine targets. SDT predicts a mere sixfold increase. Parkes, et al. con-

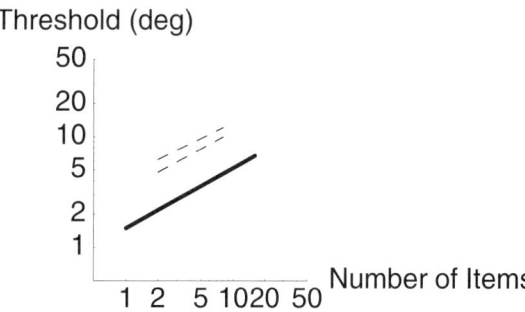

FIGURE 2.2 Square-root law. Each line shows how the threshold for orientation identification depends on the number of distracters. Threshold is derived by finding the tilt μ, which produces an accuracy of 0.75 on the psychometric function $\psi(\mu)$. The solid line is the average observer from Morgan, et al. (1998, Experiment 3). The dashed lines reflect previously unpublished results using the stimulus and observers described in Fig. 2.1. All lines have a slope near 0.5.

cluded that textural analysis was obligatory when the display items appeared in a peripheral, "crowded" array. It remains unclear why one observer made similarly inefficient use of local signals when the array was presented in the center of the visual field.

IV. FEATURE DETECTION

Although the effect is small, spatial cues appearing prior to target onset can aid spatial discrimination even when there are no distracters (Carrasco, et al., 2000; Lu and Dosher, 2000). Carrasco, et al. report the largest of these cuing effects. A peripheral pre-cue lowered the contrast required for identifying the direction of a slightly tilted Gabor by 20%. Using the steepest of the power laws fit by Bowne (1990, Fig. 2.4), this is equivalent to an 8% reduction in tilt threshold. We replicated Carrasco and colleagues' effect using the "noninformative" pre-cue technique, pioneered by Jonides (1981). In this technique, pre-cues are uncorrelated with target position; for example, when one of eight positions is cued, the target appears there on one-eighth of the trials. When the target appeared in a cued location, tilt identification was more accurate than when it did not, but this was true even when all eight locations were cued (see Fig. 2.3). Thus, precuing benefits are not necessarily a consequence of "focal" attention.

Exactly how peripheral pre-cues aid spatial discrimination remains to be determined. Lu and Dosher (2000) found evidence for both signal amplification and noise exclusion in the absence of distracters. The much larger effects found when distracters are present suggests an altogether different mechanism. We

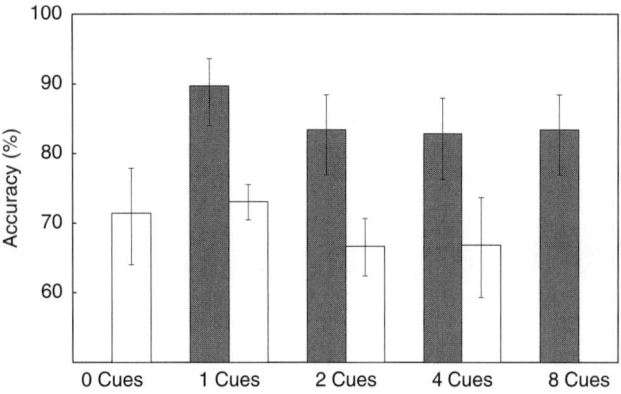

FIGURE 2.3 Precues benefit multiple locations simultaneously. white and black bars reflect orientation identification when the target appeared in uncued and precued locations, respectively. Benefits are largely independent of the number of cues, although there is a slight advantage for a single cue. Results from author J.A.S. Error bars reflect 95% confidence limits.

suggest that spatial cues help observers to more rapidly identify the most appropriate local mechanisms for spatial discrimination. Without spatial cues, observers are compelled to first analyze the scene at a more global level. In other words, focal attention can help us to identify one object only when other, irrelevant objects are present. It allows us to bypass an otherwise compulsory, texutral analysis of the visual scene.

APPENDIX

SDT assumes the orientation of each item is encoded independently, with additive gaussian noise. For simplicity, we adopt the following notation for normal density (PDF) and distribution (CDF):

$$f(x;\mu,\sigma) = \frac{1}{\sqrt{2\pi}\sigma} e^{-(x-\mu)^2/2\sigma^2} \qquad (1)$$

and

$$F(x;\mu,\sigma) = \int_{-\infty}^{x} f(u;\mu,\sigma)du. \qquad (2)$$

If distracters have zero tilt, their density and distribution can be abbreviated as

$$f_D(x) = f(x;0,\sigma) \qquad (3)$$

and

$$F_D(x) = F(x;0,\sigma), \qquad (4)$$

and we can reserve μ for the targets. Target density and distribution are thus

$$f_T(x) = f(x;\mu,\sigma) \qquad (5)$$

and

$$F_T(x) = F(x;\mu,\sigma). \qquad (6)$$

Were the observer to identify the direction of tilt, the proportion of accuarate responses would be

$$\psi = t\int_0^\infty f_T(x)[F_T(x)-F_T(-x)]^{t-1}[F_D(x)-F_D(-x)]^d dx \\ + d\int_0^\infty f_D(x)[F_D(x)-F_D(-x)]^{d-1}[F_T(x)-F_T(-x)]^t dx, \qquad (7)$$

where t is the number of targets with tilt μ, and d is the number of distracters with tilt 0.

References

Baldassi, S., and Burr, D. C. (2000). Feature-based integration of orientation signals in visual search. *Vision Res.* **40**, 1293–1300.

Baldassi, S., and Verghese, P. (2002). Comparing integration rules in visual search. *J. Vision* **2**, 559–570.

Bowne, S. F. (1990). Contrast discrimination cannot explain spatial frequency, orientation or temporal frequency resolution. *Vision Res.* **30**, 449–461.

Carrasco, M., Penpeci-Talgar, C., et al. (2000). Spatial covert attention increases contrast sensitivity across the CSF: support for signal enhancement. *Vision Res.* **40**, 1203–1215.

Jonides, J. (1981). Voluntary vs. automatic control over the mind's eye's movement. *In* "Attention and Performance IX" (J. B. Long and A. D. Badderley, Eds.), Lawrence Erlbaum, Hillsdale, NJ.

Keeble, D. R. T., Kingdom, F. A. A., et al. (1995). The detection of orientationally multimodal textures. *Vision Res.* **14**, 1991–2005.

Lu, Z. L., and Dosher, B. A. (2000). Spatial attention: different mechanisms for central and peripheral temporal precues? *J. Exp. Psychol. Hum. Percept. Perform.* **26**, 1534–1548.

Morgan, M. J., Ward, R. M., et al. (1998). Visual search for a tilted target: tests of spatial uncertainty models. *Q. J. Exp. Psychol.* **51A**, 347–370.

Palmer, J. (1994). Set-size effects in visual search: the effect of attention is independent of the stimulus for simple tasks. *Vision Res.* **34**, 1703–1721.

Palmer, J., Ames, C. T., et al. (1993). Measuring the effect of attention on simple visual search. *J. Exp. Psychol. Hum. Percept. Perform.* **19**, 108–130.

Parkes, L., Lund, J., et al. (2001). Compulsory averaging of crowded orientation signals in human vision. *Nat. Neurosci.* **4**, 739–744.

Solomon, J. A., and Morgan, M. J. (2001). Odd-men-out are poorly localized in brief exposures. *J. Vision* **1**, 9–17.

CHAPTER 3

Directed Visual Attention and the Dynamic Control of Information Flow

Charles H. Anderson, David C. Van Essen, and Bruno A. Olshausen

ABSTRACT

Visual attention is a process that directs a tiny fraction of the information arriving at primary visual cortex to high-level centers involved in visual working memory and pattern recognition. We discuss a framework for modeling this mechanism in which information flow through the visual hierarchy is regulated by dynamic control of connectivity. A key aspect of this model involves the establishment of an object-centered reference frame for visual working memory as well as object recognition. Psychophysical evidence suggests that the size of the reference frame, or region of interest, has a fixed width of about 30 resolution elements. This model is neurobiologically plausible, supported by several lines of anatomical and physiological data, and suitable for embedding into a larger computational framework for modeling of neural representations and transformations.

I. INTRODUCTION

Neural encoding of retinal images generates a barrage of information transmitted along the optic nerves that far exceeds what can actually be perceived from moment to moment. Visual attention is a process that addresses this problem by directing a tiny fraction of the information arriving at primary visual cortex to high-level centers involved in visual working memory and pattern recognition. In the information pyramid shown in Fig. 3.1, visual attention constitutes a major bottleneck. Much less than 1% of the information carried through the optic nerves reaches attentive scrutiny (see figure legend for details). Because the attended information comes from a spatially restricted region of interest (ROI), visual attention must mediate translation, scale and other invariances, so that under a variety of different poses, a given object or spatial pattern can be recognized and analyzed for how it might be manipulated.

While the concept of controlling information flow is now generally accepted (see Chapters 4, 20, 103, 105, and 109), there exist few proposals for how this is actually accomplished at the neuronal level. A computationally sound and neurobiologically plausible model of visual attention should address five broad issues, reflecting the reality that attention is an extremely sophisticated process engaging much if not most of the entire visual system. (1) What are the high-level brain structures that mediate visual pattern recognition and working memory? (2) What is the anatomical circuitry that provides ascending visual inputs to these high-level centers? (3) What are the control structures that determine where attention should be directed from moment to moment? (4) How do these control signals dynamically modulate the connections leading up to high-level centers and thereby regulate information flow? (5) How do nonattentive visual processes that operate on the whole image support attentive visual processing?

In previous publications, we presented a conceptual and computational framework for modeling visual attention as it relates to object recognition (Anderson and Van Essen, 1987; Olshausen et al., 1993, 1995; Van Essen et al., 1994). Central to this framework are the notions that (1) information flow into inferotemporal cortex is controlled dynamically across multiple hierarchical levels; (2) spatial relationships within the attended ROI are preserved explicitly; and (3) attentional control operates by dynamic modulation of connectivity weights. Here, we reassess and refine these core concepts and extend them to incorporate visual working memory as well as object recognition. We also

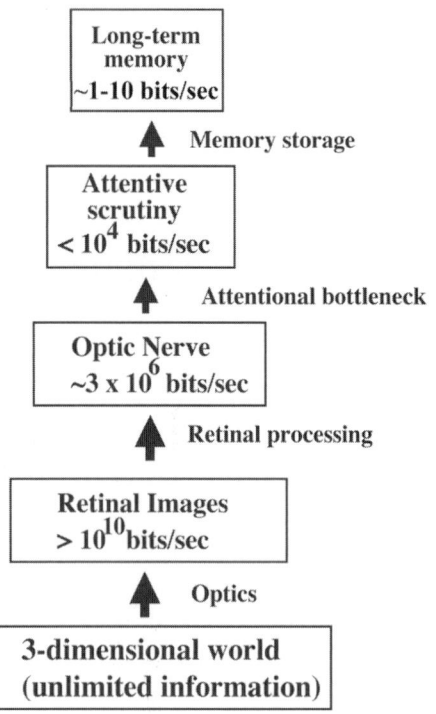

AN INFORMATION PYRAMID

FIGURE 3.1 Information pyramid for the visual system. The estimates of information available in retinal images, encoded in the optic nerves, and passed through the window of attention have been discussed by Van Essen et al. (1991) and Van Essen and Anderson (1995) and are based on information rates of 3 bits/s for each neuron (Eliasmith and Anderson, 2003). Human memory storage rates are based on estimates by Landauer (1986).

describe initial steps toward integrating attention into a broader computational framework for modeling neural representations and transformations (Eliasmith and Anderson, 2003).

II. DYNAMIC ROUTING

Dynamic control of information flow is fundamental to how both brains and modern digital computers deal with finite resources, whether it be for number crunching in a computer or for sensory processing, motor control, or cognitive processing in the brain. In computers, routing circuits constitute major components of the hardware, and a large fraction of the software is devoted to the issue of where to get information for the task at hand and where to put it after it has been processed. In the brain, dynamic reconfiguration of neural resources must be done through modulation of connectivity (see Chapter 4). We refer to this as *dynamic routing* due to the close analogy with computer routing circuits. A major difference between brains and digital computers is that memory and processing are tightly bound to one another in the brain, whereas in computers they are distinct. In essence, the brain contains numerous interlinked subsystems, each carrying out a set of subroutines for specialized computations that would he handled generically by a computer CPU.

In the visual cortex, dynamic routing is most efficiently accomplished in a hierarchical fashion, starting with local control of information flow in lower visual areas and extending to more global control at higher levels. Control signals can also operate both top-down and bottom-up. Attention is normally associated with global top-down processes that are accessible through conscious awareness, though bottom-up control signals can be regarded as involuntary attention (Yantis and Jonides, 1984). Local control interactions are not generally considered attention, although they may involve similar operations at the neural level. Overall, *attention* is inherently an imprecise term that captures only a segment of the broader and more fundamental issue of the control of information flow.

We believe that dynamic routing is essential for achieving invariant representations, both for object recognition and for working memory. An important part of achieving invariant representations is to preserve information about the variations, themselves. That is, *to interpret what has not changed, it is often important also to know what has changed.* Dynamic routing accomplishes this by remapping topographic representations from one level to the next. The control signals mediating the remapping contain information about the variations, while the remapped information itself constitutes the invariant part (see Chapter 99). This in sharp contrast to the many alternative models of object recognition that do not attempt to model the variations, but instead attempt to compute collections of loosely assembled "invariant features" (Selfridge, 1959; Fukushima, 1980; LeCun et al., 1990; Riesenhuber and Poggio, 1999). While such pandemonium-style models may be capable of simple object discriminations, we contend that they are insufficient to account for more general aspects of perception and visuomotor function in the real world. Examples include the perception of generic surfaces (a crumpled cloth); the ability not only to read handwritten text but also to recognize the idiosyncratic style of the writer; and the ability to geometrically manipulate objects in working memory and plan appropriate grasping maneuvers.

In this chapter we focus on routing circuits for translation and scale invariance, as these are fundamental requirements of any flexible vision system. Scale invariance imposes particularly stringent constraints on system design for human vision, given that we can recognize a face when it is only a foot away, where the image covers most of the visual field and occupies most of primary visual cortex (~40 cm² in both hemispheres), or when it is at a distance of 100 ft, where the image subtends about 0.4° and occupies only a few percent of primary visual cortex (~1 cm²).

III. DYNAMIC ROUTING CIRCUIT ARCHITECTURE

The basic architecture we propose for dynamic routing in the visual cortex is schematized in Fig. 3.2. The input layer at the bottom of the array contains the inputs from the retina. The diagram illustrates a spatially uniform sampling, though in fact the distribution of retinal ganglion cell inputs declines steeply with retinal eccentricity (see Olshausen et al., 1995, for a routing circuit that uses a retinal sampling lattice). The top layer represents many fewer spatial locations, reflecting the limited resolution within the ROI (see below). This upper layer may represent cortical areas involved in visual working memory and/or areas involved with object recognition. Note that no attempt is made to specify the features represented by these nodes, as there are very few neurophysiological data at present suggesting what these features should be. For now each node may be thought of as a feature vector.

The exponential increase in the spread of the receptive fields at each level and the subsampling at higher levels are essential elements of this circuit. This design minimizes the number of levels and reduces the complexity of the control, while respecting the fixed number of fan-in/fan-out connections between neurons. Many aspects of this design can be seen in the anatomy of the connections between the visual areas in the V1 to IT pathway. For example the cortical areas of V1 and V2 are approximately the same, while that of V4 is significantly smaller. Also, the width of the axonal spread increases with higher levels, and nominal receptive field sizes increase exponentially as well.

Topographic representations are remapped from one level to the next by a set of control signals that modulate the connections between each level through multiplicative interactions. Formally, the visual data signals at node μ and level l, X_μ^l, are computed by combining the visual data signals at the nodes in the layer below, X_μ^{l-1}, and the control signals at level l, C_λ^l, via

$$X_\mu^l = \sum_{\lambda,\nu} C_\lambda^l \Gamma_{\lambda\mu\nu}^l X_\nu^{l-1}. \qquad (1)$$

To see how certain geometric remappings can be accomplished by this circuit, it is helpful to rewrite Eq. (1) as

$$X_\mu^l = \sum_\nu W_{\mu\nu}^l X_\nu^{l-1}, \qquad (2)$$

$$W_{\mu\nu}^l = \sum_\lambda C_\lambda^l \Gamma_{\lambda\mu\nu}^l. \qquad (3)$$

A spatial pattern in level $l-1$ may be translated by an amount d^l and scaled by an amount α^l in level l by setting the coupling matrix as

$$W_{\mu\nu}^l = \exp\left[-\frac{1}{2}\left(\frac{\mu - \alpha^l(\nu - d^l)}{\sigma}\right)^2\right]. \qquad (4)$$

The problem of achieving a certain remapping d^l, α^l, at level l thus amounts to carving out the appropriate pattern in connection space (μ, ν). In general this may be accomplished by viewing the three-way coupling constants $\Gamma_{\lambda\mu\nu}^l$ as basis functions $\Psi_\lambda^l(\mu, \nu)$. Setting the control signals appropriately then amounts to a function approximation problem,

$$\exp\left[-\frac{1}{2}\left(\frac{\mu - \alpha^l(\nu - d^l)}{\sigma}\right)^2\right] = \sum_\lambda C_\lambda^l(\alpha^l, d^l)\Psi_\lambda^l(\mu, \nu), \qquad (5)$$

and the problem of learning amounts to finding a set of basis functions that produce good approximations for the ensemble of desired remappings.

Our previous proposal for implementing the multiplicative interactions in Eq. (1) was centered on pairwise, "and-like" interactions within dendritic trees. We now have a more general perspective based on the

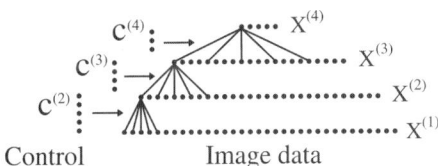

FIGURE 3.2 Dynamic routing circuit. *Top*: Connections from one level to the next are dynamically modulated by a set of control neurons. *Bottom*: Shown are two different states of effective connectivity, corresponding to different sizes and positions of the ROI. Note the size or width of the ROI in the scaled-down circuit shown in this figure is 5 nodes at the top level. A number of psychophysical studies suggest that the actual size is in the range of 20–30 sample nodes.

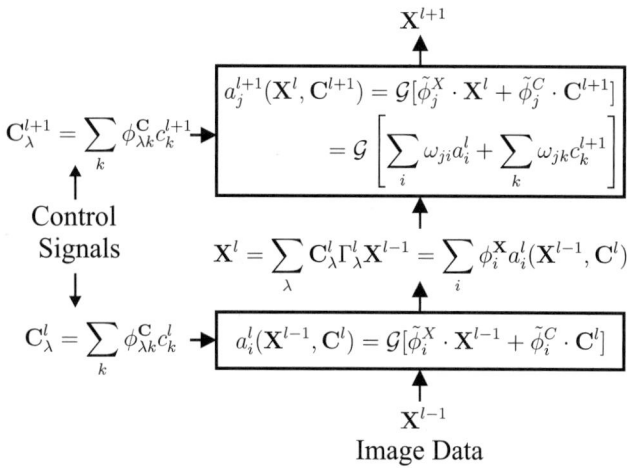

FIGURE 3.3 Neuronal circuit. The signals X^{l-1} and C^l are encoded into neural activities a_i^l using an additive projection rule. The multiplicative terms $C_\lambda^l \Gamma_\lambda^l X^{l-1}$ are then linearly decoded from this population of neurons.

The complete computation from level $l - 1$ to level $l + 1$ is illustrated in Fig. 3.3.

Equation (9), together with Fig. 3.3, illustrates that the input to the neurons in one layer consists of a weighted sum over units in the previous layer. Alternatively, the $a_i^l(X^{l-1}, C^l)$ in Eq. (9) may correspond to dendritic subunits of the level $l + 1$ neurons. In the latter case they act similarly to our previous suggestion of selective pruning in the dendrites, except that now only local summation within a nonlinear subunit is required rather than the precise pairwise multiplicative interactions in Eq. (1). Such nonlinear interactions could easily be realized by the biophysical properties of dendritic trees Mel (1994).

population coding framework described by Eliasmith and Anderson (2003). In this framework, a distinction is made between explicit neural activities and the implicit variables they represent. In the case of our routing circuit, the implicit variables are the image data signals, X_μ^l, which are represented by neural activities a_i^l in terms of a linear basis function expansion,

$$X^l = \sum_i \phi_i^X a_i^l(X^{l-1}, C^l) \approx \sum_\lambda C_\lambda^l \Gamma_\lambda^l \cdot X^{l-1}, \quad (6)$$

where the boldface variables represent vectors, and the basis vectors ϕ_i^X have the same dimensionality as X^l. The neuronal activities are computed via

$$a_i^l(X^{l-1}, C^l) = \mathcal{G}[\tilde{\phi}_i^X \cdot X^{l-1} + \tilde{\phi}_i^C \cdot C^l]. \quad (7)$$

where $\mathcal{G}[\ ]$ denotes the integrate-and-fire response nonlinearity of a neuron. The input consists of the sum of a linear projection of the data signal space at the level below and a projection of the control neuron activities. Given a population of neurons a_i^l that represent the joint space (X^{l-1}, C^l) with sufficient numbers and diversity, one can obtain the transformation described in Eq. (6). Methods for setting the parameters $\tilde{\phi}_i^X$, ϕ_i^X, $\tilde{\phi}_k^C$ and ϕ_k^C, are described in Eliasmith and Anderson (2003).

The neuronal activities corresponding to the next level are similarly computed via

$$a_i^{l+1}(X^l, C^{l+1}) = \mathcal{G}[\tilde{\phi}_i^X \cdot X^l + \tilde{\phi}_i^C \cdot C^{l+1}]. \quad (8)$$

Expanding X^l as in Eq. (6) yields

$$a_j^{l+1}[X^l, C^{l+1}] = \mathcal{G}\left[\sum_i \tilde{\phi}_j^X \phi_i^X a_i^l(X^{l-1}, C^l) + \tilde{\phi}_i^C \cdot C^{l+1}\right]. \quad (9)$$

IV. AUTONOMOUS CONTROL

The control neurons are driven in one of two modes: (1) a bottom-up mode in which they receive input from a saliency map, or (2) a top-down mode in which they are driven by multiplicatively combining the output of visual working memory with the input (see also Chapters 40, 99, 108, and 109). In both cases, the dynamics of the control neurons are governed by an energy function that measures the correlation between the input nodes X_μ^{l-1} and a template V_μ^l as coupled through the control neurons, in addition to a constraint on the control neurons that encourages control states corresponding to affine transformations. For a single stage of the routing circuit this amounts to

$$E^l = -\sum_{\lambda,\mu,\nu} V_\mu^l C_\lambda^l \Gamma_{\lambda\mu\nu}^l X_\nu^{l-1} - \sum_{\lambda,\nu} C_\lambda^l T_{\lambda\nu}^l C_\nu^l. \quad (10)$$

In the bottom-up case, V_ν^l is fixed to a gaussian blob, whereas in the top-down case the V_ν^l correspond to the outputs of an associative memory (e.g., a Hopfield network). The control neurons are then computed by performing gradient descent on Eq. (10),

$$C_\lambda^l = \sigma(u_\lambda^l)$$
$$\dot{u}_\lambda^l + \tau^{-1} u_\lambda^l = \sum_{\mu,\nu} V_\mu^l \Gamma_{\lambda\mu\nu} X_\mu^{l-1} X_\nu^{l-1} + \beta \sum_\nu T_{\lambda\nu}^l C_\nu^l, \quad (11)$$

where σ is a sigmoidal nonlinearity. The control neurons are also subject to delayed self-inhibition so that the circuit does not perseverate on any one ROI. The full control sequence proceeds as follows:

1. Set V_μ^l to a gaussian blob (focus ROI on something interesting).
2. Switch V_μ^l to the output associative memory.
3. Inhibit the current control state and return to step 1.

When an object is being recognized, then the V_μ^l evolve according to the recall dynamics of memory. Alternatively, if an object is being searched for, then the V_μ^l may be set to correspond to the object of interest, and attention will then be directed to those locations of the image containing the object. This mode would most closely correspond to "object-based attention."

V. NEUROBIOLOGICAL SUBSTRATES AND MECHANISMS

We have shown previously (Olshausen et al., 1993, 1995) that the routing circuit of Fig. 3.2 may be scaled up in a manner consistent with the anatomy of the ventral stream in visual cortex (V1 → V2 → V4 → IT). Such a circuit would comprise approximately 300,000 sample nodes (in two dimensions) at the input in V1, narrowing to approximately 1000 sample nodes in IT. The pulvinar is hypothesized to be a major source of the control signals modulating connectivity along the ventral stream, although control neurons could reside in the cortex as well (see Chapter 6).

There are two potential candidates for the top ROI level, which are not mutually exclusive. One is dedicated to the task of object recognition, which most likely resides in inferotemporal cortex. The other provides the substrate for visual working memory (VWM), which probably resides in parietal and/or frontal cortex. In VWM the image data are used for planning how to manipulate the objects in the external environment, where it is clear local spatial relationships must be made available. Several lines of psychophysical experiments suggest that VWM is limited to representing something like a region of space with a diameter of ~30 distinct spatial locations at any instant in time (Van Essen et al., 1991). A circuit like that outlined above would allow the visual system to dynamically rescale and reposition the part of the visual field that is mapped into VWM. Our current conception of VWM is that it is a distinct and separate entity from the cortical areas involved in object recognition, although the circuits for mediating both are closely related or may, in fact be the same. New objects are scrutinized using VWM, but recognition is transferred to the object recognition area after the system becomes familiar with the object. In doing so, aspects of the 30 × 30 size of VWM and details of local spatial relationships can be maintained, but need not be. Thus, rapid estimates of what objects are in the visual field may take place without these details, but when the task becomes more complex, the details of local spatial relationships become very important.

A. Physiological Evidence

The model is consistent with a number of attentional effects seen in visual cortex, most notably receptive field shifts and gain modulation. Neurons at intermediate stages of the circuit would be expected to show pronounced receptive field shifts as attention is moved to different locations in the visual field, as shown in Fig. 3.1. Such shifts are evident in a number of neurons in V4 as reported by Connor et al. (1997). However, gain changes would also be expected, because according to the population coding framework above, the neuronal activities would be subject to modulation by control neurons (Eq. 7). Such gain changes were also reported by Connor et al. (1997).

The model futher stipulates that the beginnings of an object-centered reference frame should be established at intermediate or high levels of the cortical hierarchy. Several neurophysiological studies provide strong evidence for object-centered processing in area V4. Connor et al. (1997) demonstrated that responses to a given stimulus within a cell's receptive field can be dramatically modulated according to whether attention is directed to one side or another of the receptive field. Pasupathy and Connor (2001) demonstrated that many V4 neurons encode shape characteristics within restricted subregions of a larger object. In inferotemporal cortex, there is evidence for size-invariant tuning in many neurons. It has yet to be determined whether there is explicit coding of location within the attended region in inferotemporal cortex or in frontal and parietal areas engaged in visual working memory.

Most other neurophysiological studies of attention in V4 and other visual areas have focused on different aspects of attention (e.g., the degree of enhancement produced by attention to a given stimulus). The attentional modulation reported in these studies is generally less pronounced than that found by Connor et al. (1997). Thus, although much remains to be determined about the physiological basis of attentional processing, there is solid support for the general hypothesis of object-centered processing established across multiple stages of the cortical hierarchy.

B. Experimental Predictions

One of the main predictions of the model is that there exists a separate and distinct population of neurons—control neurons—the job of which is to route information flow in the cortex. The activity of such neurons would reflect solely the attentional state rather than the contents of the ROI per se. However, at lower levels of the cortical hierarchy, the control neurons will not necessarily be tied to overt attention (as discussed previously), and so they may be difficult to correlate with behavior.

Another major prediction of the model is that invariance is mediated by dynamic routing, rather than through a feedforward filter bank of invariant features as in a pandemonium-style model. As a consequence, one should expect to find congruency effects in object recognition; i.e., recognizing an object at one position and size should make it easier to recognize a subsequent (different) object at the same position and size because the ROI would not need to be changed. While there is already some evidence for such effects in recognition, additional experiments are needed to disambiguate what exactly is being primed in the system (features at the same location vs the state of control). A related prediction is that the ROI should be limited to a fixed resolution (~30 × 30 sample nodes). Previously we have documented several lines of evidence that suggest this idea (Van Essen et al., 1991), but the current state of affairs is still inconclusive and deserves further investigation.

VI. DISCUSSION

An important attribute of any model of brain function is that it not only be able to fit the available anatomical and physiological data, but also that it is capable of explaining function. The dynamic routing model of attention was designed to solve an important problem of vision—invariant object representation. In that regard it is distinct from many other mnodels that focus mainly on emulating neural response properties or aspects of psychophysical performance measured in laboratory settings. It also differs from many other models of object recognition in that it represents both the *variations* and the *invariances* of objects, rather than simply trying to capture the invariant part as in pandemonium-style models.

Since our model was originally proposed, several other investigators have developed alternative models for achieving invariant representations that also draw on the idea of dynamic routing. Salinas and Abbott (1997) proposed a model of area V4 that uses "gain fields" to transform reference frames and is similar to our development above using the population coding framework, although their model uses direct multiplication on each neuron rather than sigmoidal non-linearities. Amit and Mascaro (2003) proposed a hierarchical model of recognition that uses a replica module to represent many different shifted or rescaled versions of the input and then switches between them. This may be seen as a special case of our routing circuit in which the control basis functions correspond to entire shifts or rescalings of a large region of the image. Arathorn has proposed a model very similar to ours that uses multiplicative gating on the inputs, and he has some very compelling demonstrations of its effectiveness in real-world object recognition tasks (see Chapter 99).

Our proposal remains a zeroth-order model in the sense that there are still many components that require better specification to offer more detailed experimental predictions. Chief among these is the choice of feature representation: What is actually being represented by the sample nodes in each area? A promising approach to this question is to use unsupervised learning procedures to discover what forms of structure are best suited for representing natural images. Exploring such procedures for learning the features, in addition to the control neuron weights, remains the subject of current investigation.

References

Amit, Y., and Mascaro, M. (2003). An integrated network for invariant visual detection and recognition. *Vision Res.* **43**, 2073–2088.

Anderson, C. H., and Van Essen, D. C. (1987). Shifter circuits: a computational strategy for dynamic aspects of visual attention. *Proc. Natl. Acad. Sci. USA* **84**, 6297–6301.

Arathorn, D. W. (2002). "Map-Seeking Circuits in Visual Cognition." Stanford Univ. Press, Palo AHo, CA.

Connor C. E., Preddie, D. G., Gallant, J. L., and Van Essen, D. C. (1997). Spatial attention effects in macaque area V4. *J. Neurosci.* **17**, 3201–3214.

Eliasmith, C., and Anderson, C. H. (2003). "Neural Engineering." MIT Press.

Fukushima, K. (1980). Neocognitron: A self-organizing neural network model for a mechanism of pattern recognition unaffected by shift in position. *Biol. Cybernet.* **36**, 193–202.

Landauer, T. K. (1986). How much do people remember? Some estimates of the quantity of learned information in long-term memory. *Cogn. Sci.* **10**, 477–493.

LeCun, Y., Boser, B., Denker, J. S., Henderson, D., Howard, R. E., Hubbard, W., and Jackel, L. D. (1990). Backpropagation applied to handwritten Zip code recognition. *Neural Comput.* **1**, 541–551.

Mel, B. W. (1994). Information processing in dendritic trees. *Neural Comput.* **6**, 1031–1085.

Olshausen, B. A., Anderson, C. H., and Van Essen, D. C. (1993). A neurobiological model of visual attention and invariant pattern recognition based on dynamic routing of information. *J. Neurosci.* **13**, 4700–4719.

Olshausen, B. A., Anderson, C. H., and Van Essen, D. C. (1995). A multiscale dynamic routing circuit for forming size- and position-invariant object representations. *J. Comput. Neurosci.* **2**, 45–62.

Pasupathy, A., and Connor, C. E. (2001). Shape representation in area V4: position-specific tuning for boundary conformation. *J. Neurophysiol.* **86**, 2505–2519.

Riesenhuber, M., and Poggio, T. (1999). Hierarchical models of object recognition in cortex. *Nat. Neurosci.* **2**, 1019–1025.

Salinas, E., and Abbott, L. F. (1997). Invariant visual responses from attentional gain fields. *J. Neurophysiol.* **77**, 3267–3272.

Selfridge, O. G. (1959). Pandemonium: a paradigm for learning. *In* "The Mechanisation of Thought Processes," pp. 511–527. HMSO, London, 1959.

Van Essen, D. C., and Anderson, C. H. (1995). Information processing strategies and pathways in the primate visual system. *In* "An Introduction to Neural and Electronic Networks," (S. Zornetzer, J. L. Davis, C. Lau, T. McKenna, Eds.), 2nd ed., pp. 45–76. Academic Press, Orlando, FL.

Van Essen, D. C., Anderson, C. H., and Olshausen, B. A. (1994). Dynamic routing strategies in sensory, motor, and cognitive processing. *In* "Large Scale Neuronal Theories of the Brain" (C. Koch and J. Davis, Eds.), MIT Press, pp. 271–299.

Van Essen, D. C., Olshausen, B., Anderson, C. H., and Gallant, J. L. (1991). Pattern recognition, attention, and information bottlenecks in the primate visual system. *In* "Proceedings, SPIE Conference on Visual Information Processing: From Neurons to Chips," Vol. 1473, pp. 17–28.

Yantis, S., and Jonides, J. (1984). Abrupt visual onsets and selective attention: Evidence from visual search. *J. Exp. Psychol. Hum. Percept. Perform.* **10**, 601–621.

CHAPTER 4

Selective Attention as an Optimal Computational Strategy

Greg Billock, Christof Koch, and Demetri Psaltis

ABSTRACT

We explore selective attention as a key conceptual inspiration from neurobiology that can motivate the design of information processing systems. In our framework, an attentional window, the "spotlight of attention," contains some reduced set of data from the environment, which is then made available to higher-order processes for planning, real-time responses, and learning. This architecture is invaluable for systems with limited computational resources. Our test bed for these ideas is the control of an articulated arm. We implemented a system that learns while behaving, guided by the attention-based content of what the higher-order logic is currently engaged in. In the early stages of learning, the higher-order computational centers are involved in every aspect of the arm's motion. The attentionally assisted learning gradually assumes responsibility for the arm's behavior at various levels (motor control, gestures, spatial, logical), freeing the resource-limited higher-order centers to spend more time problem solving.

I. THE ATTENTION–AWARENESS MODEL: AN INTRODUCTION

Computers and software have recently joined the long line of human tools inspired by biology. In this case, it is the phenomenal capabilities of biological nervous systems that intrigue and challenge us. Our desire to mimic the brain stems from the abilities it possesses, which are in so many cases superior to those we can implement today.

We here explore the extent to which attentional selection can convey functional advantages to digital machines. By attentional selection we refer to the remarkable fact—documented throughout the book—that only a very small fraction of the incoming sensory information is accessible, in a conscious or unconscious manner, to influence behavior.

Many people have speculated about consciousness and its function. According to Crick and Koch (1988; Koch, 2004), the function of conscious visual awareness in biological systems is to

"[p]roduce the best current interpretation of the visual scene in the light of past experience, either of ourselves or of our ancestors (embodied in our genes), and to make this interpretation directly available, for a sufficient time, to the parts of the brain that contemplate and plan voluntary motor output, of one sort or another, including speech."

This representation consists of a reductive transformation of the massive, real-time sensory input data. That is, the content of awareness corresponds to the state of cache memory that holds a compact version of relevant sensory data as well as recalled items. This strategy can deal with more complex scenarios and generate a strategy for action (Newman et al., 1997). This flexible, but slow, aspect of the system, is complemented by a set of very rapid and highly specialized sensorimotor modules (D. Psaltis, personal communication, 1995), "zombie agents" (Koch, 2004), that perform highly stereotyped actions (e.g., driving a car, moving the eyes, walking, running, grasping objects).

Figure 4.1 illustrates one way in which these cognitive strategies may be mapped onto a machine architecture (Billock, 2001). The sections of the diagram toward the bottom—the motor/processing modules, early processing, and error generation—reside below the level of awareness, with fast reflexes and extensive procedural memories. Selective attention and aware-

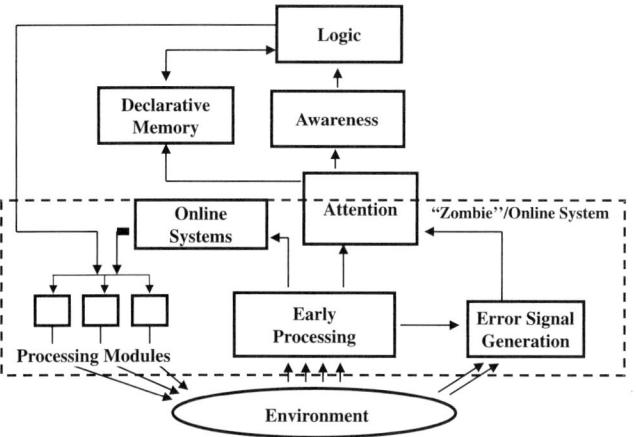

FIGURE 4.1 In this functional model of the role of attention and awareness, the pathway incorporating the attentional bottleneck operates in parallel with the faster sensorimotor agents (*zombie* systems), taking their cues from the "error signals" generated by the zombie systems.

ness are the gateways that provide preprocessed sensory data to the higher, more resource-constrained parts of the brain (the logic and planning cortices and memory). Of course, in reality many additional interconnections exist between these components. To maintain a coherent course of action, the system must be capable of alternating between volitional, top-down and reflex-level, bottom-up control.

What we would like to abstract from the biological functions of attention and awareness is a machine that can aid in performing similar tasks. We explore how, for sufficiently complex environments, using a reduced representation of the environment allows an algorithm to perform better *when under time pressure*, compared with an approach in which the entire input is represented. We would also like to understand better what advantages implementing such a bottleneck has for memory and machine learning.

II. LEARNING MOTION WITH AN ARTICULATED ARM

Machine learning is one area where we expect algorithms inspired by attentional selection strategy to outperform conventional ones. There are several ways in which attention might facilitate learning. One is during learning; if shown a single image of a car embedded in a dense background filled with other objects, the learning algorithm does not know which features belong to the object of relevance (here the car) and which ones are incidental. If attention would segment the car from the rest of the scene, however, superior performance can be obtained. This is particularly relevant to one-shot learning algorithms. The same is true during the recognition phase. Detecting the same car, say, under a different viewpoint, in a novel scene is much facilitated if an attentional selection strategy can segment the car from the background and just forward its associated features to the recognition module (see Rutishauser et al., 2004, for an illustration of this strategy).

Of course, segmentation also helps in reducing the amount of data that must be memorized, thus improving learning speed. Picking the right information to he learned and ignoring the rest is probably one of the key functions of attentional selection. Indeed, the resultant bottleneck appears to be necessary for the utilization of some kinds of memory (Naveh-Benjamin and Guez, 2000).

The test bed we use for exploring attentional learning is the control of a segmented arm moving around in a boxlike environment. It can pick up, move, and drop disks. At the most abstract level, the arm is used to solve various kinds of puzzles. The problem we explored was one of ordering various objects into target locations. This is equivalent to the Tower of Hanoi problem (Claus, 1884). In our version of this problem (see Fig. 4.2), we begin with an allotment of disks of various diameters. We assume that they have holes in their middle, that these disks are stacked in order of decreasing size (i.e., a larger disk must always be below a smaller one), and that the segmented arm can transfer the disks from one target stack to another one. The arm moves around the board and physically takes the top disk from each target and moves it to another stack, with the end goal of placing them in increasing size on a specific goal target. Various obstacles are placed on the board through which the arm cannot pass. The arm's segments can overlap as it moves. We assume that the end effector, when placed over a target, takes or releases a single disk automatically. Our problem, then, is to manipulate the joints of the arm to move its end effector between the appropriate targets in the correct order so as to solve the puzzle.

The details of the articulated arm, the playing board, and targets are shown in Fig. 4.2. For our purposes, we give the arm segments minimal dynamics involving a maximum torque and a momentum/friction decay characteristic. These force relationships are solved by the logic subsystem using a set of torque-change equations similar to those described by (Uno et al., 1989) for modeling human limb control. Initially, the system has not yet learned to drive its joints, and so must use its logic/planning functions to solve the control problem via explicit equations. The three-segment arm has a complicated inverse kinematics,

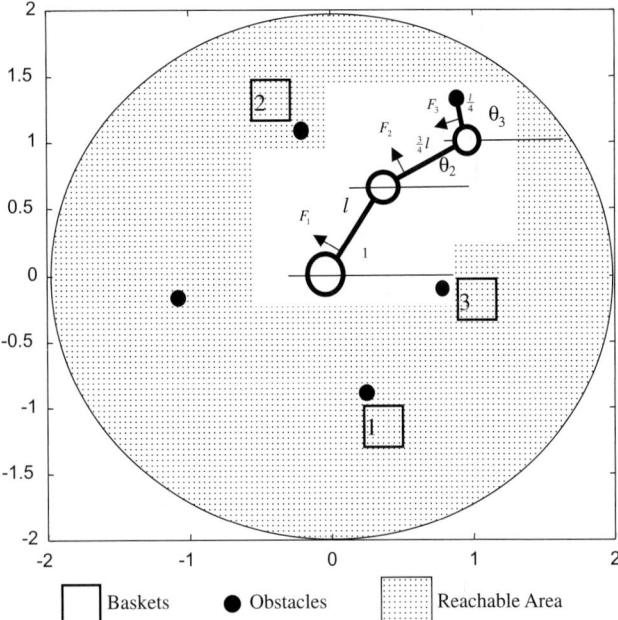

FIGURE 4.2 The Tower of Hanoi problem arranged for solution by an articulated arm. The arm must move between the marked targets without colliding with the solid obstacles. Outlined squares are the positions of the targets. Solid circles indicate obstacles around which the arm must navigate.

which requires an expensive optimization process to find the best trajectory to move from a present position to a target position. The minimum torque-change model selects a configuration out of the possible solutions that requires the minimum angular change of the arm segments to achieve. This is a very costly step in terms of required computational power, and so at first the attention of the logic unit is taken up by this low-level function.

As it does so, the reduced representation of the environment that the logic uses in solving the problem is presented to the "zombie" system for learning. Arm kinematics are learned by a neural network doing a straightforward function fit to the force curve necessary to move the arm from one angle to another. We use a three-layer neural network with four units in the hidden layer. The units use a hyperbolic tangent activation function and are fully interconnected. The input parameters are the current distance from the computed goal angle and the angular velocity of the arm segment. The output is the torque to be placed on the segment joint. The motion learning network starts its training once a sufficient number of samples (about 40 trajectories) have been collected so that it does not fall into a shallow local minimum and fail to learn the motor curves. We then use the Levenberg–Marquardt algorithm (Marquardt, 1963), which

learns rapidly over the next two dozen or so trajectories, after which it has trained enough to substantially take over arm kinematics from the logic unit.

During training, the control of the arm is shared by the network and the logic. This sharing is adjusted based on the current training error level of the network multiplied by a quality parameter that increases with the amount of training data. When the network is controlling arm motion, and fails to steer the arm to the required target, an error signal triggers the attentional mechanism, and the logic takes over and computes the kinematics, thus producing more training data which improves the network's performance as it takes over more and more the control of the arm.

This behaviorally guided learning approach produces excellent generalization: a training error of 3.46% (relative to the force given by the inverse kinematics solver) produced a test set error over the whole input field of 0.17% (both taken in the least-squares sense). For actual trajectories the test set error is still very low but is higher (around that of the training set error). This is because for actual trajectories, points tend to be more clustered in the areas of the input field where performance is critical, such as when the arm segment is close to its target location.

As basic kinematics are learned, the logic/planning system spends more time planning movements of the arm from one joint configuration to another. These motion plans are called *gestures*, such as "going around an obstacle clockwise." As a memory system for the gestural level, we use an ART-like neural network (Carpenter and Grossberg, 1987). This is an unsupervised learning model that autoclusters the trajectory data the logic presents to it during actual problem solving, and learns models for those clusters which can then be introduced into the control loop and largely replace the logic in computing gesture trajectories.

Each ART unit is associated with a linear neural network, which it uses to model the gesture parameters of the data with which it is associated. These linear networks have six inputs (a present and goal angle for each of the three arm segments) and six hidden units. Once a unit has more than three data points, it begins to train its associated neural network to model the data. This training also uses the Levenberg–Marquardt algorithm. If a new data point ruins the ability of the existing network to model the data well, an error signal is generated. The data point is then rejected and forms a new ART unit of its own (where it will compete with the existing units). If the new data point can be learned well, which usually happens, then it is incorporated into a new estimate of the mean and covariance of the unit's resonance region. The outputs of the neural network are relative coordinates for the segments of the arm to steer toward in completing the gesture. These coordinates then directly drive the kinematic level for controlling the arm joints to move the arm to that configuration.

Once an ART unit has more than six data points, it is allowed to begin to respond to the environment itself, and if the current arm parameters are within one standard deviation from a unit's center point in its input space, then that unit will be chosen to control the choice of the next trajectory path. Each unit, then, corresponds to a "gesture" that the system has learned. As the system solves puzzles, control shifts to the ART network. When no ART unit is found for the present environment, an error signal alerts the logic to calculate a new gesture trajectory. Training and operation overlap: if resonance occurs with one of the existing units, and that unit has sufficiently good performance, it is used to construct the next trajectory. If resonance occurs, but that unit fails to drive the arm successfully, an error signal causes the logic unit to compute the gesture, thus training the network. If no resonance occurs (meaning that no unit is responsible for dealing with the current state), then a new unit is created.

The consistency of the gestures permits the networks associated with the ART units to usually achieve least-squares training set errors below 10^{-3}, and frequently converge to the training threshold of 5×10^{-5} without significant overtraining. These errors are given in radians, which are target angles relative to current positions learned by the ART unit networks, and correspond to less than a tenth of a degree. On the other hand, similar gestures may not be repeated by every movement from one target to another, so it may take a while (in puzzle-solution time) for the actual arm behavior to lead to the accumulation of enough training examples for a particular unit. During solution of the first few puzzles, that is, sets of distributed targets, the logic spends a lot of the solution time (around 60%) planning gestures. After that, however, the zombie system begins to learn commonly repeated gestures and takes over the gesture planning. This reduces the total time spent in problem solution and also dramatically reduces the amount of time the logic spends "attending to" gestures to less than 10%.

The spatial sequence to follow when moving from one target to another is memorized using "declarative memory." These are basically memorized series of gestures: "to go from target 1 to target 3, first go clockwise around obstacle 2, then counterclockwise around obstacle 6, then straight on to target 3." These directions are learned as the logic solves the puzzle and continue to evolve as play proceeds. If this memory control fails, an error signal causes the logic unit to send the arm back to the originating target and try again by generating a new sequence of gestures itself. These will then replace the original sequence in memory. As it takes only one example for this memory to be useful, the declarative memory comes into play quite quickly. On the other hand, this scheme does not generalize well, except to different arrangements of the items to be sorted on the same playing board (meaning the targets and obstacles are in the same position, and only the initial arrangement of the disks to be sorted has changed).

III. ROLE OF THE ATTENTION–AWARENESS MODEL IN LEARNING

We now return to a discussion of the operation of the system in terms of the blocks of Fig. 4.1. The angular position and velocities of the arm segments and the positions of obstacles, targets, and disks form the environment. In a real-life situation a vision system might be used to extract these variables from the raw sensory data. In this environment, the job of the controller is to move the arm in the best way to solve the puzzle. The solution of the puzzle at the abstract level (i.e., which move to perform next with a view to solving the problem) always remains the province of the logic.

The different kinds of memory in the "zombie" system correspond to some varieties of memory humans employ to solve different problems. The procedural memory learns from examples to reproduce forces on the arm segments to cause desired motions. The unsupervised ART memory learns from examples of common gestures to take over motion planning. The declarative memory stores sequences of these gestures as spatial "directions" of how to move from one target to another.

The selective attention mechanism facilitates the learning process. As the system becomes more trained, the "zombie" gradually takes over control of the arm from the logic. This happens independently at the various levels as they become "reflexive" from the point of view of the logic, which then only "attends" to that level following an error signal. It spends more of its time on other parts of the problem, which then train other parts of memory.

The problem of controlling an articulated arm to solve a puzzle is one in which neither pure logic nor traditional machine learning is very good. When the logic has to plan out in detail all the motions of the arm, it can take a very long time to solve the problem. The conventional learning problem is intractable. For even quite small problems, there are dozens of dimensions in the learning problem, generating error signals is hard, and learning is very slow, if it would work at all.

This illustrates the role of the reduced representation of the "awareness window" in interfacing to different kinds of memory. In the arm example, the awareness window contains the single task that the zombie system failed to execute within acceptable error bounds and needs to be currently completed by the logic. The data in the awareness window continuously become the source of training examples for the zombie part of the system. In this way, the awareness mechanism splits up the problem into manageable "chunks." For declarative memory, the bottleneck reduces the amount of information necessary for it to learn useful patterns. For procedural memory, whether supervised or unsupervised, it assists by pruning out the information in the environment that is less relevant, reducing the dimensionality of the resulting patterns and speeding up learning. The zombie systems greatly improve the speed of overall puzzle solution: figuring out arm ballistics and computing near-optimal gestures are hard problems, and when these responses have been learned, solution times drop by an order of magnitude and more.

Figure 4.3 shows the fraction of time the system spends in the logic/planning subsystem and the training and recall of various memory subsystems. During solution of the first puzzle, the system spends almost 100% of its resources training the gesture and direction memories and only a few percent training the movement memory. However, as this is not so resource intensive, training is rather rapid once sufficient data are collected. As the system operates the fraction of time it spends training the gestures and direction sequence memories decreases, and more time is spent executing logic/planning overhead tasks (considering where to play next and so on). By the time the system is solving the fourth puzzle, resource utilization is taken up mostly by the logic/planning subsystem. Total solution time has dropped to below 20% of what was required for the first problem, and so responding to interrupts accounts for almost all of the time spent in this phase.

IV. CONCLUSIONS

Learning with the help of attentional selection that assists a logic/planning unit in training a hierarchical memory has several benefits. First, it dramatically reduces the dimensionality of the input space. The sorting problem as described has some 30 dimensions, with dependencies that are ill-suited to learning by a

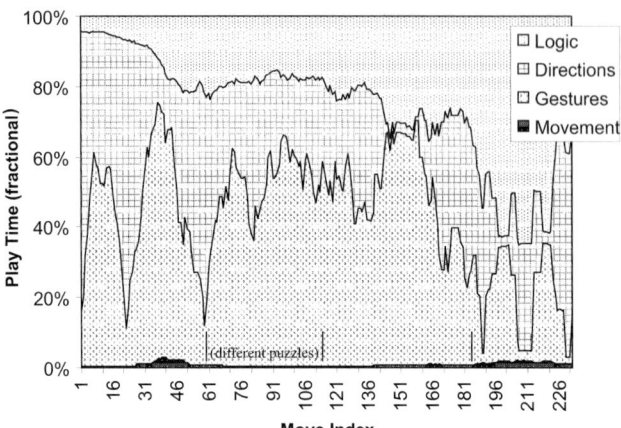

FIGURE 4.3 Fractional time spent by the system dealing with the various subsystems (movement, gesture learning and recall, directions learning and recall, and the overhead of operating the logic). The time shown is a running average over 15 moves.

traditional neural network. By segmenting the process and learning the reduced representations as used by the logic/planning subsystem to solve the problem at different levels, it becomes possible to present tractable problems to the learning algorithm.

Second, the hierarchical organization of skills enables those learned the fastest to be assumed during the learning of higher-level skills.

Third, the attentional mechanism as employed allows for cooperation between the memory subsystems and the logic/planning subsystems. When the faster network-based subsystems can respond, they do so. It is only when they make errors that the logic subsystem is aroused and spends time correcting the error.

The problems for which this approach is well-suited must satisfy at least two properties. First, they must be amenable to partial solutions. That is, an approximate solution must be initially acceptable. They cannot be of the sort where only a perfect solution is permissible. Second, they must exhibit the property that as additional information is gathered about the problem, that information becomes less and less important to the solution.

The common (although not universal) occurrence of these properties supports the argument that there is a wide class of problems, including many found in nature, whose solution is assisted by the kind of attentional selection architecture described above.

References

Billock, J. G. (2001). "Attentional Control of Complex Systems." Ph.D. thesis, California Institute of Technology.

Carpenter, G. A., and Grossberg, S. A. (1987). A massively parallel architecture for a self-organizing neural pattern recognition machine. *Computer Vision Graphics Image Process.* **37**, 54–115.

Claus, N. (1884). La Tour d'Hanoi: Jeu de Calcul. *Sci. Nat.* **1**, 127–128.
Crick, F., and Koch, C. (1998). Consciousness and neuroscience. *Cereb. Cortex* **8**, 97–107.
Garey, M. R., and Johnson, D. (1979). "Computers and Intractability: A Guide to the Theory of NP-Completeness." Freeman, San Francisco.
Koch, C. (2004). "The Quest for Consciousness: A Neurobiological Approach." Roberts, Denver, CO.
Marquardt, D. W. (1963). An algorithm for least-squares estimation of nonlinear parameters. *J. Soc. Ind. Appi. Math.* **11**, 431–441.
Naveh-Benjamin, M., and Guez, J. (2000). Effects of divided attention on encoding and retrieval processes: assessment of attentional costs and a componential analysis. *J. Exp. Psychol. Learn. Memory Cogn.* **26**, 1461–1482.
Newman, J., Baars, B. J., and Cho, S.-B. (1997). A neural global workspace model for conscious attention. *Neural Netw.* **10**, 1195–1206.
Rutishauser, U., Walther, D., Koch, C., and Perona, P. (2004). Is attention useful for object recognition? *IEEE Int. Conf. Computer Vision Pattern Recog.*, in press.
Uno, Y., Kawato, M., and Suzuki, R. (1989). Formation and control of optimal trajectory in human multijoint arm movement: minimum torque-change model. *Biol. Cybernet.* **61**, 89–101.

CHAPTER 5

Surprise: A Shortcut for Attention?

Pierre Baldi

ABSTRACT

Attention plays an essential role in the survival of an organism by enabling the real-time redirection or allocation of computing resources toward "regions" of the input field that are subjectively relevant. A particular class of relevant inputs are those that are unexpected or surprising. We propose an observer-dependent computational defintion of surprise by considering the relative entropy between the prior and the posterior distributions of the observer. Surprise requires integration over the space of models, in contrast with Shannon's entropy, which requires integration over the space of data. We show how surprise can be computed exactly in a number of discrete and continuous cases using distributions from the exponential family with conjugate priors. Surprise can be defined at multiple scales, from single synapses, to neurons, to networks, to areas and systems. Surprise provides a shortcut for relevance and a general principled way for computing centralized saliency maps in any feature space to control the deployment of attention or other information retrieval mechanisms toward the most surprising items, that is, those carrying the largest amount of information with respect to internal expectations. Because attention is a rapid mechanism likely to be driven by bottom-up cues, it could benefit from the simplicity of surprise calculation to detect mismatches between bottom-up inputs and expectations generated locally or in top-down fashion.

I. INTRODUCTION

Attention plays a fundamental role in the survival of an organism. In particular, attention enables the real-time redirection or allocation of computing resources toward particular "regions" of the input field that are relevant for the organism. A fundamental question, of course, is the algorithmic definition of what is relevant for a given organism in a given situation and at a given time. A particular class of inputs that tend to be relevant are those that are highly unusual or "surprising." Thus, here we address the problem of providing a precise algorithmic definition of surprise and speculate on its possible relationships to attentional mechanisms.

A definition of novelty or surprise, it must first of all be related to the foundations of the notion of probability, which can be approached from a frequentist or subjectivist, also called bayesian, point of view (Berger, 1985; Box and Tiao, 1992). Here we follow the bayesian approach, which has been prominent in recent years and has led to important developments in many fields (Gelman et al., 1995). The definition we propose stems directly from the bayesian foundation of probability theory, and the relation given by Bayes' theorem between the prior and posterior probabilities of an observer. The amount of surprise in the data for a given observer can be measured by the change that has taken place in going from the prior to the posterior probabilities.

A. Information and Surprise

To define surprise, we consider an "observer" with a corresponding class of hypotheses or models $\mathcal{M}$. In the subjectivist framework, degrees of belief or confidence are associated with hypotheses or models. It can be shown that under a small set of reasonable axioms, these degrees of belief can be represented by real numbers and that when rescaled to the [0, 1] interval, these degrees of confidence must obey the rules of probability and, in particular, Bayes' theorem (Cox, 1964; Jaynes, 1986; Savage, 1972). Within the bayesian framework, prior to seeing any data, the observer has a prior distribution $P(M)$ over the space $\mathcal{M}$ of models. Here the word *observer* is taken in a very general sense and could, for instance, refer to a single neuron or a neuronal circuit. The prior $P(M)$ could be hardwired or result from experience, possibly including recently observed data points. It could be generated locally and/or in top-down fashion.

If an observer has a model M for the data, associated with a prior probability $P(M)$, the arrival of a data set D leads to a reevaluation of the probability in terms of the posterior distribution:

$$P(M|D) = \frac{P(D|M)P(M)}{P(D)}. \quad (1)$$

The effect of the information contained in D is clearly to change the belief of the observer from $P(M)$ to $P(M|D)$. Thus, the surprise can be defined as the "distance" between the prior and posterior distributions: highly surprising data should correspond to substantive reevaluation of the prior and, therefore, to larger distances. Thus, a way of measuring information carried by the data D complementary to Shannon's definition of entropy,

$$I(D, M) = H(P(D|M)) = -\int_D P(D|M)\log P(D|M)dD, \quad (2)$$

(Shannon, 1948; Cover and Thomas, 1991) is to measure the distance between the prior and the posterior. To distinguish it from Shannon's communication information, we call this notion of information the surprise information or *surprise* (Baldi, 2002),

$$S(D, M) = d[P(M), P(M|D)], \quad (3)$$

where d is a distance or similarity measure. There are different ways of measuring a distance between probability distributions. In what follows, for standard well-known theoretical reasons (including invariance with respect to reparameterizations), we use the relative entropy or Kullback–Liebler (Kullback, 1968) divergence K, which is not symmetric and hence not a distance. This lack of symmetry, however, does not matter in most cases and in principle can easily be fixed by symmetrization of the divergence. The surprise then is

$$\begin{aligned}S(D, M) &= K(P(M), P(M|D)) \\ &= \int_M P(M)\log\frac{P(M)}{P(M|D)}dM \\ &= -H(P(M)) - \int P(M)\log P(M|D)dM \quad (4) \\ &= \log P(D) - \int_M P(M)\log P(D|M)dM.\end{aligned}$$

Alternatively, we can define the single model surprise by the log odd ratio

$$S(D, M) = \log\frac{P(M)}{P(M|D)}, \quad (5)$$

and the surprise by its average

$$S(D, M) = \int_M S(D, M)P(M)dM, \quad (6)$$

taken with respect to the prior distribution over the model class. In statistical mechanics terminology, the surprise can also be viewed as the free energy of the negative log posterior at a temperature $t = 1$, with respect to the prior distribution over the space of models (Baldi and Brunak, 2001).

Note that the definition of suprise addresses the "white snow" paradox of information theory according to which white snow, the most boring of all television programs, carries the largest amount of Shannon information. At the time of snow onset, the image distribution we expect and the image we perceive are very different, and therefore the snow carries a great deal of both surprise and Shannon's information. Indeed, snow may be a sign of storm, earthquake, toddler's curiosity, or military putsch. But after a few seconds, once our model of the image shifts toward a snow model of random pixels, television snow perfectly fits the prior and hence becomes boring. Because the prior and the posterior are virtually identical, snow frames carry zero surprise although megabytes of Shannon's information.

II. COMPUTATION OF SURPRISE

To be useful, the notion of suprise ought to be computable analytically, at least in simple cases. Here we consider a data set $D = \{x_1, \ldots, x_N\}$ containing N points. Surprise can be calculated exactly in a number of interesting cases. For simplicity, although this does not correspond to any restriction of the general theory, we consider here only the case of conjugate priors, where the prior and the posterior have the same functional form. In this case, to compute the surprise defined by Eq. (4), we need only compute general terms of the form

$$F(P_1, P_2) = \int P_1 \log P_2 dx, \quad (7)$$

where P_1 and P_2 have the same functional form. The surprise is then given by

$$S = F(P_1, P_1) - F(P_1, P_2), \quad (8)$$

where P_1 is the prior and P_2 is the posterior. Note also that in this case the symmetric divergence can be easily computed using $F(P_1, P_1) - F(P_1, P_2) + F(P_2, P_2) - F(P_2, P_1)$. It should also be clear that in simple cases, for instance, for certain members of the exponential family (Brown, 1986) of distributions, the posterior depends entirely on the sufficient statistics and therefore we can expect surprise also to depend only on sufficient statistics in these cases. Additional examples and mathematical details can he found in Baldi (2002).

A. Discrete Data and Dirichlet Model

Consider the case where x_i is binary. The simplest class of models for D is then $M(p)$, the first-order Markov models, with a single parameter p representing the probability of success or of emitting a 1. The conjugate prior on p is the Dirichlet prior (or beta distribution in the two-dimensional case),

$$D_1(a_1, b_1) = \frac{\Gamma(a_1+b_1)}{\Gamma(a_1)\Gamma(b_1)} x^{a_1-1}(1-x)^{b_1-1} \qquad (9)$$
$$= C_1 x^{a_1-1}(1-x)^{b_1-1},$$

with $a_1 \geq 0$, $b_1 \geq 0$, and $a_1 + b_1 > 0$. The expectation is $a_1/(a_1+b_1)$, $b_1/(a_1+b_1)$. With n sucesses in the sequence D, the posterior is a Dirichlet distribution $D_2(a_2, b_2)$ with (Baldi and Brunak, 2001)

$$a_2 = a_1 + n \quad \text{and} \quad b_2 = b_1 + (N-n). \qquad (10)$$

The surprise can be computed exactly as

$$S(D, M) = K((D_1, D_2)) = \log \frac{C_1}{C_2} + n[\Psi(a_1+b_1) - \Psi(a_1)] \\ + (N-n)[\Psi(a_1+b_1) - \Psi(b_1)], \qquad (11)$$

where Ψ is the derivative of the logarithm of the gamma function. When $N \to \infty$, and $n = pN$ with $0 < p < 1$, we have

$$S(D, M) \approx NK(p, a_1), \qquad (12)$$

where $K(p, a_1)$ represents the Kullback–Liebler divergence distance between the empirical distribution $(p, 1-p)$ and the expectation of the prior $(a_1/(a_1+b_1), b_1/(a_1+b_1))$. Thus, asymptotically surprise information grows linearly with the number of data points with a proportionality coefficient that depends on the discrepancy between the expectation of the prior and the observed distribution. The same relationship can be expected to be true in the case of a multinomial model with an arbitrary number of classes. In the case of a symmetric prior ($a_1 = b_1$), a slightly more precise approximation is provided by

$$S(D_1, D_2) \approx N\left[\sum_{k=a_1}^{2a_1-1} \frac{1}{k} - H(p)\right]. \qquad (13)$$

For instance, when $a_1 = 1$ then $R(D_1, D_2) \approx N(1 - H(p))$, and when $a_1 = 5$ then $R(D_1, D_2) \approx N[0.746 - H(p)]$.

B. Continuous Data: Unknown Mean/Known Variance

When the x_i are real, we can consider first the case of unknown mean with known variance. We have a family $M(\mu)$ of models, with a gaussian prior $G_1(\mu_1, \sigma_1^2)$. If the data have known variance σ^2, then the posterior distribution is gaussian $G_2(\mu_2, \sigma_2^2)$, with parameters given by (Gelman et al., 1995)

$$\mu_2 = \frac{\mu_1}{\sigma_1^2} + \frac{N\overline{m}}{\sigma^2} \bigg/ \frac{1}{\sigma_1^2} + \frac{N}{\sigma^2} \quad \text{and} \quad \frac{1}{\sigma_2^2} = \frac{1}{\sigma_1^2} + \frac{N}{\sigma^2}, \qquad (14)$$

where $\overline{m}$ is the observed mean. In the general case,

$$S(D, M) = KG_1, G_2) = \log \frac{\sigma}{\sqrt{\sigma^2 + N\sigma_1^2}} + N\frac{\sigma_1^2}{2\sigma^2} \\ + \frac{N^2\sigma_1^2(\mu_1-\overline{m})^2}{2\sigma^2(\sigma^2+N\sigma_1^2)} \qquad (15) \\ \approx \frac{N}{2\sigma^2}[\sigma_1^2 + (\mu_1+\overline{m})^2],$$

the approximation being valid for large N. In the special case where the prior has the same variance as the data $\sigma_1 = \sigma$, then the formula simplifies a little and yields

$$S = K(G_1, G_2) = \frac{N}{2} - \frac{1}{2}\log(N+1) + \frac{N^2(\mu_1-\overline{m})^2}{2(N+1)\sigma^2} \\ \approx \frac{N}{2\sigma^2}[\sigma^2 + (\mu_1-\overline{m})^2] \qquad (16)$$

when N is large. In any case, surprise grows linearly with N with a coefficient that is the sum of the prior variance and the square difference between the expected mean and the empirical mean scaled by the variance of the data. The cases with known mean and unknown variance, or unknown mean and unknown variance, are treated in Baldi (2002).

III. HABITUATION AND SURPRISE

There is an immediate connection between surprise and learning, or rather habituation. If we imagine that data points from a training set are presented sequentially, we can consider that the posterior distribution after the Nth point becomes the prior for the next iteration (sequential bayesian learning). In this case we can expect a form of habituation whereby, on average, surprise decreases after each iteration. As a system learns what is relevant in a data set, new data points become less and less surprising. This can be quantified precisely, at least in simple cases.

Consider, for example, a sequence of 0–1 examples $D = (d_N)$. The learner starts with a Dirichlet prior $D_0(a_0, b_0)$. With each example d_N, the learner updates its Dirichlet prior $D_N(a_N, b_N)$ into a Dirichlet posterior $D_{N+1}(a_{N+1}, b_{N+1})$ with $(a_{N+1}, b_{N+1}) = (a_N + 1, b_N)$ if $d_{N+1} = 1$, and $(a_{N+1}, b_{N+1}) = (a_N, b_N + 1)$ otherwise. When $d_{N+1} = 1$, the corresponding surprise is easily computed using the theory developed above. For simplicity, and without much loss of generality, let us assume that a_0 and b_0 are integers, so that a_N and b_N are also integers for any N. Then if $d_{N+1} = 1$, the relative surprise is

$$S(D_N, D_{N+1}) = \log \frac{a_N}{a_N + b_N} + \sum_{k=0}^{b_N-1} \frac{1}{a_N + k}, \quad (17)$$

and similarly in the case $d_{N+1} = 0$ by interchanging the role of a_N and b_N. Thus, in this case,

$$0 \leq S(D_N, D_{N+1}) \leq \frac{1}{a_N} + \log\left(1 - \frac{1}{a_N + b_N}\right). \quad (18)$$

Asymptotically we have $a_N \approx a_0 + pN$ and therefore

$$0 \leq S(D_N, D_{N+1}) \leq \frac{1-p}{pN}. \quad (19)$$

Thus, surprise decreases in time with the number of examples as $1/N$. A similar result can be obtained in the continuous case (Baldi, 2002).

IV. DISCUSSION: A SHORTCUT FOR ATTENTION

While eminently successful for the *transmission* of data, Shannon's theory of information does not address semantic and subjective dimensions of data, such as relevance and surprise. We have proposed an observer-dependent computational theory of surprise where surprise is defined by the relative entropy between the prior and the posterior distributions of an observer. As such, it is a measure of dissimilarity between the prior and posterior distributions which lies close to the axiomatic foundation of bayesian probability. Surprise is different from Shannon's information. While Shannon's definition fixes the model and varies the data, surprise fixes the data and varies the model. Surprise requires integration over the space of models, in contrast with Shannon's entropy, which requires integration over the space of data. In a number of cases, surprise can be computed analytically in terms of both exact and asymptotic computationally efficient, formulas in the discrete and continuous case. During sequential bayesian learning, habituation corresponds to a $1/N$ decay of surprise. In general, however, the computation of surprise can be expected to require Monte Carlo methods to approximate integrals over model spaces. In this respect, the computation of surprise should benefit from progress in Markov chain and other Monte Carlo methods, as well as progress in computing power.

Granted that the mathematical foundation of surprise is solid, we can speculate that surprise may he used to guide rapid attention mechanisms. With limited computing resources available relative to the volume of data, rapid identification and ranking of unusual items that require further processing to establish semantic relevance become important. Surprise provides a shortcut to relevance and a general principled way for computing centralized saliency maps in *any* feature space to control the deployment of attention or other information retrieval mechanisms toward the most surprising items, that is, those carrying the largest amount of information with respect to internal expectations. Because attention is a rapid mechanism likely to be driven by bottom-up cues, it could benefit from the simplicity of surprise calculation to detect mismatches between bottom-up inputs and expectations generated locally or in top-down fashion.

The relationship of surprise to attention then boils down to two questions: (1) Is surprise being used by biological attention mechanisms? (2) Can surprise be used as a neuroengineering principle for the design of artificial attention systems? Testing the first hypothesis in detail would require showing that probability distributions, such as priors and posteriors, are part of the biological computations carried by natural attention systems. In general, this is beyond the current state-of-the-art in neuroscience and it is unlikely that clear answers can be provided in the very near future, indirect evidence, but from psychophysical experiments may be possible.

The second question is somewhat more promising—as often the case in science, direct problems tend to be easier than reverse engineering problems—and hinge on simulating or implementing artificial attention algorithms and circuits. A positive feature of surprise as a general neural engineering design principle is the fact that it can be applied uniformly at multiple temporal and spatial scales in any feature space or sensory modality and that it can be aggregated. In principle, we can talk as well of the surprise of a synapse, a neuron, a neuronal circuit, an area, a system, or even an entire organism and, for instance, in visual, auditory, olfactory spaces. Consider, for instance, an angle detector in the [0°, 360°] range that is tuned for angles around 90°, that is, with a circular hill-shaped response curve centered at 90°. We may as well consider that this detector is computing surprise with a prior that is hill-shaped and centered at −90°.

Finally, one should not forget that Shannon's entropy, surprise, and relevance are three different facets of information that can be present in different combinations. If, while surfing the Web in search of a car, one stumbles on an old picture of Brigitte Bardot, the picture may carry a low degree of relevance, a high degree of surprise, and a small to large amount of Shannon information depending on the pixel structure. Although, despite of several attempts (Jumarie, 1990; Tishby et al., 1999), the notion of *relevance* remains the least understood, over short time scales surprise may provide an adequate approximation to relevance. Indeed, over short time scales, the overlap between relevant and surprising inputs may be large and it may also be safer for the organism not to ignore surprising inputs during early processing stages even if, on second inspection, some of these surprising inputs turn out to be irrelevant. Thus, in guiding attention, surprise may be used as a computational shortcut to relevance, but not as a perfect substitute. In this view, relevance often must await confirmation by additional and slower processes that extend beyond the realm of rapid attention.

Acknowledgments

The work of P.B. is supported by a Laurel Wilkening Faculty Innovation Award and grants from the National Institutes of Health, National Science Foundation, and Sun Microsystems at the University of California, Irvine.

References

Baldi, P. (2002). A computational theory of surprise. *In* "Information, Coding, and Mathematics" (M. Blaum, P. G. Farrell, and H. C. A. van Tilborg, Eds.). Kluwer Academic, Boston.

Baldi, P., and Brunak, S. (2001). "Bioinformatics: The Machine Learning Approach", 2nd ed. MIT Press, Cambridge, MA.

Berger, J. O. (1985). "Statistical Decision Theory and Bayesian Analysis." Springer-Verlag, New York.

Box, G. E. P., and Tiao, G. C. (1992). "Bayesian Inference in Statistical Analysis." Wiley, New York.

Brown, L. D. (1986). "Fundamentals of Statistical Exponential Families." Institute of Mathematical Statistics, Hayward, CA.

Cover, T. M., and Thomas, J. A. (1991). "Elements of Information Theory." Wiley, New York.

Cox, R. T. (1964). Probability, frequency and reasonable expectation. *Am. J. Phys.* **14**, 1–13.

Gelman, A., Carlin, J. B., Stern, H. S., and Rubin, D. B. (1995). "Bayesian Data Analysis." Chapman & Hall, London.

Jaynes, E. T. (1986). Bayesian methods: general background. *In* "Maximum Entropy and Bayesian Methods in Statistics" (J. H. Justice, Ed.), pp. 1–25. Cambridge Univ. Press, Cambridge.

Jumarie, G. (1990). "Relative Information." Springer-Verlag, New York.

Kullback, S. (1968). "Information Theory and Statistics." Dover, New York.

Savage, L. J. (1972). "The Foundations of Statistics." Dover, New York.

Shannon, C. E. (1948). A mathematical theory of communication. *Bell Syst. Tech. J.* **27**, 379–423, 623–656.

Tishby, N., Pereira, F., and Bialek, W. (1999). The information bottleneck method. *In* "Proceedings of the 37th Annual Allerton Conference on Communication, Control, and Computing", (B. Hajek and R. S. Sreenivas, Eds.), pp. 368–377. University of Illinois.

CHAPTER 6

A Heteromodal Large-Scale Network for Spatial Attention

M-Marsel Mesulam, Dana M. Small, Rik Vandenberghe, Darren R. Gitelman, and Anna C. Nobre

ABSTRACT

Hemispatial neglect is usually designated a "parietal syndrome." However, neglect can also arise after lesions in the frontal lobes, cingulate gyrus, striatum, and thalamus. These areas belong to an interconnected large-scale network subserving all aspects of spatial attention. This network helps to compile a mental representation of extrapersonal events in terms of their motivational salience, and to generate "kinetic strategies" so that the attentional focus can shift from one target to another. In the human, the left hemisphere controls attention predominantly within the contralateral right hemispace, whereas the right hemisphere controls attention in both hemispaces. As a consequence of this asymmetry, severe contralesional neglect occurs almost exclusively after right hemisphere lesions and encompasses the left side of extrapersonal space.

I. INTRODUCTION

The term *spatial attention* designates interrelated sensory, motor, and cognitive processes that collectively enable the selective allocation of neural resources to motivationally relevant parts of the environment. The spatial representation of events according to their emotional salience, their targeting for oculomotor fixation or manual grasp, the anticipatory biasing of regions where interesting events are expected to occur, and the search for objects embedded among distracters are among the manifestations of spatial attention. Certain kinds of brain damage disrupt spatial attention by interfering with the ability to attend to the opposite (contralesional) side of space. The resultant syndrome of contralesional hemispatial neglect is one of the most disabling consequences of stroke. The availability of animal models for neglect-like deficits, single-unit recordings in awake and behaving monkeys, the ability to use analogous behavioral tasks in laboratory primates and humans, and the advent of functional imaging have provided unprecedented opportunities for exploring the neurology of hemispatial neglect and spatial attention (Mesulam, 1981, 1999).

II. HEMISPATIAL NEGLECT

Any aspect of spatial attention can become impaired in patients with hemispatial neglect. Such patients may fail to dress or groom the left side of the body, copy the left side of drawings, or read the left half of sentences. Some patients ignore contralesional sensory events only in the presence of ipsilesional perceptual rivalry; others fail to make contralesional reaching movements when manually searching for an item; and still others fail to notice the contralesional side of an object, even when it is located in the ipsilesional space. Events that are not recognized at the conscious level may nonetheless exert an implicit influence on behavior; locations that are not attended during effortful search may attract attention if they contain stimuli that perceptually pop out from the background; neglect may be manifested in some sensory modalities but not others, and its severity may vary according to the emotional valence of the stimuli to be detected. This behavioral complexity of hemispatial neglect is matched by the complexity of the network subserving spatial attention and its interactions with other large-scale cognitive networks of the brain.

III. THE SPATIAL ATTENTION NETWORK

The single most common lesion site associated with neglect is located in the inferior parietal lobule. Textbooks of neurology have therefore tended to refer to hemispatial neglect as a "parietal syndrome." Clinical case studies and targeted ablations in monkeys, however, have shown that contralesional neglect and neglectlike phenomena can also arise after lesions in premotor cortex, cingulate gyrus, thalamus, and striatum. These observations, together with additional axonal transport and single-unit studies in monkeys, led to the suggestion that spatial attention is normally mediated by a parietofrontocingulate network containing subcortical components in the neostriatum, mediodorsal/pulvinar thalamic nuclei, and superior colliculus (Mesulam, 1999).

Functional imaging experiments have confirmed the existence of this network in the normal human brain (Corbetta et al., 1993; Gitelman et al., 1996; Nobre et al., 1997). According to these experiments, the parietal component of the spatial attention network appears to be centered around the intraparietal sulcus and to extend into the adjacent inferior and superior parietal lobules; the frontal component appears to be centered around the frontal eye fields (FEFs) and to extend into adjacent regions of frontal cortex; and the cingulate component appears to have an anterior sector extending into the supplementary motor area and a posterior sector extending into the retrosplenial area. This network is recruited by exogenously or endogenously triggered covert shifts of attention, visuomotor search, central attentional fixation, nonvisual manual target search, and audiospatial attention (Gitelman et al., 1996; Nobre et al., 1997; Zatorre et al., 1999). Engagement in such tasks increases the functional intercorrelation (coherence) among network components (Lipschutz et al., 2002). Recent observations have suggested that the parietotemporal junction, temporo-occipital cortex (area TO or MT+), superior temporal gyrus, insula, and inferior frontal region may also participate in spatial attention, but the role of these additional areas remains to be specified (see Mesulam, 1999, for review of the relevant literature).

It is reasonable to assume that a major function of the spatial attention network is to coordinate the top-down modulation (or control) of sensory and motor areas, where the actual work of spatial attention is ultimately executed (Yantis et al., 2002). In contrast to this "dorsal" network that directs attention to the location of behaviorally relevant extrapersonal events, a "ventral" network revolving around the fusiform gyrus mediates the selection of specific visual objects and their features for attentional targeting (Desimone, 1996). A monosynaptic pathway interconnecting the fusiform gyrus with the inferior parietal lobule has been identified in the rhesus monkey and is likely to integrate these two major streams of information processing (Mesulam et al., 1977).

IV. BRAIN–BEHAVIOR CORRELATIONS

Is there a unitary cognitive deficit underlying neglect? Are there consistent relationships between behavioral aspects of spatial attention and anatomical components of the spatial attention network? These two questions have triggered a great deal of research. Some investigators have suggested that all manifestations of neglect are based on distortions of internal representations, whereas others have proposed that they arise as a consequence of impaired intentional movements. Such unifying explanations have been difficult to reconcile with the multiple dissociations in the symptomatology of neglect. Some patients, for example, show neglect when bisecting a line but not when searching for targets, while others display the converse dissociation; some patients show sensory extinction but no distortion of figure copying while the opposite pattern is seen in others. These dissociations and the traditional tendency to link the parietal lobe with "sensory" functions and the frontal lobe with "motor" functions led to the hypothesis that lesions within the parietal component of the network would account for the perceptual aspects of neglect (such as extinction or line bisection errors), whereas lesions within the frontal component would account for the motor manifestations (such as target search failures). Despite some initial support for this contention, the bulk of evidence has failed to confirm such functional segregations within the spatial attention network (see Mesulam, 1999, for review).

There are at least three reasons why simple sensorimotor (or parietofrontal) dichotomies may have failed to provide the basis for reliable clinicopathological correlations in the neglect syndrome. First, sensory representations are necessary for guiding exploratory behaviors, and exploratory behaviors are necessary for updating representations. Second, neurons in the parietal and frontal components of the spatial attention network have motor as well as sensory fields (Andersen et al., 1997; Paré and Wurtz, 1997). Third, the intraparietal sulcus and FEFs are so strongly interconnected that damage to one induces dysfunction in the other (Chafee and Goldman-Rakic, 2000). It therefore appears that the frontal and parietal components of the spatial attention network jointly establish a level of sensorimotor integration where the boundaries between action and perception become blurred and where dysfunction in one induces dysfunction in the other.

V. INTRANETWORK SPECIALIZATIONS

The advent of functional imaging has shifted the emphasis away from global sensorimotor dichotomies. Behavioral observations show that spatial attention represents the collective outcome of numerous parallel (as in the "popout" phenomenon), serial (i.e., search-related), representational, and intentional processes, each reflecting different levels of sensorimotor integration. A major goal in functional imaging has been to understand how these cognitive computations are mapped onto the anatomical components of the spatial attention network.

The functional heterogeneity within the parietal component of the spatial attention network is in the process of being clarified, and the relevant literature contains a number of inconsistencies. Functional imaging experiments in humans show that the intraparietal sulcus and inferior parietal lobule mediate the cue-induced voluntary reorientation of spatial attention (Corbetta et al., 2000; Hopfinger et al., 2000). Other studies, however, show that the superior parietal lobule may mediate the shifting of the attentional focus, whereas the inferior parietal lobule may promote its focalization (Vandenberghe et al., 2001; Yantis et al., 2002). The slightly more ventral temporoparietal junction of the inferior parietal lobule appears to play a specific role in redirecting attention to unattended locations (Corbetta et al., 2000). The inferior parietal sulcus may also display regional specializations for the search of single versus conjoined features (Donner et al., 2002), although the possibility has been raised that such variations may reflect task difficulty and efficiency rather than the requirement for feature binding (Nobre et al., 2003). The FEF has an internal functional segregation of its own, characterized by a partially overlapping mosaic of areas mediating nonattentional eye movements, visual search, working memory, and covert shifts of attention (Courtney et al., 1998; Gitelman et al., 2002). Functional heterogeneity is also emerging within the cingulate component of the network, where an anterior sector seems to mediate processes related to cognitive control, target detection, vigilance, conflict resolution, and response selection, whereas a posterior sector seems to mediate the anticipatory allocation of attention toward locations where relevant events are expected to occur (Small et al., 2003a). Posterior cingulate activations linked to anticipatory spatial biasing are further enhanced by monetary incentive, supporting the contention that posterior cingulate neurons mediate the motivational modulation of spatial attention (Small et al., 2003b).

An intriguing question is whether the shifting of attention is encoded on the basis of a "place code" that signals the *location* of shifts or on the basis of a "vector code" that signals their *direction*, regardless of location. A functional imaging study found that the laterality of cerebral activation depended on the hemispace within which attention was being shifted rather than on the direction of the shifts within the hemispace (Corbetta et al., 1993). Clinical studies, on the other hand, suggest that unilateral brain damage interferes with contraversive shifts of attention irrespective of the hemispace or field where the shift occurs (Posner et al., 1987). This aspect of spatial attention may display regional differences in its functional anatomy. For example, the parietal component of the spatial attention network appears to mediate a vector code of attentional displacement, whereas the frontal component appears to mediate a place code (Husain et al., 2000).

VI. HEMISPHERIC ASYMMETRY

A characteristic feature of hemispatial neglect is its asymmetry. Contralesional left neglect is common, whereas contralesional right neglect is rare, even in lefthanders. This asymmetry has led to the hypothesis that the left hemisphere directs attention predominantly to the contralateral right side, whereas the right hemisphere directs attention to both sides of space. According to this model, unilateral left hemisphere lesions are not expected to cause much contralesional neglect as the ipsiversive attentional functions of the right hemisphere are likely to take over. Right hemisphere lesions, however, are expected to trigger severe left neglect because the left hemisphere has no substantial ipsiversive attentional functions. Experiments based on evoked potentials, transcranial magnetic stimulation, and functional imaging have supported this model (see Mesulam, 1999, for review). Another prediction of this model, that naturalistic settings where attention is symmetrically distributed to both sides of space would cause a greater engagement of the right hemisphere, has received preliminary support (Gitelman et al., 2002).

One of the most unexpected implications to emerge from functional imaging experiments is the possibility that different aspects of spatial attention may be associated with different patterns of asymmetry. For example, rightward asymmetry appears more pronounced in the *parietal lobe* for exogenously triggered covert shifts of attention and in the *FEFs* for visual search (Gitelman et al., 2002). Spatial attention and hemispatial neglect offer model systems for investigating the distinctively human phenomenon of functional hemispheric asymmetry. Understanding the principles of hemispheric specialization may have extensive implications for exploring factors that facilitate (or hinder) compensatory cerebral reorganization following focal brain injury.

VII. INTERNETWORK RELATIONSHIPS

None of the network components mentioned above is dedicated to spatial attention. Numerous functional imaging studies have shown that the parietal, frontal, and cingulate components of the spatial attention network are also parts of networks that mediate saccadic eye movements, working memory, temporal orientation, and motivational modulation (LaBar et al., 1998; Nobre et al., 2000; Nobre, 2001; Maddock et al., 2003). Through the mediation of these common areas, other large-scale networks can be recruited in a manner that serves the specific behavioral goals of the spatial attention task. For example, the requirement for oculomotor search co-activates the superior colliculus; motivational modulations or violations of spatial expectations elicit the co-activation of orbitofrontal and mediotemporal limbic areas; and the attentional targeting of spatial representations held in working memory co-activates prefrontal cortical areas (Nobre et al., 1999; Armony and Dolan, 2002; Gitelman et al., 2002; Small et al., 2003b; Nobre et al., 2004).

The anatomy of internetwork overlap and the physiology of how an area becomes recruited into one network versus another in a task-dependent fashion remain poorly understood. The latter process could conceivably reflect changes in the interregional correlation (i.e., effective connectivity or coherence) of neuronal signals. Tasks of spatial attention, for example, increase the correlation of inferior parietal lobule activity with activity in areas adjacent to the FEFs, whereas tasks of face perception increase the correlation between the same inferior parietal region and the visual association cortex of the fusiform gyrus (McIntosh et al., 1994).

VIII. CONCLUSIONS

Many accounts of hemispatial neglect have been published, ranging from the phenomenological to the

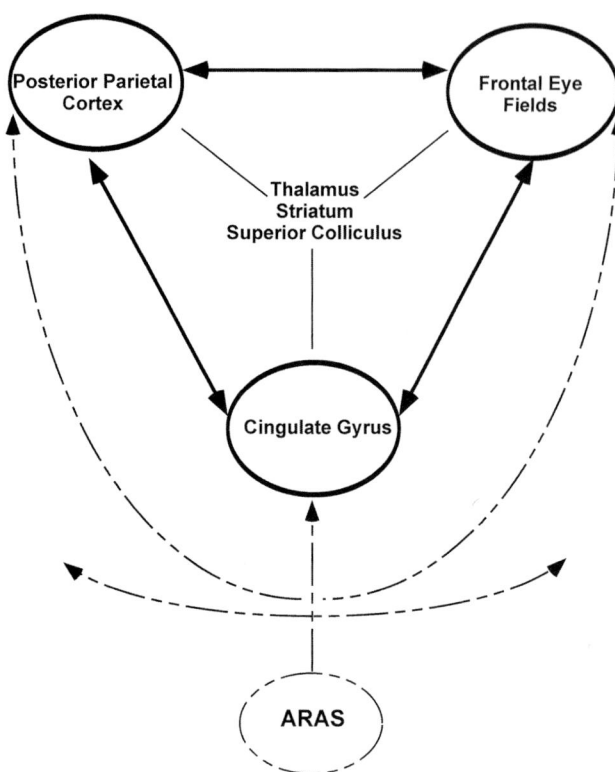

FIGURE 6.1 Large-scale network for spatial attention (Mesulam, 1999).

computational. Few, if any, have been able to account for the entire spectrum of clinical manifestations, leading some authors to question the existence of an identifiable neglect syndrome. It is becoming increasingly clear that neglect cannot be attributed to a unitary deficit of arousal, orientation, representation, or intention. Instead, it represents the collective and interactive outcome of multiple perturbations in each of these processes. As in the cases of aphasia and amnesia, neglect is a "network syndrome." It represents damage to one or more interactive components of a distributed network where each component has a different pattern of physiological and anatomical specialization.

The spatial attention network revolves around cortical epicenters in the posterior parietal cortex, FEFs, and cingulate gyrus. Each of these macroscopic components serves a dual purpose: it provides a local network for regional neural computations and also a nodal point for the linkage of distributed information. Although each component displays a relative specialization for specific behaviors, the domain of spatial attention, as a whole, is the emergent property of the entire network rather than the product of a hierarchical assembly line. Compiling internal representations, transforming extrapersonal coordinates into targets for looking or reaching, and modulating the zooming factor of spatial attention can be included among the relative specializations of the parietal component; translating representations into systematic search behaviors and scanning spatial information held in working memory seem to reflect some of the relative specializations of the FEF; and modulating spatial expectancy according to motivational valence seems to represent one of the most interesting specializations of the cingulate component. The coherent engagement of this network enables motivationally relevant extrapersonal events to become the selective targets of attentional behaviors.

Although much has been learned about the spatial attention network shown in Fig. 6.1, many crucial questions remain open. In particular, the computational features of the network, its interactions with other partially overlapping networks, the internal functional heterogeneity of its major components, and the physiological basis of right hemisphere specialization are among the major questions that will undoubtedly attract new and exciting investigations.

References

Andersen, R. A., Snyder, L. H., Bradley, D. C., and Xing, J. (1997). Multimodal representation of space in the posterior parietal cortex and its use in planning movements. *Annu. Rev. Neurosci.* **20**, 303–330.

Armony, J. L., and Dolan, R. J. (2002). Modulation of spatial attention by fear-conditioned stimuli: an event-related fMRI study. *Neuropsychologia* **40**, 817–826.

Chafee, M. V., and Goldman-Rakic, P. S. (2000). Inactivation of parietal and prefrontal cortex reveals interdependence of neural activity during memory-guided saccades. *J. Neurophysiol.* **83**, 1550–1566.

Corbetta, M., Kincade, J. M., Ollinger, J. M., McAvoy, M. P., and Shulman, G. L. (2000). Voluntary orienting is dissociated from target detection in human posterior parietal cortex. *Nat. Neurosci.* **3**, 292–297.

Corbetta, M., Miezin, F. M., Shulman, G. L., and Petersen, S. E. (1993). A PET study of visuospatial attention. *J. Neurosci.* **13**, 1202–1226.

Courtney, S. M., Petit, L., Maisog, J. M., Ungerleider, L. G., and Haxby, J. V. (1998). An area specialized for spatial working memory in human frontal cortex. *Science* **279**, 1347–1351.

Desimone, R. (1996). Neural mechanisms for visual memory and their role in attention. *Proc. Natl. Acad. Sci. (USA)* **93**, 13494–13499.

Donner, T. H., Kettermann, A., Diesch, E., Ostendorf, F., Villringer, A., and Brandt, S. A. (2002). Visual feature and conjunction searches of equal difficulty engage only partially overlapping frontoparietal networks. *NeuroImage* **15**, 16–25.

Gitelman, D. R., Alpert, N. M., Kosslyn, S. M., Daffner, K., Scinto, L., Thompson, W., and Mesulam, M.-M. (1996). Functional imaging of human right hemispheric activation for exploratory movements. *Ann. Neurol.* **39**, 174–179.

Gitelman, D. R., Parrish, T. B., Friston, K. J., and Mesulam, M.-M. (2002). Functional anatomy of visual search: regional segregations within the frontal eye fields and effective connectivity of the superior colliculus. *NeuroImage* **15**, 970–982.

Hopfinger, J. B., Buonocore, M. H., and Mangun, G. R. (2000). The neural mechanisms of top-down attentional control. *Nat. Neurosci.* **3**, 284–291.

Husain, M., Mattingley, J. B., Rorden, C., Kennard, C., and Driver, J. (2000). Distinguishing sensory and motor biases in parietal and frontal neglect. *Brain* **123**, 1643–1659.

LaBar, K., Gitelman, D. R., Parrish, T. B., Kim, Y. H., and Mesulam, M.-M. (1998). Overlap of frontoparietal activations during covert spatial attention and verbal working memory in the same set of subjects: an fMRI study. *Soc. Neurosci. Abstr.* **24**, 1896.

Lipschutz, B., Kolinsky, R., Damhaut, P., Wikler, D., and Goldman, S. (2002). Attention-dependent changes of activation and connectivity in dichotic listening. *NeuroImage* **17**, 643–656.

Maddock, R. J., Garrett, A. S., and Buonocore, M. H. (2003). Posterior cingulate cortex activation by emotional words: fMRI evidence from a valence decision task. *Hum. Brain Mapp.* **18**, 30–41.

McIntosh, A. R., Grady, C. L., Ungerleider, L. G., Haxby, J. V., Rapoport, S. I., and Horwitz, B. (1994). Network analysis of cortical visual pathways mapped with PET. *J. Neurosci.* **14**, 655–666.

Mesulam, M.-M. (1981). A cortical network for directed attention and unilateral neglect. *Ann. Neurol.* **10**, 309–325.

Mesulam, M.-M. (1999). Spatial attention and neglect: parietal, frontal, and cingulate contributions to the mental representation and attentional targeting of salient extrapersonal events. *Phil. Trans. R. Soc. B* **354**, 1325–1346.

Mesulam, M. M., Van Hoesen, G. W., Pandya, D. N., and Geschwind, N. (1977). Limbic and sensory connections of the inferior parietal lobule (area PG) in the rhesus monkey: a study with a new method for horseradish peroxidase histochemistry. *Brain Res.* **136**, 393–414.

Nobre, A. C. (2001). Orienting attention to instants in time. *Neuropsychologia* **39**, 1317–1328.

Nobre, A. C., Coull, J. T., Frith, C. D., and Mesulam, M.-M. (1999). Orbitofrontal cortex is activated during breaches of expectation in tasks of visual attention. *Nat. Neurosci.* **2**, 11–12.

Nobre, A. C., Coull, J. T., Maquet, P., Frith, C. D., Vandenberghe, R., and Mesulam, M.-M. (2004) Orienting attention to locations in perceptual versus mental representations. *J. Cogn. Neurosci* **16**, 363–373.

Nobre, A. C., Coull, J. T., Walsh, V., and Frith, C. D. (2003). Brain activations during visual search: contributions of search efficiency versus feature binding. *NeuroImage* **18**, 91–103.

Nobre, A. C., Gitelman, D. R., Dias, E. C., and Mesulam, M.-M. (2000). Covert visual spatial orienting and saccades: overlapping neural systems. *NeuroImage* **11**, 210–216.

Nobre, A. C., Sebestyen, G. N., Gitelman, D. R., Mesulam, M.-M., Frackowiak, R. S. J., and Frith, C. D. (1997). Functional localization of the system for visuospatial attention using positron emission tomography. *Brain* **120**, 515–533.

Paré, M., and Wurtz, R. H. (1997). Monkey posterior parietal cortex neurons antidromically activated from superior colliculus. *J. Neurophysiol.* **78**, 3493–3497.

Posner, M. I., Walker, J. A., Friedrich, F. A., and Rafal, R. D. (1987). How do the parietal lobes direct covert attention? *Neuropsychologia* **25**, 135–145.

Small, D. M., Gitelman, D. R., Gregory, M. D., Nobre, A. C., Parrish, T., and Mesulam, M.-M. (2003a). The posterior cingulate and medial prefrontal cortex mediate the anticipatory alloocation of spatial attention. *NeuroImage* **18**, 633–641.

Small, D. M., Nomura, E., Simmons, K., Parrish, T., Gitelman, D., and Mesulam, M.-M. (2003b). Differential effects of monetary incentives on visual spatial expectancy and disengagement: the anterior versus posterior cingulate cortex [abstract]. *NeuroImage* **19** (CD-ROM).

Vandenberghe, R., Gitelman, D. R., Parrish, T. B., and Mesulam, M.-M. (2001). Location- or feature-based targeting of peripheral attention. *NeuroImage* **14**, 37–47.

Yantis, S., Schwartzbach, J., Serences, J. T., Carlson, R. L., Steinmetz, M. A., Pekar, J. J., and Courtney, S. M. (2002). Transient neural activity in human parietal cortex during attentional shifts. *Nat. Neurosci.* **5**, 995–1002.

Zatorre, R. J., Mondor, T. A., and Evans, A. C. (1999). Auditory attention to space and frequency activates similar cerebral systems. *NeuroImage* **10**, 544–554.

CHAPTER

7

Parietal Mechanisms of Attentional Control: Locations, Features, and Objects

John T. Serences, Taosheng Liu, and Steven Yantis

ABSTRACT

Two distinct components of attentional control have been documented within subregions of parietal cortex. First, broad regions of intraparietal sulcus (IPS) and frontal eye fields (FEFs) are tonically active when attention is directed to a particular location, feature, or object in a visual scene. This tonic activity in IPS may be the source of a signal to maintain the current state of attention in visual cortex. Second, regions of superior parietal lobule, IPS, and precuneus are transiently active when attention is shifted between attentive states. This transient activity may reflect an attentional control signal that initiates abrupt changes of attentional state in sensory areas of visual cortex.

I. INTRODUCTION

Visual attention selects relevant spatial locations, perceptual features, or objects from the myriad visual stimuli in the environment (Yantis, 2000). The deployment of attention may be controlled by stimulus-driven factors, such as image-based salience, or by goal-directed (top-down) factors, such as current behavioral goals. Here we discuss the voluntary, goal-directed component of attentional control mediated by the human parietal lobes.

To organize the discussion, we adopt the biased competition model of attention (Desimone and Duncan, 1995). According to this account, subpopulations of cortical neurons that represent different aspects of the scene form a mutually inhibitory network such that when a scene contains multiple objects, they compete for representation. Attention serves to bias the competition in favor of a relevant location, feature, or object via feedback from higher attentional control centers to early sensory regions of cortex.

The biased competition model provides a useful theoretical framework without specifying the details of the component processes of attentional control. At least two distinct functional components can be identified: switching attention from one state to another, and maintaining a given attentive state. Here we review evidence gathered using functional magnetic resonance imaging (fMRI) suggesting that frontal and parietal areas exhibit unique temporal, and possibly spatial, patterns of activation during the maintenance and the switching of attention, respectively. Critically, the unique temporal signatures associated with switching attention are correlated with changes in the state of biased competition in sensory regions of visual cortex that represent spatial locations, features, and objects.

II. CUING STUDIES OF ATTENTIONAL CONTROL

One fruitful approach to investigating attentional control is inspired by the classic attentional cuing studies of Eriksen and Posner in the 1970s and 1980s (Eriksen and Hoffman, 1973; Posner, 1980). In a typical study of location-based attentional control, a centrally presented arrow cue instructs subjects to direct attention to a location in the visual periphery, without moving their eyes, in anticipation of a subsequently presented target. Functionally, cuing experiments of this type require subjects to identify the central cue, disengage attention from the cue, shift attention to the periphery, and then maintain attention in the periphery until the target appears. Neuropsychological studies suggest that these functional subcomponents can be dissociated (Posner et al., 1984); however, at the relatively coarse time scale afforded by fMRI, any changes in the blood oxygen level-dependent (BOLD) signal related to attentional control likely reflect both the initial disengage/shift operation and the tonic maintenance of attention in the periphery.

These attentional control signals can be measured by estimating changes in the BOLD signal attributable to the cue independent of the signal changes evoked by the target (Corbetta et al., 2000; Hopfinger et al., 2000) (see Chapter 11). Cue-related attentional responses are typically observed in regions of intraparietal sulcus (IPS), inferior parietal lobule (IPL), superior parietal lobule (SPL), and frontal eye fields (FEFs, near the junction of the superior frontal sulcus and the precentral gyrus; see Fig. 7.1a). Importantly, the BOLD responses in parietal and FEF regions exhibit a sustained temporal profile relative to the transient sensory response evoked by the cue in visual cortex. The sustained temporal profile suggests that parietal and FEF regions are involved in maintaining the locus of attention in the periphery and are not driven solely by the volley of sensory activity caused by the presentation of the cue (Corbetta et al., 2000; Corbetta et al., 2002). In addition, shifts of attention to one side of space in response to a central cue lead to sustained increases in the BOLD response in sensory areas of visual cortex that represent the cued location (Giesbrecht et al., 2003; Hopfinger et al., 2000; Kastner et al., 1999; Yantis et al., 2002). These spatially specific modulations in visual cortex are usually attributed to reentrant biasing signals generated in parietal and frontal cortices in response to the instruction (provided by the cue) to deploy attention. Together, the sustained temporal response profile and the ensuing modulations in spatiotopically organized regions of visual cortex support a role for parietal and FEF regions in shifting and maintaining the locus of attention to a particular spatial location.

Cuing paradigms have also been used to investigate feature-based attentional control. For example, subjects might see a central cue that instructs them to attend to a particular color or to a particular direction of motion in anticipation of the impending target display (Giesbrecht et al., 2003; Shulman et al., 1999; Vandenberghe et al., 2001b) (see Chapter 67). Although there appears to be some differentiation in the spatial pattern of activation elicited by cues to attend to

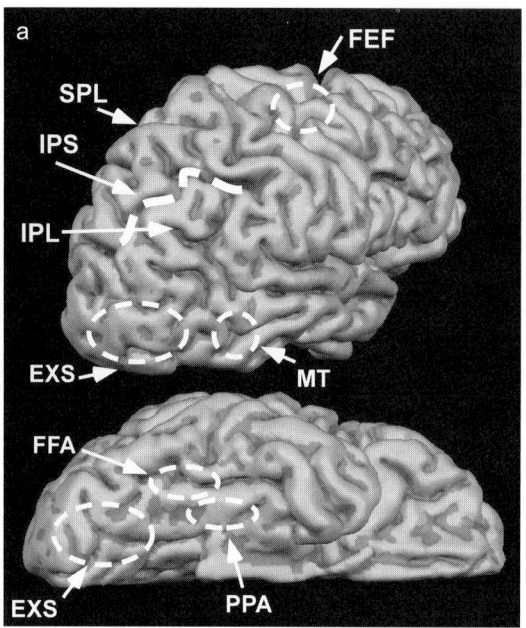

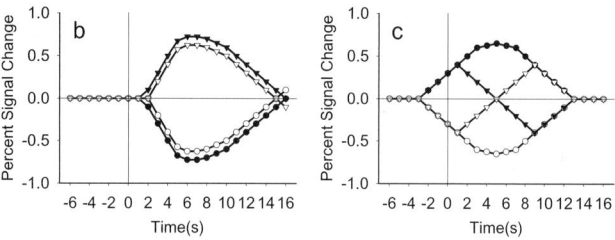

FIGURE 7.1 (a) Reconstructed cortical surface of a single subject showing areas involved in voluntary attentional control and areas in visual cortex that are modulated by attentional control signals. Top: superior–posterior view of the right hemisphere. Bottom: ventral view of the right hemisphere. In both, anterior is to the right, posterior is to the left. FEF, frontal eye fields; SPL, superior parietal lobule; IPS, intraparietal sulcus; IPL, inferior parietal lobule; EXS, extrastriate visual cortex; MT, middle temporal cortex; FFA, fusiform face area; PPA, parahippocampal place area. (b) Idealized BOLD time course from a region that is selectively more active when attention is shifted (triangles) versus when attention is maintained on the currently attention location, feature, or object (circles). (c) Idealized BOLD time course from a region whose activity is modulated by shifts of attention. As attention is shifted to the preferred stimulus, the BOLD signal increases (open triangles), and as attention is shifted to the nonpreferred stimulus, the BOLD signal decreases (closed triangles). The BOLD signal remains high or low as attention is maintained on the preferred (closed circles) and nonpreferred (open circles) stimuli, respectively.

features versus cues to attend to spatial locations, shifts of feature-based attention also lead to increased BOLD responses in IPS, IPL, SPL, and FEFs. In addition, cues to attend to a specific feature evoke a heightened BOLD response in feature-selective regions of visual cortex such as MT+ for motion when a particular direction of motion is cued (Shulman et al., 1999).

This finding suggests that activity in parietal and frontal areas may influence biased competition in visual cortex in favor of the attended feature dimensions.

III. DYNAMIC ATTENTIONAL CONTROL SIGNALS

The aforementioned studies establish that regions of parietal cortex (IPS, IPL, SPL) and frontal cortex (e.g., FEFs) play a critical role in controlling space-based and feature-based attention. In addition, the cuing paradigms used in these studies provide a method for parceling the variance in the BOLD signal related to cues and to targets, respectively, an effective approach to studying attentional control signals in isolation from target evoked responses (Corbetta et al., 2000). However, as noted earlier, shifts of attention appear to involve several cognitive operations, including disengaging attention from the cue, shifting attention to the periphery, and maintaining the new locus of attention. Cue-based studies do not permit separate measurements of these component operations.

To clarify the functional role of these brain regions in attentional control, several studies from our laboratory have been carried out to isolate BOLD signals associated with the disengage/shift operation in the domains of space-based, feature-based, and object-based attention. The general approach is to induce a highly focused attentive state at all times during the task, requiring subjects to monitor a continuously changing stimulus display for targets (Vandenberghe et al., 2001a). Targets instruct observers either to maintain attention on the currently attended aspect of the display or to shift attention to a different aspect of the display. For example, in a spatial task, "hold" cues instruct subjects to maintain attention at the currently attended location, and "shift" cues instruct them to shift attention to a different location. Therefore, any increases in the BOLD signal that are time-locked to shifts of attention must be related to the disengage and/or shift operation because maintaining the current focus of attention forms the "baseline" condition in this paradigm.

One important aspect of this experimental approach is that functional differences between different brain regions should be evident in qualitatively distinct BOLD time series. For instance, an area that is selectively involved in shifting attention between locations should show an increased BOLD response when attention is shifted as compared with when attention is maintained at the currently attended location (Fig. 7.1b). In contrast, areas whose activity reflects the current focus of attention (e.g., sensory areas that are the target of the top-down biasing signals and attention-maintenance control areas) should be selectively active when the preferred location is attended. A shift of attention to the preferred location should lead to an increase in the BOLD response, and a shift away from the preferred location should lead to a decrease in the BOLD response (Fig. 7.1c). Although this method is sensitive to dynamic changes in the BOLD signal that are specifically tied to shifts of attention, it does not reveal areas that are involved in maintaining the current locus of attention in a stimulus-independent fashion; such regions should exhibit a constant level of activity throughout the task.

A. Locations

To investigate shifts of spatial attention, observers viewed a display in which letters appeared in rapid succession (4/s) in eight locations on the screen (Yantis et al., 2002). They were to covertly attend to either the left or right central stream to detect a digit target within the attended stream. If the target was a "3", observers maintained attention at the same location; if it was a "7", observers shifted attention to the other location (targets were separated by 3–5 seconds) (see Fig. 7.2a).

Shown in Fig. 7.3a is the BOLD time series from an area in right occipital cortex that was modulated by the current locus of attention. As attention was maintained on the contralateral target stream (closed circles), the BOLD response remained high relative to when attention was directed to the ipsilateral location (open circles). A switch from the ipsilateral location to the contralateral location led to an increase in the BOLD response (open triangles); the complementary pattern was observed when attention was shifted from the contralateral to the ipsilateral side of space (closed triangles). This BOLD time series reveals dynamic changes in biased competition in regions of visual cortex that represent spatial locations as voluntary control is exerted over the current focus of attention.

Figure 7.3b shows the BOLD time series from a region in right SPL that was more active following shift versus hold events. In addition to any activity in this region related to sustaining the current locus of attention, the selective increase in activity following shift events demonstrates that the BOLD response in this region increased transiently when attention was shifted to either side of space. Functionally, this transient activity may reflect a signal to abruptly change the current state of biased competition in visual cortex, as reflected by the dynamic attentional modulations presented in Fig. 7.3a.

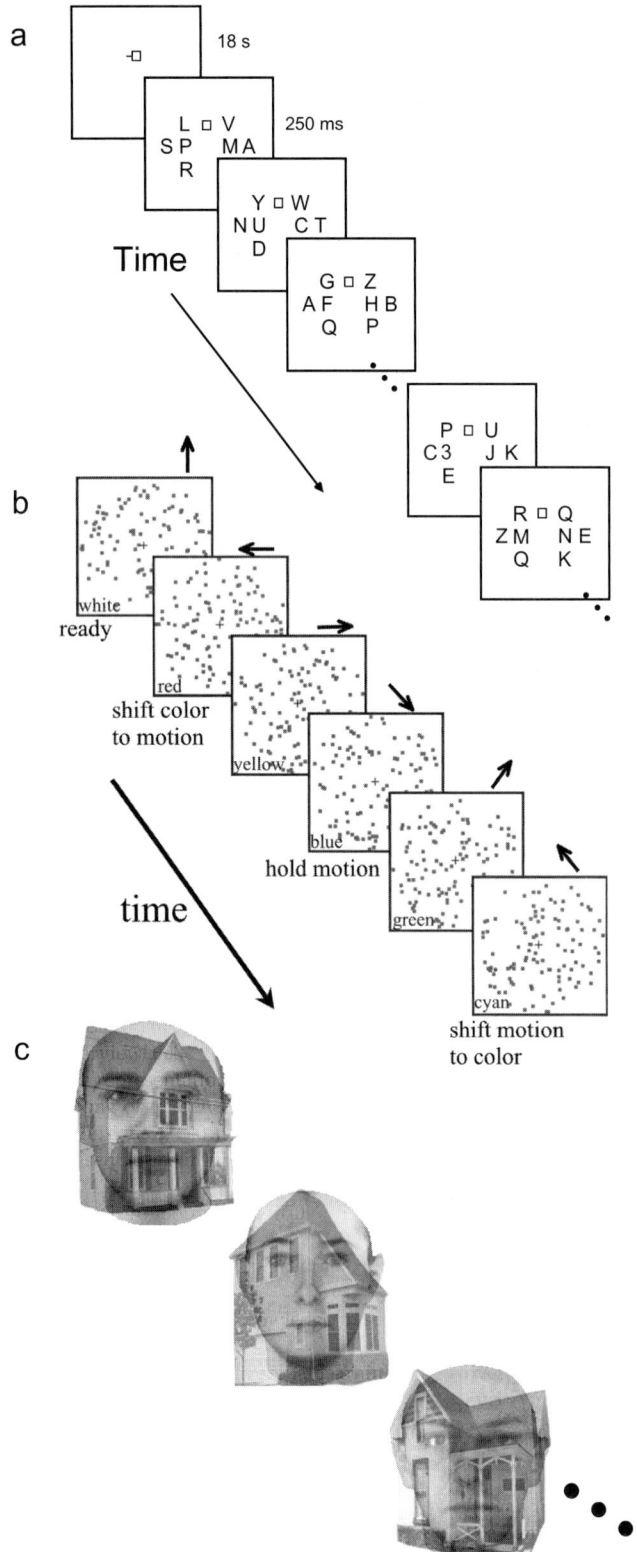

B. Features

In a subsequent study, we used a similar experimental logic to examine feature-based attentional control by examining shifts of attention between motion and color (Liu et al., 2003). Observers viewed a display containing 75% coherently moving dots; the direction of motion and the color of the dots changed once per second (six possible directions of motion and six possible colors). One color and one direction of motion instructed subjects to shift attention to the other feature dimension, whereas a different color and direction of motion instructed subjects to maintain attention on the currently attended feature (Fig. 7.2b).

Dynamic modulations in feature-selective regions of visual cortex were observed as attention was shifted between color and motion. Figure 7.3c shows the BOLD time course from the left MT+, a midlevel visual area that is selectively responsive to motion. The heightened BOLD response observed while attention was maintained on motion (closed circles) and the crossover pattern following shift events reflect the competitive advantage conferred on this region when motion is the behaviorally relevant stimulus.

In addition to the feature-selective modulations observed in visual cortex, several regions were identified in parietal cortex, including precuneus and IPS, that were more active following shifts of attention between feature dimensions. Figure 7.3d depicts the time course from a region in the precuneus that was more active when attention was shifted between motion and color. While these regions are anatomically distinct from the area of right SPL that was identified

FIGURE 7.2 Schematic of behavioral paradigms to examine attentional control. (**a**) Space-based shifts of attention. Participants fixated on the central square throughout each run and began by attending to the central stream of letters on one side (left in this example). Letters changed identity simultaneously four times per second. Hold and shift target digits (e.g., 3, 7) instructed the observer to maintain attention on the currently attended side or to shift attention to the other side. (**b**) Feature-based shifts of attention. The color of the dots and the direction of the motion changed simultaneously once every second. Subjects started a run by attending to either color or motion, and hold or shift targets (e.g., red dots, or motion to the upper left) instructed subjects to switch attention to the other feature dimension or to maintain attention on the currently attended feature. Note that the color word in each frame was not present in the experiment; it simply labels the color in which the dots were rendered. (**c**) Object-based shifts of attention. Each face spatially morphed into the subsequent face and each house spatially morphed into the subsequent house at a rate of one morph per second. Subjects started each run by attending to either faces or houses and switched or maintained attention based on prespecified face and house target stimuli.

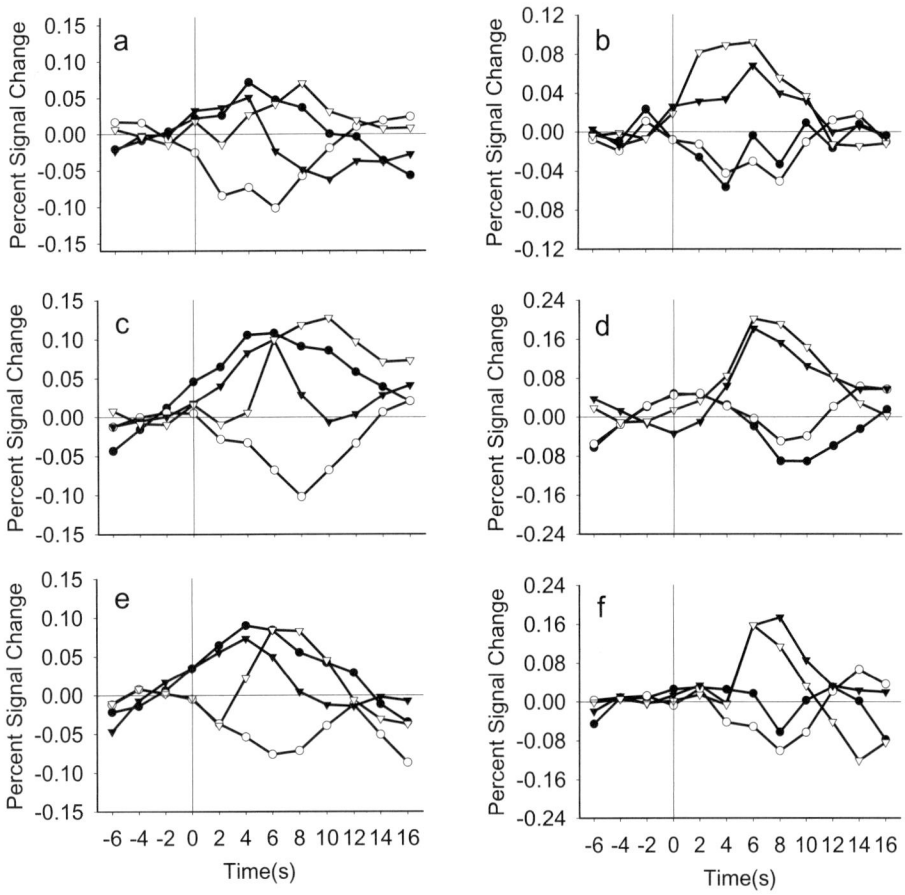

FIGURE 7.3 (a) BOLD time course from a region of right extrastriate cortex that was more active when attention was maintained (closed circles) or shifted to (open triangles) the left side of space. (b) BOLD time course from a region of right SPL that showed an increased BOLD response when attention was shifted to either side of space (open and closed triangles). (c) Time course from a region of left MT+ complex that was more active when subjects directed attention to motion. (d) Time course from a region of medial precuneus that was transiently more active when attention was shifted between feature dimensions (color and motion). (e) Time course from a region of right fusiform gyrus that was more active when attention was directed to faces as compared with houses. (f) Time course from a region of medial SPL–left IPS that was selectively more active during shifts of attention between faces and houses. For all time series: hold attention on preferred stimulus type; ●; hold nonpreferred; ○; shift preferred to nonpreferred; ▼; shift nonpreferred to preferred; ▽.

in the spatial shifting task, the medial precuneus and left IPS regions show the same pattern of transient activity tied specifically to shifts of attention between feature dimensions.

C. Objects

The majority of neuroimaging work on attentional control has focused on space- and feature-based attentional control. Although the evidence clearly indicates that locations and individual features may be selected by attention, the visual system must ultimately represent complex three-dimensional objects, not abstract locations and features. In line with this functional requirement, a growing body of evidence has revealed that attention may select unitary object representations. For example, when one aspect of an object is selected (e.g., its shape), then all other features associated with the object are also selected (e.g., color, motion) (O'Craven et al., 1999). This holds even when the attended and ignored objects are spatially superimposed.

To investigate the control of object-based attention, we used overlapping face and house stimuli and a paradigm that was similar to those used to investigate space- and feature-based attentional control (Serences et al., in press). Observers viewed a continuously changing stream of superimposed morphing faces and houses, and their task was to detect embedded targets instructing them either to shift attention to the other object stream or to maintain attention on the currently attended object stream (Fig. 7.2c). Separate functional localizer scans were used to identify regions of interest in ventral visual cortex that respond differentially to face and house stimuli (O'Craven et al., 1999). Figure 7.3e shows the BOLD time series from a region in right fusiform cortex that responds preferentially to faces. When observers were attending to faces in the attention experiment, the BOLD signal in this region remained high (closed circles). The relatively attenuated BOLD response in the region when observers attended houses and the crossover following shift events mirror the patterns observed in early visual areas during shifts of spatial attention and in feature-selective regions during shifts of feature-based attention.

Depicted in Fig. 7.3f is the BOLD time series from a region of medial SPL that extended into the left IPS, showing a greater response when attention was shifted from one object stream to the other. Anatomically, the medial SPL/left IPS activation was slightly inferior to the right SPL activation observed for shifts of spatial attention and slightly superior to the precuneus and left IPS activations observed during shifts of feature-based attention. Nevertheless, the transient BOLD response in the medial SPL/left IPS region suggests a selective role for this area in altering the state of biased competition in object-selective regions of visual cortex.

Together, these three studies of space-, feature-, and object-based attentional control reveal two key points. First, activity in subregions of parietal cortex is specifically tied to the act of shifting attention, independent of any role these regions may play in the tonic maintenance of attention at the currently attended locus. Second, this transient activity during shifts of attention is temporally correlated with dynamic changes in the state of biased competition in the regions of visual cortex that represent the attended stimuli.

IV. CONCLUSIONS

Neuroimaging research has exposed an extensive network of brain regions that subserve different aspects of attentional control. First, cuing studies show that broad regions of IPS and FEF are tonically active when subjects shift and maintain spatial attention to a peripheral location (Corbetta et al., 2000; Corbetta et al., 2002; Hopfinger et al., 2000; Vandenberghe et al., 2001a). Recent studies that use similar cuing paradigms suggest that similar anatomical regions also play a role in the deployment and maintenance of feature-based attention (Giesbrecht et al., 2003; Shulman et al., 1999). In addition, studies that focus on the transient disengage/shift function of attentional control reveal that a subset of these parietal and frontal regions are selectively more active when attention is shifted as compared to when the focus of attention is sustained on the currently attended location, feature, or object (Liu et al., 2003; Serences et al., in press; Yantis et al., 2002).

Within the studies that specifically examine shift-related activity, there is some variability in the regions of parietal cortex that are transiently active during shifts of space-, feature-, and object-based attention, respectively (Liu et al., 2003; Serences et al., in press; Yantis et al., 2002). At least two possibilities can be considered. First, there may exist functional compartments within the parietal lobes that are specialized for the control of shifts of attention within distinct perceptual domains (i.e., space, feature, object). Alternatively, the perceptual differences in the stimuli used in each of these studies may have contributed to variable patterns of activation depending on the stimulus selectivity of different parietal regions.

The big picture emerging from these studies is that different functional subcomponents of attentional control result in distinct temporal patterns in the BOLD signal (Corbetta et al., 2000; Yantis et al., 2002). This observation suggests that future experiments should address specific attentional processes (i.e., shift and maintain operations), as well as other processes that are presumably integral to exerting control over the contents of visual awareness. For example, this chapter has focused on the various attentional functions mediated by regions of parietal cortex; however, the ultimate origin of attentional control signals is most likely in prefrontal cortex, an area that is critically involved in representing current behavioral goals (Miller, 2000). One future challenge will be to use the temporal characteristics of the BOLD signal to investigate functional interactions between working memory and attention functions (e.g., Awh and Jonides, 2001; Corbetta et al., 2002), providing further insight into the various cognitive operations underlying visual experience.

Acknowledgments

This work was supported by an NSF Graduate Research Fellowship to J.T.S. and by National Institutes of Health Grant R01-DA13165 to S.Y.

References

Awh, E., and Jonides, J. (2001). Overlapping mechanisms of attention and spatial working memory. *Trends Cogn. Sci.* **5**, 119–126.

Corbetta, M., Kincade, J. M., Ollinger, J. M., McAvoy, M. P., and Shulman, G. L. (2000). Voluntary orienting is dissociated from target detection in human posterior parietal cortex. *Nat. Neurosci.* **3**, 292–297.

Corbetta, M., Kincade, J. M., and Shulman, G. L. (2002). Neural systems for visual orienting and their relationships to spatial working memory. *J. Cogn. Neurosci.* **14**, 508–523.

Desimone, R., and Duncan, J. (1995). Neural mechanisms of selective visual attention. *Annu. Rev. Neurosci.* **18**, 193–222.

Eriksen, C. W., and Hoffman, J. E. (1973). The extent of processing of noise elements during selective encoding from visual displays. *Percept. Psychophys.* **14**, 155–160.

Giesbrecht, B., Woldorff, M. G., Song, A. W., and Mangun, G. R. (2003). Neural mechanisms of top-down control during spatial and feature attention. *NeuroImage* **19**, 496–512.

Hopfinger, J. B., Buonocore, M. H., and Mangun, G. R. (2000). The neural mechanisms of top-down attentional control. *Nat. Neurosci.* **3**, 284–291.

Kastner, S., Pinsk, M. A., De Weerd, P., Desimone, R., and Ungerleider, L. G. (1999). Increased activity in human visual cortex during directed attention in the absence of visual stimulation. *Neuron* **22**, 751–761.

Liu, T., Slotnick, S. D., Serences, J. T., and Yantis, S. (2003). Cortical mechanisms of feature-based attentional control. *Cereb. Cortex* **13**, 1334–1343.

Miller, E. K. (2000). The prefrontal cortex and cognitive control. *Nat. Rev. Neurosci.* **1**, 59–65.

O'Craven, K. M., Downing, P. E., and Kanwisher, N. (1999). fMRI evidence for objects as the units of attentional selection. *Nature* **401**, 584–587.

Posner, M. I. (1980). Orienting of attention. *Q. J. Exp. Psychol.* **32**, 3–25.

Posner, M. I., Walker, J. A., Friedrich, F. J., and Rafal, R. D. (1984). Effects of parietal injury on covert orienting of attention. *J. Neurosci.* **4**, 1863–1874.

Serences, J. T., Schwarzbach, J., Golay, X., Courtney, S. M., and Yantis, S. (in press). Control of object-based attention in human cortex. *Cereb. Cortex*.

Shulman, G. L., Ollinger, J. M., Akbudak, E., Conturo, T. E., Snyder, A. Z., Petersen, S. E., and Corbetta, M. (1999). Areas involved in encoding and applying directional expectations to moving objects. *J. Neurosci.* **19**, 9480–9496.

Vandenberghe, R., Gitelman, D. R., Parrish, T. B., and Mesulam, M. M. (2001a). Functional specificity of superior parietal mediation of spatial shifting. *NeuroImage* **14**, 661–673.

Vandenberghe, R., Gitelman, D. R., Parrish, T. B., and Mesulam, M. M. (2001b). Location- or feature-based targeting of peripheral attention. *NeuroImage* **14**, 37–47.

Yantis, S. (2000). Goal-directed and stimulus-driven determinants of attentional control. *In* "Attention and Performance XVIII" (S. M. J. Driver, Ed.), pp. 73–103. MIT Press, Cambridge, MA.

Yantis, S., Schwarzbach, J., Serences, J. T., Carlson, R. L., Steinmetz, M. A., Pekar, J. J., and Courtney, S. M. (2002). Transient neural activity in human parietal cortex during spatial attention shifts. *Nat. Neurosci.* **5**, 995–1002.

CHAPTER 8

Visual Cortical Circuits and Spatial Attention

John H. Reynolds

I. INTRODUCTION

Attention has been known to play a key role in perception since the dawn of experimental psychology (James, 1890), but only recently has it become possible to describe attention in terms of its biological underpinnings. One of several fruitful approaches has been to record neuronal responses in monkeys as they perform attention-demanding tasks. This chapter focuses on the modulatory effect of attention on neurons in the visual cortex, placing emphasis on *spatial* attention, where the most detailed mechanistic understanding has been achieved. We will see that directing attention to a stimulus increases neuronal sensitivity, causing neurons to respond to an attended stimulus much as they would if its luminance had increased. This observation allows us to make contact with studies in anesthetized cats and monkeys that have provided detailed accounts of how neuronal responses vary as a function of luminance contrast. The same models that have been developed to explain contrast-dependent changes in neuronal response can also account for contrast-dependent modulation of the competive interactions that are observed when multiple stimuli appear within a neuron's receptive field. These models thus offer a relatively simple account of how the visual system selects an attended stimulus from among behaviorally irrelevant distracters.

II. SPATIAL ATTENTION: FACILITATION AND SELECTION

Psychophysical studies of spatial attention have found that when attention is directed to a location, this improves processing of stimuli appearing at that location (Posner et al., 1980). Attention increases an observer's ability to detect faint stimuli appearing at an attended location (Bashinski and Bacharach, 1980; Muller and Humphreys, 1991; Hawkins et al., 1990; Handy et al., 1996), and to discriminate features of the attended stimulus, such as its orientation (Downing, 1988; Lee et al., 1999). Attention enhances signal strength, as measured by the contrast increment required to equate accuracy in discriminating features of stimuli appearing at attended versus unattended locations (Carrasco et al., 2000; Lu and Dosher, 1998). This signal enhancement is reflected in stronger stimulus-evoked neuronal activity, as measured by scalp potentials (reviewed by Hillyard and Anllo-Vento, 1998) and brain imaging (reviewed by Pessoa et al., 2003; Yantis and Serences, 2003).

Consistent with these findings, single-unit recording studies in monkeys have also found that spatial attention can enhance responses evoked by a single stimulus appearing alone in a neuron's receptive field (Motter, 1993; Ito and Gilbert, 1999; McAdams and Maunsell, 1999). This increase in firing rate is dependent on the luminance contrast of the stimulus, as illustrated in Fig. 8.1A, which shows the responses of a single neuron in visual area V4, an extrastriate visual area at an intermediate stage of the ventral visual processing stream. A grating stimulus appeared in the neuron's receptive field at various levels of contrast, as illustrated to the right of each panel. Solid lines indicate the neuron's response when the monkey attended away from these stimuli, to detect a target far outside the receptive field. Dashed lines indicate the response evoked by the identical gratings when they were attended. Two of the gratings (5%, bottom panel, and 10%, middle panel) were too faint to evoke a response when unattended; that is, they were below the neuron's *contrast response threshold*. The third contrast (80%, top panel), exceeded the level of contrast at

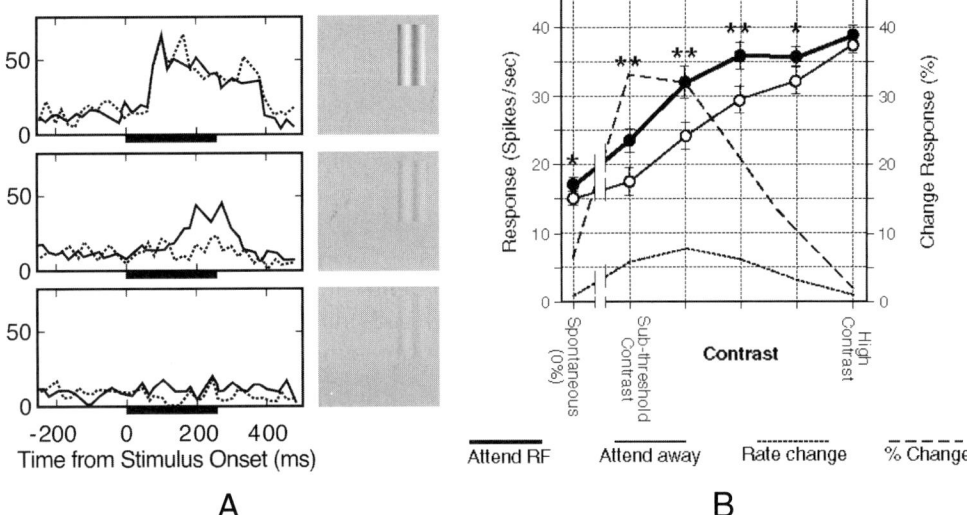

FIGURE.8.1 (A) Responses of an area V4 neuron as a function of attention and stimulus contrast. The contrast of the stimulus in the receptive field varied from 5% (bottom) to 10% (middle) to 80% (top). On any given trial attention was directed either to the location of the stimulus inside the receptive field (solid line) or to a location far away from the receptive field (dotted line). The animal's task was to detect a target grating at the attended location. Attention reduced the threshold level of contrast required to elicit a response without causing a measurable change in response at saturation contrast (80%). (B) Average responses of a population of area V4 neurons while the monkey either attended to the location of the receptive field stimulus (thick line, black circles) or else attended to a location far away from the receptive field (thin line, white circles). Luminance-modulated gratings were presented inside the receptive field at five different values of contrast, appearing along the horizontal axis, which were selected to span the dynamic range of the neuron. The leftmost point corresponds to the baseline level of activity when no stimulus was present in the receptive field, which was slightly elevated when attention was directed to the receptive field location. The dashed and dotted lines show, respectively, percentage and absolute difference in firing rate across the two attention conditions, as a function of contrast. Adapted, with permission, from the study of Reynolds, et al. (2000).

which the response saturated. Attention had no measurable effect on the response that was evoked at 5% contrast, which was well below the neuron's contrast response threshold. However, the neuron responded to the same 10% contrast stimulus when the monkey attended to its location in the receptive field. Thus, attention reduced the neuron's contrast response threshold from above 10% to a value between 5% and 10%. That is, the neuron was more sensitive to an attended stimulus than it was when the identical stimuli was unattended. Attention had no effect on the neuronal response that was elicited by the stimulus when it was presented above saturating contrast.

Figure 8.1B shows this increase in contrast sensitivity across a population of V4 neurons. For stimuli presented at contrasts just below each neuron's contrast response threshold ("subthreshold contrast"), there was a clear and significant response. The greatest increases in firing rate were observed at contrasts chosen to fall within the dynamic range of each neuron's contrast response function. Attention had no significant effect at the highest contrasts tested, which were chosen to be at or slightly above each neuron's contrast saturation point. Similar results were found for both preferred and poor stimuli, indicating that this failure to find attention effects at high contrast did not simply reflect an absolute firing rate limit, but instead reflected a leftward shift in the contrast response function. A cell-by-cell analysis revealed that for a cell to detect an unattended stimulus as reliably as it could detect an attended stimulus, the unattended stimulus needed to be half again as high in contrast as the attended stimulus. That is, under the conditions of this experiment, attention was "worth" a 51% increase in contrast.

Although attention can clearly enhance processing of a faint stimulus appearing against a blank background, an arguably more ecologically relevant purpose is served by *attentional selection*: the selection of behaviorally relevant stimuli from among distracters. Like any information-processing system, the visual system is limited in the amount of information it is able to process at each moment in time. The visual scene typically contains much more information than we can process in a single glimpse. Therefore, it is necessary to have neural mechanisms in place to select behaviorally relevant information to guide behavior.

The operation of such selection mechanisms is revealed within the extrastriate cortex. When multiple stimuli appear within a neuron's receptive field, the firing rate is characteristically determined primarily by the task-relevant stimulus. The first study to document this was carried out by Moran and Desimone (1985). They presented two stimuli within the receptive field: one that was of the neuron's preferred color and orientation and a second that was of a nonpreferred color and orientation. The monkey performed a task that required it to report the identity of one of the stimuli to earn a juice reward. The neuron's response to the pair was greater when the monkey attended to the preferred stimulus than the nonpreferred stimulus.

This fundamental observation has been replicated both in the ventral stream areas studied by Moran and Desimone (Chelazzi et al., 1993, 1998, 2001; Luck et al., 1997; Motter, 1993; Reynolds et al., 1999; Reynolds and Desimone, 2003; Sheinberg and Logothetis, 2001) and in the dorsal stream (Recanzone and Wurtz 2000; Treue and Martinez Trujillo, 1999; Treue and Maunsell, 1996; see, however, Seidemann and Newsome, 1999). This is illustrated in Fig. 8.2, which shows responses of a single V4 neuron, which was recorded by Reynolds et al. (1999). The solid black line shows the response to the preferred stimulus alone, when attention was directed away from the receptive field. The solid gray line shows the much weaker response that was elicited by a poor stimulus alone, again with attention away from the receptive field. The gray dotted line shows the response when the two unattended stimuli were presented together. The response to the pair falls between the responses elicited by either stimulus alone. When attention was directed to the preferred stimulus (dashed black line), this led to an increase in the response to the pair. When attention was directed to the poor stimulus (dashed gray line), this led to a substantial reduction in response, pushing the pair response down to a level similar to that which was observed when the poor stimulus appeared alone in the receptive field. This pattern is illustrative of what is observed in area V4, though the suppressive effect of directing attention to the poor stimulus is rarely this absolute. On average, we have found that attention moves the pair response about 70 to 80% of the way to the response that would be elicited by the attended stimulus appearing alone in the receptive field.

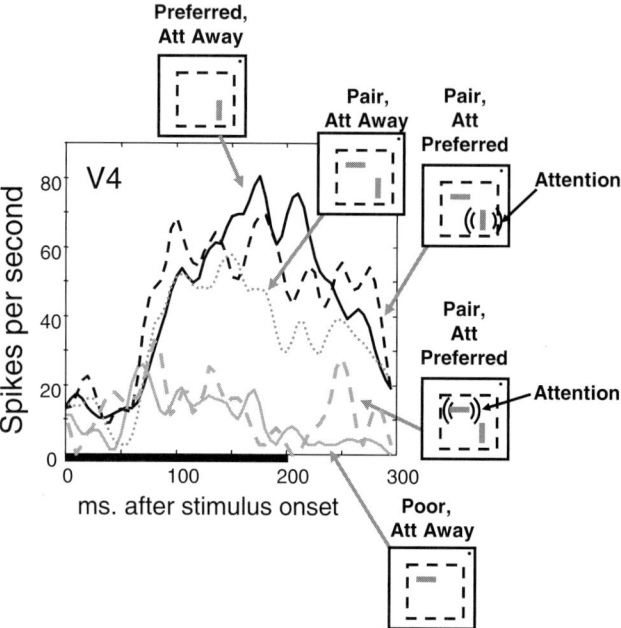

FIGURE.8.2 Effect of attention to one of two stimuli in the RF of a V4 neuron. Each line shows the response of a single V4 neuron averaged over repeated stimulus presentations, in different attentional and stimulus conditions, which are indicated by each icon. The stimuli appeared for 250 ms as indicated by the thick bar on the horizontal axis. See text for details.

Several of the above–cited studies have compared the response when attention was directed either to one of the receptive field stimuli or far away from the receptive field. Attending to the more preferred stimulus typically increases the response to the pair, but attending to the poor stimulus often *reduces* the response evoked by the pair (Chelazzi et al., 1998, 2001; Luck et al., 1997; Martínez-Trujillo and Treue, 2002; Reynolds et al., 1999; Reynolds and Desimone, 2003; Treue and Martinez Trujillo, 1999; Treue and Maunsell, 1996). This has provided support for models of attention that rely on response suppression to select one stimulus and inhibit another (Desimone and Duncan, 1995; Ferrera and Lisberger, 1995; Grossberg and Raizada, 2000; Itti and Koch, 2000; Lee et al., 1999; Niebur and Koch, 1994). These models have been used to account for observations concerning a range of topics, including the interplay between attentional selection and oculomotor control, the role of working memory in guiding attention during search, and the role of visual salience in guiding attentional selection. Given the broad explanatory power of these models, it is of interest to consider what studies in anesthetized animals have taught us about the role of response suppression in the visual cortex.

III. CONTRAST-DEPENDENT RESPONSE MODULATION IN VISUAL CORTEX

Contrast-dependent neuronal response modulation has been the focus of extensive research. The models that have been advanced to account for these modulations rely on response suppression. Here we describe three ways in which contrast has been found to modulate neuronal responses. First, cortical neuronal responses typically saturate as contrast increases, and this saturation firing rate is stimulus-dependent. This is illustrated in Fig. 8.3A. Each line shows the response of a complex cell in cat area 17 evoked by a single grating appearing at the neuron's preferred orientation (top line), a suboptimal but excitatory orientation (middle line), or its null orientation, which was slightly inhibitory (bottom line). The horizontal axis indicates the luminance contrast, so each curve is the neuron's *contrast response function* for one stimulus. The neuron did not respond to stimuli presented below a minimum level of luminance contrast. Above this threshold, the response increased over a range of contrasts that comprise the dynamic range of the contrast response function, before reaching a stimulus-dependent saturation response.

Second, increasing the contrast of a stimulus characteristically results in a multiplicative increase in the neuron's tuning curves for properties such as motion and orientation. This is illustrated in Fig. 8.3B. Each curve in this figure shows a simple cell's orientation tuning curves at different levels of contrast. The shallowest curve (open circles) reflects the cell's responses to gratings presented at 10% luminance contrast, and each higher curve reflects a doubling of contrast. This contrast-dependent multiplicative increase in the tuning curve is also thought to result from the interplay of excitation (which endows cells with their stimulus selectivity) and response suppression (which ensures that tuning is preserved as contrast increases, but keeping the responses to nonpreferred stimuli low).

The above phenomena reflect increases in response when a stimulus increases in contrast. Increasing contrast has qualitatively different effects when two stimuli appear within the receptive field. Increasing the contrast of one of them can result in increases or decreases in response, depending on the neuron's selectivity for the two stimuli. This is illustrated in Fig. 8.3C, which shows data that were recorded by Carandini et al. (1997) from a neuron in area V1 of the anesthetized macaque, when two spatially superimposed gratings, differing in orientation, appeared simultaneously within the receptive field (see also Bonds, 1989; DeAngelis et al., 1992; Morrone et al., 1982). The circles indicate mean firing rates, with shading indicating the luminance contrast of the preferred orientation grating, ranging from 0% (white) to 50% (black). Thus, the white circles indicate the response to the poor stimulus, alone. The lines show fits to the data provided by the model of Carandini et al. (1997). The contrast of the poor orientation grating is indicated on the horizontal axis.

Note that the response evoked by the poor stimulus increases with contrast, indicating that although it was poor, it was nonetheless excitatory. Despite this, increasing the contrast of the poor stimulus suppressed the response elicited by the preferred stimulus. The response evoked by the 13% contrast preferred stimulus alone (~22 spikes/s, dark gray data point, left) was strongly suppressed by the addition of the poor stimulus at 50% contrast (~8 spikes/s, dark gray data point, right). Increasing the contrast of the preferred grating had the opposite effect: to increase the response to the pair. The highest contrast preferred stimulus (black circles) was virtually immune to this suppressive effect. Thus, by increasing the contrast of one of the two stimuli, it was possible to cause it to have a dominant effect on the neuronal response.

Similar contrast-dependent response modulations have been observed when two stimuli appear at separate locations in the visual field (Reynolds and Desimone, 2003). This is illustrated in Fig. 8.4, which shows data recorded by Reynolds and Desimone (2003) from a macaque V4 neuron. The first column shows trial-by-trial spike records that illustrate the response of the neuron when a stimulus of the neuron's null orientation appeared alone in the receptive field at luminance contrasts ranging from 5% (bottom) to 80% (top). The right column shows the response elicited by a stimulus of the neuron's preferred orientation, which was presented at a fixed contrast (40%) at a separate location in the receptive field. The panels are repeated for comparison. The center column shows the response that was elicited by the pair as a function of poor stimulus contrast. The 5% contrast poor stimulus (bottom, center) has no measurable effect on the neuronal response, but at higher levels of contrast it became increasingly suppressive, and at 80% contrast (top, center) it almost entirely suppressed the response. These observations confirm earlier reports showing that a poor stimulus suppresses the response elicited by a non-superimposed preferred stimulus (Miller et al., 1993; Rolls and Tovee, 1995; see, however, Gawne and Martin, 2002).

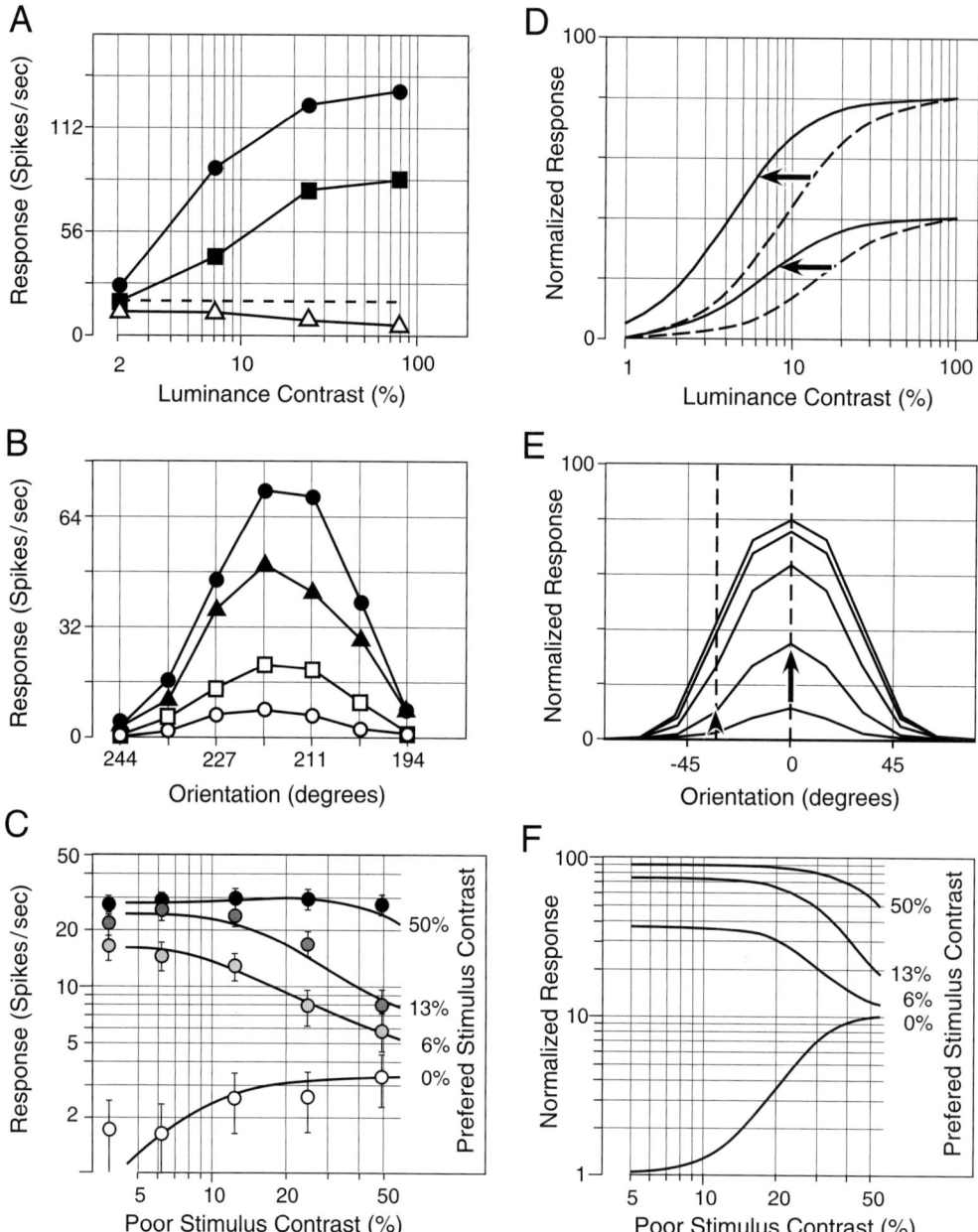

FIGURE.8.3 Contrast-dependent response modulations. (A) Contrast response functions for a stimulus of the neuron's preferred orientation (upper line), a poor but excitatory orientation (middle line), and a poor stimulus (bottom line). Adapted, with permission, from Sclar and Freeman (1982). (B) Orientation tuning curves of a second neuron, measured using a stimulus that varied in contrast from 10% (empty circles) to 80% (filled circles). Adapted, with permission, from Sclar and Freeman (1982). (C) Responses of an example neuron recorded in area V1 of the anesthetized macaque. Two spatially superimposed gratings appeared within the receptive field. One grating was of the optimal orientation for the neuron, while the second grating was of a suboptimal orientation. The preferred grating varied from 0% contrast (top row) to 50% contrast (bottom row), and the contrast of the poor grating increased from 0% (left column) to 50% (right column). Adapted, with permission, from Carandini et al., 1997 (D–F) Capacity of the contrast gain model to account for these contrast-dependent response modulations. See text for details.

IV. A LINKING HYPOTHESIS: DIRECTING SPATIAL ATTENTION TO A STIMULUS INCREASES ITS EFFECTIVE CONTRAST

These contrast-dependent response modulations closely parallel the effect of attention. First, consider the effect of elevating the contrast of a single stimulus, as illustrated in Figs. 8.4A and B. Attention causes a leftward shift in the contrast response function, enabling just-below-threshold stimuli to be detected by neurons in area V4 (Reynolds et al., 2000), as illustrated in Fig. 8.1. For stimuli within the dynamic range of the contrast response function, increasing contrast leads to a multiplicative increase in the response (Fig. 8.4B). Consistent with this, attention has been found to multiplicatively increase the orientation tuning curves of V4 (McAdams and Maunsell, 1999) and the direction of motion tuning curves of neurons in area MT (Treue and Martínez-Trujillo, 1999). Thus, if attention operates by increasing the *effective contrast* of a stimulus, this would account for elevations in response that are found when attention is directed to a single stimulus—the idea of *attentional facilitation*, described above.

Second, consider how changes in contrast alter the response when two stimuli, a preferred stimulus and a poor stimulus, appear simultaneously in the receptive field. The poor stimulus suppresses the response elicited by the preferred stimulus, and this suppression depends on the relative contrasts of the two stimuli (Fig. 8.4C). At low contrast, the poor stimulus has little or no suppressive effect, but increasing its contrast drives the pair response downward. Suppression is diminished if the preferred stimulus is elevated in contrast. This is precisely what is observed when attention is directed to one of two stimuli in the receptive field: the attended stimulus dominates the neuronal response. An appealing linking hypothesis is that attention operates by multiplying the effective contrast of the behaviorally relevant stimulus or, equivalently, increases the neuron's contrast sensitivity.

This idea is embodied in a model of attention advanced by Reynolds et al., (1999), which was conceived as a way of formalizing the biased competition model of Desimone and Duncan (1995). It assumes that attention increases the effective contrast of the attended stimulus, and it is thus referred to as the *contrast gain* model of attention. It is a functional model in that it is intended to characterize the operations that are performed by the neural circuit without committing to specific biophysical or biochemical mechanisms. It is, however, mathematically related to models that have been used to account for the contrast-dependent effects described above. It can therefore account for the same set of contrast-dependent phenomena, as documented in Figs. 8.3D–F.

The model achieves orientation selectivity as a result of tuned excitatory input, which is stronger for a preferred orientation stimulus than for a nonpreferred stimulus. Both excitatory and inhibitory inputs increase in strength with contrast, and the response at high contrast is determined by the ratio of excitatory to inhibitory input. Figure 8.3D shows the model contrast response functions for an optimal stimulus (upper dashed line) and a suboptimal but excitatory stimulus (lower dashed line). Attention is assumed to lead to increases in the strength of excitatory and inhibitory inputs activated by the attended stimulus (Reynolds et al., 1999), as would occur with an increase in the contrast of the stimulus. The effect of this change is to shift the model contrast response function to the left, as indicated by the arrows.

Figure 8.3E, which was obtained using the same set of parameters that yielded Fig. 8.3D, documents the ability of the model to exhibit approximately multiplicative increases in the orientation tuning curve with increasing contrast. The vertical lines indicate the orientations whose contrast response functions are illustrated in Fig. 8.3D. As attention yields a shift in contrast, its effect on the tuning curve is predicted to be the same as an increase in contrast: to cause a multiplicative increase in the tuning curve. This is illustrated by the upward arrows, which show the increases in response that result from a leftward shift in the contrast response function, for the two orientations whose contrast response functions are illustrated in Fig. 8.3D.

When two stimuli appear together, the effect of varying the contrast of either stimulus depends on their relative contributions of excitatory and divisive inhibitory drive. When a poor stimulus (with proportionally more inhibitory drive) is presented with a preferred stimulus, the additional inhibitory input results in a suppressed response. This response suppression can be magnified by increasing the contrast of the poor stimulus or else diminished by increasing the contrast of the preferred stimulus. Figure 8.3F illustrates the model behavior when the preferred stimulus from Fig. 8.3D appears together with a nonpreferred but excitatory stimulus also in the receptive field, at various levels of contrast. As is the case experimentally (Fig. 8.3C), the model accounts for the finding that elevating the contrast of the poor stimulus will increase its ability to suppress a fixed contrast preferred stimulus. Thus, a prediction of the model is that when two stimuli appear in the receptive field, attending to the more preferred stimulus will cause an elevation in

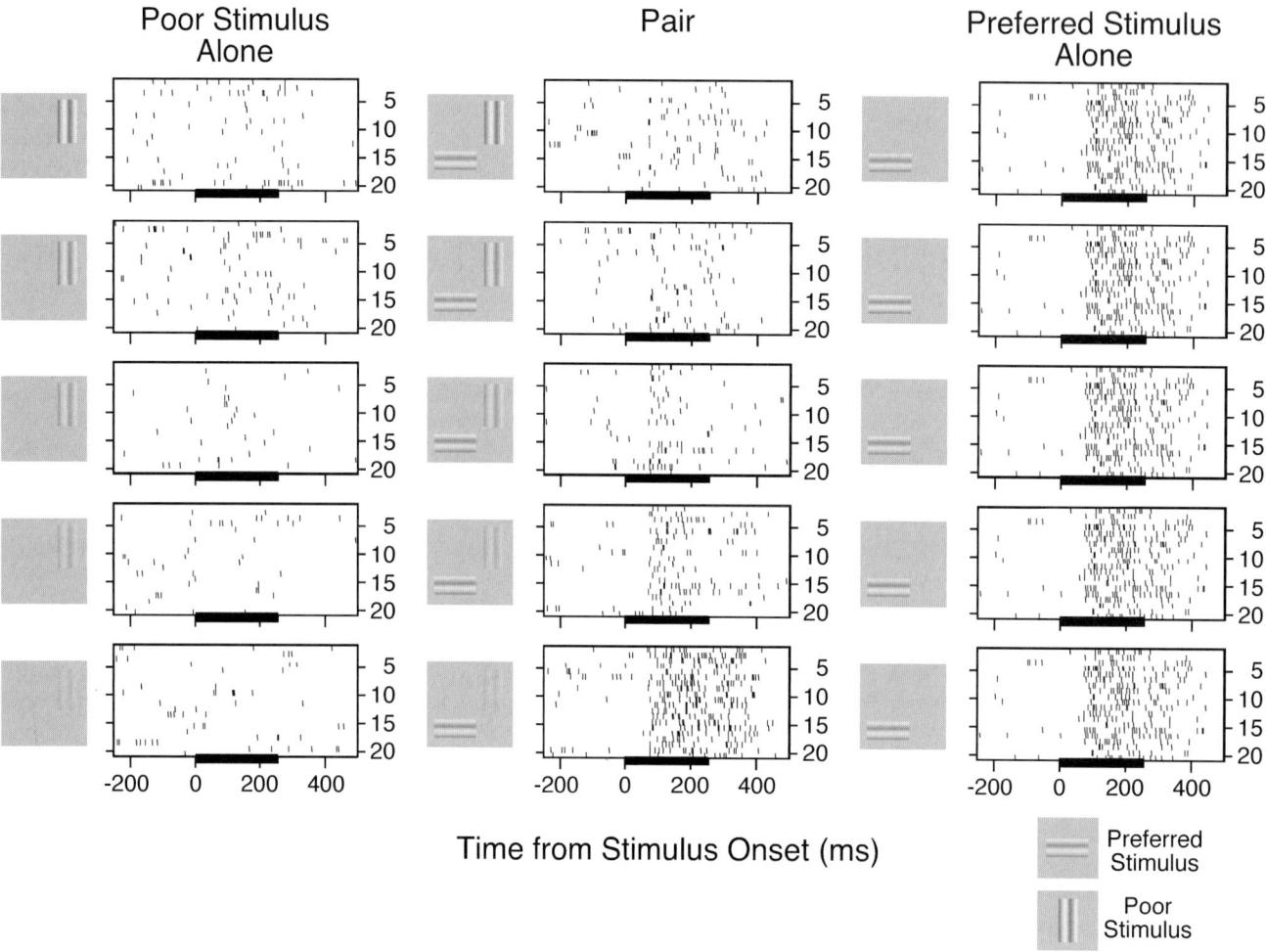

FIGURE.8.4 Increasing the contrast of a poor stimulus at one location suppresses the response elicited by a fixed contrast preferred stimulus at a second location in the receptive field of a V4 neuron. The contrast of the poor stimulus, illustrated in the first column, ranged from 5 to 80%. As indicated in the first column of raster plots, this stimulus did not elicit a clear response at any contrast. The right column shows the response elicited by the preferred stimulus, which was fixed in contrast (panels repeated down the column, for comparison). The middle column of raster plots shows the response to the pair. At low contrast (bottom), the poor stimulus had no measurable effect on the response to the preferred stimulus, but as poor stimulus contrast increased (moving up the column), it became increasingly suppressive, almost entirely suppressing the response at high contrast (top). Adapted, with permission, from Reynolds and Desimone (2003).

response, and attending to the poor stimulus will lead to a reduction in response.

V. CONCLUSIONS

Posner et al. (1980) noted that the importance of attention is its role in connecting the mental level of description used in cognitive science with the anatomical level common in neuroscience. In this chapter, we have seen an example of this integration across levels of explanation. Psychophysical measures of the effects of spatial attention on performance in detection of isolated stimuli and selection of stimuli from among distracters provide a description of attention at the mental level. Single-unit recording studies in monkeys performing attention-dependent tasks reveal direct neural correlates of attentional facilitation and selection in the extrastriate cortex. Attention-dependent increases in neuronal sensitivity and the selection of a stimulus that is presented with one or more distracters can be accounted for by a single cortical circuit model that is motivated by studies of contrast-dependent response modulation of neuronal responses. This coupling of single-unit neurophysiology and psychophysics has begun to provide a mechanistic characterization of a key cognitive process.

References

Bashinski, H. S., and Bacharach, V. R. (1980). Enhancement of perceptual sensitivity as the result of selectively attending to spatial locations. *Percept. Psychophys.* **28**, 241–248.

Bonds, A. B. (1989). Role of inhibition in the specification of orientation selectivity of cells in the cat striate cortex. *Vis. Neurosci.* **2**, 41–55.

Carandini, M., Heeger, D. J., and Movshon, J. A. (1997). Linearity and normalization in simple cells of the macaque primary visual cortex. *J. Neurosci.* **17**, 8621–8644.

Carrasco, M., Penpeci-Talgar, C., and Eckstein, M. (2000). Spatial covert attention increases contrast sensitivity across the CSF: support for signal enhancement. *Vision Res.* **40**, 1203–1215.

Chelazzi, L., Duncan, J., Miller, E. K., and Desimone, R. (1998). Responses of neurons in inferior temporal cortex during memory-guided visual search. *J. Neurophysiol.* **80**, 2918–2940.

Chelazzi, L., Miller, E. K., Duncan, J., and Desimone, R. (1993). A neural basis for visual search in inferior temporal cortex. *Nature* **363**, 345–347.

Chelazzi, L., Miller, E. K., Duncan, J., and Desimone, R. (2001). Responses of neurons in macaque area V4 during memory-guided visual search. *Cereb. Cortex.* **11**, 761–772.

DeAngelis, G. C., Robson, J. G., Ohzawa, I., and Freeman, R. D. (1992). Organization of suppression in receptive fields of neurons in cat visual cortex. *J. Neurophysiol.* **68**, 144–163.

Desimone, R., and Duncan, J. (1995). Neural mechanisms of selective visual attention. *Annu. Rev. Neurosci.* **18**, 193–222.

Downing, C. J. (1988). Expectancy and visual–spatial attention: effects on perceptual quality. *J. Exp. Psychol. Hum. Percept. Perform.* **14**, 188–202.

Ferrera, V. P., and Lisberger, S. G. (1995). Attention and target selection for smooth pursuit eye movements. *J. Neurosci.* **15**, 7472–7484.

Gawne, T. J., and Martin, J. M. (2002). Responses of primate visual cortical V4 neurons to simultaneously presented stimuli. *J. Neurophysiol.* **88**, 1128–1135.

Grossberg, S., and Raizada, R. D. (2000). Contrast-sensitive perceptual grouping and object-based attention in the laminar circuits of primary visual cortex. *Vision Res.* **40**, 1413–1432.

Handy, T. C., Kingstone, A., and Mangun, G. R. (1996). Spatial distribution of visual attention: perceptual sensitivity and response latency. *Percept. Psychophys.* **58**, 613–627.

Hawkins, H. L., Hillyard, S. A., Luck, S. J., Mouloua, M., Downing, C. J., and Woodward, D. P. (1990). Visual attention modulates signal detectability. *J. Exp. Psychol. Hum. Percept. Perform.* **16**, 802–811.

Hillyard, S. A., and Anllo-Vento, L. (1998). Event-related brain potentials in the study of visual selective attention. *Proc. Natl. Acad. Sci. USA.* **95**, 781–787.

Hoffman, J. E., and Subramaniam, B. (1995). The role of visual attention in saccadic eye movements. *Percept. Psychophys.* **57**, 787–795.

Ito, M., and Gilbert, C. D. (1999). Attention modulates contextual influences in the primary visual cortex of alert monkeys. *Neuron* **22**, 593–604.

Itti, L., and Koch, C. (2000). A saliency-based search mechanism for overt and covert shifts of visual attention. *Vision Res.* **40**, 1489–1506.

James, W. (1890). The Principles of Psychology, pp. 319–349. Henry Holt, New York.

Lee, D. K., Itti, L., Koch, C., and Braun, J. (1999). Attention activates winner-take-all competition among visual filters. *Nat. Neurosci.* **2**, 375–381.

Lu, Z. L., and Dosher, B. A. (1998). External noise distinguishes attention mechanisms. *Vision Res.* **38**, 1183–1198.

Luck, S. J., Chelazzi, L., Hillyard, S. A., and Desimone, R. (1997). Neural mechanisms of spatial selective attention in areas V1, V2, and V4 of macaque visual cortex. *J. Neurophysiol.* **77**, 24–42.

Martínez-Trujillo, J., and Treue, S. (2002). Attentional modulation strength in cortical area MT depends on stimulus contrast. *Neuron* **35**, 365–370.

McAdams, C. J., and Maunsell, J. H. (1999). Effects of attention on orientation-tuning functions of single neurons in macaque cortical area V4. *J. Neurosci.* **19**, 431–441.

Miller, E. K., Gochin, P. M., and Gross, C. G. (1993). Suppression of visual responses of neurons in inferior temporal cortex of the awake macaque by addition of a second stimulus. *Brain Res.* **616**, 25–29.

Moran, J., and Desimone, R. (1985). Selective attention gates visual processing in the extrastriate cortex. *Science* **229**, 782–784.

Morrone, M. C., Burr, D. C., and Maffei, L. (1982). Functional implications of cross-orientation inhibition of cortical visual cells: I. Neurophysiological evidence. *Proc. R. Soc. London Ser. B* **216**, 335–354.

Motter, B. C. (1993). Focal attention produces spatially selective processing in visual cortical areas V1, V2, and V4 in the presence of competing stimuli. *J. Neurophysiol.* **70**, 909–919.

Muller, H. J., and Humphreys, G. W. (1991). Luminance-increment detection: capacity-limited or not? *J. Exp. Psychol. Hum. Percept. Perform.* **17**, 107–124.

Niebur, E., and Koch, C. (1994). A model for the neuronal implementation of selective visual attention based on temporal correlation among neurons. *J. Comput. Neurosci.* **1**, 141–158.

Pessoa, L., Kastner, S., and Ungerleider, L. G. (2003). Neuroimaging studies of attention: from modulation of sensory processing to top-down control. *J. Neurosci.* **23**, 3990–3998.

Posner, M. I., Snyder, C. R., and Davidson, B. J. (1980). Attention and the detection of signals. *J. Exp. Psychol.* **109**, 160–174.

Recanzone, G. H., and Wurtz, R. H. (2000). Effects of attention on MT and MST neuronal activity during pursuit initiation. *J. Neurophysiol.* **83**, 777–790.

Reynolds, J. H., Chelazzi, L., and Desimone, R. (1999). Competitive mechanisms subserve attention in macaque areas V2 and V4. *J. Neurosci.* **19**, 1736–1753.

Reynolds, J. H., and Desimone, R. (2003). Interacting roles of attention and visual salience in V4. *Neuron* **37**, 853–863.

Reynolds, J. H., Pasternak, T., and Desimone, R. (2000). Attention increases sensitivity of V4 neurons. *Neuron* **26**, 703–714.

Rolls, E. T., and Tovee, M. J. (1995). The responses of single neurons in the temporal visual cortical areas of the macaque when more than one stimulus is present in the receptive field. *Exp. Brain Res.* **103**, 409–420.

Sclar, G., and Freeman, R. D. (1982). Orientation selectivity in the cat's striate cortex is invariant with stimulus contrast. *Exp. Brain Res.* **46**, 457–461.

Seidemann, E., and Newsome, W. T. (1999). Effect of spatial attention on the responses of area MT neurons. *J. Neurophysiol.* **81**, 1783–1794.

Sheinberg, D. L., and Logothetis, N. K. (2001). Noticing familiar objects in real world scenes: the role of temporal cortical neurons in natural vision. *J. Neurosci.* **21**, 1340–1350.

Treue, S., and Martinez Trujillo, J. C. (1999). Feature-based attention influences motion processing gain in macaque visual cortex. *Nature* **399**, 575–579.

Treue, S., and Maunsell, J. H. (1996). Attentional modulation of visual motion processing in cortical areas MT and MST. *Nature* **382**, 539–541.

Yantis, S., and Serences, J. T. (2003). Cortical mechanisms of space-based and object-based attentional control. *Curr. Opin. Neurobiol.* **13**, 187–193.

CHAPTER 9

Psychopharmacology of Human Attention

Jennifer T. Coull

ABSTRACT

Administration of drugs acting on the noradrenergic or cholinergic neuromodulatory neurotransmitter systems affects human attentional performance. The direction or magnitude of noradrenergic attentional effects, mediated primarily via the α2 receptor system, can vary according to the subject's underlying level of arousal. Functional neuroimaging studies have recently suggested the thalamus as one of the key neuroanatomical substrates for the modulatory effect of arousal on noradrenergic attentional effects. Cholinergic attentional effects can be mediated by either nicotinic or muscarinic receptor systems, although neuroimaging evidence suggests the nicotinic system may be more functionally specialized for attention, exerting its effect via the parietal cortex. In addition, attentional effects may underlie the well-known cholinergic modulation of memory and learning processes. As yet, there is insufficient evidence to propose a neurochemical specialization of noradrenaline or acetylcholine for distinct attentional processes. Future investigations may benefit from trying to dissociate these two neurotransmitter systems in terms of their neural mechanisms of action, rather than their effects on discrete attentional processes.

I. INTRODUCTION

Psychopharmacological investigation of human attention is invariably a quest for neurochemical functional specialization: which specific neurotransmitter modulates which specific attentional process. The concept of *anatomical* localization of function has been familiar for more than a century, thanks to methods such as neuropsychological investigation of brain-damaged patients and, more recently, functional neuroimaging of healthy volunteers. Neuro*chemical* specialization of cognitive function is a logically equivalent approach. It requires simply the recognition that, first, cognition (or even, in our case, attention) is a multifaceted function, with discrete and separable processes. Second, each of these processes is instantiated within a characterizable neural architecture. Finally, perturbation of different regulatory neurotransmitter systems can modulate the functioning of the architecture and, thus, functioning of the process, in potentially different ways.

A mundane analogy may help conceptualize this. Different rooms within a home serve different purposes, just as distinct brain regions (the "neural architecture") may subserve distinct cognitive functions. A perturbation to one of the regulatory control systems in the home, for example, an electrical power cut, may disrupt cooking activity in the kitchen, have no effect on sunbathing in the garden, and actually aid snoozing in the otherwise noisy living room! Similarly, reduced activity of a neurotransmitter system, due to disease or pharmacological or surgical intervention, can affect different cognitive (including attentional) processes in different ways. These effects may differ depending on the anatomic localization of each process (e.g., if a neurotransmitter does not project to the prefrontal cortex, it is unlikely to exert an effect on executive function) or the physiological mechanisms by which the neurotransmitter exerts its effect (e.g., increasing the signal-to-noise ratio of cell firing may improve selective attention but impair attentional shifting).

The noradrenergic and cholinergic neuromodulatory neurotransmitter systems are the two systems most consistently associated with the modulation of attention. This review therefore focuses on studies that manipulate one or the other of these systems. However, it is worth noting that the dopaminergic system has also been implicated in attentional processes, although mostly with those processes that require a large degree of "executive" control, such as attentional set shifting. This review is further restricted to studies of human attention, and the reader is referred to Chapter 10 for a more thorough review of the animal literature. Human psychopharmacology and animal psychopharmacology are mutually informative and coexist happily alongside one another. One cannot replace the other, and the pros and cons of each approach make this self-evident. For instance, a wide range of receptor-specific drugs are available for use in animals, and in addition, these can be injected locally into discrete brain regions to anatomically localize drug-specific cognitive effects. In humans, however, the relatively limited set of drugs available must be administered systemically, with the site of action being inferred only indirectly from task-specific effects and knowledge of existing neuropsychological or animal literature. The latter approach clearly runs the risk of producing circular arguments when trying to make the three-way link between anatomy, neurochemistry, and function. However, with the advent of functional neuroimaging techniques, such as PET and fMRI, we can now begin to make these links more confidently. Among the advantages of performing psychopharmacological functional imaging studies in humans is the ability to study the neurochemical and anatomical bases of a wider and more sophisticated range of cognitive processes than are displayed by an animal. In addition, the procedure is clearly much less invasive than animal lesion studies, and the number of hours spent training subjects is dramatically shortened.

II. THE NORADRENERGIC SYSTEM

A. Direct Effects on Attention

Electrophysiological studies in animals (see Aston-Jones et al., 1999, for a review) were among the first to suggest a role for noradrenaline (NA) in the modulation of attention by demonstrating that microiontophoretic application of NA onto monkey cortical cells selectively inhibits background firing rate and enhances the selectivity of target cell discharge. Such enhancement in the neuronal signal-to-noise ratio of evoked responses renders secondary stimuli less distracting, allowing attention to become more focused. More compellingly, Aston-Jones and colleagues (1999) recorded firing rates in monkey locus coeruleus (LC), the main noradrenergic nucleus, which gives rise to a widespread set of efferents both cortically and subcortically, during performance of a target detection "oddball" task. They showed increased phasic LC activity to targets, but not distractors, suggesting LC neurons are more active in response to "salient" stimuli. The salience of the stimulus may be determined either by virtue of being unexpected or intense (attention is captured in a "bottom-up" manner) or by becoming meaningful through training or conditioning (attention is allocated in a "top-down" manner).

In humans, manipulations of the noradrenergic α2 receptor, in particular, have shown a narrowing of the focus of selective spatial attention using the α2 anatagonist idazoxan (Smith et al., 1992) or a broadening using the α2 agonist clonidine (Coull et al., 1995). Clonidine has also been shown to facilitate the disengagement of spatial attention as measured by the Posner covert orienting of attention task, in that it diminished the cost of being invalidly cued to a target spatial location (Clark et al., 1989). This behavioral result was later confirmed in an fMRI study, which additionally demonstrated corresponding clonidine-induced attenuation of task-specific activity selectively in right superior parietal cortex (Coull et al., 2001). The same fMRI study also demonstrated clonidine-induced impairments in temporal attentional orienting (see Chapter 42 for an explanation of this task) which were associated with attenuation of task-specific activity in left prefrontal cortex and insula. The task-specific drug effects localized in right parietal versus left frontal cortex (Fig. 9.1) most likely reflected selective noradrenergic modulation of brain regions functionally specialized for temporal versus spatial attentional orienting processes, respectively, rather than general NA modulation of visual, motoric, or basic arousal processes.

B. Attentional Effects Vary According to Arousal Level

Established theories of attentional function, both psychological and neuropsychological, have emphasized the influence of underlying arousal levels on attentional function. The noradrenergic system may be critical in mediating this influence. Electrophysiological recordings in monkeys have shown that *phasic* LC firing to targets (see Section II.A) actually varies according to the level of *tonic* LC activity (equivalent to the state of arousal of the animal) and has corresponding effects on attentional performance (Aston-Jones et al., 1999). This pattern of activity follows an inverted U-shaped curve whereby too low or too high

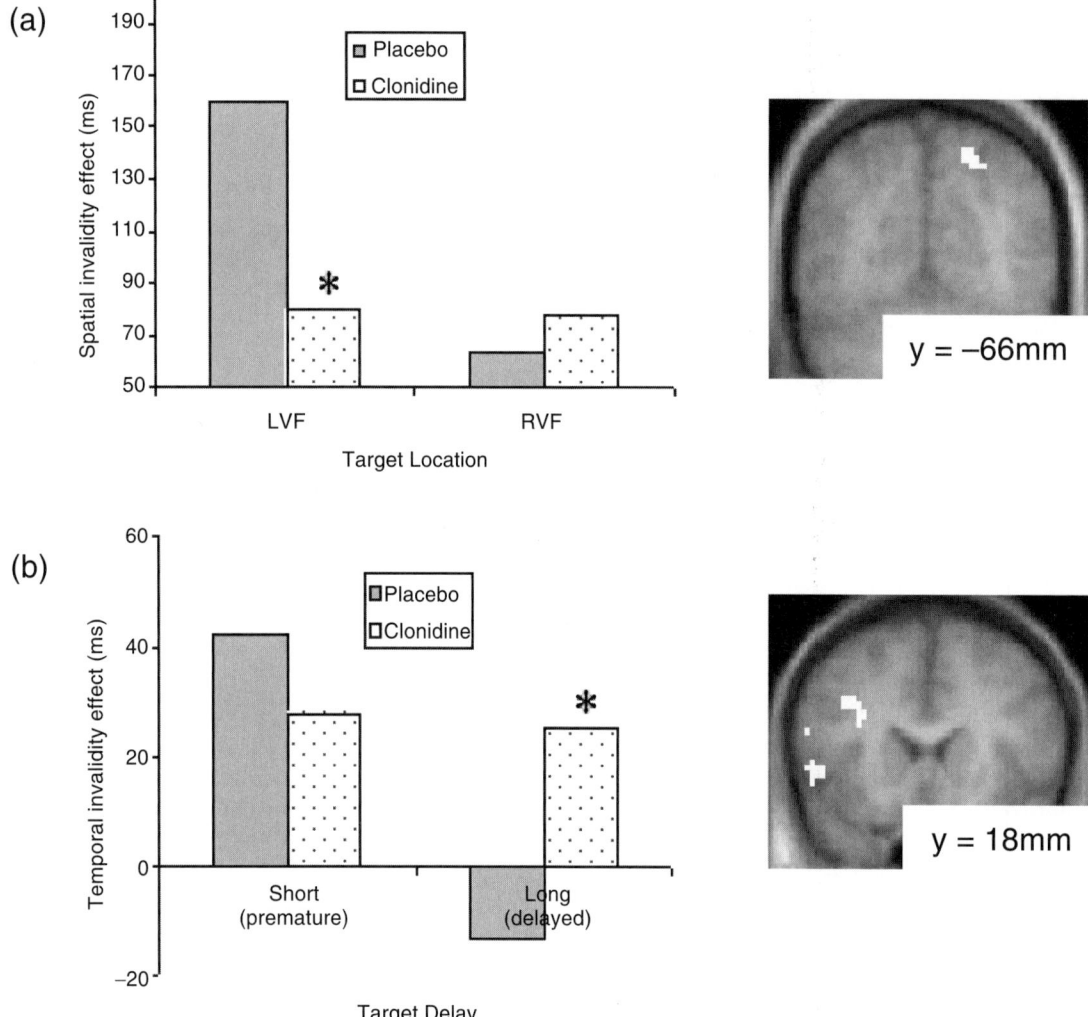

FIGURE 9.1 (a) During a covert spatial orienting task, subjects were cued to expect an upcoming target to appear in either the left (LVF) or right (RVF) visual field. Cues provided valid or invalid information, and the "spatial invalidity effect" indexed the difference in speed of target detection for invalidly minus validly cued trials. The clinical spatial neglect literature demonstrates that invalidly cued stimuli in LVF produce a larger invalidity effect than those in RVF, and are generally more sensitive to disruption. Accordingly, compared with placebo, clonidine attenuated the spatial invalidity effect for targets presented unexpectedly in LVF. Accompanying this behavioral effect was an attenuation of task-specific activity in right superior parietal cortex. (b) During a covert temporal orienting task, subjects were cued to expect an upcoming target to appear after a delay of either 600 ms (short) or 1400 ms (long). Cues provided valid or invalid information, and the "temporal invalidity effect" indexed the difference in speed of target detection for invalidly minus validly cued trials. Previous studies (see Chapter 42) have shown that the temporal invalidity effect is significant for premature targets (unexpectedly short delay) but negligible for delayed targets (unexpectedly long delay). However, compared with placebo, clonidine produced a significant temporal invalidity effect for unexpectedly delayed targets. Accompanying this behavioral effect was an attenuation of task-specific activity in left insula and frontal cortex.

a state of tonic LC firing, or arousal, alters phasic LC firing and, consequently, impairs performance. Optimal performance corresponds to an intermediate state of tonic LC discharge or arousal.

This response profile has also been associated with the effects of pharmacological manipulations of the noradrenergic system on human attentional performance. For example, Smith and Nutt (1996) reported clonidine-induced impairment in performance of a selective attention task, which was abolished by simultaneous expoure to bursts of distracting white noise. Pharmacological specificity for the α2 receptor was nicely demonstrated in this study, as the detrimental effects of clonidine on performance were abolished not only by white noise, but also by co-administration of the selective α2 antagonist idazoxan. These findings were confirmed by results from a recent fMRI study, which further demonstrates that these effects are not due simply to the sedative properties of the drug. Healthy volunteers performed a target detection task, which was presented with or without simultaneous bursts of white noise, following administration of placebo or equisedative doses of the α2 agonist dexmedotomidine or the benzodiazepine midazolam (Coull et al., 2004). Both dexmedotomidine and midazolam impaired task performance equally under "quiet" conditions. However, under "noisy" conditions, midazolam continued to impair performance, whereas the deleterious effects of dexmedotomidine were attenuated (Fig. 9.2a). In other words, the presence of loud white noise attenuated dexmedetomidine's detrimental effect on attentional performance. This behavioral effect was accompanied by a dexmedetomidine-related increase in left pulvinar activity during the noise, but not quiet, condition (Fig. 9.2b), which correlated directly with improvements in task performance. The pulvinar increase was therefore suggested to index the putative increase in phasic arousal induced by presentation of white noise which ameliorated the attentional dysfunction induced by dexmedetomidine. Our data fit neatly into the framework proposed by Aston-Jones and colleagues (1999) as phasic arousal (white noise) interacted with a state of low tonic arousal (induced by dexmedetomidine) to modify the drugs' overall effects on attentional performance. Moreover, the neural substrate of this interaction, at least for a nonspatial target detection task, was shown to be the thalamic pulvinar.

One neural mechanism suggested to underlie this effect was that pulvinar activity increased in a compensatory manner to allow the brain to cope with arousing stimuli under conditions of low tonic arousal. Such indirect, or retaliatory, effects of α2 agonists were also suggested by an analysis of a previous psychopharmacological PET study using measures of "effective connectivity" (see Chapter 75 for an explanation of this approach). Rather than examining drug-induced changes in activity *within* a single region, effective connectivity analysis allowed us to measure drug-induced changes in the functional strength of putative anatomical connections *between* discrete brain regions. In subjects performing a nonspatial sustained attention task, and as compared with placebo, clonidine increased cross-talk between brain regions implicated in attentional processing (viz. frontal and parietal cortices, thalamus, and LC) (Coull et al., 1999). There was no such increase during a rest condition (Fig. 9.3). Moreover, a significant negative correlation

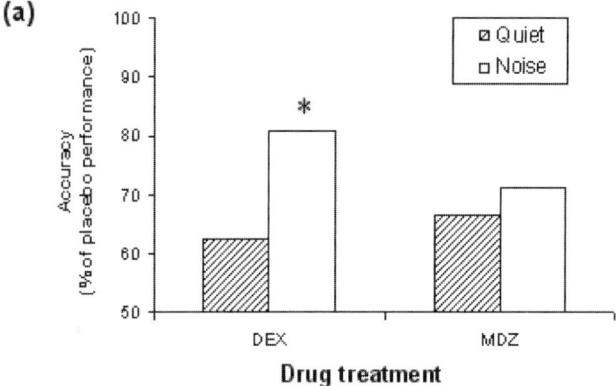

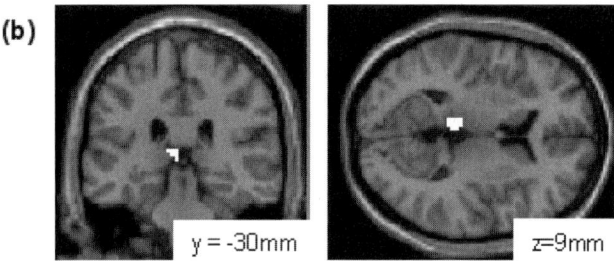

FIGURE 9.2 (a) Compared with placebo, both midazolam (MDZ) and dexmedotomidine (DEX) impaired target detection performance during quiet testing conditions ("quiet"). Bursts of white noise presented during task performance ("noise") significantly attenuated the deleterious effects of DEX on performance under quiet testing conditions, but had no effect on the deleterious effects of MDZ. Accuracy for each drug condition is represented as a percentage of placebo performance. (b) The beneficial effects of white noise on the DEX-induced attentional dysfunction correlated significantly with an increase in activity of the left pulvinar nucleus of the thalamus.

between clonidine-induced change in performance and activity in right parietal cortex was suggestive of a compensatory mechanism of brain function. The more performance deteriorated with clonidine, the more parietal activity increased. These results suggest a retaliatory response of the brain to neurochemical challenge, rather than a direct effect of the drug on the connections themselves, and illustrate the effect of clonidine on a network of areas that interact dynamically with one another to optimize behavior.

In summary, the effects of NA, or more specifically, α2 agents, on attentional performance differ according to the arousal level of the subject, such that the arousing effects of noise or novelty can attenuate, or even abolish, drug effects on performance. This noradrenergic α2 modulation of the interaction between attention and arousal may rely particularly on the neural activity of the thalamus. Furthermore, brain regions within a corticothalamic attentional network may interact dynamically with one another under states of

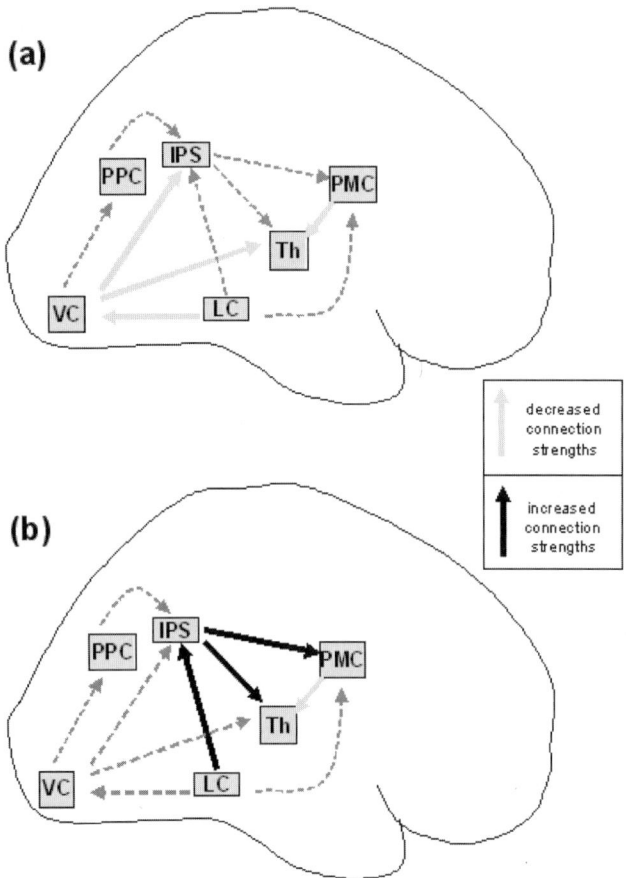

FIGURE 9.3 (a) During a rest state, functional integration between the visual cortex and three other brain regions *decreased* (solid gray arrows) following clonidine infusion. (b) During an attentional task, functional integration between the intraparietal sulcus and three other brain regions *increased* (solid black arrows) following clonidine infusion. Functional integration is measured using effective connectivity analysis, and all clonidine effects are compared with those of a placebo control. Dashed arrows represent no significant difference in connection strength between clondine and placebo conditions. LC, locus coeruleus; VC, visual cortex; PPC, posterior parietal cortex; IPS, intra-parietal sulcus; Th, thalamus; PMC, premotor cortex.

suboptimal arousal, to preserve behavioral performance following infusion of α2 agents.

III. THE CHOLINERGIC SYSTEM

A. Cholinergic Modulation of Memory and Learning: An Attentional Effect?

The cholinergic system has traditionally been implicated in processes of memory and learning, perhaps driven largely by the impact of the cholinergic hypothesis of Alzheimer's disease (AD). However, cholinergic modulation of attentional processes has increasingly been suggested to underlie these well-documented mnemonic effects. In particular, numerous experiments have shown that cholinergic modulation of memory or learning processes varies as a function of attention. For example, improvements in verbal recall induced by the cholinergic nicotinic receptor agonist nicotine were abolished by performance of a secondary attentional task (Rusted and Warburton, 1992). Conversely, deficits in verbal recall induced by the cholinergic muscarinic receptor antagonist scopolamine during performance of a dichotic listening task were significant in the attended channel only (Dunne and Hartley, 1985). Similarly, a recent fMRI study has shown that regionally specific changes in occipital cortex activity induced by repetition priming, a measure of perceptual learning, were enhanced by the anticholinesterase physostigmine, but only in attended spatial locations (Bentley et al., 2003). Several studies have even provided evidence for selective effects of cholinergic manipulations on attentional tasks, with no effect on mnemonic function (e.g., Sahakian et al., 1989).

B. Direct Effects on Attention

Converging results in rats, humans, and monkeys show that nicotine impairs spatial attentional orienting, as measured by the Posner task (e.g., Witte et al., 1997), by selectively attenuating the cost of invalid cueing (similar to the effects of the noradrenergic α2 agonist clonidine decribed in Section II.A). Scopolamine, on the other hand, impairs spatial orienting both by exacerbating the cost of being invalidly cued and by reducing the benefit of being validly cued, and recent electrophysiological measures in monkeys suggest that modulation of parietal cortex activity may underlie these attentional effects (Davidson and Marrocco, 2000). In humans, cholinergic modulation of parietal cortex has also been noted using functional neuroimaging methods in studies of both AD patients (van Dyck et al., 1997) and healthy volunteers (Mentis et al., 2001; Lawrence et al., 2002). However, both of these studies also observed cholinergic modulation of occipital cortex, with Mentis et al. (2001) suggesting that muscarinic effects (such as those produced by scopolamine) may actually be mediated via occipital, not parietal, cortex, at least for their low-level visual processing task. They further speculated that cholinergic modulation of parietal cortex activity is more likely to be a nicotinic effect.

More compellingly, distinct functional effects of cholinergic manipulations on parietal and occipital activity have recently been demonstrated using a task of sustained attention, performance of which is enhanced by administration of nicotine to both healthy volunteers (Wesnes and Warburton, 1984) and patients with mild to moderate AD (Sahakian et al., 1989). This fMRI study demonstrated that task-specific parietal cortex activity was enhanced by nicotine selectively during performance of the attentional task, while nicotine-induced increases in occipital cortex were evident during both attentional and control tasks (Lawrence et al., 2002). The authors suggest nicotinic modulation of parietal cortex underlies the drug-induced improvements in sustained attention performance per se, whereas nicotinic modulation of occipital cortex may represent an increase in the signal-to-noise ratio in stimulus-specific regions of the cortex to enhance selectivity of low-level visual processing. However, recent behavioral results (Mancuso et al., 2001) suggest that nicotinic modulation of attention may in fact be secondary to the stronger effects of nicotine on low-level sensory processing. Effective connectivity analyses of functional imaging data, similar to that described in Section II.B, may help clarify the relative contributions of parietal and occipital activity to the cholinergic modulation of attentional and visual processing.

IV. CONCLUSION

Both the noradrenergic and cholinergic neuromodulatory systems are implicated in the control of attentional processing. The effects of noradrenergic manipulations on attention may vary according to arousal levels and activity in subcortical thalamic regions. The effects of cholinergic manipulations on attention, particularly those involving the nicotinic system, may depend on a more primary modulation of visual processing and activity in a cortical parieto-occipital network. Despite these broad differences, the current evidence makes it difficult to propose a convincing theory of neurochemical specialization of attentional function for the noradrenergic and cholinergic systems that is based on attentional processes or task types (e.g., spatial attention, divided attention, selective attention). For example, drugs acting on both neurotransmitter systems have been shown to have similar effects on tests of spatial orienting and nonspatial sustained attention. In trying to dissociate the effects of these two systems on attention it may be more profitable not to examine *which* attentional process is being affected, but rather *how* this process is being differentially affected, at either the anatomical or the physiological level. We may imagine, for instance, that effects on selective attention could be mediated by either (or both) a noradrenergic frontoparietal connection or a cholinergic parieto-occipital connection.

Alternatively, both NA and ACh may enhance selective attention by increasing the signal to noise ratio, but perhaps NA does this by decreasing the noise, while ACh does it by increasing the signal. In other words, rather than trying to dissociate the attentional effects of NA and ACh in terms of functional process, it may be more informative to try and dissociate them in terms of anatomical or physiological mechanism.

Acknowledgment

J.T.C. is supported through a European Community Marie Curie Fellowship.

References

1. Aston-Jones, G., Rajkowski, J., and Cohen, J. (1999). Role of locus coeruleus in attention and behavioural flexibility. *Biol. Psychiatry* **46**, 1309–1320.
2. Bentley, P., Vuilleumier, P., Thiel, C. M., Driver, J., and Dolan, R. J. (2003). Effects of attention and emotion on repetition priming and their modulation by cholinergic enhancement. *J. Neurophysiol.* **90**, 1171–1181.
3. Clark, C. R., Geffen, G. M., and Geffen, L. B. (1989). Catecholamines and the covert orientation of attention in humans. *Neuropsychologia* **27**, 131–139.
4. Coull, J. T., Büchel, C., Friston, K. J., and Frith, C. D. (1999). Noradrenergically-mediated plasticity in a human attentional neuronal network. *NeuroImage* **10**, 705–715.
5. Coull, J. T., Jones, M. E. P., Egan, T. D., Frith, C. D., and Maze, M. (2004). Attentional effects of noradrenaline vary with arousal level: selective activation of thalamic pulvinar in humans. *Neuroimage* **22**, 315–322.
6. Coull, J. T., Nobre, A. C., and Frith, C. D. (2001). The noradrenergic α2 agonist clonidine modulates behavioural and neuroanatomical correlates of human attentional orienting and alerting. *Cereb. Cortex.* **11**, 73–84.
7. Coull, J. T., Sahakian, B. J., Middleton, H. C., Young, A. H., Park, S. B., McShane, R. H., Cowen, P. J., and Robbins, T. W. (1995). Differential effects of clonidine, haloperidol, diazepam and tryptophan depletion on focused attention and attentional search. *Psychopharmacology* **121**, 222–230.
8. Davidson, M. C., and Marrocco, R. T. (2000). Local infusion of scopolamine into intraparietal cortex slows covert orienting in rhesus monkeys. *J. Neurophysiol.* **83**, 1536–1549.
9. Dunne, M. P., and Hartley, L. R. (1985). The effects of scopolamine upon verbal memory: evidence for an attentional hypothesis. *Acta. Psychol.* **58**, 205–217.
10. Lawrence, N. S., Ross, T. J., and Stein, E. A. (2002). Cognitive mechanisms of nicotine on visual attention. *Neuron.* **36**, 539–548.
11. Mancuso, G., Lejeune, M., and Ansseau, M. (2001). Cigarette smoking and attention: processing speed or specific effects? *Psychopharmacology* **155**, 372–378.
12. Mentis, M. J., Sunderland, T., Lai, J., Connolly, C., Krasuski, J., Levine, B., Friz, J., Sobti, S., Schapiro, M., and Rapoport, S. I. (2001). Muscarinic versus nicotinic modulation of a visual task: a PET study using drug probes. *Neuropsychopharmacology* **25**, 555–564.
13. Rusted, J. M., and Warburton, D. M. (1992). Facilitation of memory by post-trial administration of nicotine: evidence for an attentional explanation. *Psychopharmacology* **108**, 452–455.

14. Sahakian, B. J., Jones, G. M. M., Levy, R., Gray, J. A., and Warburton, D. M. (1989). The effects of nicotine on attention, information processing, and short-term memory in patients with dementia of the Alzheimer type. *Br. J. Psychiatry.* **154**, 797–800.
15. Smith, A., and Nutt, D. (1996). Noradrenaline and attention lapses. *Nature* **380**, 291.
16. Smith, A. P., Wilson, S. J., Glue, P., and Nutt, D. J. (1992). The effects and after effects of the alpha-2-adrenoceptor antagonist idazoxan on mood, memory and attention in normal volunteers. *J. Psychopharmacol.* **6**, 376–381.
17. van Dyck, C. H., Lin, C. H., Robinson, R., Cellar, J., Smith, E. O., Nelson, J. C., Arnsten, A. F., and Hoffer, P. B. (1997). The acetylcholine releaser linopiridine increases parietal regional cerebral blood flow in Alzheimer's disease. *Psychopharmacology* **132**, 217–226.
18. Wesnes, K., and Warburton, D. M. (1984). Effects of scopolamine and nicotine on human rapid information processing performance. *Psychopharmacology* **82**, 147–150.
19. Witte, E. A., Davidson, M. C., and Marrocco, R. T. (1997). Effects of altering brain cholinergic activity on covert orienting of attention: comparison of monkey and human performance. *Psychopharmacology* **132**, 324–334.

CHAPTER

10

Neuropharmacology of Attention

Jean A. Milstein, Jeffrey W. Dalley, and Trevor W. Robbins

ABSTRACT

Converging evidence for the neurochemical modulation of attentional systems in animals is outlined, with the focus primarily on rat neurochemistry and behavioral pharmacology. Data are integrated from the five-choice serial reaction time task, an analog of the human continuous performance sustained attention task; a two-lever signal detection task; and convergent cross-species evidence from cued target detection tasks.

I. INTRODUCTION

The study of attention derives primarily from the fields of human cognitive psychology and neuropsychology. The animal literature is often couched in the behaviorist vocabulary of stimulus control, limiting its utility for the study of the neural and neurochemical substrates of attention. However, in simple terms, an animal must be able to make sense of its environment by processing and reacting to the stimuli within it. For example, it is adaptive for animals to monitor their environments for certain rare, but salient events (i.e., the appearance of a hawk's shadow for a typical prey animal) that require swift action. Such monitoring requires the animal to sustain attention, or vigilance, a concept used repeatedly in the human literature to mean the ability to sustain attention over a long period to detect rare events.

Similarly, like a person at a busy street corner who must selectively attend to the cars driving by, while simultaneously ignoring the traffic on the cross street, animals must differentiate and selectively attend to certain channels of information, while ignoring other irrelevant dimensions. And, assuming that the ability to process simultaneous information is limited, animals must be able to selectively switch or simultaneously monitor two streams of information and, thus, divide attention between them.

We outline converging evidence for the neurochemical modulation of attentional systems in animals, focusing primarily on rat neurochemistry and behavioral pharmacology in three complementary tasks, which tap both the frontostriatal ("anterior attentional system") and parietal ("posterior attentional system") networks (Posner and Petersen, 1990).

There are several other methods for examining attentional processes, often based on rather different theoretical approaches. These methods include prepulse inhibition of the acoustic startle response, and two other paradigms based on animal learning theory, latent inhibition and intra-/extradimensional shift discrimination learning (see Robbins, 1998, for a brief review). However, these approaches are beyond our present scope.

II. SUSTAINED ATTENTION

Sustained attention or vigilance is measured in continuous performance-type (CPT) tasks that require constant monitoring of the situation at hand. Vigilance tasks usually entail a consistent decrement in performance (i.e., a lengthening of reaction times and decrease in signal detection) with increasing time on task. Typically, sustained attention tasks fall short of the typical definitions of vigilance, but nevertheless require sustained effort to perform adequately. In one basic form of the CPT, the subject is presented with a series of letters or numbers (usually presented on a computer screen) and is asked to respond only to a certain character (the letter X, for example), which appears infrequently. In another, working memory version of the task, the individual is asked to respond to a certain character only if it is immediately preceded by a cue character (i.e., the person responds to the letter X only if the letter A immediately precedes it) (Rosvold et al., 1956).

In animals, it is quite difficult to dissociate so-called vigilance decrements in response latencies from general satiety or reduction in motivation, as the animal generally performs for a prolonged period and, thus, has been similarly rewarded for prior performance, perhaps to the point of satiety. Satiety effects can, however, be dissociated in well-designed tasks, by examining trials completed and latencies to collect food reward.

The five-choice serial reaction time task (5CSRT) is one such task that has been used extensively in the study of the neural mechanisms underlying visuospatial sustained attention in the rat (see Robbins, 2002, for a history and review of the task). At its core, this task has both continuous performance and spatial elements. The basic task requires the animal to scan a horizontal array of five nose-poke apertures in the rear of an operant chamber and respond to a brief flash of light (0.5 second) presented randomly behind one of the apertures. Responding correctly to the illuminated hole produces a food pellet in the magazine situated at the front of the chamber. Responding to an incorrect (non-illuminated) hole (commission error) or failing to respond within a prescribed time limit after stimulus presentation (omission error) results in a 5-second "time-out" punishment period, as does responding within the intertrial interval, prior to stimulus presentation (premature response). Effects of distractor stimuli (interpolated bursts of white noise) index attentional selectivity; and making the target stimuli temporally unpredictable or stimulus degradation (e.g., decreases in stimulus duration or stimulus intensity) taxes additional attentional resources. As the task requires inhibition of premature responses, it involves primarily the "anterior attentional network" including the rat frontal cortex.

Another test of visual sustained attention, which requires a forced choice between two levers ("yes" and "no") to detect a signal, has been used to measure a vigilance decrement (McGaughy and Sarter, 1995; Sarter et al., 2001). However, although this task usefully allows a signal detection analysis, the decrement does not usually occur in untreated animals. This paradigm has also been modified to study attention divided across modality (auditory and visual). Briefly, in this variant, rats are trained consecutively in visual and auditory conditional discrimination tasks before being required to perform both discriminations in the same trial block. A "frontoparietal" task, it presumably taps both attentional networks. Disadvantages of two-lever tasks include side and other response biases that make a purely attentional interpretation of performance difficult, especially when studying drugs such as the psychostimulants, which can have response-altering effects without altering attentional mechanisms per se.

III. SELECTIVE SPATIAL ATTENTION/ATTENTIONAL SHIFT

Covert/overt orienting involves attentional shifting between sensory cues (which may or may not predict the location of the subsequent target) and their associated targets. First described by Posner, this is much more of a selective attention task, and engages primarily the so-called "posterior attention network," including the parietal cortex, and is implicated in sensory neglect phenomena. Covert orienting occurs in the absence of saccades or eye movements: overt orienting occurs with intentional movements; attention to spatial location is summoned via central or peripheral cues, speeding reaction time (RT). Costs of attentional shifts can be measured by introducing cues that are contralateral to the target and, thus, shift attention away from the intended location. The cost of this attentional disengagement can be measured as the increase in RT to the invalidly cued target versus the validly cued target (validity effect). So-called "neutral" cues (cues that provide no spatial information) provide an index of the general alerting effect of the cues.

This task been used extensively in humans and in nonhuman primates, and has recently been adapted for use in the rat, there being two versions for rats, one that employs the same nine-hole box apparatus used in the five-choice task (and, thus, requires nose-poke responding) and another that employs a two-lever operant chamber (See Bushnell, 1998).

IV. NEUROCHEMICAL STUDIES OF ATTENTION

The neurochemical basis for attentional processing can be elucidated by the neuropharmacological manipulation of a variety of neurotransmitter systems. Additional comparison of the effects of these manipulations in complementary tasks should elucidate their functions.

A. Cholinergic Modulation of Attention

There is an extensive literature on the cholinergic modulation of a variety of forms of cognitive processing. The human literature has focused primarily on the role of acetylcholine (ACh) in long- and short-term memory function. However, there is a growing body of evidence for a cholinergic role in attentional processing, especially stimulus detection and response selection.

For example, ACh efflux significantly increases during performance of sustained attention tasks, including both the 5CSRT and Sarter vigilance tasks. This efflux increases in response to increases in task demand, indicating a general role in the mantainence of attention (Himmelheber et al., 2001; Dalley et al., 2001).

1. Studies of Lesions of the Cholinergic Nucleus Basalis Magnocellularis

The effects of the loss of cholinergic neurons lend support to the hypothesis that acetylcholine modulates attention. Excitotoxic lesions of the nucleus basalis magnocellularis (NBM) decrease baseline accuracy in the 5CSRT task (see Robbins, 2002). The nature of this deficit seems to be primarily attentional, as opposed to either sensory or motivational, as it was alleviated by lengthening the stimulus duration. Systemic physostigmine and nicotine treatment also corrected the deficit in the lesioned animals, but nicotine had no effect in sham-operated controls at similar doses. Increasing attentional demands, either by shortening stimulus duration or by introducing white noise, increased this performance deficit. More specific immunotoxic lesions (192 IgG-saporin lesions) decreased baseline accuracy, especially when the event rate increased.

Other impairments in the 5CSRT task include an increase in perseverative and premature responding in animals with high-dose 192 IgG-saporin lesions, interpreted as general response disinhibition. However, the effects of excitotoxic NBM lesions in cued target detection paradigms point to a specific impairment in attentional disengagement. For example, NBM lesioned monkeys showed a greater RT increase to invalid cues than control monkeys (Voytko et al., 1994). Choice accuracy, however, remained high in these animals. By contrast, rats with NBM lesions were unimpaired in both choice accuracy and RT under valid or no-cue conditions; however, they showed impairments in both accuracy and RT to invalidly cued targets (Chiba et al., 1999). Rats in the 5CSRT task can make several types of perseverative errors, including responding to the previously rewarded hole. If the lesioned animals are responding disproportionately and perseveratively to the previously rewarded hole, this may possibly indicate a specific deficit in attentional disengagement, as well as a general response inhibition.

Finally, 192 IgG-saporin lesions also affected performance of a divided attention task, in which animals were required to divide attention across the visual and auditory modalities. Rats were trained on the Sarter sustained attention task with blocks of visual alone, auditory alone, and both modalities, the last requiring divided attention. The bimodal condition was significantly more difficult for all animals as indicated by longer RTs. Response accuracy remained unaffected in the lesioned animals, but response latencies increased and were correlated with amount and extent of ACh depletion, again indicating a role for ACh under conditions of increased task demand (Turchi and Sarter, 1997).

2. Studies of the Effects of Systemic Cholinergic Agents

The effects of nicotine as a cognitive enhancer were shown by improved performance of patients with Alzheimer's disease, as well as normal elderly and young volunteers, on a rapid visual information processing continuous performance test, a demanding version of the CPT task in which the subject responds to odd or even numerical sequences; in this case, ascending triads of odd or even numbers (i.e., 2, 4, 6 or 1, 3, 5) More specifically, it increased hits and decreased RT (Sahakian et al., 1989).

Similarly, nicotine enhanced performance of the 5CSRT task, but only under specific task conditions, specifically, a low event rate, with punishment only for incorrect responses, thus removing the response-inhibition component of the task and reducing the requirement for executive control. In this version, nicotine decreased the detrimental effects of bursts of white noise on performance (Hahn et al., 2002). Also, for rats at low baseline accuracy, nicotine increased accuracy and speeded latencies, indicating a possible role in signal detection or response selection.

In the cued target detection task, nicotine facilitated response disengagement. In casual cigarette smokers, nicotine induced larger validity effects, specifically by reducing the response latency to invalidly cued targets (Murphy and Klein, 1998). There is some cross-species convergence from studies in the rat, in which although nicotine abolished the validity effect, it did so by significantly decreasing RT to invalid cues as well as speeding RTs in general (Phillips et al., 2000); Witte et al., (1997) obtained similar results in two rhesus monkeys treated with nicotine, as well as in cigarette smokers.

Whereas nicotinic receptor agonists generally appear to enhance attention and response selection, muscarinic receptor antagonists broadly have the opposite effects. For example, scopolamine impaired selective attention in the five-choice task when white noise was presented prior to stimulus onset. Similarly, whereas nicotine improved attentional shifting/response disengagement in the rat covert orienting task, scopolamine induced a slowing of RT generally, as well as a specific inability to shift from the invalid cue to make a correct response measured by RT (Phillips et al., 2000).

In general, there is converging evidence from each of the tasks described that the frontal cholinergic projections are involved in a variety of aspects of attentional processing. Aside from having disinhibitory effects similar to those observed with the stimulant drugs (which presumably act via the catecholamines), acetylcholine also facilitates switching between response sets and modulates attentional disengagement, as well as attentional performance, with increasing task demand. Note that all three of the main tasks described require rapid response selection and signal detection, as well as rapid shifts in attention. This often leads to improvements in many settings, but it may cause potential impairment in others with different demands, for example, by increasing response lability.

B. Noradrenaline

The role of noradrenaline (NA) in attentional and arousal processes has long been studied. It was one of the first transmitter systems studied in the 5CSRT task, and appears to be of special importance when contingencies change or when additional demands are placed on the system. Attentional effects are seen under specific conditions only; thus, for example, under high levels of arousal, NA appears to gate selective attention (Robbins and Everitt, 1995).

1. Studies of Lesions of the Dorsal Noradrenergic Bundle

NA can be profoundly depleted in the cerebral cortex by the injection of 6-hydroxydopamine (6-OHDA) into the dorsal noradrenergic bundle (DNAB). The effects of this depletion of cortical NA in the various tasks are as follows:

DNAB lesions impair selective attention in the 5CRST task, but not under baseline conditions. Interpolation of a white noise distractor presented just prior to stimulus onset disrupted accuracy in lesioned animals. Intra-accumbens amphetamine had similar effects, as did making stimulus onset temporally unpredictable.

By contrast, DNAB lesions have no effect on performance in the Sarter vigilance task, even under distractor conditions. However, the Sarter task, unlike the five-choice task, does not emphasize response inhibition and executive mechanisms. Additionally, five-choice deficits are only seen under very specific task demands, that is, white noise presented directly prior to stimulus presentation, and so could also reflect the timing of the distractor.

2. Studies of Adrenergic Drugs

Systemically administered adrenergic agents may have nonspecific or peripheral effects on performance that are not related to the effects of NA within the prefrontal cortex. Unfortunately, there is a considerable lack of studies using centrally administered agents.

In the 5CSRT task, agents that reduce NA function increased omission errors and reduced premature responding; accuracy also increased in the presence of these agents but only with degraded (dimmer) stimuli. Similarly, an $\alpha 1$ agonist improved choice accuracy; this effect was blocked by an adrenoreceptor antagonist, which on its own slightly impaired accuracy, but only with short stimulus durations (reviewed in Robbins, 2002).

Data from the cued target detection task also support NA as having rather nonspecific attentional alerting effects. Indeed, in two rhesus monkeys, the alerting effect apparent as a decrease in RT following a neutral cue was abolished by systemic clonidine and, to a lesser extent, by the specific $\alpha 2_a$ receptor agonist guanfacine, at doses that presumably stimulated presynaptic autoreceptors (Witte and Marrocco, 1997). By contrast, intravenous administration of clonidine decreased the cost of the invalidly cued object in normal human volunteers (Clark et al., 1989), as also seen after nicotine exposure in casual smokers.

C. Dopamine

Subcortical manipulation of dopamine (DA) tends to affect response "vigor," including motivational and motoric aspects of 5CRST performance. The indirect catecholaminergic agent *d*-amphetamine, when given systemically, increases premature responses and speeds responses without generally affecting accuracy (except for impairments at high doses). The effects of intra-accumbens amphetamine are similar, except that response latency increases, indicating a different mode of action for the response-speeding effect seen systemically. By contrast, D1 receptor agents infused into the prefrontal cortex affected accuracy of performance of the task. Specifically, for rats performing the task well at baseline, a D1 antagonist impaired performance. Conversely, for rats with a low basal level of performance, a D1 agonist improved performance.

Intra-accumbens infusions of amphetamine had no effect on performance in the Sarter sustained attention task. *cis*-Flupenthixol (a DA receptor blocker), on the other hand, impaired attentional performance, reducing both the relative number of hits and the number of correct rejections. Neither drug affected the relative number of omissions, effectively ruling out a motivational component to the behavior. Differences may be due to the fact that response inhibition is barely taxed in this task, unlike in the 5CSRT task, thereby making it less "executive" in nature.

Covert orienting of attention is also affected by dopaminergic manipulation. In humans performing the task, the anti-dopaminergic agent droperidol increased RTs in general, but reduced invalid cue cost (Clark et al., 1989). By contrast, in rhesus monkeys, administration of droperidol increased RTs to uncued targets (Marrocco and Davidson, 1998). The differences between human and monkey studies are likely a result of different methodologies; the Clark study used centrally located cues that presumably require more cognitive processing than the peripheral cues used in the monkey study. Significantly, the behavioral outcomes of DA receptor antagonism in the monkey were directly opposite those of noradrenergic receptor antagonism, which specifically abolished the alerting effect by increasing RT to neutrally cued targets.

In the rat, unilateral depletion of striatal DA produced hemispatial neglect in the covert orienting task (Ward and Brown, 1996), increasing RT on the side contralateral to the lesion, with no difference between valid and invalidly cued targets. This may reflect motor or activational effects rather than deficits in selective attention, much like the general slowing of RT seen after DA antagonism in both humans and monkeys.

Generally, dopaminergic modulation of attention seems limited to a general alerting/response preparation role as opposed to a role in selective attention per se, although D1 receptors in the prefrontal cortex appear to contribute to response selection.

V. DISCUSSION

There is converging evidence from studies using different test paradigms often adapted for comparison across species for the role of modulatory neurotransmitter systems in attention. Acetylcholine modulates attentional shifting and response selection; NA gates selective attention under higher levels of arousal (and possibly directs attention to novel situations and contingencies), and DA affects response vigor subcortically and response selection at the cortical level.

These chemically defined systems of the reticular core of the brain modulate and preferentially innervate different cortical regions and different layers in the same cortical regions. They also have different targets subcortically. Mapping out the way in which these systems interact, as well as their separate contributions to attentional function, is a major goal of future research.

Finally, for several of the test paradigms surveyed, there are only a finite number of behavioral outcomes for a given manipulation; thus, disparate drugs may cause similar behavioral effects through very different (sometimes opponent) mechanisms of action (e.g., the similar effects of cholinergic and catecholaminergic manipulations on covert orienting). The development of new tools, such as drugs selective for specific receptor subtypes as well as more selective neurotoxins and novel means of in vivo monitoring, should facilitate this functional analysis.

Acknowledgment

JM was supported by an NIH-Cambridge University Scholarship.

References

Bushnell, P. J. (1998). Behavioral approaches to the assessment of attention in animals. *Psychopharmacology* **138**, 231–259.

Chiba, A. A., Bushnell, P. J., Oshiro, W. M., and Gallagher, M. (1999). Selective removal of cholinergic neurons in the basal forebrain alters cued target detection. *NeuroReport* **10**, 3119–3123.

Clark, C. R., Geffen, G. M., and Geffen, L. B. (1989). Catecholamines and the covert orientation of attention in humans. *Neuropsychologia* **27**, 131–139.

Dalley, J. W., McGaughy, J., O'Connell, M. T., Cardinal, R. N., Levita, L., and Robbins, T. W. (2001). Distinct changes in cortical acetylcholine and noradrenaline efflux during contingent and noncontingent performance of a visual attentional task. *J. Neurosci.* **21**, 4908–4914.

Hahn, B., Shoaib, M., and Stolerman, I. P. (2002). Nicotine-induced enhancement of attention in the five-choice serial reaction time task: the influence of task demands. *Psychopharmacology* **162**, 129–137.

Himmelheber, A. M., Sarter, M., and Bruno, J. P. (2001). The effects of manipulations of attentional demand on cortical acetylcholine release. *Cogn. Brain Res.* **12**, 353–370.

Marrocco, R., and Davidson, M. C. (1998) Neurochemistry of attention "The Attentive Brain" (Parasuraman, Ed.), pp. 35–50. MIT Press, Cambridge, MA.

McGaughy, J., and Sarter, M. (1995). Behavioral vigilance in rats: task validation and effects of age, amphetamine, and benzodiazepine receptor ligands. *Psychopharmacology* **11**, 340–357.

Murphy, F. C., and Klein, R. M. (1998). The effects of nicotine on spatial and non-spatial expectancies in a covert orienting task. *Neuropsychologia* **36**, 1103–1114.

Phillips, J. M., McAlonan, K., Robb, W. G. K., and Brown, V. J. (2000). Cholinergic neurotransmission influences covert orientation of visuospatial attention in the rat. *Psychopharmacology* **150**, 112–116.

Posner, M. I., and Petersen, S. E. (1990) The attention system of the human brain. *Ann. Rev. Neurosci.* **13**, 25–42.

Robbins, T. W. (1998) The psychopharmacology and neuropsychology of attention in experimental animals. In: The Attentive Brain, ed. R. Parasuraman, Cambridge, MA: MIT Press. pp. 189–220.

Robbins, T. W. (2002). The 5-choice serial reaction time task: behavioural pharmacology and functional neurochemistry. *Psychopharmacology* **163**, 362–380.

Robbins, T. W., and Everitt, B. J. (1995). Central norepinephrine neurons and behaviour. *In* "Psychopharmacology: Fourth Generation of Progress" (F. E. Bloom and D. Kupfer, Eds.), pp. 363–372. Raven Press, New York.

Rosvold, H. E., Mirsky, A. F, Sarason, I., Bransome, E. D., and Beck, L. H. (1956). A continuous performance test of brain damage. *J. Consult. Psychol.* **20**, 343–350.

Sahakian, B. J., Jones, G., Levy, R., Warburton, G., and Gray, J. (1989). The effects of nicotine on attention, information processing, and short term memory in patients with dementia of the Alzheimer's type. *Br. J. Psychiatry* **154**, 797–800.

Sarter, M., Girens, B., and Bruno J. P. (2001) The cognitive neuroscience of sustained attention: where top-down meets bottom-up. *Brain Res. Rev.* **35**, 146–160.

Turchi, J., and Sarter, M. (1997). Cortical acetylcholine and processing capacity: effects of cortical acetylcholine deafferentation on crossmodal divided attention in rats. *Cogn. Brain Res.* **6**, 147–158.

Voytko, M. L., Olton, D. S., Richardson, R. T., Gorman, L. K., Tobin, J. R., and Price, D. L. (1994). Basal forebrain lesions in monkeys disrupt attention but not learning and memory. *J. Neurosci.* **14**, 167–186.

Ward, N. M., and Brown, V. J. (1996). Covert orienting of attention in the rat and the role of striatal dopamine. *J. Neurosci.* **16**, 3082–3088.

Witte, E. A., Davidson, M. C., and Marrocco, R. T. (1997). Effects of altering brain cholinergic activity on covert orienting of attention: comparison of monkey and human performance. *Psychopharmacology* **132**, 324–334.

Witte, E. A., and Marrocco, R. T. (1997). Alteration of brain noradrenergic activity in rhesus monkeys affects the alerting component of covert orienting. *Psychopharmacology* **132**, 315–323.

CHAPTER

11

Identifying the Neural Systems of Top-Down Attentional Control: A Meta-analytic Approach

Barry Giesbrecht and George R. Mangun

ABSTRACT

Several recent neuroimaging studies have investigated the brain systems involved in the control of voluntary orienting of selective visual attention by measuring the cortical response to attention-directing cues. Although these studies have provided strong evidence for frontal and parietal involvement in attentional control, it has proven difficult to unambiguously isolate attentional orienting responses from other cognitive operations evoked by the cue. Here we present a meta-analysis of a series of voluntary orienting studies from our laboratory. Across the studies, the only common mental operation is attentional orienting. We predicted that if regions of frontal and parietal cortex subserve top-down control of voluntary orienting specifically, then overlap among all studies should be observed. Consistent with this prediction, focal areas of superior frontal sulcus and intraparietal sulcus of both hemispheres were activated across all studies. We suggest that these subregions are critical players in the top-down control of attentional orienting.

I. INTRODUCTION

The visual environment is extremely complex. One way in which observers handle this complexity is by selectively attending to information relevant to their current goals. Understanding the selective nature of visual attention and its importance for coherent behavior has been one of the most extensively studied issues of psychology and neuroscience. Of particular interest is the notion that selective stimulus processing is mediated by the interaction between top-down executive control functions and bottom-up sensory processing systems. A key cognitive operation that is involved in this top-down interaction is voluntary covert orienting (e.g., Posner, 1980). Recent event-related fMRI studies of visual attention have focused on identifying the brain systems that support the control of voluntary covert attentional orienting (e.g., Corbetta et al., 2000; Giesbrecht et al., 2003; Hopfinger et al., 2000; Kastner et al., 1999). These fMRI studies have identified a distributed network of brain areas that support voluntary orienting, including both cortical and subcortical structures. The key cortical structures, which are the focus of the present work, are portions of superior frontal cortex, near the human homolog of the frontal eye fields (FEFs), and posterior parietal cortex (PPC), along the intraparietal sulcus (IPS).

Much of what is known about voluntary orienting comes from studies of selective attention that use the so-called cuing paradigm (Posner, 1980) (Fig. 11.1). In these studies participants are presented with a cue stimulus that directs attention to a particular location or feature (e.g., color, form, motion) to make a discrimination of a subsequently presented target stimulus that either does or does not occur at the cued location or contain the cued feature. Implication of frontal and parietal cortex in top-down control of selective attention in this task requires the dissociation of orienting responses from other cognitive operations. One approach to this dissociation, first reported by Harter et al. (1989) using electroencephalography, is to measure the cortical response to attention-directing cues in the cuing paradigm. The validity of dissociating control mechanisms from other information-processing stages via the measurement of cue-related responses rests on the logic that if an attention-

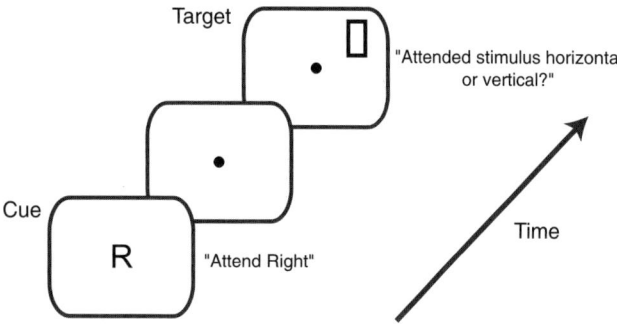

FIGURE 11.1 Schematic representation of a voluntary orienting paradigm. The cue provides an instruction to attend to a location. After a variable interval, the target stimulus is presented, for which a response is required.

directing cue engages cognitive operations that prepare the system for incoming information, then by measuring the cortical response to the cue separately from the target one can dissociate those systems that are involved in the control of orienting to task-relevant targets from those involved in selectively processing the targets themselves.

Although identifying top-down control systems by measurement of cue-related cortical activity has intuitive appeal because of its elegant simplicity, effective application of the approach is a complex endeavor. This complexity is rooted in the fact that attention-directing cues not only evoke activity in brain systems that control attentional orienting, but they also evoke activity in perceptual, cognitive, and motor stages of processing that prepare the system for the task. Within a simple cognitive framework, these stages include: (1) sensory processing of the cue, (2) extraction of an abstract/linguistic code from the cue symbol, (3) mapping of the code onto the task instruction (e.g., the arrow means attend right), (4) covertly orienting to the relevant stimulus feature or location, (5) maintaining the task instruction during the cue–target interval, and (6) preparing to respond. It is worthwhile underscoring that this framework is simple and that each of these stages may be subdivided to a more refined scale, for example, sensory processing = orientation detection + edge detection + color processing; orienting to a location = disengage + move + engage. Thus, despite the intention of using cue-related activity to identify the brain areas that subserve orienting (i.e., stage 4), areas that are activated by the cue could, in principle, support any one of or some combination of all these processing stages. Therefore, to understand the mechanisms that mediate voluntary orienting (i.e., stage 4) using the cue-related approach, one cannot simply measure activity evoked by the cue-related activity alone, but one must dissociate orienting activity from activity that is related to other sensory, motor, and cognitive operations.

The fMRI studies that have implicated frontal and parietal involvement in the control of attentional orienting via the assessment of cue-related activity have attempted to isolate the orienting response from other cognitive operations by direct comparison of cue-related activity with a reference condition. The assumption of such comparisons is that the reference condition shares many cognitive operations with the condition of interest, but ideally differs from the condition of interest only in terms of a single cognitive operation. Thus, when the condition of interest and reference condition are compared directly, the shared operations cancel or subtract out, and only those brain areas that support the cognitive operation of interest should remain. Several studies have used the subtraction approach, but in many cases the reference condition has not been ideal for isolating orienting from the other operations evoked by the cue. For instance, some studies have identified frontal and parietal top-down control systems by comparing cue-related activity with baseline activity (Corbetta et al., 2000; Hopfinger et al., 2000; Kastner et al., 1999). However, applying the cognitive framework and the subtraction approach outlined above suggests that the activity cannot be uniquely associated with voluntary orienting, but rather represents orienting and additional cognitive operations involved in processing the cue. A more complex contrast between conditions was used by Hopfinger et al. (2000), where target-related activity was used as a reference. Although this contrast controls for many basic cognitive operations, it unfortunately introduces other important differences between the cue and reference conditions, particularly in terms of response-related operations. Thus, although cue-related fMRI studies clearly indicate that portions of frontal and parietal cortex are activated in response to attention-directing voluntary orienting cues, the precise function of these areas remains unclear because the statistical comparisons have not completely dissociated the orienting response from other operations evoked by the cue.

To investigate which specific areas of frontal and parietal cortex support voluntary attentional orienting, we conducted a meta-analysis of published and unpublished attentional cuing studies from our laboratory. Cue-related contrasts from these studies were divided into three categories (see Section II). These categories differed in terms of the nature of the reference condition and, therefore, differed in terms of the cognitive operations that were revealed by the contrast. Critically, the only cognitive operation that the three categories had in common was the involvement of voluntary orienting, either to a location or to a feature (e.g., color). The analytical approach was based on the following simple logic. If the only mental operation that these comparisons have in common is the involvement of voluntary orienting, then by overlaying the activations onto a single cortical representation, those areas that are activated in all of the studies should be those areas that are involved in the control of voluntary orienting of selective attention. The results of this meta-analysis demonstrate that posterior portions of the superior frontal sulcus (SFS) and portions of IPS were common to the three categories of cue-related comparisons, suggesting that these areas are therefore critically involved in the control of voluntary orienting of selective attention.

II. METHOD

A. Details of Included Studies

Studies conducted in our laboratory were included if they were (1) published in a peer-reviewed journal or (2) presented at a scientific meeting. Each study used an attentional cuing paradigm (e.g., Fig. 11.1). In these tasks subjects were cued to attend to a location, nonspatial stimulus feature (e.g., color or global/local), or particular spatial reference frame. Table 11.1 lists the studies that were included and the details about the conditions.

B. Meta-Analysis

Cue-related contrasts were divided into three categories. The categories differed in terms of the cognitive operations that were revealed by the statistical comparison based on the cognitive framework and subtraction approach outlined above. One category, the cue versus baseline (CvsB) category, included contrasts that compared cue activity with baseline activity (as defined either by pre-cue activity or by mean level of activity across the fMRI time series). Areas revealed by this category of contrast could, in principle, support any one of the putative cognitive operations (i.e., stages 1–6). A second category, the cue versus passive (CvsP) category, included contrasts that compared activity evoked by attention-directing cues with passive cues that did not direct attention nor prepare subjects to respond. Because passive cues require sensory processing, extraction of a linguistic code, and mapping of the code onto the instruction (i.e., "do nothing") just like attention-directing cues, but not orienting- and motor-related processes, direct comparison of passive cue activity with cue-related activity should cancel out those areas that are involved in the common operations (steps 1 to 3), but leave those areas that support stages 4 to 6. Finally, the third category, referred to as the cue versus active (CvsA) category, included activations that were revealed by comparing cue-related activity with another "active" cue condition. For example, an active cue condition could be a

TABLE 11.1 Cue-Related fMRI Investigations Included in the Meta-analysis[a]

Study	Issue	Cue-related contrast	No. of foci	Classification
Giesbrecht et al., 2003	Spatial vs feature attention	Central location cue vs baseline	6	CvsB
		Central color cue vs baseline	6	CvsB
		Central location cue vs central color cue	7	CvsA
		Peripheral location cue vs baseline	7	CvsB
		Peripheral color cue vs baseline	6	CvsB
		Peripheral location cue vs peripheral color cue	5	CvsA
Kenemans et al., 2002	Feature attention	Color cue vs neutral cue	13	CvsA
Weissman et al., 2002	Global vs local attention	All cues vs passive cues	13	CvsP
Weissman et al., 2003	Global vs local attention	All cues vs passive cues	11	CvsP
Wilson et al., in press	Viewer-centered vs object-centered reference frames	Viewer-centered cues vs baseline	10	CvsB
		Object-centered cues vs baseline	13	CvsB

[a] Shown are the reference for the study, the issue addressed by the study, the specific cue-related contrast, the number of frontal and parietal foci included in the meta-analysis, and the classification for the meta-analysis based on the scheme described in Section II.
C, cue; B, baseline; P, passive; A, active.

neutral cue that does not direct subjects to orient, but does require the maintenance of a task instruction during the cue–target interval and response preparation processes. Relative to the other two categories, areas revealed by this type of contrast are likely to reflect a pure voluntary orienting response because all other operations have been roughly equated between the cue-related condition of interest and the active reference condition.

The meta-analysis was performed by selecting the coordinates of the local maxima in frontal and parietal cortex from each of the contrasts, projecting these foci onto the surface of a brain spatially normalized to the same stereotactic space (i.e., Montreal Neurological Institute), and then rendering on an inflated cortical representation (Van Essen, 2002). To compare brain activations revealed by the different contrasts, each focus was surrounded by an 8-mm radius. The application of this radius accounts for variability in the location of the foci that can be introduced by standard image processing techniques (e.g., spatial normalization, spatial smoothing) and differences in mean anatomical variability across the studies (Van Essen and Drury, 1997).

III. RESULTS

The results of 11 cue-related contrasts from five independent studies were included in the meta-analysis. There were a total of 97 frontal and parietal foci, which are shown in Fig. 11.2. Across all studies, attention-directing cues activated large portions of frontal and parietal cortex in both hemispheres. Despite the generally wide distribution of foci, several concentrations of foci can be observed. These concentrations included the IPS of both hemispheres and dorsolateral prefrontal cortex of both hemispheres including the SFS, middle frontal gyrus (MFG), and superior frontal gyrus (SFG). Other clusters of activity appeared lateralized to the left hemisphere and included the posterior aspect of the inferior frontal sulcus (IFS) spreading to the precentral sulcus (PreCS), and the medial frontal gyrus (MedFG) including the supplementary motor area (SMA).

The 11 contrasts were divided into the three categories of cue contrast described in Section II (see Table 11.1). The classification scheme resulted in 6 contrasts being assigned to the CvsB group, 2 to the CvsP group, and 3 to the CvsA group.

To identify areas of overlap among the three categories of contrasts, the anatomical location of each focus was smoothed (8 mm; see Section II) and then painted onto the inflated representation of cortex,

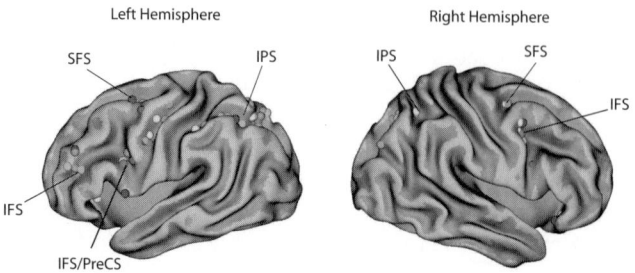

FIGURE 11.2 Foci included in the meta-analysis. Each color represents a different study and contrast (see Table 11.1). The foci are projected onto an inflated representation of the left and right hemispheres of a brain that was normalized to stereotactic space. SFS, superior frontal sulcus; IPS, intraparietal sulcus; IFS, inferior frontal sulcus; PreCS, precentral sulcus. (See color plate)

shown in Fig. 11.3. In this representation, the key areas are those common to all three contrasts and are shown in white; nonwhite areas represent CvsB (red), CvsP (green), CvsA (blue), or some combination of two of the three categories (magenta, yellow, and cyan). There were five areas that were common to all categories of cue-related contrasts. In frontal cortex, there was overlap in both hemispheres in the posterior portions of the SFS near the junction with the PreCS. In addition, there was overlap in medial prefrontal regions, including the SMA of the left hemisphere (not shown in the figure). In parietal cortex, the only area of overlap was in the IPS, bilaterally. Interestingly, the overlap in IPS appeared in both anterior and posterior portions of IPS in each hemisphere.

IV. DISCUSSION

A central issue in the study of visual attention is the identification of the neural systems that control the selective orienting of attention to relevant locations, features, and objects. Here we used a meta-analysis of several studies from our laboratory to investigate whether specific areas of frontal and parietal cortex could be related to the mental operation of attentional orienting. We approached this by searching for brain regions that were consistently activated across three different types of contrasts between cue-related brain activity and reference conditions of varying complexity. Based on a simple cognitive framework we hypothesized that because these contrasts had a single stage of processing in common, namely, attentional orienting, those areas that were commonly activated by the contrasts should be those areas that subserve the top-down control of attentional orienting. The present results demonstrated that focal areas of activation in frontal and parietal cortex were common to all

IV. DISCUSSION

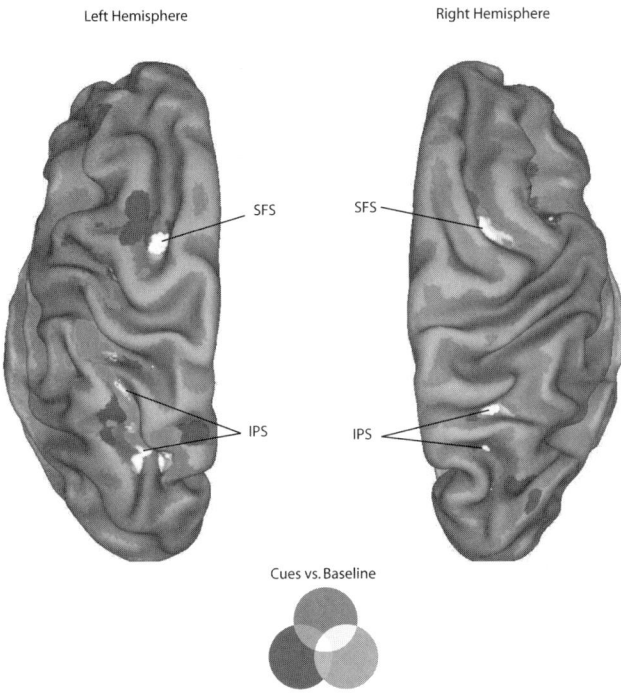

FIGURE 11.3 Anatomical overlap between the three categories of contrasts. Each focus was categorized into either the cues versus baseline (red), cues versus passive (green), or cues versus active (blue) groups, projected onto the inflated brain, smoothed (8 mm), and painted onto the surface. White represents the intersection of all three categories. Abbreviations are as in Fig. 11.2. (See color plate).

contrasts. These areas were the posterior aspect of the SFS and along the IPS of both hemispheres and the SMA of the left hemisphere. Therefore we propose that these specific regions of frontal and parietal cortex are critically involved in the top-down control of voluntary orienting of selective visual attention. Importantly, because the studies used in this analysis involved orienting of spatial and nonspatial attention, these brain regions should be considered to be related to focusing of general attentional resources in vision, rather than for a specific form of attentional orienting; we discuss this in more detail below.

While previous studies have revealed activation in distributed regions of frontal and parietal cortex by attention-directing cues (e.g., Corbetta et al., 2000; Hopfinger et al., 2000; Kastner et al., 1999), the precise control operations supported by these areas remain unclear based on the previous data alone. The uncertainty remains because the studies that have implicated posterior SFS and IPS in attentional orienting have either (1) reported large areas of activation that have included these regions as well as several neighboring regions and/or (2) revealed these regions via statistical contrasts that do not specifically isolate the mental operation of attentional orienting. Nevertheless, the results from previous studies are entirely consistent with the present result that specific regions of frontal and parietal cortex are involved in executive control. However, the meta-analytic approach adopted here offers a more precise picture of frontoparietal function in attentional control. This refinement stems from the fact that the data came from independent studies and that the different classes of cue-related activations from the different studies have only a single major operation in common, yet overlapping areas between the contrasts were still observed. These results, unlike the individually reported data, provide strong converging evidence that posterior portions of the SFS and the IPS are indeed critically involved in the top-down control of *orienting* of selective attention.

According to one prominent model of attention, orienting involves three discrete stages, disengagement, movement, and engagement of attention (Posner and Petersen, 1990). If this theoretical framework is brought to bear on the present results, then the following question can be raised: Can one be more precise about which of these three orienting operations is supported by posterior SFS and anterior IPS? When the present results are considered with other published data, there is suggestive evidence that one can indeed be more precise about the involvement of these areas in attentional orienting. For instance, evidence suggests that the disengagement of attention is subserved by temporoparietal and ventrolateral prefrontal areas (Corbetta et al., 2000; Corbetta and Shulman, 2002) and not SFS and IPS. Therefore, it is unlikely that SFS and IPS are involved in the disengagement of attention. Similarly, attentional engagement is thought to be revealed as increased neuronal excitability in attended sensory representations (e.g., Giesbrecht et al., 2003; Hopfinger et al., 2000; Kastner et al., 1999), not in the frontoparietal network. Thus, the only operation that remains is that of "moving" attention to a new focus.

Use of the term *orienting* to describe the cognitive operation subserved by posterior SFS and IPS tends to summon the idea that these areas are involved in spatial shifts of attention. Although there is little doubt that this is true, there is solid evidence that these areas are also involved in nonspatial attentional orienting. Specifically, several of the contrasts included in the present meta-analysis were from studies that cued subjects to attend to nonspatial stimulus features, such as color, or to the global or local levels of hierarchical stimuli (Giesbrecht et al., 2003; Kenemans et al., 2002; Weissman et al., 2002, 2003). Thus, these regions appear to be amodal in the sense that they generalize beyond the attended stimulus dimension. The hypothesis that portions of the frontoparietal network generalize over multiple dimensions was recently proposed by Shulman and colleagues (2002). These authors argue that amodal parts of the frontoparietal network are involved in coding and maintaining relevant information in an abstract form. The present results suggest that portions of frontal and parietal cortex are also amodal in terms of their support of attentional orienting.

V. CONCLUDING REMARKS

Previous neuroimaging and neuropsychological studies have identified a distributed network of brain areas that support visual selective attention. The meta-analysis presented here suggests that focal regions of this distributed network, specifically posterior SFS, anterior and posterior IPS, and medial frontal areas (SMA), subserve voluntary orienting that occurs in response to attention-directing cues present in the external environment. However, external sources of information are not the only mediators of attentional control processes. Indeed, multiple sources of information contribute to these control functions, including current expectations, emotions, task demands, past experience, knowledge, and arousal. Therefore, future investigations of top-down control functions must identify how these multiple sources of information mediate selective visual attention processes to reveal a precise picture of how information in our external world is represented in cortex, influences behavior, and reaches awareness.

Acknowledgments

We sincerely thank Kevin Wilson, Leon Kenemans, and Daniel Weissman for their help in the preparation of this chapter, especially the provision of stereotactic coordinates for the activation foci from the unpublished studies (i.e., Kenemans et al., 2002).

This research was supported by grants from the National Institute of Mental Health, the National Institute of Neurological Diseases and Stroke, and the Army Research Office.

References

Broadbent, D. E. (1958). "Perception and Communication" Pergamon, London.

Corbetta, M., Kincade, J. M., Ollinger, J. M., McAvoy, M. P., and Shulman, G. L. (2000). Voluntary orienting is dissociated from target detection in human posterior parietal cortex. *Nat. Neurosci.* **3**, 292–297.

Corbetta, M., and Shulman, G. L. (2002). Control of goal-directed and stimulus-driven attention in the brain. *Nat. Revi. Neurosci.* **3**, 201–215.

Giesbrecht, B., Woldorff, M. G., Song, A. W., and Mangun, G. R. (2003). Neural mechanisms of top-down control during spatial and feature attention. *NeuroImage* **19**, 496–512.

Harter, M. R., Miller, S. L., Price, N. J., LaLonde, M. E., and Keyes, A. L. (1989). Neural processes involved in directing attention. *J. Cogni. Neurosci.* **1**, 223–237.

Hopfinger, J. B., Buonocore, M. H., and Mangun, G. R. (2000). The neural mechanisms of top-down attentional control. *Nat. Neurosci.* **3**, 284–291.

Kastner, S., Pinsk, M. A., De Weerd, P., Desimone, R., and Ungerleider, L. G. (1999). Increased activity in human visual cortex during directed attention in the absence of visual stimulation. *Neuron* **22**, 751–761.

Kenemans, J. L., Grent-'t Jong, T., Giesbrecht, B., Weissman, D. H., Woldorff, M. G., and Mangun, G. R. (2002). "A Sequence of Brain activity patterns in the control of visual attention." Paper presented at the Annual Meeting of the Society for Psychophysiological Research, New York.

Posner, M. I. (1980). Orienting of attention. *Q. J. Exp. Psychol.* **32**, 3–25.

Posner, M. I., and Petersen, S. E. (1990). The attention system of the human brain. *Annu. Rev. Neurosci.* **13**, 25–42.

Shulman, G. L., d'Avossa, G., Tansy, A. P., and Corbetta, M. (2002). Two attentional processes in the parietal lobe. *Cereb. Cortex* **12**, 1124–1131.

Van Essen, D. C. (2002). Windows on the brain: the emerging role of atlases and databases in neuroscience. *Curr. Opini. Neurobiol.* **12**, 574–579.

Van Essen, D. C., and Drury, H. A. (1997). Structure and functional analyses of human cerebral cortex using a surface-based atlas. *J. Neurosci.* **17**, 7079–7102.

Weissman, D. H., Giesbrecht, B., Song, A. W., Mangun, G. R., and Woldorff, M. G. (2003). Conflict monitoring in the human anterior cingulate cortex during selective attention to global and local object features. *NeuroImage*, **19**, 1361–1368.

Weissman, D. H., Woldorff, M. G., Hazlett, C. J., and Mangun, G. R. (2002). Effects of practice on executive control investigated with fMRI. *Cogn. Brain Res.* **15**, 47–60.

Wilson, K. D., Woldorff, M. G., and Mangun, G. R. (in press). Control networks and hemispheric asymmetries in parietal cortex during attentional orienting in different spatial reference frames. *NeuroImage*.

CHAPTER 12

Attention Capture: The Interplay of Expectations, Attention, and Awareness

Michael S. Ambinder and Daniel J. Simons

ABSTRACT

The term *attention capture*, in common parlance, has two distinct connotations: (1) it is automatic or stimulus-driven and cannot be overridden by top-down control, and (2) it necessarily leads to awareness of the capturing event. Although both components are common to most intuitive definitions of attention capture, few studies have explored both aspects simultaneously. Studies examining the first component typically do not assess awareness, focusing instead on the implicit effects of a stimulus on performance (e.g., response latency). Studies examining the second component typically do not explore the automaticity or inevitability of attention shifts, focusing instead on whether or not unexpected items are explicitly detected. In this chapter, we survey the central findings from the implicit and explicit capture literatures and examine their interrelation. We discuss ways in which these literatures, when taken together, can lead to a more complete understanding of whether and how attention capture might operate in the real world. In so doing, we analyze the influence of expectations, goals, and strategies on both forms of attention capture.

There is, in the first place, an attention that we are compelled to give and are powerless to prevent... there are impressions that we cannot help attending to, that take consciousness by storm... they force their way to the focus of consciousness, whatever the obstacles that they have to overcome.
—Titchener, 1924 (original 1896 under different title), pp. 268–269

I. INTRODUCTION

Titchener describes two distinct aspects of attention capture: (1) it occurs in a stimulus-driven, automatic way that cannot be overridden by top-down control, and (2) it necessarily leads to awareness of the capturing event. Although these two components are common to most intuitive definitions of attention capture, few studies have explored both aspects simultaneously. Studies examining this first component typically do not assess awareness, focusing instead on the implicit effects of a stimulus on performance (e.g., response latency). Studies examining the second component typically do not explore the automaticity or inevitability of attention shifts, focusing instead on whether or not unexpected items are explicitly detected. In this chapter, we survey the central findings from the implicit and explicit capture literatures and examine their interrelation. We discuss ways in which these literatures, when taken together, can lead to a more complete understanding of whether and how attention capture might operate in the real world. In so doing, we analyze the influence of expectations, goals, and strategies on both forms of attention capture.

II. IMPLICIT ATTENTION CAPTURE

For preferential attention to an object, or attentional prioritization, to reflect an entirely stimulus-driven shift of attention, the critical stimulus must be *unattended* before the shift occurs. If attention were already allocated to the critical stimulus, then capture might be modulated by top-down attention sets or expectations. However, verifying that a stimulus was unattended on each trial presents an almost insurmountable problem. Consequently, most implicit capture studies do not attempt to meet the strictest form of this requirement, instead settling for a weaker variant: the critical object is not *differentially attended* or selectively processed before capture. To meet this weaker requirement, most implicit capture studies adopt the methodological strategy of making the critical stimulus *task-irrelevant*. That is, such studies attempt to eliminate any task-induced motivation or top-down strategic reason to attend to the critical feature. If the feature still receives attentional prioritization, then its influence presumably results from the nature of the stimulus itself. Most of the ongoing debates in the implicit attention capture literature focus on the extent to which tasks satisfy this irrelevance requirement: whether observers are truly "compelled to give" attention to an object in a purely bottom-up fashion or whether top-down or task-induced strategies influence prioritization.

Implicit capture tasks typically are variants of a standard visual search task, with capture measured as a change in response latency. Most studies can be classified into one of three distinct task variants with slightly different operational definitions of capture (see Fig. 12.1).

In the *additional singleton* paradigm, observers search for a unique item among a homogenous set of distracters. Singleton search of this sort is characteristically efficient, with little or no effect of the number of distracter items in the display on response latency (e.g., Treisman and Gelade, 1980). On some trials, an additional, unique singleton item appears elsewhere in

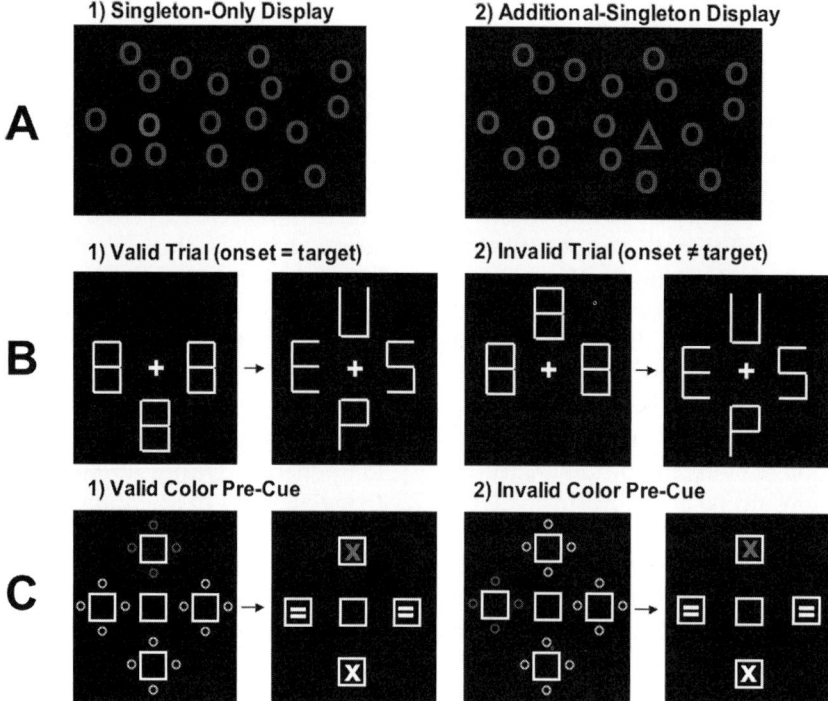

FIGURE 12.1 The three major paradigms used to study implicit attention capture. (A) Additional singleton paradigm. In the first search array, subjects search for a red circle among blue circles. In the second panel, in addition to the blue distracter circles, a blue triangle appears in the display. If this additional singleton slows search for the red circle, then it captured attention according to the logic of this task. (B) Irrelevant feature search paradigm. The initial display on both valid and invalid trials consists of a set of figure-8 premasks. These masks are then replaced by letters, and simultaneously, an additional letter appears in a location not previously occupied by a mask (an abrupt onset). Subjects try to determine whether the display contains an H or a U. If the abrupt onset captures attention, it should be searched first. When the abruptly onsetting item happens to be an H or U (a valid trial), then search should be equally efficient regardless of the number of distracters in the display. (C) Irrelevant pre-cue paradigm. A distinctive and nonpredictive cue appears in the location subsequently occupied by one of the search elements. If the features of the cue match those of the target, and the cue appears in a location subsequently occupied by a distracter, then it will slow search performance. (See color plate).

the display; this additional singleton is never the target of the search. For example, observers might search for a red circle among blue circles. On some trials, an additional blue triangle is added to the display. Attention capture is defined as a failure to ignore this additional shape singleton, resulting in slowed response latency to the color singleton target in the presence of a shape singleton. As the additional item is never the target of the search, observers should have no reason to attend to it and should thus treat it as irrelevant. In fact, they have incentive to ignore it if possible, as attending to it can only hurt their search performance. A variety of additional singletons slow responses in this task (e.g., unique colors and shapes, the abrupt appearance of an object), suggesting that many unique features are capable of capturing attention in a stimulus-driven fashion (Theeuwes, 1994). Not only are responses slowed, but the eyes often are also drawn to the additional singleton (Theeuwes et al., 1998).

This operational definition of capture, however, is not without controversy. For example, rather than reflecting stimulus-driven capture, slowed responses might result from a general bias to search for a unique item (Bacon and Egeth, 1994). If observers are set to search for any unique item, on trials with an additional singleton, they sometimes will search the wrong singleton first, leading to slowed search. In other words, if subjects adopt the strategy of searching for a singleton, then the critical additional item is relevant from the observer's perspective; subjects adopt an attention set that biases them to attend to the additional item, thereby negating the inference of stimulus-driven capture (but see Theeuwes, in press, for a counterargument). Similarly, capture may not be inevitable in this task: when they are aware of the presence of an additional singleton, younger subjects are less likely to move their eyes to it than are older subjects. Presumably the younger subjects have more effective inhibitory control of their eye movements (Kramer et al., 2000). To the extent that capture in the additional singleton paradigm can be overridden by top-down inhibition, it does not provide a measure of "attention that we are compelled to give and are powerless to prevent."

The *irrelevant feature search paradigm* is similar to the *additional singleton paradigm* with three important differences: (1) only one item in the search array is unique; (2) the unique item is the target of the search as often as it is any given distracter; and (3) the search task is more difficult, so response time increases as the number of distracters in the search array increases (i.e., serial search). The critical feature does not predict the location of the search target, so it is probabilistically irrelevant, as opposed to anti-predictive, as in the additional singleton paradigm. If the critical feature captures attention, then when it happens to be the search target, search will be relatively unaffected by the number of distracter items in the display. In contrast, when the critical feature is a distracter, search latency will increase with the number of items in the display. In this task, the sudden onset of a new object consistently captures attention; search performance is unaffected by the number of distracters when the onset happens to be the target of the search (Jonides and Yantis, 1988). Other seemingly salient features (e.g., unique color, distinct orientation, unique luminance) do not reliably capture attention (Jonides and Yantis, 1988), a finding that led to the strong conclusion that only the appearance of a new object captures attention in a stimulus-driven manner (but see Franconeri and Simons, 2003).

As in the additional singleton paradigm, stimulus-driven capture in the *irrelevant* feature search paradigm has been challenged on the grounds that the critical stimulus is actually relevant to the task. Specifically, the task induces an attention set that makes the critical stimulus relevant to the search (Gibson and Kelsey, 1998). Given that search cannot start until the onset of the search display, subjects adopt an attention set for onsets; they prioritize onsets because the appearance of the search display is relevant to their task. This attention set instills a top-down bias to prioritize onsets, leading to capture by the abrupt appearance of a new object. To the extent that top-down biases modulate performance, the task may not truly measure stimulus-driven attention capture (but see Franconeri et al., in press).

The final type of implicit capture paradigm, the *irrelevant pre-cue paradigm*, focuses on the effect of an irrelevant or anti-predictive pre-cue on search performance. Capture is inferred whenever responses are slowed by a pre-cue appearing in a location subsequently occupied by a distracter item in the search array. That is, subjects are drawn to the cue, which leads to them to focus attention on the wrong item location, thereby slowing search performance. Proponents of this paradigm typically argue that capture is modulated by top-down, task-induced expectations and goals; capture occurs only when the defining feature of the target matches the defining feature of the pre-cue (Folk et al., 1992). For example, when the search is for a uniquely colored item, subjects presumably adopt an attention set for color. Under those conditions, a color cue captures attention, but an onset cue does not. The reverse holds when the target is an abruptly onsetting item (Folk et al., 1992). In this paradigm, capture is never thought to be entirely stimulus-driven. Rather than drawing attention regardless of the goals and intentions of the perceiver, capture is contingent on the goals and expectations of the perceiver. That is, the features of the target and cue must be related to produce capture, even if the location of the cue is statistically irrelevant.

The core controversy in the implicit capture literature involves the "purity" of these paradigms. Those adopting the irrelevant pre-cue task argue that all capture effects are contingent on top-down influences. Proponents of other paradigms argue that irrelevant features can capture attention in a stimulus-driven manner. Inferences of stimulus-driven capture rest on the irrelevance of the critical feature to the task. To the extent that the irrelevance requirement is not met, then performance need not reflect stimulus-driven capture. Instead, top-down strategies and expectations could influence search performance. Given that all of these paradigms present the critical feature on many trials, they rely on the ability of the subject to treat it as irrelevant or anti-predictive. However, subjects may not be able to do so, not because the stimulus captures attention, but because they cannot distribute their attention evenly to the items in the display, something that is required if they are to treat the critical feature as irrelevant.

Two broader problems may also undermine inferences of stimulus-driven capture from implicit capture paradigms. First, the inference of stimulus-driven capture relies on a process akin to proving a null hypothesis: the tasks must definitively eliminate all possible top-down contributions to performance to claim that the stimulus was unattended and drew attention by virtue of its salience alone. The critiques noted earlier were attempts to identify top-down influences. Whether or not evidence for stimulus-driven capture in these paradigms is convincing boils down to the plausibility of the critiques of the irrelevance assumption and the extent to which a finding can withstand them. Second, the irrelevance strategy itself may be problematic. According to the irrelevance assumption, observers may well attend to the critical feature prior to capture; they just cannot preferentially attend to it. In fact, capture in the irrelevant feature search paradigm may depend on observers diffusely distributing attention to the entire display, including to the critical object (Yantis and Jonides, 1990). In a fundamental sense, in these paradigms, the critical feature must be attended before capture occurs. If so, these tasks might measure the capture of selective processing or focal attention but not the capture of attention per se, as attention was already partially devoted to the critical object.

An alternative approach to studying attention capture has attempted to avoid this controversy by eliminating the requirement that subjects treat the properties of an expected stimulus as irrelevant. In the explicit capture literature, the critical feature is presented once, unexpectedly. The primary measure of capture is whether or not subjects notice this unexpected object. Given that subjects do not expect the critical feature, the nature of the task is unlikely to induce an attention set for that feature. Of course, the relationship between the critical, unexpected object and other elements in the display may establish an attention set that affects noticing rates. In fact, expectations appear to play at least as large a role in modulating explicit capture as they do in implicit capture (Most et al., in press).

III. EXPLICIT ATTENTION CAPTURE AND INATTENTIONAL BLINDNESS

Although the perceiver's awareness of the capturing stimulus largely has been neglected in the implicit capture literature, the explicit capture literature examines whether and when unexpected stimuli capture awareness, provided attention is otherwise engaged. If an irrelevant, unexpected stimulus is noticed, then at some level, it forced its "way to the focus of consciousness." Unlike the implicit capture paradigms, explicit capture studies may well meet the strong form of the assumption that the critical stimulus was unattended prior to capture. An unexpected stimulus could not have received attention in advance of capture because it was not present in the display. Interestingly, many salient stimuli fail to reach awareness in explicit attention capture tasks, a phenomenon now known as *inattentional blindness* (Mack and Rock, 1998).

Although studies of attention capture by unexpected objects and events existed as early as the 1970s, the phenomenon of inattentional blindness and its relation to attention capture came into prominence in the late 1990s. Two primary paradigms have been used to study explicit attention capture and inattentional blindness. In one, observers are asked to make a judgment about a briefly presented stimulus (typically which arm of a cross is longer) for several trials. Then, on a critical trial, an additional object appears briefly in the display at the same time as the cross. When subjects are engaged in the cross judgment task, they frequently fail to notice the unexpected object even though it is the only other item in the display (Mack and Rock, 1998). For example, at least 25% of subjects fail to notice the appearance of an additional shape, even when it has a unique color (Mack and Rock, 1998). These findings suggest that such objects do not automatically capture attention. Rather, subjects often

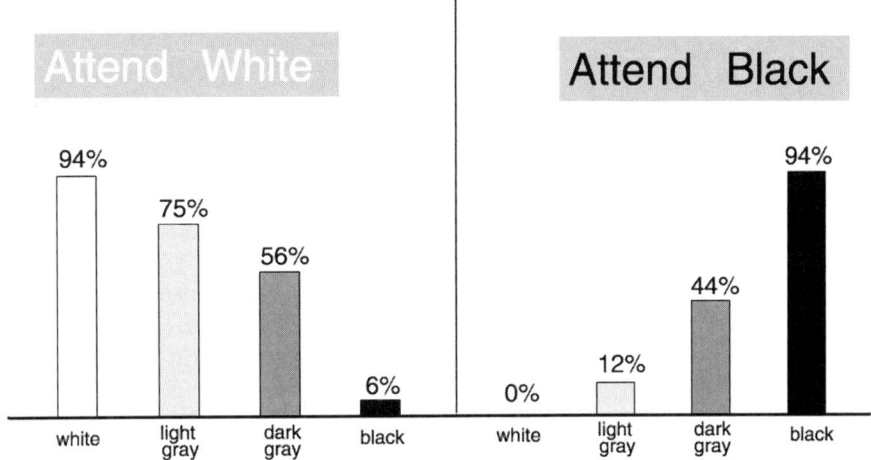

FIGURE 12.2 Percentage of observers who noticed the unexpected object. The level of noticing varied systematically as a function of the similarity of the unexpected object to the attended items. Reprinted with permission, from Most et al. (2001, Fig. 2).

are "blind" to their presence when attention is focused elsewhere.

In the other paradigm typically used to study explicit attention capture and inattentional blindness, observers engage in a sustained selective looking task in which they focus attention on some aspects of a dynamic display while ignoring others. During a critical trial, an additional, unexpected event occurs. As in the cross judgment task, the sustained event often goes unnoticed, even when it is particularly salient. In one task, subjects viewed a video in which two groups of people passed basketballs, and they counted the number of passes made by one group of people (e.g., the ones wearing white shirts). After 30 to 45 seconds, a person in a gorilla suit unexpectedly walked to the center of the display, turned to face the camera, thumped its chest, and exited on the other side of the display. Even though it was fully visible for 9 seconds, 50% of subjects failed to notice it at all (Simons and Chabris, 1999). At least under such naturalistic conditions, a new object, even a distinctive and unusual one, does not automatically capture awareness.

Interestingly, attention sets do influence the degree to which unexpected objects intrude on awareness. In one series of studies using a computer-based selective looking task, observers counted the number of times four white shapes touched the sides of a display while ignoring touches by four black shapes (Most et al., 2001). On the third trial of this task, a cross unexpectedly entered the display from the right, moved linearly across the center of the display, and exited on the left. When observers were counting the touches of the white shapes, almost all of them noticed the unexpected object when it was white and almost none noticed it when it was black. When they were focusing on the black shapes, the pattern was reversed, and intermediate shades of gray produced intermediate levels of noticing (see Fig. 12.2). Although the unexpected object was a different shape and was the only item moving in a linear path, noticing was largely determined by the similarity in luminance to the attended items. Noticing rates for an unexpected object were directly related to the dimension discriminating the attended from the ignored items. Thus, relevance seems to play a large role in explicit attention capture, just as it does in implicit attention capture. Despite its contribution to noticing rates, however, even highly distinctive and unique items often go unnoticed in this paradigm. For example, 30% of subjects failed to notice an unexpected red cross even though everything else in the display was black, white, or gray (Most et al., 2001).

Explicit attention capture is not automatic, even for salient and distinctive objects that are visible for extended periods. Distinctiveness does not guarantee awareness, and noticing rates are tied largely to the observer's primary task and to their expectations and attentional goals. If an unexpected object is similar to a selectively attended item, it often draws attention and reaches awareness. However, if it is similar to unselected items in the display, it likely will go undetected. This finding is counterintuitive: most people (including Titchener) firmly believe that salient, distinctive, and unexpected events automatically draw attention in a stimulus-driven fashion, leading to awareness.

IV. COMBINING IMPLICIT AND EXPLICIT ATTENTION CAPTURE

Although the implicit and explicit capture literatures have been relatively separate, a few studies have begun to bridge the gap. For example, a recent series of selective looking studies explored whether abrupt onsets explicitly capture attention (Most et al., in press) by contrasting the gradual appearance of an unexpected object with the abrupt appearance of the unexpected object. Rather than emerging progressively from the side of the display, the cross appeared abruptly just inside the right edge of the display and then started moving across the display as in other selective looking studies. Averaged across three experiments, noticing rates were comparable for the progressive and abrupt appearances, with less than 50% of subjects noticing either type of new object. Although the abrupt onset of a new object captures attention implicitly in most visual search paradigms, it does not automatically capture awareness.

Other studies have modified typical implicit capture paradigms to include a critical, "surprise" trial. In one study, subjects viewed a sequence of search trials with no critical features, and then, on a critical trial, the target item unexpectedly had a unique color (Gibson and Jiang, 1998). In the irrelevant feature paradigm, a unique color singleton typically does not capture attention. Consistent with these results, the unique color of the target item did not improve performance even though it was unexpected. This finding suggests that expectations did not play a crucial role in the lack of capture by a color singleton. Interestingly, subjects did show some inattentional blindness for the color feature: 37% of participants in a follow-up study did not report seeing a red singleton when asked immediately after the critical trial (Gibson and Peterson, 2001). Subjects who did remember the singleton also showed better search performance on the critical trial, presumably because they successfully divided their attention across the search display, increasing the ability of salient features to capture attention. However, the surprise color feature did not affect overall search performance in this paradigm.

Both implicit and explicit capture tasks rely on the irrelevance of the critical feature to infer stimulus-driven capture. At some level, the assumption of irrelevance will always be subject to challenge. Some arbitrary top-down expectation could always affect performance. For example, subjects might simply have a default attention set for all of the features that happen to capture attention. Given the impossibility of ruling out all such top-down biases, inference of stimulus-driven capture will have to rely on arguments for parsimony and plausibility. Even if these tasks do not determine whether some stimuli, by their very nature, impinge on performance and awareness, they may help determine which stimuli are more likely to affect performance and which are likely to reach awareness when they are unexpected. In practice, characterizing the nature of our default, top-down expectations and their interaction with stimulus properties may prove more fruitful than arguing over whether or not performance is entirely stimulus-driven.

Acknowledgment

The writing of this chapter was supported by National Institutes of Health Grant R01 MH63773-03 to D.J.S.

References

Bacon, W. F., and Egeth, H. E. (1994). Overriding stimulus-driven attentional capture. *Percept. Psychophys.* **55**, 485–496.
Folk, C. L., Remington, R. W., and Johnston, J. C. (1992). Involuntary covert orienting is contingent on attentional control settings. *J. Exp. Psychol. Hum. Percept. Perform.* **18**, 1030–1044.
Franconeri, S. L., and Simons, D. J. (2003). Moving and looming stimuli capture attention. *Percept. Psychophys.* **65**, 999–1010.
Franconeri, S. L., Simons, D. J., and Junge, J. A. (in press). Searching for stimulus-driven shifts of attention. *Psychon. Bull. Rev.*
Gibson, B. S., and Jiang, Y. (1998). Surprise! An unexpected color singleton does not capture attention in visual search. *Psychol. Sci.* **9**, 176–182.
Gibson, B. S., and Kelsey, E. M. (1998). Stimulus-driven attentional capture is contingent on attentional set for displaywide visual features. *Journal of Experimental Psychology: Human Perception and Performance*, **24**, 699–706.
Gibson, B. S., and Peterson, M. A. (2001). Inattentional blindness and attentional capture: Evidence for attention-based theories of visual salience. In B. S. Gibson, Eds. "Attraction, Distraction and Action: Multiple Perspectives on Attentional Capture" (C. L. Folk, Ed.), pp. 51–76. North-Holland, Amsterdam.
Jonides, J., and Yantis, S. (1988). Uniqueness of abrupt visual onset in capturing attention. *Percept. Psychophys.* **43**, 346–354.
Kramer, A. F., Hahn, S., Irwin, D. E., and Theeuwes, J. (2000). Age differences in the control of looking behavior: Do you know where your eyes have been? *Psychol. Sci.* **11**, 210–217.
Mack, A., and Rock, I. (1998). *Inattentional blindness.* Cambridge, MA: MIT Press.
Most, S. B., Scholl, B. J., Clifford, E., and Simons, D. J. (in press). What you see is what you set: Sustained inattentional blindness and the capture of awareness. *Psychol. Rev.*
Most, S. B., Simons, D. J., Scholl, B. J., Jimenez, R., Clifford, E., and Chabris, C. F. (2001). How not to be seen: the contribution of similarity and selective ignoring to sustained inattentional blindness. *Psychol. Sci.* **12**, 9–17.
Simons, D. J., and Chabris, C. F. (1999). Gorillas in our midst: sustained inattentional blindness for dynamic events. *Perception* **28**, 1059–1074.

Theeuwes, J. (1994). Stimulus-driven capture and attentional set: selective search for color and visual abrupt onsets. *J. Exp. Psychol. Hum. Percept. Perform.* **20**, 799–806.

Theeuwes, J. (in press). Attentional capture is independent of top-down settings. *Psychon. Bull. Rev.*

Theeuwes, J., Kramer, A. F., Hahn, S., and Irwin, D. E. (1998). Our eyes do not always go where we want them to go: capture of the eyes by new objects. *Psychol. Sci.* **9**, 379–385.

Titchener, E. B. (1924 (original 1896 under different title)). "A Textbook of Psychology." MacMillan, New York.

Treisman, A. M., and Gelade, G. (1980). A feature-integration theory of attention. *Cogn. Psychol.* **12**, 97–136.

Yantis, S., and Jonides, J. (1990). Abrupt visual onsets and selective attention: voluntary versus automatic allocation. *J. Exp. Psychol. Hum. Percept. Perform.* **16**, 121–134.

CHAPTER 13

Change Blindness

Ronald A. Rensink

ABSTRACT

Large changes that occur in clear view of an observer can become difficult to notice if made during an eye movement, blink, or other such disturbance. This *change blindness* is consistent with the proposal that focused visual attention is necessary to see change, with a change becoming difficult to notice whenever conditions prevent attention from being automatically drawn to it. The phenomenon of change blindness can provide new results on the nature of visual attention, including estimates of its capacity and the extent to which it can bind visual properties into coherent descriptions and it is also shown how the resultant characterization of attention can, in turn, provide new insights into the role that it plays in the perception of scenes and events.

I. INTRODUCTION

As observers, we have a strong impression that our visual system produces a coherent and detailed description of the world in front of us, a description, moreover, that is always stable and complete. However, various studies have shown that our ability to perceive objects and events in our visual field is far more limited than subjective experience indicates. Among the more striking phenomena in this regard is *change blindness*, the inability to notice changes that occur in clear view of the observer, even when these changes are large and the observer knows they will occur. (For a general review, see, for example, Rensink (2002).)

Change blindness has turned out to be a powerful and robust effect that can be induced in a variety of ways, such as by making the change during an eye movement, an eye blink, or a brief flash in the image. The generality of this effect indicates the involvement of mechanisms central to the way that we perceive our surroundings. The determination of these mechanisms and the way they relate to each other is far from complete. But it is clear that visual attention is critical; in particular, results indicate that focused attention is needed for the perception of change (Rensink et al., 1997). Given the strength of its effects and its tight connection with attention, change blindness appears to be a powerful way of exploring the nature of visual attention and the role it plays in our perception of the world.

A. Basic Distinctions

To avoid the confusion that often hinders investigations into the perception of change, it is useful to first make a few basic distinctions (see Rensink, 2002). One of these is the distinction between *change* and *motion*. As used here, *change* refers to the transformation of an enduring (coherent) structure over time. In contrast, *motion* refers to the temporal variation of some quantity (such as intensity or color) at a fixed point in space. Motion does not involve structure, and motion detectors do not require attention for their operation. As such, the key characteristic of focused visual attention is the creation (and perhaps maintenance) of representations capable of describing coherent spatiotemporal patterns.

Another important distinction is that between the perception of *dynamic* change (i.e., seeing a change as a dynamic visual event) and the inference of *completed* change (i.e., noticing that something has changed at some time in the past, without a phenomenological experience of anything dynamic). During the perception of dynamic change, the spatiotemporal continuity of the internal representation is maintained. In contrast, the perception of completed change does not require such continuity; in principle, it could be carried out simply by a comparison of the currently visible structure with the contents of memory, requiring at most only an intermittent application of attention.

Finally, it is also worth distinguishing between *change* and *difference*. Perception of difference is based on the lack of similarity in properties of two distinct structures. In contrast to change, difference involves no notion of temporal transformation; instead, similarity is defined via atemporal comparison. The question of whether attention is involved in the perception of difference (and perhaps of completed change) then reduces to the question of whether attention is needed for comparison.

B. Methodological Considerations

The design of any change detection experiment must provide a way to decouple change, motion, and difference. To decouple change from motion, at least two strategies are possible. First, the change can be made gradually enough that the accompanying motion signal does not draw attention (e.g., Simons et al., 2000). Second, the change can be made contingent on an event (such as a brief flash, eye movement, or occlusion) that creates a global motion signal that can swamp the localized signal associated with the change (e.g., Rensink et al., 1997).

Decoupling change from difference requires separating the effects of visual attention from the effects of long-term memory. One strategy is to have observers detect changes as soon as possible, thereby minimizing the contribution of memory. Another possibility is to have the observer respond differentially to the perception of dynamic change (which presumably relies on attention) and the inference of completed change (which presumably relies on a longer-term visual memory).

Techniques have been developed that incorporate most of these considerations into their design. Two examples are illustrated in Fig. 13.1. Figure 13.1a shows the *one-shot* paradigm, in which an image is briefly presented, followed by a brief blank or mask, and then followed by a second display, possibly containing a changed version of the first. Performance here is measured by the accuracy of change detection. Figure 13.1b shows the *flicker* paradigm, where the two displays continually alternate until the observer reports the presence or absence of the change. The measure here is the time taken to detect the change. Note that these variants correspond to the use of brief and extended displays in visual search experiments on static stimuli (see Chapter 43), with the target being a spatiotemporal pattern rather than a purely spatial one. (For a more extensive review, see Rensink (2002).)

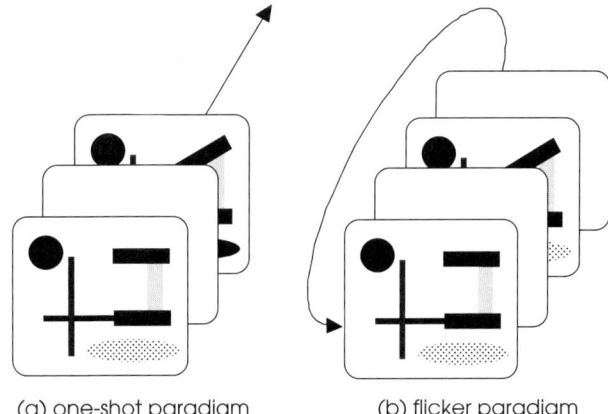

FIGURE 13.1 Examples of techniques used to induce change blindness. (a) One-shot paradigm. Here, the observer views a single alternation of displays, with a brief blank or mask between them. The task of the observer is to detect (or identify) the change; performance is measured via accuracy of response. (b) Flicker paradigm. Here, the observer views a continual cycling of displays, with a brief blank or mask after each display. The task of the observer is to detect (or identify) the change; performance is measured via response time. Both approaches can also be applied to other kinds of change, such as those made during eye movements or blinks.

II. THE NATURE OF VISUAL ATTENTION

Before examining how studies of change blindness (and its flip side, change detection) cast light on our understanding of visual attention, it is important to specify what is meant by "attention." This term can refer to several rather different things, and it is not a priori evident which would be relevant here.

As it turns out, however, all results point toward the involvement of the *focused attention* believed to bind together properties in a static item (e.g., Kahneman et al., 1992). For example, many characteristics of change detection (such as speed, capacity, and selectivity) are similar to—or at least compatible with—those of focused attention (Rensink, 2002). Also, change blindness is attenuated both for "interesting" items and for cued items (Rensink et al., 1997), both effects being consistent with what is known about the control of focused attention. Furthermore, attentional priming occurs at the location of an item seen to be changing, but not when there is no visual experience of change (Fernandez-Duque and Thornton, 2000).

A. Capacity Limits and Bottlenecks

Studies based on both one-shot and flicker paradigms demonstrate that when observers attempt to detect the *presence* of change, about four items can be attended at a time (e.g., Luck and Vogel, 1997; Rensink,

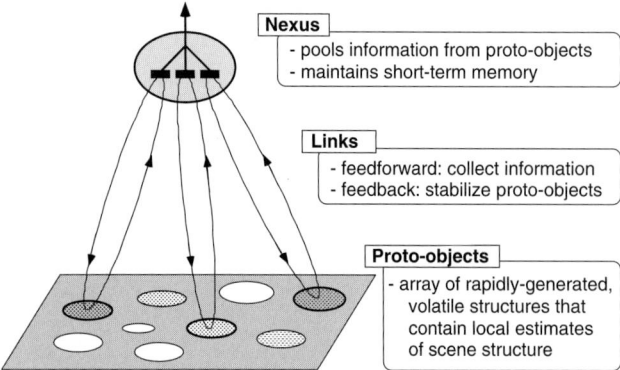

FIGURE 13.2 Pooling of attended information. Attended items are linked to a single *nexus*. (a) When searching for the presence of change, the nexus signal will either be 1 (target present) or 0 (target absent). A relatively strong signal therefore exists even when information from several links is collected. (b) When searching for the absence of change, the nexus signal will either be $n - 1$ (target present) or n (target absent). If all items are attended, n would be about 4, and the resulting signal would be quite weak; to obtain a strong signal, the nexus must collect information from only one link at a time.

2002). This is similar to the limit found for other kinds of attentional task (e.g., Pylyshyn and Storm, 1988). The extent to which separate components of attention and visual short-term memory are involved is not clear.

Interestingly, detecting the *absence* of change among a set of changing items yields a limit of one item (Rensink, 2002). One explanation for this is that information from the attended items is in some way pooled into a single collection point, or *nexus* (Fig. 13.2). If the nonchanging items do not contribute to the pooled signal, detecting a single change signal among four attended items could easily be done. But if four items were attended, it would be difficult to distinguish four changing items from three changing items + one nonchanging item; thus, to reliably detect a nonchanging target, only one item can be attended at a time. (Note that this explanation is similar to that used to explain search asymmetry, where detecting the presence of a basic property is far easier than detecting its absence (see Chapter 43).)

These limits also cast light on the nature of the bottleneck involved. If four items can be held by attention, detecting the absence of change in any of them should be easy: Simply compare each with its counterpart in the image; even if only one comparison can be made at a time, all items could eventually be compared. The finding that this is not possible indicates that the bottleneck is not the number of comparisons that can be made, but, rather, the number and nature of the representations constructed.

B. Independence of Attentional Complexes

Given that several items can be held by attention at any one time, how are the corresponding complexes[1] related to each other? It may be that each is independent of the others (Pylyshyn and Storm, 1988); on the other hand, a higher-level structure may somehow link them, imposing constraints on their operation (Rensink, 2002).

Results from change detection studies support the latter view. For example, the constraint that only one nonchange can be detected at a time would not exist if complexes were independent entities; each complex could simply be tested in turn, leading to a limit of at least four items. Additional evidence is the blindness found for switches of colors among tracked items (Saiki, 2003) and for switches of property assignments in static items containing multiple properties (Wheeler and Treisman, 2002), even when only two or three items were involved. If complexes were independent, and if properties were correctly bound to them, detection of such changes should be easy. The low level of performance actually found is compatible with some migration of properties among the attended items, a natural consequence of a pooled signal.

C. Contents of Attentional Complexes

Another issue of interest is the content of an attentional complex, that is, the number of basic properties it contains, and the amount of detail for each property. Change blindness studies indicate that this content is usually sparse, with only a small number of properties represented. For example, observers can miss large changes in an object even when it is attended, suggesting that the corresponding complex may be far from a complete representation of that object (Levin and Simons, 1997). Moreover, it appears that complexes are held in coherent form only as long as they are attended, falling apart when attention is withdrawn (Wolfe, 1999).

At least four simple properties, such as orientation, color, size, and curvature, can be simultaneously represented (Luck and Vogel, 1997), apparently via the concurrent coding of different kinds of properties

---

[1] A variety of terms have been used to describe the representational structures formed by focused attention, such as *object file* (Kahneman et al., 1992), *FINST* (Pylyshyn and Storm, 1988), and *coherence field* (Rensink, 2002). To discuss the results of various change blindness studies without respect to a particular theoretical framework, the term *complex* is used to denote the representational structure formed by attention for an item in a stimulus array, without respect to any particular theory.

(Wheeler and Treisman, 2002). Furthermore, such coding can capture not only the properties of each item, but also their parts and the structural relations between them (Carlson-Radvansky, 1999). The relation of these complexes to the elements of visual short-term memory is yet to be determined.

D. The Binding Problem

One of the more important concerns in the study of attention is the *binding problem*: how to prevent the properties (color, location, etc.) of one object representation from being erroneously assigned to another (see Chapter 24). The proposal that attended information is collected into a single nexus, that is, only one object is attended at a time, may provide a way out of this: If only one object is represented at a time, there can be no erroneous assignment of properties. Note that this solution would require the ability to construct a new complex for each object as it is needed. As such, the binding problem would be replaced by a *gating problem*: how to select the properties to be entered into the appropriate attentional complex at the appropriate moment in time.

III. VISUAL ATTENTION AND SCENE PERCEPTION

Although change blindness has provided insights into the nature of visual attention, it has also provided insights into other aspects of visual perception and the role that attention plays in them.

For example, it is believed that the initial stage of visual perception (*early vision*) involves simple visual elements created rapidly and in parallel across the visual field. Because these elements are believed to have a fleeting existence (existing only as long as light continues to enter the eye), the role of attention was sometimes seen as one of "welding" these elements into complexes that are more durable. These complexes then accumulate, providing a representation that is both dense (i.e., highly detailed) and coherent (i.e., all elements correctly bound together).

But change blindness experiments show that only a few coherent complexes exist, with relatively little detail in each; it may be that no representations exist that are both highly detailed *and* coherent. To reconcile this with the coherent, detailed picture of the world that we experience, it has been proposed that scene perception is based on a sparse, dynamic "just-in-time" system that creates object representations when (and only when) they are needed. If this coordination were done correctly, this *virtual representation* would appear to higher-level processes as if "real," that is, as if all objects simultaneously have detailed, coherent representations (Rensink, 2002).

A. Triadic Architecture

One possible implementation of a virtual representation is the *triadic architecture* (Rensink, 2002) shown in Fig. 13.3. This is composed of three systems: (1) an early system that continually generates simple visual elements, (2) an attentional system that enters a subset of these into a coherent representation of an object, and (3) a nonattentional system that determines such things as the meaning (or *gist*) of the scene and the spatial arrangement (or *layout*) of items in it.

Here, the constantly regenerating elements in the early system provide a rapid estimate of scene gist and layout. Attention is controlled both by long-term high-level factors (knowledge) and by short-term, low-level factors (salience of individual items) to create representations of the appropriate objects at the appropriate time. The objects so formed could then be used in turn as the basis of further attention guidance. As such, scene perception would involve a continually circulating flow of information between low-level representations containing retinal input and higher-level representations containing knowledge about the scene (see Chapter 33).

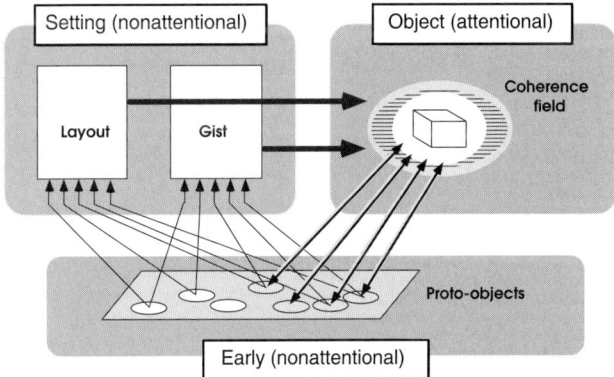

FIGURE 13.3 Triadic architecture. Thin lines indicate information flow; thick lines control. Here, visual perception is carried out by three largely independent systems: (1) an *early* system concerned with the formation of (unattended) elements rapidly and in parallel across the visual field, (2) an *object* system concerned with the formation of coherent representations (complexes) via attention, and (3) a nonattentional *setting* system that enables attentional guidance via high-level knowledge. These enable effective management of attention (and therefore conscious perception) via a combination of high- and low-level control.

B. Observer Intention

Given the dynamic nature of scene representation, perception for a given task must rely on *attentional management*, i.e., deploying attention as effectively as possible. An important factor here is the degree to which the observer expects a change, and believes that reporting it is relevant. The degree of change blindness found is much higher when the observer does not expect a change (being asked to report it afterward), although some ability to detect change still remains (Levin and Simons, 1997). This supports the view that the only properties put into coherent form (or at least compared) are those needed for the task at hand.

Another important factor is the type of change expected. Detection of orientation change is unaffected by irrelevant variations in contrast sign, again indicating that only those properties needed for the immediate task are encoded (Rensink, 2002). More generally, observers appear to be sensitive only to changes in those properties relevant to the task being carried out at the moment the change was made (see Rensink, 2002).

C. High-Level Knowledge

Attentional management is heavily dependent on the high-level knowledge of the observer. One way this can influence perception is via the particular *representations* available. For example, detection of change was better for objects learned at a specific rather than a general level (Archambault et al., 1999). This suggests that more detailed representations had been formed for the specific-level objects, with observers then taking advantage of these representations to improve their performance.

Another way that knowledge can influence perception is via a more effective *guidance* of attention. For example, a study that compared the performance of experts and nonexperts in American football found that experts could spot changes in meaning more quickly, and could attentionally scan meaningful scenes more efficiently (Werner and Thies, 2000).

D. Vision Outside the Focus of Attention

The proposal that attention is needed to see change implies that change cannot be seen outside the focus of attention. But this proposal was based on studies where observers made a volitional response; as such, the meaning of "see" must be restricted to conscious visual experience (Rensink, 2002). If other forms of response are considered, it may be that some other form of change perception is possible, perhaps mediated by the nonattentional streams proposed in the triadic architecture (see Section II.A).

Several change blindness studies report interesting results in this regard. For example, even if an observer does not consciously experience a change, their visuomotor systems can still respond to it (e.g., Bridgeman et al., 1979). Furthermore, observers without any visual experience of a change can guess its location with above-chance accuracy (Fernandez-Duque and Thornton, 2000). In such cases, however, the underlying mechanisms are poorly understood, and it is not entirely certain that the involvement of focused attention can be ruled out. Moreover, there remains the possibility that while a considerable subset of properties may be attended (and so affect subjective experience), only a subset of these may be compared on a given task (and so affect objective performance). Further work is needed to clarify these issues (see Rensink, 2002).

IV. CONCLUSIONS

Visual attention appears to be critical for the creation (and perhaps maintenance) of internal representations with a spatiotemporal coherence that in some sense matches that of the external object(s) they describe. Change blindness reflects the ability of visual attention to create (and perhaps maintain) such representational structures. Results to date on the nature and role of attention are consistent with—and in places extend—results obtained using other approaches.

Looked at more broadly, the study of change blindness is the first stage of investigation into the more general issue of the perception of organized spatiotemporal patterns, such as movements and events. Based on the results obtained so far, it is likely that the perception of such patterns will critically depend on visual attention.

References

Archambault, A., O'Donnell, C., and Schyns, P. G. (1999). Blind to object changes: when learning the same object at different levels of categorization modifies its perception. *Psychol. Sci.* **10**, 249–255.

Bridgeman, B., Lewis, S., Heit, G., and Nagle, M. (1979). Relation between Cognitive and Motor-Oriented Systems of Visual Position Perception. *J. Exp. Psychol. Hum. Percept. Perform.* **5**, 692–700.

Carlson-Radvansky, L. A. (1999). Memory for relational information across eye movements. *Percept. Psychophys.* **61**, 919–934.

Fernandez-Duque, D., and Thornton, I. M. (2000). Change detection without awareness: do explicit reports underestimate the representation of change in the visual system. *Vis. Cogn.* **7**, 323–344.

Kahneman, D., Treisman, A., and Gibbs, B. (1992). The reviewing of object files: object-specific integration of information. *Cogn. Psycho.* **24**, 175–219.

Levin, D. T., and Simons, D. J. (1997). Failure to detect changes to attended objects in motion pictures. *Psychon. Bull. Rev.* **4**, 501–506.

Luck, S. J., and Vogel, E. K. (1997). The capacity of visual working memory for features and conjunctions. *Nature*, **390**, 279–280.

Pylyshyn, Z. W., and Storm, R. W. (1988). Tracking multiple independent targets: evidence for a parallel tracking mechanism. *Spatial Vision* **3**, 179–197.

Rensink, R. A. (2002). Change detection. *Annu. Rev. Psychol.* **53**, 245–277.

Rensink, R. A., O'Regan, J. K., and Clark, J. J. (1997). To see or not to see: the need for attention to perceive changes in scenes. *Psychol. Sci.* **8**, 368–373.

Saiki, J. (2003). Feature binding in object-file representations of multiple moving items. *J. Vision* **3**, 6–21.

Simons, D. J., Franconeri, S. L., and Reimer, R. L. (2000). Change blindness in the absence of visual disruption. *Perception* **29**, 1143–1154.

Werner, S., and Thies, B. (2000). Is "change blindness" attenuated by domain-specific expertise? An expert–novice comparison of change detection in football images. *Vis. Cogn.* **7**, 163–173.

Wheeler, M. E., and Treisman, A. M. (2002). Binding in short-term visual memory. *J. Exp. Psychol. Gen.* **131**, 48–64.

Wolfe, J. M. (1999). Inattentional amnesia. *In* "Fleeting Memories" (V. Coltheart, Ed., pp. 71–94). MIT Press, Cambridge, MA.

CHAPTER 14

Development of Covert Orienting in Young Infants

John E. Richards

ABSTRACT

Adults can shift attention to different regions of space without moving the eyes, that is, covert orienting of attention. Covert orienting implies that information processing may occur for stimuli in peripheral locations. The purpose of this chapter is to review evidence that in the first 6 months of life, infants are able to shift attention throughout space covertly. These studies show that there is an increasing efficiency from birth to 6 months with which infants shift spatial attention. Some cortical areas that may be involved in the development of spatial attention are suggested.

I. INFANTS CAN SHIFT ATTENTION COVERTLY TO PERIPHERAL STIMULI

Several studies have shown that covert orienting can occur in young infants. The spatial cuing procedure developed by Posner (Posner, 1980; Posner and Cohen, 1984) (see Chapters 16, 31, and 64) was adapted by Hood (1995) to study covert orienting in infants. In Posner's procedure, the participant's fixation remained at a central location while a peripheral cue and target were presented. Two reaction time effects are used to show covert shifts of attention: response facilitation and inhibition of return (see Chapter 64). Response facilitation occurs when the cue and target occur close in time in the same location. Response slowing (inhibition of return) occurs when the cue and target are separated further in time and the cue and target are in the same location. Hood presented infants with an interesting visual pattern in the center. When the infant is fixating on this pattern, a stimulus is presented in the periphery (analogous to "cue") in addition to the central stimulus. Infants will not shift fixation from the center pattern to the peripheral pattern during the brief presentation of this peripheral stimulus (Richards, 1987, 1997, 2002). The peripheral stimulus and central stimulus are then removed, and the peripheral stimulus is presented in the periphery (analogous to "target"). The eye movement from the center position to this peripheral stimulus is the dependent measure. The target can be presented on the same side as the cue ("valid trials") or on the opposite side ("invalid trials"), cannot be presented ("no-target control"), or can be presented on a trial without the cue being presented ("neutral"). Hood and Atkinson (1991, reported in Hood, 1995; Hood, 1993) tested 3- and 6-month-old infants in this procedure. The reaction time of infants at 3 months of age for valid, invalid, and neutral trials was no different between a target presented at 200-ms stimulus onset asynchrony (SOA) or 700-ms SOA, that is, no facilitation or inhibition of return. The reaction time of the 6-month-old infants was facilitated on the 200-ms-SOA validly cued trials, relative to neutral or invalid trials. Reaction time was slowed on the 600-ms-SOA trials, that is, inhibition of return. There were no differences at either delay in the responses on the invalid or the neutral trials.

Other aspects of covert orienting have been studied in infants. Johnson and Tucker (1996) used a similar procedure for cuing the infant, and then presented bilateral stimuli in the "target" period. They found that 4- or 7-month-old infants had an increased probability of localizing the target that was ipsilateral to the cue at short delays (SOA of 133–200 ms) and facilitated reaction times to the ipsilateral target. Alternatively, the infants showed a decreased probability of localizing the ipsilateral target at a longer delay (SOA of 700 ms) and lengthened reaction times to the ipsilateral target at that delay. They did not find such an effect for

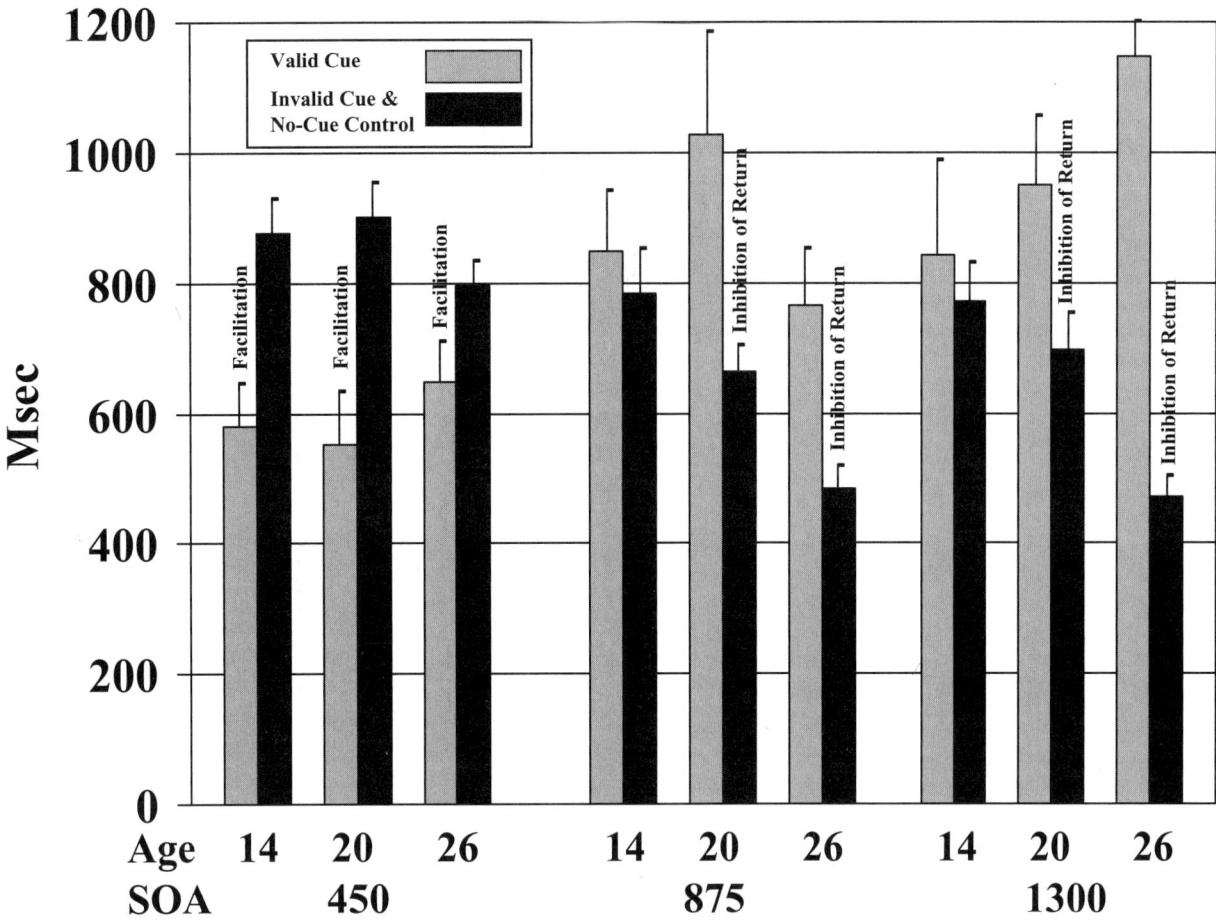

FIGURE 14.1 Latency to localize the peripheral stimulus when it was presented as a target. These figures are presented separately for the three testing ages, the three SOA conditions, and the valid cue and invalid cue/neutral trials. Response facilitation is shown by a faster response time for the valid targets than the invalid/neutral targets, and inhibition of return is shown by a slower response on the validly cued trials. Reprinted, with permission, from Richards (2000a).

2-month-old infants. Newborn infants appear to show inhibition of return following overt shifts of fixation (Valenza et al., 1994). In this procedure infants were given a peripheral cue to which they shifted fixation, and then fixation was returned to the center. A bilateral target was then presented. Newborns were more likely to look toward the target contralateral to the side to which fixation had earlier been directed. This suggests that the mechanism for inhibition of return exists at birth, but that "covert" orienting of attention does not occur at 2 to 3 months of age but does occur by 4 or 6 months.

A recent study examined covert orienting in infants aged 3, 4.5, and 6 months (14, 20, and 26 weeks) (Richards, 2000a). The spatial cuing procedure adapted for infant participants (Hood, 1995) was used. This study presented the central stimulus for 2 seconds, then the central and peripheral stimulus ("cue") together for 300ms, turned both stimuli off, and then presented the target stimulus at SOAs of 300, 875, or 1300ms. Figure 14.1 shows the latency to localize the target as a function of testing age and the spatial cuing conditions. No difference was found in the responses in the invalid and neutral conditions, so these are averaged together. The latency to localize the peripheral stimulus when it was presented as a valid target was faster than when it was presented as an invalid/neutral target at a SOA of 450ms, for all three testing ages. This indicates that response facilitation occurred at all three ages. The reaction times at the two longer SOAs (875 and 1300ms) were longer on the valid trials for the two older ages, there was no difference for the 3-month-old infants between valid and invalid/neutral trials. There was a regularly increasing amount of inhibition of return. The differences between the reaction time on the valid and invalid/neutral trials at the 1300-ms SOA were 70, 260, and 670ms for the 14-, 20-, and 26-week-olds, respectively. This study shows that facilitation of reaction time may occur earlier than found previously (Hood and Atkinson, 1991, reported in Hood, 1995; Hood, 1993; Johnson and Tucker, 1996). The results for the long SOA are consistent with the view that inhibition of return following covert orienting emerges sometime between 3 and 6 months.

II. INFANT COVERT ORIENTING OCCURS DURING CENTRAL STIMULUS ATTENTION

A difference between the spatial cuing procedure used with infants and that used with adults is the procedural step of using a central stimulus to keep infants from overtly shifting fixation toward the cue. Verbal instructions are sufficient for adults to keep fixation at the central location, but infants do not readily obey such instructions! The rationale for using this procedure is that infants are less likely to shift fixation to the onset of a peripheral stimulus when a central stimulus is present. The reason for not shifting fixation from a central to a peripheral stimulus is based on infants' attentiveness to the central stimulus (Richards, 1987, 1997, 2002). The presence of a central stimulus engages focal attention. As long as attention to the center stimulus is occurring, there is an unresponsiveness to the peripheral stimulus. Thus, when engaged in attention to the central stimulus infants will not look toward the "cue," and thus subsequent reaction time effects (response facilitation or inhibition of return) are evidence that covert orienting of attention to the cue occurred. In contrast, it is possible that infants can direct fixation to the center stimuli in the absence of active attention engagement. In this case, localization percentage of a fixed-duration peripheral stimulus is likely (Richards, 1987), or localization of a continuing stimulus occurs quickly (Richards, 1987). Focal stimulus attention results in a spatial selectivity for fixation, with fixation being directed primarily toward the location in which the attended stimulus occurs.

The effect of central stimulus attention on covert orienting was studied by Richards (2000b). In that study infants from 3 to 6 months were presented with the central stimulus and cue either simultaneously (0-second cue onset delay), delayed by 2 seconds (2 s cue onset delay), or until a significant deceleration in heart rate occurred (heart rate deceleration cue onset delay). The rationale for the heart rate deceleration condition is that heart rate deceleration in young infants is an index of the onset of sustained attention (Richards, 1987, 1997, 2002; Richards and Hunter, 1998, 2002). Figure 14.2 illustrates some results for this study. This figure shows difference scores from when the cue and the target were on the same side (valid trials) minus when the cue and target were on a different side (invalid trials) or the target was uncued (neutral trials). The results from the prior study with only a 2 second delay were replicated in this study (cf. middle portions of panels for Fig. 14.2 with Fig. 14.1). The immediate

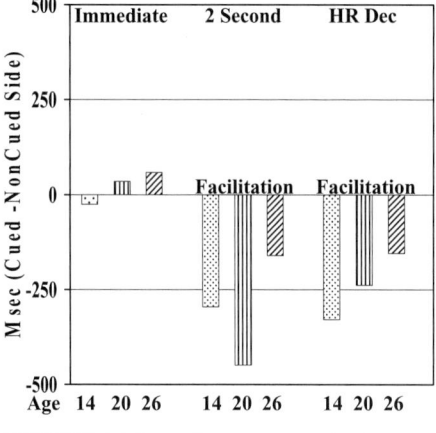
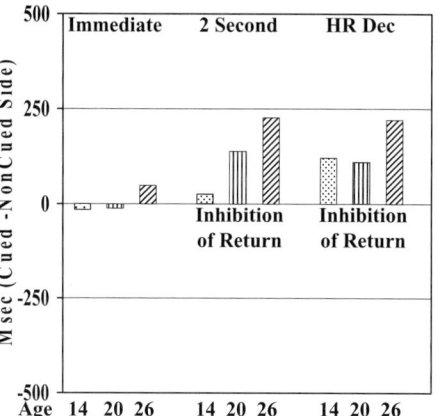

FIGURE 14.2 Difference scores for latency to localize the peripheral stimulus when it was presented as a target. The onset between the central stimulus and cue, i.e., cue onset delay, was either immediate (simultaneous), 2 seconds, or after a significant heart rate deceleration. For the 300-ms SOA, a faster reaction occurred to the validly cued target than to the invalid/neutral targets (top figures), i.e., response facilitation. For the 875 and 1300 ms SOAs, a slower reaction time occurred for the validly cued targets, i.e., inhibition of return. Reprinted, with permission, from Richards (2000b).

condition, when the central stimulus and cue were presented simultaneously, resulted in no difference between the valid and invalid/neutral trials for either SOA at any age (left-hand side of panels for Fig. 14.2). When the cue was presented contingent on the occurrence of a heart rate deceleration, indicating sustained attention was engaged, facilitation of response time and inhibition of return occurred at all three testing ages (right-hand side of panels for Fig. 14.2).

The finding that facilitation and inhibition of return occur for the heart rate deceleration condition (Fig. 14.2) indicates that attention engagement must be well underway for infants to show covert orienting of attention. This implies that this is a different task for infants than adults in that it demands focal stimulus attention be engaged in parallel with the covert orienting of attention to the cued location. The older infants apparently were able to keep processing resources on the central stimulus and shift processing resources to the peripheral cue in parallel. This resulted in continued fixation on the focal stimulus (focal stimulus attention) and processing of the stimulus location of the peripheral cue leading to inhibition of return and facilitation. The youngest infants, however, also may shift attention to the peripheral cue but not shift attention back to the center stimulus. Thus, there is a relative automatic processing of the peripheral stimulus information, a reflexive saccadic planning toward the peripheral stimulus that was inhibited, and response facilitation. However, apparently the young infants cannot shift attention from the peripheral location back to the center location and, therefore, do not show inhibition of return. These studies imply that one developmental change over this age range is an increasing flexibility in the spatial attention system resulting in the ability to orient to multiple locations in space under a wide variety of stimulus situations.

III. CORTICAL BASES OF SPATIAL ATTENTION DEVELOPMENT?

The inhibition of return effect is thought to be mediated by the superior colliculus (Rafal, 1998). It is thought that the activation of pathways in the superior colliculus responsible for fixation shifts and the inhibition of those pathways during the spatial cuing procedure result in inhibition of return (also see Chapters 16 and 64). The response facilitation that occurs at short SOAs is hypothesized to be due to the enhancement of sensory processing of information in the attended portion of visual space (Hillyard et al., 1995). This is shown elegantly in studies of selective spatial attention and the enhancements of the early components of the event-related potentials (ERP) in scalp-recorded electrical potentials (Hillyard et al., 1995) (see Chapters 84 and 85). Researchers studying infants in the spatial cuing procedure have adopted this neurophysiological perspective (Hood, 1993, 1995; Johnson and Tucker, 1996; Richards, 2000a, 2001, 2003; Richards and Hunter, 2002). The general conclusion of the "neurodevelopmental" approach is that the superior colliculus, which is relatively mature at birth, supports inhibition of return in early infancy (Hood, 1993, 1995; Valenza et al., 1994) but only for overt fixation shifts. The changes in covert attention shifts found between 3 and 6 months of age must therefore be due to cortical changes in areas such as the parietal cortex and frontal eye fields involving saccadic planning and attention shifting. This interpretation is consistent with the general view that that there is an increase in the first 6 months of life of cortical control over eye movements that occur during attention and increasing cortical control over general processes involved in attention shifting (e.g., Hood, 1995; Richards, 2003; Richards and Hunter, 1998, 2002).

Some studies have used scalp-recorded ERPs to study covert orienting in infants. The ERP represents specific events occurring in the cortex and may be useful in identifying the cortical areas involved in the development of covert orienting Richards (2000a, 2001) studied infants at 14, 20, and 26 weeks of age. The spatial cuing procedure adapted for infants was used and ERPs were measured at the beginning of the target onset, or immediately before saccade onset. Figure 14.3 (bottom) shows the ERPs changes occurring at the target onset for the occipital electrode contralateral to the target for the valid, invalid, and neutral trials. A large positive deflection occurring at about 135 ms following stimulus onset may be seen for all three testing ages (P1?). This positive potential was the same size for all three cuing conditions for the youngest ages, slightly larger for the valid condition than the other two conditions for the 20-week-old infants, and the largest for the valid condition for the 26-week-old infants. This enhanced first positive component has been found for valid trials in adults and is called the "P1 validity effect" (Hillyard et al., 1995) (see Chapters 16, 36, 84, and 85).

A second ERP component was found in these studies. Presaccadic ERP is calculated as EEG changes occurring backward in time from the onset of the saccade to the target. The presaccadic ERP reflects cortical areas involved in saccade planning (Richards, 2000a, 2001, 2003; Richards & Hunter, 2002). A presaccadic ERP component was found in frontal electrode sites about 50 ms before saccade onset, located in scalp regions contralateral to the saccade. Figure 14.3 (top

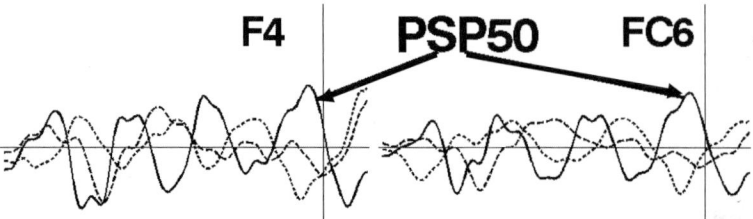

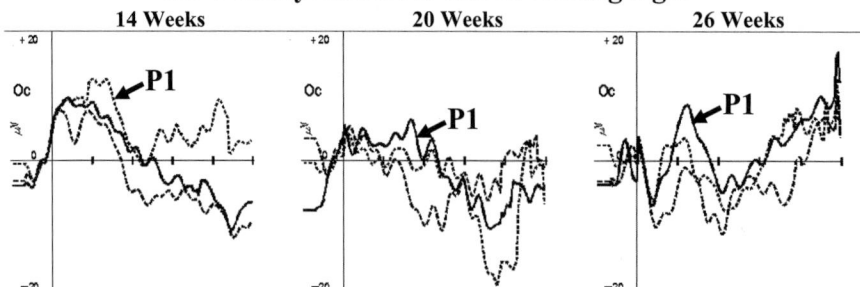

FIGURE 14.3 Top: ERP responses on the contralateral occipital electrode to the peripheral stimulus onset when it was presented as a target. The responses are presented separately for the three testing ages, and separately for the valid (solid line), invalid (small dashes), and no-cue control (long dashes) trials. The data are presented as the difference from the ERP on the no-stimulus control trial. The approximate locations of the P1 and N1 components are identified on each figure. Bottom: Presaccadic ERP component occurring about 50 ms before saccade onset. The presaccadic ERPs for F_4 and FC_6 show a large presaccadic positive ERP component that occurred about 50 ms before saccade onset for cued exogenous saccades.

figures) is an ERP recording from these electrodes (Richards, 2001). This positive component occurred primarily for saccades toward a target that occurred in the same location as the cue, that is, valid trials, and did not occur (or was smaller) for saccades toward a target that appeared in a different location than the cue (invalid trials), or for saccades toward a target that had not been preceded by a cue (neutral trials). Also, this presaccadic ERP component was absent in the youngest infants (3 months), and the amplitude of the ERP and the spread of the ERP across multiple electrode locations increased for infants at the older testing ages (4.5 and 6 months).

The cortical locations that generate these covert orienting and saccade planning effects have been examined with "equivalent current dipole analysis" ("brain electrical source analysis"; Richards, 2003; Richards and Hunter, 2002) (see Chapter 84). Figure 14.4 (top) shows a topographical scalp potential map for the presaccadic ERP component and an equivalent current dipole analysis of the presaccadic ERP. A hypothesized current dipole located in the area of the frontal eye fields generates a scalp potential map that closely corresponds to the recorded ERP. This analysis is consistent with the interpretation that the eye movements to the target in the planned location involve cortical areas that control planned eye movements (see Chapters 21 and 22). Figure 14.4 (bottom) shows some analyses of the P1 validity effect done with a 128-channel EEG system. Some of the dipoles were located in the primary visual area of the occipital cortex (Brodmann area 17; Fig. 14.4, Primary Visual Cortex). The activation of these dipoles did not show the validity effect (cf. Chapter 84). Some of the dipoles were located in the fusiform gyrus (Brodmann area 19; Fig. 14.4, Fusiform Gyrus). This area is one of the pathways from the primary visual area to the object identification areas in the temporal cortex ("ventral processing stream"). The dipoles located in the fusiform gyrus were those whose activation showed the P1 validity effect (cf. Chapters 16, 36, 84, and 85). The analysis of the cortical bases of the covert orienting effects suggests that specific brain areas may be identified that also show development and that form the basis for the changes in covert orienting seen in infants in this age range.

There are two implications of the work showing the ERP components accompanying covert orienting in young infants and their cortical bases. A first and very general implication is that brain areas involved in the control of sensory processing (e.g., P1 validity effect) and in the control of saccade planning (presaccadic

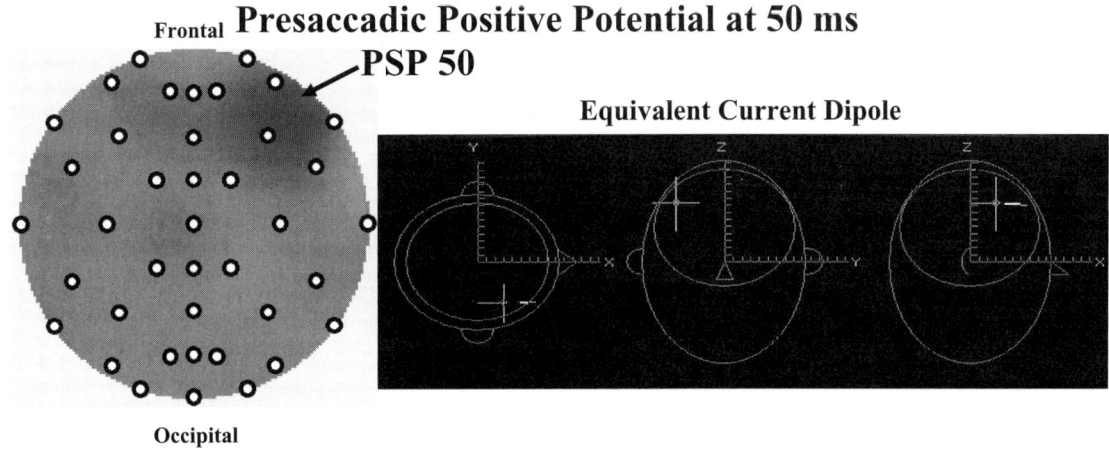

FIGURE 14.4 Top: The topographical scalp potential map is for the presaccadic ERP component (Fig. 14.3), plotted as the difference between the valid and the invalid/neutral trial saccades, plotted as if the infant were making a saccade toward the left side. The equivalent current dipole analysis resulted in a dipole located in the frontal eye fields. Bottom: Equivalent current dipole locations for individual infant participants for a component reflecting the "P1 validity effect" in a spatial cueing task. The dipole locations are plotted on a MRimage from a young child. Reprinted, with permission, from Richards (2003) and Richards and Hunter (2002). (See color plate)

ERP component) may form the basis for changes in attention to peripheral stimuli in young infants. These techniques may prove useful in identifying the cortical areas that show developmental changes that parallel the behavioral responses. Second, there are some specific implications for the development of covert orienting with these findings. Recall that infants at the earliest ages tested in these studies (14 weeks and 3 months) show response facilitation at levels comparable to those of the oldest infants. This implies that the age changes in the cortical areas involved with sensory processing (P1 validity effect, fusiform gyrus) or age changes in the cortical areas involved with saccade planning (presaccadic ERP, frontal eye fields) do not form the basis for the response facilitation. This suggests that a subcortical mechanism is responsible for this effect. A likely candidate is the activation of the pathways in the superior colliculus that are involved in peripheral saccadic eye movements (see Chapters 16 and 64). The covert orienting to the exogenous stimulus occurs without cortical involvement and is present at very early ages. Alternatively, other changes in spatial attention may be based on these cortical areas. The interpretations of the increase in inhibition of return (Figs. 14.1, 14.2) were that there was increased flexibility from 3 to 6 months in moving attention throughout space. This is accompanied in the brain by changes in the enhancement of cortical areas responsible for sensory processing (fusiform gyrus) (see Chapter 84) and the increasing sophistication of attention-based saccade planning (frontal eye fields) (see Chapters 21 and 22). These findings are consistent with "neurodevelopmental" models that posit an increasing role of the cerebral cortex in the control of attention-related eye movements (e.g., Hood, 1995; Richards, 2003; Richards and Hunter, 1998, 2002).

Acknowledgments

This research was supported by grants from the National Institute of Child Health and Human Development, R01-HD 18942, and a Major Research Instrumentation Award from the National Science Foundation, BCS-9977198.

References

Hillyard, S. A., Mangun, G. R., Woldroff, M. G., and Luck, S. J. (1995). Neural systems mediating selective attention. *In* "Cognitive Neurosciences" (M. S. Gazzaniga, Ed.), pp. 665–682. MIT Press, Cambridge, MA.

Hood, B. M. (1993). Inhibition of return produced by covert shifts of visual attention in 6-month-old infants. *Infant Behav. Dev.* **16**, 245–254.

Hood, B. M. (1995). Shifts of visual attention in the human infant: a neuroscientific approach. *Adv. Infancy Res.* **10**, 163–216.

Johnson, M. H., and Tucker, L. A. (1996). The development and temporal dynamics of spatial orienting in infants. *J. Exp. Child Psychol.* **63**, 171–188.

Posner, M. I. (1980). Orienting of attention. *Q. J. Exp. Psychol.* **32**, 3–25.

Posner, M. I., and Cohen, Y. (1984). Components of visual orienting. *In* "Attention and Performance X" (H. Bouma and D. G. Bouwhis, Eds.), pp. 531–556. Erlbaum, Hillsdale, NJ.

Rafal, R. D. (1998). The neurology of visual orienting: a pathological disintegration of development. *In* "Cognitive Neuroscience of Attention: A Developmental Perspective" (J. E. Richards, Ed.), pp. 181–218. Erlbaum, Hillsdale, NJ.

Richards, J. E. (1987). Infant visual sustained attention and respiratory sinus arrhythmia. *Child Dev.* **58**, 488–496.

Richards, J. E. (1997). Peripheral stimulus localization by infants: attention, age and individual differences in heart rate variability. *J. Exp. Psychol. Hum. Percept. Perform.* **23**, 667–680.

Richards, J. E. (2000a). Localizing the development of covert attention in infants using scalp event-related-potentials. *Dev. Psychol.* **36**, 91–108.

Richards, J. E. (2000b). The development of covert attention to peripheral targets and its relation to attention to central visual stimuli. Paper presented at the International Conference for Infancy Studies, Brighton, England, July 2000.

Richards, J. E. (2001). Cortical indices of saccade planning following covert orienting in 20-week-old infants. *Infancy* **2**, 135–157.

Richards, J. E. (2002). Attention in young infants: a developmental psychophysiological perspective. *In* "Developmental Cognitive Neuroscience" (C. A. Nelson and M. Luciana, Eds.), MIT Press, Cambridge, MA.

Richards, J. E. (2003). Development of attentional systems. *In* "The Cognitive Neuroscience of Development" (M. De Haan and M. H. Johnson, Eds.), Psychology Press, East Sussex.

Richards, J. E., and Hunter, S. K. (1998). Attention and eye movement in young infants: Neural control and development. *In* "Cognitive Neuroscience of Attention: A Developmental Perspective" (J. E. Richards, Ed.), pp. 131–162. Erlbaum, Mahway, NJ.

Richards, J. E., and Hunter, S. K. (2002). Testing neural models of the development of infant visual attention. *Dev. Psychobiol.* **40**, 226–236.

Valenza, E., Simion, F., and Umilta, C. (1994). Inhibition of return in newborn infants. *Infant Behav. Dev.* **17**, 293–302.

CHAPTER 15

Prior Entry

David I. Shore and Charles Spence

ABSTRACT

The law of prior entry was one of Titchener's seven fundamental laws of attention. According to E.B. Titchener (1908, "Lectures on the Elementary Psychology of Feeling and Attention," Macmillan, New York, p. 251): "the object of attention comes to consciousness more quickly than the objects which we are not attending to." Although research on prior entry spans more than a century, agreement on the very existence of the phenomenon has proved elusive. While some researchers are convinced of the veracity of the prior entry effect, others have argued that the empirical evidence is unconvincing. However, the last few years have seen a rapid growth of interest in the phenomenon, with a number of important confounds in previous research being identified. The latest research confirms the existence of a robust prior entry effect, identifying both perceptual and decisional components of the effect.

I. INTRODUCTION

In 1908, Titchener formalized the law of prior entry, stating that attended events are perceived prior to unattended events. While the earliest studies of prior entry tended to focus on the perceptual consequences of voluntarily (or endogenously) attending to one sensory modality versus another (see Spence et al., 2001, for a review), later research has extended the paradigm to study the consequences of both voluntary and stimulus-driven (or exogenous) *spatial* attention effects on temporal order judgments (TOJs) (e.g., Schneider and Bavelier, 2003; Shore et al., 2001; Spence et al., 2001; Stelmach and Herdman, 1991). Although prior entry effects were not always demonstrated in early work (see Spence et al., 2001, for a review), more recent studies have provided clear and robust evidence for the existence of prior entry, with most of the recent empirical debate focusing on the relative contribution of perceptual and decisional factors (such as response biases, e.g., Frey, 1990) to the effect (e.g., Schneider and Bavelier, 2003; Shore et al., 2001). The latest research has started to incorporate cognitive neuroscience techniques, such as the use of electrophysiological recordings (McDonald et al., 2003) and modeling (Schneider and Bavelier, 2003), to complement the psychophysical paradigms that are now available to determine the stages of information processing at which prior entry occurs.

II. EARLY DEVELOPMENT OF THE TASK

According to Mollon and Perkins (1996), the origins of research in the field of experimental psychology lie with the investigation of the consequences of attention for the temporal aspects of multisensory perception. They highlighted how the discovery of the personal equation in astronomy, a little more than two centuries ago, prompted early claims that attending preferentially to one or another modality could bias the relative perceived onset of sensory events (see Spence et al., 2001, for a brief historical overview). At the time, astronomers had to attend to both auditory and visual stimuli (the ticking of a clock and the movement of a star, respectively) simultaneously to determine the exact time at which a star crossed the meridian wire of their telescopes. Relatively consistent individual differences across astronomers led to the proposal of a bias, within each observer, to attend preferentially to one sensory modality or the other. Wundt's *complication experiment* represented the first attempt to bring this phenomenon into the laboratory. Observers in Wundt's studies had to indicate the relative position of a moving visual stimulus (typically an oscillating pendulum or a rotating clock arm) at the moment that an auditory event was presented. The TOJ task developed out of research on the complication situation as scientists attempted to avoid concerns related to the role of eye movements (i.e., people trying to fixate the moving stimulus) in any effects observed.

III. EARLY PRIOR ENTRY RESEARCH

Early research investigating the phenomenon of prior entry typically presented pairs of stimuli in different sensory modalities from different spatial locations, at one of a range of stimulus onset asynchronies (SOAs), while asking participants to judge which stimulus modality appeared to have been presented first. The participant's attention was biased toward one modality or the other either by means of verbal instructions or else by the use of a secondary task. The dependent measure in those experiments was the proportion of trials on which the observer reported each of the two modalities as having occurred first, as a function of the SOA that separated the onset of the two events.

Two summary statistics have typically been computed from the results of such studies: the *point of subjective simultaneity (PSS)* and the *just noticeable difference (JND)*. The PSS refers to the interval between the two stimuli at which observers report each of the two stimuli as coming first (or, in some studies, second, e.g., Frey, 1990; Shore et al., 2001) on an equal proportion of trials (i.e., 50% of the time) (see Fig. 15.1). As such, it reflects the point at which observers are maximally uncertain as to which stimulus was presented first. Conventionally, the PSS is thought to correspond to the asynchrony giving rise to the optimal perception of simultaneity (although this assumption has not, as far as we are aware, ever been tested empirically).

The phenomenon of prior entry is typically illustrated by means of a shift in the PSS as a function of the direction of a participant's attention to one or the other stimulus (no matter whether attention is directed to a particular sensory modality, as in early studies, or to a particular spatial location, as in many later studies). The magnitude of the PSS shift therefore provides a measure of how much sooner one stimulus must be presented before the other for the observer to be unable to differentiate their order (see Fig. 15.1). The divided attention condition provides a baseline against which the attention conditions can be compared. In a multisensory situation involving vision and touch, the visual stimulus must typically be presented prior to the tactile one under conditions of divided attention for the PSS: transduction is much slower on

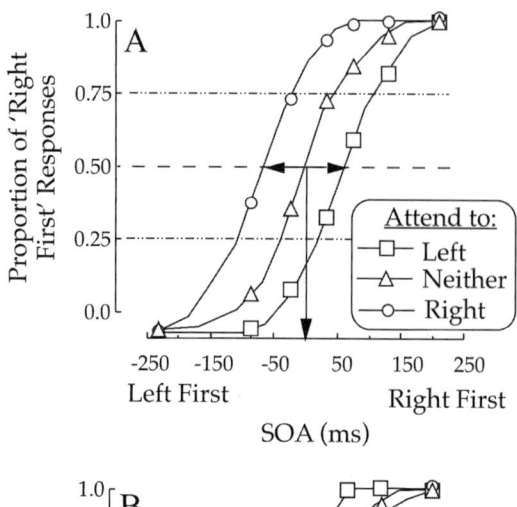

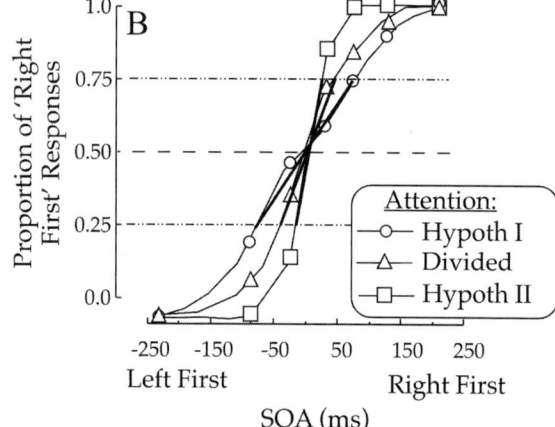

FIGURE 15.1 Hypothetical data showing the effect of attention on the PSS (A) and JND (B). The PSS represents the SOA where participants make each of the two responses equally often. The prior entry hypothesis predicts the shift in PSS shown by the double-headed arrow. When attending to the right, the stimulus on the left typically has to be presented first. Instead of directing attention to the left or right, many experiments have examined the effects of directing attention to different sensory modalities (e.g., vision, touch, or audition) with similar results. The JND is calculated as half of the SOA range between the points on the psychometric function corresponding to 75% of "right first responses" and 25% "right first responses." The JND is monotonically related to the slope of the psychometric function. Two possible effects of attention on the JND are displayed. Attention could cause a relative reduction in temporal precision resulting in a larger JND (labeled Hypoth I, shallower slope). Alternatively, attention could cause a relative increase in temporal precision resulting in a smaller JND (labeled Hypoth II, steeper slope). See text for further details.

the retina than on the skin surface. Thus, the prior entry effect is seen as an increase or reduction in this baseline (see Fig. 15.2). Critically, the baseline allows one to assess the relative benefit of attending to one modality or the other. For the visuotactile situation, attention to touch produces a much larger shift in the PSS than does attention to vision (see Spence et al., 2001, for discussions of this issue).

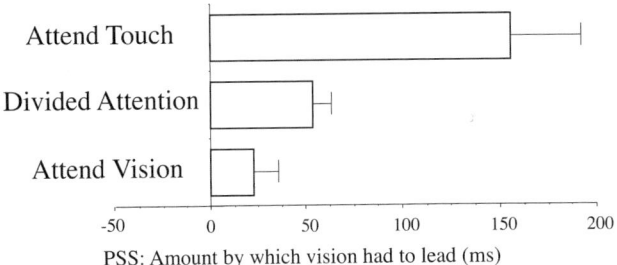

FIGURE 15.2 Data represent the shift in the PSS as a function of endogenous attention to either vision or touch in a "which side came first (left vs right)?" TOJ task. Participants were presented with a visual or tactile stimulus on the left or right and indicated which side was presented first. In the attend vision blocks, the unimodal trials were always vision–vision, and in the attend touch blocks, the unimodal trials were always touch–touch whereas in the divided attention blocks, both types of unimodal trials were presented. In all three conditions, the data presented here come from the bimodal vision–touch trials within those blocks. The PSSs were calculated from a probit analysis of the psychometric functions shown in Fig. 15.1 (see Spence et al., 2001, for details). The amount by which a visual stimulus had to be presented before a tactile stimulus was affected by the sensory modality to which observers attended. Reprinted from Fig. 6 of Spence, C., Shore, D. I., and Klein, R. M. (2001). Multisensory prior entry. *J. Exp. Psychol. Gen.* **130**, 799–832, with permission of the American Psychological Association.

The JND provides a measure of the precision with which people can make TOJs: specifically, how far apart in time two events must be presented for them to be reliably ordered in time. Historical precedent has set the 75% accuracy level as an index of reliability (Coren et al., 1999): the JND is defined as half of the temporal interval between the SOA values producing the 75 and 25% points on the psychometric function (see Fig. 15.1). Interestingly, previous research has not revealed a consistent effect of attention on JNDs. For instance, Stelmach and Herdman (1991) reported a reduction in the JND for pairs of visual stimuli presented at the attended location in their unimodal visual study (see Hypothesis II in Fig. 15.1b). By contrast, no such attentional modulation of JNDs was reported in any of Spence and colleagues' (2001) multisensory studies, regardless of whether attention was directed to a particular location or to a particular sensory modality (vision vs touch). At present, it is unclear whether this discrepancy reflects a difference between intramodal and cross-modal studies, or the influence of some other factor on performance. The third possible pattern of data (i.e., larger JNDs, Hypoth I in Fig. 15.1b) can also be predicted based on related experiments using a gap detection task (Yeshurun and Levy, 2003), where a larger gap was required at attended locations for accurate detection. That is, stimuli presented at attended locations evidenced poorer temporal precision than stimuli presented at unattended locations. Clearly, this question requires further research to better understand the relation between attention and temporal precision.

IV. SPATIAL CONFOUND IN MULTISENSORY PRIOR ENTRY RESEARCH

Much of the early work on *multisensory* prior entry presented stimuli to different sensory modalities from different spatial locations: typically, visual stimuli from a light source placed directly in front of the observer, auditory stimuli over headphones, and tactile stimuli from a stimulator placed on the forearm. To investigate the phenomenon of prior entry, an explicit (e.g., Stone, 1926) or implicit (e.g., Sternberg et al., 1971) instruction to attend to one or the other modality was also used. Such instructions may indeed have prompted observers to attend to the required modality. However, as Spence et al. (2001) point out, it is equally possible that the observers may simply have attended to the *location* from which the to-be-attended stimulus was presented. While a demonstration of prior entry under such conditions would still support an effect of attention on onset detection (i.e., a *spatial* prior entry effect), it would not necessarily support the existence of multisensory prior entry (i.e., whether attending to a particular sensory modality speeds the relative perception of stimuli in that modality).

Spence et al. (2001) controlled for this potential confound by presenting the visual and tactile stimuli in their study from either the same or different spatial locations and asking their observers to judge the relative onsets by making either an unspeeded "Which side (left vs. right) came first?" or a "Which modality (vision or touch) came first?" TOJ response. As half of the trials contained stimuli from the same location (and the remainder contained stimuli from different locations), it was possible to assess the spatial confound experimentally. A robust prior entry effect was observed (see Fig. 15.2); however, this was in the context of a clear modulatory effect of spatial location on TOJ performance (see also Spence et al., 2003).

V. RESPONSE BIAS CONFOUND

A confound that has proved more difficult to eliminate convincingly relates to the potential influence of response biases on TOJ performance. The pattern of data providing support for prior entry (see Fig. 15.1) could also reflect the influence of response biases on an observer's judgments. According to Frey (1990; see also Pashler, 1998), observers in many prior entry experiments may simply be biased to report the event to which they were instructed to attend as having the requisite quality (i.e., as coming first if instructed to report which stimulus that was presented first, or as coming second if instructed to report which stimulus was presented second). Evidence supporting the potential influence of response biases on TOJ performance comes from studies in which observers were instructed either to respond to the stimulus that they perceived first or to respond to the stimulus they perceived second (Frey, 1990; Shore et al., 2001). While these two versions of the task are, in principle, identical, Frey found a significant prior entry effect (i.e., rating the attended stimulus as having been presented earlier in time than the unattended stimulus) only when participants were instructed to respond to the stimulus they perceived first. When instructed to respond to the stimulus that came second, observers showed a highly significant effect in the reverse direction. Frey's results therefore provide a particularly dramatic demonstration of how seemingly minor differences in the way in which prior entry experiments are conducted can have dramatic consequences on the pattern of results obtained.

An *orthogonal cuing* methodology was introduced by Spence et al. (2001; Shore et al., 2001) in an attempt to reduce the contribution of response biases to TOJ performance. Participants in their prior entry experiments were required to report a stimulus attribute that was independent of the direction in which attention was manipulated. For example, judging which of two lines—horizontal or vertical—was presented first while attending to the left or right side of space (Shore et al., 2001). One cannot simply say "right" (or prepare a "right" response) when attending to the right given a horizontal or vertical response is required. Another crucial modification over previous studies was to embed the attentional manipulation within the TOJ task itself (given fears that previous null results in certain prior entry studies may have simply reflected a failure to manipulate attention effectively; see Spence et al., 2001).

The use of an orthogonal cuing design to investigate the phenomenon of prior entry should reduce the influence of any response biases on performance, because participants could not prepare the response associated with the attended stimulus in advance of its presentation (as they could have done in Frey's, 1990, study, and in the vast majority of other previous prior entry studies; see Shore et al., 2001, for a brief review). However, it can be argued that participants may still exhibit some form of second-order response bias whereby they simply use the attentional instruction to guide their selection of the appropriate response once the two stimuli have been presented (Schneider and Bavelier, 2003).

Shore et al. (2001) attempted to address this potential concern in their study of visual prior entry by comparing performance in two different orthogonal prior entry tasks, one involving a "Which came first?" task and the other a "Which came second?" task. All other aspects of the stimuli and design were identical for the two experiments. Shore et al. (2001) argued that in the "Which came second?" task, any residual response bias should work against any perceptual prior entry effect and thus reduce, or even reverse, the shift in PSS observed for stimuli on the "attended" side (cf. Frey, 1990). Consider that in the "Which came second?" task, reporting the item on the attended side would result in the proportion of "attended first" responses decreasing, producing a shift in the PSS opposite to that predicted by prior entry (see Shore et al., 2001, for a detailed description of this logic).

Shore et al. (2001) found that the prior entry effects following both the endogenous and exogenous cuing of spatial attention were significantly smaller in the "Which came second?" task than in the "Which came first?" task, consistent with there being a significant residual response bias contribution to performance, even in the orthogonal version of the prior entry task (see Fig. 15.3). The contribution of response bias was, however, much smaller than that reported in Frey's (1990) *non-orthogonal* design, thus supporting Shore and colleagues' (2001) claim that the use of an orthogonal cuing design provides an effective means of reducing the influence of response bias in the study of prior entry (see also Zampini et al., 2003). Importantly though, there was still a significant residual effect of the "attentional" manipulation on performance, once any effects of response bias had been reduced, thus demonstrating that prior entry can genuinely speed up the *perceptual* processing of attended stimuli.

Schneider and Bavelier (2003) have also explored the issue of response bias in prior entry research using a variety of cues including centrally presented deviated eye gaze cues and multiple exogenous spatially distributed cues. Their evidence for exogenous spatial cuing effects replicates the earlier results of Shore et al. (2001; see also Lupiáñez et al., 1999), whereas for endogenous spatial cues, they found a shift in PSS that they attributed solely to response bias. It should, however, be noted that Schneider and Bavelier (2003) failed to provide any independent measure of the success of their endogenous attentional manipulation. In other words, the null effect of attention from their endogenous experiment may reflect the authors'

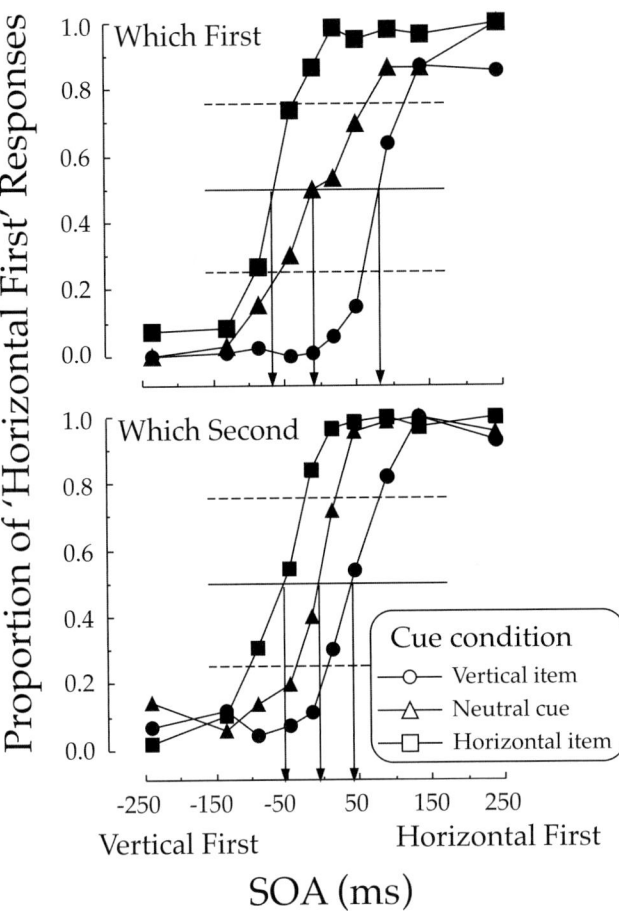

FIGURE 15.3 Average psychometric data from three observers performing a "Which orientation came first or second?" TOJ task. Attention was directed to the left or right with an exogenous cue—the thickening of the box surrounding the possible target locations. The observed shift in PSS was smaller for the "Which came second?" task than for the "Which came first?" task, confirming some residual response bias in this orthogonal cuing design (see text for further details). Reprinted from Fig. 2 of Shore, D. I., Spence, C., and Klein, R. M. (2001). Visual prior entry. *Psychol. Sci.* **12**, 205–212, with permission of the American Psychological Society.

failure to manipulate endogenous spatial attention effectively (this may be particularly problematic in TOJ tasks, which participants often find very difficult) rather than the absence of a perceptual component to the effect reported in their study. This failing makes their null results for the endogenous version of the prior entry task less than convincing. Indeed, Spence et al. (2001) had previously forwarded a similar criticism of all previous studies of the prior entry phenomenon where a null effect of prior entry had been obtained.

While the cuing effect in Shore and colleagues' (2001) intramodal visual cuing study was certainly quite small (approximately 12 ms), much larger spatial cuing effects were reported in Spence and colleagues' (2001) multisensory prior entry study (up to 120 ms when attention was directed to one sensory modality vs another) using a blocked, probability-based endogenous cue. Solely on the basis of Spence and colleagues' (2001) results, one might argue that such large effects probably reflect the combined influence of perceptual and decisional (i.e., response bias) to the PSS shifts reported, given Spence and colleagues' (2001) use of only a "Which came first?" task. Crucially, however, similar results have also been reported in subsequent work by Zampini et al. (2003) in an audio-visual version of the task, no matter whether a "Which came first?" or a "Which came second?" task was used. Given the ongoing debate in this area, it seems likely that it will prove difficult (if not impossible) to satisfy everyone that the influence of response biases has been completely removed in any given prior entry experiment. Part of the problem here may relate to the fact that virtually all published studies of prior entry have used the basic two-alternative forced-choice version of the TOJ task (though see Stone, 1926, for a ternary response version of the task in which participants were also allowed to make a simultaneous response). Future research may therefore have to develop new paradigms for studying the phenomenon of prior entry that are more resistant to the influence of response biases (cf. Pashler, 1998). One task that a number of researchers believe may successfully eliminate concerns over response bias interpretations of any prior entry effects is the simultaneous/successive judgment task (cf. Schneider and Bavelier, 2003; Zampini et al., 2003). Consider that with this task, any shift in perception would result in reporting of physically simultaneous stimuli as successive, while certain pairs of successive stimuli would actually be perceived as simultaneous. Any bias to make a simultaneous or successive response would be independent of cuing one or the other stimulus to be judged. In this regard, it is interesting to note that Zampini et al. (2003) demonstrated a small, but significant, prior entry effect for pairs of auditory and visual stimuli when the participants' attention was endogenously directed to one sensory modality or the other (mean prior entry effect of 14 ms).

VI. COGNITIVE NEUROSCIENCE APPROACHES TO THE STUDY OF PRIOR ENTRY

Event-related potentials (ERPs) may offer another way to tease apart the relative influence of response bias versus sensory contributions to the prior entry effect. Given the precise temporal resolution of ERP, it might be possible to see any sensory facilitation evidenced as a shift in peak latency for target stimuli when they are attended as compared with when another stimulus is attended. McDonald et al. (2003) recently presented some preliminary electrophysiological data using an auditory exogenous peripheral cue in a visual TOJ task ("Which light came first (red or green)?"). Interestingly, they found a reliable PSS shift (as had Lupiánez et al., 1999, in earlier behavioral work), but no latency shift, in the early sensory components of the ERP. That is, N1 and P1 were found at the same latency for attended and unattended stimuli, while there was the classic increase in amplitude observed in many previous visual attention studies. These early evoked potentials presumably originate from extrastriate cortex and are assumed to represent the initial volley of neural input into these cortical structures. McDonald et al. did not present data from the later waveforms or from nonsimultaneous conditions. However, at first glance, their results argue that the first neural volley into the cortex does not show any latency shift.

It may be possible that examination of the N2pc or P300 in future studies may reveal reliable shifts indicative of faster reentrant connections to these early processing areas. That is, attention may not speed the very first representation of the attended stimulus, but may instead facilitate later processing of that stimulus in both a feedfoward and feedback (i.e., reentrant) manner. Alternatively, of course, it is also possible that attentional effects may only be seen in later, response-related, ERP components. However, it is important to note that such a pattern of results would not conclusively argue against any perceptual facilitation of stimulus processing due to prior entry, given that it is by no means certain that the perception of temporal order is necessarily related to the relative timing of different sensory signals in the brain (i.e., temporal information may not be coded temporally in the brain).

VII. CONCLUSIONS

Much progress has been made in the last few years regarding our understanding of the phenomenon of prior entry. Successful attempts have been made to reduce, if not to eliminate entirely, the influence of response biases and other confounding factors in prior entry research (see Pashler, 1998; Schneider and Bavelier, 2003; Shore et al., 2001; Spence et al., 2001). Robust psychophysical paradigms now provide good evidence that prior entry effects can be elicited by the direction of attention to a particular sensory modality (Spence et al., 2001; Zampini et al., 2003) or spatial location (Schneider and Bavelier, 2003; Shore et al., 2001; Spence et al., 2001). While the majority of empirical research has tended to focus on the consequences of the endogenous (or voluntary) direction of attention on perception, a growing number of studies have now started to focus on the effects of exogenous (or stimulus-driven) attention effects as well (e.g., Lupiánez et al., 1999; McDonald et al., 2003; Schneider and Bavelier, 2003; Shore et al., 2001). Further progress in this area will come through the development and use of response measures that are more resistant to response bias interpretations of any prior entry effects observed (such as the simultaneous/successive response task; Schneider and Bavelier, 2003; Zampini et al., 2003). Furthermore, it is hoped that the utilization of cognitive neuroscience techniques will begin to reveal the neural underpinnings, and consequences of, the phenomenon of prior entry (e.g., McDonald et al., 2003). Modeling approaches (e.g., Schneider and Bavelier, 2003; Sternberg et al., 1971) may also become increasingly important as tools in helping us to tease apart the sensory and response-related aspects of the prior entry effect.

Acknowledgments

Both authors were funded by a Network Grant from the McDonnell–Pew Centre for Cognitive Neuroscience, University of Oxford. D.I.S. was also funded by an operating grant from the Natural Science and Engineering Council of Canada.

References

Coren, S., Ward, L. M., and Enns, J. T. (1999). "Sensation, & Perception," 5th ed. Harcourt Brace, New York.

Frey, R. D. 1990. Selective attention, event perception and the criterion of acceptability principle: evidence supporting and rejecting the doctrine of prior entry. *Hum. Move. Sci.* **9**, 481–530.

Lupiánez, J., Baddeley, R., and Spence, C. (1999). Crossmodal links in exogenous spatial attention revealed by the orthogonal temporal order judgment task. Paper presented at the 1st International Multisensory Research Conference: Crossmodal Attention and Multisensory Integration. Oxford 1st–2nd October. www.imrf.info/1999/34.

McDonald, J. J., Teder-Sälejärvi, W. A., Di Russo, F., and Hillyard, S. A. (2003). Involuntary attention to sound modulates visual temporal perception: an electrophysiological study. Paper presented at the 4th Annual International Multisensory Research Forum meeting at McMaster University, CA, 14–17 June. www.imrf.info/2003/136

Mollon, J. D., and Perkins, A. J. (1996). Errors of judgement at Greenwich in 1796. *Nature* **380**, 101–102.

Pashler, H. E. (1998). "The Psychology of Attention." MIT Press, Cambridge, MA.

Schneider, K. A., and Bavelier, D. (2003). Components of visual prior entry. *Cogn. Psychol.* **47**, 333–366.

Shore, D. I., Spence, C., and Klein, R. M. (2001). Visual prior entry. *Psychol. Sci.* **12**, 205–212.

Spence, C., Baddeley, R., Zampini, M., James, R., and Shore, D. I. (2003). Crossmodal temporal order judgments: when two locations are better than one. *Percept. Psychophys.* **65**, 318–328.

Spence, C., Shore, D. I., and Klein, R. M. (2001). Multisensory prior entry. *J. Exp. Psychol. Gen.* **130**, 799–832.

Stelmach, L. B., and Herdman, C. M. (1991). Directed attention and perception of temporal order. *J. Exp. Psychol. Hum. Percept. Perform.* **17**, 539–550.

Sternberg, S., Knoll, R. L., and Gates, B. A. (1971, November). Prior entry reexamined: effect of attentional bias on order perception. Paper presented at the meeting of the Psychonomic Society, St. Louis, MO.

Stone, S. A. (1926). Prior entry in the auditory-tactual complication. *Am. J. Psychol.* **37**, 284–287.

Titchener, E. B. (1908). "Lectures on the Elementary Psychology of Feeling and Attention." Macmillan, New York.

Yeshurun, Y., and Levy, L. (2003). Transient spatial attention degrades temporal resolution. *Psychol. Sci.* **14**, 225–231.

Zampini, M., Shore, D. I., and Spence, C. (2003). Audiovisual prior entry. Paper presented at the 4th Annual International Multisensory Research Forum meeting at McMaster University, CA, 14–17 June. www.imrf.info/2003/121

CHAPTER 16

Inhibition of Return

Raymond M. Klein and Jason Ivanoff

ABSTRACT

Using a simple model task, M. I. Posner and Y. Cohen (1984, "Attention & Performance X," pp. 531–556, Erlbaum, Hillsdale, NJ) discovered a reaction time delay following the removal of attention from a peripherally cued location that was subsequently labeled *inhibition of return* (IOR). IOR has since been well-studied in a variety of populations (e.g., normal and brain-damaged adults, infants, children, and nonhuman primates) by scientists using a variety of stimulus–response properties (e.g., visual, auditory, and tactile stimuli; static and moving displays; manual and oculomotor responses) combined with primarily behavioral or neuroscientific methods. This research strongly supports Posner and Cohen's proposal that the function of IOR is to encourage orienting toward novel components of a scene and away from already inspected components and has begun to reveal the neural mechanisms underlying IOR.

I. MODEL TASK FOR EXPLORING IOR

In a typical implementation of the model task (Fig. 16.1a) pioneered by Posner and Cohen (1984), an observer views three boxes: one at the center of the display and one on either side of fixation. A peripheral box brightens briefly, and after various delays, a target appears unpredictably (i.e., with equal probability) in one of the peripheral boxes. When a target is presented in close *temporal* proximity to the onset of the cue, the time (i.e., response time, RT) to detect, identify, or locate targets at the cued, relative to the uncued, location is usually speeded. At longer intervals, the cued and uncued performance functions cross over such that targets presented at the originally facilitated location appear to be inhibited (see Fig. 16.1b): Responding to cued targets is slower than responding to uncued targets. In Posner and Cohen's (1984) implementation of this model task, one of two procedures was used to ensure that attention did not remain at the originally cued location, but, instead, returned to fixation (or some other neutral state). First, in some experiments, attention was reflexively returned to the fixation box by brightening it after the peripheral cue. Second, in other experiments, targets were presented at fixation with higher probability than targets in the peripheral locations. It is usually not necessary to directly encourage the removal of attention from the cued location when testing normal adults who seem to spontaneously disengage attention from the cue. But, in groups with likely deficits in the anterior attention system (young children, the elderly, and schizophrenics), IOR may not be observed, or may be delayed, if one of these methods is not used (for a review, see Klein, in press). In contrast to the cuing paradigm illustrated in Fig. 16.1a, some investigators have explored IOR following responses to relevant targets (target–target paradigm) or during active search (see below).

II. BEHAVIORAL PROPERTIES OF IOR

Behavioral studies using variations on the model task illustrated above reveal that the appearance of IOR is delayed as task difficulty increases (see Klein, 2000, for a review) and that removing attention from the cued location quickly (e.g., Danziger and Kingstone, 1999) speeds the appearance of IOR. At a neural level of analysis, however, inhibition (as reflected in reduced sensory responses of neurons in the superior colliculus) seems to begin with the presentation of the cue (Klein, 2004) (see Chapter 64). It is suggested that IOR may not be observed behaviorally when it is overshadowed by strong facilitation associated with the capture of attention by the cue.

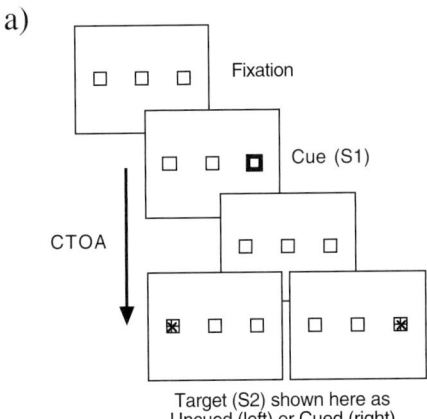

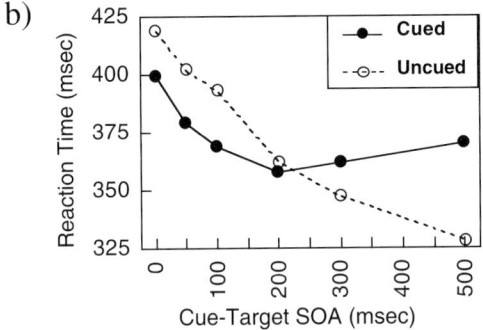

FIGURE 16.1 Inhibition of return: model task and prototypical data. (a) Sequence of events in a typical trial of the model task described in the test. A fixation display is followed by the first stimulus (S1, cue) shown here as the brightening of one of the two peripheral boxes. After varying intervals (cue–target onset asynchronies, CTOAs) from the onset of the cue a target (S2), shown here as an asterisk, is presented at the cued or uncued location. The observer's task is to make a speeded detection response as soon as the asterisk is detected. (b) Reaction time data from Posner and Cohen (1984) showing the crossover from facilition at the cued location using short CTOAs to inhibition at this location when the CTOA is longer.

Inhibition of return is not observed following the endogenous, or voluntary, allocation of attention (e.g., by means of a central arrow cue at fixation), but is observed following a peripheral cue or an endogenously prepared saccade (Posner and Cohen, 1984; Rafal et al., 1989). Together with the fact that IOR is not found following damage to midbrain structures responsible for saccadic eye movements (see below), it is generally accepted that the critical condition for *generating* IOR is activation of the oculomotor system (Taylor and Klein, 1998). Although IOR inhibits spatial orienting toward the inhibited locus or object, whether this is accomplished primarily through perceptual/attentional (Handy et al., 1999) or motoric (Ivanoff and Klein, 2001) channels appears to depend on whether eye movements are actually executed (Taylor and Klein, 2000).

IOR decreases monotonically with distance from the cued location (Bennett and Pratt, 2001; Maylor and Hockey, 1985) and may be coded in an environmental (Maylor and Hockey, 1985; Posner and Cohen, 1984) or object-based (e.g., Tipper et al., 1994) frame of reference. When an eye movement intervenes between a cue and target, environmentally coded IOR is reflected in slower performance for targets at the originally cued location, which now falls on a different region of the retina. When a cued object subsequently moves, object-coded IOR is seen in slowed responding to targets presented in the vicinity of the originally cued object, now in a new location. Not limited to the visual modality, IOR has been obtained between all pairings of vision, touch, and audition (Spence et al., 2000). Once generated, IOR can last several seconds (Samuel and Kat, 2003), and can be maintained at multiple locations as attention is exogenously shifted from one object or location to another (Snyder and Kingstone, 2000). Together, these properties make IOR highly suitable as a tagging system that could serve to facilitate foraging.

III. IOR FUNCTIONS AS A FORAGING FACILITATOR

Klein (1988) tested this proposal, with positive results, by showing that IOR operates during difficult, but not easy (popout) visual search (Fig. 16.2a). In follow-up studies (Fig. 16.2b) Klein and MacInnes (1999) monitored eye movements while participants searched complex scenes (from Martin Handford's "Where's Waldo" books). When probe targets interrupted search or were presented after the participant had temporarily stopped searching (MacInnes and Klein, 2003), saccadic reaction time to acquire the probe was a decreasing function of the probe's distance from a recently fixated region of the scene. Supporting the suggestion (cf. Tipper et al., 1994) that IOR is coded in scene coordinates when there is a rich scene, no evidence for IOR was obtained if the scene was removed when the probe was presented. Converging evidence for inhibition of reinspections during search comes from studies in which the experimenter guides oculomotor behavior by presenting a sequence of targets (Fig. 16.2c). When the observer is given a choice between targets, one presented at a new location and one presented at an old location, there is a strong bias to inspect the target at the new location (McCarley et al., 2003).

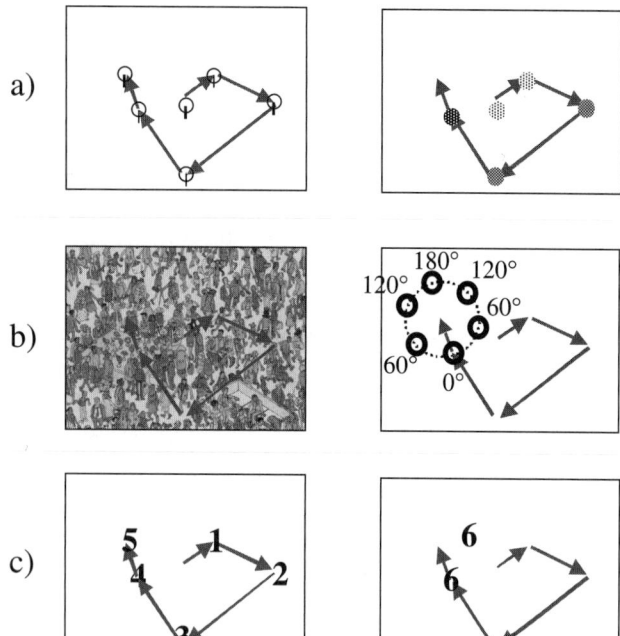

FIGURE 16.2 In three different tasks a sequence of orienting events (arrows) was followed by a probe providing evidence that orienting leaves inhibitory tags that might facilitate foraging behavior. (a) Klein (1988; see Klein, 2000, for a review) presented luminance detection probes immediately after subjects had performed a serial (illustrated here: O, target, not shown; Q, distracter) or popout search task. As indicated by the arrows, most distracters following a serial (but not following a popout) search array should have been "visited" by attention. Actual eye movements may or may not have accompanied these shifts of attention. Detection probes were presented at the locations of display items or in previously empty locations. Following serial search, probe RT was slowed for probes at previous distracter locations, whereas following popout search there was no effect of this factor. (b) In Klein and MacInnes (1999) and MacInnes and Klein (2003), the observer used eye movements to search a complex scene (a picture from the "Where's Waldo" series of books). After several saccades were made (arrows), a probe target (black disk) appeared at one of six randomly selected locations that were equi-eccentric at the time of probe presentation, one of which (marked 0°) had been previously fixated during the search episode (shown here are the six possible one-back probes), while the search array remained or was removed. Whether participants were searching for "Waldo" and the probe interrupted their search (Klein and MacInnes, 1999) or they were searching for something interesting and had stopped searching (MacInnes and Klein, 2003), saccadic reaction time increased with increases in the target's proximity to a previously fixated region, but only when the probe was presented while the scene was displayed. When the scene was removed (as in this example), saccadic reaction time was unaffected by the angular difference between the probe and a previous fixation. (c) McCarley et al. (2003) presented a sequence of targets, one at a time, in a task requiring the observer to foveate each target. The numbers here are meant to convey that the items presented in these locations were presented in sequence. During each saccade the target for the next saccade was presented. Occasionally, the observer was given a choice (the two 6's) between a new item and an old one (that remained displayed or reappeared). So long as the old item was not too old, most of the saccades made on these occasions were directed to the new item.

Although the foraging facilitation proposal might be questioned (see Chapter 43) on the grounds that IOR is too slow (as seen in Fig. 16.1) to inhibit the rapid shifts of attention (say every 50 ms) implied by the slope in serial search tasks, the time course often seen in the "model" paradigm is not necessarily pertinent. As noted above, IOR is present in the nervous system (Klein, 2004) within 50 ms of an uninformative peripheral cue and can be seen in behavior (Danziger and Kingstone, 1999) within 50 ms of a cue that elicits a rapid switch of attention (as in search). While discouraging reinspections, IOR is unlikely to completely prevent them because the number of tags may be limited and tags begin to decay after several seconds. Hence, evidence that some reinspections may occur in a difficult search task should not be asserted as evidence against a foraging facilitator role for IOR.

IV. NEURAL IMPLEMENTATION OF IOR

Several findings point to oculomotor circuitry in the superior colliculus as critical for generating IOR. For example, IOR has been observed in newborns (see Chapter 14) for whom the superior colliculus, but not cortical circuitry, is relatively fully operational. Morover, patients with damage to the superior colliculus show abnormal patterns of inhibition (Sapir et al., 1999). Yet, because the superior colliculus represents space in oculocentric coordinates, any inhibitory tags generated by it would need to be processed by cortical structures able to preserve them as the eyes or the previously attended objects move through space. Cortical involvement in object-based IOR has been demonstrated in patients whose cerebral hemispheres have been surgically separated to control severe epilepsy. They show object-based IOR within each visual field (hemisphere), but not between them (Tipper et al., 1997). Circuitry in the right intraparietal sulcus, known to mediate remapping of space in anticipation of saccades, has recently been shown to be critical for maintaining inhibitory tags in environmental coordinates when the eyes move within a scene (Sapir et al., 2004).

Among the various techniques for revealing neural activity, fruitful data on the neural correlates of IOR have been provided by ERP and single-unit recording studies. When targets are presented a half a second or more after a cue, an early ERP component, P1, thought to reflect extraction of sensory information, is reduced at the cued location (see Fig. 16.3). This reduction, also seen as an increased negativity for cued targets, extends from about 125 ms to about 250 ms. As described by Munoz and colleagues in Chapter 64, IOR

IV. NEURAL IMPLEMENTATION OF IOR

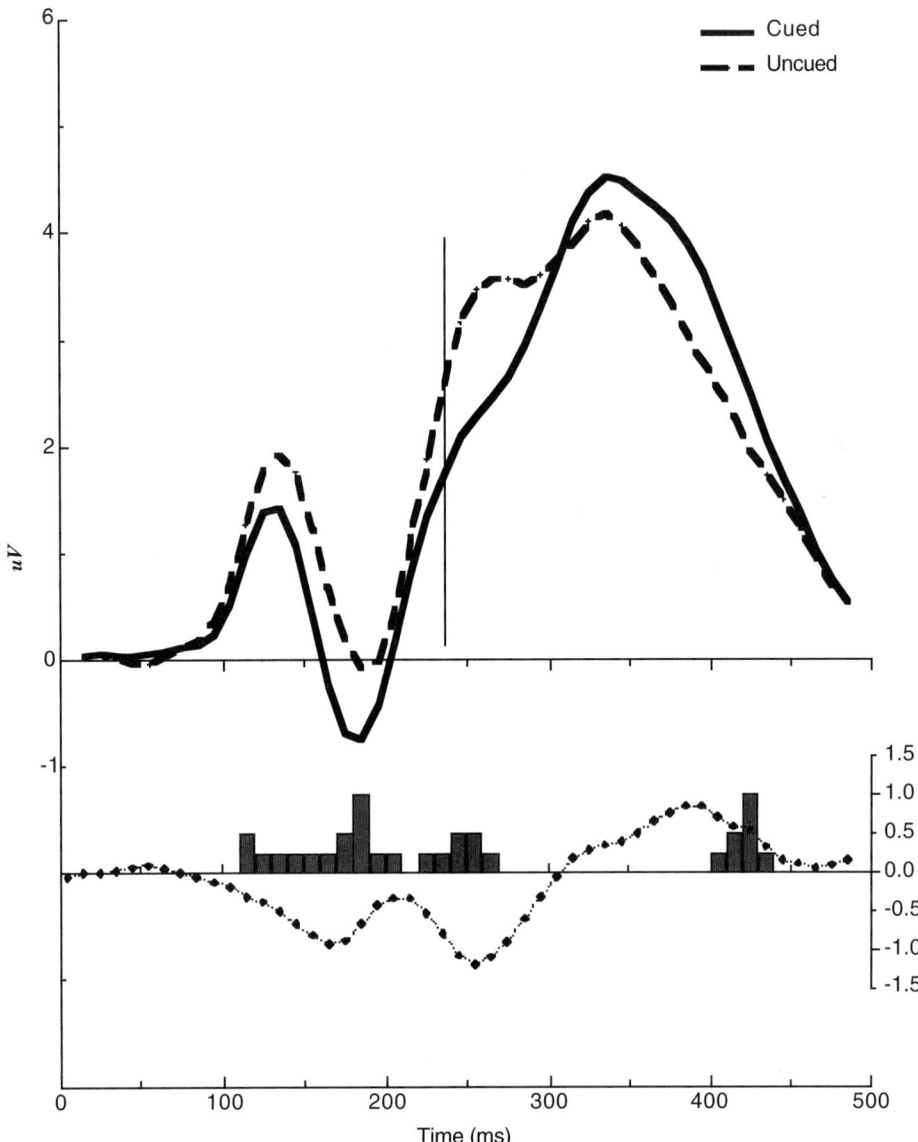

FIGURE 16.3 The data from the occipitally recorded ERP waveforms from six conditions of several studies were extracted from the original figures. When separated by hemifield of target and/or hemisphere of recording electrode, collapsed to yield two waveforms for each experiment: one for cued targets and one for uncued targets. Time 0 signals the delivery of the target. The data in the upper plot show the grand average of these waveforms (with each experiment weighted equally). The vertical line shows the point in time beyond which the data are based on only five data sets (because one study did not report ERPs beyond this point). The curve in the lower plot is the difference waveform (cued minus uncued, in microvolts) and the histogram reveals where, based on t tests with each study contributing one pair of values, the difference is significant. The heights of the bars represent P values of 0.01 (large bar), 0.025 (medium bar), and 0.05 (small bar). Note that for the waveforms positive is plotted upward. Redrawn, with permission, from Klein (in press-b).

is accompanied by reduced sensory responses of single neurons in the intermediate layers of the superior colliculus. Importantly, when electrical stimulation was delivered through the recording electrode to elicit a saccade, the latency of these electrically evoked saccades was actually faster when the cued region was stimulated. This finding demonstrates that following a cue, the cued region of the superior colliculus is not itself inhibited but, rather, is receiving inputs whose sensory responses are already reduced, possibly by circuitry in the parietal cortex (Sapir et al., 2004) or frontal eye fields.

References

Bennett, P. J., and Pratt, J. (2001). The spatial distribution of inhibition of return. *Psychological Science*, **12**, 76–80.

Danziger, S., and Kingstone, A. (1999). Unmasking the inhibition of return phenomenon. *Perception and Psychophysics*, **61**, 1024–1037.

Handy, T. C., Jha, A. P., and Mangun, G. R. (1999). Promoting novelty in vision: Inhibition of return modulates perceptual-level processing. *Psychological Science*, **10**, 157–161.

Ivanoff, J., and Klein, R. M. (2001). The presence of a nonresponding effector increases inhibition of return. *Psychon. Bull. Rev.*, **8**, 307–314.

Klein, R. M. (1988). Inhibitory tagging system facilitates visual search. *Nature* **334**, 430–431.

Klein, R. M. (2000). Inhibition of return. *Trends Cogn. Sci.* **4**, 138–147.

Klein, R. M. (in press) Volitional disengagement of attention and the timecourse of inhibition of return. *In* "Developing Individuality in the Human Brain: A Feschrift for Michael Posner" (U. Mayr, E. Awh, and S. Keele, Eds.). APA Books, Washington, DC.

Klein, R. M. (2004) Orienting of attention and inhibition of return. *In* "Handbook of Cognitive Neuroscience," (M. Gazzaniga, Ed.). pp. 545–559. 3rd ed. MIT Press, Cambridge, MA.

Klein, R. M., and MacInnes, W. J. (1999). Inhibition of return is a foraging facilitator in visual search. *Psychol. Sci.* **10**, 346–352.

MacInnes, W. J., and Klein, R. M. (2003). Inhibition of return biases orienting during the search of complex scenes. *Sci. World J.* **3**, 75–86.

Maylor, E. A., and Hockey, R. (1985). Inhibitory component of externally controlled covert orienting in visual space. *J. Exp. Psychol. Hum. Percept. Perform.* **11**, 777–787.

McCarley, J. S., Wang, R. X. F., Kramer, A. F., Irwin, D. E., and Peterson, M. S. (2003). How much memory does oculomotor search have? *Psychol. Sci.* **14**, 422–426.

Posner, M. I., and Cohen, Y. (1984). Components of visual orienting. *In* "Attention & Performance X." (H. Bouma and D. G. Bouwhuis, Eds.), pp. 531–556. Erlbaum, Hillsdale, NJ.

Rafal, R. D., Calabresi, P. A., Brennan, C. W., and Sciolto, T. K. (1989). Saccade preparation inhibits reorienting to recently attended locations. *J. Exp. Psycho. Hum. Percept. Perform.* **15**, 673–685.

Samuel, A. G., and Kat, D. (2003). Inhibition of return: a graphical meta-analysis of its timecourse, and an empirical test of its temporal and spatial properties. *Psychon. Bull. Rev.* **10**, 897–906.

Sapir, A., Soroker, N., Berger, A., and Henik, A. (1999). Inhibition of return in spatial attention: direct evidence for collicular generation. *Nat. Neurosci.* **2**, 1053–1054.

Sapir, A., Hayes, A., Henik, A., Danziger, S., and Rafal, R. (2004). Parietal lobe lesions disrupt saccadic remapping of inhibitory location tagging. *J. Cogn. Neurosci.* **16**, 503–509.

Snyder, J. J., and Kingstone, A. (2000). Inhibition of return and visual search: how many separate loci are inhibited. *Percept. Psychophys.* **62**, 452–458.

Spence, C., Lloyd, D., McGlone, F., Nicholls, M. E. R., and Driver, J. (2000). Inhibition of return is supramodal: a demonstration between all possible pairings of vision, touch, and audition. *Exp. Brain Res.* **134**, 42–48.

Taylor, T. L., and Klein. R. M. (1998). On the causes and effects of inhibition of return. *Psychon. Bull. Rev.* **5**, 625–643.

Taylor, T. L., and Klein, R. M. (2000). Visual and motor effects in inhibition of return. *Journal of Exp. Psychol. Hum. Percept. Perform.* **26**, 1639–1656.

Tipper, S. P., Rafal, R., Reuter-Lorenz, P. A., Starrveldt, Y., Ro, T., Egly, R., Danziger, S., and Weaver, B. (1997). Object based facilitation and inhibition from visual orienting in the human split brain. *J. Exp. Psycho. Hum. Percept. Perform.* **23**, 1522–1532.

Tipper, S. P., Weaver, B., Jerreat, L. M., and Burak, A. L. (1994). Object-based and environment-based inhibition of return of visual attention. *J. Exp. Psychol. Hum. Percept. Perform.* **20**, 478–499.

CHAPTER 17

Guidance of Visual Search by Preattentive Information

Jeremy M. Wolfe

ABSTRACT

When searching for one object among distracting objects, observers make use of a limited set of stimulus attributes to guide the deployment of their attention. Thus, if asked to search for ripe strawberries, an observer will use color information to guide covert attention as well as overt movements of eye and hand. While color is an unambiguous source of such guidance, there is no universally agreed list of all such sources. This chapter reviews the methods for determining if a property can guide the deployment of attention. Candidate sources of guidance are listed found in a table.

I. INTRODUCTION

Many aspects of visual processing operate in parallel. The retina, to offer an obvious example, handles input from the entire visual field at one time. Early vision continues this parallel processing. Much farther along in the stream of processing, recognizing an object must involve comparing the visible object with many stored representations in parallel. It is implausible to suppose that the visible object could be matched to all of the possible stored representations in series. However, between the early reception of visual input and the later recognition of visible objects, there is a chokepoint (perhaps more than one). We cannot recognize all of the objects in the visual field at once. At any moment in time, a limited subset of the initial parallel processing of the visual input, perhaps a single object, must be selected for delivery to parallel object recognition processes. That act of selection is not random. Information that is made available by initial parallel processing can be used to control selection. Thus, if we are looking for our red car in the parking lot, we select red cars to examine, not blue cars.

If we accept this set of propositions, several terms can be defined. Visual attention can be said to be *guided* by features (e.g., "red") of the visual stimulus. Those features can be said to be *preattentive* in the sense that the visual information about the feature must have become available before the act of selection to guide that selection. It is useful to reserve the term preattentive *features* for specific values (e.g., "red") on a preattentive *dimension* (in this case, "color"). Two types of guidance can be distinguished. If an item differs from its neighbors in a preattentive dimension, that difference, itself, can attract attention in a *"bottom-up,"* stimulus-driven manner. If the observer is told to look for targets having a specific preattentive feature, attention can be guided to items having that feature in a *"top-down"* manner. Only a limited amount of visual information is available preattentively to guide selective attention. This article briefly reviews the nature of that preattentive guidance.

II. LOOKING FOR PREATTENTIVE FEATURES

Preattentive features can be identified by several experimental tests. No one test is perfectly definitive. Ideally, a series of converging operations support the designation of a feature, though, in practice, quite a few features and dimensions are on the list on the basis of rather few data. Much evidence comes from visual search tasks where an observer looks for a target item among a number of distracting items. Typically, reaction time (RT) and/or accuracy are measured as a function of the set size (the number of items). If the target is defined by a unique preattentive feature, then the

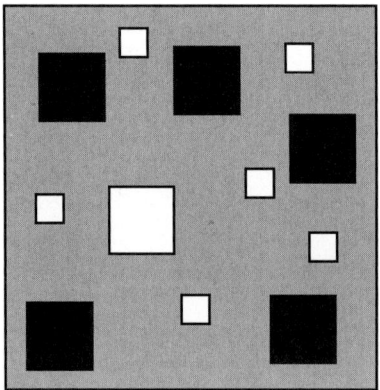

FIGURE 17.1 Find the big white target.

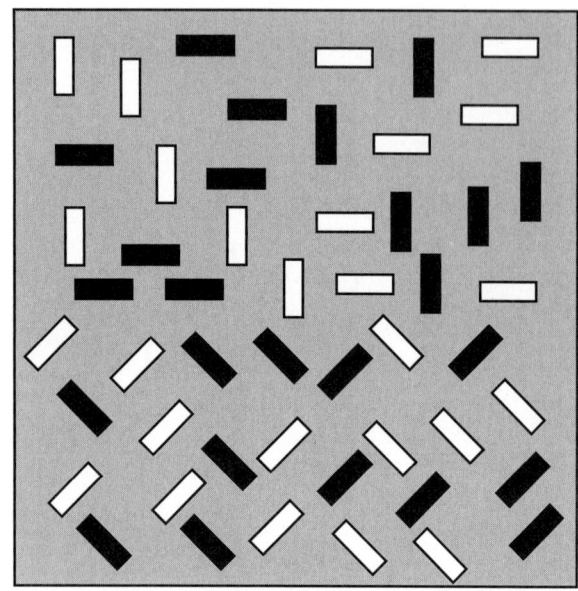

FIGURE 17.2 Where are the texture boundaries?

slope of these functions should be near zero. That is, attention should be guided to the target item with minimal interference from distracting items. Thus, search such as red among green, vertical among horizontal, moving among stationary, and big among small will typically produce RT × set size slopes near zero, indicating that color, orientation, motion, and size are dimensions that are available to guide attention (Treisman, 1988; Treisman and Gelade, 1980).

RT × set size slopes near zero are not definitive marks of a basic feature because other search tasks can produce such slopes. For example, a search for a conjunction of two features can produce highly efficient slopes if each of the two features is sufficiently salient. In Fig. 17.1, the big white target is found efficiently, not because big whiteness is a feature but because the attention is guided to white and to big and the intersection of those two sources of guidance is a good marker for the likely location of big white items (Wolfe, 1994; Wolfe et al., 1989).

What other clues to preattentive status are available if slopes alone are inadequate? As Treisman first noted, search asymmetries can be useful (Treisman and Gormican, 1988; Treisman and Souther, 1985). The presence of a feature guides attention better than its absence. For example, it is easier to find a moving stimulus among stationary than a stationary item among moving. Of course, motion is a fairly obvious feature. Search asymmetry comes into its own in ambiguous cases. For example, over a range of experiments, search for the opaque item is more efficient than search for the transparent. This suggests that opacity may be a feature and that, for purposes of preattentive guidance, transparency may merely be the absence of that feature.

In some circumstances, search asymmetries may not be evidence for the presence of a guiding feature. Suppose that a search for an item A among B or B among A is unguided. Items are sampled randomly from the display. If it takes longer to reject B's than it does to reject A's, then search for A among B will be slower than search for B among A. For example, it can be harder to search for a "smile" curve among frowns when the critical curved line is embedded in a schematic face (Suzuki and Cavanagh, 1995). This slowing of search does not have to do with the absence of a feature but with the particularities of processing facelike stimuli.

Texture segmentation is the other most useful converging operation. The difference between texture segmentation and visual search is illustrated in Fig. 17.2. It is very clear that there is a boundary, defined by a change in orientation, dividing the upper and lower parts of the figure. It is much less obvious that there is a boundary, defined by a dramatic change in conjunctive properties, between the left and right halves of the figure. A search for white vertical or black horizontal among black vertical and white horizontal would produce efficient conjunction search (similar to Fig. 17.1), but texture boundaries defined by these conjunctions are not salient (Wolfe, 1992).

III. HOW GUIDANCE WORKS

Although there is little doubt that a property like orientation can be said to be preattentive, this does not mean that any detectable difference in a preattentive dimension can be used to guide attention. The guiding signals are coarse. In orientation, for example, a vertical line will be found very efficiently among horizontal lines but not among lines tilted 5° off vertical, even though that 5° difference is well above detection threshold (Foster and Ward, 1991). Guidance is a form of a signal detection problem. The target feature provides a signal and other aspects of the display generate noise (as does intrinsic noise in the nervous system) (Verghese, 2001). A feature can guide attention if it is available preattentively *and* if it generates a signal of adequate size. This leads to a certain ambiguity. For example, search for opaque surfaces among transparent generates RT x set size functions that are on the order of 8 ms/item for target-present trials. This is much more efficient than the 20- to 30-ms slopes characterizing inefficient, unguided search, but 8 ms is not zero. The suggestion is that this stimulus property generates a relatively weak preattentive signal. Weak guidance might make search more efficient but not reduce RT x set size slopes to zero. Despite efforts to enshrine a magic value such as 10 ms/item, there is no sharp cutoff criterion for slope magnitude that allows us to declare in an unambiguous manner that a guiding feature is or is not present. The set of all search tasks produces a continuum of slopes, not a neat dichotomy between the efficiently guided (or "parallel") and the inefficiently guided or unguided ("serial") (Wolfe, 1998). Faced with weak evidence for a feature, the best remedies are to seek converging evidence (see above) or to design better stimuli.

Because the features that guide attention are properties that we can see and describe, it is tempting to think that these features behave in the same way when used by preattentive guidance as they do when they generate conscious, attended perception. This does not appear to be the case. In addition to being quite coarse, preattentive dimensions may be divided up differently from their perceptual counterparts. For example, orientation guidance may be "categorical." Preattentive orientation processing seems to discriminate between "steep," "shallow," "left," and "right," rather than the more strictly geometrical angular differences that are important in making attended discriminations (Wolfe et al., 1992).

In keeping with the view of guidance as a form of signal detection, increasing the "noise" reduces the efficiency of search in the same manner as decreasing the signal. Thus, increasing the variability of the distracter items decreases the efficiency of search. For example, 0° vertical among 20° distracters is more efficient than 0° among 20° and 40°, even though the 40° distracters are more distinct from the target. This example makes another point. There is nothing "noisy" about 20° 40° items in isolation. It is the local differences between items that seem to be the source

TABLE 17.1 Probable, Possible, and Unlikely Sources of Preattentive Guidance[a]

Probable featural dimensions
 Color
 Luminance onset (flicker)
 Luminance polarity
 Orientation
 Aspect ratio
 Size (including length and spatial frequency)
 Curvature
 Vernier offset
 Motion (possibly divided into speed and direction)
 Stereoscopic depth and tilt
 Pictorial depth cues
 Shape (probably divided into specific features like "convexity")
 Line termination
 Closure/holes (in the topological sense)
 Lighting direction (shading)
 Number
 Opacity
Possible cases
 Novelty
 Letter identity (overlearned sets)
 Shininess (gloss, luster)
 Alphanumeric category
 Threat
Probable nonfeatures
 Intersection
 Faces
 Optic flow
 Color change
 Three-dimensional volumes (e.g., geons)
 Your name

[a] Details and references available from the author.

of noise. Similarly, the "signal" is not the vertical item, in this example, but the local difference between the vertical item and surrounding items. Local differences are greater when items are closer together; as a consequence, the "signal" can increase with set size when the density of items increases with set size. In turn, this can produce negative RT x set size slopes (Bravo and Nakayama, 1992).

IV. WHAT ARE THE PREATTENTIVE FEATURES?

Of course, after the forgoing discussion, the obvious question is "What aspects of the visual stimulus provide preattentive guidance?" This short chapter does not provide the space for a full discussion with all of the several hundred relevant references. Table 17.1 lists "likely," "possible," and "unlikely" sources of guidance.

Many of the items on this list have been the subject of very limited amounts of study, sometimes only a single published paper. Consequently, it would not be a surprise if the list changed. While the details may change, the larger principle seems likely to endure. The front end of the visual system can process vast amounts of information in parallel. Later processes, like those of object recognition, can match visual input to a vast number of stored object representations, in parallel. Because it is not possible to match all of the input to all of the possible objects at the same time, selection mechanisms, guided by a set of preattentive features, govern the transfer of a limited amount of information from one massively parallel stage to the next.

References

Bravo, M., and Nakayama, K. (1992). The role of attention in different visual search tasks. *Percept. Psychophys.* **51**, 465–472.

Foster, D. H., and Ward, P. A. (1991). Horizontal–vertical filters in early vision predict anomalous line-orientation identification frequencies. *Proc. R. Soci. London Ser. B* **243**, 83–86.

Suzuki, S., and Cavanagh, P. (1995). Facial organization blocks access to low-level features: an object inferiority effect. *J. Exp. Psychol. Hum. Percept. Perform.* **21**, 901–913.

Treisman, A. (1988). Features and objects: the 14th Bartlett memorial lecture. *Q. J. Exp. Psychol.* **40A**, 201–237.

Treisman, A., and Gelade, G. (1980). A feature-integration theory of attention. *Cogn. Psychol.* **12**, 97–136.

Treisman, A., and Gormican, S. (1988). Feature analysis in early vision: evidence from search asymmetries. *Psych. Rev.* **95**, 15–48.

Treisman, A., and Souther, J. (1985). Search asymmetry: a diagnostic for preattentive processing of seperable features. *J. Exp. Psychol. Gen.* **114**, 285–310.

Verghese, P. (2001). Visual search and attention: a signal detection approach. *Neuron* **31**, 523–535.

Wolfe, J. M. (1992). "Effortless" texture segmentation and "parallel" visual search are not the same thing. *Vision Res.* **32**, 757–763.

Wolfe, J. M. (1994). Guided Search 2.0: a revised model of visual search. *Psychon. Bull. Rev.* **1**, 202–238.

Wolfe, J. M. (1998). What do 1,000,000 trials tell us about visual search? *Psychol. Sci.* **9**, 33–39.

Wolfe, J. M., Cave, K. R., and Franzel, S. L. (1989). Guided search: an alternative to the feature integration model for visual search. *J. Exp. Psychol. Hum. Percept. Perform.* **15**, 419–433.

Wolfe, J. M., Friedman-Hill, S. R., Stewart, M. I., and O'Connell, K. M. (1992). The role of categorization in visual search for orientation. *J. Exp. Psychol. Hum. Percept. Perform.* **18**, 34–49.

CHAPTER

18

The Top in Top-Down Attention

Chris Frith

ABSTRACT

The term *top-down* in the phrase *top-down control of attention* can be defined at three different levels—physiological, psychological, and phenomenological—but these definitions need not be commensurate with one another. Some effects that are "bottom-up" from a psychological point of view depend on physiological processes that are "top-down." For example, the automatic binding of the features of an attended object depends on reentrant signals arising in parietal cortex. We do not have direct top-down control of attention. Control is not exerted by choosing to attend to one stimulus rather than another, but by choosing to perform one task rather than another. Such control depends on maintaining the stimulus and response priorities defined by the task. This control uses the same processes that are engaged by traditional working memory tasks and depends on signals arising in frontal cortex. However, the "top" of this hierarchy of control is not to be found in an isolated brain. Ultimate control of what a volunteer does depends on a largely implicit mutual agreement with the experimenter.

I. DEFINITIONS OF TOP-DOWN EFFECTS

The distinction between bottom-up and top-down effects continues to be a fundamental guiding principle in accounts of attention (e.g., Corbetta and Schulman, 2002). There are, however, three different levels on which we can define what we mean by top-down. These definitions are not entirely commensurate.

1. At the physiological level the distinction is made in terms of the direction of flow of neural signals. Bottom-up signals go from primary sensory areas to secondary sensory areas and so on. Top-down signals go in the opposite direction and are sometimes referred to as reentrant or feedback signals. Anatomically, the majority of reciprocal connections between cortical areas are symmetric with respect to the cortical layers in which connections originate and terminate. This has led to the hypothesis of an anatomical hierarchy, with some connections representing forward (ascending) pathways and their reciprocal counterparts representing feedback pathways (Felleman and Van Essen, 1991). Top-down effects can be defined in terms of this anatomical hierarchy.

2. At the psychological level, the distinction is made in terms of the nature of the (stimulus processing) task being performed. Bottom-up effects are determined solely by the current stimuli, or, in other words, they are stimulus-driven. If our volunteers passively view a series of stimuli, then only bottom-up effects can operate. Their attention may be captured by a brightly colored flashing stimulus, but this is a bottom-up effect. Top-down processes require that the volunteers be in a special state prior to the appearance of the stimuli. If our volunteers are instructed "to look out for the dim red squares," then physiological responses to the dim red squares will be enhanced. This is a top-down effect induced by the instruction. In addition, there may be an increase in baseline firing in the relevant brain region *before* the stimulus appears (Driver and Frith, 2000). This is another physiological marker of a top-down effect.

3. At the phenomenological level, the distinction is made in terms of subjective experience. Top-down attention involves an act of will. There is sense of effort associated with maintaining our commitment to this task.

II. WHEN IS TOP-DOWN NOT TOP-DOWN?

It is possible to observe top-down effects at the physiological level even though there is no top-down process occurring at the psychological level. Consider the study of Macaluso et al. (2000). The volunteers' task was to detect visual targets that appeared on the left or on the right in an unpredictable manner. Tactile stimuli were also presented in the same location in space. These were irrelevant in the sense that their occurrence gave no information about the visual targets. Nevertheless, when a tactile stimulus happened to occur on the same place as a visual stimulus, there was enhanced activity in extrastriate cortex (lingual gyrus). As this is a unimodal visual area the tactile stimulus must have produced this enhancement via a top-down signal (in the physiological sense) probably emanating from multimodal areas in parietal cortex. This is not top-down in the psychological sense because the effect was automatic and entirely driven by the tactile stimulus. In multimodal interactions like this when different sensory features are bound together during attention the source of the top-down signal is likely to be in parietal cortex.

III. DO WE HAVE TOP-DOWN CONTROL OF SELECTIVE ATTENTION?

The idea that stimuli have to compete for limited perceptual resources is fundamental for our understanding of attention (e.g., Desimone and Duncan, 1995). Only a limited number of the multiple stimuli with which we are bombarded can be processed at any one time. Top-down control operates by biasing the competition in favor of a particular category of stimuli. Stimuli outwith this selected category elicit less neural activity, have less effect on behavior, and are less likely to enter consciousness. This selection of some stimuli at the expense of others is typically achieved by instructions ("look out for the dim red squares"), but is the decision to attend to one stimulus rather than another sufficient to achieve selection? Rees et al. (2001) measured the neural activity elicited in V5/MT by irrelevant visual motion signals while volunteers performed visual tasks of different levels of difficulty. During performance of the difficult task (counting the syllables in a word), activity in V5/MT was largely suppressed, but during performance of the easy task (noting whether the word was in upper- or lowercase), there was very little effect on activity in V5/MT. This result shows that, if perceptual capacity is not fully used by the primary task, then irrelevant stimuli will be processed. This effect can be seen in behavior also. Irrelevant but distracting stimuli elicit more errors during the performance of an easy than a difficult task (Lavie, 1995). These results demonstrate that we cannot simply choose to ignore irrelevant stimuli. If there is sufficient spare perceptual capacity available, then irrelevant stimuli will be processed whether we like it or not. At the psychological level, top-down control is not exerted by choosing to attend to one stimulus rather than another, but by choosing to perform one task rather than another.

IV. TOP-DOWN CONTROL AS THE MAINTENANCE OF TASK PRIORITIES

Successful task performance depends on implementing and maintaining the appropriate stimulus–response contingencies. Both these functions, implementation and maintenance, are thought to be part of an executive system instantiated in frontal cortex Shallice, T. (1988) "From neuropsychology to mental structure." Cambridge University Press. Maintenance of task priorities, in particular, is considered to be a component process of that archetypal frontal function, working memory (Baddeley, 1986). If selective attention depends on working memory, then we would expect that our ability to attend to some stimuli while ignoring others would be impaired if working memory were already engaged in some other task. This prediction was confirmed by de Fockert et al. (2001). Volunteers performed a classic working memory task in which they had to keep in mind the order of a string of digits. While they were keeping this information in mind they were presented with names (e.g., Bill Clinton) that they had to classify as politicians or popstars. At the same time irrelevant distracter faces appeared (e.g., the face of Mick Jagger). When they were performing the working memory task, the volunteers exhibited greater interference from the distracting stimuli. At the same time these irrelevant faces elicited more activity in the fusiform face area. These observations show that the system that is used to maintain items in working memory is also used to maintain the top-down biases that enable us to attend to one class of visual stimuli rather than another. In this study, engaging working memory activated a region of lateral premotor cortex (Brodmann area 8). The location of this premotor activation is almost identical to that reported in Rowe et al. (2000), where sustained activity was seen during the performance of a very different working memory task involving spatial material. We can speculate that this region is the source of the top-down signal that maintains the selective bias in visual processing areas such as fusiform gyrus. Presumably this biasing is achieved by altering the connectivity between earlier and later visual processing areas. To attend to words rather than faces, connectivity from early form processing areas to the "word form" area would be enhanced while connectivity to the face area would be attenuated. Evidence for such a mechanism in another domain comes Friston and Büchel (2000), who showed that, during attention to visual motion, connectivity between V2 and V5/MT was enhanced. It remains to be established how signals from frontal regions can modulate connectivity in this way.

V. WHAT IS AT THE TOP OF TOP-DOWN CONTROL?

We encounter a problem with the concept of top-down control if we ask what is at the top of the hierarchy or where this "top" might be located. The question seems to imply that there is a region of the brain that has outputs, but no inputs in which a homunculus resides and makes the ultimate decisions about what should be attended to (Fig. 18.1). This problem is apparent rather than real. In most studies the ultimate source of the top-down signal is the experimenter. It is she who determines where the volunteer will direct his attention. Of course, it would be possible to study "willed intention" (by analogy with "willed action") by telling the volunteer to "decide for himself" what to attend to on each trial. But such studies do not give top-down control back to the volunteer's homunculus. Instead they give the volunteer the rather more difficult task of deciding precisely what the experimenter wants him to. For example, the volunteer will probably conclude that the experimenter does not want to him to attend to the same thing on every trial. This is part of the largely implicit script (Jack and Roepstorff, 2002) agreed between the experimenter and the volunteer that determines what the volunteer will do during the experiment. Successful experiments depend on such agreements and it is the agreed script that is the ultimate source of top-down control. If we want to explore this aspect of top-down control further, then we need to study social interactions rather than isolated individuals.

FIGURE 18.1 The homunculus that sits inside Rosenberg's head. Reproduced with permission from Men in Black (Columbia Pictures, 1997).

References

Baddeley, A. D. (1986). "Working Memory." Clarendon Press, Oxford.
Corbetta, M., and Shulman, G. L. (2002). Control of goal-directed and stimulus-driven attention in the brain. *Nat. Rev. Neurosci.* **3**, 201–215.
de Fockert, J., Rees, G., Frith, C. D., and Lavie, N. (2001). The role of working memory in the control of attention. *Science* **291**, 1803.
Desimone, R., and Duncan J. (1995). Neural mechanisms of selective visual attention. *Annu. Rev. Neurosci.* **18**, 193–222.
Driver, J., and Frith, C. D. (2000). Shifting baselines in attention research. *Natu. Rev. Neurosci.* **1**, 147–148.
Felleman, D. J., and Van Essen, D. C. (1991). Distributed hierarchical processing in the primate cerebral cortex. *Cereb. Cortex* **1**, 1–47.

Friston, K. J., and Buchel, C. (2000). Attentional modulation of effective connectivity from V2 to V5 (MT) in humans. *Proc. Natl. Acad. Sci. USA* **97**, 7591–7596.

Jack, A. I., and Roepstorff, A. (2002). Introspection and cognitive brain mapping: from stimulus–response to script-report. *Trends Cogn. Sci.* **6**, 333–339.

Lavie, N. (1995). Perceptual load as a necessary condition for selective attention *J. Exp. Psychol. Hum. Percept. Perform.* **21**, 451–468.

Macaluso, E., Frith, C. D., and Driver, J. (2000). Modulation of human visual cortex by crossmodal spatial attention. *Science* **289**, 1206–1208.

Rees, G., Frith, C., and Lavie, N. (2001). Perception of irrelevant visual motion during performance of an auditory attention task. *Neuropsychologia* **39**, 937–949.

Rowe J. B., Toni, I., Josephs, O., Frackowiak, R. S., and Passingham, R. E. (2000). The prefrontal cortex: response selection or maintenance within working memory? *Science* **288**, 1656–1660.

Shallice, J. (1988) "From neuropsychology to mental structure." Cambridge University Press.

CHAPTER 19

Allocation of Attention in Three-Dimensional Space

Paul Atchley

ABSTRACT

Experiments examining shifts of attention in visual space have used largely two-dimensional representations of visual space. However, the representation of depth is of particular importance for the visual system of an active human observer, so the operation of attention in depth must also be considered. While early work was equivocal with regard to the operation of attention in depth, the present review of recent research suggests that spatial visual attention has a depth component. There is a cost for switching attention between discrete locations in depth under most conditions. Across the life span, the human attention system is generally able to attend to a region in depth selectively and to use depth to guide attentional selection of objects. Exceptions to these costs and unresolved issues are also reviewed.

I. THE PROBLEM OF ATTENTION IN THREE-DIMENSIONAL SPACE

Nearly 30 years of research has led us to the conclusion that there is a spatial extent of vision, not necessarily related to the physical receptors of the eye, within which visual information is processed and outside of which visual information is lost or less efficiently processed. What we now consider to be spatial visual attention is typically categorized in terms of one or more metaphors. One of the earliest metaphors for attention was a spotlight (e.g., Posner et al., 1980). The spotlight view depicts attention as an area within which all information is processed and outside of which information is largely ignored. Other studies have found that attention can operate from a narrow to a wide area of the visual field, leading to the suggestion that attention is like a zoom lens that can be narrowly or widely spatially focused depending on the task demands (e.g., Eriksen and Yeh, 1985), or that attention functions like a gradient with maximal processing efficiency at the center and declining efficiency at more peripheral locations (e.g., Downing and Pinker, 1985). However, two-dimensional (2D) metaphors inadequately specify how spatial visual attention must work in the real world, because observers operate in a three-dimensional (3D) environment. As metaphors of spatial attention are based on research conducted using 2D displays, the true 3D nature of the real visual environment is not adequately represented.

II. RESEARCH DEMONSTRATING ATTENTION IN DEPTH

A. Failures to Find Attention in Depth and Equivocal Studies

If visual spatial attention has a representation in the depth domain, then attending to one location in 3D space should reduce or slow the processing of objects at other locations in depth, because those objects would be further from the center of attention. The alternative is that attention is deployed differently in depth than it is in 2D space and that attention might be depth-blind (Ghiradelli and Folk, 1996; Iavecchia and Folk, 1995) or only selectively deployed in depth when objects are present in the visual field, unlike in 2D space, in which attention can be selectively focused even in an empty field.

1. Failures to Find Attention in Depth

A number of studies have indicated that attention does not operate in depth or is "depth-blind"

(Ghiradelli and Folk, 1996; Iavecchia and Folk, 1995). In the Ghiradelli and Folk study, for example, observers viewed a stereoscopic display of two empty boxes (square and diamond) that overlapped in depth and were oriented along the depth axis. Observers were cued to a location in depth (near or far box) at which a target was most likely to occur. Contrary to the notion that attention has an extent in depth, when the target appeared at a different location in depth from the cue, no cost to reaction time was found for switching attention to the new depth location.

2. Equivocal Studies

Other studies have found that attention operates in depth, but their use of real objects leads to possible alternative explanations. In a study by Downing and Pinker (1985), observers viewed two rows of lights with four lights in each row. The rows of lights were arranged in depth. The observers fixated centrally and were cued to the visual location of the target. Reaction time was slower for targets that were at a depth plane different from that of the cued location. This effect increased with increasing retinal eccentricity. Similar results with real 3D scenes have also been obtained by Gawryszewski et al. (1987). However, there are a number of potential problems with using real 3D scenes to study spatial attention. First, placement of objects at different distances from the observer creates the possibility that selection could occur on the basis of differences in object attributes (such as size or intensity in the case of Downing and Pinker) rather than on the basis of depth. Second, differences in accommodation and convergence also occur when viewing real objects at different distances. Studies using real objects contain excellent depth cues, but those cues make it difficult to disambiguate the affect of attention from perceptual differences among stimuli.

B. Studies Demonstrating Attention in Depth

Studies of focused attention in 3D space conducted using computer displays avoid many of the problems encountered using real 3D scenes (Andersen, 1990; Andersen and Kramer, 1993; Atchley et al., 1997; He and Nakayama, 1995). In these studies, depth is generally represented using stereoscopic presentation of objects. In the study by Andersen, observers viewed random-dot stereograms of centrally located horizontal or vertical bars with distracter bars placed nearby in x,y space and at various depths relative to the target bar. Display durations were less than 120 ms to avoid vergence eye shifts. Distracters were presented at various depths relative to the target. Observer performance declined the closer the distracters were to the target in depth. In the study by Atchley et al., observers viewed nonoverlapping contour stereograms of four boxes which served as target placeholders. Two boxes appeared near in depth and two appeared far in depth. The 2D distances between the locations were equivalent. A box increased in luminance to indicate the most likely target location. When this cue was invalid, the target appeared at a location next to the cued location. Target detection on invalid trials required a shift of attention in 2D space and, on trials where the depth of the target was different from that of the cue, a shift of attention in depth occurred as well. When the target was shown without distracters, there was no cost to reaction time if the target was at a different depth. However, when each location contained a distracter, it took additional time to move attention from one depth to another. This result suggests that attention has a representation in depth, but that the extent of attention in depth depends on the complexity of the visual environment. As perceptual complexity increases, the range of attention in depth decreases. In relatively simple visual environments, attention has a greater extent in depth. This result explains the null findings of previous research (e.g., Ghiradelli and Folk, 1996) in which had relatively sparse displays enabled attention to extend throughout the full extent of depth of the displays.

III. THE INTERACTION OF ATTENTION IN DEPTH AND OTHER PROCESSES

A. Depth as a Cue for Controlling Attention

The control of attention in 3D space facilitates the extraction of information of interest at a particular location in depth, as well as reducing distraction from other, irrelevant, information (Nakayama and Silverman, 1986; Theeuwes et al., 1998). For example, Theeuwes et al. found that we can use depth to guide attention to target information and to filter distracting information, such as a colored item in a field of noncolored items. In their experiments, search for a colored target of a particular shape among noncolored items of various shapes was enhanced if the observers were cued to the proper depth location of the target. Further, distracting information, such as another colored item, could be ignored partially or even completely if it was located at a different depth. In 2D space, the distracting color item would capture attention and increase the time to respond to the target. These results suggest that attention can be limited to a particular depth plane because the distracting effect was attenuated. It was also found that the degree to which distracters could be ignored was related to the degree to which the distracting information shared features with the target of visual search. Similar colored distracters at a different depth still captured attention to some degree. However, distracters presented at a different depth and with a different color than the target could be ignored completely. These finding suggest that controlling attention in 3D space can be useful for extracting critical information from the real-world visual environment, as our actual environment has a rich set of cues that allow us to recover depth. Limiting attention to a particular depth location is one way to reduce the complexity of our visual environment. The ability to conduct visual search at *relevant* depths or distances (where relevance is determined by the observer) would allow us to possibly ignore or attenuate potentially distracting information at other, less relevant, depths.

B. Attention to Objects and Surfaces That Have an Extent in Depth

Many of the objects and surfaces we encounter transcend depth. The surfaces we walk on, the walls that support our environments, and other objects often extend away from us in depth. An analysis of these surfaces based on a purely spatial approach to allocating attention to multiple depth locations would suggest that as we move our attention along these surfaces and objects, there should be some cost (in terms of the amount of time it takes to move our attention) for doing so, because we are switching attention in depth. However, there is a rich literature in visual attention that also suggests that allocating attention to objects can overcome some of the costs we observe from reallocating attention to multiple spatial locations in 2D space (e.g., Egly et al., 1994). For example, the recovery of information at two spatial locations is often found to be more rapid and to occur with fewer errors if the two locations are parts of the same object than if they are parts of different objects. Research with surfaces and objects that extend in depth (Atchley and Kramer, 2001; He and Nakayama, 1995) supports the conclusion that attention can spread in depth along surfaces, overcoming some of the costs for switching attention in depth. For example, Atchley and Kramer used a stereoscopic presentation of pipes that either extended in depth or were present at different depths (see Fig. 19.1). Observers had to detect "leaks" on the

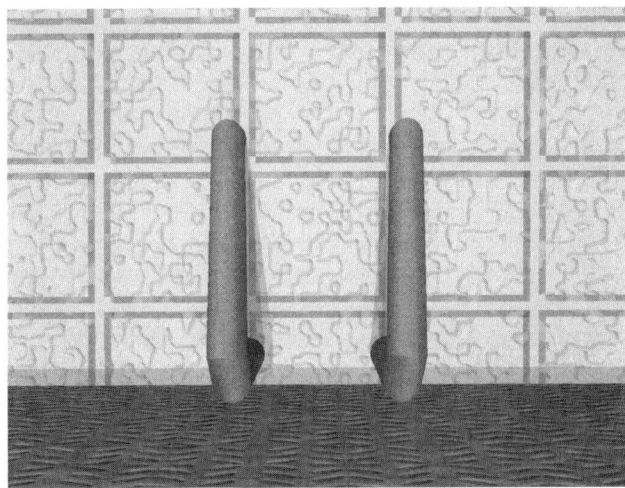

FIGURE 19.1 Example of the pipe stimuli from Atchley and Kramer (2001). In this example, the same object has an extent in depth. The near location is on the bottom, and the far location is on the top.

pipes. In the example, two targets could appear at different depths, but on the same object. Other configurations of pipes had pipes without an extent in depth, but with each pipe present at a different depth location, to examine switching attention in depth across different objects. When the leaks were on different pipes at different depths, the standard cost for switching attention in depth was observed. However, when the leaks were on the same pipe that had an extent in depth, the cost was reduced or eliminated. This supports the idea that attention can spread along surfaces, even if those surfaces extend in depth.

C. Aging and Attention in Depth

The degree to which the allocation of spatial attention to locations in 2D space remains intact with advanced age has been well established. Many studies of age-related differences for attentional tasks such as visual search, inhibitory processing, orienting of spatial attention, and dividing attention between noncontiguous spatial locations demonstrate relatively intact, though slower, visual spatial attention in older observers. This trend applies to attention in depth as well (Atchley and Kramer, 1998, 2000). With respect to reallocating attention in depth, older adults appear similar to younger adults. Older adults also retain the ability to control their attention in depth. Much like younger observers, older observers can allocate their attention to a depth plane and search this plane while excluding distracting information at another depth. The degree to which they can ignore information at other depths is related to the salience of that information and the degree to which this information can be attenuated by properties of the aging eye. In some cases, older adults are able to more effectively use attention at a particular depth plane to ignore distracting stimuli at another depth location because the aging visual system (with reductions in contrast sensitivity and acuity) serves as a "filter" by reducing the energy of the distracting stimulus. Finally, much like their younger counterparts, older adults can also use feature information, such as color, to inhibit distracters and reduce the greater slowing of response to target information that the distracters produce. Thus, attention in depth, like perception of depth, seems to be largely intact throughout the life span.

IV. UNRESOLVED ISSUES

A. Viewer-Centered versus Object-Centered Representations

One issue that has engaged the 3D attention literature has been the question of the shape of 3D attention. One possibility is that attention in depth is viewer-centered. This could take two forms. The strong form is a 3D attentional gradient that extends from the observer to the object or locus of attention. Thus, objects between the observer and attended information would be automatically within the gradient of attention. The weaker viewer-centered view is that once the gradient of attention is localized at the object or location of interest in 3D space, it then extends from that location away from the observer (Andersen, 1990; Downing and Pinker, 1985). The alternative is an object-centered account in which attention spreads from the locus of attention toward and away from the observer equivalently. Viewer-centered representations are demonstrated by an asymmetry in reaction time to stimuli (distracters or targets) located nearer versus further from the observer in 3D space. Some studies are consistent with the weaker view of attention in depth (Andersen, 1990; Atchley and Kramer, 2001). However, other studies have either the strong form of viewer-centered attention (Andersen and Kramer, 1993; Downing and Pinker, 1985; Gawryszewski et al., 1987) or have failed to find any asymmetry at all (Atchley et al., 1997; Theeuwes et al., 1998). While it is clear that attention in depth is viewer-centered, research will have to be conducted to explain why the shape seems to differ across a number of experiments.

B. What Depth Cues Support Attention in Depth?

There are numerous cues for the recovery of depth and these cues vary in efficiency with respect to their ability to specify both absolute and relative depth across different distances from the observer. For example, convergence and accommodation can indicate absolute depth, but only over a very short distance from the observer. Binocular disparity and motion parallax are much more efficient and operate across a wider range of distances than do accommodation and convergence. Pictorial depth cues such as occlusion can specify depth out to the visible horizon, but occlusion can only specify ordinal depth relations among objects. Thus, there exist cues that can support a representation for visual attention in 3D space from very near space to very far locations in space. However, with the exception of early work with real objects (Downing and Pinker, 1985; Gawryszewski et al., 1987), all work to date in 3D attention has used binocular disparity as the primary cue to depth. Since the early work *included* disparity as a cue, no work has been conducted in which disparity has not been available to support an attentional representation in depth. If attention has a representation in depth that extends beyond space immediately surrounding the observer, then the attentional representation must be supported by other depth cues, such as motion and pictorial depth cues. To date, none of these other depth cues have been examined for their ability to support the representation of attention in depth.

References

Andersen, G. J. (1990). Focused attention in three-dimensional space. *Percept. Psychophys.* **47**, 112–120.

Andersen, G. J., and Kramer, A. F. (1993). The limits of focused attention in three-dimensional space. *Percept. Psychophys.* **53**, 658–667.

Atchley, P., and Kramer, A. F. (1998). Spatial cuing in a stereoscopic display: attention remains "depth-aware" with age. *J. Geron. Psychol. Sci.* **53B**, 318–323.

Atchley, P., and Kramer, A. F. (2000). Age-related changes in the control of attention in depth. *Psychol. Aging* **15**, 78–87.

Atchley, P., and Kramer, A. F. (2001). Object-based attentional selection in three-dimensional space. *Vis. Cogn.* **8**, 1–32.

Atchley, P., Kramer, A. F., Theeuwes, J., and Andersen, G. J. (1997). Spatial cuing in a stereographic display: evidence for a depth aware attentional focus. *Psychon. Bull. Rev.* **4**, 524–529.

Downing, C., and Pinker, S. (1985). The spatial structure of visual attention. *In* "Attention of Performance XI" (M. Posner and O. Martin, Eds.), pp. 171–187. Erlbaum, Hillsdale, NJ.

Egly, R., Driver, J., and Rafal, R. D. (1994). Shifting visual attention between objects and locations: evidence from normal and parietal lesion subjects. *J. Exp. Psychol. Gen.* **123**, 161–177.

Eriksen, C. W., and Yeh, Y. (1985). Allocation of attention in the visual field. *J. Exp. Psychol. Hum. Percept. Perforin.* **11**, 583–589.

Gawryszewski, L. D. G., Riggio, L., Rizzolatti, G., and Umlità, C. (1987). Movements of attention in three spatial dimensions and the meaning of "neutral" cues. *Neuropsychologica* **25**, 19–29.

Ghiradelli, T. G., and Folk, C. L. (1996). Spatial cueing in a stereoscopic display: evidence for a "depth blind" attentional spotlight. *Psychon. Bull. Rev.* **3**, 81–86.

He, Z. J., and Nakayama, K. (1995). Visual attention to surfaces in three-dimensional space. *Proc. Natl. Acad. Sci. USA* **92**, 11155–11159.

Iavecchia, H. P., and Folk, C. L. (1995). Shifting visual attention in stereographic displays: a timecourse analysis. *Hum. Factors* **36**, 606–618.

Nakayama, K., and Silverman, G. H. (1986). Serial and parallel processing of visual feature conjunctions. *Nature* **320**, 264–265.

Posner, M., Snyder, C., and Davidson, B. (1980). Attention and the detection of signals. *J. Exp. Psychol. Gen.* **109**, 160–174.

Theeuwes, J., Atchley, P., and Kramer, A. F. (1998). Attentional control within three-dimensional space. *Journal of Experimental Psychology: Human Perception & Performance*, **24**, 1476–1485.

CHAPTER 20

Covert Attention and Saccadic Eye Movements

John M. Findlay

ABSTRACT

In normal vision the eyes make overt saccadic eye movements several times each second. We have a good understanding of how saccadic targets are selected, particularly in visual search and in reading. Visual processing is enhanced at the saccade destination before the movement itself occurs, and this peripheral preview assists the smooth uptake of visual information. Thus, covert attention to locations in space operates as a supplementary process to assist active vision. Situations outside the laboratory in which covert attention acts as a substitute process without eye movements occurring are unusual.

I. INTRODUCTION

A traditional approach to vision emphasizes the neural processes that operate on information derived from the retinal image. Covert visual attention is often considered as one of these processes, operating to select specific information. One familiar metaphor for covert attention is that it operates like a mental spotlight, selecting a region of peripheral vision away from the fovea. In this article, I propose that a better understanding of visual attention can be obtained by adopting a rather different starting point. An elaboration of the ideas can be found in Findlay and Walker (1999) and Findlay and Gilchrist (2003).

II. VISUAL INFORMATION EXTRACTION NORMALLY INVOLVES OVERT ATTENTION (EYE MOVEMENTS)

Human vision is characterized by the presence of a fovea. The fovea is both the retinal region of highest visual acuity and the reference point for a unique visual direction. The visual axis defines the gaze location, the region of the visual field selected for high-acuity analysis. In most visual activities, a new location is selected several times each second. The fovea is redirected to the new location with an efficient steplike movement of the eyes, termed a *saccadic eye movement* or *saccade*. The gaze rests for a brief fixation period at this new location before the next saccade in the sequence is made. These saccades and fixations are almost always automatic. We are aware neither of the processes involved in the selection of new gaze locations, nor of the massive changes that occur in retinal stimulation when the eyes move.

These saccadic eye movements constitute an *overt* process for directing visual attention. We normally attend where we are looking. When studied in detail, a striking finding emerges. Visual information at the location of the saccade destination is subject to enhanced visual processing in the brief period *before* the eye movement itself occurs. Deubel and Schneider (1996) asked observers to look at a display containing items at three visual locations centered at 5° eccentricity. One of the items was designated as the saccade target and the task was to discriminate a visual probe appearing briefly while the saccade was in preparation, disappearing again before the eyes reached the target. Discrimination was highest at the designated saccade target and reduced for material positioned even as little as 1° distant. As discussed in Section V, this improved processing assists vision through the phenomenon of *preview advantage*. If, through technical laboratory manipulations of a visual display, the visual information at a saccade target is not available prior to the generation of the saccade, vision works less efficiently.

III. COVERT ATTENTION OPERATES IN CONJUNCTION WITH SACCADIC EYE MOVEMENTS

Enhanced visual processing at a peripheral location is a hallmark of *covert* visual attention, attending while looking elsewhere. Hence one account of the phenomenon described in the previous paragraph is to say that attention is moved covertly to the saccade target before the eyes move. It might then be claimed that covert attention is primary, with the eyes following. However, there is a danger here in letting the tail wag the dog by shifting the emphasis to the covert processes and forgetting the significance of the overt movements. Much writing about visual attention has adopted this emphasis and takes little account of eye mobility. This is surely misguided. In what circumstances would it make sense for covert attention to operate without being linked to the overt scanning process? Outside the laboratory, it is very unusual to attend without looking. Although situations are described in Section VI where attending covertly without moving the eyes is worthwhile, such situations are comparatively rare and it seems more plausible to suppose that covert attention co-evolved with eye scanning to support active vision.

It would be easy to formulate a separate function for covert attention if the attentional spotlight could be redirected at a greater speed than could the eyeball. Early work, in particular that derived from the classic analysis of visual search by Treisman and Gelade (1980), supported the suggestion that in a search task involving a number of discrete items, the spotlight could be moved between items at a rate as fast as 30 ms/item. This figure was based on the finding that the average time to complete a search increased incrementally at this rate as the number of distracter items was increased. However, alternative approaches to measure the rate at which covert attention can be redeployed have found a much slower rate (Egeth and Yantis, 1997), and alternative accounts can be provided of the search rates that originally gave rise to the proposal for rapid scanning (Eckstein, 1998). There is therefore no good evidence that the spotlight can move any faster than the eyes themselves can move.

The same conclusion has come from studies of visual search in which searchers are free to move their eyes. Assuming covert attention could be scanned at a rapid rate, it would be possible that several covert attentional movements occurred within the duration of each overt eye fixation (typically around 250 ms). If that were the case, some predictions could be made (e.g., saccades directed to the search target would occur following shorter than normal fixation periods). Results from several studies (summarized in Findlay and Gilchrist, 2001) have failed to uphold these predictions.

IV. SACCADIC TARGET SELECTION INVOLVES A SALIENCE MAP

An alternative proposal of how visual material is processed within each eye fixation is that retinal information is analyzed in a spatially parallel way to create a *salience map*. This concept refers to a hypothesized two-dimensional representation of the visual field in which a single property, salience, is represented at each point. A concrete realization might be a two-dimensional neural net, mapping the visual field, with neural activity at each point corresponding to the salience at that point. Selection of a unique location for the saccade requires some form of winner-take-all process. For example, the next saccade target might be the point of maximum salience at the instant, toward the end of the fixation, when the saccade is triggered (see Chapter 65). The salience map is derived from visual stimulation on the retina, processed according to task demands, and is also in important respects dependent on the structure of the visual system. Thus, in a visual search task, salience depends on the *similarity* of material to the visual search target and on the *proximity* of items to the current point of fixation (Findlay and Gilchrist, 2001; Motter and Belky, 1998). Visual search can be studied readily in primates, and elucidating the neural correlates of these selection processes is a topic of intense current interest (see, Chapters 21 and 22).

V. PLANNING AHEAD IN THE SACCADIC SYSTEM

This is only part of the fascinating story about how vision operates actively. Selection and analysis of the saccade target can also commence earlier than the immediately preceding fixation. Following the presentation of a search display, an erroneous first saccade is often followed very rapidly by a second saccadic movement (McPeek et al., 2000). These second saccades are often directed accurately at the target, even following fixations too brief for new visual analysis to contribute. The destination of the second saccade must therefore have been decided using *preview* information carried over the intervening first saccade. The analysis of McPeek et al. (2000) suggests that information is used in this way very frequently.

A similar use of preview information has been intensively studied when individuals make saccadic eye movements during the course of reading (Rayner, 1998). During the reading process, the great majority of saccades take the eyes in a left to right scan across each successive line of text. Using computer-controlled technology, it is possible to modify the text displayed on a VDU screen in a *gaze-contingent* way, that is, dependent on where the viewer's gaze is directed. It thus becomes possible to control the extent and nature of the material available in the upcoming text to the right of where the eyes are directed. Research with gaze-contingent displays demonstrates a *preview advantage*; reading is faster when preview material is available than when it is not. The advantage comes about for two reasons. First, information obtained from preview is integrated with information obtained when the word is subsequently fixated, reducing the time needed for the latter fixational process. Second, the preview can affect saccade targeting; in particular, it may allow sufficient information to be obtained for an upcoming word to be skipped and not fixated (Gautier et al., 2000).

On this account, covert attention is a contributor to the *active vision cycle* of saccades and fixations (Findlay and Gilchrist, 2003), but does not, so to speak, have an independent life of its own. A critical question then becomes to what extent the process of forward planning for saccades might echo that believed to occur for the execution of individual saccades and, through a further winner-take-all process, take the form of a mental spotlight that selects a specific location from a salience map. Two recent contributions suggest that such a selection might be present. First, a detailed theoretical account of the reading process (Reichle et al., 2003) operates on the basis of a word-by-word attentional selection that is decoupled from saccade selection. Second, on the basis of an ingenious physiological study, Bisley and Goldberg (2003) claim that intention (defined as a plan to move the eyes) can be maintained at a location when attention (defined as selection of a location in the peripheral visual field for enhanced processing) is momentarily directed elsewhere. The crucial finding is that enhanced visual processing at the location for attention is accompanied by reciprocal diminished performance at the intended saccade location.

VI. WHEN MIGHT COVERT ATTENTION OPERATE WITHOUT THE EYES MOVING?

Are there situations in which covert attention is used to acquire information from the visual periphery in the absence of overt movements of the eyes? Outside the laboratory, the situations where this occurs would seem to be social ones, for example, where actually moving the eyes would provide a cue that the individual wishes to conceal. Deception about intentions in competitive situations such as sports activities might lead to use of covert attention, as might the conventions of social activities such as conversation in which eye gaze can signal turn taking. In reviewing this area, Argyle and Cook (1976) noted that procedures for using gaze and covert attention separately are explicitly taught to infants in some cultures. Overall, it seems reasonable to conclude that this type of behavior is the exception and, in most circumstances of normal vision, covert attention and overt attention are intimately linked in the way described in this chapter.

References

Argyle, M., and Cook, M. (1976). "Gaze and Mutual Gaze." Cambridge Univ. Press, Cambridge.
Bisley, J. W., and Goldberg, M. E. (2003). Neuronal activity in the lateral intraparietal area and spatial attention. *Science* **299**, 81–86.
Deubel, H., and Schneider, W. X. (1996). Saccade target selection and object recognition: evidence for a common attentional mechanism. *Vision Res.* **36**, 1827–1837.
Eckstein, M. P. (1998). The lower visual search efficiency for conjunctions is due to noise and not serial attentional processing. *Psychol. Sci.* **9**, 111–118.
Egeth, H. E., and Yantis, S. (1997). Visual attention: control, representation, and time course. *Annu. Rev. Psychol.* **48**, 269–297.
Findlay, J. M., and Gilchrist, I. D. (2001). Visual attention: the active vision perspective. *In* "Vision and Attention" (M. Jenkin and L. R. Harris, Eds.) pp. 83–103. Springer-Verlag, New York.
Findlay, J. M., and Gilchrist, I. D. (2003). "Active Vision: The Psychology of Looking and Seeing." Oxford Univ. Press, Oxford.
Findlay, J. M., and Walker, R. (1999). A model of saccadic eye movement generation based on parallel processing and competitive inhibition. *Behav. Brain Sci.* **22**, 661–721.
Gautier, V., O'Regan, J. K., and Le Gargasson, J. F. (2000). "The-skipping" revisited in French: programming saccades to skip the article "les". *Vision Res.* **40**, 2517–2531.
McPeek, R. M., Skavenski, A. A., and Nakayama, K. (2000). Concurrent processing of saccades in visual search. *Vision Res.* **40**, 2499–2516.
Motter, B. C., and Belky, E. J. (1998). The guidance of eye movements during active visual search. *Vision Res.* **38**, 1805–1815.
Rayner, K. (1998). Eye movements in reading and information processing: 20 years of research. *Psychol. Bull.* **124**, 372–422.
Recihle, E. D., Rayner, K., and Pollatsek, A. (2003). The E-Z reader model of eye movement control in reading: comparisons to other models. *Behav. Brain Sci.* **26**, 445–518.
Treisman, A. M., and Gelade, G. (1980). A feature integration theory of attention. *Cogn. Psychol.* **12**, 97–136.

CHAPTER 21

Prefrontal Selection and Control of Covert and Overt Orienting

Narcisse P. Bichot and Jeffrey D. Schall

ABSTRACT

Recent research has provided new insights into the neural processes that select the target for and control the production of a shift of gaze. Being a key node in the network that subserves visual processing and saccade production, the frontal eye field has been an effective area in which to monitor these processes. Certain neurons in the frontal eye field signal the location of conspicuous or meaningful stimuli that may be the targets for saccades. Other neurons control whether and when gaze shifts. The existence of distinct neural processes for visual selection and saccade production is necessary to explain the flexibility of visually guided behavior.

I. INTRODUCTION

Vision is an active process; the eyes move constantly so that images of objects of interest project onto the fovea which provides high-acuity vision. As gaze can be directed only at one location at a time, gaze shifts are typically preceded by the attentional selection of objects that are behaviorally relevant. Although the relation between attention and eye movements is not obligatory (i.e., one can attend to an object without making an eye movement to it, as it is in the case of many studies of covert attention), recent behavioral research indicates that a common visual selection mechanism governs both covert and overt orienting (see Chapter 20). Neurophysiological studies, including those involving functional brain imaging, have also supported the view that attention and eye movements are functionally related (see Chapter 6). Altogether, these results suggest that covert orienting may be little more than a state of visual selection without activating motor circuitry to produce overt orienting. This conclusion is consistent with the premotor theory of attention proposing that the attentional system has evolved as part of the motor systems and is part of the premotor processing of the brain (see Chapters 20 and 31).

By virtue of its connections and physiological properties, the frontal eye field (FEF), located on the rostral bank of the arcuate sulcus in macaque monkeys, is ideally positioned to transform the outcome of visual processing into a command to move the eyes (reviewed by Schall, 1997) (see also Chapter 22). Consequently, FEF has two aspects, one sensory and one motor.

Most research on FEF has emphasized its motor aspect, and its involvement in the generation of saccadic eye movements is unequivocal (reviewed by Schall, 1997). Surprisingly, the role of FEF in visual selection has been relatively ignored despite equally compelling evidence for the visual aspect of this area. About half of the neurons in FEF have visual responses that are mediated by massive converging input from extrastriate visual areas of both the dorsal (or "where") and ventral (or "what") streams (see Chapter 22) that have been implicated in visual selection (reviewed by Desimone and Duncan, 1995; Maunsell, 1995). In fact, FEF appears to be one of the highest points of convergence of the dorsal and ventral visual information streams in the brain.

Here, we review evidence showing that visually responsive neurons in FEF signal the overall behavioral relevance of stimuli, whether relevance is derived from the intrinsic properties of the stimuli or from the viewer's knowledge and goals. We also review evidence showing that movement neurons are more directly involved in whether or when a saccade will be generated.

II. VISUAL SELECTION INVOLVING CONSPICUOUS STIMULI

A. Simple Popout Visual Search

We studied the involvement of FEF in selection during visual search tasks that have commonly been used in psychophysical studies of attention (see Chapter 17). The simplest visual search involves the detection of a target stimulus that differs in one or more visual attributes from neighboring stimuli. Such stimuli are said to "pop out" and are likely to be attended and fixated. Neural correlates of visual selection during popout search in FEF have been reviewed (see Chapter 22). Briefly, the initial activity of most visually responsive neurons in FEF did not discriminate whether the target or distracters of the search array appeared in their receptive field, consistent with the fact that visual responses in FEF are not selective for stimulus properties. However, before saccades were generated, the activity of FEF neurons evolved to signal the location of the popout target regardless of the visual feature that distinguished it from distracters. Furthermore, the time at which the target was discriminated did not predict the variability of saccade latencies, and the selection of the popout stimulus was not contingent on saccade planning. These results, along with characteristics of neural modulation during a search-step visual search task, suggest that visually responsive neurons in FEF are more involved in signaling the behavioral relevance of a stimulus than in initiating a saccade to that stimulus, per se. To test this hypothesis, we monitored activity in FEF during a variety of visual search tasks in which the behavioral relevance of stimuli derived from both bottom-up and top-down factors was manipulated.

B. Feature Expectancy during Popout Visual Search

The observations above regarding neural modulation in FEF during popout search were made in monkeys trained to detect the oddball target in complementary visual search arrays. We found that both monkeys' search strategy and the concomitant neural selection process in FEF can be modified profoundly by cognitive strategies (Bichot et al., 1996). While monkeys trained to search for a popout regardless of the particular feature that defines it generalized a strategy of shifting their gaze according to visual conspicuity, monkeys given exclusive experience with one visual search array adopted a strategy of ignoring stimuli with the learned distracter feature, even if those same stimuli became the popout stimulus in the complementary visual search array presented occasionally (Fig. 21.1A). In monkeys using this strategy, about half of FEF neurons exhibited a suppressed response to the learned distracter as soon as they responded (Fig. 21.1A). In other words, FEF neurons exhibited an apparent feature selectivity in their initial response unlike what had been observed before in this area. This study shows how cognitive strategies can dramatically affect the behavioral relevance of stimuli and their underlying neural representation in FEF even during a simple popout visual search.

C. Priming during Popout Visual Search

Although striking, the effect described above developed after extensive training with a given search configuration leading to the observed bias. However, the performance of both humans and monkeys during popout search is also affected by visual experience on a much shorter time scale (McPeek et al., 1999; Bichot and Schall, 2002). This robust effect, known as *perceptual priming*, manifests itself as an improvement of behavioral performance with the repetition of stimulus features over consecutive trials (Fig. 21.1B).

We found that the properties of the neural selection signals in FEF could fully explain changes in behavior due to priming (Bichot and Schall, 2002). Neurons discriminated the target from distracters earlier and better with repetition of stimulus features, corresponding to improvements in saccade latency and accuracy, respectively (Fig. 21.1B). Furthermore, consistent with earlier psychophysical explorations of priming of popout, the improvement in the neural target selection signals consisted of both target enhancement and distracter suppression. Moreover, the time course of neural discrimination in FEF accounted for the increase in saccade latencies associated with the location-based inhibition of return when target position repeated across consecutive trials. These results show adjustments of the neural selection process in FEF mirroring the dynamic changes in performance across trials due to sequential regularities in display properties.

D. Effect of a Singleton Distracter during Popout Visual Search

In some instances, the item sought by a viewer in one feature dimension may not be the most salient item in the image. The presence of such singleton distracters has been shown to disrupt performance (see Chapter 69). If the target selection signals in FEF represent the saccade goal only, the representation of such a singleton distracter should not differ from that of other distracters on trials in which the viewer correctly

II. VISUAL SELECTION INVOLVING CONSPICUOUS STIMULI 119

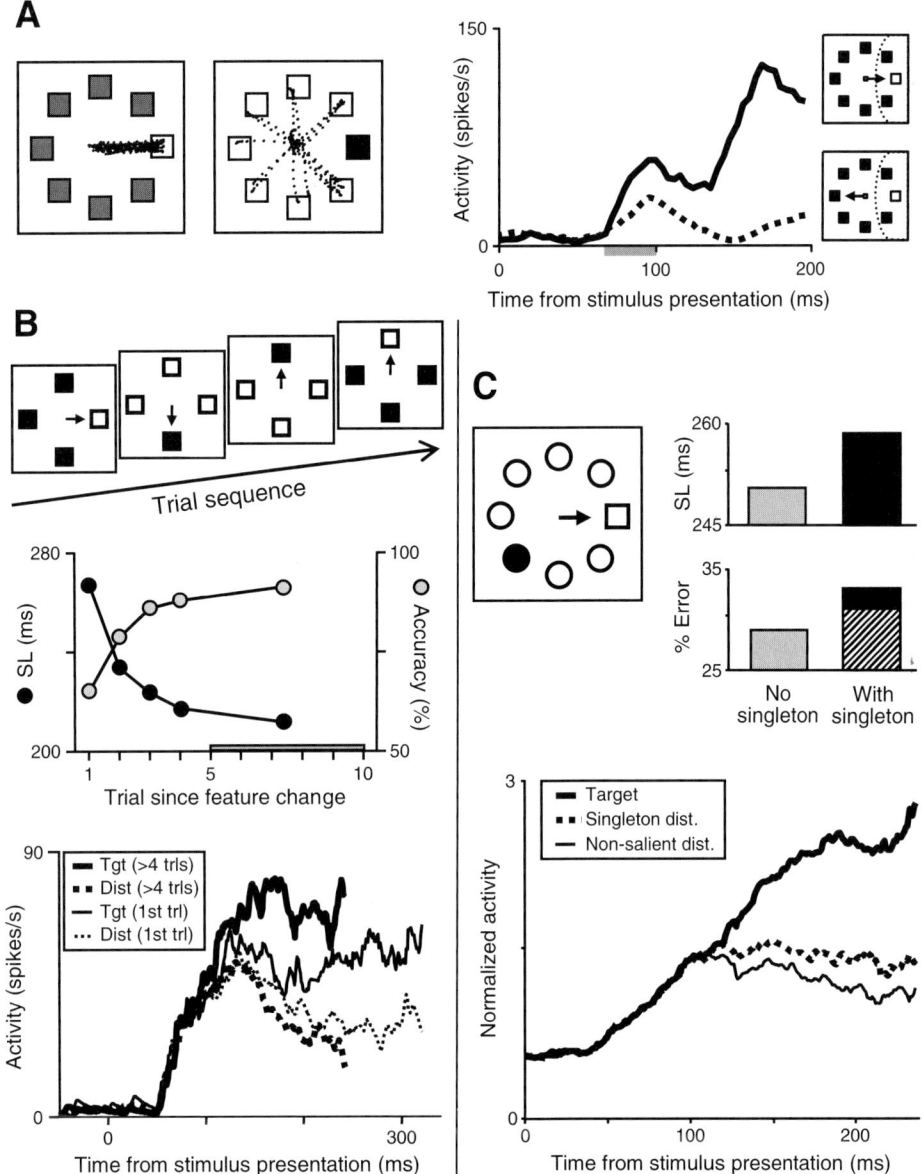

FIGURE 21.1 Performance and neural modulation in FEF during visual search tasks involving conspicuous stimuli. **(A)** Effect of feature expectation during a color popout search. Left: Eye movements of one monkey during search with the learned search array (left) and during viewing of the complementary search array (right). Right: Response of one FEF visuomovement neuron while this monkey performed search with the learned search array. Spike density functions during correctly performed trials when the target was in the neurons's response field (solid line) and when distracters were in its response field (dashed line) are shown superimposed. **(B)** Top: Task display. The arrow illustrates the saccade to the target. In this sequence, the second trial represents the first trial after a feature switch; the next trial represents the second trial after the feature switch; and the last trial represents another first trial after a feature switch Middle: Mean saccade latency and accuracy as a function of trial number since feature switch. Bottom: Response of one FEF visuomovement neuron to the target (thin solid line) and to distracters (thin dotted) during trials immediately following a feature switch, and to the target (thick solid) and distracters (thick dotted) during trials more than four removed from the feature switch. **(C)** Effect of a color singleton distracter during a shape popout search. Top left: Task display. Top right: Mean saccade latency and error rate during the shape search task with (black bars) and without (gray bars) the color singleton. The proportion of saccades to the color singleton are represented by the white stripes. Bottom: Pooled normalized response of a population of FEF visuomovement neurons during correctly performed feature search trials with a singleton distracter. Responses to the target (thick solid line), to the singleton distracter (dashed line), and to the nonsalient distracters (thin solid line) are shown superimposed.

SECTION I. FOUNDATIONS

locates the less salient oddball target. On the other hand, if activity in FEF represents a visual salience map resulting from visual selection, the added behavioral significance of the singleton distracter should manifest itself in that map.

Monkeys performing a search task for a shape singleton exhibited slower saccade latencies and increased error rates when the search array included a color singleton distracter to be ignored (Bichot et al., 2001a) (Fig. 21.1C). During correctly performed popout search trials with a singleton distracter, neural modulation in FEF reflected the behavioral significance of all stimulus types. After an initial period of nonselective visually related responses, neurons exhibited the highest activation for the popout shape target in their response field. More importantly, they responded to the singleton distracter significantly more strongly than to nonsalient distracters in their response field. Thus, results in this search extend the argument that FEF neurons signal behavioral significance rather than the saccade goal only.

E. Pro- and Antisaccade Popout Visual Search

Although singleton distracters tend to attract attention more or less automatically, the study by Bichot et al. (1996) showed that, with experience, monkeys can learn to ignore stimuli of the learned distracter color even if they become the popout in a search display. However, the way in which the feature bias developed in that study precluded further investigation of the dissociation between bottom-up attentional capture by the singleton and top-down selection of distracters in the complementary search array. Such a manipulation was recently implemented by Sato and Schall (2003), who trained monkeys to perform an antisaccade popout search task in which reward was contingent on making a saccade to the distracter located diametrically opposite the singleton. They found two types of neurons: one type initially selected the singleton and then usually responded preferentially to the anti-saccade endpoint (i.e., the saccade goal), while the other type selected only the anti-saccade endpoint. The modulation pattern of the first type of neuron resembles closely the expected pattern of covert orienting that must take place in this task: allocation of attention to the singleton both because of its conspicuity and because of its condition cuing attribute, followed by the selection of the saccade endpoint. Interestingly, however, although neurons of the first type tended to exhibit more visual- than saccade-related activity during a memory-guided saccade task, and neurons of the second type showed a somewhat opposite trend, these two populations of neurons could not be distinguished significantly on a visual-to-movement type of continuum.

III. VISUAL SELECTION BASED ON KNOWLEDGE

Altogether, the studies described thus far support the hypothesis that neural modulation in FEF represents the allocation of attention to stimuli in terms of their significance, making FEF an effective area in which to monitor covert orienting. However, in all of those studies the search guiding feature was an oddball item, while in many real-world situations, an object of interest cannot be located based on conspicuity, and a memory of that object is required to locate it (e.g., "searching for a face in the crowd"). An analogous situation is obtained during a conjunction search where the target is defined by one combination of possible features, and distracters are formed by other possible combinations (Fig. 21.2A). Many studies have shown a dichotomy between feature (or popout) search that appears to be effortless because performance is not affected by the number of distracters in the display, and conjunction search that appears more attentionally demanding because performance worsens to some degree with increasing number of distracters. This dichotomy has played an important role in the development of theories of visual attention, including the guided search model postulating that stimuli during conjunction search can be processed in parallel to identify those with the desired features (see Chapter 17).

To the extent that attention and eye movements are functionally related, such a parallel search strategy predicts that subjects would be more likely to shift gaze to a distracter that shares a target feature (i.e., "similar distracter") than to a distracter that has no features in common with the target (i.e., "opposite distracter"). We have confirmed this prediction in monkeys searching for a target defined by the combination of color and shape (Bichot and Schall, 1999) (Fig. 21.2B). Interestingly, when target properties remained the same within an experimental session, but changed across experimental sessions, we also found evidence that the history of target properties affected behavior, revealed by an increased tendency of errant saccades to land on the distracter that was the search target during the previous session (Fig. 21.2B). Further evidence for a target template has been obtained during singleton search in trials with no target (Sato et al., 2003).

Recordings in FEF during conjunction search further supported the hypothesis that this area

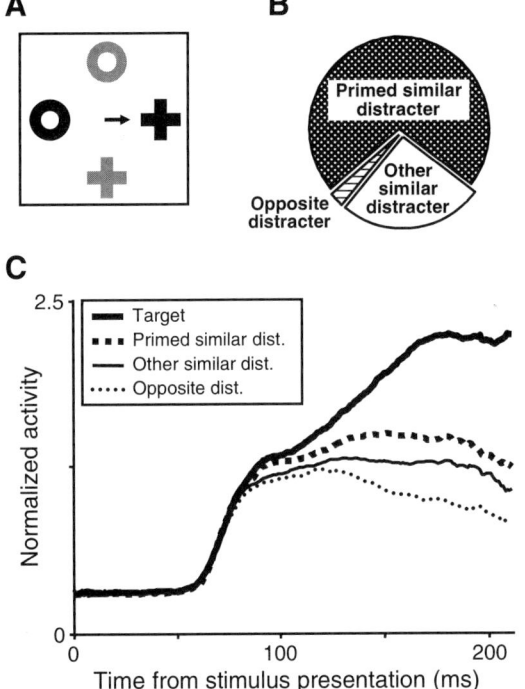

FIGURE 21.2 Gaze behavior and neural modulation in FEF during conjunction visual search. (A) Conjunction search display. The arrow illustrates the saccade to the target. (B) Probability of shifting gaze to each of the distracter types during error trials in sessions in which the target on the previous session was a similar distracter. A similar distracter refers to a distracter that shares a target feature (i.e., same color or same shape); a primed distracter refers to a distracter that was the search target during the previous session. (C) Normalized response of a population of FEF visuomovement neurons during correctly performed conjunction search trials (i.e., first and only saccade was made to the target) in the same sessions contributing to the plot of incidence of saccades to distracters during error trials (B). The responses to the target (thick solid line), to the primed similar distracter (thick dashed), to the unprimed similar distracter (thin solid), and to the opposite distracter (thin dotted) are shown superimposed.

represents a visual salience map. After an initially nonselective response, FEF neurons not only discriminated the target from distracters, but also discriminated among the distracters based on their visual similarity to the target and the history of target properties across sessions (Fig. 21.2C). In other words, while the highest activation was associated with the target, the responses to distracters similar to the target and to distracters primed by virtue of being the target of the previous session were relatively enhanced. Thus, this study shows that neural modulation in FEF reflects a variety of top-down influences, and predicts covert and overt orienting patterns during a complex visual search. Furthermore, because the above similarity- and priming-based modulations were observed during correctly performed trials in which a single saccade was made to the target, these results rule out the possibility that the selection signal in FEF represents nothing more than a command to move the eyes.

IV. CONTROL OF OVERT ORIENTING

A. When to Move the Eyes: Predicting Reaction Time

We have shown across a wide range of visual search tasks and efficiency that the time at which visually responsive FEF neurons discriminate the target predicts accurately mean saccade latencies (e.g., Bichot et al., 2001b; Bichot and Schall, 2002), occurring usually around 80 ms before the average time of saccade initiation. However, as mentioned earlier, even during a simple popout search task, the discrimination time of these neurons does not predict the variability of saccade latencies within a search condition (i.e., latency on individual trials) (see Chapter 22).

In contrast, during a simple detection task, the stochastic variability in the rate at which movement-related neural activity in FEF grows toward a constant threshold predicts and generates the distribution of saccade latencies (Hanes and Schall, 1996). In other words, saccades were initiated if and only if the activity of movement neurons in FEF reached a given threshold, with activity growing toward that threshold faster on short-latency trials and slower on long-latency trials (Fig. 21.3A).

B. Whether to Move the Eyes: Predicting Response Generation

As reviewed earlier, the pattern of modulation of visually responsive FEF neurons does predict the shift of gaze to the target and even to distracters as a function of their behavioral significance during visual search tasks (e.g., Bichot and Schall, 1999; Bichot et al., 2001a). However, the ability to predict where they eyes may move as a function of covert orienting across the visual map does not translate into the ability to predict whether or not the eyes will move to a given stimulus. In fact, as mentioned earlier, visually responsive FEF neurons select the oddball of a search array even if a saccade to it is not programmed or executed (see Chapter 22).

Hanes et al. (1998) further investigated the role of different types of FEF neurons in gaze production using a countermanding paradigm. In this task, monkeys made a saccade to a peripheral target presented alone during a majority of trials. However, on a fraction of trials, the reappearance of the central

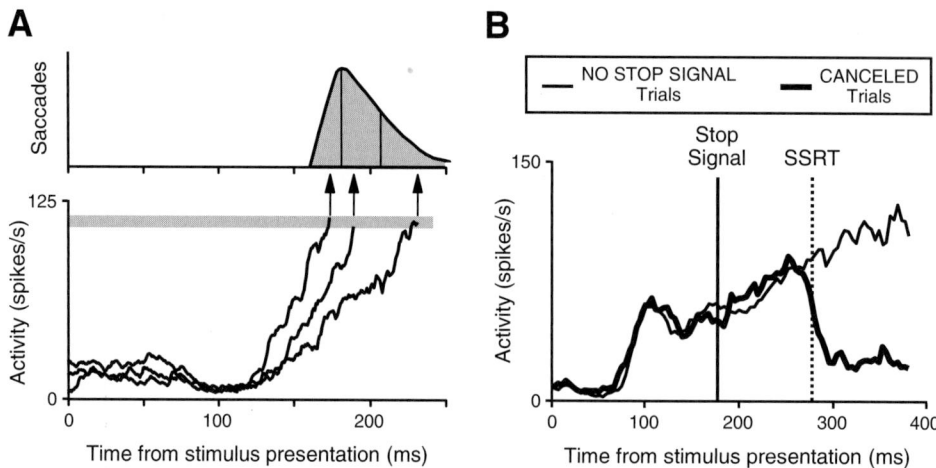

FIGURE 21.3 Relationship between FEF movement-related activity and saccade production. **(A)** Activity of one FEF movement neuron (bottom) for saccades in three different ranges of latency (distribution shown on top). Note that the level of activity reached (gray bar) in all saccade latency ranges is fairly constant. **(B)** Activity of one FEF movement neuron during the countermanding task. The response during trials in which no stop signal was presented and the monkey made a single saccade to the peripheral target (thin line) and the response during trials in which a stop signal was presented and the monkey successfully inhibited the saccade to the target (thick line) are shown superimposed. The average time of the stop signal (solid vertical line) and the average time needed to cancel the planned saccade (stop signal reaction time or SSRT) (dotted vertical line) are indicated.

fixation spot (stop signal) instructed the monkeys to withhold the saccade to the peripheral target. Based on a race model, they estimated the time needed to cancel the planned saccade, and found that movement-related activity decayed in response to the stop signal within that time (Fig. 21.3B). Although less common, fixation-related neurons also exhibited gaze control signals by reactivating in response to the stop signal within the estimated time to cancel the saccade. In contrast, visual-related neurons did not exhibit modulation that could control the production of the gaze shift.

V. DISCUSSION

The diversity of individual neurons reveals a dissociation of covert and overt orienting. Attention is allocated covertly when visual neurons in a circuit including FEF exhibit differential activity. Gaze shifts overtly when a different pool of neurons in the circuit including FEF exhibit a sufficient degree of activation.

We have reviewed evidence for this dual role of FEF, one more closely related to visual selection and the other more closely related to saccade production. Across a variety of visual search tasks, we have shown evidence that visually responsive FEF neurons appear to represent a visual salience map (see Chapters 65, 90, and 94) in which the behavioral significance of stimuli derived from both visual attributes and goal-driven biases is reflected. These neurons predict gaze patterns (to both the target and distracters), overall accuracy, and mean saccade latencies, but their activity does not control either the precise time at which saccades are initiated or whether or not the eyes move. The control in FEF of whether and when saccades will be executed is instead mediated by movement-related neurons (see also Chapter 22).

These studies therefore suggest that FEF should also be considered as part of the visual pathway responsible for the selection of stimuli, whether covertly or overtly. This view is consistent with the findings of brain imaging studies that many areas, including FEF, are activated during both covert and overt orienting (see Chapter 6). Thus, the evidence reviewed here demonstrates that FEF is an effective area in which to monitor covert and overt orienting. However, we should also consider that the connections of FEF with other visual areas are topographically organized and reciprocal (see Chapter 22), and therefore, selection signals in FEF can in turn modulate activity in these areas. This idea is consistent with recent evidence from microstimulation of FEF suggesting that feedback from this area can improve monkeys' performance (Moore and Fallah, 2001) and enhance the responses of extrastriate visual neurons (Moore and Armstrong, 2003) during spatial attention tasks.

Finally, it should be noted that, although this review focused on FEF, the kinds of neural signals described here have or likely will be observed in other functionally related structures such as the superior colliculus and the posterior parietal cortex.

References

Bichot, N. P., and Schall, J. D. (1999). Effects of similarity and history on neural mechanisms of visual selection. *Nat. Neurosci.* **2**, 549–554.

Bichot, N. P., and Schall, J. D. (2002). Priming in macaque frontal cortex during popout visual search: feature-based facilitation and location-based inhibition of return. *J. Neurosci.* **22**, 4675–4685.

Bichot, N. P., Rao, S. C., and Schall, J. D. (2001a). Continuous processing in macaque frontal cortex during visual search. *Neuropsychologia* **39**, 972–982.

Bichot, N. P., Schall, J. D., and Thompson, K. G. (1996). Visual feature selectivity in frontal eye fields induced by experience in mature macaques. *Nature* **381**, 697–699.

Bichot, N. P., Thompson, K. G., Rao, S. C., and Schall, J. D. (2001b). Reliability of macaque frontal eye field neurons signaling saccade targets during visual search. *J. Neurosci.* **21**, 713–725.

Desimone, R., and Duncan, J. (1995). Neural mechanisms of selective visual attention. *Annu. Rev. Neurosci.* **18**, 193–222.

Hanes, D. P., Patterson, W. F., and Schall, J. D. (1998). Role of frontal eye fields in countermanding saccades: visual, movement, and fixation activity. *J. Neurophysiol.* **79**, 817–834.

Hanes, D. P., and Schall, J. D. (1996). Neural control of voluntary movement initiation. *Science* **274**, 427–430.

Maunsell, J. H. R. (1995). The brain's visual world: representation of visual targets in cerebral cortex. *Science* **270**, 764–769.

McPeek, R. M., Maljkovic, V., and Nakayama, K. (1999). Saccades require focal attention and are facilitated by a short-term memory system. *Vision Res.* **39**, 1555–1566.

Moore, T., and Armstrong, K. M. (2003). Selective gating of visual signals by microstimulation of frontal cortex. *Nature* **421**, 370–373.

Moore, T., and Fallah, M. (2001). Control of eye movements and spatial attention. *Proc. Natl. Acad. Sci. USA* **98**, 1273–1276.

Sato, T. R., and Schall, J. D. (2003). Effects of stimulus–response compatibility on neural selection in frontal eye field. *Neuron* **38**, 637–648.

Sato, T. R., Watanabe, K., Thompson, K. G., and Schall, J. D. (2003). Effect of target-distractor similarity on FEF visual selection in the absence of the target. *Exp. Brain Res.* **151**, 356–363.

Schall, J. D. (1997). Visuomotor areas of the frontal lobe. *In* "Cerebral Cortex," Vol. 12: "Extrastriate Cortex of Primates" (K. Rockland, A. Peters, and J. Kaas, Eds.), pp. 527–638. Plenum, New York.

CHAPTER

22

Dissociation of Selection from Saccade Programming

Kirk G. Thompson

ABSTRACT

In monkeys performing visual search tasks, frontal eye field visual neurons discriminate the singleton target from distracters in a search array before a saccade to that target is made. Recently, experiments have been done to determine whether the selection process corresponds to visual selection or to the preparation of the next saccade. The evidence indicates that the selective activity observed in the frontal eye field is an explicit representation of the visual stimulus and not an obligatory saccade command. Distinct neural processes for visual selection and saccade production provide flexibility in visuomotor transformations. The visual selection in the frontal eye field corresponds to the allocation of covert spatial attention that can subsequently be used to guide overt saccades and other visually guided behaviors.

I. INTRODUCTION

We are faced with an ever-changing visual world with a multitude of objects simultaneously vying for our attention. Although we usually look at the things that "catch" our attention (see Chapter 20), we are considerably limited in that we can direct our gaze toward only one narrowly focused region of the visual world at a time. To overcome this limitation, we inspect complex visual scenes using sequences of fast ballistic eye movements, called *saccades*, to direct the fovea to interesting or "attention-grabbing" locations for further visual processing. Visual processing takes place during intersaccadic intervals while gaze is held steady. During each period of fixation, at least two selection processes must take place: a visual selection process by which the coordinates of one object or location is identified, and a motor selection process by which the parameters of the next saccade are programmed. During active scanning, the cycle of visual encoding, target selection, and saccade execution occurs about three to five times per second.

Behavioral studies have shown that before a saccade, visual attention is obligatorily directed to the saccade target (Kowler et al., 1995; Deubel and Schneider, 1996). This suggests that saccade target selection and visual spatial attention are implemented via a common mechanism (see Chapters 20 and 31). However, the reverse is not the case. The deployment of attention toward a peripheral location does not necessarily lead to a saccade. Numerous studies have shown that attention can be directed covertly to specific objects or locations in the visual field without saccades (Kinchla, 1992; Egeth and Yantis, 1997) (see Chapter 73). Therefore, visual selection and motor selection processes seem to be organized independently. This chapter reviews recent neurophysiological evidence obtained from the frontal eye field of monkeys performing visual search tasks that identifies a visual selection process that precedes saccades but is only loosely related to saccade production. Instead, the selective activity in FEF is more closely related to visual and behavioral salience (see Chapter 94) and likely guides both covert and overt orienting (see Chapter 21).

II. THE FRONTAL EYE FIELD

A. Role in Saccade Production

The frontal eye field (FEF) is located in the rostral bank of the arcuate sulcus in the prefrontal cortex of macaques and is usually regarded as part of the

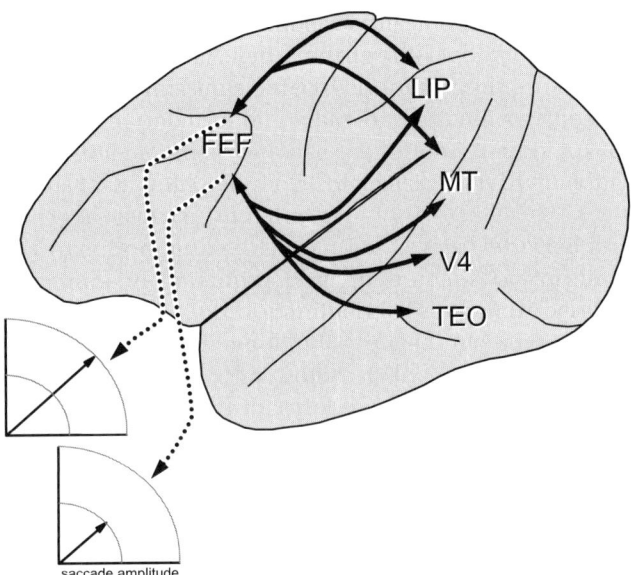

FIGURE 22.1 Connectivity between FEF and visual cortex. FEF is reciprocally connected with multiple extrastriate cortical areas and sends a saccade command to the brainstem. FEF is topographically organized. Ventrolateral FEF is connected with foveal representations in visual cortex and produces shorter-amplitude saccades. Dorsomedial FEF is connected with more peripheral visual field representations in visual cortex and produces longer-amplitude saccades.

oculomotor system (Fig. 22.1). The evidence for this is beyond dispute. Low-intensity electrical stimulation of FEF elicits saccades that are topographically organized (Bruce and Goldberg, 1985), and reversible inactivation of FEF prevents saccades (Sommer and Tehovnik, 1997). Two types of neurons found in FEF are directly related to gaze. Movement neurons are active before and during saccades, and fixation neurons are active while gaze is held steady. Movement and fixation neurons are located in layer 5 and innervate the superior colliculus and parts of the neural circuit in the brainstem that generates saccades (Segraves, 1997). The function of FEF in gaze control has been reviewed (Hanes et al., 1998).

B. Role in Visual Processing

FEF is also a part of the visual system (reviewed in Schall et al., 2003). FEF is reciprocally connected with both the dorsal and ventral visual processing streams, and these connections are topographically organized (Fig. 22.1) (Schall et al., 1995b). The central visual field representations of retinotopically organized areas such as V4, TEO, and MT, as well as areas that overrepresent the central field (e.g., caudal TE), project to the ventrolateral portion of FEF. This part of FEF produces short-amplitude saccades. The peripheral field representation of retinotopically organized areas, as well as areas that overrepresent the peripheral visual field (e.g., PO and MSTd), project to the dorsomedial part of FEF. This part of FEF produces larger-amplitude saccades. Roughly half of FEF neurons have visual responses with spatially defined receptive fields. Unlike neurons found in extrastriate visual cortex (see Chapter 49), FEF visual neurons do not exhibit selectivity for specific features (e.g., color, shape, motion), but instead exhibit selective activation that is related to visual salience (see Chapter 21).

III. TARGET SELECTION IN FEF

The visual search paradigm has been used extensively to investigate visual selection and attention (see Chapters 17, 65, 69, and 90). In a visual search task, multiple objects are presented, and from among them a target is identified. The task of finding the target among the many objects can be made easy or hard by manipulating target/nontarget similarity. The term *popout* refers to the search condition in which the target can be distinguished from distracters with seemingly no effort, such as a red spot among several green spots (see Chapters 34 and 68). To investigate how the brain selects targets for visually guided saccades, recordings of neural activity were made in the FEF of monkeys trained to shift gaze to an oddball target stimulus in a popout search array (Schall et al., 1995a). Most visually responsive neurons in FEF responded initially indiscriminately to the target or the distracter of the search array in their receptive field. However, before a saccade shifted gaze to the target, a discrimination process transpired by which most visually responsive neurons in FEF ultimately signaled the location of the oddball target stimulus. Thus, instead of representing specific features of objects, visual neurons in FEF represent the behavioral relevance of objects. But is FEF selection an explicit visual selection signal, or is it related exclusively to saccade production? The next section reviews a series of recent experiments designed to distinguish between these two possibilities.

IV. DISSOCIATION OF TARGET SELECTION FROM SACCADE PRODUCTION

A. Dissociation in Time

Following the observation that the activity of FEF neurons identifies the target for the upcoming saccade, an analysis was carried out to determine when the

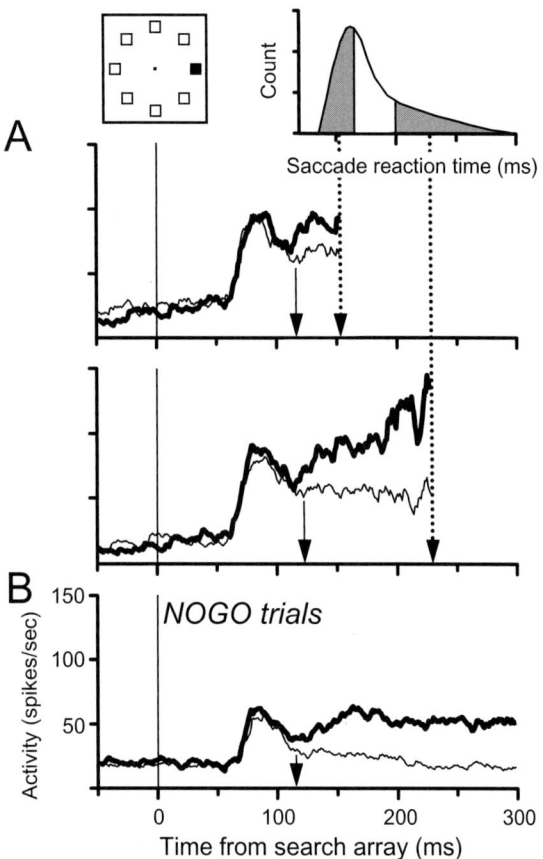

FIGURE 22.2 Activity of an FEF visual neuron following presentation of a popout search array. Each plot shows the average activation when the oddball stimulus appeared in the neuron's receptive field (thick lines), and when distracters appeared in the receptive field (thin line). (A) The top two plots show the average activity, grouped according to reaction time, during a block of trials in which the monkey made a saccade to the oddball. The ranges of reaction times selected for analysis are indicated by the shaded regions of the reaction time distribution in the upper plot. As reaction time increases (dotted arrows), the time of target selection does not (solid arrows). (B) The bottom plot shows the activity of the same neuron during a block of trials in which the monkey was instructed to withhold a saccade while passively viewing the same search array. When no saccade was produced, the overall activation was attenuated, but the same selection process was observed. Thus, selection was not contingent on saccade production. Adapted from Thompson et al. (1996, 1997).

selection was accomplished in relation to the time of the presentation search array and the time of the saccade (Thompson et al., 1996). It was found that FEF visual neurons discriminate the target from the distracters in a popout search array at a fairly constant interval after search array presentation (Fig. 22.2A). In other words, the time at which the target of the saccade was identified did not predict when the eyes moved. This finding demonstrated that, in an easy popout search task, a relatively constant period is taken to identify the saccade target, and the variability in saccade reaction time is due to variability in the time needed to prepare and execute the eye movement.

Subsequent experiments found that when the search task is made difficult by making the target similar to the distracters, the monkey's reaction time increases and the time taken by FEF neurons to select the target increases (Sato et al., 2001). This increase in time taken for FEF neurons to locate targets is accompanied by an increase in the variability of the selection time across trials, and accounts for a larger fraction of the variability in the monkey's reaction time. This occurs because the production of an accurate saccade cannot proceed until the target is located. However, when a monkey's reaction time increases as a result of interference in motor preparation processes, as what occurs in an a popout search task in which the target changes locations (see the next section), there is not a concomitant increase in the duration or variability of time taken by FEF neurons to select the oddball target. These findings indicate that the time of target discrimination in FEF marks the conclusion of the visual selection stage of processing.

B. Dissociation in Saccade Choice

The question of whether the visual selection in FEF necessarily leads to a saccade to the selected location was addressed in an experiment in which the target of a search array changes location before the monkey can make a saccade (Murthy et al., 2001). In the search-step task, most trials were identical to the popout search task in which the monkey was rewarded for making a single saccade to the oddball, but on about a third of the trials the target unexpectedly switched places with a distracter. The monkeys' task was to shift gaze to the new target location to receive a juice reward. The timing of the target step was adjusted so that the monkeys were able to compensate for the target step on about half the trials. On the remaining trials the monkeys did not compensate for the target step and made a saccade to the original target location and were not rewarded. The question to be answered was: Does the selective activation track the location of the popout target or does the selective activation predict the monkey's impending eye movement? The answer is that the selective activity reflects the location of the visual stimulus, not the monkey's behavior. This was most evident on those trials in which the first stimulus presented in the receptive was a distracter and then changed to the target.

Figure 22.3 shows the typical results obtained from FEF visual neurons during the search step task when the distracter was initially presented in the receptive

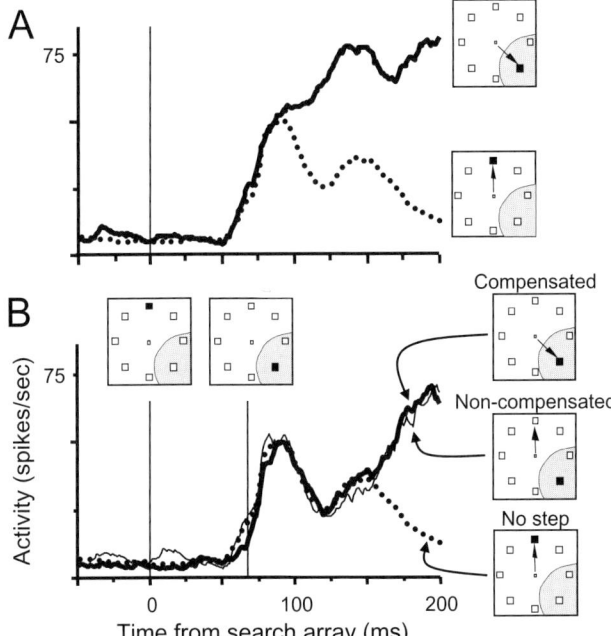

FIGURE 22.3 Activity of an FEF visual neuron during the search-step task. Most of the trials were identical to the popout search task shown in Fig. 22.2. But on about one-third of the trials, the oddball target and a distracter switched places before the monkey could make a saccade to it. The monkey was rewarded for making a saccade to the new target location. The timing was adjusted so that the monkey was successful on about half of the trials (compensated trials). On the remaining trials, the monkey made the first saccade to the original target location (noncompensated trials) and was not rewarded. The most revealing trials were those that began with a distracter in the receptive field (shaded region in search array panels). Initially, the activity on step trials (solid lines) was identical to that on no step trials (dotted line). About 130 ms following the target's appearance in the receptive field, the activity on both compensated (thick line) and noncompensated (thin line) trials grew equally strongly. The activity selected the location of the oddball stimulus, not the location of the saccade target. Reprinted, with permission, from Murthy et al. (2001).

field. The early phase of the response was identical to the response elicited by a distracter during trials in which the target does not step; the early nonselective visual activity was followed by decrease in activation. This is because the stimulus conditions were identical prior to the target step. After the target appeared in the receptive field, the activation grew substantially. The critical point is that this growth of activity occurred not only on trials in which the saccade was into the neuron's receptive field, but also on trials in which the saccade was to the original target location outside the neuron's receptive field. In other words, the activity of this neuron represented accurately the new target location regardless of what saccade was made. This is strong evidence that the selection process in FEF visual neurons is distinct from immediate saccade production. However, following an incorrect saccade to the first target location, the monkeys usually made a second saccade to the new target location. This behavior introduces the possibility that the selective activation can be used to guide the corrective saccade and, therefore, may be a form of saccade planning. The next section describes two experiments that address this question.

C. Visual Selection without Saccades

Does visual selection require saccade execution? To answer this question, the same monkeys that were trained to make a saccade to the oddball in the popout search experiment (Schall et al., 1995a; Thompson et al., 1996) were also trained in a NOGO version of the visual search task (Fig. 22.2B). The monkeys were rewarded for maintaining fixation while passively viewing the visual search array (Thompson et al., 1997). In NOGO search, FEF visual neurons still discriminated the oddball from distracters at the same time and to the same degree as when a gaze shift was produced in the search task in which saccades were made. However, although no saccade was made to the oddball location after the trial was over, the possibility exists that saccades were being planned but suppressed during the trial. This is a notable concern because these monkeys were highly trained to make saccades to the oddball in the GO version of the search task. Nevertheless, we can conclude that the visual selection observed in FEF does not require saccade execution.

The question of the necessity of saccade programming for visual selection was recently addressed in a manual version of the popout search task (Biscoe and Thompson, 2003) (Fig. 22.4). In monkeys that were trained only to fixate (i.e., they were never rewarded for making a saccade to any object other than a fixation spot), recordings were made from visual and movement neurons while they reported the location of the oddball using a manual lever turn. Neurons were classified as having visual- and/or movement-related modulation based on the activity during a memory-guided saccade task to a flashed fixation spot. During the lever search task most visually responsive neurons exhibited a selective response for the oddball stimulus before the manual lever response. This selection usually disappeared after the lever turn but before the search array disappeared. There are two reasons why we are confident that saccades were not being planned during this task. First, monkeys had ample opportunity and were free to shift gaze to the oddball following the reward for turning the lever but before the search array was removed, but they did not. Instead,

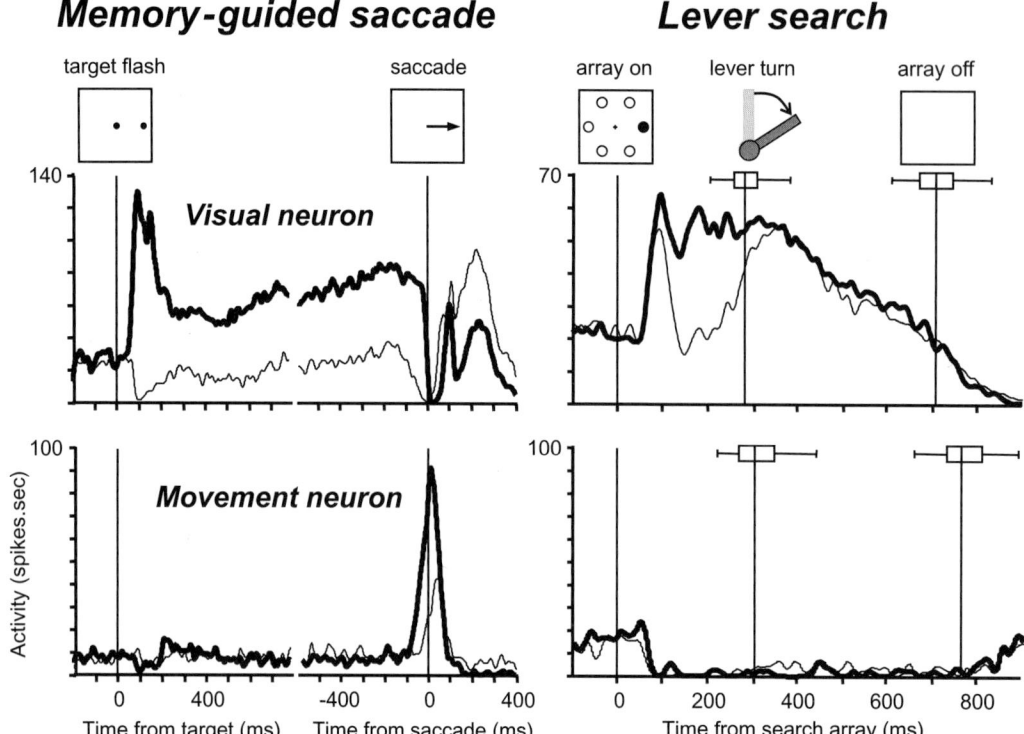

FIGURE 22.4 Activity of an FEF visual neuron (top) and an FEF movement neuron (bottom) recorded during blocks of memory-guided saccade trials (left) and blocks of popout search trials in which the monkey indicated oddball location with a lever turn (right). In the memory-guided saccade task, the monkey was required to hold gaze on a central spot for 800 to 1200 ms after a peripheral spot flashed for 50 ms. After the central spot disappeared, the monkey was rewarded for making a saccade to the remembered peripheral location. This task is useful for identifying neurons with visual (top)- and movement (bottom)-related activity. The activity is aligned on target presentation and saccade initiation. In the "lever search" task the monkey was required to maintain fixation on the central spot and was rewarded 100 ms after indicating, with a lever turn, which side of the fixation spot the oddball appeared. Even though the search array remained on for about 400 ms after the reward for the correct lever turn, the monkey did not tend to look at the oddball. The box-whisker plots indicate the median (vertical line) and range of lever reaction times and subsequent search array offset times. Before the lever turn, the visual neuron exhibited target selection, but the movement neuron was silent. Even though there was no evidence of saccade planning in either the monkey's behavior or the activity of saccade-related neurons, visual selection occurred in FEF visual neurons.

during this time they consistently maintained fixation on the central spot and returned the lever to the central position to begin the next trial. Second, the activity of movement neurons was suppressed during the lever search task. There was no evidence of covert saccade planning in neurons with saccade-related activity. Therefore, we conclude that the visual selection signal in FEF is an explicit representation of visual salience and is independent of saccade production.

V. DISCUSSION

It can be argued that visual selection is the most important function of the visual system. The visual selection process identifies which object to act on. But once an object is selected, what is to be done with it? Look at it or not? Reach for it or not? Pursue it or avoid it? Turn the lever left or right? Given the complexity of the world we live in and our limitations in being able to make only one action of a kind at a time, separate visual and motor selection processes are a necessity. The dissociation of visual selection from saccade production in FEF highlights this flexibility in stimulus–response mappings.

Orienting saccadic eye movements and covert spatial attention probably co-evolved to support active vision (see Chapter 20). Covert spatial attention without eye movements improves perceptual abilities (see Chapters 73 and 81), and modulates feature-specific activity in visual cortex (see Chapters 49 and 80). Interestingly, a set of experiments by Moore and colleagues showed that electrical stimulation of FEF in monkeys improves perceptual ability (Moore and Fallah, 2001) and modulates visual related activity in extrastriate visual cortex (Moore and Armstrong, 2003). In addition, functional imaging studies have shown that the human homolog of FEF is involved in both overt and covert shifts of attention (Corbetta and Shulman, 2002). These findings, combined with the evidence reviewed in this chapter, lead to the conclusion that, in addition to providing a signal for guiding action, the selective activation of FEF visual neurons corresponds to the allocation of visual spatial attention and can influence ongoing visual processing (see Chapters 18, 25, and 50).

References

Biscoe, K. L., and Thompson, K. G. (2003). Visual selection without saccade planning in the frontal eye field of macaques. *In* "Soc. Neurosci. Abstr., Program No. 180.7, Abstracts" Viewer/Itinerary Planner. CD-ROM. Washington, DC.

Bruce, C. J., and Goldberg, M. E. (1985). Primate frontal eye fields: I. Single neurons discharging before saccades. *J. Neurophysiol.* **53**, 603–635.

Corbetta, M., and Shulman, G. L. (2002). Control of goal-directed and stimulus-driven attention in the brain. *Nat. Rev. Neurosci.* **3**, 201–215.

Deubel, H., and Schneider, W. (1996). Saccade target selection and object recognition: evidence for a common attentional mechanism. *Vision Res.* **36**, 1827–1837.

Egeth, H. E., and Yantis, S. (1997). Visual attention: control, representation, and time course. *Annu. Rev. Psychol.* **48**, 269–297.

Hanes, D. P., Patterson, W. F., and Schall, J. D. (1998). The role of frontal eye field in countermanding saccades: visual, movement, and fixation activity. *J. Neurophysiol.* **79**, 817–834.

Kinchla, R. A. (1992). Attention. *Annu. Rev. Psychol.* **43**, 711–743.

Kowler, E., Anderson, E., Dosher, B., and Blaser E. (1995). The role of attention in the programming of saccades. *Vision Res.* **35**, 1897–1916.

Moore, T., and Armstrong, K. M. (2003). Selective gating of visual signals by microstimulation of frontal cortex. *Nature* **421**, 370–373.

Moore, T., and Fallah, M. (2001). Control of eye movements and spatial attention. *Proc. Natl. Acad. Sci. USA* **98**, 1273–1276.

Murthy, A., Thompson, K. G., and Schall, J. D. (2001). Dynamic dissociation of visual selection from saccade programming in frontal eye field. *J. Neurophysiol.* **86**, 2634–2637.

Sato, T., Murthy, A., Thompson, K. G., and Schall, J. D. (2001). Search efficiency but not response interference affects visual selection in frontal eye field. *Neuron* **30**, 583–591.

Schall, J. D., Hanes, D. P., Thompson, K. G., and King, D. J. (1995a). Saccade target selection in frontal eye field of macaques: I. Visual and premovement activation. *J. Neurosci.* **15**, 6905–6918.

Schall, J. D., Morel, A., King, D. J., and Bullier, J. (1995b). Topography of visual cortical afferents to frontal eye field in macaque: convergence and segregation of processing streams. *J. Neurosci.* **15**, 4464–4487.

Schall, J. D., Thompson, K. G., Bichot, N. P., Murthy, A., and Sato T. R. (2003). Visual processing in the macaque frontal eye. *In* "The Primate Visual System" (J. Kass and C. Collins, Eds.), CRC Press, Boca Raton, FL.

Segraves, M. A. (1992). Activity of monkey frontal eye field neurons projecting to oculomotor regions of the pons. *J. Neurophysiol.* **68**, 1967–1985.

Sommer, M. A., and Tehovnik, E. J. (1997). Reversible inactivation of macaque frontal eye field. *Exp. Brain Res.* **116**, 229–249.

Thompson, K. G., Bichot, N. P., and Schall, J. D. (1997). Dissociation of target selection from saccade planning in macaque frontal eye field. *J. Neurophysiol.* **77**, 1046–1050.

Thompson, K. G., Hanes, D. P., Bichot, N. P., and Schall, J. D. (1996). Perceptual and motor processing stages identified in the activity of macaque frontal eye field neurons during visual search. *J. Neurophysiol.* **76**, 4040–4055.

CHAPTER

23

Space- and Object-Based Attention

Michael C. Mozer, Shaun P. Vecera

ABSTRACT

Behavioral studies of visual attention have suggested two complementary modes of selection. In a space-based mode, locations in the visual field are selected; in an object-based mode, organized chunks of visual information—roughly, objects—are selected, even if the objects overlap in space or are spatially discontinuous. Although the two modes are distinct, they can operate in concert to influence the allocation of attention. This chapter presents key experimental results on space- and object-based attention and their interaction, and sketches a theoretical framework in which the two attentional modes can be unified. This chapter also discusses alternative notions of object-based attention, from perceptual grouping of low-level features in a retinotopic reference frame to construction of structural descriptions, and argues that the data are consistent with the former—a simple, low-level mechanism.

I. TWO MODES OF ATTENTIONAL SELECTION

Behavioral studies of visual attention have suggested two distinct and complementary modes of selection, one involving space and the other objects. In a space-based mode, stimuli are selected by location in the visual field (e.g., Ericksen and Hoffman, 1973; Posner, 1980). Evidence for this mode has come from a variety of sources, including spatial pre-cuing tasks, in which an abrupt luminance change (a *cue*) summons attention to a region in space. Observers are faster to detect or identify a subsequent target that appears at the cued location than one that appears at an uncued location (Posner, 1980). The space-based mode of attention has given rise to the attention-as-a-spotlight metaphor, in which attention acts as a beam to illuminate a contiguous region of the visual field. More recently, a zoom-lens metaphor has been suggested (Eriksen and Yeh, 1985), consistent with the finding that the region of space selected by attention can vary in size.

In contrast to the space-based mode, evidence has also been found for an object-based mode in which attention is directed to organized chunks of visual information corresponding to an object or a coherent form in the environment, even if objects overlap in space or are spatially discontinuous. All visual features of an attended object are processed concurrently, and features of an attended object are processed faster and more accurately than features of other objects. In one well-known task (Duncan, 1984), observers view two overlapping objects, a box and a line. Each object varies on two feature dimensions: the box is short or tall and has a gap on its left or right side; the line is dotted or dashed and tilts to the left or right. Observers are instructed to report pairs of features. Observers are more accurate at reporting two features of the same object (e.g., the height and side of gap of the box) than two features that belong to different objects (e.g., the height of the box and the tilt of the line). The cost in accuracy cannot be attributed to spatial factors, because the two objects overlap in space; rather, the cost must be attributed to the switching of attention from one object to the other. Indeed, Vecera and Farah (1994) have shown that no additional cost is incurred if the two objects are separated in space, suggesting that spatial factors are not at play in the object-based deployment of attention.

Some studies have shown that both spatial and object factors can simultaneously influence the allocation of attention. Egly et al. (1994) presented displays containing two rectangles (Fig. 23.1a). One end of one of the rectangles is cued with a brief flicker (Fig. 23.1b); a target then appears, and observers make a key-press response to the appearance of the target. The target

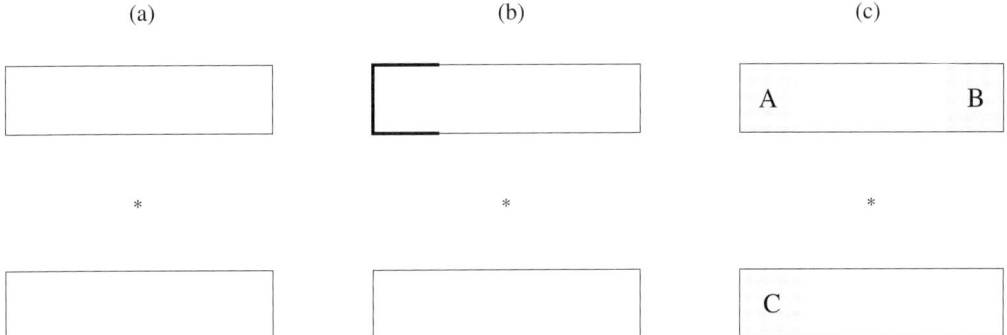

FIGURE 23.1 The Egly et al. (1994) experiment: (a) start of trial, (b) cue, (c) possible target locations.

appears either at the cued location, at the uncued end of the cued object, or in the uncued object (locations A, B, and C in Fig. 23.1c). Observers show a cue validity effect: detecting a target at the cued location is fastest. This result can be interpreted in terms of ordinary space-based attention. However, although the distance from the cued location to the target is the same for the noncued target locations, observers are faster to detect targets at the uncued end of the cued object (B) than those in the uncued object (C), indicating that the cue summoned attention to the entire cued object. This object-based effect is nonetheless modulated by spatial proximity: If the two objects are moved close together, the object effect is reduced in magnitude (Vecera, 1994). Further interactions between space- and object-based attention have been found via demonstrations that object-based effects occur only within the focus of spatial attention (Lavie and Driver, 1996), and the outputs of preattentive grouping processes influence the allocations of spatial attention (e.g., Baylis and Driver, 1992; Kramer and Jacobson, 1991; Logan, 1996).

II. CLARIFYING THE NOTION OF OBJECT-BASED ATTENTION

The phrase *object based* is ambiguous, and a lack of clarity as to its intended meaning has resulted in some confusion in the literature. "Object-based" can be a descriptive term for experimental results: An object-based effect is observed in any experimental study in which attentional allocation or performance depends not merely on the location of an object in space, but on the extent, shape, or movement of the object itself. "Object based" can also be a characterization of processes and internal representations. Object-based representations arise from processes that use object-based frames of reference to transform visual features to achieve partial or complete view invariance. Object-based effects do not require object-based representations or frames of reference (Mozer, 2002; Vecera, 1994).

The distinction between object-based effects and object-based representations does not entirely remove the ambiguity in the phrase "object based." One can conceive of a continuum of senses in which attentional mechanisms and representations might be considered object based. Examples of at least four alternatives can be found in the literature. Ordered from weakest to strongest notions of object based, these alternatives are as follows. (See Driver (1999) for a similar enumeration.)

1. *Grouping in a viewer-based frame* (Grossberg and Raizada, 2000; Mozer et al., 1992; Vecera, 1994; Vecera and Farah, 1994). Attention might act to select the set of locations in which visual features of an object are present. The resulting segmentation has been referred to as a *grouped array representation* (Vecera, 1994), because visual features are coded in a viewer-centered (e.g., retinotopic) array of locations, and labeled to indicate their grouping (Fig. 23.2). The segmentation can be achieved via heuristics, such as the Gestalt grouping principles, or might exploit low-order statistical regularities in visual scenes (Mozer et al., 1992). Whole-object knowledge is not required, nor are object-based frames of reference.

2. *Grouping and determination of principal axis* (Driver, 1999). In addition to performing segmentation in a viewer-based frame, attentional processes might also determine the principal axis of an object: the axis of symmetry or elongation. Using the axis to establish a partial frame of reference, such as an up–down direction, visual features could be reinterpreted with respect to the partial frame. For example, the shape in Fig. 23.3a evokes a principal axis from which the midline of the shape can be

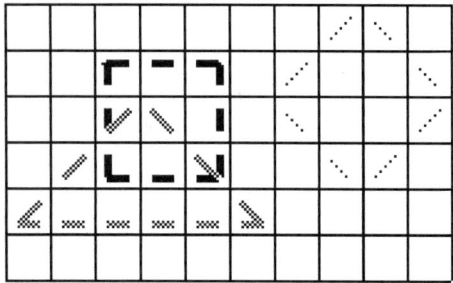

FIGURE 23.2 Illustration of the grouped array representation. Grouping of the visual features is indicated by their shading.

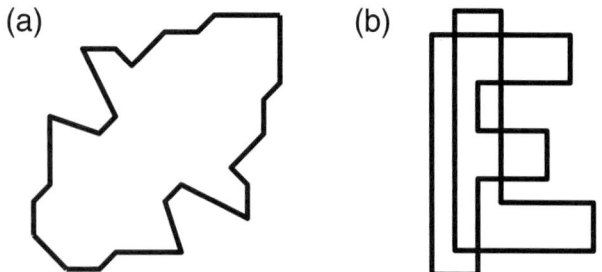

FIGURE 23.3 (a) A shape and its principal axis. (b) Sample stimulus from Vecera and Farah (1997).

determined, and the left–right position of visual features is then determined with respect to the midline, although the specification of which direction is "left" and which is "right" can be derived from the viewer-based frame.

3. *Grouping and determination of an object-based frame of reference* (Marr and Nishihara, 1978). Attentional processes might determine not only the up–down direction of an object, but also its left–right and front–back direction, allowing for the establishment of a full-blown object-based frame of reference. It is difficult to imagine that this process could operate in a purely bottom-up fashion; most likely contact with object representations stored in long-term visual memory would be necessary.

4. *Grouping and determination of a structural description* (Biederman, 1987). For complex, articulated objects, attention might operate on a structural description of an object—a description that decomposes an object or scene into its parts, and that characterizes the relationships among the parts in terms of multiple allocentric frames of reference. Attention would then act to select a subtree of the structural description.

These alternatives vary on two dimensions: the reference frame in which object-based attention operates—egocentric for (1) and (2), allocentric for (3) and (4)—and on the degree of interaction with object knowledge required for selection—from weak, low-order statistics for (1) to high-order object knowledge for (4). The key issue is the degree to which object-based attentional effects require the explicit computation of object properties, such as a principal axis, frame of reference, or structural description, and the degree to which the data mandate such computations. All four alternatives have been invoked in the literature to explain object-based attentional effects.

III. EVIDENCE POINTS TO ONE ACCOUNT OF OBJECT-BASED ATTENTIONAL SELECTION

In terms of the computations required, the grouped array account of object-based selection is the simplest and can operate at the earliest point in the visual processing stream, and hence should be preferred on the grounds of parsimony. Mozer (2002) and Vecera (1994) have argued that a variety of experimental results seeming to call for more complex accounts of object-based attention can in fact be explained in terms of the grouped array account, and object-based representations and contact with object knowledge are not required. Although the existence of multiple attentional processes raises the possibility of multiple types of object-based selection, the grouped array account can explain most, if not all, of the object-based attention literature.

A complete account of object-based attention in terms of grouped arrays requires an understanding of the cues that determine grouping and, hence, that specify objects. The Gestalt cues of similarity, closure, and connectedness are all grouping cues that can influence the allocation of attention. For example, Kramer and Jacobson (1991) reported that a target that was physically connected to adjacent flanking items was attended as a single unit or group; a target that was not connected to the flankers could be selectively attended with little influence from the surrounding flankers (also see Baylis and Driver, 1992). Thus, grouping cues, including grouping, by similarity, connectedness, and good continuation, can determine which stimuli or visual features are attended simultaneously.

Grouping cues need not be primitive and innate; they might also be learned and influenced by familiarity with a visual environment. For example, Vecera and Farah (1997) presented displays consisting of two overlapping outline block letters, in either the upright or the inverted position (Fig. 23.3b). Observers were asked to determine whether an "X" in the display was contained in one of the forms or in neither of the forms. Response times were faster for upright displays than inverted displays, suggesting that familiarity with upright letter forms was at play in the allocation of object-based attention. This result seems to argue against an account that relied solely on Gestalt grouping cues for segmentation, but might nonetheless be explained by the existence of adaptive grouping mechanisms that exploit low-order statistical regularities in the environment (Zemel et al., 2002).

IV. RELATIONSHIP BETWEEN SPACE-BASED AND OBJECT-BASED ATTENTION

Although many grouping cues have been identified, the relationship between these cues, which direct attention to objects, and the factors directing attention to locations in space has proved elusive. Initially, evidence for both space-based and object-based attentional selection led to a debate about whether selection was object based or space based. The current consensus is that both of these attentional modes coexist in the visual system and may influence one another. Despite this emerging consensus, many studies continue to address the relationship between these two modes of selection as if one mode of selection is more important than the other. For example, Lavie and Driver (1996) suggested a space-then-object account by demonstrating that object-based effects occur only within the focus of spatial attention. Other theorists have argued for an object-then-space relationship in which that the outputs of preattentive grouping processes influence the allocation of spatial attention (e.g., Baylis and Driver, 1992; Kramer and Jacobson, 1991; Logan, 1996).

Neither of these accounts is completely satisfactory. For example, data exist that appear inconsistent with the object-then-space account. Mack et al. (1992) demonstrated that if spatial attention is occupied at fixation by a visually demanding discrimination, grouping fails. If grouping occurred before attention, then occupying spatial attention with a demanding task should not impair object-based grouping. Similarly, a space-then-object account has difficulty explaining results from object-based attention studies. In many studies (e.g., Kramer and Jacobson, 1991), observers are instructed to perform a discrimination on a centrally presented target item. With such instructions, spatial attention should be focused centrally, and grouping outside this central region should not influence observers' responses to this central target.

The alternative to accounts supposing primacy of either space-based or object-based attention is an interactive account in which space- and object-based attentional processing operate simultaneously, each one helping to guide the other. However, interactive accounts face a serious computational problem. Object attention requires a search to partition visual features into objects; spatial attention requires a search for salient locations in the visual field. Each of these searches entails distinct and possibly conflicting computational objectives and, hence, incompatible solutions. Consequently, the reality of interactive accounts is that they are tricky to implement: The solutions reached are often suboptimal, where each search converges but the two outcomes are inconsistent with one another and each is suboptimal within its own domain (Hinton and Lang, 1985; Mozer et al., 1992).

Thus, a significant challenge lies ahead to unify mechanisms of space- and object-based attention. The grouped array view of object-based attention provides one key insight toward a coherent theory, via its proposal of a common substrate for the two varieties of attention: a topographic, viewer-based representation of space, often referred to in the attentional modeling literature as a saliency map. Another key insight concerns the role of strategic control. Because one form of attention does not always dominate over the other, it is likely that task demands and stimulus structure influence the relative contribution of each form of attention. Thus, a complete theory of attention requires claims concerning the processes by which the flexible attentional system is configured to operate for a given task in a given environment.

V. TOWARD A UNIFIED THEORY OF SPACE- AND OBJECT-BASED ATTENTION

Rather than viewing space-based and object-based attention as two qualitatively different forms of attention, unification is possible by conceptualizing attentional processing as fundamentally aimed at grouping related locations in the visual field. One can think of space-based and object-based processes as providing weak constraints concerning which locations belong together: Space-based attention suggests that adjacent, contiguous locations be grouped; object-based attention suggests that locations containing visual features likely to belong to the same object be grouped. The attentional state is then determined by a constraint-satisfaction search that attempts to identify groupings that are consistent with as many of the suggestions as possible. Thus, the operation of attention is viewed as a single search, not two searches with distinct goals; this unification is possible via a shared representation of space.

This view suggests a weaker role for grouping processes than is ordinarily considered. Grouping processes can be heuristic and spatially local and can operate on multiple dimensions (e.g., color, shape) independently, and the global attentional state results from resolving the assorted grouping constraints with the space-based constraints.

Given the contribution of constraints from many different processes converging in attentional selection, the weighting of constraints becomes a key issue. Space-based attentional states result when space-based constraints dominate; object-based attentional states result when object-based constraints dominate. Because the data suggest the nature of the task and the stimulus display can influence which form of attention dominates, it seems natural to suggest that the weighting of constraints is under strategic control. One particularly elegant, computationally motivated form of control might involve selection of task- and environment-specific weightings that yield optimal performance, e.g., minimal response time or maximal accuracy. For example, reinforcement learning (Sutton and Barto, 1998) might be used to fine-tune the operation of the attentional system to achieve optimal performance. Beyond the virtue of integrating space-based and object-based attention, this perspective has the additional potential virtue of explaining integration of the various and diverse Gestalt grouping cues in determination of the attentional state.

References

Baylis, G. C., and Driver, J. (1992). Visual parsing and response competition: the effect of grouping factors. *Percept. Psychophys.* **51**, 145–162.

Biederman, I. (1987). Recognition-by-components: a theory of human image understanding. *Psychol. Rev.* **94**, 115–147.

Driver, J. (1999). Egocentric and object-based visual neglect. *In* "The Hippocampal and Parietal Foundations of Spatial Cognition" (N. Burgess, K. J. Jeffery, and J. O. Keefe, Eds.), pp. 67–89. Oxford Univ. Press, New York.

Driver, J., Baylis, G. C., Goodrich, S. J., and Rafal, R. D. (1994). Axis-based neglect of visual shapes. *Neuropsychologia* **32**, 1353–1365.

Duncan, J. (1984). Selective attention and the organization of visual information. *J. Exp. Psychol. Gen.* **113**, 501–517.

Egly, R., Driver, J., and Rafal, R. D. (1994). Shifting visual attention between objects and locations: evidence from normal and parietal lesion subjects. *J. Exp. Psychol. Gen.* **123**, 161–177.

Eriksen, C. W., and Hoffman, J. E. (1973). The extent of processing of noise elements during selective encoding from visual displays. *Percept. Psychophys.* **14**, 155–160.

Eriksen, C. W., and Yeh, Y.-Y. (1985). Allocation of attention in the visual field. *J. Exp. Psychol. Hum. Percept. Perform.* **11**, 583–597.

Grossberg, S., and Raizada, R. D. S. (2000). Contrast-sensitive perceptual grouping and object-based attention in the laminar circuits of primary visual cortex. *Vision Res.* **40**, 1413–1432.

Hinton, G. E., and Lang, K. J. (1985). Shape recognition and illusory conjunctions. *In* "Ninth Annual Joint Conference on Artificial Intelligence," pp. 252–259. Morgan–Kaufmann, Los Altos, CA.

Kramer, A. F., and Jacobson, A. (1991). Perceptual organization and focused attention: the role of objects and proximity in visual processing. *Percept. Psychophys.* **50**, 267–284.

Lavie, N., and Driver, J. (1996). On the spatial extent of attention in object-based visual selection. *Percept. Psychophys.* **58**, 1238–1251.

Logan, G. D. (1996). The CODE theory of visual attention: an integration of space-based and object-based attention. *Psychol. Rev.* **103**, 603–649.

Mack, A., Tang, B., Tuma, R., Kahn, S., and Rock, I. (1992). Perceptual organization and attention. *Cogn. Psychol.* **24**, 475–501.

Marr, D., and Nishihara, H. K. (1978). Representation and recognition of the spatial organization of three dimensional structure. *Proc. R. Soc. London Ser. B* **200**, 269–294.

Mozer, M. C. (2002). Frames of reference in unilateral neglect and visual perception: a computational perspective. *Psychol. Rev.* **109**, 156–185.

Mozer, M. C., Zemel, R. S., Behrmann, M., and Williams, C. K. (1992). Learning to segment images using dynamic feature binding. *Neural Comput.* **4**, 650–665.

Posner, M. I. (1980). Orienting of attention. *Q. J. Exp. Psychol.* **32**, 3–25.

Sutton, R. S., and Barto, A. G. (1998). "Reinforcement Learning." MIT Press, Cambridge, MA.

Vecera, S. P. (1994). Grouped locations and object-based attention: comment on Egly, Driver, and Rafal (1994). *J. Exp. Psychol. Gen.* **123**, 316–320.

Vecera, S. P., and Farah, M. J. (1994). Does visual attention select objects or locations? *J. Exp. Psychol. Gen.* **123**, 146–160.

Vecera, S. P., and Farah, M. J. (1997). Is visual image segmentation a bottom-up or an interactive process? *Percept. Psychophys.* **59**, 1280–1296.

Zemel, R. S., Behrmann, M., and Mozer, M. C. (2002). Experience-dependent perceptual grouping and object-based attention. *J. Exp. Psychol. Hum. Percept. Perform.* **28**, 202–217.

CHAPTER

24

Attention and Binding

Lynn C. Robertson

ABSTRACT

The "binding problem" arose from neurobiological investigations demonstrating different cortical areas of increased neural activity in response to different features of a visual stimulus (e.g., color, motion, shape). Consistently, neuropsychological evidence with humans collected over the prior century suggested that the perception of certain features could be disrupted without disrupting others. For instance, motion, color, shape, size, or location can be uniquely affected with damage to different areas of the human brain. This observation poses the question of what mechanisms bind these features together to account for perception of a unified world. Recent work in cognitive neuroscience has implicated spatial attentional functions associated with parietal lobes in this binding process. The findings were predicted by feature integration theory and show that features can be detected without spatial attention while conjoining two features together seems to requires it. This chapter provides an overview of the evidence and issues.

I. WHAT MUST BE BOUND?

There are many binding problems in perception (Treisman, 1996), but they all basically revolve around the question of how distributed information is integrated to form the coherent world we see. How are the parts of an object bound together to form the whole? How are properties such as color and texture bound to the correct shapes? How are objects bound to their locations? These are but a few of the binding problems in perception that have generated scientific interest.

As with other perceptual and cognitive phenomena, binding problems can be approached from different levels of analysis. For instance, if one asks how an object is bound to a location, a hypothesis could be stated in any of the following ways: (1) Synchronous neural firing across separate cortical areas sends signals upstream to merge with mechanisms that increase or decrease the gain of these signals, (2) Executive functions of the frontal lobe that receive inputs from parietal "where" and temporal "what" processing streams bind what and where information together, (3) Individual feature maps, each coding different features such as color, shape, and size are bound together in a master map of locations. The first is a description at the cellular level, the second at the functional level, and the third at the cognitive level. Whatever the level of focus, there is a binding problem to be solved. The challenge is to find consistency across levels, and one caveat is that arguments against a binding problem at one level do not necessarily explain away a binding problem at the other (e.g., evidence that color-tuned neurons are also orientation-tuned does not negate the neuropsychological evidence that orientation and color perception can be disrupted independently). One must be careful not to discard a problem that exists at one level with conflicting data from another. However, when evidence from different levels converges, much progress can be made.

Whatever level is chosen in addressing a perceptual binding problem, the common theme is that independent nodes (whether they be neurons, assemblies of neurons, cortical areas, or basic perceptual features) must be coordinated to account for an organized and spatially unified view of the world. Binding is required to integrate different modalities, different action systems, different features, as well as binding objects to locations and parts to wholes. A binding problem that is not often targeted for study concerns how the integration of different *spatial* representations from distributed areas of the brain is accomplished. For it is binding of this information into a unified spatial map that provides the spatial locations that attention can selectively enhance or inhibit, the perception of spatial relationships between individuated objects, and the spatial relations between different reference frames (e.g., environment, viewer, object, scene) (see Robertson, 2004).

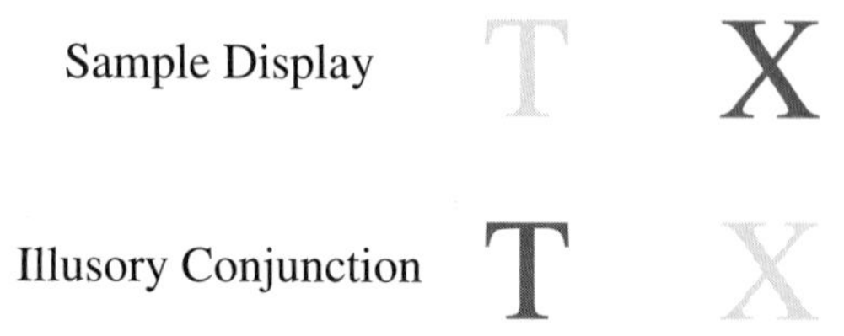

II. IS THERE A BINDING PROBLEM?

Thoughtful and sound arguments have been made against binding as a problem in perception, but there remains the difficulty of dealing with the neuropsychological evidence. Perception breaks down in systematic ways. Although visual agnosias come in various forms and severity, one surprising thing about them is how specific they can be (see Behrmann, 2001). Local elements or parts of an object can become unbound from their global form and be seen floating or misplaced. Objects can be seen without a location, and locations can be seen in the absence of objects. Motion can disappear, producing frozen rivers in summer, or can be added, producing houses that move. Sizes or shapes can change, and perceiving the orientation of a line through vision alone can become nearly impossible. These are real perceptual binding problems to the individuals who experience them. Shapes must be correctly bound to locations, and color and motion must be correctly bound to shape. Parts must be bound to each other to produce meaningful wholes, and so on.

A particular binding problem that has generated much recent interest is that of binding properties of objects such as color, orientation, motion, and shape to form say the perception of an elongated, red wagon moving uphill. There is no one cell or assembly of cells that has yet been found to respond to this exact configuration, and perhaps it should not be surprising that features are distributed in the visual system. Despite the fact that they are delivered to sensation together, they appear to be rapidly channeled to different parts of the brain. Consistently, perception of different features can be disturbed by damage to different brain areas.

But the question at hand is about binding itself, and the neuropsychological evidence has again shown that a particular type of binding, namely, surface feature binding, can be affected by damage to areas outside those that are thought to initially register the basic features in a display. A particularly striking example is that of R.M., a patient with Balint's syndrome produced by bilateral parietal damage. R.M. was able to detect single features such as color and shape in a multi-item array relatively well (Robertson et al., 1997), but he made as many as 38% binding errors

FIGURE 24.1 Example of illusory conjunctions. The sample display contains two shape features (T and X) and two color features (yellow and red). Illusory conjunctions occur when the colors and shapes are incorrectly bound in perception as represented in the lower part of the figure. (See color plate)

(known as illusory conjunctions, or ICs) in displays with only two colored letters (Fig. 24.1). He incorrectly bound color and letter in perception and did so whether the two-letter display was presented for 500ms or 10 seconds. R.M. also perceived ICs with motion and shape (Bernstein and Robertson, 1998) and size and shape (Friedman-Hill et al., 1995). Notably, these binding errors occurred only when the two forms were presented simultaneously in different locations but not when they were presented sequentially in the same location. For R.M., binding represented a spatial problem in perception and not a temporal one. Consistently, it was nearly impossible for him to find a target defined by the conjunction of two features in a cluttered visual search array.

Studies with healthy normal perceivers using functional imaging techniques have verified the importance of the parietal lobes (especially the superior parietal lobe and inferior parietal sulcus) in perceptually binding surface features together (Corbetta et al., 1995; Donner et al., 2002; Shafritz et al., 2002). For instance, searching for a single feature such as color or motion does not necessarily activate parietal areas, while searching for a conjunction of the two features does. Moreover, transient disruption of parietal function in healthy humans by transcranial magnetic stimulation (TMS) disrupts the ability to search for a conjunction of features but does not disrupt single-feature search (Walsh and Cowey, 1998).

In sum, neuropsychological data verify that the parietal lobe is necessary for binding surface features normally in perception, and functional imaging and TMS evidence with normal perceivers support this conclusion. However, it is probably not binding per se that is involved, but rather the need for spatial attention (or, more specifically, spatial representations necessary to guide attention) to bind together features registered in various cortical maps in other brain areas.

III. ATTENTION AND BINDING

As early as 1980, Treisman proposed feature integration theory (FIT) to account for her findings that reaction times increased linearly across set size when searching for a conjunction of two features, but were unaffected by set size when searching for a unique feature (Treisman and Gelade, 1980). She proposed that conjunction search required a serial scan, whereas feature search could be done in parallel. Later, Treisman and Schmidt (1982) reported ICs in normal observers when target stimuli were presented outside the focus of attention. Single features continued to be registered correctly but required spatial attention to be properly bound. The original theory predicted that in a cluttered array, unique features could be detected without spatial attention, while spatial attention was engaged when searching for a conjunction.

This theory produced an avalanche of research, especially using visual search methods. Many of the experiments addressed the serial versus parallel proposal of FIT and often focused on whether these were distinct modes of processing or one mode lying on a continuum. Others used visual search methods to determine the basic features that could be detected without attention or the extent to which the stimulus was preattentively organized. Several experiments focused on the special role space was presumed to play in the theory. But at its heart, FIT has always been about binding. How are basic features perceptually integrated to form the objects we see?

Two major theories have been proposed to account for the difficulty of conjunction search and the occurrence of ICs: FIT and a theory based on competition between feature maps, the contents of which are weighted or "biased" by top-down control (Desimone and Duncan, 1995). In the biased competition model, binding is the result of saliency within and between feature maps and reflects competition for the receptive fields of neurons in specialized cortical areas (e.g., color, shape). In this model, attention selects the features that are relevant for the task, but no special role is assigned to spatial attention in binding. Instead, binding results from the global activation of cells via both bottom-up and top-down influences. However, the theory as presently formed lacks a clear explanation of why parietal lesions affect conjunction but not feature search and increase ICs. It seems that parietal lobe damage would leave color/shape binding relatively intact (as areas that encode these features within the ventral cortical stream remain functional) but disrupt object/location binding. Yet both types of binding are affected. When spatial attention is disrupted (either directly or indirectly by affecting the spatial representations required to guide it), binding errors appear between features registered in more ventral areas that should theoretically continue to directly interact. Where and what streams interact early and in important ways to support integration and to solve at least some binding problems (Robertson, 2003, 2004)?

IV. IMPLICIT AND EXPLICIT BINDING

A. Implicit Binding of Features and Objects

The majority of research concerned with feature binding has focused on what we consciously perceive. However, if attention is a critical part of the binding process then evidence for implicit binding might indicate reconsidering the role of attention. Rather than perceptual binding resulting as a function of spatial attention, attention may select from among already bound items. To some extent, we already know this is the case. Detecting two elements on the same object is easier than detecting two elements on different objects, presumably due to object-based selection (Duncan, 1984).

Work like that with R.M. established that spatial attentional functions, at least those associated with parietal lobes, are not required to bind together various lines, edges, T-junctions, and so on, to form a correctly perceived shape (Robertson et al., 1997). Despite extensive damage to both parietal lobes, an object's shape could be correctly perceived. Even the recognition of faces and reading of single words appeared to remain intact. However, the ability to volitionally select an object and to localize it (binding objects to locations) all but disappeared. (Bilateral parietal damage also produces simultanagnosia, the inability to see more than one object at a time.) Without spatial attention objects cannot be selected volitionally, and coherent objects appear to pop in and out of awareness in a random manner. Critically, parietal damage also disrupts binding of surface features to an object, presumably because the co-location of features is difficult to disambiguate without spatial attention.

It is clear from the ICs made by normal perceivers and by patients with Balint's syndrome that objects do not necessarily come into view bound with all their features intact (e.g., color and form). But at early sensory processing stages, the features are co-located (i.e., they project to identical portions of the retina which are transferred to isomorphic spatial maps in V1). But do these early bindings influence performance before attention is engaged?

Evidence against preattentive binding was reported by Lavie (1997), who showed that discriminating foveal conjunction targets in the presence of flanking conjunction distracters was slower when attention was broadly distributed than when it was narrowly focused on the target location. Conjunctions outside the window of attention did not interfere with identification of a foveal conjunction target. Conversely, simple features interfered with target detection under both attentional conditions.

Other evidence from neuropsychology suggests that preattentive binding can influence performance. For instance, patient R.M. with Balint's syndrome showed Stroop interference from colored words placed in unattended locations (Wocjialick and Kanwisher, 1998), although this effect was highly attenuated and practically disappeared when perceived brightness was equated. Other evidence comes from studies with normal perceivers. Goldsmith (1998) showed that the rate of visual search for a target letter overlapping a nontarget letter was more efficient when a relevant color was attached to the target, suggesting that the color feature was bound to the target before detection.

Results from studies using negative priming procedures, as well as perceptual phenomena such as the McCollough effect (where color and line orientation are perceptually bound), have also been invoked to support preattentive binding. However, in both cases attention is needed, either at the time the prime is processed (negative priming) or to examine the inducing stimulus (McCollough effect), so the degree to which these findings support preattentive binding outside attention as opposed to binding influences from a memory source is not clear.

In sum, the evidence is mixed, but support for the influence of some type of preattentively bound conjunctions appears to be growing. However, even if further data support a preattentive feature binding process, there remains the question of why ICs appear when attention is diverted or when spatial attention becomes dysfunctional. Surface features appear to require spatial attention to be bound normally to their objects in awareness.

B. Implicit Binding of Objects to Location

Another example of preattentive binding again comes from the study of patients with Balint's syndrome. As noted earlier, these patients mostly see only one object in a display at a time, although they cannot report where it is. They are aware that other "things" must exist, but perceptual organization of more than one object is not available to perceptual awareness. Explicitly binding an object to a location and spatially relating it to other objects are nearly impossible without explicit spatial information. So patients with Balint's syndrome seem to offer the potential for understanding what can be perceived without a master map of locations, and the answer is: not much. In fact, one may wonder how even a single object can be perceived without becoming unbound in the absence of spatial information. The answer is unknown, but the phenomenon is a demonstration of complex binding in the absence of attentional selection (although see Chapter 47).

Despite the almost complete loss of perceptual information outside the object that captures attention in these patients, there is evidence that binding items to locations can be accomplished implicitly. When shown a global form created from the spatial arrangement of local forms (Fig. 24.2), they almost always report the local and not the global shape. Even under forced choice conditions the global shape is not reported much above chance. Yet, despite the nearly complete inability to perceive the global form, reaction time measures demonstrate that the global configuration is processed, because it interferes with response to the local letter (Karnath et al., 2000). For this to occur, the spatial configuration of the local elements must be coded properly. Yet perceptual organization of the stimulus as a whole does not enter awareness, and attentional selection of the global shape is nearly impossible. These data complement others collected with patients with Balint's syndrome showing that implicit spatial information remains intact (Robertson et al., 1997). Nonetheless this again raises the question of how and why an implicitly bound visual stimulus (in this case, each shape to its location) becomes

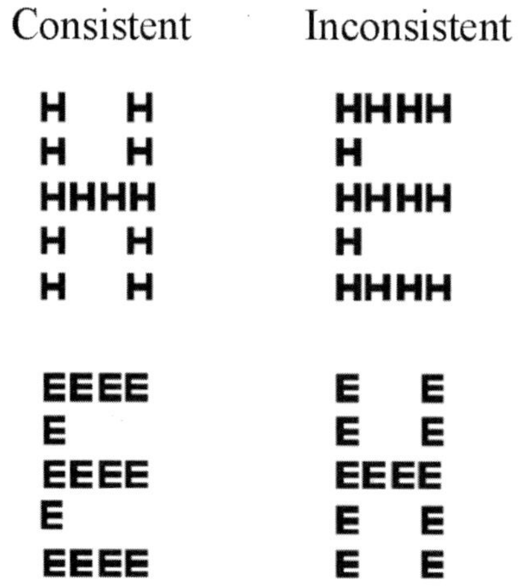

```
     Consistent        Inconsistent

      H   H              HHHH
      H   H              H
      HHHH               HHHH
      H   H              H
      H   H              HHHH


      EEEE               E   E
      E                  E   E
      EEEE               EEEE
      E                  E   E
      EEEE               E   E
```

FIGURE 24.2 Example of hierarchically organized stimuli. The local letters are spatially arranged to form global letters that are consistent or inconsistent with the local form.

unbound in perception when spatial attention (or the spatial representations that guide it) is not available.

V. SUMMARY

The role of attention in binding has evolved from the relatively simple question of whether attention is involved at all to how it is involved, under what conditions, and to what extent. Contrary to arguments against binding as a problem, the neuropsychological evidence has shown that it is a real problem that can manifest in everyday life for people with parietal lesions. This cortical area has been most often associated with spatial attention and spatial representation. The study of these patients has confirmed a special role for spatial attention in binding surface features to objects and objects to locations. As usual, there is more to know than is known, especially as it relates to binding parts to wholes and why implicitly bound representations are not sufficient for explicit awareness.

References

Behrmann, M. (Ed.) (2001). "Handbook of Neuropsychology." Vol. 4: "Disorders of Visual Behavior." Elsevier Press, Amsterdam.

Bernstein, L. J., and Robertson, L. C. (1998). Independence between illusory conjunctions of color and motion with shape following bilateral parietal lesions. *Psychol. Sci.* **9**, 167–175.

Corbetta, M., Shulman, G. L., Miezen, F. M., and Petersen, S. E. (1995). Superior parietal cortex activation during spatial attention shifts and visual feature conjunction. *Science* **270**, 802–805.

Desimone, R., and Duncan, J. (1995). Neural mechanisms of selective visual attention. *Annu. Rev. Neurosci.* **18**, 193–222.

Donner, T.H., Kettermann, A., Diesch, E., Ostendorf, F., Villringer, A., and Brandt, S. A. (2002). Visual feature and conjunction searches of equal difficulty engage only partially overlapping frontoparietal networks. *NeuroImage* **15**, 16–25.

Duncan, J. (1984). Selective attention and the organization of visual information. *J. Exp. Psychol. Gen.* **113**, 501–517.

Friedman-Hill, S., Robertson, L. C., and Treisman, A. (1995). Parietal contributions to visual feature binding: evidence from a patient with bilateral lesions. *Science* **269**, 853–855.

Goldsmith, M. (1998). What's in a location? Comparing object-based and space-based models of feature integration in visual search. *J. Exp. Psychol. Gen.* **12**, 189–219.

Karnath, H.-O., Ferber, S., Rorden, C., and Driver, J. (2000). The fate of global information in dorsal simultanagnosia. *Neurocase* **6**, 295–306.

Lavie, N. (1997). Visual feature integration and focused attention: Response competition form multiple distracter features. *Percept. Psychophys.* **59**, 543–556.

Robertson, L. C. (2003). Binding, spatial attention and perceptual awareness. *Nat. Neurosci. Rev.* **4**, 93–102.

Robertson, L. C. (2004). "Space, Objects, Minds & Brains." Psychology Press, New York.

Robertson, L. C., Treisman, A., Friedman-Hill, S., and Grabowecky, M. (1997). The interaction of spatial and object pathways: evidence from Balint's syndrome. *J. Cogn. Neurosci.* **9**, 295–317.

Shafritz, K. M., Gore, J. C., and Marois, R. (2002). The role of the parietal cortex in visual feature binding. *Proc. Nat. Acad. Sci. USA* **99**, 10917–10922.

Treisman, A. (1996). The binding problem. *Curr. Opin. Neurobiol.* **6**, 171–178.

Treisman, A., and Gelade, G. (1980). A feature integration theory of attention. *Cogn. Psychol.* **12**, 97–136.

Treisman, A., and Schmidt, H. (1982). Illusory conjunctions in the perception of objects. *Cogn. Psychol.* **14**, 107–141.

Walsh, V., and Cowey, A. (1998). Magnetic stimulation studies of visual cognition. *Trends Cogn. Sci.* **2**, 202–238.

Wocjialick, E., and Kanwisher, N. (1998). Implicit but not explicit feature binding in a Balint's patient. *Vis. Cogn.* **5**, 157–181.

CHAPTER

25

Top-Down Facilitation of Visual Object Recognition

Moshe Bar

ABSTRACT

Cortical processing required for object recognition is traditionally thought to be propagating serially along a hierarchy of visual areas that analyze increasingly complex information. Recent efforts gradually promote the involvement of top-down facilitation in cortical functions, but how such processing is initiated remains a puzzle. A specific mechanism for the activation of top-down facilitation during visual object recognition is described here. This mechanism is triggered by the rapid projection of a partially analyzed version of the input image (i.e., a blurred image) from early visual areas directly to the prefrontal cortex. The information that is activated in the prefrontal cortex as a result is backprojected to "object-related" regions in the temporal cortex to produce expectations about the most likely interpretation of the input image. This top-down process facilitates recognition by substantially limiting the number of object representations that need to be considered. The relation between top-down processes that facilitate recognition and top-down processes that exert attentional influence is discussed.

I. INTRODUCTION

Attention plays a major role in guiding our visual inquiry of the environment. For example, attentional influences may enhance selective processing of task-relevant attributes and dictate where to look next and what aspects of a visual scene should remain represented in long-term memory. Ample evidence, as apparent in this book, indicates that attentional control relies on top-down processes. Such top-down mechanisms help prioritize the importance of various aspects of the input, therefore directing the subsequent effects of attention. But top-down processes may also facilitate the actual perception of a stimulus, regardless of attention-related considerations. This top-down facilitation and the top-down processes that guide attention may be independent, and they have different time courses and possibly different cortical origin, but they may nevertheless interact and inform each other. Top-down facilitation of perception, primarily in the domain of visual object recognition, is the focus of this chapter.

Anatomical studies have shown that connections between the visual areas in the ventral pathway are ascending as well as descending. Nonetheless, the majority of the relevant research has concentrated on bottom-up analysis, where the visual input is analyzed in a hierarchy of cortical regions. In such a model, which emphasizes the role of forward connections, an object is recognized only after the last visual area in the pathway for object recognition has received and analyzed all the required input from earlier areas. However, the massive parallel and feedback connections that are known to exist (Ungerieider et al., 1989; Bullier and Nowak, 1995; Rempel-Clower and Barbas, 2000) indicate that top-down mechanisms may also play a central role in visual processing (Ullman, 1995).

Several previous theories have promoted the involvement of top-down analysis in cortical mechanisms (e.g., Grossberg, 1980; Kosslyn, 1994; Ullman, 1995). In these models, the bottom-up flow can be seen as analogous to the thoroughly studied hierarchical processing, from V1 to IT. The top-down process, however, is assumed to represent multiple possible interpretations of the input and it is less clear what initiates this process. Specifically, how can these high-level representations be activated before the input has been fully analyzed? I propose a detailed mechanism that mediates the cortical triggering of top-down processing during object recognition. In addition to providing a concrete mechanism for key aspects of those previous models (Grossberg, 1980; Kosslyn, 1994; Ullman, 1995), the present proposal accounts for several established properties of visual object recognition and it is further used to produce novel predictions. In the next sections I elaborate on this mechanism, propose a cortical origin for top-down facilitation, and discuss the relationship between top-down processing and attentional control.

II. ACTIVATING THE TOP FROM THE BOTTOM

Object recognition is typically accomplished within less than 200 ms from stimulus onset (not including response time). To start even earlier than that, top-down processing should rely on quick mechanisms that might use partially processed information. Within the proposed framework, low spatial frequencies in the image are extracted quickly and projected from early visual areas directly to the prefrontal cortex. This projection is considerably faster than the detailed bottom-up analysis and, therefore, is predicted to use especially quick anatomical connections. A possible cortical route to mediate this rapid projection is the *magnocellular* pathway, which is known to transfer low-spatial-frequency information, and to do so early and rapidly (e.g., Shapley, 1990; Merigan and Maunsell, 1993; Bullier and Nowak, 1995).

Different spatial frequencies convey different information about the appearance of a stimulus. High spatial frequencies represent fine spatial changes in the image (e.g., edges), and generally correspond to featural information and details. Low spatial frequencies, on the other hand, represent global information about the shape (e.g., general orientation and proportions). A key observation is that such global information is typically sufficient for activating a relatively small set of probable candidate interpretations of the input (i.e., "initial guesses"). Figure 25.1 depicts an example of a low-frequency image and the three possible interpretations that may be activated as an initial guess based exclusively on this blurred input. Although they all share the same global appearance with the target object, most of the representations activated by the low frequencies may be of irrelevant objects. Nevertheless, such initial activation significantly reduces the number of candidate object representations that need to be considered. When the input representation is associated with one of the candidates, recognition is accomplished and the other "initial guesses" are suppressed.

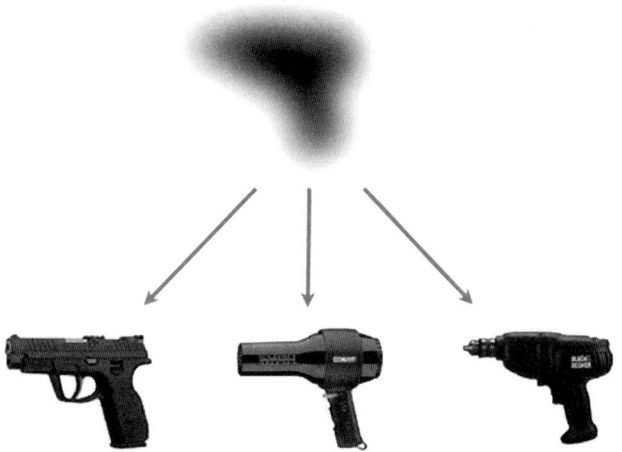

FIGURE 25.1 "Initial guesses" activated by a low-spatial-frequency image. Although the blurred picture cannot be recognized with high certainty, there is only a small set of possibilities that one can produce about its identity. Therefore, such representations, which are minimal in the amount of details that they convey, are sufficient for activating a typically accurate set of alternatives that need to be considered.

Indeed, two recent neurophysiological studies (Sugase et al., 1999; Tamura and Tanaka, 2001) provide support for the notion that IT initially responds to low-spatial-frequency information before it receives high-spatial-frequency information. In addition, psychophysical and physiological experiments with stimuli ranging from gratings (e.g., DeValois and DeValois, 1988) to complex scenes (Schyns and Oliva, 1994) indicate that observers perceive the low-spatial-frequency components considerably earlier than they perceive the high frequencies. A gradually increasing perception of details is also suggested by *global precedence* (i.e., the coarse-to-fine perception of information) (Navon, 1977) and by the reaction time pattern observed in classification studies of objects at different levels of detail and specificity (Rosch et al., 1976). Finally, global and local information seem to be represented differently by the left and right hemispheres (e.g., Robertson and Ivry, 2000). A different time course of response between the left and the right hemispheres may therefore give rise to the gradual coarse-to-fine perception. In summary, these data suggest that coarse information is perceived earlier than the fine details and is therefore processed and propagated faster in the cortex. It is proposed here that the low frequencies are projected rapidly and directly from early visual areas to initiate a top-down process in the prefrontal cortex (PFC). Specifically, this rapid projection activates in the PFC information that, when subsequently backprojected to the temporal cortex, activates there a small set of expectations about the possible identity of the input object.

III. THE CORTICAL ORIGIN OF TOP-DOWN FACILITATION

While the PFC is traditionally not considered as part of the visual system, several studies suggest its involvement in visual recognition (e.g., Bachevalier and Mishkin, 1986; Wilson et al., 1993; Freedman, et al., 2001). Here I discuss support for the specific proposal that the PFC is a likely origin of top-down facilitation in object recognition.

Anatomical data indicate the general existence of cortical and subcortical bypass projections. Specifically, the magnocellular pathway has direct connections from as early as visual area V2 to the dorsolateral PFC and from the ventral area V4 directly to the ventrolateral PFC (Rempel-Clower and Barbas, 2000). Therefore, these data demonstrate the existence of the neural infrastructure required for the rapid projection of low spatial frequencies from early visual cortex to the PFC.

Given our current incomplete understanding of the functional subdivisions of the PFC, as well as the lack of data that will allow defining the human homologs of the PFC regions known from the monkey, an attempt to specify the PFC modules involved in object recognition has to be considered with caution. Nevertheless, two regions seem particularly suitable for triggering top-down facilitation in object recognition: the ventrolateral PFC and the orbital PFC. They are adjacent to each other and massively interconnected. Within the present proposal, these regions constitute a network where a projection of a low-spatial-frequency image is translated into information that subsequently allows the activation of the most likely candidate object representations in IT.

The ventrolateral PFC has been shown to be involved in visual object analysis in monkeys (Wilson et al., 1993; Freedman et al., 2001). In humans, the inferior frontal gyrus within the PFC, in particular, is widely believed to be an integral part in semantic analysis of words and pictures (e.g., Petersen et al., 1989; Brewer et al., 1998; Smith, 1999). The second candidate, the orbital PFC, is the prefrontal region with the strongest connections to the inferior temporal cortex (Cavada et al., 2000). It has been shown to be involved in the analysis of visual information in numerous studies (e.g., Voytko, 1985; Meunier et al., 1997; Frey and Petrides, 2000). Furthermore, activity in the orbital region has been associated with guessing and hypothesis testing, as well as with the generation of expectations (e.g., Bechara et al., 1996; Frith and Dolan, 1997; Petrides et al., 2002), all of which are in agreement with the role attributed here to this region as a source of top-down predictions.

Responding based on rapidly available rudimentary information can facilitate perception and action by focusing cortical processing. Specifically, an early activation of knowledge related to the input image has obvious advantages for producing expectations about the environment. This is particularly pronounced in situations such as danger, where the brain needs to analyze the image of an imminent threat as quickly as possible, to infer the possibility of danger and immediately share this information with other areas relevant to our subsequent decision and action. Indeed, the orbitofrontal PFC is situated optimally for mediating the recognition of danger and controlling a subsequent response; it is part of an orbitofrontal–amygdala–IT triad (Morecraft et al., 1992; Ghashghaei and Barbas, 2002) and it has massive and reciprocal connections with other limbic areas, the autonomic motor system through the thalamus, and premotor areas (Amaral and Price, 1984; Cavada et al., 2000; Ghashghaei and Barbas, 2002).

In the present model the PFC is proposed to participate in the facilitation of object recognition in general, and not exclusively in danger-related recognition. Therefore, an important question is what information about visual objects is represented in the PFC.

IV. OBJECT REPRESENTATIONS IN THE PREFRONTAL CORTEX?

It is not clear what is the nature of the PFC representations that participate in early top-down facilitation of object recognition. A related issue of interest is what information exactly is projected from the PFC to activate accurate expectations in IT. For example, do object-related representations in the PFC contain the same visual information as the object representations in IT, or are they more abstract? On the one hand, the PFC has been shown repeatedly to represent and analyze semantic and abstract information. On the other hand, accumulating reports suggest the existence of visual representations in the PFC (e.g., Wilson et al., 1993; Goldman-Rakic, 1995; Smith, 1999; Frey and Petrides, 2000).

It is proposed here that the PFC translates the rapid projection of sensory information into pointer-like signals that activate the corresponding visual representations in IT. Although one can only speculate on the exact nature of such a mechanism at the moment, this alternative seems plausible given the work of Miyashita and his colleagues (Tomita et al., 1999; Miyashita and Hayashi, 2000), demonstrating the existence of prefrontal signals that trigger memory retrieval of visual object representations in IT.

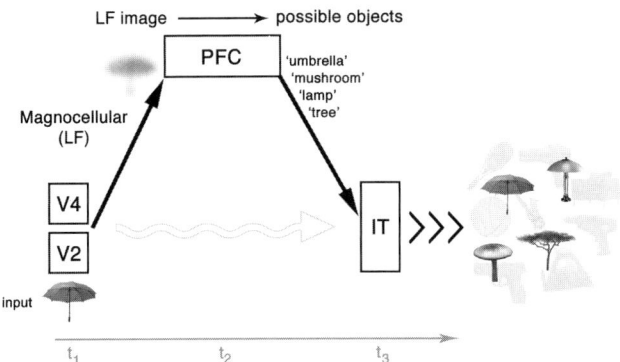

FIGURE 25.2 Schematic illustration of the proposed model. The triggering of top-down facilitation advances in the temporal order t_1, t_2, t_3. Afterward, information transfers in all these pathways in a reciprocal manner. For the sake of simplicity, this figure shows only a subset of areas and connections; additional but unrelated functions that are mediated by these anatomical structures are not discussed.

Therefore, it is possible that PFC representations of objects are not as detailed as those in IT, but they are sufficient to activate expectations based on coarse information. Specifically, the PFC region that participates in top-down facilitation of recognition may contain a "look-up table" associating a low-spatial-frequency image with several alternative objects, whose actual representations reside in the temporal cortex. For example, the blurred image in Fig. 25.1 will be stored in the PFC together with information that will subsequently allow the activation of "gun," "hairdryer," and "drill" in IT. This account implies that the PFC does represent visual objects, but these representations contain information only about very coarse physical properties (e.g., global shape) and therefore are economical. In other words, objects are represented in the PFC in a much degraded manner compared with IT object representations. Furthermore, each PFC representation may be considered a prototype representing a category of multiple objects, all of which look alike when blurred. A summary of this proposed model is depicted in Fig. 25.2.

This representation of coarse information seems powerful enough for mediating complex semantic analysis such as categorization. A recent report demonstrating the critical role of the PFC in representing categories of visual objects (Freedman et al., 2001) is in direct agreement with the present proposal. The findings of this study suggest that PFC representations distinguish between objects belonging to different categories, while not emphasizing detailed information regarding different exemplars within the same category. Exemplar-specific detailed shape information has been shown to be represented instead in IT (Op de Beeck et al., 2001). In other words, the PFC represents relatively coarse visual information that can mediate between-category decisions (Bar, 2003).

V. PREDICTIONS AND OPEN QUESTIONS

Several testable predictions stem from the proposal described here:

1. Object recognition entails an early peak of activation in the PFC, related to the activation of an "initial guess."
2. Early recognition-related activity in the PFC is determined by the low-spatial-frequency content of the input image.
3. The top-down process is suggested to facilitate object recognition by reducing the number of the possibilities that need to be considered. Therefore, an absence of the top-down process is predicted to slow object recognition (i.e., prolong reaction times), but not to preclude successful recognition. In other words, object recognition in this case will rely solely on the bottom-up process. Studies of patients with lesions in the frontal cortex (e.g., Greenlee et al., 1997) support this prediction.
4. Although top-down facilitation is suggested to be generally an integral part of recognition, its role and magnitude can nevertheless be modulated by attention as determined by task demands. In other words, when object recognition is relatively trivial, where objects are presented clearly and in isolation (usually on a computer screen in the laboratory), bottom-up recognition may be sufficiently efficient as to not depend on top-down facilitation. In our environment, however, we see objects with clutter, where they need to be segmented from their background, they may be occluded, have interfering shading, seen from accidental viewing angles, and so on, and in those cases, our system could benefit from additional sources and thus rely more on top-down facilitation.

Three important questions remain open and their resolution in the future is critical. First, how do early top-down facilitation and attentional control interact? For example, could processes that guide attention also be activated by coarse information such as low spatial frequencies? Can attention modulate the contribution of individual components in the proposed model?

Second, it is not clear what is the exact nature of the object-related representations in the PFC. Several types of representations, with a varying degree of visual specificity, may be capable of mediating top-down facilitation. Of specific interest is the interaction between those representations and the mechanism that generates expectations based on low-spatial-frequency images. A possible model has been proposed (Fig. 25.2), but it will have to be tested in the future.

Finally, what is the role of the magnocellular projection that is known to exist from early visual cortex to IT? This projection may mediate the integration of the bottom-up and top-down processes. Alternatively, that similar magnocellular information is projected from early visual cortex simultaneously both to IT and to PFC has been used to propose an expansion of the present model to further account for context-based facilitation in recognition (Bar, 2004).

It is important to note that not all models of object recognition are predicted to gain equally from a top-down process. Naturally, the benefit from a top-down rapid projection is less obvious in models that are inherently based on a bottom-up progression of cortical analysis (Marr, 1982; Biederman, 1987). The underlying principle in these cases involves direct "indexing" from visual primitives to a specific object representation. In other words, there is no explicit process of searching for a match, and therefore, there is no search space that can be reduced by a rapid "initial guess." Nevertheless, it is conceivable that sensitizing the correct identity in advance may be beneficial for its subsequent activation regardless of the specific components of a certain recognition model. Furthermore, early activation of an "initial guess" based only on coarse information will be certainly helpful in extreme situations, where an immediate reaction may be necessary.

Finally, it is proposed that object recognition involves a co-activation of multiple possible representations of the input object. However, we subjectively perceive objects in our environment as possessing singular identities. In other words, we typically become aware of an identity only after the subserving cortical processes have reached a sufficient level of certainty and converged onto a single interpretation. The awareness of an object's identity has been suggested to be associated with a gradually increasing, rather than abruptly changing, cortical activity (Bar et al., 2001), and the time before awareness may be the time during which the most likely interpretation is selected.

Acknowledgments

This chapter is based on Bar (2003). I thank S. Ullman, H. Barbas, B. Rosen, A. Shmuel, and R. Tootell for helpful discussions, and E. Aminoff and K. Kassam for technical assistance. This work was supported by NINDS R01NS44319-01, by the MIND Institute, and by the James S. McDonnell Foundation – 21st Century Science Research Award #21002039.

References

Amaral, D. G., and Price, J. L. (1984). Amygdalo-cortical projections in the monkey (*Macaca fascicularis*). *J. Comp. Neurol.* **230**, 465–496.

Bachevalier, J., and Mishkin, M. (1986). Visual recognition impairment follows ventromedial but not dorsolateral prefrontal lesions in monkeys. *Behav. Brain Res.* **20**, 249–261.

Bar, M. (2003). A cortical mechanism for triggering top-down facilitation in visual object recognition. *J. Cogn. Neurosci.* **15**, 600–609.

Bar, M., (2004). Visual Objects in context. Nature Reviews: Neuroscience **5**, 619–629.

Bar, M., Tootell, R., et al. (2001). "Cortical mechanisms of explicit visual object recognition." *Neuron* **29**, 529–535.

Bechara, A., Tranel, D., et al. (1996). Failure to respond autonomically to anticipated future outcomes following damage to prefrontal cortex. *Cereb. Cortex* **6**, 215–225.

Biederman, I. (1987). Recognition-by-components: a theory of human image understanding. *Psychol. Rev.* **94**, 115–117.

Brewer, J. B., Zhao, Z., et al. (1998). Making memories: brain activity that predicts how well visual experience will be remembered. *Science* **281**, 1185–1187.

Bullier, J., and Nowak, L. G. (1995). Parallel versus serial processing: new vistas on the distributed organization of the visual system. *Curr. Opin. Neurobiol.* **5**, 497–503.

Cavada, C., Company, T., et al. (2000). The anatomical connections of the macaque monkey orbitofrontal cortex: a review. *Cereb. Cortex* **10**, 220–242.

DeValois, R. L., and DeValois, K. K. (1988). Spatial Vision. Oxford Science, New York.

Freedman, D. J., Riesenhuber, M., et al. (2001). Categorical representation of visual stimuli in the primate prefrontal cortex [see comments.]. *Science* **291**, 312–316.

Frey, S., and Petrides, M. (2000). Orbitofrontal cortex: a key prefrontal region for encoding information. *Proc. Nal. Acad. Sci. USA* **97**, 8723–8727.

Frith, C., and Dolan, R. J. (1997). Brain mechanisms associated with top-down processes in perception. *Phil. Trans. R. Soc. London Ser. B* **352**, 1221–1230.

Ghashghaei, H., and Barbas, H. (2002). Pathways for emotion: interactions of prefrontal and anterior temporal pathways in the amygdala of the rhesus monkey. *Neuroscience* **115**, 1261–1279.

Goldman-Rakic, P. S. (1995). Architecture of the prefrontal cortex and the central executive. *Ann. NY Acad. Sci.* **769**, 71–83.

Greenlee, M. W., Koessler, M., et al. (1997). Visual discrimination and short-term memory for random patterns in patients with a focal cortical lesion. *Cereb. Cortex* **7**, 253–267.

Grossberg, S. (1980). How does a brain build a cognitive code? *Psychol. Rev.* **87**, 1–51.

Kosslyn, S. M. (1994). "Image and Brain." MIT Press, Cambridge, MA.

Marr, D. (1982). "Vision: A Computational Investigation into the Human Representation and Processing of Visual Information." Freeman, San Francisco, CA.

Merigan, W. H., and Maunsell, J. H. (1993). How parallel are the primate visual pathways? *Annu. Rev. Neurosci.* **16**, 369–402.

Meunier, M., Bachevalier, J., et al. (1997). Effects of orbital frontal and anterior cingulate lesions on object and spatial memory in rhesus monkeys. *Neuropsychologia* **35**, 999–1015.

Miyashita, Y., and Hayashi, T. (2000). Neural representation of visual objects: encoding and top-down activation. *Curr. Opin. Neurobiol.* **10**, 187–194.

Morecraft, R. J., Geula, C., et al. (1992). Cytoarchitecture and neural afferents of orbitofrontal cortex in the brain of the monkey. *J. Comp. Neurol.* **323**, 341–358.

Navon, D. (1977). Forest before trees: the precedence of global features in visual perception. *Cogn. Psychol.* **9**, 1–32.

Nowak, L. G., and Bullier, J. (1997). The timing of information transfer in the visual system. *In* "Cerebral Cortex: Extrastriate Cortex in Primate" (K. Rockland, J. Kaas, and A. Peters, Eds.), vol. 21, pp. 205–241. Plenum, New York.

Op de Beeck, H., Wagemans, J., et al. (2001). Inferotemporal neurons represent low-dimensional configurations of parameterized shapes. *Nat. Neurosci.* **4**, 1244–1252.

Petersen, S. E., Fox, P. T., et al. (1989). Positron emission tomographic studies of the processing of single words. *J. Cogn. Neurosci.* **1**, 153–170.

Petrides, M., Alivisatos, B., et al. (2002). Differential activation of the human orbital, mid-ventrolateral, and mid-dorsolateral prefrontal cortex during the processing of visual stimuli. *Proc. Natl. Acad. Sci. USA* **99**, 5649–5654.

Rempel-Clower, N. L., and Barbas, H. (2000). The laminar pattern of connections between prefrontal and anterior temporal cortices in the rhesus monkey is related to cortical structure and function. *Cereb. Cortex* **10**, 851–865.

Robertson, L. C., and Ivry, R. (2000). Hemispheric asymmetries: attention to visual and auditory primitives. *Curr. Directions Psychol. Sci.* **9**, 59–63.

Rosch, E., Mervis, C., et al. (1976). Basic objects in natural categories. *Cogn. Psychol.* **8**, 382–439.

Schyns, P. G., and Oliva, A. (1994). From blobs to boundary edges: evidence for time- and spatial-dependent scene recognition. *Psychol. Sci.* **5**, 195–200.

Shapley, P. (1990). Visual sensitivity and parallel retinocortical channels. *Annu. Rev. Psychol.* **41**, 635–658.

Smith, E. E. (1999). Storage and executive processes in the frontal lobes. *Science* **283**, 1657–1661.

Sugase, Y., Yamane, S., et al. (1999). Global and fine information coded by single neurons in the temporal visual cortex. *Nature* **400**, 869–873.

Tamura, H., and Tanaka, K. (2001). Visual response properties of cells in the ventral and dorsal parts of the macaque inferotemporal cortex. *Cereb. Cortex* **11**, 384–399.

Tomita, H., Ohbayashi, M., et al. (1999). Top-down from prefrontal cortex in executive control of memory retrieval. *Nature* **401**, 699–703.

Ullman, S. (1995). Sequence seeking and counter streams: a computational model for bidirectional information flow in the visual cortex. *Cereb. Cortex* **1**, 1–11.

Ungerleider, I. G., Gaffan, D., et al. (1989). Projections from inferior temporal cortex to prefrontal cortex via the unicate fascicle in rhesus monkeys. *Exp. Brain Res.* **76**, 473–484.

Voytko, M. L. (1985). Cooling orbital frontal cortex disrupts matching-to-sample and visual discrimination learning in monkeys. *Physiol. Psychol.* **13**, 219–229.

Wilson, F. A. W., O'Scalaidhe, S. P., et al. (1993). Dissociation of object and spatial processing domains in primate prefrontal cortex. *Science* **260**, 1955–1958.

CHAPTER 26

Spatial Processing of Environmental Representations

James R. Brockmole and Ranxiao Frances Wang

ABSTRACT

In everyday life, we accomplish tasks that require the storage and access of mental representations of many familiar locations. Humans store this environmental knowledge in a series of representations in memory. We discuss recent studies on how humans attend to and process multiple environmental representations to reason and act within the spatial world. These studies suggest that people access one environment at a time, and they automatically update their relationship to their immediate environment, but not to remote environments. Navigation across environments involves reorienting to upcoming environments at certain spatial regions and dropping the old environment from the processing. This selective processing sheds light on the nature and structure of human environmental representations.

I. INTRODUCTION

Throughout this book, attention and selective processing have been shown to play an important role in various visually based cognitive tasks, such as visual search, change detection, object identification, scene processing, and learning. In studies on the nature and structure of spatial representations in humans, however, the role of attention and selective processing has received little consideration. As a result, traditional models of environmental representations in humans largely ignore issues of information processing load, a critical oversight given the vastness of our world. In this chapter, we discuss the limitations of human spatial processing ability, show how traditional models of environmental knowledge encounter difficulty when attention and processing limitations are considered, and describe an alternative view of spatial processing that is consistent with these limitations.

II. THE HIERARCHICAL MODEL OF ENVIRONMENTAL REPRESENTATIONS

Humans live in amazingly complicated environments. On a daily basis, we encounter environments that span miles and include natural and artificial structures, as well as subjectively imposed divisions among areas such as neighborhoods, cities, and states. Given this complexity, humans cannot directly perceive the entire extent of their world from a single vantage point, yet exhibit an impressive ability to reason about, and navigate within, the spatial world. We can give directions to our house to a friend or describe the layout of our kitchen while sitting in our living room. We can turn from one direction to face another without losing track of our bearings with respect to regions of space beyond our field of view. We can navigate from one place to another without being able to see our desired destination.

To perform these tasks, humans must store, recall, and transform information about the world in memory. This information is referred to as an environmental representation. Environmental representations must encode environments at different scales. For example, people have spatial knowledge of small places such as the arrangement of objects on their desk or the location of furniture in a room as well as large places such as the arrangement of rooms in a building, or of buildings in a city. According to the hierarchical network model, the representation of an environment is composed of a number of distinct units, which encode information at different levels of detail.

Moreover, these units are organized into a systematic, hierarchical structure. For example, humans maintain separate representations of a room, the building the room is in, the block the building is on, and so forth. Within this network, the spatial relationships among locations in the world are explicitly encoded in memory only if the locations are encoded within the same representation (e.g., McNamara, 1986; Stevens and Coupe, 1978). When making spatial judgments about locations encoded in different representations, information from each representation and the corresponding superordinate representations must be accessed and combined to compute a solution (Hirtle and Jonides, 1985; McNamara 1986; McNamara et al., 1989; Stevens and Coupe, 1978; Taylor and Tversky, 1992; Wilton, 1979).

The hierarchical network model is intuitively appealing and explains several phenomena well. For example, when people judge the directional relationship between items encoded in separate units, they typically bias their judgments in accordance to the relationship shared by superordinate units. For example, Reno, Nevada, is overwhelmingly judged to be northeast of San Diego, California, although Reno is actually northwest. According to the hierarchical model, the relationship shared by Reno and San Diego is not encoded in memory but must be computed by combining the subordinate knowledge that Nevada is east of California and the subordinate knowledge that San Diego is in Southern California and Reno is in northern Nevada, which results in the biased judgment (Stevens and Coupe, 1978). In addition, within-unit distances are typically underestimated while across-unit distances are often overestimated (Hirtle and Jonides, 1985; McNamara, 1986), and spatial judgments about targets in a single representation are made faster than across-representation judgments regardless of the distance between the locations (McNamara, 1986; Wilton, 1979). These findings suggest that locations within a single unit are "mentally closer" than those in different units.

There is one aspect of the hierarchical network model that has traditionally been overlooked, however. To infer spatial relationships based on two pieces of information, they need to be available at the same time. That is, one needs to know that A is south of B and B is east of C simultaneously to infer that A is southeast of C. In other words, information from multiple units and levels in a hierarchical network may need to be accessed simultaneously to allow inferences about spatial relationships across units. Compared with the vastness of our world and the infinite capacity of long-term memory, however, human information processing abilities are comparatively limited; for example, the capacity of visual working memory is often estimated to be only four to six items (e.g., Irwin, 1992; Luck and Vogel, 1997). Thus, it is an empirical question whether multiple levels of a hierarchical structure can be accessed simultaneously.

III. ACCESSING MULTIPLE ENVIRONMENTAL REPRESENTATIONS

To examine the issue of whether people are able to access spatial knowledge of nested environments (e.g., a building and a room in the building) at the same time, Brockmole and Wang (2002; see also Brockmole and Wang, 2003) employed a task-set switching paradigm where, across successive trials, participants made spatial judgments about objects drawn from the same environment or from different environments. If people can access representations of both environments at the same time, then participants should be able to switch between representations freely without a cost in performance. If, however, only one environment can be accessed at a time, then switching between environments would be associated with a temporal cost as subjects deactivate the old representation and activate the new representation. In one experiment, subjects evaluated target locations in a building and in an office within that building; in another experiment, locations on a college campus and a building within that campus were evaluated.

Consistent with the sequential access account, participants required additional time to judge spatial relationships immediately following a switch in the probed environment, an effect that could not be attributed to switching between two semantic categories (objects in a room environment versus rooms in a building environment). Strikingly, the effect also occurred despite the fact that each participant had at least 2 years of daily experience in the tested environments (see Fig. 26.1).

These results suggest that representations of nested environments are not accessed simultaneously, but rather, they are accessed one at a time. When a switch of representation is required, the inactive representation must be retrieved, a process that takes time. These findings put constraints on how spatial inference may occur in the hierarchical network model. That is, spatial inference across representations cannot be the simple combination of two or more concurrently accessed representations at different hierarchical levels. Either other mechanisms are needed to mediate such a combination process, or these environments, though familiar, are not represented in an integrated hierarchical network.

position relative to different environments during navigation. Almost all participants (93%) were unable to identify the location of the union while in our laboratory. Instead, participants walked a median distance of 55 m before they could. This distance corresponded to a point at which the participants were actually outside of the building (but could not physically see the union). Surprisingly, at this point, 80% of participants could not identify the location of the laboratory from which they just came! Instead, they walked a median distance of 13 m more before they could. This point corresponded to a place where an entrance to the building could be observed. Even at this position, however, 87% of participants could not orient the map of the room with respect to geography and had to continue a median distance of 22 m more before they could. This point corresponded to a position at which an entrance to the laboratory could be observed.

To examine whether the location changes that were needed to respond to each query resulted from time to access information from the relevant environmental representations or the need to reach certain spatial regions before the relevant environmental knowledge could be accessed, the experiment was replicated with half of the participants walking at a fast speed, and half at normal pace while response time was recorded. Slow and fast walkers differed in response time, with fast walkers responding sooner than slow walkers. However, the distance traversed by each group was the same. These results suggested that certain spatial regions, not time, were needed to access the appropriate information from memory: fast walkers simply arrived at those critical locations more quickly.

VI. PROCESSING OF ENVIRONMENTAL REPRESENTATIONS

Together, these studies suggest that people have separate representations for different environments, which are not organized into an integrated network such as a hierarchy, as the dominant view of spatial representations argues. The traditional hierarchical network model may require simultaneous access to information from multiple representations to make spatial inferences. Moreover, an integrated network of spatial relations, such as a hierarchical network, allows spatial updating that occurs within one unit to be generalized to other units. Thus, updating does not need to operate on each unit individually and should not be "capacity limited."

Human performance fails to support these predictions, however. When completing tasks that require knowledge of locations in different environments, humans engage in a representation switching process. Only one representation can be accessed and processed at a time. Thus, only a navigator's relationship with respect to the active representation is computed. When the representation must be switched, a new set of targets and relationships need to be derived and updated. This comes at the cost of retaining information about one's relationship to other environments, even those locations from which one just came. Finally, the determination of when and which representation to access can be cued by perceived locations in the environment as one moves.

Attending to locations in the immediate environment or the upcoming environment for these computations is most appropriate given nonequivalence in the utility of all locations for this process. For example, upcoming targets serve a more functional role in guiding action. Once one passes through an area of space, continued orientation to that area is not necessary for progress toward the goal location. Furthermore, distant targets cannot be updated as accurately as near targets (the further away a location is, the more difficult it is to ascertain an accurate vector between oneself and that target). Thus, assuming spatial updating is a limited-capacity process, then it is no surprise that targets have to be constantly dropped and reintroduced at given spatial locations where cues (certain landmarks, etc.) can be found to activate those representations.

If people only keep track of their relationship to a limited set of targets in the overall environment, for example, their current environment and the environment they are approaching, they may have very poor knowledge about their spatial relationship to many other environments that they are not actively processing. Wang and Brockmole (2003a) showed that people indeed are constantly disoriented relative to remote environments even when they are perfectly oriented to the immediate surroundings. Why do we rarely have the subjective experience of disorientation or "being lost" despite the fact that we are almost always "disoriented" relative to many environments? One possibility is that the sense of being oriented is like the sense of visual richness we perceive. Despite limitations on how much visual detail can be attended and retained in online visual memory, as shown in the research of change blindness (for reviews, see Rensink, 2002, and Simons, 2000), we experience a rich, detailed visual world as long as we have perceptual access to the visual details of those things that fall within the scope of attention. Similarly, we may experience perfect sense of orientation as long as we know our relationship to the immediate surroundings, despite our poor knowledge of our relationship to many, or even most, parts of the world.

VII. CONCLUSIONS

In summary, our environment is parsed into a series of independent representations that can only be accessed one at a time. Thus, when people keep track of their relationship to one environment, they do not necessarily update their relationship to other environments. Navigation across nested environments involves updating one's relationship to the upcoming environments, and changes in the active representation can be cued by locations in the environment that are perceived during motion.

These conclusions are inconsistent with integrated network models such as a hierarchical network model, which predicts that one may need to access information from multiple representations to make spatial inferences and, by the same token, that they may maintain knowledge of their relationship to multiple environments concurrently. Environmental representations appear to be fragmented in nature, and the mechanisms by which they are processed have a limited computational capacity and therefore do not apply to all environments simultaneously. This emerging incoherence underscores the importance of understanding the limitations in how environmental representations are processed to develop a more complete view of the nature and structure of spatial representations. Whether the hierarchical network model can be revised to account for the current findings or if new models of environmental representations need to be devised constitutes an important avenue for future research.

References

Amorim, M. A., and Stucchi, N. (1997). Viewer- and object-centered mental explorations of an imagined environment are not equivalent. *Cogn. Brain Res.* **5**, 229–239.

Brockmole, J. R., and Wang, R. F. (2002). Switching between environmental representations in memory. *Cognition* **83**, 295–316.

Brockmole, J. R., and Wang, R. F (2003). Changing perspective within and across environments. *Cognition* **87**, B59–B67.

Farrell, M. J., and Robertson, I. H. (1998). Mental rotation and the automatic updating of body centered spatial relationships. *J. Exp. Psychol. Learn. Mem. Cogn.* **24**, 227–233.

Hirtle, S. C., and Jonides, J. (1985). Evidence of hierarchies in cognitive maps. *Mem. Cogn.* **13**, 208–217.

Irwin, D. E. (1992). Memory for position and identity across eye movements. *J. Exp. Psychol. Learn. Mem. Cogn.* **18**, 307–317.

Luck, S. J., and Vogel, E. K. (1997). The capacity of visual working memory for features and conjunctions. *Nature* **390**, 279–281.

McNamara, T. P. (1986). Mental representations of spatial judgments. *Cogn. Psychol.* **18**, 87–121.

McNamara, T. P., Hardy, J. K., and Hirtle, S. C. (1989). Subjective hierarchies in spatial memory. *J. Exp. Psychol. Learn. Mem. Cogn.* **15**, 211–227.

Rensink, R. A. (2002). Change detection. *Annu. Rev. Psychol.* **53**, 245–277.

Rieser, J. J. (1989). Access to knowledge of spatial structure at novel points of observation. *J. Exp. Psychol. Learn. Mem. Cogn.* **15**, 1157–1165.

Simons, D. J. (2000). Current approaches to change blindness. *Visual Cogn.* **7**, 1–15.

Stevens, A., and Coupe, P. (1978). Distortions in judged spatial relations. *Cogn. Psychol.* **10**, 422–437.

Taylor, H. A., and Tversky, B. (1992). Descriptions and depictions of environments. *Mem. Cogn.* **20**, 483–496.

Wang, R. F., and Brockmole, J. R. (2003a). Simultaneous spatial updating in nested environments. *Psychon. Bull. Rev.* **10**, 981–986.

Wang, R. F., and Brockmole, J. R. (2003b). Human navigation in nested environments. *J. Exp. Psychol. Learn. Mem. Cogn.* **29**, 398–404.

Wilton, R. N. (1979). Knowledge of spatial relations: the specification of information used in inferences. *Q. J. Exp. Psychol.* **31**, 133–146.

CHAPTER 27

Decision and Attention

Andrei Gorea and Dov Sagi

ABSTRACT

In real life, individuals are faced with more than one perceptual event on which they have to make distinct decisions. It is shown that for a range of such multistimulus environments, decision behavior departs from optimality in the sense that subjects do not set their decision criteria in accordance with the requirements of each individual event. This behavior is explained in terms of a unified internal representation of the multistimulus environment, presumably resulting from the relaxation of attention to the critical dimension associated with each stimulus. Exceptions are observed for cross-modal (audiovisual) stimulations and for stimuli showing sensory interference. It is proposed that decision behavior and the selection process required to segment sensory objects are intimately related. Response criterion interaction may account for phenomena such as extinction and may be the substrate of a number of contextual effects.

I. INTRODUCTION

Despite the progressive and by now practically completed withdrawal of introspective thought from the field of experimental psychology, or perhaps because of it, a little introspection may guide the reader of this chapter. After all, every research report on attention that saves itself a formal definition of this concept—the overwhelming majority of attentional studies—relies on its intuitive understanding. Intuition tells us that making a decision is by essence an attentional state, that it requires pondering and hence involves an effortful, intentional, voluntary (terms willy-nilly equivalent to free will) component to be contrasted with a default, "relaxed," freewheeling behavior. This being said and inasmuch as intention, volition, free will, and perhaps mental effort remain purely intuitive, ill-defined terms, attention itself, at least its endogenous aspect, stands out as an equally indefinite concept.

The work to be described here bears on human decisional behavior and, more specifically, on a subject's capacity to deal with a number of *decision criteria* when faced with an equal number of distinct events. Technically speaking, this is an exploration into the *psychophysics of decision*. The basic results show that when confronted with a range of equally likely but different-strength events, subjects use a unique decision criterion although optimal behavior requires the use of criteria proportional to the stimuli strengths (see below). We refer to this behavior as to *criterion attraction* and we interpret it as the consequence of a *unitary internal representation* (UIR) of the (physically) distinct events, at least along the dimension under study. *Criterion attraction* (and its underlying UIR) is not observed under all the experimental conditions considered. A survey of the ensemble of our data led us to conclude that the UIR is the instantiation of a default, nonattentional state whereby a number of stimulus dimensions are merged together, hence sparing the observer the sustained effort of keeping them apart in exchange for a negligible loss in performance.

II. HISTORICAL BACKGROUND

By the mid-19th century, Fechner (1860) asked (and answered) the question of how to measure sensation; he thereby founded *psychophysics*. Some 50 years later, the Gestalt school put forward the study of shape perception relying on what could be called *conceptual observation*. Another 50 years later, a group of what one might call "neo-psychophysicists" set forth signal detection theory (SDT) (see Green and Swets, 1966), a set of principles and psychophysical tools by means of which sensation and decision were associated in an unbreakable conceptual tandem. This may have sounded as a more than 50-year-deferred echo to von Helmholtz's (1856–1867) view of perception as an "unconscious inference" process subject to perpetual decision making.

SDT quite rapidly became and has remained the main (and perhaps the only) general framework for measuring and modeling sensory/perceptual processes and for separating them from the decision/subjective part involved in the detection of an event and in its discrimination from other events. Still, SDT allows the measurement of sensation/perception while ignoring the decision parameter, provided that the subject's decisional state remains unchanged. Because such a "simplification" comes along with notable savings in experimental time, it has been extensively used to quantify a great deal of auditory, visual, and attentional processes to the expense of the study of the associated decision processes.[1] Although some recent neurophysiological work has revived interest in the process of perceptual and motor decisions (Schall, 2001), psychophysics and the cognitive sciences as a whole have not followed up. This chapter is such an enterprise.

III. PRECIS OF SIGNAL DETECTION THEORY

The cornerstone of SDT is the distinction it makes between *sensitivity* and *decision criterion*. The former is meant to characterize the processing efficiency of the underlying sensory system, and it increases with stimulus strength. The latter is regarded as the manifestation of a subjective operation whereby individuals decide on (as opposed to react reflexively to) the occurrence of an event based on factors such as expectation and payoff, in addition to its strength. To do so, individuals need to have some knowledge of the internal response distributions evoked by this event or its absence.

SDT (see Fig. 27.1a) posits that the activity of a sensory system along an arbitrary *sensory continuum* (or processing dimension) is non-null, even in the absence of a stimulus, and that this "reference" or *noise* (N) activity distributes normally over time. The mean and standard deviation, σ, of the N distribution are arbitrarily set to 0 and 1, respectively (standardized normal distribution). More generally, N designs the internal activity evoked by a reference condition (including the absence of any stimulus) against which the observer has to detect any stimulation change or *signal* (S). Depending on whether this change is referred to no stimulation at all or to some invariant reference stimulus, the task is coined *detection* or *discrimination*, respectively. The mean internal response difference between S and N normalized by σ is the SDT sensitivity index, d'. In other words, d' is defined as the S-to-N ratio.

In practice, the observer is randomly presented with N and S trials, each type in a given proportion P so that $P_N + P_S = 1$. The observer is faced with a binary choice: S (i.e., a "yes" response) or N (a "no" response). The conditional measured proportions $p(Yes|N)$, $p(Yes|S)$, $p(No|N)$, and $p(No|S)$ are referred to as *false alarms* (FA), *hits* (H), *correct rejections* (CR), and *misses*, respectively. Referring FA and H rates to the standardized normal distribution yields their standardized scores $z(FA)$ and $z(H)$, that is, their abscissa values (measured in σ units) with respect to the mean of N and of S, respectively. The sensitivity index d' is given by $z(H) - z(FA)$.

SDT posits that there is a point along the sensory continuum where larger and smaller internal response values entail "yes" and "no" responses, respectively. This frontier is called the *criterion*, $c = [z(H) + z(FA)]/2$, typically measured with reference to an unbiased response strategy. If the criterion is measured with respect to the mean of the N distribution, it is referred to as the *absolute criterion*, $c' = z(FA)$. In SDT, c (or c') is optimal (i.e., maximizes the percentage of correct responses) when its corresponding *likelihood ratio*, $\beta = p_S(z = c')/p_N(z = c')$ (read the ratio of the probabilities of the internal responses evoked by the signal and the noise at $z = c'$) is equal to P_N/P_S. Experimental results show (Green and Swets, 1966; Gorea and Sagi, 2000) that human observers are close to optimality only for p_N/p_S ratios close to 1 and are reluctant to unbalance their "yes" and "no" responses in proportion to the N and S probabilities when they are too different. It emerges from the account above that decision is an unconstrained process, fully controlled by the observer's strategy subjected to "free will." In a multistimulus environment, such a process requires distinct internal representations of the different sensory "objects" liable to a decision/response. SDT implicitly assumes that the simultaneous construction of these internal representations is possible and that, as a consequence, a subject's decisional behavior should be the same in single and multidecisional tasks. However, this implication has never been tested and the rules

---

[1] Of course, what is today abusively referred to as *cognitive* (in opposition to psychophysical) *studies* offers an equivalent modeling based on alternative behavioral techniques involving mostly response time and magnitude estimation techniques. Although these techniques are not discussed here, it should be noted that data collected with any of them are also related to the SDT format analysis: they depend on both sensory and decision parameters.

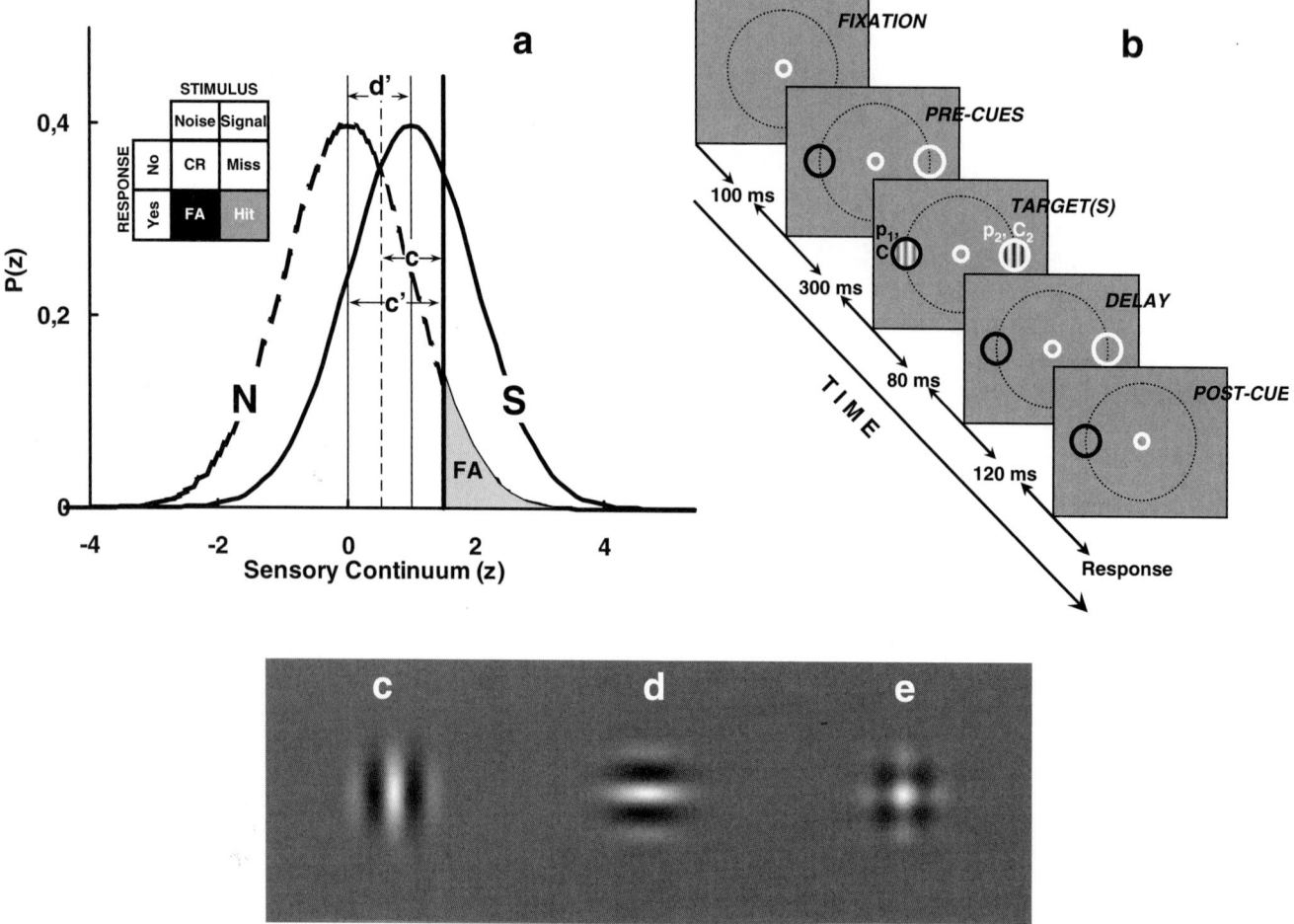

FIGURE 27.1 (a) Standard signal detection theory (SDT) framework for the dual-criterion experiments. Gaussian functions describe the probability density, $p(z)$, of the internal response distributions (in standard z scores, abscissa) for the noise [N: $p_N(z)$] alone (dashed curve) and for the signal + noise [S: $p_S(z)$]. Thin vertical lines show their means with sensitivity ($d' = zH - zFA$, with zH and zFA the z scores for the observed correct target detectia, Hit, and False Alarm rates) being the distance between these means ($d' = 1$ in this case) measured in units of the noise standard deviation, σ_N, and assuming that N and S are normally distributed with $\sigma = \sigma_N = \sigma_S$. The "absolute" criterion is defined as $c' = -zFA$. Defined in this way, criteria are independent of the univariance assumption (i.e., $\sigma_S = \sigma_N$), because they depend on the N distribution only. The corresponding values of the likelihood ratio criterion, $\beta = p_S(z = c')/p_N(z = c')$, characterize observers' response bias independently of d'. Error rate is minimized when $\beta = P_N/P_S$, (with P_N and P_S the a priori N and S probabilities) but experimental results show that observers adopt a more conservative behavior with β's closer to one (Green and Swets, 1966). The vertical dashed and continuous heavy lines show optimal criterion for $P_S = 0.5$ and $P_S = 0.25$, respectively. The shaded area denotes the FA rate for the latter case. (b) One trial sequence with two unequal contrast stimuli, a condition referred to as dual-different. The color of the pre-cues is systematically associated with a given contrast (and probability of occurrence) so that observers have full knowledge of the stimulus to be presented in each circle. The post-cue specifies the stimulus/location to be reported on (partial report paradigm). Dual-same (i.e., equal contrast stimuli) and Single (only one and the same stimulus presented at a time) conditions yield response criteria close to those predicted by standard SDT, whereas the dual-different condition yields a unique response criterion, contrary to SDT predictions. (c, d, e) Respectively a vertical and horizontal Gabor patch and their superposition in a "plaid." The segregation of the two plaid components is rather difficult.

governing the putative interference of multiple criteria remain unknown.

The experiments described here focus on a subject's capacity to handle multiple internal representations. Their outcome points to a strong limitation of this capacity with its implications on decision behavior.

The observed constraint in handling multiple internal representations may be thought of as being rooted in the selection process required to segment sensory objects. Hence, the present query is susceptible of throwing new light on the attentional process as a whole.

IV. CRITERION ATTRACTION AND ITS INTERPRETATION, THE *UNIQUE INTERNAL REPRESENTATION*

A. The Basic Phenomenon

A few years ago we presented data supporting the notion that in a multidecision visual task involving different-strength (i.e., contrast), noninterfering stimuli (Gabor patches) (Figs. 27.1c,d) presented simultaneously (Fig. 27.1b), observers behave nonoptimally, in the sense that for equally likely signals, they adopt, contrary to SDT prediction, one single decision criterion despite the fact that they have full knowledge of the stimulus properties, that is, of their contrasts and occurrence probabilities (Gorea and Sagi, 1999, 2000). Observers do use distinct criteria when the different targets appear with different probabilities; however, these criteria were found to be strongly biased toward each other, as if mutually attracted. Figure 27.2 illustrates the optimal and unique criterion behavior for two equally likely signals within the standard SDT format (Fig. 27.2a) and, as actually measured (idealized representation), under single, dual-same, and dual-different conditions (Fig. 27.2b; see caption). While stimuli of different strengths should yield distinct response criteria whether presented in isolation (single condition) or simultaneously (dual-different), the latter condition yields a *unique criterion* (UC). Interestingly, this is the case only if observers are asked to judge each of the two stimuli within the same experimental block. If, instead, they are always asked to judge one and the same stimulus out of the two, their response behavior shifts back to optimal (in the SDT sense), suggesting that the UC is a consequence of mixing *decisions* about stimuli rather than just mixing stimuli. Indeed, in agreement with SDT, these decisional interactions do not affect sensory aspects of

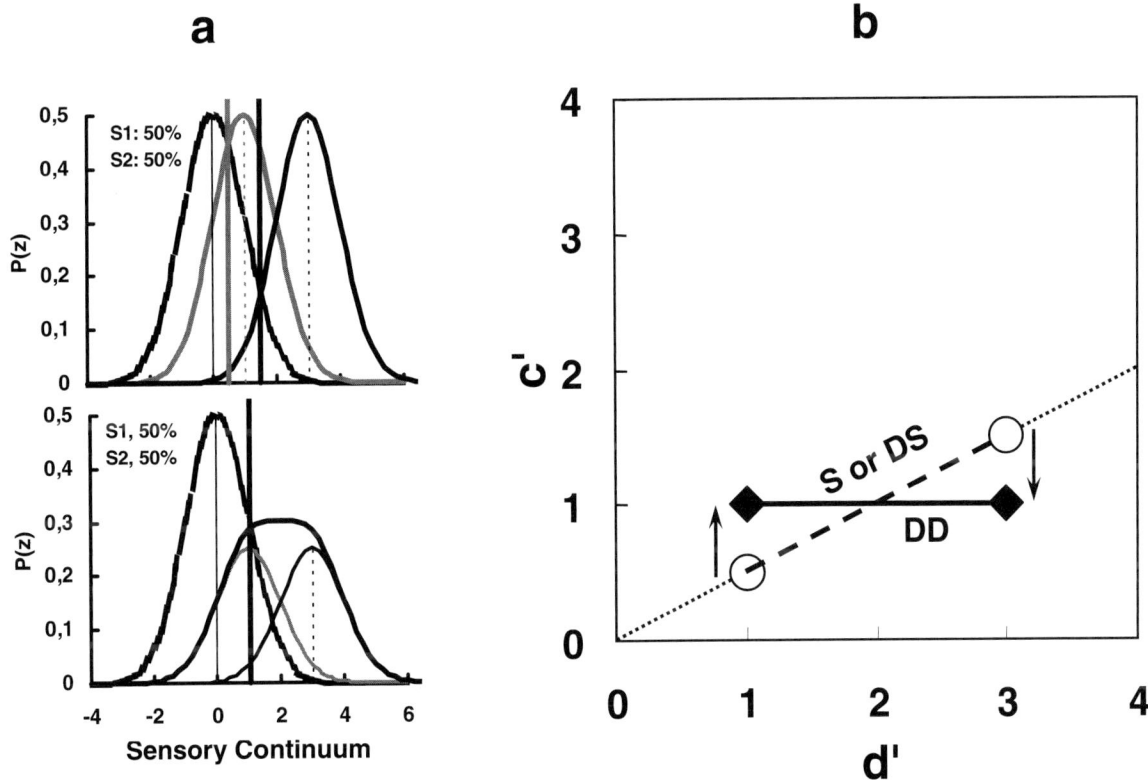

FIGURE 27.2 Optimal (top panel in (a) and the circles linked by the dashed line in (b)) and nonoptimal (the unique criterion predicted by the unique internal distribution model of Gorea and Sagi (2000) (bottom panel in (a) and diamonds linked by the continuous line in (b)) absolute criteria (c') for two known, equal-probability but different d' stimuli. Optimal behavior was observed under conditions where only one stimulus was presented within an experimental block (Single, S, case) or where two identical d' stimuli were mixed in one block (dual-same, DS, case). The unique criterion was observed when two different d' stimuli were mixed in one block (dual-different, DD, case) despite the fact that pre-cues indicated the type of stimulus to be presented on a trial-by-trial basis. SDT predicts optimal behavior in all cases. Arrows in (b) indicate the criterion shifts for the low- and high-d' stimuli.

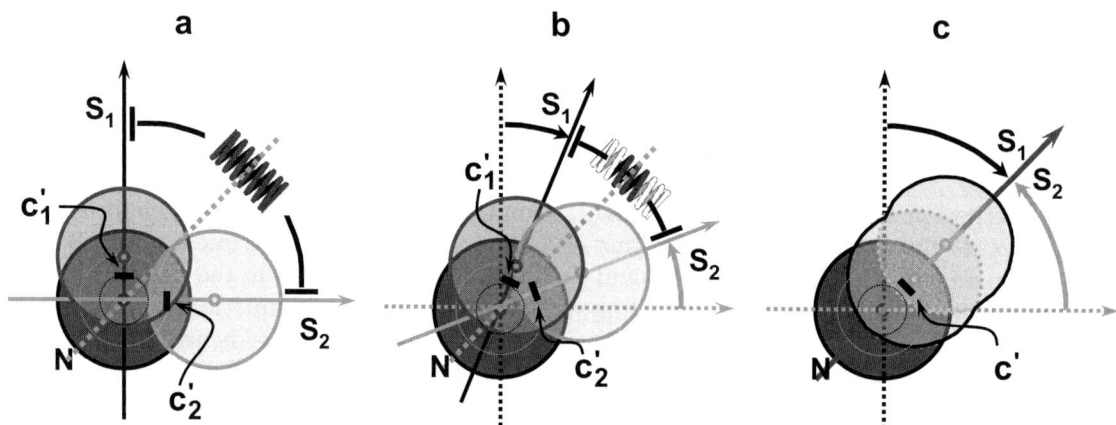

FIGURE 27.3 Three "attentional states" yielding noise (N, disk centered at the origin)- and stimulus (S_1 and S_2)-evoked internal representations (remaining two disks) given that S_1 and S_2 do not interfere at the sensory level. Internal representations, decision axes, and response criteria (c') are (**a**) independent (attention is ON), (**b**) partially mixed (attention fluctuates), or (**c**) entirely merged (attention is OFF). The intersections of the decision axes with the dotted concentric circles within the noise (N) internal distributions are the optimal (in the SDT sense) locations of the criteria. Attention is represented as an expander spring that keeps the decision axes apart.

account does not prejudge the limitation of such resources as it is perfectly conceivable that, had they been asked to also discriminate the two signals, observers could have done so without dropping their detection d' (Watson and Robson, 1981). The point here is that, as long as *not* measured, the discrimination performance should logically be blank and normal otherwise. This conceptual conundrum appears to prevent a direct test of the logical implication of the UC. It may, however, inspire future theoretical and empirical apprehensions of the attentional fact.

Attentional effects have always been measured and interpreted as they pertain to performance, that is, to sensitivity. At the same time, *selective* (endogenous) attention unequivocally refers to a fact of choice or strategy. Nonetheless, attentional effects, in general, and selective attention effects, in particular, have not been referred as yet to shifts of the *decision/response criterion*. The typical way of evading this obvious temptation was to expose the fact that selective attention (hence choice criterion) modulates the number of potential noise sources and, ultimately, sensitivity (i.e., the signal-to-noise ratio) (Lu and Dosher, 1998). This inclination has been formalized in the framework of an *uncertainty theory* (Green and Swets, 1966; Pelli, 1981). SDT's postulate that sensitivity is independent of the response criterion, together with the general consensus on what is meant by selective attention, apparently discouraged an experimental and theoretical effort focused on the attention–decision tandem. The present studies are but the shy beginning of such an enterprise. They suggest at least one new definition of attention (i.e., *a decision preparatory state*) and link it to the notion of stimulus dimensionality as it is determined by the task and to the dimensionality of the task itself as it is (consciously or not) formulated by the observer. This conceptual chain lends itself to a future expansion of the notion of sensory *objecthood* (see Han et al., 2003) as it relates to decision and attention processes.

VI. CONCLUSION

The preceding discussion of the implications of decisional behavior on the attentional process remains by and large speculative. We believe nonetheless that it may well be integrated within the wider conceptual spectrum of older (for a review, see Boring, 1942) and more recent (e.g., Gilchrist et al., 1999; Adelson, 2000) context theories (including Gestalt, adaptation-level, and other anchoring effects). Most importantly, we believe it offers for the first time, as far as we know, a means of linking perception, decision, and attention into a conceptually united whole. It is our opinion that in the last decade, perception and attention investigations have lacked more than ever new conceptual frameworks. The present review is a modest contribution to that avail.

References

Adelson, E. H. (2000). Lightness Perception and Lightness Illusions. In *The New Cognitive Neurosciences*, 2nd ed., M. Gazzaniga, ed. Cambridge, MA: MIT Press (pp. 339–351).

Boring, E. G. (1942). "Sensation and Perception in the History of Experimental Psychology." Appleton–Century, New York.

Fechner, G. T. (1860). "Elemente der Psychophysik." Breitkopf & Härtel, Leipzig.

Gilchrist, A., Kossyfidis, C., Bonato, F., Agostini, T., Cataliotti, J., Li, X., Spehar, B., Annan, V., Economou, E. (1999). An anchoring theory of lightness perception. *Psychological Review*, **106**, 795–834.

Gorea, A., and Sagi, D. (1999). Explorations into the psychophysics of decision: criterion attraction. *Invest. Ophthalmol. Vis. Sci. Suppl.* **40**, S796.

Gorea, A., and Sagi, D. (2000). Failure to handle more than one internal representation in visual detection tasks. *Proc. Nat. Acad. Sci. USA* **97**, 12380–12384.

Gorea, A., and Sagi, D. (2001). Disentangling signal from noise in visual contrast discrimination. *Nat. Neurosci.* **4**, 146–1150.

Gorea, A., and Sagi, D. (2002). Natural extinction: a criterion shift phenomenon. *Vis. Cogn.* **9**, 913–936.

Gorea, A., and Sagi, D. (2003). Selective attention as the substrate of optimal decision behavior in multistimulus environments. *Perception* Suppl. **32**, 5.

Green, D. M., and Swets, J. A. (1966). "Signal Detection Theory." Wiley, New York.

Halligan, P. W., and Marshall, J. C. (1998). Visuospatial neglect: the ultimate deconstruction? *Brain Cogn.* **37**, 419–438.

Han, S., Dosher, B. A., and Lu, Z.-L. (2003). Object attention revisited: identifying mechanisms and boundary conditions. *Psychol. Sci.* **14**, 598–604.

Lu, Z., and Dosher, B. A. (1998). External noise distinguishes attention mechanisms. *Vision Res.* **38**, 1183–1198.

Pelli, D. (1981). Uncertainty explains many aspects of visual contrast detection and discrimination. *J. Opt. Soc. Am. A* **2**, 1508–1532.

Schall, J. D. (2001). Neural basis of deciding, choosing and acting. *Nat. Rev. Neurosc.* **2**, 33–42.

von Helmholtz, H. L. F. (1856–1867). "Handbuch der physiologischen Optik." L. Voss, Hamburg, 3rd ed. (1909–1911). "Treatise of Physiological Optics" (J. P. C. Southall, Transl.), Dover, New York, 1962.

Watson, A. B., and Robson, J. (1981). Discrimination at threshold: labelled detectors in human vision. *Vision Res.* **21**, 1115–1122.

CHAPTER

28

Visual Attention and Emotional Perception

Luiz Pessoa and Leslie G. Ungerleider

ABSTRACT

Over the past 25 years, a great deal has been learned about the neural mechanisms of visual attention. Converging evidence from single-cell recording studies in monkeys and neuroimaging and event-related potential studies in humans have shown that the processing of attended information is enhanced relative to the processing of unattended information. At the same time, there is increasing evidence that, in general, visual processing requires attention. We review several key issues at the interface between visual attention and emotional perception, namely, the perception of emotional-laden stimuli, such as a picture of a fearful face. We review findings that suggest that attention is also required for the processing of emotion-laden faces, a category of stimuli previously proposed to be processed "automatically." We propose that, when attentional resources have not been exhausted, emotional stimuli can bias competition for processing resources. A likely source of biasing signals for emotional stimuli is the amygdala, a key structure involved in the processing of valence.

I. INTRODUCTION

Because the processing capacity of the visual system is limited, selective attention to one part of the visual field comes at the cost of neglecting other parts. Thus, several investigators have proposed that there is *competition* for neural resources (see Chapter 107). One instance of this proposal is the *biased competition model* of attention (Desimone and Duncan, 1995). According to this model, the competition among stimuli for neural representation, which occurs within visual cortex itself, can be biased in several ways. One way is by bottom-up sensory-driven mechanisms, such as stimulus salience. For example, stimuli that are colorful or of high contrast will be at a competitive advantage. But, another way is by attentional top-down feedback, which is generated in areas outside the visual cortex. For example, directed attention to a particular location in space facilitates processing of stimuli presented at that location. Stimuli surviving the competition for neural representation will have further access to memory systems for mnemonic encoding and retrieval and to motor systems for guiding action and behavior. At the neural level, an important consequence of attention is to enhance the influence of behaviorally relevant stimuli at the expense of irrelevant ones, providing a mechanism for the filtering of distracting information in cluttered visual scenes (Kastner and Ungerleider, 2000) (see Chapter 50).

II. ATTENTION IS NEEDED TO PROCESS VISUAL STIMULI

An implicit prediction of the biased competition model is that only items that survive the competition for neural representation in visual processing areas will impact on subsequent memory and motor systems. A related, but stronger, proposal has been advanced by Lavie (1995), who has suggested that the extent to which unattended objects are processed depends on the available capacity of the visual system. If, for example, the processing load of a target task exhausts available capacity, then stimuli irrelevant to that task would not be processed at all.

Hence, perceptually such stimuli may not even reach awareness.

Consistent with this idea, psychophysical studies in the past decade have demonstrated that processing outside the focus of attention is attenuated and may be eliminated under some conditions. Rock and colleagues (1992) showed that even the simplest visual tasks are compromised when attention is taken up elsewhere, a phenomenon they termed *inattentional blindness*. Further, in a striking demonstration, Joseph and colleagues (1997) showed that so-called "preattentive" tasks, such as orientation popout, require attention to be successfully performed. The necessity of attention for perception is perhaps most compellingly illustrated by "change blindness" studies (see Chapter 13), in which subjects may miss even very large changes in complex scenes, provided the changes are not associated with stimulus transients that capture attention.

But what is the fate of unattended stimuli? In extrastriate areas V2 and V4, single-cell studies in monkeys have shown that when an effective stimulus and ineffective stimulus are placed within a neuron's receptive field, spatially directed attention to the effective stimulus results in a response similar to that elicited by the effective stimulus presented alone (Reynolds et al., 1999). Remarkably, spatially directed attention to the ineffective stimulus results in a response similar to that elicited by the ineffective stimulus when presented alone. In essence, it is as if the unattended stimulus, be it the effective or the ineffective one, were not in the receptive field. These findings suggest that, at the neural level, responses evoked by unattended items may be eliminated.

Such an interpretation is consistent with fMRI studies demonstrating that the stimulus-evoked fMRI response is essentially abolished when subjects are engaged in a competing task with high attentional load (see also Chapter 71). In one study, Rees and colleagues (1997) showed that moving stimuli did not elicit fMRI activation in area MT when subjects performed a concurrent, highly demanding linguistic task. In a related study, Rees and colleagues (1999) showed that activations associated with words were not elicited when subjects performed a concurrent, highly demanding object working memory task. Thus, like the processing of visual motion, even word processing seems to require attention, contrary to claims for full automaticity.

III. IS ATTENTION NECESSARY FOR THE PROCESSING OF EMOTION-LADEN FACES?

A major exception to the critical role of attention may be in the neural processing of emotion-laden stimuli, which are reported to be processed automatically, namely, without attention (Vuilleumier et al., 2001; Ohman, 2002). For example, subjects exhibit fast, involuntary autonomic responses to emotional stimuli, such as aversive pictures and faces with fearful expressions. Other behavioral studies suggest that such autonomic responses to facial expressions not only occur "automatically," but may even take place without conscious awareness (Ohman, 2002). This conclusion is also supported by imaging studies of the neural processing of emotional stimuli in the amygdala, a structure that is known to be important in emotion, particularly the processing of fear (Aggleton, 2000). Such studies report that the amygdala is activated not only when normal subjects view fearful faces, but even when these stimuli are masked and subjects appear to be unaware of their occurrence. Using the backward masking paradigms developed by Ohman and colleagues, Whalen and colleagues (1998) showed that fMRI signals in the amygdala were significantly larger during viewing of masked, fearful faces than during the viewing of masked, happy faces. In another study, Morris and colleagues (1998) combined backward masking with classic conditioning to investigate responses to perceived and nonperceived angry faces. Although the participants never reported seeing the masked, angry stimuli, the contrast of conditioned and nonconditioned masked, angry faces activated the right amygdala.

The view has thus emerged that the amygdala is specialized for the fast detection of emotionally relevant stimuli in the environment, and that this can occur without attention and even without conscious awareness. If this were indeed the case, amygdala activity would reflect an obligatory response independent of the locus of spatial attention. Vuilleumier and colleagues (2001) tested this prediction in an fMRI study in which subjects fixated a central cue and matched either two faces or two houses presented eccentrically. Both fearful and neutral faces were used. As in earlier studies, activity in the fusiform gyrus, which is known to respond strongly to faces, was modulated by attention. At the same time, Vuilleumier and colleagues failed to see evidence that attention modulated responses in the amygdala, regardless of stimulus valence. Not surprisingly, these results were interpreted as further evidence for obligatory activation of the amygdala by negative stimuli.

IV. A STRONG TEST OF AUTOMATIC AMYGDALA ACTIVATION

In a recent study, we tested the alternative possibility, namely, the neural processing of stimuli with emotional content is not automatic and instead requires some degree of attention, similar to the processing of neutral stimuli (Pessoa et al., 2002). We hypothesized that the failure to modulate the processing of emotional stimuli by attention in previous studies was due to a failure to fully engage attention by a competing task. In other words, activation in the amygdala by emotional stimuli should resemble activation in MT to moving stimuli; if the competing task is of high load, activation should be reduced or absent. We therefore employed fMRI and measured activations in the amygdala and other brain regions that responded differentially to faces with emotional expressions compared with neutral faces and then examined how those responses were modulated by attention.

We measured fMRI responses evoked by pictures of faces with fearful, happy, or neutral expressions when attention was focused on them (attended condition), and compared the responses evoked by the same stimuli when attention was directed to oriented bars (unattended condition). In designing our bar orientation task, we chose one that was sufficiently demanding to exhaust all attentional resources on that task and leave little or none available to focus on the faces, even though they were viewed foveally during the bar orientation task. We found that compared with unattended faces, attended faces evoked significantly greater activations bilaterally in the amygdala for all facial expressions (Fig. 28.1A). Importantly, there was a significant interaction between stimulus valence and attention. That is, the differential response to stimulus valence was observed *only* in the attended condition (Fig. 28.1B). Moreover, for the unattended condition, responses to all stimulus types were equivalent and not significantly different from zero. Thus, amygdala responses to emotional stimuli are not automatic and instead require attention.

Our findings are in direct contrast to those by Vuilleumier and colleagues who failed to see evidence that attention modulated responses in the amygdala, regardless of stimulus valence (Vuilleumier et al., 2001). What is the explanation for their negative findings? The most likely explanation is that the attentional manipulation in the Vuilleumier et al. study was not as effective as ours. For example, behavioral performance for the bar orientation task in our study and that for house matching in the Vuilleumier et al. study were 64 and 86% correct, respectively, indicating that our competing task was a more demanding one (see Pessoa et al., 2002, for further discussion).

To summarize, contrary to the prevailing view, we found that the amygdala responded differentially to faces with emotional content only when sufficient attentional resources were available to process those faces. Indeed, when all attentional resources were consumed by another task, responses to faces were eliminated, consistent with Lavie's (1995) proposal that if the processing load of a target task exhausts available capacity, stimuli irrelevant to that task will not be

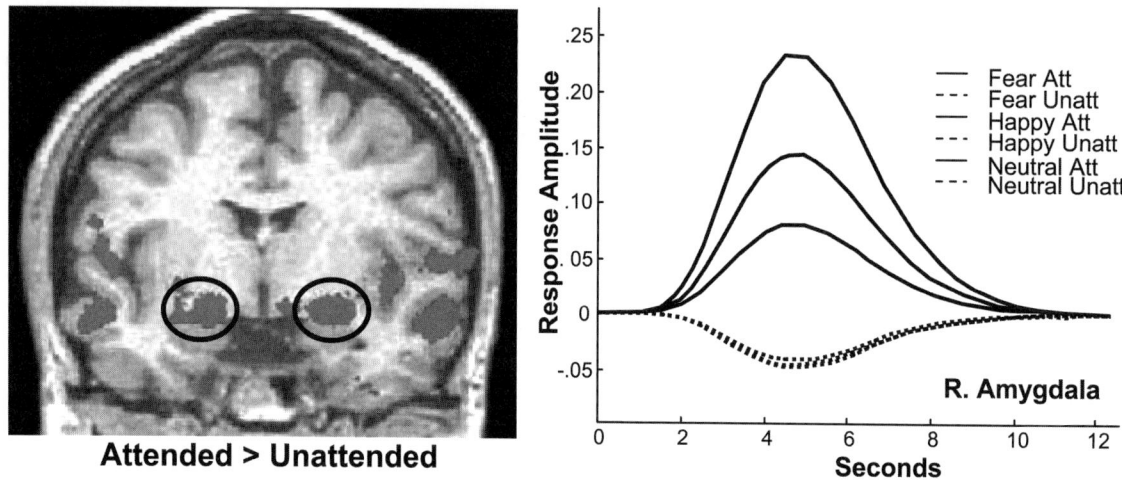

FIGURE 28.1 Attention is required for the processing of stimulus valence. Left: Attended faces compared with unattended faces evoked significantly greater activations for all facial expressions. Circles indicate the region of the amygdala. Right: Estimated responses for the right amygdala region of interest as a function of attention and valence.

CHAPTER 29

The Difference between Visual Attention and Awareness: A Cognitive Neuroscience Perspective

Victor A.F. Lamme

ABSTRACT

Visual awareness is often equated to what is in the focus of attention. There are a number of reasons from the psychological or theoretical perspective to distinguish between the two, but perhaps the clearest argument to separate attention from awareness can be made from the neural perspective. In neural terms, definitions of visual attention and awareness can be formulated that clearly distinguish between the two, yet explain why attention and awareness are so intricately related. Attention can be defined as the convolution of sensory processing with long- and short-term memory. Awareness, on the other hand, depends critically on recurrent processing. By combining the two, we understand why we seem only capable of conscious reports about what is in the focus of attention. At the same time, we must conclude, however, that we also have phenomenal experience of what is outside that focus.

I. TWO FORMS OF SELECTION: IDENTICAL OR NOT?

Fundamental to the study of conscious experiences is the assumption that they are selective; we are not aware of everything we lay our eyes on. From the neuroscience perspective this implies that there is neural activity that produces conscious experience and neural activity that does not. Hence the search for the neural correlate of consciousness (NCC) (Crick and Koch, 1998). Another type of selection is attention; some sensory inputs are processed faster or deeper than others, and thus become more readily available for action, memory, or thought (Egeth and Yantis, 1997). It seems only natural that the two forms of selection share properties, and some have even argued that they are identical: we are aware of what is in the focus of attention (O'Regan and Noe, 2001). This would strongly invalidate the search for the NCC. However, a strong argument can be made to distinguish between the two. This argument has components from the fields of experimental psychology, theoretical modeling, and neuroscience, which taken together explain (1) why attention and awareness should be considered very different and (2) why the two seem so intricately related (reviewed in Lamme, 2003, 2000). Here, I focus on the neuroscience part of the argument, because it is the strongest. Moreover, complex issues can be put in a comprehensive framework and counterintuitive conclusions can be drawn.

II. STARTING POINTS: PROCESSING AND MEMORY

Although many issues in neural processing are not fully understood, we do understand some at a principal level. We know that the senses transduce physical information from the outside into neural activity. We know that neurons integrate inputs at their dendrites, resulting in an output at the axon. These neurons are embedded in an intricate network in which we can identify nuclei, areas, pathways, modules, and so on. In essence this is the basis of what we call sensory processing. The immense complexity of the anatomical connections between neurons renders the true nature of how successive neurons transfer information rather difficult to study. For that reason, many details, and also many fundamental properties, of how a sensory input is translated into a motor output still evade us. But still, we can in principle imagine how the brain might do that. If we start from our understanding of the reflex arch, it is imaginable how we can build sophisticated input–output mappings from this principle. We may have to include very complex computational concepts, such as recurrent processes, synchrony, and modulatory influences, of which we still understand very little, but there is no real explanatory gap there, only a lot of work still to do.

A similar reasoning can be applied to memory. The synapses that mediate processing are plastic, so that the transfer of information may be modified. In this way, preceding events may induce changes to the network, and this is what we call memory. Again, the issue is highly complex, both with respect to the "preceding events" and with respect to the "changes in the network." In principle, any preceding event that results in neural processing will have an influence on the network. Ongoing activity changes, some activity starts to reverberate in the network, and we enter the domain of working memory. Some events may induce more lasting changes in synaptic transfer (like long-term potentiation, LTP), which eventually may result in an anatomical consolidation of the changes in synaptic transfer. Now we are in the domain of long-term memory. This is not to say that all aspects of memory and its neural basis are understood, but again we have at least some idea of basic principles.

III. ATTENTION = PROCESSING × MEMORY

Combining the core concepts of sensory processing and memory may be sufficient to explain visual attention (Desimone, 1996, 1998). Attention is a selection process where some inputs are processed faster, better, or deeper than others, so that they have a better chance of producing or influencing a behavioral response or of being memorized. Attention induces increased and synchronous neuronal activity of those neurons processing the attended stimuli and increased activity in parietal and frontal regions of the brain (Desimone and Duncan, 1995; Fries et al., 2001). The increased neural activity is in principle sufficient to explain why the associated stimuli are processed faster, deeper, and so on. The main problem lies in explaining what brings the enhanced activity about.

Attention may be grabbed externally (Egeth and Yantis, 1997). Some stimuli are simply processed more efficiently than others. These stimuli we call salient. A bright stimulus will win from a dark one, a moving from a stationary, a foveal from a peripheral, and so on. This is due mainly to the properties of the adult processing network, shaped by genetics and visual experience. In other words, saliency reflects how long-term memory has shaped and modified sensory processing.

But preceding stimuli may subtly change these properties. Imagine a stimulus entering the system and another stimulus following within 100 ms or so. If the two stimuli share properties (such as retinal position), it is understandable why processing of the second stimulus will be more efficient than of a similar stimulus not preceded by the first stimulus. The first stimulus will have "paved the way" in the sense that neurons are already activated above threshold, and this activity may persevere for some time (Fig. 29.1). This is a typical attentional priming situation (Dehaene et al., 1998) (Fig. 29.1). More specifically, processing of the first stimulus has led to a short-term memory trace (in this example, maybe better called a sensory or iconic memory trace), and processing of the second stimulus is influenced by this trace. Also, inhibitory influences from the first stimulus are possible, for example, at other locations or at later times, when neural activity rebounds, resulting in inhibition of return (see Chapter 16) (Egeth and Yantis, 1997).

With endogenous attention, the situation becomes more complex, but not fundamentally different. Now, an external event, such as an abstract cue, has to be translated into something akin to the "paving of the way" described above. Parts of the brain that extract the meaning of the cue, and that are able to relate this to current needs and goals, must preactivate or otherwise facilitate the appropriate sensory pathways, mostly via corticocortical feedback or subcortical routes. Regions that are able to do so will be at the interface between sensory and motor representations (parietal cortex), or will be where sensory, motivational, and internal milieu information meets (prefrontal cortex). Such top-down paving requires more time, yet it is more flexible than bottom-up types of attention, and provides higher-level control.

This, however, is not to say that something like intention or free will has to be incorporated into the idea. What we may experience as free will or intention is, in this simplified scheme, nothing more than a combination of current and past inputs that operate on the current state of the network. We do not need anything else than the combination of sensory processing, the processing of internal milieu variables, and short- and long-term memory to explain why a particular brain at a particular moment in time is inclined to favor one stimulus over another (Desimone, 1996).

So in summary, I think that all forms of attentional selection can be explained at the fundamental level as a convolution of sensory processing with short- and long-term memory, even though, again, many details still need to be worked out. Therefore, I strongly argue that attentional selection is not a priori associated with visual awareness. We can imagine all the operations described above to occur in brains (or machines for that matter) without any phenomenal experience. We do not need to have an explanation for phenomenal experience to understand attention.

IV. VISUAL AWARENESS = RECURRENT PROCESSING

What remains to be found, then, is a similar core understanding of phenomenal experience. We know that neural (including cortical) activation does not necessarily lead to awareness. Hence the search for the NCC: where it is investigated, what kind of neural activity is (and what kind is not) capable of producing awareness. Elsewhere (Lamme et al., 2000), I have argued that a strictly localizationist approach in the search for the visual NCC will be barren; there is no region in the brain whose activation automatically leads to visual awareness. The NCC is not anatomically defined, but functionally; some type of neural activity leads to awareness, while other types do not. With respect to that question, I have made a strong point of distinguishing between the so-called feedforward sweep (FFS) and recurrent processing (RP) (Lamme, 2000; Lamme and Roelfsema, 2000).

The FFS is defined as the earliest activation of cells in successive areas of the cortical hierarchy. Typically, V1 starts to respond 40 ms after stimulus onset, and higher, extrastriate areas respond at successively

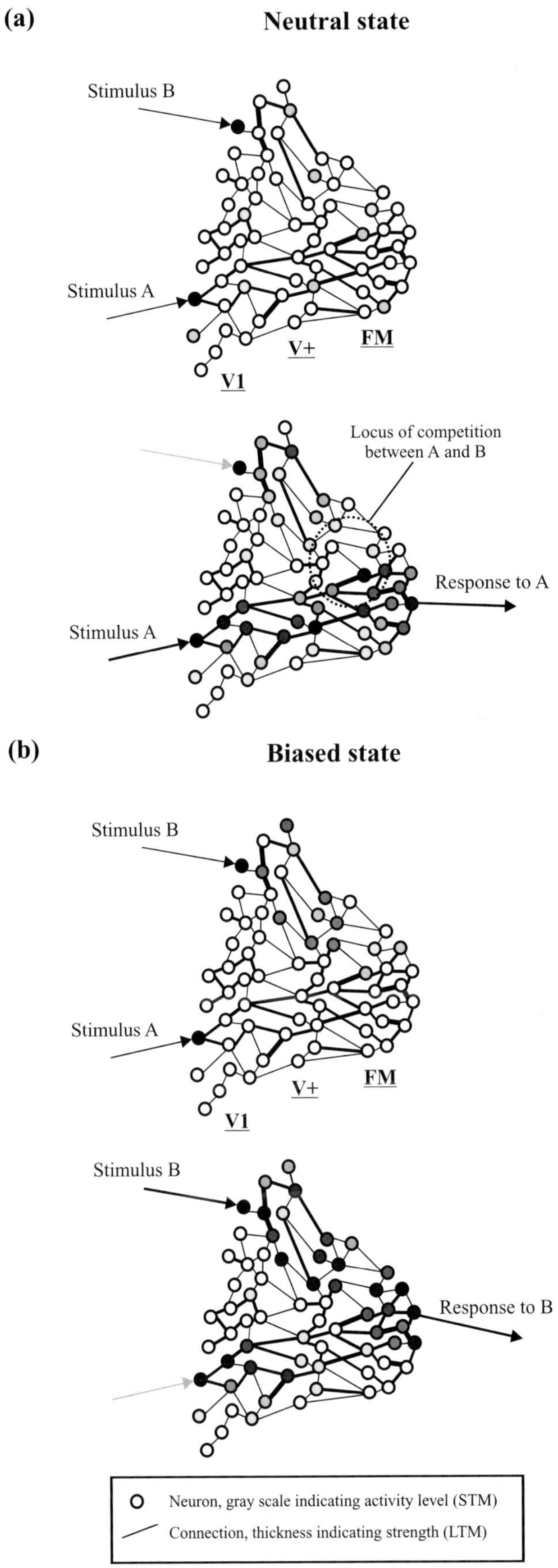

FIGURE 29.1 **Attentional selection is a convolution of memory and processing.** Selection is necessary when two stimuli (A, B) reach the brain, yet only one response is possible. Competition, typically at the level of the extrastriate areas, prevents all inputs to reach output areas of the brain. Depending on the state of the brain when stimuli arrive, either of the two outputs may be selected. (**a**) Neutral state, where stimulus A is processed more efficiently, that is, better matches stored synaptic weights: stimulus A is more salient, and the associated neural activity is stronger or more synchronous (darker dots). (**b**) Biased state, where the processing of a previous stimulus has left a short-term trace of activity (light gray dots). Now, processing of stimulus B toward a response is favored. Thus, attentional selection results from the convolution of the processing of current inputs with long- and short-term memory.

increasing latencies. At about 80 ms most visual areas are activated; at 120 ms visual activation can be found in all cortical areas, including motor cortex. Surprisingly, these early responses already fully express the receptive field (RF) tuning properties of cells, even complex ones like face selectivity in area IT. Feedforward connections are apparently capable of generating sophisticated RF tuning properties and thus extracting high-level information, which could lead to categorization and selective behavioral responses (for references, see Lamme and Roelfsema 2000).

As soon as the FFS has reached an area, recurrent interactions between neurons within that area and neurons that have been activated earlier at lower levels may start. These interactions are mediated by horizontal connections and feedback–feedforward circuits between and within areas. They are expressed as modulatory influences from beyond the classic, feedforward, RF (Lamme and Spekreijse, 2000; Albright and Stoner, 2002).

The hypothesis I put forward is that the feedforward activation of whatever area in the brain is not sufficient for visual awareness. Even when high-level areas in temporal, parietal, or frontal cortex are reached, this in itself does not lead to visual awareness, that is, is unconscious. Recurrent interactions between areas, most notably between V1 and extrastriate areas, are necessary to become aware of the visual input. Some important observations can be made about the relation between FFS, RP, and visual awareness in support of that idea (for references, see Lamme, 2003).

1. Backward masking renders a visual stimulus invisible by presenting a second stimulus shortly (e.g., 40 ms) after the first. The masked stimulus, even though invisible, still evokes selective feedforward activation in visual and nonvisual areas as widespread as V1, IT, FEF, and motor cortex. Neurophysiological manifestations of recurrent interactions are, however, suppressed by backward masking.
2. With transcranial magnetic stimulation (TMS), the ongoing activity in a particular brain region can be shortly disrupted. Applying TMS to early visual areas at a latency far beyond the FFS still renders stimuli invisible. Also, TMS over the motion-selective area MT induces motion sensations, unless V1 activity is disrupted at a later moment. MT is higher in the visual hierarchy than V1; this implies that feedback from MT to V1 is necessary for motion awareness.
3. Feedforward activation of neurons can still be recorded in anesthetized animals, with RF tuning properties that hardly differ from those in the awake animal. Manifestations of recurrent processing, in particular those contextual modulations that express aspects of perceptual organization, are, however, reduced or fully suppressed under anesthesia.
4. Feedforward activation of neurons in V1 is not affected when stimuli are reported as not seen by animals engaged in a figure–ground detection task. A neural correlate of figure–ground segregation, probably mediated by recurrent interactions between V1 and extrastriate areas and present when stimuli are seen, is, however, fully suppressed when stimuli are not seen.

This has led me and others (see Lamme, 2000, 2003) to conclude that visual processing mediated by the FFS, however sophisticated, is not accompanied by awareness. Recurrent interactions are necessary for visual awareness to arise (Fig. 29.2).

V. AWARENESS × ATTENTIONAL SELECTION: THREE STAGES OF PROCESSING

We may now have a look at what happens when the proposed neural mechanism of visual awareness (recurrent processing) interacts with the mechanism of attention (processing × memory). Suppose a visual scene is presented to the eyes. The feedforward sweep reaches V1 at a latency of about 40 ms. If multiple stimuli are presented, these are almost all represented at this stage. Next (60–80 ms), this information is fed forward to the extrastriate areas. At these intermediate levels, there is already some competition between multiple stimuli, in particular when they are close by. Not all stimuli can be processed in full by the receptive fields, which grow larger and larger going upstream in the visual cortical hierarchy. This results in crowding phenomena. Attentional selection (in one way or another, see above) may resolve this competition (Desimone, 1998). In the end, only a few stimuli reach the highest levels, up to and including areas in executive space. This whole feedforward event evolves very rapidly (within ~120 ms) and is hypothesized to be fully unconscious. Feedforward activation alone may, in certain circumstances, result in a behavioral response (or modify ongoing behavior), but if it does, it will be a reflexlike action that is fully unconsciously initiated (which is not to say that we may not become aware of it later).

Meanwhile, the early visual areas have started to engage in recurrent interactions, mediated by horizontal and feedback connections. By means of these

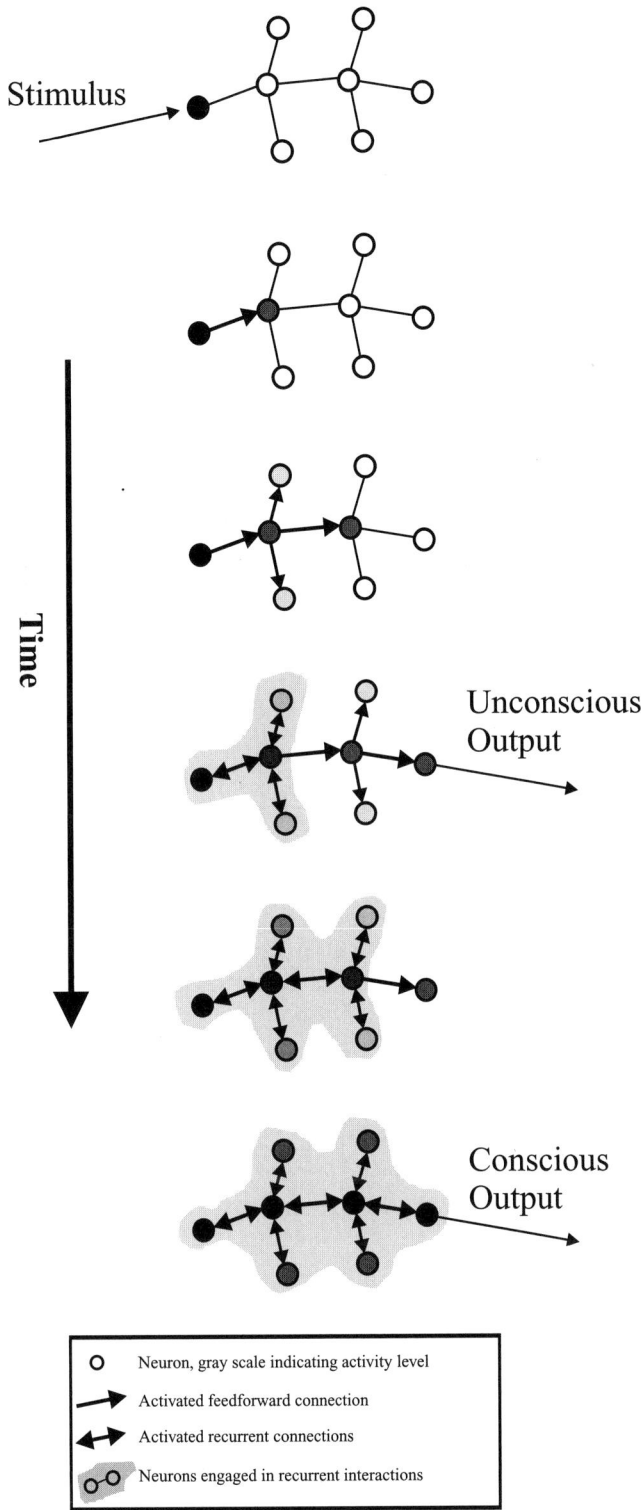

recurrent interactions, visual features are related to each other, binding and segregation may occur, and perceptual organization evolves. This is what produces awareness. Without these recurrent interactions there is no awareness at all. Because at low levels there is relatively little competition between stimuli (unless they are close by), groups of recurrent interactions representing multiple stimuli are possible. This may occur for many items in a scene.

When these recurrent interactions grow more and more widespread, and eventually include areas in executive or mnemonic space (frontal, prefrontal, temporal cortex), the visual information is put into the context of the systems' current needs, goals, and full history. There is considerable competition, however, for interaction with these higher levels. Attentional selection during the feedforward sweep will already have predisposed some interactions over others. Alternatively, this selection may operate at the recurrent interactions themselves. In any case, only a limited number of recurrent groups can span the range from visual to more frontal (parietal, temporal) areas. Therefore, these more widespread recurrent interactions are limited to a few items in the scene. On the other hand, the recurrent process is less "superficial" in the sense that stimuli are processed more deeply; more behavioral and mnemonic context is added to the stimuli than in the case of more low-level recurrent interactions, limited to the visual areas only.

We are thus able to discern at least three fundamentally different types of processing: (1) feedforward processing, influenced by attentional selection, but unconscious for reasons outlined above; (2) recurrent processing of a restricted nature, limited to, say, visual areas; (3) widespread recurrent processing of information that involves many regions of the brain and has passed the attentional bottleneck between sensory and executive areas. It is clear that widespread recurrent interactions should correspond to a stage of processing that we could call conscious. Here, a selected part of the information is embedded in mnemonic and behavioral context. Information thus processed is

FIGURE 29.2 **Conscious visual experience requires recurrent processing.** Feedforward processing rapidly activates (unidirectional arrows) successive levels of processing (from left to right), potentially leading to a reflexlike unconscious output or modification of behavior, based on basic ("hard-wired") categorizations and stimulus–response associations. Recurrent processing, mediated by horizontal or feedback connections (bidirectional arrows), lags behind this feedforward sweep (unless parallel feedforward sweeps exist, one being slower than the other; see Bullier, 2001). Recurrent procesing mediates more complex stimulus–response associations, and is required for visual awareness or for a conscious response.

available for conscious access and can be reported about. Related theories refer to such a state as *resonant* (Grossberg, 1999) or as having reached *global workspace* (Dehaene and Naccache, 2001). But what exactly is the nature of information that has achieved local (say only visual) recurrent embedding? Or to put it in another way: what span should a state of recurrent resonance have to call it conscious?

VI. A CASE FOR PHENOMENAL AWARENESS

There are many properties of stimuli that never reach consciousness, not even when attended. Yet these invisible stimuli or attributes activate neurons. Examples are high temporal and spatial frequencies, anticorrelated disparity, physical wavelength (instead of color), crowded or masked stimuli, or the nondominant patterns during perceptual rivalry (for references see Lamme, 2003). Also, fully attended stimuli may occasionally not be perceived, suggesting that sensory processing does not necessarily always complete to a perceptual stage (Super et al., 2001). The reason these properties do not reach awareness is that they are extracted during the feedforward sweep.

During the feedforward sweep, information has been extracted, but this information has not yet interacted. Interaction between the distributed information requires recurrent interactions. Visual recurrent processing goes beyond initial feature detection and may be the neural correlate of binding or perceptual organization: features are tentatively bound, surfaces are defined, figure–ground relationships may be established. Others have called this stage *midlevel vision* or the *2.5D sketch* (Nakayama et al., 1995). There is strong evidence that this stage is indeed a manifestation of recurrent interactions between early visual areas (Lamme and Spekreijse, 2000). Moreover, this stage is hardly susceptible to attentional bottlenecks or top-down control. On the other hand, it is the first stage to which we have conscious access. Ingenious stereo display experiments have shown very elegantly that vision (more specifically, motion perception, object recognition, and visual search) is based on this surface representation stage, and that it is the basis of our phenomenal experience (Nakayama et al., 1995). In my view, recurrent processing limited to the visual areas (Fig. 29.3) forms the neural basis of this stage, a stage where perceptual organization occurs, and that we should call phenomenally conscious.

To report about these percepts, however, the information has to become globally recurrent, has to reach access awareness. But that is not to say that we have no phenomenal experience of this information. We do, and that gives us the rich experience of vision we have. This experience is not an illusion, as change blindness (Simons and Levin, 1997) or inattentional blindness (Mack and Rock, 1998) experiments might suggest (O'Regan and Noe, 2001). Those experiments reveal the attentional bottleneck between experience and report or between experience and the storage of experiences (Landman, 2003), not the limitations of experience itself. A similar distinction between phenomenal and access awareness has been made by Block (1996) on philosophical and theoretical grounds. Also in the domain of memory we find a similar distinction. Working memory, the limited-capacity yet stable storage of information, has clear similarities to access awareness, and may even share neural mechanisms. Iconic memory, the large-capacity yet fleeting form of memory we have of a scene in its entirety, may be linked to phenomenal awareness. Both are forms of memory, however, and in that sense different from awareness, which is present only when stimuli are there.

VII. CONCLUSIONS

From the cognitive neuroscience perspective a clear distinction can be made between attention and awareness. Attentional selection is how sensorimotor processing is modified by the current state of the neural network, shaped by genetic factors, experience, and recent events (memory). Phenomenal experience has a different origin, which is the recurrent interaction between groups of neurons. Depending on the extent to which recurrent interactions between visual areas incorporate interactions with action- or memory-related areas, awareness evolves from phenomenal to access awareness. Whether this occurs depends on attentional selection mechanisms, via influences on both the feedforward sweep and recurrent interactions. Other mechanisms, however, determine whether neurons will engage in recurrent interactions at all and, thus, whether processing will go from an unconscious to a conscious state.

Conscious stimuli have reached a level of processing beyond initial feature detection, where at least an initial coherent perceptual interpretation of the scene is achieved. Whether at this stage the binding problem, in all its diversity, has been solved is not clear yet. The binding of some features of an object, such as its color and shape, may require attention, while other feature combinations are detected preattentively. So it may be that the conscious level, before attention has been allocated, consists of tentatively bound features and sur-

VII. CONCLUSIONS

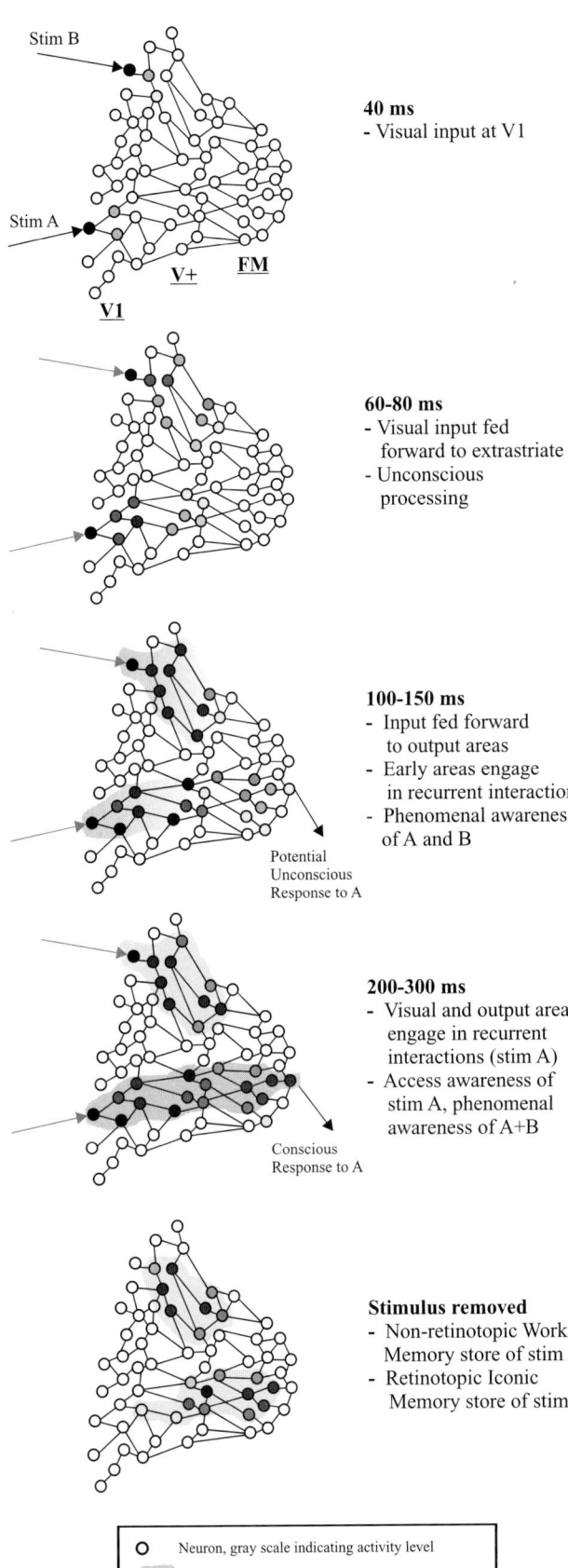

FIGURE 29.3 **Phenomenal versus access awareness.** The interaction between recurrent processing (Fig. 29.2) and mechanisms of attentional selection (Fig. 29.1) is shown. As in Fig. 29.2, competition between the neural representations of multiple stimuli (stimulus A, B) may prevent the feedforward transfer from V1 to the executive areas (FM, frontal or motor regions) of all but a few stimuli (in this case A). At lower levels (V1, V+, extrastriate areas), however, simultaneous representations (of both A and B) may exist. Either way, feedforward activation (gray dots) of both selected (i.e., attended) and not selected inputs is unconscious, even though it may trigger or modify behavior. Meanwhile, neurons in activated regions may start to engage in recurrent interactions, which is accompanied by increased activity or synchronous firing (dots enclosed by gray shading). This produces *phenomenal* awareness of the visual inputs (and iconic memory after removal of the stimulus). Some of these recurrent interactions grow more widespread than others, and may even incorporate high-level, executive, or planning regions, depending on attentional selection, in part already established during the feedforward sweep. Stimuli associated with these widespread interactions reach *access* awareness, and may be stored in nonretinotopic working memory after removal of the stimulus.

faces, akin to midlevel vision or the 2.5D sketch. There is a clear distinction, however, with unconscious stages, where individual features, even features that are never perceived, are represented.

The hypothesis forms a core understanding of the different forms of awareness, in the same spirit as we have core understandings of sensorimotor transformations, memory, and attentional selection. Again, many things still need to be worked out, the most important being of course to explain why recurrent interactions are necessary for phenomenal experience to arise, and how we go from such a neural process to the phenomenon of mental experience. In that sense, the "hard problem" remains as hard as it was. With this theory, however, we may have a better sense of what to look for.

References

Albright, T. D., and Stoner, G. R. (2002). Contextual influences on visual processing. *Annu. Rev. Neurosci.* **25**, 339–379.

Block, N. (1996). How can we find the neural correlate of consciousness. *Trends Neurosci.* **19**, 456–459.

Bullier, J. (2001) Integrated model of visual processing. *Brain Research Reviews*, **36**, 96–107.

Crick, F., and Koch, C. (1998). Consciousness and neuroscience. *Cereb. Cortex* **8**, 97–107.

Dehaene, S., and Naccache, L. (2001). Towards a cognitive neuroscience of consciousness: basic evidence and a workspace gramework. *Cognition* **79**, 1–37.

Dehaene, S., et al. (1998). Imaging unconscious semantic priming. *Nature* **395**, 597–600.

Desimone, R. (1996). Neural mechanisms for visual memory and their role in attention. *Proc. Nat. Acad. Sci. USA* **93**, 13494–13499.

Desimone, R. (1998). Visual attention mediated by biased competition in extrastriate visual cortex. *Philos. Trans. R. Soc. London Ser. B* **353**, 1245–1255.

Desimone, R., and Duncan, J. (1995). Neural mechanisms of selective visual attention. *Annu. Rev. Neurosci.* **18**, 193–222.

Egeth, H. E., and Yantis, S. (1997). Visual attention: control, representation, and time course. *Annu. Rev. Psychol.* **48**, 269–297.

Enns, J. T., and Di Lollo, V. (2000). What's new in visual masking? *Trends Cogn. Sci.* **4**, 345–352.

Fries, P., et al. (2001). Modulation of oscillatory neuronal synchronization by selective visual attention. *Science* **291**, 1560–1563.

Grossberg, S. (1999). The link between brain learning, attention and consciousness. *Consciousness Cogn.* **8**, 1–44.

Lamme, V. A. F. (2000). Neural mechanisms of visual awareness: a linking proposition. *Brain Mind* **1**, 385–406.

Lamme, V. A. F. (2003). Why visual attention and awareness are different. *Trends Cogn. Sci.* **7**, 12–18.

Lamme, V. A. F., and Roelfsema, P. R. (2000). The distinct modes of vision offered by feedforward and recurrent processing. *Trends Neurosci.* **23**, 571–579.

Lamme, V. A. F., and Spekreijse, H. (2000). Modulations of primary visual cortex activity representing attentive and conscious scene perception. *Front. Biosci.* **5**, D232–D243.

Lamme, V. A. F., et al. (2000). The role of primary visual cortex (V1) in visual awareness. *Vision Res.* **40**, 1507–1521.

Landman, R., et al. (2003). Large capacity storage of integrated objects before change blindness. *Vision Res.* **43**, 149–164.

Mack, A., and Rock, I. (1998). "Inattentional Blindness." MIT Press, Cambridge, MA.

Nakayama, K., He, Z. J., and Shimojo, S. (1995). Visual surface representation: a critical link between lower-level and higher level vision. *In* "Invitation to Cognitive Science" (S. M. Kosslyn and D. N. Osherson, Eds.), pp. 1–70. MIT Press, Cambridge, MA.

O'Regan, J. K., and Noe, A. (2001). A sensorimotor account of vision and visual consciousness. *Behav. Brain Sci.* **24**, 939–1031.

Simons, D.J., and Levin, D.T. (1997). Change blindness. *Trends Cogn. Sci.* **1**, 261–267.

Super, H., et al. (2001). Two distinct modes of sensory processing observed in monkey primary visual cortex (V1). *Nat. Neurosci.* **4**, 304–310.

CHAPTER

30

Reaching Affects Saccade Trajectories

Steven P. Tipper

ABSTRACT

When a task requires accurate reaching, attention to a location activates the motor circuits controlling saccades and manual reaches. These actions involve separate neural systems for the control of eye and hand movements. However, I believe, first, that the selection processes acting on neural population codes within these systems are similar; and second, that there is cross-talk between these systems to guide coherent behavior. The present study shows that the path the eye takes as it saccades to a target is determined by inhibitory selection processes acting on population codes. Furthermore, the path of the saccade is influenced by whether a reach toward the target is also produced. This hand–eye interaction effect is interpreted as the influence of a hand-centered frame used in reaching on the spatial frame of reference required for the saccade.

I. INTRODUCTION

A number of properties of visuomotor systems are to be discussed in this chapter. First, actions such as saccades and manual reaches are represented in population codes. That is, a range of cells are active when, for example, a reach or saccade (e.g., Georgopoulos, 1990; Georgopoulos et al., 1984) is made in a particular direction. The direction of the action is determined by the population vector. Second, a number of actions can be simultaneously represented (e.g., Cisek and Kalsask, 2002; Goldberg and Segraves, 1987; Snyder et al., 1997), and therefore inhibitory selection mechanisms are necessary to select the appropriate action at the appropriate time (e.g., Tipper et al., 1992). And third, different action systems require different frames of reference. Thus, for the hand to be directed to a target, the distance and direction between the hand and target must be coded in a hand-centered frame (e.g., Kalaska et al., 1998; Tipper et al., 1992); whereas for a saccade to the same target, the location of the stimulus must be encoded relative to the fovea.

Evidence for the first two of these ideas emerges from a number of studies. For example, Tipper et al. (1997) required subjects to reach for a target in the presence of irrelevant distracters. They argued that both target and distracter object were automatically encoded and competed for the control of action. This competition took place in networks containing populations of activated neurons. For example, in Fig. 30.1A the direction of the reach for a target is represented. In Fig. 30.1B the direction of reach for a competing distracter is represented. In Fig. 30.1C, the overall neural activity is represented. Two things are of note: First, many cells are common to both reaches; and second, without selection mechanisms acting on this representation, action would be directed between the two objects.

The properties of such overlapping neural populations (Fig. 30.1C) have recently been investigated by Treue et al. (2000) in MT motion cells. They have demonstrated that the underlying populations (Figs. 30.1A, B) that produce the complex population codes of Fig. 30.1C can be recovered (see Tipper et al., 2001, for discussion). Therefore, selection can be achieved by inhibition of the population code activated by the distracter stimulus, shown in Fig. 30.1B (see Tipper et al., 1997 for details). The level of inhibition is determined by the initial activation state of the distracter via a reactive feedback system (e.g., Houghton and Tipper, 1999). Thus, more powerful distracters receive greater levels of self-inhibition feeding back onto themselves. The effect of such selection processes is to alter the path of the hand as it reaches to the target. This is because the inhibition of cells common to both reaches (cells 10 to 13 in this example) changes the overall population profile of the cells encoding the

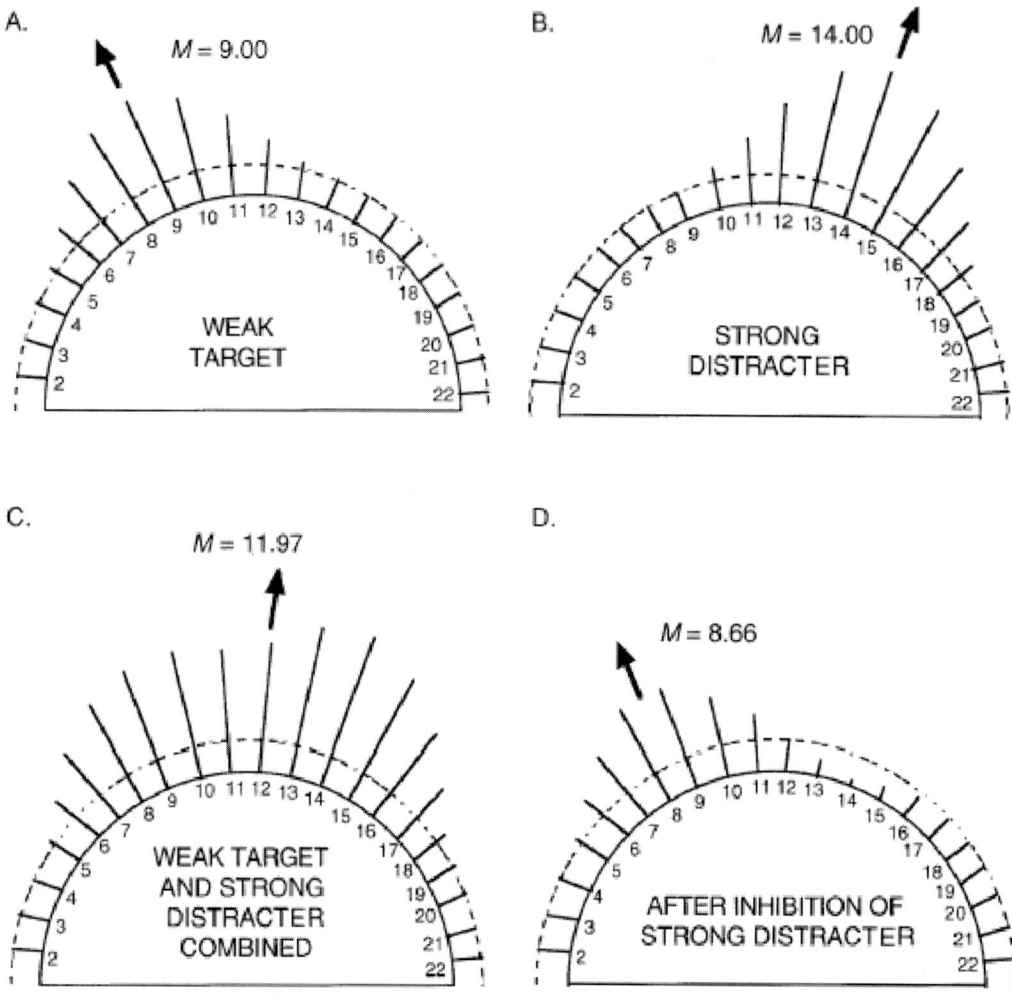

FIGURE 30.1 Neural activity of a population of cells coding reach direction under four different conditions. Each spoke represents a cell responding to a specific reach direction, and is numbered; the length of the spoke represents level of activity. Baseline activity is shown by the dotted semicircle around the population. The population vector (and the reach direction) is shown by the direction of the arrow. (A) Coding of a weak target. Activity is normally distributed around the mean in direction 9, and the spokes are relatively short. (B) Coding of a distracter, with activity centered on direction 14, and with long spokes, denoting activity greater than that of the target. (C) Sum of distribution of activity in (A) and (B), showing that the direction of a reach falls between the target and distracter. (D) Inhibition has been applied to the distribution in (C). The inhibition is normally distributed, centered on the direction of the distracter, and the activity of some cells has been pushed below baseline. The population vector is now, broadly speaking, toward the target, but the reach deviates slightly to the side away from the distracter (compare A and D).

target reach. This is shown in Fig. 30.1D, in which severe levels of inhibition acting on the distracter population of Fig. 30.1B have resulted in a population vector that, although successfully approaching the target, veers slightly away from the direction of the distracter.

In such reaching tasks we think that the level of inhibition is determined by the distance of the distracter from the reaching hand. Those distracters that are close to and ipsilateral with the hand appear to produce a strong competing response and, hence, require high levels of inhibition to prevent inappropriate reaches. This level of inhibition results in the hand veering away. In contrast, distracters further from the hand are expected to provoke much weaker actions, and therefore little or no reactive inhibition is evoked, and hence small deviations toward the distracter can be observed (see Tipper et al., 1997).

II. SACCADE POPULATION CODE SELECTION

We propose that exactly the same selection processes take place in the oculomotor system. For example, multiple saccades can be activated simultaneously (Morrison, 1984), which will entail inhibitory selection mechanisms. Like reaches, saccade directions are also represented in population codes (e.g., Georgopoulos, 1990; Sparks et al., 1990). When two saccades have been evoked by external stimuli, then selection from population codes can be expected to result in changes to the path of the saccade. Sheliga and colleagues (1994) have indeed demonstrated changes to saccade paths. Subjects were required to attend to an imperative stimulus to analyze a cue informing them of the eye movement to be made. According to the premotor theory put forward by Sheliga and colleagues, even though a saccade to the imperative stimulus was never made, the act of attending evoked a saccade in that direction that was not carried out, which affected the speed and direction of the subsequent saccade. We propose that to prevent the inappropriate action, the saccade toward the cue must be suppressed. This means that when a subsequent saccade is made to the target, the suppression of cells overlapping the population coding the imperative stimulus changes the population vector and the saccade veers away from the location of the cue.

Such saccade deviations are robust effects, being observed in a range of studies. A consistent result that has repeatedly been observed is that saccade deviations are greater when the attended imperative cue is in the same visual field (VF) as the subsequent saccade (e.g., Sheliga et al., 1994). Overlapping populations of neural activity can predict this asymmetry. Thus, consider Fig. 30.2. Subjects are fixating the center of the display and attend to one of the four light-emitting diode (LED) cues. If the cue is flashed green, a rapid saccade is made to the target at the top of the display. In contrast, if the cue flashes red no response is to be produced. When attending to the top right LED, in accordance with the findings of Sheliga et al. (1994), we would predict that saccade deviations away from this LED cue will be greater than when attending to the bottom right LED. This is because the population of neurons encoding the saccade to the top right LED and the target at the top of the board overlap far more extensively than the populations encoding the bottom right LED and target saccades, which, from the fixation point, are in quite different directions. Therefore, assuming that all LEDs have equivalent salience, simply because there is more overlap between com-

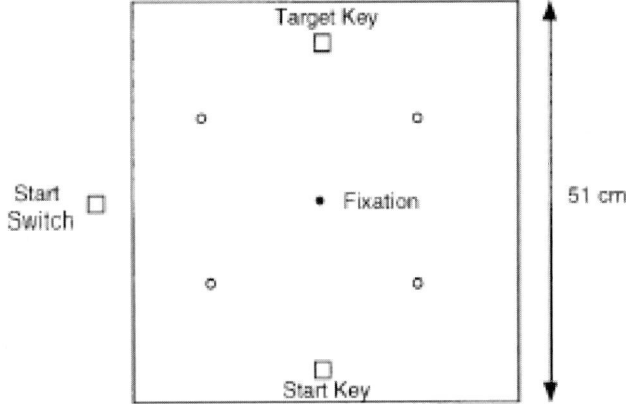

FIGURE 30.2 Overhead view of a stimulus board with start key at the front and a target at the far side of the board, these were 39 cm apart. Pairs of light-emitting diodes (LEDs) are shown embedded on either side of the board. These LEDs were 23 cm apart, and 9 cm (near hand/lower VF) and 30 cm (far from hand/upper VF) from the start key. In the center of the board is a fixation point; to the left is a start switch, which was 7 cm from the edge of the board.

peting neural populations in the target and top right LED condition, there is greater opportunity for inhibition to change the population vector.

In summary, when saccades are made from the central fixation point to the target in the upper VF, deviations away from LED cues will be larger when they are also in the upper VF as compared with the LEDs in the lower VF. We now ask whether the pattern of saccade deviations is affected when subjects also reach to the target. If reach and saccade systems are independent modules, then the saccade deviation pattern should remain unchanged. However, if there is cross-talk between these motor systems, then saccade deviations will be changed when reaches are required.

III. INTERACTIONS BETWEEN EYE & HAND

Recall that when reaching to targets, distracters are encoded in hand-centered frames (e.g., Tipper et al., 1997). Generally those distracters that are nearer and ipsilateral to the responding hand win the race for the control of action and, hence, interfere more. In the task described in Fig. 30.2, in one condition subjects are required to reach with the right hand from the start key at the bottom of the board to the target at the top of the board. In this situation, the near-right LED is the most potent stimulus in a hand-centered frame.

Therefore if the activation state in the reaching system can directly influence states in the saccade system, then a specific result should be observed. That is, the near-right LED typically has little effect when only saccades to the target are produced. However, when reaches to the target are also initiated, this near-right LED will be much more potent, and will now influence saccades by causing them to deviate away from itself.

The experimental setup is shown in Fig. 30.2. In separate blocks of trials subjects either attended to the LEDs in the upper VF or those in the lower VF. The procedure for the reaching task will be described first. Participants used the index or middle finger of the right hand for the task. Initially the start key at the front of the board was depressed and the blue dot in the center of the stimulus board was fixated. A 250-ms tone then signaled to subjects whether the right (300 Hz) or left (800 Hz) cue would flash in either the upper or lower VF, depending on block. Subjects were requested to shift covert attention to the LED while maintaining fixation. There was a fixed interval of 1250 ms after the tone. The LED cue that had been indicated then flashed either red or green for 100 ms, and eye movement recordings were triggered. If the cue was green, the subject was to make a visually guided reach to press the target key. The saccade from fixation point to target passed through an arc of approximately 14°. If the cue was red, the subject was to keep holding down the start key to begin the next trial, and to maintain fixation.

In the saccade-only task, the procedure was similar except for the following modifications: Participants initiated trials by pressing and releasing a switch that was placed to the left of the workspace and level with the fixation point, with one of the fingers of the left hand, keeping the right hand in their lap. Subjects were requested to look at the target as fast as possible if the LED cue flashed green, but maintain fixation if it flashed red (see Tipper et al., 2001, for details).

The saccade trajectories (recorded via an EOG Biopac system) are illustrated in Fig. 30.3. Consider Figs. 30.3C and D. This is the situation where only saccades to the target are produced, and the findings of Sheliga et al. (1994) are clearly replicated. That is, saccades deviate away from LEDs when the LED and target are in the same VF (upper in this case: Fig. 30.3D). In sharp contrast, when the LEDs are in the VF opposite (lower) the saccade target, no significant deviations are observed (Fig. 30.3C).

As discussed above, one explanation for these results is in terms of overlap of population codes. When LED and target are in the same VF, there is much greater overlap in their population codes and, hence, greater possibility for population vector change due to inhibition of the saccade evoked by the LED. Whereas when the LED and saccade target are in opposite VFs, fewer cells overlap, and hence there is reduced ability for inhibition of one population code to change the population vector of the other.

Now consider Figs. 30.3A and B, which illustrate saccade trajectories when subjects reach for the target. In Fig. 30.3B it can be seen that the saccade deviations away from upper VF LEDs is similar to the situation where only saccades are produced (Fig. 30.3D). Therefore, reach to a target has little effect on saccade deviations when attending to stimuli in the same VF as the target.

In sharp contrast, reaching has a significant effect on saccades when attending to cues in the lower VF. Precisely as predicted, it appears that when reaching to targets with the right hand, the near-right LED has a particularly potent effect on saccade trajectory (Fig. 30.3A, dotted line), which is not evident when only saccades are produced (Fig. 30.3C).

A further point of interest is that the potency of the left versus right LED seems to depend on whether a reach is produced. In Figs. 30.3A and B, the bar graphs show the amount of deviation from a straight line. It is clear that there is more deviation from LEDs on the right (dotted pattern) than to LEDs on the left (black pattern). This is predicted by hand-centered frames, because the LEDs on the right are ipsilateral to the responding right hand and, hence, are more potent.

In contrast, when only a saccade is required, there is a tendency for LEDs on the left (black pattern) to have more potent effects on saccade trajectories than those on the right (dotted pattern, Figs. 30.3C, D). Initially we thought that this might be because the left hand on the left side of the stimulus board started trials in the saccade-only condition and, hence, was another situation where hand-centered representations were affecting saccades. However, recently, we have undertaken further studies and found this left bias in the saccade deviation effects when hand position was more carefully controlled. Hence, this hemisphere difference in the effect of attention on saccade deviation seems to be a real effect that may be worthy of future study.

IV. CONCLUSIONS

These results support our notion of cross-talk between neural systems encoding eye and hand movements. This appears to be a complex interaction, being determined by the behavioral goals of the task: one frame of reference can become dominant and affect the other. Hence, when subjects are only making a saccade to targets, the pattern of results accurately reflects this

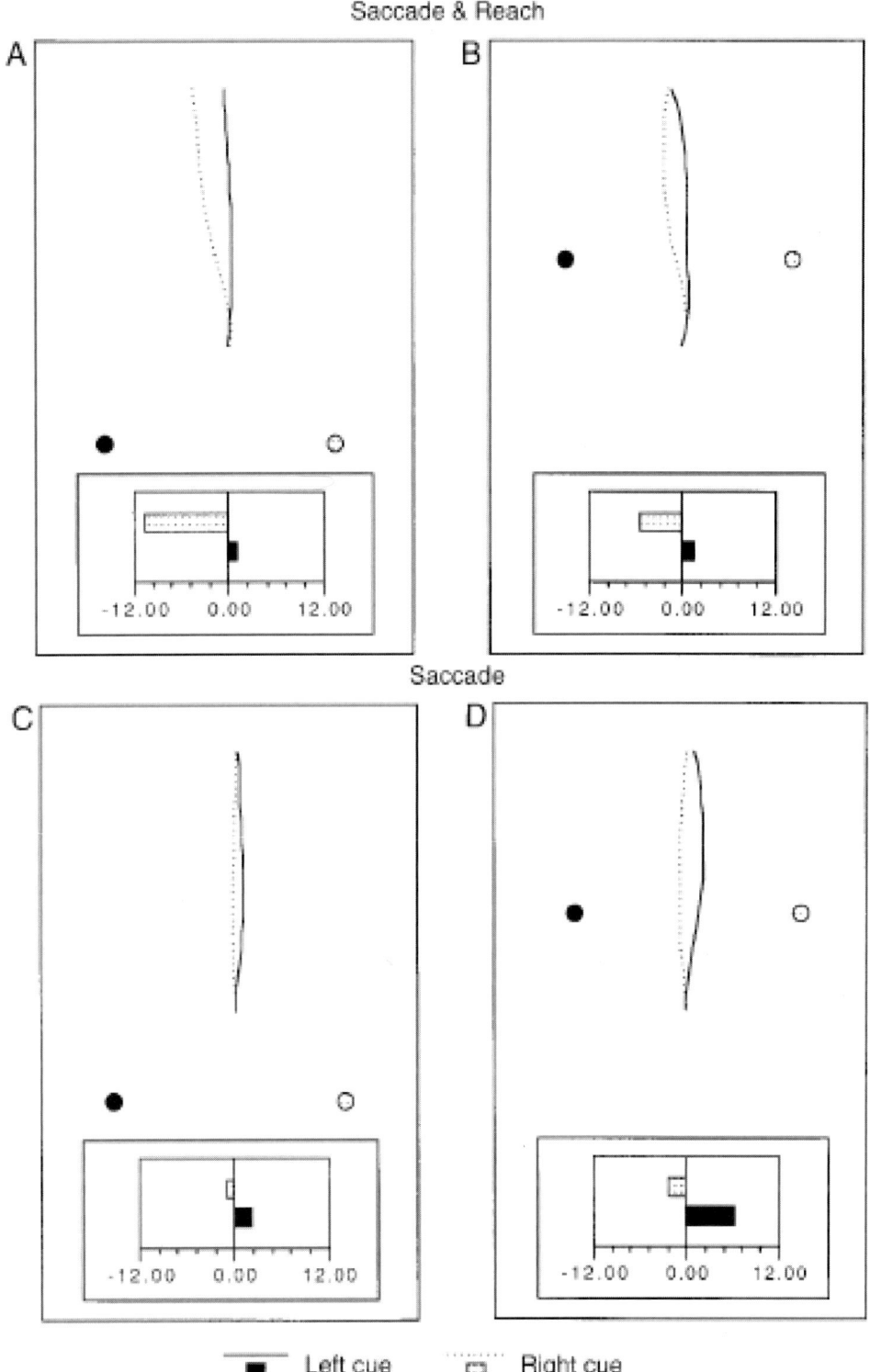

FIGURE 30.3 Saccades when participants also reach (A, B) or when they saccade only (C, D), and when cues are presented in the upper VF (B, D) or lower VF (A, C). The approximate locations of left (solid) and right (dotted) LED cues are represented by the circles. The lines show the mean path of saccades from their origin in the center of each panel to the target toward the top of each panel, after a cue on either the left (solid) or the right (dotted). In the lower part of each panel a bar graph shows the mean deviation of each saccade trajectory from its origin to the left (−) or right (+) in approximate millimeters.

system. As illustrated in Fig. 30.3, attention to an LED in the same VF (upper) as the subsequent saccade produces a much larger deviation effect (Fig. 30.3D) than when attention is oriented to the opposite (lower) VF (Fig. 30.3C). In contrast, when the task requires a manual reach to the target with the right hand, the near-right LED has a significant effect on saccade trajectory, as predicted by a hand-centered frame of reference.

Further work is necessary to provide physiological evidence for such interactions between eye and hand movement systems; and formal computational models are necessary to make explicit the processes involved. These models will have to take into account notions of population codes and the properties of behavior that emerge from them when selecting information for action, such as path deviation effects (e.g., Houghton and Tipper, 1999). They will also have to describe possible mechanisms of cross-talk. Concerning the latter, one of the simplest notions could be spreading activation between eye and hand representations. Thus, the highly activated representation of the LED near and ipsilateral to the hand in the manual system while reaching could spread to the representation of the same LED in the oculomotor system. The result of this increase in the salience of this LED near the hand in the oculomotor system would counter the sparser overlap of neural populations when saccades are evoked by stimuli in the opposite VFs.

Such notions of cross-talk must of course be confirmed via single-cell recording techniques, which would examine whether oculomotor activity in areas such as frontal eye fields, superior colliculus, and lateral intraparietal vary depending on whether a reach to a target is also required. Such an interaction could perhaps take place within the common coordinate frame of the posterior parietal lobe proposed by Andersen et al. (1998). This common code enables quite different forms of spatial representation (e.g., eye-centered, head-centered, limb-centered) to communicate within a common frame. That is, feedback from salience maps in limb-centered representations (e.g., dorsal premotor cortex—Pmd) into common eye-centered codes could change the salience of stimuli. Thus, the activity of cells encoding a saccade to an LED close to the hand might increase when a reach is required as compared with only when saccades are made. Indeed, such cross-talk could take place in parallel within multiple neural structures such as supplementary eye fields, cerebellum, and the parietal lobe, where eye and hand movements are jointly coded.

Acknowledgment

This work was supported by the Biotechnology and Biological Sciences Research Council (Reference S07727).

References

Andersen, R. A., Snyder, L. H., Batista, A. P., Bueno, C. A., and Cohen, Y. E. (1998). Posterior parietal areas specialized for eye movements (LIP) and reach (PRR) using a common coordinate frame. In "Sensory Guidance of Movement", Novartis Foundation Symposium 218, pp. 109–128. Wiley, Chichester.

Cisek, P., and Kalaska, J. F. (2002). Simultaneous encoding of multiple reach direction in dorsal premotor cortex. *J. Neurophysiol.* **87**, 1149–1154.

Georgopoulos, A. P. (1990). Neural coding of the direction of reaching and a comparison with saccadic eye movements. *Cold Spring Harbor Symp. Quant. Biol.* **55**, 849–859.

Georgopoulos, A. P., Kalaska, J. F., Crutcher, M. D., Caminiti, R., and Massey, J. T. (1984). The representation of movement direction in the motor cortex: single cell and population studies. In "Dynamic Aspects of Neocortical Function" (G. M. Edelman, Ed.), pp. 501–515. Wiley.

Goldberg, M. E., and Segraves, M. A. (1987). Visuospatial and motor attention in the monkey. *Neuropsychologia* **25**, 107–118.

Houghton, G., and Tipper, S. P. (1999). Attention and the control of action: an investigation of the effects of selection on population coding of hand and eye movement. In "Proceedings of the 5th Neural Computational and Psychological Workshop" (D Heinke, G. W. Humphreys, and A. Olson, Eds.), pp. 283–298. Springer-Verlag., Berlin/New York.

Kalaska, J. F., Sergio, L. E., and Cisek, P. (1998). Cortical control of whole-arm motor tasks. In "Sensory Guidance of Movement" Novartis Foundation Symposium 218, pp. 176–201. Wiley, Chichester.

Morrison, R. E. (1984). Manipulation of stimulus onset delay in reading: evidence for parallel programming of saccades. *J. Exp. Psychol. Hum. Percept. Perform.* **10**, 667–682.

Sheliga, B. M., Riggio, L., and Rizzolatti, G. (1994). Orienting of attention and eye movements. *Exp. Brain Res.* **98**, 507–522.

Snyder, L. H., Batista, A., and Andersen, R. A. (1997). Coding of intention in the posterior parietal cortex. *Nature* **386**, 167–170.

Sparks, D. L., Lee, C., and Rohrer, W. H. (1990). Population coding of the direction, amplitude, and velocity of saccadic eye movements by neurons in the superior colliculus. *Cold Spring Harbor Symp. Quant. Biol.* **55**, 805–811.

Tipper, S. P., Howard, L. A., and Jackson, S. R. (1997). Selective reaching to grasp: Evidence for distracter interference effects. *Vis. Cogn.* **4**, 1–38.

Tipper, S. P., Howard, L. A., and Paul, M. (2001). Reaching affects saccade trajectories. *Exp. Brain Res.* **136**, 241–249.

Tipper, S. P., Lortie, C., and Baylis, G. C. (1992). Selective reaching: evidence for action-centered attention. *J. Exp. Psychol. Hum. Percept. Perform.* **18**, 891–905.

Treue, S., Hol, K., and Rauber, H.-J. (2000). Seeing multiple directions of motion: physiology and psychophysics. *Nat. Neurosc.* **3**, 270–276.

CHAPTER
31

The Premotor Theory of Attention

Laila Craighero, Giacomo Rizzolatti

ABSTRACT

Traditionally, attention was conceived as a cognitive mechanism subserved by specific, dedicated anatomical centers independent of those involved in data processing and action execution. Attention mechanism was seen either as unitary or as formed by two or more independent anatomical circuits. The premotor theory of attention challenges the notion of attention systems separated from those for sensorimotor integration. In contrast, it proposes that spatial attention results from an activation of the same circuits that program eye movements as well as other motor activities. It maintains that spatial attention differs from movement execution in the degree of activation of the circuits coding the representation of action in space, rather than in the activation of dedicated systems. Evidence coming from a variety of sources—behavioral, neurophysiological, brain imaging—are reviewed, demonstrating that spatial attention can indeed be "reduced" to activation of sensorimotor circuits.

I. INTRODUCTION

The ability to detect visual stimuli in space is enhanced when the observer knows in advance the location of the incoming stimulus. The term *spatial attention* describes this phenomenon. Although, normally, eye and body movements accompany attention allocation, there is ample evidence that attention may act in their absence.

Until the introduction of brain imaging techniques, the dominant view on how attention is organized was that of a unitary, dedicated system, anatomically separated from the circuits underlying sensorimotor transformations ("classic theory of attention"). According to this theory, attention acts as a control system, increasing the efficiency of the basic sensorimotor systems. The idea of a *single* attention system was, however, abandoned when brain imaging studies showed that different brain circuits become active according to the attentional tasks individuals were required to execute. A modified version of classic theory was then advanced. This new version proposed that two different control systems exist: a posterior, parietal system subserving spatial attention, and an anterior one involved in attention recruitment and control of brain areas to perform complex cognitive tasks (Posner and Dehaene, 1994).

This new formulation eliminates the notion of attention as a global, unitary phenomenon. Yet it maintains the central tenet of the classic theory: attention is still conceived as a supramodal activity, independent of the basic sensorimotor circuits.

Two series of experimental findings, one coming from lesion experiments, the other from psychological studies, already challenged this tenet in the eighties.

The lesion studies concerned the possible localization of the putative spatial attention control system. Experiments carried out in monkeys showed that, in contrast to the classic theory prediction, there was no control center for attention in space. According to the site of the lesion, monkeys showed either attention deficit (neglect) for space outside their reach ("far" space) or attention deficit for peripersonal and personal space. The lesions producing these deficits were located in frontal eye field (FEF, area 8) and in ventral premotor cortex (area 6), respectively (Rizzolatti et al., 1983). Similar dissociations were subsequently reported in humans with neglect following lesion of parietal and frontal lobe (see Berti and Rizzolatti, 2002). The conclusion of these studies was that neglect is not a consequence of the destruction of an attention control system, but results from lesions of the same anatomical circuits that control action in space.

The validity of the classic theory of attention was also challenged by the discovery of the so-called "meridian effect." This effect was demonstrated in a series of experiments based on the "Posner paradigm." In these experiments a visual display was presented consisting of a central box and a row of peripheral boxes on each side of the central box, either vertically or horizontally oriented. Subjects were instructed to maintain fixation on the central box, to direct attention to a cued box, and to press a switch as fast as possible at the occurrence of the imperative stimulus. The cue indicated that the most likely location of the incoming stimulus was a digit placed near the central box. The imperative stimulus was a small geometrical shape that appeared at the center of one of the peripheral boxes.

It is well known that in the Posner paradigm, reaction times to stimuli presented at the cued location (valid trials) are faster than reaction times to stimuli at an uncued location (invalid trials). This finding was replicated. In addition, it was found that when attention had to move from one hemifield to another, crossing either the horizontal or vertical visual field meridian, there was a cost of 20 to 25 ms ("meridian effect") (see Rizzolatti et al., 1987)

According to the classic theory of spatial attention, when the target is shown at an uncued box, attention has to move from the cued box to the target, determining a lengthening of reaction times. The lengthening is greater when the invalid stimulus is far from the cued box than when it is close to it, because the costs depends on the distance attention has to cover. Classic theory, however, does not predict and is unable to explain why attention should take more time to cross the visual field meridians than to cover the same distance within one visual hemifield. If attention is a control system independent of basic anatomical and physiological circuits, why should anatomical landmarks such as the principal meridians of the visual field delay its action?

II. THE PREMOTOR THEORY OF ATTENTION

The incapacity of classic theory of attention to account for these findings led to formulation of a new theory called the "premotor theory of attention." The fundamental claim of this theory is that attention does not result from, nor require a control system separated from sensorimotor circuits. Attention derives from activation of the same circuits that, under other conditions, determine perception and motor activity (Rizzolatti et al., 1987).

The premotor theory of attention originally formulated by Rizzolatti and co-workers (1987) was subsequently expanded by Rizzolatti and Craighero (1998). According to them, spatial attention derives from an endogenous or exogenous activation of brain maps that transforms spatial information into movement representations. Activation of these maps determines both an increase in the motor readiness to respond to a specific space sector and a facilitation in processing stimuli coming from that space sector. A fundamental assumption is that in action organization there is a stage in which motor programs are set, but not executed. The brain activity of this stage is what determines the cognitive state that introspectively we call attention. Although spatial attention can be produced by any map that codes space, the strong development in primates of foveal vision and of neural mechanisms related to foveation gives a central role in attention to those maps that code space for eye movements.

The "meridian effect," which the classic theory of attention was unable to account for, is easily explained by the premotor theory. As soon as the location of an impending imperative stimulus is predicted, a motor program is prepared toward the expected location. This program specifies the direction and the amplitude of the future saccade. The presence of a motor program has two consequences: the location of the expected stimulus becomes salient with respect to all other locations via backward connections, and stimuli appearing in that location are responded to faster. This is true both when the required response is a saccade or another arbitrary response (e.g., hand movement). If the imperative stimulus appears in a noncued location, the response can be emitted only when a new motor program is made toward this location. Thus, the invalid response is delayed both because the expected location is not facilitated and because a time-consuming change in the saccade program takes place before emission of the response. This conceptual framework easily explains the meridian effect. Because the goal-directed saccades are prepared first setting the saccade direction and then the saccade amplitude, changes in the direction require a radical modification in the oculomotor program, whereas changes in the amplitude require only a readjustment of a preexisting program. Thus, when the amplitude of the attention movement has to be modified without changing the basic direction parameters, only an adjustment of the motor program is needed. In contrast, when the target appears in the hemifield opposite that containing the cued location, then also the direction of the programmed movement has to be modified. This process is time-consuming because it requires construction of a new program involving (if executed) different movements. This radical program change is at the origin of the meridian effect.

III. EVIDENCE IN FAVOR OF THE PREMOTOR THEORY OF ATTENTION

Strong evidence in favor of the premotor theory of attention was provided by a series of experiments in which ocular saccades were used as the variable measured during an attention task.

It is well established that when two simultaneous or closely consecutive stimuli activate the oculomotor system, there is response interference. From these findings it follows that if spatial attention is based on an activation of oculomotor circuits, as claimed by the premotor theory of attention, attention allocation should influence overt oculomotor responses. In contrast, there is no reason why this should occur if spatial attention does not depend on the oculomotor system.

Sheliga and co-workers tested this issue in a series of elegant experiments (e.g., Sheliga et al., 1995, see for review Rizzolatti and Craighero, 1998). The essence of their experiments was the following. Normal participants were instructed to pay attention to a given spatial location and to perform a predetermined vertical or horizontal saccade at the presentation of an imperative stimulus. Results showed that the trajectory of ocular saccades in response to visual or acoustical imperative stimuli deviates according to the location of attention. The deviation increases as the attentional task becomes more difficult. It is obvious that if spatial attention were independent of oculomotor programming, ocular saccades should have not been influenced by attention location.

A similar close link between spatial attention and eye movements was found by Kustov and Robinson (1996). These authors trained macaque monkeys to pay attention to different spatial locations, and while the monkey performed the task, they electrically stimulated the superior colliculus. They showed that when attention is allocated to a given space position, there is a shift of the saccade trajectory evoked by the electrical stimulation with respect to the saccade trajectory evoked by the same electrical stimulation with no attention allocation. The shift was present when attention was allocated as a consequence of an endogenous or an exogenous cue presentation. Furthermore, and most importantly, the saccade shift was also present when the monkey was instructed to make a manual response and to keep its eyes still, following the imperative stimulus presentation. This latter finding demonstrates that a shift of attention even without any request of subsequent eye movements determines a change in the excitability of the oculomotor system. This is exactly what the premotor theory of attention predicted.

The existence of a possible causal relation between the capacity to perform a saccade and the capacity to orient attention in space was recently investigated by studying normal individuals with their eyes rotated in the orbit. Behavioral testing was based on the Posner paradigm. Participants were studied in two different monocular conditions. In the "frontal" condition, they performed the experiment sitting in front of the computer screen, whereas in the "rotated" session they were rotated 40°, either clockwise or counterclockwise with respect to it. In both conditions the participants were instructed to maintain their gaze on a fixation point. The tested eye was the one closer to the screen (see Fig. 31.1). Note that, while in the frontal condition, both eyes were able to move toward the temporal and nasal visual hemifields, in the rotated condition, the tested eye could not move, for mechanical reasons, toward the temporal hemifield. The tested hypothesis was whether in this condition, the impossibility of making temporally directed eye movements would also affect the participants' capacity to orient temporally attention. Results showed that this was the case. In the frontal condition participants were (not surprisingly) able to correctly deploy attention to all locations. In contrast, in the rotated condition they could allocate attention only to the nasal visual hemifield. No difference in reaction times between valid and invalid trials was observed when the stimulus appeared in the temporal visual hemifield (Craighero et al., 2004). These data clearly indicate a causal link between eye movements and attention. When the former cannot be executed, the latter also cannot be allocated.

The premotor theory of attention predicts that the setting of an oculomotor program not only makes the individual ready to respond, but also facilitates, through backward connections, the processing of incoming stimuli spatially congruent with the motor program (see above). In a recent, brilliant electrophysiological experiment, Moore and Fallah (2001) reported evidence supporting this prediction. They trained two macaque monkeys to make manual responses to signal the transient dimming of a peripheral visual target in the presence of flashing distracters. They then tested the effects of FEF microstimulation on the monkeys' performance. More specifically, they determined first the position in space to which suprathreshold microstimulation shifted the direction of gaze. In this way they defined the motor field of the stimulated site. They then tested the effects of *subthreshold* microstimulation of the FEF on monkeys' performance when the target was placed inside and outside the motor field. Results showed that stimulation of the FEF without the occurrence of eye movements facilitates attention to the target stimulus (lower

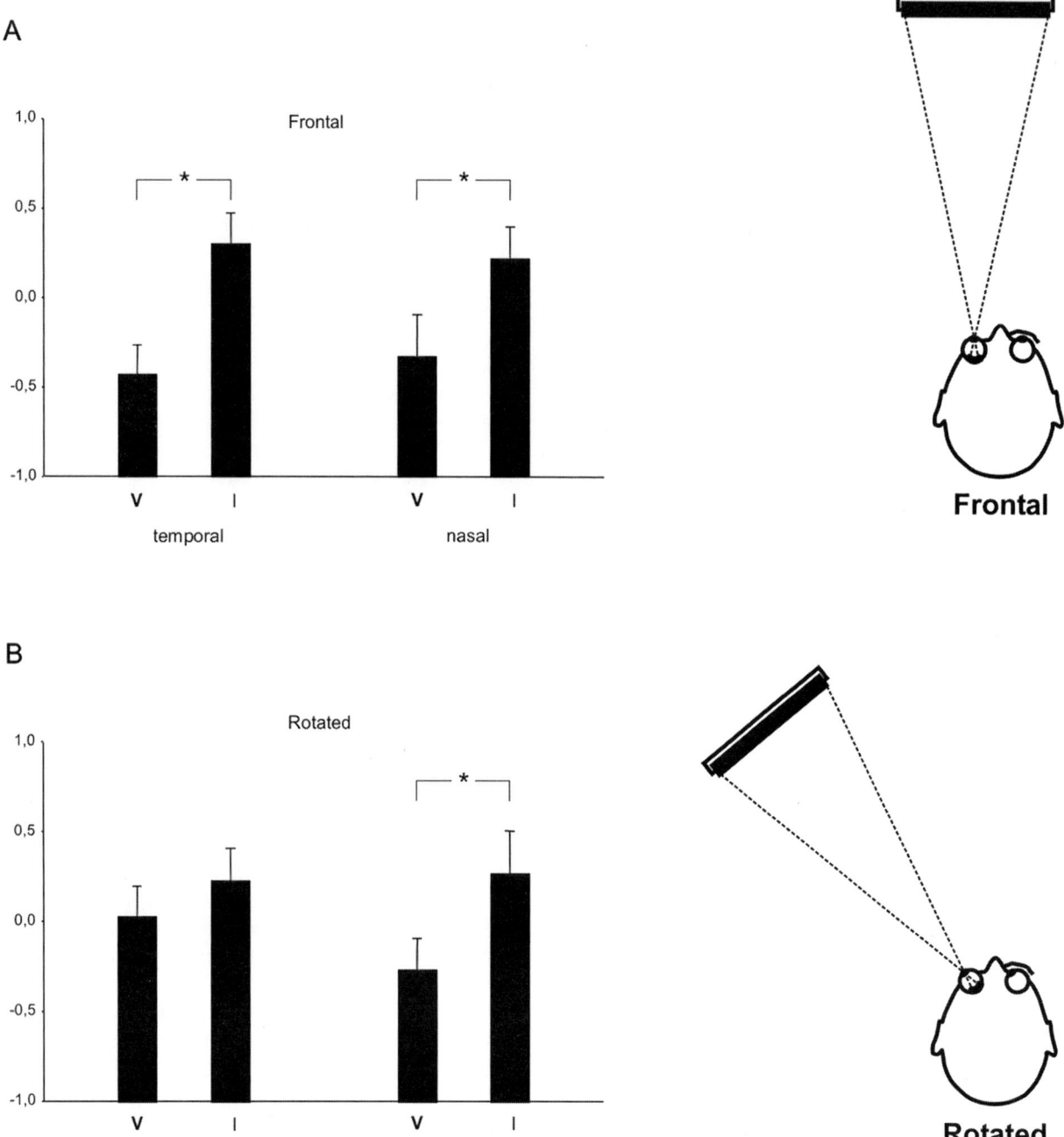

FIGURE 31.1 Influence of eye position on attention orienting. Right: Schematic illustration of frontal and rotated conditions. Left eye tested. Left: Standardized mean reaction time and SE to stimulus presentation during frontal (A) and rotated (B) conditions. Data for the attended (V, valid) and unattended (I, invalid) locations in both temporal and nasal hemifields are shown. Asterisks indicate the presence of a statistically significant difference between valid and invalid trials ($P < 0.05$). Ordinates, z- scores. Modified, with permission, from Craighero et al. (2004).

psychophysical threshold level), but exclusively when the target stimulus was located in the motor field. These data provide direct evidence that oculomotor signaling determines the allocation of spatial attention, as predicted by the premotor theory of attention.

IV. ACTIVATION OF OCULOMOTOR AREAS DURING ORIENTING OF SPATIAL ATTENTION: BRAIN IMAGING STUDIES

Brain imaging studies were very influential in demystifying the notion that, in the brain, there is a dedicated system for selective attention independent of the basic sensorimotor circuit. They also showed a marked similarity between the activations obtained during spatial attention and during eye movement tasks. Corbetta and co-workers (1998) studied normal participants during a task involving covert shifts of attention to peripheral visual stimuli and during a task involving both attentional and saccadic shifts to the same stimuli. Results showed overlapping activation in both the parietal and frontal lobes in the two conditions. Nobre et al. (2000) compared brain areas activated in tasks of covert visuospatial orienting and in tasks requiring large and repetitive saccades toward peripheral stimuli. Results showed that the two tasks activated highly overlapping neural systems. No system of distinct brain areas was activated exclusively by the covert attention or by the oculomotor task.

More recently, Beauchamp et al. (2001) reexamined the issue by comparing overt shifts of spatial attention (shifts of attention with saccadic eye movements) and covert shifts of spatial attention (shifts of attention without eye movements) using identical tasks and stimuli located at similar distance from fixation. Results showed that both overt and covert shifts of visuospatial attention induce activation of the frontal cortex region where human FEFs are located, of the cortex of the intraparietal sulcus in the correspondence of area LIP (an area involved in the control of eye movements), and in lateral occipital cortex. Overt shifts of attention elicited more neural activity than did covert shifts, reflecting additional activity associated with saccade execution. The authors concluded that overt and covert attentional shifts are subserved by the same network of areas.

V. NON-OCULOMOTOR ATTENTION

In everyday life most of our actions in space are preceded by foveation. This gives the oculomotor system a special, central position in spatial attention. There are, however, some conditions in which we select stimuli in space not to look at them, but to act on them. In these cases spatial attention should depend basically on circuits other than those controlling eye movements. Probably the best evidence in favor of spatial attention not related to eye movements is that deriving from experiments conducted by Tipper et al. (1992). They studied, in normal participants, the effect of an irrelevant stimulus located either within the arm trajectory necessary to execute a pointing response or outside it. The result showed that an interference effect was present only when the distracter was located in the trajectory necessary to execute a pointing response. Control experiments indicated that the effect was not due to a purely visual representation of the stimuli or to spatial attention related to eye movements. Rather, the organization of the arm–hand movement determined a change in the attentional relevance of stimuli close to the hand or far from it.

Recently, studies were carried out to assess whether the premotor theory of attention could be extended from space to objects. Objects are represented in both the ventral and dorsal streams. However, whereas processing in the ventral stream is responsible for perceptual and cognitive representations of the visual characteristics of objects and their significance, processing in the dorsal stream underlies the organization of the appropriate object-related hand movements. What happens when individuals prepare to grasp a specific object? Would the motor preparation for acting determine perceptual salience of those stimuli that are congruent with the prepared motor program? If the premotor theory of attention is valid for hand movement, motor preparation should cause, in addition to greater readiness to respond, also facilitation in processing stimuli congruent in size and shape with those for which the grasping program was prepared.

This problem was addressed by Craighero et al. (1999). Normal human participants were required to prepare to grasp a bar and then to execute the action as fast as possible on presentation of a visual stimulus. On the basis of the degree of sharing of their intrinsic properties with those of the to-be-grasped bar, visual stimuli were categorized as "congruent" or "incongruent." Results showed that grasping reaction times to congruent visual stimuli were faster than reaction times to incongruent ones. These data indicate that preparation to act on an object produces faster processing of stimuli congruent with that object. The same facilitation was also present when, after the preparation of hand grasping, participants were instructed to inhibit the prepared grasping movement and to respond with a different motor effector.

These results and others (e.g., Tucker and Ellis, 1998) show that there is a clear parallelism between the facilitation resulting from the preparation of a hand action and that resulting from oculomotor programming. In the latter case, facilitation favors a specific spatial location; in the former, it facilitates objects with intrinsic properties similar to those for which the motor program was prepared. The premotor attentional mechanism appears, therefore, to be a general mechanism valid for attention in space (with the distinction between oculomotor and reaching), as well as for attention to objects. There is no need for hypothetical supramodal mechanisms, separate from the sensorimotor circuits.

Acknowledgments

This study was supported by European Grants IST-2000-29689 and IST-2001-35282 and by MIUR (Ministero Istruzione Università e Ricerca).

References

Beauchamp, M. S., Petit, L., Ellmore, T. M., Ingeholm, J., and Haxby J. V. (2001). A parametric fMRI study of overt and covert shifts of visuospatial attention. *NeuroImage* **14**, 310–321.

Berti, A., and Rizzolatti, G. (2002). Coding near and far space. *In* "The Cognitive and Neural Bases of Spatial Neglect" (H-O Karnath, A. D. Milner, and G. Vallar, Eds.), pp. 119–130. Oxford University Press, Oxford.

Corbetta, M., Akbudak, E., Conturo, T. E., Snyder, A. Z., Ollinger, J. M., Drury, H. A., Linenweber, M. R., Petersen, S. E., Raichle, M. E., Van Essen, D. C., and Shulman, G. L. (1998). A common network of functional areas for attention and eye movements. *Neuron* **21**, 761–773.

Craighero, L., Fadiga, L., Rizzolatti, G., and Umiltà, C. (1999). Action for perception: a motor–visual attentional effect. *J. Exp. Psychol. Hum. Percept. Perform.* **25**, 1673–1692.

Craighero L., Nascimben M., and Fadiga L. (2004). Eye position affects orienting of visuospatial attention. *Curr. Biol.* **14**, 331–333.

Kustov, A. A., and Robinson, D. L. (1996). Shared neural control of attentional shifts and eye movements. *Nature* **384**, 74–77.

Moore, T., and Fallah, M. (2001). Control of eye movements and spatial attention. *Proc. Natl. Acad. Sci. USA* **98**, 1273–1276.

Nobre, A. C., Gitelman, D. R., Dias, E. C., and Mesulam, M. M. (2000). Covert visual spatial orienting and saccades: overlapping neural systems. *NeuroImage* **11**, 210–216.

Posner, M. I., and Dehaene, S. (1994). Attentional networks. *Trends Neurosci.* **17**, 75–79.

Rizzolatti, G., and Craighero, L. (1998). Spatial attention: Mechanisms and theories. *In* "Advances in Psychological Science," Vol. 2: "Biological and Cognitive Aspects" (M. Sabourin, F. Craik, and M. Robert, Eds.), pp. 171–198. Psychology Press, East Sussex.

Rizzolatti, G., Matelli, M., and Pavesi, G. (1983). Deficit in attention and movement following the removal of postarcuate (area 6) and prearcuate (area 8) cortex in monkey. *Brain* **106**, 655–673.

Rizzolatti, G., Riggio, L., Dascola, I., and Umiltà, C. (1987). Reorienting attention across the horizontal and vertical meridians: evidence in favor of a premotor theory of attention. *Neuropsychologia* **25**, 31–40.

Rizzolatti, G., Riggio, L., and Sheliga, B. M. (1994). Space and selective attention. *In* "Attention and Performance XV" (C. Umiltà and M. Moscovitch, Eds.), pp. 231–265. MIT Press, Cambridge, MA.

Sheliga, B. M., Riggio, L., and Rizzolatti, G. (1995). Spatial attention and eye movements. *Exp. Brain Res.* **98**, 507–522.

Tipper, S. P., Lortie, C., and Baylis, G. C. (1992). Selective reaching: evidence for action-centered attention. *J. Exp. Psychol. Hum. Percept. Perform.* **23**, 823–844.

Tucker, M., and Ellis, R. (1998). On the relations between seen objects and components of potential actions. *J. Exp. Psychol. Hum. Percept. Perform.* **24**, 830–846.

CHAPTER

32

Cross-Modal Consequences of Human Spatial Attention

Jon Driver, Martin Eimer and Emiliano Macaluso

ABSTRACT

Traditionally selective attention was studied separately for each sensory modality, but an increasing number of multisensory studies have now examined whether directing attention for a task in one particular sensory modality may have consequences for processing in other modalities also. For cases of *spatially* directed covert attention, many such cross-modal consequences of attentional selection have now been shown. Attending toward a particular location for a task in one modality often produces spatially corresponding performance consequences for other modalities also, typically with better performance at the location attended for a task in one modality being found in other modalities also. Such cross-modal effects typically remain in register with respect to external space across changes in posture (e.g., eye, head, or hand movements) that realign the receptors from different senses. ERP and fMRI studies have begun to uncover the neural correlates of such cross-modal spatial attention effects. A key finding is that the direction of spatial attention for a task in one modality (e.g., within touch) can modulate sensory-specific processing for another modality (e.g., vision), to produce cross-modal influences at stages that would traditionally be considered "unimodal." For instance, tactile spatial attention can affect occipital cortex activations in response to visual stimuli as measured with fMRI, and also P1 and N1 components in visual evoked potentials recorded at the scalp. The cross-modal results we review appear consistent with spatial attention being directed at multimodal levels of representation, which may then exert influences on unimodal representations via backprojections.

I. INTRODUCTION

Selective attention exists in every sensory modality, although traditional attention studies have typically considered only a single sensory modality at a time. More recently there has been much interest in possibly multisensory or cross-modal aspects of attention (e.g. Chapter 89). Here we focus specifically on *spatial* interactions in attention between different sensory modalities, which have been the most fully researched and documented to date, perhaps because space may serve as a common "language" for exchange of information between different senses (though time might do so also). There is now much evidence for cross-modal spatial interactions in human spatial attention, from both behavioral and neurobiological studies (e.g., see Spence and Driver, 2004, for a collected volume of reviews by various authors). Here, given the space constraints, we focus primarily on our own studies.

II. BEHAVIORAL EVIDENCE FOR CROSS-MODAL EFFECTS OF SPATIAL ATTENTION

Driver and Spence (1994) found behaviorally that in dual-task situations, auditory and visual streams from a common spatial location could be judged more efficiently than two streams from different locations. This suggested that it may be difficult to attend covertly to separate locations in different modalities. Spence et al. (2000b) found that visual distracters were harder to ignore if coming from the same location as auditory targets (see also Pavani et al., 2001, for analogous results involving visual–tactile situations). This

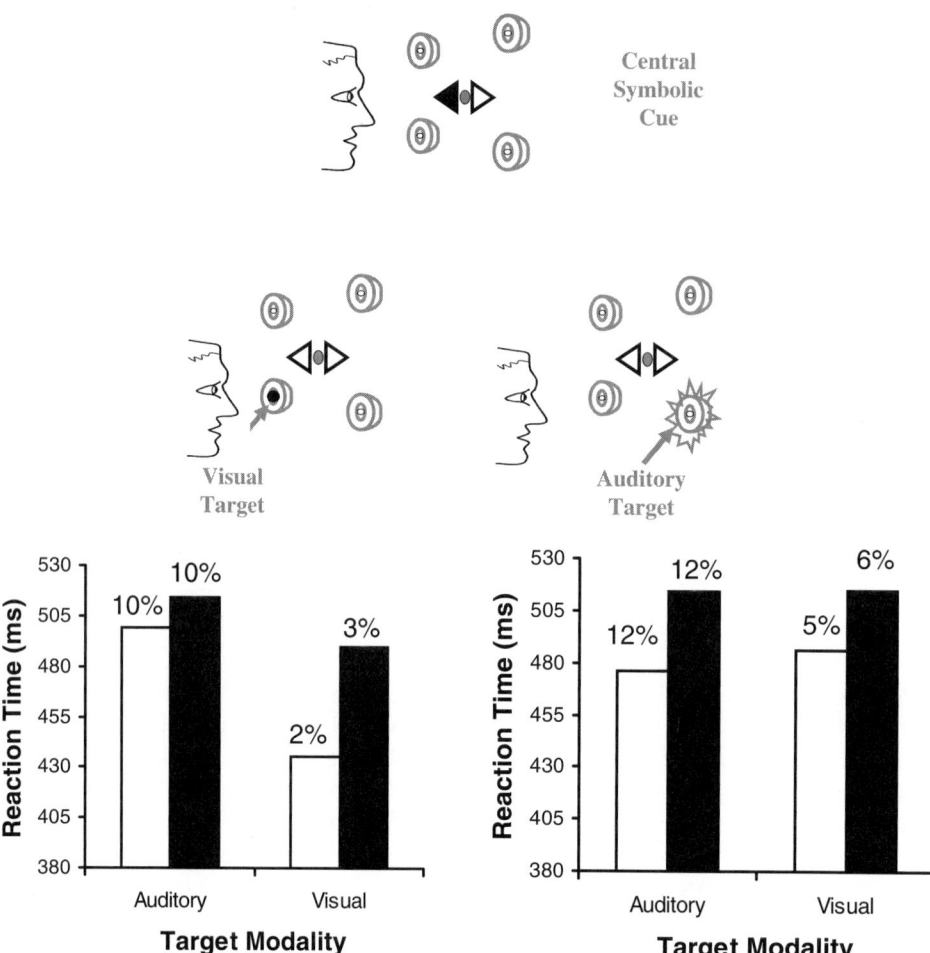

FIGURE 32.1 (a) Cartooned experimental setup in Spence and Driver (1996), plus example results. Participants kept eyes and head straight ahead throughout. A central symbolic cue (schematized here by central visual arrows) indicated the highly likely target side for the most common ("primary") modality at the start of each trial. In different experiments, either audition or vision served as the primary modality, but on some trials targets could be presented instead in the rarer "secondary" modality (see main text). After the central cue, a single target appeared in just one sensory modality on each trial, with auditory and visual trials randomly intermingled. A target in the primary modality appeared on the cued side on 87.5% of trials (as shown here for the visual target appearing at bottom left of (a); but any target within the rarer secondary modality was twice as likely to appear on the uncued side (as shown for the auditory target at bottom right of (a)). Participants judged whether each target came from an upper or lower location, regardless of its side or modality. (b, c) The results showed faster judgments (and better or equivalent accuracy, see percentage error scores above each bar) for targets in the primary modality (i.e., vision in (b) and audition in (c)) when appearing on the cued side (empty bars), where they were most likely, than on the uncued side (filled bars). Critically, this advantage for the cued side also applied for targets in the secondary modality (i.e., audition in (b) and vision in (c)), even though targets in this modality were now themselves actually more likely on the *uncued* side. This suggests that a strong spatial expectancy for one (primary) modality can lead to spatially correspondingly attention effects in another (secondary) modality.

suggested the corollary that it can be difficult *not* to attend to the same location in different modalities.

Spence and Driver (1996) reported that directing covert attention endogenously to one particular location, in anticipation of *auditory* targets that were strongly expected there, led to better *visual* as well as auditory performance at that location (even when the visual targets themselves were twice as likely elsewhere) (Fig. 32.1). Analogously, directing covert attention to one location in anticipation of a visual target that was strongly expected there benefited auditory as well as visual performance at that location, even when the spatial probabilities within audition did not encourage this. Furthermore, when subjects explicitly attempted to direct their auditory attention to one side but concurrently to direct their visual attention to the other side, effects of spatial attention on performance were significantly reduced. Spence et al. (2000a) found analogous interactions in endogenous covert spatial attention between vision and touch.

Cross-modal spatial interactions have also now been documented behaviorally for so-called "exogenous" (or stimulus-driven) cross-modal spatial "cuing," as when a spatially nonpredictive cue event in one modality leads to better judgments for a target in another modality if this appears in the same location at around the same time, as compared with a target appearing elsewhere (e.g., Spence et al., 1998; see Spence et al., 2004, for a recent comprehensive review). Exogenous cross-modal spatial cuing effects of this sort have now been reported for all possible pairings of cue and subsequent target involving vision, hearing, and touch, although the exact outcome does depend on specifics of the behavioral paradigm (see Spence et al., 2004). Such cross-modal spatial cuing effects can influence perceptual sensitivity (i.e., d') in the target modality, for at least some cases. For instance, McDonald et al. (2000) found greater sensitivity for visual detection of a masked target at the location of a shortly preceding sound.

In many of these cross-modal spatial "cuing" studies, a spatially nonpredictive cue in one modality enhances discrimination performance for a subsequent target nearby in another modality *even though the cue itself carries no information whatsoever* about the property that must be judged for the target modality. For instance, a sound can enhance judgments of the presence versus absence of a visual event nearby (McDonald et al., 2000); and a visual event can enhance judgments of whether a tactile vibration nearby has low versus high frequency (Gray and Tan, 2002), or whether it contains one versus two pulses (Spence et al., 1998), and so on. This type of result is therefore somewhat different from those often studied under the general heading of "multisensory integration," in which two or more sensory modalities provide *redundant* information about the same external property (as when seen lip movements provide additional information about heard speech sounds, e.g., Bertelson and De Gelder, 2004). In cases of cross-modal spatial cuing, the spatial cue can influence judgments about a target property that is not available to the cue modality. This suggests that when a sudden event in one sensory modality makes a particular location salient, other sensory modalities (extracting their own stimulus properties) may also start to focus their specialist processing on that same location (see Driver and Spence, 2000; Driver et al., 2004)

III. ERP EVIDENCE FOR SENSORY EFFECTS OF CROSS-MODAL SPATIAL ATTENTION

Several studies have now examined whether ERPs for particular sensory events, presented in any one of several possible modalities, may show modulations that reflect cross-modal links in covert spatial attention. This might reveal the level at which cross-modal attentional influences can arise, for instance, whether early, sensory-specific components are affected, as often found with ERPs in traditional studies of unimodal spatial attention. Work on this issue has typically involved cross-modal extensions of designs previously established in ERP work on unimodal endogenous spatial attention. In a prototypical study (e.g., Hillyard et al., 1984; Eimer and Schräger, 1998; Teder-Sälejärvi et al., 1999; Eimer and Driver, 2000), a series of sensory events are presented one at a time, with each appearing unpredictably at one or another specific location (e.g., on the left or right) and also unpredictably in one or another sensory modality. The task requires endogenous monitoring of just one location or side, in just one modality, to detect occasional deviant stimuli (e.g., longer than "standard" events) with an overt response, while maintaining central fixation and head. Any stimuli at other locations in the target modality, or at any location whatsoever in other modalities, can simply be ignored. But ERPs can of course be recorded for these task-irrelevant stimuli as well as for any task-relevant stimuli.

The well-known result from many traditional unimodal studies with such designs (e.g., for the visual modality see Mangun and Hillyard, 1991; Eimer, 1994; Hopfinger et al., 2000) is that the amplitude of relatively early ERP components (e.g., visual P1 and N1, peaking at around 90–120 and 160–200 ms poststimulus) is often larger for a given stimulus when covert spatial attention is directed toward it, than when the same stimulus is ignored with attention being directed elsewhere (see Figs. 32.2a and c for examples of this for visual ERPs recorded from occipital electrodes).

Possible *cross-modal* ERP effects of spatial attention can also be addressed with this type of design, simply by examining ERPs for stimuli in other task-relevant modalities, as a function of their location relative to the direction of attention for the task-relevant modality. For instance, one might present not only visual events but also (separately) auditory events on the left or right, one at a time, in an unpredictable intermingled sequence of events in different modalities and at different locations, while requiring subjects to monitor only one side in only one modality for occasional deviants. Such designs have now revealed cross-modal

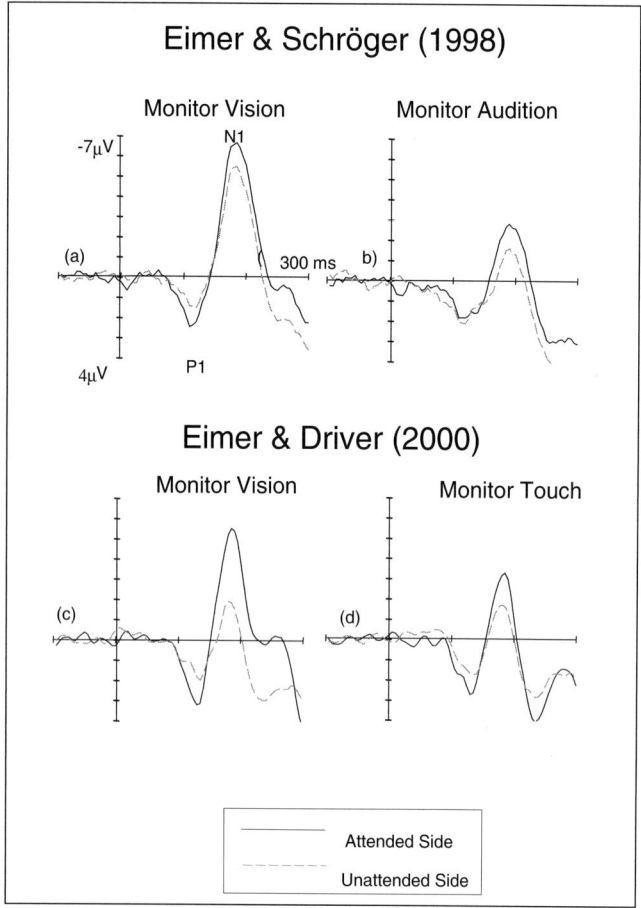

FIGURE 32.2 Grand-averaged event-related potentials (ERPs) at occipital electrodes (OL/OR) contralateral to visual stimuli at attended locations (solid lines) and unattended locations (dashed lines). (a, b) Data from Eimer and Schröger (1998): audiovisual links. (c, d) Data from Eimer and Driver (2000): tactile–visual links. Waveforms shown separately for conditions in which attention was directed toward one side to monitor visual events (a, c) or to monitor auditory (b) or tactile (d) events. Three types of effects on ERP amplitudes can be seen. First, visual ERPs are larger when vision is task-relevant (a, c) versus irrelevant, reflecting intermodal attention. Second, P1 and N1 components are larger for visual stimuli on the side currently attended for visual monitoring (solid vs dashed lines within (a) and (c)), reflecting unimodal spatial attention. Third, the visual N1 component (and also P1 for (d)) is also larger for visual stimuli on the side currently attended for monitoring in a nonvisual modality (solid vs dashed lines within (b) and (d)), reflecting cross-modal spatial attention.

effects of endogenous spatial attention on ERPs in several studies (see Eimer, 2001, 2004; Eimer and Driver, 2001, for reviews). Figure 32.2b provides one example, from a study by Eimer and Schröger (1998), where ERPs in response to task-irrelevant *visual* events were modulated as a function of which side was attended for an *auditory* monitoring task (see also Hillyard et al., 1984; Teder-Sälejärvi et al., 1999). Note that this particular cross-modal ERP effect influenced the amplitude of the visual N1 component, traditionally considered to be a sensory-specific component, with cross-modal attentional modulation arising within around 150 ms poststimulus in Fig. 32.2b.

Figure 32.2d provides another example of visual ERPs being influenced cross-modally, now by which side was attended covertly for a purely *tactile* task on one or the other *hand*. In this situation (Eimer and Driver, 2000), the amplitude of both the visual N1 component and the earlier P1 component was affected cross-modally, by which hand/side was monitored covertly for occasional tactile deviants. Such early ERP components (e.g., starting around 100 ms poststimulus for the P1 in Fig. 32.2d) are traditionally regarded as reflecting modality-specific sensory processing (probably within extrastriate visual cortex for the visual P1; see Mangun, 1995). But the recent cross-modal studies show that they can be modulated by where attention is directed for a *different* modality, here touch (Fig. 32.2d). Such ERP results are consistent with the behavioral evidence (e.g., Fig. 32.1) for cross-modal links in endogenous spatial attention that was discussed earlier (e.g., Spence and Driver, 1996; Spence et al., 2000a). But the ERP data go further in suggesting that relatively early stages of sensory processing can be affected by cross-modal interactions in spatial attention, probably affecting brain areas that would traditionally be regarded as "unimodal" (e.g., visual cortex).

Analogous effects have also been obtained for *auditory* rather than visual ERPs, with the descending flank of the auditory N1 showing amplitude modulation not only as a function of where endogenous spatial attention is directed for unimodal *auditory* tasks, but also as a function of which side is covertly attended for *visual* or *tactile* monitoring tasks (e.g., Eimer and Schröger, 1998; Teder-Sälejärvi et al., 1999; Eimer et al., 2002). Somatosensory ERPs can also be affected by the direction of endogenous attention for tasks in other modalities (e.g., vision), under some conditions (see Eimer and Driver, 2000).

Finally, some analogous ERP results have also been observed for *exogenous* cross-modal spatial cuing situations, where attention is captured by a salient cue stimulus rather than being directed endogenously to a particular location for a task. Again, the cross-modal effects can modulate ERP components that would usually be considered modality specific. For instance, McDonald and Ward (2000) reported that preceding a visual stimulus with a sound on the same rather than opposite side led to negative-going visual ERP differences, while McDonald et al. (2001) found an analogous effect within 120 to 140 ms into the ERPs for auditory targets following a shortly preceding visual cue on the same of opposite side. Similarly, Kennett et al. (2001) observed that a spatially nonpredictive, task-irrelevant tactile cue to one hand could modulate ERPs to a visual event presented near one or the other hand shortly afterward, with an enhanced visual N1 for stimuli at the same location as the preceding tactile cue. Kennett et al. also examined any impact on this cross-modal effect of changing the current position of the hands in external space, manipulating posture by crossing the hands over, so that the left hand now lay in the right hemifield and vice versa. They observed that the spatial cuing effect on the visual N1 crossed over correspondingly, so that with the left hand placed in the right visual field, an unseen touch on it now boosted the N1 to a subsequent right rather than left visual event (and vice versa for the crossed right hand), a result opposite that found with uncrossed hands (see also Eimer et al., 2001). The common location of events from different modalities in external space thus appears more important for cross-modal spatial attention than their initial anatomical projections to one or the other hemisphere. A similar "remapping" of the cross-modal spatial cuing effect between touch and vision with changes in hand posture is found behaviorally, even when the hands and arms are occluded from view (Kennett et al., 2002), implying a possible role for proprioception in modulating spatial interactions between vision and touch. This suggests that tactile events might influence visual processing via communicating brain areas that also receive postural information (see also Spence et al., 2000b; Eimer et al., 2001, for related results from tactile–visual studies of cross-modal endogenous rather than exogenous spatial attention).

IV. NEUROIMAGING EVIDENCE FOR MODULATION OF SENSORY CORTEX BY CROSS-MODAL SPATIAL ATTENTION

PET and fMRI studies have provided further evidence that attending to a particular location or side to perform a task in one sensory modality (or because of a sudden stimulus appearing there in that modality) can have implications for other modalities also. This can affect sensory brain areas that would traditionally be regarded as unimodal. For instance, areas of visual cortex may be affected by the location attended for a purely tactile task. Macaluso et al. (2000a) presented bilateral streams of stimuli in either vision or touch (i.e., unimodal stimulation only, at any one time). Unseen tactile stimuli were delivered to the hands, which were uncrossed, and any visual events were near the hands. The task was to attend covertly to just one side, to discriminate stimuli (single- versus double-pulse) on that side only, in the currently stimulated modality which was blocked. Activations for attending to the left minus attending to the right, and vice versa, were examined, separately for visual and for tactile blocks. Critically, this study also examined any effects of attending to one particular side versus the other that were common across both touch and vision ("multimodal" spatial attention effects in this sense), versus those that were found only for one of the two modalities (henceforth "unimodal" spatial attention activations).

As expected given previous studies of spatial attention for just vision or just touch, unimodal effects of spatial attention were found in contralateral posterior occipital cortex specifically for vision and in contralateral somatosensory cortex specifically for touch. Importantly, some multimodal effects of spatial attention (i.e., common to both vision and touch) were also detected (Macaluso et al., 2000a). These included the intraparietal sulcus contralateral to the attended side, a brain region that seems likely to be a multisensory area, given electrophysiological studies in monkeys showing both visual and tactile receptive fields for neurons in this general region (e.g., Duhamel et al., 1998), plus some our own fMRI work (Macaluso and Driver, 2001) confirming that intraparietal regions can be activated in humans by passive stimulation on the contralateral side in either vision or touch.

More importantly for present purposes, Macaluso et al. (2000a) also found *multimodal* effects of spatial attention in brain areas that would traditionally be regarded as *unimodal* visual areas (thus raising some potential analogies with the cross-modal effects on early visual ERP components discussed above). Lateral and superior occipital regions were more active when attending to the contralateral side for a visual task, but also when attending contralaterally for a tactile task. Thus, the direction of attention for a tactile task modulated activity in areas of visual cortex, in apparent analogy to the ERP findings reviewed above, which had shown that selecting one side for a tactile task can modulate visual ERPs.

In a follow-up study, Macaluso et al. (2002a) implemented a similar design, but now stimulating both modalities at the same time, on both sides. Thus, there were now always four concurrent streams of stimuli in total: streams on the left and right in both touch and vision. The task was again to discriminate stimuli (double vs single pulses) on just one side (left or right) in just one modality (vision or touch). Distracters in the irrelevant modality could now be incongruent with the concurrent target (e.g., a double pulse when the target

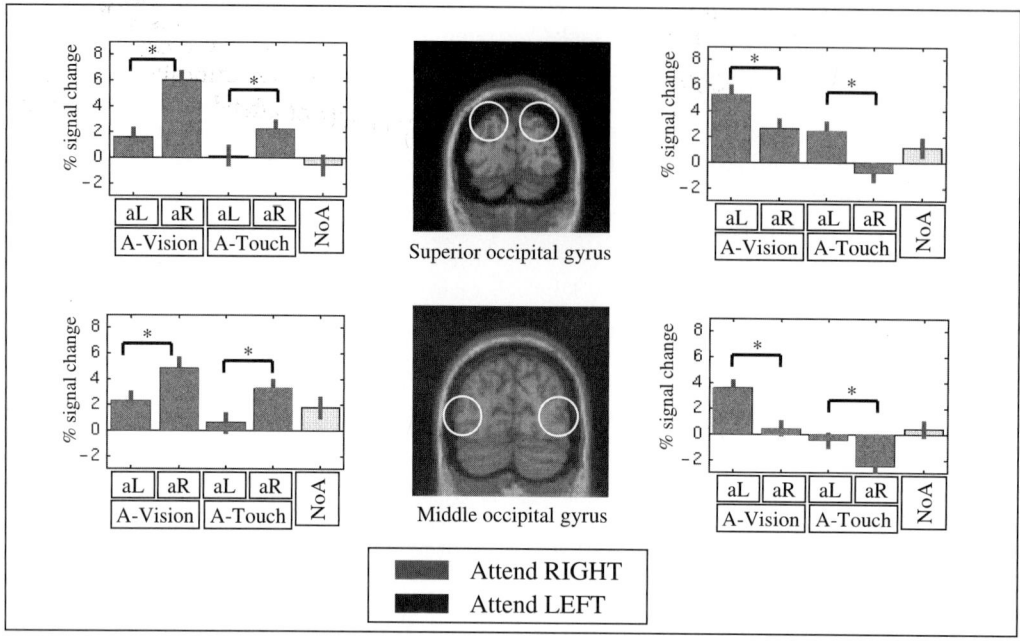

FIGURE 32.3 Multimodal attentional modulation in visual cortex when attending covertly to one or the other hemifield, during bimodal and bilateral visuotactile stimulation. Plots refer to the voxel at the maxima of the activated cluster, always found contralateral to the attended side. Note that these occipital areas displayed differential activity depending on the attended side not only for attend-vision blocks (bars 1 and 2 in each plot), but also during blocks of tactile attention (see bars 3 and 4 in each plot). Activities are expressed as percentage signal change compared with a baseline condition without any peripheral stimulation. The NoA condition (fifth bar in each graph) refers to brain activity in the presence of bilateral, bimodal stimulation, but with attention directed to a central task performed on the fixation point (see Macaluso et al., 2002a, for details). Coronal sections are taken at $y = -90$ (top slice) and $y = -80$ (bottom slice). aL and aR indicate the attended side. (*$P < 0.05$). Adapted, with permission, from Macaluso et al. (2002a). (See color plate)

was single). Hence the task now strongly motivated subjects to try and ignore the currently irrelevant modality. This was implemented to preclude any strategic incentive for shifting attention within that modality to the same side as for the task-relevant modality. But once again, multimodal effects of spatial attention (i.e., observed both when attending vision and when attending touch) were found in the anterior intraparietal sulcus, contralateral to the attended side. More importantly, *multimodal* effects of spatial attention were also again found in areas that would traditionally be considered as *unimodal* visual cortex, in the middle occipital gyrus and a portion of the superior occipital gyrus (see Fig. 32.3). These "visual" areas also showed a main effect of attended modality, being most active overall during the visual task. But nevertheless, they still showed increased activation when attending contralaterally for the tactile task (see plots in Fig. 32.3), when the visual stimuli could serve only as distracters.

These two neuroimaging studies (Macaluso et al., 2000a, 2002a) suggest that the direction of spatial attention for a *tactile* task can modulate activity not only in somatosensory and multimodal (intraparietal) cortex, but also in *visual* cortex, suggesting that anatomically remote brain areas specializing in processing for different senses can come to focus on a common spatial location. While these studies were concerned with endogenous spatial attention, further work with event-related fMRI has shown that exogenous (stimulus-driven) cross-modal spatial cuing can also reveal spatially specific tactile influences on visual cortex.

Macaluso et al. (2000b) presented single visual targets randomly in the left or right hemifield, unpredictably with or without a concurrent touch on one hand (right hand near the right-visual-field target location for one group of subjects, see schematic at top of Fig. 32.4A; left hand near the left-visual-field target for the other group of subjects, see top of Fig. 32.4B). They examined whether multimodal stimulation at the same external location would lead to an enhanced response from contralateral visual cortex, as compared with purely visual stimulation, or with the addition of touch at a different location. A region in lingual gyrus (plus fusiform) showed exactly such a pattern (see bottom of Fig. 32.4). The response to a contralateral visual

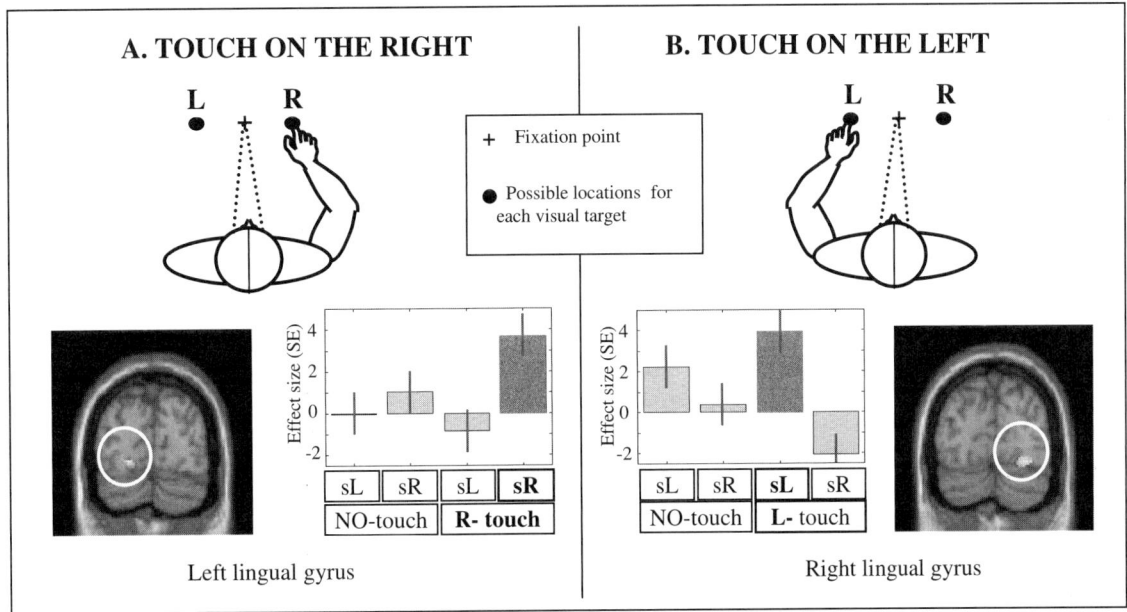

FIGURE 32.4 Effect of multisensory spatial correspondence on visual responses in occipital cortex (see Macaluso et al., 2000b). For each group (A: group receiving tactile stimulations to the right hand, B: group with touch to the left hand), we cartoon the spatial relation between possible visual and tactile input (top), and plot brain activity from the critical maxima for four experimental conditions: sl, stimulate left visual field; sR, stimulate right visual field; in the presence of touch (on left or right hand) or the absence of touch. The sections ($y = -82$ and $y = -80$, for left and right hemisphere activation, respectively) show the anatomical location of the cross-modal interaction (highest activity for contralateral visual stimulation with concurrent touch at same location) observed in the hemisphere contralateral to the location where spatially congruent multi-modal stimulation could be presented. The signal plots show the amplification of visual responses when vision and touch were stimulated at the same contralateral location (see red bars in both graphs). Effect sizes are expressed in standard error (SE) units. For display purposes, the anatomical sections shown were thresholded at P-uncorrected $= 0.05$. (See color plate.)

stimulus was boosted here by adding a concurrent tactile event at the same external location, even though this region of visual cortex did not respond to the presence of tactile stimulation per se (i.e., did not show any overall main effect of adding tactile stimulation, but only an interaction that depended on the added touch being at the same location as the visual target to which this area responded, namely, in the contralateral hemifield); see the plots at bottom of Fig. 32.4.

In this study, tactile and visual stimuli from the same external location (e.g., right hand with right-visual-field target, or left hand with left-visual-field target) would project initially to the same contralateral hemisphere, and hence, a simple account in terms of "hemispheric competition" (Kinsbourne, 1970) might be invoked to explain the cross-modal enhancement. But in a follow-up study by Macaluso et al. (2002b), the direction of gaze was varied (as occurs continuously in daily life), so that the unseen right hand on which any tactile stimulation was delivered could now fall in either the right or left visual field (see schematic of experimental setup in Fig. 32.5A). The critical result was that touch on the right hand boosted response for a concurrent right-visual-field target (in left lingual/fusiform gyri) when it lay near that target; but boosted response for a concurrent *left*-visual-field target instead (in right lingual/fusiform gyri) when the stimulated right hand now lay in the left visual field (see Fig. 32.5B). This reversal of which hemisphere exhibited a boost in visual activity by right-hand stimulation arose even when the hand was occluded and unseen (as for the data shown in Fig. 32.5B). This suggests that eye position signals can modulate spatial interactions between touch and vision, and can even reverse which hemisphere is affected.

One implication from these fMRI studies of exogenous spatial cuing effects (Macaluso et al., 2000b, 2002b) is that spatial information can be shared between remote brain areas, as the location of a tactile event can boost responses in visual cortex to a concurrent visual target at the same external location. A further analysis of "functional connectivity" between brain areas in the fMRI data tested for regions that

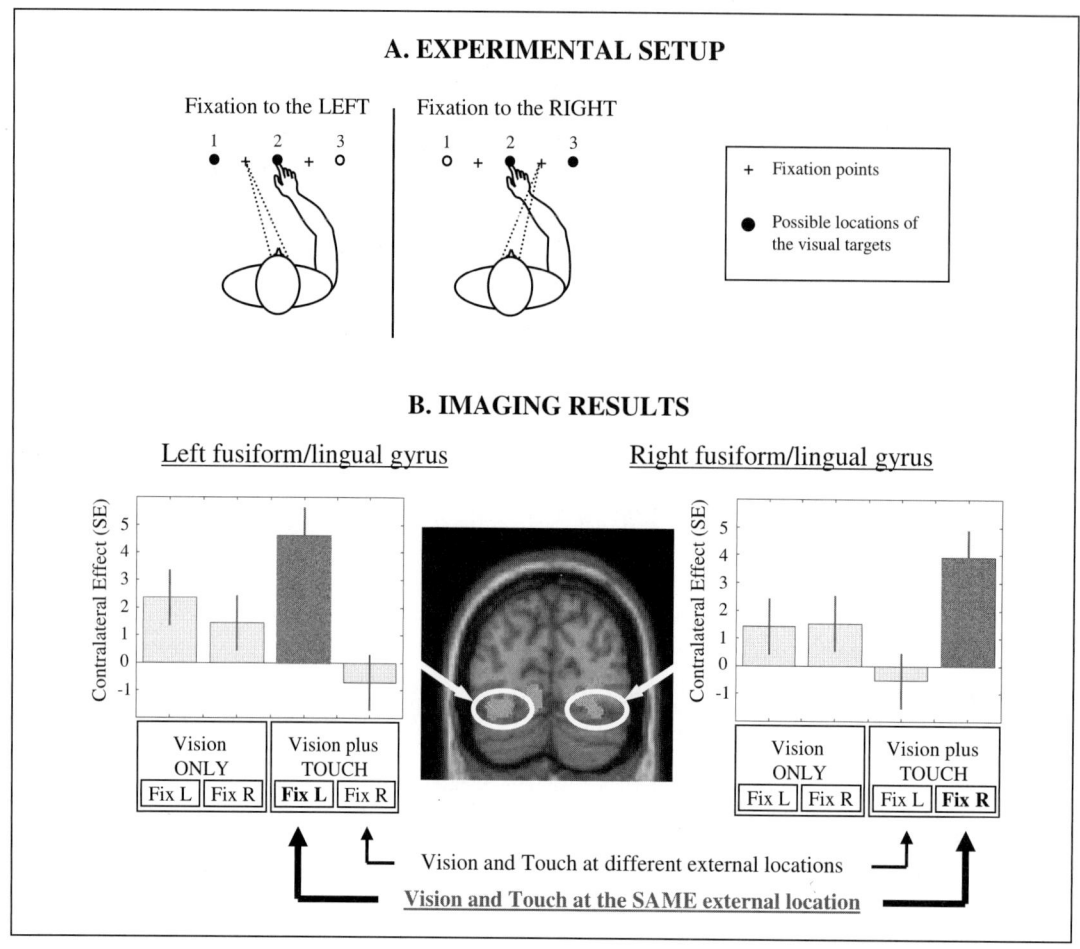

FIGURE 32.5 Gaze direction modulates cross-modal spatial interactions in visual cortex (Macaluso et al., 2002b). (A) Cartoon of experimental setup (possible direction of gaze, possible location of visual targets, and position of unseen right hand). In different blocks, participants fixated either leftward or rightward of the centrally placed right hand. During leftward fixation visual stimuli could be delivered in either position 1 or position 2. When a visual stimulus in position 2 was coupled with tactile stimulation of the right hand, the multimodal stimulation was spatially congruent. However, both stimuli projected to the left hemisphere, so any amplification of visual responses during leftward fixation might be due to intrahemispheric effects, rather than the spatial relation of the stimuli in external space. Including the rightward fixation conditions allowed us to distinguish these alternatives. During rightward fixation, multimodal spatial congruency occurred for simultaneous stimulation of the *left* visual field plus *right*-hand touch (both stimuli in position 2), with the two stimuli now projecting to different hemispheres. (B) Activity in ventral visual cortex for the different stimulation conditions and directions of gaze. Signal plots show the activity in the two hemispheres expressed as the difference between visual stimulation of the contralateral-minus-ipsilateral visual hemifield (i.e., contralateral effect, in standard error units (SE)). Maximal effects of contralateral-minus-ipsilateral visual stimulation were detected when the visual target was coupled with touch at the same external location (see red bars), in the hemisphere contralateral to the *external* location of the multimodal event (i.e., left hemisphere for leftward fixation, left panel; right hemisphere for rightward fixation, right panel). Coronal section taken at $y = -78$; section threshold set to P-uncorrected = 0.01. (See color plate)

showed stronger coupling during multimodal stimulation at the same rather than at different external locations. This found that lingual gyrus contralateral to the visual target showed stronger coupling with somatosensory cortex contralateral to the stimulated hand, and with inferior parietal/superior temporal cortex, specifically for the situation with multimodal spatial congruence. Somatosensory cortex might thus influence visual cortex via parietal regions that receive postural information, such as current eye position, which may be important for updating the spatial register of vision and touch across different postures that realign then, in accord with the effects shown in Fig. 32.5.

Macaluso and Driver (2001), Eimer and Driver (2001), and Driver et al. (2004) review further evidence from both fMRI and ERP studies that focused on the possible "control processes" that may direct spatial attention to produce the cross-modal effects described here. They suggest that such control processes for directing spatial attention may operate at multimodal or even at "supramodal" levels of spatial representation, which take current posture into account. If attended locations are indeed typically selected at multimodal levels of spatial representation, such selection may then inevitably have multisensory consequences, although this would raise an apparent paradox. On the one hand, spatial locations would be selected attentionally at multimodal levels of representation; but, on the other hand, as documented here, we find that the resulting cross-modal effects can influence sensory components and brain areas that would traditionally be considered "unimodal" (e.g., visual cortex and visual evoked potentials). This apparent paradox can be resolved if backprojection influences arise from multisensory brain regions, where attentional selection may arise, to affect unimodal brain areas, where some of the cross-modal influence can become manifest.

Acknowledgments

Our research was supported by grants from the Medical Research Council and the Wellcome Trust. We thank our many colleagues, especially Charles Spence for cross-modal collaborations and Geraint Rees for patience and encouragement.

References

Bertelson, P., and De Gelder, B. (2004). The psychology of multimodal perception. *In* "Crossmodal Space and Crossmodal Attention" (C. Spence and J. Driver, Eds.), pp. 141–177. Oxford University Press, Oxford.

Driver, J., Eimer, M., Macaluso, E., and Van Velzen, J. (2004). Neurobiology of human spatial attention: Modulation, generation and integration. *In* "Functional Neuroimaging of Visual Cognition: Attention and Performance XX" (N. Kanwisher and J. Duncan, Eds.), pp. 267–300. Oxford University Press, Oxford.

Driver, J., and Spence, C. J. (1994). Spatial synergies between auditory and visual attention. In C. Umilta & M. Moscovitch (eds), Attention and Performance: Vol XV (pp. 311–331). Cambridge, MA: MIT press.

Driver, J., and Spence, C. (2000). Multisensory perception: beyond modularity and convergence. *Curr. Biol.* **10**, R731–R735.

Duhamel, J. R., Colby, C. L., and Goldberg, M. E. (1998). Ventral intraparietal area of the macaque: congruent visual and somatic response properties. *J. Neurophysiol.* **79**, 126–136.

Eason, R. G. (1981). Visual evoked potential correlates of early neural filtering during selective attention. *Bull. Psychon. Soc.* **18**, 203–206.

Eason, R., Harter, M., and White, C. (1969). Effects of attention and arousal on visually evoked cortical potentials and reaction time in man. *Physiol. Behav.* **4**, 283–289.

Eimer, M. (1994). "Sensory gating" as a mechanism for visual-spatial orienting: electrophysiological evidence from trial-by-trial cueing experiments. *Percept. Psychophys.* **55**, 667–675.

Eimer, M. (2001). Crossmodal links in spatial attention between vision, audition, and touch: Evidence from event-related brain potentials. *Neuropsychologia* **39**, 1292–1303.

Eimer, M. (2004). Electrophsiology of human crossmodal spatial attention. *In* "Crossmodal Space and Crossmodal Attention" (C. Spence and J. Driver, Eds.), pp. 221–245. Oxford University Press, Oxford.

Eimer, M., Cockburn, D., Smedley, B., and Driver, J. (2001). Crossmodal links in endogenous spatial attention are mediated by common external locations: Evidence from event-related potentials. *Exp. Brain Res.* **139**, 398–411.

Eimer, M., and Driver, J. (2000). An event-related brain potential study of cross-modal links in spatial attention between vision and touch. *Psychophysiology* **37**, 697–705.

Eimer, M., and Driver, J. (2001). Crossmodal links in endogenous and exogenous spatial attention: Evidence from event-related brain potential studies. *Neurosci. Biobehav. Rev.* **25**, 497–511.

Eimer, M., and Schräger, E. (1998). ERP effects of intermodal attention and cross-modal links in spatial attention. *Psychophysiology* **35**, 313–327.

Eimer, M., Van Velzen, J., and Driver, J. (2002). Crossmodal interactions between audition, touch, and vision in endogenous spatial attention: ERP evidence on preparatory states and sensory modulations. *J. Cogn. Neurosci.* **14**, 254–271.

Gray, R., and Tan, H. Z. (2002). Dynamic and predictive links between touch and vision. *Exp. Brain Res.* **145**, 50–55.

Heinze, H. J., Mangun, G. R., Burchert, W., Hinrichs, H., Scholz, M., Munte, T. F., Gos, A., Scherg, M., Johannes, S., and Hundeshagen, H. (1994). Combined spatial and temporal imaging of brain activity during visual selective attention in humans. *Nature* **372**, 543–546.

Hillyard, S. A., Simpson, G. V., Woods, D. L., Van Voorhis, S., and Münte, T. F. (1984). Event-related brain potentials and selective attention to different modalities. *In* "Cortical Integration" (F. Reinoso-Suarez and C. Ajmone-Marsan, Eds.), pp. 395–414. Raven Press, New York.

Hopfinger, J. B., Jha, A. P., Hopf, J. M., Girelli, M., and Mangun, G. R. (2000). Electrophysiological and neuroimaging studies of voluntary and reflexive attention. *In* "Attention and Performance XVII" (S. Monsell and J. Driver, Eds.), pp. 125–153. MIT Press, Cambridge, MA.

Kennett, S., Eimer, M., Spence, C., and Driver, J. (2001). Tactile–visual links in exogenous spatial attention under different postures: convergent evidence from psychophysics and ERPs. *J. Cogn. Neurosci.* **13**, 462–478.

Kennett, S., Spence, C., and Driver, J. (2002). Visuo-tactile links in covert exogenous spatial attention remap across changes in unseen hand posture. *Percept. Psychophys.* **64**, 1083–1094.

Kinsbourne, M. (1970). A model for the mechanism of unilateral neglect of space. *Trans. Am. Neurol. Assoc.* **95**, 143–146.

Luck, S. J., Woodman, G. F., and Vogel, E. K. (2000). Event-related potential studies of attention. *Trends Cogn. Sci.* **4**, 432–440.

Macaluso, E., and Driver, J. (2001). Spatial attention and crossmodal interactions between vision and touch. *Neuropsychologia* **39**, 1304–1316.

Macaluso, E., Frith, C., and Driver, J. (2000a). Selective spatial attention in vision and touch: unimodal and multimodal mechanisms revealed by PET. *J. Neurophysiol.* **83**, 3062–3075.

Macaluso, E., Frith, C. D., and Driver, J. (2000b). Modulation of human visual cortex by crossmodal spatial attention. *Science* **289**, 1206–1208.

Macaluso, E., Frith, C. D., and Driver, J. (2002a). Directing attention to locations and to sensory modalities: multiple levels of selective processing revealed with PET. *Cereb. Cortex* **12**, 357–368.

Macaluso, E., Frith, C. D., and Driver, J. (2002b). Crossmodal spatial influences of touch on extrastriate visual areas take current gaze direction into account. *Neuron* **34**, 647–658.

Mangun, G. R. (1995). The neural mechanisms of visual selective attention. *Psychophysiology* **32**, 4–18.

Mangun, G. R., and Hillyard, S. A. (1991). Modulations of sensory-evoked brain potentials indicate changes in perceptual processing during visual-spatial priming. *J. Exp. Psychol. Hum. Percept. Perform.* **17**, 1057–1074.

McDonald, J. J., Teder-Salejarvi, W. A., and Hillyard, S. A. (2000). Involuntary orienting to sound improves visual perception. *Nature* **407**, 906–908.

McDonald, J. J., Teder-Sälejärvi, W. A., Heraldez, D., and Hillyard, S. A. (2001). Electrophysiological evidence for the "missing link" in crossmodal attention. *Can. J. Exp. Psychol.* **55**, 143–151.

McDonald, J. J., and Ward, L. M. (2000). Involuntary listening aids seeing: evidence from human electrophysiology. *Psychol. Sci.* **11**, 167–171.

Pavani, F., Spence, C., and Driver, J. (2001). Visual capture of touch: out-of-the-body experiences with rubber gloves. *Psychol. Sci.* **11**, 353–359.

Spence, C., and Driver, J. (1996). Audiovisual links in endogenous covert spatial attention. *J. Exp. Psychol. Hum. Percept. Perform.* **22**, 1005–1030.

Spence, C., and Driver, J. (Eds.) (2004). "Crossmodal Space and Crossmodal Attention". Oxford University Press, Oxford.

Spence, C., McDonald, J., and Driver, J. (2004). Exogenous spatial-cuing studies of human crossmodal attention and multisensory integration. *In* "Crossmodal Space and Crossmodal Attention" (C. Spence and J. Driver, Eds.), pp. 277–320. Oxford University Press, Oxford.

Spence, C., Nicholls, M. E., Gillespie, N., and Driver, J. (1998). Crossmodal links in exogenous covert spatial orienting between touch, audition, and vision. *Percept. Psychophys.* **60**, 544–557.

Spence, C., Pavani, F., and Driver, J. (2000a). Crossmodal links between vision and touch in covert endogenous spatial attention. *J. Exp. Psychol. Hum. Percept. Perform.* **26**, 1298–1319.

Spence, C. J., Ranson, J., and Driver, J. (2000b). Crossmodal selective attention: on the difficulty of ignoring sounds at the locus of visual attention. *Percept. Psychophys.* **62**, 410–424.

Teder-Sälejärvi, W. A., Münte, T. F., Sperlich, F.-J., and Hillyard, S. A. (1999). Intra-modal and cross-modal spatial attention to auditory and visual stimuli: an event-related brain potential (ERP) study. *Cogni. Brain Res.* **8**, 327–343.

Van Voorhis, S. T., and Hillyard, S. A. (1977). Visual evoked potentials and selective attention to points in space. *Percept. Psychophys.* **22**, 54–62.

CHAPTER

33

Attention and Scene Understanding

Vidhya Navalpakkam, Michael Arbib and Laurent Itti

ABSTRACT

This chapter presents a simplified, introductory view of how visual attention may contribute to and integrate within the broader framework of visual scene understanding. Several key components are identified which cooperate with attention during the analysis of complex dynamic visual inputs, namely, rapid computation of scene gist and layout, localized object recognition and tracking at attended locations, working memory that holds a representation of currently relevant targets, and long-term memory of known world entities and their interrelationships. Evidence from neurobiology and psychophysics is provided to support the proposed architecture.

I. INTRODUCTION

Primates, including humans, use focal visual attention and rapid eye movements to analyze complex visual inputs in real time, in a manner that highly depends on current behavioral priorities and goals. A striking example of how a verbally communicated task specification may dramatically affect attentional deployment and eye movements was provided by the pioneering experiments of Yarbus (1967). Using an eye-tracking device, Yarbus recorded the scanpaths of eye movements executed by human observers while analyzing a single photograph under various task instructions (Fig. 33.1). Given the unique visual stimulus used in these experiments, the highly variable spatiotemporal characteristics of the measured eye movements for different task specifications exemplify the extent to which behavioral goals may affect eye movements and scene analysis.

Subsequent eye-tracking experiments during spoken sentence comprehension have further demonstrated how the interplay between task demands and active vision is often reciprocal. For instance, Tanenhaus et al. (1995) tracked eye movements while subjects received ambiguous verbal instructions about manipulating objects lying on a table in front of them. They demonstrated not only that tasks influenced eye movements, but also that visual context influenced spoken word recognition and mediated syntactic processing, when the ambiguous verbal instructions could be resolved through visual analysis of the objects present in the scene.

Building computational architectures that can replicate the interplay between task demands specified at a symbolic level (e.g., through verbally delivered instructions) and scene contents captured by an array of photoreceptors is a challenging task. We review several key components and achitectures that have attacked this problem, and we explore in particular the involvment of focal visual attention during goal-oriented scene understanding.

II. BASIC COMPONENTS OF SCENE UNDERSTANDING

A recent overview of a computational architecture for visual processing in the primate brain was provided by Rensink (2000), and is used as a starting point for the present analysis. In Rensink's triadic architecture, low-level visual features are computed in parallel over the entire visual field, up to a level of complexity termed proto-objects (an intermediary between simple features such as edges and corners, and sophisticated object representations). One branch of subsequent processing is concerned with the computation of the so-called setting, which includes a fairly crude semantic analysis of the nature of the scene (its gist, for example, whether it is a busy street, kitchen, or beach; and its coarse spatial layout. See Chapter 41). In the other branch, attention selects

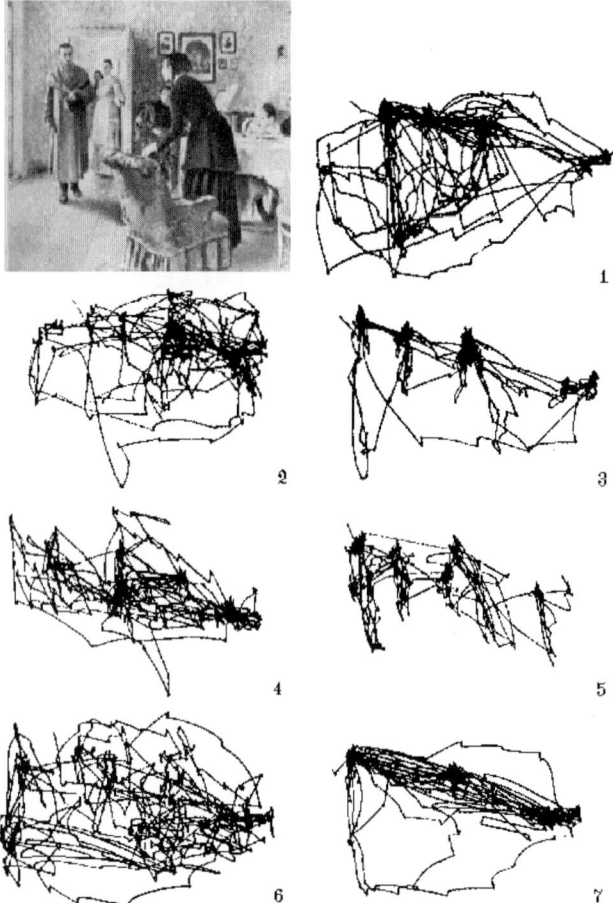

FIGURE 33.1 Stimulus (top-left) and corresponding eye movement traced and recorded from human observers under seven task specifications: 1) free examination of the picture, 2) estimate the material circumstances of the family in the picture, 3) give the ages of the people, 4) surmise what the family had been doing before the arrival of the "unexpected visitor," 5) remember the clothes worn by the people, 6) remember the position of the people and objects in the room, and 7) estimate how long the "unexpected visitor" had been away from the family. (From Yarbus, 1967.)

Visual attention has been often compared to a virtual spotlight through which our brain sees the world (see Chapter 61), and shifts of attention have been classified into several types based on whether or not they involve eye movements (overt vs. covert), and are guided primarily by scene features or volition (bottom-up vs. top-down) (for review, see Itti and Koch, 2001, Chapter 94). The first explicit biologically plausible computational architecture for controlling bottom-up attention was proposed by Koch and Ullman (1985) (also see Didday and Arbib, 1975). In their model, several feature maps (such as color, orientation, and intensity) are computed in parallel across the visual field (Treisman and Gelade, 1980) and are combined into a single salience map. Then, a selection process sequentially deploys attention to locations in decreasing order of their salience. We here assume a similar architecture for the attentional branch of visual processing and explore how it can be enhanced by modeling the influence of task on attention. The choice of features which may guide attention bottom-up has been extensively studied in the visual search literature (see Chapter 17).

At the early stages of visual processing, task modulates neural activity by enhancing the responses of neurons tuned to the location and features of a stimulus (Chapter 50, 51, 70, 73 and many others in this book). In addition, psychophysics experiments have shown that knowledge of the target contributes to an amplification of its salience; for example, white vertical lines become more salient if we are looking for them (Blaser *et al.*, 1999). A recent study even shows that better knowledge of the target leads to faster search (e.g., seeing an exact picture of the target is better than seeing a picture of the same semantic type or category as the target (Kenner and Wolfe, 2003). These studies demonstrate the effects of biasing for features of the target. Other experiments (e.g., Treisman and Gelade, 1980) have shown that searching for feature conjunctions (e.g., color × orientation conjunction search: find a red-vertical item among red-horizontal and green-vertical items) is slower than "pop-out" (e.g., find a green item among red items). These observations impose constraints on the possible biasing mechanisms and eliminate the possibility of generating new composite features on the fly (as a combination of simple features).

A popular model to account for top-down feature biasing and visual search behavior is *Guided Search* (Wolfe, 1994). This model has the same basic architecture as proposed by Koch and Ullman (1985), but, in addition, achieves feature-based biasing by weighing feature maps in a top-down manner. For example, with the task of detecting a red bar, the red-sensitive feature

a small spatial portion of the visual inputs and transiently binds the volatile proto-objects into coherent representations of attended objects. Attended objects are then processed in further detail, yielding the recognition of their identity and attributes. Figure 33.2 builds upon and extends this purely visual architecture.

As the present review focuses on computational modeling of the interaction between cognitively represented task demands and this simplified view of visual processing, we refer the reader to several articles in this volume that explore the putative components of the triadic architecture in further details (e.g., Chapters 41, 94, and 96).

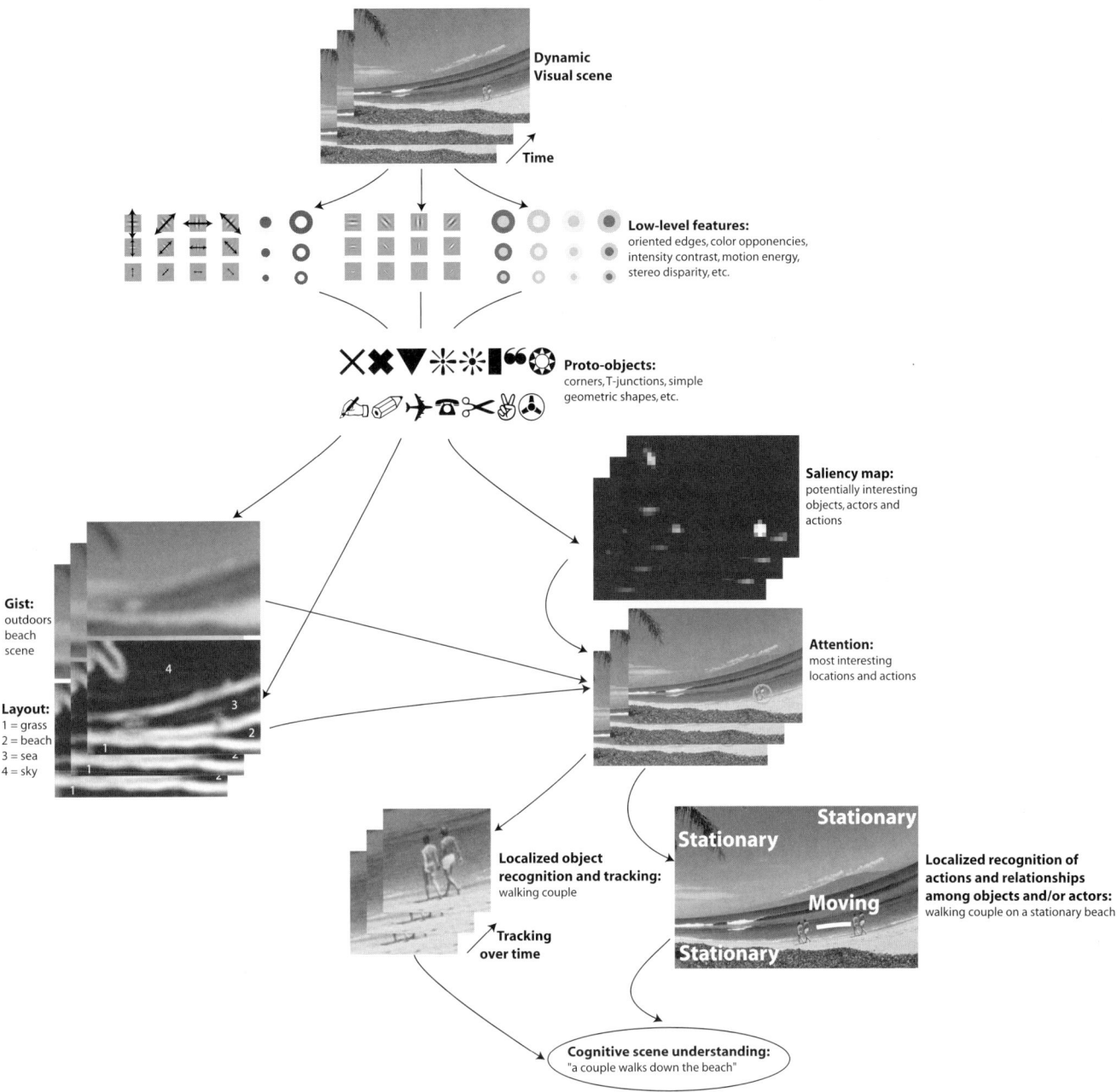

FIGURE 33.2 The beginning of an architecture for complex dynamic visual scene understanding, starting from the visual end. This diagram augments the triadic architecture proposed by Rensink (2000), which identified three key components of visual processing: preattentive processing up to a proto-object representation (top), identification of the setting (scene gist and layout; left), and attentional vision including detailed object recognition within the spatially circumscribed focus of attention (right). In this figure, we have extended Rensink's architecture to include a saliency map to guide attention bottom-up toward salient image locations, and action recognition in dynamic scenes. The following figure explores in more detail the interplay between the visual processing components depicted here and symbolic decriptions of tasks and visual objects of current behavioral interest.

map gains more weight, hence making the red bar more salient. One question that remains is how to optimally set the relative feature weights so as to maximize the detectability of a set of behaviorally relevant targets among clutter (Navalpakkam and Itti, 2004).

Given a saliency map, several models have been proposed to select the next attended location, including various forms of winner-take-all (maximum-selector) algorithms (see Chapters 90, 91, 92, and 94).

Having selected the focus of attention, it is important to recognize the entity at that scene location. Many recognition models have been proposed that can be classified based on factors including the choice of basic primitives (e.g., Gabor jets, geometric primitives like geons, image patches or blobs, and view-tuned units), the process of matching (e.g., self-organizing dynamic link matching, probabilistic matching), and other factors (for review, see Chapter 46).

Recognition is followed by the problem of memorization of visual information. A popular theory, the *object file theory of trans-saccadic memory* (Irwin, 1992), posits that when attention is directed to an object, the visual features and location information are bound into an object file (Kahneman and Treisman, 1984) that is maintained in visual short-term memory across saccades. Psychophysics experiments have further shown that up to three or four object files may be retained in memory (Irwin, 1992). Studies investigating the neural substrates of working memory in primates and humans suggest that the frontal and extrastriate cortices may both be functionally and anatomically separated into a "what" memory for storing the visual features of the stimuli, and a "where" memory for storing spatial information (Wilson *et al.*, 1993).

To memorize the location of objects, we here extend the earlier hypothesis of a salience map (Koch and Ullman, 1985) to propose a two-dimensional topographic *task-relevance map* (TRM) that encodes the task relevance of scene entities. Our motivation for maintaining various maps stems from biological evidence. Single-unit recordings in the visual system of the macaque indicate the existence of a number of distinct maps of the visual environment that appear to encode the salience and/or the behavioral significance of targets. Such maps have been found in the superior colliculus, the inferior and lateral subdivisions of the pulvinar, the frontal-eye fields, and areas within the intraparietal sulcus (see Chapters 21 and 22). Since these neurons are found in different parts of the brain that specialize in different functions, we hypothesize that they may encode different types of salience: the posterior parietal cortex may encode a visual salience map, while the prefrontal cortex may encode a top-down task-relevance map, and the final eye movements may be generated by integrating information across the visual salience map and task-relevance map to form an attention guidance map possibly stored in the superior colliculus.

Our analysis so far has focused on the attentional pathway. As shown in Figure 33.2, nonattentional pathways also play an important role; in particular, rapid identification of the gist (semantic category) of a scene is very useful in determining scene context, and is known to guide eye movements (Chapter 41). It is computed rapidly within the first 150 ms of scene onset, and the neural correlate of this computation is still unknown (Chapter 131). Recently, Torralba (Chapter 96) proposed holistic representation of the scene based on spatial envelope properties (such as openness, naturalness etc.) that bypasses the analysis of component objects and represents the scene as a single identity. This approach formalizes the gist as a vector of contextual features. By processing several annotated scenes, these authors learned the relationship between the scene context and categories of objects that can occur, including object properties such as locations, size, or scale, and used it to focus attention on likely target locations. This provides a good starting point for modeling the role of gist in guiding attention. Since the gist is computed rapidly, it can serve as an initial guide to attention. But subsequently, our proposed TRM that is continuously updated may serve as a better guide. For instance, in dynamic scenes such as traffic scenes where the environment is continuously changing and the targets such as cars and pedestrians are moving around, the gist may remain unchanged and hence, it may not be so useful, except as an initial guide.

The use of gist in guiding attention to likely target locations motivates knowledge-based approaches to modeling eye movements, in contrast to image-based approaches. One such famous approach is the scan-path theory, which proposes that attention is mostly guided in a top-down manner based on an internal model of the scene (Chapter 48). Computer vision models have employed a similar approach to recognize objects. For example, Rybak et al. (Chapter 105) recognize objects by explicitly replaying a sequence of eye movements and matching the expected features at each fixation with the image features. (also see Chapters 91, 105, 109).

To summarize, we have motivated the basic architectural components which we believe are crucial for scene understanding. Ours is certainly not the first attempt to address this problem. For example, one of the finest examples of real-time scene analysis systems is *The Visual Translator* (VITRA) (Herzog and Wazinski, 1994), a computer vision system that generates real-time verbal commentaries while watching a televised soccer game. Their low-level visual system recognizes and tracks all visible objects from an overhead (bird's eye) camera view, and creates a geometric representation of the perceived scene (the 22 players, the field, and the goal locations). This intermediate representation is then analyzed by series of Bayesian belief networks, which evaluate spatial relations, recognize interesting motion events, and incrementally recognize plans and intentions. The model includes an abstract, nonvisual notion of salience that characterizes each recognized event on the basis of recency, frequency, complexity, importance for the game, and other factors. The system finally generates a verbal commentary, which typically starts as soon as the beginning of an event has been recognized but may be interjected if highly salient events occur before the current sentence has been completed. While this system delivers very impressive results in the specific application domain considered, due to its computational complexity it is restricted to one highly structured environment and one specific task, and cannot be extended to a general scene understanding model. Indeed, unlike humans who selectively perceive the relevant objects in the scene, VITRA attends to and continuously monitors *all* objects and attempts to simultaneously recognize *all* known actions. The approach proposed here differs from VITRA not only in that there is nothing in our model that commits it to a specific environment or task. In addition, we only memorize those objects and events that we expect to be relevant to the task at hand, thus saving enormously on computation complexity.

III. DISCUSSION: SUMMARY OF A PUTATIVE FUNCTIONAL ARCHITECTURE

In this section, we present a summary of an architecture which can be understood in four phases (figure 33.3). Partial implementation and testing of this architecture has been developed elsewhere (Navalpakkam and Itti, 2004), so that here we focus mainly on reviewing the components and their interplay during active goal-oriented scene understanding.

A. Phase 1: Eyes Closed

In the first phase known as the "eyes-closed" phase, the symbolic working memory (WM) is initialized by the user with a task definition in the form of keywords and their relevance (any number greater than baseline 1.0). Given the relevant keywords in symbolic WM, volitional effects such as "look at the center of the scene" could be achieved by allowing the symbolic WM to bias the TRM so that the center of the scene becomes relevant and everything else is irrelevant (but our current implementation has not explored this yet). For more complex tasks such as "who is doing what to whom," the symbolic WM requires prior knowledge and, hence, seeks the aid of the symbolic long-term memory (LTM). For example, to find what the man in the scene is eating, prior knowledge about eating being a mouth and hand-related action, and being related to food items helps us guide attention toward mouth or hand and determine the food item. Using such prior knowledge, the symbolic WM parses the task and determines the task-relevant targets and how they are related to each other. Our implementation (Navalpakkam and Itti, 2004) explores this mechanism using a simple hand-coded symbolic knowledge base to describe long-term knowledge about objects, actors, and actions. Next, it determines the current most task-relevant target as the desired target. To detect the desired target in the scene, the visual WM retrieves the learned visual representation of the target from the visual LTM and biases the low-level visual system with the target's features.

B. Phase 2: Computing

In the second phase, known as the "computing" phase, the eyes are open, and the visual system receives the input scene. The low-level visual system that is biased by the target's features computes the biased salience map. Apart from such feature-based attention, spatial attention may be used to focus on likely target locations (e.g., gist and layout may be used to bias the TRM to focus on relevant locations but this is not implemented yet). Since we are interested in attending to locations that are salient and relevant, the biased salience and task-relevance maps are combined by taking a pointwise product to form the attention-guidance map (AGM). To select the focus of attention, we deploy a winner-take-all competition that chooses the most active location in the AGM. It is important to note that there is no intelligence in this selection and all the intelligence of the model lies in the WM.

C. Phase 3: Attending

In the third phase known as the "attending" phase, the low-level features or prototype objects are bound into a spatiotemporal structure or a midlevel representation called the "coherence field" (a midlevel representation of an object formed by grouping prototype objects into a spatiotemporal structure that exhibits coherence in space and time). Spatial coherence implies that the prototype objects at the different locations belong to the same object, and temporal coherence implies that different representations across time refer to the same object (Rensink, 2000); in our implementation, this step simply extracts a vector of visual features at the attended location. The object recognition module determines the identity of the entity at the currently attended location, and the symbolic WM

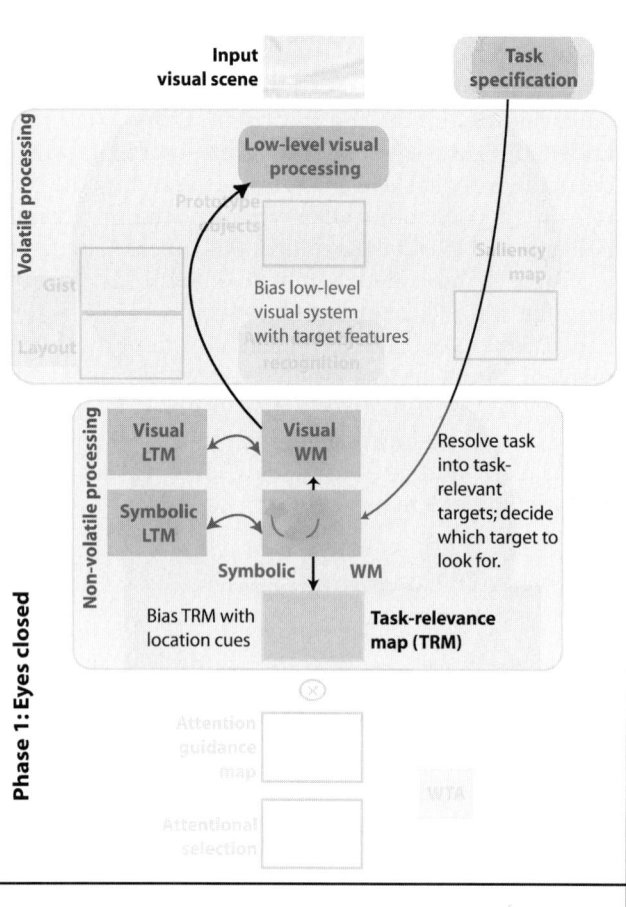
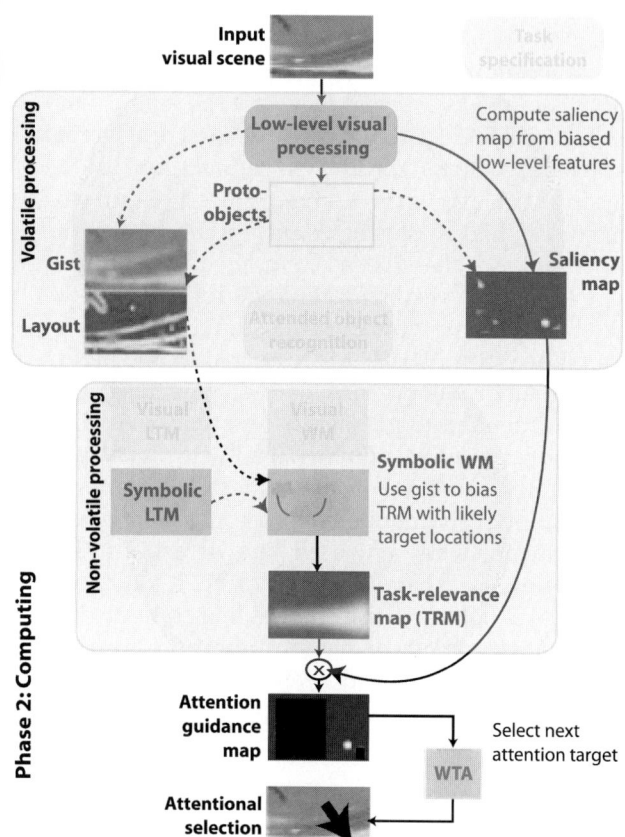
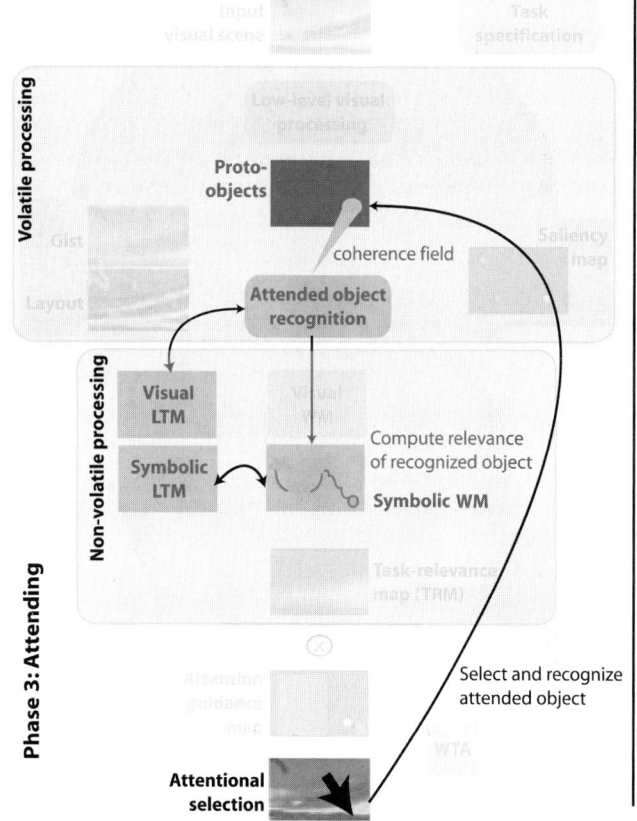
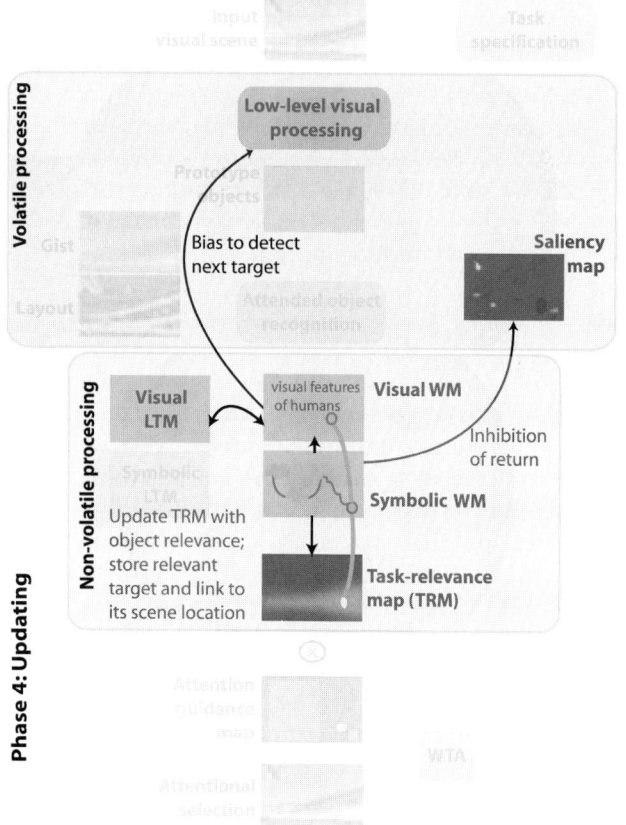

FIGURE 33.3 Phase 1 (top left): Eyes closed, Phase 2 (top right): Computing, Phase 3 (bottom left): Attending, Phase 4 (bottom right): Updating. Please refer to section 33.2 for details about each phase. All four panels represent the same model; however, to enable easy comparison of the different phases, we have highlighted the components that are active in each phase and faded those that are inactive. Dashed lines indicate parts that have not been implemented yet. Following Rensink's (2000) terminology, volatile processing stages refer to those that are under constant flux and regenerate as the input changes.

◄─────────

estimates the task relevance of the recognized entity (Navalpakkam and Itti, 2004).

D. Phase 4: Updating

In the final phase known as the "updating" phase, the WM updates its state (e.g., records that it has found the man's hand). It updates the TRM by recording the relevance of the currently attended location. The estimated relevance may influence attention in several ways. For instance, it may affect the duration of fixation (not implemented). If the relevance of the entity is less than the baseline 1.0, it is marked as irrelevant in the TRM, and hence will be ignored by preventing future fixations on it (e.g., a chair is irrelevant when we are trying to find what the man is eating. Hence, if we see a chair, we ignore it). If it is somewhat relevant (e.g., man's eyes), it may be used to guide attention to a more relevant target by means of directed attention shifts (e.g., look down to find the man's mouth or hand; not implemented). Also, if it is relevant (e.g., man's hand), a detailed representation of the scene entity may be created for further scrutiny (e.g., a spatiotemporal structure for tracking the hand; not implemented). The WM also inhibits the current focus of attention from continuously demanding attention (inhibition of return in SM). Then, the symbolic WM determines the next most task-relevant target, retrieves the target's learned visual representation from visual LTM, and uses it to bias the low-level visual system.

This completes one iteration. The computing, attending, and updating phases repeat until the task is complete. Upon completion, the TRM shows all task-relevant locations, and the symbolic WM contains all task-relevant targets.

IV. CONCLUSION

We have presented a brief overview of how some of the crucial components of primate vision may interact during active goal-oriented scene understanding. From the rather partial and sketchy figure proposed here, it is clear that much systems-level, integrative research will be required to further piece together more focused and localized studies of the neural subsystems involved.

References

Blaser, E., Sperling, G., and Lu, Z. L. (1999). Measuring the amplification of attention. *Proc. Natl. Acad. Sci. USA*, **96**(20), 11681–11686.

Didday, R. L., and Arbib, M. A. (1975). Eye Movements and Visual Perception: A "Two Visual System" Model. *Int. J. Man-Machine Studies*, **7**, 547–569.

Herzog, G., and Wazinski, P. (1994). VIsual TRAnslator: Linking Perceptions and Natural Language Descriptions. *Artificial Intelligence Review*, **8**(2–3), 175–187.

Irwin, D. E. (1992). Memory for position and identity across eye movements. *Journal of Experimental Psychology: Learning, Memory, and Cognition*, **18**, 307–317.

Itti, L., and Koch, C. (2001). Computational Modeling of Visual Attention. *Nature Reviews Neuroscience*, **2**(3), 194–203.

Kahneman, D., and Treisman, A. (1984). Changing views of attention and auto-maticity. Pages 29–61 of: Parasuraman, D., Davies, R., and Beatty, J. (eds), *Varieties of attention*. New York, NY: Academic.

Kenner, N., and Wolfe, J. M. (2003). An exact picture of your target guides visual search better than any other representation [abstract]. *Journal of Vision*, **3**(9), 230a.

Koch, C., and Ullman, S. (1985). Shifts in selective visual attention: towards the underlying neural cicuitry. *Hum Neurobiol*, **4**(4), 219–27.

Navalpakkam, V., and Itti, L. submitted. Modeling the influence of task on attention. *Vision Research*.

Rensink, R. A. (2000). The dynamic representation of scenes. *Visual Cognition*, **7**, 17–42.

Tanenhaus, M. K., Spivey-Knowlton, M. J., Eberhard, K. M., and Sedivy, J. C. (1995). Integration of visual and linguistic information in spoken language comprehension. *Science*, **268**(5217), 1632–1634.

Treisman, A. M., and Gelade, G. (1980). A feature-integration theory of attention. *Cognit. Psychol.*, **12**(1), 97–136.

Wilson, F. A., O'Scalaidhe, S. P., and Goldman-Rakic, P. S. (1993). Dissociation of object and spatial processing domains in primate prefrontal cortex. *Science*, **260**, 1955–1958.

Wolfe, J. M. (1994). Guided search 2.0: a revised model of visual search. *Psyonomic Bulletin and Review*, **1**(2), 202–238.

Yarbus, A. (1967). *Eye Movements and Vision*. Plenum Press, New York.

› SECTION II

FUNCTIONS

CHAPTER

34

Visual Search and Popout in Infancy

Scott A. Adler

ABSTRACT

Attentional selection enables efficient information processing by selecting relevant information and filtering out irrelevant information. For infants, attentional selection is critical because they are first constructing a knowledge base of the world. Visual search and texture segregation studies with adults have indicated that items that contain a unique perceptual feature automatically and selectively capture attention (i.e., popout), irrespective of the number of distracter items. In contrast, items that lack a unique perceptual feature are detected by serially allocating attention, resulting in longer search times as the number of distracters increases. Presumably, popout is due to an early preattentive processing stage that decomposes items in parallel into their primitive perceptual features, whereas serial search is accomplished by a later attentive processing stage. Research addressing infants' exhibition of visual search, popout, the functioning of the underlying attentive mechanisms, and development of the possible neural foundations is reviewed.

I. INTRODUCTION

Our world is populated by many visual objects and events, all of which are potential targets for our attention and processing resources. Because our processing resources are limited, we cannot attend to all possible objects and events simultaneously, but instead we allocate attention selectively to individual items. For infants, whose visual and mental processing was once described as "assailed by eyes, ears, nose, skin and entrails at once, feels it all as one great blooming, buzzing confusion" (James, 1890, p. 488) and who are first constructing a knowledge base, it is particularly necessary to possess mechanisms of selective attention. Thus, infants, as do adults, need to filter the information impinging on their sensory mechanisms to construct an organized knowledge base.

II. SEARCH, SEGREGATION, POPOUT, AND SELECTION MECHANISMS IN ADULTS

In adults, one particular set of mechanisms of attentional selectivity that have been investigated is preattentive versus attentive mechanisms, with the primary paradigms for their investigation being visual search (Treisman and Gelade, 1980) and texture segregation (Julesz, 1981). The functioning of these mechanisms has been formalized in two influential theories of adults' visual information processing in which an initial "preattentive" mechanism decomposes all stimuli in the visual array in parallel into their basic perceptual features (Julesz, 1981; Treisman and Gelade, 1980). The basic perceptual features have been hypothesized to include elongated blobs, orientations, width and length, size, color, motion, and elongated blob terminators—a list that agrees well with the properties that physiological evidence suggests are processed in parallel by the early visual system (Deco et al., 2002) (see Chapter 93). A later "attentive" mechanism selectively focuses processing resources serially to individual stimuli for the purpose of binding the features into a unified object percept and for object recognition (Julesz, 1981; Treisman and Gelade, 1980) (see Chapters 24, 65, and 90).

Behaviorally, these mechanisms have been explored with visual search and texture segregation paradigms and the phenomenon of popout. Popout can be described simply as the situation in which stimuli that are defined by a unique perceptual feature automatically and selectively capture attention (Treisman and Gelade, 1980) (see Chapters 17 and 68). That is, the preattentive mechanism decomposes stimuli into their basic features and when a feature map indicates a stimulus unique for that property, then attentive processes are selectively allocated to that stimulus location. Consequently, regardless of the number of stimuli in the array, the time it takes to detect the stimulus with the unique feature remains relatively stable. In contrast, when a stimulus does not consist of a single unique identifying feature but is defined by a unique combination of features, it does not popout. Instead, the attentive mechanism allocates processing resources to each stimulus to detect the stimulus in the array with the unique conjunction of features (Treisman and Gelade, 1980) (see Chapters 17 and 43). Consequently, the time it takes to detect the stimulus with the unique conjunction increases as the number of stimuli in the array increases.

III. FEATURE SEARCH, SEGREGATION, AND POPOUT IN INFANCY

If "popout" is due to the early processing of features that are considered to be the building blocks of perception (Julesz, 1981; Treisman and Gelade, 1980), then popout might be evident early in development. Unfortunately, it is not possible to conduct manual reaction time studies with young infants as with adults, because infants do not have the motor control to press buttons to indicate detection. Instead, researchers have harnessed aspects of infants' existing behavior, including their looking (e.g., preferential looking, habituation–dishabituation, and novelty preference paradigms) and kicking (e.g., mobile conjugate reinforcement paradigm) behaviors, to assess their attentional and perceptual capacities. None of these paradigms, however, provides measures (e.g., reaction time) comparable to those obtained with adults, and they measure visual search and popout over larger time scales (seconds and even minutes) than assessed with adults, thereby making comparison of the underlying attentional and perceptual mechanisms difficult, as is discussed later in this chapter.

One of the first studies to suggest that popout occurs in early infancy was reported by Salapatek (1975), who used a preferential looking paradigm and presented infants with a patch of unique stimuli embedded in a field of dissimilar stimuli, for example, squares in horizontal lines (or vice versa). He found that 3-month-olds always oriented to the unique stimuli, whereas 2-month-olds did not, suggesting that 3-month-olds' but not 2-month-olds' attention was captured by the unique patch. This suggests that the mechanism responsible for popout might develop around 3 months of age.

In the last 10 years or so, there has been considerable interest in this prospect and numerous studies have investigated popout in infants. In 1992, motivated by Salapatek's findings, two studies further investigated the development of popout of discrepant patches or textures (Atkinson and Braddick, 1992; Sireteanu and Rieth, 1992). Atkinson and Braddick found, in relative agreement with Salapatek (1975), that 4-month-old infants, but not 2- to 3-month-olds, oriented to a patch of oriented lines embedded in a texture of orthogonally oriented lines, suggesting that the ability to exhibit popout on the basis of orientation differences does develop around 3 months of age. However, when the patch of lines differed from the surrounding texture on the basis of size (and luminance), even the youngest infants exhibited popout.

Sireteanu and Rieth (1992) also found that infants as young as 2 months preferentially oriented toward the discrepant patch when it was defined by size, suggesting that it popped out. In contrast to Atkinson and Braddick (1992), however, Sireteanu and Rieth found that a discrepant patch defined by orientation was not preferentially oriented until approximately 12 months of age. The discrepancy between the two studies in the age at which popout of orientation-defined patches is exhibited may be due to methodological differences, including whether the homogeneous and discrepant areas of the texture were presented in two separate fields (Sireteanu and Rieth, 1992) or in one large field (Atkinson and Braddick, 1992), whether the oriented lines were the same length (Sireteanu and Rieth, 1992) or varied in length (Atkinson and Braddick, 1992), and whether the textures contained 16 lines (Sireteanu and Rieth, 1992) or 38 lines (Atkinson and Braddick, 1992). Regardless, these studies demonstrate that popout and the segregation of textures, at least those defined by size or luminance differences, is evident in infants as young as 2 months of age.

A number of other studies using different stimuli and paradigms have demonstrated popout in 3-month-old infants (Adler et al., 1998; Colombo et al., 1995; Rovee-Collier et al., 1992). Rovee-Collier et al. (1992), for example, used as a basis the finding by Julesz (1981) that a patch of +'s pops out for adults when it is embedded in a surrounding texture of L's, presumably because the +'s contain the unique perceptual feature of the line crossing. Rovee-Collier et al. used the mobile conjugate reinforcement paradigm in which 3-month-olds were trained to kick to move an overhead seven-block crib mobile that displayed either L's or +'s on every block side and then tested them 24 hours later with a mobile that consisted of either a single unique L block among six + blocks or a single unique block of +'s among six blocks of L's. Previous research with this paradigm has indicated that infants will exhibit recognition of the test mobile only if the information presented on the blocks match (i.e., is familiar) the information infants remember were on the blocks during training; otherwise they exhibit discrimination. Rovee-Collier et al. found that infants' recognition performance was controlled by the familiarity or novelty of the unique characters on the single block regardless of whether the characters on the surrounding blocks were familiar or novel. For example, infants trained with a mobile that consisted of all L's and tested 24 hours later with a mobile that consisted of a single + among L's discriminate the test mobile even though it was predominately identical to the mobile with which they were trained. Thus, the single, unique + block controlled infants' recognition performance at test. This suggests that the unique character popped out from amid the surrounding dissimilar characters, similar to the findings of popout in visual search and texture segregation studies with adults.

Another fundamental outcome of preattentively decomposing stimuli into their basic features is that a search asymmetry exists in popout (Treisman and Gelade, 1980). That is, a stimulus that contains a unique feature will pop out from amid stimuli in which that feature is absent, whereas a stimulus in which that feature is absent will not pop out from stimuli in which that feature is present. For example, in adults, C's will popout from amid O's but not vice versa, presumably because the C's contain line terminators which are absent in the O's. Adler et al. (1998) and Colombo et al. (1995) have separately found evidence that suggests that search asymmetries are even exhibited in early infancy.

Adler et al. (1998), using the mobile conjugate reinforcement paradigm (described above), trained 3-month-olds to kick to move a mobile that displayed either R's (feature-present) or P's (feature-absent). Twenty-four hours later, infants were tested with either a single P among R's or a single R among P's. Results indicated that recognition or discrimination was determined by the feature-present stimulus (R) regardless of whether it was the target (R among P's) or distracter (P among R's), suggesting that it popped out when it was the target whereas the feature-absent P failed to pop out when it was the target (see Table 34.1). For example, infants trained with a P mobile and tested 24 hours later with a mobile that consisted either of a single R among P's or of a single P among R's discriminate the test mobile because their recognition performance is controlled by the R, which does not match the P with which they were trained. This occurs because infants' attention was allocated to the feature-present stimulus in both cases, apparently popping out in the first case and receiving attentional allocation in the second due to the feature-absent stimulus failing to pop out.

Colombo et al. (1995), using a preferential looking task, presented 3- to 4-month-old infants with a homogeneous array of either O's or Q's paired with a heterogeneous array of a single Q among O's (feature-present condition) or with a single O among Q's (feature-absent condition). Results indicated the infants preferentially oriented their attention to the heterogeneous array only in the feature-present condition but not in the feature-absent condition. This asymmetry in infants' attentional allocation suggested that the feature-present target (Q) popped out from amid the feature-absent distracters (O's), whereas the feature-absent target (O) did not pop out from amid

TABLE 34.1 Summary of Results from Adler et al. (1998), Who Investigated the Perceptual Asymmetry of Popout in 3-Month-Old Infants

Train mobile	Test mobile	FAM or NOV	Recognition	Popout?
Feature-absent popout target				
R	Single P Among R's	NOV	Yes	No
P	Single P Among R's	FAM	No	No
Feature-present popout target				
P	Single R Among P's	NOV	No	Yes
R	Single R Among P's	FAM	Yes	Yes

Note. Shown are the training mobile, test mobile, familiarity (FAM) or novelty (NOV), and feature absence versus presence of the single popout target on the test mobile, whether recognition (or discrimination) was exhibited and, consequently, whether popout was indicated. From Adler, Inslicht, Rovee-Collier, and Gerhardstein, 1998.

Source. Reprinted, with permission, from Adler et al. (1998).

the feature-present distracters (Q's). Together, these two studies suggest that search asymmetries and the relevant preattentive and attentive mechanisms outlined in theories of adults' visual information processing are also evident in early infancy.

IV. PARALLEL SEARCH IN INFANTS

Collectively, the studies described above and others seem to indicate that very young infants, as young as 3 months, exhibit the phenomenon of "popout." This would further suggest that the preattentive mechanisms for selectively allocating early visual processing resources are functioning in early infancy. However, there are a couple of issues that have yet to be resolved by the infant "popout" studies that would provide definitive evidence for popout and preattentive processing in infancy. First, in adults, popout typically occurs on the order of milliseconds (e.g., Treisman and Gelade, 1980). In the infant studies, however, the paradigms used assess popout during test phases that last seconds (Colombo et al., 1995) and even minutes (Adler et al., 1998; Rovee-Collier et al., 1992). Thus, sufficient time exists in these studies for infants' behavior to be due to serial attentive processes rather than to preattentive popout.

A second issue that has yet to be addressed in the infant studies is that a key function of preattentive processing is that items in the visual world are processed in parallel. Quantitatively, in terms of popout, this means that the time to detect a target remains relatively unchanged as the number of distracter items increases. This contrasts with the effect of not presenting a popout target but only presenting distracter items, in which case the time to detect increases as the number of distracters increases. No infant study, due to the fact the paradigms used to date do not measure reaction time, has yet to examine whether processing of items in a visual array and popout occur in parallel, that is, that detection time is unaffected by the number of distracters.

Consequently, due to these two limitations of the infant studies, protracted test phase for assessing popout and failure to test for set size effects, whether infants actually exhibit the phenomenon of popout and a functioning preattentive mechanism as found in adults has yet to definitively demonstrated. Recently, Orprecio and Adler (2003) measured the latency of 3-month-old infants' saccadic eye movements to visual arrays that either contained a popout target or did not. Because infants' eye movements to stimuli occur on the order of milliseconds, Orprecio and Adler reasoned that such a measure would be more comparable in timing to the reaction time measure used to assess popout in adults and would be able to assess the effect of set size on detection time. Consequently, measuring infants' eye movements would allow for a more comparable assessment of pop-out in infants and be able to measure the functioning of the parallel preattentive mechanism (see Chapters 20 and 65).

Specifically, Orprecio and Adler (2003) presented infants with visual arrays in which the popout target (+) was either present or absent and in which the number of distracter items (L's) varied (see Fig. 34.1), and infants' latency to make an eye movement to the target in the target-present condition and to one of the distracters in the target-absent condition was measured. Furthermore, to assess whether popout occurred in parallel, the effect of increasing set sizes on infants' eye movement latencies was measured. The results indicated that infants' saccade latencies remained unchanged in the target-present conditions as set size

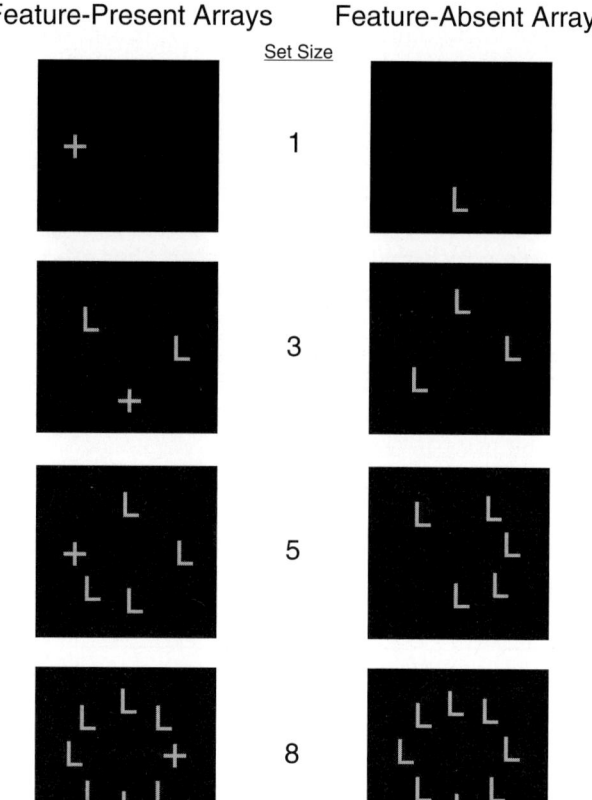

FIGURE 34.1 Example of the visual search arrays used by Orprecio and Adler (2003). Shown are target-present and target-absent search arrays with set sizes of 1, 3, 5, and 8. The stimuli in the array shown to infants were actually red in color. On those trials when a target was present it could randomly occur in either the 3, 6, 9, or 12 o'clock location, and the distracters randomly occurred in any of the remaining locations.

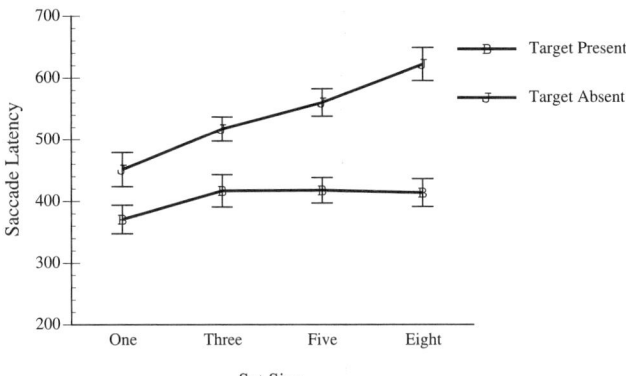

FIGURE 34.2 Infants' saccade latencies to visual search arrays when the target (+) was present among distracters (L's) and when it was absent as a function of the array set size. Results indicate popout and a parallel search function when the target was present and a serial search function when the target was absent.

increased from 1 to 3, 5, and 8 items, whereas their saccade latencies increased linearly in the target-absent conditions as set size increased (see Fig. 34.2). These results are identical to what is typically found in studies of visual search in adults (Treisman and Gelade, 1980) (see Chapter 65). Moreover, they indicate that infants as young as 3 months of age do exhibit "popout" on a millisecond scale, that it is unaffected by the number of distracters, and that it is likely due to a functioning parallel preattentive processing mechanism.

V. CONJUNCTION SEARCH IN INFANTS

The two-stage models of early perceptual processing (Julesz, 1981; Treisman and Gelade, 1980) posit that when an item is not defined by a single unique feature but is defined by the conjunction of two or more features, its detection does not occur via the parallel preattentive mechanism (first stage) and, consequently, it does not pop out. Instead the item defined by a conjunction of features is detected by an effortful serial search conducted by the attentive mechanism (second stage). As a result, detection time increases as the number of distracter items increases (Treisman and Gelade, 1980). In early infancy, in contrast to feature search and popout, there have been few studies investigating conjunction search.

In one study, Bhatt et al. (1999) used the novelty preference paradigm to compare the ability of 5.5-month-old infants to detect a discrepant texture (red X's) embedded in a surrounding texture (blue X's and green O's) when they differed by a single unique feature (red) versus their ability to detect a discrepant texture (blue O's among blue X's and green O's) when it was defined by a conjunction of features (blue and O). Bhatt et al. familiarized infants to texture arrays consisting solely of distracter items and tested infants with the familiarized texture paired with a texture array that contained the discrepant embedded texture. They found that infants exhibited a novelty preference when the embedded texture was defined by a single unique feature but not when it was defined by a conjunction of features. This finding suggests that 5.5-month-old infants exhibit preattentive popout but are unable to conduct an attentive serial search. Another study by Gerhardstein and Rovee-Collier (2002), which measured 1- to 3-year-olds' reaction time to touch the location of the target item on a computer screen, found that these older infants and children were able to detect the target when it was defined by a conjunction of features, and their reaction time as a function of the number of distracters was consistent with a serial attentive search. Together, these studies indicate that, unlike preattentive processing and popout, which are exhibited as early as 3 months of age, attentive processing and serial search may not be available until the end of the first year.

VI. DEVELOPMENTAL ASPECTS OF THE NEUROBIOLOGY OF PREATTENTIVE AND ATTENTIVE MECHANISMS

Together, the infant studies indicate that preattentive mechanisms and popout are present in young infants, whereas attentive mechanisms and serial search may come online near the end of the first year. This provides developmental support for the hypothesis that preattentive and attentive mechanisms are distinct processing modules. If they are distinct mechanisms, perhaps they are also distinct neurally, in which case the developmental trends for the exhibition of preattentive and attentive processing in the visual search situation should mirror developmental trends in the maturation of neural mechanisms.

Johnson (1993) has proposed a theory of neural development, particulary as it relates to visual attention, in which attentional processing and behavior in the first few months of life is controlled primarily by the subcortical superior colliculus pathways, but around 3 months of age the cortical pathway through the frontal eye fields (FEFs) comes online. Considering that substantive research has established that neurons in FEFs show activity during presentation of a popout search array that discriminates the popout target from the distracters even in the absence of eye movements (Thompson et al., 1997) (see Chapters 21 and 22) and that the areas of the visual cortex from which it receives projections process the primitive perceptual units (Deco et al., 2002) (see Chapter 93), it is consistent that popout, which relies on the processing of these primitive perceptual units, would be evident at 3 months of age.

Johnson (1993) further proposes that around 6 months of age, there is rapid development of the parietal cortex, which has been hypothesized to be involved in the ability to disengage from stimuli, a capacity that would seem to be critical to serially search through the stimuli in a visual search array (see Chapters 7 and 24). Consequently, the ability to serially and attentively search an array may be significantly hampered until the parietal cortex is more functionally mature at 6 months, perhaps accounting for why 5.5-month-olds were not able to search for a conjunction target but 1- to 3-year-olds could.

VII. CONCLUSIONS

A key factor in efficiently processing information from our world is the ability to select relevant information while filtering out irrelevant information. Selection and filtering of information are even more critical for the infant who is first constructing his or her knowledge base about the world. One set of mechanisms that has been proposed to accomplish this attentional selection is preattentive versus attentive mechanisms. These mechanisms are manifested behaviorally in adults to visual search arrays as popout versus serial search. In this chapter we reviewed developmental research that indicated that popout is exhibited as early as 3 months of age, whereas serial search is not exhibited before approximately 6 months of age. This developmental trend is consistent with theories of neural development in which the FEF pathway, which has been hypothesized to be involved in popout, is functionally mature by 3 months of age, and the parietal cortex, which is involved in attentional disengagement and therefore may be involved in serial search, is functionally mature around 6 months of age. Regardless, the primitive preattentive mechanism for automatic attentional selection of unique stimuli in the visual world is available very early in infancy, whereas the more intentional attentive mechanism is not available until the second half of the first year.

References

Adler, S. A., Inslicht, S., Rovee-Collier, C., and Gerhardstein, P. C. (1998). Perceptual asymmetry and memory retrieval in 3-month-old infants. *Infant Behav. Dev.* **21**, 253–272.

Atkinson, J., and Braddick, O. (1992). Visual segmentation of oriented textures by infants. *Behav. Brain Res.* **49**, 123–131.

Bhatt, R. S., Bertin, E., and Gilbert, J. (1999). Discrepancy detection and developmental changes in attentional engagement in infancy. *Infant Behav. Dev.* **22**, 197–219.

Colombo, J., Ryther, J. S., Frick, J. E., and Gifford, J. J. (1995). Visual popout in infants: evidence for preattentive search in 3- and 4-month-olds. *Psychon. Bull. Rev.* **2**, 266–268.

Deco, G., Pollatos, O., and Zihl, J. (2002). The time course of selective visual attention: theory and experiments. *Vision Res.* **42**, 2925–2945.

Gerhardstein, P., and Rovee-Collier, C. (2002). The development of visual search in infants and very young children. *J. Exp. Child Psychol.* **81**, 194–215.

James, W. (1890). "Principles of psychology," Vol. 1. Henry Holt, New York.

Johnson, M. H. (1993). The development of visual attention: a cognitive neuroscience perspective. *In* "Brain Development and Cognition: A Reader" (M. H. Johnson, Y. Munakata, and R. O. Gilmore, Eds.), pp. 134–150. Blackwell, Oxford.

Julesz, B. (1981). A theory of preattentive texture discrimination based on first-order statistics of textons. *Biol. Cybernet.* **41**, 131–138.

Orprecio, J., and Adler, S. A. (2003). Visual popout in infancy: effects of set-size on the latency of their eye movements. *J. Vision* **3**, 725a.

Rovee-Collier, C., Hankins, E., and Bhatt, R. (1992). Textons, visual pop-out effects, and object recognition in infancy. *J. Exp. Psychol. Gen.* **121**, 435–445.

Salapatek, P. (1975). Pattern perception in early infancy. *In* "Infant Perception: From Sensation to Cognition" (L. B. Cohen and P. Salapatek, Eds.), Vol. 1, pp. 133–248. Academic Press, New York.

Sireteanu, R., and Rieth, C. (1992). Texture segregation in infants and children. *Behav. Brain Res.* **49**, 133–139.

Thompson, K. G., Bichot, N. P., and Schall, J. D. (1997). Dissociation of target selection from saccade planning in macaque frontal eye field. *J. Neurophysiol.* **77**, 1046–1050.

Treisman, A., and Gelade, G. (1980). A feature-integration theory of attention. *Cogn. Psychol.* **12**, 97–106.

CHAPTER

35

Attention in Conditioning

Peter Dayan

ABSTRACT

We consider two ideal learner models of some simple paradigms in animal conditioning. The models exhibit phenomena of selective attention, preferring to allocate learning to some stimuli over others according to their relative uncertainties, or preferring to believe the predictions made by some stimuli over others according to their relative reliabilities. These characteristics are statistically normative rather than being the consequence of competition for any restricted resource. We also describe some recent evidence as to the neural substrate of competition for learning; competition for making predictions is less well understood.

I. INTRODUCTION

Various paradigms in animal conditioning (Dickinson, 1980), and most influentially that of blocking (Kamin, 1969), suggest that learning is not always accorded equally when many simultaneously presented neutral stimuli (called conditioned stimuli, CSs, such as lights and tones) could potentially be linked with an outcome such as a reward or punishment (called the unconditioned stimulus, US, such as food, water, or electrical stimulation). Theories eagerly embraced ideas of selective attention when faced with this evidence. Perhaps, the notion went, the psychological processor responsible for plasticity might have only a limited capacity, forcing CSs and USs to compete for access and thus the allocation of learning. Prominent suggestions (not all of which are strongly tied to the notion of limited capacity) were that USs that are *already* well predicted by CSs that are present should be denied learning (Rescorla and Wagner, 1972) and/or that CSs that either are not *predictive* (Mackintosh, 1975) or whose predictions are already completely *certain* (Pearce and Hall, 1980) should not have their associations adapt. Grossberg (1982; see Chapter 107) advocated the idea that CSs (and USs) are in constant competition with one another, both to make predictions and for learning, supported by the strength and nature of their existing predictions. Sophisticated behavioral neuroscience methods have been used to test the psychological and neural veracity of these models (see Holland, 1997; Holland and Gallagher, 1999), leading to a rich understanding of the role of various nuclei of the amygdala and neuromodulatory systems (particularly acetylcholine).

Although one route to competition and selection indeed lies in the existence of a limited, shared resource, there are other reasons why stimuli might compete. Indeed, since most conditioning experiments involve at most a handful of stimuli, it is strange to think of a learning mechanism so handicapped that it could not consider them all simultaneously. Rather, as in other theoretical treatments of selective attention (see also Chapter 4), it is important to consider the *statistical* basis of competition. Perhaps less learning is accorded some stimuli because it is statistically *normative* for this to be so; that is, in accordance with an ideal learner (the plastic counterpart of an ideal observer).

In this contribution, we review recent theoretical approaches along these lines (Dayan, Kakade and Montague, 2000). Two nearly opposite forms of selection are identified. One, which arises from a standard statistical learning device called a Kalman filter (Anderson and Moore, 1979), allocates learning to stimuli in proportion to relative uncertainty. The other considers the task of making predictions in the light of stimuli with different reliabilities. In neither case is there a limited capacity processor—rather, the roots of competition lie in statistical optimality. There are inferential counterparts to the effects on learning (Dayan and Yu, 2003), but they are not critical for modeling

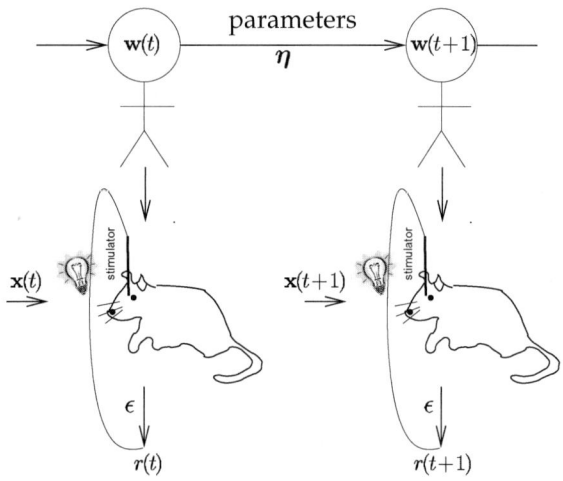

FIGURE 35.1. Statistical treatment of the ideal learner. On trial t, the ideal learner (here shown as a rat) experiences a relationship $\mathbf{w}(t)$ between conditioned stimuli $\mathbf{x}(t)$ (often lights or tones) and an unconditioned stimulus $r(t)$ (here, electrical stimulation). This relationship may be noisy, governed by ε. According to the learner's model, the experimenter's (stick figure) choice of parameters $\mathbf{w}(t)$ changes over time in an orderly, though possibly random, way (governed by η). The learner has to maintain an estimate of $\mathbf{w}(t)$. After Sutton (1992).

effects in conditioning and we do not discuss them here.

Normative statistical models of learning are based on the two simple ideas of ignorance and information. The learner (i.e., the subject) is assumed to be at least partially ignorant about some aspect of the world, but to receive information about it from observations or experiments. Figure 35.1 illustrates this for conditioning (Sutton, 1992). Here, the subject faces repeated experiments in an environment (the Skinner box). The experimenter (the stick figure) controls the relationship between CSs (quantified as $\mathbf{x}(t) = \{x_1(t), x_2(t), \ldots\}$, one component for each CS) and the US (quantified as $r(t)$) according to a set of parameters $\mathbf{w}(t)$, which may change over time (as in reversals, say). The learner is somewhat or wholly ignorant of the values of the parameters. Individual experiments provide the learner with partial and noisy information about $\mathbf{w}(t)$, and its task is to make inferences or predictions about $\mathbf{w}(t)$ on the basis of the data provided. Different ways of parameterizing the relationship between the CSs and the US lead to different models for learning; indeed, the two aspects of selective attention that we consider employ two different parameterizations.

In Section II, we consider selection in the context of learning; in Section III, we consider selection in the context of prediction.

II. ATTENTION IN LEARNING

In order to turn the qualitative model of Fig. 35.1 into a quantitative model of learning, it is necessary to specify how $\mathbf{w}(t)$ parameterizes the relationship between the CSs $\mathbf{x}(t)$ and the reward $r(t)$, and also how the learner might imagine the parameters to change over time (based on η). Of course, just as with a standard ideal observer model, we are not assuming that these parameters or relationships can be found explicitly in the learner; merely that they license appropriate inferences.

The Kalman filter offers just such a specification. According to this, the unconditioned stimulus on trial t is noisily, but linearly, related to the CSs (for succinctness, subsequent references to (t) are dropped unless otherwise noted):

$$r = x \cdot \mathbf{w} + \varepsilon = \sum_i x_i w_i + \varepsilon \qquad (1)$$

where ε is a Gaussian random variable, with mean 0 and variance τ^2, and $x_i(t) \in \{0, 1\}$ indicates whether or not stimulus i is present on trial t. That is, the net US should be the sum of the prediction *weights* associated with all the stimuli that are present on the trial, barring random factors. Of course, if there are many stimuli, then each experiment only provides one piece of information about a whole collection of weights, and therefore data must be accumulated over multiple trials to form estimates of $\mathbf{w}$. Noise reinforces this problem.

According to the simplest version of the Kalman filter, the weights themselves are assumed to drift slowly over time in a random way (this is formally called a random walk):

$$\mathbf{w}(t+1) = \mathbf{w}(t) + \eta(t) \qquad (2)$$

where $\eta(t)$ is a Gaussian random variable with mean 0 and covariance matrix $\mathcal{T}$. Typically, we assume that there are no particular correlations in the way the weights change, so $\mathcal{T} = \mathcal{I}$ is just the identity matrix. That is, we assume that the experimenter changes the weights slowly, stochastically, and independently. It could also be, for instance, that the change in w_i is coupled to the presence of the associated CS on a trial, so that the uncertainty about stimuli that are not presented does not grow without bound.

Let us assume that the subject has learned about τ^2 and $\mathcal{T}$ from past experience, so it is only ignorant about $\mathbf{w}$. Formally, its exact ignorance should be expressed in terms of a probability distribution $\mathcal{P}[\mathbf{w}(t)|\mathcal{D}]$, where $\mathcal{D}$ contains all its observations. Under the many simplifications described above, it turns out that this distribution is itself Gaussian, characterized exactly by a mean $\overline{\mathbf{w}}(t)$ and a covariance matrix $\Sigma(t)$. Based on the new observation $(\mathbf{x}(t), r(t))$ (and remembering our convention about dropping (t) where not confusing), it should first calculate the prediction error, based on the previous predictions:

$$\delta = r - \mathbf{x} \cdot \overline{\mathbf{w}} \qquad (3)$$

and then update the mean and covariance according to:

$$\overline{\mathbf{w}}(t+1) = \overline{\mathbf{w}} + \frac{\Sigma \mathbf{x}}{\mathbf{x}^T \Sigma \mathbf{x} + \tau^2} \delta \qquad (4)$$

and

$$\Sigma(t+1) = \Sigma + \mathcal{T} - \frac{\Sigma \mathbf{x} \mathbf{x}^T \Sigma}{\mathbf{x}^T \Sigma \mathbf{x} + \tau^2} \qquad (5)$$

The update for the mean is particularly easy to understand if we make the approximation of ignoring the correlations between the weights (Sutton, 1992). This means that we write Σ as a diagonal matrix, with entries σ_i^2, which indicate exactly how uncertain the subject is about w_i, given all the experiments that have been observed. In this case, Eq. (4) reduces to

$$\overline{w}_i(t+1) = \overline{w}_i + \frac{\sigma_i^2}{\Sigma_j \sigma_j^2 + \tau^2} \delta \qquad (6)$$

where we assume that the ith CS is provided on trial t, ($x_i = 1$), and the sum over j is over all the CSs that are presented. The rule thus says that the prediction error δ is shared among all the CSs that are present in proportion to their uncertainties (σ_i^2), taking account of the noise τ^2. The point about τ^2 is that if each observation is noisy (τ^2 is large), so there is only a weak relationship between the reward that is delivered and the CSs, then large prediction errors will be the norm and should not lead to substantial changes in any weights. On the other hand, if observations are not noisy (τ^2 is small), then any prediction error should be accounted for and thus the stimulus about which the subject is most uncertain should change its prediction the most.

If we consider $\sigma_i^2 / (\Sigma_j \sigma_j^2 + \tau^2)$ as the *associability* of stimulus i, then learning rule (6) has a quite similar effect to the suggestion of Pearce and Hall (1980) that associabilities of stimuli should be determined by the unpredictability or uncertainty of their consequences. This is, of course, exactly the interpretation of σ_i^2. What it adds, apart from its statistical normativity, is how associabilities compete, which is the link to selective attention and also the effect of the observational noise τ^2. The rule differs from Pearce and Hall's (1980) suggestion in the form of δ in Eq. (3). The net prediction of the reward comes from adding up contributions from each CS that is present ($\mathbf{x} \cdot \overline{\mathbf{w}}$), and so the prediction error associated with stimulus i depends directly on the weights $\overline{w}_j$ of the other stimuli x_j that are

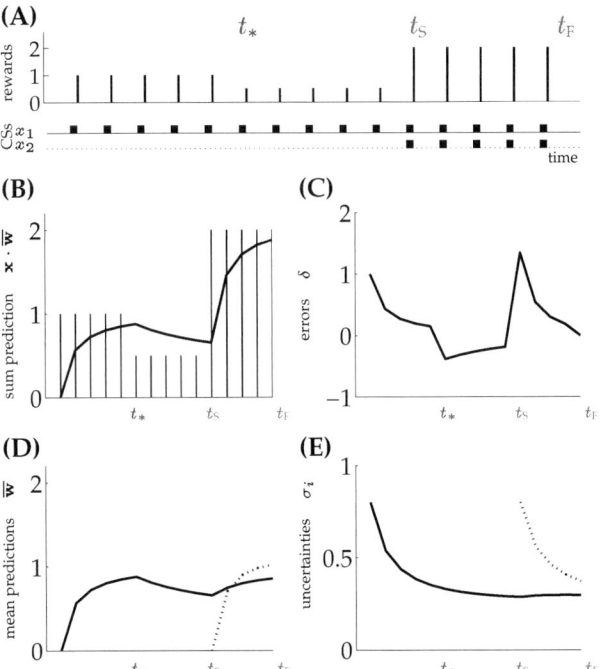

FIGURE 35.2. Associability and learning using the approximate form of Eq. (6) and assuming that the uncertainty σ_2 does not grow when CS_2 is not presented. (A) A set of 15 learning trials, with two CSs (CS_2 shown by the dotted line) and rewards varying between 0.5 and 2 units. (B) Net prediction made by the Kalman filter model (curve), overlaid by the rewards (sticks). Learning is slow between t_* and t_s because the associability of CS_1 is low even though the reward has changed. (C) Prediction error δ over the learning trials. (D) Mean predictions $\overline{w}_1$ (solid) and $\overline{w}_2$ (dotted) over trials. At t_s, more learning is accorded to CS_2, because its predictions are more uncertain ($\sigma_2 > \sigma_1$). (E) Uncertainties σ_1 (solid) and σ_2 (dotted) over trials. Modified from Dayan, Kakade, and Montague (2000).

present. In this respect, the learning rule is a form of Rescorla and Wagner (1972) or delta rule. Thus, it combines in a precise way learning based on prediction error and associability.

Figure 35.2 illustrates this proposal. Figure 35.2A shows a set of 15 learning trials, with a reward r varying between $\frac{1}{2}$ and 2 and with two potentially predictive CSs, one being shown on every trial and the second only introduced when the reward goes to $r = 2$ (on trial t_s). Figure 35.2B shows the total prediction $\mathbf{x} \cdot \overline{\mathbf{w}}$ made by the Kalman filter model over trials (continuous curve), with the actual rewards (sticks) overlaid. Learning is rapid up to t_*, when the reward changes without a change in the stimuli, and is then rapid again at t_s. Figure 35.2C indicates the prediction errors δ underlying the learning; the change in circumstance at t_* and t_s are clearly evident. Figure 35.2D and E shows what underlies the predictions, with solid and dotted lines showing means (D; $\overline{w}_i$) and standard deviations (E; σ_i) for CS_1 and CS_2, respectively. Because the uncertainty about CS_2 is greater than that for CS_1

at t_s, it attracts a far greater share of the learning. In this simple model, uncertainties change monotonically with the number of trials; more sophisticated models exhibit richer behaviors.

Holland, Gallagher, and their colleagues (see Holland, 1997; Holland and Gallagher, 1999) performed a series of experiments in rats, probing the neural substrate of the control over learning exerted by associabilities. The central nucleus of the amygdala, the hippocampus, the parietal cortex, and various aspects of the basal forebrain cholinergic system are all involved in mediating increases and decreases in associability consequent on making animals more or less uncertain about the consequences of CSs. We would therefore suggest that these areas are involved in implementing uncertainties σ_i^2 and their competition as in $\sigma_i^2/(\Sigma_j \sigma_j^2 + \tau^2)$, that is, one component of learning rule (6). Furthermore, these experiments have shown that if all the structures associated with manipulating attention are damaged, then learning still occurs in a way suggestive of a critical role for δ. This argues for the other component of learning rule (6) and is in support of one crucial aspect of the Rescorla-Wagner rule against the standard view of the Pearce-Hall rule. Unfortunately, an experiment entirely akin to that shown in Fig. 35.2, with directly competing uncertainties (on trial t_s and beyond) has not yet been carried out using methods that would elucidate its neural basis.

Compared with the approximate learning rule in Eq. (6), the normative learning rule in Eq. (4) employs a full covariance matrix Σ governing the uncertainty. Dayan and Kakade (2001) suggested that this characteristic is important to model conditioning paradigms such as backwards blocking, which indicate another form of apparently competitive interaction between CSs. In backwards blocking, subjects first learn information indicating the *sum* of the weights for two CSs, say that $w_1 + w_2 = 1$, and only subsequently learn information about one of the weights by itself, say $w_1 = 1$. The consequence of the second stage of learning is that the association of the other stimulus (CS_2) is weakened, a sort of intertemporal competition between the stimuli. This is counterintuitive for regular conditioning models because the association of the second CS is changed even when that stimulus is not presented. In the Kalman filter model, this arises because the first stage of learning establishes $w_1 = w_2 = \frac{1}{2}$, but with a negative correlation between the two CSs, $\Sigma_{12} < 0$ because, if w_1 were a little bigger than $\frac{1}{2}$, then w_2 must be a little less than $\frac{1}{2}$ in order that $w_1 + w_2 = 1$. Given this negative correlation, Eq. (4) arranges that $\overline{w}_2$ decreases during the second set of trials, even though the second CS is not presented.

Of course, all the various effects on learning that we have just been discussing depend on the uncertainties Σ taking appropriate values. The intuition underlying the two leftmost terms of Eq. (5) are that uncertainty about the weights *grows* according to the drift ($\mathcal{T}$) but shrinks according to the observation. The observation specifies information according to how the particular stimuli presented relate to the current set of uncertainties and the overall noise in the observation. This rule is rather different from the one that Pearce and Hall (1980) suggested. Unfortunately, there is rather little psychological or neural evidence about the way the uncertainties actually change.

Of the two main assumptions in the model, Eq. (2) appears far less reasonable than Eq. (1), at least for the bulk of conditioning experiments that involve coordinated, punctated changes in all the parameters (usually between blocks of trials) rather than slow and independent changes. Some studies have focused on modifying this assumption, and, in particular, in considering the role of norepinephrine in reporting coordinated changes (Yu and Dayan, 2003) as a counterpart to acetylcholine's apparent involvement with uncertainties. However, there is little evidence as to the interaction of norepinephrine and acetylcholine in conditioning.

III. ATTENTION IN PREDICTION

The previous section discussed a version of the general statistical model of learning in Fig. 35.1 in which the predictions associated with all the stimuli present are simply summed ($\mathbf{x} \cdot \overline{\mathbf{w}}$). A natural alternative is to consider the different CSs as making competing predictions about the US, just like competing weather forecasters or stock market analysts. Two factors control the weight or significance accorded to a predictor in this competition, uncertainty, as discussed in the previous section, and reliability. Just as some weather forecasters are more reliable than others, some CSs could be more reliable than others and therefore have more attention paid to their predictions. However, on a particular occasion, a normally reliable weather forecaster might have based her forecast on only little information and therefore be uncertain. In this case, her forecast should be comparatively discounted. This is, of course, another form of selective attention.

In terms of the ideal learner, the implication of competitive prediction comes in terms of changing Eq. (1). Two different statistical models have been suggested for this form of attention, one (Kruschke, 2001) motivated by Mackintosh's (1975) suggestion about the competitive allocation of learning (though *not* prediction) and based on the mixture of experts architecture and the other (Dayan and Long, 1998) motivated by a conditioning paradigm called downwards unblocking (which invalidates standard versions of learning rules based either on associabilities or on prediction errors) and grounded in Bayesian statistical ideas about combining the predictions of multiple experts of different reliabilities. In either version, what results is a predictive distribution over the reward whose mean is of the form

$$\hat{r} = \sum_i \pi_i \overline{w}_i \quad (7)$$

where the sum is over all the stimuli that are present on a trial, and the π_i are the result of competition between those stimuli. The π_i evidently instantiate a form of selective attention. In a good approximation to the reliability model, we can define

$$\rho_i = \frac{1}{\gamma_i^2 + \sigma_i^2} \quad (8)$$

whereupon the competition terms are

$$\pi_i = \frac{\rho_i x_i}{\sum_j \rho_j x_j} \quad (9)$$

again just for those stimuli that are present. Here the γ_i^2 terms indicate how far τ might be expected to stray from w_i, and, if $\sigma_i^2 = 0$, then ρ_i is called the precision of the estimate w_i. The opposite relationship between this form of selective attention and the selective attention for learning is clear in the inverse way that σ_i^2 enters Eq. (6) compared with Eq. (8) via (9).

Although competitive combination is favored by learning paradigms such as downwards unblocking, other experiments on more general forms of inhibitory learning suggest that additive combination as in Eq. (1) is also important. Furthermore, few studies have probed the neural basis of this form of selective attention.

Grossberg's (1982) model also involves competition in making predictions. An important aspect of his model is that stimuli that make large predictions of USs are favored in the competition. By contrast in the model of Eq. (9), *what* is predicted and the *reliability* with which it is predicted are two independent quantities. It is not yet experimentally clear whether animals suffer from the inferential fallacies that would result from the former view.

IV. DISCUSSION

We have considered two aspects of selective attention in an ideal learner model of animal conditioning. In the first (Eq. 6), *learning* is competitively allocated among the stimuli that are present in proportion to the relative degree to which the learner is uncertain about their associations. In the second (Eqs. 7, 8, and 9), *predictions* are made on the basis of the relative reliabilities of the stimuli. Even in this simple model, reliability is separately parameterized from uncertainty; however, uncertain predictors are also discounted. Most important, in neither case does selective attention arise from anything akin to a melancholy consequence of a resource limitation. Rather, aspects of an underlying model make it statistically optimal to downweight stimuli.

The actual instantiations of the general framework of Fig. 35.1 are woefully simple and have yet to be normatively combined. Nevertheless, there are strong hints of a close relationsbip with behavioral psychological data on the involvement of cholinergic (and potentially noradrenergic) systems in attentional modulation and fair detail as to various of the neural pathways involved.

Acknowledgments

I am very grateful to Sham Kakade, Read Montague, Rich Sutton, Theresa Long, and Angela Yu, my collaborators on parts of this work. I am also grateful to the Centre for Theoretical Biology, Peking University, for their generous hospitality Funding was from the Gatsby Charitable Foundation.

References

Anderson, B. D. O., and Moore, J. R. (1979). "Optimal Filtering." Prentice-Hall, Englewood Cliffs, NJ.

Dayan, P., and Kakade, S. (2001). Explaining away in weight space. In "Advances in Neural Information Processing Systems 12" (T. K. Leen, T. G. Dietterich, and V. Tresp, Eds.), MIT Press, Cambridge MA. pp. 451–457.

Dayan, P., Kakade, S., and Montague, P. R. (2000). Learning and selective attention. *Nat. Neurosci.* **3**, 1218–1223.

Dayan, P., and Long, T. (1998). Statistical models of conditioning. In "Advances in Neural Information Processing Systems 10" (M. I. Jordan, M. Kearns, and S. A. Solla, Eds.), MIT Press, Cambridge, MA. pp. 117–123.

Dayan, P., and Yu, A. J. (2003). Uncertainty and learning. *IETE J. Res.* **49**, 171–182.

Dickinson, A. (1980). "Contemporary Animal Learning Theory." Cambridge University Press, Cambridge, UK.

Grossberg, S. (1982). Processing of expected and unexpected events during conditioning and attention: A psychophysiological theory. *Psychol. Rev.* **89**, 529–572.

Holland, P. C. (1997). Brain mechanisms for changes in processing of conditioned stimuli in Pavlovian conditioning: Implications for behavior theory. *Anim. Learning Behav.* **25**, 373–399.

Holland, P. C., and Gallagher, M. (1999). Amygdala circuitry in attentional and representational processes. *Trends Cogn. Sci.* **3**, 65–73.

Kamin, L. J. (1969). Selective association and conditioning. *In* "Fundamental Issues in Associative Learning" (N. J. Mackintosh and W. K. Honig, Eds.), pp. 42–64. Dalhousie University Press, Halifax, Canada.

Kruschke, J. K. (2001). Toward a unified model of attention in associative learning. *J. Math. Psychol.* **45**, 812–863.

Mackintosh, N. J. (1975). A theory of attention: Variations in the associability of stimuli with reinforcement. *Psychol. Rev.* **82**, 276–298.

Pearce, J. M., and Hall, G. (1980). A model for Pavlovian learning: Variation in the effectiveness of conditioned but not unconditioned stimuli. *Psychol. Rev.* **87**, 532–552.

Rescorla, R. A., and Wagner, A. R. (1972). In "Classical Conditioning II: Current Research and Theory" (A. H. Black and W. F. Prokasy, Eds.), pp. 64–69. Appleton-Century-Crofts, New York.

Sutton, R. (1992). Gain adaptation beats least squares? *In* "Proceedings of the 7th Yale Workshop on Adaptive and Learning Systems", pp. 161–166. New Haven, CT.

Yu, A. J., and Dayan, P. (2003). Expected and unexpected uncertainty: ACh and NE in the neocortex. *In* "Advances in Neural Information Processing" (S. Becher, S. Throm, U. Chermayer, Eds.), pp. 157–164. MIT Press, Cambridge, MA.

CHAPTER

36

Electrophysiology of Reflexive Attention

Joseph B. Hopfinger

ABSTRACT

Reflexive attention is quickly and effortlessly captured by salient events in the world, in contrast to the slower, effortful processes of voluntary attention. Despite extensive research on the neural mechanisms of voluntary attention, however, comparatively little is known about the neural dynamics of reflexive attentional capture. Recent studies using scalp-recorded event-related brain potentials (ERPs) are providing new insights into the brain mechanisms of reflexive attention. ERP studies have shown that reflexive attention initially enhances neural processing of subsequent stimuli at an early sensory level (as indexed by the P1 ERP component) as well as at a later higher-order stage of processing (as indexed by the P300 component). These effects are transient, however, and are soon replaced by a relative inhibition in sensory processing. Electrophysiological studies are thus providing new information about the dynamic consequences of reflexive attentional capture that is critical for a complete understanding of the neural computations and mechanisms underlying human attention and perception.

I. INTRODUCTION

The ability of attention mechanisms to alter neural processing is a vital mechanism allowing us to effectively concentrate on important stimuli in an environment filled with numerous potentially distracting and irrelevant stimuli. Two distinct types of attention may be engaged in such situations: voluntary and reflexive attention. Voluntary attention provides us with a means to purposefully select those items or spatial locations to which we wish to attend. Although providing needed flexibility, this top-down voluntary control requires effort and cognitive resources and is relatively slow to engage compared to the reflexive attention system. Reflexive attention is a bottom-up system, driven by the physical properties and dynamic salience of items in the world. The reflexive system is quick to engage, but does not typically persist for long, unless voluntary attention mechanisms are subsequently engaged at that location. In fact, reflexive attention acting in isolation quickly dissipates from the attended location and is replaced by a relative inhibition. Specifically, there is a slowing in behavioral responses at the previously attended location, a phenomenon termed inhibition of return (IOR; Posner and Cohen, 1984; see Klein, 2000 for comprehensive review). In addition to acting on different time scales, reflexive and voluntary attention are thought to be distinct mechanisms, controlled by partially different neural systems (Rafal, 1996). In order to fully understand the effects of these attention systems on human perception and action, a complete understanding of both of these systems is required. Although the larger goal is an understanding of how these systems interact to affect our ultimate perceptions of the world, a necessary intermediate step is to understand these systems when working in relative isolation.

Previous research in humans and nonhuman primates has provided compelling evidence that voluntary visual spatial attention enhances cortical processing, starting at relatively early levels in the visual system (e.g., Moran and Desimone, 1985; Mangun, 1995). Much less is known about the neural mechanisms of reflexive attention and its effects on sensory processing. This chapter reviews recent evidence using event-related brain potentials (ERPs) to understand the neural basis of reflexive attention. Given the very different behavioral effects that result depending on the length of time separating the attention-capturing event and subsequent target stimuli (i.e., cue-to-target interstimulus interval, ISI), this chapter is divided into separate sections for the ERP effects seen at short ISIs versus the effects seen at long ISIs.

II. SHORT ISI EFFECTS

A typical paradigm to study reflexive attention involves preceding the relevant target stimulus with a brief peripheral flash (cue) that is not predictive of the upcoming location or type of target. When a short interval (less than 300 ms) separates the cue and target stimuli, responses to the target are typically faster and more accurate at the cued location, despite the fact that participants know the target is just as likely to occur at an uncued location (e.g., Jonides, 1981; Muller and Rabbitt, 1989). Recently, we (Hopfinger and Mangun, 1998) recorded ERPs from human participants engaged in a similar task, in order to investigate the effects that a nonpredictive cue has on the neural processing of subsequent visual stimuli. The cue in our study was the disappearance, and then rapid reappearance, of four dots on one side of a fixation cross (Fig. 36.1). In this experiment, targets could follow the cue at either a short (<250 ms) or long (>550 ms) interval; the results from the long cue-target interval are described later in this chapter. The target was a vertical bar, and participants performed a discrimination task, pressing one button for a tall bar (2.3 deg × 0.69 deg) or a second button for a short bar (1.8 deg × .069 deg).

Our results provide evidence supporting the idea that reflexive attention is able to affect neural activity at multiple levels of processing. The earliest effect of reflexive attention was a significant enhancement of the lateral-occipital, visually evoked P1 component (Fig. 36.2A). Topographic voltage maps across the scalp showed that the distribution of activity over the lateral occipital electrodes was highly similar for cued- and uncued-location targets during the latency range of the P1 with the principal difference being the amplitude, or strength, of the P1 component. This suggests that the effect we observed on the P1 was an enhancement of the neural activity that produces the P1

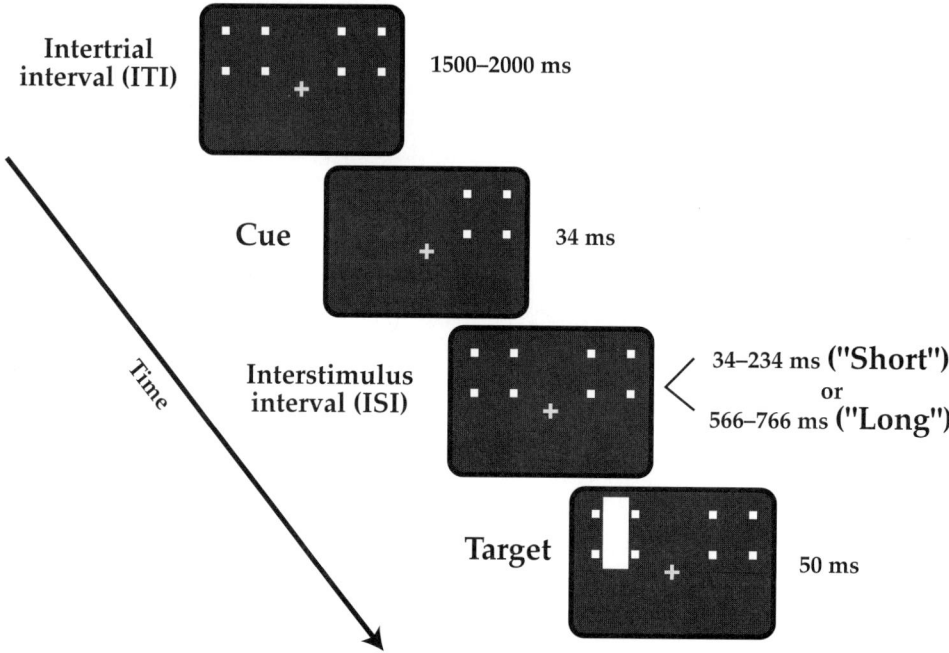

FIGURE 36.1 Example of stimulus display showing a trial with a target occurring at a cued location. The cue was a 34-ms offset and then reappearance of the four dots on one side of fixation (a left cue is shown here). The cue was nonpredictive of target location and target type. The cue-to-target ISI was randomly varied over a short (34- to 234-ms) or long (566- to 766-ms) interval; each interval occurred equally often. The target was a vertical bar presented for 50 ms. Figure adapted with permission from Hopfinger & Mangun (1998), *Psychological Science*, **9**, 441–447. © Blackwell Publishing.

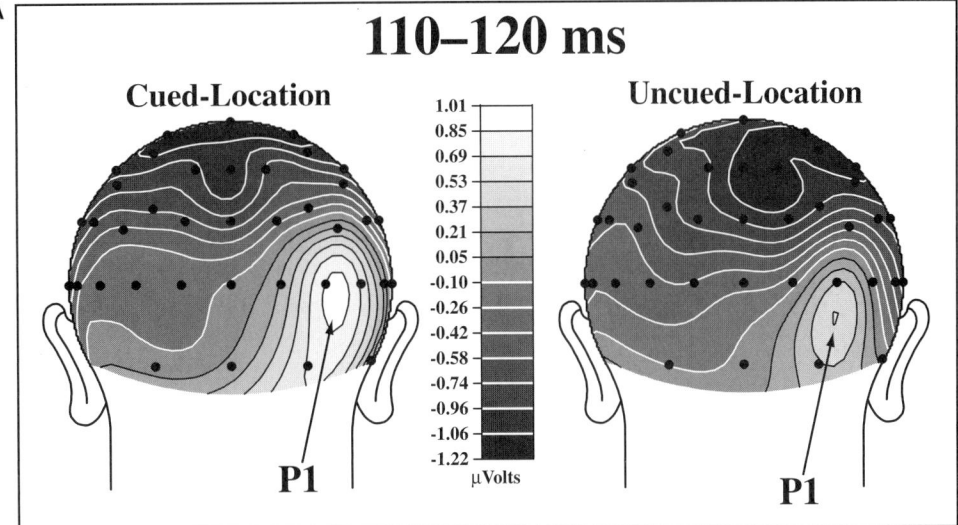
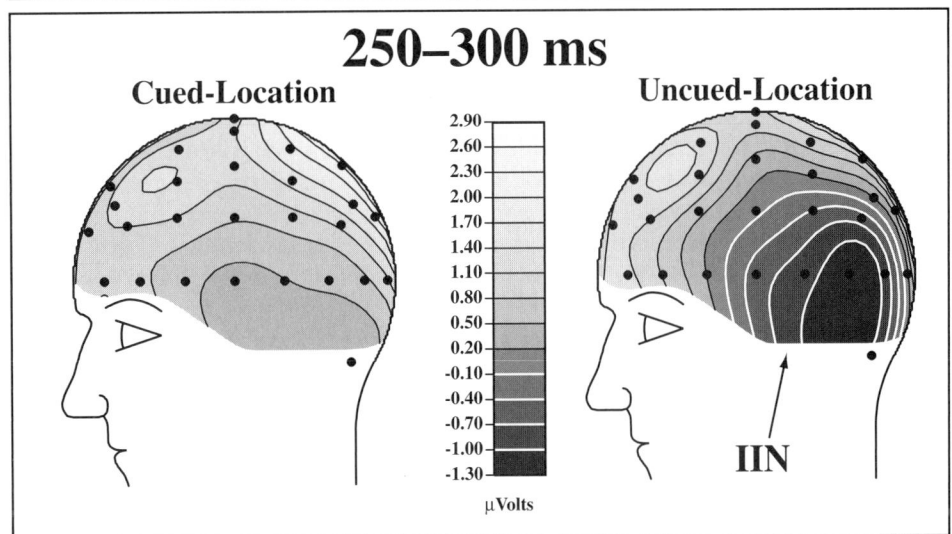
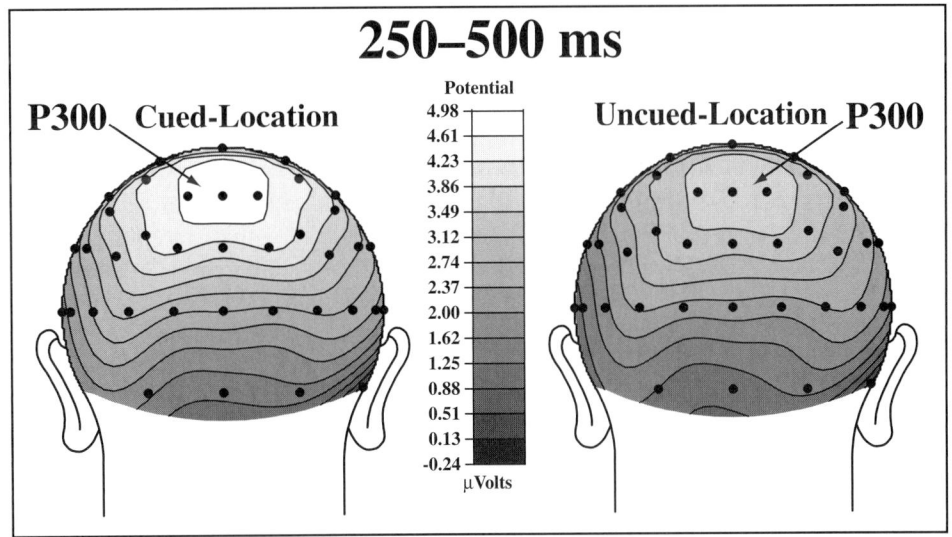

FIGURE 36.2 Short cue-to-target ISI.
Scalp topographic voltage maps for short ISI targets, collapsed over contralateral and ipsilateral scalp sites. The left scalp hemisphere of each map represents the ipsilateral hemisphere (data from the left hemisphere for left visual field targets combined with data from the right hemisphere for right visual field targets), whereas the right scalp hemisphere of each map represents the contralateral hemisphere. The small black dots on each topographic map indicate the location of the electrodes.
A. Voltage maps shown from a back view of the head, for the time period corresponding to the peak of the P1 component (110–120 ms). At these short ISIs, the cued-location targets (left map) produced a significantly enhanced P1 component relative to uncued-location targets (right map).
B. Side view of the head at 250–300 ms latency, showing the ipsilateral activity. The ipsilateral invalid negativity (IIN) refers to the voltage negativity (negative voltages shown as dark gray with white topography lines) that is clearly present only for uncued-location targets.
C. Maps of the time period corresponding to the peak of the P300 component (250–300 ms). The distribution of activity across the scalp is similar for cued versus uncued location targets, but the amplitude of the P300 is significantly larger for cued-location compared to uncued-location targets.

component and was not due to a separate overlapping activity from a different neural source underlying a different process. The P1 component is known to be a visually-evoked sensory ERP component generated in extrastriate visual cortex, and numerous studies have found that this stage of processing can be modulated by voluntary attention (e.g., Heinze et al., 1994; Mangun, 1995; Martinez et al., 1999). Our results showed that reflexive attention mechanisms can modulate visual processing at this same early stage. These results suggest that one aspect of reflexive attention is to act as a sensory gain control, modulating the strength of neural processing at an early sensory level.

In addition to affecting early sensory processing, reflexive attention also enhances neural processing at later stages. Specifically, we found that cued location targets generated an enhanced P300 component compared to uncued location targets (Fig. 36.2C). The P300 (posterior P3b) is an ERP component thought to index context updating and stimulus relevance or importance (Johnson, 1998). Our results suggest that reflexive attention mechanisms work not only at early sensory levels to increase the gain of sensory processing, but also act at higher order stages of processing, briefly tagging those locations as being potentially more important.

A. Attentional Load and Reflexive Attention Effects

Studies of voluntary attention have shown that task and perceptual difficulty may determine the stage of neural processing at which voluntary attention acts (e.g., Handy and Mangun, 2000). Our initial experiment just described used a challenging discrimination task, raising the possibility that our observation of attention effects at multiple stages of processing may have been due in part to the high attentional load of the task and not simply an automatic consequence of sensory-driven reflexive attentional capture. Therefore, in a second study (Hopfinger and Mangun, 2001), we asked participants to perform a simple detection task, using the identical stimuli and task parameters as in the original study. If the enhancements of cortical processing we found in our initial study were due solely to reflexive attention automatically triggered by the abrupt onset cues, then the easier task should not change the pattern of results. On the other hand, if task difficulty were to modulate the effects we found earlier, it would suggest that the effects we previously attributed to simple reflexive processing were actually due to an interaction of reflexive attention and general top-down expectancies concerning the current task goals.

The results of this second study revealed that the effects were independent of the difficulty of the task being performed. Specifically, the P1 and P300 were significantly enhanced for cued- versus uncued-location targets in the simple detection task (Hopfinger and Mangun, 2001), just as they were in the difficult discrimination task (Hopfinger and Mangun, 1998). Together, the two studies provide evidence that the processing enhancements are indeed involuntary reflexive effects. Therefore, although voluntary attention effects are dependent on attention load (e.g., Handy and Mangun, 2000), reflexive attention effects are quickly engaged by salient physical stimuli regardless of the task being performed.

B. Potential ERP Marker of Disengagement and Reorienting

If attentional resources are captured at the cued location, then an uncued-location target should require a disengagement from the cued location and an extra orienting of attention to the new location of the target stimulus. Cued-location targets would not require any further orienting of attention or disengagement of attention because attention would already be at the cued location. Consistent with this, we found that uncued-location (invalid) targets generated a distinct negativity over ipsilateral posterior scalp sites between 200 and 300 ms that was not present for cued-location targets. We labeled this aspect of the waveform the ipsilateral invalid negativity (IIN), and we found it to be maximal over lateral posterior scalp sites (Fig. 36.2B). Although future electrical-dipole modeling of this component will provide a better indication of the neural generators, the scalp distribution is consistent with a generator at the temporal parietal junction, a region previously implicated in the processes of disengaging and reorienting attention (reviewed in Rafal, 1996). Similar to the P1 and P300 enhancements, the IIN was evoked in both simple detection and difficult discrimination experiments at the short cue-to-target ISIs.

III. LONG ISI EFFECTS

As described in the introduction, a unique and potentially critical aspect of reflexive attention is that at long intervals between an attention-capturing cue and a subsequent target, participants are actually slower and less accurate to respond to targets at the cued location. This IOR effect is also modulated by the nature of the task being performed. Specifically, IOR is commonly found in simple detection tasks, but is less commonly found in discrimination tasks (reviewed in Klein, 2000). IOR has been investigated extensively with behavioral measures, but only recently have physiological studies begun to examine the neural underpinnings of IOR. A major issue in the investigation of IOR has been whether it reflects an attention effect on early sensory processing, a motor-programming effect, or a combination of early and late effects. Because ERPs provide a direct real-time measure of neural processing, this method of investigation can provide critical new data in this debate.

In our initial discrimination experiment (Hopfinger and Mangun, 1998), we found that all the enhancements of processing present at short ISIs were absent at the longer ISI. Even though the long ISI was only slightly (i.e., less than one-half second) longer than the short ISI, there was no evidence at the long ISIs for the generation of the IIN component for either target type (Fig. 36.3B), and the P3 showed no difference in amplitude or latency between cued- and uncued-location targets (Fig. 36.3C). The facilitation of the P1 component was also absent and, furthermore, was replaced by a relative inhibition. Specifically, at the long ISI cued-location targets produced significantly reduced amplitude P1 components compared to uncued-location targets (Fig. 36.3A). This relative inhibition at the cued location is similar to IOR and suggests that IOR may be produced in part by early sensory modulations. In our initial discrimination experiment, however, there was no significant difference in manual response times to the cued- versus uncued-location targets. A 1999 study by McDonald and colleagues, however, found a reduction in P1 amplitude along with the typical pattern of IOR in behavior, suggesting that the early sensory inhibition may be leading to IOR.

In our simple detection task experiment (Hopfinger and Mangun, 2001), we again found that the effects of reflexive attention at short ISIs were no longer present at long ISIs. There was no evidence of an IIN, and the enhancements of the P1 and P300 were no longer present at the long ISI. With this simple detection task, participants were significantly slower to respond to targets at the cued location (at the long ISI) relative to the uncued location—the typical IOR pattern of behavioral responses. Although there was a trend for a reduction of the P1 amplitude at the cued location, this difference did not reach significance. Therefore, across these studies it appears that an inhibitory effect on early perceptual processing may occur as a consequence of reflexive attention, but it may not wholly account for the IOR phenomenon.

IV. THE EFFECTS OF CROSS-MODAL ATTENTIONAL CAPTURE ON CORTICAL VISUAL PROCESSING

In addition to visual stimuli capturing attention, stimuli in other sensory modalities also capture attention reflexively, and physiological studies have begun to investigate the neural basis of this phenomenon. McDonald and Ward (2000) recorded ERPs during a visual detection task in which visual targets were preceded by nonpredictive auditory sounds. When the time interval between the auditory cue and visual target was short (100–300 ms), an enhanced negativity for cued-location targets versus uncued-location targets was observed over the contralateral scalp at a latency of 200–300 ms. This experiment provides evidence that reflexive attention, captured by auditory stimuli, can affect visual sensory processing. However, this cross-modal attentional capture did not modulate sensory processing as early as did the visual attentional capture described previously.

V. CONCLUSION

Mechanisms of reflexive attention can have powerful effects on our perceptions and our ability to act effectively and efficiently on items in our world. Results from electrophysiological studies are shedding new light on the complex and dynamic neural mechanisms of attentional capture. Rapidly engaged reflexive attention mechanisms result in an initial facilitation of early sensory processing, as well as an enhancement of later stages of context updating. These effects are fleeting, however, and are soon replaced by a relative inhibition of sensory processing for cued-location stimuli. Our results suggest that these are relatively automatic processes, engaged in both simple detection and difficult discrimination tasks. Ongoing investigations into the neural architecture of reflexive attention, combined with studies of voluntary attention, promise to significantly enhance our understanding of the attention systems of the human brain and their effects on perception, action, and cognition.

Acknowledgments

This research was supported by funding from the NSF and the NIMH. I thank Marty Woldorff for consultation on portions of the analyses, Ron Mangun and Jeff Maxwell for assistance with the experiments described here, and Geraint Rees and Jennifer Schaaf for very helpful comments on earlier versions of this manuscript.

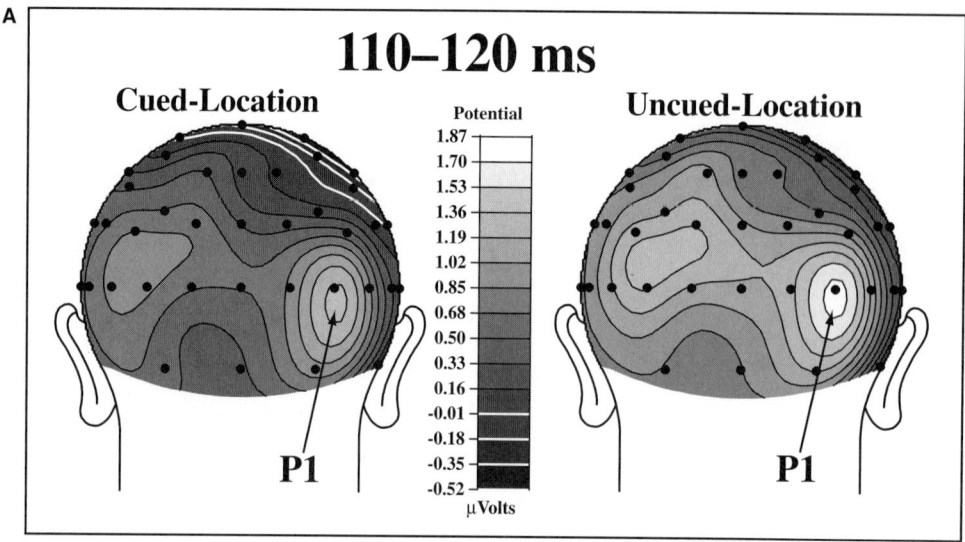
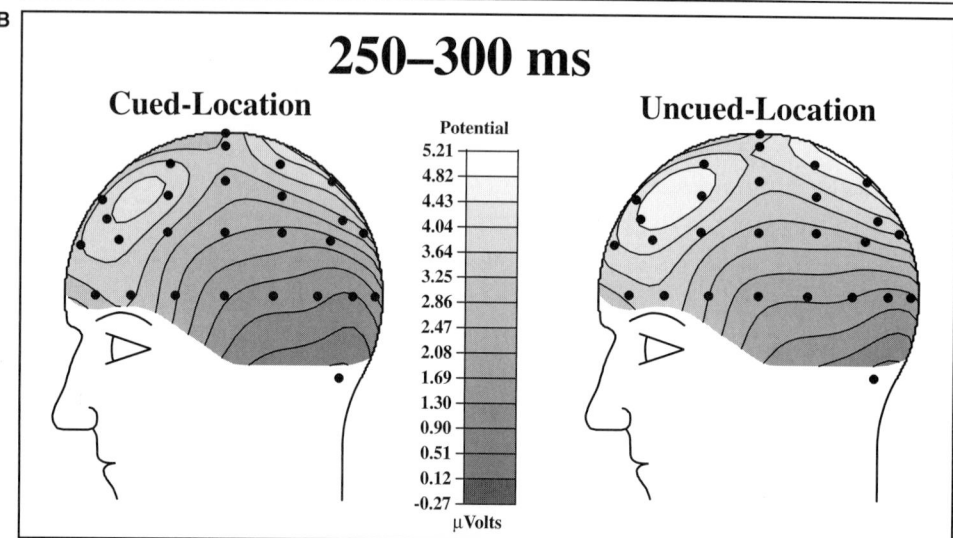
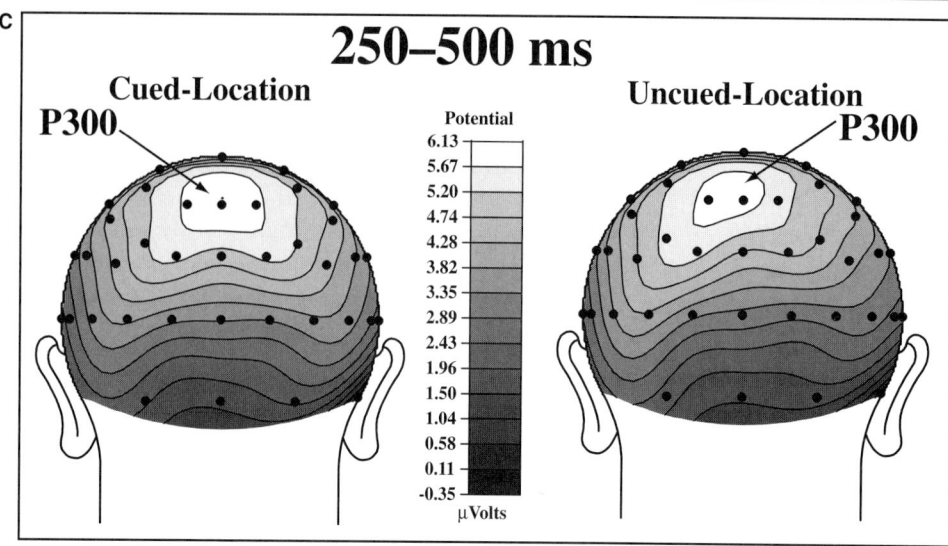

FIGURE 36.3 Same as Figure 36.2, but for targets at the long cue-to-target ISIs.
A. Scalp topographic voltage maps of the time period corresponding to the peak of the P1 component (110–120 ms). At these long ISIs, the cued-location targets (left map) produced a significantly reduced P1 component relative to uncued-location targets (right map). Again, the distribution of activity was very similar, with the main difference being the strength of the P1.
B. Side view of the head, showing ipsilateral activity during the latency range (250–300 ms) at which a clear IIN was found for short ISI targets. No IIN or negativity during this latency range was found over ipsilateral temporal-occipital-parietal regions at these long cue-to-target ISIs for either cued- or uncued-location targets.
C. Topographic voltage maps of the time period corresponding to the peak of the P300 component (250–300 ms). The P300 was not significantly different for cued- versus uncued-location targets at the long ISI. More important, the lack of a difference in the P300, despite a significant inhibition of the P1 component, argues that the P300 amplitude is not simply a function of the strength of early sensory processing.

References

Handy, T. C., and Mangun, G. R. (2000). Attention and spatial selection: Electrophysiological evidence for modulation by perceptual load. *Perception Psychophys.* **62**, 175–186.

Heinze, H. J., Mangun, G. R., Burchert, W., Hinrichs, H., Sholz, M., Munte, T. F., Gos, A., Scherg, M., Johannes, S., Hundeshagen, H., Gazzaniga, M. S., and Hillyard, S. A. (1994). Combined spatial and temporal imaging of brain activity during selective attention in humans. *Nature* **372**, 543–546.

Hopfinger, J. B., and Mangun, G. R. (1998). Reflexive attention modulates processing of visual stimuli in human extrastriate cortex. *Psychol. Sci.* **9**, 441–447.

Hopfinger, J. B., and Mangun, G. R. (2001). Tracking the influence of reflexive attention on sensory and cognitive processing. *Cogn. Affective Behav. Neurosci.* **1**, 56–65.

Johnson, R. (1988). The amplitude of the P300 component of the event-related potential: Review and synthesis. *Adv. Psychophysiol.* **3**, 69–137.

Jonides, J. (1981). Voluntary versus automatic control over the mind's eye movement. *In* "Attention and Performance IX" (J. B. Long and A. D. Baddeley, Eds.), pp. 187–203. Erlbaum, Hillsdale, NJ.

Klein, R. M. (2000). Inhibition of return. *Trends Cogn. Sci.* **4**, 138–147.

Mangun, G. R. (1995). Neural mechanisms of visual selective attention. *Psychophysiol.* **32**, 4–18.

Martinez, A., Anllo-Vento, L., Sereno, M. I., Frank, L. R., Buxton, R. B., Dubowitz, D. J., Wong, E. C., Heinze, H. J., and Hillyard, S. A. (1999). Involvement of striate and extrastriate visual cortical areas in spatial selective attention. *Nature Neurosci.* **2**, 364–369.

McDonald, J. J., and Ward, L. M. (2000). Involuntary listening aids seeing: Evidence from human electrophysiology. *Psychol. Sci.* **11**, 167–171.

McDonald, J. J., Ward, L. M., and Kiehl, K. A. (1999). An event-related brain potential study of inhibition of return. *Perception Psychophys.* **61**, 1411–1423.

Moran, J., and Desimone, R. (1985). Selective attention gates visual processing in the extrastriate cortex. *Science* **229**, 782–784.

Müller, H. J., and Rabbitt, P. M. (1989). Reflexive and voluntary orienting of attention: Time course of activation and resistance to interruption. *J. Exp. Psychol.: Hum. Perception Performance* **15**, 315–330.

Posner, M. I., and Cohen, Y. (1984). Components of visual orienting. *In* "Attention & Performance X" (H. Bouma and D. Bouwhis, Eds.), pp. 531–556. Erlbaum, Hillsdale, NJ.

Rafal, R. (1996). Visual attention: Converging operations from neurology and psychology. *In* "Converging Operations in the Study of Visual Selective Attention" (A. F. Kramer, M. G. H. Coles, and G. D. Logan, Eds.), pp. 139–192. American Psychological Association, Washington, D.C.

CHAPTER 37

Natural Scene Statistics and Salient Visual Features

Christoph Zetzsche

ABSTRACT

The control of attention and of the associated saccadic fixations is to a large degree determined by the image content, that is, by salient local features in the image. The nature of this salient features and of the underlying strategy for their selection can be investigated within the context of natural scene statistics. For this, saccadic eye movements of human observers are recorded for a variety of natural and artificial test images, and the statistical properties of the fixated image regions are analyzed. The second-order statistics indicate that regions with higher spatial variance have a significantly higher probability to be fixated. Local contrast is therefore a salient feature. It was difficult to find additional differences in the local power spectra that can yield unequivocal information about a preference for specific local form properties. Differences that can be more easily interpreted as structural properties are derived from an investigation with higher-order statistics (bispectral density). The results indicate that nonredundant, intrinsically two-dimensional image features such as curved lines and edges, occlusions, isolated spots, and so on, play an important role in the saccadic selection process. Such features cannot be extracted by the classic spatial-frequency selective filter mechanisms but require nonlinear AND-like interactions between frequency components as possibly provided by end-stopping or hypercomplex properties related to the extraclassical receptive field.

I. INTRODUCTION

The analysis of complex objects or scenes by a sequence of rapid saccadic eye movements is a proto- typical example of efficient sensorimotor information processing. In general, this efficiency can be regarded to result from an evolutionary adaptation to the regularities of the terrestrial environment. A well-known example for such an adaptation is the optimized structure of receptive fields (for review see Simoncelli and Olshausen, 2001; Zetzsche and Krieger, 2001). Although adaptation takes place at various structural levels and on different time scales, there is a common underlying principle: the optimized allocation of restricted neural processing resources to environmental data of high complexity. This principle seems also to be the basis of the efficient selection strategy of the saccadic system. The major part of the information-processing machinery is devoted to the small area of the fovea centralis, whereas the remainder, orders of magnitude larger in area, is served by a much coarser representation with limited computational potential. Such a strategy can only work without risking overlooking important information because efficient mechanisms can rapidly redirect the foveal processing resources to the important locations within the visual scene.

Here we investigate this selection process with statistical information theory. Our key hypothesis is that the foveated image regions are selected on the basis of the statistical redundancies of natural images: The least redundant (i.e., most informative) features are more often fixated than the more redundant parts. For our investigations, we have first obtained an empirical database by recording the eye movements of human observers on a variety of artificial and natural images (Sec. IIA). From these data, we derive informations about the selection criteria and about the visual mechanisms that are involved in the computation of the salient image locations. For this, we investigate the differences between the statistical properties of those image regions that were selected by fixations, as opposed to the average statistics of randomly selected regions (Secs. IIB–C). The results of this statistical analysis are then used to derive design criteria for non-linear saliency-detectors in early vision (Sec. III).

II. STATISTICS OF FIXATED REGIONS

A. Methods

Images have been chosen from seven different topical classes (portraits, landscapes, technics, architecture, computer graphics, polygons, and textured polygons) to enable a differentiated analysis and to avoid any bias on the relevant image features. Every class contained seven images; hence, a total number of 49 images has been investigated. All images were grey-level images of size 512×512. Detailed descriptions of the experimental setup and procedures, the statistical measurements, and the simulations can be found in Zetzsche et al. (1998) and Krieger et al. (2000).

From Fig. 37.1 it becomes apparent that the locations of saccadic eye movements are not randomly distributed across natural scenes, although there is a substantial degree of variability in the data. Small regions around the measured fixation points of the subjects were extracted for statistical analysis (cf. Fig. 37.1). For the statistical analysis we used the seven images shown in Fig. 37.1. Several types of statistics were computed for these regions and compared to the average statistics, as obtained from randomly selected regions. Differences in the statistics should indicate which image features are preferred in the selection process.

B. Analysis of Spatial Variance

In a first step, we measured the signal variability σ^2_{eye} of fixated image regions and compared it to the variance σ^2_{rand} of regions, which had been chosen by a random number generator:

$$\frac{\sigma^2_{eye}}{\sigma^2_{rand}} = \frac{\sum_{\vec{x}_{eye}} \sum_{\vec{x} \in W} \left(u(\vec{x}_{eye} - \vec{x}) - \overline{u}\right)^2}{\sum_{\vec{x}_{rand}} \sum_{\vec{x} \in W} (u(\vec{x}_{rand} - \vec{x}) - \overline{u})^2} \quad (1)$$

Here $u(\vec{x})$ is the image intensity at $\vec{x}$, and $\overline{u}$ is the mean intensity value within a local window W, which in our analysis spanned $1/8$ of the total image size. The vectors $\vec{x}_{eye}$ and $\vec{x}_{rand}$ indicate the center of fixated and randomly selected image regions, respectively.

For most images, the regions selected by eye movements had a significantly higher amount of variance. The only exception was the lake scene in Fig. 37.1, where the variance was approximately equal for both the randomly selected and the fixated regions. For the whole set of images, we obtained a variance ratio of $\sigma^2_{eye}/\sigma^2_{rand} = 1.35$ (Zetzsche et al., 1998).

C. Second-Order Statistics

The analysis of spatial variance is only a crude approximation of the statistical dependencies within a

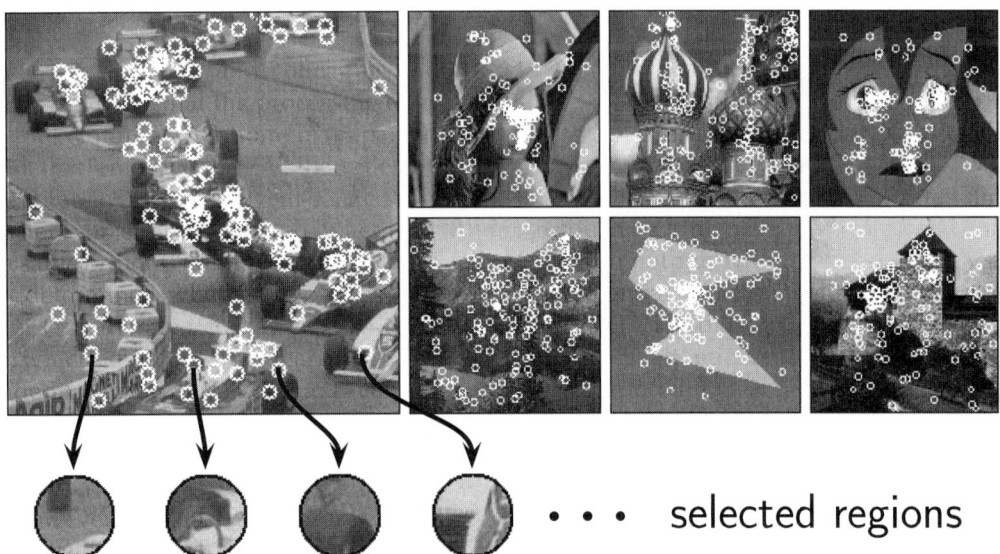

FIGURE 37.1 Test images used in our experiments. The small circles indicate positions of saccadic fixations. (From Krieger et al., 2000)

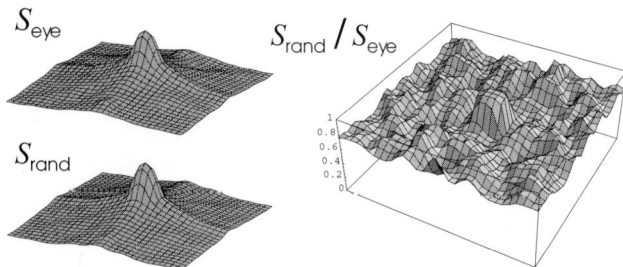

FIGURE 37.2 Power spectra of image regions selected by eye fixations (upper left) and by a random number generator (lower left). The two power spectra are very similar in their basic shape. Their ratio S_{rand}/S_{eye} (right side) is consistently below unity, which is in agreement with our analysis of the variance. In addition, there may exist a small whitening tendency (the ratio may be closer to 1 for low frequencies than for high frequencies). (Adapted from Zetzsche et al., 1998)

local window, and a more detailed characterization may be obtained by the autocorrelation function or its frequency domain counterpart, the power spectrum. In Fig. 37.2, we compare the power spectrum S_{eye} of fixated image regions with the power spectrum S_{rand} of the randomly chosen regions. It is difficult to find systematic differences because the two power spectra have a very similar basic shape. Their ratio S_{rand}/S_{eye} shows considerable fluctuations, but it is consistently smaller than 1, which is in agreement with the higher variance in the fixated regions. In addition, there could exist a systematic tendency that $S_{rand}/S_{eye} \approx 1$ at low frequencies and $S_{rand}/S_{eye} < 1$ at medium and high frequencies. This would imply that for higher spatial frequencies the fixated image regions contribute more power than the randomly chosen regions and would thus be consistent with a small whitening (decorrelation) tendency, in agreement with the slightly steeper decay of the autocorrelation function of the fixated regions reported by other authors (see Discusion).

D. Higher-Order Statistics

One measure for higher-order dependencies is provided by the third-order cumulant c_3^u, which basically calculates the expectation of an image that has been multiplied by two shifted copies of itself. For a zero-mean stationary random process $\{u(\vec{x})\}$, the third-order cumulant c_3^u is defined as:

$$c_3^u(\vec{x}_1, \vec{x}_2) = E[u(\vec{x}) \cdot u(\vec{x} + \vec{x}_1) \cdot u(\vec{x} + \vec{x}_2)] \quad (2)$$

As for autocorrelations, higher-order statistics are often more conveniently investigated in the frequency domain. The Fourier transform of c_3^u is known as the bispectrum C_3^u:

$$C_3^u(\vec{f}_1, \vec{f}_2) = F\{c_3^u(\vec{x}_1, \vec{x}_2)\} \quad (3)$$

Alternatively, the Fourier-Stieltjes representation of $\{u(\vec{x})\}$ offers the possibility to express the bispectrum directly in terms of the frequency components $dU(\vec{f})$ (Nikias and Petropulu, 1993):

$$E[dU(\vec{f}_1) \cdot dU(\vec{f}_2) \cdot dU(\vec{f}_3)]$$
$$= \begin{cases} \int C_3^u(\vec{f}_1, \vec{f}_2) \cdot d\vec{f}_1 d\vec{f}_2 & \text{if } \vec{f}_3 = \vec{f}_1 + \vec{f}_2 \\ 0 & \text{otherwise} \end{cases} \quad (4)$$

From this equation it becomes apparent that the bispectrum measures the statistical dependencies between three frequency components, the sum of which equals zero. A direct computation in the frequency domain can also be derived from this notation (Nikias and Petropulu, 1993).

In Fig. 37.3, we compare the bispectrum $C_3^{U_{eye}}$ ($[f_{x1}, f_{y1}]$, $[f_{x2}, f_{y2}]$) of the fixated image regions to the bispectrum $C_3^{U_{rand}}$($[f_{x1}, f_{y1}]$, $[f_{x2}, f_{y2}]$) of randomly chosen areas (Fig. 37.3A1 vs. 37.3A2). The two bispectra show a clear difference in their basic shape: For the randomly chosen regions we obtain a starlike shape of the bispectrum, which indicates strong statistical dependencies between frequency components aligned to one another with respect to orientation, $\text{atan}(f_{y1}/f_{x1}) \approx \text{atan}(f_{y2}/f_{x2})$. This is consistent with our systematic investigations of the bispectra of a large set of natural images (Krieger et al., 1997). In contrast, the bispectrum of fixated regions exhibits sections with more circular structures, thus indicating a strong simultaneous contribution from frequency components of *different* orientations, $\text{atan}(f_{y1}/f_{x1}) \neq \text{atan}(f_{y2}/f_{x2})$.

III. SALIENCY DETECTORS

A. The Concept of Intrinsic Dimensionality

These statistical observations fit well with the idea of an information-theoretic hierarchy in terms of *intrinsic dimensionality* (Zetzsche and Barth, 1990; Zetzsche et al., 1993). The concept of intrinsic dimensionality relates the degrees of freedom provided by a signal domain to the degrees of freedom actually used by a given signal and thus provides a hierarchy of local image signals in terms of different degrees of redundancy:

- *i0D signals* are constant. That is, $u(x, y) = $ constant within a local window.
- *i1D signals* can locally be approximated by a function of only one variable. That is, $u(x, y) = u(ax + by)$. Examples are straight lines and edges. Sinusoidal gratings, the eigenfunctions of linear systems, are also members of this class.

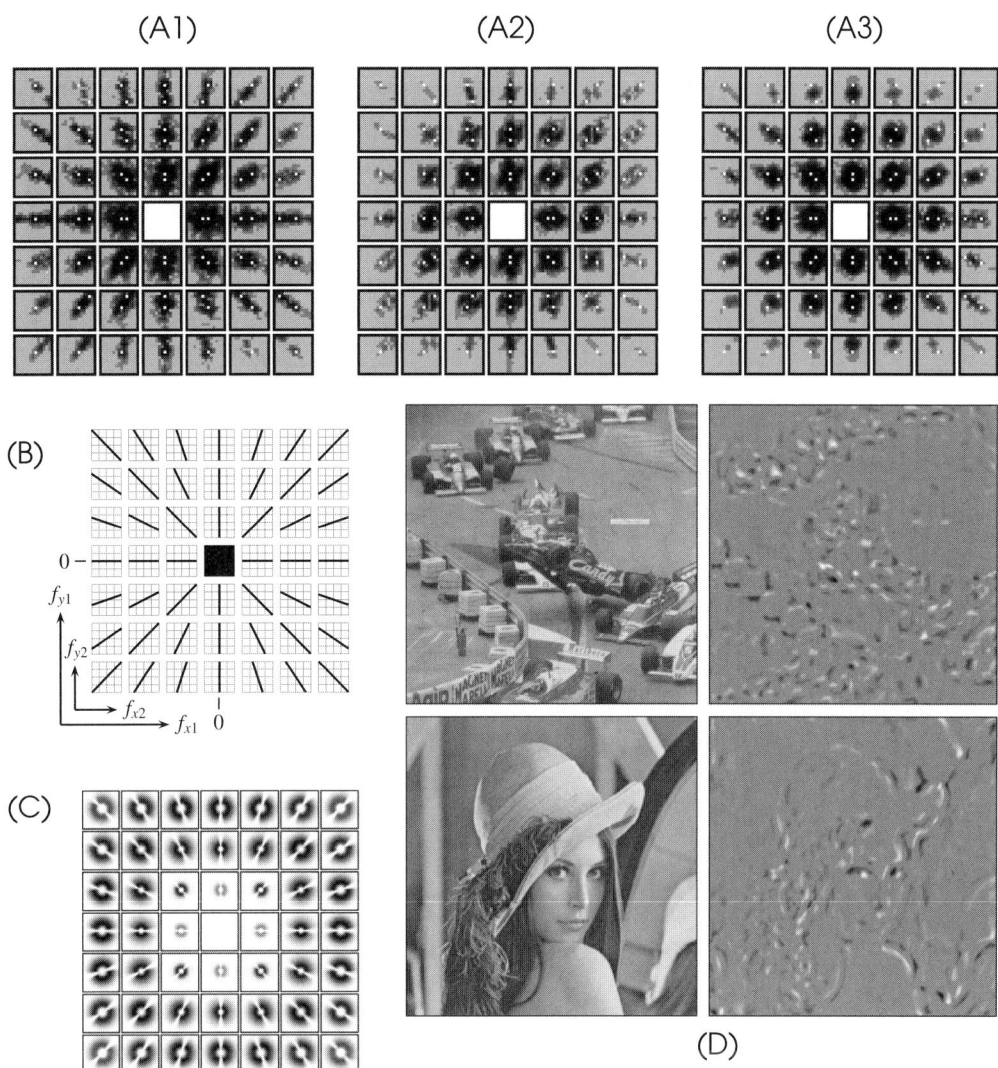

FIGURE 37.3 (A1–A3) Bispectra of image regions selected by different criteria. Shown are several sections of the four-dimensional bispectral magnitude $|C_3^U([f_{x1}, f_{y1}], [f_{x2}, f_{y2}])|$ with $[f_{x1}, f_{y1}]$ = constant. The coordinate axes are indicated in (B). The iso-orientation regions of the four-dimensional bispectrum where the frequency components are aligned in orientation, $\mathrm{atan}(f_{y1}/f_{x1}) \approx \mathrm{atan}(f_{y2}/f_{x2})$, are marked by lines in (B). (A1) $C_3^{U_{\mathrm{rand}}}$-selection by a random number generator. The bispectral energy is concentrated along the iso-orientation lines. (A2) $C_3^{U_{\mathrm{eye}}}$-selection by saccadic eye movements. The bispectral energy of the fixated image regions is distributed more circularly around the void points $[f_{x1}, f_{y1}] = [0, 0]$, $[f_{x2}, f_{y2}] = [0, 0]$, and $[f_{x1} + f_{x2}, f_{y1} + f_{y2}] = [0, 0]$. This indicates that the eye (as opposed to the random number generator) selects more often image regions with strong statistical dependencies between frequency components of *different* orientation. (A3) $C_3^{U_{i2D}}$-selection by an *i*2D operator. Note the similarity to the bispectrum of the fixated image regions shown in (A2). (C) The nonlinear transfer function of an *i*2D-selective operator, which can block *i*1D signals (the iso-orientation lines in (B) are the stopbands) and can enhance cross-orientation, combinations. (D) Features extracted by *i*2D–selective operators. If the selection of local image regions is determined by an *i*2D-selective operator, the bispectrum shown in (A3) results. (Adapted from Zetzsche et al., 1998 and Krieger et al., 2000)

- *i2D signals* are neither *i0D* nor *i1D*. Examples are corners, junctions, curved lines and curved edges. These local signals are characterized by the fact that they contain local frequency components with $\mathrm{atan}(f_{y1}/f_{x1}) \neq \mathrm{atan}(f_{y2}/f_{x2})$.

This hierarchy seems well suited for the description of the selection process: *i*0D signals are least frequently fixated, *i*1D signals are sometimes fixated, and *i*2D signals are the most attractive targets. That *i*0D signals are of low relevance is indicated by the fact that they have a very small variance, whereas our statistical results show that regions with high variance are preferred. Straight edges and other i1D signals have a higher variance, and hence an increased chance of getting fixated, but they are definitely not as relevant as the i2D signals. This is shown by our statistical measurements via bispectra, which demonstrate that signals containing local frequency components with $\mathrm{atan}(f_{y1}/f_{x1}) \neq \mathrm{atan}(f_{y2}/f_{x2})$ are preferred to i1D signals with $\mathrm{atan}(f_{y1}/f_{x1}) \approx \mathrm{atan}(f_{y2}/f_{x2})$. We therefore suggest that i2D signals provide significant information about complex scenes and that i2D-selective detectors are important members of the class of saliency detectors. However, unlike classical filters such i2D-selective detectors are highly nonlinear devices, thus requiring special techniques for their modeling and functional analysis (Zetzsche and Barth, 1990).

B. i2D-Selective Operators

Suitable i2D-selective operators can be developed by a $\Sigma\Pi$-approach (Zetzsche and Barth, 1990) or by the Volterra-Wiener expansion of nonlinear functionals (Krieger and Zetzsche, 1996). The Volterra series relates the input $u_1(\vec{x})$ of a nonlinear shift-invariant system to its output $u_2(\vec{x})$ in the following way (Schetzen, 1989):

$$u_2(\vec{x}) = h_0 + \int h_1(\vec{x}_1) \cdot u_1(\vec{x} - \vec{x}_1) \cdot d\vec{x}_1 \\ + \int h_2(\vec{x}_1, \vec{x}_2) \cdot u_1(\vec{x} - \vec{x}_1) \cdot u_1(\vec{x} - \vec{x}_2) \cdot d\vec{x}_1 d\vec{x}_2 + \cdots \quad (5)$$

The quadratic part of Eq. (5) may be expressed in the frequency domain as

$$U_2(\vec{f}) = \int \tilde{U}_2(\vec{f}_1, \vec{f} - \vec{f}_1) \cdot d\vec{f}_1 \quad (6)$$

where

$$\tilde{U}_2(\vec{f}_1, \vec{f}_2) = H_2(\vec{f}_1, \vec{f}_2) \cdot U_1(\vec{f}_1) \cdot U_1(\vec{f}_2) \quad (7)$$

is the expanded output spectrum and $H_2(\vec{f}_1, \vec{f}_2)$ is the Fourier transform of the second-order Volterra kernel $h_2(\vec{x}_1, \vec{x}_2)$. Note that Eq. (7) may be regarded as the weighting of an AND-like conjunction between frequency components.

A necessary and sufficient condition for a quadratic Volterra operator to be i2D selective (i.e., insensitive to i0D and i1D signals) is given by (Krieger and Zetzsche, 1996):

$$H_2([f_{x1}, f_{y1}], [f_{x2}, f_{y2}]) = 0 \quad \forall f_{x1} \cdot f_{y2} = f_{y1} \cdot f_{x2} \quad (8)$$

Systems constrained by this condition block signals with aligned frequency components $\mathrm{atan}(f_{y1}/f_{x1}) =$ $\mathrm{atan}(f_{y2}/f_{x2})$. These forbidden zones for an i2D-selective Volterra kernel H_2 ($[f_{x1}, f_{y1}]$, $[f_{x2}, f_{y2}]$) are indicated as black lines in Fig. 37.3B. An example for a quadratic system with its symmetric kernel vanishing in the forbidden zones is shown in Fig. 37.3C and the application of an i2D-selective operator to two test images is illustrated in Fig. 37.3D. The regions that are selected by such operators are sparsely distributed in the images, but carry a substantial amount of information about the images.

In order to test our hypothesis that i2D signals are preferentially fixated, we have computed the bispectrum $C_3^{U_{i2D}}$ of the image regions that are selected by a nonlinear i2D-selective saliency detector (Fig. 37.3A3). As can be seen from the circular shape of this bispectrum, the image regions extracted by i2D operators exhibit the same strong statistical dependencies between frequency components of different orientations as measured for the fixated regions (bispectrum $C_3^{U_{eye}}$. Altogether this suggests that i2D signals provide significant information about complex scenes, that i2D-selective detectors are interesting candidates for the set of saliency detectors, and that they can explain a substantial part of the saccadic selection process.

IV. DISCUSSION

The findings in Sec. II indicate that the saccadic selection system avoids the fixation of image regions with little structural content (low variance), such as regions with approximately constant luminance (i0D signals). Similar results on the variance have been reported by Mannan et al. (1996), Reinagl and Zador (1999) and in Chapter 39. However, such a basic distinction of structure versus nonstructure cannot differentiate between different types of structure and thus can not provide any clues with respect to *form properties* of the salient features. We thus applied the standard second-order analysis of the data via power spectra, but could not obtain an unequivocal result. There could exist a tendency toward a flattening of the spectrum for the fixated regions (i.e., the selection process may cause a sort of whitening effect), but our data are too noisy to allow definite conclusions about these small differences. Measurements of spatial second-order correlations by other authors yield a reduction of the local correlations by about 0.05 in the fixated regions (Reinagl and Zador, 1999), and it has been shown that this difference is indeed statistically significant (see Chapter 39).

However, it is not easy to derive definite conclusions from these data with respect to the nature of the salient image features that are involved in the selection process. In particular, the role of the variance effects and the role of the second-order correlations cannot be easily separated. First, we have to rule out a simple variance-based model in which images are composed of two sorts of regions, spatially extended low-variance regions and more confined high-variance regions, with otherwise identical correlation structure. Theoretically, this could easily cause a steeper decay of the measured two-point correlation function. But even if this possibility could be excluded, it remains difficult to improve the explanation of the selection process that is already offered by the variance effect. That luminance changes, like edges or lines, have a higher fixation probability than smooth regions is already explained by the variance. A more differentiated explanation could be obtained, for example, by using the second-order structure to determine subclasses that have a similar variance but different fixation probabilities. However, this would require a more sophisticated analysis of the data.

Higher-order spectra (bispectra) make it easier to find meaningful structural differences. We observed that the bispectrum of natural images is dominated by the presence of statistical dependencies between frequency components with aligned orientations, and this is consistent with the frequent occurrence of locally straight edges or lines (i1D signals). The bispectrum of the fixated image regions shows a different pattern. Rather than being concentrated on frequency components with aligned orientations, the bispectral energy is more or less evenly distributed on the possible combinations. This implies that the saccadic selection system avoids image regions, which are dominated by a single oriented structure, and selects instead preferably i2D signals. These are regions that contain different orientations, such as corners, curved contours, occlusions, T-junctions, and also pop-out configurations, such as a region in which one short line segment is surrounded by segments having a different orientation. It can be shown that the least redundant i2D information captures indeed the most relevant image structure because it is basically sufficient for a reconstruction of the original image signal (Zetzsche et al., 1993). Further arguments for the relevance of i2D signals for eye-movement control and object recognition can be found in Krieger et al. (2000).

A further important question is how the statistical observations can be used to deduce which neural mechanisms are involved in the control of attentional shifts via saccadic fixations. The observed variance effect fits well with the basic operation of lateral inhibition because neurons with center-surround antagonism suppress the response to constant regions (i0D signals) and enhance the response to varying signals, such as luminance transitions. Again, it is difficult to find separate mechanisms that make specific use of the second-order correlation structure. However, for the higher-order spectra, it is easier to find the neural substrate in form of nonlinear i2D-selective cells whose properties are denoted as hypercomplex or end-stopped (Orban, 1984), or are attributed to the nonclassical receptive field (Allman et al., 1985). In primates, approximately one-half of the neurons in area V2 of the visual cortex exhibit such a preference to i2D stimuli (Orban, 1984). The class of i2D operators described in Sec. IV can provide functional models of these nonlinear neurons. It is interesting to note that effects of a nonclassical receptive field arise naturally from this nonlinear approach (Zetzsche and Krieger, 2001) and that there exist close relations to model approaches that are based on nonlinear center-surround interactions (see Chapters 38, 93 and 94). For example, there is a close relation between iterated center-surround interactions and i2D selectivity (Barth and Zetzsche, 1998).

However, although it may well be possible to identify nearly all potentially interesting locations within a complex scene by the use of a specific set of saliency detectors, it would usually not be an optimal strategy to systematically scan all these potential fixation regions. Rather, it must be possible to select a subset out of this preselection, and it must also be possible to fixate points outside the preselected set (note, however, that untrained subjects have great difficulties in making a saccadic eye movement to a prescribed position in space if they are confronted with a smooth surface of constant luminance). This implies that any model that is based solely on bottom-up processes can only provide an approximate statistical prediction of the fixations by human observers, and indeed it has been shown that top-down influences may have a substantial influence on the pattern of saccadic fixations (see, e.g., Chapters 18, 25, 33 and 48). In an interdisciplinary project, we have hence investigated how the bottom-up/low-level feature extraction process can be combined with top-down influences from a higher-level knowledge-based scheme into an integrated architecture (Zetzsche et al., 1998; Chapter 109). In the philosophy underlying this architecture, the exploration of a scene by saccadic eye movements corresponds to a sequence of "best questions to ask." The current results suggest that these best questions are partially determined by the statistical regularities of the natural environment.

Acknowledgement

Study was supported by DFG-SFB 462/B5.

References

Allman, J., Miezin, F., and MeGuiness, E. (1985). Stimulus specific responses from beyond the classical receptive field: Neurophysiological mechanisms for local-global comparisons in visual neurons. *Annu. Rev. Neurosci.* **8**, 407–430.

Barth, E., and Zetzsche, C. (1998). Endstopped operators based on iterated nonlinear center-surround inhibition. *In* "Human Vision and Electronic Image Processing" (B. Rogowitz and T. Papathomas, Eds.), pp. 67–78. WA. SPIE, Bellingham.

Krieger, G., Rentschler, I., Hauske, G., Schill, K., and Zetzsche, C. (2000). Object and scene analysis by saccadic eye-movements: an investigation with higher-order statistics. *Spatial Vis.* **13**, 201–214.

Krieger, G., and Zetzsche, C. (1996). Nonlinear image operators for the evaluation of local intrinsic dimensionality. *IEEE Trans. Image Processing* **5**, 1026–1042.

Krieger, G., Zetzsche, C., and Barth, E. (1997). Higher-order statistics of natural images and their exploitation by operators selective to intrinsic dimensionality. *In* "Proceedings of the IEEE Signal Processing Workshop on Higher-Order Statistics," Vol. PRO800S, pp. 147–151. IEEE Computer Society, Los Alamitos, CA.

Mannan, S., Ruddock, K., and Wooding, D. (1996). The relationship between the locations of spatial features and those of fixations made during visual examination of briefly presented images. *Spatial Vis.* **10**, 65–188.

Nikias, C. L., and Petropulu, A. P. (1993). "Higher-Order Spectral-Analysis: A Nonlinear Signal Processing Framework." Prentice-Hall, Englewood Cliffs, NJ.

Orban, G. A. (1984). "Neuronal Operations in the Visual Cortex." Springer, Heidelberg.

Reinagl, P., and Zador, A. M. (1999). Natural scene statistics at the centre of gaze. *Network: Comput. Neural Syst.* **10**, 341–350.

Schetzen, M. (1989). "The Volterra and Wiener Theories of Nonlinear Systems." Rev. ed. Krieger's, Malabar, FL.

Simoncelli, E., and Olshausen, B. (2001). Natural image statistics and neural representation. *Annu. Rev. Neurosci.* **24**, 1193–1216.

Zetzsche, C., and Barth, E. (1990). Fundamental limits of linear filters in the visual processing of two-dimensional signals. *Vis. Res.* **30**, 1111–1117.

Zetzsche, C., Barth, E., and Wegmann, B. (1993). The importance of intrinsically two-dimensional image features in biological vision and picture coding. *In* "Digital Images and Human Vision" (A. Watson, Ed.), pp. 109–138. MIT Press, Cambridge, MA.

Zetzsche, C., and Krieger, G. (2001). Nonlinear mechanisms and higher-order statistics in biological vision and electronic image processing: Review and perspectives. *J. Electronic Imaging* **10**, 56–99.

Zetzsche, C., Schill, K., Deubel, H., Krieger, G., Umkehrer, E., and Beinlich, S. (1998). Investigation of a sensorimotor system for saccadic scene analysis: An integrated approach. *In* "From Animal to Animats" (R. Pfeifer, B. Blumenberg, J. Meyer, and S. Wilson, Eds.), pp. 120–126. MIT Press, Cambridge, MA.

CHAPTER 38

Salience of Feature Contrast

Hans-Christoph Nothdurft

ABSTRACT

The paper introduces salience from feature contrast and reviews perceptual studies undertaken to measure its properties and to identify the underlying neural mechanisms. A strong link is made to contextual modulation in area V1. If the response variations there observed are related to salience, then some of the phenomena seen in single-cell recordings should also be evident in salience perception. This was found for temporal and spatial properties of contextual modulation and for the summation of saliency effects from feature contrast in different dimensions.

When we look around a scene, there likely are objects that are more conspicuous than others (see Chapter 39); objects that pop out (Chapters 17 and 34) and attract our gaze and attention (Chapters 69, 94, and 96). This paper addresses this phenomenon and summarizes studies undertaken to quantify the properties of salience and to relate it to single cell responses in the brain (see also Chapters 45 and 93).

I. MEASURES OF SALIENCE

Various methods have been used to measure the different salience of objects. Salient and nonsalient targets were presented in different scenes and the time needed for observers to detect them was measured. It turned out that observers could detect salient targets very fast even when these were embedded in a large number of nonsalient items, but that it took quite a while to detect a nonsalient target, in particular when it was presented among a large number of other items (Treisman, 1985).

Address: Christoph Nothdurft, MPI Biophysical Chemistry, 37070 Göttingen, Germany

In other studies, stimulus patterns were presented for short periods of time and observers were asked to indicate whether or not a salient item was present. To make the task less dependent on an observer's willingness to say "yes" when salience is not yet strong, we may instead use a two-alternative forced-choice task and, for example, ask observers to indicate whether the salient element is on the left or right side of the fixation point, but this task would require observers to locate the target in addition to detecting it. We also may ask observers to compare two targets and tell us which one is more salient. If one of the two targets is taken from a series of fixed-reference items of graded salience, the method should give a quantitative estimate of the other target's salience (Nothdurft, 1993a; cf. Fig. 38.2).

II. PHENOMENOLOGY OF SALIENCE

Although salience is an immediate perceptual impression, neither its properties nor its neural basis is yet fully understood. Several studies have suggested that certain features are salient and pop out if they rarely occur in a pattern (Treisman, 1985; Chapter 17). Other work has shown that differences between items (e.g., orientation differences) appear as salient; this led to the postulation of a special role of feature contrast for visual salience (Nothdurft, 1993b).

The difference is illustrated in Fig. 38.1. A single object (a vertical line) in an otherwise empty world is salient and immediately attracts attention (Fig. 38.1A). However, if the line is surrounded by other similar lines, it is not salient anymore (Fig. 38.1B). In order to make it stand out, it would be necessary to make it different from the other lines, for example, by increasing its luminance or changing its orientation relative to

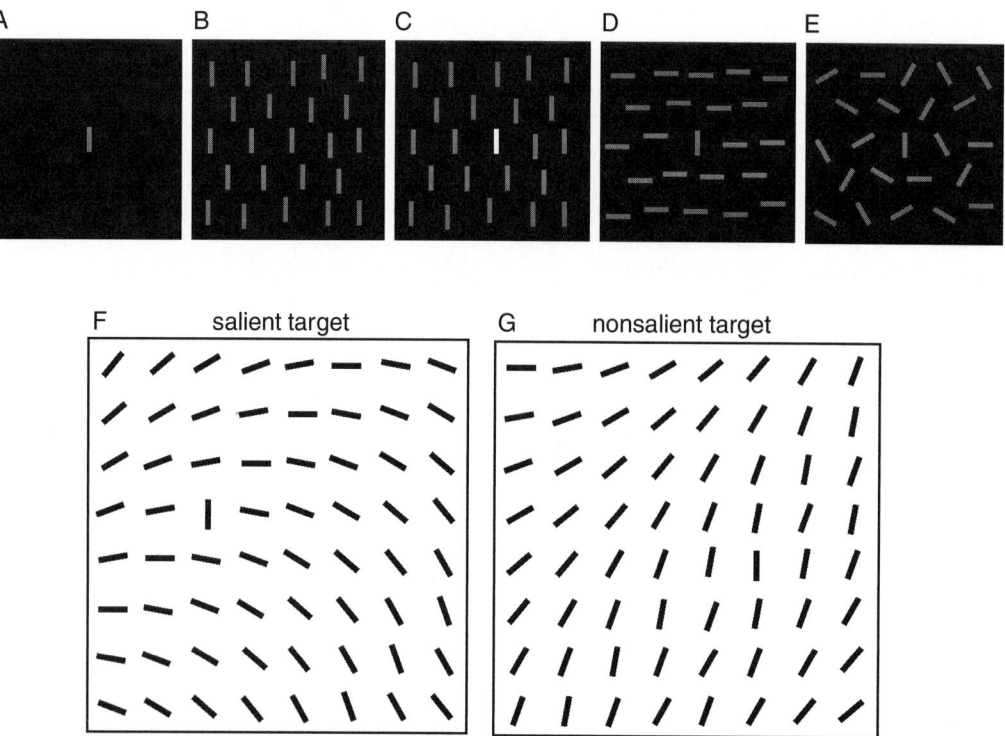

FIGURE 38.1 Examples of salient targets. A single vertical line in an empty field is salient (A) but loses its salience when surrounded by similar lines (B). Salience comes back when the target line differs from lines in the surround (C–D) unless the surround itself is variable (E). (F–G) Targets (vertical lines) with or without local orientation contrast do or do not appear as salient.

that of its neighbors (Fig. 38.1C–D). This indicates that the salience of a target is not based (alone) on the presence or absence of one particular feature but also requires the target to be different from nearby items. Several differences have been found to produce salience (e.g., differences in luminance, color, orientation, spatial frequency, motion, and depth), and the general term to describe this property is *local feature contrast*. Interestingly, if the neighboring items themselves vary in their properties, even the locally contrasting line will lose its salience (Fig. 38.1E). Thus, global variations also matter. One possible explanation is that neighbors, if displaying large features that contrast to *their* neighbors, themselves become salient; hence, the presumed target is not more salient than these.

Salience from feature contrast suggests an underlying interactive process between targets and neighbors and between the neighboring items themselves. But it should be stressed that even a single item can be salient; thus, feature contrast and the interaction between items is not a prerequisite for salience (but it should be recognized that even a single item does resemble luminance contrast to the empty regions around it).

III. POPOUT FROM FEATURE CONTRAST

The variation of salience with local feature contrast, on the one hand, and with global feature variation, on the other hand, was quantitatively studied in texture-like line arrays (Nothdurft, 1993a, 1993b). It was shown that targets pop out when their local feature contrast (in orientation, motion, or color) is sufficiently large and when the global feature variation is small (Fig. 38.1F–G). For example, a local orientation difference of 10–15 deg to the homogeneous line orientation in the surround is sufficient to make the target pop out, but global variations in the surround diminish target salience; a difference of approximately 35 deg between neighboring lines will cancel the salience of a line at even maximum orientation contrast (90 deg). Similar data were obtained for the direction of motion: A difference of only 10 deg let a moving dot pop out from homogeneously moving dots in the surround, but an overall direction shift of 70 deg between neighboring dots hindered the popout of a dot moving even in the opposite direction to that of its neighbors. Direct measurements of target salience confirmed this relationship; salience increases (nonlinearly) with local feature contrast of the target and decreases with increased variation of items in the surround (Nothdurft, 1993a).

While the general dependence of salience on (high) local feature contrast and (not too high) global feature variation is obvious and holds for a number of feature dimensions, exact investigations of the underlying spatial parameters are still rare. It should be intuitively clear that feature variations far off from the target only weakly affect its salience (e.g., the distracters in Fig. 38.1B do not seem to affect the salience of the target in Fig. 38.1A).

IV. NEUROBIOLOGY OF SALIENCE

Let us consider how salience in the examples of Fig. 38.1 might be encoded in the visual system. The line in Fig. 38.1A will evoke vigorous responses from neurons representing this orientation in this region of the visual field, whereas neurons representing the empty regions of the scene will be quiet. Thus, the salience of the target will be accompanied by an increased firing of neurons. By contrast, the identical lines in Fig. 38.1B will probably generate very similar and globally indistinguishable responses, corresponding to the missing salience of any of these lines. Only when one line is made brighter, or the surround turned orthogonal, will mean responses to this line increase, according to the increased salience of this line. Note, however, that there is a difference between these two modifications. A brighter line (Fig. 38.1C) might already evoke stronger responses in the retina and the LGN, whereas turning the surround orthogonal (Fig. 38.1D) should merely affect responses of cortical cells.

It is important to note that salience is not related to response differences of individual neurons as such. This is obvious from the fact that much smaller orientation differences can be discriminated (and hence must evoke different responses in the brain) than those required to become salient. The just-noticeable difference in orientation is on the order of 1–2 deg or less, whereas salience from orientation contrast starts to become notable for differences of 5–10 deg. This indicates that salience from orientation contrast is based on different mechanisms to those providing orientation discrimination.

Several recent studies suggest that salience from feature contrast is reflected in response differences between populations of cells in area V1. Some neurons in V1 respond with different strength to a stimulus presented with or without feature contrast to neighboring stimuli. The phenomenon, known as contextual modulation, has been reported in quite a number of single-cell studies (e.g., Knierim and Van Essen, 1992; Lamme, 1995; Sillito et al., 1995; Zipser et al., 1996; Lee et al., 1998; Kastner et al., 1999; Nothdurft et al., 1999), which have led to precise models of salience encoding in area V1 (Li, 2002; Chapter 93).

V. CONTEXTUAL MODULATION IN AREA V1

Comparison of Fig. 38.1B and D shows that a line with feature contrast appears as salient, whereas a line within a homogeneous array of similar lines does not. Single-cell recordings in monkeys and cats have revealed that these different stimulus conditions do indeed produce responses of different strength; this was found for differences in orientation (as depicted in Fig. 38.1D), motion, and several other properties. Apparently, the surround modulates the responses to a stimulus over the receptive field. Two possible sources of this modulation have been discussed: lateral interaction between neurons in the same area and feedback from subsequent areas. The modulation seems to be primarily inhibitory (although facilitating effects have also been reported) and comes from regions around the classical receptive field, the area in which visual stimulation directly activates the cell. Two components of contextual modulation should be distinguished: nonspecific (type I) inhibition, which is apparently produced by any stimulus in this region and probably modulated by luminance contrast but not stimulus form, and specific (type II) inhibition, which is produced by features similar to those presented in the center of the receptive field. Both components were found in the responses of cells in area V1 (e.g., Knierim and Van Essen, 1992; Kastner et al., 1999; Nothdurft et al., 1999). Nonspecific suppression modulates cell responses so that a line surrounded by other lines evokes a smaller response than a single line (reflecting salience differences as illustrated in Fig. 38.1A vs. B, D). Specific suppression produces response differences that depend on the form (or motion) of items in the surround: Stimuli surrounded by similar items evoke smaller responses than stimuli surrounded by contrasting items (reflecting the salience differences illustrated in Fig. 38.1B vs. C, D). While nonspecific suppression is frequently seen in V1 cells (although not always reaching the level of significance), specific suppression seems to occur only in a subpopulation of cells. Specific contextual modulation was originally observed in the cell responses of alert monkeys, first for motion (Allman et al., 1985) and later also for orientation (Knierim and Van Essen, 1992), but has meanwhile also been found in slightly anesthetized animals (Nothdurft et al., 1999; Kastner et al., 1999) and for a large variety of features (Zipser et al., 1996).

The link to salience is compelling. Due to contextual modulation, the population response to the target in the different conditions of Fig. 38.1 vary in qualitatively the same way as target salience. This suggests that salience is reflected in the (population) response of neurons in area V1.

VI. STUDYING PROPERTIES OF CONTEXTUAL MODULATION IN SALIENCE PERCEPTION

How can we attempt to ensure this conjecture? One approach is to study properties of salience from feature contrast that are predicted by properties of contextual modulation in single cells but have not yet been looked at in perception. In a series of psychophysical experiments, I have investigated three such aspects; these data are reviewed in the following section.

A. Salience from Orientation Contrast Should Be Delayed against Salience from Onset or Offset

Independent of whether contextual suppression of type II (suppression by similar features in the surround) are due to lateral inhibition within area V1 or to feedback modulation from subsequent areas, the modulation effects should be delayed, and indeed are, against the response to the target stimulus itself. Knierim and Van Essen (1992) and Nothdurft et al. (1999) found delays of approximately 40 ms for the population response to a single line but delays of approximately 60 ms to the onset of response differences caused by similar versus contrasting surrounds. After stimulus onset, the population response to target and nontargets in Fig. 38.1D will first be identical and only start to differentiate when the specific suppression from the similar surround has built up. Thus, when we compare the dynamics of saliency effects that are caused by the pure onset of the target (that is, from luminance contrast) with those from orientation (or motion) contrast, the latter should establish with a delay.

This was indeed found; subjects could detect saliency effects from target onset (cf. Fig. 38.1A) faster than saliency effects from orientation differences such as in Fig. 38.1D (Nothdurft, 2000b). It was also found that the temporal modulation of saliency effects produced by orientation or motion contrast (contextual modulation type II) could not be resolved up to the same frequencies at which salience from luminance or color contrast was resolved (type I modulation). This is consistent with the slower development of type II modulation.

B. Target Salience Should Be Modulated by the Distance to Neighboring Items

Contextual modulation only occurs over a limited distance (Knierim and Van Essen, 1992; Lee et al., 1998; Nothdurft et al., 1999). Thus, if salience from feature contrast is based on the contextual modulation seen in V1, it should diminish when the spacing between targets and surrounding lines is increased. In the extreme, when suppression from neighboring items cannot reach the target site and hence cannot modulate the responses there, saliency effects from feature contrast should disappear.

This was confirmed in an experiment in which target salience was measured against a series of luminance targets of different salience (Fig. 38.2; Nothdurft, 2000c). Salience from orientation contrast was maximal for a line spacing slightly above line length and diminished with increased spacing down to virtually no salience for spacings beyond 2–4 deg. These values were obtained for 0.7-deg-long targets at an eccentricity of 3.5–7 deg; the measures should vary with target location as cortical magnification and, hence, the size of receptive fields, the range of cortical interactions, and the corresponding visual angle vary with eccentricity.

There is another interesting result in these data. It is well documented that different visual areas are specialized for the analysis of different stimulus attributes. Whereas neurons in area MT are highly sensitive to movement and movement direction, neurons in area V4 and area IT seem to represent spatial properties of the stimulus. If contextual modulation were based on feedback from these areas, and thus on different feedback for the various feature dimensions, we should expect some variations in the range over which contextual modulation affects salience. This was indeed the case. Salience from motion contrast was found up to wider line spacing than salience from orientation contrast. Were contextual modulation based on lateral inhibition in area V1 rather than on feedback from subsequent areas, sensitivity profiles for different saliency effects should be identical. Saliency effects from luminance contrast, which are likely encoded already in retina and the LGN, did not strongly vary with line spacing (Fig. 38.2).

C. Contextual Modulation in Different Feature Domains Should Add in the Perception of Salience

Single-cell recordings in the cat have shown that specific contextual modulation is found for both orientation and motion contrast, but not always in the

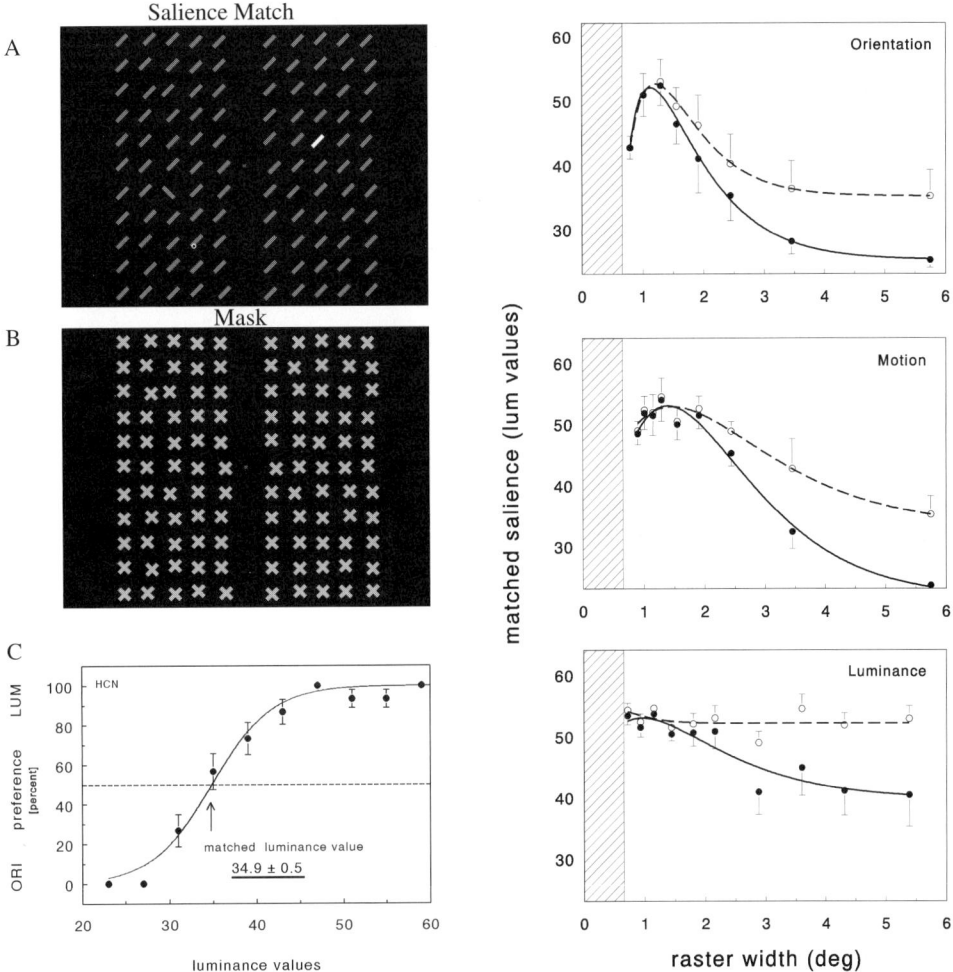

FIGURE 38.2 Salience of feature contrast depends on line density. The left half of the figure illustrates how salience is measured. The right half shows mean data of five subjects when the salience of a contrasting target was measured in line arrays of various raster widths.

Left: The salience matching task (from Nothdurft, 2000a). (A) Example of a stimulus pattern; (B) mask; (C) data and analysis. Subjects saw texture patterns with two salient lines (A), one on each side of the fixation spot, and were asked to indicate which target was more salient. Test targets, randomly assigned to one side of the screen, were lines orthogonal to the surround (as shown here) or moving in a different direction, or brighter lines. Reference targets, presented on the other side of the screen, were lines brighter than background elements but otherwise identical to them; different luminance levels were used to estimate the relative salience of the test target. Data were summed over 25–50 presentations of each individual target pair (C). Finally, sigmoid curves were fitted to the data to estimate the 50% value that then was taken as the salience-matched luminance value of the test target (arrow). The luminance value of background lines was 23.

Right: Mean salience profiles for targets defined by orientation, motion, or luminance contrast (modified from Nothdurft, 2000c). Each data point represents the average of salience-matched luminance values (found as illustrated in the bottom graph on the left) of five subjects; bars indicate SEM. Dashed curves and open circles plot the actual measurements; continuous curves and filled circles give the salience profiles after baseline saliency effects were subtracted. Baseline saliency effects represent the salience of sparsely arranged lines and were obtained from measurements with similar patterns without target feature contrast. Hatched areas mark the length of texture lines (0.7 deg). Salience was maximal for raster widths of 1–2 deg and decreased toward denser and sparser line arrangements, except for luminance contrast. Different modulation profiles for orientation, motion, and luminance contrast indicate differences in the spatial range of modulatory effects in these dimensions and hence the likely involvement of distinct neural mechanisms.

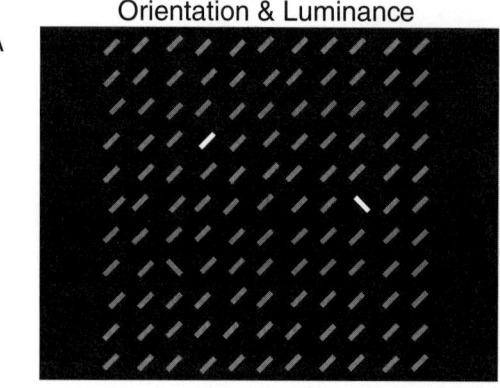

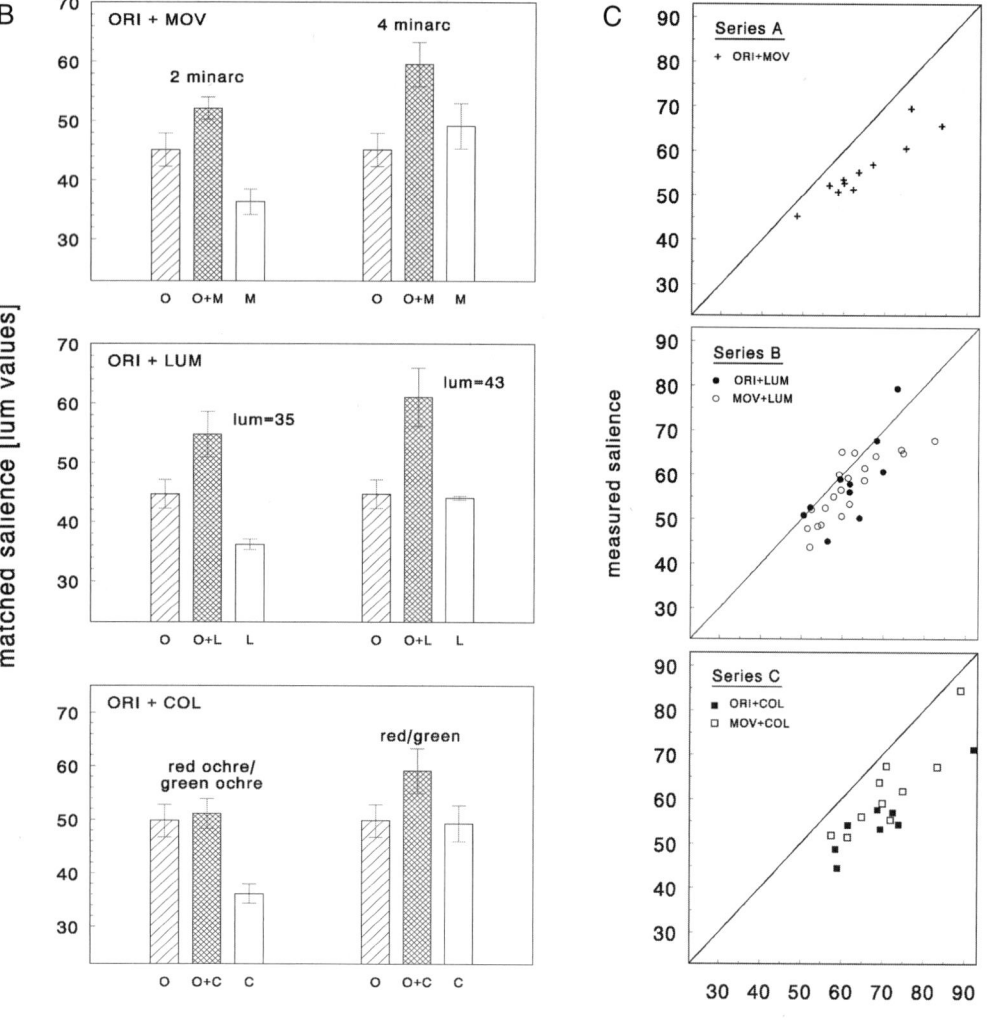

FIGURE 38.3 Partial summation of saliency effects (modified from Nothdurft, 2000a). (A) Illustration of summation effects between salience of orientation contrast and salience of luminance contrast. The target with combined saliency effects appears as most salient (and can, for example, be seen farther in periphery than the other two targets). (B) Summation of saliency effects in orientation and one other dimension (MOV or M, motion; LUM or L, luminance; COL or C, color); averaged data of five subjects; error bars indicate SEM. Measurements were made for maximum orientation contrast (90 deg) and two different levels of the second feature contrast, as indicated. Bar triplets indicate the measured salience of individual saliency effects (outer bars) and of the combination of both saliency effects together (center bar). Saliency effects nearly sum for combinations of orientation and luminance contrast, add less strongly for orientation and motion, and show only little summation for orientation and color contrast. The deviations from linearity are summarized in (C). For all tested combinations, measured salience of the combination (ordinates) is plotted against the sum of component saliency effects (abscissa). Each point represents data from one subject on a single combination. Perfect summation (indicating largely independent mechanisms) should make data points fall on the midline, but real measurements deviated considerably from that line (though less so for combinations with luminance contrast).

same neurons (Kastner et al., 1999). That is, modulation in these two dimensions is partly independent. If salience is based on response variations due to these effects, saliency from orientation contrast and saliency from motion contrast should also be partly independent.

This was studied by measuring the summation of saliency effects produced by different types of feature contrast (Nothdurft, 2000a). Indeed, a target that popped out from either orientation or motion contrast was less salient than a target that differed in both dimensions from its neighbors (Fig. 38.3). However, although different saliency effects added, they generally did not add linearly; that is, the various contrast effects were not entirely independent but interacted to some degree. The smallest amount of overlap (maximal independence) was found for saliency effects produced by luminance contrast and saliency effects produced by orientation or motion contrast. The largest overlap (minimal independence) was found for orientation and small differences in color. Altogether, the results indicate that saliency effects from feature contrast in one dimension are not entirely independent of saliency effects from feature contrast in another dimension. A different approach to the interaction of feature domains has recently reported increased summation at low feature contrast (Meinhardt and Persike, 2003).

All three studies confirmed predictions made from contextual modulation in single cells in the perception of visual salience. They thus support the model that salience is already reflected in response differences of cells in area V1. However, the various saliency effects differed in their properties; the properties of luminance contrast were quite distinct from those of orientation or motion contrast. This underlines that area V1 is not the only place where salience is probably encoded; in particular, salience of luminance contrast is already encoded earlier. The data thus seem to support a distributed representation of salience (cf. Chapter 45).

VII. THE ROLE OF SALIENCE AND FEATURE CONTRAST IN VISION

Why would salience from feature contrast be at all important? The continuous flow and large amount of visual information in vision makes it necessary to select and concentrate on important things. Important *a priori* are objects that must be identified and distinguished for their impetus on a subject's behavior and contours that help to segment such objects from background. Real objects differ in luminance, color, or texture from the surrounding world and often appear at another distance in depth or move differently to the background. Given the large variations of illumination under sunshine and moonlight, with and without clouds, object segmentation based on fixed luminance or color properties would make visually oriented life very complicated. Surface textures also vary with illumination; in addition, their structure depends on the angle at which the object is looked at. If salience is based on feature contrast, contours along a (texture) orientation difference do remain salient irrespective of the angle under which they are viewed. Were salience instead based on the analysis of fixed orientation features, many branches of processing would be necessary to handle the various occurrences of salience from the same object under various viewing conditions. It would be far more economical to select and enhance, nonspecifically, any local differences in a scene and only later, or maybe even in parallel, analyze them for behavioral relevance.

References

Allman, J., Miezin, F., and McGuinness, E. L. (1985). Direction- and velocity-specific responses from beyond the classical receptive field in the middle temporal visual area (MT). *Perception* **14**, 105–126.

Kastner, S., Nothdurft, H. C., and Pigarev, I. N. (1999). Neuronal responses to orientation and motion contrast in cat striate cortex. *Vis. Neurosci.* **16**, 587–600.

Knierim, J. J., and Van Essen, D. C. (1992). Neuronal responses to static texture patterns in area V1 of the alert macaque monkey. *J. Neurophysiol.* **67**, 961–980.

Lamme, V. A. F. (1995). The neurophysiology of figure-ground segregation in primary visual cortex. *J. Neurosci.* **15**, 1605–1615.

Lee, T. S., Mumford, D., Romero, R., and Lamme, V. A. F. (1998). The role of the primary visual cortex in higher level vision. *Vis. Res.* **38**, 2429–2454.

Li, Z. (2002). A salience map in primary visual cortex. *Trends Cogn. Sci.* **6**, 9–16.

Meinhardt, G., and Persike, M. (2003). Strength of feature contrast mediates interaction among feature domains. *Spatial Vis.* **16**, 459–478.

Nothdurft, H. C. (1993a). The conspicuousness of orientation and motion contrast. *Spatial Vis.* **7**, 341–363.

Nothdurft, H. C. (1993b). The role of features in preattentive vision: Comparison of orientation, motion, and color cues. *Vis. Res.* **33**, 1937–1958.

Nothdurft, H. C. (2000a). Salience from feature contrast: Additivity across dimensions. *Vis. Res.* **40**, 1183–1201.

Nothdurft, H. C. (2000b). Salience from feature contrast: Temporal properties of saliency mechanisms. *Vis. Res.* **40**, 2421–2435.

Nothdurft, H. C. (2000c). Salience from feature contrast: Variations with texture density. *Vis. Res.* **40**, 3181–3200.

Nothdurft, H. C., Gallant, J. L., and Van Essen, D. C. (1999). Response modulation by texture surround in primate area V1: Correlates of "popout" under anesthesia. *Vis. Neurosci.* **16**, 15–34.

Sillito, A. M., Grieve, K. L., Jones, H. E., Cudeiro, J., and Davis, J. (1995). Visual cortical mechanisms detecting focal orientation discontinuities. *Nature* **378**, 492–496.

Treisman, A. (1985). Preattentive processing in vision. *Comput. Vis. Graphics Image Processing* **31**, 156–177.

Zipser, K., Lamme, V. A. F., and Schiller, P. H. (1996). Contextual modulation in primary visual cortex. *J. Neurosci.* **16**, 7376–7389.

CHAPTER 39

Stimulus-Driven Guidance of Visual Attention in Natural Scenes

Derrick J. Parkhurst and Ernst Niebur

ABSTRACT

A large body of research indicates that the focus of attention and eye movements are guided by bottom-up, stimulus-driven mechanisms of visual attention. This research has primarily used simple experimental tasks and simple visual displays in order to maximize experimental control. In this chapter, we discuss recent work that reexamines bottom-up guidance of attention by measuring eye movements made by observers viewing complex, natural scenes. The results of this research indicate that under natural viewing conditions, attention is indeed guided by stimulus-driven mechanisms.

I. INTRODUCTION

The primate visual system receives an enormous amount of information as input and, rather than attempting to fully process all this information, portions of the input are selected for detailed processing while the remaining information is left relatively unprocessed. Two classes of attentional mechanisms control this selection process. Bottom-up selection involves fast, and often compulsory, stimulus-driven mechanisms. That is to say, computational resources are allocated to particular parts of the visual input, based on the properties of that input. For example, attention is preferentially allocated to unique features, abrupt onsets, and the appearance of new perceptual objects. Top-down selection is typically slower and governed by the observer's expectations, intentions, or memory. For example, observers can volitionally select objects or regions of space for detailed processing. Note, however, that intentionality is not a necessary component of top-down selection because familiar scene contexts (Chapter 40) and semantic associations (Moores et al., 2003) can influence attention, even in the absence of the explicit knowledge of the observer.

The majority of the research on visual attention has used relatively simple experimental paradigms designed to obtain a high degree of experimental control. Simplified visual stimuli, for example, visual search arrays consisting of colored bars of varied orientations, are nearly always used in conjunction with visual search tasks (see Chapter 17). This research has been extremely valuable. However, it is not clear whether the principles of attentional guidance gleaned from this research generalize to more complex stimuli. Thus, the results obtained should be validated using paradigms that use natural scenes and natural tasks.

The traditional measures of attention (e.g., those inferred through reaction times or error rates) are not easily determined for natural viewing conditions. However, important insights into the allocation of attention can be derived by examining the way in which people make eye movements. The logic of this approach rests on the assumption that eye movements and attention are correlated. This assumption is a reasonable one given that both eye movements and attention are related to the selection of the most important parts of the visual input. Although the locations of the focus of attention and the center of gaze can be dissociated, psychophysical evidence indicates that focal attention at the location of a pending eye movement is a necessary precursor for that movement (for review, see Chapter 20).

II. EYE MOVEMENTS IN NATURAL SCENES

Over the years, a number of studies have examined eye movements recorded from observers' viewing complex natural scenes and doing a variety of different tasks. For the most part, these studies have made qualitative claims about the relationship between eye movements and stimulus properties. Some of the earliest of these studies indicated that observers preferentially look at people and faces, although this result depended heavily on the task of the observers (Yarbus, 1967). Later, more quantitative analyses indicated that observers look at regions that are deemed to be informative (Antes, 1976). It is only recently that extensive quantitative analyses have begun to be used to examine the relationship between stimulus properties and eye movements. This approach has become feasible given rapidly improving eye tracking techniques and readily available computational resources for image processing. The majority of these quantitative studies have shown a significant correlation between stimulus features and eye movements (Mannan et al., 1996; Reinagel and Zador, 1999; Krieger et al., 2000) in free-viewing paradigms. This suggests that image features guide attention in a bottom-up fashion under natural conditions.

To further quantify the relationship between stimulus features and eye movements, we recently recorded eye movements from participants viewing a variety of natural and artificial scenes including home interiors, fractals, natural landscapes, and city scenes (Parkhurst and Niebur, 2003). Participants were told to free-view the images and that the only requirement was that they look around in the images. Presented in Fig. 39.1 are examples of the natural landscapes used in the experiment accompanied by the quantitative results of our eye movement analyses. We focus on the results obtained using the natural landscapes database because they are characteristic of the pattern of results obtained with other image databases.

We began by examining the relationship between local contrast and the observed fixation locations. To accomplish this, image patches were extracted from the images at the observed fixation locations and contrast was calculated as the standard deviation of pixel intensities within each patch. We refer to this ensemble of extracted images patches as the participant-selected ensemble. If contrast attracts attention, we expect the contrast in the participant-selected ensemble to be greater than that expected by chance factors alone. To test this prediction, we first estimated the contrast expected by chance using the average contrast in each of two additional image ensembles, a uniformly selected ensemble and an image-shuffled ensemble. The uniformly selected ensemble was created by extracting patches from random locations in each image. The contrast obtained using this ensemble serves as an estimate of the average image statistics of the images. The image-shuffled ensemble is used to control for the fact that participants may not sample the images uniformly; for instance, subjects typically show a bias to fixate central locations. This ensemble is created by extracting image patches at the observed fixation locations (tending to be central) but using a shuffled image database. Although this procedure equates the distribution of fixation locations in the participant-selected and image-shuffled databases, using it as a baseline comparison may be overly conservative given that the central fixation bias is likely to be due, at least in part, to a greater presence of interesting stimulus features near the center of the images.

As can be seen in Fig. 39.1B, the average contrast for the participant-selected ensemble (dashed line, circle) is always significantly greater than that obtained with the uniformly selected ensemble (solid line, square) or the image-shuffled ensemble (solid line, triangle). This result indicates that regions of high contrast tend to be fixated under natural viewing conditions and suggests that attention is guided to regions of high contrast in a bottom-up fashion.

The question of why attention should be drawn to regions of high contrast presents itself. One answer is that these regions may tend to be more informative for accomplishing behaviorally relevant tasks, for example, searching for and recognizing objects in a natural scene. A simple, purely stimulus-based, measure of the informativeness of a region is the correlation between local pixel intensities. Although it is well known that local intensities in images of natural scenes tend to be correlated due to common lighting, the degree of correlation can vary dramatically across different locations in a scene. For example, correlation will be low for regions that contain luminance discontinuities, such as edges, whereas the correlation is high for uniform regions, such as surfaces. Note that this measure of correlation differs from a measure of contrast in that the structure of the image patch is important. Whereas low contrast necessarily implies a high correlation between the intensity at different locations, high contrast can cooccur both with low correlation (e.g., with a random noise stimulus) or with high correlation (e.g., with a sine wave, a checkerboard pattern, or more complex patterns).

To explore the dependence of eye movements on the structure of the scene, we used the two-point correlation function. It is defined between the points at the center of each patch (i.e., at the observed fixation locations) and the points at a given distance from the center. The correlations obtained for each of the three image patch ensembles are shown in Fig. 39.1C. As expected for natural scenes, the correlations are highest for short distances and monotonically decrease with increasing distances. More important, the corre-

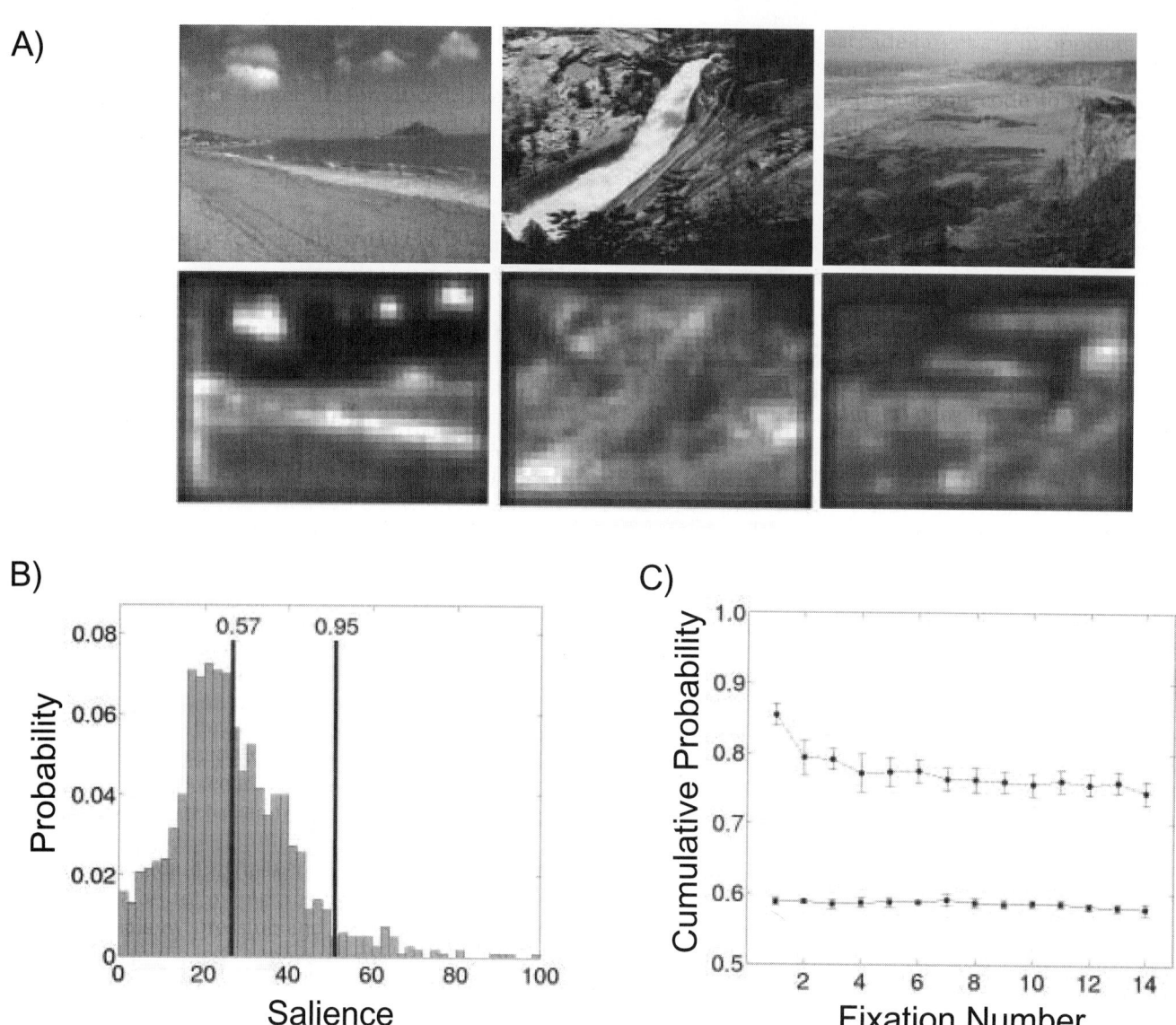

FIGURE 39.1 (A) Example natural landscapes. (B) Average contrast in the participant-selected ensemble (dashed line, circle), uniformly selected ensemble (solid line, square) and image-shuffled ensemble (solid line, triangle) all as a function of image patch size. (C) Two-point correlation using a 4-deg radius patch as a function of the distance from fixation. Same symbols as in (B). Error bars represent plus/minus one standard error of the mean taken across random permutations. (See color plate)

lations observed with the participant-selected image ensemble tends to be less than those observed for the uniformly selected or image-shuffled ensembles. This indicates that it is not regions of high contrast *per se* that attract fixation but rather regions that show an especially low correlation. Regions of low correlation have the highest information content and thus fixating these regions is an ideal strategy to gain information about the stimulus. These results suggest that bottom-up attention guides eye movements in order to maximize information about the stimulus.

III. STIMULUS SALIENCE IN NATURAL SCENES

A number of stimulus properties in addition to luminance attract attention in a bottom-up fashion including color, orientation, and motion. Under controlled conditions, the conspicuousness of a stimulus, also referred to as its salience, depends strongly on local feature contrast (see Chapter 38). For example, consider a red circle on a neutral gray background. The circle is salient because it is the only object in the display. However, surround this circle with a number of identical red circles and the salience of the original circle is greatly reduced. If we instead surround the red circle by green circles, its salience is greatly enhanced. The salience of a stimulus is related to the contrast between its features and the features of its neighbors. Functionally speaking, salience is determined by the visual uniqueness of a stimulus within the context of a scene, and local feature contrast is one plausible way to calculate the degree of uniqueness of a stimulus.

The question arises whether this analysis of salience, derived from simple displays, generalizes to natural scenes. For example, if the circles are, instead, red and green apples on display at a local market, will attention still be a function of stimulus salience? To answer this question, we might consider investigating the ability of a variety of different stimulus features to guide attention, as we did for luminance contrast in the previous section. However, given the ongoing disagreement over the fundamental set of stimulus features that attract attention in a bottom-up fashion (see Chapter 17), this approach is difficult and at best tedious. In lieu of such an approach, we decided to implement a biologically motivated computational model of the primate visual system and use this model to quantify stimulus salience (Parkhurst et al., 2002). The design of this model, its representations and algorithms, were based on what has been learned about the primate visual system from a large number of neurophysiological and neuroanatomical studies (see Niebur and Koch, 1996; Itti et al., 1998; Parkhurst, 2002, for more details about the model). In this way, the representation of stimulus salience is derived from a single, neurally plausible implementation rather than from a potentially large battery of psychophysical studies. This computational implementation allows us to explicitly quantify stimulus salience and predict attentional allocation in complex natural scenes, our primary interest.

The model takes a photograph of a natural scene as input and processes it in three parallel feature channels, representing luminance, color, and orientation across a range of spatial scales. This processing results in a set of topographic center-surround feature maps. To derive an estimate of stimulus salience, these feature maps are combined across scales and feature channels to form a saliency map (Koch and Ullman, 1985). The saliency map indicates the most salient, or visually unique, regions in the scene. A number of natural scenes and their respective salience maps are shown in Fig. 39.2A.

We reasoned that if attention is indeed a function of stimulus salience under natural viewing conditions, there should be a correspondence between fixation locations and the salience of the stimuli at those locations. To test this logic, we examined the stimulus salience, as determined by our model, at the fixation locations observed in the free-viewing paradigm described in the previous section. We used the following procedure to quantify the correspondence between stimulus salience and fixations. First, the salience at each fixation location was extracted from the relevant salience map and compared to the overall distribution of salience in that map. Note that the distribution of salience in any given salience map is often positively skewed. This is because there can be only a small number of very salient locations in a scene; otherwise these locations, being no longer unique, would no longer be salient. An example distribution is shown in Fig. 39.2B. Next, the probability of finding a salience value less than or equal to that extracted from the observed fixation location was calculated. This is a cumulative probabihiy and is equal to 1.0 if the location of maximum salience is fixated. The cumulative probability expected by chance factors alone, in other words the value that is expected if fixation locations are chosen at random, is the cumulative probability for the average salience value. Example cumulative probabilities for a single fixation are shown as dark bars in Fig. 39.2B. Given the positive skew of the salience distributions, the cumulative probability expected by chance factors alone is approximately 0.6, on average.

We found that the average cumulative probability for the observed fixations significantly exceeded the cumulative probabilities expected by chance. This is shown in Fig. 39.2C, which indicates that there is a significant correspondence between fixation locations and stimulus salience. We also found that the largest effect is seen for fixations made just after stimulus onset (Koch and Ullman, 1985). This is consistent with the time course of top-down attentional mechanisms that are known to have a slower onset than bottom-up mechanisms. These results support the conclusion that attention is indeed guided by bottom-up mechanisms nuder natural viewing conditions. Furthermore, given the magnitude of the effect, bottom-up mechanisms can play an important role in determining the guidance of attention.

IV. CONCLUSION

In order to deal with the complexity of natural scenes, the visual system must select a small portion of the scene to process in detail, leaving the remainder of the scene for processing at a later time, or not at all. Both bottom-up attentional mechanisms, which are dependent on the stimulus, and top-down attentional

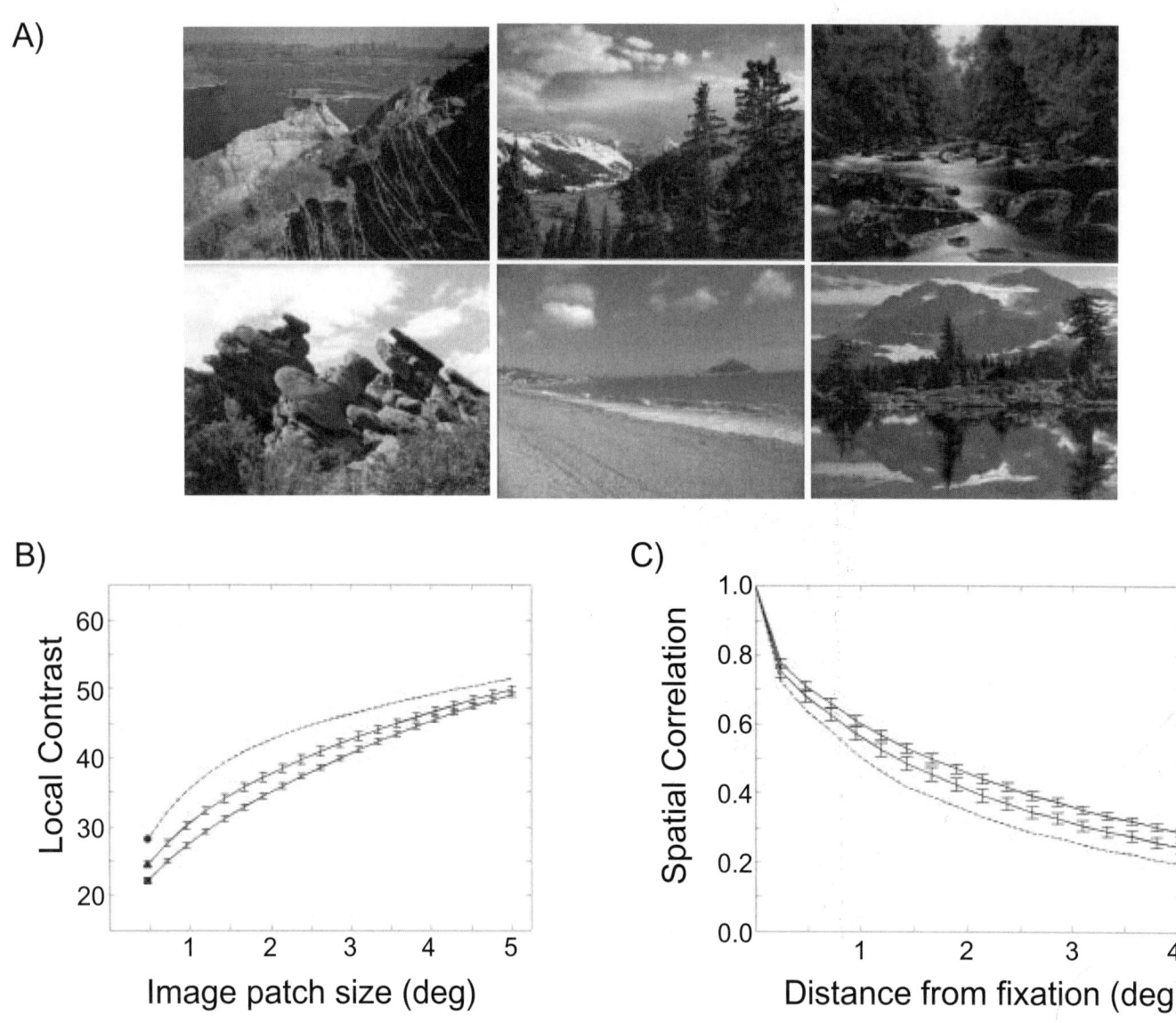

FIGURE 39.2 (A) Example natural landscapes and their respective salience maps. (B) The distribution of salience in an example salience map is shown with the salience expected by chance factors alone (cumulative probability = 0.57) and the salience obtained for a single example fixation (cumulative probability = 0.95). (C) The cumulative probabilities at the points of fixation (dashed line; circle) and expected by chance (solid line; square) are shown as a function of fixation number after stimulus onset. Error bars represent plus/minus one standard error of the mean taken across participants. (See color plate)

mechanismns, which are dependent on the viewer, contribute to this selection process.

In this chapter we described how we quantified the role of bottom-up attentional guidance using eye movements obtained from participants viewing natural scenes. Because these studies were observational rather than experimental in nature, there is the possibility that an unobserved variable not related to stimulus salience could account for the results that we obtained. We argue that this is not likely to be the case given the converging evidence in support of our conclusion from observational, computational, and recent experimental studies (see Parkhurst & Niebur, 2004).

These studies are just the beginning of our investigation into attentional allocation under natural conditions, and a number of questions remain open. For example, whereas free-viewing of scenes probably captures the way in which we view a scene when we are free from task constraints, how is attentional allocation determined when a complex task needs to be performed? Other open questions include how attentional allocation is affected by episodic memory (e.g., having viewed a scene before; see Chapter 40) and semantic memory (e.g., the gist of a scene; see Chapter 41). More generally, how can the understanding of bottom-up and top-down influences be integrated into a common framework?

We argue that computational modeling of visual processing will be an important tool to answer these questions given the inherent difficulty of studying attentional allocation in natural scenes. A number of important insights into attentional guidance in natural scenes have come from modeling approaches (e.g., see Chapters 65 and 96). Computational models allow for explicit implementations of conceptual hypotheses and can make quantitailve predictions for complex, natural stimuli. However, it is important that the modeling be integrated with behavioral and neuroscientific approaches to achieve its full potential.

Acknowledgments

This work was supported by NSF through a CAREER award to EN. Derrick Parkhurst was also supported by a NIH-NEI postdoctoral training fellowship.

References

Antes, J. R. (1976). The time course of picture viewing. *J. Exper. Psychol.* **103**, 62–70.

Itti, L., Niebur, E., and Koch, C. (1998). A model of saliency-based fast visual attention for rapid scene analysis. *IEEE Trans. Pattern Anal. Machine Intell.* **20**, 1254–1259.

Koch, C., and Ullman, S. (1985). Shifts in selective visual attention: Towards the underlying neural circuitry. *Human Neurobiol.* **4**, 219–227.

Krieger, G., Rentschler, I., Hauske, G., Schill, K., and Zetzsche, C. (2000). Object and scene analysis by saccadic eye-movements: An investigation with higher-order statistics. *Spatial Vis.* **13**, 201–214.

Mannan, S. K., Ruddock, K. H., and Wooding, D. S. (1996). The relationship between the locations of spatial features and those fixations made during visual examination of briefly presented images. *Spatial Vis.* **10**, 165–188.

Moores, E., Laiti, L., and Chelazzi, L. (2003). Associate knowledge controls deployment of visual selective attention. *Nature Neurosci.* **6**, 182–189.

Niebur, E., and Koch, C. (1996). Control of selective visual attention: Modeling the "where" pathway. *In* "Advances in Neural Information Processing Systems" (D. S. Touretzky, M. C. Mozer, and M. E. Hasselmo, eds.), Vol. 8, pp. 802–808. MIT Press, Cambridge, MA.

Parkhurst, D. (2002). "Selective Attention in Natural Vision: Using Computational Models to Quantify Stimulus-Driven Attentional Allocation." Unpublished Ph.D. iss., Johns Hopkins University, Baltimore, MD.

Parkhurst, D., Law, K., and Niebur, E. (2002). Modeling the role of salience in the allocation of overt visual selective attention. *Vis. Res.* **42**, 107–123.

Parkhurst, D. J., and Niebur, E. (2003). Scene content selected by active vision. *Spatial Vis.* **6**, 125–154.

Parkhurst, D. J., and Niebur, E. (2004). Texture contrast attracts overt attention in natural scenes. *Eur. J. Neurosci.* **19**, 783–789.

Reinagel, P., and Zador, A. M. (1999). Natural scene statistics at the center of gaze. *Network: Comput. Neural Syst.* **10**, 341–350.

Yarbus, A. (1967). "Eye Movements and Vision." Plenum Press, New York.

CHAPTER 40

Contextual Guidance of Visual Attention

Marvin M. Chun

ABSTRACT

Visual context guides what to expect and where to look in complex scenes. Such contextual guidance of attention and eye movements is possible because the content and layout of visual scenes are not entirely random but, rather, they are structured in a way that resembles other scenes that the observer may have viewed in the past. The rich, global context of a scene allows observers to match an incoming scene with an identical or similar one encountered before, allowing observers to interact with it more efficiently. Contextual cueing refers to the process of how observers encode scene contexts and how such learning serves to guide attention to important task-relevant objects. Studies using the contextual cueing task show that observers are sensitive to stable spatial layout information, how object shapes correlate, and how objects move about in predictable ways. Other experiments reveal the general importance of statistical learning in guiding visual processing.

I. INTRODUCTION

A large number of factors influence what people attend to. Salient visual features, such as flashing lights, draw attention (Chapters 17, 37, 39, and 93). However, unique, salient visual features do not necessarily signal the most important objects within complex displays or scenes (e.g., blinking neon ad signs while driving through fast city traffic; for discussion on other limitations of purely data-driven, unbounded, visual search, see Chapter 1). Human observers can also focus their search based on their behavioral goals (see Chapters 11 and 18), such as limiting their gaze to yellow vehicles while trying to hail a taxicab. However, observers do not always view the visual world with such explicit targets in mind.

One factor that exists in almost every viewing situation is the global meaning or gist of a scene (Chapter 41). People can easily recognize an office scene as an office and a farm scene as a farm. Scene meaning or gist constrains what to expect and where to fixate key objects in scenes, allowing the observer to efficiently find a stapler in an office or a tractor on a farm (Chapter 96).

Although no one would contest that scene gist plays a crucial role in guiding attention and eye movements, gist is difficult to study because natural scenes are too complex and diverse to be precisely defined or controlled in the laboratory. Hence, some researchers have turned to the visual search task that presents artificial, well-controlled scenes to capture important principles governing how global scene factors may guide attention.

II. THE CONTEXTUAL CUEING TASK

Scene gist is presented in the form of global visual context, namely the rich mosaic of features, surfaces, and objects that make up each scene. A key feature of scenes is that they contain regularities that are useful for vision. The spatial layout of surfaces and objects tends to remain stable from one moment to the next. Objects themselves tend to cooccur such that the presence of a stapler is correlated with paper, whereas the presence of a barn is correlated with hay (see also Chapter 25). Finally, there are regularities in how objects move about. With some driving experience, traffic patterns of other cars tend to be somewhat predictable; this is also true of patterns in many ball sports. These regularities are presented to observers in the form of visual context, such that sensitivity to such useful contextual information may help constrain visual processing in adaptive ways.

Chun and Jiang (1998) proposed that top-down guidance of attention based on such regularities allows visual context to guide visual attention. This is referred to as contextual cueing (see also Chapters 91, 99, and 107). Reflecting the different types of regularities that exist in scenes, there are various forms of contextual cueing tasks. The tasks are all based on visual search, requiring subjects to quickly detect targets embedded among distractors. Faster reaction times reflect efficient deployment of attention to the target and its location.

A. Spatial Contextual Cueing

Spatial contextual cueing reveals how regularities in spatial layout information may guide search to a target location. Consider the speedometer in a car's instrumentation panel. The spatial configuration of gauges and lights on the dashboard is invariant, so the position of the speedometer is fixed relative to its spatial context (the dashboard). Sensitivity to these regularities allows the driver to quickly locate relevant information as needed.

To study this using a search task, Chun and Jiang (1998) had observers search for a rotated T target among L distractors, a task that requires focal attention to detect the target. Spatial visual context was defined by the spatial layout of the distractors around the T target (see Fig. 40.1). To make these invariant, a set of search displays was simply repeated across blocks of trials throughout the experiment. Within repeated displays, the target appeared in fixed locations relative to its surrounding spatial context. In other words, the global visual context was predictive of target location. If subjects are sensitive to spatial context information and their repetitions, their search should be faster for these old displays relative to newly generated displays. Figure 40.2 shows the typical result. Search performance was faster for old displays than for new displays after approximately five repetitions—the contextual cueing effect. The learning is so robust that it lasts at least 1 week; subjects showed immediate contextual cueing to old displays within the first block of trials in a second session conducted 1 week after initial exposure, as shown in Fig. 40.2 (Chun and Jiang, 2003). Other studies have generalized the contextual cueing effect to 3D displays (Kawahara, 2003) and pseudorealistic scene displays (Chua and Chun, 2003). It is also somewhat robust to distortions to the global visual context as long as local context around the target is preserved (Olson and Chun, 2002).

Interestingly, this spatial contextual cueing effect can occur without conscious awareness of the repetitions (Chun and Jiang, 1998, 2003). When asked to explicitly discriminate old versus new displays in a surprise recognition test at the end of the search task, subjects were at chance levels. Even when they were simply asked to guess which quadrant of the display was most likely to contain the target given a certain display (where the target was replaced by a distractor),

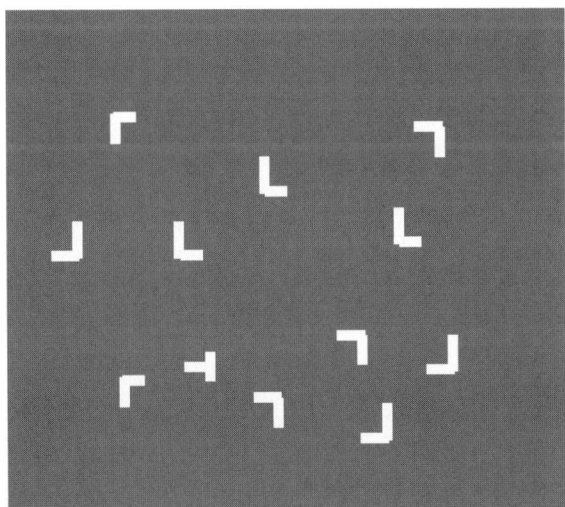

FIGURE 40.1 The configuration of rotated L distractors form a spatial context around the rotated T target to be searched (Chun and Jiang, 1998, 2003).

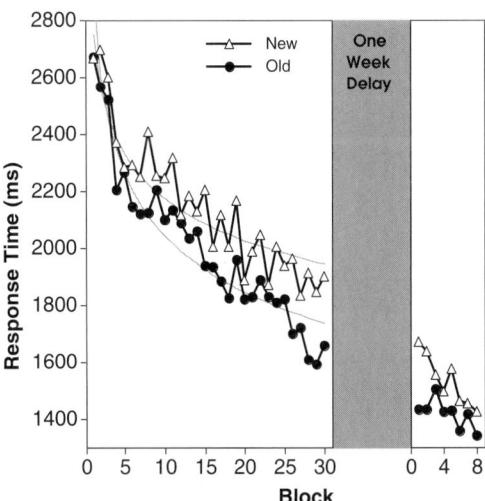

FIGURE 40.2 Targets appearing in repeated (old) scenes were detected more quickly than in new scenes. The learning persisted up to at least 1 week (Chun and Jiang, 2003, experiment 3).

subjects performed no better on the old displays that they had just viewed 24 times each.

In everyday viewing, attention serves to guide foveal vision to targets of interest (see Chapter 20), so it is useful to note that a number of studies have revealed contextual cueing of eye movements as well. Fewer eye movements were needed to foveate a target in old displays than new displays in human subjects (Peterson and Kramer, 2001a). Moreover, the effects of context were powerful enough to even override the distracting effects of abrupt onsets (Peterson and Kramer, 2001b).

B. Object Contextual Cueing

Not only is the spatial layout of the visual environment stable, but there are regularities in the types of objects an observer may expect to encounter in a scene. In other words, objects tend to cooccur. If so, observers should be sensitive to these correlations. To study this, Chun and Jiang (1999) asked subjects to search for a novel target shape appearing among other novel distractor shapes. The target was specified as the single item in a display that was symmetric around a vertical axis, whereas the distractors were objects that were symmetric around axes rotated away from vertical (see Fig. 40.3). In old trials, repeated across blocks throughout the experiment, the target shape was correlated with the distractor shapes such that sensitivity to the correlation should allow the distractor shapes in the shape context to cue the target shape. In the new displays, target and distractor shapes were also repeated across blocks, but they were not correlated. The spatial position and layout of all of the items were always randomized. The results showed that subjects are indeed sensitive to shape correlations because performance was faster in old displays than in new displays.

This behavioral demonstration of object contextual cueing reveals an important property of neural processing. Namely, neurons in the temporal cortex not only encode the shapes of objects but also the associations between them (Miyashita, 1988). The sensitivity to correlations between different cooccurring objects helps the visual system encode regularities that exist in the visual environment.

C. Temporal Contextual Cueing

The visual world is not static like a picture but dynamic like a movie. Thus, in addition to the static properties of spatial layout and object identity, dynamic properties in the form of motion and sequential order information form an important context for visual processing. Such temporal context helps the observer to predict upcoming events (see Chapter 42), just as static context helps constrain the position and identity of objects in complex scenes.

To present a common example, regularities in how players move on a basketball court or football field help players predict how a play will unfold over the next few seconds. Intelligent successful players enjoy great field sense, based on their ability to read the dynamic field and determine what will happen next. Such field sense presumably derives from perceptual experience, and so it can be mimicked in the laboratory. In a task that required the detection of a T moving around the computer screen among L distractors that moved around in different trajectories as well, subjects were faster when the target trajectory was correlated with background distractor trajectories, compared to when they were not correlated (Chun and Jiang, 1999).

Even when motion cues were not present, having a predictable sequence of visual events was sufficient to produce temporal contextual cueing (Olson and Chun, 2001). Letter targets were embedded in a rapid stream of other letters. On some trials, the serial position of the letter target in the sequence was predictable by the identity of the distractor letter sequence appearing before it. Subjects were unconsciously sensitive to this predictive relation and showed faster detection performance.

III. STATISTICAL LEARNING

The contextual cueing studies just described show how predictive context information guides visual attention and eye movements to targets in complex scenes with multiple objects. Although it is well

FIGURE 40.3 Search for the single vertically symmetric novel shape can be facilitated when it is correlated with the distractor shapes (Chun and Jiang, 1999).

known that context information guides visual processing, the value of the contextual cueing studies lies in the effort to identify *how* context can be defined and *how* contextual information is learned.

More broadly speaking, mechanisms for contextual guidance of attention take advantage of two important insights. One fundamental issue concerns a useful property of the environment: The visual world is highly structured. Features are not random, and they do not change radically over time. The visual world is stable such that regularities in the visual input can be consulted to guide visual processing.

Second, neural mechanisms have an extraordinary ability to encode these stable properties of the visual environment. Visual neurons exhibit tuning properties that are optimized to respond to recurring features in the visual input. In humans, memory mechanisms such as the hippocampus and other structures in the medial temporal lobe participate in vision so that behaviors such as search may benefit from past experience (Chun and Phelps, 1999). Contextual cueing studies reveal that the visual system is exquisitely sensitive to regularities in the environment, and it will encode and retrieve such information when it is predictive of an ongoing visual behavior such as search (this memory should be distinguished from within-trial memory, which is under debate; see Chapters 16 and 43). Thus, when a spatial layout of objects is stable, the visual system will not treat that scene as if it were new when seen on a different occasion. Likewise, knowing that certain objects go together helps generate expectancies and tunes the system to novel features. Finally, encoding regularities in how objects move about reduces the processing burden of tracking those objects. This is one of the many reasons why driving becomes easier with experience.

Thus, more broadly speaking, contextual cueing represents an example of statistical learning, a field that addresses how cognitive and perceptual systems encode important regularities in the environment. Observers take advantage of the predictive nature of such regularities in a variety of situations. One basic early finding was that people were faster at detecting targets appearing in probable locations. That is, if a target appeared more often in one location than another, people were better at detecting it at the more frequently presented location (Shaw and Shaw, 1978). An interesting extension of this work is that such location probability effects occur even without conscious visual awareness. Patients with hemispatial neglect lack awareness of visual information appearing in the contralesional neglected field, but they show sensitivity to target presentation probabilities in their neglected field (Geng and Behrmann, 2002).

Observers are sensitive to more complex statistical properties as well, and the mechanisms for doing so appear to be in place early in development. Adult and even infant observers appear to encode the joint occurrence of geometric shapes, as well as the conditional probabilities between shapes (Fiser and Aslin, 2001, 2002). Thus, statistical learning allows observers to acquire knowledge of which features go together for objects and which objects go together in scenes.

IV. CLOSING COMMENTS

Contextual guidance of attention demonstrates how top-down information derived from perceptual experience facilitates vision. The visual environment is highly structured, such that the observer can interact with the complex input more efficiently based on sensitivity and memory for statistical regularities in how objects are spatially arrayed, how their shapes covary with one another, and how they change and move about. These regularities are typically presented to observers in the form of visual context, which guides attention to important objects within scenes.

References

Chua, K.-P., and Chun, M. M. (2003). Implicit spatial learning is viewpoint-dependent. *Perception Psychophys.* **65**, 72–80.

Chun, M. M., and Jiang, Y. (1998). Contextual cueing: Implicit learning and memory of visual context guides spatial attention. *Cogn. Psychol.* **36**, 28–71.

Chun, M. M., and Jiang, Y. (1999). Top-down attentional guidance based on implicit learning of visual covariation. *Psychol. Sci.* **10**, 360–365.

Chun, M. M., and Jiang, Y. (2003). Implicit, long-term spatial contextual memory. *J. Exper. Psychol.: Learning, Memory, Cogn.* **29**, 224–234.

Chun, M. M., and Phelps, E. A. (1999). Memory deficits for implicit contextual information in amnesic subjects with hippocampal damage. *Nature Neurosci.* **2**, 844–847.

Fiser, J., and Aslin, R. N. (2001). Unsupervised statistical learning of higher-order spatial structures from visual scenes. *Psychol. Sci.* **12**, 499–504.

Fiser, J., and Aslin, R. N. (2002). Statistical learning of new visual feature combinations by infants. *Proc. Nat. Acad. Sci. U.S.A.* **99**, 15250–15251.

Geng, J. J., and Behrmann, M. (2002). Probability cuing of target location facilitates visual search implicitly in normal participants and patients with hemispatial neglect. *Psychol. Sci.* **13**, 520–525.

Kawahara, J. (2003). Contextual cueing in 3-D layouts defined by binocular disparity. *Vis. Cogn.* **10**, 837–852.

Lee, D., and Quessy, S. (2003). Visual search is facilitated by scene and sequence familiarity in rhesus monkeys. *Vis. Res.* **43**, 1455–1463.

Miyashita, Y. (1988). Neuronal correlate of visual associative long-term memory in the primate temporal cortex. *Nature* **335**, 817–820.

Olson, I. R., and Chun, M. M. (2001). Temporal contextual cueing of visual attention. *J. Exper. Psychol.: Learning, Memory, Cogn.* **27**, 1299–1313.

Olson, I. R., and Chun, M. M. (2002). Perceptual constraints on implicit learning of spatial context. *Vis. Cogn.* **9**, 273–302.

Peterson, M. S., and Kramer, A. F. (2001a). Attentional guidance of the eyes by contextual information and abrupt onsets. *Perception Psychophys.* **63**, 1239–1249.

Peterson, M. S., and Kramer, A. F. (2001b). Contextual cueing reduces interference from task-irrelevant onset distractors. *Vis. Cogn.* **8**, 843–859.

Shaw, M. L., and Shaw, P. (1977). Optimal allocation of cognitive resources to spatial locations. *J. Exper. Psychol.: Hum. Perception Performance* **3**, 201–211.

CHAPTER

41

Gist of the Scene

Aude Oliva

ABSTRACT

Studies in scene perception have shown that observers recognize a real-world scene at a single glance. During this expeditious process of seeing, the visual system forms a spatial representation of the outside world that is rich enough to grasp the meaning of the scene, to recognize a few objects and other salient information in the image, and to facilitate object detection and the deployment of attention. This representation refers to the *gist* of a scene, which includes all levels of processing, from low-level features (e.g., color, spatial frequencies) to intermediate image properties (e.g., surface, volume) and high-level information (e.g., objects, activation of semantic knowledge). Therefore, *gist* can be studied at both perceptual and conceptual levels.

I. WHAT IS THE "GIST OF A SCENE"?

With just a glance at a complex real-world scene, an observer can comprehend a variety of perceptual and semantic information. The phenomenal experience of understanding everything at once, regardless of the visual complexity of the scene, can be experienced while watching television and flipping rapidly through the channels: with a mere glimpse of each picture, observers can grasp each one's meaning (a politician, a car chase, the news, cartoons, etc.) independently of the clutter and the variety of details. This is referred to as the *gist* of a scene (Friedman, 1979; Potter, 1976).

Behavioral studies have shown that observers can recognize the basic-level category of the scene (e.g., a street; Potter, 1976), its spatial layout (e.g., a street with tall vertical blocks on both sides (Schyns and Oliva, 1994), as well as other global structural information (e.g., a large volume in perspective) in less than 100 msec. Observers may also remember a few objects (e.g., a red car and green car), the context in which they appeared (e.g., parked on the side) and other low-level characteristics of regions that were particularly salient (see Chapter 39).

Concurrent with gist development is the automatic activation of a framework of semantic information, including scripts (e.g., the actions occurring in a scene; Friedman, 1979), scene-related knowledge (e.g., the typicality or familiarity of a particular scene), as well as predictions of which objects are likely to be found in the environment. The gist benefits object detection mechanisms almost instantaneously, as well as attention deployment and gaze control in cluttered scenes (Oliva, Torralba, Castelhano, and Henderson, 2003; Torralba and Oliva, 2003 (see Chapter 96)).

II. THE NATURE OF THE GIST

Because gist includes all levels of visual information—ranging from low-level features (e.g., color, contours) to intermediate (e.g., shapes, texture regions) and high-level information (e.g., activation of semantic knowledge)—it can be represented at both perceptual and conceptual levels. Perceptual gist refers to the structural representation of a scene built during perception. Conceptual gist includes the semantic information that is inferred while viewing a scene or shortly after the scene has disappeared from view. Conceptual gist is enriched and modified as the perceptual information bubbles up from early stages of visual processing.

A. Conceptual Gist

The pioneering work of Mary Potter (1976; Potter and Levy, 1969) described the conceptual information that observers are able to quickly comprehend from a picture. In their original study, Potter and Levy (1969) allowed observers a single glance at a series of meaningful images before testing their memory of these images. When presented alone for 100 msec, each image was easily remembered. However, when embedded in a Rapid Serial Visual Presentation paradigm (e.g., RSVP pace of 125 msec exposure per image), performances deteriorated to the level of chance. In a second study (Potter, 1976), observers were cued ahead of time about the possible appearance of a picture in the RSVP stream (the cue consisted of a picture, or a short verbal description of the picture, "a picnic at the beach") and were asked to detect it. Results improved drastically, with performances of detection reaching 60 percent and 80 percent, respectively, at a pace of 125 and 250 msec per picture (Potter, 1976, Experiment 1) as compared with 12 percent and 30 percent in the control recognition memory task. Together with other experimental evidence (for reviews, see Intraub, 1999; Potter, 1999), the results obtained by Mary Potter demonstrated that during a rapid sequential visual presentation, the processing of a new picture might disrupt the consolidation in short-term memory of the previous picture. Within 100 msec, a picture is indeed instantly understood, and observers seem to comprehend a lot of visual information, but a delay of a few hundreds msec is required for the picture to be consolidated in memory. When consolidated, conceptual gist can be represented as a verbal description of the scene image, including that which was perceived and inferred.

B. Perceptual Gist

Ascertaining the perceptual content of the gist involves determining the image properties (e.g., spatial frequency, color, texture) that provide a structural representation of a scene (see, for example, Fig. 41.3). By manipulating the availability of these image properties (filtering out the edges of an image, for example), as well as imposing task constraints (e.g., duration of exposure, level of categorization, attentional demands), one can determine empirically the information needed to build a structural perceptual gist, and, therefore, enable scene identification.

Oliva and Schyns (1997, 2000; Schyns and Oliva, 1994) showed that a coarse description of the input scene (oriented blobs in a particular spatial organization at a resolution as low as 4 cycles per image) would initiate recognition before the identity of the objects was processed. Similarly, the structure of a scene can be assembled rapidly from a layout of parts (spatial arrangement of simple volumetric forms like *geons*; Biederman, 1995) or a coarse layout description of texture density (Figure 41.3; Torralba and Oliva, 2002). Common to these representations is their holistic nature: the structure of the scene is inferred, with no need to represent the shape or meaning of the objects.

In a series of articles, Oliva and collaborators (Oliva and Schyns, 1997, 2000; Schyns and Oliva, 1994, 1999; Oliva and Torralba, 2001; Torralba and Oliva, 2002, 2003) studied the image properties that enable an efficient categorization of a real-world scene. These properties included spatial frequency orientations and scales, color, and texture density. In their original study, Schyns and Oliva (1994) aimed to determine the role of spatial frequency scales for rapid scene categorization tasks. They contrasted information from blobs and edges by presenting participants with ambiguous visual stimuli (termed *hybrids*) combining the low spatial frequency (LSF) of one image with the high spatial frequency (HSF) of another. Figure 41.1 illustrates complementary hybrid stimuli: low spatial scale conveys information related to the spatial arrangement of oriented blobs, while high spatial scale conveys details, surfaces, and object contours. Two experiments revealed that hybrids were preferentially categorized in a *coarse-before-fine* sequence. In a categorization task, participants were briefly shown a hybrid scene (prime image, e.g., Fig. 41.1*a*) and asked to match the picture with a subsequent normal image (target). If the prime were the hybrid of Fig. 41.1*a*, the subsequent target image could be a city scene, matching the LSF components of the hybrid, or a hallway scene, matching the HSF components. Brief (30 msec) presentations of prime hybrids elicited matching based on their coarse structures (*city* in Fig. 41.1*a*). Longer (150 msec) presentation of the same hybrid elicited the opposite matching based on fine structures (*hallway* in Fig. 41.1*a*). This effect was reproduced in a categorization task in which an animated sequence of two hybrids (Fig. 41.1*a–b*) was preferentially categorized according to a *coarse-to-fine* sequence (*city*, 67% of coarse-to-fine interpretation), although the animation simultaneously presented the *fine-to-coarse* sequence of another scene (*hallway*, 29% of fine-to-coarse interpretations).

Because different spatial frequency scales transmit different information about the scene image (blobs preferentially convey scene spatial layout information, and edges convey surfaces and density of texture regions), an identical image may be quickly perceived at a scale that would optimize the information required to resolve a specific task: a diagnostic scale. Consequently, the allocation of attention to a specific spatial scale might determine what type of visual information enters the perceptual gist. Indeed, related experiments with hybrid stimuli (Oliva and Schyns, 1997) demonstrated a mandatory registration of

FIGURE 41.1 (*a*) A hybrid image representing a hallway scene in high spatial frequency (HSF, 24 cycles per image) and a city scene in low spatial frequency (LSF, 8 cycles per image). If you squint, blink, or defocus, the city scene should replace the hallway scene. (If this demonstration fails, step back from the image until your perception changes.) (*b*) The complementary hybrid scene, showing a city scene in HSF and the hallway scene in LSF (from Schyns and Oliva, 1994; Oliva and Schyns, 1997).

multiple spatial scales, but a flexible use of the information (blobs versus edges) proved most relevant in resolving a categorization task. In Experiment 2 of Oliva and Schyns (1997), the authors initially had two groups of observers (the LSF and HSF group) view, for 150 msec each, a set of hybrids that were only meaningful at one scale (either LSF, or HSF), the other scale being structured noise. The rationale was that these stimuli would sensitize categorization processes to seek scene cues on a diagnostic scale (either LSF, or HSF depending on the training). The testing phase consisted of presenting the two groups of observers with the same hybrid images, where the two scales were both meaningful (as in Fig. 41.1). With the observers unaware of the presence of two meaningful scenes in the hybrids, the two groups categorized the same images orthogonally: the group sensitized to the LSF scale categorized 73 percent of hybrids according to their LSF components (*city* in Fig. 41.1*a*), while HSF participants categorized 72 percent of the same stimuli on the basis of their HSF (*hallway* in Fig. 41.1*a*). These results demonstrated that attention could determine which information pertaining to spatial scale would be used during fast scene identification, leaving participants consciously blind to the unattended scale.

To complement these data, results from a third and a fourth experiment (Oliva and Schyns, 1997) suggested that, nevertheless, some covert processing of the neglected spatial frequency scale seemed to have taken place: two scene representations may be simultaneously activated within the duration of a brief glance, with the covertly processed HSF scene influencing the overtly processed LSF scene.

In another study, Oliva and Schyns (2000) evaluated the role of color in express recognition of scene gist. They did a comparison between performances of fast categorization of scenes with their normal natural colors and scenes with their colors transformed (Fig. 41.2). A scene was presented for 150 msec, and participants were asked to identify it as quickly as they could (e.g., street, bedroom, forest). The rationale was that when the color of a surface is unrelated to the meaning of a scene (Fig. 41.2*a*), as is the case in most man-made environments, color information is not necessarily used by high-level cognitive processes. Consequently, color manipulation should not necessarily affect fast scene recognition. However, when color is a feature diagnostic of the meaning of a scene, as is the case in most natural environments, altering color information should impair recognition (Fig. 41.2*b*). Results demonstrated that color influences fast scene recognition when it is *diagnostic* of the scene category: the addition of normal color to a grey-level scene accelerates its recognition, whereas the addition of

FIGURE 41.2 (*a*) Examples of color-nondiagnostic scenes: surface color is unrelated to meaning, (*b*) Examples of color-diagnostic scenes: surface color is related to the meaning of the object or region, (*c*) Illustration of color and luminance layout information available at 2, 4, and 8 cycles per image, (*d*) Performances of scene recognition as a function of cycles per image, for color and grey-level images (from Oliva and Schyns, 2000). (See color plate)

abnormal color impedes it. Additional experiments showed that the spatial layout of color blobs at a scale as coarse as 2 to 4 cycles per image (Fig. 41.2c) still mediates fast scene identification, while the grey-level counterparts do not provide enough structural cues for recognition before 6 to 8 cycles per image (Fig. 41.2c and d).

Together, these empirical results suggest that a reliable perceptual gist may be structured quickly based on coarse spatial scale information (from 4 to 8 cycles per image). At this resolution, enough structural cues are provided to allow the identification of the scene, although the local identity of objects may not be recovered. The flexible usage of spatial scale during scene perception encourages a soft-wiring view of the selection of image properties, with task constraints guiding the attentional selection of spatial scales. If a scene is unknown and must be categorized very quickly, highly salient, though uncertain, information may be more efficient for an initial rough estimate of the scene's gist. However, if one already knows what the content of the scene might be before seeing it, the informational constraints of a fast verification task may lead to a selection of fine edges before coarse blobs (Schyns and Oliva, 1994). By first attending to the coarse scale, the visual system can quickly get a rough estimate of the input to activate the conceptual part of the gist (scene schemas in memory). Attending to the fine information allows local contours to be bound together providing a refinement or refutation of the raw estimate. This view encourages a holistic mechanism during the perceptual gist elaboration that may be independent of object localization and identification (Oliva and Torralba, 2001).

III. A HOLISTIC REPRESENTATION OF GIST

In order to satisfy the main requirement of the gist—to deliver a structural summary that is meaningful enough to permit image identification—most proposals of gist content have included spatial layout information. For instance, an estimation of the spatial layout of a scene may be built upon volumetric forms termed *geons* (Biederman, 1987, 1995), spatial arrangements of blobs of different contrasts and colors (Schyns and Oliva, 1994; Oliva and Schyns, 2000), or a representation of the coarse layout of principal contours and texture density (Oliva and Torralba, 2001, Fig. 41.3). Any description related to the diagnostic structure of a scene category (e.g., urban zones are vertically structured, forests are textured) is a likely candidate for perceptual gist.

One prominent view of scene recognition is based on the idea that a scene is built as a collection of objects. This notion has been influenced by seminal approaches in computational vision that have depicted visual processing as a hierarchical organization of modules of increasing complexity (edges, surfaces, objects), with the highest level, object identification, eventually initiating scene schema activation (Marr, 1982). However, empirical results suggest that a scene may be initially processed as a single entity and that segmentation of the scene in objects operates at a later stage during gist formation. For example, speed and accuracy in scene recognition are not affected by the quantity of objects in a scene (for a review, see Biederman, 1995), and recognition can be achieved equally well even when object information is degraded so much that objects cannot be locally recovered (Schyns and Oliva, 1994). If a scene is initially processed as a single entity, then what is the nature of this entity? An alternative approach to gist representation (Oliva and Torralba, 2001) takes advantage of the regularities found in the statistical distribution of image properties when considering a specific scene category (e.g., a highway must afford speed, so ground is a flat surface stretching to the horizon). Along these lines, perceptual and conceptual representations of gist could be initiated without processing object information. To this end, Oliva and Torralba (2001; Torralba and Oliva, 2002, 2003) reasoned that, since a scene is arranged in three-dimensional space, fast recognition could be based on image properties that are diagnostic of the space the scene subtends. The authors found that eight perceptual dimensions capture most of the three-dimensional structures of real-world scenes (naturalness, openness, perspective or expansion, size or roughness, ruggedness, mean depth, symmetry, and complexity). They observed that scenes with similar perceptual dimensions shared the same semantic category. In particular, scenes given the same basic-level name (e.g., street, beach) by observers tend to cluster within the same region of a multidimensional space in which the axes are the perceptual properties.

 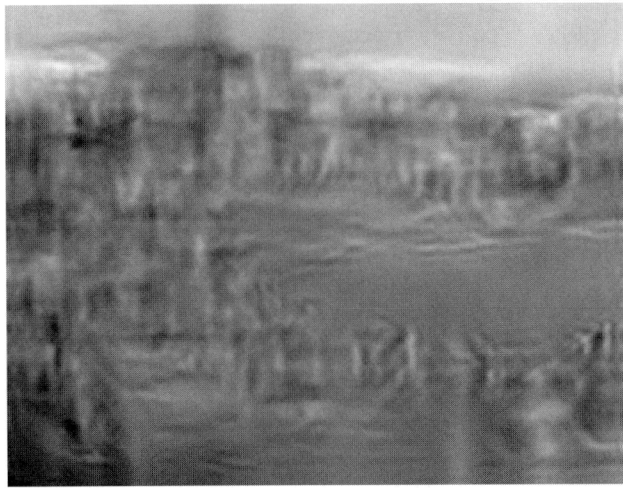

FIGURE 41.3 Illustration of a scene gist representation that conserves sufficient structural cues to infer the probable category of the scene. This global scene representation is used to determine spatial envelope properties of a scene. The information preserved by this global representation is illustrated on the right-hand image: it represents a sketch version of the original scene, computed by coercing noise images to have the same global features as the left scene (Torralba and Oliva, 2002). The scene sketch corresponds to an "unbound" spatial layout representation of contours, texture density and colors in the original scene picture. (See color plate)

The information about these perceptual dimensions can be extracted from the image using simple combinations of linear filters of the sort thought to exist in the early visual system. All together, the spatial perceptual dimensions form the *spatial envelope* of a scene—a holistic representation that does not require the use of objects as an intermediate representation and, therefore, is not being based on stages of region and object segmentation. For example, the spatial envelope of a forest scene could be described as "an enclosed natural environment, in the range of 100 meters, vertically structured in the background (trees) and connected to a textured horizontal and rough surface (grass)."

IV. CONCLUSION

In the attempt to explain how the brain represents the gist of a scene, the part-based approach (Marr, 1982; Biederman, 1987) depicts access to scene meaning as the last step within a hierarchical organization of modules of visual processing with increasing complexity (edges, surfaces, objects, scene). The "geon" theory put forth by Irving Biederman (1987, 1995) suggests that fast scene understanding could be achieved via a representation of the arrangement of simple volumetric forms from which the identity of the individual objects and scenes can be inferred. Alternatively, a holistic-based approach (*spatial envelope* theory; see Oliva and Torralba, 2001) constructs a meaningful representation of scene gist directly from the low-level features pool, without binding contours to form surfaces, or surfaces to form objects. Both approaches are based on the biological evidence that mechanisms of visual encoding are grounded in a multiscale and multiorientation representation. They differ in the number of steps, modules and binding procedures required to achieve high-level scene categorization.

Models of the mandatory role of scene gist in object detection tasks (see Chapter 96) and its ability to predict the zones where attention might be allocated during object search have been proposed in the computational domain (Oliva, Torralba, Castelhano, and Henderson, 2003; Torralba, 2003). However, whether the neural correlate of the gist of a visual scene is a final product representation hosted in a dedicated cortical region (Epstein and Kanwisher, 1998), or a representation already formed in earlier cortical areas, or even a distributed topological representation at several cortical and functional levels—all are intriguing possibilities that remain to be explored.

References

Biederman, I. (1987). Recognition-by-components: A theory of human image interpretation. *Psychological Review* **94**, 115–148.

Biederman, I. (1995). Visual object recognition. *In* "An Invitation to Cognitive Science: Visual Cognition" (2nd edition). (M. Kosslyn and D. N. Osherson, Eds.), vol. **2**, 121–165.

Epstein, R., and Kanwisher, N. (1998). A cortical representation of the local visual environment. *Nature* **392**, 598–601.

Friedman, A. (1979). Framing pictures: the role of knowledge in automatized encoding and memory for gist. *Journal for Experimental Psychology: General* **108**, 316–355.

Intraub, H. (1999). Understanding and remembering briefly glimpsed pictures: implications for visual scanning and memory. *Fleeting Memories: Cognition of Brief Visual Stimuli*, V. Coltheart (ed.), 47–70.

Marr, D. (1982). *Vision*. W. H., Freeman, San Francisco, CA.

Oliva, A., and Schyns, P. G. (1997). Coarse blobs or fine edges? Evidence that information diagnosticity changes the perception of complex visual stimuli. *Cognitive Psychology* **34**, 72–107.

Oliva, A., and Schyns, P. G. (2000). Diagnostic colors mediate scene recognition. *Cognitive Psychology* **41**, 176–210.

Oliva, A., and Torralba, A. (2001). Modeling the shape of the scene: a holistic representation of the spatial envelope. *International Journal in Computer Vision* **42**, 145–175.

Oliva, A., Torralba, A., Castelhano, M. S., and Henderson, J. M. (2003). Top-down control of visual attention in object detection. *Proceedings of the IEEE International Conference on Image Processing*, Vol. I, 253–256.

Potter, M. C. (1976). Short-term conceptual memory for pictures. *Journal of Experimental Psychology: Human Learning and Memory* **2**, 509–522.

Potter, M. C. (1999). Understanding sentences and scenes: the role of conceptual short-term memory. *In* "Fleeting Memories: Cognition of Brief Visual Stimuli," (V. Coltheart, ed.), 13–46.

Potter, M. C., and Levy, E. I. (1969). Recognition memory for a rapid sequence of pictures. *Journal of Experimental Psychology* **81**, 10–15.

Schyns, P. G., and Oliva, A. (1994). From blobs to boundary edges: Evidence for time- and spatial-scale-dependent scene recognition. *Psychological Science* **5**, 195–200.

Torralba, A., and Oliva, A. (2002). Depth estimation from image structure. *IEEE Pattern Analysis and Machine Intelligence* **24**, 1226–1238.

Torralba, A., and Oliva, A. (2003). Statistics of natural images categories. *Network: Computation in Neural Systems* **14**, 391–412.

CHAPTER

42

Temporal Orienting of Attention

Ivan C. Griffin and Anna C. Nobre

ABSTRACT

Intuition suggests that knowing when something is going to happen would help us to focus our resources at that expected point in time and enhance our behavior. Recent experiments have validated this notion, and have begun to reveal the neural systems and mechanisms involved in the temporal orienting of attention. These studies indicate that we are able to use time information flexibly to orient attention selectively to different time intervals. This is achieved via a left-hemisphere weighted parietal-frontal system. Temporal orienting capitalizes on modulation of motor-related mechanisms. Temporal orienting, when contrasted to spatial orienting, illustrates the flexibility of attentional functions in the human brain. Furthermore, undetected effects of temporal expectancies may be pervasive in behavioral and neuroscientific experiments.

I. INTRODUCTION

Temporal information can be valuable in aiding selection in our dynamic world. For example, the approach of the football striker toward the penalty area may provide critical information about the time for the goalkeeper to act.

From the warning-signal literature, it has long been known that behavioral performance can be improved when target stimuli occur at a constant and predictable time after a cue (see Niemi and Näätänen, 1981, for review). However, this literature did not indicate whether the preparatory processes underlying performance are constrained by a buildup with a set time course, or under flexible cognitive control—that is, whether we can use information about expected time intervals to orient attention flexibly to *any* point in time at which a stimulus is expected, to optimize behavior.

II. USING TEMPORAL INFORMATION TO ORIENT ATTENTION

A. Behavioral Studies

Griffin et al. (2001) and Nobre (2001) reported a series of experiments in which a symbolic cue provided information about *when* a stimulus would occur, in an analogous fashion to the way subjects are cued as to where a stimulus will appear in spatial orienting studies (e.g., Posner, 1980). Subjects were cued to expect a stimulus after a particular time interval (short or long), with the cues predicting the time interval correctly on the majority of trials (75 to 80%, valid cue), but occasionally predicting the time interval incorrectly (invalid cue). Stimuli were presented foveally, so there was no spatial information available to guide selection. Many versions of this basic task were run, which manipulated several variables: the physical form of the stimuli, the predicted cue-target intervals, and the judgment and response required to the target.

Overall it was found that reaction times (RTs) to detect or discriminate a target were facilitated when the target occurred at the correctly predicted time interval (see figure 42.1). This shows that the cues were effective in generating temporal expectancies that could modulate some aspect of target processing, and enhance behavioral performance.

These effects of temporal orienting were evident regardless of the specific stimuli used, with similar results being obtained when circles were used as cues and crosses as targets, and vice versa (see Nobre, 2001). Behavioral advantages for validly cued targets were robust over different cued time intervals (300/700, 600/1200, 600/1400 ms). This indicated that orienting attention in the temporal domain is under flexible control, rather than having a fixed time-course. It was consistently found that the behavioral advantages of valid cueing were greater for targets appearing at short

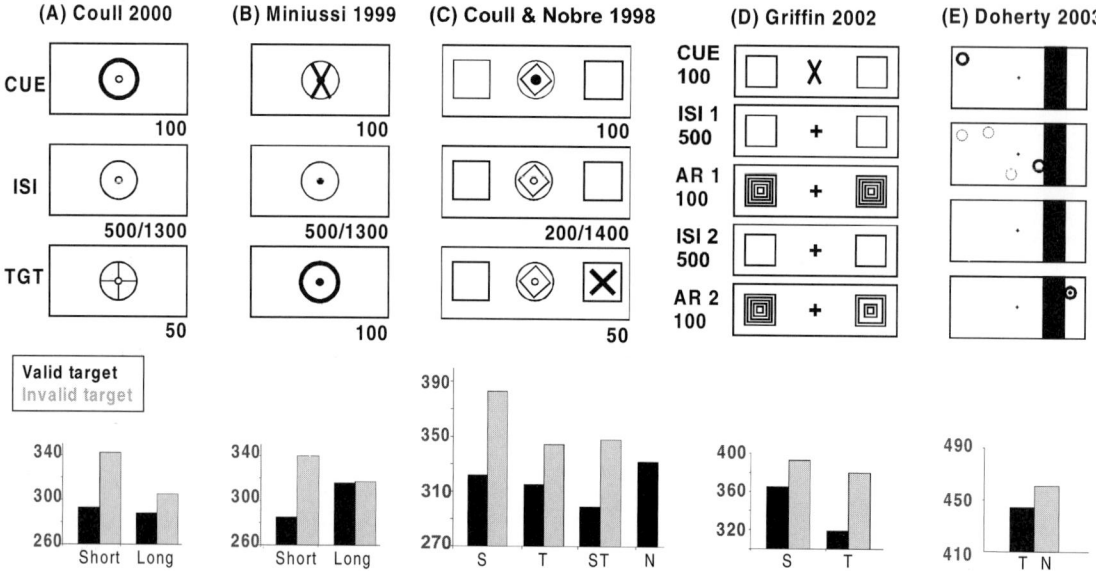

FIGURE 42.1 Behavioral tasks and reaction-time results in representative temporal orienting experiments. In all diagrams time flows from top to bottom. In cued temporal orienting tasks (A–D), stimulus durations and intervals are indicated in the corresponding frames. Tasks A and B were purely temporal-orienting experiments, using foveal stimuli, where a central symbolic cue indicated the likely interval for target appearance. Graphs plot the reaction times for the target appearing at the short or long interval when it was predicted correctly (valid) or incorrectly (invalid). Tasks C and D compared temporal versus spatial orienting. Symbolic cues indicated either the likely temporal interval (T) or spatial location (S) for target appearance. In task C, cues could also provide combined information about both temporal interval and spatial location (ST) or provide no information (N). Graphs plot the reaction times for targets separated by the type of information provided by the cue (S, T, ST, N). In task E, a ball moved across the screen from left to right in either constant (550 ms/step) or random (200–900 ms/step) temporal steps and passed under an occluding strip. Constant movement afforded the online buildup of a temporal expectancy about the reappearance of the ball (T), whereas random movement afforded no temporal expectancy (N). The graph plots reaction times for detection of target reappearance with (T) or without (N) temporal expectancies.

intervals. The relevant variable determining temporal orienting therefore seems to be the conditional probability that the target will appear given that it has not yet occurred (also known as the "hazard function"). In most tasks used, the target was guaranteed to occur at the long interval if it had not appeared at the short interval. The hazard function for long-interval targets was equivalent, regardless of which interval the subjects were cued to attend; and subjects had time to reorient their attention to the long interval. Comparison between valid and neutral cues (see Griffin et al., 2001) showed that in addition to RT costs being cued to the wrong time interval, there was also a facilitation of responses to stimuli that appeared at the predicted time relative to neutral trials, where there was no temporal expectancy.

Experiments requiring subjects to make easy or difficult perceptual discriminations of the target stimuli, and to make a choice response accordingly, also demonstrated the advantages of temporal orienting (Griffin et al., 2001). This shows that these effects are not determined solely by preparation of a specific motor response. However, the cueing effects were smaller in the discrimination than the detection tasks, suggesting that planning of specific motor responses may contribute to some extent. Experiments using delayed or unspeeded responses, and different types of response (e.g., verbal, saccades) should help clarify the extent to which temporal orienting is dependent upon both general and specific aspects of motor preparation.

In everyday situations, temporal expectancies tend to be built from the properties of dynamic stimuli, such as the speed and acceleration of a moving ball. Rarely are there symbolic cues giving explicit information about when a static stimulus is likely to appear. To capture temporal orienting in a more naturalistic condition, we developed a new paradigm in which subjects observed a ball that moved across the screen and passed "under" an occluding strip (Doherty et al., 2003). The ball moved at constant or temporally unpredictable intervals. When the ball moved at a constant pace, this enabled subjects to build up an expectancy about when the ball would reappear from behind the occluding strip. Upon the ball's reappearance, subjects were required to discriminate the presence of a black dot in its center (50% probability). Temporal expectancies about when the ball would reappear produced significant behavioral advantages compared to when there was no expectation (unpredictable timings of ball movements). This shows that the behavioral advantages of temporal orienting are robust across different tasks, including more natural tasks where temporal expectancies are provided by implicit information in the temporal predictability of a dynamic moving stimulus.

Overall, these behavioral experiments show that there is no fixed optimal foreperiod between cue and target stimuli. Rather, we are able to use time information flexibly to orient attention selectively to different time intervals, enhancing behavioral performance. It would be interesting to investigate whether temporal expectancies can overcome temporal constraints of selective attention and information processing, such as inhibition of return, attentional blink, and psychological refractory period.

Hemodynamic and electrophysiological studies have started to investigate the neural system that controls orienting attention toward specific time intervals, and the mechanisms by which it influences stimulus processing.

B. Neural Substrates and Mechanisms of Temporal Orienting

Coull et al. (2000) investigated the neuroanatomical substrates of temporal orienting using event-related functional magnetic resonance imaging (efMRI) to monitor brain activity during a temporal orienting task, similar to those described above, with foveal stimuli. Inner or outer circles predicted (80% validity) either a short (600 ms) or long (1400 ms) interval before the target stimulus (an upright cross) appeared. Behavioral advantages of temporal orienting were observed, again being greater for targets occurring at the short time interval. Comparison between valid and invalid trials revealed the effects of temporal orienting, since during invalid trials temporal attention is shifted from one time interval to the other. Temporal orienting activated inferior parietal, and inferior premotor and prefrontal areas in the left hemisphere predominantly. This pattern of activation is similar to that seen during motor preparation and motor attention (Rushworth et al., 2003), and raises the possibility that orienting attention in time is closely linked with motor preparation and motor attention.

To investigate the online modulation of brain activity by temporal orienting, Miniussi et al. (1999) recorded event-related potentials (ERPs) while subjects performed a temporal orienting task. Central cues predicted the appearance of a central target stimulus after a short (600 ms) or long (1400 ms) time interval. Analysis of ERPs to the target stimuli revealed the modulatory effect of temporal orienting on stimulus processing. Like the behavioral effects, ERP differences were restricted to targets appearing at the short interval. Unlike spatial orienting (see Mangun, 1995), temporal orienting had no effect on the visual-evoked potentials (VEPs) reflecting early activity in extrastriate visual cortex. Temporal attention modulated later ERP components, including the amplitude of the N2 potential, and amplitude and latency of the P300 potential. These later potentials have been linked to decisions and response preparation (e.g., Mangun, 1995). The absence of VEP modulation, and modulation on later, response-related potentials suggests that temporal orienting exerts its effects via different mechanisms than spatial orienting, tuning later, decision or response-related variables. However, one limitation of this study was that unlike spatial attention studies, the task used bright, transient foveal stimuli. Processing of foveal stimuli is already greatly enhanced by the visual system, so there may be no room to enhance visual processing further, and therefore no modulation of VEPs during this task.

Miniussi et al. (1999) also analyzed ERPs elicited by the cue stimuli, to investigate the neural dynamics during the control of temporal orienting. Directing attention to the short or long interval modulated the CNV component (contingent negative variation) of the ERP signal, which is linked to expectancies and motor preparation (e.g., Macar et al., 1999). Nobre (2001) reported that dipole localization of this CNV modulation revealed its likely source to be near the supplementary or presupplementary motor area, the same cortical area found to be active during manipulation of short versus long intervals during temporal orienting by event-related fMRI (Coull et al., 2000).

Together these studies suggest that temporal orienting involves contributions from a network of frontal and parietal areas with similar left-hemisphere lateralization and areas of activation to motor attention and preparation. ERP evidence also suggests that temporal orienting involves modulation of late potentials related to expectancies and response preparation, rather than the early perceptual enhancements seen with spatial attention. However, the use of foveal stimuli in all of the temporal-orienting experiments described above precludes direct comparisons between temporal and spatial orienting systems. Comparisons between temporal and spatial orienting were thus conducted in tasks using identical peripheral stimuli. This would also help investigation of the more general issue of whether there is a single attentional orienting system in the human brain, independent of stimulus dimension, or whether different systems are involved, depending on the nature of the information available.

III. COMPARISON OF TEMPORAL AND SPATIAL ORIENTING

Coull and Nobre (1998) used positron-emission tomography and fMRI to compare directly the neural systems involved in orienting attention to spatial locations and time intervals (see figure 42.2). Symbolic central stimuli cued subjects (80% validity) either to the spatial location (left, right) or temporal interval (300, 1500 ms) of a peripheral target stimulus. Behavioral advantages were observed in both conditions, showing that the benefits of orienting attention in time are not restricted to foveal stimuli.

There was considerable overlap of activation between spatial and temporal orienting conditions in frontal and parietal regions compared to a low-level resting baseline. However, when identifying areas specifically involved in focused spatial versus temporal orienting, there was a hemispheric asymmetry within this common frontal-parietal system. Spatial orienting preferentially activated the right posterior parietal cortex, whereas temporal orienting was associated with preferential activation of left inferior parietal and inferior lateral premotor cortex, as seen also by Coull et al. (2000). Once again, this suggests that areas involved in motor preparation and motor attention are involved in temporal orienting. At a more general level, the findings show that whereas parietal-frontal sensorimotor circuits may contribute to attentional orienting in general, the specific functionally specialized areas involved depend on the type of expectancy subjects have (see Nobre, 2001b).

Griffin et al. (2002) used ERPs to compare directly the modulation of stimulus processing by spatial and temporal attention (see figure 42.3). Central symbolic cues predicted (75% validity) either the spatial location (left, right) or temporal interval (600, 1200 ms) of an upcoming target stimulus. In order to minimize the automatic capture of attention by the sudden appearance of a peripheral stimulus, bilateral peripheral stimulus arrays appeared at the two time intervals following the cue. One target stimulus appeared in every trial, at one of the locations in the array, at one of the two time intervals. In this way, each condition was identical apart from the information carried by the cueing stimulus. Also, the distribution of nontargets in

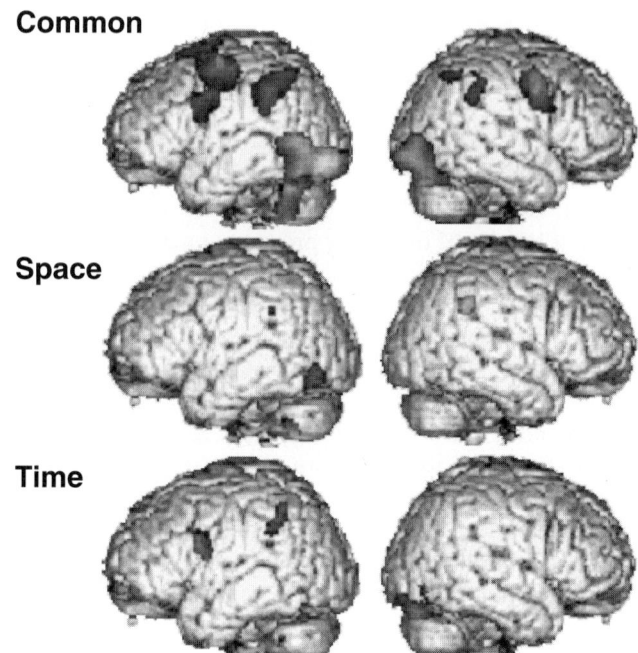

FIGURE 42.2 Brain-imaging results from Coull and Nobre (1998), who compared orienting attention to spatial locations versus temporal intervals. PET brain activations are shown superimposed upon a representative brain volume with standardized anatomy, from left and right lateral perspectives. The top row shows common areas activated in both spatial and temporal orienting conditions, which occurred in a distributed network of parietal, frontal and occipital areas. The next two rows show selective frontal and parietal activations for spatial- and temporal-orienting, respectively. (See color plate).

both space and time minimized exogenous shifts of attention. Behavioral benefits of valid cueing were present in both spatial and temporal orienting conditions, even in these tightly controlled conditions, where purely endogenous attentional mechanisms were isolated. ERPs to the identical, nontarget stimulus arrays revealed that the behavioral benefits of spatial and temporal attention were due to different modulatory mechanisms. Spatial orienting affected the amplitude of early visual P1 and N1 components, as has been seen before (e.g., Mangun, 1995). Temporal orienting involved a different pattern of ERP modulation, enhancing later potentials linked to decisions and responses, such as the N2 and P300, as seen by Miniussi et al. (1999). However, there was some evidence that temporal orienting could sometimes modulate visual processing reflected by the N1 ERP component, though in a different and more diffuse manner than spatial orienting. A second experiment used unilateral target stimuli, which occurred with

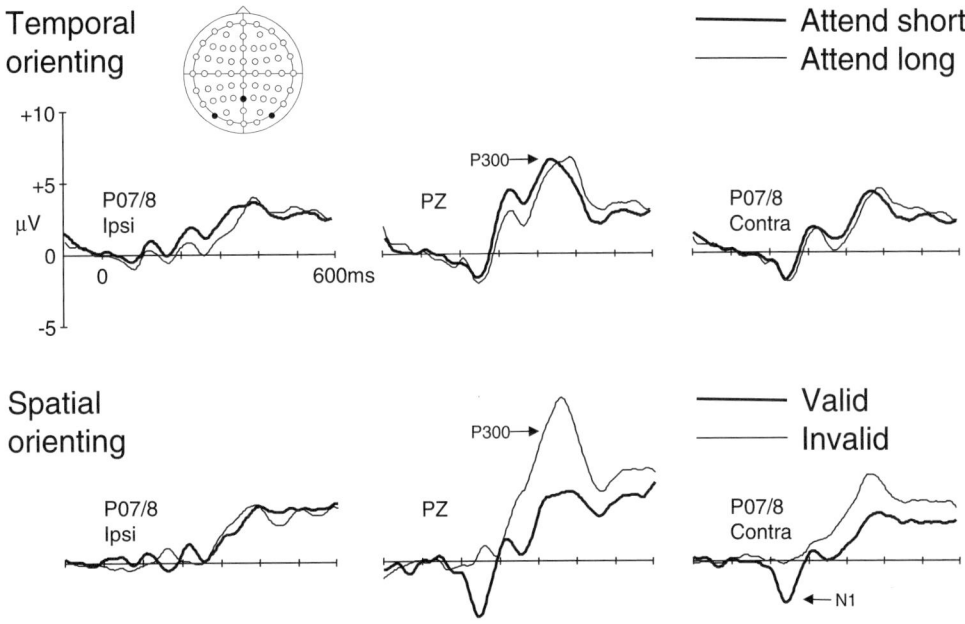

FIGURE 42.3 Grand-averaged ERPs elicited by unilateral standard stimuli during temporal orienting (top) and spatial orienting (bottom) in Experiment 2 by Griffin et al. (2002), at parietal-occipital electrodes (shaded black on montage). Waveforms are shown at the electrode location depending on whether the stimulus was ipsilateral (Ipsi) or contralateral (Contra) to the electrode. The ERPs showed distinct patterns of modulation by temporal and spatial orienting. Temporal orienting had no significant effects on early VEPs, but modulated the latency of the P300 component. Spatial orienting modulated the amplitude of the visual N1 component at contralateral posterior electrodes and the amplitude of the P300.

equal probability at only one of the two time intervals, and at the left or right of fixation (similar in principle to the experiment by Coull and Nobre, 1998). Once again there was a distinct pattern of modulation of stimulus processing by spatial and temporal orienting, showing that these effects were not driven by differential probabilities of stimulus occurrence at different spatial locations or time intervals.

Together these studies show that although both spatial and temporal orienting can enhance behavioral performance and often to the same degree, this does not necessarily mean that they recruit the same brain areas or use the same modulatory mechanisms. Comparison of spatial and temporal orienting suggests that there is not one ubiquitous dedicated attentional system in the human brain. Rather, the optimization of behavior by attentional orienting can be achieved via different functionally specialized brain regions and modulatory mechanisms, depending on the information that is guiding selection.

To date, temporal orienting has had more pronounced effects on motor-related processing, whereas spatial orienting has influenced perceptual analysis. It must be noted, however, that temporal-orienting tasks used so far have required speeded motor responses and have involved relatively low perceptual demands. In tasks emphasizing perceptual factors, temporal orienting may also influence perceptual analysis.

IV. IMPLICATIONS OF TEMPORAL EXPECTANCIES FOR COGNITIVE RESEARCH

The robust effects of temporal orienting warrant greater regard for temporal factors in cognitive research in general and attention research in particular. The effects of temporal expectancies have been typically disregarded, but may be insidious in experiments of many types. For example, the use of fixed or a narrow band of intervals between stimuli is very common and may inadvertently engender temporal expectancies that could determine or interact with the experimental factors of interest.

Milliken et al. (2003) investigated spatial-orienting effects at different cue-target intervals in a behavioral experiment in humans. In addition to giving a spatial-orienting cue (left, right) at the start of each trial, they varied the proportion of trials, with the target occurring at each interval (100, 500, 900 ms), so subjects could also develop a temporal expectancy of when the target was likely to occur. They found effects of both spatial and temporal orienting, and an interaction between the two factors when subjects were required to discriminate targets at the short interval. That is, the effects of spatial orienting were larger at the short interval when subjects had a greater temporal expectancy of the target appearing at that interval. Although the interaction was only in this one condition, this does show that the engagement of spatial attention can be affected by factors unrelated to spatial predictability of cues (temporal expectancy), and raises a more general point about the possible role of temporal expectancies in other findings.

Naccache et al. (2002) explored the effects of temporal expectancies on unconscious masked priming of number stimuli. They noted that in most masked priming experiments fixed temporal proximity between primes and targets could lead to temporal orienting to the expected time of target occurrence. To test the role of temporal orienting directly, they manipulated the temporal certainty between prime and target. Indeed, temporal expectancies were found to be necessary for repetition priming and response-congruity effects to occur. In addition to reinforcing the notion of the pervasiveness of temporal expectancies in cognitive paradigms, their findings show in particular that temporal orienting can influence unconscious processing of meaningful stimuli.

Ghose and Maunsell (2002) investigated the role of temporal expectancies in the modulation of neuronal firing rates by spatial attention in visual area V4. Monkeys were trained to respond to stimulus changes at a cued location. The behaviorally relevant probability that such a change would occur at a point in time, given that it had not yet occurred, was the same in each trial. It was found that the modulation of neuronal firing rates by spatial attention changed on a moment-to-moment basis depending on the probability of the stimulus change occurring at that instant in time. This has important implications for a number of reasons. First, given that the timing of behaviorally relevant events within a task affects the timing of attentional modulation, it is difficult to make inferences based on the latency of any attentional modulations observed, as they may be a reflection of task timing, rather than an underlying aspect of attention. Second, studies using the average modulation over a whole trial may miss out important aspects of any modulation that is occurring.

The fact that modulation by other forms of attention or priming can be influenced by temporal expectancies has strong implications for considering results from other paradigms. For example, most neuroimaging studies entail temporal expectancies, with stimuli occurring at predictable intervals. It is thus possible that aspects of the modulations observed are a reflection of temporal expectancies about upcoming stimuli, rather than the topic of interest.

V. SUMMARY AND CONCLUSIONS

In summary, several experiments have demonstrated that we are able to use predictive information about when stimuli will occur to enhance behavioral performance. This process appears to be under flexible cognitive control. Temporal orienting, at least in speeded response tasks, involves a parietal-frontal sensorimotor network of brain areas related to motor attention, with different lateralization and to some extent localization than the network controlling spatial orienting. The modulatory effects of temporal attention occur mainly at later, response and decision-related stages of processing. Whereas there are some consequences of temporal orienting during perceptual processing, the modulations are probably not dependent on stimulus location in contrast to the effects of spatial orienting. This shows that there is not one general attentional network and mechanism of action in the brain, regardless of what stimulus attributes are being expected. Instead, different areas and modulatory processes can be recruited flexibly in order to optimize behavior, depending on what type of information is guiding attentional selection. Finally, the widespread influence of temporal expectancies on behavioral performance as well as neuronal firing rates in visual areas indicates the importance of considering temporal factors when interpreting results from attentional paradigms.

References

Coull, J. T., Frith, C. D., Buchel, C., and Nobre, A. C. (2000). Orienting attention in time: behavioral and neuroanatomical distinction between exogenous and endogenous shifts. *Neuropsychologia* **38**, 808–819.

Coull, J. T., and Nobre, A. C. (1998). Where and when to pay attention: the neural systems for directing attention to spatial locations and to time intervals as revealed by both PET and fMRI. *Journal of Neuroscience* **18**, 7426–7435.

Doherty, J., Griffin, I. C., Rao, A., and Nobre, A. C. (2003). Spatial versus temporal expectancies derived from object movement differentially modulate stimulus processing. Program No. 228.10. *Abstract Viewer/Itinerary Planner*. Washington, D.C.: Society for Neuroscience. Online.

Ghose, G. M., and Maunsell, J. H. R. (2002). Attentional modulation in visual cortex depends on task timing. *Nature* **419**, 616–620.

Griffin, I. C., Miniussi, C., and Nobre, A. C. (2001). Orienting attention in time. *Frontiers in Bioscience* **6**, 660–671.

Griffin, I. C., Miniussi, C., and Nobre, A. C. (2002). Multiple mechanisms of selective attention: differential modulation of stimulus processing by attention to space or time. *Neuropsychologia* **40**, 2325–2340.

Macar, F., Vidal, F., and Casini, L. (1999). The supplementary motor area in motor and sensory timing: evidence from slow brain potential changes. *Experimental Brain Research* **125**, 339–350.

Mangun, G. R. (1995). Neural mechanisms of visual selective attention. *Psychophysiology* **32**, 4–18.

Milliken, B., Lupianez, J., Roberts, M., and Stevanovski, B. (2003). Orienting in space and time: Joint contributions to exogenous spatial cueing effects. *Psychonomic Bulletin and Review* **10**, 877–883.

Miniussi, C., Wilding, E. L., Coull, J. T., and Nobre, A. C. (1999). Orienting attention in time. Modulation of brain potentials. *Brain* **122**, 1507–1518.

Naccache, L., Blandin, E., and Dehaene, S. (2002). Unconscious masked priming depends on temporal attention. *Psychological Science* **13**, 416–424.

Niemi, P., and Näätänen, R. (1981). Foreperiod and simple reaction time. *Psychological Bulletin* **89**, 133–162.

Nobre, A. C. (2001a). Orienting attention to instants in time. *Neuropsychologia* **39**, 1317–1328.

Nobre, A. C. (2001b). The attentive homunculus: Now you see it, now you don't. *Neuroscience and Biobehavioral Reviews* **25**, 477–496.

Posner, M. I. (1980). Orienting of attention. *Quarterly Journal of Experimental Psychology* **32**, 3–25.

Rushworth, M. F., Johansen-Berg, H., Goebel, S. M., and Devlin, J. T. (2003). The left parietal and premotor cortices: motor attention and selection. *Neuroimage* **20**, Suppl 1, S89–100.

CHAPTER 43

Visual Search: The Role of Memory for Rejected Distractors

Todd S. Horowitz

ABSTRACT

Serial models of attentional deployment in visual search have traditionally assumed sampling without replacement. Each item in a search array was thought to be selected only once, and rejected distractors were never revisited. This efficient search pattern requires some form of high-capacity memory for every deployment of attention. Several investigators suggested that inhibition of return served to implement this memory (see Chapter 16). Recently, this assumption has been challenged on several fronts. Experiments using the randomized search paradigm, search for multiple targets, and attentional reaction time methods suggest that search often uses sampling with replacement. Studies of eye movements have yielded a range of findings, from perfect memory to substantial evidence for resampling. Simple strategies can reduce the probability of resampling without positing high-capacity memory. Such strategies carry a cost in terms of speed and may explain the oculomotor data. Memory for rejected distractors in visual search is likely to be of limited capacity, similar to other aspects of visual memory.

I. SAMPLING IN VISUAL SEARCH

How is attention deployed during search of a cluttered scene? We know that the salience of objects in the present drives attention, whether salience derives from intrinsic properties (bottom-up, see Chapters 38, 39, 43, 93), or task relevance (top-down, see Chapters 12, 14, 18, 40, 96). Furthermore, it has long been assumed that past attentional deployments determine future deployments, in that attention is assumed to avoid objects that have already been searched. In the language of probability theory, visual attention is supposed to use sampling without replacement from the search array. We can use a simple urn model to illustrate the difference between sampling with replacement and sampling without replacement. Imagine we are looking for the red marble in an urn that contains 1 red marble and 20 blue marbles. Using sampling without replacement, we would set aside each marble after determining its color. If we were sampling with replacement, however, we would return the marble to the urn before pulling the next marble, thus making it likely that we would pull the same marble from the urn more than once. Sampling without replacement is clearly more efficient, which presumably explains why it was the default assumption for theories of search.

A. The Standard Model

This assumption is explicit in a wide range of search models that employ a serial component (see Chapter 43). Despite their differences, all of these models state that attended items are never resampled. Moreover, the assumption is implicit in many other papers. For instance, the routine claim that a serial model should produce a 2:1 ratio of target absent to present slopes is based on this assumption. Therefore, we will call the sampling without replacement option the Standard Model.

Under the Standard Model, as each distractor is rejected, it is somehow "marked" or inhibited. Consequently, the probability of detecting the target in a given epoch increases during the course of the search. For example, assume a standard search task in which there is one target among d distractors. Further assume that all items are equally salient. In the first epoch, the probability of detecting the target is $1/d$. In the second epoch, the probability is $1/(d-1)$, and so forth. Thus, the "hazard function" rises exponentially during the course of the search, reaching 1.0 in the dth epoch.

Models that do not employ a serial attentional component are formally equivalent to the Standard Model for present purposes if they produce an increasing hazard function.

The time to find targets using sampling without replacement is governed by the negative hypergeometrical distribution, and the average number of samples $E(s)$ to find a target in a display containing t targets and d distractors is given by Eq. (43.1):

$$E(S) = \frac{t+d+1}{t+1} \quad (43.1)$$

The common claim that "half of the items are searched before the target is found" derives from this equation. If we set $t = 1$, and N = set size $(d + t)$ $E(S) = (N + 1)/2$; on average, $(N + 1)/2$ items will be sampled before the target is found.

B. The Amnesic Model

The obvious alternative to the Standard Model is a model based on sampling with replacement. In such a model, items would be selected without regard to whether they had been selected before. Free recall, for instance, is widely held to proceed by sampling with replacement from memory. Under sampling with replacement, only the current salience would determine the probability of selection. If all objects were equally salient, selection would be essentially random. In the human factors literature (Arani, Karwan, and Drury, 1984; Chan and Courtney, 1998), this has been termed the *random search model*, as opposed to the *systematic search model*, which corresponds to what we have been calling the Standard Model here. However, we prefer the term *amnesic search*, which emphasizes the memoryless character of the model.

In contrast to the Standard Model, amnesic models produce flat hazard functions; the probability of finding the target is constant over time. In the case of a single target among d distractors, in each epoch the model has a $1/d$ probability of detecting the target. As a result, there is a small but finite probability that the target will never be attended.

Sampling with replacement is governed by the negative binomial distribution, and the number of samples required to find a target in a display containing t targets and d distractors is given by Eq. (43.2):

$$E(S) = 1 + \frac{d}{t} \quad (43.2)$$

In the case of a single target, $E(S) = d + 1$, so the number of samples needed to find the target is equal to the set size. Note that this does not mean that every item in the array is examined; this would be an exhaustive, memory-driven search. Instead, it is likely that some items would be examined several times, while other items might not be examined at all.

C. Other Roles for Memory in Visual Search

In statistics, a process using sampling with replacement is said to be "memoryless" because the state of the system (here the focus of attention) at any given time is not affected by previous states of the system. When we say that visual search is memoryless (Horowitz and Wolfe, 1998), we do not mean that no memory is involved in the process of search. On the contrary, memory is known to be critical to search. Memory processes and search processes interact on many levels. These include working memory (see Chapter 100), as well as implicit contextual guidance of visual attention (Chapter 40). The claim that search is memoryless is strictly an argument about whether the deployment of attention is determined by the history of previous deployments (or, to put it more broadly, about the form of the hazard function). Note that this view is also not incompatible with the idea that information about distractors is acquired during search. Subjects may have an accurate memory for the characteristics of distractors, yet still be unable to prevent attention from returning to them.

At the same time, memory-driven search does require some mechanism to prevent attention from returning to rejected distractors. The dominant hypothesis in the literature is that inhibition of return (see Chapter 16) serves this purpose. One could postulate other mechanisms more like conventional memory systems. Whatever "memory" structure is held to underlie sampling without replacement, it must have a high capacity to keep track of all distractor locations in a standard visual search experiment.

II. EMPIRICAL TESTS OF THE STANDARD MODEL

Researchers have studied the question of memory-driven versus amnesic search (aka systematic versus random search strategies) since Krendel and Wodinsky (1960). However, until recently, this debate has taken place within the applied literature and has had little impact on the development of mainstream cognitive theories of search. Recent work in the basic vision science literature was sparked by the development of the randomized search procedure by Horowitz and Wolfe (1998).

A. Randomized Search

One strategy to determine whether subjects use sampling with or without replacement under normal circumstances is to devise conditions under which the sampling strategy is known and to compare performance in such a forced condition to performance under standard conditions. The logic behind the randomized search procedure is to force subjects to use sampling with replacement. In the randomized search procedure, each trial consists of a series of search frames. The stimuli (targets and distractors) are held constant across frames. In the *dynamic* condition, however, the positions of all items are randomly shuffled (within certain constraints, see below) from frame to frame, while in the *static* condition, locations are also held constant, so that there is in fact no stimulus change from frame to frame. In the static condition, subjects are free to use either sampling without replacement or sampling with replacement. In the dynamic condition, however, sampling without replacement is impossible because there is no way for the visual system to know which distractors in the current frame correspond to distractors rejected on previous frames. Therefore, on each frame, search starts anew.

Predictions for the standard and amnesic models can be derived in a straightforward fashion from Eqs. (43.1–43.2). Under both models, performance in the dynamic condition is governed by Eq. (43.2). If there are n items in the search array, subjects have to examine an average of n items before finding the target. In the static case, the standard model holds that subjects can sample without replacement, which is much more efficient. To be precise, subjects need to examine only $(n + 1)/2$ items on average. Thus, the RT × set size slope in the dynamic condition should be steeper than that in the static condition by a ratio of $2n:(n + 1)$, or roughly 2:1. According to the amnesic model, however, subjects in the static condition are using sampling with replacement anyway, so there should be no difference in search efficiency (RT × set size slope) between the two conditions. Of course, the two conditions are not identical from a stimulus point of view. The dynamic condition by definition contains a variety of dynamic events, onsets and offsets, which may affect RT. However, these factors should have their influence on the nonsearch aspects of the task, and show up in the intercept of the RT × set size function. Thus, the critical datum is the slope ratio.

Horowitz and Wolfe (1998) found that the slopes for static and dynamic conditions were not reliably different from each other in a range of conditions. These included versions intended to thwart "sit and wait" strategies where the subject might fixate at one location and wait for the randomly replotted target to appear at fixation (von Muhlenen, Muller, and Muller, 2003). The basic finding was replicated in a number of labs (Gibson, Li, Skow, Salvagni, and Cooke, 2000; Kristjánsson, 2000). Kristjánsson (2000) produced a notable failure to replicate when he used large set sizes. However, his stimulus configuration may have led to more masking in the dynamic condition. Horowitz and Wolfe (2003) repeated the experiment with large set sizes under conditions intended to reduce masking in the dynamic condition. As before, they found that dynamic and static slopes were essentially identical. Moreover, RT distributions were similar across conditions. This should not be the case if static and dynamic search used different sampling mechanisms.

Although these data argue against the standard model, Gibson et al. (2000) used a similar method to show that, while subjects may not use memory for rejected distractors to guide subsequent search, they remember the location of a target when searching for multiple targets (see also Horowitz and Wolfe, 2001; Takeda, in press).

Although the substantial similarity between dynamic and static search is impressive, the randomized search method has a number of drawbacks. First, as noted earlier, the two conditions can never be equated for stimulus quality, so performance will never be equalized on all measures. Second, the argument for amnesic search, at least, relies on accepting the null hypothesis. Third, the logic of the method will fail if the search can be completed during a single frame. Fourth, experiments must be designed in order to thwart subjects from adopting different search strategies in the two conditions. In particular, if the target location is selected randomly on every frame, subjects may simply monitor a single location or set of locations and wait for the target to come to them. Horowitz and Wolfe attempted to circumvent this problem by restricting the locations at which the target could appear, though this may not have been entirely successful (von Muhlenen et al., 2003). Finally, subjects appear to have difficulty deciding when to give up on target-absent trials in dynamic conditions, so switching to a forced-choice design (i.e., one of two targets is always present) will significantly reduce dynamic error rates (Horowitz and Wolfe, 2003).

B. Multiple-Target Search

In the classic visual search task, where subjects have to find a single target, both the standard model and the amnesic model make similar predictions for the central tendency of RTs. (Distributions are a different matter, see next section.) However, when subjects must find multiple targets during a single trial, the two models make quite different predictions. Compare Eq. (43.1) and Eq. (43.2). As each target is found, the value of t in the denominator of both equations decreases. However, the value of d in the numerator of Eq. (43.1) also decreases as distractors are marked off, so that the interval between finding targets is constant for the Standard Model. In contrast, the value of d in Eq. (43.2) is constant, so that in the amnesic model, the intervals increase nonlinearly.

Of course, measuring multiple RTs during a single trial is problematic, so Horowitz and Wolfe (2001) devised a method in which subjects are asked to report as soon as they know that there are at least n targets in a display containing m targets. Both n and m are varied across trials (n across blocks in this case). The RT × n function for constant m is a proxy for the RTs to m successive targets.

Horowitz and Wolfe, using alphanumeric stimuli (digit targets among letters), found that this function is highly accelerated, disconfirming the standard model. However, Takeda (in press), noted that the search rate may depend on n, possibly due to memory load. The interaction of working memory (see Chapter 102), and visual search is not entirely clear.

C. Cumulative Distribution Functions

Since the Standard Model and the amnesic model have different hazard functions (section I), they also have different cumulative distribution functions (CDFs). Sampling without replacement will yield linear CDFs, which reach 1.0. Sampling with replacement will produce exponential CDFs, which approach but never quite reach 1.0, since there is always a chance that the target will not be located (Krendel and Wodinsky, 1960). Visual search studies in the mainstream cognitive literature rarely report CDFs. However, fitting CDFs has been a popular method to distinguish between random and systematic search in the human factors literature. A number of papers have reported exponential CDFs in visual search (e.g., Chan and Courtney, 1998), though linear CDFs have also been reported.

As Chan and Courtney (1998) noted, RT CDFs have two components: the search time (which may be distributed linearly or exponentially), and a nonsearch time (including response time), which is normally distributed. They found that search RTs are well described by the ex-Gaussian distribution, in which a normal distribution (representing nonsearch time) is convolved with an exponential (representing search time).

Horowitz, Wolfe, and Alvarez (see discussion in Horowitz and Wolfe, 2003) attempted to measure the search CDF directly, eliminating the nonsearch component. Subjects searched for a mirror-reversed letter and reported its color. The colors of all items changed at some time T during the search. The dependent variable was the probability $p(T)$ that the subject reported the initial color of the target. Plotting $p(T)$ against T yielded the CDF for the search process. This function was exponential, supporting an amnesic account of search.

III. MEMORY IN THE OCULOMOTOR DOMAIN

Although the issue is not settled, the balance of the data suggest that covert deployments of attention are not guided by memory for prior deployments. What about overt deployments of the eyes? As we noted above (see Chapter 16) IOR has been proposed as an inhibitory tagging mechanism to support sampling without replacement in search, and has been tightly linked to the oculomotor system. Therefore, one might expect that eye movements would be memory-driven. Indeed, eye movements tend to be biased away from the previous fixation locations (see Chapter 16). On the other hand, Horowitz and Wolfe (2003) employed the randomized search method using small stimuli that required fixation and found no difference between the slopes of static and dynamic functions. Measurements of CDFs for oculomotor search support a memoryless account (Scinto, Pillalamarri, and Karsh, 1986).

The most obvious way to assess whether eye movements sample a visual search stimulus with or without replacement might be to record fixations and see whether or not items are revisited. A number of researchers have taken just such a straightforward approach. Peterson et al. (2001) reported that the hazard function to fixate a search target in an oculomotor search task was increasing rather than flat, indicating as their title stated that "visual search has memory," at least in the oculomotor domain. The question is not as easily resolved as one might think. Gilchrist and Harvey (2000) pointed out that much depends on the algorithm used to determine which fixations are counted as refixations. They arrived at a figure of 50 percent refixations for their task, but suggested that this may have been an overestimation. Moreover, the search tasks used in these experiments seem to be particularly inefficient and may have encouraged subjects to use an ordered pattern of eye movements; "reading" a display is one example.

IV. THE COST OF SYSTEMATIC SEARCH?

Why don't all searches use a systematic pattern of attentional deployments? This could provide the benefits of the standard model without the demand for perfect memory. If sampling without replacement could be accomplished by shifting attention or the eyes in a predetermined scanpath, then the only memory required would be memory for which direction to move in. You are using such a scanpath in reading this article. Why don't subjects simply scan search arrays left-to-right, in an expanding spiral, or some other efficient search pattern?

Wolfe, Alvarez, and Horowitz (2000) have demonstrated that when subjects are required to search systematically (e.g., clockwise around a circular array), search rates are slow (·300 ms/item), compared to control conditions in which search strategy is unconstrained. Although systematic search is efficient in terms of the number of samples required, the cost in terms of search rate appears to be too great for displays in which rapid, covert deployments of attention are possible. When search rate is limited to the slower rate of overt deployments of the eyes, systematic search may become a viable option.

V. LIMITED-CAPACITY MEMORY?

Memoryless search and the standard model are really two ends of a continuum. It is possible that there is some, imperfect memory for rejected distractors. The experiments that falsify perfect sampling without replacement do not falsify limited or fallible memory systems. Indeed, a successful limited-capacity model might resolve many controversies in this area. Two versions of this solution have been proposed so far. Horowitz and Wolfe (2001) introduced a buffer model in which attention was prevented from orienting to the c most recently examined items, but the remaining arrays items were sampled randomly, even those that had been sampled in the past. After each item was attended, whichever item had been attended $c + 1$ samples back was dropped off the stack, and the current item was added. For the first c samples, this model followed the Standard Model and afterward behaved as an amnesic model with the set size reduced by c items. This compromise model was a better fit to their data. The average estimate of c was three items for their subjects. (Note that substantially larger estimates were obtained by Takeda, in press.) A similar model was used to fit CDFs by Horowitz, Wolfe, and Alvarez (see discussion in Horowitz and Wolfe, 2003), who arrived at an estimate of one to two items. A similar account of memory for overt shifts of attention was proposed by McCarley et al. (2003), who observed a buffer size of three to four items. Moreover, recent work suggests that inhibition of return can be measured for the last five to six attended items.

An alternative conception is the variable memory model of Arani, Karwan, and Drury (1984). Instead of allocating a fixed capacity, Arani et al. proposed that subjects might fail either to encode or to retrieve attended locations, and that re-fixation or re-attending might occur through such forgetting.

VI. GENERAL CONCLUSIONS

It seems likely that the Standard Model is wrong. Current research is developing a more nuanced and precise understanding of the degree to which rejected distractors are remembered in visual search and the mechanisms that might underlie such a memory.

References

Arani, T., Karwan, M. H., and Drury, C. G. (1984). A variable-memory model of search. *Human Factors* **26**, 631–639.

Chan, A. H., and Courtney, A. J. (1998). Revising and validating the random search model for competitive search. *Perceptual & Motor Skills* **87**, 251–260.

Gibson, B. S., Li, L., Skow, E., Salvagni, K., and Cooke, L. (2000). Memory-based tagging of targets during visual search for one versus two identical targets. *Psychological Science* **11**, 324–328.

Gilchrist, I. D., and Harvey, M. (2000). Refixation frequency and memory mechanisms in visual search. *Current Biology* **10**, 1209–1212.

Horowitz, T. S., and Wolfe, J. M. (1998). Visual search has no memory. *Nature* **394**, 575–577.

Horowitz, T. S., and Wolfe, J. M. (2001). Search for multiple targets: remember the targets, forget the search. *Perception & Psychophysics* **63**, 272–285.

Horowitz, T. S., and Wolfe, J. M. (2003). Memory for rejected distractors in visual search? *Visual Cognition* **10**, 257–298.

Krendel, S. E., and Wodinsky, J. (1960). Search in an unstructured visual field. *Journal of the Optical Society of America* **50**, 562–568.

Kristjánsson, A. (2000). In search of remembrance: Evidence for memory in visual search. *Psychological Science* **11**, 328–332.

McCarley, J. S., Wang, R. F., Kramer, A., Irwin, D. E., and Peterson, M. S. (2003). How much memory does oculomotor search have? *Psychological Science* **14**, 422–426.

Peterson, M. S., Kramer, A. F., Wang, R. F., Irwin, D. E., and McCarley, J. S. (2001). Visual search has memory. *Psychological Science* **12**, 287–292.

Scinto, L. F. M., Pillalamarri, R., and Karsh, R. (1986). Cognitive strategies for visual search. *Acta Psychologica* **62**, 263–292.

Takeda, Y. (in press). Search for multiple targets: Evidence for memory-based control of attention. *Psychonomic Bulletin & Review*.

von Muhlenen, A., Muller, H. J., and Muller, D. (2003). Sit-and-wait strategies in dynamic visual search. *Psychological Science* **14**, 309–314.

Wolfe, J. M., Alvarez, G. A., and Horowitz, T. S. (2000). Attention is fast but volition is slow. *Nature* **406**, 691.

CHAPTER 44

The Neuropsychology of Visual Feature Binding

Glyn W. Humphreys and M. Jane Riddoch

ABSTRACT

We review evidence from the neuropsychology of feature binding that reveals that (a) binding processes are mediated by the parietal lobe and (b) several binding processes may take place at different stages of vision. The relations between neuropsychological disorders of binding and synesthesia are discussed.

There is a good deal of evidence from electrophysiological studies in nonhuman primates, and from brain imaging and neuropsychological studies of humans, that many attributes of the visual world are processed independently and in parallel. For example, separate areas of cortex are specialized for processing color and motion (see Zeki, 1993), and selective lesions of these areas can lead to isolated problems in color and motion perception (Heywood and Cowey, 1999; Heywood and Zihl, 1999). In order for a coherent perceptual representation of the world to be coded, there may need to be some binding process that enables the different properties of objects to be integrated correctly.

Two main accounts have been offered to explain this binding process. One proposal, supported by physiological data on neuronal firing patterns, is that binding depends on the temporal synchrony of neuronal firing between features that belong to the same object (e.g., Singer and Gray, 1995). The temporal signal provided by neurons responsive to different features being active together indicates that the features "belong together." The second proposal, expressed in Feature Integration Theory (FIT; e.g., Treisman, 1998), is that features are bound by spatial attention. By attending to the location where the stimuli fall, the activation levels for these features will be increased and those for unattended features suppressed (cf. Moran and Desimone, 1985). This solves the "binding problem" because it prevents bindings between the features of the attended object and those of unattended objects. According to FIT, participants may make "binding errors" (perceiving incorrect conjunctions of properties) if they are prevented from attending appropriately to stimuli. There is evidence that these "illusory conjunctions" occur in vision (e.g., incorrect combinations of color and shape; Treisman and Schmidt, 1982) and even for incorrect bindings between a stimulus and its modality of occurrence (a felt texture being reported as being seen; Cinel, Humphreys, and Poli, 2002). There is also evidence that binding is dependent on intact spatial coding. The neuropsychological disorder of Balint's syndrome (Balint, 1909), found after bilateral damage to parietal cortex, is associated with poor visual localization. Patients with Balint's syndrome can make large numbers of illusory conjunctions even when stimuli are presented for prolonged durations (Friedman-Hill, Robertson, and Treisman, 1995; Humphreys et al., 2000). Similar results have been found in patients with unilateral pulvinar lesions, especially when stimuli are mislocalized (Ward et al., 2002).

Although both the temporal and spatial binding accounts stress the operation of a single binding process, there is some suggestion that several forms of binding operate, probably modulated by different brain regions. For example, studies with normal participants show that illusory conjunctions are greater if the miscombined elements fall within a single group (Prinzmetal, 1981; Prinzmetal, Presti, and Rho, 1986), suggesting that elements may be grouped along one dimension before being "bound" to their other properties. Similarly, patients with Balint's syndrome can be influenced by grouping between form elements, even though they miscombine forms and colors from different stimuli (Humphreys et al., 2000). Form-based grouping may take place within ventral areas of cortex concerned with pattern recognition. In contrast, binding across feature dimensions such as form and color may rely on interactions between ventral areas of cortex and more dorsal areas (e.g., posterior parietal cortex) that provided a spatial register of visual features. Damage to these dorsal regions, found in patients with Balint's syndrome, may lead to a selective disturbance in binding across visual dimensions, though binding within the form dimension is relatively spared. This is consistent with such patients frequently being able to recognize single objects. The distinction between binding between form elements, and the separate binding of those form elements to surface properties of stimuli (such as color), is also made by some computational models of visual processing (e.g., Grossberg and Mingolla, 1985).

Although patients with Balint's syndrome are abnormally poor at binding properties such as the shape and color of stimuli, some data indicate that this problem is most apparent when the patients have to explicitly report on the relations between the different visual properties. When patients have to report only on a single property (e.g., they have to name a color and not report the word that the color is carried by), they can still be affected by the other property present. For example, Wojcuilik and Kanwisher (1998) showed that their patient was slow to name a color carried by an incongruent color word (a Stroop effect), even though he was at chance at explicitly reporting the relation between the word and its color. One reason for this difference between (poor) explicit report of the relations between visual features and at least some form of implicit coding may be that explicit report, in particular, depends on top-down feedback from later processes, to stabilize the codes formed in early visual areas (Hochstein and Ahissar, 2002). This stabilization process is disrupted after parietal damage.

There is also some neuropsychological evidence for temporal binding in vision. Goodrich and Ward (1997) first reported the phenomenon of anti-extinction, when a patient could not detect a single stimulus presented in their impaired (contralesional) field, but could report the same item if another stimulus appeared on the spared (ipsilesional) side. Humphreys et al. (2002) showed that anti-extinction was sensitive to the two stimuli appearing as onsets, and that it was stronger when the stimuli appeared together relative to when they were staggered in time. They suggested that anti-extinction could reflect temporal grouping between visual onsets.

When an item in the impaired field groups with an item in the spared field, the two stimuli may be selected as a single object, allowing even impoverished perceptual information from the contralateral item to be reported.

In neuropsychological studies, we can witness the breakdown of normal binding processes, which can be mimicked in normal observers by limiting their attention to displays. In the syndrome of synesthesia, however, we may observe the operation of abnormal binding processes, formed between properties that would not usually have a necessary association. People who report synesthesia may always see a number in a particular color (even when it is presented in a different color), or they may associate a smell with a given sound (Robertson, 2003). These unusual bindings are typically consistent within an individual and may reflect long-term connections between properties coded in different brain areas (both within and across modalities). On this view, synesthesia is the result of too-active a binding process, perhaps reflecting increased connectivity across neural areas. In contrast, disorders such as Balint's syndrome can reflect a failure to establish active bindings, particularly between separate visual dimensions.

References

Balint, R. (1909). Seelenlähmung des Schauens, optische Ataxie, räumliche Störung der Aufmerksamkeit. *Monatsschr. Für Psychiat. Und Neurol.* **25**, 51–81.

Chinel, C., Humphreys, G. W., and Poli, R. (2002). Cross-modal illusory conjunctions between vision and touch. *J. Exp. Psych.: Hunt. Perc. and Perf.* **28**, 1243–1266.

Friedman-Hill, S. R., Robertson, L. C., and Treisman, A. (1995). Parietal contributions to visual feature binding: evidence from a patient with bilateral lesions. *Science* **269**, 853–855.

Goodrich, S., and Ward, R. (1997). Anti-extinction following unilateral parietal damage. *Cog. Neuropsych.* **14**, 595–612.

Grossberg, S., and Mingolla, E. (1985). Neural dynamics of form perception: Boundary completion, illusory figures and neon color spreading. *Psych. Rev.* **92**, 173–211.

Heywood, C. A., and Cowey, A. (1999). Cerebral achromatopsia. *In* The cognitive neuropsychology of vision. (G. W. Humphreys Ed.), London: Psychology Press.

Heywood, C. A., and Zihl, J. (1999). Motion blindness. *In* The cognitive neuropsychology of vision. (G. W. Humphreys Ed.), London: Psychology Press.

Hochstein, S., and Ahissar, M. (2002). View from the top: Hierarchies and reverse hierarchies in the visual system. *Neuron* **36**, 791–804.

Humphreys, G. W., Cinel, C., Wolfe, J., Olson, A., and Klempen, N. (2000). Fractionating the binding process: Neuropsychological evidence distinguishing binding of form from binding of surface features. *Vis. Res.* **40**, 1569–1596.

Humphreys, G. W., Riddoch, M. J, Nys, G., and Heinke, D. (2002). Unconscious transient binding by time: Neuropsychological evidence from anti-extinction. *Cog. Neuropsych.* **19**, 361–380.

Moran, J., and Desimone, R. (1985). Selective attention gates visual processing in the extrastriate cortex. *Science* **229**, 782–784.

Prinzmetal, W. (1981). Principles of feature integration. *Perc. & Psychophys.* **30**, 330–340.

Prinzmetal, W., Presti, D. E., and Rho, S. H. (1986). Does attention affect visual feature integration? *J. Exp. Psych.: Hum. Perc. & Perf.* **12**, 361–370.

Robertson, L. C. (2003). Binding, spatial attention and perceptual awareness. *Nature Rev. Neurosci.* **4**(2), 93–102, February.

Robertson, L. C., Treisman, A., Friedman-Hill, S., and Grabowecky, M. (1997). The interaction of spatial and object pathways: Evidence from Balint's syndrome. *J. Cog. Neurosci.* **9**, 295–317.

Singer, W., and Gray, C. M. (1995). Visual feature integration and the temporal correlation hypothesis. *Ann. Rev. Neurosci.* **18**, 555–586.

Treisman, A. (1998). Feature binding, attention, and object perception. *Phil. Trans. R. Soc. Lond. B* **353**, 1295–1306.

Treisman, A., and Schmidt, H. (1982). Illusory conjunctions in the perception of objects. *Cognitive Psychology*, **14**, 107–141.

Ward, R., Danziger, S., Owen, V., and Rafal, R. D. (2002). Deficits in spatial coding and feature binding following damage to spatiotopic maps in the human pulvinar. *Nature Neurosci.* **5**, 99–100.

Wojciulik, E., and Kanwisher, N. (1998). Implicit but not explicit feature binding in a Balint's patient. *Vis. Cog.* **5**, 157–181.

Zeki, S. (1993). *A vision of the brain.* Oxford: Oxford University Press.

CHAPTER

45

Visual Saliency and Spike Timing in the Ventral Visual Pathway

Rufin VanRullen

ABSTRACT

I argue against the view that "bottom-up" saliency necessarily recruits the attentional system prior to object recognition. A number of visual processing tasks are clearly performed too fast for such a costly strategy to be employed. Rather, visual attention could simply act by biasing a "saliency-based" object recognition system. At any given level of the ventral pathway, the most activated cells of the neural population simply represent the most salient locations. The notion of saliency itself grows increasingly complex throughout the system, based mostly on luminance contrast until information reaches visual cortex, gradually incorporating information about features such as orientation or color in primary visual cortex and early extrastriate areas, and finally the identity and behavioral relevance of objects in temporal cortex and beyond. Under these conditions, the object that dominates perception, that is, the object yielding not only the strongest, but also the first selective neural response, is by definition the one whose features are most "salient"—without the need for any external saliency map.

I. INTRODUCTION

It is classically assumed that saliency information is systematically extracted from the visual scene and explicitly represented in our visual system in the form of a "saliency map" (see Chapter 94). When an intrinsically salient object is presented, its neural representation in the saliency map automatically directs "bottom-up" attention to its location. This in turn allows the object to be further processed in the ventral visual stream and eventually recognized or categorized. This hypothesis of a saliency map has proven to be a rather effective way of directing attention to regions of interest in the visual scene.

One problem with this classical view is that a number of seemingly complex visual recognition tasks are performed so rapidly that this type of processing strategy would simply not be viable. From rapid serial visual presentation or rapid visual categorization paradigms (Thorpe et al., 1996) it is usually concluded that a complex natural scene can be categorized based on semantic aspects such as the presence or absence of a target object (animal, vehicle) in around 150 ms. Given the number of processing stages involved, this remarkable speed makes it highly unlikely that such processing could rely on a first pass through a saliency map followed by a shift of spatial attention, finally leading to object recognition. Rather, if saliency information is used at all in such extreme cases, it must be extracted during object recognition itself. In addition, recent evidence suggests that rapid visual categorization of natural scenes can even be performed under conditions where focal attention is tied elsewhere (Li et al., 2002), a result that constitutes a strong challenge for a theory based on bottom-up recruitment of spatial attention prior to object recognition.

Here I argue that saliency is in fact represented, albeit implicitly, in neural responses throughout the ventral visual stream. I propose that saliency could serve as an organizational principle in this hierarchy of visual areas: the visual system must represent those features that are *potentially salient* for the observer. These not only include luminance, orientation, color or disparity contrast, transient events, but also object identity and/or behavioral relevance, and depend on the spatial as well as the temporal context.

Before developing this theory, I will first attempt to provide a working definition of saliency. Finally, I will speculate that spike timing is certainly the most appropriate carrier of saliency information in the ventral visual stream.

II. A "BEHAVIORAL" DEFINITION OF VISUAL SALIENCY

Defining visual saliency is a challenging problem, since it is a volatile concept that changes with the observer and its particular goals. A reasonable approach is to simply enumerate the different possible factors that can participate in determining saliency.

Luminance contrast appears to be the primary variable on which saliency computation is based (Fig. 45.1a,b). It is also the first type of information extracted by our visual systems in the retina. At higher levels of processing in the visual cortex, other feature dimensions are encoded and participate in defining visual saliency. Among these are edge or line orientation (Fig. 45.1c), color, motion, and stereo disparity. One major observation—and possibly the only generalization at this point—is that in each case the relevant variable is not the amplitude of visual signals along a particular feature dimension, but the *contrast* between this amplitude at a given point and at the surrounding locations. In other words, the saliency of a visual object depends on the *context* in which the object is presented. This generalization holds for visual "features" at a much higher level of complexity: semantic contrast, for example (in space, Fig. 45.1d, or in time, Fig. 45.1e) can also boost the saliency of an object. As we move to higher levels of representation, the notion of saliency grows to include the identity and behavioral relevance of objects (such as whether they are dangerous or appetizing; Fig. 45.1f), and is biased in a top-down fashion by voluntary shifts of attention (Fig. 45.1g).

A critical aspect of saliency is that the different dimensions listed so far are not mutually exclusive but generally all interact and contribute simultaneously

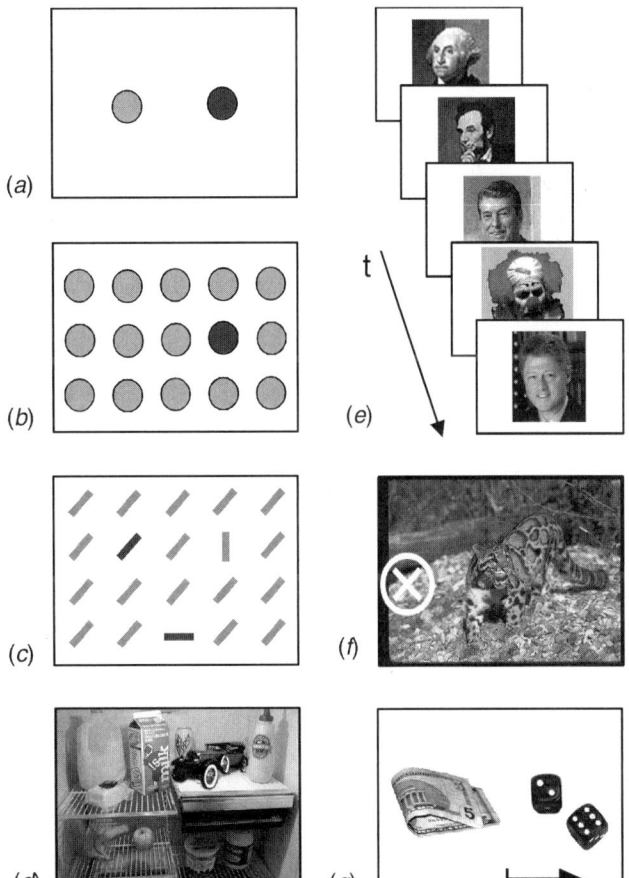

FIGURE 45.1 Saliency is a property of objects or locations, but can also depend strongly on the context (spatial, temporal, or semantic) of the scene. (*a*) Contrast (luminance contrast in particular) is the primary determinant of saliency. In this rectangle, both disks are salient, but the one on the right is more readily perceived because the local contrast is stronger at that location. (*b*) Even when based on luminance contrast, saliency depends on the spatial context: placing similar disks in the background diminishes the saliency of the disk on the left, while the other one remains salient. This type of luminance-based saliency can be derived from the retinal ganglion cells' or the lateral geniculate cells' response. (*c*) Other features (such as orientation or color) or feature contrasts can determine saliency. The vertical bar on the right appears salient because it differs in orientation from the other bars. This feature-based saliency (which can be derived from neural responses in primary visual cortex and beyond) adds up to the one described previously, so that a conjunction of luminance and orientation contrast appears more salient than either of them separately (see the horizontal dark bar at the bottom). (*d*) A "semantic" contrast can also render an object more salient. In this case the automobile (which is not necessarily the most contrasted object in the scene) is rendered salient by its presence in a semantically incongruous context: the interior of a refrigerator. Note that the difference in spatial scale between the car and the other objects in the scene also participates in making it conspicuous. (*e*) A semantic contrast in time—rather than in space—has a similar effect. The picture of the clown in a sequence of pictures of former presidents is made salient by this semantic discrepancy. (*f*) The behavioral relevance of objects is also a critical factor in determining their saliency. To any human observer the predator is clearly the most relevant object in this scene. However, a system only taking into account the physical properties of the scene (Chapter 94) would consider the location indicated by the cross as the most salient. (*g*) Top-down effects, finally, can bias this race for saliency. The luminance contrast of the dice is higher than that of the 5-dollar bill. As mentioned previously, the behavioral relevance attached to "money" can potentially revert this saliency ranking. Voluntarily drawing attention to the dice (as if cued by the arrow, or as in a gambling context) could in turn render them more salient.

to the representation of saliency (Chapter 38). For example, the conjunction of luminance contrast and orientation contrast is more salient than either of them separately (Fig. 45.1c). These interactions between different feature dimensions are not disorganized but follow a particular hierarchy, such that an increase in luminance contrast (the "base" of this hierarchy) will systematically enhance the saliency of the corresponding location (Chapter 38), even when top-down priorities request that this location should be ignored (see Chapter 8).

Saliency is usually associated with bottom-up properties of the scene, while attentional modulation is generally referred to in the context of top-down control of visual processing. In other words, attentional effects are thought to mimic saliency, so that the two might be indistinguishable at the single-cell level (see Chapter 8), yet one is actually driven by intrinsic properties of the scene (i.e., exogenous), while the other one is voluntarily initiated (i.e., endogenous). Note, however, that the boundary between bottom-up and top-down components of visual saliency might not be as clear-cut as this distinction would suggest (e.g., is it because of bottom-up or top-down effects that we are startled by an unexpected behaviorally relevant object such as the face of the predator in Fig. 45.1f?). Both low-level (e.g., contrast, color) and higher-level (e.g., object identity) features of a stimulus can be made salient by either bottom-up or top-down effects. How do these interacting processes map onto the neuronal architecture of the visual cortex?

III. VISUAL SALIENCY IN THE VENTRAL VISUAL PATHWAY

The visual system has been divided according to anatomical, neuropsychological, functional and electrophysiological evidence into two main streams (Ungerleider and Mishkin, 1982; Morel and Bullier, 1990): the dorsal stream where information about motion and spatial relationships among objects is represented (and containing one potential candidate for a saliency map; Colby and Goldberg, 1999), and the ventral stream concerned primarily with the identity or the category of objects. It is supposedly in the ventral stream that spatial attention, controlled by the external saliency map, must gate visual processing. However, I will argue here that because saliency is an intrinsic property of the objects that are processed within the ventral stream, saliency information is necessarily present in these neuronal receptive fields and can act to effectively gate the relevant aspects of the visual scene.

Along the ventral pathway the specific visual properties and features to which cells are selective become more and more complex. For simplicity, let us focus here on the features that participate in 2D shape recognition (and keep in mind that similar mechanisms must exist for color, disparity or motion). As mentioned previously, the first feature dimension extracted by the visual system in the retina and also present in the LGN is luminance contrast. In the primary visual cortex, neurons use this input to build selectivity for line or edge orientation and sometimes display a certain degree of invariance to retinotopic position (i.e., complex cells). Further down the line neurons respond to more abstract (e.g., illusory) contours and figure/ground boundaries in V2, and to complex geometric patterns in V4. Selectivity for the identity and category of complex objects or their components arises in the posterior part of the inferotemporal cortex (PIT) and is refined as visual information advances to the anterior part (AIT). Typically, neurons in IT respond (sometimes in a viewpoint invariant manner) to meaningful objects, in particular those with obvious biological relevance such as faces (Tanaka, 1996). IT is thus often considered as the end-point of the ventral stream hierarchy. Yet one should not overlook the fact that IT is heavily interconnected with medial temporal lobe structures and the prefrontal cortex (Morel and Bullier, 1990). The parahippocampal complex (PHC) of the medial temporal lobe (i.e., perirhinal and entorhinal cortices) is thought to maintain a memory of task-related information such as the identity or category of a target (Suzuki, 1996). The prefrontal cortex could mediate even higher-level visual information related to the current decision or to longer-term behavioral plans (Crick and Koch, 1998; Wallis et al., 2001).

This hierarchy, represented in Fig. 45.2, is widely taken as evidence for a functional architecture in which, in a sequence of relatively small computational steps, visual areas extract from their afferents increasingly complex features of the stimulus ("feature-detection" theory; see Barlow, 1972). At the last levels such "features" are by construction complex enough to represent object identity or category. Firing in these cells thus corresponds to object recognition. However, because the models must satisfy the mutually exclusive constraints of (i) achieving position and size invariance and (ii) minimizing interference from multiple objects in a receptive field, they must restrict processing to a certain location using an attentional "gating," implemented (in most cases) by external feedback modulation from the saliency map.

While this external feedback mechanism is a reasonable assumption for top-down attentional modulation, where only higher-level areas possess the

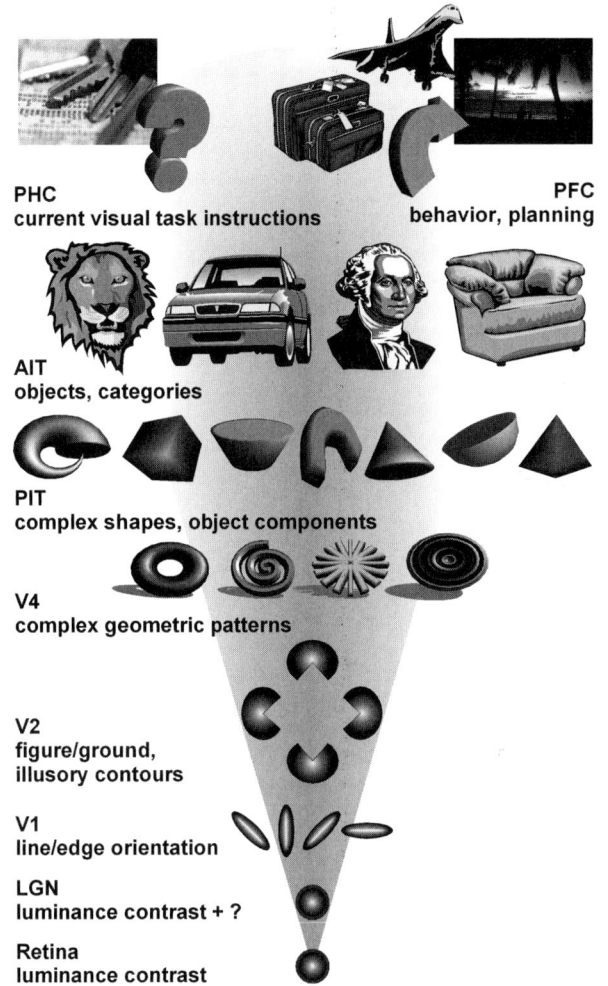

FIGURE 45.2 Feature-detection hierarchy, or saliency-based object recognition system? The hierarchical organization of the ventral visual stream (here limited to aspects of 2D shape recognition) is usually interpreted in terms of a "feature-detection" framework: complex selectivities are built step by step from simpler ones. From contrast and edges in the retina and V1, neurons can build receptive fields selective to complex shapes and finally objects and categories in inferotemporal cortex. However, processing does not end in IT. Cells in its efferent structures can represent task-related information (e.g., "where are my keys?") in the parahippocampal complex (PHC), and behavioral relevance or longer-term plans (e.g., "I must find my suitcases before I can leave for vacation") in prefrontal cortex (PFC). The ultimate output of the ventral stream might not be simply "object recognition," but this pathway could be more generally directed to determining the relevance of visual objects. This hierarchy can thus also be interpreted in terms of saliency extraction rather than feature-detection, since saliency can be thought of as the "exogenous" component of behavioral relevance. In particular, because of center-surround receptive fields, lateral interactions and other nonlinear competition mechanisms, neurons at a given level $\mathcal{L}$ (say, V2) not only represent the "feature" specifically assigned to that level of the feature-detection hierarchy (e.g., figure-ground boundary), but also encode the saliency of their inputs in terms of the "features" assigned to levels 1 to $\mathcal{L}$-1 (e.g., contrast *and* orientation).

information that should modulate visual neurons' responses, it would seem computationally absurd in the case of bottom-up attention, since the features that can make an object salient are already represented within the ventral pathway. Is it possible to make use of this information and compute saliency in parallel with the integration of afferents leading to feature (or object) recognition in a receptive field?

Upon examining Fig. 45.2, one realizes that the ultimate goal of the ventral visual pathway might not be simply object recognition, but more importantly to make judgments about the behavioral relevance of stimuli. Why did our visual systems choose, through evolution, to represent a particular feature or object rather than another? Surely because it is *informative* in some sense, that is, *potentially salient* and of relevance to the organism as a whole. As proposed in section II, saliency can be defined hierarchically, from purely exogenous factors (e.g., stimulus contrast, orientation) to more endogenous ones (e.g., task relevance). In other words, and to extend (rather than replace) the feature-detection theory, the hierarchical structure of the ventral visual pathway also reflects the fact that saliency is a constructed notion based on a hierarchy of feature contrasts.

Consider a given level $\mathcal{L}$ of the ventral hierarchy, at which a certain feature dimension is explicitly represented (e.g., figure-ground boundaries in V2). In a feature-detection context, neuronal receptive fields build their selectivities upon those of the preceding level $\mathcal{L}$-1 (e.g., orientation). But because of the center-surround organization of these receptive fields, of lateral interactions at the preceding level and possibly other types of nonlinear competition mechanisms (e.g., Max-like operation (see Chapter 46); see also the next section for an alternative nonlinear mechanism implemented in the temporal domain), the neurons at level $\mathcal{L}$ also respond to local contrast in terms of the feature dimension represented at level $\mathcal{L}$-1 (e.g., orientation contrast). By induction (remember that the first dimension extracted in the retina is not luminance per se but luminance contrast), the receptive fields at level $\mathcal{L}$ actually contain information about local contrasts in *all* feature dimensions from level 1 to $\mathcal{L}$-1 (e.g., luminance *and* orientation contrast), that is, about the "saliency" of the stimulus in their receptive fields. In this scheme, the notion of saliency itself depends on the processing level considered, and, for example, could potentially include semantic contrast or behavioral relevance in higher levels of the hierarchy such as the prefrontal cortex. What is important here is that the activation of a neuron in the ventral pathway not only reflects the "match" between its preferred feature and the stimulus in its receptive field (i.e., what is predicted by the feature-detection theory), but also the global saliency at that location. The most activated neuron of a population corresponds *by construction* to the most salient location at that stage. This result is obtained without the need for an explicit saliency map and could explain how humans are able to detect target objects (e.g., animals, vehicles) in natural scenes in only 150 ms (Thorpe et al., 1996), without the need for focal attention (Li et al., 2002).

IV. SPIKE TIMING AND VISUAL SALIENCY

How could such a processing scheme be implemented by neurons in the visual system? Essentially, a multitude of nonlinear operations that result in an enhancement of activity at one salient location compared to the surrounding locations (e.g., center-surround filtering, max or "softmax" operations) could be compatible with this hypothesis, and the previous sections make no assumption as to the type of coding mechanism used. Here, I would like to suggest that using relative spike timing as an indicator of saliency is a particularly efficient way of implementing this strategy.

One property of biological neurons, often overlooked by both theoretical and experimental neuroscientists, is that the level of postsynaptic excitation (i.e., the net activation) of a neuron translates not only into changes of spike firing rate, but also of spike firing latency. Consider a population of neurons encoding the local luminance contrast in their receptive fields. As shown in Fig. 45.3a, the relative firing times (i.e., the specific *order* of firing) in the first wave of spikes generated by this population in response to a given stimulus pattern carry most of the information present in the input. This information can be accessed by efferent neurons if we assume that they possess some (even vague) form of prior knowledge of the statistics of input ordering in their receptive field, and apply this knowledge by progressively decreasing their sensitivity as they receive more and more input spikes (for example, this could be achieved using fast shunting inhibition). This desensitization will constitute the nonlinear mechanism on which saliency extraction is based. Under these conditions the spatiotemporal firing pattern in this wave of spikes can be very effectively decoded, even when only the first 1 to 2 percent of spikes in the wave are taken into account (VanRullen and Thorpe, 2001).

At the following processing level (e.g., orientation detection), the most activated cell (whether this activation comes from feature detection or saliency computation) will also be the first one to fire. In other words, this population coding scheme can be reproduced in a cascade, and used to build efficient hierarchical object recognition systems compatible with the time constraints observed experimentally (VanRullen et al., 1998). How does spatial routing of information occur in such systems?

As mentioned in the previous section, neurons at a given level not only represent a particular feature dimension but also the local saliency in their receptive field. Let us go back to the simple example (Fig. 45.3) of a population of neurons encoding luminance contrast at different spatial scales, and take the point of view of a neuron (e.g., orientation-selective) reading out the firing order of cells in this population. In order to understand what this neuron "perceives," we can reconstruct (in the context of wavelet theory; Mallat, 1989) the visual information in its receptive field using the same decoding scheme as the neuron itself uses (i.e., we simply desensitize the reconstruction process with each incoming spike). As is apparent in Fig. 45.3b, the most salient (i.e., contrasted) locations of the input naturally dominate the reconstructed receptive field, even at the very beginning of the reconstruction process: indeed, they are by definition represented by the first spikes generated in the input population. If we increase the input saliency of a particular region (by enhancing local contrast at each input spatial scale using Gaussian-distributed weights), the receptive field now seems to "shrink" or "shift" around that location (given the same number of input spikes; Fig. 45.3c). The "active" region of the receptive field virtually follows the salient input location. In addition, surrounding locations have a decreased effect on this receptive field, even though they were not affected by the contrast modulation. In other words, salient information is effectively "gated" inside the receptive field (see Chapter 3, and Salinas & Abbott, 1997, for a related demonstration using endogenous attentional control). This is a consequence of the neuron's sensitivity behaving as a limited resource, as the "juice" of saliency that can be spread over the field or recruited at particular locations. In image processing jargon, it realizes an automatic normalization of inputs inside a receptive field.

Going back to the bigger picture, we should insist on two points. First, this saliency-controlled gating of information based on relative spike timing can, of course, take place at any level within the ventral stream. Just as proposed in the previous section, at each stage neurons respond to a particular feature dimension *and* necessarily to input saliency (defined by previous feature dimensions) in their receptive fields. Therefore, here again the output of the system

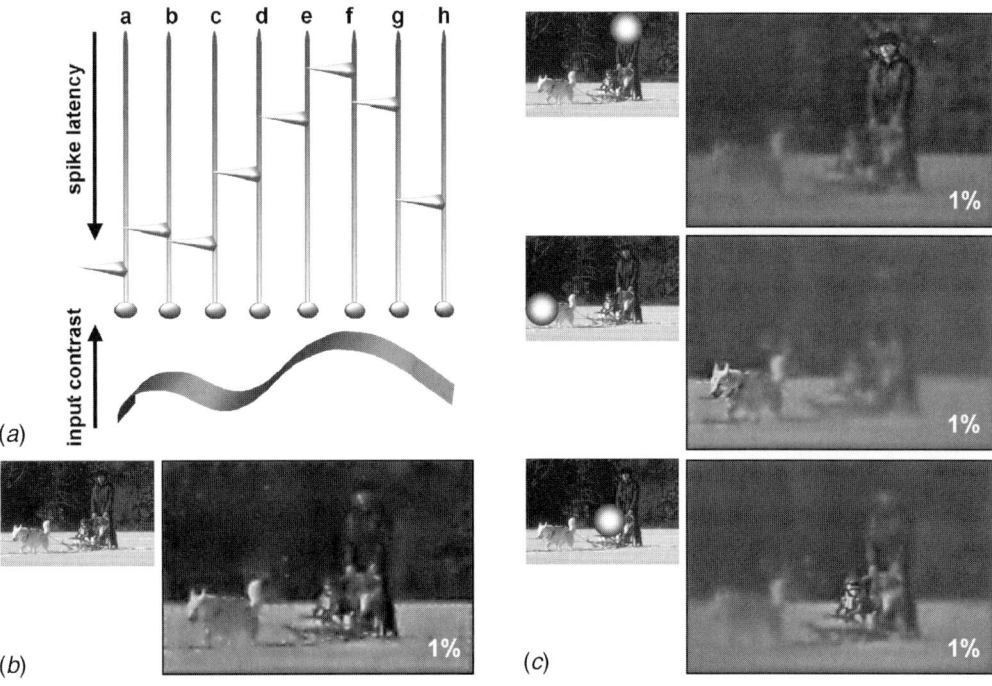

FIGURE 45.3 Using spike timing to implement dynamic routing of salient information. (*a*) An input signal to a neural population is converted into a spatiotemporal wave of action potentials: the most activated neurons fire earlier. Considering the order of spikes inside this wave is a computationally efficient coding scheme. (*b*) This temporal information can be decoded by assuming that a target neuron becomes desensitized as it receives more and more inputs: the neuron would have "learned" the statistics associated with input ordering. Suppose that the (reduced) image on the left is presented to a population of contrast-selective cells (at 2 polarities, ON- and OFF-center, and at different spatial scales). The larger image represents a reconstruction of what an efferent neuron "perceives" in its receptive field, after only 1 percent of the input neurons have fired, if it is sensitive only to the order of these incoming spikes. The most salient locations of the original image appear first in this reconstruction. (*c*) When the saliency of a given region (indicated by the circle in each reduced image) is increased by modulating (with Gaussian-distributed weights) the input contrast in that region (and at different spatial scales), the reconstruction is biased toward the salient location: not only does this region appear more contrasted, but the rest of the receptive field (which was not directly affected by the saliency manipulation) seems to "fade away." The receptive field has "shrunk" or "shifted" to the salient location, because the desensitization mechanism effectively acts as an automatic normalization process. The same result can be obtained if we assume that spatial top-down attention acts to decrease firing latencies in a certain region of the receptive field (VanRullen and Thorpe, 1999). Everything happens in fact as if the receptive field contained a fixed amount of a certain "resource" (i.e., the efferent neuron's sensitivity) that can be spread across the field or focused at particular locations, depending on input saliency and attentional factors.

represents *by construction* the most salient or behaviorally relevant object. In this case, however, this result is obtained on the basis of the feedforward propagation of the very first wave of spikes generated by each neuron: such a system is certainly the fastest imaginable for such processing. Second, all the model properties described here in the case of input ("bottom-up" or "exogenous") saliency are still valid in the case of "endogenous" saliency (i.e., top-down spatial attention), if we assume that spatial attention acts by decreasing firing latencies in the attended region, therefore mimicking an increase in local saliency (VanRullen and Thorpe, 1999). At the cellular level, the two components of saliency will even be indistinguishable, a phenomenon observed in electrophysiological recordings of visual neurons (see Chapter 8).

V. DISCUSSION

Even though attempts have been made to explain saliency in different terms, the saliency map hypothesis is still the most widely used in current models of attentional control. Desimone and Duncan (1995) certainly opened the way to alternative thinking with their biased competition framework, but did not get clearly involved in the "saliency map" debate. Zhaoping Li proposed that input saliency could be encoded implicitly in feature detectors of the primary visual cortex (see Chapter 93). It is shown here that a similar idea can in fact be extended to the entire ventral visual stream. In particular, building up visual representations on a hierarchy of feature contrasts allows the system to utilize a definition of saliency that extends beyond pure bottom-up (exogenous) aspects and includes behavioral or task-relevance.

This is not to say, however, that the saliency map is a useless idea. For example, the present model only ensures that the most salient object(s) will be detected in the first feedforward pass through the system, but does not provide any control mechanism for subsequently scanning the next most salient locations in a crowded scene. A saliency map network is still one of the best known ways to achieve this, and it is fairly possible to imagine the two ideas working together in a single model (see Chapter 98).

References

Barlow, H. B. (1972). Single units and sensation: a neuron doctrine for perceptual psychology? *Perception* **1**(4), 371–394.

Colby, C. L., and Goldberg, M. E. (1999). Space and attention in parietal cortex. *Annu. Rev. Neurosci.* **22**, 319–349.

Crick, F., and Koch, C. (1998). Consciousness and neuroscience. *Cereb. Cortex* **8**(2), 97–107.

Desimone, R., and Duncan, J. (1995). Neural mechanisms of selective visual attention. *Annual. Review. Neurosci.* **18**, 193–222.

Li, F. F., VanRullen, R., Koch, C., and Perona, P. (2002). Rapid natural scene categorization in the near absence of attention. *Proc. Natl. Acad. Sci. USA* **99**(14), 9596–9601.

Mallat, S. G. (1989). A theory for multiresolution signal decomposition: The wavelet representation. *Inst. Electrical Electronics Engrs. Trans. on Pattern Analysis and Machine Intelligence.* **11**, 674–693.

Morel, A., and Bullier, J. (1990). Anatomical segregation of two cortical visual pathways in the macaque monkey. *Vis. Neurosci.* **4**(6), 555–578.

Salinas, E., and Abbott, L. F. (1997). Invariant visual responses from attentional gain fields. *J. Neurophysiol.* **77**(6), 3267–3272.

Suzuki, W. A. (1996). The anatomy, physiology and functions of the perirhinal cortex. *Curr. Opin. Neurobiol.* **6**(2), 179–186.

Tanaka, K. (1996). Inferotemporal cortex and object vision. *Annu. Rev. Neurosci.* **19**, 109–139.

Thorpe, S. J., Fize, D., and Marlot, C. (1996). Speed of processing in the human visual system. *Nature* **381**, 520–522.

Ungerleider, L. G., and Mishkin, M. (1982). Two cortical visual systems. In "Analysis of Visual Behavior" (D. J. Ingle, & M. A. Goodale, and R. J. W. Mansfield Eds.), MIT Press, Cambridge, MA: 549–586.

VanRullen, R., Gautrais J., Delorme A., and Thorpe, S. J. (1998). Face processing using one spike per neuron. *Biosystems* **48**(1–3), 229–239.

VanRullen, R., and Thorpe, S. J. (1999). Spatial attention in asynchronous neural networks. *Neurocomputing* **26–27**, 911–918.

VanRullen, R., and Thorpe, S. J. (2001). Rate coding versus temporal order coding: what the retinal ganglion cells tell the visual cortex. *Neural Comput* **13**(6), 1255–1283.

Wallis, J. D., Anderson, K. C., and Miller, E. K. (2001). Single neurons in prefrontal cortex encode abstract rules. *Nature* **411**(6840), 953–956.

CHAPTER

46

Object Recognition in Cortex: Neural Mechanisms, and Possible Roles for Attention

Maximilian Riesenhuber

ABSTRACT

The primate visual system can rapidly and with great accuracy recognize a large number of diverse objects in cluttered scenes under widely varying viewing conditions. Recent data (Thorpe et al., 1996; Li et al., 2002) have suggested that complex object recognition tasks can be performed in one feedforward pass without the need for attention, providing strong constraints for models of object recognition in cortex. I will review a Standard Model that is an extension of the original model of simple and complex cells of Hubel and Wiesel. Despite its simplicity, this feedforward model can already explain a number of experimental findings, and has been shown to be able to perform object detection in natural images. Moreover, the model can be extended in a straightforward way to investigate how "top-down" attention can modulate "bottom-up" processing to improve its performance. This leads to constraints on the scenarios in which attention can aid object recognition, and to experimental predictions on how attention and feedforward processing might interact.

I. INTRODUCTION

Object recognition is a fundamental cognitive task essential for survival, for example, to detect predators or to discriminate food from nonfood. Despite the apparent ease with which the visual system performs object recognition, it is a very complex computational task requiring a quantitative tradeoff between invariance to certain object transformations on the one hand, and specificity for individual objects on the other. For instance, object recognition needs to be invariant across huge variations in the appearance of objects such as faces, due to viewpoint, illumination, or occlusions, At the same time, the system needs to maintain specificity, that is, the ability to discriminate between different faces.

How does the brain perform object recognition, and what is the role of attention in this process? The experimental data paint a complex picture: Even very complicated visual tasks, such as determining whether an arbitrary natural scene contains an animal, can be performed in the absence of attention. However, other tasks that would appear to be "simpler" such as discriminating bisected colored discs from their mirror images seem to require attention (Li et al., 2002).

In this chapter, I will first review some basic experimental data on object recognition in cortex which motivate a simple computational model that can be viewed as an extension of the original simple-to-complex cell scheme of Hubel and Wiesel. The model can perform object recognition in cluttered natural scenes in one feedforward pass, in agreement with the experimental data. I will then discuss how the model can be extended to incorporate attentional effects, and under what conditions attention can aid object recognition, leading to predictions for experiments.

II. OBJECT RECOGNITION IN CORTEX: SOME EXPERIMENTAL RESULTS

Object recognition in cortex is thought to be mediated by a hierarchy of brain areas called the ventral visual stream (Ungerleider and Haxby, 1994) extending from primary visual cortex (V1) to inferior temporal cortex (IT). IT in turn provides input to prefrontal cortex (PFC) which appears to play a crucial role in linking perception to action. Starting from *simple cells* in primary visual cortex, V1, with small receptive fields that respond preferably to oriented bars, neurons along the ventral stream show an increase in receptive field size as well as in the complexity of their preferred stimuli (Kobatake and Tanaka, 1994). At the top of the ventral stream, in anterior inferotemporal cortex (AIT), cells are tuned to complex stimuli such as faces and other relevant stimuli from the monkey's environment (Logothetis and Sheinberg, 1996). A hallmark of these IT cells is the robustness of their firing to stimulus transformations such as scale and position changes (Logothetis and Sheinberg, 1996). In addition, as these and other studies have shown, most neurons show specificity for a certain object view or lighting condition (so-called view-tuned neurons), while some neurons show view-invariant tuning (view-invariant/object-tuned neurons). The tuning of the view-tuned and object-tuned cells in AIT can be modified by visual experience (for references, see Riesenhuber and Poggio, 2002). Recent fMRI data have shown a similar pattern of tuning properties for the lateral occipital cortex (LOC), a brain region in human visual cortex central to object recognition and believed to be the homologue of monkey area IT (Grill-Spector et al., 2001).

ERP experiments have established that the visual system is able to perform even complex recognition tasks such as object detection in natural images within 150 ms (Thorpe et al, 1996), which is on the order of the latency of neurons in prefrontal cortex, close to the site of the measured ERP effect. Further experiments have shown that such detection tasks can be performed in the absence of attention (Li et al., 2002), and in parallel for two images (Rousselet et al., 2002). These results point to a feedforward account of object recognition in cortex—at least for object detection tasks in natural images—in which recognition is achieved in one processing pass through the ventral visual stream.

III. THE "STANDARD MODEL"

The data described in the previous section motivate a Standard Model of visual processing in cortex, which reflects in its general structure the average belief of many visual physiologists and cognitive scientists. We have provided a quantitative computational implementation (Riesenhuber and Poggio, 1999b) that demonstrates the feasibility of the basic architecture and allows us to integrate experimental data in a rigorous framework and make quantitative predictions for new experiments. The model reflects the general organization of visual cortex in a series of layers from V1 to IT to PFC. From the point of view of invariance properties, it consists of a sequence of two main modules based on two key ideas. The first module, shown schematically in the inset in Fig. 46.1, leads to model units showing the same scale and position invariance properties as view-tuned IT neurons (Riesenhuber and Poggio, 2002; Logothetis and Sheinberg, 1996; Riesenhuber and Poggio, 1999b). Computationally, this is accomplished by a scheme consisting of a hierarchy of just two operations: (1) a "MAX" operation, and (2) a "template match" operation.

In detail, the model proposes that a MAX pooling function, in which a cell's output is determined by its strongest afferent, provides invariance to scaling and translation as well as robustness to clutter while maintaining feature specificity. To illustrate the idea of MAX pooling, consider the example of simple and complex cells in primary visual cortex: Both simple and complex cells respond to bars of a certain orientation, but while simple cells have separate on and off regions (i.e., responding to light and dark bars, respectively) and small receptive fields, complex cells have larger receptive fields with overlapping on/off regions. In the model, complex cells (C1 units in Fig. 46.1) increase translation invariance by performing a MAX pooling operation over simple cells (S1) tuned to the same feature but at different positions (and phase). Besides increasing translation (and scale) invariance, the MAX pooling function is also advantageous for object recognition in clutter: By design, the MAX operation only selects the strongest input to a cell, and the response is not affected by the presence of other objects that activate other afferents to a lesser degree (but might cause strong activation of the afferents to another cell, e.g., one tuned to a different orientation). Thus, the MAX operation provides a biologically plausible mechanism to perform invariant object recognition in clutter (see Riesenhuber and Poggio, 1999a) without the need for a separate segmentation process or special neural circuits to reroute the visual input to a standard reference frame, which are challenged by the timing constraints for object detection imposed by the experimental data, in particular for visual scenes containing more than a single object (but see Chapter 3).

Although oriented edges are a good model for the preferred features of V1 neurons, neurons in higher areas along the ventral stream are tuned to more complex shapes. This is achieved in the model by the other basic neural mechanism, a "template match" operation in which feature complexity is increased by

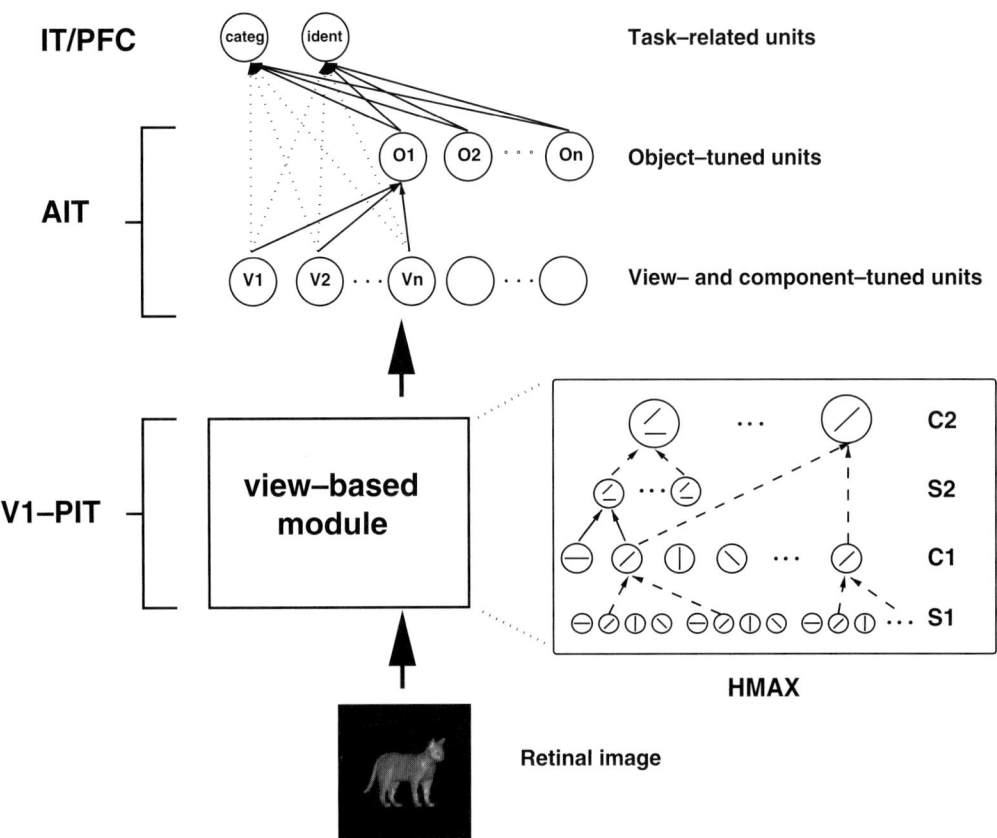

FIGURE 46.1 Sketch of the "Standard Model" of the recognition architecture in cortex. It combines and extends several recent models and effectively summarizes many experimental findings. The view-based module shown in the inset is an hierarchical extension of Hubel and Wiesel's classical paradigm of building complex cells from simple cells. The circuitry consists of a hierarchy of layers leading to greater specificity and greater invariance by using two different types of mechanisms (a MAX pooling mechanism (dashed lines), to increase invariance, and a template match operation (solid lines), to increase feature specificity, see text). The output of the view-based module is represented by view-tuned model units V_n that exhibit tight tuning to rotation in depth (and illumination, and other object-dependent transformations such as facial expression, etc. but are tolerant to scaling and translation of their preferred object view. Invariance to rotation in depth (or other object-specific transformations) is obtained by combining in a learning module several view-tuned units V_n tuned to different views (or differently transformed versions) of the same object (Poggio and Edelman, 1990), creating view-invariant (object-tuned) units O_n. These, as well as the view-tuned units, can then serve as input to task modules that learn to perform different visual tasks such as identification or object categorization. They consist of same generic learning circuitry but are trained with appropriate sets of examples to perform specific tasks. The stages up to the object-centered units probably encompass V1 to anterior IT (AIT). The last stage of task dependent modules may be localized in AIT or prefrontal cortex (PFC). For more information on the model, including source code, see http://riesenhuberlab.neuro.georgetown.edu/hmax. Modified from (Riesenhuber and Poggio, 2003).

combining simpler features into more complex ones (e.g., from the C1 to S2 and C2 to VTU levels in Fig. 46.1). This operation is performed at different levels in the hierarchy to build increasingly complex features while maintaining invariance.

In the second part of the architecture, arbitrary transformations can be learned by interpolating between multiple examples, that is, different view-tuned neurons, leading to neural circuits performing specific tasks. The key idea here is that interpolation and generalization can be obtained by simple networks that learn from a set of examples, that is, input–output pairs (Poggio and Girosi, 1990). In this case, inputs are views, and the outputs are the parameters of interest such as the label of the object or its pose or expression (for a face). The weights from the view-tuned units to the output are learned from the set of examples (see Riesenhuber and Poggio, 2002). In principle, two networks sharing the same VTU input units but with different weights (from the VTUs to the respective output units) could be trained to perform different tasks such as pose estimation, view-invariant recognition, or categorization.

Despite its simplicity, our implementation of this Standard Model of object recognition in cortex has turned out to explain a number of experimental results (for a recent review, see Riesenhuber and Poggio, 2002), and make predictions for new experiments. Importantly, there is now evidence from physiology for the MAX pooling prediction, in comnplex cells in V1 (Lampl et al., 2004) as well as V4 (Gawne and Martin, 2002). Also, recent data from an experiment in which monkeys were trained to categorize "cat" and "dog" stimuli followed by recordings from the animals' IT and PFC support the model prediction of a shape-based but object-class specific representation (in this case for "cat"/"dog"-like shapes) that provides input to task-specific circuits, in this case trained on the categorization task (in IT and PFC, respectively, see Freedman et al., 2003).

Although the Standard Model is purely feedforward and thus in principle fast enough to explain the data of Thorpe et al. (1996), the question is whether such a simple model can indeed perform real-world object recognition tasks. Recent work (Serre et al., 2002; Louie, 2003) has provided some very encouraging results: The feedforward architecture of Fig. 46.1 can detect objects (in this case, faces) in natural images, at a level comparable or even superior to state-of-the-art machine vision systems (Fig. 46.2). Key to the model's success on this difficult task is the learning of a set of object class-specific features, at the S2 level in the model, roughly corresponding to V4 cells in cortex. In combination with the MAX pooling operation, this specialized set of features allows the system to isolate the relevant features from the surrounding clutter, without the need for a separate segmentation step. Moreover, the specialized object class representation also greatly simplifies the complexity of the learning problem, permitting the use of a simple linear classifier (see curve for "Kmeans classifier" in Fig. 46.2) similar to the architecture shown in Fig. 46.1.

The observed difference in the attentional demands of different recognition tasks (Li et al., 2002) could thus be related to the "naturalness" of the objects involved: If the visual system has learned optimized features of intermediate complexity for familiar objects (like animals in natural scenes), this would facilitate detection of these objects in a single feedforward processing pass even in the presence of clutter. In the case of unfamiliar or artificial objects, such as bisected colored disks (Li et al., 2002), for which there are no specialized features, however, the higher level of interference caused by the simultaneous presence of other objects

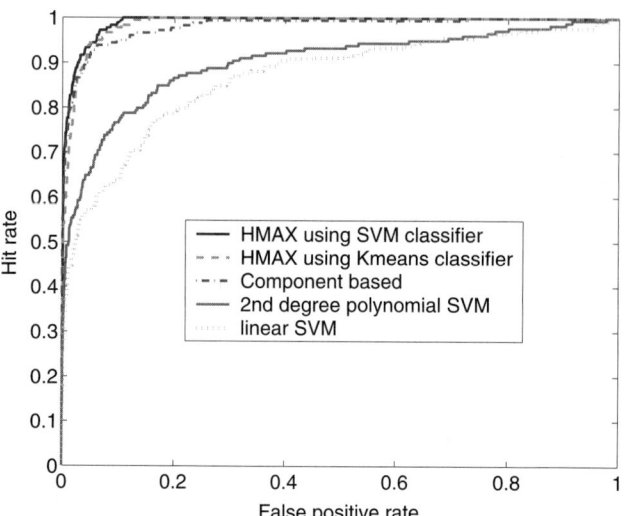

FIGURE 46.2 The feedforward model of object recognition in cortex can perform face detection in natural images (faces were a subset of the CMU PIE database, nonfaces were selections from natural scenes selected as "face-like" by an LDA classifer, see Louie, 2003) at a level comparable to that of one of the best available machine vision face detection systems ("Component-based", Heisele et al., 2002). The figure shows ROC curves of the biological system with feature learning ("HMAX") and different machine vision face detection systems. The HMAX system was like the one shown in Fig. 46.1, with the difference that S2 features were learned from a set of faces images (Serre et al., 2002). This object-class specific feature set enabled the system to robustly detect faces in cluttered images. For details, see Louie, 2003.

is more likely to prevent "attentionless" recognition of the target objects (i.e., in one feedforward pass through the ventral stream), as observed in the experiment (Li et al., 2002). In section IV we will discuss possible mechanisms of how attention could modulate processing in the recognition system to improve performance in such cases.

A. Limitations of the Feedforward Approach

As described in the previous section, the purely feedforward Standard Model is already able to perform very complex object recognition tasks. However, the feedforward architecture has some limitations:

- There is experimental evidence that the visual system can exploit information about target location (spatial cues, e.g., Posner, 1980) to enhance processing at the location of interest. The model cannot explain such top-down effects.
- There are situations where the feedforward system is overwhelmed and cannot correctly detect the object of interest, for instance, when the target appears together with a number of other objects

and the surrounding clutter interferes with the representation of the target object (see Tsotsos, 1990; Riesenhuber and Poggio, 1999a). This is the case in some visual search tasks where the visual system appears to resort to a serial approach to sequentially process different parts of the visual input.

IV. EXTENDING THE FEEDFORWARD SYSTEM: ROLES FOR TOP-DOWN ATTENTIONAL AND TASK-DEPENDENT MODULATIONS

A. Spatial Cueing

It is straightforward to incorporate task-relevant information in form of a spatial cue into the framework of the feedforward model by appropriately modulating the pooling range of units performing a MAX operation. In this way, spatial attention could enhance signals from the region of interest and suppress input from nonrelevant parts of the visual field (see Fig. 46.3). This is compatible with reports from physiology that show that the receptive fields of neurons in V4 can constrict around the location of interest (Luck et al., 1997), and similar observations of enhancement of processing for the region of interest and suppression elsewhere in fMRI experiments (see Chapter 50). Regarding the underlying neural mechanisms, recent data suggest that deploying spatial attention to a region that includes the receptive field of a particular neuron causes a leftward shift of that neuron's contrast response curve (see Chapter 8). Thus, focusing attention on a particular region in space would be equivalent to raising the effective contrast of that part of the input (and conversely, nonattended regions would be expected to show a lowered effective contrast). In the framework of the model, such a modulation of effective contrast directly reduces the interference caused by nonattended regions, as high-contrast stimuli cause higher responses that are more likely to win the MAX competition and thus determine the response of pooling units along the pathway and ultimately of view-tuned IT neurons (Riesenhuber and Poggio, 1999a). This parallel between attentional modulation and contrast is also very appealing since it directly relates to the notion of "salience" at the heart of popular models of attentional selection (Itti and Koch, 2000) (see Chapter 94).

B. Nonspatial Cueing

The case of nonspatial cueing is not as straightforward as the spatial case, however. While a spatial signal can be translated into a modulatory signal for cells at all levels of the processing hierarchy depending on the overlap of a neuron's receptive field with the extent of the "spotlight" of attention, it is not clear how a nonspatial (e.g., object-level cue, such as "look for a face") can be translated into response modulations of neurons tuned to different features along the ventral pathway to selectively improve detection of the object of interest. For instance, if the goal is to detect "a face," there is a multitude of potential target objects, and it is not clear which neurons should be modulated and in which way to increase the detectability of any face versus nonfaces. Consider the simplest case: Assume the target is a particular face in a particular pose (lighting condition, expression, etc.), and that further there is a particular view-tuned cell in IT tuned to this exact face (a so-called grandmother cell), and the target face is declared "detected" if the activation of this VTU exceeds a certain threshold. How should afferent neurons tuned to simpler features in lower processing levels, for example in V4, be modulated to improve the system's selectivity (i.e., to improve detection without an increase in false alarms) for the target object? If the VTU is tuned to a characteristic distributed activation pattern over its afferents (with high and low activations, depending on which features are present in the face and to what degree), then how should those afferents be modulated, in particular in the absence of information about target contrast? Increasing the afferents' gain might change their response to the target object in such a way that the resulting activation pattern over the afferents could actually be less optimal to activate the VTU than the unmodulated activation pattern (for supporting simulation results, see Schneider and Riesenhuber, 2004). The situation is even more problematic in the more general case where object identity is encoded by a population of view-tuned units tuned to, for example, different faces (Young and Yamane, 1992). Here, the same V4 neuron can provide input to different VTUs and conceivably receive top-down signals from more than one IT cell. How should the possibly different top-down inputs be combined to modulate the V4 neuron?

These computational arguments concerning the conceptual simplicity of spatial cueing on the one hand and the difficulties associated with nonspatial cueing on the other are compatible with reports from electrophysiology that suggest that featural and spatial attentional effects are qualitatively different. In particular, while spatial effects appear to occur close to response onset, nonspatial modulations appear to have a latency of at least 150 ms (Motter, 1994; Hillyard and Anllo-Vento, 1998). A possible interpretation of these data

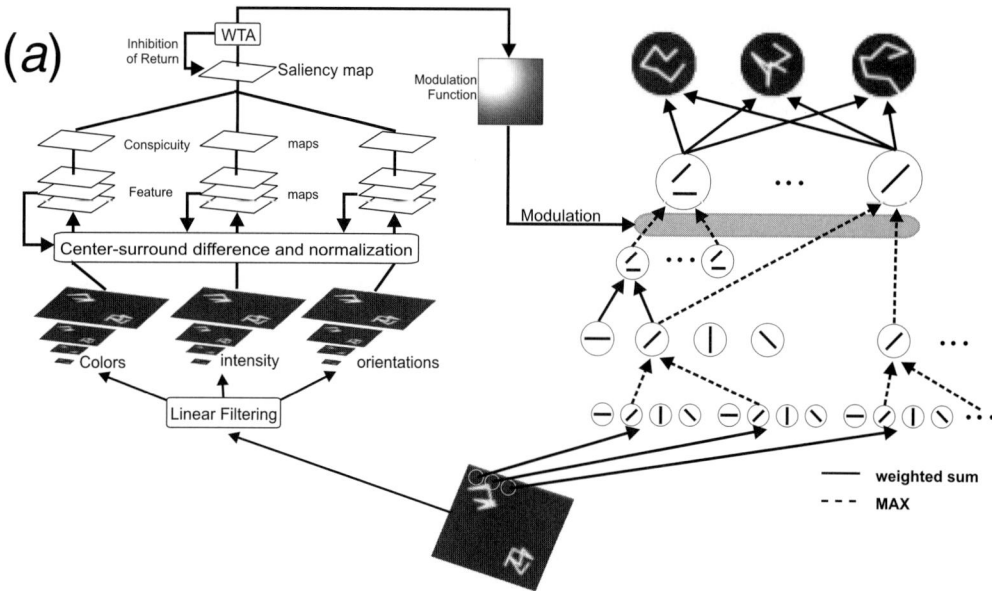

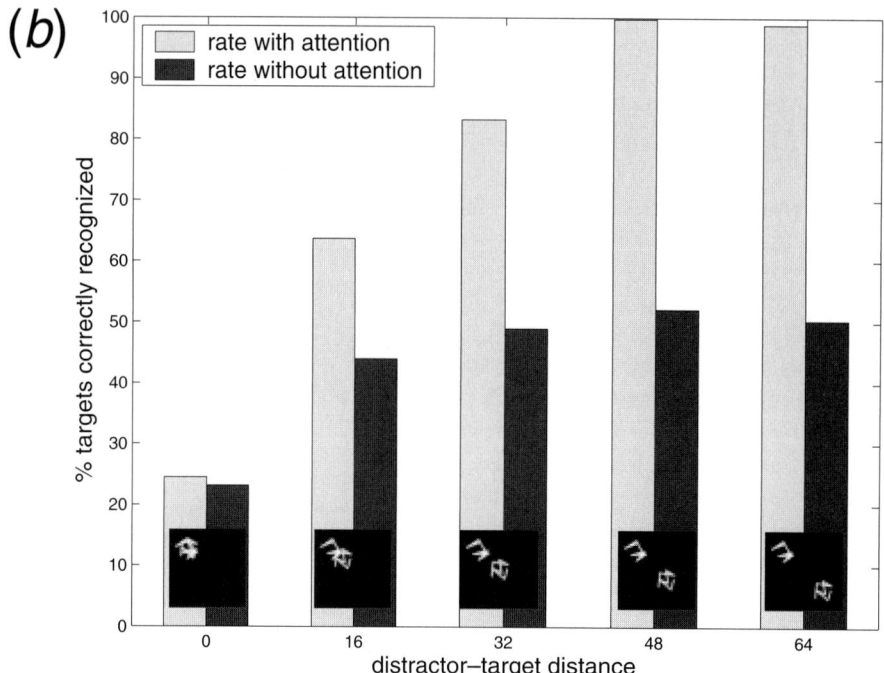

FIGURE 46.3 Coupling of the Saliency Map model of attention (Itti and Koch, 2000) and the model of object recognition (Riesenhuber and Poggio, 1999b). (*a*) Sketch of the integrated system (Walther et al., 2002): The saliency map (left) provides a modulatory signal to C2 model units (corresponding to V4 neurons in cortex) to modulate their pooling range, causing their receptive fields to focus around the location of interest selected by the saliency map. (*b*) Recognition results with and without attentional modulation, as a function of target and distractor stimulus separation. While a purely feedforward analysis of the image yields only poor performance, iterative piecewise analysis of the image through attentional modulation of the spatial extent of receptive fields dramatically improves recognition performance. From Riesenhuber and Poggio, 2003.

could be that nonspatial attention only sets in *after* an initial feedforward pass through the visual system, when the precise shape, contrast, etc. of the target and the activation it evokes at the different processing stages would be known. However, this would suggest very different roles for spatial and featural attention, at least for the case of rapidly presented images (when processing is limited mainly to a feedforward pass): While spatial cues could aid recognition as it is possible to "tune" the system before stimulus onset, information about features would not be able to enhance performance of the initial feedforward processing pass, but might serve to, for example, "highlight" instances of the target object in the visual field, as in a saliency map (Motter, 1994; Mazer and Gallant, 2003) to inform other processes such as eye movements to potential targets (see also section V).

If the target object appears together with a number of similar distractor objects that interfere with the representation of the target object so much that it cannot be recognized in a feedforward pass anymore (see Fig. 46.3b, dark bars) one computational strategy for the visual system is to "divide and conquer" to reduce the influence of clutter and detect the target by sequentially analyzing parts of the image. This piecemeal approach could use the same mechanism underlying spatial attention described above, that is, spatial modulation of the pooling range of neurons along the processing hierarchy, but now controlled not by external spatial cues but by, for instance, a "saliency map" (Itti and Koch, 2000) (see Chapter 94) as shown in Fig. 46.3 (Walther et al., 2002). It is an interesting question whether this saliency map is based solely on bottom-up factors such as orientation or intensity contrast, or whether there are task-specific components to saliency (that would serve to, e.g., increase the salience of purple regions when looking for Barney, the Dinosaur). Also, given the feature-based modulations of V4 neurons and their possible role in object recognition described in section IV, it is interesting to ask whether more complex features, like those represented by V4 neurons (Kobatake and Tanaka, 1994) are integrated into the saliency map. Clearly, investigating this link between attention and recognition should be a priority for future studies.

V. SUMMARY AND CONCLUSIONS

"Basic" object recognition tasks such as object detection in natural images can be understood to a first approximation as resulting from a single feedforward pass through the processing hierarchy of the ventral visual stream in cortex from primary visual cortex, V1, to inferotemporal cortex, IT. A hierarchical computational model of the ventral stream based on just two operations, a MAX pooling function to increase tolerance to stimulus translation and scaling and a template match operation to increase feature complexity, can provide an explanation for the shape tuning and invariance properties of view-tuned cells in IT and explain how the ventral stream can perform object detection in complex natural scenes. The model makes very few assumptions. For instance, model unit responses have no dynamics, and there are no lateral interactions, oscillations, or synchronous ensembles of units. This does not mean such mechanisms do not play a role in vision. However, the simulation results show that they are not necessary to explain the relevant data on object recognition.

Not surprisingly, simulations show that there are situations where such a simple feedforward system breaks down, for instance, in the case of visual clutter when the receptive field of a model IT unit contains several interfering objects. Similar effects are observed in natural vision, and it will be interesting to compare the conditions under which feedforward vision fails in the model and in the experiment. As the modeling studies on face detection demonstrate (Serre et al., 2002; Louie, 2003), familiarity with objects from the target class is expected to play a crucial role: If a subject is well trained on a certain object class such that specific intermediate representations have been learned, then interference caused by simultaneously presented distractor objects and thus attentional demands for this task should decrease. Reports that some initally "serial" tasks can become "parallel" with practice (Sireteanu and Rettenbach, 1995), and that well-practiced object recognition tasks (such as animal detection in natural scenes) do not seem to require attention, whereas more artificial ones (like discriminating bisected colored disks) do (Li et al., 2002) are compatible with this hypothesis.

The challenge to perform object recognition also in more difficult situations when the feedforward system fails, together with experimental data that show spatial cueing effects in behavior as well as attention-related modulations of processing observed in physiology and fMRI, has motivated extensions of the basic model to incorporate spatial modulations of processing to reduce the effect of clutter. These modulations can be based on explicit information about the target (for instance, in the form of a spatial cue), or could possibly be driven more indirectly by stimulus saliency (Itti and Koch, 2000; Walther et al., 2002) as in serial visual search. In the case of spatial attention, information about target location permits a "tuning" of the visual system prior to stimulus exposure that can improve the performance of the feedforward system.

From a computational point of view, featural attention, where the system is given information about the shape of a target but not its location, seems to be fundamentally different (Schneider and Riesenhuber, 2004): While the translation of information about target location into a neuronal modulation pattern is straightforward based on the match of the location of a cell's receptive field with the region of interest, and can be applied to any cell regardless of its position in the processing hierarchy, information about complex target objects is difficult to translate into appropriate attentional modulations of simpler feature detectors in lower levels of the hierarchy. Nevertheless, object recognition tasks that leave sufficient time to perform iterations of feedforward and feedback processing might profit from featural attention. In these cases, an initial feedforward pass could provide hypotheses about possible targets, which could then provide specific top-down signals to influence lower levels of processing, possibly explaining observed effects of featural attention on task performance in some experiments (Rossi and Paradiso, 1995; Blaser et al., 1999; Lee et al., 1999). Alternatively, these results might suggest the intriguing and computationally feasible hypothesis that pre-stimulus "featural tuning" of feedforward processing is possible for the basic visual features that neurons at the lower processing levels are tuned to, such as color and orientation. Such modulations of processing could consist in a sharpening of tuning curves (possibly paired with an increase in gain akin to the aforementioned increase in effective contrast in the spatial case) (Lee et al., 1999) of those neurons that are directly tuned to the target stimulus. Opportunities abound for interesting hypothesis-driven experiments seeking to clarify the role of attention in object recognition.

Acknowledgments

Thanks to Tommy Poggio for discussions, to Thomas Serre for creating Fig. 46.2, and to Dirk Walther for creating Fig. 46.3.

References

Blaser, E., Sperling, G., and Lu, Z. L. (1999). Measuring the amplification of attention. *Proc. Nat. Acad. Sci. USA* **96**, 11681–11686.

Freedman, D. J., Riesenhuber, M., Poggio, T., and Miller, E. K. (2003). Comparison of primate prefrontal and inferior temporal cortices during visual categorization. *J. Neurosci.* **23**, 5235–5246.

Gawne, T. J., and Martin, J. M. (2002). Responses of primate visual cortical V4 neurons to simultaneously presented stimuli. *J. Neurophys.* **88**, 1128–1135.

Grill-Spector, K., Kourtzi, Z., and Kanwisher, N. (2001). The lateral occipital complex and its role in object recognition. *Vis. Res.* **41**, 1409–1422.

Heisele, B., Serre, T., Pontil, M., Vetter, T., and Poggio, T. (2002). Categorization by learning and combining object parts. In *Advances in Neural Information Processing Systems*, vol. 14.

Hillyard, S. A., and Anllo-Vento, L. (1998). Event-related brain potentials in the study of visual selective attention. *Proc. Nat. Acad. Sci. USA* **95**, 781–787.

Itti, L., and Koch, C. (2000). A saliency-based search mechanism for overt and covert shifts of visual attention. *Vis. Res.* **40**, 1489–1506.

Kobatake, E., and Tanaka, K. (1994). Neuronal selectivities to complex object features in the ventral visual pathway of the macaque cerebral cortex. *J. Neurophys.* **71**, 856–867.

Lampl, I., Ferster, D., Poggio, T., and Riesenhuber, M. (2004). Intracellular Measurements of Spatial Integration and the MAX Operation in Complex Cells of the Cat Primary Visual Cortex. *Journal of Neurophysiology* **92**, 2704–2713.

Lee, D. K., Itti, L., Koch, C., and Braun, J. (1999). Attention activates winner-take-all competition among visual filters. *Nat. Neurosci.* **2**, 375–381.

Li, F. F., van Rullen, R., Koch, C., and Perona, P. (2002). Rapid natural scene categorization in the near absence of attention. *Proc. Nat. Acad. Sci. USA* **99**, 9596–9601.

Logothetis, N. K., and Sheinberg, D. L. (1996). Visual object recognition. *Ann. Rev. Neurosci.* **19**, 577–621.

Louie, J. (2003). A biological model of object recognition with feature learning. Master's thesis, MIT, Cambridge, MA.

Luck, S. J., Chelazzi, L., Hillyard, S. A., and Desimone, R. (1997). Neural mechanisms of spatial selective attention in areas V1, V2, and V4 of macaque visual cortex. *J. Neurophys.* **77**, 24–42.

Maser, J. A., and Gallant, J. L. (2003). Goal-related activity in V4 during free viewing visual search. Evidence for a ventral stream visual salience map. *Neuron* **40**, 1241–1250.

Motter, B. C. (1994). Neural correlates of feature selective memory and pop-out in extrastriate area V4. *J. Neurosci.* **14**, 2190–2199.

Poggio, T., and Edelman, S. (1990). A network that learns to recognize 3D objects. *Nature* **343**, 263–266.

Poggio, T., and Girosi, F. (1990). Networks for approximation and learning. *Proc. IEEE* **78**(9), 1481–1497.

Posner, M. I. Orienting of attention. (1980). *Quart. J. Exp. Psych.* **32**, 3–25.

Riesenhuber, M., and Poggio, T. (1999). Are cortical models really bound by the "Binding Problem"? *Neuron* **24**, 87–93.

Riesenhuber, M., and Poggio, T. (1999). Hierarchical models of object recognition in cortex. *Nat. Neurosci.* **2**, 1019–1025.

Riesenhuber, M., and Poggio, T. (2002). Neural mechanisms of object recognition. *Curr. Op. Neurobiol* **12**, 162–168.

Riesenhuber, M., and Poggio, T. (2003). How Visual Cortex Recognizes Objects: The Tale of the Standard Model. In: *The Visual Neurosciences* (Eds. L. M. Chalupa and J. S. Werner), MIT Press, Cambridge, MA, 1640–1653.

Rossi, A. F., and Paradiso, M. A. (1995). Feature-specific effects of selective visual attention. *Vis. Res.* **35**, 621–634.

Rousselet, G. A., Fabre-Thorpe, M., and Thorpe, S. J. (2002). Parallel processing in high-level categorization of natural images. *Nat. Neurosci.* **5**, 629–630.

Schneider, R., and Riesenhuber, M. (2004). On the difficulty of feature-based attentional modulations in visual object recognition: A modeling study. Technical Report AI Memo 2004-004, CBCL paper 235, MIT AI Lab and CBCL, Cambridge, MA.

Serre, T., Riesenhuber, M., Louie, J., and Poggio, T. (2002). On the role of object-specific features for real world object recognition.

In H. H. Buelthoff, S.-W. Lee, T. Poggio, and C. Wallraven, (eds.). *Proceedings of BMCV2002*, volume 2525 of *Lecture Notes in Computer Science*, New York, Springer.

Sireteanu, R., and Rettenbach, R. (1995). Perceptual learning in visual search: fast, enduring, but non-specific. *Vis. Res.* **35**, 2037–2043.

Thorpe, S. J., Fize, D., and Marlot, C. (1996). Speed of processing in the human visual system. *Nature* **381**, 520–522.

Tsotsos, J. K. (1990). Analyzing vision at the complexity level. *Behav. Brain Sci.* **13**, 423–469.

Ungerleider, L. G., and Haxby, J. V. (1994). "What" and "where" in the human brain. *Curr. Op. Neurobiol.* **4**, 157–165.

Walther, D., Itti, L., Riesenhuber, M., Poggio, T., and Koch, C. (2002). Attentional selection for object recognition—a gentle way. In H. H. Buelthoff, S.-W. Lee, Poggio, T., and Wallraven, C. (eds.), *Proceedings of BMCV2002*, volume 2525 of *Lecture Notes in Computer Science*, New York, Springer.

Young, M. P., and Yamane, S. (1992). Sparse population coding of faces in the inferotemporal cortex. *Science* **256**, 1327–1331.

CHAPTER
47

Binding Contour Segments into Spatially Extended Objects

Pieter R. Roelfsema and Henk Spekreijse

ABSTRACT

As a first step on the way to image comprehension, the visual system groups together contours that belong to the same object and segregates them from contours that belong to different objects and the background. Here we outline a simple algorithm that explains how contours are grouped together. We demonstrate that this algorithm is implemented in the visual cortex, where an enhancement of neuronal firing rates spreads through the network of corticocortical connections until the neuronal responses to all contours of a single object are enhanced. Thereby, these contours are grouped together and segregated from the rest of the image. In the primary visual cortex of monkeys that carry out a contour grouping task, neuronal responses evoked by all contours that belong to a target object are indeed stronger than responses evoked by contours of other objects. This firing rate enhancement provides a correlate of visual attention. Thus, at a psychological level of description, attention gradually spreads across the contours of a single perceptual object until they have all been bound into a coherent representation.

I. INTRODUCTION

Psychological theories usually subdivide visual processing into a pre-attentive and an attentive stage (e.g., Treisman and Gelade, 1980). Pre-attentive vision extracts elementary visual features in parallel across the entire image. This stage can, to a first approximation, be identified with processing that is carried by feedforward connections that propagate information from the retina through the LGN and primary visual cortex to higher areas of the visual cortex (Lamme and Roelfsema, 2000). After an image has been presented, the feedforward connections rapidly activate neurons in all the areas of the visual cortex within 120 ms. This "feedforward sweep" extracts remarkably complex features, such as the shape of animals or cars (Thorpe, Fize, and Marlot, 1996), but there are also a number of tasks that remain unsolved in this pre-attentive phase.

Many contour grouping tasks are, for example, not yet solved when the highest visual areas are activated by the feedforward sweep (Jolicoeur, Ullman, and MacKay, 1986, 1991; Roelfsema, Scholte, and Spekreijse, 1999). At this point in time, there are no further downstream areas to be activated by the feedforward sweep, and only a recirculation of activity within the visual areas by recurrent connections can account for the additional processing time. This recurrent processing presumably corresponds to the engagement of visual attention. Recurrent connections modulate neuronal responses, but unlike the feedforward connections, they usually do not *drive* the cells directly. There are several classes of recurrent connections. First, feedback connections mediate the influences from higher to lower processing levels (Felleman and van Essen, 1991). Second, lateral connections provide interactions between areas at the same hierarchical level. Third, horizontal connections allow interactions between neurons in different columns of the same visual area (Gilbert, 1992).

An important function that is attributed to recurrent connections, and therefore to attentive vision, is to establish relations between the features that have been extracted by pre-attentive vision. The visual system has to group features together that belong to a single perceptual object, and to segregate them from features that belong to other objects, and the background. In this chapter, we will describe how attention binds contours into coherent object representations at an algorithmic, neurophysiological, as well as a psychological level of description.

II. AN ALGORITHM FOR GROUPING OF CONNECTED IMAGE ELEMENTS

It is unlikely that the visual cortex reserves feature detectors for the identification of any arbitrary shape (von der Malsburg, 1986). We may safely assume, for example, that you, the reader, have never seen some of the shapes that are depicted in Fig. 47.1a and b. It is also unlikely that you will see them again in your lifetime, unless you decide to reread this chapter. There are actually more of these arbitrary pixel configurations than there are neurons in the visual cortex (a genuine combinatorial explosion). The implication is that these shapes have to be encoded by ensembles of neurons, with individual neurons tuned to the shape of simpler object fragments. If there are multiple shapes, as in Fig. 47.1, multiple ensembles are activated at the same time. This leads to a "binding problem" since information about whether fragments belong to the same or different objects is not available. We can nevertheless identify all pixels that belong to one object if we are required to do so. It is easy to check, for example, whether the two gray pixels of Fig. 47.1a and b are on the same or on different objects. We solve this task by applying the grouping criterion of connectedness. Pixels that are connected to each other are grouped by our visual system. This is also useful in everyday vision, since connected image components usually belong to the same object. Here we outline an algorithm that is implemented in the primate visual cortex to evaluate grouping criteria such as connectedness, thereby binding the components of single objects into a coherent representation.

The proposed algorithm can, in principle, be implemented in a single retinotopic area, such as the primary visual cortex (area V1). It is illustrated in Fig. 47.1c, where there is a neuron for every pixel. Some of the neurons have a pixel in their receptive field, and these cells are activated by feedforward connections (gray circles). Other neurons remain silent (white circles). The algorithm uses a label that is spread through the network of recurrent (here horizontal) connections. The pattern of activity induced by feedforward connections constrains the label spreading process, since the label can only spread among neurons with a pixel in their receptive field. We may therefore distinguish between two types of connections. First, there are *enabled* connections, which can spread the label, since they have an active cell on both

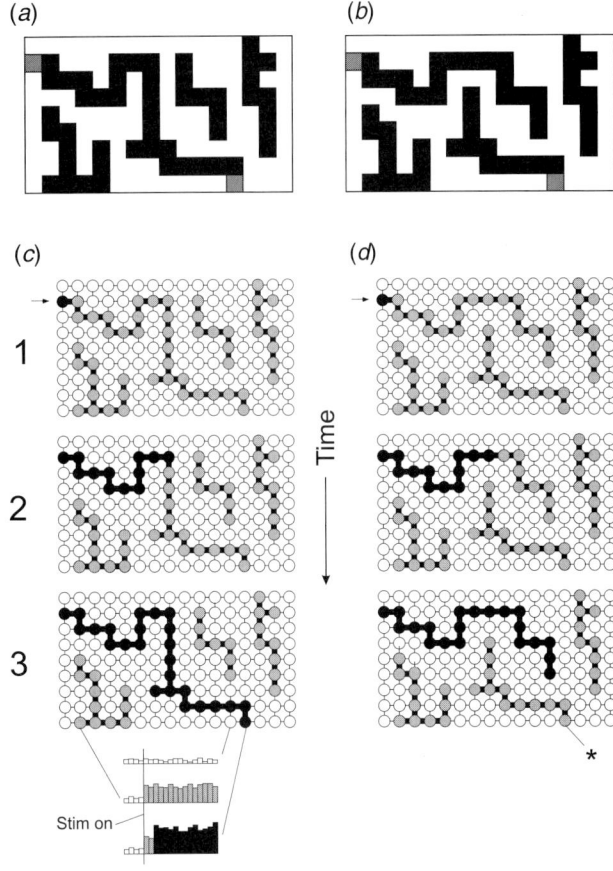

FIGURE 47.1 An algorithm for the detection of connected image components. a,b, Two images. In (a) the two gray squares are connected to each other, but in (b) they are not. (c) (d), Retinotopic area with a neuron for each pixel. Neurons that are activated by feedforward connections have been indicated in gray. The other, white cells are not activated. A label (black) spreads in time through the network of enabled connections (thick connections). Connections between silent cells are disabled (thin lines). Inset below, neurons not activated by feedforward connections have some spontaneous activity (white). Neurons driven by feedforward connections have an intermediate level of activity (gray). Neurons that, in addition, receive the label have an enhanced response (black). Star (*), neuron that receives the label in case of the left, but not in case of the right image (modified from Roelfsema and Singer, 1998).

sides (thick lines in Fig. 47.1c,d). We call the subset of connections that is enabled in this way the *interaction skeleton* (Roelfsema, Lamme, and Spekreijse, 2000). Second, there are *disabled* connections (thin lines) with at least on one side an inactive cell. The algorithm starts by labeling one of the active neurons (black in Fig. 47.1c-1). This label will be spread in time across the enabled connections to all neurons that respond to the pixels connected to the pixel at which the process is initiated. When a small change is made in the image, by shifting a single pixel (compare Fig. 47.1a and b) the pattern of feedforward activity changes, and therefore the subset of enabled connections changes too (compare Fig. 47.1c and d). Again, the label will be spread to neurons that respond to all pixels connected to the pixel at which label spreading starts. Thus, the label is selectively distributed to neurons that respond to pixels connected to each other, and it may therefore be used to group them into a coherent representation.

An essential question is if and how this algorithm is implemented in the visual cortex. First we need to know the identity of the label. We will demonstrate below that the label is an enhancement of neuronal firing rates (as indicated in the PSTHs below Fig. 47.1c). Thus, neurons that have a pixel in their receptive field are activated by the feedforward connections, but neurons that also receive the label have an even stronger response.

The model illustrates many of the differences between pre-attentive and attentive vision. In the pre-attentive phase, neurons with a pixel in their receptive field are activated in parallel through feedforward connections from the retina. In this phase neurons are only influenced by information inside their receptive fields. The induced activity pattern enables a subset of the recurrent connections, the interaction skeleton, which implicitly groups together pixels that are connected to each other. But these groupings are not yet explicit, that is, evident to neurons in other areas of the visual cortex (gray bins in the PSTHs below Fig. 47.1c). To make the groupings explicit, the rate enhancement (black bins) has to be spread through the interaction skeleton. This is a serial process that takes time and that belongs to the domain of attentive vision. During the attentive phase, neurons become sensitive to information that is outside their receptive field. The neuron indicated by a star in Fig. 47.1d, for example, does not enhance its response, whereas it does so in Fig. 47.1c, although the bottom-up input coming from its receptive field is identical for both stimuli.

III. THE NEUROPHYSIOLOGY OF CONTOUR GROUPING

To investigate if and how this algorithm is implemented in the visual cortex, we trained monkeys in a task that requires grouping of contours into elongated curves (Roelfsema, Lamme, and Spekreijse, 1998). The task is shown in Fig. 47.2a. When the monkeys looked at a fixation point, two curves and two circles appeared on the screen. One of these circles was connected to the fixation point by a curve that will be called the target curve (T in Fig. 47.2a). The other circle was connected to a distractor curve (D), not connected to the fixation point. The monkey maintained fixation until the fixation point was removed (usually after 600 ms) and was then required to make an eye movement to the circle at the other end of the target curve. To solve this task, and to locate the circle at the end of the target curve, the monkeys had to group all contours of this curve into a coherent representation. A small change in the vicinity of the fixation point interchanged target and distractor curve (stimuli I and II in Fig. 47.2b). A change at another location introduced an intersection between the two curves (stimuli III and IV).

We recorded from a number of electrodes, which had been chronically implanted in the primary visual cortex (area V1) during this task. In the example of Fig. 47.2, we recorded from five groups of V1 neurons. Their receptive fields are shown in Fig. 47.2b as rectangles. Some receptive fields fell on the target curve, and others fell on the distractor curve. The strength of the neuronal responses depended on whether the receptive fields fell on the target or distractor curve. The receptive fields of neurons that enhanced their responses are indicated in gray, and white receptive fields belong to neurons that did not enhance their response. Note that contours belonging to the target curve evoked stronger responses than contours of the distractor curve in each of the four stimulus configurations, and even if the two curves crossed each other.

Figure 47.2c illustrates the time course of neuronal responses at one of the recording sites. The initial responses evoked by the target and distractor curve occurred at a latency of about 40 ms and were of equal magnitude for stimulus I and II. These initial responses are determined by what is inside the receptive field, and this did not differ between these stimuli. The initial response of these neurons was somewhat larger if there was an intersection (stimulus III and IV), which can be explained by a slightly different shape of the contour in the receptive field. After the initial response, the neurons started to distinguish between target and distractor curve. A response enhancement to the target curve (gray area between the two responses) emerged after about 150 ms and was maintained until the monkey responded by making an eye movement. In this task, contours at locations that are remote from the receptive field determine whether it is on the target or distractor curve. Information about these remote contours can only arrive at the cell through recurrent connections, such as horizontal connections between neurons in area V1, and feedback connections from higher visual areas. It is likely that the relatively long latency of the response enhancement reflects the length of these routes.

These results are in accordance with the proposed algorithm (Fig. 47.1). Initial neuronal responses are

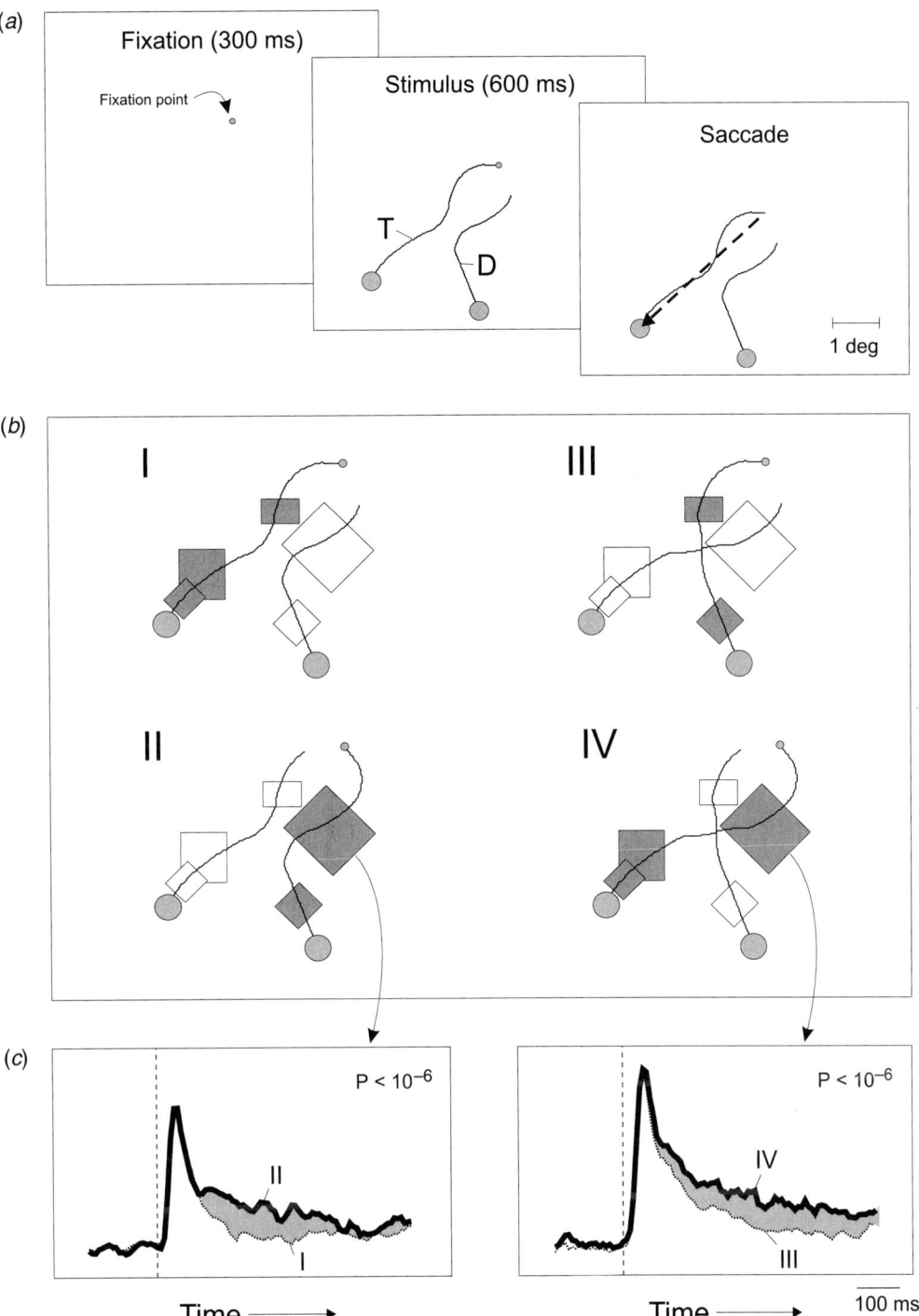

FIGURE 47.2 Activity in area V1 during contour grouping. (*a*) Grouping task. The monkey has to look at a fixation point. Two curves and two circles appear on the screen, and the monkey has to group contours of the target curve (T) that is connected to the fixation point in order to locate a larger circle on the other end of this curve. The second curve is a distractor (D). When the fixation point disappears, the monkey makes an eye movement to a circle at the other end of the target curve. (*b*) Receptive fields of five recorded groups of neurons in area V1 relative to the curves. Receptive fields of neurons that enhanced their response are shown as gray rectangles, whereas receptive fields of neurons that did not are shown as white rectangles. (*c*) The initial response at the indicated recording site does not yet distinguish between target and distractor curve. After about 150 ms, however, the response to the target curve is enhanced. Thick line, response to the target curve. Thin dashed line, response to the distractor curve. Gray region between responses indicates the response enhancement. (modified from Roelfsema et al., 1998).

determined by the feedforward connections. It takes some more time before the recurrent connections label the neuronal responses to all contours that belong to a single object. An essential aspect of the model is that these recurrent connections should be enabled only if they interconnect neurons that are activated by feedforward connections. This assumption is also supported by a number of studies showing that influences from outside the receptive field are particularly strong for V1 neurons that are well driven by the visual stimulus, and much weaker for cells that receive little bottom-up activation (Knierim and Van Essen, 1992; Kapadia, Ito, Gilbert, and Westheimer, 1995; Zipser, Lamme, and Schiller, 1996).

In the task of Fig. 47.2, connectedness is not the only available grouping cue. Another cue is colinearity, since contours that are colinear to each other usually belong to the same curve (Kellman and Shipley, 1991; Sigman, Cecchi, Gilbert, and Magnasco, 2001). At intersections between the curves, colinearity is actually the only cue that is useful, since at these locations both curves are connected to each other. We would have to include orientation selective neurons in our model to account for the specificity of contour grouping at intersections. In a more elaborate model, horizontal connections should interconnect neurons tuned to colinear configurations. Anatomical studies show that horizontal connections in the primary visual cortex indeed selectively interconnect neurons that are tuned to colinear contours (Bosking, Zhang, Schofield, and Fitzpatrick, 1997; Schmidt, Goebel, Löwel, and Singer, 1997). Further grouping criteria can also be accommodated in an equivalent way, as has been outlined elsewhere (Roelfsema et al., 2000).

In a recent study, we manipulated the difficulty of contour grouping by making it hard for the monkeys to discriminate intersections from nonintersections (Roelfsema and Spekreijse, 2001). Now monkeys made errors on some trials with the more difficult configurations. Thus, on these trials the monkeys erroneously grouped together contour segments that actually belonged to different curves. The errors were reflected by an altered pattern of the rate enhancement, since now the contours that were wrongly grouped were labeled by an enhanced response. Thus, it is possible to monitor the animals' interpretation of a visual image by recording from the primary visual cortex.

IV. THE PSYCHOLOGY OF CONTOUR GROUPING

The neurophysiological data indicate that contours that are grouped together are labeled by an enhanced neuronal response in the visual cortex. Such response enhancements, which have also been observed in a variety of other tasks (Chelazzi, Miller, Duncan, and Desimone, 1993; Maunsell, 1995), are usually attributed to visual attention. We therefore set out to investigate the role of attention in our contour grouping task. We tested human observers, who saw one of the eight stimuli of Fig. 47.3a on each trial. The stimuli were composed of two curves that could intersect each other. Subjects had to maintain fixation, and press a button in their left hand if a left circle (L in Fig. 47.3a) was connected to the fixation point, and a button in their right hand if the other circle (R) was connected. We measured reaction times, which turned out to depend strongly on the number of intersections. Each intersection increased reaction time by at least 100 ms, which is a first indication that attention is involved in this task (Roelfsema et al., 1999). Our physiological data provide a straightforward prediction about the involvement of attention. If the response enhancement observed in area V1 reflects attention, then attention should be directed to all the contours of the target curve.

To measure attention during contour grouping, we added a secondary task (Scholte, Spekreijse, and Roelfsema, 2001). Colors were briefly displayed on segments of the target or distractor curve (white segments in Fig. 47.3a). If a contour is attended, then its color should be reported more reliably than if it is not. Thus, we can use the subjects' performance in the secondary task to distinguish between attended and unattended contours. As expected, colors on the target curve were reported much more reliably than colors on the distractor curve. This shows that subjects indeed direct their attention to the contours that need to be grouped together. We recently measured the time course of attention, by presenting the colors at different times during the contour grouping task (Houtkamp, Spekreijse, and Roelfsema, 2003). Percentages in Fig. 47.3b indicate performance in the secondary color-report task. At an early interval, 0–150 ms after stimulus appearance, colors on segments of the target curve close to the fixation point were reported more reliably than colors on the distractor curve (Fig. 47.3b, left). At an intermediate time interval, performance also improved at the middle contour segments, and started to improve at the lower contour segments (Fig. 47.3b, middle). At a late interval, just before the subjects responded by pressing the button in their left or right hand, color performance was improved at all contour segments of the target curve (Fig. 47.3b, right). Thus, attention gradually spreads over the target curve, starting at the fixation point. It spreads from attended contours to contours that are colinear and

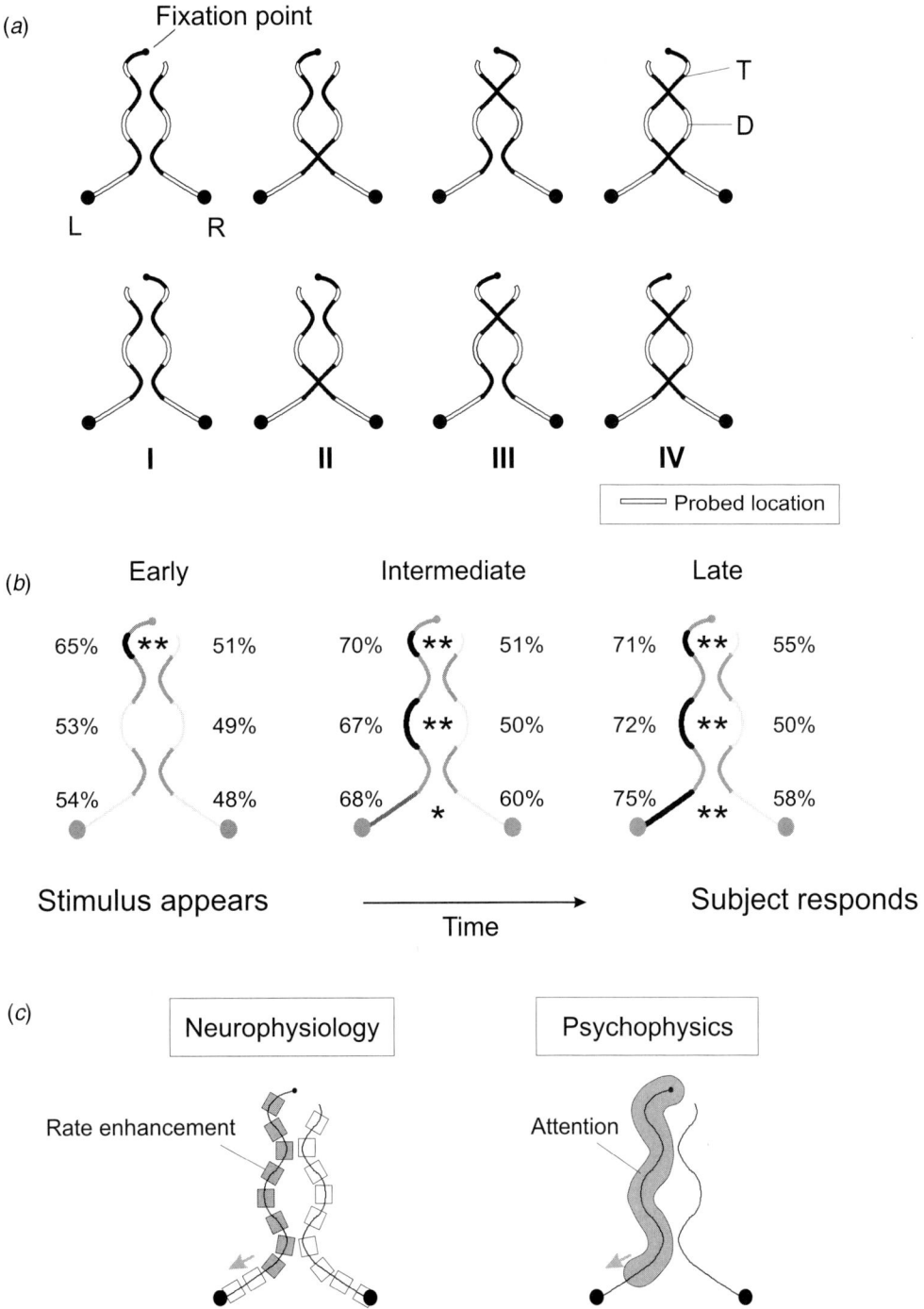

FIGURE 47.3 The spatial profile of attention during contour grouping. (a) Eight stimuli, one of these was presented on each trial. The subjects had to maintain fixation on the fixation point, while they had to identify a larger circle that was connected to this point by a target curve (T). They pressed a button in their left hand if the fixation point was connected to the left circle (L) and a button in their right hand if it was connected to the right circle (R). There also was a distractor curve (D) that could be ignored. The subject had a secondary task to report colors of contour segments indicated in white. (b) Percentages indicate performance in the secondary, color-report task. Within 150 ms after stimulus presentation, colors at the upper contour segments of the target curve (black) are reported more reliably than those of the distractor curve. A similar benefit for colors on the target curve occurs for the middle and lower segments, albeit at a later point in time. Thus, attention spread gradually from the fixation point across all segments of the target curve. Significance of the difference in performance between contours of the target and distractor curve is indicated by *($P < 0.05$) and **($P < 0.01$). (c) Neuronal responses to contours that have to be grouped together are enhanced. There is a close correspondence with the psychophysical results, which indicate that attention is directed to the contours that are grouped. We call this the "sausage of attention". (modified from Houtkamp et al., 2003).

connected to them (see also Kramer and Jacobson, 1991), until the whole curve is labeled by visual attention. The hypothesis that attention thereby groups these contour segments into a coherent representation partially agrees with the feature integration theory (see Chapter 24) (Treisman and Gelade, 1980). In the feature integration theory, attention groups different features of the same object such as colors and orientations. Our theory generalizes the feature-integration theory to the domain of contour grouping on the basis of colinearity and connectedness.

V. CONCLUSIONS

We have outlined an algorithm for the grouping of connected image elements that is implemented in the visual cortex. It proposes distinct roles for pre-attentive and attentive vision (Lamme and Roelfsema, 2000). In the pre-attentive phase, feedforward connections generate an initial activity pattern across the many visual areas. This activity pattern enables a subset of recurrent connections, the interaction skeleton, so that neurons responding to image elements of the same object are linked. The grouping criterion of connectedness is implemented by the selectivity of the recurrent connections, which only link neurons with receptive fields that are close together. Other grouping criteria, such as colinearity, can be implemented equivalently by interconnecting neurons that respond to colinear or co-circular contours (Bosking et al., 1997; Schmidt et al., 1997; Sigman et al., 2001). The linkage of neurons is not yet explicit during this pre-attentive phase; that is, it cannot yet be read out by neurons in other cortical areas, since they do not "see" the pattern of enabled connections. Attentive vision has to make these linkages explicit. In the attentive phase, a response enhancement gradually spreads though the network of enabled connections, until the image elements of a single object have been labeled (Fig. 37.3c).

Although our neurophysiological data were obtained in the primary visual cortex, a number of arguments make it unlikely that this area solves all contour grouping tasks on its own. First, Jolicoeur et al. (1991) showed that the speed of contour grouping depends on the distance between a target curve and surrounding distractors. The speed (measured in deg/s) is highest if the target curve and the distractors are widely separated. This suggests that contour grouping is implemented at multiple spatial scales, which presumably correspond to different visual areas. Receptive fields are large in higher visual areas, and the rate enhancement can cross long distances in the visual field with a few synapses (Roelfsema et al., 2000). If the curves come close together, however, these large receptive fields will fall on the target as well as on the distractor curve, which prohibits the selective labeling of a single curve. In that case, the spread of the rate enhancement has to be carried out in lower visual areas where receptive fields are smaller, at the cost of a decreased grouping speed. Second, other cues such as motion coherence or similarity of color can speed up contour grouping (Nothdurft, 1993), which implies that areas where neurons are tuned to these additional grouping cues can be involved. Moreover, contours of familiar objects are grouped more rapidly than contours of less familiar objects (Vecera and Farah, 1997). This can also be explained by the involvement of higher visual areas that are involved in object recognition, such as the inferotemporal cortex. Here neurons are tuned to shapes and enhance their response if attention is directed to their preferred shape (Chelazzi et al., 1993). When neurons that are tuned to the shape of a familiar target object enhance their response, they may feed back to lower visual areas to enhance the response of neurons with receptive fields on the target object. If the shape is unfamiliar, however, its contours can also be grouped together, but only on the basis of low-level grouping cues, such as colinearity and connectedness. In that case there will be little help from high-level areas, which results in a decreased grouping speed.

We conclude that grouping criteria, such as connectedness, are evaluated by spreading a enhanced firing rate through a network of enabled corticocortical connections. This enhanced firing rate is a correlate of visual attention. Attentive vision thus flexibly binds image fragments into coherent object representations, even if these fragments belong to objects that have never been seen before.

Acknowledgments

P.R.R. was supported by a grant of the McDonnell-Pew foundation and a grant of the HFSP. We thank Roos Houtkamp for helpful comments on a draft of the manuscript.

References

Bosking, W. H., Zhang, Y., Schofield, B., and Fitzpatrick, D. (1997). Orientation selectivity and the arrangement of horizontal connections in tree shrew striate cortex. *Journal of Neuroscience* **15**, 2112–2127.

Chelazzi, L., Miller, E. K., Duncan, J., and Desimone, R. (1993). A neural basis for visual search in inferior temporal cortex. *Nature* **363**, 345–347.

Felleman, D. J., and Van Essen, D. C. (1991). Distributed hierarchical processing in the primate cerebral cortex. *Cereb. Cortex* **1**, 1–47.

Gilbert, C. D. (1992). Horizontal integration and cortical dynamics. *Neuron* **9**, 1–13.

Houtkamp, R., Spekreijse, H., and Roelfsema, P. R. (2003). The cause of delays in contour integration. *Perception & Psychophysics*, **65**, 1136–1144.

Jolicoeur, P., Ullman, S., and MacKay, M. (1986). Curve tracing: a possible basic operation in the perception of spatial relations. *Memory & Cognition* **14**, 129–140.

Jolicoeur, P., Ullman, S., and MacKay, M. (1991). Visual curve tracing properties. *Journal of Experimental Psychology: Human Perception and Performance* **17**, 997–1022.

Kapadia, M. K., Ito, M., Gilbert, C. D., and Westheimer, G. (1995). Improvement in visual sensitivity by changes in local context: parallel studies in human observers and in V1 of alert monkeys. *Neuron* **15**, 843–856.

Kellman, P. J., and Shipley, T. F. (1991). A theory of visual interpolation in object perception. *Cognitive Psychology* **23**, 141–221.

Knierim, J. J., and Van Essen, D. C. (1992). Neuronal responses to static texture patterns in area V1 of the alert macaque monkey. *Journal of Neurophysiology* **67**, 961–980.

Kramer, A. F., and Jacobson, A. (1991). Perceptual organization and focused attention: the role of objects and proximity in visual processing. *Perception & Psychophysics* **50**, 267–284.

Maunsell, J. H. R. (1995). The brain's visual world: representation of visual targets in cerebral cortex. *Science* **270**, 764–769.

Lamme, V. A. F., and Roelfsema, P. R. (2000). The distinct modes of vision offered by feedforward and recurrent processing. *Trends Neurosci.* **23**, 571–579.

Nothdurft, H. C. (1993). The role of features in preattentive vision: comparison of orientation, motion and color cues. *Vision Reseach* **33**, 1937–1958.

Roelfsema, P. R., Lamme, V. A. F., and Spekreijse, H. (1998). Object-based attention in the primary visual cortex of the macaque monkey. *Nature* **395**, 376–381.

Roelfsema, P. R., Lamme, V. A. F., and Spekreijse, H. (2000). The implementation of visual routines. *Vision Research* **40**, 1385–1411.

Roelfsema, P. R., Scholte, H. S., and Spekreijse, H. (1999). Temporal constraints on the grouping of contour segments into spatially extended objects. *Vision Research* **39**, 1509–1529.

Roelfsema, P. R., and Singer, W. (1998). Detecting connectedness. *Cereb. Cortex* **8**, 385–396.

Roelfsema, P. R., and Spekreijse, H. (2001). The representation of erroneously perceived stimuli in the primary visual cortex. *Neuron* **31**, 853–863.

Schmidt, K. E., Goebel, R., Löwel, S., and Singer, W. (1997). The perceptual grouping criterion of colinearity is reflected by anisotropies of connections in the primary visual cortex. *Eur. J. Neurosci.* **9**, 1083–1089.

Scholte, H. S., Spekreijse, H., and Roelfsema, P. R. (2001). The spatial profile of visual attention in mental curve tracing. *Vision Research* **41**, 2569–2580.

Sigman, M., Cecchi, G. A., Gilbert, C. D., and Magnasco, M. O. (2001). On a common circle: natural scenes and Gestalt rules. *Proceedings of the National Academy of Sciences of the U.S.A.* **98**, 1935–1940.

Thorpe, S., Fize, D., and Marlot, C. (1996). Speed of processing in the human visual system. *Nature* **381**, 520–522.

Treisman, A. M., and Gelade, G. (1980). A feature-integration theory of attention. *Cognitive Psychology* **12**, 97–136.

Vecera, S. P., and Farah, M. J. (1997). Is visual image segmentation a bottom-up or an interactive process? *Perception & Psychophysics* **59**, 1280–1296.

von der Malsburg, C. (1986). Am I thinking Assemblies? *In* G. Palm and A. Aertsen (eds.), Springer Verlag, Berlin. 161–175.

Zipser, K., Lamme, V. A. F., and Schiller, P. H. (1996). Contextual modulation in primary visual cortex. *Journal of Neuroscience* **16**, 7376–7389.

CHAPTER 48

Scanpath Theory, Attention, and Image Processing Algorithms for Predicting Human Eye Fixations

Claudio M. Privitera and Lawrence W. Stark

ABSTACT

Attention has two strong links to the top-down scanpath theory. The initial intentional conscious organization of the perceptual cognitive-spatial model that directs vision must occur at a high level of cortical processing. Then, the directions of the scanpath eye movements controlling successive attentional foveations toward the informative regions-of-interest enable matching and checking of high-resolution detail from bottom-up sensory signals.

The scanpath theory firstly takes into account the dichotomy between low-resolution peripheral vision and high-resolution foveal vision; this dualism necessitates an important role for eye movements. It further develops that *representation* over distributed modules of the cortex that generate the complex model of perception as an active process. Finally, the scanpath controls eye movement foveations that are the usual and natural means of directing visual attention.

Understanding the role of the attention shifts–eye movenients scanpath in human vision is an important step toward the achieving of independent and automatic image processing in the computer vision community. We have demonstrated that a small and manageable collection of image processing algorithms, experimentally selected and then combined together can serve in a task such as predicting human eye fixations.

I. SCANPATH AND ATTENTION

The **scanpath theory** outlines how a top-down spatial-cognitive model can control active eye movements (EMs) and visual perception. Philosophers had long speculated about perception and how we see in our mind's eye, but little scientific evidence was adduced until the scanpath sequence of eye movements enabled an approach to the problem. The scanpath sequence, itself consists of alternating saccadic EMs and fixations that enable the active-looking paradigm. The controlling top-down (TD) model has its roots in philosophy, where, for 2000 years, thoughtful scholars have considered how could it be possible that inner awareness or conscious perception, most likely not all iconic, can check perception by iconic matching to bottom-up (BU) physical signals arriving to the brain via peripheral nerves and sensory organs.

Eye movements are an essential part of vision because of the dual nature of the visual system: (1) the fovea, a narrow field, about 1/2 to 2 degrees, of high-resolution vision (Fig. 48.1, lower right panel); and (2) the periphery, a very wide field, about 180 degrees, of low-resolution vision, sensitive to motion and flicker (Fig. 48.1, upper right panel). Eye movements (experimentally recorded, Fig. 48.1, middle left panel) must carry the fovea to fixate each part of a scene (Fig. 48.1, middle left panel) or picture or page of reading matter to be processed with high resolution. We call these foveal fixations, hROIs—human Regions-of-Interest. An illusion of clarity exists, that we "see" the entire visual field with high resolution, exists, but this cannot be true (Fig. 48.1, bottom right panel).

Human Regions-of-Interest (hROIs) are explained according to the scanpath theory of Noton and Stark (1971a,b) based on long-standing philosophical views

FIGURE 48.1 Experimentally recorded eye movements and hROIs (human Regions-of-Interest).

and recent, neurological studies. The scanpath theory proposes that a TD internal cognitive model of what we "see" not only controls our vision, but also drives the sequences of rapid eye movements and fixations, or glances, that so efficiently travel over a scene or picture of interest. These scanpath sequences are idiosyncratic to the subject and to the picture. The contrary belief is that features of the external world control eye fixations and vision in a BU mode by impinging on the retina and sending signals to the brain.

Experiments have shown that when we look at ambiguous pictures, patterns of eye movement change with the mental image we have of the ambiguous figure (Stark and Ellis, 1981). When we engage in visual imagery, looking at a blank screen and visualizing a previously seen figure, our scanpath eye movements are similar whether viewing the figure or the blank screen. This provides strong evidence that the internal cognitive model, and not the external world (since this is absent in visual imagery), drives the scanpath (Brandt and Stark, 1997). Recent evidence presented by Stark et al. (2001) uses string editing distances to measure the similarity and dissimilarity between scanpaths. Also, studies of visual search indicate that a primitive form of precognitive spatial model controls a searchpath sequence of eye movements.

A **TD cognitive-spatial model** of the image in the mind's eye (Fig. 48.2, upper right) contains a knowledge representation (*What*, upper left) of the elements of the background setting of the picture, of the loci (*Where*, 2nd row, right), and of the sequences (*When*, 2nd row, left) and controls active looking to each of these hROIs by the experimentally found EM scanpath (lower right). EMs and attentional shifts are very closely linked. When the fovea is placed on an object, BU visual signals in log polar retinotopic iconic form arrive via retina and lateral geniculate nuclei to layers 4 and 5 of the visual cortex, area 17 (3rd row, BU icons). These icons are matched in sequence with the cognitive-spatial representations projected TD from other cortical areas to layers 1, 2, and 3 (3rd row, TD icons). Matching confirmation of these hypothesized icons generated by the TD model would thus support the overall percept. Of course, the match-

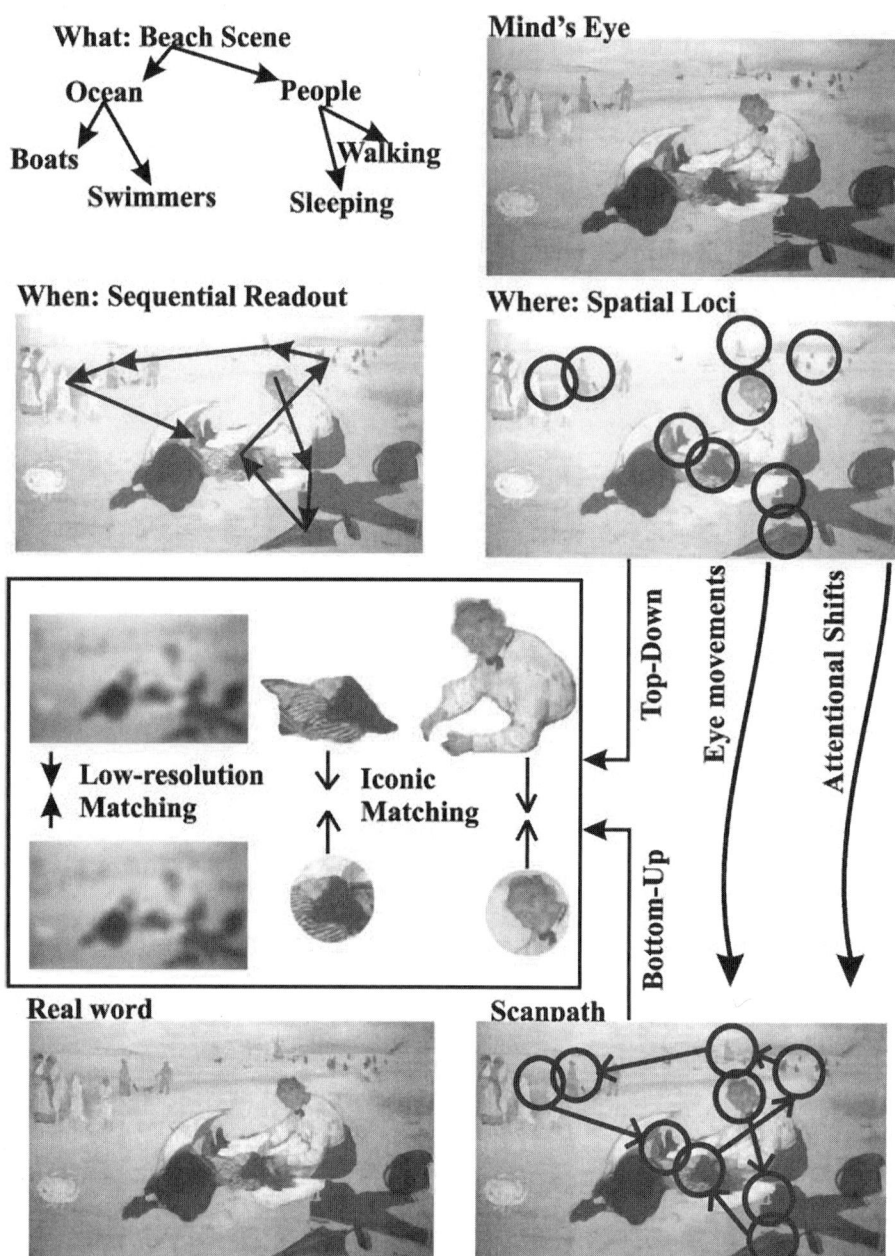

FIGURE 48.2 Human Regions-of-Interest, hROIs (circled regions, bottom right panel), are explained according to the scanpath theory of Noton and Stark, based on long-standing philosophical views and recent neurological studies. A top-down (TD) cognitive model of the external world is composed of *What*, *Where*, and *When* information. This model controls the scanpath shifts of attention (eye movements sequence) and the iconic matching in the visual cortex between foveated bottom-up (BU), visual signals and the TD model cognitive representation; all together they yield visual perception. (The authors thank Dr. Chernyak for an earlier version of this figure.) (See color plate).

ing of low-resolution peripheral (salient) regions is important to inform the planning control and helps to make the saccades more accurate and sure in their generation.

As a laboratory curiosity, **attention shifts** can be obtained without eye movements, yet at a high cost in several ways. A lower resolution is available only at retinal eccentricities away from the fovea. Also, motor instabilities occur when attempting the fixing of eye position away from the intended position as studied in Zeevi et al. (1979). There is also no advantage in speed. The information obtained with successive shifts of attention has the same quarter-second brain processing requirements as do successive foveations (Stark and Krischer, 1988). This is likely due to generation and matching interactions of the TD and BU flows of visual signals.

II. SALIENCY PREDICTS INFORMATIVENESS

During active looking, fixations, thus attention, are directed toward the most **informative regions in the visual scene**. A Bayesian conditional probability framework was used to model this TD mechanism in Chernyak and Stark (2001) and Schill et al. (2001).

Can saliency predict informativeness? In other words, is it possible to predict EM fixations using bottom-up image processing algorithms or saliency operators? We considered several sources of operators in Privitera and Stark (2000). (1) Using introspection as a metaphorical approach to top-down human vision, one might come up with operators such as those that identify regions in an image that have localized projections or concentric features. (2) Considering early-stage vision as described in the pioneering neurophysiology of J. Lettvin, and D. Hubel and T. Weisel, one might be led to operators that identify regions with lines or edges, center-surround operators, spot or bug detectors, or orientation elements. (3) Canonical operators based on some mathematical algorithms could be chosen; examples are structural operators such as laplacians, Fourier series and wavelets. (4) Other possible candidates are statistical operators such as entropy or generative operators like fractals. (5) Multiple adaptive matched filters experimentally derived from EM fixation loci can be constructed and incorporated in this operator collection.

Each operator gives emphasis to a different **bottom-up visual saliency** when applied to an image and correspondingly defines its own energy map. Local maxima in the output energy map define peaks of saliency, and the corresponding loci are possible candidates to be selected as the algorithmic Regions-of-Interest, or aROIs.

The procedure for defining aROIs is divided into three basic stages. First, the saliency operator is applied to the image, and the resulting energy map (which spatially coincides with the input image) is used to identify an initial, large collection of saliency local maxima. Maxima are then clustered in a final number of eccentric clusters of maxima; for each cluster, the highest maximum identifies the aROI. This three-stage procedure is similar to the classical winner-take-all sequential selection widely proposed in the literature, yet it encapsulates a few important advantages as the clustering of the local maxima is a parameter-free process that does not depend, for example, on the size of the image and on the scale of aROIs (Privitera and Stark, 2000).

Although we cannot here try to match the complex cognitive-perceptual functions of the human viewer, in many cases we have demonstrated (in Privitera and Stark, 2000) that a small and manageable collection of image processing algorithms can define **aROIs, which indeed identify those hROIs** chosen by the human scanpath EM fixations driven by TD guidance. Ten different algorithms were studied in Privitera and Stark (2000), symmetry transform, wavelet transform, two different center-surround quasi-receptive field masks, orientation contrast, edge intensity per unit area, entropy, graylevel Michaelson contrast, discrete cosine transform, and laplacian of Gaussian.

Thus, automatic picture analysis based on human vision could be an essential element in many computer vision application that require selective attention. An example provided in Privitera and Stark (2003) is for planetary and geological exploration.

References

Brandt, S. A., and Stark, L. W. (1997). Spontaneous eye movements dnring visual imagery reflect the content of the visual scene. *J. Cognitive Neuroscience* **9**(1), 27–38.

Chernyak, D. A., and Stark, L. W. (2001). Top-down guided eye movements. *IEEE Trans. SMC-B* **31**(4), 514–522.

Noton, D., and Stark, L. W. (1971a). Eye movements and visual perception. *Scientific American* **224**, 34–43.

Noton, D., and Stark, U W. (1971b). Scanpaths in eye movements during pattern perception. *Science* pp. 308–311.

Privitera, C. M., and Stark, L. W. (2000). Algorithms for defining visual region-of-interest: comparison with eye fixations. *IEEE Trans. PAMI* **22**(9), 970–981.

Privitera, C. M., and Stark, L. W. (2003) Human-vision-based selection of image processing algorithms for planetary exploration. *IEEE Trans. IP* **12**(8), 917–923.

Schill, K., Umkehrer, E., Beinlich, S., Krieger, G., and Zetzsche, C. (2001). Scene analysis with saccadic eye movements: Top-down and bottom-up modeling. *J. Elect. Imaging* **10**(1), 152–160.

Stark, L. W., and Ellis, S. R. (1981). Scanpath revised: cognitive models direct active looking. In D. Fisher (ed.), "Eye movements: Cognition and Visual Perception", Lawrence Erlbaum Associates, Hillside, NJ, 193–226.

Stark, L. W., and Krischer, C. C. (1988). Reading with and without eye movements. In "Brain and Reading". Stockton Press, New York.

Stark, L. W., Privitera, C. M., Yang, H., Azzariti, M., Ho, Y. F., Blackmon, T., and Chernyak, D. (2001). Representation of human vision in the brain: How does human perception recognize images? *J. Elect. Imaging* **10**(1), 123–151.

Zeevi, Y. Y., Pehi, E., and Stark, L. W. (1979). Study of eccentric with secondary visual feedback. *J. Opt. Soc. Am.* **69**(5), 669–675.

CHAPTER 49

The Feature Similarity Gain Model of Attention: Unifying Multiplicative Effects of Spatial and Feature-based Attention

Julio C. Martínez-Trujillo and Stefan Treue

ABSTRACT

Attention modulates neuronal responses throughout the visual cortex of primates. For direction-selective neurons in cortical area MT, this modulation is multiplicative; that is, it changes a neuron's response to all motion directions by the same factor without affecting the neuron's selectivity. The intensity of the modulation and whether it results in an increase or decrease of cells' responses are determined by the attended stimulus location (spatial attention) and the attended stimulus feature (feature-based attention). Rather than representing distinct attentional effects, spatial and feature-based attention can be unified in a "feature similarity gain model" in which attended locations and features interact with a cell's selectivity to determine the attentional modulation of information processing in the cortex.

Attention modulates the responses of neurons in the visual cortex of primates (Moran and Desimone, 1985). We studied such attentional modulation in the middle temporal area (MT) of nonhuman primates in which a large proportion of cells are direction-selective; that is, they respond strongly to only a subset of directions of visual motion. Direction-selective neurons in area MT play a central role in visual motion perception (Newsome et al., 1985) and have been linked directly to psychophysical performance in motion tasks (Britten et al., 1996). Their most characteristic feature, their direction tuning curves (i.e., the bell-shaped response profile as a function of stimulus direction), correlates well with psychophysical thresholds of motion perception (Snowden et al., 1992).

I. MEASURING THE EFFECTS OF SPATIAL ATTENTION IN AREA MT

We recorded the responses of 131 neurons in area MT of two macaques to moving random dot patterns (RDP) presented on a computer monitor positioned in front of the animals while they attended or ignored such stimuli. One RDP was placed inside the receptive field (RF) of the neuron being recorded, and a second pattern, moving in the same direction, was placed in the opposite visual hemifield (Fig. 49.1a). On a given trial, using a spatial cue (i.e., a stationary RDP appearing at the beginning of the trial), the animal's attention was directed to either the pattern located inside the RF (attending *inside*) or to the pattern located outside the RF (attending *outside*). The animal was required to report the occurrence of a transient direction change in the cued/target pattern while ignoring an identical change in the other pattern (Treue and Martínez-Trujillo, 1999). In both the attending *inside* (upper panel) and the attending *outside* (lower panel) conditions, we derived the neuron's tuning curve by recording the responses to 12 different motion directions spaced every 30 degrees and fitting them with a Gaussian function (see equation in Fig. 49.1b and Treue and Martínez-Trujillo, 1999).

I. MEASURING THE EFFECTS OF SPATIAL ATTENTION IN AREA MT

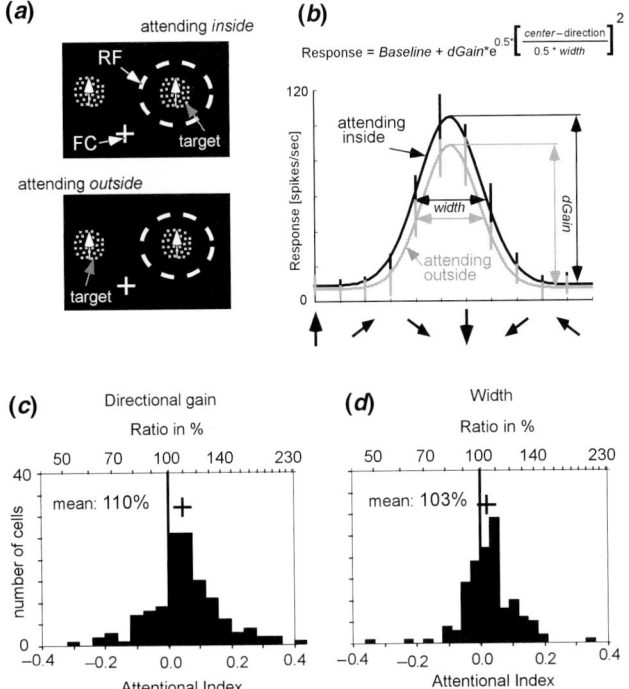

FIGURE 49.1 Measuring the effects of spatial attention. (a) One moving RDP was presented inside the neuron's RF (dashed circle), while the other one was presented about the same distance from the fixation point in the opposite hemifield. In a given trial, both RDPs moved in the same of 12 possible directions (white arrows). In the upper panel, the target was the pattern located inside the RF, while in the lower panel the target was the pattern located outside. (FC: fixation cross, RF: receptive field) (b) Cell example. The upper black curve represents the tuning curve in the attending *inside* and the gray curve in the attending *outside* condition. The former curve shows a visible increase in height (dGain) gain with almost no change in *width* in the former relative to the latter condition. The equation represents the Gauss function used to fit the responses. (c, d) Histograms showing the influence of attention on the tuning curve parameters dGain (c) and width (d) across the sample of 131 MT neurons. Binning is according to the attentional modulation index values (see text). The top scale displays the AMI converted into response ratios. The left histogram shows a clear shift to the right, with an average index of 0.05 (marked by the cross, where the horizontal arms span the 95% confidence interval of the mean), indicating that attention increases the height of the tuning curves on average (geometric mean) by about 10 percent. The right histogram shows no shift to the left, demonstrating that attention does not sharpen the tuning curves. Rather, we find a small, nonsignificant increase in the width of the tuning curves (average increase: 4%, $p > 0.05$ in paired t-test).

At least three main outcomes for this experiment are possible: (1) spatial attention might have no effect on neuronal responses—that is, the tuning curve profiles do not change as a function of whether attention is directed inside or outside the RF of the neurons; (2) spatial attention might enhance the sensory gain of neurons—that is, increase the response to all attended stimuli by the same proportion (multiplicative modulation) leaving neurons' selectivity for motion direction (i.e., the width of the tuning curve) unchanged (McAdams and Maunsell, 1999); and (3) spatial attention might increase a neuron's response only for attended stimuli within its RF moving in the preferred and similar directions, increasing the neurons selectivity for that direction through a nonmultiplicative sharpening of the tuning curve (Spitzer et al., 1988).

Figure 49.1b shows results from an example neuron. For this cell the tuning curve in the attending *inside* condition (black line) is a scaled version of the tuning curve in the attending *outside* condition (gray line). Essentially, the two curves differ in their height (dGain), but their *width* is very similar. Thus, for this neuron spatial attention multiplied the responses to different motion directions by the same factor (multiplicative modulation), leaving the neuron's selectivity for motion direction unchanged. We repeated the same analysis for each recorded neuron and computed an attentional modulation index (AMI) between the tuning curves parameters in the two conditions (i.e., AMI = [parameter in attending *inside*—parameters in attending *outside*] / [parameter in attending *inside* + parameters in attending *outside*]) (Treue and Martínez-Trujillo, 1999).

Figure 49.1c,d shows histograms of the AMI distributions for the parameters dGain and *width* across all our cells. On average, the dGain was about 10 percent larger in the attending *inside* relative to the attending outside condition (Fig. 49.1c). Although differences in the *width* of the curves were not significant, there was a slight nonsignificant increase (~4%) in the former relative to the latter condition (Fig. 49.1d). This increase in the height of the tuning curve in the absence of narrowing indicates that spatial attention has the same effect across the whole tuning curve; that is, it increases the responses through a multiplicative modulation.

The absence of a sharpening of the tuning curves agrees with a report of multiplicative changes in the tuning curves of neurons in area V4 of the ventral pathway with attention (McAdams and Maunsell, 1999) suggesting that attention acts similarly in both the dorsal and ventral pathways. A recent study attempting to model psychophysical orientation discrimination performance in dual-task attentional paradigms has suggested that the observed performance can only be accounted for by models that implement sharpening of tuning curves with attention (Itti et al., 1998). Since we have found no indication for such a sharpening, further studies will be necessary to understand the reasons behind this discrepancy (Martínez-Trujillo and Treue, 2004). It should be noted, though, that the attentional enhancement we observe does support better stimulus discriminability even without tuning sharpening by increasing the slope of the tuning curve (McAdams and Maunsell, 1999).

II. MEASURING THE EFFECTS OF FEATURE-BASED ATTENTION IN AREA MT

Psychophysical studies have suggested that attention can also be selectively allocated to stimuli that match a particular nonspatial feature (i.e., color, motion direction) (Lankheet et al., 1995; Duncan and Nimmo-Smith, 1996; Valdes-Sosa et al., 1998). To test for such effects in area MT, we designed the experiment illustrated in Fig. 49.2a. The stimulus configuration resembled the one in Fig. 49.1a. However, in this case the stimulus inside the RF was never attended and always moved in a given cell's preferred direction, and the stimulus outside the RF—always attended— could move either in the same preferred direction (attend-*preferred*, upper panel) or in the opposite (attend-*null*) direction (lower panel). This allowed switching the attended feature (direction) without changing the attended location (always outside the RF) and without changing the stimulus inside the RF. Note that changing the direction of the stimulus outside the RF (e.g., preferred or null) had no effect on the responses when that stimulus was behaviorally irrelevant—that is, when the animal was attending inside the RF or when simply fixating (control experiment, data not shown).

We recorded the responses from the same 131 direction-selective cells in area MT of two macaques and, for each individual neuron, computed the AMI ([response in attend-*preferred*—response in attend-*null*] / [response in attend-*preferred* + response in attend-*null*]). The histogram in Fig. 49.2a shows the distribution of the AMI across all cells. Attending to the preferred direction caused an average increase in response of about 13 percent relative to attending to the null direction. This is not an effect of spatial attention since the attended location was the same in both conditions. Rather, it represents a neural correlate of attention to a stimulus feature (i.e., motion direction). To determine whether this effect was due to an enhancement of sensory responses when the preferred direction was attended and/or a suppression of responses when the null direction was attended, we compared the responses in both attentional conditions against responses evoked in trials in which none of the stimuli was attended and the animal was performing a different task (detecting a luminance change) at the fixation point. We found that the attentional modulation is a combination of response enhancement

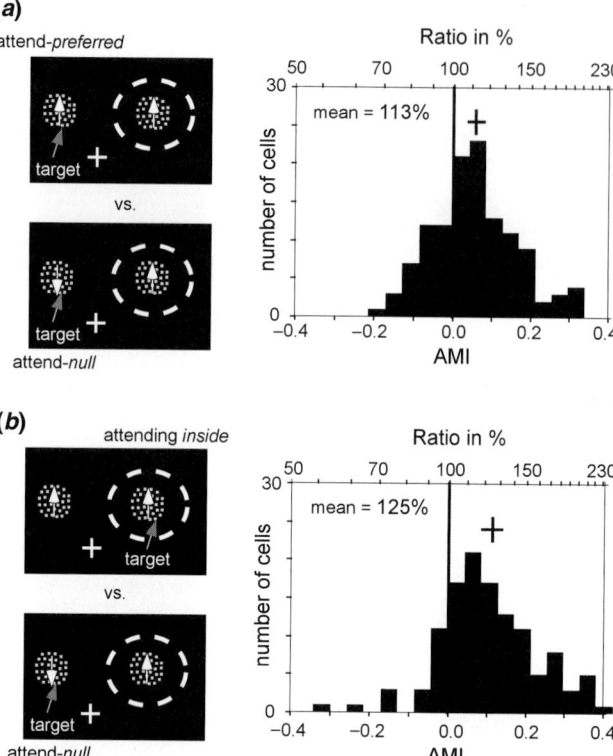

FIGURE 49.2 Feature-based attention acting alone and in combination with spatial attention. (*a*) Measuring feature-based attention. Left: The RDP inside the RF always moved in the cell's preferred direction (upward pointing arrow), and the stimulus outside moved in either the same (upper panel) or the opposite (null) direction (lower panel). Right: Histogram of the AMI between the responses of 131 MT neurons in the attend-preferred and attend-null conditions. The symbols and axes are the same as in Fig. 49.1c, d. The histogram is shifted to the right, (mean shift 13%), indicating an increased response in the attend-*preferred* relative to the attend-*null* condition. (*b*) Combined effects of feature and space-based attention. Right: Histogram of the AMI between the responses of the same 131 neurons in the attending *inside* (upper panel on the left) and attend-*null* (lower panel on the left) conditions. The histogram is shifted to the right, indicating a mean increase in response of ~25 percent in the former relative to the latter condition.

when the preferred direction was attended (mean enhancement of ~5%) and response suppression when the null direction was attended (mean suppression of ~6%). Thus, attending to a given direction enhances the responses of neurons whose preferred direction aligns with the attended direction and reduces the responses of those neurons preferring the opposite direction. This influence is far reaching since our stimuli were as much as 20 degrees apart and in opposite visual hemifields.

Though nonspatial, feature-based modulation of sensory responses has been observed in imaging studies (O'Craven et al., 1997; Beauchamp et al., 1997) and psychophysical paradigms (Lankheet et al., 1995;

Duncan and Nimmo-Smith, 1996; Valdes-Sosa et al., 1998), our study provided the first unambiguous single-cell correlate. Previous studies of attention did not discard the possibility that the modulation itself was based on the location of a given stimulus (Motter, 1994) or have confounded a change in the attended feature with a simultaneous change in attended location (McAdams and Maunsell, 1999). Subsequently, McAdams and Maunsell reported feature-based attentional modulation in area V4 of similar magnitude as the effects we have observed in area MT (McAdams and Maunsell, 2000); an fMRI study (Saenz, 2002) demonstrated feature-based attention in human visual cortex using a very similar design as the one we used.

Having demonstrated space- and feature-based attentional modulations of about equal size, we determined how these two effects interact. Figure 49.2b compares responses when the animals attended to the preferred direction inside the RF (attending *inside*) and when they attended to the null direction outside the RF (attend-*null*). The average attentional modulation between these conditions (25%) was approximately the sum of the shifts shown in Figs. 49.1c and Figs. 49.2a, demonstrating that space- and feature-based attentional effects can be additively combined. Comparing the two attentional conditions against responses when neither of the stimuli was behaviorally significant shows that the attentional modulation was a combination of the suppressive effect of switching attention to the null direction outside the RF (~6% suppression) and the enhancing effect of directing attention into the RF onto the preferred direction (~15% enhancement).

III. FEATURE SIMILARITY GAIN CAN EXPLAIN ATTENTIONAL EFFECTS

We have demonstrated physiological correlates of space- and feature-based attention in area MT of macaques, and we have further shown that these are additive processes that have a multiplicative effect on the response of neurons. This effect resembles changes to a neuron's sensory gain and thus can be mimicked by sensory effects, such as reducing a stimulus's luminance contrast. Similarly to our multiplicative effect, sensory effects do not change the tuning width of direction-selective neurons, suggesting that response modulation based on attentional and sensory aspects employ common mechanisms (Martínez-Trujillo and Treue, 2002).

Desimone and Duncan (1995) have proposed a "biased competition model of attention" in which attention influences the competition between two stimuli for access to a given cell in favor of the attended stimulus. This is achieved by increasing the strength of the signal coming from the population of input cells activated by the attended stimulus (Reynolds et al., 1999). Given the model's emphasis on competition between two stimuli within the RF of a given neuron, the attentional effects outside the RF and the summation of spatial and nonspatial effects when attention is switched into the RF that contains only one stimulus do not seem to be predicted by the model. We suggest that spatial and nonspatial attentional effects can be unified in a "feature similarity gain model," in which the up or down regulation of the gain of a sensory neuron reflects the similarity of the currently attended features and the sensory selectivity of the neuron along all target dimensions. Thus, this up or down regulation will also affect neurons whose RFs do not include the attended stimulus location. The relevant attended features include the spatial location, the direction of motion, and presumably others. Correspondingly, the sensory selectivity of the neuron includes the location of the RF (or of the smaller RFs of its input neurons), the preferred direction of motion, and presumably other preferred features. This model does not only provides a good account of other physiological studies of attention that did not use competing stimuli inside the receptive field (McAdams and Maunsell, 1999; Seidemann and Newsome, 1999) but also incorporates the suggestion of psychophysical and imaging studies as well as other modeling attempts that nonspatial stimulus features can be the basis of attentional effects.

In summary, the feature similarity gain model of attention provides a unified explanation for attentional effects on physiological signals that are based either on the attended location or on attended nonspatial features of visual stimuli. Further studies are needed to establish the link between the feature similarity gain modulation of physiological signals and the effects of attention on behavioral performance.

References

Beauchamp, M. S., Cox, R. W., and DeYoe, E. A. (1997). Graded effects of spatial and featural attention on human area MT and associated motion processing areas. *J. Neurophysiol.* **78**, 516–520.

Britten, K. H., Newsome, W. T., Shadlen, M. N., Celebrini, S., and Movshon, J. A. (1996). A relationship between behavioral choice and the visual responses of neurons in macaque MT. *Vis. Neurosci.* **13**, 87–100.

Desimone, R., and Duncan, J. (1995). Neural mechanisms of selective visual attention. *Annu. Revi. of Neurosci.* **18**, 193–222.

Duncan, J., and Nimmo-Smith, I. (1996). Objects and attributes in divided attention: Surface and boundary systems. *Percept. & Psychophy.* **58**, 1076–1084.

Itti, L., Braun, J., Lee, D. K., et al. (1998). Attentional modulation of human pattern discrimination psychophysics reproduced by a quantitative model. *Neural Information Processing Systems* (NIPS).

Lankheet, M. J. M., and Verstraten, F. A. J. (1995). Attentional modulation of adaptation to two-component transparent motion. *Vision Rese.* **35**, 1401–1412.

Martínez-Trujillo, J. C., and Treue, S. (2002). Attention changes apparent stimulus contrast in primate visual cortex. *Neuron* **35**, 365–370.

Martínez-Trujillo, J. C., and Treue, S. (2004). Feature-based attention increases the selectivity of population responses in primate visual cortex. *Current Biol.* **14**, 1–20.

McAdams, C. J., and Maunsell, J. H. R. (1999). Effects of attention on orientation tuning functions of single neurons in macaque cortical area V4. *J. Neurosci.* **19**, 431–441.

McAdams, C. J., and Maunsell, J. H. (2000). Attention to both space and feature modulates neuronal responses in macaque area V4. *J. Neurophysiol.* **83**(3), 1751–1755.

Moran, J., and Desimone, R. (1985). Selective attention gates visual processing in the extrastriate cortex. *Science* **229**, 782–784.

Motter, B. C. (1994). Neural correlates of attentive selection for color or luminance in extrastriate area V4. *J. Neurosci.* **14**, 2178–2189.

Newsome, W. T., Wurtz, R. H., Drsteler, M. R., and Mikami, A. (1985). Deficits in visual motion processing following ibotenic acid lesions of the middle temporal visual area of the macaque monkey. *J. Neurosci.* **5**, 825–840.

O'Craven, K. M., Rosen, B. R., Kwong, K. K., Treisman, A., and Savoy, R. L. (1997). Voluntary attention modulates fMRI activity in human MT-MST. *Neuron* **18**, 591–598.

Reynolds, J. H., Chelazzi, L., and Desimone, R. (1999). Competitive mechanisms subserve attention in macaque areas V2 and V4. *J. Neurosci.* **19**, 1736–1753.

Saenz, M., Buracas, G. T., and Boynton, G. M. (2002). Global effects of feature-based attention in human visisual cortex. *Nature Neurosci.* **5**, 631–632.

Seidemann, E., and Newsome, W. T. (1999). Effect of spatial attention on the responses of area MT neurons. *J. Neurophysiol.* **81**, 1783–1794.

Snowden, R. J., Treue, S., and Andersen, R. A. (1992). The response of neurons in areas V1 and MT of the alert rhesus monkey to moving random dot patterns. *Experimental Brain Research* **88**, 389–400.

Spitzer, H., Desimone, R., and Moran, J. (1988). Increased attention enhances both behavioral and neuronal performance. *Science* **240**, 338–340.

Treue, S., and Martínez-Trujillo, J. C. (1999). Feature-based attention influences motion processing gain in macaque visual cortex. *Nature* **399**, 575–579.

Valdes-Sosa, M., Bobes, M. A., Rodriguez, V., et al. (1998). Switching attention without shifting the spotlight: Object-based attentional modulation of brain potentials. *J. of Cog. Neurosci.* **10**, 137–151.

CHAPTER

50

Biasing Competition in Human Visual Cortex

Sabine Kastner and Diane M. Beck

ABSTRACT

A typical scene contains many different objects that compete for neural representation owing to the limited processing capacity of the visual system. At the neural level, competition among multiple stimuli is evidenced by the mutual suppression of their visually evoked responses and occurs most strongly at the level of the receptive field. The competition among multiple objects can be biased by both bottom-up sensory-driven mechanisms and top-down influences, such as selective attention. Functional brain imaging studies reveal that biasing signals due to selective attention can modulate neural activity in visual cortex not only in the presence but also in the absence of visual stimulation.

I. LIMITED PROCESSING CAPACITY AND COMPETITION

Natural visual scenes are cluttered and contain many different objects that cannot all be processed at the same time. The limited processing capacity of the visual system becomes immediately evident when we view the many words that fill this page. Obviously, we are unable to process them all at once. One limitation is that the more distant words from the current point of fixation fall outside of our range of acuity. However, even if we restricted our example to those words that fall within a small window around the center of gaze, in which we could resolve all the letters, we would still find that we could only grab on to a few words at a time. That is, our visual experience is very often limited to a subset of the information available at any point in time. This limited processing capacity has led researchers to propose that multiple items in a visual scene compete for representation at the neural level (Desimone and Duncan, 1995). The winner of the competition then becomes accessible to further processing.

How is the competition resolved, or what determines the winner? Our example above may have suggested one mechanism. When attempting to see as many words as possible, while fixating a point on the page, we focus selectively on particular words, allowing them to reach awareness. Hence, selective attention may be one means by which the competition among multiple items for neural representation can be influenced. Such a mechanism would constitute a top-down bias on competition, but similar effects may occur for bottom-up mechanisms, such as visual salience. Consider, for example, if one of the words on this page had been printed in red ink. One would immediately notice its presence. It would have won the competition by virtue of physical characteristics rather than through the deliberate top-down allocation of attention.

In this chapter, we will review evidence for competition among multiple visual stimuli for neural representation and mechanisms of spatially directed attention that operate in visual cortical areas and appear to bias competition through top-down feedback in favor of the selected stimulus.

II. A NEURAL BASIS FOR COMPETITION AMONG MULTIPLE STIMULI

What is the neural basis for competition among multiple objects in the visual field? In single-cell physiology studies, neural responses to a single visual stimulus presented alone in a neuron's receptive field (RF) were compared to the responses evoked by the same stimulus when a second one was presented simultaneously within the same RF. The responses to the paired stimuli were found to be smaller than the sum of the responses evoked by each stimulus individually, and turned out to be a weighted average of the individual responses (Reynolds et al., 1999). This result suggests that multiple stimuli present at the same time within a neuron's RF are not processed independently, but interact with each other in a mutually suppressive way, indicating competition for neural representation. Competitive interactions among multiple stimuli present at the same time in the visual field have been found in several areas, including V2, V4, MT, MST, and IT.

Based on hypotheses derived from these monkey physiology studies, we investigated competitive interactions among multiple stimuli in the human cortex using fMRI (Kastner et al., 1998, 2001). Colorful visual stimuli, which optimally activate ventral visual cortex, were presented in four nearby locations to the periphery of the visual field, while subjects maintained fixation. Fixation was ensured by having subjects count the occurrences of Ts or Ls at fixation, which is an attentionally demanding task. The stimuli were presented under two different presentation conditions, sequential and simultaneous. In the sequential presentation condition, a single stimulus appeared in one of the four locations, then another appeared in a different location, and so on, until each of the four stimuli had been presented in the different locations. In the simultaneous presentation condition, the same four stimuli appeared in the same four locations, but they were presented together. Thus, integrated over time, the physical stimulation parameters were identical in each of the four locations in the two presentation conditions. However, suppressive (competitive) interactions among stimuli within RFs could take place only in the simultaneous, not in the sequential presentation, condition. Based on the results from monkey physiology, we predicted that the fMRI signals would be smaller during the simultaneous than during the sequential presentation condition owing to the presumed mutual suppression induced by the competitively interacting stimuli.

Consistent activations evoked by visual presentations as compared to blank periods were found in areas V1, V2/VP, V4, TEO, V3A, and MT, which were determined on the basis of retinotopic mapping. Briefly, areas V1, V2, and V3/VP were identified by the alternating representations of the vertical and horizontal meridians, which form the borders of these areas. Areas V4, TEO, V3A, and MT were identified by their characteristic upper and lower visual field topography (for further details, see Kastner et al., 2001). As predicted by our hypothesis, simultaneous presentations evoked weaker responses than sequential presentations in all activated visual areas. The response differences were smallest in V1 and increased in magnitude toward ventral extrastriate areas V4 (Fig. 50.1a) and TEO, and dorsal extrastriate areas V3A and MT. This increase in magnitude of the suppression effects across visual areas suggested that the competitive interactions were scaled to the increase in RF size of neurons within these areas. That is, the small RFs of neurons in V1 and V2 would encompass only a small portion of the visual display, whereas the larger RFs of neurons in V4, TEO, V3A, and MT would encompass all four stimuli. Therefore, suppressive interactions among the stimuli within RFs could take place most effectively in these more anterior extrastriate visual areas. In V1 and V2, it is likely that surround inhibition from regions outside the classical RF contributed to the small sensory suppression effects. To rule out the possibility that the differential responses evoked by the two presentation conditions reflected differences in the rate of transient stimulus onsets, suppressive interactions were also demonstrated in a control experiment, in which the presentation rate was kept constant (see Kastner et al., 1998, 2001).

The idea that suppressive interactions are scaled to RF size was tested directly in a second study, in which

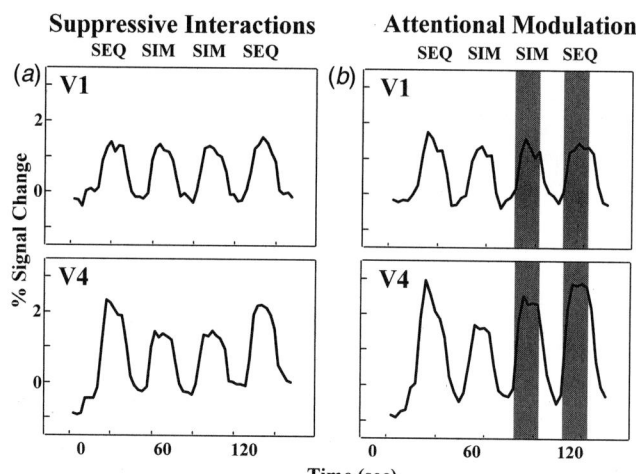

FIGURE 50.1 Competitive interactions and attentional modulation in visual cortex. (a) Suppressive interactions in V1 and V4. Simultaneously presented stimuli evoked less activity than sequentially presented stimuli in V4, but not in V1, suggesting that suppressive interactions were scaled to the RF size of neurons in visual cortex. (b) Attentional modulation of suppressive interactions. The suppression effect in V4 was replicated in the unattended condition of this experiment, when the subjects' attention was directed away from the stimulus display (unshaded time series). Spatially directed attention (shaded time series) increased responses to simultaneously presented stimuli to a larger degree than to sequentially presented ones in V4. Adapted from Kastner et al. (1998).

the spatial separation among the stimuli was increased (Kastner et al., 2001). According to the RF hypothesis, the magnitude of the suppressive interactions should be inversely related to the degree of spatial separation among the stimuli. In agreement with this idea, separating the stimuli by 4 degrees abolished suppressive interactions in V2, reduced them in V4, but did not affect them in TEO. Separating the stimuli by 6 degrees led to a further reduction of suppression effects in V4, but again had no effect in TEO. This is shown for a single subject in Fig. 50.2. Simultaneously presented stimuli induced strong suppressive interactions in area V4 with a 2 × 2 degree display, but not with a 7 × 7 degree display, whereas the differences in display size did not affect the activity evoked in area TEO. These results confirmed our hypothesis that competitive interactions occur mainly at the level of the RF. Furthermore, by systematically varying the spatial separation among the stimuli and measuring the magnitude of suppressive interactions, we estimated the average RF sizes at an eccentricity of about 5 degrees to be less than 2 degrees in V1, in the range of 2 to 4 degrees in V2, about 6 degrees in V4, and more than 6 degrees but still confined to a quadrant in TEO. These numbers may underestimate RF sizes in the human visual cortex owing to additional suppressive influences from beyond the RF, which cannot be distinguished from interactions within RFs in our experimental paradigm. It was striking, however, that these estimates of RF sizes in human visual cortex as determined on the basis of hemodynamic responses are similar to those measured in the homologous visual areas of monkeys as defined at the level of single cells (e.g., Desimone & Ungerleider, 1989).

In summary, these fMRI studies have begun to establish a neural basis for competition among multiple stimuli present at the same time in the visual field. Importantly, the degree to which this competition occurs appears to critically depend on the RF sizes of neurons across visual cortical areas. This has important implications for the operations of spatial selective attention in visual cortex, as will be described in the next section.

III. EVIDENCE FOR AN ATTENTIONAL TOP-DOWN BIAS IN VISUAL CORTEX

A. Filtering of Unwanted Information

In single-cell recording studies, it has been demonstrated that spatially directed attention can influence the competition among multiple stimuli in favor of one of the stimuli by modulating competitive interactions. When a monkey directed attention to one of two competing stimuli within a RF, the responses in extrastriate areas V2, V4, and MT were as large as those to that stimulus presented alone, thereby eliminating the suppressive influence of the competing stimulus (Reynolds et al., 1999; Recanzone & Wurtz, 2000). The attentional effects were less pronounced when the second stimulus was presented outside the RF, suggesting that competition for processing resources within visual cortical areas takes place most strongly at the level of the RF. These findings imply that attention may resolve the competition among multiple stimuli by counteracting the suppressive influences of nearby stimuli, thereby enhancing information processing at the attended location. This may be an important mechanism by which attention filters out information from nearby distracters (Desimone & Duncan, 1995).

A similar mechanism appears to operate in the human visual cortex (Kastner et al., 1998). We studied the effects of spatially directed attention on multiple competing visual stimuli in a variation of the paradigm, described in the last section. In addition to the

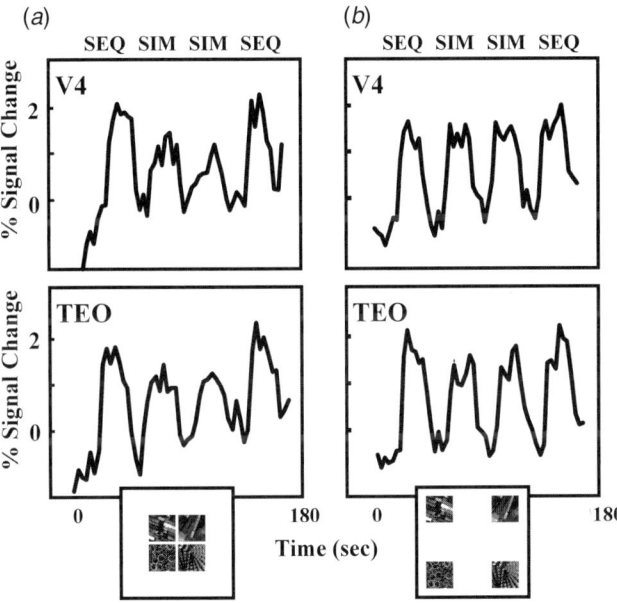

FIGURE 50.2 Modulation of sensory suppression in areas V4 and TEO. Time series of fMRI signals in V4 and TEO with display sizes of 2 × 2 degrees (a) and 7 × 7 degrees (b). Displays consisting of four stimuli were presented within the same quadrant and in the sequential and simultaneous conditions. Data are from a single subject. When stimuli were presented with the 2 × 2 degrees display, response differences to sequentially and simultaneously presented stimuli were found in V4 and TEO. When stimuli were presented with the 7 × 7 degrees display, the response differences were abolished in V4 but unchanged in TEO.

two different presentation conditions, sequential and simultaneous, two different attentional conditions were tested, attended and unattended. During the unattended condition, attention was directed away from the peripheral visual display by having subjects count letters at fixation. In the attended condition, subjects were instructed to attend covertly to the peripheral stimulus location closest to fixation in the display and to count the occurrences of one of the four stimuli. Based on the results from monkey physiology, we predicted that attention should reduce suppressive interactions among stimuli. Thus, responses evoked by the competing, simultaneously presented stimuli should be enhanced more strongly than responses evoked by the noncompeting sequentially presented stimuli.

Directing attention to the location closest to fixation in the display enhanced activity to sequentially and to simultaneously presented stimuli in extrastriate areas V2/VP, V4 (Fig. 1B), TEO, V3A, and MT, with increasing effects from early to later stages of visual processing; attentional effects were absent in V1 (Fig. 50.1b). In accordance with our prediction, directed attention led to greater increases of fMRI signals to simultaneously presented stimuli than to sequentially presented stimuli in areas V4 (Fig. 50.1b) and TEO. The magnitude of the attentional effect scaled with the magnitude of the suppressive interactions among stimuli, with the strongest reduction of suppression occurring in ventral extrastriate areas V4 (Fig. 50.1b) and TEO, suggesting that the effects scaled with RF size. These findings support the idea that directed attention enhances information processing of stimuli at the attended location by counteracting suppression induced by nearby stimuli, which compete for limited processing resources, thereby filtering out unwanted information from nearby distracters.

This filter mechanism is compatible with the idea that directed attention to a stimulus may cause the RF to shrink around the attended stimulus, thereby leaving the unattended stimuli in nearby locations outside the RF (Moran and Desimone, 1985). Given that the magnitude of suppressive interactions scaled with RF size in our fMRI studies (Kastner et al., 2001), we estimated the RF sizes in V4 and TEO during directed attention to the display. The reduced suppressive interactions in V4 and TEO during directed attention were similar in magnitude to the suppressive interactions obtained in area V2 in the unattended condition. Hence, it is possible that directed attention caused a constriction of RFs in V4 and TEO from 4/8 degrees to about 2 degrees, thereby presumably enhancing spatial resolution. This interpretation is compatible with behavioral studies showing that spatial attention improves acuity (Yeshurun and Carrasco, 1998).

In contrast to ventral extrastriate areas, in dorsal extrastriate areas V3A and MT, spatially directed attention led to comparable increases of activity for sequentially and simultaneously presented stimuli, indicating that the spatial filter mechanism did not operate within these areas. Because we used visual stimuli that activated ventral areas more effectively than dorsal areas, this finding suggests that the spatial filtering of unwanted information depends not only on RF size but also on the selectivity of neural populations to process preferred stimulus features.

In summary, areas at intermediate levels of visual processing such as V4 and TEO appear to be important sites for the filtering of unwanted information by counteracting competitive interactions among stimuli at the level of the RF. This notion has also been supported by studies in a patient with an isolated V4 lesion and in monkeys with lesions of areas V4 and TEO (DeWeerd et al., 1999; Gallant, Shoup, and Mazer, 2000). In these studies, subjects performed an orientation discrimination of a grating stimulus in the absence and in the presence of surrounding distracter stimuli. Significant performance deficits were observed in the distracter-present but not in the distracter-absent condition, suggesting a deficit in the efficacy of the filtering of distracter information.

B. Increases of Baseline Activity

The previous experiment suggests that competition among multiple stimuli within a RF can be biased by selective attention. There is evidence that attentional biasing signals can be obtained not only for the modulation of visually driven activity, but also in the absence of any visual stimulation whatsoever. Single-cell recording studies have shown that spontaneous (baseline) firing rates were 30 to 40 percent higher for neurons in areas V2 and V4 when the animal was cued to attend covertly to a location within the neuron's RF before the stimulus was presented there—that is, in the absence of visual stimulation (Luck et al., 1997). This increased baseline activity, termed the *baseline shift*, has been interpreted as a direct demonstration of a top-down signal that feeds back from higher order to lower order areas. In the latter areas, this feedback signal appears to bias neurons representing the attended location, thereby favoring stimuli that will appear there at the expense of those appearing at unattended locations. Thus, stimuli at attended locations are biased to "win" the competition for processing resources in a winner-take-all fashion (Itti and Koch, 2001).

We studied attentional biasing signals in the human visual cortex in the absence of visual stimulation by adding a third experimental condition to the design used to investigate sensory suppressive interactions and their modulation by attention (Kastner et al., 1999). In addition to the two visual presentation conditions, sequential and simultaneous, and the two attentional conditions, unattended and attended, an expectation period preceding the attended presentations was introduced. The expectation period, during which subjects were required to direct attention covertly to the target location and instructed to expect the occurrences of the stimulus presentations, was initiated by a marker presented briefly next to the fixation point 11 seconds before the onset of the stimuli. In this way, the effects of attention in the presence (ATT in Fig. 50.3) and absence (EXP in Fig. 50.3) of visual stimulation could be studied.

We found that, during the expectation period preceding the attended presentations, regions within visual areas with a representation of the attended location were activated. This activity was related to directing attention to the target location in the absence of visual stimulation. Notably, the increase in activity during expectation was topographically specific, inasmuch as it was only seen in areas with a spatial representation of the attended location. As illustrated for area V4 in Fig. 50.3b, the fMRI signals increased during the expectation period (light shaded epochs in the figure), before any stimuli were present on the screen. This increase of baseline activity was followed by a further increase of activity evoked by the onset of the stimulus presentations (dark shaded epochs in the figure). The baseline increase was found in all visual areas with a representation of the attended location. It was strongest in V4, but was also seen in early visual areas. It is noteworthy that baseline increases were found in V1 (Fig. 50.3a), even though no significant attentional modulation of visually evoked activity was seen in this area. This dissociation suggests either that different mechanisms underlie the effects of attention on visually evoked activity and on baseline activity, as suggested by Luck et al. (1997), or that some of the attentional effects previously reported with visual stimulation in V1 actually derive from sustained shifts in baseline activity rather than increases in the stimulus-evoked response per se. Importantly, the increase in baseline activity in V1 has also been found to depend on the expected task difficulty. Ress and colleagues (2001) showed that increases in baseline activity were stronger when subjects expected a visual pattern that was difficult to discriminate compared to a pattern that was easy to discriminate. In areas that preferentially process a particular stimulus feature (e.g., color or motion), increases in baseline activity were shown to be stronger during the expectation of a preferred compared to a nonpreferred stimulus feature (e.g., Chawla et al., 2000).

The baseline increases found in human visual cortex (Kastner et al., 1999; Ress et al., 2000; Chawla et al., 2000) may be subserved by increases in spontaneous firing rate similar to those found in the single-cell recording studies (Luck et al., 1997), but summed over large populations of neurons. The increases evoked by directing attention to a target location in anticipation of a behaviorally relevant stimulus at that attended location are thus likely to reflect a top-down feedback bias in favor of the attended location in human visual cortex.

In summary, neural activity in visual cortex is modulated by spatially directed attention. Biasing signals due to spatial attention affect neural processing in several ways. These include, as reviewed in this chapter, the filtering of unwanted information by counteracting the suppression induced by nearby distracters, and the biasing of signals in favor of an attended location by increases of baseline activity in the absence of visual stimulation.

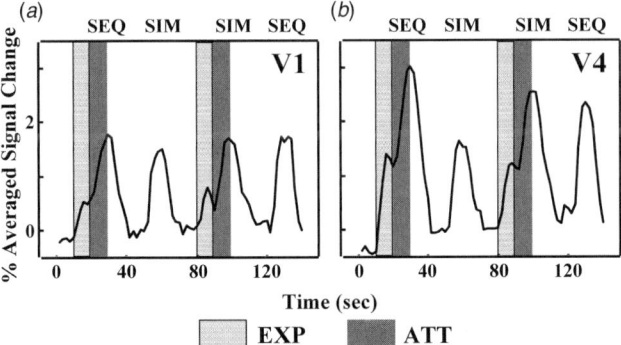

FIGURE 50.3 Increases of baseline activity in the absence of visual stimulation. Time series of fMRI signals in areas V1 (a) and V4 (b). Directing attention to a peripheral target location in the absence of visual stimulation led to an increase of baseline activity (light shaded blocks), which was followed by a further increase after the onset of the stimuli (dark shaded blocks). Baseline increases were found in both striate and extrastriate visual cortex. In striate cortex, no attentional effects on visually evoked activity were found.

References

Chawla, D., Rees, G., and Friston, K. J. (1999). The physiological basis of attentional modulation in extrastriate visual areas. *Nat. Neurosci.* **2**, 671–676.

Desimone, R., and Duncan, J. (1995). Neural mechanisms of selective visual attention. *Annu. Rev. of Neurosci.* **18**, 193–222.

Desimone, R., and Ungerleider, L. G. (1989). Neural mechanisms of visual processing in monkeys. *In* "Handbook of Neuropsychology" (F. Boller and J. Grafman, eds.), Vol. 2. Elsevier, Amsterdam.

De Weerd, P., Peralta, M. R. III., Desimone, R., and Ungerleider, L. G. (1999). Loss of attentional stimulus selection after extrastriate cortical lesions in macaques. *Nat. Neurosci.* **2**, 753–758.

Gallant, J. L., Shoup, R. E., and Mazer, J. A. (2000). A human extrastriate area functionally homologous to macaque V4. *Neuron* **27**, 227–235.

Itti, L., and Koch, C. (2001). Computational modeling of visual attention. *Nat. Rev. Neurosci.* **2**, 194–203.

Kastner, S., De Weerd, P., Desimone, R., and Ungerleider, L. G. (1998). Mechanisms of directed attention in the human extrastriate cortex as revealed by functional MRI. *Science* **282**, 108–111.

Kastner, S., De Weerd, P., Pinsk, M. A., Elizondo, M. I., Desimone, R., and Ungerleider, L. G. (2001). Modulation of sensory suppression: Implications for receptive field sizes in the human visual cortex. *J. Neurophysiol.* **86**, 1398–1411.

Kastner, S., Pinsk, M. A., De Weerd, P., Desimone, R., and Ungerleider, L. G. (1999). Increased activity in human visual cortex during directed attention in the absence of visual stimulation. *Neuron* **22**, 751–761.

Luck, S. J., Chelazzi, L., Hillyard, S. A., and Desimone, R. (1997). Neural mechanisms of spatial selective attention in areas V1, V2, and V4 of macaque visual cortex. *J. Neurophysiol.* **77**, 24–42.

Moran, J., and Desimone, R. (1985). Selective attention gates visual processing in the extrastriate cortex. *Science* **229**, 782–784.

Recanzone, G. H., and Wurtz, R. H. (2000). Effects of attention on MT and MST neuronal activity during pursuit initiation. *J. Neurophysiol.* **83**, 777–790.

Ress, D., Backus, B. T., and Heeger, D. J. (2000). Activity in primary visual cortex predicts performance in a visual detection task. *Nat. Neurosci.* **3**, 940–945.

Reynolds, J. H., Chelazzi, L., and Desimone, R. (1999). Competitive mechanisms subserve attention in macaque areas V2 and V4. *J. Neurosci.* **19**, 1736–1753.

Yeshurun, Y., and Carrasco, M. (1998). Attention improves or impairs visual performance by enhancing spatial resolution. *Nature* **396**, 72–75.

CHAPTER

51

Nonsensory Signals in Early Visual Cortex

David Ress and David J. Heeger

ABSTRACT

We performed two sets of experiments that revealed nonsensory signals in early visual cortex. The first experiment concerned visual attention. Activity in early visual cortex was measured using functional magnetic resonance imaging (fMRI) while subjects performed a challenging contrast-detection task. We observed a large, stimulus-independent response in early visual cortex, the *base* response. The base response showed three attributes that are characteristic of visual attention: it depended on task difficulty; it was spatially selective; and most significant, the trial-to-trial variability in the base response predicted behavioral performance on the task. The second set of experiments concerned perception. We modified the behavioral protocol to control the effects of the base response and thereby reliably observe the smaller perceptual signals. Subjects now attempted to detect the presence of slight contrast increments added to a high-contrast background pattern. Behavioral responses were recorded so that the corresponding cortical activity could be grouped into the usual signal detection categories: hits, false alarms, misses, and correct rejects. The cortical activity ranked as: hits ≈ false alarms > correct rejects ≈ misses. Thus, the activity in early visual cortex corresponded to the subjects' percepts, even when that percept was the opposite of what was physically presented in the stimulus. Together, the results suggest that early visual areas do more than encode raw sensory signals: they also participate in processing activities that correspond to visual attention and perception.

I. NEURAL CORRELATES OF VISUAL ATTENTION

Our ability to perform a visual discrimination task improves when we are cued to attend, without moving our eyes, toward the relevant stimulus (Posner, 1980; Pashler, 1998). Shifts in attention are correlated with systematic changes in cortical activity that have been measured in a number of cortical areas using a variety of methods. It is widely believed that top-down projections from prefrontal and/or parietal areas provide the attentional control signals that modulate visual processing (see e.g. Chapter 7). These top-down control signals modulate activity in early visual cortical areas including V1. These neural correlates of attention are spatially selective and depend on task difficulty. (Also see Chapter 61, 84)

Although there is considerable evidence that attention modulates of activity in visual cortex, only recently have there been any demonstrations of a link between attention-related increases in neural activity and improvements in behavioral performance (Ress, Backus et al., 2000; Moore and Fallah, 2001; Bisley and Goldberg, 2003). In one of these recent studies, we used fMRI to measure activity in human visual cortex using during a challenging contrast-detection task. The cortical activity measured during these experiments was dominated by signals with three characteristics of visual attention: they varied with task difficulty; they were spatially selective; and their magnitude was contingent with performance accuracy (Ress, Backus et al., 2000).

Subjects viewed a uniform gray field and continuously fixated a small high-contrast mark at its center

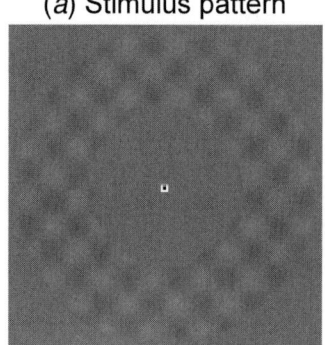

(a) Stimulus pattern

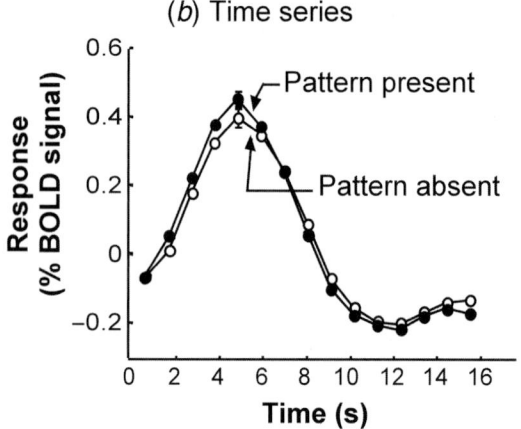

(b) Time series

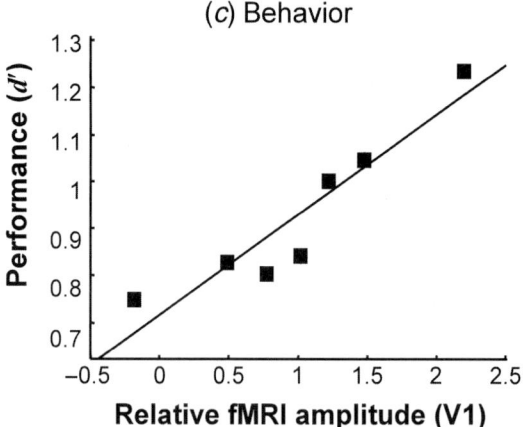

(c) Behavior

FIGURE 51.1 Neuronal correlates of visual attention in area V1. (a) Stimulus was a barely perceptible flickering checkerboard presented in an eccentric annulus. (b) A stimulus independent base response was visible on every trial. (c) Subject performance on the detection task was contingent on the magnitude of the base response.

(Fig. 51.1a) while lying in the bore of the MR scanner. Once every 20 s, a short auditory tone cued subjects that a new trial was beginning. On half the trials (randomly interleaved), a low-contrast pattern was presented briefly in a peripheral annulus around the fixation mark; on the other trials, no pattern was presented, and the display remained uniformly gray. Subjects pressed one of two buttons to indicate whether they believed the pattern was present. fMRI data were collected during several hundred trials for each of three subjects. The trials were categorized according to whether the stimulus pattern was present or absent and whether the subject responded correctly or incorrectly.

We observed a large response in early visual cortex (areas V1, V2, and V3) on every trial, even in the absence of a visual stimulus (Fig. 51.1b), evoked by the presentation of the auditory cue at the beginning of each trial. This cue-related *base response* was very similar for pattern-present and pattern-absent trials. This base response exhibited the signature characteristics of visual attention: it depended on task difficulty; and it was retinotopically selective, that is, evident only in the subregion of visual cortex corresponding to the representation of the stimulus aperture. Most importantly, the cue-related base response quantitatively predicted the subjects' pattern-detection performance: when activity was greater, the subject was more likely to correctly discern the presence or absence of the pattern (Fig. 51.1c).

We were surprised by the large magnitude of the cue-related base response. The result appeared to disagree with a number of electrophysiological studies for which baseline firing rates did not increase with attention (e.g., McAdams and Maunsell, 1999). Moreover, the studies that did find baseline increases in extrastriate cortex failed to find them in V1 (e.g., Luck, Chelazzi, et al., 1997). There are a number of possible explanations for the discrepancy between those results and our results (along with Kastner, Pinsk et al., 1999) that do indicate baseline increases in V1. First, because cortical neurons generally have low baseline firing rates, the responses of many neurons must be recorded to obtain statistically significant results. Therefore, these effects may be better revealed by fMRI measurements that reflect the activity of a large population of neurons. Second, the fMRI measurement may reflect subthreshold activity (e.g., due to excitation from distant inputs in the frontal or parietal lobes) that would be invisible to extracellular electrodes. Third, our data suggest that reliable and robust increases in baseline firing rates may be evident only when performing a particularly demanding task. The studies that failed to find baseline increases may have been using tasks that did not place sufficiently high demands on spatial attention. Fourth, the monkeys in those experiments may have been so highly trained that the usual attentional mechanisms were no longer needed to perform the task; training can have a critical effect on attentional signals (Ito and Gilbert, 1999). Fifth, small shifts in eye position, equal in size to the V1 receptive fields, present a difficulty for the electrophysiology experiments. Small shifts in eye position do not present a difficulty in the fMRI experiments because they have a negligible effect on measurements of pooled neuronal activity. Sixth, there may be a genuine species difference.

To understand how the base response could lead to improved performance, we can divide the hypothetical mechanisms mediating the base response into sensory enhancement and template enhancement categories. The base response could operate via either or both of these mechanisms.

Sensory enhancement mechanisms improve performance by increasing the signal-to-noise ratio of relevant sensory signals in early visual cortex. Signal detection theory offers a framework for understanding how a boost in the relevant neuronal signals can lead to improved performance (e.g., Palmer, Verghese et al., 2000). Accuracy is improved by boosting the relevant neuronal signals (e.g., from neurons with receptive fields that overlap the stimulus aperture) relative to other signals (e.g., from neurons with receptive fields outside the stimulus aperture), which contribute only noise.

Template-enhancing mechanisms improve performance by increasing the efficacy of how the sensory signals are compared to an internal representation of the visual target. For example, the base response might correspond to the processing of a visual working-memory representation of the target as it is compared with the incoming visual signals to perform the detection task. Errors would then occur on trials when the working-memory processing within the retinotopic representation of the stimulus aperture is weak (see Chapter 83).

II. NEURAL CORRELATES OF VISUAL PERCEPTION

For more than 30 years, psychophysical studies of visual pattern discrimination have paralleled research on the neurophysiological response properties of neurons in the visual cortex. The prevailing view has been that psychophysical judgments about visual patterns are limited by neuronal signals in early visual cortical areas. Signal detection theory has provided the theoretical framework for linking psychophysics and physiology.

The relationship between psychophysics and neurophysiology, as predicted by signal detection theory, can be illustrated in the context of the contrast detection task that we used in the experiments described below. On each trial, subjects were presented with one of two stimuli, either a background presented alone or the same background with a low-contrast target pattern superimposed (Fig. 51.2a). Subjects pressed one of two buttons to indicate whether they believed the target was present or absent. Logically, there are four possible outcomes on a given trial: hits, when the observer correctly responds "yes" on a target-present trial; correct rejects, when the observer correctly responds "no" on a target-absent trial; false alarms, when the observer erroneously responds "yes" on a target-absent trial; and misses, when the observer erroneously responds "no" on a target-present trial. Nearly all neurons in early visual cortex increase their activity monotonically with contrast, so the target-present stimulus will, on average, evoke slightly greater neuronal activity than the target-absent stimulus. Neuronal responses are, however, variable from one trial to the next, even when physically identical stimuli are presented repeatedly (Dean 1981; Shadlen and Newsome 1998). This variability in neuronal responses implies that the target-present stimulus can sometimes evoke less activity than the target-absent stimulus (Fig. 51.2b, overlap between the two probability distributions).

A simple model of the decision process is that observers respond "yes" when the neuronal activity exceeds a fixed criterion (Fig. 51.2b, the vertical line), and otherwise respond "no." This criterion divides the

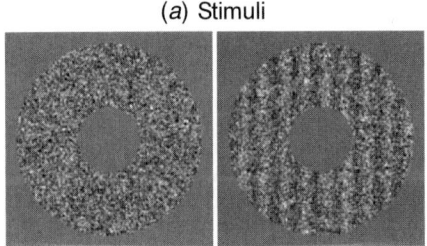

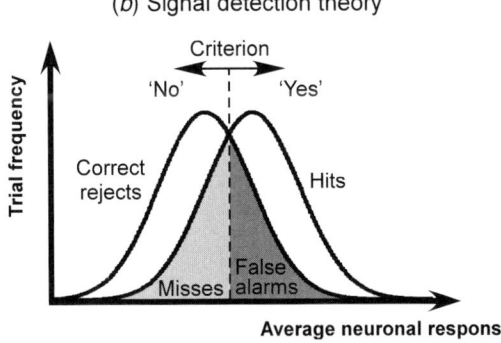

FIGURE 51.2 The visual perception experiment. (a) Subjects has to detect the presence or absence of a low-contrast vertical grating superimposed on a high-contrast noise background. Orientation and spatial phase of the target gratings were randomized. (b) Signal detection theory predicts that the average neuronal response on a great number of such detection trials should rank as correct rejects < misses < false alarms < hits.

two response distributions into four parts corresponding to the four possible outcomes. According to this model, we would expect the cortical activity averaged over many neurons and many trials of each trial type to rank as follows: responses to hits > false alarms > misses > correct rejects. This prediction is intuitive for the trials when the subject responds correctly (hits > correct rejects); cortical activity should be greater when the target contrast pattern is physically present in the stimulus. The prediction is counterintuitive for the error trials (false alarms > misses); cortical activity should now follow the subject's percept, the opposite of what is physically presented in the stimulus. The experiments described below were designed to test this prediction.

The experimental procedure was similar to the first set of experiments but modified to control the trial-to-trial variability of the base response. There were two critical changes. First, spatial uncertainty was eliminated by changing the task to contrast discrimination rather than detection. On most of the trials, only a high-contrast background pattern was presented; on the remaining (~1/6, randomly interleaved) trials, a low-contrast target grating was added to the background. Second, the pace of the experiment was greatly increased, with discrimination trials now occurring once every 2 s, thus encouraging subjects to continuously maintain their visual attention on the task. Because the target pattern was presented infrequently, most of the trials (~70%) corresponded to correct rejects. Hence, fMRI activity levels associated with correct reject trials were taken as a baseline, and we calculated the differential activity associated with hits, misses, and false alarms.

The observed ranking of cortical activity was hits ≈ false alarms > correct rejects ≈ misses. This was evident in the fMRI time series acquired from individual visual cortical areas in individual subjects (Fig. 51.3a). When the data were then combined across all subjects, the same ranking (hits ≈ false alarms > correct rejects ≈ misses) was again clearly evident (Fig. 51.3b). The differences between hits/false alarms and misses/correct rejects were highly statistically significant ($p \ll 0.001$). These effects were retinotopically specific, suggesting that they are related to the percepts themselves as opposed to a nonspecific effect (e.g., arousal) associated with "yes" responses. Of particular interest are the responses to the error trials (false alarms and misses) because they dissociate the percept from the physical presence or absence of the target. Critically, we found that activity was larger on average for false alarm trials than for misses. In other words, the activity in early visual cortex corresponded to the subjects' percepts even when that percept was the opposite of what was

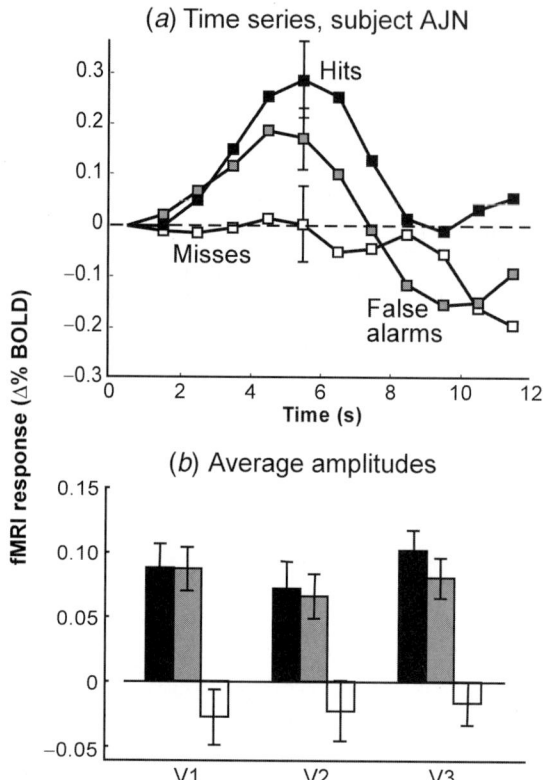

FIGURE 51.3 fMRI results from the visual perception experiment: (a) time series from a single subject; (b) response amplitudes averaged across all subjects.

physically presented in the stimulus, consistent with observers making perceptual decisions based on activity in these early visual areas.

These results are largely, but not entirely, consistent with the predictions of signal detection theory. We did not observe response amplitudes to hits > false alarms nor misses > correct rejects. There are at least two possible explanations for these discrepancies. First, cortical activity during our task may have consisted of an immediate response to the stimulus and a later signal associated with the percept (as reported in some single-unit electrophysiology experiments cited below). Because of the sluggishness of the hemodynamic response, our fMRI measurements would have yielded a superposition of the bottom-up sensory inputs and top-down percept-related signals. Second, there may be a residual performance-related effect, similar to what was discussed above in our first experiment. There was, in fact, a significant trend for response amplitudes to misses < correct rejects, consistent with the results described above in which errors were associated with a lapse in attention and a decreased base response, but completely opposite to what is predicted by signal detection theory.

Signal detection in humans has been previously studied with event-related potentials. Some of these experiments used stimuli analogous to ours, either auditory or visual targets masked by a distractor stimulus or noise. Particular attention has been paid to a transient evoked potential called the P3 or P300 (Hillyard, Squires et al., 1971; Squires, Squires et al., 1975; Hunter, Turner et al., 2001). Under auditory stimulus conditions analogous to our experiments, the magnitude of the P3 exhibited a dependence on trial type similar to what we have observed with fMRI (Squires, Squires et al., 1975). One interpretation of the P3 is that it reflects a working-memory representation used in the process of stimulus discrimination. Another common interpretation of the P3 is that it reflects an oddball effect, a response to the detection of infrequent targets. Further experiments will have to be performed to elucidate the relationship between the P3 and our fMRI measurements.

Although many previous fMRI experiments have found that a number of brain areas show greater responses during correct target detection (hits > misses, hits > correct rejects), none of these experiments were designed to measure the responses to false alarms, nor did they assess the retinotopic specificity of the results, making a meaningful comparison to our results difficult. One recent fMRI study reported a result that is similar to our own, but measured in an area of visual cortex that is believed to subserve object recognition (Grill-Spector and Kanwisher, 2001).

Threshold-level signal detection has also been studied previously by measuring single- and multi-unit activity in monkey cortex. One series of experiments demonstrated that threshold-level visual-motion signals are coded by the spike rates of neurons in cortical area MT, an area specialized for visual motion (Parker and Newsome, 1998). Our results suggest an analogous process in human visual cortex, as early as V1, for threshold-level visual contrast signals. A second line of research used backward masking to affect the detection of a visual target while recording from macaque frontal eye-field neurons (Thompson and Schall, 1999; Thompson and Schall, 2000; see also Chapter 21). The initial responses of these neurons (<100 msec after stimulus onset) corresponded to the actual presence or absence of the target, while later activity (100–300 msec after stimulus onset) corresponded to the monkey's behavioral report such that responses to false-alarms were greater than responses to misses. A third series of experiments recorded multi-unit activity in V1 while monkeys performed a figure-ground discrimination task (Super, Spekreijse et al., 2001) OR (see Chapter 83). Once again, two phases of responses were evident: an early response (<90 ms) that was directly related to the stimulus and a later response (100–240 ms) that was stronger when the monkey correctly performed the task (hits > misses). Our results might also have been caused by an immediate response to the stimulus and a later signal associated with the percept.

III. CONCLUSIONS

The results presented here, together with many other recent experimental results, indicate a profound transformation in our view of cortical processing in early visual areas. The earlier viewpoint construed early visual cortex processing mostly as a set of feed-forward decoding operations, extracting salient visual features such as contrast, orientation, color, and motion, from several incoming sensory data streams. Presumably, these decoded visual features were then processed more fully in other "higher" cortical areas. In contrast, the emerging body of data reviewed here suggests that early visual cortex actively participates in higher-level functions such as attention and perception, and that higher cortical areas are involved in the processing in which the visual features are first extracted. The mechanisms by which bottom-up sensory signals interact with top-down control and feedback signals in early visual cortex during visually guided behavior will be the grist of much new research in vision science.

References

Bisley, J. W., and Goldberg, M. E. (2003). Neuronal activity in the lateral intraparietal area and spatial attention. *Science* **299**, 81–86.

Dean, A. F. (1981). The variability of discharge of simple cells in the cat striate cortex. *Experimental Brain Research* **44**, 437–440.

Grill-Spector, K., and Kanwisher, N. (2001). The functional organization of human ventral temporal cortex is based on stimulus selectivity not recognition task. Soc. for Neurosci. 31st annual meeting.

Hillyard, S. A., Squires, K. C., et al. (1971). Evoked potential correlates of auditory signal detection. *Science* **172**, 1357–1360.

Hunter, M., Turner, A., et al. (2001). Visual signal detection measured by event-related potentials. *Brain. Cogn.* **46**, 342–356.

Ito, M., and Gilbert, C. D. (1999). Attention modulates contextual influences in the primary visual cortex of alert monkeys. *Neuron.* **22**, 593–604.

Kastner, S., Pinsk, M. A., et al. (1999). Increased activity in human visual cortex during directed attention in the absence of visual stimulation. *Neuron.* **22**, 751–761.

Luck, S. J., Chelazzi, L., et al. (1997). Neural mechanisms of spatial selective attention in areas V1, V2, and V4 of macaque visual cortex. *J. Neurophysiol.* **77**, 24–42.

McAdams, C. J., and Maunsell, J. H. R. (1999). Effects of attention on orientation-tuning functions of single neurons in macaque cortical area V4. *J. Neurosci.* **19**, 431–441.

Moore, T., and Fallah, M. (2001). Control of eye movements and spatial attention. *Proc. Natl. Acad. Sci. U S A* **98**, 1273–1276.

Palmer, J., Verghese, P., et al. (2000). The psychophysics of visual search. *Vision Res.* **40**, 1227–1268.

Parker, A. J., and Newsome, W. T. (1998). Sense and the single neuron: probing the physiology of perception. *Annu. Rev. Neurosci.* **21**, 227–277.

Pashler, H. E. (1998). The Psychology of Attention. MIT Press, Cambridge, MA.

Posner, M. I. (1980). Orienting of attention. *Q. J. Exp. Psychol.* **32**, 3–25.

Ress, D., Backus, B. T., et al. (2000). Activity in primary visual cortex predicts performance in a visual detection task. *Nat. Neurosci.* **3**, 940–945.

Shadlen, M. N., and Newsome, W. T. (1998). The variable discharge of cortical neurons: implications for connectivity, computation, and information coding. *J. Neurosci.* **18**, 3870–3896.

Squires, K. C., Squires, N. K., et al. (1975). Vertex evoked potentials in a rating-scale detection task: relation to signal probability. *Behav. Biol.* **13**, 21–34.

Super, H., Spekreijse, H., et al. (2001). Two distinct modes of sensory processing observed in monkey primary visual cortex (V1). *Nat. Neurosci.* **4**, 304–310.

Thompson, K. G., and Schall, J. D. (1999). The detection of visual signals by macaque frontal eye field during masking. *Nat. Neurosci.* **2**, 283–288.

Thompson, K. G., and Schall, J. D. (2000). Antecedents and correlates of visual detection and awareness in macaque prefrontal cortex. *Vision Res.* **40**, 1523–1538.

CHAPTER

52

Effects of Attention on Auditory Perceptual Organization

Robert P. Carlyon and Rhodri Cusack

ABSTRACT

This chapter summarizes the evidence on the effects of attention on auditory perceptual organization, with particular reference to the phenomenon of auditory streaming. Four approaches are described, and the limitations of each discussed. It is concluded that, although some streaming can take place without subjects devoting all their attention to the sounds, attention can have a profound effect on auditory perceptual organization.

I. INTRODUCTION

The extent to which perceptual organization is influenced by attention has recently been the subject of research in both the visual and auditory modalities (Visual: see Kovacs, this volume. Auditory: Bregman and Campbell, 1971; Sussman et al., 1998; Sussman et al., 1999; Carlyon et al., 2001a; Macken et al., 2003; Cusack et al., in press). The answer may shed light on the stages of neural processing involved in forming perceptual organization, and, at the same time, inform models of sensory processes. For example, many aspects of "auditory streaming" (e.g., Bregman and Campbell, 1971; van Noorden, 1975; Anstis and Saida, 1985) can effectively be modeled by automatic processes, in which the role of attention is restricted to selecting from competing representations at the output of low-level neural mechanisms (Beauvois and Meddis, 1991). However, evidence that attention can affect the streaming process itself (Carlyon et al., 2001a; Carlyon et al., 2001b; Cusack et al., in press) suggests that these models will have to be recast, either in terms of cortical processes or with more peripheral structures receiving input from those neural pathways involved in attention (e.g., McCabe and Denham, 1997).

A challenge facing researchers in all modalities is that of measuring the perceptual organization of a stimulus to which the participant is not fully attending. The challenge arises because, in order to test the effects of attending to the stimulus, one must divert attention away from it in some conditions. Consequently, subjective reports of the perceived organization may either be unreliable (for example, not all attributes of the stimulus may be encoded in short-term memory) or cause attention to be reoriented to the stimulus. In audition, research into the effects of attention on perceptual organization has focused not on the grouping of simultaneous frequency components into the same or different objects, but on the sequential organization of successive sounds ("auditory streaming"). Here we briefly describe auditory streaming and then review the different approaches that have been adopted when studying its modulation by attention.

II. AUDITORY STREAMING

The ability to organize sounds into separate streams is important in everyday situations when a listener needs to track the voice of one speaker in the presence of an interfering talker, and has also been widely used by composers of polyphonic music. It is commonly studied using simplified sequences of pure tones, like the one shown in Fig. 52.1 (van Noorden, 1975). This stimulus consists of a sequence of tones of frequencies A and B, in a series of repeating triplets "ABA-ABA-ABA..." (the dashes represent silent intervals). When the frequencies of A and B are similar, or the presentation rate is slow, subjects typically report hearing all

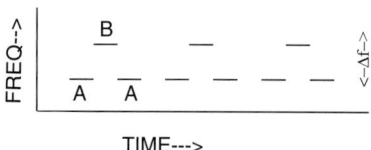

FIGURE 52.1 Example of a tone sequence used to study auditory streaming. The sequence, invented by van Noorden (1975), consists of repeating triplets of tone. The frequency of the "B" tones differs from that of the "A" tones by an amount Δf.

the tones in a single auditory stream, with a galloping rhythm. In contrast, when the A and B frequencies are far apart, and/or the presentation rate is fast, the "A" and "B" tones stream apart, and the percept of the galloping rhythm is lost. Furthermore, when subjects attend to a long sequence and monitor what they hear throughout, two further interesting findings emerge. First, for a wide range of intermediate frequency separations and rates, the percept flips between one and two streams. Second, superimposed on this flipping, the tendency to report hearing two streams increases markedly over the first 5 to 10 seconds. As we shall see in section V, this buildup of streaming has proved extremely useful in the study of the effects of attention on auditory streaming

III. ELECTROPHYSIOLOGICAL MEASURES

A conceptually straightforward way of avoiding the problems associated with subjective reports, described in section I, is to perform an EEG experiment. One measure, the "mismatch negativity (MMN)," lends itself to this approach because it can be obtained when subjects are not attending to the stimulus, and are instead reading a book or watching a silent movie. It is measured in a paradigm where subjects receive repeated presentation of a common standard sound, mixed with a small proportion (usually <20%) of "deviants." (For example, the deviants might be quieter than the standards or have a different frequency.) The MMN takes the form of a negative wave, with a latency of 150 to 200 ms, observed when the response to the standards is subtracted from that to the deviants. Although the MMN is thought to have multiple generators, its major source has been shown to be in auditory areas along the supratemporal plane.

A. Streaming Outside the Focus of Attention

Some MMN experiments have led to the conclusion that streaming can occur when participants are not listening to the to-be-streamed sounds. Sussman et al. (1999) played participants a tone sequence that alternated regularly between a high-frequency and a low-frequency range, with the tones in each range playing a simple melody. On a minority of trials, they altered the order of tones in the low range and argued that this should elicit an MMN only when the interfering high-frequency tones were pulled into a separate auditory stream. They observed an MMN when the stimulus onset asynchrony (SOA) between successive tones was 100 ms, at which behavioral experiments suggest that streaming should occur. In contrast, when the SOA was 750 ms, for which streaming should be much weaker, no MMN was observed. They attributed the difference in MMN between the two conditions to differences in streaming, which, as subjects were reading a book, they described as occurring outside the focus of attention. However, it should be noted that even when the low tones were omitted, the MMN to the high tone deviants was very weak at the 750-ms SOA, indicating that the observed effects of repetition rate may not have been entirely due to its effects on streaming.

Yabe et al. (2001) used MEG to measure the magnetic MMN ("MMNm") produced by the omission of an occasional high (3000 Hz, "H") tone in a sequence of alternating high and low ("L") tones (i.e., HLHLHL_LHL). Previous work had suggested that stimulus omissions only elicit an MMN when the SOA between successive tones is less than 160 to 170 ms. Using an SOA of 125 ms, they observed an MMN when the lower tone frequency was 2800 Hz, but not when it was 500 Hz. They attributed the lack of an MMN in the latter condition to the high and low tones being pulled into separate auditory streams, so that the SOA in the high stream was now too long (250 ms) to elicit the MMN. Their finding provides a clear demonstration that the omission MMN operates on a frequency selective basis, and, as described in section II, so does auditory streaming. However, a demonstration that the MMN difference is *due* to streaming would, we believe, require one to show that it varies with other parameters known to affect streaming. Indeed, the results of Shinozaki et al. (2000) show that size of the frequency separation between pairs of tone sequences can affect the MMN, even when participants hear all pairs of sequences as a single stream.

B. Manipulating Attention

One general limitation of the approach described above is that, although subjects were instructed to ignore the sounds, one can never be sure that they did not "sneak a little listen". For this reason, researchers are careful to conclude that some streaming occurs outside the *focus* of attention. An alternative approach is to manipulate the amount of attention paid to the sounds and see whether this affects the MMN response. This was adopted by Sussmann et al., using a paradigm similar to that in their 1989 experiment described above. They used an SOA of 500ms and observed an MMN only when subjects were attending to the tones; they concluded that, for this stimulus, attention did indeed modulate streaming.

IV. INDIRECT EFFECTS OF STREAMING ON COMPETING TASKS

Another way of measuring the perceptual organization of unattended stimuli has been developed by Jones and his colleagues, who measured the effects of streaming on a secondary task—namely, serial recall of visually presented letters (Jones et al., 1995; Jones et al., 1999; Macken et al., 2003). Performance on this task can be disrupted by sequences of sounds that subjects are instructed to ignore, provided that the sounds change over time. For example, two tones that alternate in frequency disrupt performance, whereas a single repeated tone does not. Importantly, this "irrelevant sound effect (ISE)" appears to be sensitive to streaming, so that when a sequence splits into two streams, each of which is heard as a repetition of a single sound, its deleterious effect on visual recall is reduced. Furthermore, this effect has been observed when streaming is produced by a number of different manipulations, including differences in frequency separation, in perceived location, and in repetition rate (for a summary see Macken et al., 2003). For example, Macken et al. (2003) visually presented letters accompanied by a sequence of tones that alternated at a low, medium, or high rate. When the alternation rate increased from low to medium, the number of errors increased, consistent with the increased number of changes in the (ignored) auditory sequence. However, when the rate was increased further, the percentage of errors dropped, consistent with the two tones now falling into separate auditory streams, each containing a repeating single tone.

A strong advantage of using the ISE to measure auditory streaming is that it provides a *performance* measure of the effect. Hence, any effects that one observes cannot be attributed to nonperceptual factors such as response biases. However, it too suffers from some limitations when it comes to studying the effects of attention on auditory streaming. An important problem is that it is hard to be sure that the task of remembering visually presented letters is sufficiently demanding to prevent subjects from attending to the sounds. Furthermore, because attention is not explicitly manipulated, one does not know whether any streaming observed would have been greater if subjects had been paying attention to the sounds. Essentially, these experiments lead to the important conclusion that *some* streaming can occur without *full* attention, but do not show whether, or by how much, streaming can be affected by attention.

V. MANIPULATING ATTENTION DURING THE BUILDUP OF AUDITORY STREAMING

A third approach, developed by Carlyon et al. (2001a), exploited the fact that the initial percept of a sequence of tones as a single stream can turn into a two-stream percept after several seconds. In their first experiment they presented a 20-sec sequence of repeating ABA-triplets (see section II and Figure 52.1) to subjects' left ears, and manipulated attention during the first 10sec of the sequence. In a baseline condition, they simply asked subjects to monitor how many streams they heard throughout the sequence by switching between two response buttons marked "1 stream" and "2 streams." In the remaining two conditions, a series of noise bursts was presented to the opposite ear during the first 10sec of the sequence. For one of these, subjects were instructed to perform a demanding task on the noise bursts for the first 10sec, and then switch their attention to the tones and start making streaming judgements. The question was, when they did this, whether the streaming would be the same as if they had been attending to the tones all along, or more like the tones at the very start of an attended sequence. The results, shown in Fig. 52.2*a*, reveal that the latter prediction was correct; when attention was diverted to the tones (circles), the buildup of streaming was much less than in the baseline condition (triangles). A third condition (squares) showed that the noise bursts had no effect when subjects ignored them and attended to the tones throughout.

An important feature of Carlyon et al.'s paradigm, which differs from the MMN and ISE experiments, is that it requires subjects to make direct reports of what they hear. It overcomes the problems of such reports themselves affecting attention by manipulating attention during the *first* half of a sequence but measuring streaming during the *second* half. However, the use of a subjective measure does make it possible that, in Carlyon et al.'s first experiment, a form of response bias was operating. Specifically, subjects may have decided to always respond "1 stream" whenever they

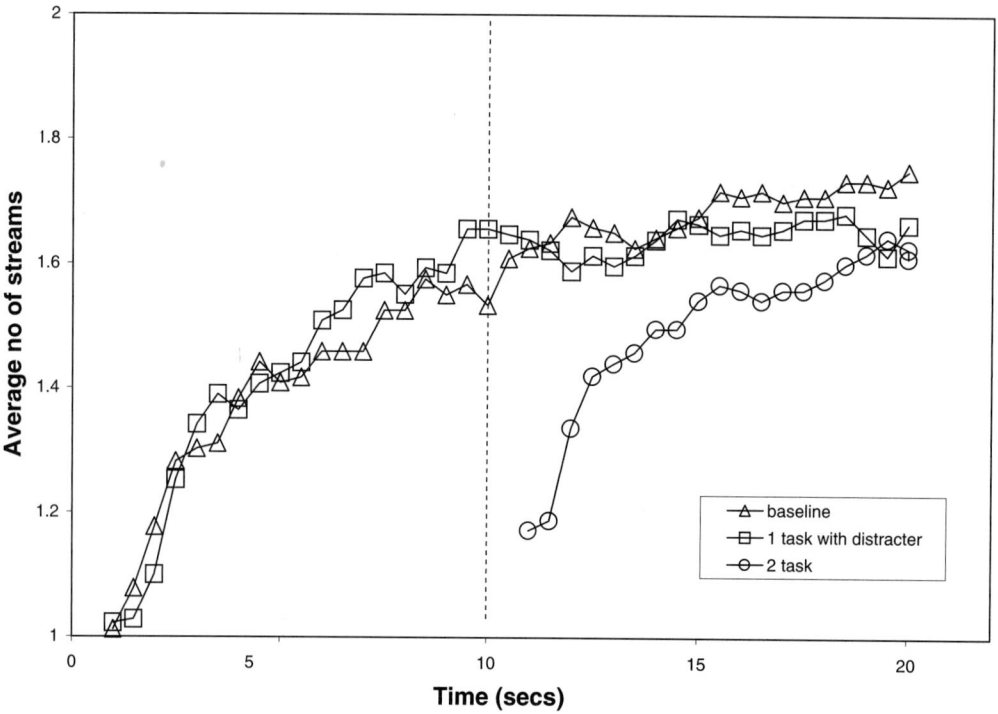

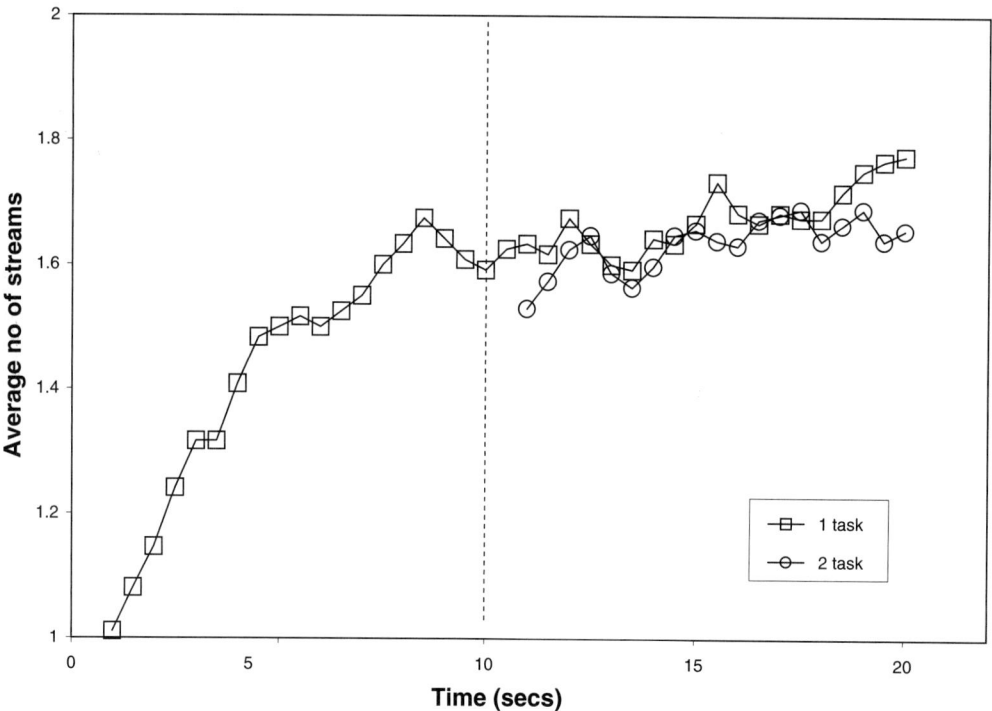

FIGURE 52.2 Average number of streams, plotted as a function of time, for (a) experiment 1 and (b) experiment 2 of the study by Carlyon et al. (2001a).

started to make streaming judgments about a sequence. To control for this, they performed a second experiment with no noise bursts and in which the tones were amplitude modulated (AM), with the rate of modulation switching every few seconds. In one condition, subjects were told to detect the switches in AM rate for the first 10 seconds, and then start making the streaming judgments. These judgments (Fig. 52.2b, circles) resembled those after 10 sec of a baseline condition where subjects made streaming judgments throughout (Fig. 52.2b, squares). Carlyon et al. noted that this result was inconsistent with a response bias and concluded that streaming does build up provided one is attending to the tones, even if one is performing a different type of task on them. Further experiments, using slightly different paradigms, have lent support to this conclusion (Carlyon et al., 2001b; Cusack et al., 2004). These experiments also showed that the buildup of streaming can be reduced by attending to sounds in the same ear, or even to visual stimuli.

Although the experiments by Carlyon and colleagues indicate a clear influence of attention on streaming, they do leave one important question unanswered. This is that we only know what subjects have reported after they have switched attention back to the tones. It may be that instead of diverting attention really preventing the buildup of streaming, the act of switching attention back to a sequence "resets" streaming. Indeed, Cusack et al. have suggested that the default organization might always be for a sequence to be heard as two streams, except when attention is directed toward it—either by instructions to the listener or simply because it has just been turned on. According to this explanation, the "buildup" of streaming observed in numerous experiments should instead be thought of as recovery from an active attentional influence. What does seem clear is that attention can have a profound effect on auditory perceptual organization, either by preventing a gradual buildup or by an active resetting mechanism.

VI. THE HIERARCHICAL DECOMPOSITION MODEL

Cusack et al. (2004) further investigated the effect of attention on the buildup of auditory streaming by examining the domain over which auditory streaming was prevented when attention was directed elsewhere. They found that, rather than just those sounds in unattended locations or frequency regions failing to show a buildup of streaming, *any* stream that was outside the focus of attention failed to show further fragmentation. This result was interpreted in terms of a *hierarchical decomposition model*, illustrated in Fig. 52.3. In this example of a real-world auditory scene, initial (perhaps automatic) grouping allows the listener to focus attention on (e.g.) the music, at the expense of the speech and traffic noise. This attentional focusing then causes the internal representation of the music to fragment, and several further streams become available for attention. If one of these is then attended, further fragmentation takes place and so on. As shown by Cusack et al. in a further experiment, if attention is withdrawn from a sequence, its grouping is then reset: in our example, if attention is switched to the traffic, the fragmentation of the music hierarchy will collapse. The advantages of such hierarchical decomposition are that on arrival in the auditory scene the listener is not bombarded with tiny components of all the sounds, and that any limited capacity processes involved in grouping are not wasted by the organization of unattended parts of the auditory scene.

VII. NEUROPSYCHOLOGICAL APPROACH

Only one study has adopted a neuropsychological approach to the effects of attention on auditory streaming. Carlyon et al. (2001a) studied patients with unilateral neglect, who had been diagnosed on the basis of having attentional deficits toward the left side of visual space. They asked patients to report how many streams they heard when 6-second sequences of alternating tones (Fig. 52.1) were presented to their left or right ears. The amount of stream segregation for sequences presented to their left ears was reduced compared both to the same sequences presented to their right ears and to sequences presented to either ear of healthy or brain-damaged controls. Carlyon et al. suggested that the reduced streaming in the left ears of neglect patients was caused by damage to cortical attentional processes. However, their patients had a variety of lesions, and so it is not possible to specify which areas of cortex are important for the effect.

VIII. CONCLUSIONS

Measuring the perceptual organization of an unattended stimulus, without requiring the participant to attend to it, stretches the imagination of the cognitive neuroscientist to the limit. We have described four approaches to this problem, each of which has its own limitations. Nevertheless, two broad principles emerge. First, the ISE experiments of Jones and colleagues allow us to conclude that some streaming can

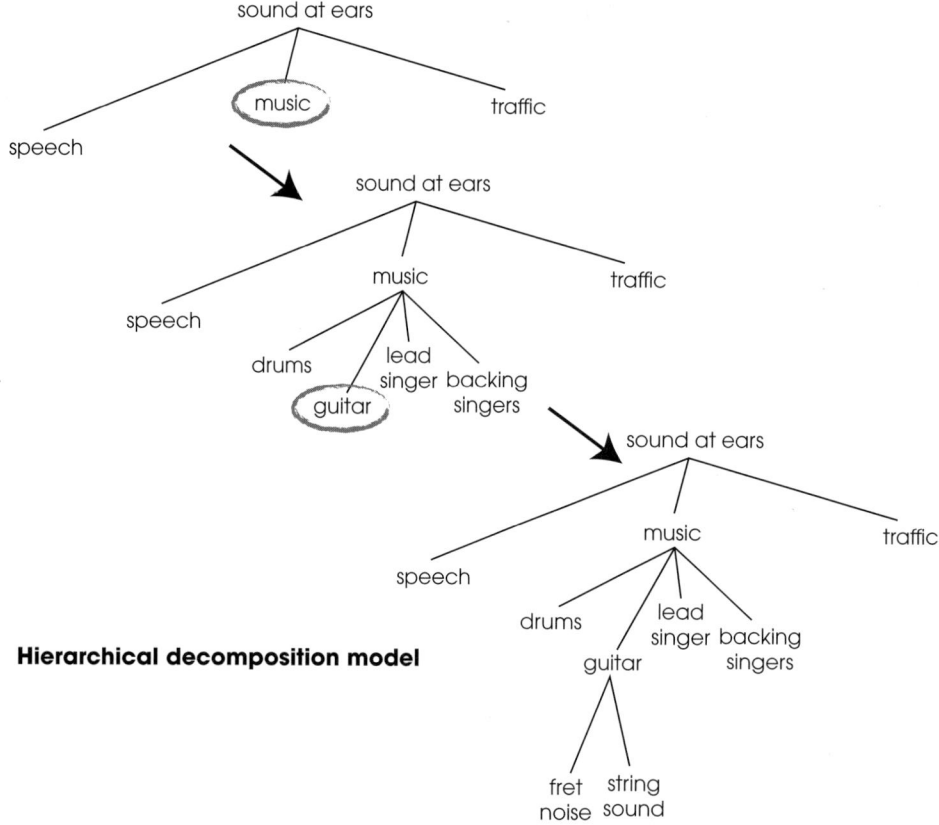

FIGURE 52.3 An example of the interaction between selective attention and grouping as proposed by the hierarchical decomposition model (Cusack et al., 2004).

take place without full attention. Second, Sussman et al.'s (1998) MMN experiment, combined with Carlyon et al.'s studies of the buildup of streaming, clearly indicate that attention can affect streaming. Crucially, those results show that that attention can have a profound effect on the streaming process per se, rather than simply allowing subjects to choose from the outputs of a purely automatic streaming mechanism (van Noorden, 1975).

References

Anstis, S., and Saida, S. (1985). Adaptation to auditory streaming of frequency-modulated tones. *Journal of Experimental Psychology: Human Perception and Performance* **11**, 257–272.

Beauvois, M. W., and Meddis, R. (1991). A computer model of auditory stream segregation. *Quarterly Journal of Experimental Psychology* **43A**, 517–542.

Bregman, A. S., and Campbell, J. (1971). Primary auditory stream segregation and perception of order in rapid sequences of tones. *Journal of Experimental Psychology* **89**, 244–249.

Carlyon, R. P., Cusack, R., Foxton, J. M., and Robertson, I. H. (2001a). Effects of attention and unilateral neglect on auditory stream segregation. *Journal of Experimental Psychology: Human Perception and Performance* **27**, 115–127.

Carlyon, R. P., Plack, C. J., and Cusack, R. (2001b). Cross-modal and cognitive influences on the build-up of auditory streaming. *Brit. J. Audiol.* **35**, 139–140.

Cusack, R., Deeks, J., Aikman, G., and Carlyon, R. P. (2004). Effects of location, frequency region, and time course of selective attention on auditory scene analysis. *Journal of Experimental Psychology: Human Perception and Performance* **30**, 643–656.

Jones, D. M., Alford, D., Bridges, A., Tremblay, S., and Macken, W. J. (1995). Organizational factors selective attention: The interplay of acoustic distinctiveness and auditory streaming in the irrelevant sound effect. *Journal of Experimental Psychology: Learning, Memory, and Cognition* **25**, 464–473.

Jones, D. M., Alford, D., Bridges, A., Tremblay, S., and Macken, W. J. (1999). Organizational factors in selective attention: The interplay of acoustic distinctiveness and auditory streaming in the irrelevant sound effect. *Journal of Experimental Psychology: Learning, Memory, and Cognition* **25**, 464–473.

Macken, W. J., Tremblay, S., Houghton, R. J., Nicholls, A. P., and Jones, D. M. (2003). Does auditory streaming require attention? Evidence from attentional selectivity in short-term memory. *Journal of Experimental Psychology: Human Perception and Performance* **29**, 43–51.

McCabe, S. L., and Denham, M. J. (1997). A model of auditory streaming. *J. Acoust. Soc. Am.* **101**, 1611–1621.

Shinozaki, N., Yabe, H., Sato, Y., Sutoh, T., Hiruma, T., Nashida, T., and Kaneko, S. (2000). Mismatch negativity (MMN) reveals sound grouping in the human brain. *Neuroreport* **11**, 1597–1601.

Sussman, E., Ritter, W., and Jr., H. G. V. (1998). Attention affects the organization of auditory input associated with the mismatch negativity system. *Brain Research* **789**, 130–138.

Sussman, E., Ritter, W., and Vaughan, H. G., Jr. (1999). An investigation of the auditory streaming effect using event-related potentials. *Psychophysiology* **36**, 22–34.

van Noorden, L. P. A. S. (1975). Temporal coherence in the perception of tone sequences. Ph.D. Eindhoven University of Technology.

Yabe, H., Winkler, I., Czigler, I., Koyama, S., Kakigi, R., Sutoh, T., Hiruma, T., and Kaneko, S. (2001). Organizing sound sequences in the human brain: the interplay of auditory streaming and temporal integration. *Brain Research* **897**, 222–227.

Attention in Language

Andriy Myachykov and Michael I. Posner

ABSTRACT

Attention plays an important role in critical aspects of the use of grammar and lexicon in human discourse. We explore methods for the activation of these operations during the choice of an adequate syntactic structure, referential control, and other grammatical operations, during the processing of word and sentence meaning and in the skill of reading. Neuroimaging has suggested separate systems for syntactic and semantic processing and has provided some details on the anatomy of attentional systems related to semantic analysis.

I. INTRODUCTION

The neurobiology of attention and most models discussed in this volume have quite naturally involved rather simple forms of repetitive behavior that can be studied relatively easily by experimental methods and are common to both humans and other animals. From this perspective, the role of attention in language may seem an odd topic. Sentences are hardly ever repeated, and grammar is a complex set of operations within each language. Nonetheless, it is important to understand how top-down control is exercised over species-specific higher mental activities that are involved in aspects of language such as grammar, semantics, or reading.

Spoken language is a good example of a species-specific behavior of human beings. The ability to read represents a high-level skill, closely coordinated with spoken language, but learned only by a subset of the human community. The role of attention in reading provides an entry into the many complex but arbitrary skills that human beings can learn to perform.

Developments in neuroimaging make it possible to examine the anatomy, circuitry, and plasticity of brain networks that are present, at least in full form, only in the human species. However, prior to imaging, it is important to have tasks that activate selectively key aspects of grammar and semantics. In this chapter we consider empirical methods and results that trace the role of attention in processing language and report some imaging experiments that have examined the mechanisms involved.

II. ATTENTION AND THE STRUCTURE OF DISCOURSE

Two aspects of discourse grammar are usually ascribed to attentional mechanisms:

1. Assignment of syntactic roles in a clause.
2. Referential choice and resolution of ambiguous reference in discourse.

A. Assignment of Syntactic Roles

A good example of the research in this area is a set of experimental studies by R. Tomlin (1997), in which he demonstrated that the choice of syntactic subject position in English narrative might result from the direction of attention. In a set of experiments, Tomlin used a computer animation program called the "Fish Film" (Fig. 53.1). Subjects viewed and described an unfolding engagement of the two fish in real time. In each trial one of the two fish was visually cued (see arrow in Fig. 53.1) attracting the subjects' attention. The critical part of each trial is the *dynamic event*, which is described by the subject. In their description of the dynamic event, assignment of the syntactic subject is varied depending on the visual cue. The dynamic event was the eating of one of the fish by the other. When the dark fish was cued and was then eaten by the light fish, the subjects said, "the dark fish was eaten by the light fish." When the light fish was cued, the very same dynamic event was described as "the light fish ate the dark fish." Attention to the cue influenced the choice of syntactic subject of the sentence and the grammatical voice, depending on which stimulus was cued. The effect demonstrated by Tomlin was robust: the cued stimulus appeared as the subject of the sentence in over 90 percent of the trials. Further analyses showed that these results hold as well for other films and languages other than English.

Tomlin's approach was extended to the study of the descriptions of static events (Forrest, 1996). L. Forrest manipulated subjects' attention to one of the quadrants of the display with the help of a visual cue and showed that it influenced the narrative. (see Fig. 53.2). These results show how the choice of syntactic subject is determined by the direction of attention.

B. Reference and Anaphora

Engaging in discourse requires being able to properly identify and track objects, events, or people that the discourse concerns. Directing attention to the entities in oral or written text usually involves reference by use of a pronoun or a noun phrase. The choice of a referent to indicate an entity previously mentioned in discourse is sometimes called *anaphora resolution*. Consider the sentence:

(1) **A researcher** has written **a new book** recently and (Ø) managed to publish **it**, too.

What aspects of the antecedents "a book" and "a researcher" influence the consequent choice of the referring anaphoric pronoun "it" for the former and not mentioning the referent ("zero anaphora"), for the later? Usually, both zero and a pronoun anaphora are chosen if the antecedent has been mentioned relatively recently. In this case, a common assumption is that the pronoun's antecedent is a *salient* entity in the text. More salient entities are accessed more easily and therefore are marked with anaphoric expressions of a lesser semantic content, for example, a definite pronoun or a zero anaphora. Less salient entities require more semantically specific forms of reference, for example, a full noun phrase.

Salience is often described as a combination of the concept's degree of activation in the working memory and manipulations of the focus of attention. In Tomlin's research, described above, subject assignment was cued by an external event. However, in comprehension of discourse, the most common way of subject assignment is to use linguistic cues to focus attention. For example, when the antecedent is the subject of the first clause, then the anaphoric pronoun will also assume the position of the subject in the second clause. Another example of a linguistic cue is that the first mentioned entity has a greater chance of being referred to by a pronoun rather than a nonphrase (see Garnham, 1999 for a review).

Both of these linguistic cues have received substantial support. For example, in sentence completion studies it was found that subject antecedents are indeed prefered for the reference in the second clause. However, the preference of the antecedent may also depend on other factors. For example, implicit causal-

FIGURE 53.1 Fish Film.

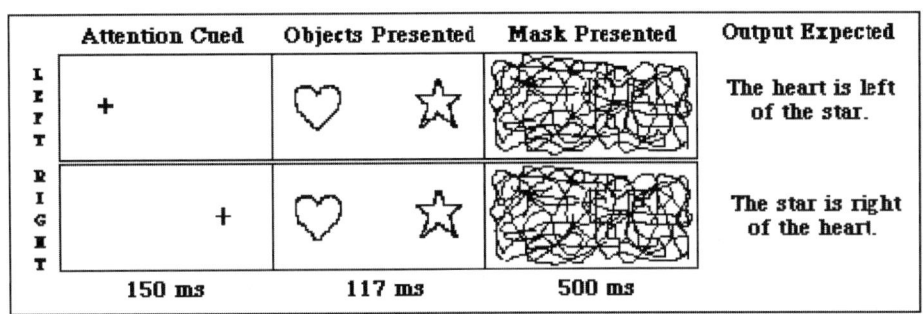

FIGURE 53.2 Example of stimuli from Forrest (1996).

ity of the verb may bias subjects' preference for object of the first clause to be refered to by the anaphora. (Garnham, 1987; 1999 for a review).

A more complicated explanation of discourse focus is a computational model known as *centering theory* (e.g., Grosz et al., 1995). This model hypothesizes the existence of two sets of discourse centers. The *forward-looking center* concerns salience of the discourse entities. The *backward-looking center* is related to references to these entities. Forward and backward centers provide a mechanism of establishing the coherence of the discourse. Different factors affect the salience (or "preference") of forward-looking centers. The most frequently mentioned are surface position, grammatical role, and being a backward-looking center of the utterance.

Experimental work in the centering theory shows that pronouns referring to more preferred characters are understood more quickly than those that refer to less preferred ones. Studies also show that sentences are read faster when backward-looking center is a pronoun than when it is a noun, and subjects judge these sentences as more coherent. Also, in reading short texts consisting of three sentences, processing time is reduced when the third sentence pronoun refers to the second sentence antecedent than when it refers to an antecedent in the first sentence (e.g., Gordon et al., 1993).

The choice of anaphoric expression also depends on the maintenance of the referent in memory. Although other linguistic factors may contribute, this issue is closely related to distance separating the anaphora from its antecedent. One task of psycholinguists is to trace the change of linguistic forms dynamically as the referential distance varies. Support for the role of distance comes from studies of the frequency of different forms of reference in written passages. The results of one such study are illustrated in Table 53.1 (adapted from Givón, 1995).

A new entity is usually introduced by means of an indefinite noun or a proper name. After that, the rule is: the larger the textual distance, the more likely one is to choose a noun phrase in order to refer to the antecedent; the shorter the distance, the more likely it will be a pronoun (presented or implied as a zero anaphora). If the referent is an anaphoric pronoun, the cognitive search for an antecedent within a file is started. This scan is very likely to be short-ranged, localized within a couple of preceding clauses, highly automatic, and routine. Finally, a definite noun demands a long-ranged, less automatized, and more attentionaly demanding search of an antecedent (Givón, 1995).

Psycholinguists also use experimental methods to gain evidence supporting the role of referential distance. For example, when subjects are asked questions concerning the referents of the pronoun in the last sentence of the passage they had just read, the number of errors tends to increase with increasing distance between the pronoun and its antecedent. Another way is to measure the time it takes to read a target sentence. In the standard variant of this task, subjects are given a segment of discourse in which the antecedent introduced in the first sentence is later referred to by means of different linguistic forms at different textual distances. One result is that it takes less time to read and understand the sentence that contains a pronominal referent if its antecedent is at a shorter distance, and therefore salient and highly activated (Garnham, 1987).

Priming of lexical decisions can also be used to trace the degree and time course of activation in referential choice. Subjects read the sentence and then decide whether a target word (usually one of the two possible antecedents was in the sentence they have just read. The experimental evidence is that in the case of two possible candidate antecedents, the pronoun actually activates both of them, but context serves to direct attention to the correct antecedent (Garnham, 1987).

By measuring eye movements, one can examine the time course of anaphoric resolution in more detail. The time and location of the fixations are both shown to reflect the referential distance between the target pronoun and its antecedent (see Stevenson, 1996, for a review). Eye movements may also be of help when trying to disambiguate a sentence in the course of comprehension. When linguistic cues are missing or ambiguous, eye movements may be triggered to resolve the ambiguity and attempt to gather the missing information (Tanenhaus et al., 1995).

Syntactic and semantic processing have been studied by neuroimaging. Violations of semantic or syntactic structure of the sentence are commonly used in studies recording scalp electrical potentials (ERP). These violations give rise to quite distinct wave forms.

TABLE 53.1 Referential Distance and the Choice of Linguistic Form

Construction	Referential distance in # of clauses
1. Zero anaphora	1.0
2. Unstressed pronoun	1.0
3. Stressed pronoun	2.5
4. Definite noun	7.0
5. Definite noun with a modifier	10.0

Semantic violations produce a negative wave with a peak latency of 400ms (N400). (See Kutas and Van Petten, 1994, for a review). In the cases of structural or syntactic violations, an early left anterior negativity (e.g., Osterhout and Mobley, 1995) and/or a late positive wave with a peak at 600ms (P600) (e.g., Hagoort et al., 1993) are frequently observed.

Hahne and Friederici (1999) hypothesized that the early left anterior negativity is a highly automatic process, whereas the P600 involves more attention. In a study aimed at testing this hypothesis, the proportion of correct sentences and sentences with structure violations was varied. Incorrect sentences might appear to be either of a low (20% violation) or a high (80% violation) proportion. The early left negativity was elicited under both conditions. P600 was elicited only for a low proportion of incorrect sentences. These results support the idea that early left negativity is an automated first-pass phrase parsing, whereas P600 relates to a second-pass parsing that requires a larger allocation of subjects' attention.

A limited number of reports have used fMRI in order to identify brain regions associated with syntactic and semantic processes. One study (Newman et al., 2001), found that syntactic violations elicit significantly greater activation in superior frontal cortex. Semantic anomalies result in greater activation in inferior frontal and temporoparietal regions, both of which are considered in more detail in the next section.

III. ATTENTION AND SEMANTICS

A. Word Association

A common task for studying the brain systems involved in semantic processing is to ask subjects to give the use of a common noun (e.g., hammer). In a typical version of this task for imaging, the subjects are shown a series of 40 simple nouns (Raichle et al., 1994). In the experimental condition, they indicate the use of each noun (for example, hammer → pound). In the control condition, they simply read the word aloud. The difference in activation between the two tasks illustrates what happens in the brain when subjects are required to develop a very simple thought, in this case how to use a hammer. Results illustrate that the anatomy of this high-level cognitive activity is similar enough among individuals to produce focal average activations that are both statistically significant and reproducible.

One area that is more strongly activated when generating the use of a word is on the midline of the frontal lobe in the anterior cingulate gyrus. Two additional areas of cortical activation that are greater in the generate condition are in the left lateral prefrontal cortex and posterior temporal cortex. Both of these areas have been shown to be involved in many tasks dealing with processing the meaning of words or sentences.

To examine the time course of these activations, it is possible to use a large number of scalp electrodes to obtain an ERP, which is the scalp signature of the generators found active in imaging studies (Abdullaev and Posner, 1998). When subjects think of the use of a noun, there is an area of positive electrical activity over frontal electrodes starting about 150msec after the word appears. This early electrical activity is generated by the large area of activation in the anterior cingulate. A left prefrontal area (anterior to and overlapping classical Broca's area) begins to show activity about 200msec after the word occurs. At first this area was called semantic because it was more active in the semantic task than in reading aloud. However, because this lateral area was often involved in working memory and its time course was early, it is more likely that this lateral frontal area is related to operations such as holding the lexical item in a brief store during the time needed to which to look up the associated word use.

The left posterior brain area found to be more active during the processing of the meaning of visual words did not appear until a much later time (500msec). This activity is near the classical Wernicke's area, lesions of which are known to produce a loss of understanding of meaningful speech. These findings suggest this area is important in finding the semantic association. There is also evidence of the transfer of information from left frontal electrodes to the posterior area at about 450 millisec into the task. Since the response time for this task was about 1,100msec, this would leave time for the generation of related associations needed to solve the task.

Practice on a single list of words reduces the activation in both the anterior cingulate and lateral cortical areas (Raichle et al., 1994). Thus, the very same task, when it is highly over learned, avoids the circuits involved in thought and relies upon an entirely different circuitry.

These studies provide a start in understanding the functional roles of different brain areas in carrying out executive control. The medial frontal area appears most related to the executive attention network and is active when there is conflict among stimuli and responses. It may be serving as a monitor of conflict, but it is possible that it plays other roles as well. The lateral prefrontal area seems to be important in holding in mind the information relevant to the task.

Even when a single item is presented, it may still be necessary to hold it in some temporary area while other brain areas retrieve information relevant to the response. Together these two areas are needed to solve nearly any problem, which depends upon retrieval of stored information (Duncan et al., 2000). Both of these areas could be said to be related to attention, or one might identify only the medial area with attention and the lateral area with working memory. In either case they begin to give us a handle on how the brain parses high-level tasks into individual operations that are carried out in separate parts of the network.

B. Reading

The ability of fluent adult reading seems to rely upon two areas of posterior cortex that in fluent readers automatically organize visual words. The visual word form area lies within the visual system in the fusiform gyrus of the left hemisphere. In skilled adult readers, this area appears to package the letters of words into visual units. This system allows the reader to avoid scanning the individual letters. A second area that lies close to auditory cortex allows the fluent reader to retrieve a phonological word unit from the visual input. Children with difficulties related to the word form can read slowly but lack fluency, while those with phonological difficulties cannot sound out new words and are limited to reading familiar words and guessing at those they do not know. Although in a logical sense the visual word form precedes the phonological system, there is evidence that the visual word form may develop later, and the visual and phonological system function in close interaction (Harm & Seidenberg, 1999).

Little or no attention is needed for these systems to operate in the skilled reader, but attention is needed for these systems to develop properly. Children with difficulty in learning to read employ the executive attention network (including both lateral and medial frontal areas) to compensate for insufficient development of the more automatic posterior visual word form and phonological systems (Shaywitz et al., 1999). The word form system seems to develop relatively late. While one can read without an efficient word form system, it apparently improves the speed and fluency of reading. Studies show that at age 10, activation of the visual word form system requires familiar words that children have already learned. However, in skilled adults the visual word form is activated by letter strings never before seen, provided that the strings obey the rules of the written language and thus are orthographically regular (Posner & McCandliss, 1999).

References

Abdullaev, Y. G., and Posner, M. I. (1998). Event-related brain potential imaging of semantic encoding during processing single words. *Neuroimage* **7**, 1–13.

Duncan, J., Seitz, R. J., Kolodny, J., Bor, D., Herzog, H., Ahmed, A., Newell, F. N., and Emslie, H. (2000). A neural basis for general intelligence. *Science* **289**, 457–460.

Forrest, L. B. (1996). Discourse goals and attentional processes in sentence production: the dynamic construal of events. *In* "Conceptual Structure, Discourse and Language." (A. E. Goldberg, Ed.), Stanford, CA, CSLI Publications, 149–162.

Garnham, A. (1987). Understanding anaphora. *In* "Progress in the Psychology of Language." (A. W. Allis, Ed.), Vol. 3. Lawrence Erlbaum Associates, London, 253–300.

Garnham, A. (1999). Reference and anaphora. *In* "Language Processing." (S. Garrod and M. J. Pickering, Eds.), Psychology Press, Hove, England, 335–362.

Givón, T. (1995). Coherence in text versus coherence in mind. *In* "Coherence in spontaneous text." (M. A. Gernsbacher and T. Givon, Eds.), TSL#31. John Benjamins, Amsterdam, 59–115.

Gordon, P. C., Grosz, B. J., and Gilliom, L. A. (1993). Pronouns, names, and the centering of attention in discourse. *Cognitive Science* **17**, 311–347.

Grosz, B. J., Joshi, A. K., and Weinstein, S. (1995). Centering: a framework for modelling the local coherence of discourse. *Computational Linguistics* **21**, 203–225.

Hagoort, P., Brown, C., and Groothusen, J. (1993). The syntactic positive shift as an ERP measure of syntactic processing. *Language and Cognitive Processes* **8**, 439–484.

Hahne, A., and Frederici, A. D. (1999). Electrophysiological evidence for two steps in syntactic analysis: early automatic and late controlled processes. *Journal of Cognitive Neuroscience* **11**, 194–205.

Harm, M. W., and Seidenberg, M. S. (1999). Phonology, reading acquisition, and dyslexia: Insights from connectionist models. *Psychological Review* **106**, 491–528.

Kutas, M., and Van Petten, C. (1994). Psycholinguistics electrified: Event-related brain potential investigations. *In* "Handbook of Psycholinguistics." (M. Gernsbacher, Ed.), Academic Press, New York, 83–143.

Newman, A. J., Pancheva, R., Ozawa, K., Neville, H. J., and Ullman, M. T. (2001). An event-related fMRI study of syntactic and semantic violations. *Journal of Psycholinguist Resources* **30**, 339–364.

Osterhout, L., and Mobley, L. A. (1995). Event-related brain potentials elicited by failure to agree. *Journal of Memory and Language* **34**, 739–773.

Posner, M. I., and McCandliss, B. D. (1999). Brain circuitry during reading. *In* "Converging methods for understanding reading and dyslexia." (R. Klein and P. McMullen, Eds.), MIT Press, Cambridge, 305–337.

Raichle, M. E., Fiez, J. A., Videen, T. O., McCleod, A. M. K., Pardo, J. V., Fox, P. T., and Petersen, S. E. (1994). Practice-related changes in the human brain: functional anatomy during nonmotor learning. *Cerebral Cortex* **4**, 8–26.

Shaywitz, S. E., Shaywitz, B. A., Pugh, K. R., Fulbright, R. K., Constable, R. T., Mencl, W. E., Shankweiler, D. P., Liberman, A. M., Skudlarski, P., Fletcher, J. M., Katz, L., Marchione, K. E., Lacadie, C., Gatenby, C., and Gore, J. C. (1998). Functional disruption in the organization of the brain for reading in dyslexia. *Proceedings of the National Academy of Sciences of the United States of America* **95**, 2636–2641.

Stevenson, R. J. (1996). Mental models, propositions and the comprehension of pronouns. *In* "Mental Models in Cognitive Science." (J. Oakhill and A. Garnham, Eds.), Psychology Press, Hove, England, 53–76.

Tanenhaus, M. K., Spivey-Knowlton, M. J., Eberhard, K. M., and Sedivy, J. E. (1995). Integration of visual and linguistic information in spoken language comprehension. *Science* **268**, 1632–1634.

Tomlin, R. S. (1997). Mapping conceptual representations into linguistic representations: The role of attention in grammar. *In* "Language and Conceptualization" (J. Nuyts and E. Pederson.), 162–189.

CHAPTER 54

Attention and Spatial Language

Laura A. Carlson and Gordon D. Logan

ABSTRACT

This chapter summarizes recent research on the role of attention in spatial language. Spatial language is a domain that focuses on the comprehension and production of spatial descriptions of objects and relations in an environment. First, we establish that attention is necessary for the apprehension of spatial language. Second, we describe a computational model of the apprehension of spatial language, and examine how attention plays a role within its constituent steps. We conclude by underscoring the importance of looking at the interface between attention and spatial language, and point out areas for future work.

I. INTRODUCTION

Spatial language is a domain that focuses on spatial descriptions of objects and their relations in a given environment. A prominent focus of work in this area has been on describing the end-products of apprehension, including what people understand an utterance to mean or what utterance they choose to produce. However, a smaller body of work within this domain has focused on the processes by which spatial descriptions are produced and the representations upon which these processes operate. Within this focus, attention plays a critical role, serving as a mechanism that establishes correspondences among various representations via a set of processes that operate on the objects and their relations. The goal of the current article is to review the work that articulates the role of attention in spatial language. There are two main sections. First, we overview research showing that attention is necessary for the apprehension of spatial language. Second, we describe how attention plays its role, grounding our inquiry in a particular computational model (Logan and Sadler, 1996) of spatial language use. We conclude by underscoring the importance of examining processes and representations within the domain of spatial language, and indicating areas in need of research for further elucidating the role of attention within spatial language.

II. ATTENTION IS NECESSARY FOR APPREHENDING SPATIAL RELATIONS

A common type of spatial description indicates the location of one object (the located object) by spatially relating it to a second object (the reference object), as in "The coffee mug is in front of the centerpiece of flowers." Logan (1994, 1995) suggested that attention is involved in the apprehension of such descriptions. For example, attention may be initially allocated to the reference object and then moved in a specified direction (i.e., front) to the located object. An important first step in verifying this claim is to empirically demonstrate that attention is necessarily involved during apprehension. To assess this proposal, Logan (1994) had participants perform a visual search task in which the target was defined by the spatial relation between two stimuli (a dash and a plus), and the distractors were defined by a different spatial relation (for material relating to visual search, see Chapters 34, 49, 65, and 90). For example, participants would be instructed to search for the "dash above plus" in a display containing a target that was dash-above-plus amid distractors that were pluses-above-dashes. The logic was that if attention were required to process the spatial relations between the elements, then attention would have to be allocated to each set of stimuli in turn. This would be similar to the classic conjunction search in which participants would have to attend to each stimulus in order to combine multiple features, in contrast to the classic feature search in which participants could allocate attention across stimuli in the display. Thus, the diagnostic of whether attention was required for computing spatial relations was a linearly increasing function relating search time to display size with a substantial slope. In contrast, if attention were not required, search time should be relatively constant across display sizes, with the function relating search to display size having a slope near zero.

Across a series of experiments, Logan (1994) found that response time increased linearly with a sizable slope when the spatial relations among the elements defined the target. This pattern of results was obtained for both vertical (above/below) and horizontal (left/right) relations. Recently, Moore, Elsinger, and Lleras (2001) extended this effect to spatial relations that are defined in depth (i.e., front). Logan (1994) also showed that it held over extensive practice and was obtained when a picture stimulus rather than a linguistic description was used to define the target. In contrast to the increasing slope pattern, search was relatively flat when the target was defined by an individual element (plus or dash), indicating that it was the spatial relation between the elements that required attention rather than identification of the individual elements. Moreover, cueing the target by displaying one of its elements in a unique color facilitated search, as compared to cueing a distractor by coloring one of its elements, based on the idea that the unique color would automatically draw attention. This pattern of data is consistent with the idea that attention must be allocated to the target during the computation of the spatial relation; when attention is already there (due to the unique color), detection of the target can occur more quickly than when attention is initially allocated to a distractor.

Converging evidence for the necessity of attention in the computation of spatial relations has been obtained within the flicker paradigm (Rosielle, Crabb, and Cooper, 2002; for more on this paradigm, see Chapter 13). Rosielle et al. were interested in whether the participants encoded categorical relations (such as one object above another object) in addition to metric relations (the direction and distance between the objects). To assess this, they showed participants alternating scenes of objects that were separated by a brief masked interval. The task was to press a button when a change across the displays was detected. The critical comparison was between consecutive displays that differed in the metric relations between two of the objects while preserving the categorical relation (i.e., one object would be above a second object in both displays, but the distance between the objects would vary across displays), or that differed in both metric relations and categorical relations (i.e., one object would be above a second object in one display but to the side-of the second object in the second display). Importantly, the degree of change in the metric relations was the same for both types of change. Rosielle et al. reasoned that if categorical relations were encoded between objects, then a change in the categorical and metric relations should be detected faster than a change in the metric relations alone. This pattern was obtained, demonstrating that the relations among objects are encoded categorically during scene viewing. Attentional allocation to the changed objects is implicated in this result, given evidence within the flicker paradigm that attention modulates change detection (Rensink, O'Regan, and Clark, 1997; see also Chapter 13). However, a stronger test of whether the categorical encoding of the relations between the objects requires attention was provided by adding a precue to some displays that indicated with 100 percent validity the quadrant in which a change (metric or categorical) would take place. If categorical relations are encoded only after attention is allocated to a given object in the scene, then allocating attention via the precue should not impact the size of the categorical encoding advantage. That is, detection of a change in categorical and metric relations should always be facilitated relative to detection of a change in metric relations, regardless of the presence of the cue. The results showed a main effect of cue, such that detection of both types of changes was facilitated in the presence of a cue. However, for both cue and no cue trials, detection of a categorical and metric change was significantly faster than detection of a metric change, and the size of this difference did not vary as a function of cue. The fact that a cue that draws attention to the objects had no impact on the effect indicates that attention was already allocated to the objects during the computation of the spatial relation.

In summary, two different paradigms provide converging evidence for the necessary role of attention in the computation of spatial relations. The following section more closely examines the particular role of attention, grounded within Logan and Sadler's (1996) computational model of the apprehension of spatial relations.

III. THE ROLE OF ATTENTION IN APPREHENDING SPATIAL RELATIONS

A. A Computational Model of the Apprehension of Spatial Language

Consider again the utterance, "The coffee mug is in front of the centerpiece of flowers" and the claim that apprehension of the utterance involved allocating attention to the reference object, and then moving attention in the "front" direction to the located object (Logan, 1994, 1995). Given the evidence that attention is necessary to apprehend this spatial description, the next step is to specify the manner by which attention is allocated to the objects and how it is moved in a particular direction, and to empirically verify these roles.

A good place to start is with Logan and Sadler's (1996) computational model of the apprehension of spatial language in which attention coordinates various representations and directs processes that operate upon these representations. The representations include a perceptual representation that is an analog array derived from a given display or scene in the environment, a linguistic representation that dictates the spatial description, and a conceptual representation that serves as an interface between the perceptual and linguistic representations (see also Jackendoff, 1996). For example, imagine pairing our utterance, "The coffee mug is in front of the centerpiece of flowers," with a scene showing an assortment of objects on a kitchen table, including a coffee cup, a centerpiece of flowers, and two candlesticks. The perceptual token for coffee cup must be linked to the linguistic label "coffee cup" via an intermediate symbol within the conceptual representation, a process termed spatial indexing (Logan, 1994, 1995). More generally, spatial indexing involves selection of the relevant objects from the scene as dictated by the linguistic utterance (i.e, the coffee cup) from among a set of distractor objects (i.e, the candlesticks).

Two intermediate representations are also involved in linking the linguistic spatial term onto the appropriate region of space around the reference object within the conceptual representation (i.e., linking "front" to the front of the centerpiece). These representations are reference frames and spatial templates. In their most general form, reference frames consist of sets of coordinate axes that are defined by the setting of a number of parameters including *orientation and direction*, which indicate a dimension (i.e., vertical or horizontal) and an end point (i.e., above or below); *scale*, which indicates a distance between the objects; and *origin*, which corresponds to where in space the reference frame is imposed. Typically, the reference frame is imposed on the reference object, thereby dividing the surrounding space into regions. These regions are referred to as *spatial templates*. Spatial templates encapsulate how well a given point in space corresponds to a given relation with respect to the origin of the reference frame, with each template divided into three general regions: good, acceptable, and bad. Fig. 54.1 presents the spatial template associated with the term *above* based on data obtained by Logan and Sadler (1996). Specifically, they placed one stimulus

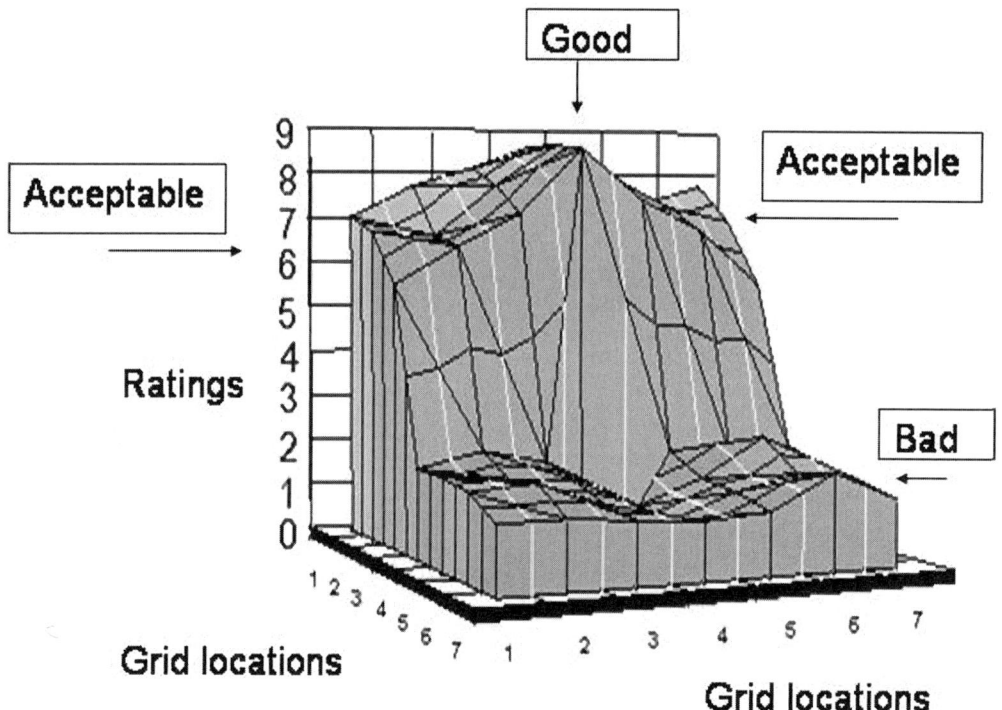

FIGURE 54.1 Illustration of a spatial template for "above." Data from Logan and Sadler (1996).

(an O) in the center of a 7 × 7 grid, at location (4, 4). Across trials, they placed another stimulus (an X) at each point within the grid, and collected an acceptability rating at each point that corresponded to how well the sentence "X is above O" described the spatial configuration of the X and the O. Plotted in the X and Y dimensions are the grid locations. In the Z dimension are the mean acceptability ratings. As indicated in the figure, the good region corresponds to the peak in the ratings, extending along the above end point of the vertical axis of a reference frame imposed on the reference object (the O). The acceptable region flanks the good region and is associated with acceptable uses of the spatial term, extending outward from the vertical axis at this end point. The bad region corresponds to locations on the opposite end point of the vertical axis. The processes by which the parameters of the reference frame are set, and the spatial templates are constructed, are known collectively as assigning directions to space. A final process involves computing the goodness of fit of the located object with respect to good, acceptable, and bad regions across various spatial templates, and comparing a given template in which the object appears in a good or acceptable region (depending on some threshold), with the spatial term provided in the linguistic description.

In sum, within the Logan and Sadler (1996) model, there are three constituent steps: spatially indexing the relevant objects, assigning directions to space, and computing and comparing the spatial relation. The order of these steps is not fixed but may vary as a function of the task and communicative goal (Logan, 1994, 1995). As reviewed next, research has observed effects of attention within the processes of spatially indexing and assigning directions to space.

B. Attention and Spatially Indexing the Reference and Located Objects

The effects of attention as a mechanism of selection have been observed within the process of spatially indexing (for more on attentional selection, see Chapters 22, 90, and 94). Specifically, Carlson and Logan (2001) presented participants with displays containing reference objects and located objects, and asked them to verify as quickly as possible whether an accompanying sentence was an acceptable description of the display. On some trials, the displays contained an additional distractor object. When the distractor was present, its placement was manipulated so that on some trials it stood in the same spatial relationship with respect to the reference object as the located object. For example, stimuli could be vertically aligned, with the distractor on top, the located object in the middle, and the reference object on the bottom. In this case, both the distractor and the located object are above the reference object. On other trials, the distractor would be in a different spatial relationship with the reference object than the located object. For example, stimuli could be vertically aligned, with the located object on top, the reference object in the middle, and the distractor on the bottom. In this case, the located object is above the reference object, and the distractor is below the reference object. Logan and Compton (1996) had found that the presence of the distractor within a display slowed performance. However, it was unclear in their study within which constituent step (spatially indexing, assigning directions to space, computing the spatial relation) the distractor was having its influence. If the distractor impacted the ability to spatially index the objects, causing difficulty in the selection of the relevant objects from the scene, then one might expect to observe a cost in response time for distractor present trials relative to distractor absent trials. However, this cost should not be impacted by whether the distractor was placed in the same spatial relation with respect to the reference object as the located object. In contrast, if the spatial relation between the distractor and the reference object were calculated along with the spatial relation between the reference object and the located object, and the distractor had its impact in terms of forcing a selection between these two viable relations, then when the distractor was in the same spatial relation as the located object, selection should be more difficult because both instantiate the same spatial relation, as compared to when the distractor was in a different spatial relation. If this were the case, one would expect to see a cost in response time for distractor present relative to distractor absent trials. However, this cost should be influenced by the placement of the distractor, being present when the distractor was placed in the same spatial relation but not when it was in a different spatial relation.

Carlson and Logan (2001) observed an effect of distractor presence, with significantly slower response times to trials in which the distractor was present, relative to when it was absent. However, there was no difference in this cost as a function of distractor placement; rather, there was significant slowing regardless of whether the distractor instantiated the same relation. This result suggests that the distractor had its impact in both the Carlson and Logan (2001) study and in the Logan and Compton (1996) study by making it more difficult to select the relevant objects from the display. This implicates the role of attention in spatial indexing.

C. Attention and Selecting a Reference Frame

Not only do the reference and located objects need to be spatially indexed and identified, but the linguistic spatial term must also be assigned to the relevant region of space around the reference object. This is accomplished via the selection and setting of a reference frame. Logan (1995) presented an extensive series of experiments demonstrating that reference frames are required for directing attention from a cue to a target, analogous to the example of directing attention from a reference object (centerpiece of flowers) in a given direction to a located object (coffee mug). His diagnostic of the use of a reference frame was a particular order of responding to various spatial relations. Specifically, following work by Franklin and Tversky (1990), within a spatial framework analysis, access to the vertical above/below axis is facilitated relative to access to the horizontal left/right axis. Thus, differential access to particular locations in space is presumed to reflect use of a reference frame. For example, in one experiment, Logan presented participants with displays of colored dots, preceded by a cue that indicated a given spatial relation (above, below, left, right). The task of the participants was to indicate the color of the target referred to by the cue. If processing the cue required the allocation of attention via a reference frame, then responding to the color of the target in the above/below positions in response to the vertical cues should be faster than responding to the color of the target in the left/right positions in response to the horizontal cues. Logan observed this pattern, and extended the result to cues defined either with respect to the viewer or with respect to an internal feature of the display, and at various locations within the display, indicating that reference frames can be translated and rotated into different orientations. In contrast, when participants had to report the color of a target that was indicated by a cue that didn't require a reference frame (i.e., with a digit arbitrarily mapped onto a location), the facilitation for the vertical axis over the horizontal axis was not observed.

Additional evidence for the role of attention in selecting a reference frame comes from situations in which use of a spatial term is ambiguous because it can be defined by different types of reference frames (for a review, see Carlson, 2000). For example, the term *above* can be defined with respect to environmental features (such as gravity), with respect to the orientation of the reference object (such as above the top of a trashcan that is overturned), or with respect to the orientation of speaker (e.g., above one's head if one is reclining). Carlson-Radvansky and Irwin (1994) showed that initially multiple reference frames were automatically activated and competed to assign a direction to a spatial term, consistent with a broader literature in spatial cognition that illustrates multiple encoding of locations within diverse reference frames (for example, see Chapter 30). In the case of spatial language, the competition among reference frames is resolved by selecting one frame to define the spatial term. This makes sense for language because the goal of the speaker uttering a spatial description is to convey a particular location. This selection process involves inhibition of the nonselected frames, as demonstrated within a negative priming paradigm (Carlson-Radvansky and Jiang, 1998). This is consistent with the literature in visual attention which argues that selection is facilitated not only by enhancing a target but also by inhibiting distractors (for review, see Fox, 1995; more generally, see Chapters 16 and 43).

D. Attention and Constructing a Spatial Template

In addition to selecting a reference frame and applying it to the reference object, apprehension also involves defining particular regions of space in terms of spatial templates. It has been observed that the shape of a spatial template and its regions is generally preserved across types of reference and located objects, across various spatial terms, and across different types of reference frames. This suggests a common set of mechanisms that underlie its construction across the diverse contexts. Regier and Carlson (2001) present a computational model of spatial language in which spatial templates emerge from the use of two general mechanisms: attention and vector sum coding of direction. Specifically, according to the attention vector-sum (AVS) model, an attentional beam is anchored at a point on the reference object, and radiates out to encompass the located object. This is shown in Fig. 54.2*a*, with the rectangle as the reference object and the small circle as the located object. Attentional strength is maximal at the focus of the beam and drops off with distance. As a result, some parts of the reference object receive more attention than others. In addition, the direction of the located object with respect to the reference object is represented as a population of vectors that project from each point along the reference object to the located object, as illustrated in Fig. 54.2*b*. The attentional beam and the vector sum are combined by weighting each vector as a function of the amount of attention being allocated to its point on the reference object, as reflected in Fig. 54.2*c* by the different lengths of the vectors. The resulting weighted vectors are then summed to create an overall direction (Fig. 54.2*d*), and its alignment with respect to a reference direction

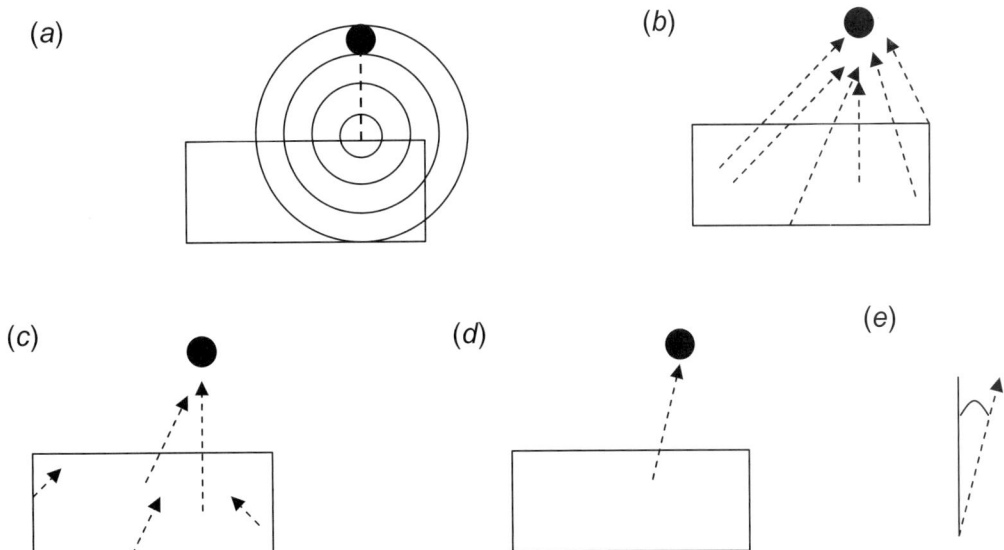

FIGURE 54.2 Illustration of the Attention-Vector Sum model. Figure from Regier and Carlson (2001). Grounding spatial language in perception: An empirical and computational investigation *Journal of Experimental Psychology: General*, 130, 273–298. Copyright © 2001 by the American Psychological Association. Reprinted with permission. See text for panel descriptions.

(such as upright vertical for "above") is measured (Fig. 54.2e). In general, perfect alignment with the reference direction corresponds to the best use of a spatial term, corresponding to the good region on a spatial template (see Fig. 54.1). Acceptability drops off in a linear fashion with increasing deviations from the reference axis, corresponding to the acceptable regions of a spatial template. Finally, there is a cutoff, below which the term is not considered acceptable, regardless of the vector sum, corresponding to the bad region. Regier and Carlson (2001) provide a formal presentation of the model and demonstrate that it outperforms various competing models both quantitatively in terms of the fit to empirical data and qualitatively in terms of its output illustrating the effects of interest.

In summary, this section has provided empirical support for the role of attention in two of Logan and Sadler's (1996) constituent steps in apprehending a spatial description: spatially indexing and assigning directions to space, both with respect to selecting and accessing a reference frame, and constructing a spatial template.

IV. CONCLUSION

We have reviewed evidence for the role of attention in spatial language, documenting both that attention is necessary for computing the spatial relation and that it has effects within many of the constituent processes of apprehension, including spatial indexing, selecting and defining a reference frame, and constructing a spatial template. Future work will address how attention mediates the computation of the goodness of fit of the located object within the various spatial template. Logan and Sadler (1996) suggest that there is an iteration through multiple templates. However, the manner (parallel or serial) and possible order in which the templates are consulted, the stopping function associated with the iteration, and the decision rule for determining an acceptable relation are all currently undetermined. More generally, examining how spatial language interfaces with other more general cognitive processes such as attention represents a promising direction for future work because it encourages a focus on mechanism and representation, and allows one to step beyond looking at spatial language as a set of constrained isolated utterances, but rather as a reflection of a more wholistic general cognitive system.

Acknowledgments

Address correspondence to either Laura Carlson, Department of Psychology, University of Notre Dame, Notre Dame, IN 46556, LCarlson@nd.edu or Gordon Logan, Department of Psychology, Vanderbilt University, Nashville, TN, 37203, Gordon.logan@vanderbilt.edu. This work was supported by National Science Foundation Grants BCS 0133202 and BCS 0218507.

References

Carlson, L. A. (2000). Selecting a reference frame. *Spatial Cognition and Computation* **1**, 365–379.

Carlson, L. A., and Logan, G. D. (2001). Using spatial terms to select an object. *Memory & Cognition* **29**, 883–892.

Carlson-Radvansky, L. A., and Irwin, D. E. (1994). Reference frame activation during spatial term assignment. *Journal of Memory and Language* **33**, 646–671.

Carlson-Radvansky, L. A., and Jiang, Y. (1998). Inhibition accompanies reference frame selection. *Psychological Science* **9**, 386–391.

Fox, E. (1995). Negative priming from ignored distractors in visual selection: A review. *Psychonomic Bulletin & Review* **2**, 145–173.

Franklin, N., and Tversky, B. (1990). Searching imagined environments. *Journal of Experimental Psychology: General* **119**, 63–76.

Jackendoff, R. (1996). The architecture of the linguistic-spatial interface. *In* "Language and Space" (P. Bloom, M. A. Peterson, L. Nadel, and M. Garrett, Eds.), MIT Press, Cambridge, MA, 1–30.

Logan, G. D. (1994). Spatial attention and the apprehension of spatial relations. *Journal of Experimental Psychology: Human Perception and Performance* **20**, 1015–1036.

Logan, G. D. (1995). Linguistic and conceptual control of visual spatial attention. *Cognitive Psychology* **28**, 103–174.

Logan, G. D., and Compton, B. J. (1996). Distance and distraction effects in the apprehension of spatial relations. *Journal of Experimental Psychology: Human Perception and Performance* **22**, 159–172.

Logan, G. D., and Sadler, D. D. (1996). A computational analysis of the apprehension of spatial relations. *In* "Language and Space" (P. Bloom, M. A. Peterson, L. Nadel, and M. Garrett, Eds.), MIT Press, Cambridge, MA, 493–529.

Moore, C. M., Elsinger, C. L., and Lleras, A. (2001). Visual attention and the apprehension of spatial relations: The case of depth. *Perception & Psychophysics* **63**, 595–606.

Regier, T., and Carlson, L. A. (2001). Grounding spatial language in perception: An Empirical and Computational Investigation. *Journal of Experimental Psychology: General* **130**, 273–298.

Rensink, R. A., O'Regan, J. K., and Clark, J. J. (1997). To see or not to see: The need for attention to perceive changes in scenes. *Psychological Science* **8**, 368–373.

Rosielle, L. J., Crabb, B. T., and Cooper, E. E. (2002). Attentional coding of categorical relations in scene perception: Evidence from the flicker paradigm. *Psychonomic Bulletin & Review* **9**, 319–326.

CHAPTER 55

The Sustained Attention to Response Test (SART)

Tom Manly and Ian H. Robertson

ABSTRACT

Oops! If you have ever discovered your keys in the fridge, or dialed 9 to get an outside line from home, you will be aware that action doesn't necessarily stop when our mind is elsewhere. Since William James's seminal discussion of habit, such "slips of action," in which routine responses are triggered despite being inappropriate to current intentions, have informed a distinction between relatively automatic response production and executive or supervisory control. The increased difficulties in this respect experienced by brain-injured patients with dysexecutive deficits— together with the results of functional imaging studies—point to the importance of the prefrontal cortex in mediating these capacities. Such problems, which generally occur under the complex and multiple demands of everyday life, are difficult to capture using typical controlled, standardized forms of assessment. The Sustained Attention to Response Test (SART) was designed to form a simple, controlled and replicable measure of such lapses. In the test, participants view a computer monitor on which a random series of single digits are presented at the regular rate of 1 per 1.15 seconds. The task is to press a single response key following each presentation with the exception of a nominated no-go digit, to which no response should be made. The repetitive and unselective responding required for 89 percent of presentations was designed to lull participants into a rather absentminded "stimulus-press-stimulus-press" mode—a tendency that they should aim to resist if they are to avoid inadvertently responding on the rare and unpredictable no-go trials. In the standard version of the test, 225 digits are presented in a continuous sequence over 4.3 minutes.

I. VALIDITY

A. Ecological and Clinical Validity

Robertson et al. (1997) administered the SART to a large group of traumatically brain-injured patients (in whom frontal damage is probable) and matched controls. Error of commission rates on the SART were related to brain injury severity. Errors were also correlated to the frequency of everyday attentional lapses as indexed by the Cognitive Failures Questionnaire (CFQ) in both the patient and control groups. The association between the SART and CFQ was robust when the same participants were retested two years later, suggesting that the test is picking up a relatively stable aspect of individual differences (Manly et al., 1999). The SART has also been shown to be sensitive to everyday attentional deficits following right hemisphere stroke (Manly et al., in press) and, in a paper and pencil variant (the "Walk Don't Walk" subtest of the Test of Everyday Attention for Children; TEA-Ch) to childhood attention deficit hyperactivity disorder (Manly et al., 2002). In line with previous findings on dysexecutive-type errors in sleepy participants, errors on the SART shows a clear circadian pattern in healthy adults (Manly et al., 2001).

B. What Does It Measure?

The SART requires participants to withhold a response, and it is possible that the test is better conceived as a "response inhibition" test than as a measure of sustained executive control. Results of relevance to this issue include the positive relationship between commission error rates and the length of intervals between the occurrence of no-go targets

(Manly et al., 1999). This suggests that individuals find it increasingly difficult to self-sustain anticipatory control with the passage of time between no-go trials. It has also been shown that periodic auditory cues to "pay attention" are associated with significant improvements in performance in adult (Manly et al., in press) and child populations. As these cues had no predictive value for upcoming events in the task, it again suggests a strong role for maintained top-down control in determining the success of response inhibition on relevant trials.

II. PSYCHOMETRIC PROPERTIES

The test has been shown to be reliable over retest in the short (one week: Robertson et al., 1997) and the longer term (two years: Manly et al., 1999). Ceiling levels of performance are rare in healthy volunteers (mean commission error rates = 6.36 (SD = 4.36).

One difficulty in interpreting error rates in isolation concerns the negative relationship between speed-of-go responses and commission errors on no-go trials that are common in such tasks. Although differences in response speed have been unable to account for the clinical group differences reported above, or indeed the positive effects of periodic reminders to pay attention, they nevertheless can be a confound in interpreting individual's scores. In an attempt to reduce both within- and between-subject variability in reaction time, an auditory "press" cue was introduced at a regular interval after the onset of each trial, with participants being instructed to press in-time with this click. Interestingly, although this manipulation was associated with a significant reduction in RT variability, error rates on the task remained highly correlated with the standard version. This suggests that the speed differences observed in the standard version may themselves reflect wandering attention.

III. NEURAL BASIS OF SART PERFORMANCE

Recent work using fMRI (O'Connor, Robertson, Manly, Hevenor, and Levine, 2003) has shown that, in comparison to a control task of simply pressing for every digit presented, the SART produces significant activation, including in the right dorsolateral prefrontal cortex. This activation is significantly attenuated when participants are provided with periodic reminders to "pay attention." This suggests a role for this region in monitoring and maintaining performance against attentional drift, in addition to any response inhibition demands.

References

Manly, T., Anderson, V., Nimmo-Smith, I., Turner, A., Watson, P. C., and Robertson, I. H. (2002). The differential assessment of children's attention: The Test of Everyday Attention for Children (TEA-Ch), normative sample and ADHD performance. *Child Psychology and Psychiatry* **42**, 1065–1081.

Manly, T., Davison, B., Heutink, J., Galloway, M., and Robertson, I. (2000). Not enough time or not enough attention?: Speed, error and self-maintained control in the Sustained Attention to Response Test (SART). *Clinical Neuropsychological Assessment* **3**, 167–177.

Manly, T., Heutink, J., Davison, B., Gaynord, B., Greenfield, E., Parr, A., Ridgeway, V., and Robertson, I. H. in press. An electronic knot in the handkerchief: "Content free cueing" and the maintenance of attentive control. *Neuropsychological Rehabilitation*.

Manly, T., Lewis, G. H., Robertson, I. H., Watson, P. C., and Datta, A. K. (2001). Coffee in the cornflakes: Time-of-day, routine response control and subjective sleepiness. *Neuropsychologia* **40**, 747–758.

Manly, T., Robertson, I. H., Galloway, M., and Hawkins, K. (1999). The absent mind: Further investigations of sustained attention to response. *Neuropsychologia* **37**, 661–670.

O'Connor, C., Robertson, I. H., Manly, T., Hevenor, S., and Levine, B. (2003). Endogenous versus exogenous engagement of sustained attention: An fMRI study. *Clinical Neuropsychologist* **17**, 117.

Robertson, I. H., Manly, T., Andrade, J., Baddeley, B. T., and Yiend, J. (1997). 'Oops!': Performance correlates of everyday attentional failures in traumatic brain injured and normal subjects. *Neuropsychologia* **35**, 747–758.

CHAPTER 56

ERP Measures of Multiple Attention Deficits Following Prefrontal Damage

Leon Y. Deouell and Robert T. Knight

ABSTRACT

Maintaining a goal-directed behavior requires selectively attending to a subset of the sensory input at the expense of the rest of the input. At the same time, a surveillance mechanism must be in operation, so that deviant or novel events may bring about reorientation of attention and avoidance of potential hazards. Event-related potential (ERP) studies in patients with lateral prefrontal damage, due mainly to stroke, reveal deficits in all of these components of the attentional system. These deficits may explain some pervasive neuropsychological impairment in these patients.

I. INTRODUCTION

Maintaining a goal-directed behavior requires attention, that is, biased processing of a subset of sensory input, motor planning, and thought streams. Multiple goals (from survival to hitting a speeding tennis ball) are present at any one time, requiring a control system that will dynamically prioritize attention allocation based on changing internal goals. While enhancing one stream of input and suppressing another ("voluntary" or "endogenous" attention), the control system needs to allow for changes in the environment, as well as changes in internal drives to interfere with ongoing behavior as needed in a rapid and flexible manner ("involuntary" or "reflexive" attention). Consequently, this control system needs to interact with multiple sensory regions of the brain, as well as with motor output regions involved in orientation, locomotion, and speech. Extensive research using event-related brain potentials (ERPs) established scalp-recorded signatures of voluntary and involuntary attentional processes (see Chapter 84), which may serve as probes into mechanisms of attention in brain-damaged patients. Together with behavioral studies, these measures support critical involvement of the lateral prefrontal (LPFC) region in multiple components of attention control.

II. SELECTIVE ATTENTION

A. Inhibition

Selectively attending to an auditory stream, to a visual location or a visual object, and maintaining the goal of behavior requires not only enhancement of processing of some sensory stimuli or motor plans, but also suppression, or inhibition, of competitive stimuli or plans. A "biased competition" model suggests that excitatory signals to neurons result in inhibition of nearby task-irrelevant neurons, resulting in a sharpening of the attentional focus (see Chapter 50). Patients with lateral prefrontal (LPFC) damage are frequently unable to suppress inappropriate reaction to objects and events in their environment (so-called environmental dependency syndrome (Lhermitte, 1986) or to suppress prepotent responses (Barcelo and Knight, 2002). Whereas some of these effects may be at the level of response selection, ERP studies of these patients suggests that LPFC is essential for normal suppression of information processing at the level of unimodal sensory cortices (Fig. 56.1).

Knight and co-workers delivered auditory stimuli (clicks or tone bursts) or brief electric shocks to the median nerve to patients with damage to LPFC and to patients with comparably sized lesions in the temporoparietal junction or the lateral parietal cortex, as well as to age-matched controls. Both types of stimuli were distractors, not relevant to the task. Lesions in the

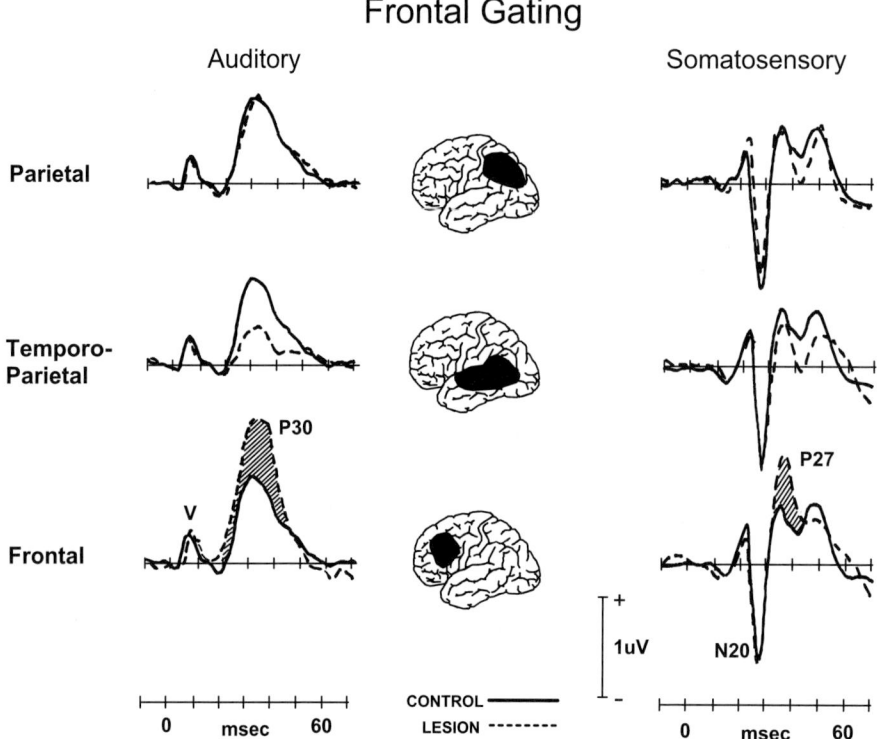

FIGURE 56.1 Primary cortical auditory and somatosensory event-related potentials for controls (solid line) and patients (dashed line) with focal damage in the lateral parietal cortex (top, $n = 8$), temporal-parietal junction (middle, $n = 13$), or dorsolateral prefrontal cortex (bottom, $n = 13$). Reconstructions of the center of damage in each patient group are shown in the middle. Somatosensory stimuli were square-wave pulses of 0.15 ms duration delivered to the median nerve at the wrist. Auditory stimuli were clicks delivered at a rate of 13/s at intensity levels of 50 dB HL. Prefrontal damage resulted in a selective increase in the amplitude of the P27 primary somatosensory response and P30 primary auditory response (hatched area). Data from Knight et al., 1988.

posterior association cortex, sparing primary sensory regions, had no effects on the amplitudes or latencies of the primary cortical evoked responses. Lesions invading either the primary auditory or somatosensory cortex reduced early latency (20 to 40 ms) evoked responses generated in these regions (Knight, Scabini, Woods, and Clayworth, 1988). In a distinct contrast, LPFC damage resulted in enhanced amplitude of both the primary auditory and somatosensory evoked responses generated from 20 to 40 ms poststimulation (Chao and Knight, 1998; Knight, Scabini, and Woods, 1989; Yamaguchi and Knight, 1990). Spinal cord and brainstem potentials were unaffected by prefrontal damage, indicating that amplitude enhancement of primary cortical responses was due to abnormalities in either prefrontal-thalamic connectivity or in direct interactions between LPFC and sensory cortex.

The failure to inhibit irrelevant sensory input indeed affects the patients' performance. Chao and Knight (Chao and Knight, 1995) showed that patients with LPFC lesions are impaired in an auditory delayed-match-to-sample task when the delay period is filled with irrelevant tone pips. In the same task, the irrelevant tones elicited abnormally augmented primary auditory potentials (Na, Pa) in the patients, and the performance decrement directly correlated with the Pa enhancement (Chao and Knight, 1998). Similarly, in patients with right prefrontal damage, sounds presented to an unattended ear in a dichotic paradigm abnormally reduced the attentional enhancement (see next section) of subsequent to-be-attended sounds (Woods and Knight, 1986). In agreement with these findings, patients with prefrontal lesions show reduced or even reversed negative priming effects (Metzler and Parkin, 2000).

B. Facilitation

The LPFC seems to be essential not only for suppression of irrelevant information, but also for facilitation of relevant inputs. A series of ERP studies in patients with LPFC damage (centered on Brodmann's

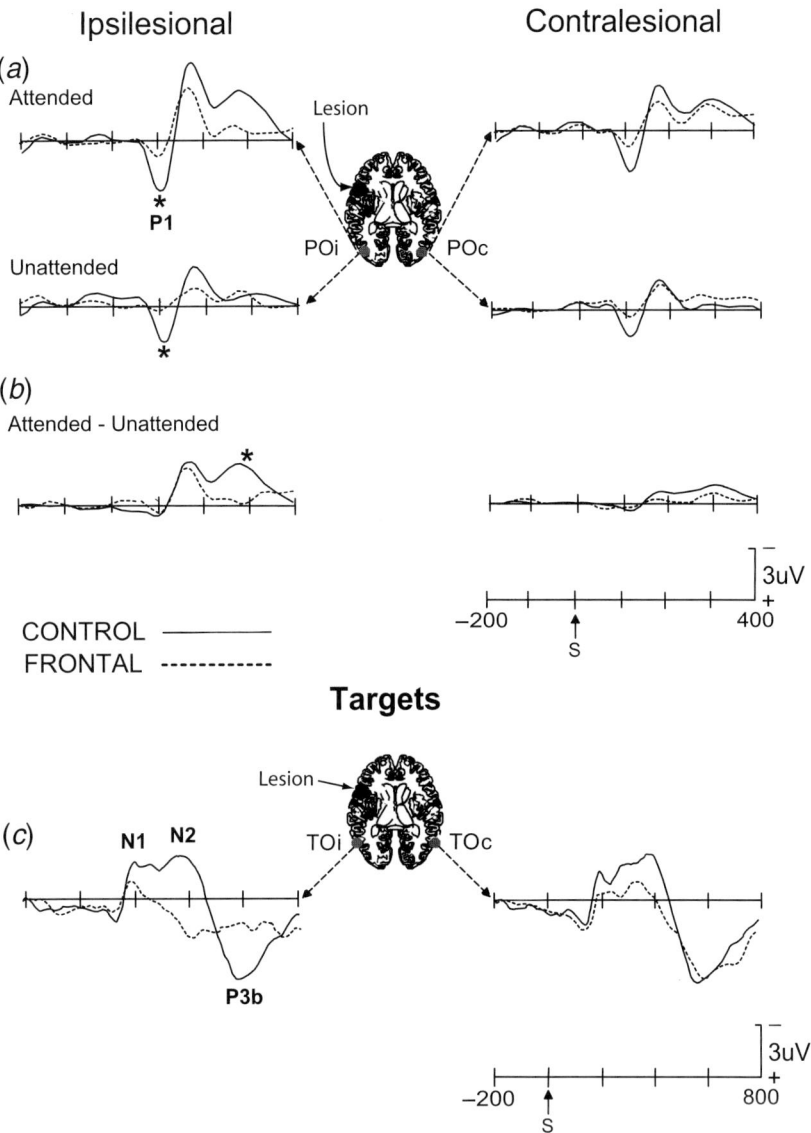

FIGURE 56.2 Visual ERPs in healthy controls and in patients with unilateral right and left lateral prefrontal damage. Patients fixated centrally and were required to detect inverted triangles in a series of upright triangles and novel stimuli presented to the left and right of fixation. In separate blocks, subjects directed attention either to the ipsilesional hemifield or to the contralesional hemifield. All lesions are presented on the left hemisphere. POc/POi—Parieto-Occipital recording sites contralateral or ipsilateral to the lesion, respectively. TOc/TOi—Temporo-Occipital recording sites contralateral or ipsilateral to the lesion, respectively. (a) Responses to attended and unattended nontargets. Note reduction in P1 amplitude for contralesional stimuli regardless of attention. (b) Significant differences in attention effects (attended minus unattended) between patients and controls start around 200 ms post-onset. (c) Responses to targets. Reduced amplitude of N2-P3b. Data from Barcelo and Knight, 2000, and Yago et al., 2001).

areas 9 and 46) suggests that human LPFC regulates extrastriate neural activity through three distinct mechanisms: (1) through a tonic excitatory influence on ipsilateral posterior areas, affecting attended and nonattended sensory inputs alike; (2) by enhancement of extrastriate cortex response to attended information; (3) by a phasic excitatory influence on ipsilateral posterior areas response to correctly perceived task-relevant stimuli (targets). In these experiments, the patients were asked to detect targets (inverted triangles) in a series of distractors (upright triangles). We refer to this paradigm below as the triangles task.

1. Tonic Excitatory Modulation

When the stimuli in the triangles task were presented at fixation, the patients showed a diminished extrastriate N1 (at 170 ms) component relative to controls (Knight, 1997). A similar reduction of N1 was observed in another experiment using visual word stimuli (Swick, 1998). Even more revealing were conditions in which targets and nontargets were lateralized and could appear in either visual field. In one of these conditions, the patients were instructed to attend to all stimuli regardless of their location (Barcelo, Suwazono, and Knight, 2000; Yago and Knight, 2000). The extrastriate P1, immediately preceding the N1 component, was reduced for all stimuli presented to the contralesional hemifield (Fig. 56.2a, b). Directing patients' attention to one of the fields at a time, revealed that this reduction of P1 over ipsilesional extrastriate cortex was attention independent; it was observed for attended and not attended contralateral stimuli alike (Yago and Knight, 2000). Thus, lateral prefrontal cortex seems to exert a modulatory tonic facilitation over processing at ipsilateral extrastriate cortex. A similar effect of reduced association cortex activity ipsilateral to LPFC damage has been reported for auditory-induced N1 potentials (Chao and Knight, 1998).

2. Attentional Enhancement of a Selected Channel

Directing attention to one hemifield in a target detection task normally elicits a sustained negativity starting around 100 ms for all stimuli presented at this hemifield ("Selection Negativity," or "Negative Difference"; see Chapter 84). When attention was directed to the contralesional visual field of LPFC patients, this attention effect on extrastriate cortex was normal in the first 200 milliseconds but significantly disrupted thereafter (Yago and Knight, 2000; Fig. 56.2b). This suggests that LPFC facilitation effect begins around 200 ms after stimulus onset, whereas other cortical areas are responsible for attention-dependent regulation of extrastriate cortex in the first 200 milliseconds. It is conceivable that inferior parietal cortex is responsible for the early reflexive component of attention, whereas LPFC is responsible for more controlled and sustained aspects of visual attention beginning later on. The effect of LPFC lesion on the selection negativity is not limited to visual areas. In an auditory selective attention task, prefrontal patients generated reduced selection negativity as well (Knight, Hillyard, Woods, and Neville, 1981). Notably, this diminished attention effect depended on the side of the lesion. Patients with left hemisphere lesions showed a mildly reduced attention effect regardless of which ear was attended. In contrast, patients with right prefrontal damage showed reduced effect mainly when the contralateral ear was to be attended. Posterior association cortex lesions in the temporoparietal junction had comparable effects on the selection negativity regardless of ear of stimulation (Woods, Knight, and Scabini, 1993).

3. Target-Related Phasic Activation

Target detection in the visual oddball paradigm described above is normally associated with a prominent late response, starting at 200 ms and continuing for the next 500 ms, including the N2-P3b complex. Since this component is elicited only by detected targets, it is considered to be a manifestation of top-down effects. Following LPFC lesions, the N2, a component that likely reflects postselection processing of the target in the inferior temporal lobe, was not observable over the lesioned hemisphere, in response to targets in either visual field (Barcelo et al., 2000; Fig. 56.2c). The P3b was reduced over the temporo-occipital electrodes, but not at parietal sites, attesting to the fact that the P3b most likely reflects multiple distinct cortical processes related to target detection (Soltani and Knight, 2000). The patient's performance in this study, producing more errors, was concordant with this electrophysiological evidence of impaired top-down effects following LPFC lesion. A spatially limited reduction of target P3 was recently reported also by Daffner et al. (Daffner et al., 2003), although in their group of LPFC patients the main reduction of target P3 was reported to be in anterior electrodes. Further investigations will be needed to clarify the source of these differences.

III. NOVELTY AND DEVIANCE DETECTION AND INVOLUNTARY ATTENTION SHIFT

Being able to selectively attend to a subset of the sensory stream and suppress another subset of the input relies on the concomitant operation of a surveillance mechanism allowing for salient events to penetrate and shift attention, so that the organism may avoid hazards or take advantage of opportunities. Such events may be slight perturbations of an established sensory regularity (such an intensity decrement in background noise) or grossly unexpected ("novel") events such as screeching brakes sound. Automatic responses to novel sounds and visual stimuli are typified by a frontoparietal response consisting of the novelty P3a component, peaking around 300 ms following the onset of the triggering stimulus. The P3a is considered a marker of attention orienting, elicited by a distributed multimodal corticolimbic system. Automatic response to slight changes in acoustic regularity outside the focus of attention bears the electrophysiological signature of the "mismatch negativity" (MMN) (Näätänen 1990), a frontal negativity accompanied by lower temporal positivity with a peak latency of 100 to 250 ms following the deviation. The MMN presumably reflects an "error signal," generated automatically in the secondary auditory cortex by a neural mechanism comparing a perceived stimulus to a sensory "memory trace" formed by the regular stimuli, or the process of updating of the existing model of the environment. This signal may be a trigger for ensuing shift of attention toward the deviant event. Both the MMN and the novelty P3a components were shown to be reduced in patients with PFC lesions.

A. Deviance Detection

In paradigms designed to study the MMN, patient and control subjects watched silent movies or were engaged with a visual reaction time paradigm, and were instructed to ignore series of repetitive sounds. Infrequently, the regularity of the sound stream was broken by the occurrence of a deviant stimulus. The MMN in response to pitch and pattern changes was reduced in patients with PFC lesions, indicating an early deficit in automatic detection of deviance outside the focus of attention (Alain, Woods, and Knight, 1998; Alho, Woods, Algazi, Knight, and Näätänen, 1994). Whereas temporoparietal lesions caused an MMN reduction mainly to contralesional stimuli, LPFC lesions elicited comparable deficits regardless of stimulus side in one study (Alain et al., 1998), and more to ipsilesional sounds in another (Alho et al., 1994). In addition, there was a tendency for right prefrontal lesions to be associated with larger MMN reductions than left frontal lesions, while no such asymmetry was found for the temporoparietal lesions (Alain et al., 1998). These results suggest different contributions of the temporoparietal region and the prefrontal region to the MMN (cf. Deouell, Bentin, and Giard, 1998). Further research will be needed to determine whether the reduced MMN following prefrontal damage reflects a weakened memory trace of the previous regularity due to disinhibition of irrelevant information (see above), reduced frontal facilitation of a comparator mechanism in the secondary auditory cortex which is the main generator of scalp MMN, a failure to initiate an attention switch following the detection of the change, or damage to a postulated frontal generator of MMN. In addition, it is still unclear whether the effect of LPFC damage is independent of the dimension of the regularity that is disturbed (e.g., spatial location, pitch or more abstract dimensions of the sound; cf. Deouell, Bentin, and Soroker, 2000).

B. Novelty Detection

Novelty P3a responses generated over prefrontal scalp sites to unexpected novel stimuli are reduced by prefrontal lesions, with reductions observed throughout the lesioned hemisphere (Knight, 1997). Comparable P3a decrements have been observed in the auditory (Knight, 1984), visual (Daffner et al., 2000; Knight, 1997), and somatosensory (Yamaguchi and Knight, 1991) modalities in humans with prefrontal damage. Reductions appear to be more severe after right prefrontal damage. In most studies of novelty P3a, the novel stimuli are completely task irrelevant, and the response is considered to reflect an "involuntary" orienting response toward the salient event. Accordingly, galvanic skin response (GSR), a peripheral marker of the orienting response, is also reduced by damage to the prefrontal as well as posterior association cortex (Tranel and Damasio, 1994). Nevertheless, LPFC damage caused a dramatic reduction in P3a amplitude even in a paradigm in which the novels where relevant to one of two tasks the patients were involved with (Daffner et al., 2003). In this paradigm, patients had to self-pace a succession of visual stimuli, while also looking for a designated (nonnovel) target. LPFC patients not only showed a significantly attenuated P3a response to novel visual stimuli, but also spent less time than controls looking at novels, suggesting a more general decrement of "novelty seeking." Taken together, these findings converge with both clinical observations and animal experimentation supporting a critical role of prefrontal structures in the processing of novel stimuli, probably as part as a prefrontal-hippocampal network (Kimble, Bagshaw, and Pribram, 1965; Knight and Nakada, 1998).

C. Conclusion

The behavior of patients with LPFC lesions suggests an interruption of attentional control at multiple levels. Thus, in everyday life patients may inadequately respond to environmental stimuli, get disproportionally distracted by irrelevant information, and are unable to focus on a task on the one hand and unable to flexibly shift from one goal to another (perseveration) on the other hand. The data from electrophysiology reveals the role of LPFC in automatic processes such as maintaining an adequate level of responsiveness to all stimuli, change detection, and novelty processing, as well as controlled processes such as inhibition of irrelevant information, enhancement of processing of relevant information, and phasic responses to targets. Moreover, the electrophysiological data reveals that LPFC implements some of these functions through interaction with remote regions of the brain, including unimodal sensory cortices and the hippocampus. These effects are evident early in the course of processing.

The fact that LPFC lesions impair the response to novel stimuli and to unattended changes in the sensory environment seems to be at odds with the finding that processing of irrelevant information is disinhibited at primary sensory cortices. One might have predicted that disinhibition and failure to maintain attentional focus would enhance the response to irrelevant novels and deviants making the patient reorient attention inappropriately. One reason for the reduction of responses to novels and deviants may be a reduction of the signal-to-noise ratio in the system, owing to elevated noise levels. In other words, the relative salience, or novelty, of a given stimulus is reduced when all stimuli are processed and perhaps attract some attentional resources owing to disinhibition of early sensory flow. Alternatively, or concomitantly, this combination of results suggest that deviance detection (as exhibited by the MMN) and novelty processing (as reflected by the P3a) are active processes requiring input from the LPFC, rather than purely local processes within sensory areas. Thus, lesions to the LPFC can at the same time cause disinhibition but still prevent adequate responses to novelty.

References

Alain, C., Woods, D. L., and Knight, R. T. (1998). A distributed cortical network for auditory sensory memory in humans. *Brain Res.* **812**, 23–37.

Alho, K., Woods, D. L., Algazi, A., Knight, R. T., and Näätänen, R. (1994). Lesions of frontal cortex diminish the auditory mismatch negativity. *Electroencephalography and Clinical Neurophysiology* **91**, 353–362.

Barcelo, F., and Knight, R. T. (2002). Both random and perseverative errors underlie WCST deficits in prefrontal patients. *Neuropsychologia* **40**, 349–356.

Barcelo, F., Suwazono, S., and Knight, R. T. (2000). Prefrontal modulation of visual processing in humans. *Nat. Neurosci.* **3**, 399–403.

Chao, L. L., and Knight, R. T. (1995). Human prefrontal lesions increase distractibility to irrelevant sensory inputs. *Neuroreport* **6**, 1605–1610.

Chao, L. L., and Knight, R. T. (1998). Contribution of human prefrontal cortex to delay performance. *J. Cogn. Neurosci.* **10**, 167–177.

Daffner, K. R., Mesulam, M. M., Scinto, L. F. M., Acar, D., Calvo, V., Faust, R., et al. (2000). The central role of the prefrontal cortex in directing attention to novel events. *Brain* **123**, 927–939.

Daffner, K. R., Scinto, L. F. M., Weitzman, A. M., Faust, R., Rentz, D. M., Budson, A. E., et al. (2003). Frontal and parietal components of a cerebral network mediating voluntary attention to novel events. *J. Cogn. Neurosci.* **15**, 294–313.

Deouell, L. Y., Bentin, S., and Giard, M. H. (1998). Mismatch negativity in dichotic listening: evidence for interhemispheric differences and multiple generators. *Psychophysiology* **35**, 355–365.

Deouell, L. Y., Bentin, S., and Soroker, N. (2000). Electrophysiological evidence for an early (pre-attentive) information processing deficit in patients with right hemisphere damage and unilateral neglect. *Brain* **123**, 353–365.

Kimble, D. P., Bagshaw, M. H., and Pribram, K. H. (1965). The GSR of monkeys during orienting and habituation after selective partial ablations of the cingulate and frontal cortex. *Neuropsychology* **3**, 121–128.

Knight, R. T. (1984). Decreased response to novel stimuli after prefrontal lesions in man. *Electroencephalogr. Clin. Neurophysiol.* **59**, 9–20.

Knight, R. T. (1997). Distributed cortical network for visual attention. *J. Cogn. Neurosci.* **9**, 75–91.

Knight, R. T., Hillyard, S. A., Woods, D. L., and Neville, H. J. (1981). The effects of frontal cortex lesions on event-related potentials during auditory selective attention. *Electroencephalogr. Clin. Neurophysiol.* **52**, 571–582.

Knight, R. T., and Nakada, T. (1998). Cortico-limbic circuits and novelty: a review of EEG and blood flow data. *Rev. Neurosci.* **9**, 57–70.

Knight, R. T., Scabini, D., and Woods, D. L. (1989). Prefrontal cortex gating of auditory transmission in humans. *Brain Res.* **504**, 338–342.

Knight, R. T., Scabini, D., Woods, D. L., and Clayworth, C. (1988). The effects of lesions of superior temporal gyrus and inferior parietal lobe on temporal and vertex components of the human AEP. *Electroencephalogr. Clin. Neurophysiol.* **70**, 499–509.

Lhermitte, F. (1986). Human autonomy and the frontal lobes. Part II: Patient behavior in complex and social situations: the "environmental dependency syndrome." *Ann. Neurol.* **19**, 335–343.

Metzler, C., and Parkin, A. J. (2000). Reversed negative priming following frontal lobe lesions. *Neuropsychologia* **38**, 363–379.

Näätänen, R. (1990). The role of attention in auditory information processing as revealed by event-related potentials and other brain measures of cognitive function. *Behav. and Brain Sci.* **13**, 201–288.

Soltani, M., and Knight, R. T. (2000). Neural origins of the P300. *Crit. Rev. Neurobiol.* **14**, 199–224.

Swick, D. (1998). Effects of prefrontal lesions on lexical processing and repetition priming: an ERP study. *Cogn. Brain Res.* **7**, 143–157.

Tranel, D., and Damasio, H. (1994). Neuroanatomical correlates of electrodermal skin conductance responses. *Psychophysiology* **31**, 427–438.

Woods, D. L., and Knight, R. T. (1986). Electrophysiologic evidence of increased distractibility after dorsolateral prefrontal lesions. *Neurology* **36**, 212–216.

Woods, D. L., Knight, R. T., and Scabini, D. (1993). Anatomical substrates of auditory selective attention: behavioral and electrophysiological effects of posterior association cortex lesions. *Brain Res. Cogn. Brain Res.* **1**, 227–240.

Yago, E., and Knight, R. T. (2000). Tonic and phasic prefrontal modulation of extrastriate processing during visual attention. Paper presented at the Society for Neuroscience.

Yamaguchi, S., and Knight, R. T. (1990). Gating of somatosensory input by human prefrontal cortex. *Brain Res.* **521**, 281–288.

Yamaguchi, S., and Knight, R. T. (1991). Anterior and posterior association cortex contributions to the somatosensory P300. *J. Neurosci.* **11**, 2039–2054.

CHAPTER 57

Nonspatially Lateralized Mechanisms in Hemispatial Neglect

Masud Husain

ABSTRACT

Hemispatial neglect is a common disabling disorder following right-hemisphere stroke. Although much work has focused on the lateralized components of neglect, recent investigations have also revealed deficits that are not spatially lateralized and are present in both the good (ipsilesional) side and the neglected (contralesional) side of space. These findings are consistent with new observations from functional imaging, human neuropsychological and monkey electrophysiological studies that also suggest nonspatially lateralized functions for the regions damaged in neglect. Understanding the interaction between spatially lateralized and nonlateralized mechanisms may provide important insights into the neglect syndrome, the normal functions of brain structures commonly damaged in neglect patients, and the development of treatments for the condition.

I. INTRODUCTION

The neglect syndrome is the most common disorder of visual perception following stroke, with approximately two-thirds of right-hemisphere cases affected acutely. These patients fail to orientate toward or detect items on their contralesional left side, even though they are not blind to stimuli on that side (see Chapter 58). Neglect is most often associated with damage to the right inferior parietal lobe (IPL) (Mort et al., 2003), although focal lesions involving the right inferior frontal lobe may also lead to a similar syndrome (Fig. 57.1). Similarly, focal damage to subcortical regions including the basal ganglia, thalamus, or deep white matter may produce neglect, perhaps because of cortical hypoperfusion or disconnection. Typically, however, patients have large lesions spanning many of these cortical and subcortical regions, consistent with the view that neglect represents a heterogeneous disorder, with different patients suffering different combinations of component deficit. Note that severe and enduring neglect is not common in patients with left-hemisphere damage, perhaps because of the different abilities of the two hemispheres to orient attention (see Chapter 59).

II. SPATIALLY LATERALIZED DEFICITS MAY NOT BE ENOUGH

Understandably, much of the research into neglect has focused on its spatially lateralized presentation, and investigators have proposed that the syndrome arises from lateralized deficits that are worse toward one side of space, for example, in directing spatial attention, limb or eye movements, or representing space. But although such lateralized mechanisms are undoubtedly critical, they might not—on their own—fully explain the neglect behavior of all patients (Husain and Rorden, 2003; Robertson, 2001). Consider the performance of a right-hemisphere neglect patient on a cancellation task in which he has to mark all the targets he can find among distractor elements (Fig. 57.2). A spatially lateralized attentional bias that makes the rightmost items effectively more salient than leftward ones might explain why such a patient starts to search for items on the right. But how does it *alone* prevent him from moving on toward the left after he has marked targets on the right, and when given unlimited time? Although the relative perceived salience of marked items to the right could repeatedly draw a patient's attention back, this seems unlikely to be the full explanation. Erasing the targets (so that they

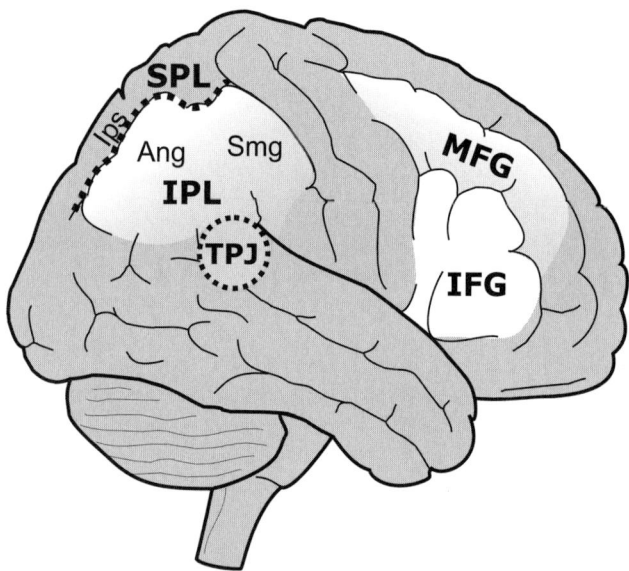

FIGURE 57.1 Lateral view of right hemisphere demonstrating the superior parietal lobe (SPL), intraparietal sulcus (IPS) and the inferior parietal lobe (IPL), which consists of the angular gyrus (Ang) and supramarginal gyrus (Smg). Also marked are the temporoparietal junction (TPJ), inferior frontal gyrus (IFG), and middle frontal gyrus (MFG).

FIGURE 57.2 Typical performance by a patient with severe left-sided neglect on a cancellation task. Note how the patient has found targets on only the right side of the sheet, neglecting those to the left.

are no longer salient) rather than marking them does improve search performance, but patients still continue to demonstrate neglect.

Similarly, although a directional motor deficit can lead to prolonged reaction times for leftward movements, patients can nevertheless make movements to the left. Indeed, on visual search tasks they make just as many leftward as rightward saccades. So why do they not eventually find targets to the left? A loss of the representation of left space, regardless of the nature of that representation (Pouget and Driver, 2000), might offer a simpler explanation because one would not expect patients to search a part of space they fail to represent. But patients who do not search toward the left of cluttered cancellation tests nevertheless find the same items to the left if they are presented individually in uncluttered displays, so objects in that part of space can be represented by these patients.

These observations suggest that *spatially lateralized* attentional, motoric and representational models have difficulties in fully explaining the performance of unilateral neglect patients on a standard test such as cancellation. Spatially lateralized models also have difficulties in explaining the behavior of patients with *bilateral* posterior cortical damage who have Balint's syndrome (Rafal, 2001). Individuals with this condition do not necessarily demonstrate a lateralized bias (presumably because both hemispheres are affected), but nevertheless suffer a profound disorder of visual perception. They report seeing only one thing at a time (simultanagnosia), even if objects such as line drawings are superimposed over each other at one location in space. They appear to have a severe limitation in their visual processing capacity. In addition, they have a greatly constricted effective field of vision, so that they are unaware of all but the most salient stimuli in the periphery of *either* the left or right visual field. These findings suggest that it might be worth considering nonspatially lateralized contributions to neglect.

III. NONSPATIALLY LATERALIZED DEFICITS MIGHT BE IMPORTANT

Recent investigations have revealed deficits in neglect patients that are not necessarily worse toward one side of space, that is, are nonspatially lateralized (Husain and Rorden, 2003; Robertson, 2001). These nonlateralized components might combine with spatially lateralized mechanisms to exacerbate the severity of neglect. Moreover, such nonlateralized mechanisms need not be specific to the neglect syndrome; they can occur separately in patients who do not have neglect. Thus findings from other branches of cognitive neuroscience that have hitherto been considered unrelated to neglect may be important in understanding the syndrome. Furthermore, brain regions that are typically associated with neglect and are considered to have spatial functions, such as the IPL, also have been found to have nonlateralized functions. An understanding of this functional anatomy provides a new way to view the mechanisms driving neglect, with different combinations of spatially lateralized and nonlateralized impairment occurring in individual patients, depending upon the extent and location of brain damage. Finally, nonspatially lateralized mechanisms, when combined with lateralized components, might reduce the potential for recovery from neglect and therefore are potentially important targets for treatment.

IV. SELECTIVE ATTENTION CAPACITY

The visual system has a limited capacity, so only stimuli that are salient or potentially important for the task at hand capture or drive our attention. One way to probe the capacity of the visual system is to measure the time course of attentional processing. When we identify a visual object, our ability to detect a second object is impaired if it appears within 400ms of the first. This phenomenon, termed the *attentional blink* (see Chapter 63) provides a measure of the temporal dynamics of nonlateralized selective attention: the time taken by the visual system to identify a visual stimulus before it is free to detect a subsequent stimulus.

Patients with hemispatial neglect have a more severe and protracted attentional blink than healthy individuals (Husain et al., 1997). When neglect patients identify a target letter in a stream of letters presented successively at central fixation, their ability to detect a subsequent target letter is impaired for more than 1200ms. Thus, they show a deficit in selective attention on this nonspatially lateralized task. And this impairment cannot be attributed to a difficulty in sustaining attention throughout a trial because detection of the second target actually improves as the interval between the first and second target increases.

Moreover, patients' ability to detect only a single target on a control task is excellent regardless of whether the target is presented early or late in a trial.

Just as patients with Balint's syndrome appear to have a profound capacity limit in visual processing (perceiving only one item at a time), patients with neglect appear to demonstrate a severe capacity limit when tested with this attentional blink paradigm. Importantly, their deficit on this nonspatially lateralized task correlates with the severity of spatial neglect. Recent work shows that the impairment might be anatomically specific rather than neglect-specific (Shapiro et al., 2002), for it can occur in patients with posterior lesions involving the inferior parietal lobe and superior temporal gyrus who do *not* have neglect, but not in patients with more dorsal lesions involving the superior parietal lobe (SPL). Auditory studies, too, have begun to find evidence for a nonlateralized deficit in selective attention, with neglect patients found to encounter difficulty on a task that requires comparisons between brief successive *central* sounds (Cusack et al., 2000).

Recent investigations have also reported impairments on *both* sides of space (Battelli et al., 2003; Duncan et al., 1999). One of these investigations attempted to measure the capacity of visual attention using the visual attention model developed by Bundesen for normal human observers. Neglect patients were found to have significantly reduced capacity for encoding stimuli presented transiently in *either* the left or right visual field (Duncan et al., 1999).

Functional imaging studies have identified specific locations around the intraparietal sulcus and frontal cortex (Fig. 57.3) as being associated with non-spatially lateralized visual processing, e.g., on the attentional blink task (Husain and Rorden, 2003). These regions, or their connections, are also often damaged in neglect patients, and this may explain the severe reduction in

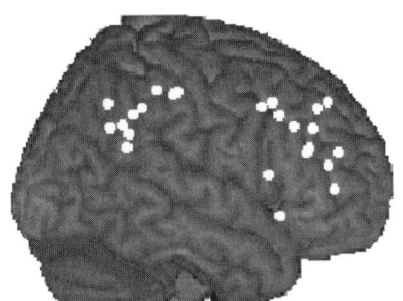
(*a*) Sustained attention

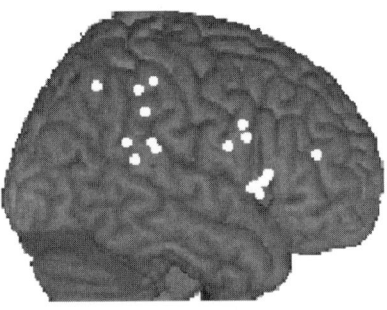
(*b*) Salience detection

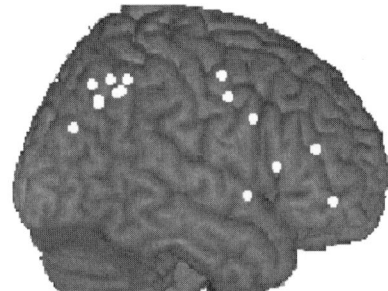
(*c*) Selective attention at fixation

FIGURE 57.3 Meta-analysis plotting centroids of activation in functional imaging tasks that have investigated nonspatially lateralized sustained attention (*a*), salience detection using the "oddball task" (*b*) and visual selective attention at fixation, for example, using the attentional blink paradigm (*c*).

non-lateralised capacity and prolongation of visual processing observed in these individuals. When combined with any lateralised deficits, these impairments would be expected to exacerbate spatial neglect, by prolonging the time spent in processing stimuli on the right at the expense of those on the left.

V. NONSPATIALLY LATERALIZED SUSTAINED ATTENTION

In addition to selective attention capacity, the ability to *maintain* attention over periods of time is important. Individuals who cannot be vigilant miss critical events (see Chapter 55). For patients with brain lesions, sustained attention may be an important factor in determining prognosis for recovery. It is often tested using tasks that require observers to respond to transient, relatively infrequent stimuli that occur randomly over a protracted period of time. In their simplest and purest forms, these paradigms present observers with only one form of stimulus, and the task is to detect that stimulus without having to discriminate it from distractors. Importantly, these tasks do not necessarily require stimuli to be presented at different locations in space.

Human lesion studies have implicated the right frontal lobe as a key site in mediating sustained attention. More recently, sustained attention deficits have also been reported in patients with posterior lesions using a visual vigilance task. In all these investigations, the patients were *not* noted to show signs of unilateral neglect at the time of testing. Thus, deficits of sustained attention may occur independently of neglect. On the other hand, patients with neglect can show impaired sustained attention on tasks that do not use spatially lateralized stimuli, or that present stimuli only in the "good," ipsilesional side of space.

Right-hemisphere stroke patients with neglect performed significantly worse on a tone-counting task than stroke patients without neglect (Robertson et al., 1997). Moreover, there was a strong correlation between severity of neglect and degree of impairment on the nonspatially lateralized sustained attention task. The deficit in sustained attention also proved to be an independent predictor of performance on clinical measures of visual neglect, such as cancellation. Other studies have shown that patients with persistent spatial neglect continue to be impaired on sustained attention tasks, whereas patients who recover from neglect show improvements in sustained attention.

In healthy individuals, the results of several functional imaging studies point to locations in the right inferior frontal lobe and predominantly right midfrontal lobe (Fig. 57.3) as being critical for sustained visual attention (Husain and Rorden, 2003). Damage to these regions might be necessary to produce deficits in sustained attention after stroke, and these areas are also often involved in neglect patients. In such individuals, a lateral spatial bias (initially directing attention to the right) combined with a nonlateralized deficit in sustained attention might exacerbate leftward neglect, with patients experiencing difficulty in their ability to continue searching from their initial rightward starting position.

VI. DETECTING SALIENCE OVER SPACE AND TIME

Whereas sustained attention tasks probe the ability to maintain attention over time, other paradigms have been developed to investigate brain mechanisms involved in detecting salient stimuli among distractors. Neuronal recording studies have found activity related to the *spatial location* of salient stimuli in monkey parietal cortex. Remarkably, some neurones in the inferior parietal cortex respond only when a particular stimulus is salient among distractor objects, but not when the same stimulus acts as a distractor in a different visual array (Constantinidis and Steinmetz, 2001). Such studies have led to the view that the neural representation in the posterior parietal cortex may be very sparse, with only salient or behaviorally relevant stimuli being encoded there.

In humans, too, there is evidence of such a representation of stimulus salience, using the "oddball" paradigm (see chapter 56). Typically, a sequence of standard stimuli is presented *over time*. Embedded in this stream are infrequent target stimuli (to which the subject has to make a response), familiar nontargets (which do not require a response), or completely novel nontarget stimuli (which, again, do not require a response). ERP (evoked response potential) studies reveal a consistent posterior P3 or P300 positive waveform ~400ms after either targets *or* novel nontargets, but not after familiar nontargets. Thus, the P3 might be a marker of salient stimuli that draw attention, regardless of whether the stimulus is a target or a novel nontarget (although some investigators have distinguished between an earlier P3a component in response to novelty and a slightly later P3b wave evoked by a target). Unilateral lesions of the temporoparietal junction eliminate P3a and P3b components in *both* cerebral hemispheres, indicating that this region might be at least one critical site for their generation. There is also evidence of prefrontal involvement in salience encoding, as unilateral lesions of lateral prefrontal cortex also reduce P3b amplitude bilaterally over posterior regions.

Functional imaging researchers have also adopted the "oddball" paradigm. The results of their studies point to a key role for the temporoparietal junction (TPJ) region, as well as inferior frontal regions (Fig. 57.3), predominantly in the right hemisphere (Husain and Rorden, 2003). Both the posterior and anterior brain regions implicated in detecting salience are damaged in many patients with neglect.

VII. TRANS-SACCADIC SPATIAL WORKING MEMORY

A common observation is that patients with neglect often re-inspect items they have already looked at. For example, when right-hemisphere neglect patients perform a cancellation task, some of them return repeatedly to re-examine rightward targets they have already found. Although the relative perceived salience of items on the right, compared to those on the left, might be one explanation for such behavior, a failure to keep track of object locations across saccadic eye movements might be an important independent contributing factor in some patients.

To examine this issue, a paradigm to probe memory for previously inspected spatial locations during visual search has been developed (Husain et al., 2001). Subjects' eye movements are recorded while they attempt to find targets among distractor items on a multitarget search task. They are asked to click on a response button only when they find a new target—one they have not fixated before. Using this system, it has been confirmed that many neglect patients refixate targets they have previously found. Critically, analysis of the click responses also shows that many neglect patients mistake targets they have previously found for new discoveries, indicating that they do not remember having inspected them. Note that this impairment is actually on the right, or "good," side of space for these patients.

Importantly, high-resolution MRI shows that this type of behavior is associated with specific lesion sites near the right intraparietal sulcus and inferior frontal gyrus. Functional imaging studies, using double- or triple-step saccades in healthy individuals also reveal a critical role of specific regions within inferior parietal and frontal lobes in keeping track of spatial locations across saccades (Heide et al., 2001). Note that such an impairment is not neglect-specific but rather appears to be related to lesion location, since parietal patients without neglect may also be impaired on such a task.

Across neglect patients, the severity of the working memory deficit correlates with the degree of hemispatial neglect. The greater the difficulty in keeping track of previously inspected locations to the right, the more likely that patients do not shift their search toward the left. Furthermore, the trans-saccadic spatial working memory deficit is not spatially lateralized, as neglect patients are impaired even when required to keep track of locations in a vertical array. The combination of a lateral bias plus a (nonlateralized) failure to remember which locations have already been searched offers a new explanation for the inability of these patients to direct their search to the contralesional side.

VIII. COMBINING NONLATERALIZED AND SPATIALLY LATERALIZED IMPAIRMENTS

The nonspatially lateralized mechanisms implicated in neglect correlate well with functional imaging evidence that inferior parietal and lateral frontal cortices are involved in these functions in the normal brain. Within parietal and frontal cortex there might be regional specialization for these nonlateralized processes. Different nonlateralized functions appear to activate different parts of the parietal cortex, extending from the intraparietal sulcus down to the IPL and TPJ (Husain and Rorden, 2003).

In addition, there is some evidence that spatially lateralized and nonlateralized functions might be anatomically segregated. Several functional imaging studies show lateralized attentional mechanisms localized in the SPL, whereas the nonlateralized ones cluster in the IPL. Regions within the intraparietal sulcus (between the SPL and IPL) appear to participate in both lateralized and nonspatially lateralized functions. This specialization within parietal cortex might explain why different patients show different symptoms, depending on the extent and location of their lesion. It is possible that the combination of spatially lateralized and nonlateralized deficits leads to persistent neglect. According to this model, although SPL lesions (or their connections) might lead to a lateralized bias they would not lead to lasting neglect. By contrast, damage to the IPL (including the intraparietal sulcus and TPJ) or its connections would lead to severe and protracted neglect, with both lateralized and nonlateralized deficits.

Of course, it is possible that the distinction between spatially lateralized and nonlateralized processes is not strictly dichotomous, at least for some processes underlying neglect. Indeed, some investigators have argued that a left-right spatial gradient in, for example, the attentional blink might underlie part of the lateralized bias in neglect. Further work will be required to investigate this possibility. What is clear, however, is that studying lateralized and nonlateralized component deficits and their interaction may be an important way to improve our understanding of neglect. It may also be important in directing future attempts at treatment.

Currently, there is no established treatment for neglect. Conventional methods that have targeted the spatially lateralized deficit have been singularly unsuccessful. For successful treatment *both* lateralized and nonlateralized mechanisms might need to be targeted. For example, Robertson and colleagues found that neglect patients became aware of left stimuli more than half a second later than those on the right. Remarkably, this spatial bias was abolished when patients heard a central auditory warning sound, suggesting not only that a tonic deficit in sustained attention normally contributes to neglect, but also that improving this deficit might also improve the spatial bias (Robertson et al., 1998). In addition to behavioral methods, pharmacological interventions directed toward specific nonlateralized components are likely to be the next step.

For impairments of sustained attention, research in animals and healthy humans has suggested that the noradrenergic or cholinergic systems might be potential pharmacological targets. By contrast, the dopaminergic system might be an important target for treatments aimed at improving trans-saccadic spatial working memory. Neurones in monkey frontal cortex, like those in the intraparietal sulcus, encode the remembered locations of saccadic targets. Both their activity and the monkeys' memory of saccadic targets are modulated by dopamine D1-receptor agents. Some clinical studies have used the dopamine agonist bromocriptine and reported both favorable and adverse effects on neglect. The variability in response might be due to the heterogeneity of the component deficits in the patients studied, as well as the fact that bromocriptine acts mainly at dopamine D2-receptors. Future dopaminergic treatment might need to be aimed selectively at D1 receptors specifically in patients with a trans-saccadic spatial working memory deficit.

Several behavioral and pharmacological strategies, perhaps in combination, are eventually likely to be deployed in the treatment of the neglect syndrome. However, for such interventions to be successful a better understanding of the component deficits underlying the disorder—both lateralized and nonspatially lateralized—will be essential.

Acknowledgments

This research is funded by a grant to MH from the Wellcome Trust. My thanks to Jon Driver, Paresh Malhotra, and Andy Parton for all their help.

References

Battelli, L., Cavanagh, P., Martini, P., and Barton, J. J. (2003). Bilateral deficits of transient visual attention in right parietal patients. *Brain* **126**, 2164–2174.

Constantinidis, C., and Steinmetz, M. A. (2001). Neuronal responses in area 7a to multiple-stimulus displays: I. neurons encode the location of the salient stimulus. *Cerebral Cortex* **11**, 581–591.

Cusack, R., Carlyon, R. P., and Robertson, I. H. (2000). Neglect between but not within auditory objects. *J. of Cog. Neurosci.* **12**, 1056–1065.

Duncan, J., Bundesen, C., Olson, A., Humphreys, G., Chavda, S., and Shibuya, H. (1999). Systematic analysis of deficits in visual attention. *J. of Exper. Psychol: General* **128**, 450–478.

Heide, W., Binkofski, F., Seitz, R. J., Posse, S., Nitschke, M. F., Freund, H. J., et al. (2001). Activation of frontoparietal cortices during memorized triple-step sequences of saccadic eye movements: an fMRI study. *European J. of Neurosci.* **13**, 1177–1189.

Husain, M., Mannan, S., Hodgson, T., Wejciulik, E., Driver, J., and Kennard, C. (2001). Impaired spatial working memory across saccades contributes to abnormal search in parietal neglect. *Brain* **124**, 941–952.

Husain, M., and Rorden, C. (2003). Non-spatially lateralized mechanisms in hemispatial neglect. *Nat. Rev. Neurosci.* **4**, 26–36.

Husain, M., Shapiro, K., Martin, J., and Kennard, C. (1997). Abnormal temporal dynamics of visual attention in spatial neglect patients. *Nature* **385**, 154–156.

Mort, D. J., Malhotra, P., Mannan, S. K., Rorden, C., Pambakian, A., Kennard, C., et al. (2003). The anatomy of visual neglect. *Brain* **126**, 1986–1997.

Pouget, A., and Driver, J. (2000). Relating unilateral neglect to the neural coding of space. *Curr. Opin. Neurobiol.* **10**, 242–249.

Rafal, R. (2001). Balint's syndrome. *In* M. Behrmann, (ed.) Handbook of Neuropsychology. Vol 4. Elsevier: Amsterdam, 121–141.

Robertson, I. H. (2001). Do we need the "lateral" in unilateral neglect? Spatially nonselective attention deficits in unilateral neglect and their implications for rehabilitation. *Neuroimage* **14**, S85–S90.

Robertson, I. H., Manly, T., Beschin, N., Daini, R., Haeske-Dewick, H., Homberg, V., et al. (1997). Auditory sustained attention is a marker of unilateral spatial neglect. *Neuropsychologia* **35**, 1527–1532.

Robertson, I. H., Mattingley, J. B., Rorden, C., and Driver, J. (1998). Phasic alerting of neglect patients overcomes their spatial deficit in visual awareness. *Nature* **395**, 169–172.

Shapiro, K., Hillstrom, A., and Husain, M. (2002). Control of visuotemporal attention by inferior parietal and superior temporal cortex. *Curr. Biol.* **12**, 1320–1325.

CHAPTER 58

Visual Extinction and Hemispatial Neglect after Brain Damage: Neurophysiological Basis of Residual Processing

Patrik Vuilleumier

ABSTRACT

Visual extinction, associated with unilateral spatial neglect after parietal lesions, is characterized by unawareness of contralesional stimuli in the presence of competing ipsilesional stimuli. Behavioral studies suggest that extinguished stimuli are still processed without attention or without awareness. fMRI and evoked-potentials studies in patients with neglect and extinction indicate that unseen visual stimuli can activate striate and extrastriate areas in ventral pathways, without affording conscious perception. By contrast, awareness is associated with greater activity in a distributed network at early and later stages of visual processing, particularly parietal and frontal areas. Retinotopic mapping of intact visual cortex also shows an abnormal bias in activations of early cortical areas of the damaged hemisphere, increasing from V1 to higher areas. These results suggest that residual processing in extinction may occur at relatively early stages of visual analysis but need to be integrated with, and to receive feedback modulation from, concomitant processing in dorsal stream in order to reach awareness.

I. BEHAVIORAL AND ANATOMICAL ASPECTS

Hemispatial neglect is a frequent neurological disorder, involving a failure to perceive and respond to stimuli located on the side of space opposite to a focal brain lesion, not attributable to either sensory or motor deficits (for a review, see Driver and Vuilleumier, 2001a). Most often, neglect follows right-hemisphere damage and hence concerns stimuli toward the left side. Although neglect may arise in a variety of tasks and potentially implicate several functional components, including nonspatial deficits (see Husain, Chapter 57), a core symptom is reflected by the phenomenon of perceptual extinction that can typically be observed when the patients have no basic sensory losses. Such patients may perceive a stimulus presented alone on the left side, but remain unaware of the same stimulus at the same location when presented with a simultaneous stimulus on the right side. The ipsilesional event thus seems to extinguish awareness of concomitant information toward the contralesional side. Extinction on double simultaneous stimulation may sometimes arise within the intact right visual field (e.g., for the leftmost of two stimuli) and increase with perceptual difficulty of the task. Extinction can occur not only in vision but also in any other modality, or between stimuli simultaneously presented in different modalities. However, this chapter will focus on the visual domain.

The phenomenon of extinction demonstrates that awareness of contralesional information can be abolished by the presence of competing inputs, even though processing of contralesional information is possible otherwise. This can be considered as a pathological, spatially biased exaggeration of perceptual failures exhibited by healthy subjects under conditions of restricted attention (Driver et al., 2001a). At the neuronal level, such failures may also be related to single-cell recordings in the monkey showing competitive suppression between concurrent stimuli within the receptive fields, with spatial attention required to resolve competition in favor of a single behaviorally-relevant stimulus (see Desimone, this volume, Chapter 8). Neuropsychological studies of extinction therefore provide important insights into the cerebral mechanisms underlying attentional selection and awareness.

Clinically, extinction often persists after recovery of more florid neglect symptoms. Some patients show neglect in spatial tasks without extinction on double simultaneous stimulation, or vice versa, but such dissociations might reflect different attentional demands in the tasks used (Cocchini, Cubelli, DellaSala, and Beschin, 1999). Whereas severe and long-lasting neglect is commonly associated with damage in inferior posterior parietal cortex, a distinct role of the superior temporal gyrus and temporoparietal junction has recently been proposed for unilateral spatial neglect and extinction on double stimulation, respectively (Karnath, Himmelbach, and Kuker, 2003). Similar but more transient deficits can also be seen after frontal and subcortical (posterior thalamic and basal ganglia) lesions. All these brain areas are implicated in attentional processes in healthy subjects (see Mesulam, this volume, chapter 6), supporting the idea that deficits in attention may play a crucial role in neglect and extinction. By contrast, basic sensory pathways (e.g., from retina to primary visual cortex and higher-level perceptual machinery of the visual system) are often spared by the lesion in these patients. Accordingly, recent behavioral findings and functional imaging studies converge to indicate that these anatomically intact sensory pathways can still process incoming stimuli, yet fail to afford conscious perception due to damage in nonsensory brain regions (e.g., in parietal and frontal lobe) whose functions have been related to spatial attention, working memory, and/or action planning.

II. IMPLICIT RESIDUAL PROCESSING

Considerable information can be extracted from contralesional stimuli despite spatial neglect or extinction. Such residual processing has been demonstrated by a variety of indirect measures where the patient's behavior is influenced by stimuli in contralesional space, even when these stimuli are neglected or extinguished from conscious awareness. For example, patients who fail to detect contralesional visual stimuli during bilateral simultaneous presentations may nonetheless guess above chance whether the unseen stimulus was the same or different as the one seen on the ipsilesional side, even when pairs of objects are semantically but not physically similar (Berti et al., 1992). Extinguished stimuli can also produce semantic priming effects in object or word classification tasks, as well as Stroop or flanker interference effects. These findings suggest that unconscious processing may proceed up to stages where the visual system can extract not only simple features but also abstract categorical information. In addition, visual neglect or extinction can be modulated by intrinsic stimulus properties (e.g., meaningfulness, affective value), or by "gestalt" grouping principles that link two bilateral events into a single object (e.g., collinearity, illusory contours, closure). This also suggests that some visual processes responsible for extracting such properties and organizing visual inputs into coherent objects are not disrupted in these patients.

Residual processing in extinction and neglect therefore seems fairly more sophisticated than "blindsight" observed in patients with lesions of primary visual cortex (Weiskrantz, 2004; Whatham et al., 2003). In these patients, a variety of unconscious visual abilities have been documented, but typically concern elementary processes related to stimulus detection and visuomotor acts that can be subserved by subcortical visual pathways.

By contrast, it has been hypothesized that residual processing in extinction and neglect might be mediated by intact pathways in occipital and temporal cortex that normally code for visual features and semantics. In support of this hypothesis, visual grouping by illusory contours may occur in neglect patients with intact occipital cortex, but not in those whose damage extends into lateral occipital areas (Vuilleumier, Valenza, and Landis, 2001b). However, this does not imply that residual processing in anatomically intact areas is normal or equivalent to the processing occurring within the same areas during conscious recognition. Such residual processing may exhibit some similarities but also differences with the processing of unattended stimuli in healthy subjects (Fuentes and Humphreys, 1996). Furthermore, theoretical controversies have arisen concerning whether the neural bases of unconscious versus conscious processing involve impaired access of information from lower perceptual modules to some critical higher-level stages in the cognitive system; impaired integration between separate modules; or partial degradation and subthreshold activation of perceptual representations. Recent studies using functional imaging and electrophysiology measures in neglect patients have begun to answer these questions by identifying specific patterns of neural activity associated with impaired perceptual awareness and residual processing abilities after parietal damage.

III. FUNCTIONAL NEUROIMAGING

Early imaging studies used SPECT and PET techniques in patients to show that decreased cerebral blood flow in the temporoparietal junction best correlate with neglect symptoms, even with lesions situated elsewhere (e.g., subcortically). These remote effects were attributed to functional disconnections between the lesioned areas and intact brain regions (diaschisis). Although some changes might also reflect decreased vascular supply in the context of acute stroke, decreased blood flow was also observed in distant territories, including right ventral occipitotemporal areas and anterior cingulate cortex (Leibovitch et al., 1999). More recently, fMRI with event-related measures has allowed us to investigate cerebral activity in neglect patients during specific task conditions, including extinction. These new data provide more direct evidence for altered neural activity within structurally intact areas, beyond the local destruction of brain tissue.

Several event-related fMRI studies specifically examined the neural correlates of visual extinction, during both unconscious and conscious processing (Rees et al., 2002; Rees et al., 2000; Vuilleumier et al., 2002; Vuilleumier et al., 2001a). In these studies, patients with right parietal damage were shown faces and other stimuli (e.g., simple shapes or houses) in their right, left, or both visual fields (in unpredictable order). Faces were used because previous imaging work in healthy people has delineated specific neural substrates for face perception within occipitotemporal regions, which are typically spared in neglect patients. Comparing fMRI responses on bilateral trials where a left-sided face was extinguished from awareness with trials where only a unilateral right stimulus was presented (i.e., same awareness but different inputs in contralesional field) revealed a consistent pattern of residual unconscious processing across different patients and different tasks (Rees et al., 2000; Vuilleumier et al., 2002; Vuilleumier et al., 2001a). Although escaping awareness, extinguished faces in the left field were found to activate primary visual cortex (V1) and posterior temporal areas of the damaged right hemisphere (Fig. 58.1). Some activation was also found in face-selective regions of fusiform cortex when comparing extinction of faces versus houses (Rees et al., 2000) and in limbic regions remote from visual cortex (e.g., amygdala and orbitofrontal cortex) when comparing extinction of emotional versus neutral faces (Vuilleumier et al., 2002). These results corroborate the hypothesis that activation of primary visual cortex and ventral visual pathways might mediate residual processing abilities in extinction. They also indicate that such activation, by itself, is insufficient to support conscious awareness of stimulus properties that are presumably represented in these areas (e.g., facial traits and expressions).

Event-related fMRI in the same patients (Rees et al., 2002; Vuilleumier et al., 2002; Vuilleumier et al., 2001a) also revealed differential activations on those occasional bilateral trials where a left-side face was perceived, rather than extinguished (i.e., comparing identical bilateral stimuli but different awareness). Greater responses were found in occipital and temporal cortex of the damaged hemisphere, including V1 and fusiform areas that were already activated by extinguished stimuli, plus in parietal and frontal cortex of the intact left hemisphere, and in bilateral cuneus regions (Fig. 58.1). Activations in left parietal cortex were relatively symmetric to lesioned areas in the right hemisphere. Thus, neural responses associated with awareness were not restricted to early visual areas (as during extinction) but now extended to a broad neural network. Furthermore, an analysis of "functional connectivity" showed increased coupling of visual areas with left parietal and left frontal cortex when the left stimulus was detected, as opposed to extinguished (Vuilleumier et al., 2001a).

These data suggest not only that awareness may depend on the access of visual inputs to a critical network in the brain, but also that some interaction within the network may enhance neural responses to these inputs within visual areas, perhaps through reentrant feedback mechanisms (see Walsh, this volume, Chapter 7). Thus, while V1 and ventral temporal cortex are activated by extinguished stimuli, activation is greater when stimuli are seen and elicit concomitant responses in frontoparietal areas. This may accord with the latter areas controlling spatial attention in the normal brain by enhancing activity in early sensory areas (see Kastner, this volume, Chapter 72). Enhancement of visual of visual processing and of stimulus representation in early visual areas by top-down attention has also been observed at the neuronal level in monkey neurophysiology. Right-hemisphere lesions in neglect patients would presumably compromise such modulatory projections for inputs on the contralesional side of space, resulting in their perceptual extinction when there is competition between simultaneous stimulations. This would also be consistent with event-related potentials obtained in the patients (Driver, Vuilleumier, Eimer, and Rees, 2001b;

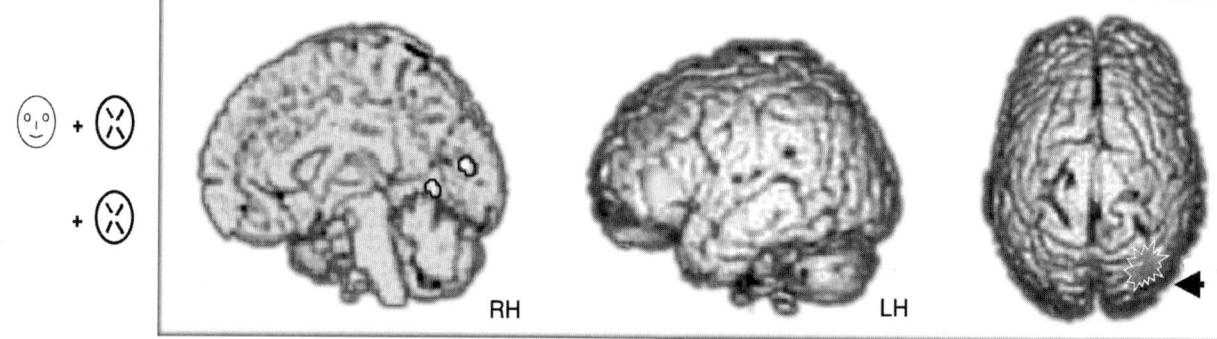

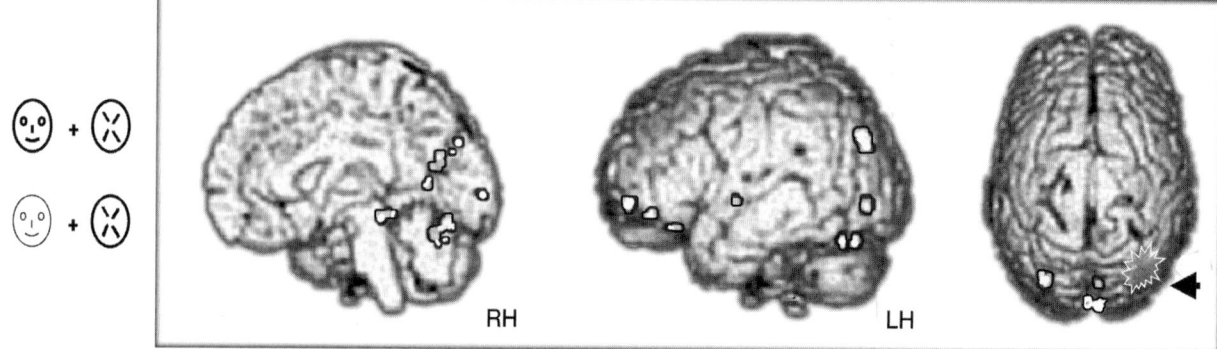

FIGURE 58.1 Neural correlates of residual processing versus conscious awareness in extinction, shown by event-related fMRI in a patient with a chronic right parietal stroke. Lesion location is highlighted by a jagged white contour (black arrow), and activations are displayed as white blobs with black outlines. (*a*) Bilateral events (left face + right shape) where the patient was unaware of the left stimulus, compared with unilateral right stimuli alone, produced residual activation restricted to visual areas of the contralateral occipital and temporal cortex. (*b*) The same bilateral events where the patient correctly detected both stimuli (left face + right shape), compared to left extinction, produced increased activation in early visual areas as well as bilateral fusiform cortex and cuneus, left parietal cortex (homologous to the right lesion), and left frontal cortex.

Marzi et al., 2000; Vuilleumier et al., 2001a), showing that early visual components P1 and N1 (generated in extrastriate cortex around 100 ms poststimulus onset) are present but reduced over the damaged hemisphere for perceived unilateral stimuli, with further suppression on bilateral trials with extinction (Marzi et al., 2000), and with complete abolition of later responses during extinction (Vuilleumier et al., 2001a), such as P200 and P300 (likely generated by a network of higher-level areas in frontal, parietal, and hippocampal regions)

Moreover, a spatial bias can affect the activity of early visual cortex independent of sensory inputs. This is suggested by an additional analysis of our fMRI data in a right parietal patient (from Vuilleumier et al., 2001a) during trials where visual stimuli were expected but not presented (i.e., catch trials with a central fixation cross only, randomly interspersed among trials with right or left peripheral stimuli). Event-related activation was measured for these single central prompts in both the right and left visual cortex, as defined by selective responses to unilateral contralateral stimuli. Such activation was significantly lower in the right than left visual cortex (Fig. 58.2), even without any lateralized stimulus. This may correspond to an abnormal "baseline shift" of activity in the visual cortex of the damaged hemisphere, similar to the retinotopic effect of directed spatial attention during preparatory states in healthy subjects (see Kastner, this volume, Chapter 72). This asymmetry may reflect unbalanced top-down influences due to unilateral parietal lesion, which will then bias the processing of incoming inputs in favor of the ipsilesional side.

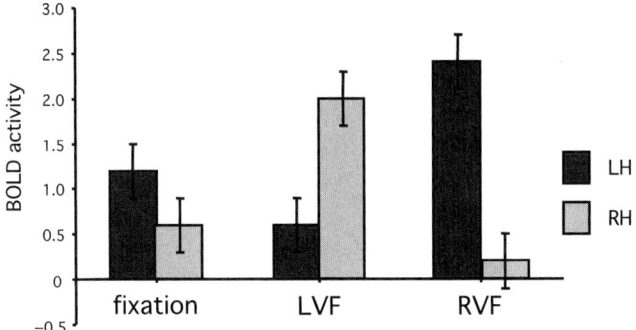

FIGURE 58.2 Event-related activation in visual cortex shown by fMRI for central and peripheral visual events, in the same patient with right parietal stroke and left neglect as in Fig. 58.1. Activity was averaged (across all trials from eight scanning blocks) for occipital clusters of the right (RH) and left hemisphere (LH), where responses were greater for peripheral stimuli in the left (LVF) or right visual field (RVF), respectively. Critically, for a central event consisting of a fixation prompt alone ("catch trials" without peripheral stimuli), activation was lesser in the right than left occipital cortex ($t(7) = 4.62$, $p < 0.05$), suggesting an abnormal bias in "baseline shifts" induced by preparatory spatial attention.

Such an abnormal spatial bias in visual cortex may become worse in these patients when increasing demands on spatial attention. Behaviorally, it is known that tasks requiring more focal attention may indeed aggravate contralesional neglect or extinction (e.g., difficult target discrimination). This is consistent with preliminary findings from a new fMRI study in right parietal patients with left extinction, where visual cortical responses were measured during two different attentional conditions. Checkerboard stimuli were presented in each hemifield, either during epochs of passive fixation on a central stream of coloured stimuli (without any task) or during epochs of an active discrimination task with higher attentional load (e.g., responding to pre-specified colour targets). Increased focusing of attention on the central discrimination task significantly reduced responses to contralesional visual stimuli in the damaged right hemisphere (Fig. 58.3). These decreases occurred even in early retinotopic visual areas (defined beforehand by standard mapping procedure), including V1 and progressively increasing along the ventral visual pathways toward higher areas, such as V4 and beyond. No decreases occurred in the intact left hemisphere under these conditions. In normals, cortical responses to peripheral visual stimuli can also be reduced by central load but during a more difficult task, bilaterally and symmetrically (Schwartz et al., 2004). These results provide direct evidence for a pathological bias in visual responses arising during directed attention in parietal patients, and reveal the damaging functional consequences of such lesions on anatomically intact visual pathways.

IV. CONCLUSION

By combining behavioral and functional neuroimaging approaches, recent studies of spatial neglect and extinction have begun not only to unveil the neurobiological bases of these disorders, but also help to understand fundamental aspects of attentional mechanisms in the brain, converging with studies in healthy people and nonhuman primates. Right parietal damage may suppress awareness of contralesional visual stimuli, even though ventral occipitotemporal pathways remain intact and capable of considerable residual activity. The lesion may disrupt critical interactions with dorsal pathways involved in action planning and working memory, which appear necessary to select and amplify the representation of visual information established in ventral pathways, perhaps through feedback mechanisms. Pathological biases in such interactions might weaken contralesional visual signals and result in their extinction from awareness. Moreover, whereas unconscious processing of extinguished stimuli is mainly confined to ventral visual pathways, awareness correlates with increased activity in a distributed network, involving frontoparietal areas in the intact hemisphere. Future studies are needed to determine whether such increases are a direct cause, rather than one of the consequences, of visual stimuli reaching conscious awareness.

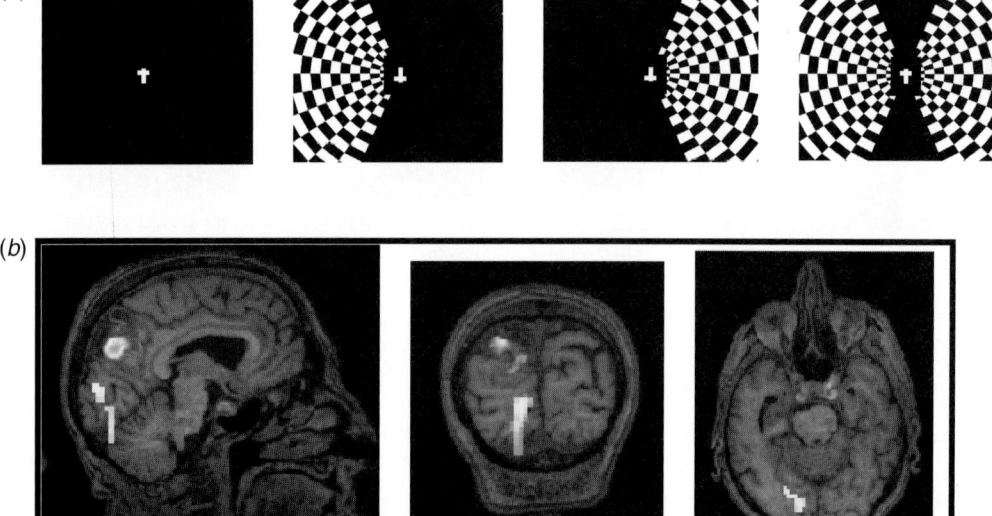

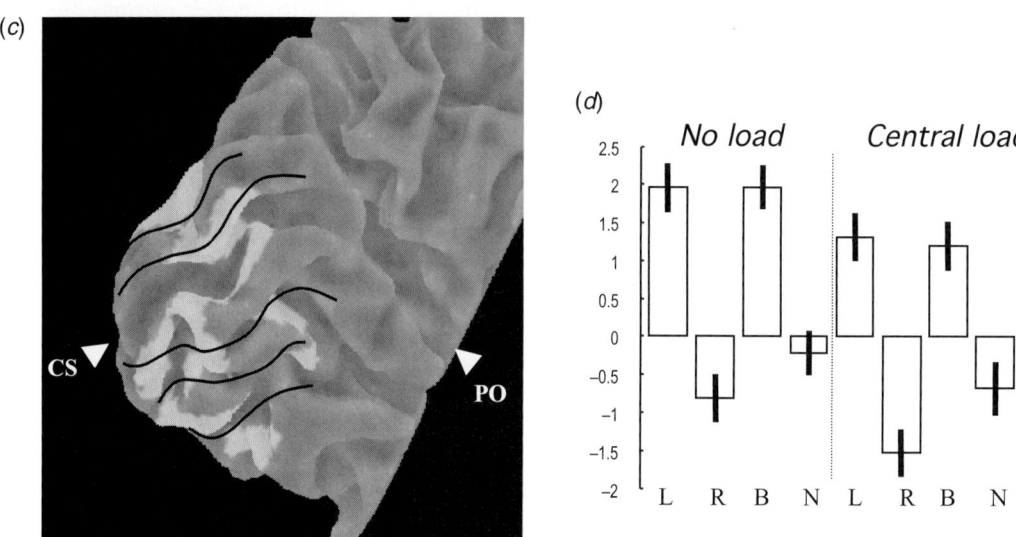

FIGURE 58.3 Reduced response of the right visual cortex during increased attentional load at fixation, as shown by fMRI in a patient with right parietal stroke. Lesion location is indicated by a white arrow. (a) Checkerboard were flashed in the right, left, or both hemifields, or in none, while the patient performed one of two tasks: either passive fixation on rapid central visual stream of letters "T" (low load) or discrimination of a prespecified target colour among the letter sequence (high load). (b) Decreased responses to visual stimuli during high versus low load (depicted by white blobs with black outlines) were found in occipital cortex of the damaged right hemisphere. (c) Peaks of decreases, overlapped on a 3D reconstruction of the segmented cortical surface, showing that central load occurred in early retinotopic areas (as schematically delineated here based on separate mapping scans), including area V1 along the calcarine sulcus (CS). (d) Average activity across different blocks for each stimulation and load condition (R = checkerboard in right visual field; L = checkerboard to the left; B = bilateral stimulation; N = no stimulation). PO = parieto-occipital sulcus.

References

Berti, A., Allport, A., Driver, J., Dienes, Z., Oxbury, J., and Oxbury, S. (1992). Level of processing for stimuli in an "extinguished" visual field. *Neuropsychologia* **30**, 403–415.

Cocchini, G., Cubelli, R., DellaSala, S., and Beschin, N. (1999). Neglect without extinction. *Cortex* **35**, 285–313.

Driver, J., and Vuilleumier, P. (2001a). Perceptual awareness and its loss in unilateral neglect and extinction. *Cognition* **79**, 39–88.

Driver, J., Vuilleumier, P., Eimer, M., and Rees, G. (2001b). Functional MRI and evoked potential correlates of conscious and unconscious vision in parietal extinction patients. *Neuroimage* **14**, 68–75.

Fuentes, L. J., and Humphreys, G. W. (1996). On the processing of "extinguished" stimuli in unilateral visual neglect: An approach using negative priming. *Cognitive Neuropsychol.* **13**, 111–136.

Karnath, H. O., Himmelbach, M., and Kuker, W. (2003). The cortical substrate of visual extinction. *Neuroreport* **14**, 437–442.

Leibovitch, F. S., Black, S. E., Caldwell, C. B., McIntosh, A. R., Ehrlich, L. E., and Szalai, J. P. (1999). Brain SPECT imaging and left hemispatial neglect covaried using partial least squares: the Sunnybrook Stroke study. *Hum. Brain Mapp.* **7**, 244–253.

Marzi, C., Girelli, M., Miniussi, C., Smania, N., and Maravita, A. (2000). Electrophysiological correlates of conscious vision: Evidence from unilateral extinction. *J. of Cog. Neurosci.* **12**, 869–877.

Maunsell J. H, and Cook, E. P. (2002). The role of attention in visual processing. *Philos. Trans. R. Soc. Lond. B. Biol. Sci.* **29**, 357: 1063–1072.

Rees, G., Wojciulik, E., Clarke, K., Husain, M., Frith, C., and Driver, J. (2002). Neural correlates of conscious and unconscious vision in parietal extinction. *Neurocase* **8**, 387–393.

Rees, G., Wojciulik, E., Clarke, K., Husain, M., Frith, C. D., and Driver, J. (2000). Unconscious activation of visual cortex in the damaged right hemisphere of a parietal patient with extinction. *Brain* **123**, 1624–1633.

Schwartz, S., Vuilleumier, P., Hutton, C., Maravita, A., Driver, J., and Dolan, R. J. (2004). Attentional Load and Sensory Competition in Human Vision: Modulation of fMRI Responses by Load at Fixation during Task-irrelevant Stimulation in the Peripheral Visual Field. *Cereb. Cortex.* Sep 30.

Vuilleumier, P., Armony, J., Clarke, K., Husain, M., Driver, J., and Dolan, R. (2002). Neural response to emotional faces with and without awareness: Event-related fMRI in a parietal patient with visual extinction and spatial neglect. *Neuropsychologia* **40**, 2156–2166.

Vuilleumier, P., Sagiv, N., Hazeltine, E., Poldrack, R., Swick, D., Rafal, R., and Gabrieli, J. (2001a). Neural fate of seen and unseen faces in unilateral spatial neglect: A combined event-related fMRI and ERP study of visual extinction. *Proc. Nat. Acad. Sci. USA* **98**, 3495–3500.

Vuilleumier, P., Valenza, N., and Landis, T. (2001b). Explicit and implicit perception of illusory contours in unilateral spatial neglect: behavioural and anatomical correlates of preattentive grouping mechanisms. *Neuropsychologia* **39**, 597–610.

Weiskrantz L. (2004). Roots of blindsight. *Prog. Brain Res.* **144**, 229–241.

Whatham A. R., Vuilleumier, P., Landis, T., and Safran, A. B. (2003). Visual consciousness in health and disease. *Neurol. Clin.* Aug; **21**, 647–686.

CHAPTER 59

Attention in Split-Brain Patients

Todd C. Handy and Michael S. Gazzaniga

ABSTRACT

Attention can take many different forms in human information processing, yet one question central to all domains of attention research concerns understanding how our attentional capacities are controlled. Accordingly, the study of attentional functioning in split-brain patients has long focused on investigating the degree to which attention is independently controlled in the two cerebral hemispheres. In particular, severing the corpus callosum disrupts the ability of the two hemispheres to directly communicate, and as a result, the hemispheres can no longer coordinate their operations in a strategic and organized manner. As a result, each hemisphere in the split-brain patient is thus believed to function in a relatively independent manner, a condition that is ideal for examining the laterality of attentional control processes. For example, if a split-brain patient is given a visual search task to perform, at least three different results may be obtained, each of which is informative regarding the implementation of attentional control processes in cortex. If each hemisphere can independently search the display elements in the contralateral visual hemifield, it would suggest that the control processes mediating visual search exist in parallel within each hemisphere. Conversely, if only one of the two hemispheres shows an ability to perform visual search, it would support the proposal that control processes are lateralized to that hemisphere. Finally, if the split-brain patient is unable to perform the visual search task, it would be consistent with the idea that attentional processes underlying visual search are distributed—or integrated—across the two cerebral hemispheres.

I. INTRODUCTION

In the following chapter we review four specific areas of attention research in split-brain patients that have been germane to understanding the cortical implementation of attentional control processes: (1) the orienting of visual attention to locations in visual space, (2) the control of attention during visual search, (3) the allocation of attentional resources in dual-task situations, and (4) comparing top-down/endogenous/volitional versus bottom-up/exogenous/automatic influences on attentional orienting. In common to all of these research domains has been the question of whether the two cerebral hemispheres show evidence of parallel or asymmetric attentional capacities. When taken together, the collective evidence suggests that the hemispheric control of attentional processes varies as a function of the specific form of attention involved and the neural processes mediating its control.

II. SPATIAL ORIENTING

One of the dominant paradigms for visual attention research has been to ask participants to voluntarily dissociate the focus of their attention from the focus of their gaze. When attention is covertly oriented in this manner, manual responses to stimuli in the attended location are faster and more accurate relative to the responses to stimuli falling outside the focus of gaze—data which has suggested that attention enhances or facilitates the processing of stimuli in attended locations of space. Given this ability to orient attention to nonfoveal locations, researchers have long been interested in understanding how the control of spatial orienting is integrated between the two cerebral hemispheres. Specifically, each hemisphere receives direct sensory input from the contralateral visual hemifield, indicating that a unified representation of visual space requires assimilating information initially projected to separate halves of the brain. If so, what attentional orienting capacities remain intact within each hemisphere when the corpus callosum is severed, and what capacities are disrupted when the two hemispheres can no longer directly communicate?

To address such questions, Reuter-Lorenz and Fendrich (1990) tested the prediction that, due to the necessary integration of information between the two hemispheres, orienting attention *between* the two lateral visual hemifields would have a cost in split-brain patients that would not be observed when these patients were required to orient attention within a single hemifield. Two patients with complete callosotomies were cued to covertly orient their attention to a location in either the left or right visual hemifield. On most trials a target was then presented in the cued location. However, on a small proportion of trials the target appeared at a location that wasn't cued (or attended). Target responses on these latter trials were then analyzed as a function of whether or not the target was presented in the same visual hemifield as the cued (or attended) location. In both patients, Reuter-Lorenz and Fendrich (1990) found that there was a significant increase in the reaction time to uncued targets that occurred in the hemifield opposite of the attended location, relative to uncued targets presented in the same hemifield as the attended location. This suggested that there was a cost uniquely associated with redirecting attention across the vertical meridian of visual space, in comparison to when attention was redirected to a new location within the same hemifield. Interestingly, a third patient who performed the task who had had only a partial callosotomy—leaving the posterior third of the corpus callosum intact—failed to show this same cost of redirecting attention across the vertical meridian. This finding supported the conclusion that normal interhemispheric integration of visuospatial attentional orienting is mediated by the posterior region of the corpus callosum.

Researchers have also studied split-brain patients as a means of investigating what spatial orienting capacities exists within each hemisphere. In other words, if we can covertly orient our attention to nonfoveal locations within the visual field, is each hemisphere equally adept at performing this function, or are there systematic asymmetries in each hemisphere's orienting abilities? Mangun and colleagues examined this issue by asking split-brain patients to covertly orient their attention to the location of a peripheral visual cue that was in either the left or right visual field and that predicted the most likely location of an upcoming target (Mangun et al., 1994). Of interest was comparing the reaction times to these targets as a function of the target's visual field and whether or not the target's location had been cued (or was attended). For targets in the left visual field, reaction times were faster for cued relative to uncued targets—a pattern suggesting that the right cerebral hemisphere oriented attention to the left visual field when that location was cued. In contrast, the reaction times to targets in the right visual field were unaffected by the cuing condition, and further, they were equally as fast as the cued (or attended) targets in the left visual field. Mangun et al. (1994) interpreted these data as supporting the hypothesis that in the split-brain patients, the left cerebral hemisphere oriented attention to the right visual field on every trial, independent of whether or not a cue was presented in that visual field.

Taken together, the results of the study by Mangun et al. (1994) were thus consistent with the proposal that whereas the right cerebral hemisphere can orient attention to the left or right visual field depending on the cuing condition, the left cerebral hemisphere is strongly biased toward maintaining covert attention in the right visual field. Importantly, as Mangun et al. (1994) point out, this hemispheric asymmetry in attentional orienting capacities may help explain why lesions of the right hemisphere typically produce more dramatic and consistent deficits in left visual field awareness, relative to the effects of left hemisphere lesions on right visual field awareness—if the right hemisphere is damaged, the intact left hemisphere will only be paying attention to the right visual field, but if the left hemisphere is damaged, the intact right hemisphere can still orient attention to either visual field.

III. VISUAL SEARCH

Given our capacity for orienting attention to locations in visual space, a critical use for this capacity concerns identifying an object of interest that may be embedded within a cluttered display containing a variety of "distractor" objects. When confronted with such situations the focus or "spotlight" of attention may be swept around the visual scene in a systematic fashion, moving from item to item until the target—or object of interest—is identified. In other words, models of covert visual search posit that search tasks are predicated on strategically controlling a unitary attentional focus that is used to serially scan the multi element array. Accordingly, a key issue arising in this literature has been understanding the degree to which search abilities depend on integrating functions between the two cerebral hemispheres, or whether each hemisphere has the capacity to control search independently within the contralateral visual hemifield.

Luck and colleagues investigated this issue by asking callosotomy patients to perform a visual search task that held the number of display elements constant, but varied whether they were all contained within a single lateral visual hemifield, or whether they were dispersed across both visual hemifields (Luck, Hillyard, Mangun, and Gazzaniga, 1989). The experimental design was based on the notion that as the number of nontarget distractors is increased in a visual search display, search becomes less efficient and a linear increase is observed in the amount of time it takes to identify the target owing to the increase in the number of items (or distractors) that will likely need to be scanned before the target is identified. Luck et al. (1989) reasoned that if efficient search depends on integrating functions between the two hemispheres, search times for split-brain patients should be unaffected by whether or not all the elements in the display were contained within a single visual hemifield. On the other hand, if each hemisphere of the split-brain patient can perform search independently, then the study predicted search times should be reduced when display elements were equally distributed between the two lateral visual hemifields. Why? In this situation, not only would each hemisphere be simultaneously performing a search within the contralateral hemifield, but the effective number of items in each search would be halved, relative to when all the items were in a single hemifield.

Luck et al. (1989) found evidence consistent with the hypothesis that each hemisphere can in fact perform visual search independent of the other hemisphere. When display elements were dispersed across the entire visual field, the search time was significantly reduced relative to the time taken to find the target when all display elements were in a single lateral hemifield. These data thus supported the conclusion that visual search is a process that can be performed in parallel by the surgically separated cerebral hemispheres. Interestingly, a similar conclusion was made by Mangun et al. (1994) regarding the orienting of attention in response to a spatial cue. In particular, Mangun et al. (1994) reported that each cerebral hemisphere in a split-brain patient could be simultaneously cued to orient attention to the contralateral visual hemifield. In conjunction with the data reported by Luck et al. (1989; see also Luck et al., 1994), the collective evidence thus supports the proposal that in split-brain patients, each hemisphere is in control of its own independent attentional focus—"spotlights" that can be oriented in a parallel in response to either spatial cues or the demands of a visual search task. However, as discussed above, the right cerebral hemisphere may be unique in its ability to orient attention to the ipsilateral visual hemifield. Given this understanding of visuospatial attentional capacities, we now turn to more general models of attentional processing that have been examined with split-brain patients.

IV. RESOURCES AND DUAL-TASK PERFORMANCE

A central theme in most models of human information processing is that it is "capacity limited," in that there is a finite limit on how much information we can processes—or how many things we can simultaneously do—at any given moment. In brief, performing a perceptual, cognitive, or motor task is assumed to consume some amount of attentional "resources," which exist in the brain as a fixed-volume "pool" upon which all processing operations draw. One's ability to perform a task thus depends on the efficiency of the processing strategy adopted and the percentage of attentional resource capacity (or volume) allocated to the task. As a consequence, if a single task being performed becomes more difficult or if multiple tasks are engaged simultaneously, task performance will necessarily decline at that point where resource demand exceeds the available supply. Given this basic theoretical framework, a critical question of interest has been whether a separate "pool" of attentional resources exists within each cerebral hemisphere, or whether the two hemispheres draw upon a common resource capacity.

As one approach to resolving this issue, Holtzman and Gazzaniga (1982) asked a split-brain patient to perform a working memory task that involved presenting a set of three geometric figures (e.g., triangle or circle) to each cerebral hemisphere, while varying as an independent variable whether each hemisphere received an "easy" or "hard" version of the task. In the easy condition the three figures seen by the given hemisphere were the same (e.g., three circles), while in the hard condition the three figures were all different (e.g., a circle, triangle, and square). After each hemisphere saw its "test" set of three figures to memorize, the memory performance of one of the two hemispheres was tested by asking it whether a subsequent target figure matched one of the three figures in the memorized "test" set. Holtzman and Gazzaniga (1982) predicted that if each hemisphere draws upon its own resource capacity, the performance of the task in one cerebral hemisphere should be unaffected by the difficulty level of the task performed concurrently by the other cerebral hemisphere. Alternatively, if the two hemispheres share a common pool of attentional resources, it predicted that task performance in the two hemispheres should interact, such that performance in one hemisphere should be worse when the other hemisphere is performing the hard relative to easy version of the task. Consistent with this latter possibility, the results revealed that the speed of the target responses was slower, and the accuracy of the responses (old vs.

new figure) was lower when the hemisphere not responding on that trial was performing the difficult version of the task, relative to the easy version.

The data of Holtzman and Gazzaniga (1982) thus support the hypothesis that the two cerebral hemispheres draw on a common pool of attentional resources. Interestingly, evidence in favor of this hypothesis was obtained even though the two hemispheres in the split-brain patient have no ability to share information via direct callosal connections. This suggests that resources may either reside in subcortical structures or may be transferable between hemispheres via subcortical connections, and further, that resources are not specialized for the specific processing operations that are uniquely lateralized within each hemisphere.

V. TOP-DOWN VS. BOTTOM-UP CONTROL

In addition to resource models, another general way in which attention can be conceptualized is in terms of how it is controlled. On the one hand, top-down attentional control concerns situations where a voluntary or willful decision is made regarding where to orient attention or what tasks to pay attention to—such as deciding to listen to one speaker in a crowded and noisy room while ignoring all other voices that may be impinging upon the auditory system. On the other hand, bottom-up attentional control refers to conditions where a stimulus automatically draws our attention to it in a reflexive or automatic manner—such as occurs when our attention is drawn to a flashing ad on the margin of web page, or the flashing lights of a police car up ahead in the distance. Although in each of these control conditions attention is ultimately drawn to a stimulus of interest, the central idea is that the underlying neural mechanism by which attention is influenced differs in the two situations. In short, bottom-up control is believed to be mediated by subcortical structures that can rapidly orient our attention to salient stimulus events, while top-down control appears to be a slower-acting influence on attention that involves mediation by prefrontal cortical areas (e.g., Desimone and Duncan, 1995).

Given this dichotomy between top-down and bottom-up influences on attention, growing evidence from split-brain patients has been consistent with the hypothesis that each cerebral hemisphere can orient attention independently in response to bottom-up attentional cues, but that when attention is controlled by volitional decisions or top-down influences, the hemispheres appear to compete for control of a unitary focus. For example, Kingstone (1995) directly compared visuospatial attentional orienting to the left or right visual field in a split-brain patient under two conditions: either under top-down control in response to a spatial cue that predicted the most likely location of an upcoming target, or under bottom-up control in response to the brief flashing of a stimulus in the visual periphery that was unpredictive of an upcoming target. In the former condition, reaction times to targets were found to be faster on "cued" relative to "neutral" trials (or trials where both target locations were cued simultaneously). Conversely, in the latter condition, reaction times to targets were equivalent on "cued" and "neutral" trials—and importantly, both of these reaction times were faster relative to "uncued" trials, where a target was presented to a hemisphere not expecting a target. Kingstone (1995) reasoned that this lack of a difference in reaction times between "cued" and "neutral" trials reflected an ability of each hemisphere to reflexively orient independently, an independence not possible under volitional orienting conditions.

More recent split-brain studies have dissociated between different forms of reflexive orienting within the cerebral hemispheres. In particular, there is a strong tendency to reflexively orient one's attention to the location where someone else is looking. However, Kingstone, Friesen, and Gazzaniga (2000) have reported that in split-brain patients, only the hemisphere dominant for face processing will reflexively orient attention to the location of eye gaze direction. This finding suggested that a dissociation exists between reflexive orienting associated with brief flashes in the visual periphery (e.g., as studied by Kingstone, 1995) and reflexive orienting to eye gaze. A similar conclusion was reached by Ristic, Friesen, and Kingstone (2002), who replicated the hemispheric asymmetry in eye-gaze orienting in split-brain patients while demonstrating that both hemispheres will reflexively orient to the location indicated by a non-spatially predictive arrow cue. Taken together, the collective data thus suggest not only differences between the hemispheres in how they can volitionally orient to locations in space, but differences as well in how they may—or may not—reflexively orient to events in the visual world.

VI. CONCLUSIONS

The study of attention in split-brain patients has been critical to the effort to understand what attentional capacities reside within each of the cerebral hemispheres and how those capacities are integrated.

Within this context, the proposal that hemispheric independence depends on whether top-down or bottom-up influences are mediating attentional control has the advantage of being consistent with the putative neural systems involved (see Enns and Kingstone, 1997; Kingstone et al., 1997). The cerebral hemispheres may be labile to independent control via bottom-up influences because these influences presumably derive from subcortical inputs. In this sense, independence may arise when the hemispheres themselves are not directly responsible for deciding how or where to orient attention. Conversely, the hemispheres may compete for attentional control when voluntary or conscious decisions are made regarding what to do with attention, situations where the hemispheres are directly involved in the top-down decision-making process. If so, the idea that the attentional focus is unitary within split-brain patients may only be an illusion of competition for conscious attentional control. Under conditions where both hemispheres are in voluntary agreement as to what to do—for example, performing a visual search task (e.g., Luck et al., 1989, 1994)—then the hemispheres may be equally free to operate independently as they would be under strict bottom-up control.

References

Desimone, R., and Duncan, J. (1995). Neural mechanisms of selective visual attention. *Ann. Rev. Neurosci.* **18**, 193–222.

Enns, J., and Kingstone, A. (1997). Hemispheric coordination of spatial attention. *In* "Cerebral symmetries in sensory and perceptual processing." (S. Christman, Ed.), Elsevier, New York, 197–231.

Holtzman, J. D., and Gazzaniga, M, S. (1982). Dual task interactions due exclusively to limits in processing resources. *Science* **218**, 1325–1327.

Kingstone, A. (1995). Covert orienting and the cerebral hemispheres: Solution to a paradox? *Cog. Neurosci. Soc. Abs.* **2**, 21. San Francisco, CA.

Kingstone, A., Friesen, C. K., and Gazzaniga, M. S. (2000). Reflexive joint attention depends on lateralized cortical connections. *Psychol. Sci.* **11**, 159–166.

Kingstone, A., Grabowecky, M., Mangun, G. R., Valsangkar-Smyth, M., and Gazzaniga, M. S. (1997). Paying attention to the brain. The study of selective visual attention in cognitive neuroscience. *In* "Attention, Development, and Psychopathology" (J. Burak and J. Enns, Eds.), Guilford, New York, 263–287.

Luck, S. J., Hillyard, S. A., Mangun, G. R., and Gazzaniga, M. S. (1989). Independent hemispheric attentional systems mediate visual search in split-brain patients. *Nature* **342**, 543–545.

Luck, S. J., Hillyard, S. A., Mangun, G. R., and Gazzaniga, M. S. (1994). Independent attentional scanning in the separated hemispheres of split-brain patients. *J. Cog. Neurosci.* **6**, 84–91.

Mangun, G. R., Luck, S. J., Plager, R., Loftus, W., Hillyard, S. A., Handy, T. C., Clark, V. P., and Gazzaniga, M. S. (1994). Monitoring the visual world: Hemispheric asymmetries and subcortical processes in attention. *J. Cog. Neurosci.* **6**, 267–275.

Reuter-Lorenz, P. A., and Fendrich, R. (1990). Orienting attention across the vertical meridian: Evidence from callosotomy patients. *J. Cog. Neurosci.* **2**, 232–239.

Ristic, J., Friesen, C. K., and Kingstone, A. (2002). Are eyes special? It depends on how you look at it. *Psychonomic Bul. & Rev.* **9**, 507–513.

CHAPTER 60

Divided Attention in the Normal and the Split Brain: Chronometry and Imaging

Marco Iacoboni

ABSTRACT

Divided attention is the ability to integrate in parallel multiple stimuli. A relevant experimental effect that has been studied for almost a century is the redundant target effect. When multiple copies of the same stimulus are presented to subjects, in choice, go no-go, and even a simple reaction time task, reaction times (RT) tend to be faster, compared to RT to a single copy of the stimulus. Paradoxically, this effect is larger in split-brain patients when two stimuli are presented in the two opposite hemifields. Recent RT and imaging studies reviewed here suggest that cortico-subcortical interactions between the superior colliculus and the extrastriate cortex, which are modulated by the corpus callosum, are reflected in different levels of activation in dorsal premotor cortex during divided attention tasks and can account for the paradoxical facilitation observed in split-brain patients.

I. INTRODUCTION

To achieve a flexible and adaptive behavior, we must coordinate our activities with the surrounding world. This requires an efficient processing of the large number of stimuli that we receive. Often, we must process stimuli in parallel and hopefully integrate these processes in a unitary behavior. The study of the human capacity to deal with multiple stimuli is certainly daunting. Psychologists and neuroscientists have devised a variety of clever paradigms to study cognitive and neural mechanisms of parallel processing and divided attention. The paradigm I would like to focus here is called Redundant Target Effect (RTE) (Fig. 60.1). Here, response times to the detection of multiple copies of the same stimulus are compared to response times to the detection of a single copy of the stimulus (Todd, 1912). Typically, response times to multiple copies of the stimulus are faster than response times to a single stimulus. This difference in response times is called the Redundancy Gain (RG).

II. ACCOUNTS OF REDUNDANCY GAIN

Two main accounts of RG have been provided. One account argues that RG occurs because multiple copies of the stimulus initiate multiple independent processes (Raab, 1962). The sum of the probability that each of these multiple independent processes reaches the threshold for motor response is higher than the probability that a single process reaches the threshold for motor response. This model is called *race model* because an effective analogy comes from horse races: if you take a series of horse races, the average time of the winners is shorter than the average time of each horse participating in the races. An alternative account is called co-activation model (Miller, 1982). This model assumes that multiple copies of the stimulus initiate processes that reinforce each other, rather than running in parallel as in the race model. The co-activation of these processes determines RG. Miller (Miller, 1982) proposed a now widely adopted approach to test these two alternative models. It turns out that one can calculate the upper boundary of probability summation. If RG exceeds this boundary (this is called "violation of race models"), then race models cannot account for RG and a co-activation must have occurred. This is calculated by using the cumulative distribution function (CDF) of the response times to single copies of the stimulus. The sum of these CDF determines the upper boundary of probability summation (Fig. 60.2)

A specific version of the RTE paradigm has recently generated several chronometric studies in normal subjects and in patients with callosal lesions, mostly because of some seemingly paradoxical results. Moreover, this research has also generated some electrophysiological and brain imaging studies of the neural correlates of RG. This article will review these recent results and will discuss a model that may provide a unitary interpretation of the main empirical findings.

III. PARADOXICAL INTERHEMISPHERIC RG INCREASE IN THE SPLIT BRAIN

Reaction times (RTs) to lateralized flashes that are presented simultaneously and bilaterally to both visual hemifields are faster than RTs to a single flash presented unilaterally, even when responses are made with the hand ipsilateral to the single lateralized flash. In normal subjects, however, this interhemispheric RG does not usually yield violation of race models. Paradoxically, in the split brain the interhemispheric RG is typically much larger than in normal subjects, and typically yields violation of race models (Reuter-Lorenz et al., 1995). This evidence had been taken as suggesting that the weak interhemispheric RG effect is mediated by an inhibitory role of the corpus callosum, the largest commissure of the human brain connecting the two cerebral hemispheres (Reuter-Lorenz et al., 1995). The lack of the corpus callosum in split-brain patients who underwent callosotomy or in patients with callosal agenesis would then result in a release from callosal inhibition and a large interhemispheric RG. In keeping with this prediction, reports of interhemispheric RG violating race models in both callosotomy and callosal agenesis patients are now available (Corballis, 1998; Corballis et al., 2002; Iacoboni et al., 2000). Moreover, callosotomy and acallosal patients showing large interhemispheric RG violating race model do not show race model violation when the two stimuli are presented to the same visual hemifield (Corballis et al., 2002). Taken together, this evidence seems to suggest a critical role of callosal inhibition in the reduced interhemispheric RG observed in normal subjects.

There is also evidence, however, that does not fit with this simple model. First, some patients lacking the corpus callosum do show RG that does not yield violation of race models (Corballis et al., 2004). Second, normal subjects presented with two lateralized stimuli, one of which is below threshold for conscious detection, do show a RG violating race models (Savazzi and Marzi, 2002). Both types of evidence suggest that a simple inhibitory role of the corpus callosum cannot account for the paradoxically larger interhemispheric RG observed in some callosotomy and acallosal patients.

FIGURE 60.1 The lateralized version of RTE tasks. One lateralized flash is presented to the right, to the left, or to both visual fields (redundant targets condition).

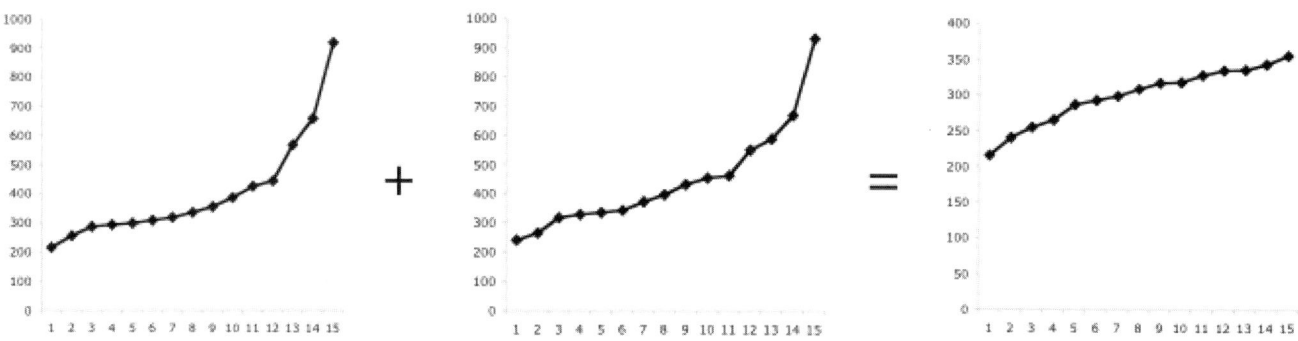

FIGURE 60.2 The sum of the CDF of each stimulus represents the upper boundary of race models (the rightmost CDF in the figure). When the CDF of redundant targets goes below this upper boundary, a violation of race models occurs.

A possible role of subcortical structures in this phenomenon is suggested by the observation of RG in patients with hemispherectomy (Tomaiuolo et al., 1997). Among subcortical structures, the superior colliculus is a likely candidate structure mediating this effect. At single-cell level, collicular neurons are known to show multiplicative effects when two stimuli are presented simultaneously in their receptive fields (Stein and Meredith, 1993). Moreover, when normal subjects were tested using three different types of motor responses, vocal, manual, and saccadic, saccadic responses yielded the largest RG violating race models (Hughes et al., 1994), suggesting that the superior colliculus, strongly associated with oculomotor behavior, is a major locus of RG. Finally, RG violating race models disappear in split-brain patients when the stimuli used are equiluminant with the background (Corballis, 1998), an experimental condition that should restrict processing to the cortical parvocellular system. However, RGs violating race models do not depend on symmetric location of the stimuli in the two visual fields (Roser and Corballis, 2002), thus making it unlikely that the superior colliculus, which is organized in a retinotopic fashion, is the only responsible structure for the effect.

If the corpus callosum, the major cortical commissure of the human brain, and the superior colliculus, a subcortical structure, are both implicated in RG, but neither one seems sufficient to produce the effect, then it is possible that cortico-subcortical interactions play a major role in producing and modulating RG. In the largest series of patients with callosal lesions studied so far on RG, it was found that patients with interhemispheric transfer time around 20 msec or longer, regardless of their callosal pathology, had large RGs violating race models, whereas patients with interhemispheric transfer time shorter than 15 msec had RGs not violating race models (Iacoboni et al., 2000). Note that in the normal brain interhemispheric transfer time is estimated around 4 msec (Marzi et al., 1991). How do we explain this result? If one considers the oscillatory patterns of cortical activity in the gamma band and the essential role of the corpus callosum in it (Munk et al., 1995), and if one takes into account that oscillatory systems can be phase-locked only if the conduction delay between them is less than one-third of the duration of the oscillatory cycle (Konig and Schillen, 1991), then long interhemispheric transfer times would interfere with phase-locked interhemispheric oscillations. This would result in asynchronous cortical activity that, summed over time, would produce a larger cortical input over the superior colliculus, producing a stronger reentrant signal from the colliculus back to extrastriate areas. Anatomically, the extrastriate cortex connects to frontal areas. Thus, the greater extrastriate activity would then generate stronger premotor activation, producing RG.

Recent functional Magnetic Resonance Imaging (fMRI) data support this model. Two patients with callosal agenesis were studied with fMRI (Iacoboni et al., 2000). One patient (J.L.) had long interhemispheric transfer time and large RG violating race models, and the other patient (M.M.) had short interhemispheric transfer time and RG not violating race models. The fMRI study demonstrated extrastriate activation in J.L. but not in M.M. when brain activity during detection of two simultaneous lateralized light flashes was compared to brain activity during a control task (Iacoboni et al., 2000). Taken together, the chronometric and imaging data in normal subjects and split-brain patients suggest that the paradoxically larger RG observed in the split brain compared to the normal brain is not simply due to removal of callosal inhibition, but rather to cortico-subcortical interactions between extrastriate cortical areas and the superior colliculus. In these interactions, the role of the corpus callosum would be to synchronize cortical activity that regulates collicular activity. Recent data on partial callosotomy patients support this conclusion (Corballis et al., 2004). Anterior callosal sections are associated with normal RG not violating race models, whereas posterior callosal sections, severing visual callosal fibers, are associated with large RG violating race models. However, an open question remains: what are the functional and neural correlates of the RG observed in normal subjects with intact corpus callosum? The next section of the chapter will discuss findings relevant to this issue.

IV. THE FUNCTIONAL AND NEURAL LOCUS OF RG IN THE NORMAL BRAIN

Recent studies using electrical scalp recordings and brain imaging techniques based on blood flow have investigated the neural locus of RG in the normal brain. However, before discussing these studies, it is useful to address a series of behavioral experiments that investigated the functional locus of the effect. In principle, the effect can occur at a sensory level, at a central, cognitive, decisional level, or at a motor level. When RG to multimodal (auditory and visual) stimuli is compared to RG to unimodal (typically, both visual) stimuli, RG to multimodal stimuli is much larger than RG to unimodal stimuli (Miller, 1982). This evidence suggests that the effect does not occur at an early sensory level. In fact, even within the visual modality, RG is larger for cross-dimensional tasks (for instance, color and shape). Some evidence suggests that the functional locus of the effect is at a late motoric stage. Intermanual reaction time differences during bimanual responses decrease in trials with redundant targets (Diederich and Colonius, 1987). Morever, response force increases when responding to redundant targets (Giray and Ulrich, 1993). However, some evidence does suggest that the locus of the effect is not at the very late stage of motor execution.

A study of the effect applied to latencies of single cells in primary motor cortex of macaques performing the task shows that the effect occurs before the neural level of primary motor cortex (Miller et al., 2001). Furthermore, in a paradigm in which subjects are asked to refrain from responding when presented with stop signals, redundant stop signals are more effective than single-stop signals in inhibiting a motor response in normal subjects. This suggests that RG occurs before motor plans are actually executed (Cavina-Pratesi et al., 2001).

Two electrical scalp-recording studies have used the redundant target paradigm (Miniussi et al., 1998; Murray et al., 2001). Both studies provide evidence for a relatively early locus of RG. Even though slightly different paradigms were used and slightly different recording and processing techniques were adopted, both studies provide evidence that the earliest detectable site of RG is at the extrastriate level. Obviously, electrical scalp recordings do not yield precise cortical localization, so it is difficult to establish, on the basis of these studies, whether the observed effect originates in the occipital, in the posterior temporal, or in the inferior parietal cortex. A good spatial localization is, however, provided by fMRI. The only study on RG in normal subjects that adopted fMRI has provided results somewhat different from the results reported in the electrical scalp recording studies. Blood oxygenation level dependent (BOLD) fMRI signal was shown to increase in three cortical areas when trials with redundant targets were compared to trials with single targets (Iacoboni and Zaidel, 2003). These three regions were the left and the right dorsal premotor cortex and the right intra-parietal sulcus. This latter activation may be compatible with the observation obtained by the electrical scalp-recording studies. However, in the fMRI study the right intraparietal area demonstrated similar activity for redundant targets and unilateral left visual hemifield targets, thus suggesting that this area may simply reflect generic attentional processing directed to the contralateral left visual hemifield, rather than real RG. In contrast, the two dorsal premotor activated areas clearly demonstrate increased signal for redundant targets compared to unilateral targets in both left and right visual hemifields. In keeping with these findings, one of the two electrical scalp-recording studies also shows evidence of RG in central electrodes (Miniussi et al., 1998).

How do we explain the seemingly different results obtained by electrical scalp recordings and fMRI? It is possible that the fMRI study better detected local processing at the premotor level, whereas the electrical scalp-recording studies were able to detect neuronal output from posterior areas that was sent to dorsal premotor regions. Parietal and premotor areas are strongly interconnected in the primate brain and play a major role in several other aspects of attentional behavior. Thus, seemingly different results may be explained by the different sensitivity of the techniques used, to different aspects of cortical processing. Taken together, the evidence from electrical scalp recordings and fMRI suggests that RG in the normal brain likely occurs in parieto-premotor networks. It is also possible that activation within this large network likely percolates from posterior to anterior regions on the basis of task characteristics, stimulus type, and cognitive strategies adopted during redundant target paradigms. The definition of the factors that determine specific activations in specific sites of the network will be the next major question to be addressed by imaging studies adopting the redundant target paradigm.

Acknowledgments

Supported, in part, by the Brain Mapping Medical Research Organization, Brain Mapping Support Foundation, Pierson-Lovelace Foundation, The Ahmanson Foundation, Tamkin Foundation, Jennifer Jones-Simon Foundation, Capital Group Companies Charitable Foundation, Robson Family, William M. and Linda R. Dietel Philanthropic Fund at the Northern Piedmont Community Foundation, Northstar Fund, the National Center for Research Resources grants RR12169, RR13642 and RR08655, and NIH grant NS-20187.

References

Cavina-Pratesi, C., Bricolo, E., Prior, M., and Marzi, C. A. (2001). Redundancy gain in the stop-signal paradigm: implications for the locus of coactivation in simple reaction time. *J. Exp. Psychol.: Hum. Percept. Perform.* **27**, 932–941.

Corballis, M. C. (1998). Interhemispheric neural summation in the absence of the corpus callosum. *Brain* **121**, 1795–1807.

Corballis, M. C., Corballis, P. M., and Fabri, M. (2004). Redundancy gain in simple reaction time following partial and complete callosotomy. *Neuropsychologia* **42**, 71–81.

Corballis, M. C., Hamm, J. P., Barnett, K. J., and Corballis, P. M. (2002). Paradoxical interhemispheric summation in the split brain. *J. Cogn. Neurosci.* **14**, 1151–1157.

Diederich, A., and Colonius, H. (1987). Intersensory facilitation in the motor component? A reaction time analysis. *Psychol.: Res.* **49**, 23–29.

Giray, M., and Ulrich, R. (1993). Motor coactivation revealed by response force in divided and focused attention. *J. Exp. Psychol.: Hum. Perc. Perform.* **19**, 1278–1291.

Hughes, H. C., Reuter-Lorenz, P. A., Nozawa, G., and Fendrich, R. (1994). Visual-auditory interactions in sensorimotor processing: saccades versus manual responses. *J. Exp. Psychol.: Hum. Perc. Perf.* **20**, 131–153.

Iacoboni, M., and Zaidel, E. (2003). Interhemispheric visuo-motor integration in humans: the effect of redundant targets. *Eur. J. Neurosci.* **17**, 1981–1986.

Iacoboni, M., Ptito, A., Weekes, N. Y., and Zaidel, E. (2000). Parallel visuomotor processing in the split brain: cortico-subcortical interactions. *Brain* **123** (Pt 4), 759–769.

Konig, P., and Schillen, T. B. (1991). Stimulus-dependent assembly formation of oscillatory responses. I. Synchronization. *Neural Comput.* **3**, 155–166.

Marzi, C. A., Bisiacchi, P., and Nicoletti, R. (1991). Is interhemispheric transfer of visuomotor information asymmetric? Evidence from a meta-analysis. *Neuropsychologia* **29**, 1163–1177.

Miller, J. (1982). Divided attention: Evidence for coactivation with redundant signals. *Cogn. Psychol.* **14**, 247–279.

Miller, J., Ulrich, R., and Lamarre, Y. (2001). Locus of the redundant-signals effect in bimodal divided attention: a neurophysiological analysis. *Percept. Psychophys.* **63**, 555–562.

Miniussi, C., Girelli, M., and Marzi, C. A. (1998). Neural site of the redundant target effect: electrophysiological evidence. *J. Cogn. Neurosci.* **10**, 216–230.

Munk, M. H. J., Nowak, L. G., Nelson, J. I., and Bullier, J. (1995). Structural basis of cortical synchronization. II. Effects of cortical lesions. *J. Neurophysiol.* **74**, 2401–2414.

Murray, M. M., Foxe, J. J., Higgins, B. A., Javitt, D. C., and Schroeder, C. E. (2001). Visuo-spatial neural response interactions in early cortical processing during a simple reaction time task: a high-density electrical mapping study. *Neuropsychologia* **39**, 828–844.

Raab, D. H. (1962). Statistical facilitation of simple reaction times. *Trans. N.Y. Acad. Sci.* **24**, 574–590.

Reuter-Lorenz, P. A., Nozawa, G., Gazzaniga, M. S., and Hughes, H. C. (1995). Fate of neglected targets: a chronometric analysis of redundant target effects in the bisected brain. *J. Exp. Psychol.: Hum. Perc. Perf.* **21**, 211–230.

Roser, M., and Corballis, M. C. (2002). Interhemispheric neural summation in the split brain with symmetrical and asymmetrical displays. *Neuropsychologia* **40**, 1300–1312.

Savazzi, S., and Marzi, C. A. (2002). Speeding up reaction time with invisible stimuli. *Curr. Biol.* **12**, 403–407.

Stein, B. E., and Meredith, M. A. (1993). The merging of the senses. Cambridge, MA, MIT Press.

Todd, J. W. (1912). Reaction to multiple stimuli. The Science Press. New York.

Tomaiuolo, F., Ptito, M., Marzi, C. A., Paus, T., and Ptito, A. (1997). Blindsight in hemispherectomized patients as revealed by spatial summation across the vertical meridian. *Brain* **120**, 795–803.

SECTION III

MECHANISMS

COLOR PLATE

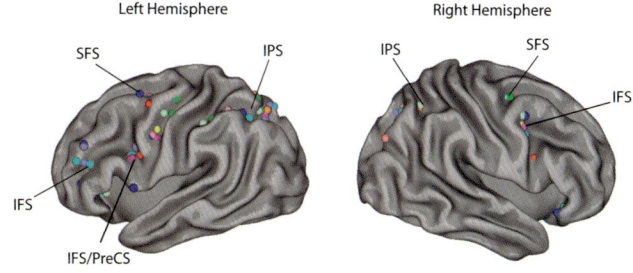

FIGURE 11.2 Foci included in the meta-analysis. Each color represents a different study and contrast (see Table 11.1). The foci are projected onto an inflated representation of the left and right hemispheres of a brain that was normalized to stereotactic space. SFS, superior frontal sulcus; IPS, intraparietal sulcus; IFS, inferior frontal sulcus; PreCS, precentral sulcus.

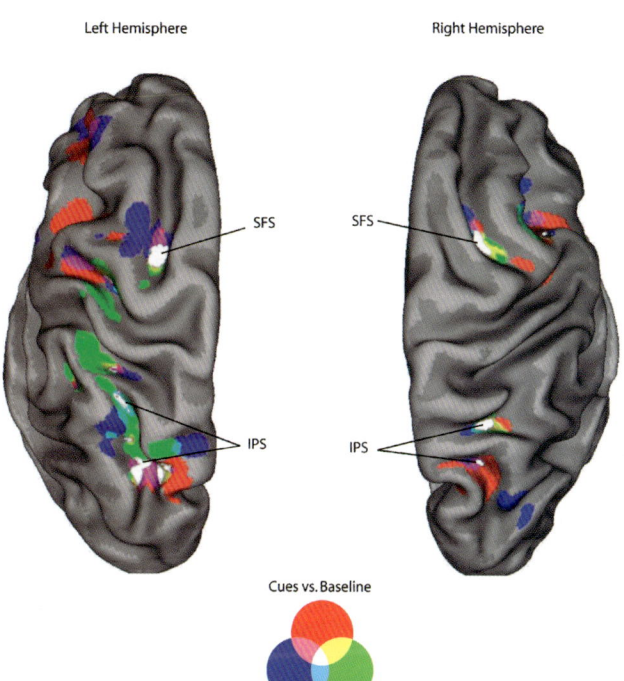

FIGURE 11.3 Anatomical overlap between the three categories of contrasts. Each focus was categorized into either the cues versus baseline (red), cues versus passive (green), or cues versus active (blue) groups, projected onto the inflated brain, smoothed (8 mm), and painted onto the surface. White represents the intersection of all three categories. Abbreviations are as in Fig. 11.2.

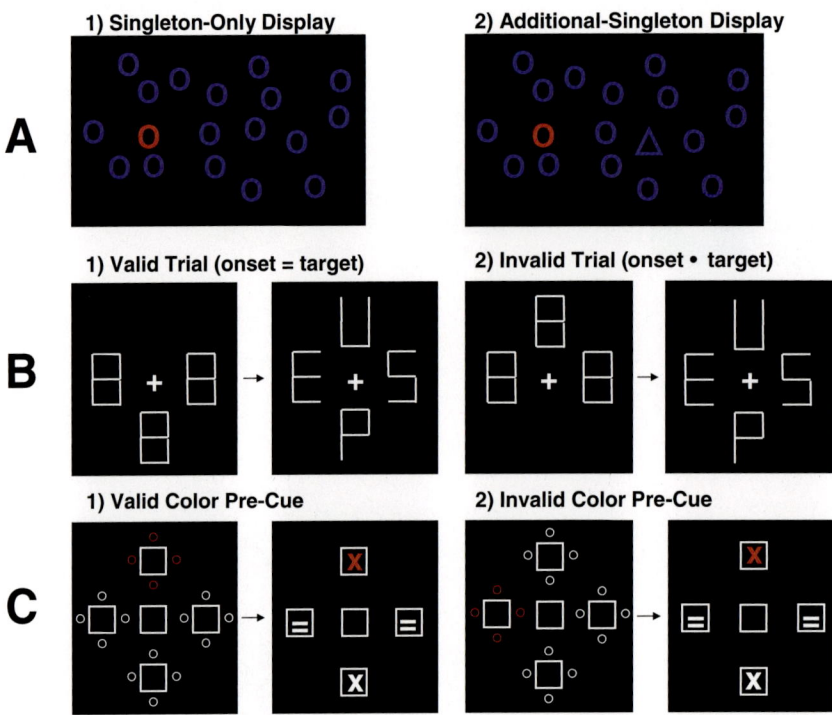

FIGURE 12.1 The three major paradigms used to study implicit attention capture. (A) Additional singleton paradigm. In the first search array, subjects search for a red circle among blue circles. In the second panel, in addition to the blue distracter circles, a blue triangle appears in the display. If this additional singleton slows search for the red circle, then it captured attention according to the logic of this task. (B) Irrelevant feature search paradigm. The initial display on both valid and invalid trials consists of a set of figure-8 premasks. These masks are then replaced by letters, and simultaneously, an additional letter appears in a location not previously occupied by a mask (an abrupt onset). Subjects try to determine whether the display contains an H or a U. If the abrupt onset captures attention, it should be searched first. When the abruptly onsetting item happens to be an H or U (a valid trial), then search should be equally efficient regardless of the number of distracters in the display. (C) Irrelevant pre-cue paradigm. A distinctive and nonpredictive cue appears in the location subsequently occupied by one of the search elements. If the features of the cue match those of the target, and the cue appears in a location subsequently occupied by a distracter, then it will slow search performance.

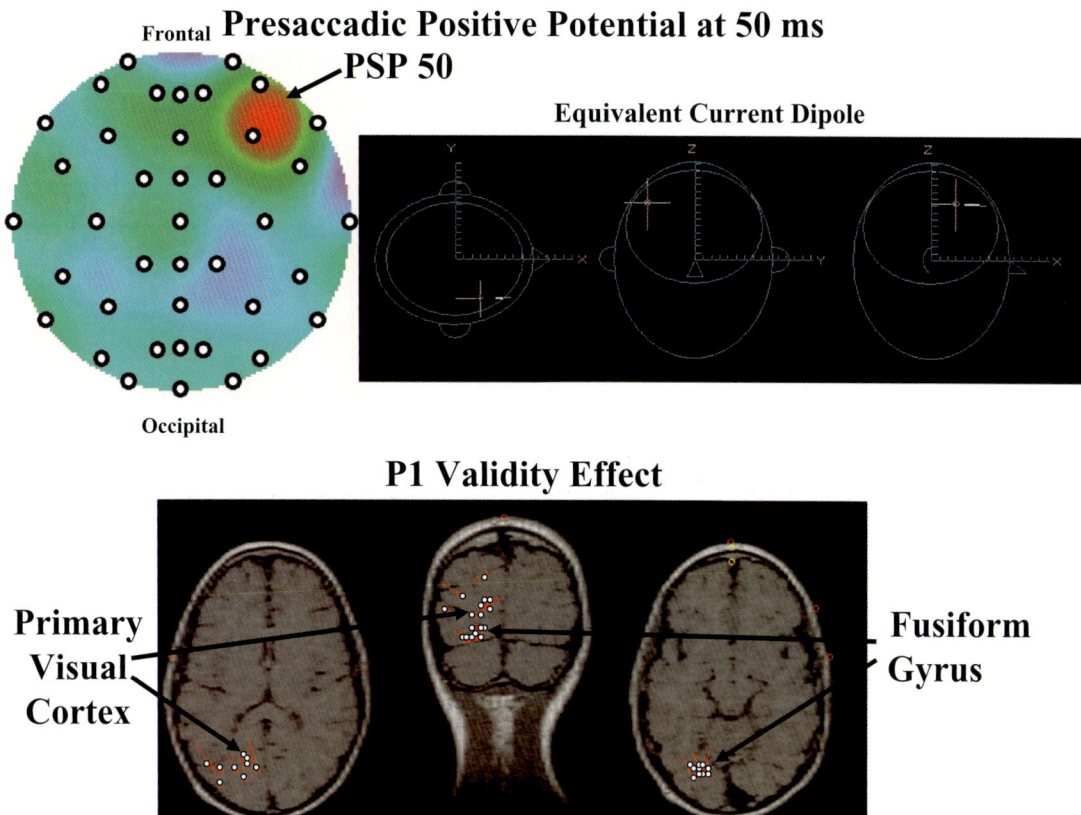

FIGURE 14.4 Top: The topographical scalp potential map is for the presaccadic ERP component (Fig. 14.3), plotted as the difference between the valid and the invalid/neutral trial saccades, plotted as if the infant were making a saccade toward the left side. The equivalent current dipole analysis resulted in a dipole located in the frontal eye fields. Bottom: Equivalent current dipole locations for individual infant participants for a component reflecting the "P1 validity effect" in a spatial cueing task. The dipole locations are plotted on a MRimage from a young child. Reprinted, with permission, from Richards (2003) and Richards and Hunter (2002).

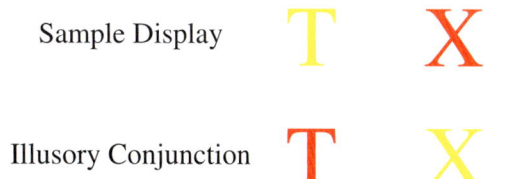

FIGURE 24.1 Example of illusory conjunctions. The sample display contains two shape features (T and X) and two color features (yellow and red). Illusory conjunctions occur when the colors and shapes are incorrectly bound in perception as represented in the lower part of the figure.

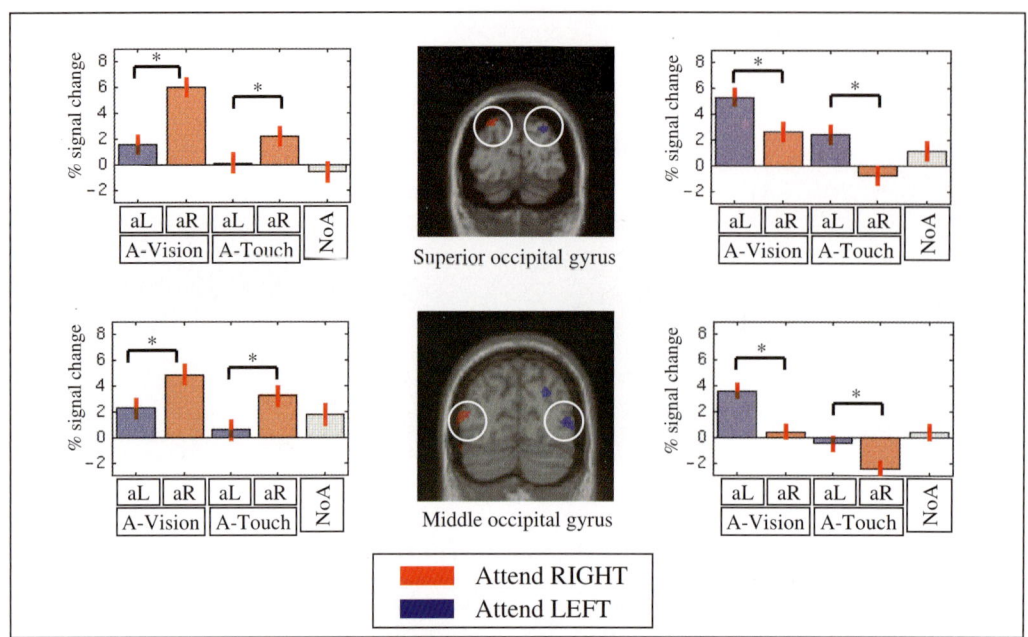

FIGURE 32.3 Multimodal attentional modulation in visual cortex when attending covertly to one or the other hemifield, during bimodal and bilateral visuotactile stimulation. Plots refer to the voxel at the maxima of the activated cluster, always found contralateral to the attended side. Note that these occipital areas displayed differential activity depending on the attended side not only for attend-vision blocks (bars 1 and 2 in each plot), but also during blocks of tactile attention (see bars 3 and 4 in each plot). Activities are expressed as percentage signal change compared with a baseline condition without any peripheral stimulation. The NoA condition (fifth bar in each graph) refers to brain activity in the presence of bilateral, bimodal stimulation, but with attention directed to a central task performed on the fixation point (see Macaluso et al., 2002a, for details). Coronal sections are taken at $y = -90$ (top slice) and $y = -80$ (bottom slice). aL and aR indicate the attended side. (*$P < 0.05$). Adapted, with permission, from Macaluso et al. (2002a).

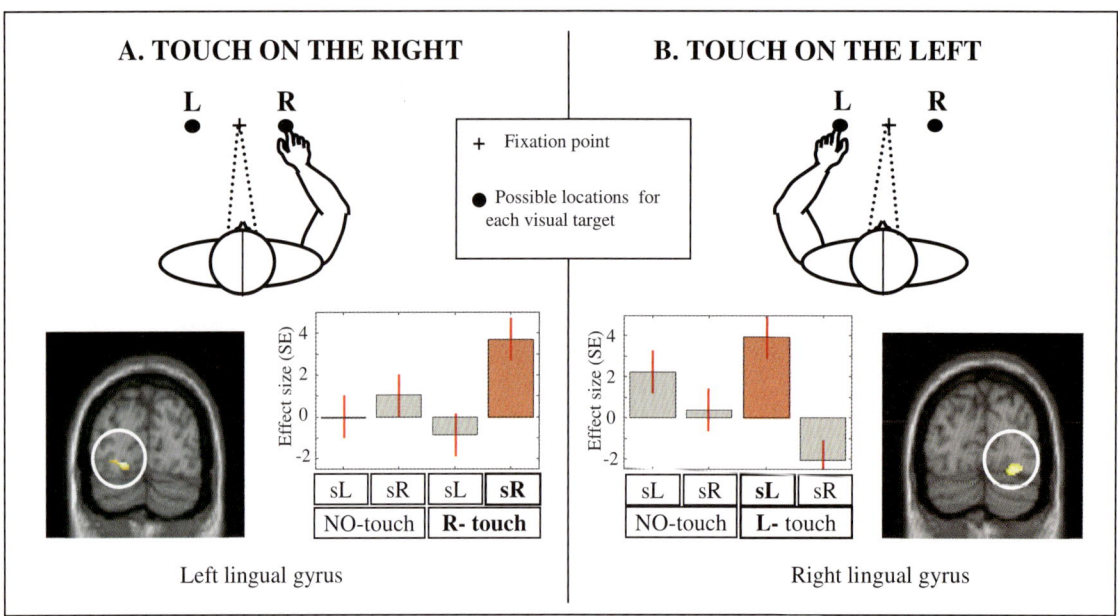

FIGURE 32.4 Effect of multisensory spatial correspondence on visual responses in occipital cortex (see Macaluso et al., 2000b). For each group (A: group receiving tactile stimulations to the right hand, B: group with touch to the left hand), we cartoon the spatial relation between possible visual and tactile input (top), and plot brain activity from the critical maxima for four experimental conditions: sL, stimulate left visual field; sR, stimulate right visual field; in the presence of touch (on left or right hand) or the absence of touch. The sections ($y = -82$ and $y = -80$, for left and right hemisphere activation, respectively) show the anatomical location of the cross-modal interaction (highest activity for contralateral visual stimulation with concurrent touch at same location) observed in the hemisphere contralateral to the location where spatially congruent multimodal stimulation could be presented. The signal plots show the amplification of visual responses when vision and touch were stimulated at the same contralateral location (see red bars in both graphs). Effect sizes are expressed in standard error (SE) units. For display purposes, the anatomical sections shown were thresholded at P-uncorrected = 0.05.

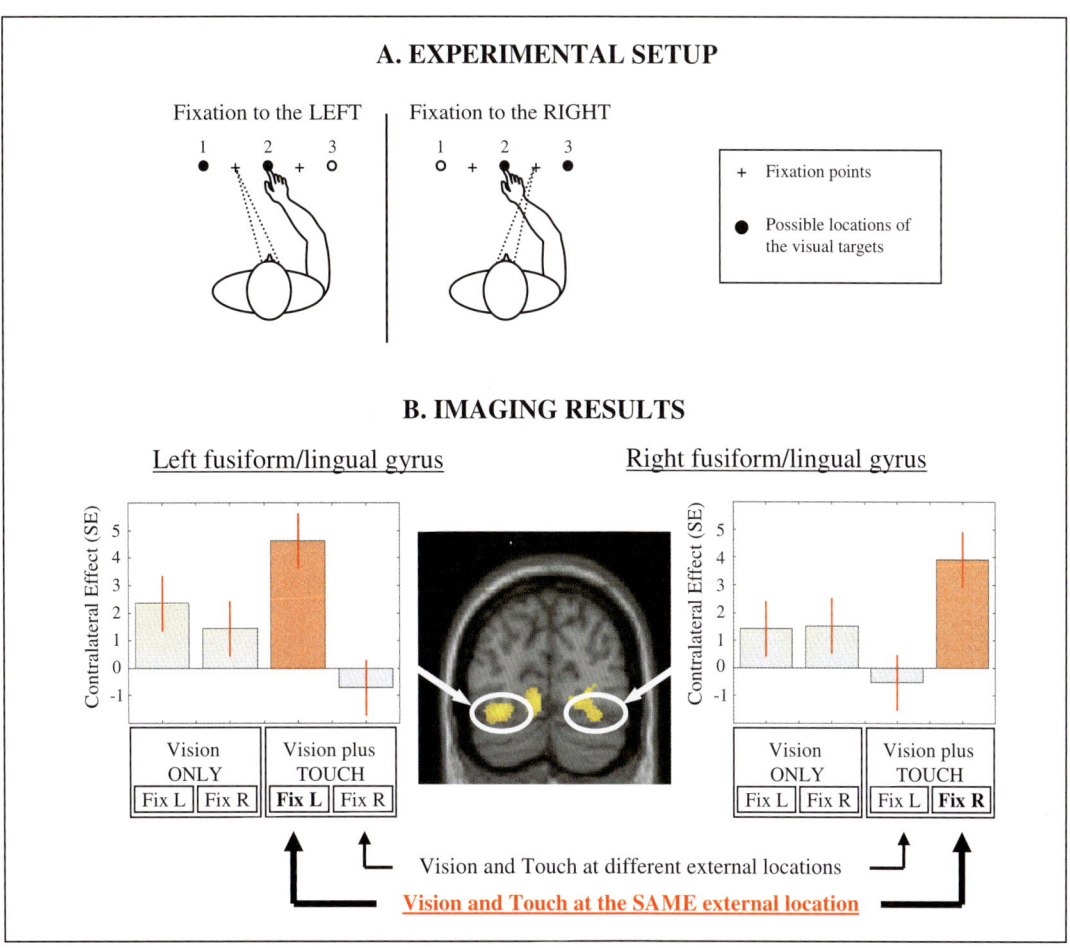

FIGURE 32.5 Gaze direction modulates cross-modal spatial interactions in visual cortex (Macaluso et al., 2002b). (A) Cartoon of experimental setup (possible direction of gaze, possible location of visual targets, and position of unseen right hand). In different blocks, participants fixated either leftward or rightward of the centrally placed right hand. During leftward fixation visual stimuli could be delivered in either position 1 or position 2. When a visual stimulus in position 2 was coupled with tactile stimulation of the right hand, the multimodal stimulation was spatially congruent. However, both stimuli projected to the left hemisphere, so any amplification of visual responses during leftward fixation might be due to intrahemispheric effects, rather than the spatial relation of the stimuli in external space. Including the rightward fixation conditions allowed us to distinguish these alternatives. During rightward fixation, multimodal spatial congruency occurred for simultaneous stimulation of the *left* visual field plus *right*-hand touch (both stimuli in position 2), with the two stimuli now projecting to different hemispheres. (B) Activity in ventral visual cortex for the different stimulation conditions and directions of gaze. Signal plots show the activity in the two hemispheres expressed as the difference between visual stimulation of the contralateral-minus-ipsilateral visual hemifield (i.e., contralateral effect, in standard error units (SE)). Maximal effects of contralateral-minus-ipsilateral visual stimulation were detected when the visual target was coupled with touch at the same external location (see red bars), in the hemisphere contralateral to the *external* location of the multimodal event (i.e., left hemisphere for leftward fixation, left panel; right hemisphere for rightward fixation, right panel). Coronal section taken at $y = -78$; section threshold set to P-uncorrected $= 0.01$.

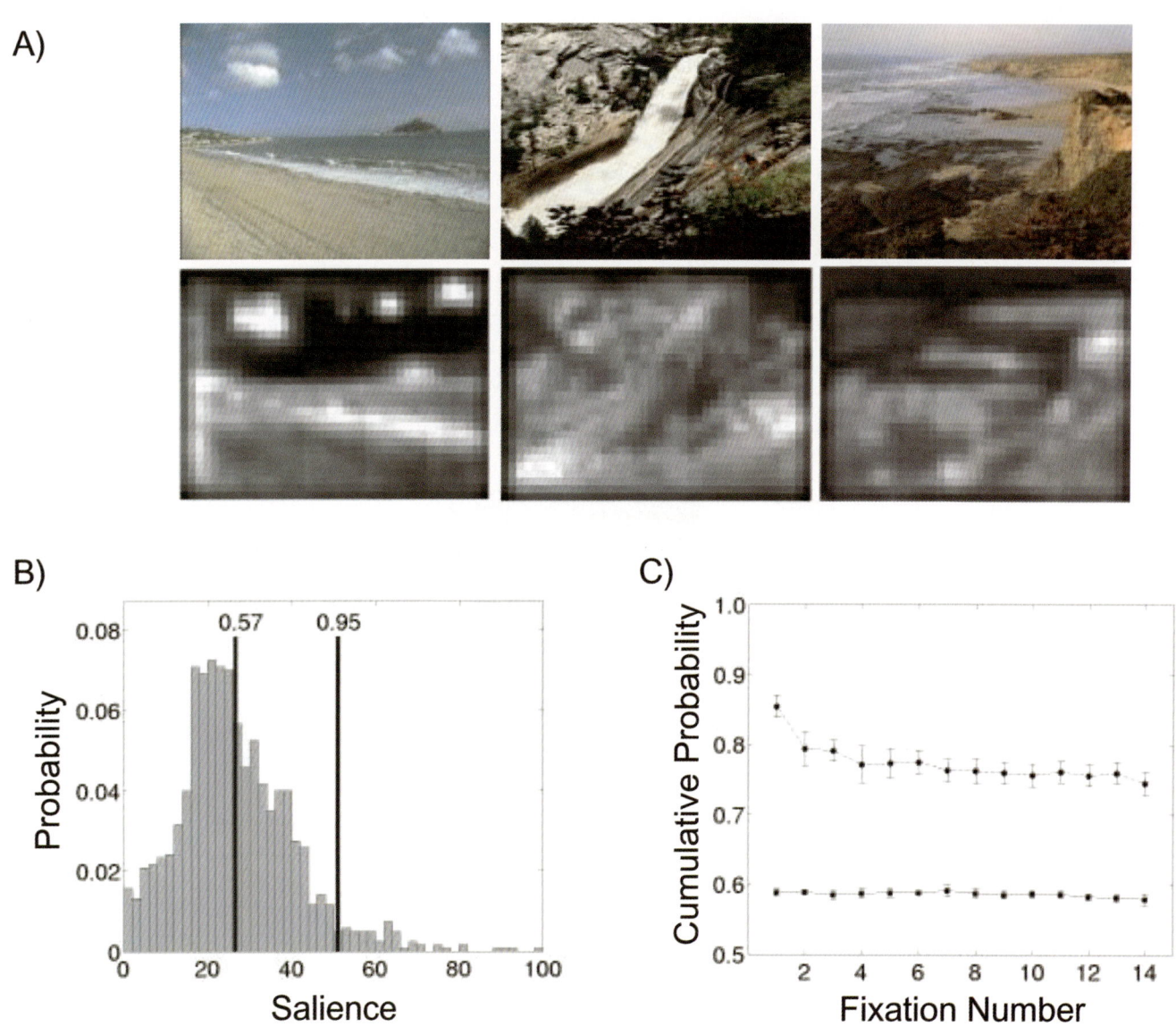

FIGURE 39.1 (A) Example natural landscapes. (B) Average contrast in the participant-selected ensemble (dashed line, circle), uniformly selected ensemble (solid line, square) and image-shuffled ensemble (solid line, triangle) all as a function of image patch size. (C) Two-point correlation using a 4-deg radius patch as a function of the distance from fixation. Same symbols as in (B). Error bars represent plus/minus one standard error of the mean taken across random permutations.

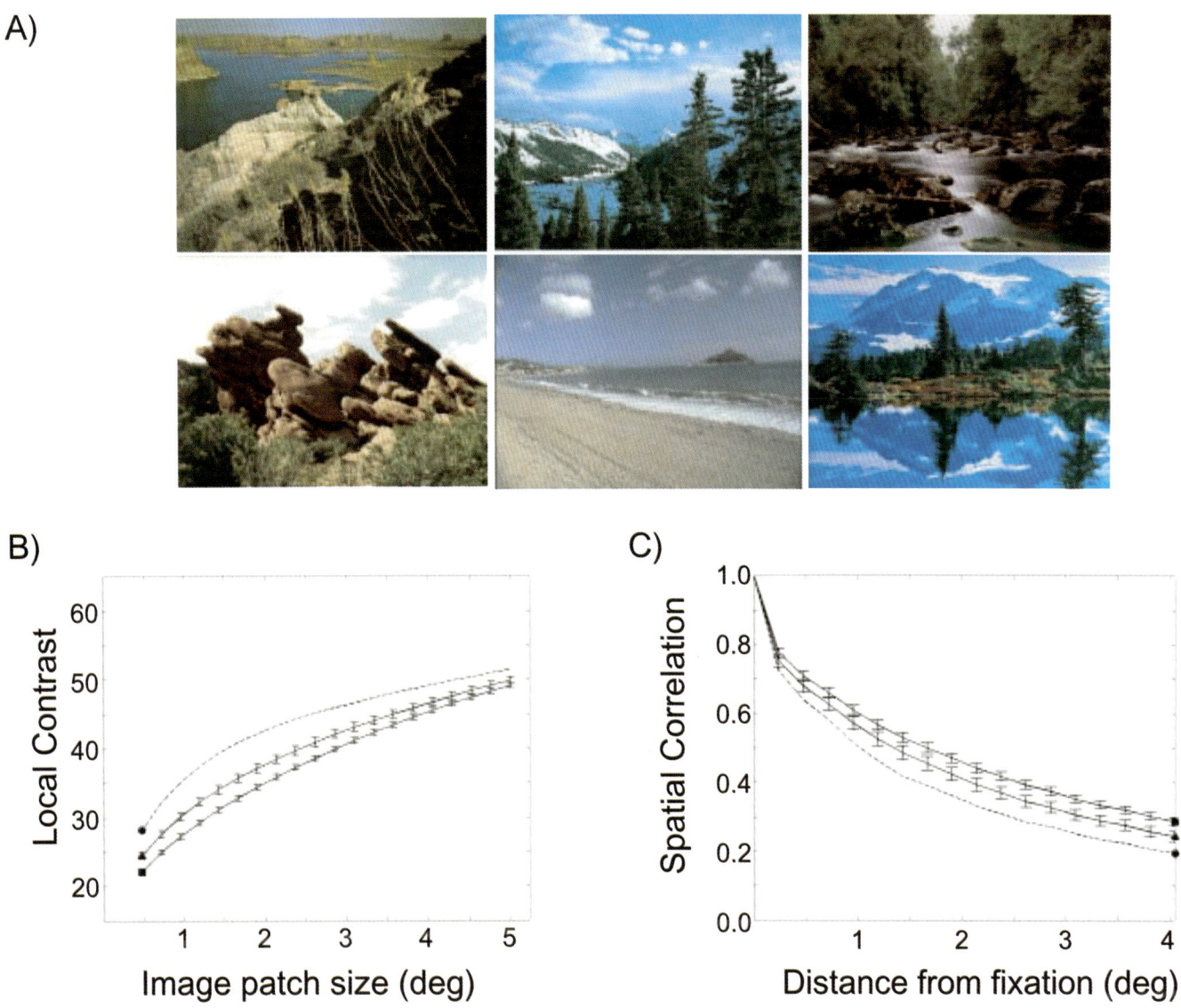

FIGURE 39.2 (A) Example natural landscapes and their respective salience maps. (B) The distribution of salience in an example salience map is shown with the salience expected by chance factors alone (cumulative probability = 0.57) and the salience obtained for a single example fixation (cumulative probability = 0.95). (C) The cumulative probabilities at the points of fixation (dashed line; circle) and expected by chance (solid line; square) are shown as a function of fixation number after stimulus onset. Error bars represent plus/minus one standard error of the mean taken across participants.

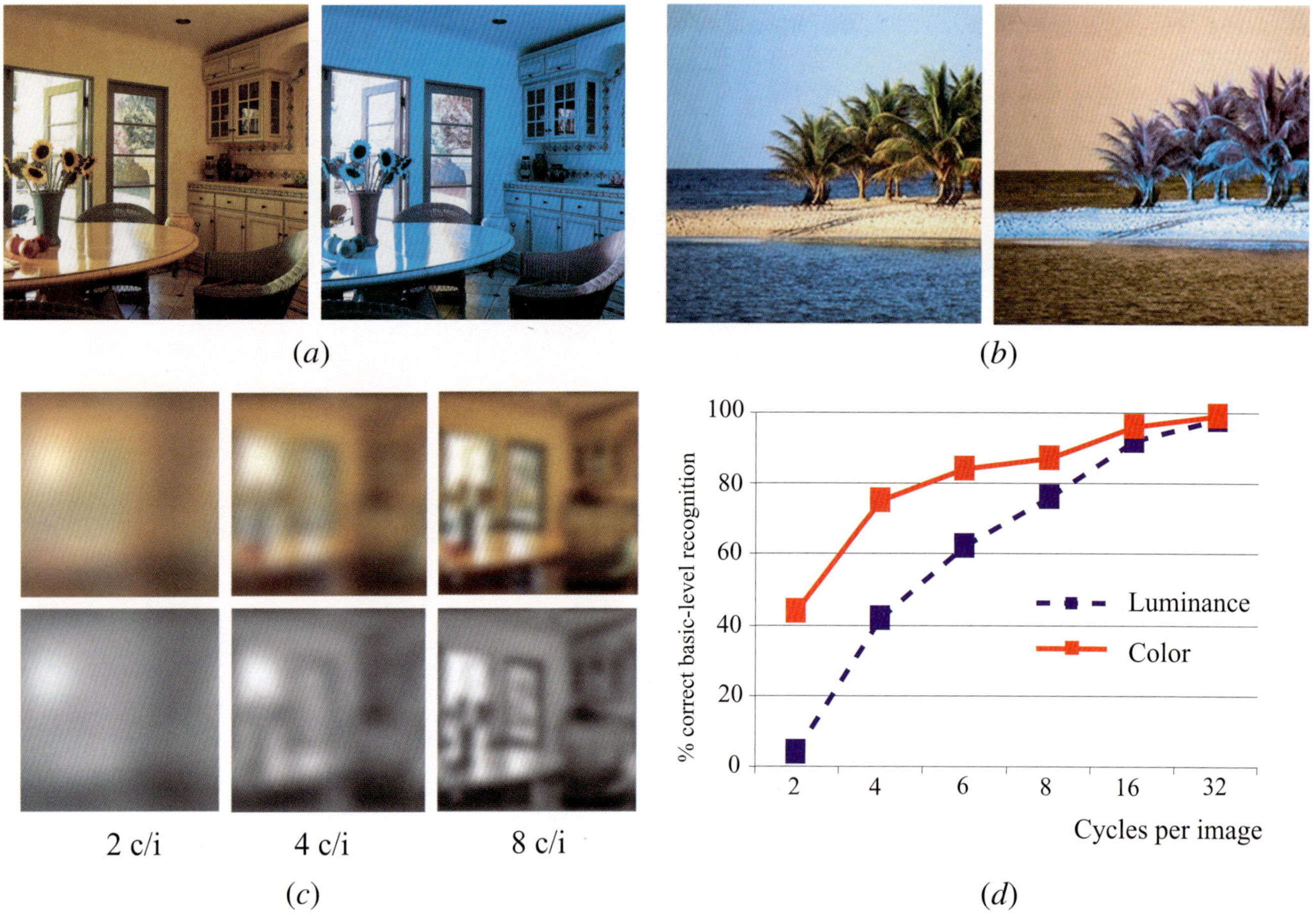

FIGURE 41.2 (a) Examples of color-nondiagnostic scenes: surface color is unrelated to meaning, (b) Examples of color-diagnostic scenes: surface color is related to the meaning of the object or region, (c) Illustration of color and luminance layout information available at 2, 4, and 8 cycles per image, (d) Performances of scene recognition as a function of cycles per image, for color and grey-level images (from Oliva and Schyns, 2000). (see color plate)

FIGURE 41.3 Illustration of a scene gist representation that conserves sufficient structural cues to infer the probable category of the scene. This global scene representation is used to determine spatial envelope properties of a scene. The information preserved by this global representation is illustrated on the right-hand image: it represents a sketch version of the original scene, computed by coercing noise images to have the same global features as the left scene (Torralba and Oliva, 2002). The scene sketch corresponds to an "unbound" spatial layout representation of contours, texture density and colors in the original scene picture.

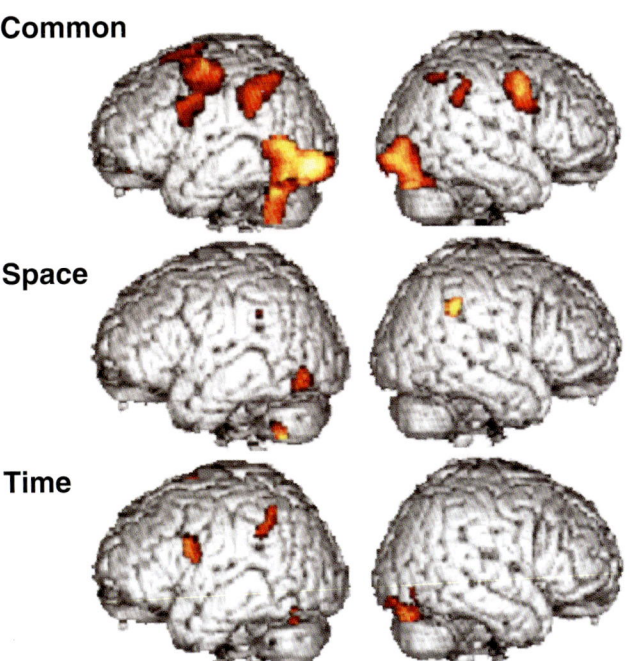

FIGURE 42.2 Brain-imaging results from Coull and Nobre (1998), who compared orienting attention to spatial locations versus temporal intervals. PET brain activations are shown superimposed upon a representative brain volume with standardized anatomy, from left and right lateral perspectives. The top row shows common areas activated in both spatial and temporal orienting conditions, which occurred in a distributed network of parietal, frontal and occipital areas. The next two rows show selective frontal and parietal activations for spatial- and temporal-orienting, respectively.

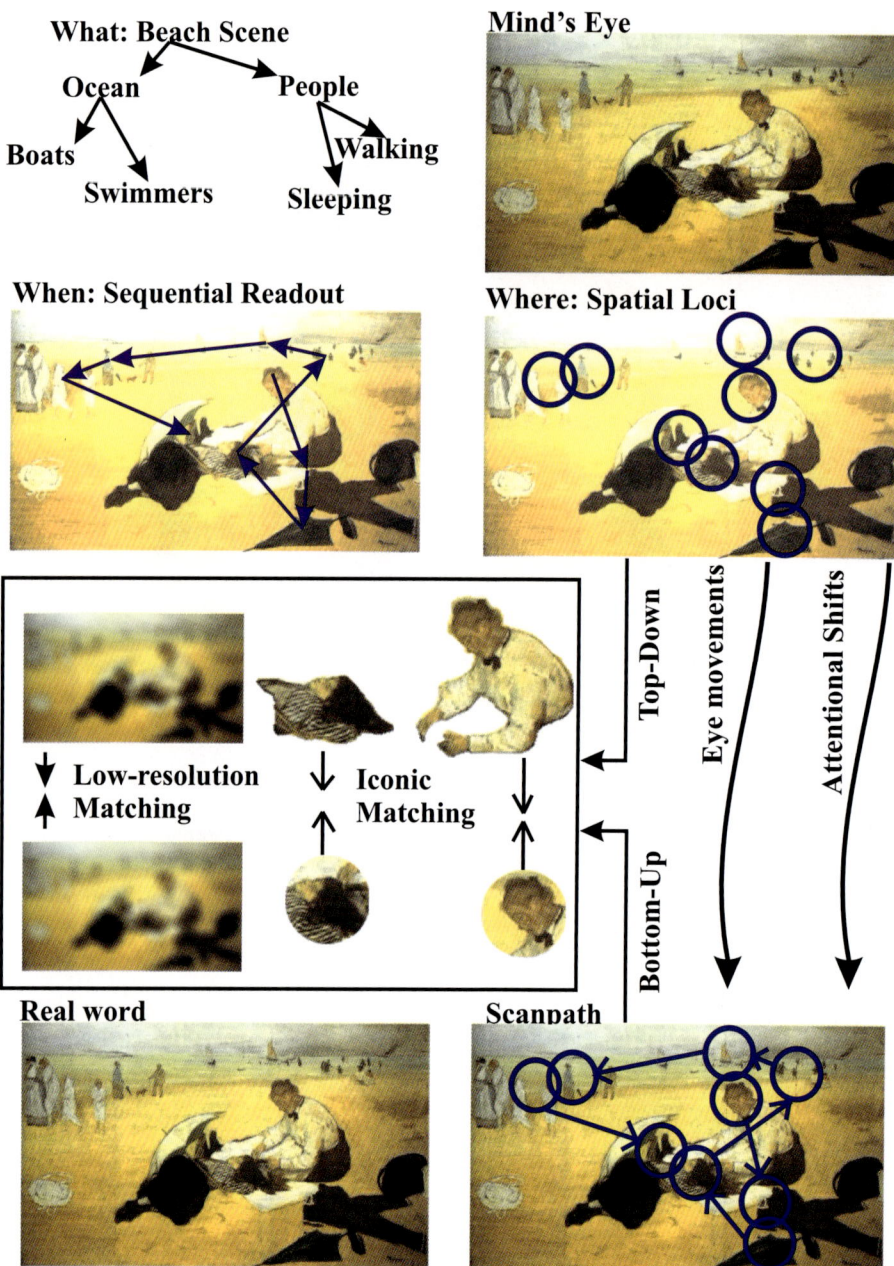

FIGURE 48.2 Human Regions-of-Interest, hROIs (circled regions, bottom right panel), are explained according to the scanpath theory of Noton and Stark, based on long-standing philosophical views and recent neurological studies. A top-down (TD) cognitive model of the external world is composed of *What*, *Where*, and *When* information. This model controls the scanpath shifts of attention (eye movements sequence) and the iconic matching in the visual cortex between foveated bottom-up (BU), visual signals and the TD model cognitive representation; all together they yield visual perception. (The authors thank Dr. Chernyak for an earlier version of this figure.) (See color plate).

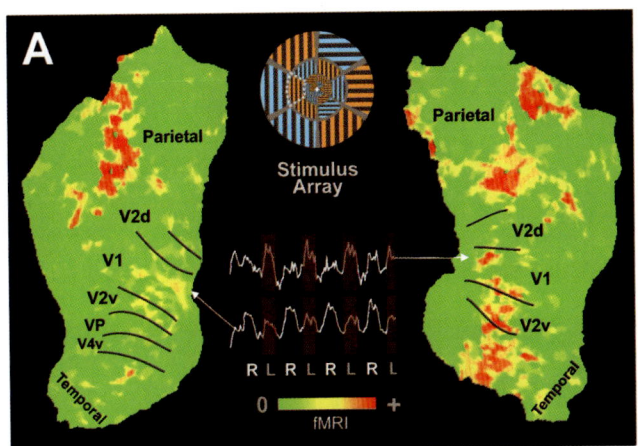

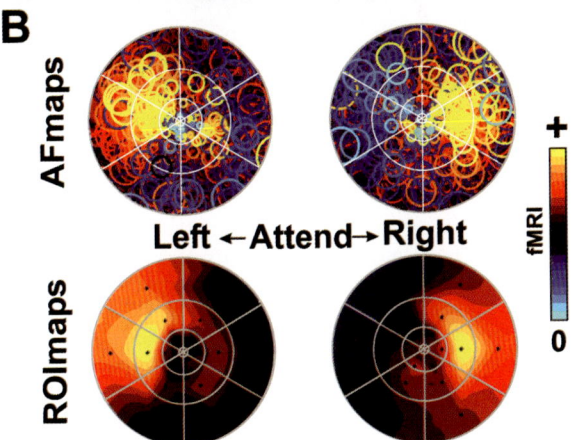

FIGURE 61.1 Topography of attention-related activation for single targets. **A.** Flat maps of parieto-occipital cortex showing pattern of activation (yellow/red patches) in left and right hemispheres when observer attended to middle left target in array of colored (blue/orange) and striped (vertical/horizontal) segments (top, center). Attended segment marked by white dotted line that was not present in actual stimulus. Timecourse graphs at middle, center illustrate the fMRI responses for regions of interest in right V1 (upper graph) and left V1 (lower graph). Epochs (10 sec long) in which subject attend to left (L, red bands) or right (R) middle segment were separated by epochs in which the subject passively viewed the fixation marker (not marked in diagram). Scale at bottom shows color scheme representing amplitude of attention effects. V1, V2d, V2v, VP, V4v indicate human visual areas as established in DeYoe et al., 1996. **B.** Attentional field maps (AFmaps) and region-of-interest maps (ROImaps) showing the topography of attentional activation "back-projected" from cortex onto diagrams of the visual field. Each circle in the AFmaps represents a visually active voxel in medial occipital cortex that responded best to visual stimuli at that location. The size of the circle indicates the potential error in positioning the symbol. The color of each circle codes the magnitude of the attention-related effects observed in the corresponding voxel. More intense activation is indicated by lighter circles. Outlines of the 18 stimulus segments are overlaid in gray on the AFmaps and ROImaps for reference.

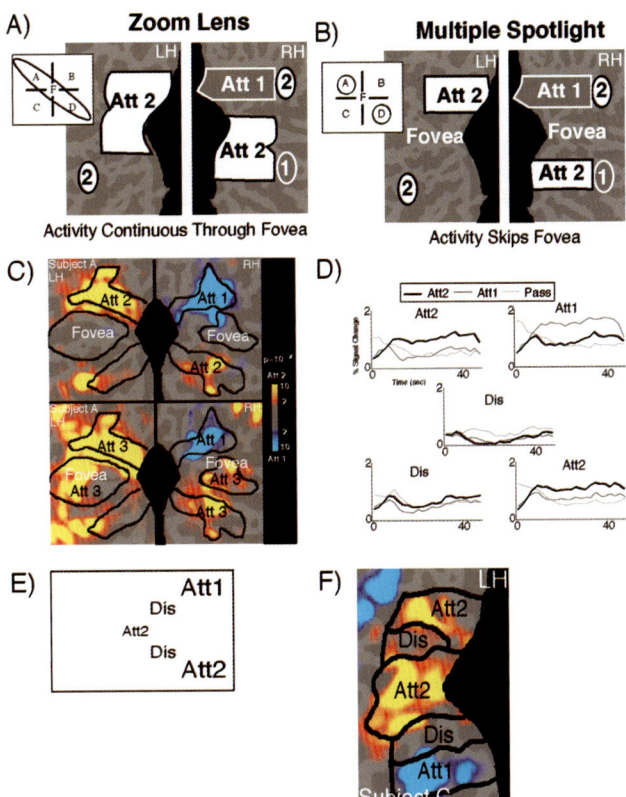

FIGURE 62.3 Multiple spotlights of spatial attention. **A** and **B**. Predicted activation patterns for Attend2 vs. Attend1 for the two competing hypotheses. **C**. Activation patterns for Attend2 vs. Attend1, for two subjects, supporting the multiple spotlight hypothesis. **D**. Time course data averaged across all subjects. **E**. Spatial layout for hemifield experiment. **F**. Activation maps for one subject in the right hemifield configuration comparing Attend2 vs. Attend1.

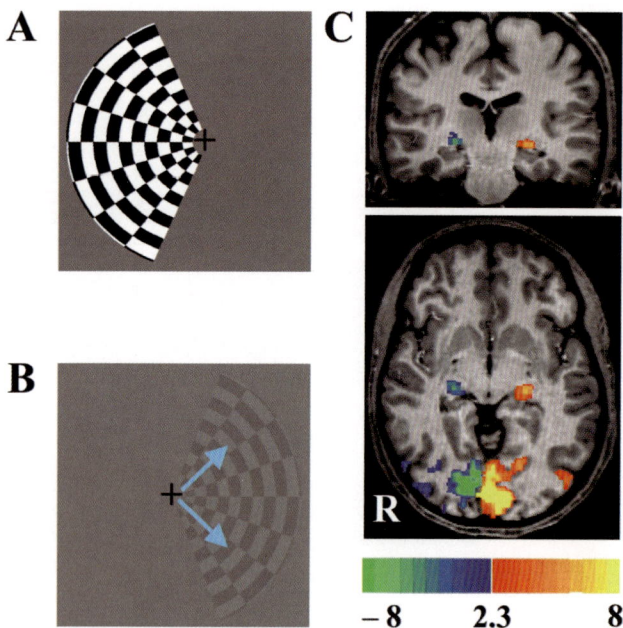

FIGURE 72.1 Visual stimuli and LGN activation. **A, B.** High- or low-contrast checkerboard stimuli were presented to the left or right of a central fixation (+) point. In Experiments 1 and 3, subjects covertly directed attention to a checkerboard arc (blue arrows) and detected randomly occurring luminance changes along that arc. The detection of luminance changes in low-contrast and high-contrast checkerboard stimuli was not matched for task difficulty. **C.** Axial slice showing activations of the right (green-blue) and left (yellow-orange) LGN and visual cortex evoked by the checkerboard stimuli. Activations in the thalamus were restricted to the LGN; no activations within the pulvinar were obtained. Scale indicates Z score values of activations in colored regions. R = right hemisphere. (Adapted from O'Connor et al., 2002.)

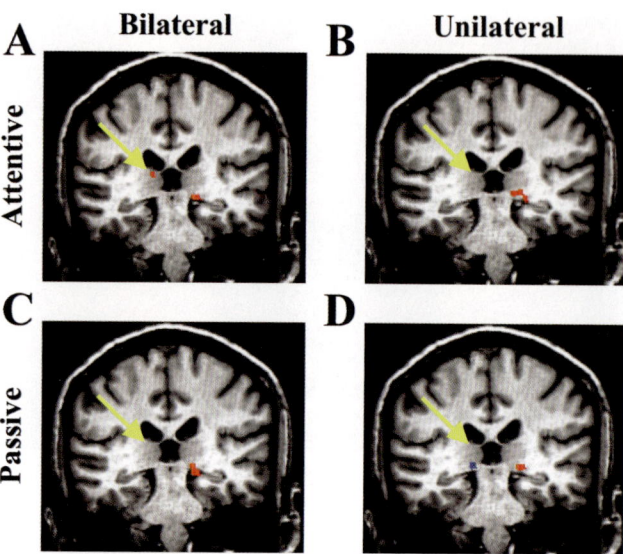

FIGURE 72.4 Functional activation of the pulvinar under different conditions. Results are shown from one subject tested in four different experiments. **A.** Bilateral attentive. The arrow indicates functional activation of the right pulvinar. The subject directed attention to a checkerboard stimulus presented simultaneously to the right and left visual hemifield. **B.** Unilateral attentive. No functional activation was found in the pulvinar when the stimuli were presented unilaterally and in alternation to the right and left hemifield while subjects directed attention to the checkerboards. **C.** Bilateral passive. No functional activation was obtained in the pulvinar when the stimuli were presented bilaterally and passively viewed by the subjects while maintaining fixation. **D.** Unilateral passive. No functional activation was found in the pulvinar in Experiment 1, in which checkerboard stimuli were presented in alternation to the right or left hemifield. Other conventions as in Fig. 72.1. (Adapted from Kastner et al., 2004.)

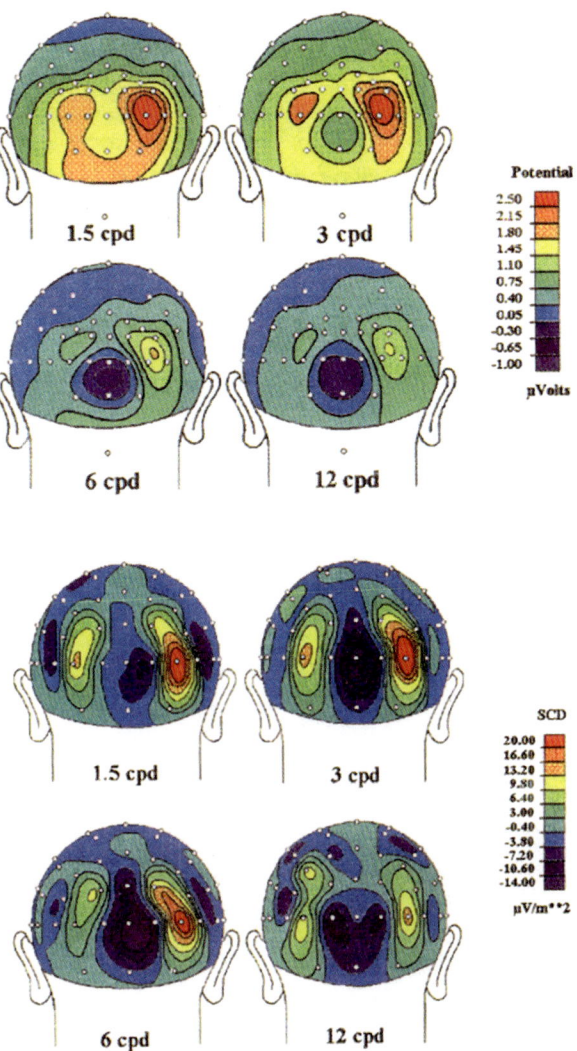

FIGURE 82.1 Grand-average voltage (top) and Scalp Current Density (SCD) maps (bottom) of brain activity recorded in response to four different luminance-modulated gratings going from a low (1.5 cycles per degree, cpd) up to a high (12 cpd) spatial frequency, presented in the central visual field during passive viewing. Note the topographic dissociation between the pattern of response to high vs. low frequency gratings, reflecting the equivalent distinction between the dorsal/ventral type of activation. Reprinted from *Brain Topography* **9**; A. M. Proverbio, A. Zani, and C. Avella; Differential activation of multiple current sources of foveal VEPs as a function of spatial frequency, pp. 59–69. Copyright (1996) with permission from Kluwer Academic/Plenum Publishers and authors.

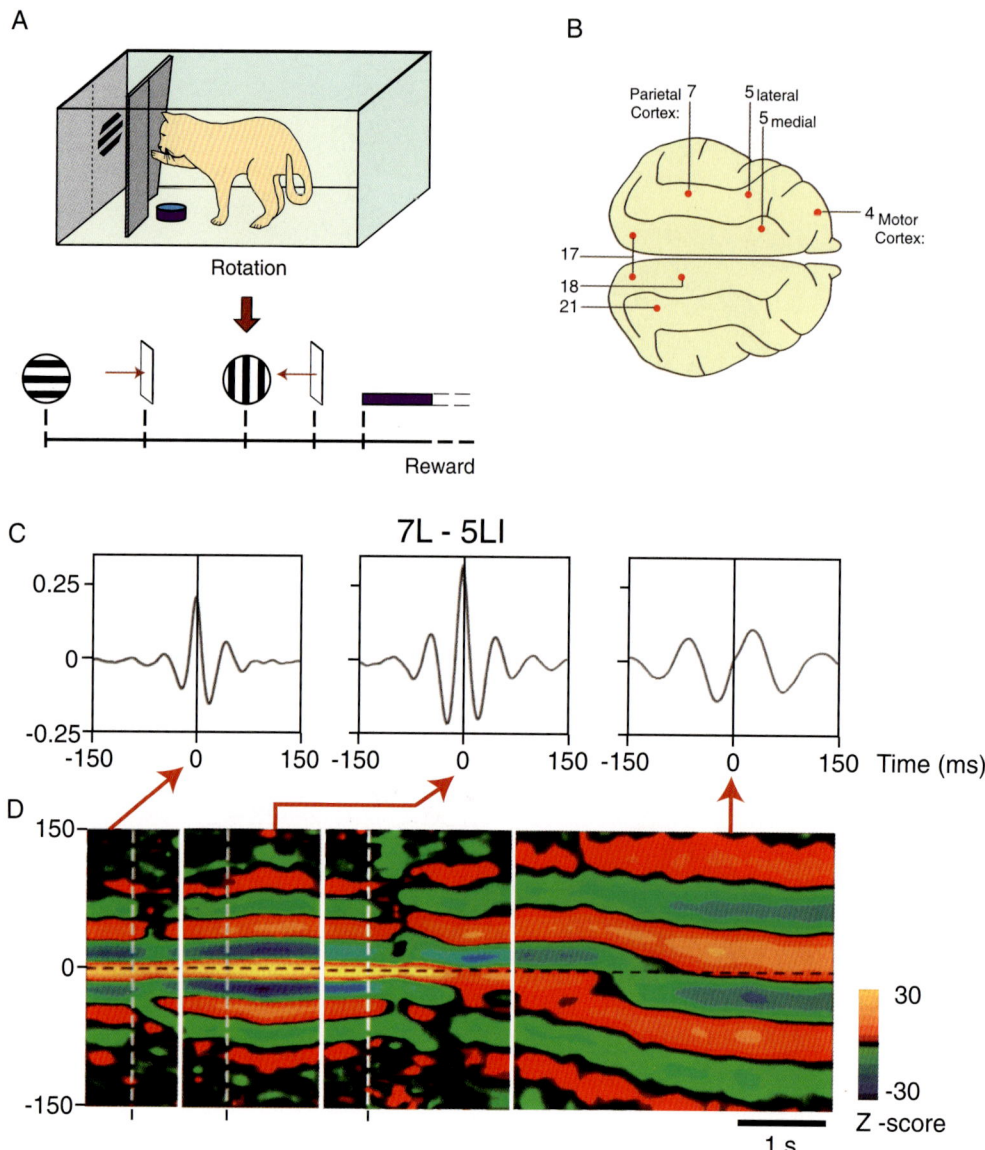

FIGURE 87.2 Expectation-related synchrony in visual cortex in awake cats during a visuomotor coordination task (Roelfsema et al., 1997). **A.** Cats had to respond to a visual stimulus by pressing and holding a transparent door as they waited for a rotation of the stimulus. **B.** Local field potentials (LFPs) were recorded from visual, parietal, and motor cortex. **C.** Cross-correlograms between responses recorded from electrodes 7 and 5 during various phases of the trial. Left: prior to target presentation; Middle: after target presentation; Right: during consumption of reward. **D.** The contour plot shows how the temporal correlation between parietal areas 5 and 7 evolves over time. Abscissa: time relative to stimulus appearance. Zero on the ordinate indicates synchrony between the recording sites. Red–yellow, positive correlation; Green–blue, negative correlation. The high-amplitude zero peak shows that synchrony occurs before the appearance of the stimulus (**C**, left), presumably reflecting expectancy, and increases during attentive visuomotor processing (**C**, middle). Synchrony breaks down completely after the cat has responded to rotation of the pattern by releasing the door (**C**, right). (Adapted from Engel et al., 2001).

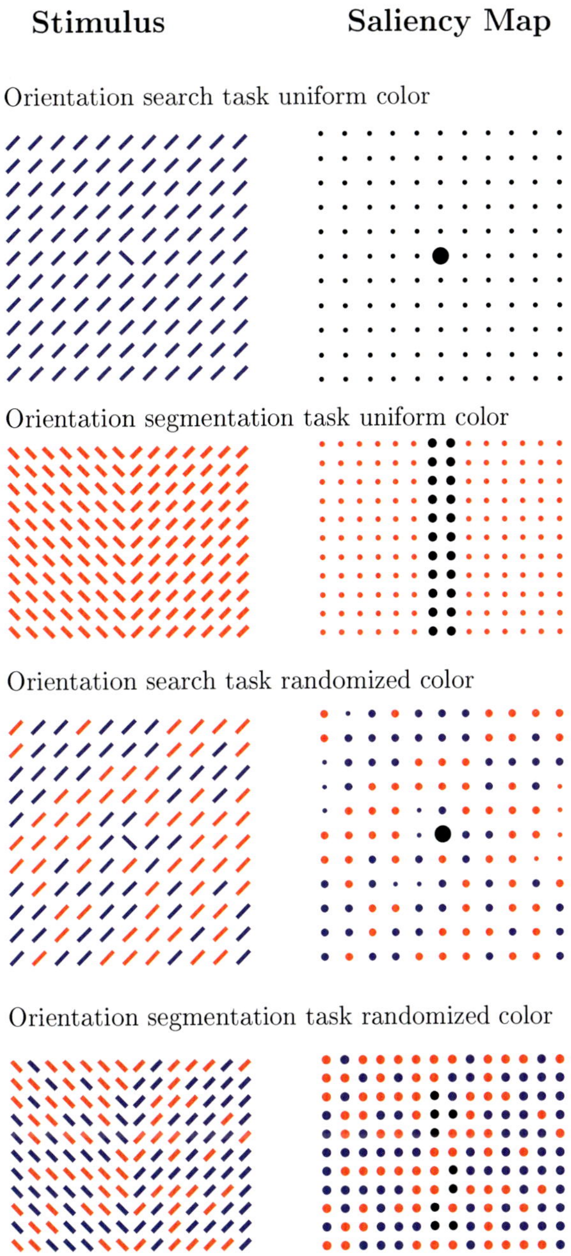

Stimulus   Saliency Map

Orientation search task uniform color

Orientation segmentation task uniform color

Orientation search task randomized color

Orientation segmentation task randomized color

FIGURE 93.3 Color inference in orientation feature-based tasks. The two random colors of the stimulus bars are visualized in the grey scale images as black and white bars. Each bar evokes responses from color-tuned cells *and* from orientation-tuned cells. Randomizing the colors of the stimulus bars increases the response levels (and variations) in the color tuned cells, submerging the responses from the orientation tuned cells to the texture border but not that to the search target. Hence, the texture segmentation task, but not the search task, is impaired. Note that in uniform color stimuli, the saliency of the target is much higher than that of the texture border against their respective background.

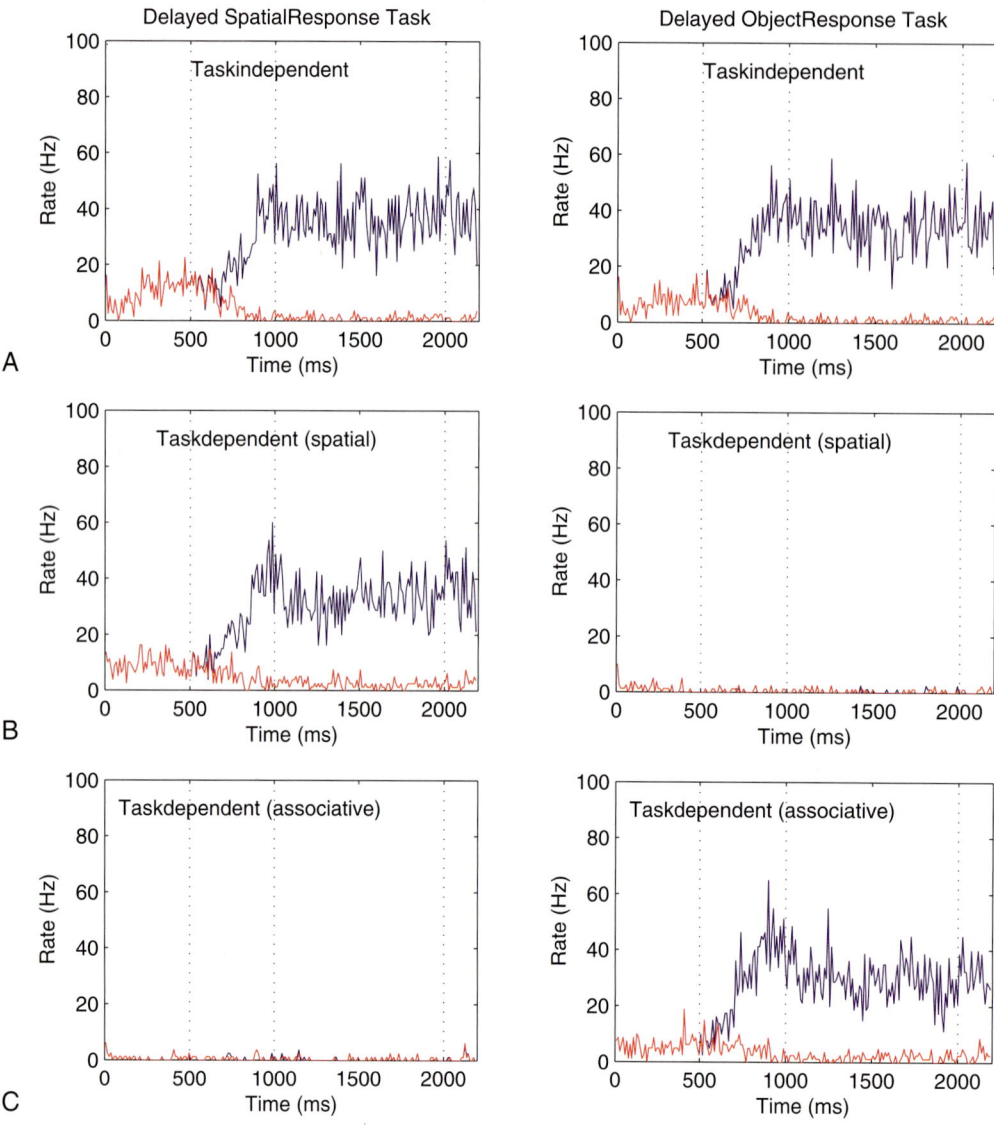

FIGURE 100.2 Simulation results corresponding to the experimental paradigm of Asaad et al. (2000). The temporal evolution of the averaged population activity for three neural pools during the execution of the object-response (associative) and delayed spatial response task. Cue, response, and selective context-specific associative activity is explicitly maintained during the short-term-memory-related delay period in the period 1000–2000 ms. The stimulus was shown in the period 500–1000 ms. The left panel corresponds to the delayed spatial response task condition and the right panel to the object-response (associative) task condition. Each picture plots two curves corresponding to the two possible responses (dark corresponds to L and light to R). (A) the premotor pool L, which was response direction-selective in both tasks. (B) the intermediate spatial pool S1-L, which was direction-selective (to the L, dark curve) in only the delayed spatial response task. (C) the intermediate associative pool O1-L, which was direction-selective in only the object-response (associative) task.

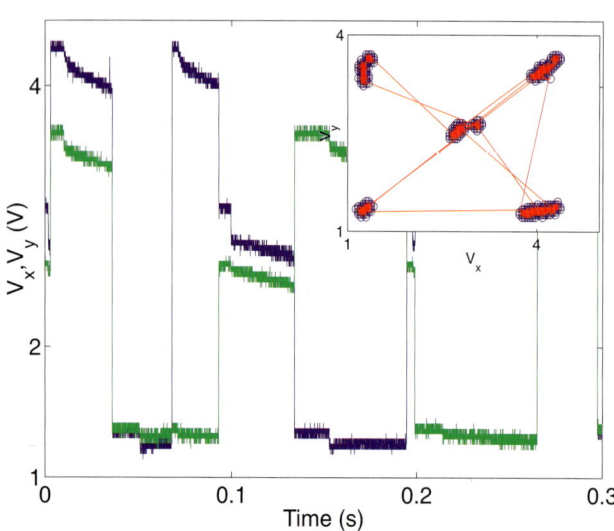

FIGURE 103.3 Output of the P2V x circuit and y circuit of the selective attention chip in response to a test stimulus exciting four corners of the input array at a rate of 30 Hz and a central cell at a rate of 50 Hz. The inset shows the x output trace versus the y output trace.

CHAPTER
61

Neurophysiological Correlates of the Attentional Spotlight

Edgar A DeYoe and Julie Brefczynski

ABSTRACT

Though it has long been known that visual attention is spatially selective, only recently has it been possible to view the topography of attentional effects as mapped within the brain and as mapped within the observer's field of view. Functional MRI was used to identify and map cortical activation as an observer shifted attention to verbally cued targets within a field of similar distracters without moving the eyes. The resulting pattern of attention-related activation was strongest at the cortical sites representing the retinotopic location of the target but also spread to nearby cortical loci and weakly to some sites in the opposite hemifield. Thus the locus of attention was mapped accurately in retinotopic coordinates within occipital visual cortex. Averaged across subjects, the attentional topography resembled a gradient model (Downing and Pinker, 1985), though individual subjects could show consistent variations on this pattern. Together, the results provide an increasingly detailed picture of the neurophysiological mechanisms underlying visuospatial attention.

I. VISUAL ATTENTION IS SPATIALLY SELECTIVE

The ability to shift visual attention to an object or location in space at will or reflexively when a novel event occurs is such an integral part of our ongoing perceptual experience that we rarely notice its operation. If we do become aware of the operation of attention it is usually so closely linked to the control of eye movements that we tend to assume that they are one and the same. However, it is relatively easy to demonstrate to oneself that the linkage with eye movements is not always necessary. For example, an experienced basketball player can mislead an opponent by looking in one direction but covertly attending to a teammate in another direction who, seconds later, becomes the recipient of a pass. Perhaps the earliest reported scientific demonstration of this ability to shift attention without moving the eyes was in Herman von Helmholtz's extraordinary Treatise on Physiological Optics (Helmholtz, 1910), in which he described the ability of an observer to report the identity of letters in one portion of a large array during the briefest of flashes from an electric spark. He noted that the observer's eyes could be fixed on a small pinhole, yet if cued before the flash to attend to, say, the upper right letters, he would be unable to report letters at other locations. This elegant experiment not only demonstrated the separability of attention and eye movements, but also demonstrated that the effects of attention are, as noted by Posner years later (Posner et al., 1980), like a "spotlight that enhances the efficiency of detection of events within its beam." Later investigators extended the concept that visual attention has a topographic distribution in visual space that can vary in extent and may be graded in efficacy around the attended locus (Downing and Pinker, 1985; Eriksen and St. James, 1986). Neurophysiological studies with trained animals have provided insight into the neural mechanisms underlying visuospatial attention (Maunsell et al., 1991; Desimone and Duncan, 1995) and, together with more recent psychophysical studies, have resulted in a number of influential theories directly, or indirectly, concerning the topography of attention (Duncan, 1984; Desimone and Duncan, 1995; LaBerge, 1995; Shih and Sperling, 1996; Dosher & Lu, 2000). This article describes current insights into the neurophysiological basis of visuospatial attention as revealed by neuroimaging of human subjects performing attention-demanding tasks.

II. NEUROIMAGING THE SPATIAL TOPOGRAPHY OF ATTENTION

Though various properties of attention have been revealed by studies since Helmholtz's time, a complete understanding of the detailed topography of attention has been lacking. A breakthrough in this respect has been achieved using functional magnetic resonance imaging (fMRI) to observe the patterns of brain activation when human observers perform attention-demanding tasks. FMRI provides images of localized changes in blood flow and oxygenation that are closely coupled to changes in neuronal activity evoked by sensory, motor, or cognitive events (DeYoe et al., 1994). Using this technique, it has been possible to show that visual attention directed to a specific location or object within the field of view results in increased brain activation at sites in the visual cortex that correspond to the neural representation of those locations or objects (Tootell et al., 1998; Brefczynski and DeYoe, 1999; Gandhi et al., 1999; Somers et al., 1999).

A. Topography for a Single Target

Figure 61.1 illustrates the results of one of these experiments (Brefczynski and DeYoe, 1999) in which the observer was asked to attend to one of 18 segments within a dartboard-like array of segments that were colored (blue/orange) and striped (horizontal/vertical). (In Fig. 61.1A, the attended segment is marked by a dotted white outline that was not present in the actual stimulus. Total width of the stimulus array was 56°.) Every two seconds, the pattern of colors and stripe orientations changed randomly throughout the array, so that there was nothing physically unique about any of the segments including the target segment. The subject's task was to report on the color/orientation combination of the target segment on each 2 sec. trial while ignoring the patterns in the other distracter segments. It is known that to perform such a task well, the observer must accurately direct attention to the target segment (Treisman and Gelade, 1980), thereby ensuring adequate attentional control. Figure 61.1A shows the resulting pattern of brain activation when the observer attended to the middle segment on the left. The maps of computationally unfolded and flattened occipito-parietal cortex illustrate the topography of the attention-related activation. The dark lines on the flat maps mark the approximate boundaries of V1 and several extrastriate visual areas. Note that the attention-related activation involved virtually all known occipital visual areas including primary visual cortex, V1, and extended into the temporal and parietal lobes as well. The strongest activation, indicated by the darkest patches, was found in the right occipital cortex (contralateral to the attended target) and bilaterally in parietal cortex. It is important to emphasize that this topographic specificity was not associated with the stimulus array itself, but rather, was linked to attention directed to a particular target location. In other words, it was a reflection of a purely mental event having no physical correlate in the stimulus itself. When the subject was cued to switch attention to the right middle target, the patterns in the left and right hemispheres switched, with the strongest activation now in the left hemisphere. This lateralized effect was also evident in the graphs shown in the middle of Figure 61.1A, which illustrate the average timecourse of the fMRI signal for two regions of interest in left and right primary visual cortex, V1. Again, the strongest activation was always associated with attention directed to the contralateral target location (the flat maps show the pattern of activation during the epochs overlaid with gray bars). However, the timecourse graphs also show that the fMRI signal returned to baseline or was slightly suppressed in the passive fixation epochs separating each left and right attention epoch. Relative to these fixation epochs, attention directed to, say, the left target location also evoked a weak response in some locations in the ipsilateral hemisphere, appearing in the timecourse graph as a lower amplitude peak and as light gray patches in the left hemisphere flat map. Although the topography of this weak ipsilateral activation varied across subjects, it tended to be consistent within the same individual.

Although it is informative to view the topography of attentional effects as they exist in the cortex, this view does not readily reveal how the attentional effects are distributed relative to the observer's field of view. Fortunately, however, all occipital visual areas are retinotopically organized, and it is possible to use fMRI to map this retinotopic organization individually for each observer (DeYoe et al., 1996). Using such retinotopic information, it is then possible to translate the pattern of cortical activation into a schematic diagram of the visual field, thereby obtaining a picture of the attentional activation as it would appear to the observer. Such a diagram is termed an attentional field map (AFmap). AFmaps for the attend-left and attend-right conditions pooled across a sample of five subjects are illustrated in Figure 61.1B. In these maps, the activity of each responsive cortical voxel is represented by

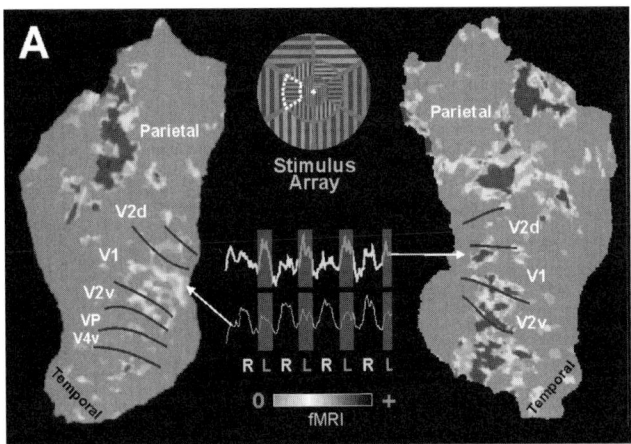

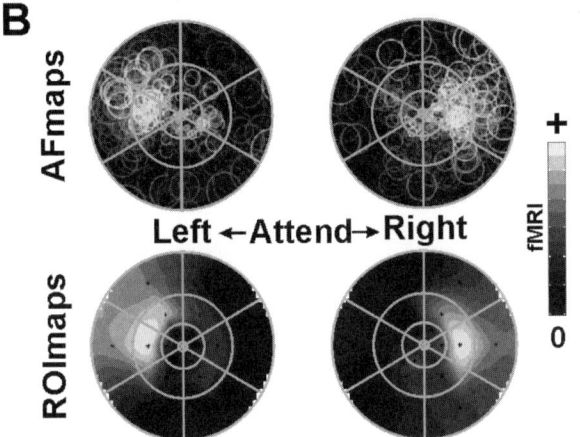

FIGURE 61.1 Topography of attention-related activation for single targets. **A.** Flat maps of parieto-occipital cortex showing pattern of activation (gray patches) in left and right hemispheres when observer attended to middle left target in array of colored (blue/orange) and striped (vertical/horizontal) segments (top, center). Attended segment marked by white dotted line that was not present in actual stimulus. Timecourse graphs at middle, center illustrate the fMRI responses for regions of interest in right V1 (upper graph) and left V1 (lower graph). Epochs (10 sec long) in which subject attend to left (L, gray bands) or right (R) middle segment were separated by epochs in which the subject passively viewed the fixation marker (not marked in diagram). Scale at bottom shows color scheme representing amplitude of attention effects. V1, V2d, V2v, VP, V4v indicate human visual areas as established in DeYoe et al., 1996. **B.** Attentional field maps (AFmaps) and region-of-interest maps (ROImaps) showing the topography of attentional activation "back-projected" from cortex onto diagrams of the visual field. Each circle in the AFmaps represents a visually active voxel in medial occipital cortex that responded best to visual stimuli at that location. The size of the circle indicates the potential error in positioning the symbol. The brightness of each circle codes the magnitude of the attention-related effects observed in the corresponding voxel. More intense activation is indicated by lighter circles. Outlines of the 18 stimulus segments are overlaid in gray on the AFmaps and ROImaps for reference. (See color plate.)

a circular symbol placed at the retinotopic location coded by that voxel. The size of each circle shows the potential error in estimating this retinotopic location and the brightness of each circle shows the intensity of the attention-related activation. All the circles together show the complete retinotopic pattern of attention-related activity. In other words, the AFmap is a picture of the attentional "window" itself.

The average AFmaps in Figure 61.1B show that the strongest attention-related activation (light colored circles) was associated with the attended target location but spread to nearby segments and weakly to some locations in the opposite hemifield. (For reference, the outlines of the stimulus segments have been superimposed in gray on the AFmaps.) To test the significance of differences in attentional effects across the target and distracter locations, the data points within each segment were pooled and a mean attentional effect was computed for each segment as a whole. The resulting values were plotted as individual points at the center position of each segment on a new diagram of the visual field. These data points were then fit with a smooth surface and viewed as a contour diagram. Such ROImaps are illustrated at the bottom of Figure 61.1B. A repeated measures ANOVA verified that the differences in attentional effect across segment locations was significant and that this pattern changed significantly when the observer switched attention from left to right target locations.

B. The Locus of Attention is Represented in Retinotopic Coordinates

The results of the preceding experiment indicated that directing attention to a single target produces enhanced fMRI signals at the corresponding retinotopic location in occipital visual cortex. If so, then shifting attention successively from one target to another at increasing visual field eccentricity should produce a corresponding shift of activation in the cortex, but how accurately? This was tested in a second experiment for which the results are illustrated in Figure 61.2. In this case, the subject was verbally cued every 10 sec. to attend to successive target locations extending rightward along the horizontal meridian as depicted schematically in the left column of Figure 61.2. Again, the task was to identify the color/orientation combination at the currently attended location. However, the stimulus array now consisted of four rings of six segments each (total width of the array was 56°). The spacing and size of the stimulus segments were scaled in proportion to eccentricity in order to roughly account for cortical magnification (expanded representation of the fovea relative to the periphery). In other words, shifts of attention to segments farther and farther from fixation would be expected to result in roughly equal shifts of activation in the visual cortex. The experiment was analyzed as a temporal phase-mapped design in which the activation associated with each shift of attention was uniquely identified by its timing relative to the start of the shift sequence.

As illustrated in the left half of Figure 61.2, shifting attention from one target location to the next did indeed produce attention-related activation (dark patches with light outlines marked 1–4 in the correlation maps) that shifted roughly equal steps away from the occipital pole (as viewed in a parasagittal brain section through occipital cortex). The multiple dorsoventral patches in each panel correspond to simultaneous attentional activation in different visual areas. To test the accuracy of the shifts in attentional foci, a second experiment was performed in which just the target segments themselves were presented in the absence of distracters (right half of Figure 61.2). In this case, the activations produced by the targets themselves were mapped in order to identify the true retinotopic locations at which they were represented in the cortex. If the attention-related foci were accurately associated with each target location, then the attentional activations and the stimulus-driven activations

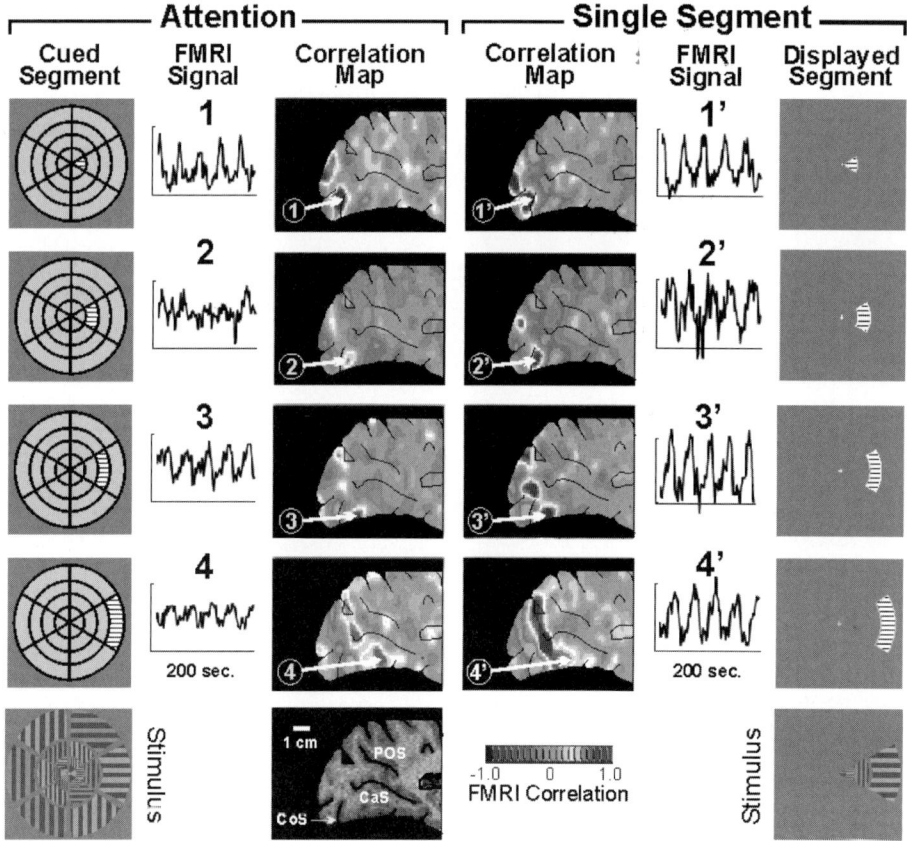

FIGURE 61.2 Retinotopic attentional modulation compared to activation evoked by cued targets alone. **Cued Segment** (left column) shows schematic sequence of target segments cued for attentional scrutiny. **Stimulus** array (bottom left) is composed of 4 rings of 6 colored (blue/orange) and striped (vertical/horizontal) segments subtending 54°. **FMRI Signal** (left) shows response modulation of individual voxels at sites indicated on adjacent **Correlation Maps**. Temporal phase shift of the signal at each site identifies the corresponding locus of attention. Correlation maps show sites where timing of modulation was positively correlated (dark patches with white outlines) with the timing of attentional shifts. **Displayed Segment** (right column) shows schematic sequence of single segments presented during otherwise identical control experiment. Composite of single segments shown in **Stimulus** (bottom right), **FMRI Signal** and **Correlation Map** on right show results of the control experiment. **Structural MRI** (bottom) is a parasagittal section (13.6 mm left of midline) through the occipital lobe in the same plane as the correlation maps. Sulcal landmarks: CaS—calcarine sulcus, CoS—Collateral sulcus, POS—parieto-occipital sulcus. (Adapted from Brefczynski and DeYoe, 1999.)

should match. They did, exceptionally well. In a temporal phase mapped design, retinotopic location is coded as time delay, that is, temporal phase. By constructing a scattergram of the temporal phase of the attentional activation versus the temporal phase of the isolated target activation for each voxel, we could quantitatively compare the retinotopic accuracy of the match between the two data sets. Figure 61.3 shows the results of this analysis for the eccentricity experiment just described and for an analogous experiment in which attention was shifted from segment to segment at different clock positions within a single ring of targets. This latter experiment tested retinotopic accuracy along a polar angle dimension (retinotopy being represented in polar coordinates). In both cases, the correlation between the attentional retinotopy and the stimulus target retinotopy was exceptionally high (r = 0.87–0.97). This was true for voxels from medial occipital cortex (primarily V1 and V2) and also for voxels from ventral occipital cortex (VP [also known as V3 v], V4v).

The results of these experiments show that attention directed to a specific target location in the visual field produced multiple foci of cortical enhancement in occipital visual cortex. The positions of these foci within the cortex corresponded precisely with the cortical representations of the attended target, as if presented in isolation. When the focus of attention shifted, the cortical enhancement shifted in precise correspondence. Since this study used a dense array of visual targets filling the field of view, the resulting focal activation reflected the spatial characteristics of the attentional modulation itself, not a "filtering" of the modulation through a pattern of spatially isolated stimulus features. In sum, these results not only demonstrated a physiological correlate of spatial attention but also showed that it is accurately described using a retinotopic metric.

It is important to note that these findings do not necessarily imply that the perceived attentional "spotlight" must be two-dimensional, or that it corresponds to an undifferentiated region of space. Within a given retinotopic zone of cortex, distinct populations of cells representing different distances, objects, features, or surfaces may be selectively modulated depending on the attentional task. Thus, it is possible for attentional effects to be both retinotopic and "object-based" (Duncan, 1984), especially for two-dimensional displays such as those used in this study.

III. IMPLICATIONS FOR THEORIES OF SPATIAL ATTENTION

A. Neurophysiological Insights

The results reviewed here provide several important insights into the brain mechanisms responsible for mediating the effects of directed visual attention. First, it is clear that attention-related modulation of brain activity can be very widespread and probably can affect virtually any occipital visual area. This includes the primary visual cortex V1 and, as other studies have shown (Beauchamp et al., 1997), areas such as the middle temporal visual area, MT, that originally had been thought to be immune to attentional modulation. Despite this wide distribution, the attentional effects can be very spatially accurate. This seeming paradox is accounted for by the presence of multiple visual areas that each have a more or less complete retinotopic representation of the visual field. Moreover, using the attentional field map (AFMap) technology just described, it is now possible to obtain a brain-based picture of the "window" of attention or, more precisely, of the topographic distribution of attentional effects as they would appear in the observer's field of view. Averaged across individuals, this topography appears relatively simple, with the strongest effects associated with the attended target location but with notable spread of activation to adjacent distracters

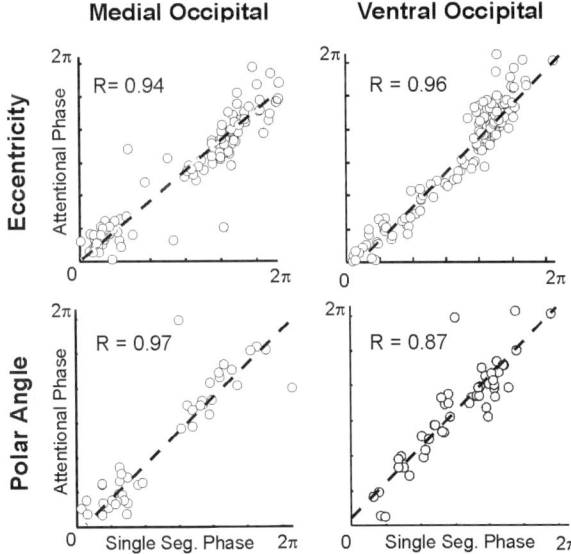

FIGURE 61.3 Comparison of visual field topography (coded by temporal phase of FMRI response) for attentional foci (y-axis) versus single segments (x-axis). Each circle represents a single responsive voxel. Data are pooled across subjects. **Left column**: Medial occipital cortex consisting primarily of V1 and V2. **Right column**: Ventral occipito-temporal cortex within and surrounding the collateral sulcus. **Top row**: Shifts in eccentricity. **Bottom row**: Shifts in polar angle. (Adapted from Brefczynski and DeYoe, 1999.)

and, weakly, into the opposite hemifield. It is noteworthy that suppression of the fMRI signal in distracter locations near the target was not a prominent feature of the results. Clear suppression occasionally can be seen for peripheral locations in the opposite hemifield (not evident in the grayscale images shown here). Thus, although it is possible that fMRI activation in nontarget regions might reflect the enhanced activity of local inhibitory neurons, there is little or no extant evidence to support such an interpretation (Logothetis, 2003). The relatively simple picture of the attentional "window" depicted in Figure 61.1B may partly reflect the common practice of averaging data across multiple individuals. For a single observer, the pattern may not be quite so simple, perhaps reflecting both a universal topographic pattern as well as consistent, but idiosyncratic, variations of this "theme."

1. Implications for Theories of Attention

The average topography reflected in the AFmaps shown earlier, is roughly consistent with a gradient model of attentional topography (Downing and Pinker, 1985) in which the attentional effects are strongest at the attended location and decrease gradually at more distant locations, especially peripherally. Although the gradient may appear steeper at locations closer to the center of gaze than the target, this may be a product of cortical magnification. Within the cortex, the steepness of central and peripheral gradients may be more similar than they appear in the visual field.

At this juncture, it is not clear how the attentional topography revealed in the AFmaps relates to object-based theories of attention (Duncan, 1984). If the attended target in our stimulus array can be considered a single object then, taken at face value, the AFmaps seem to indicate that attention is not strictly "confined" within the boundaries of the object, as might be predicted by object-based theories. However, given that fMRI activation reflects a spatial and temporal summation of neural activity that may obscure details at the level of single neurons, it would be premature to assert that this stands as an indictment of object-based attention. Also, there is mounting evidence that attention can reflect both object- and space-based models (Yantis and Serences, 2003), perhaps depending on the nature of the visual environment and the task at hand. Whether this can be revealed with fMRI remains a future challenge. In any event, the work reviewed here demonstrates that neuroimaging can provide unique new tools to explore these issues and to gain further insight into the neural basis of cognitive attributes such as attention.

References

Beauchamp, M. S., Cox, R. W., and DeYoe, E. A. (1997). Graded effects of spatial and featural attention in human area MT and associated motion processing areas. *Journal of Neurophysiology* **78**, 516–520.

Brefczynski, J. A., and DeYoe, E. A. (1999). A physiological correlate of the "spotlight" of visual attention. *Nature Neuroscience* **2**, 370–374.

Desimone, R., and Duncan, J. (1995). Neural Mechanisms of Selective Visual Attention. *Annual Review of Neuroscience* **18**, 193–222.

DeYoe, E. A., Bandettini, P., Neitz, J., Miller, D., and Winans, P. (1994). Functional magnetic resonance imaging (FMRI) of the human brain. *Journal of Neuroscience Methods* **54**, 171–187.

DeYoe, E. A., Carman, G., Bandettini, P., Glickman, S., Wieser, J., Cox, R., Miller, D., and Neitz, J. (1996). Mapping striate and extrastriate visual areas in human cerebral cortex. *Proceedings of the National Academy of Sciences—USA* **93**, 2382–2386.

Dosher, B. A., and Lu, Z. L. (2000). Mechanisms of perceptual attention in precuing of location. *Vision. Res.* **40**, 1269–1292.

Downing, C. J., and Pinker, S., Eds. (1985). The spatial structure of visual attention. *Attention and Performance*. Erlbaum, Hillsdale, NJ.

Duncan, J. (1984). Selective attention and the organization of visual information. *Journal of Experimental Psychology: General* **113**, 501–517.

Eriksen, C. W., and St. James, J. D. (1986). Visual attention within and around the field of focal attention: a zoom lens model. *Perception and Psychophysics* **40**, 225–240.

Gandhi, S. P., Heeger, D. J., and Boynton, G. M. (1999). Spatial attention affects brain activity in human primary visual cortex. *Proceedings of the National Academy of Sciences* **96**, 3314–3319.

Helmholtz, H. (1910). *Treatise on Physiological Optics*. Dover Publications, Inc., New York.

LaBerge, D. (1995). *Attentional Processing: The Brain's Art of Mindfulness*. Harvard University Press, Cambridge, MA.

Logothetis, N. K. (2003). The underpinnings of the BOLD functional magnetic resonance imaging signal. *Journal of Neuroscience* **23**, 3963–3971.

Maunsell, J. H. R., Sclar, G., Nealey, T. A., and DePriest, D. D. (1991). Extraretinal representations in area V4 in the macaque monkey. *Visual Neuroscience* **7**, 561–573.

Posner, M. I., Snyder, C. R. R., and Davidson, B. J. (1980). Attention and the detection of signals. *Journal of Experimental Psychology* **109**, 160–174.

Shih, S-I., and Sperling, G. (1996). Is there feature-based attentional selection in visual search? *Journal of Experimental Psychology: Human Perception and Performance* **22**, 758–779.

Somers, D. C., Dale, A. M., Seiffert, A. E., and Tootell, R. B. (1999). Functional MRI reveals spatially specific attentional modulation in human primary visual cortex. *Proceedings of the National Academy of Science USA* **96**, 1663–1668.

Tootell, R. B. H., Hadjikhani, N., Hall, E. K., Marrett, S., Vanduffel, W., Vaughan, J. T., and Dale, A. M. (1998). The retinotopy of spatial attention. *Neuron.* **21**, 1409–1422.

Treisman, A. M., and Gelade, G. (1980). A feature integration theory of attention. *Cognitive Psychology* **12**, 97–136.

Yantis, S., and Serences, J. T. (2003). Cortical mechanisms of space-based and object-based attentional control. *Curr. Opin. Neurobiol.* **13**, 187–193.

CHAPTER

62

Spatially-Specific Attentional Modulation Revealed by fMRI

David C. Somers and Stephanie A. McMains

ABSTRACT

Functional MRI and visual psychophysics were employed to investigate space-based attentional selection mechanisms in human occipital cortex. Our 1999 findings, along with nearly simultaneous findings from other laboratories, demonstrated that spatially specific attention could operate robustly even in primary visual cortex. Spatial deployment of attention operates in a "push-pull" fashion, both increasing responses at attended locations and decreasing responses at nonattended locations. Parametric studies suggest that spatial attention acts primarily as an "additive bias" signal, whose amplitude is largely independent of stimulus strength. Finally, the spatial window of attention, as reflected in retinotopic cortical activation, is highly flexible and it may be split into multiple "spotlights," if the task so demands.

I. INTRODUCTION

Complex visual environments bombard our eyes with more information than our cognitive systems can act on at one time. Although the retina performs massively parallel processing, visual cognition operates on no more than a few items at once. Given these limitations, what we do or do not perceive is largely determined by attentional mechanisms that select information for enhanced cognitive processing. William James proposed that visual attention acts like a spotlight, selecting a single region of visual space, while largely neglecting the rest of the visual field. The "spotlight" model, further refined by Eriksen (Eriksen and St. James, 1986) and others as a "zoom lens" mechanism that selects one contiguous, convex spatial window, has successfully addressed a wealth of psychophysical data. This article summarizes functional MRI-based research explorations of the mechanisms of spatial attention in human occipital cortex, including primary visual cortex.

II. ATTENTIONAL MODULATION IN AREA VI AND OTHER VISUAL CORTICAL AREAS

Our initial fMRI experiments began with the question of how visual cortical processing of a moving stimulus is influenced by attention (Somers et al., 1999a). These experiments depended critically on the ability to distinguish the visual cortical representations of different parts of the visual field (e.g., (Sereno et al., 1995) see Figure 62.1D). The cortical representation of the center of the visual field (or fovea) occupies a large swath of cortex in the center of the flattened occipital cortex patches (see Figure 62.1C). More peripheral eccentricities are represented as horizontal bands above and below the central region. The stimulus display was configured to exploit the eccentricity bias of the cortical map. A two-part stimulus display was utilized (see Figure 62.1A). It consisted of an annular region containing a grating pattern, which rotated either clockwise or counterclockwise on a given trial, and a central disk in which letters were displayed in a rapid serial visual presentation (RSVP) format. In this annulus-disk configuration, the RSVP letters would drive the central eccentricity band, and the motion annulus would drive the horizontal bands above and below the central region.

Subjects held central fixation on this two-part display and directed their attention to one portion or the other. Eye position measurements performed in the

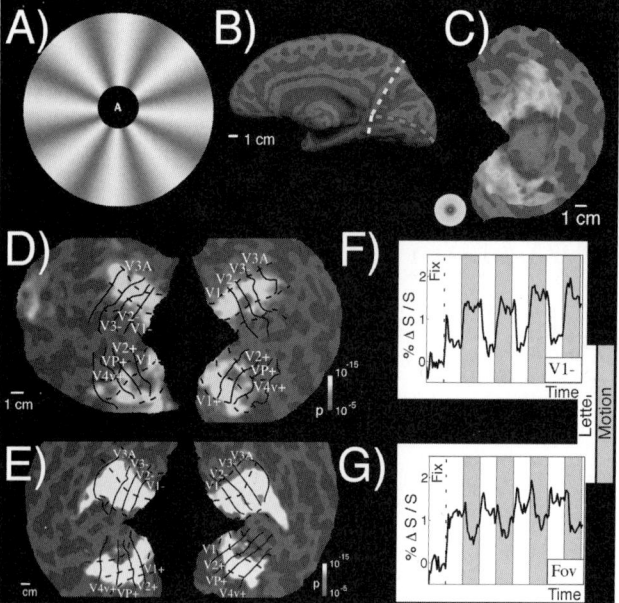

FIGURE 62.1 Attentional modulation of striate and extrastriate visual cortex. **A**. Visual stimuli were composed of an annulus with rotating radial wedge patters and a central target that was either a fixation point or single letters in an RSVP stream. **B**. The medial view of a mathematically inflated cortical surface revealing the buried sulci. The posterior portion is cut off and cut along the calcarine sulcus. The resulting patch is flattened for data visualization. **C**. Functional mapping of visual eccentricity, with the foveal representation in the center of the flattened patch. **D** and **E**. Patterns of statistically significant increased activation for attend extrafoveal motion vs. attend foveal letters for both hemispheres for two subjects, extending across all labeled retinotopic areas. **F** and **G**. Time course data for the fovea and peripherial V1, averaged across subjects for attend foveal letters vs. attend extrafoveal motion.

scanner confirmed that subjects could hold central fixation during these experiments. Subjects performed alternating blocks of trials in which they either judged the rotation direction of the motion annulus or identified five consecutive letters appearing in the central RSVP stream. The central task was designed to be highly demanding in order to strongly draw attention away from the motion annulus. Comparison of fMRI activation between the two conditions revealed spatially specific attentional modulations across all early visual cortical areas (V1, V2, V3, VP, V3A, V4v). When attention was directed to the motion annulus, activation increased in the iso-eccentricity bands corresponding to the cortical representation of the annulus (see Figure 62.1D, E, F). When attention was directed to the fovea, fMRI activation increased in the cortical representation of the fovea (see Figure 62.1G). These findings were also confirmed with a comparison of equivalent tasks, in which the RSVP stream was replaced by a motion disk stimulus that rotated independently of the annulus.

Two features of these results were remarkable. First, the attentional modulations closely followed the iso-eccentricity bands and thus attentional effects were strongly spatially selective. Perhaps the more surprising observation was that attention robustly modulated primary visual cortex (see Figure 62.1F). These attentional modulations in area V1 were as large as 1% signal change in some subjects. Overall, the attentional modulations alone comprised 60% of the combined stimulus plus attention activation (compared to passive viewing of a blank fixation display). This implies that the attentional modulation amplitude was of the same order as the pure stimulus drive activation and was perhaps even greater than the stimulus drive. The conventional wisdom had held that primary visual cortex was a "preprocessing" stage that was "cognitively impenetrable." The seminal primate electrophysiology studies of Moran and Desimone in 1985 had observed attentional modulations in area V4, but not V1. As late as 1998, a leading research group predicted that attention could not operate in area V1. Our findings flew in the face of this conventional wisdom and were supported by nearly simultaneous reports of V1 attention from other fMRI laboratories using widely different stimuli and task paradigms (see Chapter 61) (Tootell et al., 1998; Watanabe et al., 1998; Gandhi et al., 1999; Martinez et al., 1999). This wave of studies resulted in the quick acceptance of the V1 attention finding. Thus, the V1 attention result is an example of a result that was widely accepted on the basis of human fMRI data before it was embraced on the basis of primate physiology. Motter (Motter, 1993) had reported V1 attentional effects in monkeys, but this finding was questioned due to concerns that the effects might be due to small eye movement artifacts. The large areas of cortex modulated in the fMRI studies could not be explained away as an eye movement effect. Ironically, the relatively coarse spatial resolution of fMRI was an important asset in establishing this finding.

III. SELECTION MECHANISMS OF SPATIAL ATTENTION

Further experiments were aimed at determining whether the observed attentional modulations reflected an enhancement of activity at the attended locations, suppression of activity at the unattended locations, or both (Somers et al., 1999a). In the initial experiments this issue could not be resolved since there was a direct comparison of attention directed to two spatially complementary regions. This issue was addressed by performing experiments in which each attentional state (attend center, attend periphery) was paired with a condition in which the stimulus was passively viewed. The passive viewing condition represents the attentional baseline condition. Each experiment makes complementary predictions about the modulations to be observed in the two spatial regions. In the comparison of Attend Periphery vs. Passive Viewing, modulations observed in the periphery would be interpreted as attentional enhancements, whereas modulations observed in the fovea region would be interpreted as attentional suppressions. Conversely, in the "Attend Center" vs. Passive Viewing comparison, peripheral modulations would reflect suppression and central modulations would reflect enhancement. In both sets of experiments, attentional modulations were observed in both spatial regions (in antiphase relationship), implying that spatial attention acts in a "push-pull" manner, increasing responses at the cortical representations of the attended locations and diminishing responses at the nonattended cortical representations (see Figure 62.2A) (see Chapters 71 and 50).

Many of the stimulus response properties of visual cortical neurons have been extremely well characterized in electrophysiology studies. Our next question was as follows: By what computational means does attention modulate these stimulus responses? Does attention have a multiplicative effect, an additive effect, or some other form of influence on stimulus responses? We performed parametric fMRI studies in order to reveal these computational mechanisms. Contrast was the stimulus property varied. Contrast Response Functions (CRFs) have been well-established in psychophysics and single-unit physiology. As stimulus contrast increases, responses increase monotonically over a broad range of contrasts. Boynton et al. (1996) extended this CRF result to fMRI. In our experiments (Somers et al., 1999b), we constructed two fMRI CRFs using seven different contrast values, each with attention covertly directed to the stimulus and with attention directed away from the stimulus in alternate blocks. Comparison of these two CRFs will reveal how attention interacts with stimulus contrast responses (see Figure 62.2C).

Three hypotheses were proposed. Attention *might* multiplicatively "gate" stimulus responses. The multiplicative model predicts that the effects of attention should increase as the stimulus drive (and contrast) increase. Attention *might* provide an additive bias signal to the attended locations. The additive model predicts that attention effects should be constant across

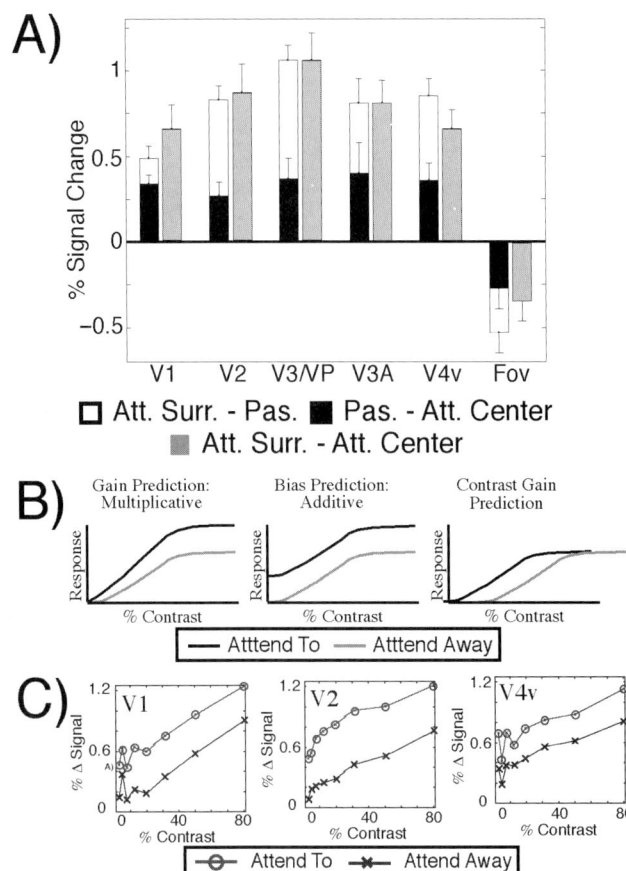

FIGURE 62.2 Mechanisms of attention. **A**. Average attentional modulation amplitudes for attend conditions vs. passive viewing, demonstrating the push-pull manner of spatial attention. **B**. Three hypothesis for how attention might interact with stimulus contrast responses. **C**. Contrast response functions for three visual areas, with and without attention, supporting the additive bias model.

all stimulus drive (and contrast) levels. Attention *might* increase the contrast gain, making a stimulus effectively appear as if it were at a higher contrast. This predicts that attention should shift the CRF to the left and that attentional effects should be strongest for the middle range of stimulus drives/contrasts (see Figure 62.2B).

In these experiments, the "with attention" and "without attention" CRFs were observed to differ primarily by a constant factor across all stimulus contrasts (see Figure 62.2C). This result supports the additive bias model prediction and is inconsistent with the multiplicative gain model. In addition to the dominant additive effect, the attentional effects exhibit a modest peak at the middle contrast ranges, consistent with some weak contribution from a contrast gain mechanism. However, since fMRI CRFs do not saturate at high contrasts (while single unit CRFs do saturate), the fMRI results are not well described by a leftward shift of the CRF or by the contrast gain modulation hypothesis.

The additive bias finding is somewhat surprising. Adding a constant factor to a response actually *decreases* the signal-to-noise ratio at that location. If perceptual decisions were computed locally this would be a destructive influence of attention. However, if perceptual decisions were computed upstream in a higher cortical area and different locations had to compete for access to this circuitry then an additive signal would be a constructive influence as it would bias the competition toward that location. This is highly consistent with the predictions of the 'Biased Competition' model of attention (Desimone, 1998) (see Chapter 50). The finding of an "attentional pedestal" of activity that is independent of stimulus drive (see Chapter 51) appears consistent with Kastner et al.'s fMRI finding that the expectation of the appearance of a stimulus can yield increased activation even in the absence of a stimulus (see Chapter 50). However, these findings differ substantially from a primate study of attentional effects on CRFs in area V4 (Reynolds et al., 2000), which reported that the primary effect was a leftward shift of the CRF1 (contrast gain hypothesis) (see Chapter 70). A small additive effect was also reported. It is difficult to reconcile these two sets of findings in different species other than to suggest that single-unit recordings reflect activation from a tiny minority of the neurons reflected in the fMRI recordings. Perhaps, a small set of highly responsive neurons exhibit robust contrast gain modulation effects under directed attention, whereas a much larger population of more weakly responsive neurons exhibits an additive pedestal influence of spatial attention. Without a more complete understanding of the workings of the neural code, it is difficult to know which neural population plays a greater role in constructing perception. Hopefully, future work will better reconcile the differences between human fMRI and primate electrophysiology studies.

IV. FLEXIBLE WINDOWS OF SPATIAL ATTENTION

William James suggested that spatial attention acts like a "spotlight," selecting what is in its beam and filtering out what falls outside the spotlight. This "spotlight" metaphor was further developed into the "zoom lens" model (Eriksen and St. James, 1986). The zoom lens model suggests that spatial attention is flexible in that it may be narrowed or expanded (and distorted) to suit task demands. However, the zoom lens model also constrains attention to act over a single contiguous region of space. This model appears too limiting in that some tasks may require attention to separated regions of space. Other tasks may benefit from excluding strong distracting stimuli from the window of selection. Some psychophysical studies have suggested that the window of spatial attention may indeed be more flexible than the zoom lens model suggests (e.g., Awh and Pashler, 2000).

We performed a set of experiments that test this hypothesis and demonstrate that spatial attention can be split into multiple distinct "spotlights" of activation in area V1 and other human occipital cortical regions (McMains and Somers, 2004). Since fMRI studies are limited by their temporal resolution and since spatial attention is capable of moving as fast as every 200–500 ms (e.g., Weichselgartner and Sperling, 1987), it is extremely difficult to rule out the movement of spatial attention as a factor in the pattern of activation observed if one relies purely on the fMRI data. A central feature of these experiments was to employ a psychophysical task that excluded the possibility that spatial attention was rapidly switching between locations of interest.

Subjects were required to compare the identity of targets simultaneously displayed in two separated locations. The targets appeared in RSVP streams and thus were visible only briefly (173 ms) and were masked immediately by another stimulus. The targets were digits appearing in RSVP streams of letters (use of letter targets did not change the psychophysical findings) and there were no other overt cues to the appearance of a target. Subjects were required to report whether the target digits were the same or different. This task cannot be performed at above chance levels unless both targets are identified. These two RSVP streams were imbedded in a display consisting of five RSVP streams, a central one surrounded by four equidistant peripheral streams, one per visual field quadrant (see Figure 62.3A). The central RSVP stream lay directly in between the two attended RSVP streams, which were placed in opposing visual field quadrants. In order to make the central stream highly distracting, only digits appeared in this stream. If information appearing in the central stream were selected along with the information in the two peripheral streams of interest, it likely would interfere with performance of the digit comparison task. Thus for this task, it was advantageous to avoid selecting the central region while selecting the two peripheral streams. Psychophysical performance in this task was investigated by parametrically varying the letter presentation duration of the RSVP streams. Threshold level performance ($d' = 1$) was observed at letter durations of 67 ms. This rate is much faster than the minimum estimates for

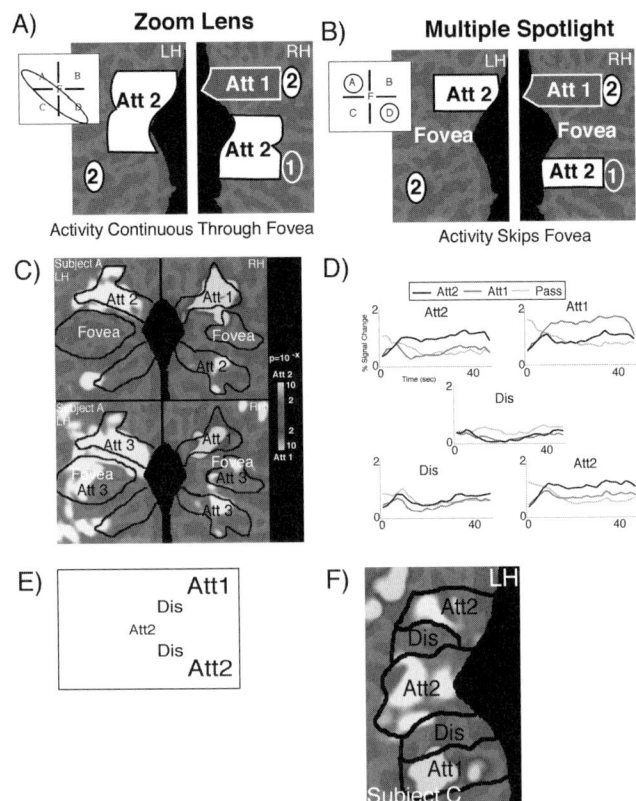

FIGURE 62.3 Multiple spotlights of spatial attention. **A** and **B**. Predicted activation patterns for Attend2 vs. Attend1 for the two competing hypotheses. **C**. Activation patterns for Attend2 vs. Attend1, for two subjects, supporting the multiple spotlight hypothesis. **D**. Time course data averaged across all subjects. **E**. Spatial layout for hemifield experiment. **F**. Activation maps for one subject in the right hemifield configuration comparing Attend2 vs. Attend1. (See color plate)

how quickly spatial attention can select a target, deploy to a new location, and select a second target (200–500 ms). In addition to the Attend2 task, an Attend1 condition was included in which attention was covertly directed only to a (third) peripheral RSVP stream, while viewing the same display as that used in the Attend2 condition. In the Attend1 task subjects had to identify digits appearing in the stream and report whether they matched a predefined target digit. Although the Attend1 task was somewhat easier than the Attend2 task, performance differences were far less than would be predicted if the Attend2 task were done by purely serial processing. The threshold in the Attend1 task was 59 ms, only 8 ms shorter than for the Attend2 task. These modest task performance differences were observed across all RSVP speeds of interest. Thus the psychophysical results demonstrate that spatial attention was simultaneously deployed to the two RSVP streams of interest.

Having thus ruled out the "rapid switching" hypothesis for this task, two hypotheses then remained that could explain selection of the two peripheral streams. The zoom lens model predicts that attention stretches to select the two RSVP streams and necessarily also selects the distracting central RSVP stream (Figure 62.3A). The alternate hypothesis suggests that the spatial window of attention can be split into two distinct spotlights that select the two RSVP streams of interest while filtering out the central stream (Figure 62.3B). This question was addressed using fMRI of occipital lobe activation during task performance. Comparison of Attend2 vs. Attend1 activation (see Figure 62.3C, D) revealed two hotspots of activation at the visual cortical representations of the two Attend2 streams. Critically, the cortical representation of the central RSVP stream was not activated (nor was it activated in comparisons with the passive viewing condition). This result demonstrates that the window of spatial attention may be split to select multiple distinct regions. This pattern was observed for all subjects. Group analysis revealed that this split attentional spotlight effect occurred in all "early" visual cortical areas, including primary visual cortex. As expected, the Attend1 condition (vs. Attend2) revealed a single hotspot of activation at the cortical representation of the attended RSVP stream (see Figure 62.3C, D).

We repeated these experiments using a different spatial configuration of RSVP streams (see Figure 62.3E). All streams were placed in one hemifield and the two RSVP streams of interest in the Attend2 task were placed in one visual field quadrant—one at the fovea and one in the periphery. The critical distracter was placed in between at a mid-eccentricity. The Attend1 RSVP stream was placed in the other visual quadrant. The fMRI results again revealed two hotspots of activation corresponding to the locations of the two targets, with a sparing of the intermediate region (see Figure 62.3F).

Our results demonstrate that attention may be split between targets located in opposite hemifields and thus split across different cortical hemispheres. We also observed splitting of the spotlight within a visual quadrant and within a cortical hemisphere. The finding that the foveal representation may either be included or excluded as one focus of attention demonstrates that attention may be divided between overt and covert targets or divided between two covertly monitored targets. In these regards, the ability to divide spatial attention exhibits remarkable flexibility.

In hindsight, the requirement that spatial attention be a serial process seems unnecessary (see Chapter 97). The capacity for parallel processing is an implicit feature of spatial representations. Once spatial attention was demonstrated to operate in early visual cortical areas (see Chapter 61) (Tootell et al., 1998; Watanabe et al., 1998; Gandhi et al., 1999; Martinez et al., 1999; Somers et al., 1999a), the general requirement for a single spotlight of spatial attention was ready to fall. These cortical areas contain primarily neurons with spatially limited receptive fields (RFs), so multiple targets need not compete for resources at these stages, provided that targets are far enough apart. Even in higher visual cortical areas, where attention shrinks down the normally large RFs to isolate single targets (Desimone, 1998), attentional selection may divide the pool of relevant neurons between multiple targets rather than assigning them in a "winner-take-all" fashion to one target location. This strategy might result in some cost for attending to multiple targets but would not necessarily result in a complete failure to select some locations. A modestly parallel spatial attention system of this form appears much simpler to implement physiologically than does a rapidly switching serial spatial attention system.

V. SUMMARY

fMRI is a powerful technique for revealing how spatial attention is deployed across the cortical representations of the visual field. fMRI has helped to reveal that spatial attention may have robust effects in primary visual cortex and that attention may be divided between multiple, distinct, spatial windows or "spotlights." Parametric fMRI studies hold the promise of a greater understanding of the computational mechanisms of attention. Our studies revealed an additive push-pull bias signal that is consistent with the Biased Competition hypothesis (Desimone, 1998).

References

Awh, E., and Pashler, H. (2000). Evidence for split attentional foci. *J. Exp. Psychol. Hum. Percept. Perform.* **26**, 834–846.
Boynton, G. M., Engel, S. A., Glover, G. H., and Heeger, D. J. (1996). Linear systems analysis of functional magnetic resonance imaging in human V1. *J. Neurosci.* **16**, 4207–4221.
Desimone, R. (1998). Visual attention mediated by biased competition in extrastriate visual cortex. *Philos. Trans. R. Soc. Lond. B. Biol. Sci.* **353**, 1245–1255.
Eriksen, C. W., and St. James, J. D. (1986). Visual attention within and around the field of focal attention: a zoom lens model. *Percept. Psychophys.* **40**, 225–240.
Gandhi, S. P., Heeger, D. J., and Boynton, G. M. (1999). Spatial attention affects brain activity in human primary visual cortex. *Proc. Natl. Acad. Sci. USA* **96**, 3314–3319.
Martinez, A., Anllo-Vento, L., Sereno, M. I., Frank, L. R., Buxton, R. B., Dubowitz, D. J., Wong, E. C., Hinrichs, H., Heinze, R. I., and Hillyard, S. A. (1999). Involvement of striate and extrastriate visual cortical areas in spatial attention. *Nat. Neurosci.* **2**, 364–369.
McMains, S. A., and Somers, D. C. (2004). Multiple spotlights of attentional selection in human visual cortex. *Neuron.* **42**, 677–686.
Motter, B. C. (1993). Focal attention produces spatially selective processing in visual cortical areas V1, V2, and V4 in the presence of competing stimuli. *J. Neurophysiol.* **70**, 909–919.
Reynolds, J. H., Pasternak, T., Desimone, R. (2000). Attention increases sensitivity of V4 neurons. *Neuron.* **26**, 703–714.
Sereno, M. I., Dale, A. M., Reppas, J. B., Kwong, K. K., Belliveau, J. W., Brady, T. J., Rosen, B. R., and Tootell, R. B. (1995). Borders of multiple visual areas in humans revealed by functional magnetic resonance imaging. *Science.* **268**, 889–893.
Somers, D. C., Dale, A. M., Seiffert, A. E., and Tootell, R. B. (1999a). Functional MRI reveals spatially specific attentional modulation in human primary visual cortex. *Proc. Natl. Acad. Sci. USA* **96**, 1663–1668.
Somers, D. C., Seiffert, A. E., Dale, A. M., and Tootell, R. B. (1999b). Effects of task difficulty and stimulus contrast on attentional modulation in human striate and extrastriate cortex revealed by fMRI. In: *Society for Neuroscience*, 6.10.
Tootell, R. B., Hadjikhani, N., Hall, E. K., Marrett, S., Vanduffel, W., Vaughan, J. T., and Dale, A. M. (1998). The retinotopy of visual spatial attention. *Neuron.* **21**, 1409–1422.
Watanabe, T., Harner, A. M., Miyauchi, S., Sasaki, Y., Nielsen, M., Palomo, D., and Mukai, I. (1998). Task-dependent influences of attention on the activation of human primary visual cortex. *Proc. Natl. Acad. Sci. USA* **95**, 11489–11492.
Weichselgartner, E., and Sperling, U. (1987). Dynamics of automatic and controlled visual attention. *Science.* **238**, 778–780.

CHAPTER

63

The Neural Basis of the Attentional Blink

René Marois

ABSTRACT

A fundamental characteristic of visual attention is its severely limited processing capacity: We have great difficulty in attending to more than one visual event at a time. This limitation is clearly illustrated by the attentional blink (AB) paradigm: When subjects search for two targets in a rapid serial visual presentation (RSVP) of distracter items, they are severely impaired at detecting the second of the two targets when it is presented within 500 ms of the first target. Imaging and electrophysiological studies of the attentional blink suggest that the bottleneck of information processing revealed by the AB is not located in visual cortex, but instead primarily resides in a network of fronto-parietal regions previously implicated in visuo-spatial attention. Brain lesions studies are consistent with these findings, but also suggest that additional cortical and subcortical regions may be involved as well. Overall, these findings are consistent with a late, post-perceptual locus for the bottleneck of information processing for the attentional blink, and further suggest that the AB may be a multicomponent deficit arising from the contribution of a multitude of brain regions.

The human brain is heralded for its complexity and enormous processing capacity: Its billions of neurons and trillions of synaptic connections can process and exchange massive amounts of information over distant extents of brain tissue in the matter of milliseconds. Correspondingly, early stages of visual information processing are endowed with massive parallel processing capacities. Yet, for all their neurocomputational sophistication, humans show severe capacity limits in the number of objects they can attend to or hold in mind, or in the number of tasks they can perform at once. Thus, from early sensory stages with seemingly unlimited capacities, information flow hits capacity-limited processing stages on its way to response execution.

I. THE ATTENTIONAL BLINK

The division between capacity-limited and capacity-unlimited stages of information processing is a prominent feature of virtually all cognitive models of visual attention (e.g., Duncan, 1980; Chun and Potter, 1995). In these two-stage models, the early stage permits the rapid, initial categorization of the visual world, whereas later attention-demanding, capacity-limited stages are necessary for the conscious report of, and action upon, the stimuli. The dual nature of perception is illustrated by the attentional blink (AB) paradigm: When subjects search for two targets presented in a rapid serial visual display of distracter items, they are severely impaired at detecting the second of the two targets when it is presented within 500 ms of the first target (see Fig. 63.1) (Raymond et al., 1992; Chun and Potter, 1995). The deficit with the second target (T2) is a result of attending to the first target (T1): subjects have no difficulty in reporting T2 when only it is required to be detected (Raymond et al., 1992; Chun and Potter, 1995). Thus, T2 report is slowed and demanding only when attention is engaged in processing a previously presented target (T1). The observation that a single target can easily be detected in an RSVP whereas detection of the second of two targets is drastically more difficult supports a two-stage model of the AB, in which an initial stage of rapid representation and categorization of visual events is followed by the slow, capacity-limited and attention-demanding consolidation of these events for conscious report (Chun and Potter, 1995).

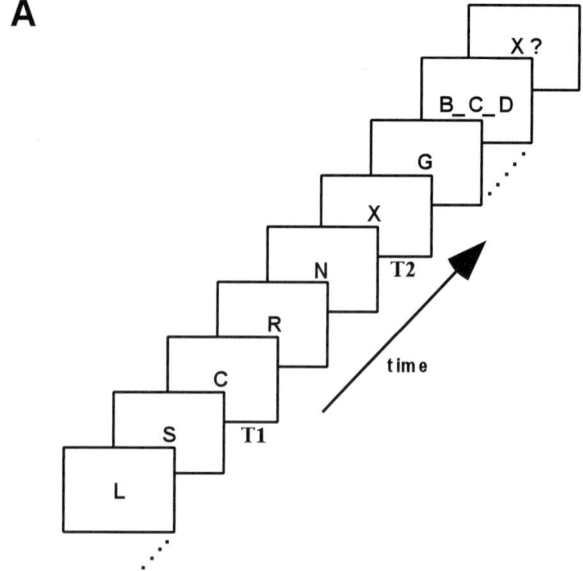

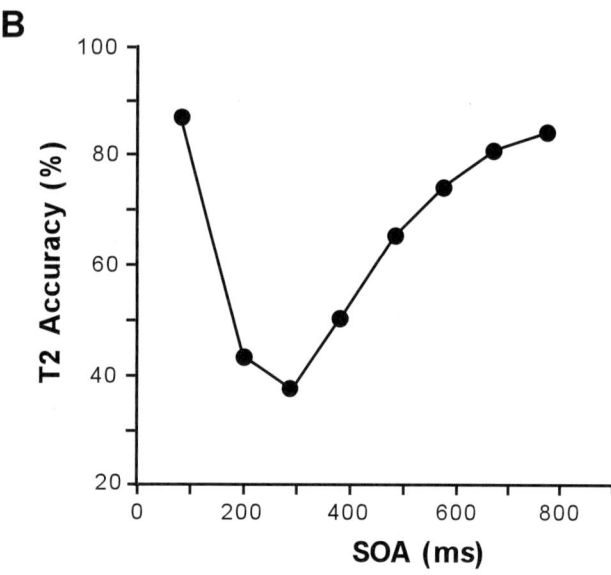

FIGURE 63.1 The Attentional Blink paradigm. A) Standard Trial Design. Subjects search for two targets (T1: letter B, C, or D; T2: presence or absence of X) presented among distracters, and respond at the end of each trial. B) Standard T2 performance by T1-T2 SOA for trials with correctly identified T1.

Although there is little doubt that the AB demonstrates a limitation in attentional processing, there is considerably more debate regarding the nature of this capacity-limited process (Shapiro et al., 1997a). Although some models suggest that it reveals the necessary time an item must be attended before other items can be attended, several others propose instead that it is related to the limited capacity of short-term memory, with some theories ascribing the limitation at the stage of encoding or consolidating items into short-term memory, and yet others assigning it to interference between items during retrieval from short-term memory. Importantly, not all these models are mutually exclusive, and it is certainly conceivable that the AB results from more than one capacity-limited process (see Section V).

II. THE NEURAL BASIS OF THE AB BOTTLENECK

Although theories may differ about the nature of the AB bottleneck, they are consistent in placing this bottleneck at a relatively late stage of information processing (Shapiro et al., 1997a). In keeping with this notion, event-related potential (ERP) studies indicate that visually presented target (T2) words that are not explicitly perceived by the subjects in an attentional blink paradigm are nonetheless processed up to semantic identity by the brain (Luck et al., 1996). This finding is further supported by behavioral studies demonstrating that undetected T2 words can still induce semantic priming (Shapiro et al., 1997b). Although these behavioral and electrophysiological studies clearly show that missed T2 items are nevertheless processed, they reveal little about the neural locus of the AB bottleneck. Such information, however, can be acquired with functional magnetic resonance imaging (fMRI).

An initial series of fMRI experiments sought to localize the AB bottleneck by measuring brain activity associated with manipulations known to modulate the attentional blink (Marois et al., 2000). Since the attentional blink is caused by the processing demands of T1 (Raymond et al., 1992; Chun and Potter, 1995), these studies focused on T1 in the absence of T2 confounds. In addition, the experimental design took advantage of the fact that the AB critically depends on the perceptual interference generated by distracter items, especially the one immediately following T1 (Raymond et al. 1992; Chun and Potter, 1995; Marois et al., 2000). Identifying a target among competing distracters not only requires attention, but in an RSVP attention must be deployed before T1 is perceptually erased by the subsequent, masking distracters. As a result of the high attentional demands imposed on T1 processing by interfering distracters, attention cannot be redeployed immediately for processing of a second target, thereby preventing the latter from entering awareness (Chun and Potter, 1995; Marois et al., 2000).

According to this AB model, increased T1 interference not only leads to bigger attentional blinks (Raymond et al., 1992; Chun and Potter, 1995); but it should also lead to enhanced activation in brain regions involved in attentional resolution of T1 identification. These brain regions can therefore be isolated by comparing two conditions that differ only by the degree to which T1 is masked by temporally adjacent distractors. When such manipulation was carried out in the scanner, increased target interference was associated with activation of the intra-parietal sulcus (IPS), inferior lateral frontal, and anterior cingulate cortex (Marois et al., 2000), predominantly in the right hemisphere (see Fig. 63.2). This activation pattern was also observed in a second experiment that varied the degree of target-distracter discriminability (Marois et al., 2000), another manipulation previously shown to affect the magnitude of the AB (Chun and Potter, 1995). Furthermore, similar results are also obtained with a spatial form of attentional blink in which target-distracter interference is induced, not by distracters that temporally precede or follow the first target, but by presenting the distracters simultaneously with, but spatially adjacent to, T1. Such spatial interference paradigm not only induced a powerful AB, but also recruited the same parieto-frontal network observed with the classical temporal interference paradigm (Marois et al., 2000).

Taken together, these findings clearly indicate that T1 manipulations shown to affect the magnitude of the attentional blink recruit a parieto-frontal network of brain regions, and therefore point to this network as a candidate neural locus of the capacity-limited process underlying the AB deficit. The localization of the AB bottleneck in a parieto-frontal network is also highly consistent with behavioral and electrophysiological studies suggesting that it occurs at a late, post-perceptual stage of information processing (Luck et al., 1996; Shapiro et al., 1997b). Interestingly, the frontal and parietal cortical areas activated in the AB fMRI studies are similar to the brain regions involved in visuo-spatial attention (Corbetta and Schulman, 2002). Based on this convergence of findings, it has been proposed that the neural network involved in visuo-spatial attention may form the capacity-limited bottleneck of visual information processing revealed by the AB (Marois et al., 2000; Chun and Marois, 2002).

III. NEURAL FATE OF T2

The initial fMRI study by Marois and colleagues (2000) aimed at identifying candidate neural loci of the AB bottleneck by focusing on T1 processing. Complementarily, a recent study (Marois et al., 2004) examined neural processing of T2. Both behavioral (Shapiro et al., 1997b) and electrophysiological (Luck et al., 1996) data suggest that T2 is processed by the brain even when it is not detected by the subjects. These findings predict that a substantial extent of the visual information processing pathway should be activated by T2 regardless of whether T2 is consciously perceived or not. Furthermore, prior imaging data (Marois et al., 2000) suggest that the fronto-parietal cortex should be recruited only or primarily when the T2 target is consciously reported, that is, when the target reaches the slow, capacity-limited and attention-demanding consolidation stage. These predictions were tested in an fMRI experiment (Marois et al., 2004) in which subjects searched first for a target face (T1) and then for a target scene (T2) presented among an RSVP of scrambled scenes (see Fig. 63.3A). Since scenes activate a canonical region of the visual cortex—the parahippocampal place area (PPA)—that is largely insensitive to faces (Epstein and Kanwisher, 1998), this experimental design permits the assessment of the response to T2 (scenes) in a brain region supporting high-level visual categorization uncontaminated by the processing of T1 (faces).

Behaviorally, the experimental paradigm leads to a robust attentional blink for scenes (Marois et al., 2004). Yet, even though the T2 scene target frequently went undetected by the subjects, it activated the PPA (see Fig. 63.3B). Moreover, this activation was amplified when the stimulus was consciously perceived. These results suggest that activation of a visual cortex area subserving high-level categorization can still take place even in the absence of visual awareness, as predicted from the behavioral and ERP studies (Luck et al., 1996; Shapiro et al., 1997b). However, the enhanced activation observed with conscious scene perception suggests that the PPA's role may not be limited to automatic stimulus categorization, as this signal amplification may be a necessary component of scene awareness.

In contrast to the PPA, the lateral frontal cortex was activated only when scenes were successfully

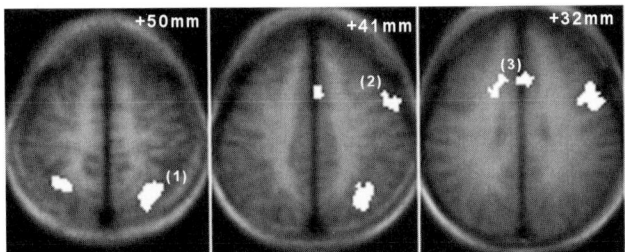

FIGURE 63.2 Brain activations associated with temporal masking interference manipulations shown to affect the magnitude of the Attentional Blink. Increase in target-distracter interference recruits the intra-parietal sulcus (1), lateral frontal (2), and anterior cingulate (3) cortex (from Marois et al., 2000).

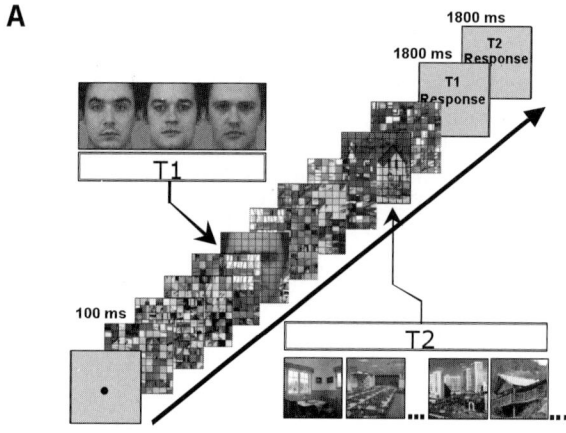

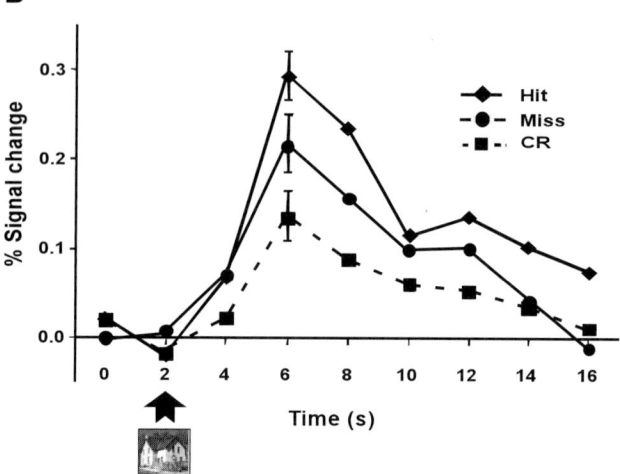

FIGURE 63.3 T2-related activation in the parahippocampal place area (PPA). (A) Experimental Design. Subjects searched for a face target (T1) and a scene target (T2) presented in an RSVP of scrambled distracter scenes. Insets show the three face targets and examples of both indoor and outdoor scene targets. (B) PPA activation timecourses with T2 Hit, Miss, and Correct Rejection (CR) trials (from Marois et al., 2004).

reported, whereas the response of the intra-parietal cortex did not clearly distinguish between T2 detections and misses, although it showed a trend for prolonged activation with T2 hits relative to T2 misses and correct rejections (Marois et al., 2004).

Taken together, these results generally support the proposal (Marois et al., 2000) that the bottleneck of information processing revealed by the attentional blink is not primarily located in visual cortex, but in a fronto-parietal cortical network. These findings are also consistent with the two-stage model of the attentional blink, which proposes that stimuli are initially characterized and registered at an early stage of visual information processing, followed by a second, capacity-limited attention-demanding stage required for conscious report (Chun and Potter, 1995). Specifically, the imaging data suggest that stimulus categorization in visual cortex may correspond to the first stage of Chun and Potter's two-stage model, in which the input is registered in an efficient, preconscious manner, and the lateral frontal cortex may be associated with the attention-demanding, capacity-limited second stage of information processing (Chun and Marois, 2002). As for the intra-parietal cortex, its specific contribution to the AB has not yet been fully resolved. On the one hand, the IPS seems to have a limited role in conscious target report (Marois et al., 2004). On the other, T1 manipulations of distracter interference that affect the magnitude of the AB modulate IPS activation (Marois et al., 2000). It is conceivable that the role of the intra-parietal cortex primarily consists in resolving perceptual (target-mask) interference (Marois et al., 2000) or, more broadly speaking, in controlling the distribution of attention among visual events, rather than in explicit target perception *per se*.

In sum, the neuroimaging studies suggest that activation of the visual cortex is not sufficient for conscious target perception in the AB, which would necessitate the recruitment of the frontal cortex, and possibly of the parietal cortex as well. Evidently, the explicit perception of a visual stimulus is likely to result from the interaction between a sensory representation of the stimulus in visual cortex and a fronto-parietal attentional network necessary to consolidate that stimulus for full report (Beck et al., 2001).

IV. EFFECTS OF REAL AND VIRTUAL BRAIN LESIONS ON THE ATTENTIONAL BLINK

Another approach to understanding the neural basis of the AB consists in investigating how it is affected by brain lesions. In the first of such studies, patients with hemispatial neglect following lesions in inferior parietal cortex, inferior frontal cortex, or basal ganglia exhibited more pronounced ABs than patients without neglect following lesions in other parietal, frontal, or subcortical regions (Husain et al., 1997). Importantly, there were no differences in performance between the two patient groups when they were required only to identify T2, demonstrating that the exacerbated AB in the neglect group was not a result of general cognitive impairment. These findings provide a novel insight into the syndrome of neglect, for they suggest that it is a disorder of directing attention in time as well as in space. Although these results also suggest that the attentional blink and neglect arise from similar neural mechanisms, more recent brain lesion studies have shown that AB abnormalities can occur independently of neglect (Rizzo et al., 2001; Shapiro et al., 2002): Patients without neglect but with focal brain lesions in inferior parietal cortex (Shapiro et al., 2002), and perhaps occipito-temporal and frontal cortex as well (Rizzo *et al.*, 2001), can still display abnormally pronounced ABs. Importantly, although these studies suggest that damage to several brain regions may affect the AB, this behavioral impairment exhibits neuroanatomical specificity since lesions to other cortical regions, such as the superior parietal cortex, do not affect the AB (Shapiro et al., 2002).

In general, the brain lesion studies accord with the imaging findings of a fronto-parietal locus for the AB bottleneck (Marois et al., 2000; 2004). In particular, both neuropsychological and fMRI approaches highlight the lateral frontal cortex as an important neural substrate of the attentional blink. Although imaging and neuropsychological data also implicate the parietal cortex in the AB, the precise localization and nature of its involvement has not yet been fully resolved. Imaging studies suggest that the intra-parietal cortex may be involved in attentional resolution of target-distracter interference rather than in target detection *per se* (Marois et al., 2000; 2004); brain lesion studies suggest that the inferior, but not superior, parietal cortex contribute to the AB (Shapiro et al., 2002). Since the IPS separates the inferior and superior parietal cortex, it is unclear how this region was affected in the brain lesion experiments. However, virtual lesion studies provide converging evidence for a role of the intra-parietal cortex in the attentional blink: Transcranial magnetic stimulation (TMS) over the IPS modulates the magnitude of the AB (Cooper et al., 2004) although, somewhat surprisingly, by lessening the AB deficit. While these results support the involvement of the IPS in the AB, it is worth emphasizing that they do not preclude other parietal regions, notably the temporo-parietal junction, from contributing to other aspects of the AB (Shapiro et al., 2002; but see Marois et al., 2004), such as target detection.

V. CONCLUSION

Cognitive models of the AB propose that the attentional blink can be accounted for by a two-stage process (Chun and Potter, 1995). In this model, unattended visual stimuli are initially processed to the level of identification, but capacity-limited attentional processes are needed to bring these visual representations into a state that can be consciously reported. Behavioral, electrophysiological, and imaging data generally support this model (Luck et al., 1996; Shapiro et al., 1997b; Marois et al., 2000; 2004). Visual cortical areas involved in high-level categorization/identification can still be activated by T2 targets that are not explicitly perceived (Marois et al., 2004), suggesting that the first stage of the dual-stage model includes even relatively high-level visual cortex. Furthermore, both brain lesion and imaging data place the second, bottleneck stage in a multitude of brain regions that include, but most likely are not limited to, the frontal and parietal cortex. This notion in turn suggests that the AB bottleneck may be a neural network property rather than emanating from a single brain region.

The fMRI and neuropsychological findings may not be easily pegged into a ventral parietal-frontal network subserving stimulus-driven attention, or a dorsal parietal-frontal network subserving goal-driven attention (Corbetta and Schulman, 2002), since there is evidence that brain regions from both of these networks are involved in the AB (e.g., intra-parietal cortex, inferior lateral frontal cortex). Rather, the notion that several brain regions contribute to the attentional blink is consistent with the view that the AB might result from processing deficits at multiple, late stages of information processing, perhaps involving both top-down and bottom-up attentional mechanisms. For instance, impairments in target-distracter interference resolution, in reorienting of attention (from T1 to T2), or in target consolidation into working memory would all be expected to affect the attentional blink. Importantly, because all these processes evidently occur after initial stimulus categorization, a multibottleneck view is still consistent with a refined version of two-stage models (Duncan, 1980; Chun and Potter, 1995) in which the second, capacity-limited stage is comprised of several substages. Viewed from this multibottleneck perspective, it would not be surprising if a plethora of brain regions were associated with the AB.

Regardless of how many neural processes may turn out to be associated with the AB, the fact that the brain regions most consistently associated with the attentional blink appear to be the same regions subserving visual attention (Corbetta and Schulman, 2002) provides neural evidence that the attentional blink is not only aptly named, but that it can also be employed to yield fundamental insights into the nature of our all-too-costly attentional limitations.

References

Beck, D. M., Rees, G., Frith, C. D., and Lavie, N. (2001). Neural correlates of change detection and change blindness. *Nat. Neurosci.* **4**, 645–650.

Chun, M. M., and Marois, R. (2002). The dark side of visual attention. *Curr. Opin. Neurobiol.* **12**, 184–189.

Chun, M. M., and Potter, M. C. (1995). A two-stage model for multiple target detection in rapid serial visual presentation. *J. Exp. Psychol. Hum. Percept. Perform.* **21**, 109–127.

Cooper, A. C. G., Humphreys, G. W., Hulleman, J., Praamstra, P., and Georgeson, M. (2004). Transcranial magnetic stimulation to right parietal cortex modifies the attentional blink. *Exp. Brain Res.* **155**, 24–29.

Corbetta, M., and Schulman, G. L. (2002). Control of goal-directed and stimulus-driven attention in the brain. *Nat. Rev. Neurosci.* **3**, 201–215.

Duncan, J. (1980). The locus of interference in the perception of simultaneous stimuli. *Psychol. Rev.* **87**, 272–300.

Epstein, R., and Kanwisher, N. (1998). A cortical representation of the local visual environment. *Nature* **392**, 598–601.

Husain, M., Shapiro, K., Martin, J., and Kennard, C. (1997). Abnormal temporal dynamics of visual attention in spatial neglect patients. *Nature* **385**, 154–156.

Luck, S. J., Vogel, E. K., and Shapiro, K. L. (1996). Word meanings can be accessed but not reported during the attentional blink. *Nature* **383**, 616–618.

Marois, R., Chun, M. M., and Gore, J. C. (2000). Neural correlates of the attentional blink. *Neuron* **28**, 299–308.

Marois, R., Yi, D. J., and Chun, M. M. (2004). The Neural Fate of Consciously Perceived and Missed Events in the Attentional Blink. *Neuron* **41**, 465–472.

Raymond, J. E., Shapiro, K. L., and Arnell, K. M. (1992). Temporary suppression of visual processing in an RSVP task: An attentional blink? *J. Exp. Psychol. Hum. Percept. Perform* **18**, 849–860.

Rizzo, M., Akutsu, H., and Dawson, J. (2001). Increased attentional blink after focal cerebral lesions. *Neurology* **57**, 795–800.

Shapiro, K. L., Arnell, K. M., and Raymond, J. E. (1997a). The attentional blink. *Trends Cog. Sci.* **1**, 291–296.

Shapiro, K., Driver, J., Ward, R., and Sorensen, R. E. (1997b). Priming from the attentional blink: A failure to extract visual tokens but not visual types. *Psychol. Sci.* **8**, 95–100.

Shapiro, K., Hillstrom, A. P., and Husain, M. (2002). Control of Visuotemporal Attention by Inferior Parietal and Superior Temporal Cortex. *Curr. Biol.* **12**, 1320–1325.

CHAPTER

64

Neurophysiological Correlates of the Reflexive Orienting of Spatial Attention

Jillian H. Fecteau, Andrew H. Bell, Michael C. Dorris, and Douglas P. Munoz

ABSTRACT

There are two reflexive biases in orienting attention that assist in the exploration of the visual scene. A distinct object will draw, or *capture*, spatial attention to its locus. After this object has been inspected (and deemed irrelevant), *inhibition of return* prevents its repeated inspection. Here, we describe the neurophysiological correlates of these biases in orienting attention. In the superior colliculus, both originate from changes in sensory processing—the capture of attention is linked to a strong neural representation of a visual target, whereas inhibition of return is associated with a weak representation of this target. We describe how changes in this sensory signal may produce changes in behavior and can explain the typical and anomalous findings associated with these biases in reflexively orienting attention.

Figure 64.1a illustrates the cue-target task that elicits both the capture of attention and inhibition of return (Posner and Cohen, 1984). In this task, a flash of light in peripheral visual field (the cue) is followed by a second visual stimulus (the target) that appears at the same or opposite location as the cue. The cue serves as the distinct, or salient, object. Responding to the target probes the consequences of the cue on orienting attention toward a new object (i.e., the target).

Manipulating the time between the onset of the cue and target (cue-target onset asynchrony, CTOA) produces a cross-over interaction, as shown in the mean correct saccadic reaction time of 16 human (see Fig. 64.1b, left) and 2 monkey (right) observers. This may be seen more clearly in the subtraction plot illustrated in Fig. 64.1c (positive values signify the advantage in saccadic reaction time when the cue and target appear at the same location, whereas negative values signify the advantage in saccadic reaction time when the cue and target appear at opposite locations). At the short CTOAs, the participants responded faster when the cue and target appeared at the same location. This represents the capture of attention—the compulsion that observers have to inspect abrupt changes in the visual scene (Jonides, 1981; see Chapter 69). At longer CTOAs, the participants responded more slowly when the cue and target appeared at the same location. This represents inhibition of return (see Chapter 16), the tendency of observers to favor new locations in the visual scene, as opposed to previously inspected locations (see also Posner and Cohen, 1984; Posner et al., 1985). Although both humans and monkeys produced similar patterns of behavior, the timing of the cross-over was shifted forward for monkeys; this originates from the repeated testing of monkeys on this task (Dorris et al., 2002).

As the monkeys performed this task, we monitored the activity of visuomotor neurons in the intermediate layers of the superior colliculus (SC). The SC is of particular interest when exploring the neurophysiological basis of reflexive orienting because it may produce inhibition of return (Posner et al., 1985; Sapir et al., 1999). In addition, its visuomotor neurons produce three distinct neural signals that could carry these biases in orienting attention. Each neuron discharges in response to the appearance of a visual stimulus in and the initiation of a saccade generated to its response field (see Munoz et al., 2000; Munoz and Fecteau, 2002 for reviews). Moreover, visuomotor neurons can have low-frequency activity that is modulated by attention, motor preparation, and target selection (e.g., Glimcher

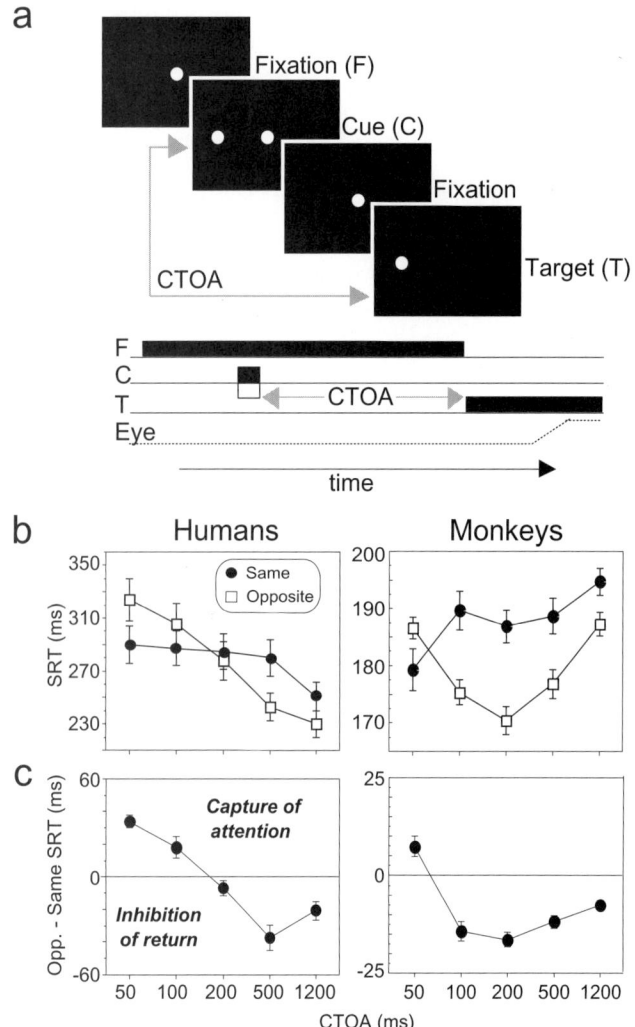

FIGURE 64.1 a. An illustration of the cue-target paradigm used in these studies. Observers fixated a central fixation marker during the presentation of a peripheral cue. A target (T) appeared (fixation marker disappeared simultaneously) and the monkey generated an eye movement to the target's location. The cue and target could appear at the same or opposite locations. The time between cue and target onset (cue-target onset asynchrony: CTOA) was manipulated. b. Mean correct saccadic reaction times for human (left) and monkey (right) observers when the target appeared at the *same* (filled circle) and *opposite* (squared) locations as the cue across the CTOAs tested (adapted from Fecteau et al., in submission). c. Subtraction plot showing the differences between same and opposite cueing conditions for both groups of participants.

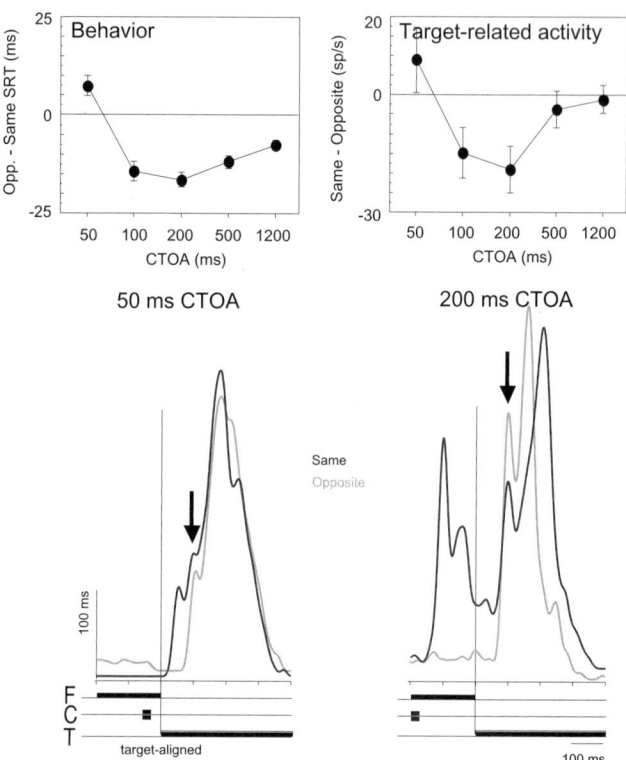

FIGURE 64.2 a. Differences in saccadic reaction time (left) and the corresponding changes in the target-related response (right). b. Representative example of the activity of a single neuron at the 50 ms CTOA when the capture of attention was obtained in behavior. c. Representative example of the activity of a single neuron at the 200 ms CTOA when inhibition of return was obtained in behavior (adapted from Fecteau et al., 2003).

and Sparks, 1992; Basso and Wurtz, 1997; Dorris and Munoz, 1998).

Of these three candidates, only the sensory response to the target (peak target-related activity; see Fig. 64.2a, right) mimicked the changes observed in orienting behavior (left). Stronger target-related activity was obtained at the 50 ms CTOA, whereas weaker target-related activity was obtained at longer CTOAs (≥100 ms) when the cue and target appeared at the same location.

In addition, a strong correlation was obtained between target-related activity and saccadic reaction time on a trial-by-trial basis for each neuron, with stronger target-related activity associated with shorter reaction times and weaker target-related activity linked to longer reaction times (Bell et al., 2004; Dorris et al. 2002; Fecteau et al., 2004). A similar relationship was not observed for other neurophysiological signals (pretarget activity or for saccade-related activity) and behavior (Fecteau et al., 2004).

Demonstrating that changes in target-related activity link to the capture of attention and inhibition of return does not reveal what is responsible for these changes. To provide some insight into this issue, we show the activity of two representative neurons. At the 50 ms CTOA (see Fig. 64.2b), target-related activity was stronger because the appearance of the cue was still eliciting a response in the neuron (black trace). So, when the target-related signal entered (see Fig. 64.2b, black arrow), it summed with the residual activity from the cue, enhancing its peak compared to when the cue appeared at the opposite location (gray trace, Bell et al., 2002; Fecteau et al., 2004). By contrast, at the 200 ms CTOA (see Fig. 64.2c), target-related activity was attenuated when the cue appeared at the same location as the target (black trace) than when it appeared at the opposite location (gray trace; Dorris et al. 2002; Bell et al., 2004; Fecteau et al., 2004).

One intriguing feature of the pattern of neural activity obtained at the 200 ms CTOA was that the neuron was more excitable after the appearance of the cue (neural activity was elevated before the visual target was registered by the neuron; black trace above gray trace). This effect, observable in most visuomotor neurons in our sample, is somewhat counterintuitive because elevated pretarget activity is associated with faster, not slower, reaction times (Dorris and Munoz, 1998).

This observation forces us to reevaluate the role of the SC in the active generation of inhibition of return. Previously, it was hypothesized that the SC produces inhibition of return because it is eliminated after the SC is lesioned (Posner et al., 1985; Sapir et al., 1999). However, our neurophysiological data indicate that the SC is facilitated after the appearance of the cue (evidenced through elevated pretarget activity) and that the *incoming* target-related response is weakened. If this interpretation is accurate, then saccades initiated through other means should be facilitated when generated to the same location as the cue.

To test this hypothesis, we compared saccadic reaction times in two conditions: when monkeys initiated a saccade to the visual target (75% of the trials; Fig. 64.3a) and when weak electrical stimulation drove a saccade to the same location (25% of the trials; Fig. 64.3b). When using electrical stimulation, the time required to evoke the saccade depends on the level of neural excitability at the time of stimulation (Munoz et al., 2000). In line with our expectations, electrical stimulation produced the opposite pattern of saccadic reaction times (see Fig. 64.3b), than did visual targets (see Fig. 64.3a). This indicates that the incoming target-related signal, as opposed to active inhibition within the SC, is responsible for inhibition of return (Dorris et al., 2002).

At what point along the sensory-to-motor axis does the weakening of the target-related response occur? Although we cannot answer this directly, we have obtained evidence indicating that it originates early on when using this cue-target task. The SC consists of two functionally distinct subregions—the superficial and the intermediate layers. Up to now, we have described changes in neural activity in the intermediate layers, which receives its inputs from widespread cortical areas, including the frontal eye fields and the lateral intraparietal area, and subcortical areas, including regions of the basal ganglia and thalamus (reviewed in Munoz and Fecteau, 2002). By contrast, the superficial layers receive input from visual areas early in sensory processing, such as V1 and the retina (e.g., Lui et al., 1995; Pollack and Hickey, 1979).

Comparing these functional layers revealed that weaker target-related activity was observed in both (Dorris et al., 2002; Fecteau and Munoz, in submission). This indicates that the attenuation of this target-related signal occurs early (i.e., as early in processing as V1) and is propagated throughout the entire oculomotor network and, perhaps, the entire dorsal visual stream (Fecteau et al., 2004; Robinson et al., 1995). Indeed, weak target-related activity has been observed during similar tasks in the lateral intraparietal area (Robinson et al., 1995) and area 7a, as well (Constandtinidis and Steinmetz, 2001; Steinmetz et al., 1994). Area 7a is not strictly a member of the oculomotor network (Andersen et al., 1990).

Taken together then, the changes in reaction times associated with the capture of attention and inhibition of return correlate with changes in the neural salience of the visual target. Both may be viewed as the interactions of sensory stimuli in the cue-target task. Strong target-related activity, linked to the capture of attention, originates from the summation of cue- and target-related activities. Weak target-related activity, linked to inhibition of return, originates at an early stage in sensory processing. Similar findings have been demonstrated in human electrophysiology studies (see Chapter 36). This raises an interesting question—what happens when the sensory signals linked to the cue and target do *not* interact?

In the intermediate and deep layers of the superior colliculus, sensory input from several modalities impinges upon its neurons, including audition (Wallace et al., 1996; Bell et al., 2001). However, unlike the robust activity elicited by visual stimuli, auditory stimuli elicit weaker responses that are registered much sooner after the presentation of the auditory stimulus (Wallace et al., 1996; Bell et al., 2001). These features have important consequences with regard to the capture of attention. Because the neural activity linked to the auditory stimulus will have passed before the neural activity linked to the visual target enters, there is little opportunity for these signals to sum and produce the capture of attention. Regarding the mechanisms responsible for inhibition of return, cross-modal interactions between auditory and visual stimuli occur at a later stage along this sensory-to-motor axis compared to the superficial layers of the

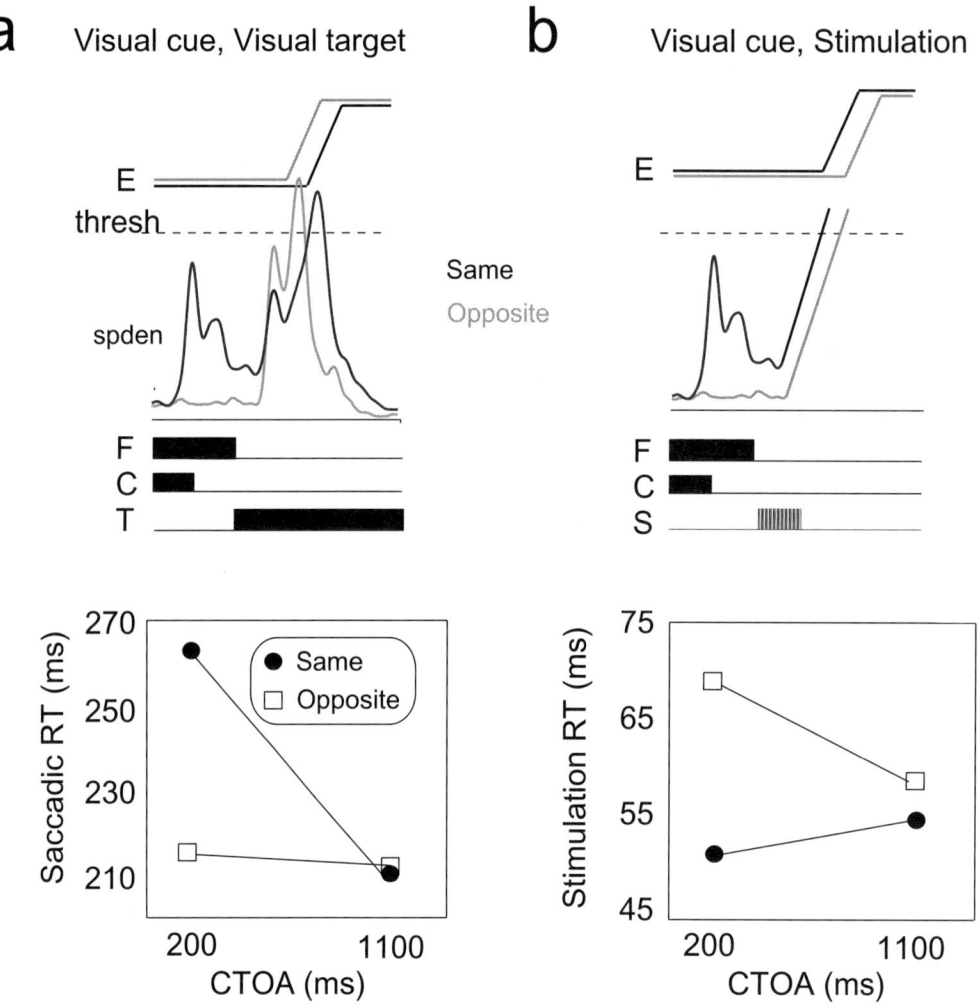

FIGURE 64.3 **a.** *Top.* Representative example of a single neuron when a visual target drives the saccadic response. Illustration shows that a theoretical saccadic threshold (dotted line) is achieved sooner when the cue appeared at the opposite location. **b.** Hypothetical example of what may happen when electrical microstimulation is applied to the SC instead of presenting a visual target, in which elevated excitability leads to saccadic threshold being achieved sooner. *Bottom.* Saccadic reaction times obtained when a visual target guided the response (left) and when microstimulation drove the response (right).

SC that show strong attenuation of the target-related response. Therefore, it is quite possible that presentation of an auditory cue will not alter the neural activity linked to the visual target for inhibition of return.

Recently, we have confirmed this hypothesis (Bell et al., 2004). Using a cue-target paradigm to contrast the consequences of auditory and visual cues on a visual target, auditory cues did not modify the behavior of the monkeys and did not alter the peak target-related activity of sensorimotor neurons in the SC.

All told, both the capture of attention and inhibition of return are associated with changes in a sensory signal or the neural salience of the target—stronger target-related activity is linked to the capture of attention and weaker target-related activity is linked to inhibition of return (see also Chapter 36).

What do these findings mean for our understanding of the reflexive orienting of spatial attention?

Attention describes a phenomenological experience (James, 1890)—the spotlight that illuminates (brings to awareness) items that are within its beam (Posner, 1980). By many accounts, spatial attention is a stand-alone cognitive ability (e.g., Bisley and Goldberg, 2003; Colby and Goldberg, 1999; Posner, 1980; Posner and diGirolamo, 2000 (see Chapter 22). Strong support for this claim comes from the consistency with which attentional phenomena are observed across different task manipulations—in particular, the apparent independence of attentional phenomena from the effector (hand versus eye) used to respond (Posner and diGirolamo, 2000; see Chapter 22).

With regard to the capture of attention and inhibition of return, careful scrutiny of the role of effector shows important *inconsistencies* when hand and eye responses are compared, though. For one, the crossover from the capture of attention to inhibition of return occurs sooner when a saccade is initiated to the locus of the target than when a manual button press is used instead (Briand et al., 2000). A more dramatic difference is obtained when retinotopic and environmental spatial reference frames are dissociated with a smooth pursuit eye movement (Abrams and Pratt, 2000). In this case, inhibition of return is encoded in an environmental reference frame for manual responses, but in a retinotopic reference frame for saccadic responses. Taken together, these findings suggest that both the capture of attention and inhibition of return are represented differently in the manual and oculomotor neural networks.

Showing that the capture of attention and inhibition of return link to changes in a visual signal makes it easy to explain how these biases appear to influence a single attentional network in most instances, but change with the effector used to respond in others. In the dorsal visual pathway, visual information is used to guide action (Milner and Goodale, 1995). Changes in the strength, or salience, of this visual signal will change how efficiently or quickly it can be translated into motor acts. This would produce the changes in reaction times that we associate with the capture of attention and inhibition of return. The widespread distribution of visual information explains the congruity associated with the capture of attention and inhibition of return in most studies. However, because different networks use this information to generate a response explains the important differences that have been observed across effectors. For instance, visual information is encoded in different coordinate frames for different effectors (Colby and Goldberg, 1999), potentially explaining why inhibition of return is encoded in different reference frames for manual and saccadic responses (Abrams and Pratt, 2000).

This description explains how changes in the strength of a visual signal could give rise to behaviors we associate with the capture of attention and inhibition of return. Does this mean that these effects produced with the cue-target task tell us more about the processes involved in initiating actions than the reflexive orienting of spatial attention (see also Chapters 20 and 31)?

Albeit one possible interpretation, it alone cannot explain our unified percept of the spotlight. For instance, our phenomenological experience of the capture of attention does not feel more fleeting when responding with the eye than with the hand. (Indeed, our phenomenological experiences can be dissociated from the actions we produce altogether. See Castiello et al., 1991; Kramer et al., 2000.) If similar visual signals are distributed widely across the brain, this information will produce different consequences depending on the network(s) expressing and using this information. Perhaps, in addition to gaining access to saccadic and manual systems that produce the required response, this visual information may modify the neural processes that give rise to the phenomenological experience of the spotlight as well. Of course the future may favor another interpretation. Nonetheless, the accumulated evidence to date casts doubt on the view that these reflexive biases in orienting attention originate from and/or influence only one attentional network in the brain.

References

Abrams, R. A., and Pratt, J. (2000). Oculocentric coding of inhibited eye movements to recently attended locations. *J. Exp. Psychol. Hum. Percept. and Perform.* **26**, 776–88.

Andersen, R. A., Asanuma, C., Essick, G., and Siegel, R. M. (1990). Corticocortical connections of anatomically and physiologically defined subdivisions within the inferior parietal lobule. *J. Comp. Neurol.* **296**, 65–113.

Basso, M., and Wurtz, R. (1997). Modulation of neural activity by target uncertainty. *Nature* **389**, 66–69.

Bell, A. H., Corneil, B. D., Meredith, M. A., and Munoz, D. P. (2001). The influence of stimulus properties on multisensory processing in the awake primate superior colliculus *Can. J. Exp. Psychol.*, **55**, 125–134.

Bell, A. H., Fecteau, J. H., and Munoz, D. P. (1994) Using auditory and visual stimuli to investigate the behavioral and neuronal consequences of reflexive covert orienting. *J. Neurophysiol.* **91**, 2172–2184.

Bisley, J. W., and Goldberg, M. E. (1998). Neuronal activity in the lateral intraparietal area and spatial attention. *Science* **299**, 54–56.

Briand, K. A., Larrison, A. L., and Sereno, A. B. (2000). Inhibition of return in manual and saccadic response systems. *Percept Psychophys* **62**, 1512–1524.

Castiello, U., Paulignan, Y., and Jeannerod, M. (1991). Temporal dissociation of motor responses and subjective awareness. A study in normal subjects. *Brain* **114**, 2639–2655.

Colby, C. L., and Goldberg, M. E. (1999). Space and attention in parietal cortex. *Ann. Rev. Neurosci.* **22**, 319–349

Constandtinidis, C., and Steinmetz, M. A. (2001). Neuronal responses in area 7a to multiple stimulus displays: II. responses are suppressed at the cued location. *Cereb. Cortex.* **11**, 592–597.

Dorris, M. C., and Munoz, D. P. (1998). Saccadic probability influences motor preparation signals and time to saccadic initiation. *J. Neurosci.* **18**, 7015–7026.

Dorris, M. C., Klein, R. M., Everling, S., and Munoz, D. P. (2002). Contribution of the primate superior colliculus to inhibition of return. *J. Cog. Neurosci* **14**, 1256–1263.

Fecteau, J. H., Bell, A. H., and Munoz, D. P. (2004). Neural correlates of the automatic and goal-driven biases in orienting spatial attention. *J Neurophysiol.* **92**, 1728–1737.

Fecteau, J. H., and Munoz, D. P. (in submission) Correlates of capture of attention and inhibition of return across stages of visual processing.

Glimcher, P. W., and Sparks, D. L. (1992). Movement selection in advance of action in the superior colliculus. *Nature* **355**, 542–545.

Goodale and Milner. (1995). *A Visual Brain in Action.* Oxford: Oxford University Press.

Gottlieb, J., Kusunoki, M., and Goldberg, M. E. (1998). The representation of visual salience in monkey parietal cortex. *Nature* **391**, 481–484.

James, W. (1890). *The Principles of Psychology.* In C.D. Green, *Classics in the History of Psychology.*
http://psychclassics.asu.edu/James/Principles/index.htm.

Jonides, J. (1981). Voluntary vs. Automatic control over the mind's eye's movement. In J. B. Long and A. D. Baddeley (Eds.) Attention and Performance IX. Lawrence Erlbaum Associates, Hillsdale, N.J.

Kramer, A. F., Hahn, S., Irwin, D. E., Theeuwes, J. (2000) Age differences in the control of looking behavior: do you know where your eyes have been? *Psychol. Sci.* **11**, 210–217.

Lui, F., Gregory, K. M., Blanks, R. H., Giolli, R. A. (1995). Projections from visual areas of the cerebral cortex to pretectal nuclear complex, terminal accessory optic nuclei, and superior colliculus in macaque monkey. *J. Comp. Neurol.* **363**, 439–460.

Munoz, D. P., Dorris, M. C., Paré, M., Everling, S. (2000). On your mark, get set: brainstem circuitry underlying saccadic initiation. *Can. J. Physiol. Pharmacol.* **78**, 934–944.

Munoz, D. P., and Fecteau, J. H. (2002). Vying for dominance: Dynamic interactions control visual fixation and saccadic initiation in the superior colliculus. *Prog. Brain Res.* **140**, 3–19.

Pollack, J. G., and Hickey, T. L. (1979). The distribution of retinocollicular axon terminals in rhesus monkey. *J. Comp. Neurol.* **185**, 587–602.

Posner, M. I., and Cohen, Y. (1984). Components of visual orienting. In H. Bouma and D. G. Bouwhuis (Eds.), *Attention and performance X* (pp. 531–556). Erlbaum Hillsdale, NJ.

Posner, M. I. (1980). Orienting of spatial attention. *Quarterly Journal of experimental Psychology* **32**, 2–27.

Posner, M. I., Rafal, R. D., Choate, L. S., and Vaughan, J. (1985). Inhibition of return: neural basis and fucntion. *Cog. Neuropsychology* **2**, 211–228.

Posner, M. I., and diGirolamo, G. J. (2000). Attention in Cognitive Neuroscience: An Overview. In The New Cognitive Neurosciences, 2nd ed., M. S. Gazzaniga, ed., 623–631. MIT Press, USA.

Robinson, D. L., Bowman, E. M., and Kertzman, C. (1995). Covert orienting of attention in macaques. II. Contributions of the parietal cortex. *J. Neurophys.* **74**, 698–712.

Sapir, A., Soroker, N., Berger, A., and Henik, A. (1999). Inhibition of return in spatial attention: direct evidence for collicular generation. *Nat. Neurosci.* **2**, 1053–1054.

Steinmetz, M. A., Connor, C. E., Constantinidis, C., and McLaughlin, J. R. (1994). Covert attention suppresses neuronal responses in area 7a of the the posterior parietal cortex. *J. Neurophysiol.* **72**, 1020–1023.

Wallace, M. T., Wilkinson, L. K., and Stein, B. E. (1996). Representation and integration of multiple sensory inputs in primate superior colliculus. *J. Neurophysiol.* **76**, 1246–1266.

CHAPTER 65

Specifying the Components of Attention in a Visual Search Task

Gregory J. Zelinsky

ABSTRACT

Although commonly treated as a unitary process, attention is more likely a collection of task-related but separable operations. Three components of attention (set, selection, and movement) are identified and defined within the context of a computationally explicit model of eye movements during visual search. The model compares filter-based representations of the target and search displays to derive a salience map indicating likely target candidates in a scene. Eye position is defined as the centroid of activity on this saliency map. As this map is thresholded over time, the changing centroid produces a sequence of movements that eventually cause simulated gaze to become aligned with the target. By adopting a more computational language and making explicit the underlying operations of the task, visual search, a behavior that has long been hobbled to the concept of attention, can be well described without appeal to an abstracted attention theory.

I. A THREE-STEP PLAN FOR BEATING ATTENTION ADDICTION

Consider the following armchair experiment. Think back to the last time you used the word *attention*. Now imagine snipping out *attention* from this sentence and substituting the word *processing*. If you find that you can make this substitution without a significant loss of underlying meaning, then your reference to the term *attention* probably lacked precision. Moreover, if you find yourself repeatedly invoking the concept of attention and repeatedly failing this 'processing test', then you may be suffering from 'attention addiction'.

I suggest the following three-step plan for beating this addiction to the attention concept and remedying the systemic misuse of the word. As is often the case when confronting an addition, Step 1 is to acknowledge the problem. We must learn to recognize when we are using the concept of attention as an abstraction, and then stop it. The processing test is a valuable tool in this effort, and I recommend integrating it into our everyday thinking. If a concept as vague as *processing* can simply be substituted for the word *attention*, then the word is probably not being used in reference to a well-defined cognitive operation. Step 2 requires us to identify the individual components of attention. Perhaps the reason we gravitate toward abstract usages of the word is because attention itself is not a unitary concept but rather a collection of related processes and operations. If this is indeed the case, we need to identify these operations so that they can be referred to directly and not simply lumped under one abstract term. Having identified the components of attention, Step 3 asks that we isolate these individual components in our behavioral tasks and begin addressing them specifically in our experiments and theories. This step is likely the most painful as it involves some recalibration of our thinking. Rather than thinking of attention as a unitary thing that can be manipulated or measured, we must drop down a level in our thinking and start being more specific as to the individual component of attention that we are attempting to study.

II. SEARCHING FOR ATTENTIONAL COMPONENTS IN A VISUAL SEARCH TASK

This chapter applies these three steps to the topic of attention in visual search. With regard to Step 1, the attention and search literatures have been closely intertwined for the past two decades, and this has created ample opportunity for abstract attentional concepts to permeate our theories of search. Part of this problem can be traced back to our choice of dependent measures in a search task. The visual search literature has relied disproportionately on manual RTs as a dependent measure, and has even built from these RT studies elaborate theories of how attention is thought to shift as we search for visual targets (see Chapter 17 for a review). RTs, however, do not adequately constrain search theory, meaning that there are multiple ways to go about explaining the same set of search data (Townsend, 1976). This problem is most pronounced when it comes to hypothesizing attentional movements in space over time. It is difficult to know whether a 600 ms RT corresponds to three 200 ms search movements or thirty 20 ms movements (Zelinsky and Sheinberg, 1997). Search theory has attempted to infer these spatio-temporal dynamics from RT × Set Size functions, but here again there exists the potential for radically different interpretations of the same data (e.g., Wolfe, 1994; Eckstein, 1998). Because there exists no machine that can track the hypothetical spotlight of attention as it jumps from one display item to the next (see Chapter 61 for a step in this direction), directly observable data regarding the individual search movement simply does not exist.

The task of adding specificity to our language, therefore, starts with the choice of dependent measure, and for this reason I will focus here on the topic of eye movements during a visual search task. Until someone actually builds that machine to measure the covert search movement, describing overt search behavior seems the next best thing (Deubel and Schneider, 1996). Unlike the hypothetical attention movement, one thing we know for sure about eye movements is that they really exist—they can actually be measured with an eye tracker. An eye movement analysis also provides an enormous amount of detailed spatio-temporal data, the exact sort of data needed to properly constrain a detailed theory of search. To the extent that attention and eye movements are coupled (see Chapters 20 and 22), an oculomotor measure provides: (1) the spatial x,y coordinates of specific locations visited by attention during search, (2) the temporal order of these search movements, and (3) an estimate of how long attention remained at each of these locations before shifting to another region of the display. Although there may also be attentional search movements that are not revealed by an eye movement analysis, with these data it becomes possible to at least assign attention to specific locations and times in the search process (Zelinsky, Rao, Hayhoe, and Ballard, 1997).

Step 2, identifying the components of attention, has been the goal of many valiant efforts over the years (see Pashler, 1998, for a review), but standing in the way of these efforts have been two obstacles. First, the task of defining attentional components is complicated by the myriad of paradigms used to study attention and the fact that each may invoke different attentional operations. Whereas components **a** and **b** may seem important in a Stroop task, components **c** and **d** may seem more relevant to visual search. In the interest of specificity, it is therefore probably best to identify the components of attention within the context of individual tasks. Second, the ease with which attentional components can be defined is directly proportional to the explicitness of existing attention theory. Purely descriptive theories breed abstraction. They make it seem correct and proper to assume that attention has no components; that it is a unitary process that can be directly manipulated and measured. However, with the steady retirement of these theories and the adoption of a more mathematical (e.g., Tsotsos et al., 1995) and computational language (e.g., Itti and Koch, 2001), we are forced to confront these assumptions as we make explicit the details of the operations previously ascribed to attention. This forced specificity is perhaps the least appreciated benefit of a computational model of behavior. Computers hate vague and imprecise descriptions. A necessary first step toward implementing a computer model is therefore the decomposition of the behavior in question into well-defined elemental operations. It is this focus on elemental processes that makes computational modeling a valuable tool for identifying and defining the components of attention.

My goal here is not to attempt a definitive list of the attentional components, but rather to identify a handful of these components in a visual search task and to make explicit their operation within the context of a computational model. Three components will be considered, two of which are borrowed directly from Posner and Boies' (1971) treatment of this topic. The first component, alertness or preparation, corresponds to the 'get ready' usage of the word attention. If you are stopped at a traffic light and have prepared your perceptual machinery to see green, this preparation will allow you to respond a bit faster to the light change compared to someone who is casually listening to the radio or talking on the cell phone. This component also is discussed in the context of attentional control settings; the idea that we can adjust internal parameters to prepare for a particular task or stimulus. The second component is selection. When someone tells you to "pay attention", they are asking you to selectively process them and to temporally ignore everything else. Selection is considered by many to be the real meat of the attention concept, but one that is arguably impotent when studied in isolation from other components. The third component listed by Posner and Boies was capacity, but this is not so much a component or computation as it is one of potentially many constraints that might be imposed on a computation, so it really does not belong in this list. I will instead substitute for this component the idea of movement. This is intended to correspond to someone asking "how are you attending to something?", such as when shifting your processing focus while actively viewing a picture or scene. Included in this component are both the duration of individual object inspections as well as the trajectory or scanpath taken by the processing focus as it shifts from object to object or region to region in a scene (see Chapter 48 for a more detailed discussion of scanpaths).

Step 3 in weaning ourselves from abstract attentional thinking requires that we recalibrate our theories to address the individual components identified in Step 2. The remainder of this chapter demonstrates how this can be accomplished in the context of a visual search task.

III. A NEUROCOMPUTATIONAL MODEL OF EYE MOVEMENTS DURING VISUAL SEARCH

A. Model Overview

The computational approach described here builds on the Rao, Zelinsky, Hayhoe, and Ballard (2002) model of visual search (see also Chapter 91), but differs from this model in its selection and saccade initiation routines. Briefly, visual patterns are represented by collections of biologically inspired oriented Gaussian derivative filter responses at multiple spatial scales and color channels—a representation that I have previously referred to as a BOLAR (bank of linear analyzer responses) vector (Zelinsky, 2003). These vectors are used to quantify visual information in the target and search displays, in effect creating a featural signature uniquely identifying a particular target in a search image. To obtain this featural signature I use a 4-level pyramid representation starting with two types of spatial filters at the topmost level. These filters, roughly corresponding to edge and bar detectors in V1, were created by taking the first and second derivatives of a three-dimensional Gaussian. Orientation is represented in the second level, with the two filters oriented at 0°, 45°, 90°, and 135° before being applied to an image. These eight filters are then duplicated at three spatial scales in the third level; 7×7 pixels, 15×15 pixels, and 31×31 pixels. Finally, this feature vector is divided into three opponent-process color channels in the fourth level. The first channel actually performs an achromatic analysis of the image in which the R, G, and B pixel values are averaged. The other two channels are chromatic, with the second coding R-G and the third coding B-Y, where Y is the average of R and G. It is this representation of 72 filter responses that comprises a BOLAR vector.

To represent the stimuli used in each search trial, a single BOLAR vector is found for the midpoint of the target pattern (see Fig. 65.1a), and this vector is correlated with a BOLAR vector derived for every point in the search image (see Fig. 65.1b). The matches from this comparison operation are plotted in Fig. 65.1c as a *saliency map* (Koch and Ullman, 1985; see also Itti and Koch, 2000; Parkhurst, Law, and Niebur, 2002; and Chapters 94 and 39), with a brighter point indicating a better match. As expected, the brightest points in this saliency map correspond to those points defining the dinosaur pattern in the search image. However, note that there are also many nontarget regions represented in Fig. 65.1c. In fact, literally thousands of nontarget points in the Fig. 65.1b search scene bear some visual similarity to the target and therefore appear as points of activation on the saliency map. These partial matches, just by virtue of their enormous number, play a crucial role in this model of search. If you take the centroid of activity on the saliency map, the point obtained will be near the center of the image rather than on the target—similar to where human observers initially directed their gaze in the search displays (see the scanpath illustrated in Fig. 65.1b). This centroid computation is essential to the model, which effectively defines the current locus of gaze during search as the weighted centroid of the search saliency map.

The centroid computation, however, characterizes only the initial saccade landing position of observers—it says nothing about the subsequent eye movements that bring gaze finally to the target. Because these eye movements occur over time, to account for these behaviors it is necessary to build time into the model. To do this, a moving threshold is used to exclude points from the saliency map over time, pruning off those points that are least similar to the target. When this threshold is very low, even very poor matches will appear on the saliency map

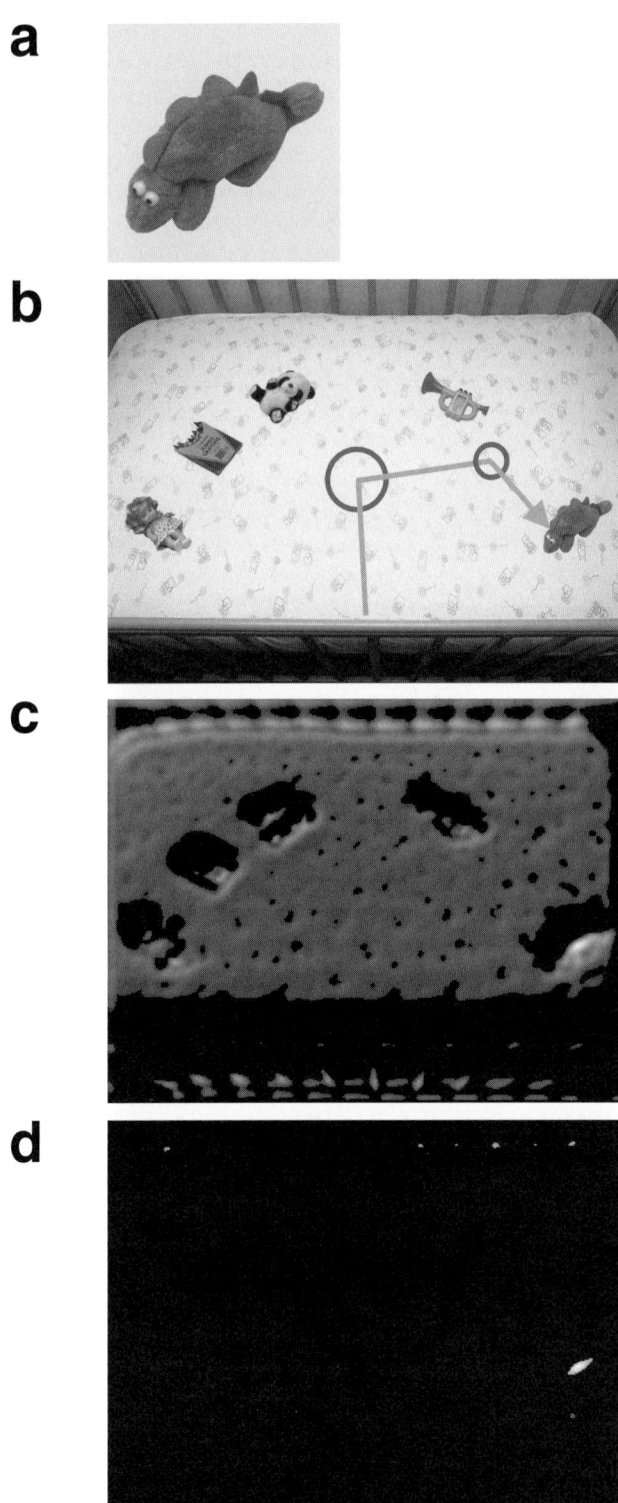

FIGURE 65.1 Samples of search stimuli and saliency maps. **a.** Search target. **b.** Search scene, with a representative eye movement scanpath superimposed over the image. Fixation duration is indicated by the diameter of the grey rings. **c.** A saliency map obtained by correlating the vector of filter responses defining the target with the vector array defining the search scene. **d.** The same saliency map after a thresholding process pruned off all but the best matching points.

and contribute to the centroid computation. However, as this threshold increases, points are removed from the saliency map until, eventually, only the most salient points remain—presumably those corresponding to the target (see Fig. 65.1d). The alignment of gaze with the target is therefore accomplished by: (1) a thresholding process that removes nontarget activity from the saliency map over time, and (2) the model's definition of gaze position as the spatial average of the saliency map activity.

Left unconstrained, this dynamic would produce an inordinately large number of eye movements, with each increment of the saliency map threshold resulting in a slightly different centroid and a new gaze position. To thwart this unrealistic behavior, the model implements a distance threshold (the large dashed ring superimposed over the saliency map in Fig. 65.2a) around the current fixation position. The function of this threshold is to impose a minimum amplitude on the following saccade. Immediately after each new fixation, the geometric center of this ring will correspond to the weighted centroid of the saliency map (the central white dot in Fig. 65.2a). However, as the saliency map threshold increases, points will be excluded from the saliency map and this will change the map's centroid, moving it closer to the target (see Fig. 65.2b). Eventually, the dynamically changing centroid will exceed the distance threshold and at this time the model will make an eye movement (see Fig. 65.2c). The number of saliency threshold increments needed for this event to occur is then used to derive a duration estimate for the preceding fixation (the solid grey ring in Fig. 65.2c). After this gaze shift, a new distance threshold will be set around the new fixation position and the process will repeat. Search terminates when the hotspot of the saliency map is included within the distance threshold, a condition corresponding to gaze becoming aligned with the search target (see Fig. 65.2d).

B. Identifying Attentional Components in the Model

With this brief overview of the model behind us, let us now return to the questions of how, and where, it implements the components of attention identified in Step 2. Recall that the first of these components was preparation or set. When we are searching for an object, such as the dinosaur in Fig. 65.1a, we are able to create an attentional set for this target that will help guide our search to that pattern in the image (see Fig. 65.1b). This preparatory component of attention is distributed over two operations in the model. First and foremost, the creation of attentional set is a byproduct of target definition. In effect, the 72-dimensional BOLAR representation of the target is identifying those features that are relevant to the search task and for which an attentional set should be constructed. Second, the use of this attentional set to guide search is accomplished by the correlation operation comparing the BOLAR target vector to the BOLAR representation of the search image. It is this operation that produces the target-related bias on the saliency map illustrated in Fig. 65.1c and ultimately results in the direction of gaze to the target. To the extent that preparation or set is a component of attention, then Fig. 65.1c illustrates where attention is located in this model—right in the actual saliency map representation.

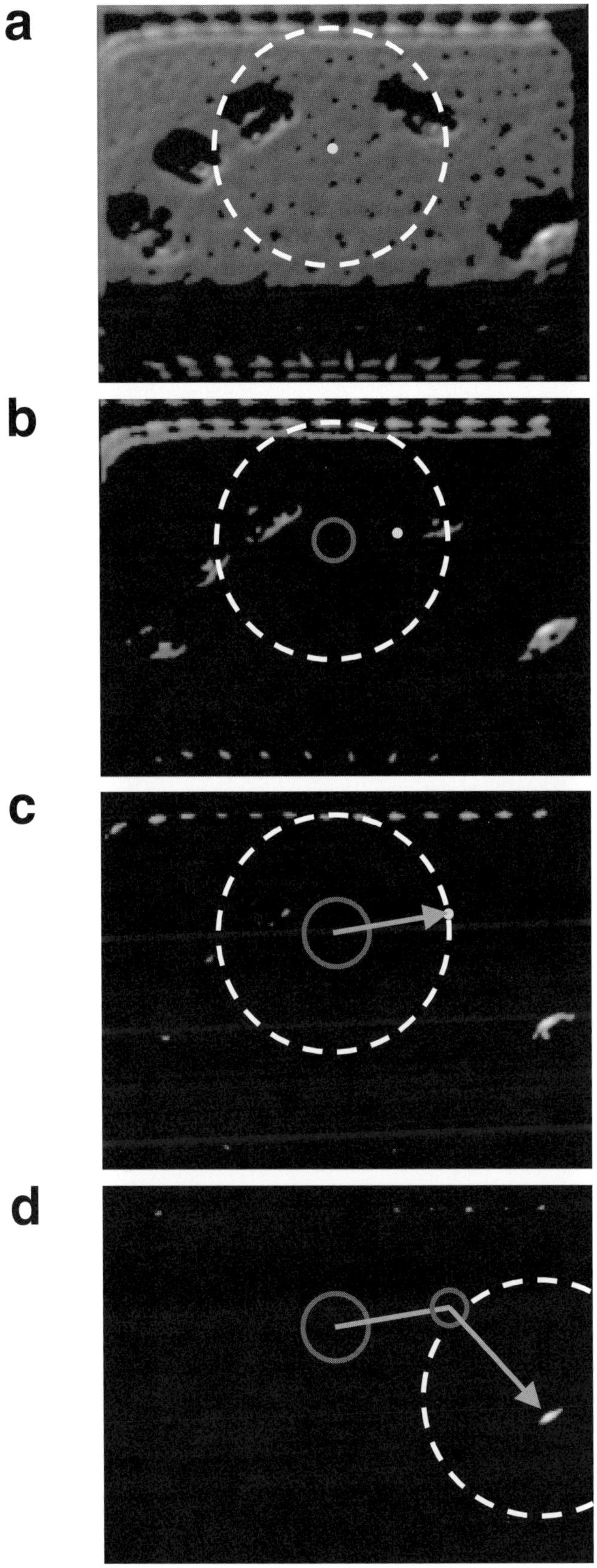

The second component of attention, selection, is a bit more obvious in the model and is accomplished by incrementing the saliency map threshold over time. Initially, the model is in a nonselective state with activity distributed broadly over the saliency map (see Fig. 65.2a). However, by incrementing the saliency map threshold, the model gradually isolates or selects the hottest region of map activation, in this case the region corresponding to the target (see Fig. 65.2d). According to the model, the selection process is therefore extended in time, with all of the eye movements occurring during search happening before the final selection of the target. For the purpose of the current discussion, this also means that attention is identified in this model by the selective pruning of activity from the saliency map.

The third component of attention, its movement in space and time, is distributed over several operations in the model but principally is embodied by the minimum distance threshold. It is this threshold, working together with the saliency map and the selection process, that determines when eye movements will occur and where they will be directed in a scene (see Fig. 65.2, c–d). To the extent that these spatiotemporal dynamics are yet another component of attention, then here, too, is where attention is located in this model.

FIGURE 65.2 An example of the target selection process and the accompanying pattern of simulated eye movements. **a.** Initial saliency map shown with the weighted centroid (white dot) and the minimum distance threshold (dashed ring). **b.** An intermediate saliency map following several increments of the saliency map threshold. The diameter of the solid ring reflects the number of threshold increments (i.e., simulated fixation duration). Note that gaze has not yet shifted because the centroid is still contained with the minimum distance threshold. **c.** The initial saccade latency and eye movement. **d.** Final saliency map reflecting the alignment of simulated gaze with the target.

IV. CONCLUSION

In relating attention to these search processes, my goal was to highlight the fact that attention is not a unitary concept. At the very least, attention is a trinity of components—and probably many more. Given that the term *attention* can refer to fundamentally different operations, we need to take more care in how we use this term in our language. We need to resist the temptation to treat attention as if it were a unitary thing that can be manipulated and measured, and work instead toward better specifying the individual component of attention that we wish to address.

Adopting this more specific language may not be easy, but it will become so with practice. This transition can be made less painful by tailoring our dependent measures to specific attentional components and by embracing a more computational perspective. Computational models have very little tolerance for vague and poorly defined constructs, and so should we. The rewards gained by adopting a computational perspective more than offset the technical hurdles that may need to be crossed as one recalibrates to this more specific mode of thinking. The added clarity of concept alone makes this effort worthwhile. As indicated by the many titles in this volume, such a computational systems approach to attention is alive and well, and gaining widespread acceptance (for examples, see Chapters 3, 90, 97, 98, and 1). As attention researchers continue to embrace this perspective in the years to come, we may one day soon find ourselves even wondering why we need to invoke the abstracted concept of attention at all in our descriptions of these dynamical systems.

In some sense, we are outgrowing the concept of attention. We no longer need it to describe behavior. We can do better. We now have the behavioral and computational tools to be more precise in our language and as we use these tools we may find that the concept of attention, much like the concept of ether in ancient Greek philosophy, will fade from our scientific vocabulary—it will be pushed back into the vernacular where it belongs.

References

Deubel, H., and Schneider, W. (1996). Saccade target selection and object recognition: Evidence for a common attentional mechanism. *Vision Research* **36**, 1827–1837.

Eckstein, M. (1998). The lower visual search efficiency for conjunctions is due to noise and not serial attentional processing. *Psychological Science* **9**, 111–118.

Itti, L., and Koch, C. (2001). Computational Modeling of Visual Attention. *Nature Reviews Neuroscience* **2**, 194–203.

Itti, L., and Koch, C. (2000). A saliency-based search mechanism for overt and covert shifts of visual attention. *Vision Research* **40**, 1489–1506.

Koch, C., and Ullman, S. (1985). Shifts of selective visual attention: Toward the underlying neural circuitry. *Human Neurobiology* **4**, 219–227.

Parkhurst, D., Law, K., and Niebur, E. (2002). Modeling the role of salience in the allocation of overt visual selective attention. *Vision Research* **42**, 107–123.

Pashler, H. (1998). *The Psychology of Attention*. MIT Press; Cambridge, MA.

Posner, M., and Boies, S. (1971). Components of attention. *Psychological Review* **78**, 391–408.

Rao, R., Zelinsky, G., Hayhoe, M., and Ballard, D. (2002). Eye movements in iconic visual search. *Vision Research* **42**, 1447–1463.

Townsend, J. (1976). Serial and within-stage independent parallel model equivalence on the minimum completion time. *Journal of Mathematical Psychology* **14**, 219–238.

Tsotsos, J., Culhane, S., Wai, W., Lai, Y., Davis, N., and Nuflo, F. (1995). Modeling visual attention via selective tuning. *Artificial Intelligence* **78**, 507–545.

Wolfe, J. (1994). Guided search 2.0: A revised model of visual search. *Psychonomic Bulletin and Review* **1**, 202–238.

Zelinsky, G. (2003). Detecting changes between real-world objects using spatio-chromatic filters. *Psychonomic Bulletin & Review* **10**, 533–555.

Zelinsky, G., Rao, R., Hayhoe, M., and Ballard, D. (1997). Eye movements reveal the spatio-temporal dynamics of visual search. *Psychological Science* **8**, 448–453.

Zelinsky, G., and Sheinberg, D. (1997). Eye movements during parallel–serial visual search. *Journal of Experimental Psychology: Human Perception and Performance* **23**, 244–262.

CHAPTER

66

Neural Evidence for Object-based Attention

Kathleen M. O'Craven

ABSTRACT

An object-based view of attention holds that when one aspect of a visual scene is attended, the object to which that attribute "belongs" is attended in its entirety, and all attributes of that object will receive enhanced processing. This differs from the more traditional spatial attention view, which likens attention to a spotlight or zoom lens moving from location to location and enhancing processing of whatever falls within its beam. The experiments described here give neural evidence for the existence of object-based attention. When subjects attend to one attribute of a scene consisting of two superimposed transparent objects, fMRI activity is increased for the attended attribute and for the unattended attribute of the attended object, relative to the attributes of the unattended object at the same spatial location.

I. OBJECT-BASED ATTENTION

At any given moment, our visual system (and indeed every sensory system) takes in far more information than can be fully processed. Selective attention allows an individual to choose certain subsets of that information to receive additional processing. Selective attention can act in a voluntary, top-down manner, in which the individual intentionally decides to allocate attention to some subset of the information; or in an involuntary, bottom-up manner, in which attention is simply drawn to a part of the stimulus by external factors (such as motion or salience). But what is the "unit" to which attention is allocated? Traditional theories of attention have used space as the basic unit, describing how attention is directed to a particular region of space (a particular retinal location), and that the features present in that location then receive additional processing including binding them together. A prime example of this is Treisman's Feature Integration Theory (1980). Popular analogies compare attention to a spotlight or a zoom lens, moving across space and then focusing on a particular portion of the visual field, enhancing the view of the visual "stuff" that is present at that location. Feature Integration Theory describes attention moving from location to location in a search array, sequentially binding the features at each location. Although this way of thinking about attention is very popular, there also has been evidence that attention can sometimes be allocated to the features themselves, as opposed to the locations they occupy. Wolfe's (1994) Guided Search Model suggests that the search through locations can be guided by using features as a unit of selection (see Chapter 17). Yet another view, and the focus of this chapter, is object-based attention, which holds that attention is allocated to objects, rather than to their component features or the locations they occupy. John Duncan (1984) popularized this idea with an intriguing behavioral study. Subjects made speeded perceptual decisions about the features of a stimulus consisting of two overlapping objects, a C and a diagonal line (see Fig. 66.1a). When subjects had to make two decisions about the same object (e.g., orientation and dottedness of the diagonal line), they performed better than when they had to make decisions about two features of different objects (e.g., orientation of the line and gap location of the C). In fact, subjects were just as good at answering two questions about a single object as they were at answering one. Duncan explained this in terms of attention being allocated to the object, such that when attending to one feature of the object the other features of that

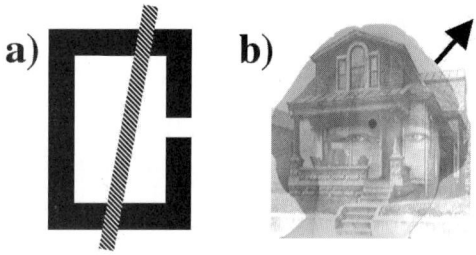

FIGURE 66.1 **Experimental Stimuli. a.** A composite stimulus similar to that used by Duncan (1984) to demonstrate object-based attention. Subjects were better at responding to two features of the same object (e.g., orientation and dottedness of the line) than to two features of separate objects (e.g., orientation of the line and gap location of the C). **b.** Composite stimulus used by O'Craven (1999), composed of a transparent face plus house. On each trial, either the face or the house oscillated in one of four motion directions. Demo available at http://psych.utoronto.ca/~ocraven/objattentiondemo.html.

object also received enhanced processing. Essentially, the additional features "come along for free" when attention is allocated to some aspect of the object. Spatial attention could not account for these results because the two objects were overlapping in space, occupying (approximately) the same location. This can be interpreted in the context of biased competition models of attention (see Chapter 50). Although appealing, this idea remained unresolved for many years. Additional behavioral experiments added support to the idea (Baylis and Driver, 1993; Egly et al., 1994; see Scholl, 2001 for a review), but it continued to be controversial. In most cases, including typical natural scenes and simplified lab stimulus setups, space and objects co-occur—multiple objects occur in separate locations. It is only with carefully constructed stimuli, such as the overlapping stimuli shown here, that we are able to completely tease apart the differential effects of space and objects.

II. fMRI OF OBJECT-BASED ATTENTION

With the advent of cognitive neuroimaging techniques, we gained a powerful new way of exploring the nature of attentional processing. A rather large collection of studies has emerged over the past decade showing that attention can modulate the neural response in many visual areas. For example, when subjects view a stimulus consisting of both moving and stationary dots, motion processing area MT+ is more active when they attend to the moving dots than when they attend to the stationary dots, even when the stimulus remains identical (O'Craven et al., 1997). Similar effects can be seen in the Fusiform Face Area (FFA) when attention is directed to faces (Wojciulik et al., 1998). Attentional modulation has now been observed in many visual areas (and indeed other sensory areas as well), including even primary visual cortex (see Chapter 62). We used fMRI to look for neural evidence that attention can be allocated to objects (O'Craven, Downing, and Kanwisher, 1999). In our study, subjects viewed a series of stimuli consisting of two transparently overlapping objects, a face and a house (see Fig. 66.1b), in which one of the objects was moving and one was stationary. While the subjects attended to either the face, the house, or the motion, we measured neural activity in three extrastriate visual areas that respond selectively to these three attributes: the Fusiform Face Area, the Parahippocampal Place Area, and MT+, respectively. These areas were functionally defined, for each individual subject, using separate independent localizer scans. We interpreted the magnitude of the fMRI response in each area as an indicator of the degree of processing for that region's preferred stimulus attribute. In separate blocks, subjects performed a one-back task on either face identity, house identity, or direction of motion, to direct attention to one of these three attributes. These three tasks were performed on both sets of face-house stimuli: those in which the faces moved (oscillating along a path 10% as long as the image) and the houses were stationary, and the reverse, for a total of six conditions. All six conditions contain the same three stimulus attributes (face, house, motion) in the same location. If attention truly were allocated only to locations in retinal space, a strong claim would be that no attentional modulation would be observed between these conditions, since all attributes occur at the same location. However, an object-based view of attention would predict that attention to a single attribute would have differential effects on the three brain areas. It would of course predict increased activity in the region responding to the attended attribute (as would a featural account). But critically, it would also predict increased activity in other attributes of an object to which attention was allocated. For example, while performing the Face Attention task, an object-based attention view would predict increased activity in FFA for either stimulus type (face moving or house moving) *and increased activity in MT+ when the face was moving* (face and motion part of same object) *but not when the house was moving* (face and motion in same location but on two separate objects). This follows directly from Duncan's original description—when one attribute of an object is attended, all attributes of that object receive enhanced processing, whereas attributes of other objects (even in the same location) do not. In this study, we essentially get to perform three separate experiments with the same dataset, by looking at activity in the three extrastriate brain regions during the six different attention conditions. For each region, we asked: 1) Does the level of neural activity in the region depend on whether the subject is attending to the effective stimulus for that region? 2) Does it depend on the *unattended attributes* of the attended items?

A. Behavioral Results

Subjects responded with a button press whenever the specified attribute for that block repeated. For example, during an "Attend Face" block, subjects pressed a button if the face on a given trial was the same as the face on the previous trial. Repetitions occurred on 12% of the trials. Accuracy on the task was between 60% and 80% for each of the six conditions, with no significant differences (possibly due to the low number of repeated stimuli). Exact accuracy is unimportant, but these numbers confirm the subjects' reports that the task was do-able (accuracy was above chance) but very challenging (not at ceiling).

B. fMRI Results

Timecourse data from each subject's individual FFA, PPA, and MT+, as defined functionally from the independent localizer scans, were averaged across seven subjects and 6 to 8 runs per subject (see Fig. 66.2). We use the magnitude of the fMRI response in each area as a measure of the degree of processing of that area's preferred stimulus. All six conditions produced highly significant increases in FFA, PPA, and MT+ relative to the fixation baseline. This is to be expected since all conditions included faces, houses, and motion. Significant standard attentional modulation was also observed in all three extrastriate regions. The Fusiform Face Area was more active during attention to faces than during attention to houses or motion. The Parahippocampal Place Area was more active during attention to houses than to faces or motion. And MT+ was more active during attention to direction of motion than to faces or houses. The most intriguing finding, however, is that in each case, we observed not just increased activation to the attended attribute, but also to the other attribute of the attended object. These key comparisons are shown by the thick straight lines in the graph. Specifically: 1) In Fusiform Face Area (see Fig. 66.2a) when comparing the two blocks in which the subject monitored for repetitions in the direction of motion, there was increased FFA activity when the moving item (target) was a face than when it was a house. In the Parahippocampal Place Area (see Fig. 66.2b) the opposite pattern was observed: for the two "Attend Motion" blocks, PPA activity was higher when the face was stationary and the house was moving, than for the other stimulus type. Finally, in MT+ (see Fig. 66.2c), we compare the pairs of blocks during which subjects attended to faces and to houses. For the Face Task, there was more MT+ activation when the face was moving, but for the House Task, there was more MT+ activation when the house was moving. In all four predicted comparisons, an unattended attribute that was bound to the attended attribute, as part of the same object, received enhanced processing relative to when the same stimulus attribute was present in the same location, but as part of a separate unattended object. This provides strong neural evidence for object-based attention. When attention is directed to an attribute of one object in a multi-object display, all attributes of that object receive enhanced processing relative to attributes of other objects, even at the same location.

C. Tightening the Argument with Event-Related fMRI

But could these results be due to the fact that motion is "special," attracting attention to itself automatically regardless of whether it is bound to an attended object? Or, might the blocked nature of this experiment, in which subjects performed the same attention task on the same stimulus type for a series of 14 consecutive trials (28 seconds), encourage subjects to use the expected cooccurrence of the cues to aid in their task performance? To exclude these possible explanations of the results, a companion study was conducted using event-related fMRI. Stimuli were as described earlier, except that the stationary item on each trial was displaced up, down, left, or right of fixation. Trials in which faces were moving were intermixed randomly with trials in which houses were moving. In this version, subjects performed a one-back task either on the direction of motion, as before, or on the direction of displacement of the stationary item. Attention was never explicitly directed to the faces or houses, and indeed their identities were completely irrelevant to the task throughout the experiment. Because trial types were intermixed, face and house attributes gave no useful cues at all about the relevant dimensions (direction of motion or displacement). Examining the resulting event-related timecourses (see Fig. 66.3) again reveals enhanced processing of the unattended attribute of the attended object, compared to the

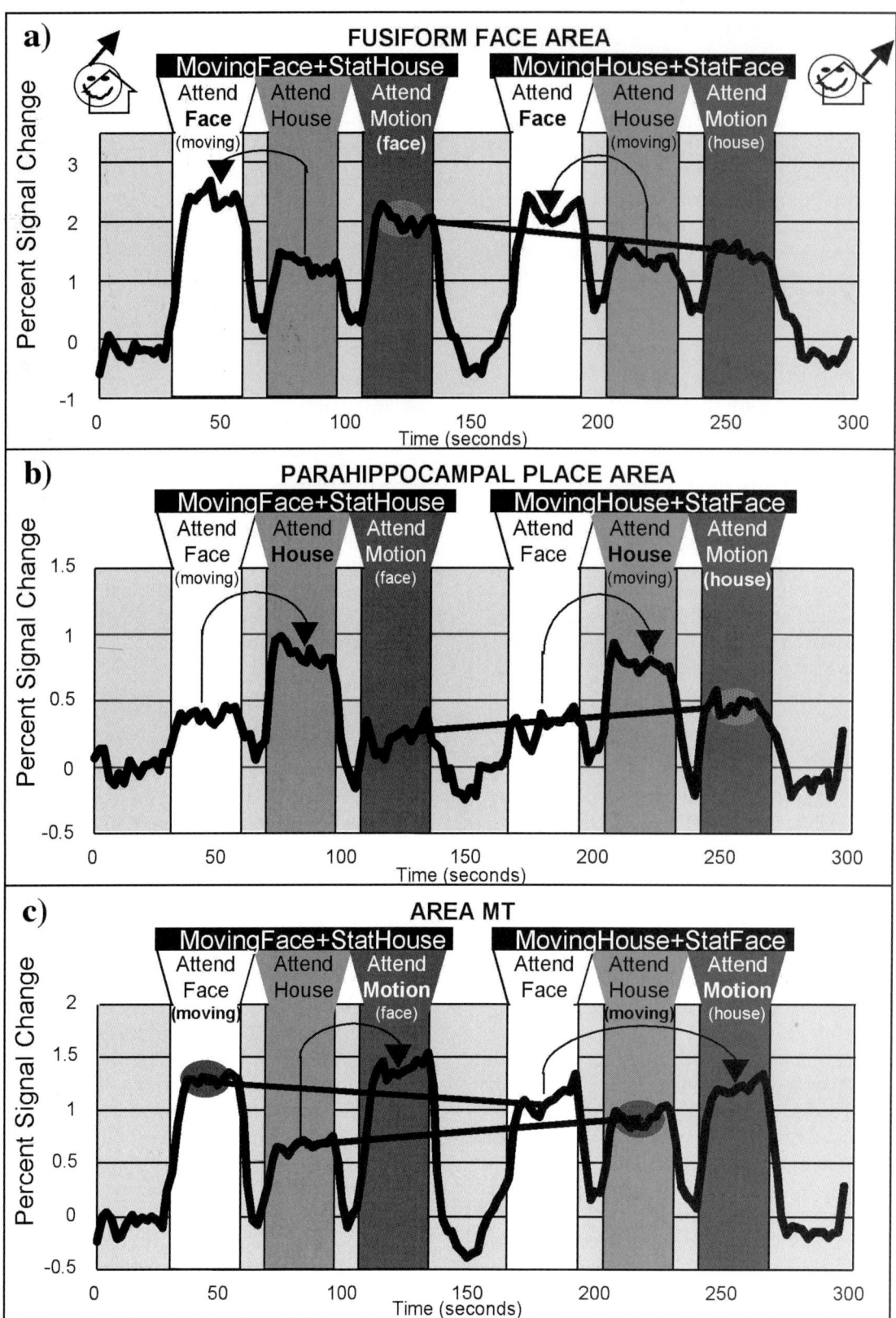

FIGURE 66.2 Timecourses of activation in (a) FFA, (b) PPA, and (c) MT+. For the first three blocks the stimulus was a moving face superimposed on a stationary house; the last three blocks were a moving house plus stationary face. Subjects attended to faces (white bars), houses (gray bars), or direction of motion (dark gray bars). Thin rounded lines designate the comparisons for "standard" attentional modulation (which would not occur if attention were allocated purely spatially). Thicker straight lines indicate key comparisons for assessing the processing of the irrelevant attribute of the attended object, with the circle identifying the epoch predicted to have greater activation if attention is object-based. Predictions were upheld in 8/8 comparisons (all $p < .05$ or greater).

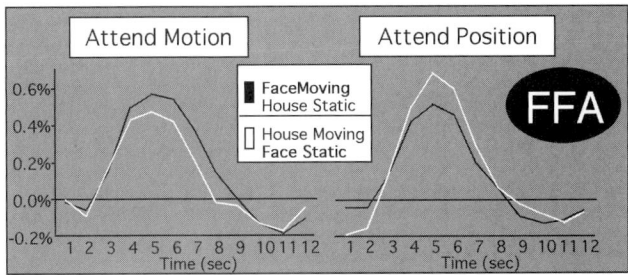

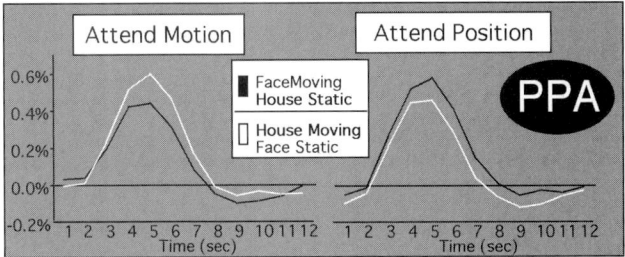

FIGURE 66.3 Event-related fMRI timecourses showing responses in FFA and PPA during the attend-motion and attend-position (stationary) tasks, for the two stimulus types (black = house + moving face, white = face + moving house). In each case, activation is stronger for the unattended attribute of the attended object than for the corresponding attribute in the unattended object.

unattended object at the same location. Even when faces are completely task irrelevant, the FFA is more active when the attended attribute—position or motion—is bound to the face than when the other attribute is bound to the face. The analogous results were seen in the PPA.

III. DISCUSSION AND RELATED STUDIES

We thus conclude with confidence that attention is attributed not just to spatial locations or to features, but to objects. Which unit will dominate will depend on the setting in which the stimulus occurs, and the task at hand. The addition of ERP techniques (Schoenfeld et al., 2003) adds critical information about the temporal parameters of these effects, demonstrating that processing of irrelevant features is enhanced quite soon (~50ms) after the features are first registered. This suggests the effects are not merely due to feedback from subsequent processing, which would be expected to occur much later.

Many neuroimaging studies have demonstrated spatial attention effects. Although most of these could be interpreted as object-based, because the different locations typically are occupied by separate objects, current studies are expected to demonstrate that object-based attention works together with spatial attention, and that the visual system uses whichever one is most appropriate for a particular visual stimulus (see also Avidan et al., 2003). This claim is supported by several findings that show an impact of objects on attentional modulation, though they do not address the fate of unattended attributes of attended objects. Activation in retinotopic cortex is greater to a region bounded by an object than to the same region when unbounded (Arrington et al., 2000). Furthermore, activation to an uncued location is stronger at a location that is part of the same object as a cued location, than at an equidistant uncued location on a separate object (Muller and Kleinschmidt, 2003). Interestingly, effects of object membership have even been demonstrated at the single unit level (Roelfsema et al., 1998); V1 cells of monkeys viewing a stimulus containing two curvy irregular lines respond more strongly to the line segments in their receptive fields that belong to the target line (defined as connected to the fixation point, outside the receptive field) than to a distractor line, even if the two lines intersect. Finally, Serences et al. (2004) have explored the control mechanisms for object-based attention, showing that posterior parietal and frontal regions were transiently active when attention was shifted between spatially superimposed perceptual objects based on those shown in Fig. 66.1b.

References

Arrington, C. M., Carr, T. H., Mayer, A. R., and Rao, S. M. (2000). Neural mechanisms of visual attention: Object-Based selection of a region in space. *Journal of Cognitive Neuroscience* **12** Supplement 2, 106–117.

Avidan, G., Levy, I., Hendler, T., Zohary, E., and Malach, R. (2003). Spatial vs. object specific attention in high-order visual areas. *Neuroimage* **19**, 308–318.

Baylis, G. C., and Driver, J. (1993). Visual attention and objects: Evidence for a hierarchical coding of location. *Journal of Experimental Psychology: Human Perception and Performance* **19**, 451–470.

Duncan, J. (1984). Selective attention and the organization of visual information. *J Exp Psychol Gen.* **113**, 501–517.

Egly, R., Driver, J., and Rafal, R. D. (1994). Shifting visual attention between objects and locations: Evidence from normal and parietal lesion subjects. *Journal of Experimental Psychology: General* **123**, 161–177.

Muller, N. G., and Kleinschmidt, A. (2003). Dynamic interaction of object- and space-based attention in retinotopic visual areas. *J Neurosci.* **23**, 9812–9816.

O'Craven, K. M., Downing, P. E., and Kanwisher, N. (1999). fMRI evidence for objects as the units of attentional selection. *Nature* **401**, 584–587.

O'Craven, K. M., Rosen, B. R., Kwong, K. K., Treisman, A., and Savoy, R. L. (1997). Voluntary attention modulates fMRI activity in human MT-MST. *Neuron* **18**, 591–598.

Schoenfeld, M. A., Tempelmann, C., Martinez, A., Hopf, J. M., Sattler, C., Heinze, H. J., and Hillyard, S. A. (2003). Dynamics of feature binding during object-selective attention. *Proc Natl Acad Sci USA* 2003 Sep 30 100, 11806–11811.

Scholl, B. J. (2001). Objects and attention: the state of the art. *Cognition* 80, 1–46.

Triesman, A., and Gelade, G. (1980). A feature-integration theory of attention. *Cognitive Psychology* 12, 97–136.

Wolfe, J. M. (1994). Guided search 2.0: A revised model of visual search. *Psychonomic Bulletin & Review* 1, 202–238.

Roelfsema, P. R., Lamme, V. A. F., and Spekreijse, H. (1998). Object-based attention in the primary visual cortex of the macaque monkey. *Nature* 395, 376–381.

Serences, J. T., Schwarzbach, J., Courtney, S. M., Golay, X., and Yantis. S. (2004). Control of Object-based Attention in Human Cortex. *Cereb Cortex* (in press).

CHAPTER

67

Location- or Feature-based Targeting of Spatial Attention

R. Vandenberghe

ABSTRACT

Two of the most widely used paradigms for selective attention are spatial cueing and visual feature search. Although in their classical form these two paradigms differ in several respects, one essential distinction lies in the selection criterion, location versus feature. Using event-related fMRI we contrasted location- or feature-based targeting of spatial attention. Each event consisted of a cue followed by an array of four stimuli, one in each quadrant, same-color stimuli appearing in the same hemifield. The spatial or feature cue directed attention to either the two left-sided or the two right-sided stimuli, because of their location or their color, respectively. The right inferior parietal lobule, at a location almost identical to the lesion site implicated in spatial neglect, was more active during spatial cueing than during feature cueing trials, indicative of a role in selecting peripheral stimuli cued on the basis of their location, both for ipsi- and contralateral hemispace.

I. INTRODUCTION

Two of the most fertile paradigms in the interdisciplinary study of selective attention are spatial cueing and visual search (e.g., see Chapter 17). During spatial cueing, a spatial selection criterion allows subjects to constrain the spatial focus of attention beforehand. During visual search, a feature selection criterion requires subjects to search for items bearing the relevant feature regardless of their location. Location and features of an object are not symmetrical properties, and spatial factors play an important role in certain types of visual search, too.

In healthy subjects spatial cueing and visual search activate many areas in common, including the intraparietal sulcus, anterior cingulate, premotor cortex, and right middle frontal gyrus. Comparison between studies of spatial cueing and visual search suggests that the main neuroanatomical difference lies in occipital cortex. In the case of spatial cueing, attentive selection of a location enhances activity in contralateral ventral occipital areas (Heinze et al., 1994). In the case of visual search, attentive selection of a feature enhances activity in occipital areas that are specialized for that feature (Corbetta et al., 1991). Despite these anatomical differences in occipital cortex, the neuronal mechanisms of attentional enhancement may be very similar between spatial and feature-selective attention and can be described by a single functional model (Treue and Maunsell, 1996). Patient lesion data also indicate a close neuroanatomical overlap: A focal right parietotemporal brain lesion can impair spatial cueing together with visual search for both simple features and conjunctions (Sprenger et al., 2002). A previous electrophysiological study compared feature and spatial cueing directly and indicated a high degree of overlap between the two paradigms (Luck and Cirelli, 1998).

In this event-related fMRI study (Vandenberghe et al., 2001b) we compared spatial cueing and visual search directly. The essential difference we were interested in was the selection criterion: We wanted to test whether specific areas are involved in selecting stimuli on the basis of their location compared to selection on the basis of a feature, that is, color. Otherwise we matched the two trial types as closely as possible. In both trial types subjects had to fixate the center of the screen. The stimulus array during the test phase consisted of four stimuli, one in each visual quadrant, two of which were relevant. The target criterion was identical between the two conditions (same-different shape judgement).

II. METHODS

Nineteen right-handed subjects participated in the main fMRI study and eight and nine of them also participated in control experiments 1 and 2, respectively.

The experiment was conducted in different runs, a run being defined as a continuous series of 228 volume acquisitions. Throughout each run a central fixation point was present and subjects were instructed to maintain fixation (on-line eye movement control). An event of interest was a trial consisting of a cue (100 ms duration), a delay, and an array (150 ms duration). The cue directed attentive selection trial-by-trial. The selection criterion could be feature- or location-based. During the cue phase of feature cueing trials (see Fig. 67.1A) the fixation point turned either blue or red. This directed the subject's attention to either the two blue or the two red stimuli in the upcoming array. During the spatial cueing trials (see Fig. 67.1B) two empty squares (0.54 deg side length, 0.06 deg line thickness) appeared either in left or in right hemifield, one in the upper and one in the lower quadrant, at 4.10 deg eccentricity. This directed the subject's attention to either the two left-field or the two right-field stimuli in the upcoming array. After a 200 ms delay we presented an array of four peripheral stimuli (1 deg height), one in each quadrant at 4.10 deg eccentricity. Two of these stimuli were red and two were blue, same-color stimuli occurring in the same hemifield (see Fig. 67.1). Subjects held a button in their right hand and were instructed to press the key when the two cued stimuli were identical in shape (either both + or both ×) and not to respond when their shapes differed (one + and one ×). Event-onset asynchrony was 2260 ms.

In control experiment 1 we compared central spatial cueing and peripheral spatial cueing trials. In a central spatial cueing trial (see Fig. 67.1C) a green arrowhead briefly covered the fixation point during the cue phase. It pointed either to the left or to the right. This directed the subject's attention to either the two left-field or the two right-field stimuli in the upcoming array. This was contrasted with peripheral spatial cueing trials identical to those studied in the main experiment (see Fig. 67.1B).

In control experiment 2 we assessed the sensory effects of the cues. During a trial the peripheral spatial cue stimulus or the central feature cue stimulus was presented for 100 msec without subsequent presentation of a stimulus array. The intertrial interval was 2260 msec. Subjects were informed that no arrays

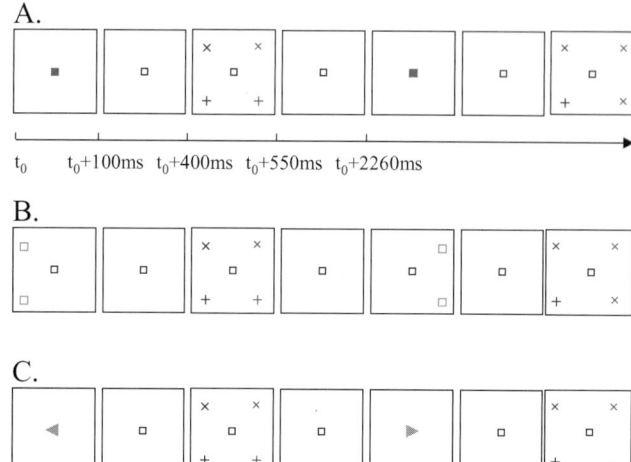

FIGURE 67.1 The stimuli are rendered in black and white. The actual stimuli were colored. The color of the stimulus array in the third and the seventh column is as follows: The two left-sided stimuli are blue, the two right-sided stimuli red. In all conditions and trials the stimuli to the left were identical in color (both red or both blue) as were the stimuli to the right (both red or both blue) and the color differed between the left-sided and the right-sided stimuli. **A.** Central feature cueing. A trial consisted of a cue, a delay, and a four-stimulus array. Subjects were instructed to press a button when the two cued stimuli were identical in shape. In the first trial the central fixation point turns blue in the cue phase, in the array the two stimuli to the left are blue and the two stimuli to the right, red. Subjects would not have to press for the first trial since the cue is blue and the blue stimuli differ in shape. In the second trial the central fixation point turns red; they would have to respond since the cue directed their attention to the red stimuli and the red stimuli were identical in shape. **B.** Peripheral spatial cueing. Here, subjects would have to press for the second trial since the two peripheral boxes appeared to the right and the two right-sided stimuli in the array were identical in shape. **C.** Central spatial cueing (first control experiment). Here, subjects would have to press for the second trial since the arrow pointed to the right and the two right-sided stimuli were identical in shape.

would come up and were told to fixate the center of the screen and view the stimuli passively.

The data were analysed using SPM99 and a random-effects analysis.

III. RESULTS

Reaction times and accuracies were analysed using a two-way ANOVA with these factors: cue type (spatial versus feature) and direction of attention (leftward or rightward). In the main experiment peripheral spatial cueing trials were associated with shorter reaction times ($F(1,17)$ 71.6, $P < 0.0001$) and a higher percentage correct responses ($F(1,17)$ 6.9, $P < 0.005$) compared to central feature cueing trials. Left-sided attention also was associated with shorter reaction times ($F(1,17)$ 15.1, $P < 0.0001$). In control experiment

1, peripheral spatial cueing trials were associated with a higher percentage of correct responses (F(1,17) 6.9, P < 0.01) than central spatial cueing trials. There was no significant reaction time difference at this stimulus-onset asynchrony of 300 msec.

Neuroanatomically, there was considerable overlap between the activation maps for different cue types and cued locations. The overlap included the intraparietal sulcus, premotor regions, anterior cingulate, and right middle frontal gyrus. These areas have been previously activated across a wide variety of different attention experiments and will not be discussed any further. There were a few notable differences, too, between the spatial cueing and the feature cueing trials (see Fig. 67.2).

The right inferior parietal lobule (60–51 24, Z 4.21, extent (ext.) 76, $P < 0.01$) showed significantly higher activity when attentional selection was based on stimulus location rather than stimulus feature (see Fig. 67.2A). This activation was centered around the ascending branch of the superior temporal sulcus and corresponded to the angular gyrus. The position of the spatial cue, central or peripheral, did not influence the inferior parietal activation, even when the threshold was lowered to uncorrected $P < 0.05$ (control experiment 1). There was no difference in the sensory response to the two types of cue stimuli in the right inferior parietal region (uncorr. $P < 0.05$) (control experiment 2).

The cingulate sulcus (−6 12 45, Z 4.24, ext. 96) and the left frontal operculum (−51,12,0, Z 4.10, ext. 112, $P < 0.005$) showed significantly higher activity during central feature cueing than during peripheral spatial cueing (see Fig. 67.2B) and also during central spatial cueing compared to peripheral spatial cueing (control experiment 1). Broadly speaking these activations might relate to the higher need for attentional control processes due to the requirement for translation (or verbalization) of the meaning of the central cue to an attentional set.

In agreement with earlier studies, activity during left-sided compared to right-sided attention was significantly higher in the right fusiform, lingual, and middle occipital gyrus, and vice versa for right-sided attention. This was true for both spatial cueing and feature cueing trials when the relevant stimuli occurred to the contralateral side. During feature cueing in comparison to spatial cueing we did not find stronger activation in any ventral occipitotemporal areas.

IV. DISCUSSION

We will discuss two distinct spatial effects: One is related to the type of cue, spatial versus feature, and localizes to the right angular gyrus, the second is related to the direction of attention, left versus right, and is situated in occipital cortex.

In the spatial cueing trials the right angular gyrus was more active than in the feature-cueing trials. Activity did not depend on whether the spatial cue was peripheral or central (control experiment 1), or on the direction of attention, leftward or rightward. The right angular gyrus has displayed selective activation only in a limited number of functional imaging studies of visuospatial attention (e.g., Corbetta et al., 2000). This is surprising given the close association between right inferior parietal lesions and neglect (Mort et al., 2003) (see Chapter 57). This contrasts with the more dorsally located intraparietal sulcus, where activation is found almost invariably in attention-requiring functional imaging paradigms (see Chapter 11). The inferior parietal activation could be brought out in the current study only through comparison with the feature cueing trials. When peripheral spatial cueing trials were compared to baseline, no activation was seen. The contrast between spatial and feature cueing isolates the cognitive component of location-based

A. Peripheral spatial cue minus central feature cue

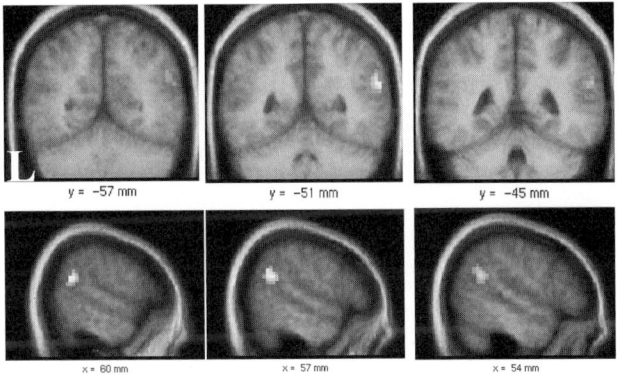

B. Central feature cue minus peripheral spatial cue

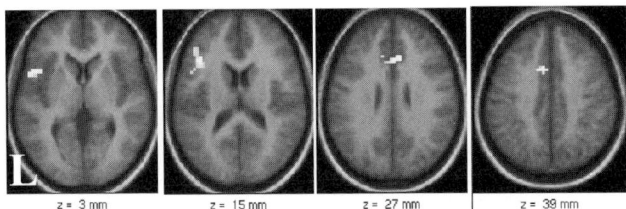

FIGURE 67.2 **A.** Peripheral spatial cueing trials minus central feature cueing trials. Projection upon the averaged brain MRI. Height threshold $P < 0.001$ uncorrected and cluster-level threshold corr. $P < 0.05$. Transverse, coronal and sagittal sections. **B.** Central feature cueing trials minus peripheral spatial cueing trials.

selection as opposed to feature-based selection. A recent event-related fMRI study of a classical spatial cueing paradigm (Corbetta et al., 2000) has found inferior parietal activation in response to a central spatial cue as well as to the subsequent target stimulus. Activation was stronger for invalidly cued than for validly cued targets (Corbetta et al., 2000). Together with the current findings this indicates that the right angular gyrus is involved in processes related to the development and revision of spatial expectancies. Which specific processes? According to a recent model proposed by Corbetta and Shulman (2002) the right inferior parietal lobule fulfills a role in detecting new spatial events. Attentional shifts in response to invalidly cued trials (Corbetta et al., 2000) and automatic capture of attention by luminant stimuli of sudden onset (Yantis and Jonides, 1990) constitute special instances of this function. This could explain the inferior parietal activation obtained in the main experiment. According to this model one would expect this area to be more active during peripheral spatial cueing than during central spatial cueing but this prediction was not born out by our control experiment of eight subjects where we compared these two conditions directly. We specifically probed the inferior parietal area and did not find a difference even at a very low threshold. As an alternative explanation, we propose that the inferior parietal lobule may be involved in locking (re-engaging) and maintaining the spatial focus of attention, in accordance with its involvement in spatial working memory. Higher activity during the central or peripheral spatial compared to the feature cueing trials could then be explained by the longer period of maintained focal attention during the delay period (Vandenberghe et al., 2001b; Nobre et al., 2004). We propose that the inferior and the superior parietal regions together realize the constant alternation between periods of maintained attention and transient shifts, respectively, that characterizes the way in which we explore our visual environment (Yarbus, 1967). According to this model, the right angular gyrus is involved in locking (re-engagement) and maintenance of the spatial focus of attention and the superior parietal lobule in spatial shifts of attention, both exogenous (Vandenberghe et al., 2001a) and endogenous (see Chapter 7).

The right angular gyrus activation during spatial cueing did not depend on the direction of attention, in agreement with previous studies (Corbetta and Shulman, 2002) (see Chapter 57). In contrast, ventral occipitotemporal activity depended strongly on the direction of attention and not on the type of cue, location- or feature-based. The direction-sensitive attentional occipital effects in this study are related to the location of the relevant stimuli and not to the spatial nature of the cue. This can be understood within the biased-competition model (Desimone and Duncan, 1995) that postulates that the response to a relevant object is enhanced in distributed brain areas that are involved in processing many different aspects of the relevant object.

During feature cueing in comparison to spatial cueing we did not find stronger activation in any ventral occipitotemporal areas. This is distinct from results of previous neuroimaging studies of working memory comparing spatial and feature cueing where retention periods are usually several seconds (Courtney et al., 1998). One possible explanation is that the stimulus onset asynchrony was only 300 ms, so that subjects did not need to hold any color information on-line for a prolonged period of time. Alternatively, the feature-cue related attentional enhancement may have overlapped spatially as well as temporally with the fMRI response evoked by the colored stimulus array (see Chapter 84).

To conclude, we describe two distinct spatial-attentional effects. One well-known effect (Heinze et al., 1994) is related to the position of the attended stimuli in left or right hemifield regardless of cue type and is situated in the ventral occipital cortex contralateral to the direction of attention. The second effect is related to the type of cue regardless of direction of attention. This effect localizes to an area that causes neglect when damaged (Mort et al. 2003): the right angular gyrus. Our findings highlight the critical role of the right angular gyrus in development and maintainenance of a spatial focus of attention when the deployment mode is relatively focal and targeting is based on stimulus location.

References

Corbetta, M., Miezin, F., Dobmeyer, S., Shulman, G., and Petersen, S. (1991). Selective and divided attention during visual discriminations of shape and color and speed: functional anatomy by positron emission tomography. *J. Neurosci.* **11**, 2383–2402.

Corbetta, M., Kincade, J., Ollinger, J., McAvoy, M., and Shulman, C. (2000). Voluntary orienting is dissociated from target detection in human posterior parietal cortex. *Nature Neurosci.* **3**, 292–297.

Corbetta, M., and Shulman, G. (2002). Control of goal-directed and stimulus-driven attention in the brain. *Nature Review Neuroscience* **3**, 201–215.

Courtney, S., Petit, L., Maisog, J., Ungerleider, L., and Haxby, J. (1998). An area specialized for spatial working memory in human frontal cortex. *Science* **279**, 1347–1351.

Desimone, R., and Duncan, J. (1995). Neural mechanisms of selective visual attention. *Ann. Rev. Neurosci.* **18**, 193–222.

Heinze, H., Mangun, G., Burchert, W., Hinrichs, H., Scholz, M., Munte, T., Gos, A., Scherg, M., Johannes, S., and Hundeshagen, H. (1994). Combined spatial and temporal imaging of brain activity during visual selective attention in humans. *Nature* **372**, 543–546.

Luck, S., and Girelli, M. (1998). Electrophysiological approaches to the study of selective attention in the human brain. *In* "The attentive brain" (R. Parasuraman, Ed.,) 71–94. MIT Press, Cambridge, MA.

Mort, D., Malhotra, P., Mannan, S., Rorden, C., Pamhakian, A., Kennard, C., and Husain, M. (2003). The anatomy of visual neglect. *Brain* **126**, 1986–1997.

Nobre, A., Coull, J., Maquet, P., Frith, C., Vandenberghe, R., and Mesulam, M. (2004). Orienting attention to locations in perceptual versus mental representations. *J. Cogn. Neurosci.* in press.

Sprenger, A., Kompf, D., and Heide, W. (2002). Visual search in patients with left visual hemineglect. *Progress in Brain Research* **140**, 395–416.

Treue, S., and Maunsell, J. (1996). Attentional modulation of visual motion processing in cortical areas MT and MST. *Nature* **382**, 539–541.

Vandenberghe, R., Gitelman, D., Parrish, T., and Mesulam, M. (2001a). Functional specificity of superior parietal mediation of spatial shifting. *NeuroImage* **14**, 661–673.

Vandenberghe, H., Gitelman, D., Parrish, T., and Mesulam, M. (2001b). Location- or feature-based targeting of peripheral attention. *NeuroImage* **14**, 34–47.

Yantis, S., and Jonides, J. (1990). Abrupt visual onsets and selective attention: voluntary versus automatic allocation. *J. Exp. Psychol. Hum. Percept Perform* **16**:121–134.

Yarbus, A. (1967). *Eye movements and vision.* Plenum Press, New York.

CHAPTER 68

Dimension-based Attention in Pop-out Search

Joseph Krummenacher and Hermann J. Müller

ABSTRACT

Visual objects that differ from distracting objects in a single salient attribute can be discerned in a seemingly effortless, automatic, fashion. Phenomenally, they seem to pop out of the background (pop-out effect). However, recent work, reviewed in this article, has provided evidence that pop-out is modulated by dimension-based attentional weighting processes: Pop-out is based on dimension-specific feature contrast/saliency signals. The computation of these signals, and/or their integration by superordinate mechanisms, may be primed, in bottom-up manner by the repeated presentation of targets defined within the same dimension, and top-down modulated based on knowledge of the upcoming target-defining dimension. Saliency signals are computed in parallel in multiple dimensions, and their integration into a super-ordinate (overall) saliency representation, that guides the allocation of focal attention, is spatially specific and subject to the same modulatory effects mentioned earlier.

I. VISUAL SEARCH FOR POP-OUT TARGETS: AUTOMATIC OR ATTENTIONALLY MODULATED?

A large share of the time we spend interacting with the environment is dedicated to searching for information required for the control of ongoing mental and/or behavioral activities. Information processing traditionally has been conceptualized as involving automatic processes that transform information in a predetermined fashion, and controlled processes that are subject to attentional (capacity) limitations (e.g., Shiffrin and Schneider, 1977).

Mechanisms that underlie the extraction of basic features in the visual field, as well as the computation of feature contrast or saliency signals (see Chapter 38), are thought to work in a spatially parallel, automatic fashion (see Chapters 11 and 94) prior to the allocation of limited-capacity (focal) attention (e.g., Koch and Ullman, 1985; Treisman, 1988; Treisman and Gelade, 1980; Wolfe, 1994). Preattentive, automatic processing commonly is invoked in explaining a striking phenomenon in visual search: If a searched-for (target) object differs in one salient feature from nearby task-irrelevant (distractor) objects, its presence in the field can be discerned very rapidly and effortlessly. For example, a red apple is detected quickly in a box of green apples (of similar size and form to the red apple). Phenomenally, it seems to pop out of the box by virtue of its differing color. Pop-out can even be observed when the object to be searched for is not known in advance.

Until recently, relatively little research has been directed to investigating the mechanisms underlying visual pop-out. It was simply assumed (or, rather, asserted) that the critical information is extracted automatically; that is, the stages of feature extraction and feature contrast/saliency computation are themselves not modulable by attentional processes (e.g., Theeuwes, 1991). The following sections will consider a number of recent (convergent) experimental approaches that have reexamined the role of feature- and dimension-specific processes underlying pop-out in visual search. This work has provided evidence that pop-out is subject to dimension-based processing limitations: Pop-out is based on dimension-specific feature contrast/saliency signals. The computation of these signals, and/or their integration by superordinate mechanisms, may be primed, in bottom-up manner, by the repeated presentation of targets defined within the same dimension (providing a form of implicit short-term memory for the critical dimension), and top-down modulated based on knowledge of the upcoming target-defining dimension. Saliency signals are computed in parallel in multiple dimensions, and their integration into a superordinate (overall) saliency representation, which guides the allocation of focal attention, is spatially specific and subject to the bottom-up and top-down modulatory effects already mentioned.

II. DIMENSION-SPECIFIC EFFECTS IN POP-OUT SEARCH

A. Dimension-Specific Target Definition Affects Search Performance across Trials

There are various accounts of how salient features in a visual field can be detected. Two of the most influential ones are Treisman's (1988; Treisman and Gelade, 1980) Feature Integration Theory (FIT) and Wolfe's (1994) Guided Search (GS) model. FIT proposes that topologically organized maps of feature detectors register the presence of a particular feature at a given location, in parallel across the visual field. The signals generated by different feature detectors are integrated by focal attention, which, by being directed to a particular object location on a master map, transfers the object's various defining features, represented by the detectors at the corresponding feature map locations, to an object-based short-term memory for further recognition processes. Since focal attention can be allocated to only one location/object at a time, feature integration is serial in nature. In the case of pop-out search, however, a differing (unique) feature is present at only one location in one feature map. If the critical, target-defining feature is known, the spatially parallel detection of any activity within this map will indicate target presence, and responses can be made very quickly, without involving feature integration.

GS provides an alternative account, which separates the computation of feature contrast/saliency signals from the extraction of the features itself. GS assumes that maps of saliency signals are computed in parallel for all visual dimensions (brightness, color, motion, [stereoscopic] depth, orientation, etc.) and that these dimension-specific saliency signals are then integrated, in an unweighted fashion, by the units in an overall-saliency map, which guides the allocation of focal attention. Importantly, active saliency units just signal the presence of feature contrast at the corresponding field locations, rather than conveying information as to the precise feature identities.

It has long been known that search performance is improved by observers knowing the featural definition of upcoming targets, and early versions of FIT and GS were modified to provide an explanation for this top-down effect. FIT proposed that the coding of features that are not target-defining (but distractor-defining) is inhibited; that is, the processing capacity allocated to their detectors is set to zero. In contrast, GS proposed that target-defining features are assigned more processing capacity, thus expediting the computation of the saliency signals in dimensions likely to contain a target. That is, both accounts propose top-down mechanisms for modulating visual feature processing.

The purely top-down position was challenged by a series of studies by Maljkovic and Nakayama (1994, 2000). They used a singleton search paradigm in which the precise featural definition of the target varied randomly across trials (e.g., the target, a color singleton, being red among green distractors on some trials and green among red distractors on other trials). Maljkovic and Nakayama demonstrated that target detection on a given trial, n, was affected by the feature value of the target on the preceding trial, $n-1$ (e.g., detection was expedited if a red target followed a red target, as compared to a blue target), which they referred to as priming of pop-out. Explanations based on top-down processes were excluded by the use of the singleton paradigm, with target definition varying unpredictably across trials; even when the sequence of target definitions was made perfectly predictable (i.e., changing every two trials) and observers were informed of this, they were found to be quite unable to use this knowledge to modify the priming in top-down fashion.

A similar effect of the previous target definition on search performance on a given trial was reported by Müller et al. (1995). Their observers were presented with two conditions, with either all targets presented in a trial block defined in one dimension (orientation: right-tilted, left-tilted, and horizontal small gray bars within an array of small gray vertical distractors; *within-dimension* search) or with targets defined, variably across trials, in one of the three possible dimensions (orientation: right-tilted; color: black; size: large; *cross-dimension* search). The observers' task was to indicate whether a target was present or absent. Comparison of the search reaction times (RTs) to the right-tilted orientation target presented in both conditions revealed RT costs, of some 60 ms, for cross-dimension relative to within-dimension search (see also Treisman, 1988). Similar to Maljkovic and Nakayama (1994, 2000), Müller et al. (1995) took this result as evidence for a bottom-up modulation of search performance. However, since Müller et al. (1995) used dimensionally different targets, they accordingly emphasized the importance of the dimensional change for the RT costs that they had observed (in contrast to the feature change costs reported by Maljkovic and Nakayama (1994, 2000)).

The relative contributions of the dimensional and featural definitions of the target on trial *n-1* on the detection of the targets on trial *n* was investigated by Found and Müller (1996). Their observers were presented with singleton targets that were either color-defined, red or blue, or orientation-defined, tilted to the left or the right (among green, vertical distractors), with dimensional and featural target definition varying randomly across trials. Observers simply had to discern the presence of a singleton object in the display. Found and Müller (1996) observed prolonged RTs for trials on which the target was defined in a different dimension to that on the preceding trial (e.g., a red [color] target preceded by a right-tilted [orientation] target), but only slightly (if at all) slowed RTs for trials on which the target was defined by a different feature within the same dimension as the target on the preceding trial (e.g., a red target preceded by a blue target). Found and Müller (1996) concluded that the RT slowing reflects a cost incurred by (time-consuming) processes associated with dimensional, rather than featural, changes in target definitions.

In their *dimension weighting* account, Müller and his colleagues suggested that the efficiency of dimension-specific (feature contrast/saliency) coding processes is modulated by the allocation of a limited-capacity resource: attentional weight (Duncan and Humphreys, 1989). In within-dimension search, when the target is consistently defined in one dimension, (almost) all weight is allocated to the coding mechanisms in this dimension, making target detection in this dimension fast. In contrast, in cross-dimension search, when the target dimension changes from one trial to the next, the critical dimension needs to be determined in a time-consuming, and possibly serial (Treisman, 1988), process, which involves the reallocation of attentional weight to the new target-defining dimension. The dimensional weight pattern established in this way persists across time, producing inter-trial facilitation (priming) if the target-defining dimension is repeated and costs if the critical dimension changes.

Taking together the findings described so far, the following picture emerges: Search performance is affected by the defining attributes of the preceding target(s); this inter-trial effect is implicit in nature—in that observers, when asked to report the identity of the target on the previous trial, perform at chance level for both dimensional and featural identity (see Müller et al., 2004; see also Maljkovic and Nakayama, 2000); and it is largely dimension-based. The divergence between the results of Maljkovic and Nakayama (2000) and of Müller et al. (2004) (featural and dimensional inter-trial effects, respectively) may have a number of reasons. One is that the color dimension, for which Maljkovic and Nakayama (2000) found feature-specific inter-trial effects, may be special: it might consist of a number of subdimensions, corresponding to the cardinal color channels, that may themselves be attentionally weighted (Found and Müller, 1996). Another reason is that, on feature change trials, Maljkovic and Nakayama (2000) reversed both the target and the distractor definitions relative to the preceding trial. Assuming that the computation of dimension-specific feature contrast adapts to the background features (based on the assumption of background stability across moments of time; see Krummenacher and Müller, 2004), feature-specific effects would be expected to be more pronounced under Maljkovic and Nakayama's (2000) conditions than under the constant-background conditions used by Found and Müller (1996).

B. Dimensional Processing is Top-Down Modulable

The dimensional weighting and inter-trial (priming) effects described earlier may be entirely stimulus-driven (i.e., controlled by the inter-trial history of target definitions; see Maljkovic and Nakayama, 1994). Alternatively, they may be top-down modulable, if not entirely top-down controllable. In order to investigate this issue, Müller et al. (2003) presented observers, on a trial-by-trial basis, with a symbolic (verbal) pre-cue indicating the dimension in which the upcoming target was highly likely to be defined. Two types of cue were used, specifying either the target dimension (e.g., color, orientation) or feature (e.g., red, blue, left-tilted, right-tilted). The results revealed effects of largely dimension-specific costs and benefits, with both dimension and feature cues. That is, with valid cues, dimension-based RT-benefits were evident irrespective of whether the cue was dimensional or featural in nature. For example, the processing of a blue target was expedited even if the (feature) cue predicted a red target; moreover, detection RTs were expedited even on trials on which the target was defined within the predicted dimension, but by a feature that was used extremely rarely throughout the whole experiment (i.e., rarer than the standard feature within the alternative dimension). Similarly, dimensionally, but not featurally, invalid cues gave rise to RT costs. Finally, both valid and invalid cues reduced the dimension-specific effect of the target on the preceding trial (stimulus-driven inter-trial effect), relative to a neutral-cueing condition (i.e., a cue with chance level validity) in which the upcoming target was unpredictable. This pattern of results suggests that feature contrast coding is modulable by top-down mechanisms, and that the modulation is dimension-specific in nature. However, residual inter-trial effects (e.g., costs associated with a dimension change) remained evident even when the cues were 100% predictive of the target dimension. This indicates that the top-down modulation cannot completely overcome stimulus-driven (bottom-up) priming effects.

C. Integration of Dimensional Feature Contrast Signals

All objects encountered in the visual world comprise multiple visual attributes. Given the costs associated with changes in the dimensional definition of targets in visual search, how does the visual system deal with singleton objects that differ from distractor objects by two (or more) features in two (or more) dimensions (e.g., a green watermelon in a box of red apples)? That is, are the feature contrast/saliency signals generated in the separate dimensional modules conveyed to the overall-saliency map in a serial, all-or-nothing, fashion (e.g., Treisman, 1988) or in a parallel fashion? And, if the latter, is there a parallel race of separate signals for activating an overall-saliency unit (with the unit being activated by either one or the other signal), or is the overall-saliency unit co-activated by both dimensional feature contrast signals in parallel? Assuming that the activation of an overall-saliency unit determines both focal attention and detection responses to the target, the answer to these questions has important implications for the conceptualization of how responses to target objects in the visual environment are controlled.

One way to decide between the various hypotheses is to examine visual search for dimensionally redundant targets; that is, targets that differ in two or more feature dimensions from the distractors (e.g., in color and orientation, such as a red and right-tilted target bar among green, vertical distractor bars). All accounts predict mean RT redundancy gains; that is, on average, faster RTs when a singleton target is dimensionally redundantly defined compared to when it is singly defined (in the example, a red vertical bar and, respectively, a green right-tilted bar). The reason is that, with two (redundant) dimensional feature contrast signals, faster RTs would simply result from the fact that, statistically, one of the two signals is computed at an earlier time than when only one signal is computed (with the speed of signal computation in one dimension being independent of whether or not a signal is computed in the other dimension). However, analysis of the RT distributions can provide evidence that permits a decision among the three alternatives. If dimension-based feature contrast signals are processed separately, whether serially or in terms of a parallel race, the fastest RTs to redundantly defined targets should not be faster than the fastest RTs to the singly defined targets (in the example, a red vertical bar and, respectively, a green right-tilted bar). The reason is that, for example, in a competitive race of separate dimensional feature contrast signals (at one location) for activating of an overall-saliency unit (representing that location), the race is won by that signal that is computed fastest, which is independent of how many signals (two or only one) are computed. In contrast, if dimension-specific saliency signals coactivate a common overall-saliency unit, the fastest RTs to dimensionally redundant targets should be faster than the fastest RTs to singly defined targets, because the overall-saliency unit would exceed its threshold activation at a relatively earlier point in time.

Miller's (1982) race model inequality (RMI) provides a solution to the problem of possible statistical confounds (see earlier). When the entire distributions of RTs are analyzed (rather than just mean RTs), a form of redundancy gain may be revealed that is inconsistent with any strictly serial model. Miller (1982) demonstrated that all models that assume that each target produces an independent, separate activation must satisfy the following inequality:

$$P(RT < t | T_1 \& T_2) \leq P(RT < t | T_1) + P(RT < t | T_2), \quad (1)$$

where t is the time since display onset and T_1 and T_2 are targets 1 and 2. Importantly, this inequality entails that the fastest RTs to displays with redundant targets be no faster than RTs to displays with single targets; however, fast RTs may occur more often with redundant targets. Violations of this inequality constitute evidence against serial and parallel-race accounts, and in favor of parallel-coactive accounts.

Krummenacher et al. (2001, 2002) applied this procedure to cross-dimension search in experiments with a variable number of dimensions in which a single target was defined; for example, color-only or orientation-only (singly defined targets, e.g., a red target or a right-tilted target), or color and orientation simultaneously (redundantly defined target, e.g., a red

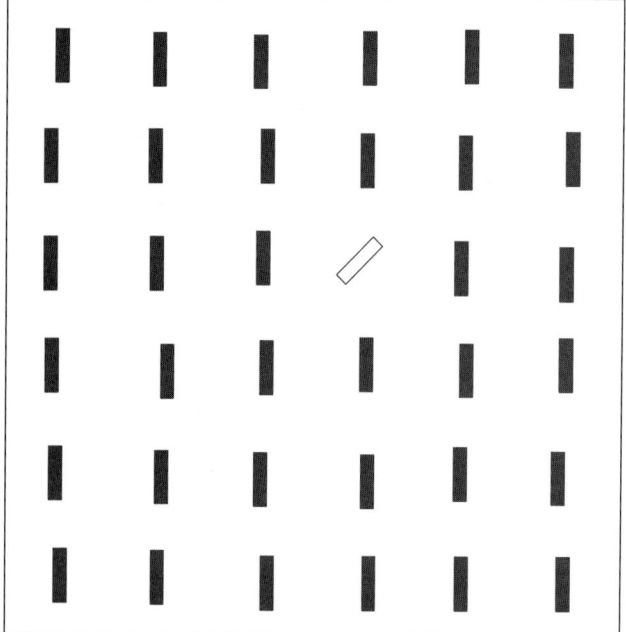

FIGURE 68.1 Illustration of a visual search display with a redundant target that differs from the distractors in both color and orientation, as used by Krummenacher et al. (2001, 2002).

right-tilted target). It was then examined, by testing for violations of the RMI, whether only one dimension (dimension-specific feature contrast signal) at a time can activate a response-relevant (e.g., master map) representation, or whether there is co-activation from multiple dimensions.

The results revealed that not only were RTs to redundantly defined targets on average faster than RTs to singly defined targets (mean RT redundancy gain), but also that the fastest RTs to redundantly defined targets were faster than the fastest RTs to singly defined targets, violating the RMI. The second finding constitutes strong evidence in favor of dimension-specific feature contrast signals coactivating, or being integrated by, a common (response-relevant) overall-saliency unit. The implication is that cross-dimension search for singleton feature targets does indeed proceed in parallel in multiple dimensions (e.g., Müller et al., 1995; see also Mordkoff and Yantis, 1993), rather than serially, in only one dimension at a time (e.g., Treisman, 1988).

A number of other findings of Krummenacher et al. (2001, 2002) are important: First, there was no evidence of coactivation when there were dual (redundant) targets at separate locations, both defined within the same dimension (e.g., a red and a blue color target), consistent with Mordkoff and Yantis (1993). Second, when there were dual (redundant) targets defined in different dimensions (e.g., a red color target and a 45°-tilted orientation target), evidence for coactivation was found only when the two targets were spatially adjacent—though the evidence tended to be weaker compared to when there was only a single, dimensionally redundant target (i.e., two-dimensional feature contrast signals at the same location). This pattern of effects is consistent with the idea that overall-saliency units integrate saliency signals only from separate dimensions and that the integration is spatially specific. Third, evidence of dimensionally redundant signal integration was found at both focally attended (spatially pre-cued) and unattended (uncued) field locations, suggesting that the signal integration occurs at a preselective stage of processing. Fourth, signal integration was found to be modulated by inter-trial effects, in particular: the evidence for coactivation was stronger for the second, relative to the first, of two repeated redundantly defined targets, relative to the second and, respectively, first of two repeated singly defined targets—consistent with idea that the first redundant target adjusts the relative weights assigned to the two target-defining dimensions so as to optimize dimensionally redundant target detection.

Taken together, these findings are consistent with the dimension-weighting account. This account assumes that the limited attentional weight resource is allocated to the various target-defining dimensions in a fashion that allows for the most efficient processing of the next (expected) target object. Redundancy gains and violations of the RMI suggest that attentional weight is allocated to different dimensions simultaneously, though not necessarily in equal proportions.

III. THE FUNCTIONAL ROLE OF DIMENSION-BASED INTER-TRIAL MEMORY AND REDUNDANCY GAINS

Traditional accounts of the efficient detection of singleton feature targets in visual search (e.g., Wolfe et al., 1989) have assumed that pop-out arises at a stage of processing characterized by automatic mechanisms of feature contrast/saliency computation, independent of limited-capacity resources and impenetrable by top-down control. At variance with this notion, search efficiency on a given trial is influenced by the dimensional definition of a target object on the preceding trial(s), suggesting attentional modulation of mechanisms that code dimension-based feature contrast/saliency signals.

According to the dimension-weighting account, attentional weight is allocated to feature dimensions in such a way as to achieve the most efficient processing of expected target objects within the visual environment. On this account, there is a limit to the total attentional weight available to be allocated at any one time to the various dimensions of the target object, with potential target-defining dimensions (i.e., dimensions in which the target might differ from nontarget objects) being assigned weight in accordance with their instructed importance (top-down) and their variability across trials (bottom-up). The greater the weight allocated to a particular dimension, the faster can the presence of a target defined in that dimension be discerned.

The dimension-weighting account assumes that the modulation of dimension-based coding occurs at an early, preselective stage of visual processing. This implies that observers do not need to be consciously aware of the dimensional and featural identity of a target the presence of which they were able to discern and respond to (Müller et al., 2004). That is, under conditions in which explicit knowledge of target identity is not required to perform the task, observers' responses can be based on preselective feature contrast signals.

An alternative, response-based account of dimension-based effects, especially coactivation effects, has been proposed by Cohen and his colleagues in their dimension-action model (DA; Cohen and Feintuch, 2002). This model assumes that target (and distractor) features are registered in dimensionally organized feature maps, with feature-specific response decision units pooling activity across separate maps. The intradimensional response devices are mutually inhibitory, so that one response must win the competition to be transferred to the central response execution stage. With multiple stimuli in the display, multiple (incompatible) response units may be activated in parallel (e.g., in the flanker paradigm, by the central target and the flanking stimuli, respectively). To resolve the ensuing competition, spatial attention must be focused on the task-critical stimulus for its associated response to win the competition. These postselective stages of processing, according the DA model, give rise to dimension-based effects.

Further research is needed to decide between the two accounts. It remains possible, however, that dimension-based processing limitations may operate at both preselective (perceptual) and postselective (response-related) stages of visual processing.

References

Cohen, A., and Feintuch, U. (2002). The dimensional-action system: A distinct visual system. *In* (W. Prinz and B. Hommel, Eds.), *Attention and performance XIX: Common mechanisms in perception and action*, 587–608. Oxford University Press, Oxford, England.

Duncan, J., and Humphreys, G. W. (1989). Visual search and stimulus similarity. *Psychological Review* **96**, 433–458.

Found, A., and Müller, H. J. (1996). Searching for unknown feature targets on more than one dimension: Investigating a "dimension weighting" account. *Perception & Psychophysics* **58**, 88–101.

Koch, C., and Ullman, S. (1985). Shifts in selective visual attention: Towards the underlying neural circuitry. *Human Neurobiology* **4**, 219–227.

Krummenacher, J., and Müller, H. J. (2004). Dimension-based priming and redundancy gain effects in visual singleton search under variable background conditions. Unpublished manuscript.

Krummenacher, J., Müller, H. J., and Heller, D. (2001). Visual search for dimensionally redundant pop-out targets: Evidence for parallel-coactive processing of dimensions. *Perception & Psychophysics* **63**, 901–917.

Krummenacher, J., Müller, H. J., and Heller, D. (2002). Visual search for dimensionally redundant pop-out targets: Parallel-coactive processing of dimensions is location-specific. *Journal of Experimental Psychology: Human Perception & Performance* **28**, 1303–1322.

Maljkovic, V., and Nakayama, K. (1994). Priming of pop-out: I. Role of features. *Memory & Cognition* **22**, 657–672.

Maljkovic, V., and Nakayama, K. (2000). Priming of pop-out: III. A short-term implicit memory system beneficial for rapid target selection. *Visual Cognition* **7**, 571–595.

Miller, J. (1982). Divided attention: Evidence for coactivation with redundant signals. *Cognitive Psychology* **14**, 247–279.

Mordkoff, J. T., and Yantis, S. (1993). Dividing attention between color and shape: evidence of coactivation. *Perception & Psychophysics* **53**, 357–366.

Müller, H. J., Heller, D., and Ziegler, J. (1995). Visual search for singleton feature targets within and across feature dimensions. *Perception & Psychophysics* **57**, 1–17.

Müller, H. J., Krummenacher, J., and Heller, D. (2004). Dimension-specific intertrial facilitation in visual search for pop-out targets: Evidence for a top-down modulable visual short-term memory effect. *Visual Cognition.* **11**, 577–602.

Müller, H. J., Reimann, B., and Krummenacher, J. (2003). Visual search for singleton feature targets across dimensions: Stimulus- and expectancy-driven effects in dimensional weighting. *Journal of Experimental Psychology: Human Perception & Performance* **29**, 1021–1035.

Shiffrin, R. M., and Schneider, W. (1977). Controlled and automatic information processing: II. Perceptual learning, automatic attending, and a general theory. *Psychological Review* **84**, 127–190.

Theeuwes, J. (1991). Cross-dimensional perceptual selectivity. *Perception & Psychophysics* **50**, 184–193.

Treisman, A. (1988). Features and objects. The fourteenth Bartlett Memorial Lecture. *Quarterly Journal of Experimental Psychology* **40A**, 201–236.

Treisman, A. M., and Gelade, G. (1980). A feature-integration theory of attention. *Cognitive Psychology* **12**, 97–136.

Wolfe, J. M., Cave, K. R., and Franzel, S. L. (1989). Guided search: An alternative to the feature integration model for visual search. *Journal of Experimental Psychology: Human Perception & Performance* **15**, 419–433.

Wolfe, J. M. (1994). Guided search 2.0: A revised model of visual search. *Psychonomic Bulletin & Review* **1**, 202–238.

CHAPTER

69

Irrelevant Singletons Capture Attention

Jan Theeuwes

ABSTRACT

When searching for a particular target, irrelevant distractors that happen to be more salient than the target will capture our attention against our intentions. We show that (1) spatial attention is first directed to the location of the most salient object in the field; (2) after attention is captured by the location of the salient singleton, this location becomes inhibited to allow disengagement of attention; (3) disengagement of attention from an inhibited location can proceed very fast (within 150 ms); (4) salient singletons may not only capture our attention but may also capture our eyes. We conclude that the visual system encodes differences in features in parallel across the visual field irrespective of the top-down goals of the observer and attention is directed automatically to the location of the highest saliency.

I. INTRODUCTION

A crucial research question is the extent to which we are able to exert attentional control over what we select from the visual environment. Selection may either be controlled by the properties of the stimulus field or by intentions of the observer. When we intentionally select only those objects and events needed for our current tasks, selection is said to occur in a voluntary, goal-directed manner. When, irrespective of our goals and beliefs, specific properties present in the visual field determine what we select, this selection is said to occur in an involuntary, stimulus-driven manner. It is as if attention is captured by the stimulus (e.g., Yantis, 2000). Often such attentional capture is accompanied by an exogenous saccade to the location of the object or event, a phenomenon termed *oculomotor capture* (Theeuwes et al., 1998).

When confronted with a display in which one element is unique in a basic visual dimension (such as a red element surrounded by green elements), one is able to immediately detect this element without effort. Search time to determine whether such a salient target is present or not is independent of the number of elements in the display. Elements that pop out from the display are referred to as *feature singletons* or simply *singletons*. Typically, search functions with slopes of less than 5 or 6 ms per item are considered to reflect parallel search (e.g., Treisman and Gormican, 1988). Such a "pop-out effect" is used as a diagnostic for determining whether visual information is available at the preattentive parallel level, which is assumed to segment the visual field into objects and background. Although the preattentive parallel stage can segment the visual field, it is assumed that a second stage of focused attention is necessary before one can act upon segmented objects.

Given the observation that upon presentation, a feature singleton is detected immediately, one may believe that feature singletons receive attentional priority independent of the intentions of the observer. In other words, when searching for a prespecified target (such as a red circle between green circles) one may argue that attention is captured in a bottom-up way by the uniquely colored element. However, this argument is not necessarily incorrect because the element that pops out is the target one is looking for and thus possibly part of the observer's top-down attentional set. To investigate whether salient singletons capture attention in a purely stimulus driven manner Theeuwes (1991b, 1992, 1994) developed a paradigm referred to as the *additional singleton* paradigm (e.g.,

Simons, 2000). In this visual search task, two feature singletons are simultaneously present in the visual field; one of the singletons is the target one is looking for while the other is a distracter that has to be ignored.

II. FEATURE SINGLETONS CAPTURE ATTENTION

The logic underlying the irrelevant singleton task is simple: Participants were asked to perform a visual search task, and one item in the search display is a unique salient feature singleton that is completely unrelated and completely irrelevant to the search task. The feature singleton is never the search item, and should therefore play no role in the observer's attentional set. This condition is compared to a condition in which an irrelevant featural singleton is not present. Figure 69.1 presents an example of typical stimulus displays and results (from Theeuwes, 1992). Participants with displays consisting of colored circles or diamonds appear on the circumference of an imaginary circle. Line segments of different orientations appear in the circles and diamonds. Participants had to determine the orientation of the line segment appearing in the target shape. The target shape that participants searched for was a singleton because it was the only diamond present in the display. In the distracter condition, an irrelevant color distractor singleton was also present in the display. Time to find the shape singleton increased when an irrelevant color singleton was present (i.e., one of the circles was red).

Even though participants had a clear top-down set to search for the shape singleton (i.e., the single green diamond), the presence of an irrelevant singleton (i.e., the single red circle) caused interference. It should be noted that search was performed by preattentive parallel search as the search slopes were basically flat (see Fig. 69.1). In subsequent experiments, it was shown that selection depended on the relative saliency of the stimulus attributes: When the color singleton was made less salient than the shape singleton (by reducing the color difference between the target and the nontarget elements), the shape singleton interfered with search for the color singleton while the color singleton no longer interfered with the search for the shape singleton.

Theeuwes (1991b, 1992, 1994, 1996) explained the increase in search time in conditions in which an irrelevant singleton was present in terms of attentional capture. Because attention was automatically captured by the distracter singleton (the most salient element in the display), it took longer before attention could be redirected to the location of the target singleton and a response could be emitted. Given the observation that selectivity depended completely on the relative saliency of target and distracter singleton, it was suggested that early visual preattentive parallel processing is driven only by bottom-up factors such as saliency. Irrespective of the top-down goals of the observer, spatial attention is automatically and involuntarily captured by the most salient singleton. The shift of spatial attention to the location of the singleton implies that the singleton is selected for further processing. If this singleton is the target, a response is made. If it is not the target, attention is directed to the next most salient singleton. The initial shift of attention to the most salient singleton is thought to be the result of relatively inflexible, hardwired mechanisms that are triggered by the presence of a feature difference signals. It is assumed that at each location in the visual field a local feature contrast is calculated that represents how different that object is within a particular primitive feature dimension (e.g., color, shape, movement, etc.).

The notion suggested by Theeuwes is similar to that of Koch and Ullman (1985), who introduced the notion of a saliency map to accomplish preattentive selection. This map is a two-dimension topographical map that encodes the saliency of objects in their visual environ-

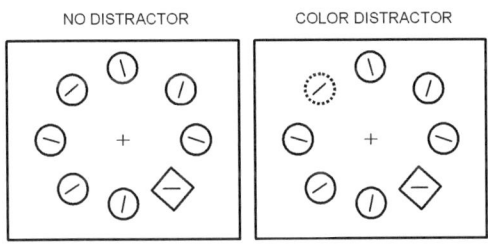

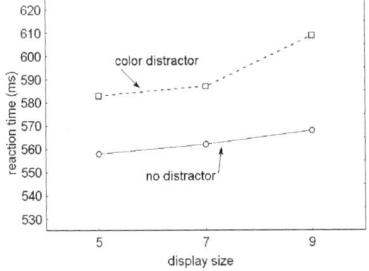

FIGURE 69.1 Stimuli and data from Theeuwes (1992). Top: A vertical and horizontal bar appeared within a green diamond (the other bars were oblique). Top left: all circles are green (solid lines); top right: one of the circles is red (dashed lines). Participants were to report the orientation of the bar in the diamond. Bottom: The presence of the irrelevant color singleton slowed responses to the bar (from Theeuwes, 1992; Experiment 1).

ment. Neurons in this map compete among each other, giving rise to a single winning location (winner take all) that contains the most salient element. The saliency map is the result of preattentive parallel encoding across the visual field calculating differences in simple visual features such as intensity, contrast, color, and orientation. Focused spatial attention simply scans the locations of decreasing activation (saliency). If a location is inhibited, the next salient location will receive spatial attention (see also Itti and Koch, 2001; Sagi and Julesz, 1985; Nothdurft, 2000).

It has been argued that preattentive computation simply calculates differences in features within dimensions resulting in a pattern of activations at different locations (Theeuwes, 1992, 1994). Given the previous example (see Fig. 69.1), at the location of the red singleton a large difference signal arises because the singleton differs from all other nontargets within the color dimension. At the location of the circle singleton, a large difference signal arises because the circle differs from all other elements within the shape dimension. Focal attention is automatically and unintentionally shifted to locations in the display containing large local feature differences, regardless of the dimension in which this feature difference occurs. The source of the preattentively calculated difference signal (whether it is caused by a color singleton or a shape singleton) is not available preattentively and can be determined only after focused attention has moved to the location of the difference signal. In other words, an observer knows whether the singleton was the target only after selecting the location with the large difference signal. In this view, the saliency of the singleton, and not its identity, its color, its shape, its brightness, and so on, will determine which element captures attention. Obviously, given this model, selection operates purely bottom-up irrespective of the task demands. The automatic shifts of attention are considered to be the result of relatively inflexible, hardwired mechanisms that are triggered by the presence of these difference signal interrupts. It is assumed that the parallel process can perform only a *local-mismatch* detection followed by a serial stage in which the most mismatching areas are selected for further analysis.

A. Spatial Attention Is Focused on the Most Active (Winner) Location

The increase in reaction time for those conditions in which the irrelevant singleton was present has been explained in terms of attentional capture: focused attention moved to the location of the salient singleton before it could move to the location of the (less salient) target singleton. However, one may argue that the presence of an irrelevant singleton increased RT not because of an additional shift of spatial attention to the distracter location but because the preattentive computation of the target feature took longer in the presence of a distracter singleton. In other words, simply because another irrelevant singleton was present, computing the target feature singleton and directing attention to it may take more time than when no such irrelevant singleton was present. Note that this view does *not* entail a shift of spatial focused attention to the most active location.

To address this issue Theeuwes (1996) used the same additional singleton search task but changed the elements inside the search items from line segments to large letter characters. When the letter R was presented in the shape singleton (i.e, the diamond) participants had to respond with their right hand. When the letter L was presented in the shape singleton participants responded with the left hand. Importantly, Theeuwes (1996) manipulated the congruency of the character at the location of the irrelevant distracter (see Fig. 69.2). In half of the trials the character at the distracter location was associated with the same response as was required by the target (e.g., an R in the target singleton and an R in the distractor singleton); on the other half it was the opposite of what was required by the target (e.g., an R in the target singleton and an L in the distractor singleton).

Theeuwes (1996) reasoned that if focal attention never went to the location of the irrelevant singleton, the identity of its character would have no effect on responding. However, a clear congruency effect was found (see Fig. 69.2), which provided evidence that before a response was given, attention was at the location of the irrelevant singleton. This finding is completely in line with the notion that *spatial* focused attention is directed automatically to the most salient location in the scene.

B. Attentional Selection and Inhibition of Return

The saliency map guides the attentional focus to the most salient location in the scene. But how can one prevent attention from permanently focusing onto the most active location? Within their model Itti and Koch (2001) suggested that after a location is attended it will be suppressed allowing the winner-take-all network to focus on the next most salient location. The phenomenon labeled as inhibition of return (Posner and Cohen, 1984) represents such an inhibitory tagging mechanism: after attention has been

II. FEATURE SINGLETONS CAPTURE ATTENTION

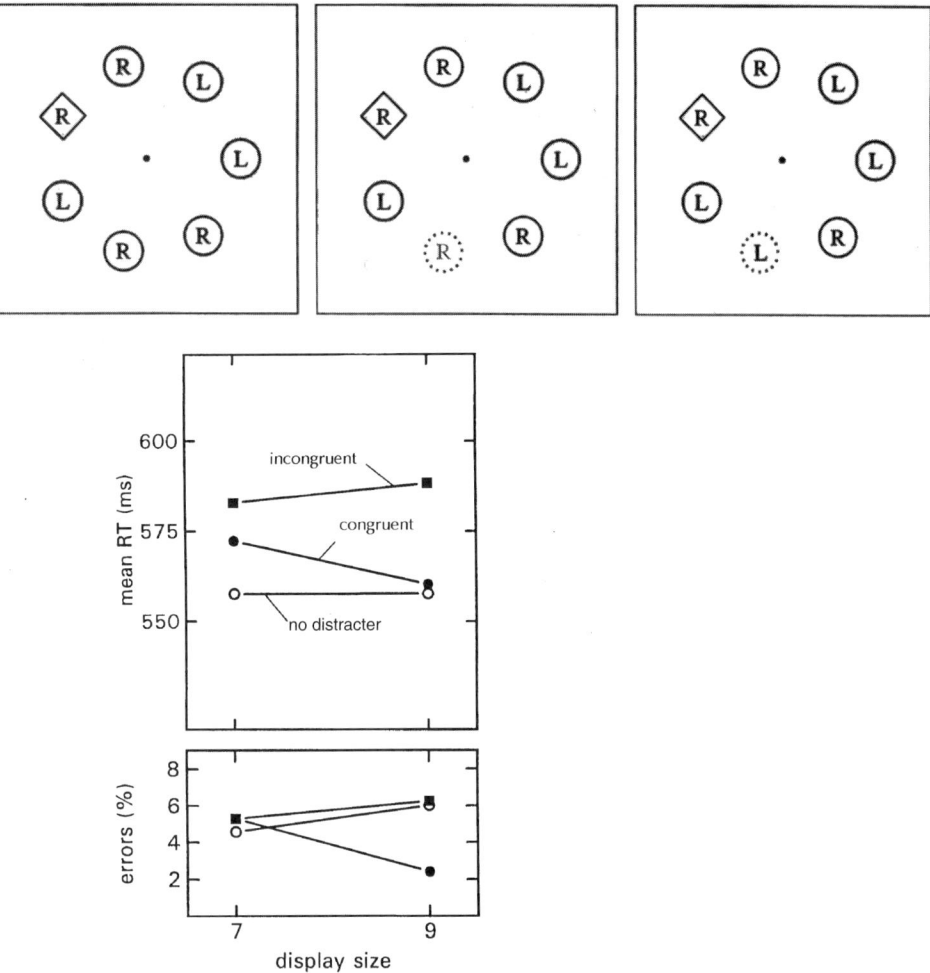

FIGURE 69.2 Stimuli and data from Theeuwes (1996). Top: Sample stimulus display (display size 7). In the no-distractor condition (top left) the green diamond shape appears among green circles. In the congruent condition (top middle), the letter inside the green diamond target shape (in this case the letter R) is identical to the letter inside the red circle distracter. In the incongruent condition (top right), the letter inside the green diamond target shape is different from the letter inside the red circle distracter. Solid lines indicate green and dotted lines indicate red. Bottom: The incongruent condition is significantly slower than the congruent condition suggesting that the letter at the location of the to-be-ignored singleton was processed (from Theeuwes, 1996; Experiment 1).

shifted to a location in space, there is delayed responding to stimuli subsequently displayed at that location (see Klein [2000] for a review). Theeuwes and Godijn (2002) provided direct evidence for the occurence IOR at a salient distracter singleton location: reaction times for stimuli presented at the location of a salient singleton were higher than for stimuli presented at a nonsalient distracter location. Based on the agreement that IOR is the result of an early, reflexive, involuntary orienting system (e.g., Posner and Cohen, 1984), Theeuwes and Godijn (2002) argued that these results could only be interpreted as evidence for exogenous bottom-up attentional capture.

C. Speed of Disengaging Spatial Attention

The bottom-up model assumes that attention is directed automatically to the most salient location regardless of whether it is a target or distracter (Theeuwes, 1992, 1994). The preattentive mechanism that extracts visual features calculates difference signals; only after focused attention has been directed to such a salient location, the exact feature becomes available (e.g., it is a color singleton, shape singleton, etc.). Therefore, top-down knowledge cannot guide the focused attentional selection mechanism. Knowing for example that the target is a red and squared cannot help selection but only processes that occur after an element has been selected for further processing. One crucial question is how long attention stays focused at a salient location when it is clear that the location attended does not contain the target one is looking for.

Theeuwes et al. (2000) showed that once attention is directed to the location of an irrelevant singleton it takes only a very brief time to disengage attention from that location. Theeuwes et al. (2000) used a visual search task similar to that of Theeuwes (1992; see Fig. 69.1) in which participants searched for a shape singleton. Prior to the presentation of the target display, at different times (50, 100, 150, 200, 250, and 300 ms) a color singleton was presented. Theeuwes et al. (2000) showed that the presence of an irrelevant salient distractor had an effect only when the target and distracter singleton were presented in close succession (at times of 50 and 100 ms). When the distracter singleton was presented considerable time (times of 150 to 300 ms) before the presentation of the target singleton, there was no evidence that attention was focused on the salient distracter. These results indicate that bottom-up attention is always directed to the most salient singleton; however, when it is clear that the element that captured attention in a bottom-up way is not the target, within 150 ms attention may be redirected to another location. Note that top-down processing plays a crucial role in the speed with which one is able to disengage attention. Thus, if the singleton that captures attention in a bottom-up way resembles the target one is looking for, the speed of disengagement may be much slower. As noted earlier, in order to disengage attention one may require the suppression of the (features at the) attended location.

III. ATTENTION AND EYE MOVEMENTS

It is generally agreed that there is an obligatory and selective coupling between saccade execution and visual attention to one common target object (e.g., Deubel and Schneider, 1996). When a saccade is programmed, attention precedes the eyes to the saccade target location. To determine whether salient singletons not only capture attention but also eye movements, Theeuwes et al. (1998) developed a paradigm referred to as the oculomotor capture paradigm, which uses the same logic as the additional singleton paradigm. Instead of inferring capture on the basis of a slowed response to the target, capture is reflected by an inappropriate eye movement toward the irrelevant item. Participants had the explicit instruction to search and make a saccadic eye movement toward the only gray element in the display. On some trials, an irrelevant singleton (an abrupt onset) was added to the display. Participants knew the onset was irrelevant and also knew that they had to ignore it. The condition in which a to-be-ignored onset was presented somewhere in the visual field was compared to a control condition in which there was no onset added to the display. The results showed that when no onset was added to the display, observers made saccades that generally went directly to the uniquely colored circle. However, in those trials in which an onset was added to the display, in about 30 to 40 percent of the trials the eye went in the direction of the onset, stopped briefly, and then went on to the target. Figure 69.3 shows the results. The graphs on the left side depict the control condition without the onset; the graphs on the right side depict the condition in which an onset was presented. Note that in the condition with the onset, the eye often went to the distracter. This occurred even when the onset appeared at a side opposite to that of the target circle (see Fig. 69.3 bottom panels: an abrupt onset is presented at 150 degrees separation). These findings suggest that salient singletons may not only capture our attention but may also capture our eyes.

IV. CONCLUSION

In conclusion, there is strong evidence that salient singletons capture spatial attention in a purely bottom-up way. The most parsimonious interpretation of the current findings is that top-down control during early preattentive segmentation is not possible. Selectivity is determined by the saliency of objects in the visual field; that is, the most salient singleton gets attention first. After this location is inhibited the next salient location will receive attention.

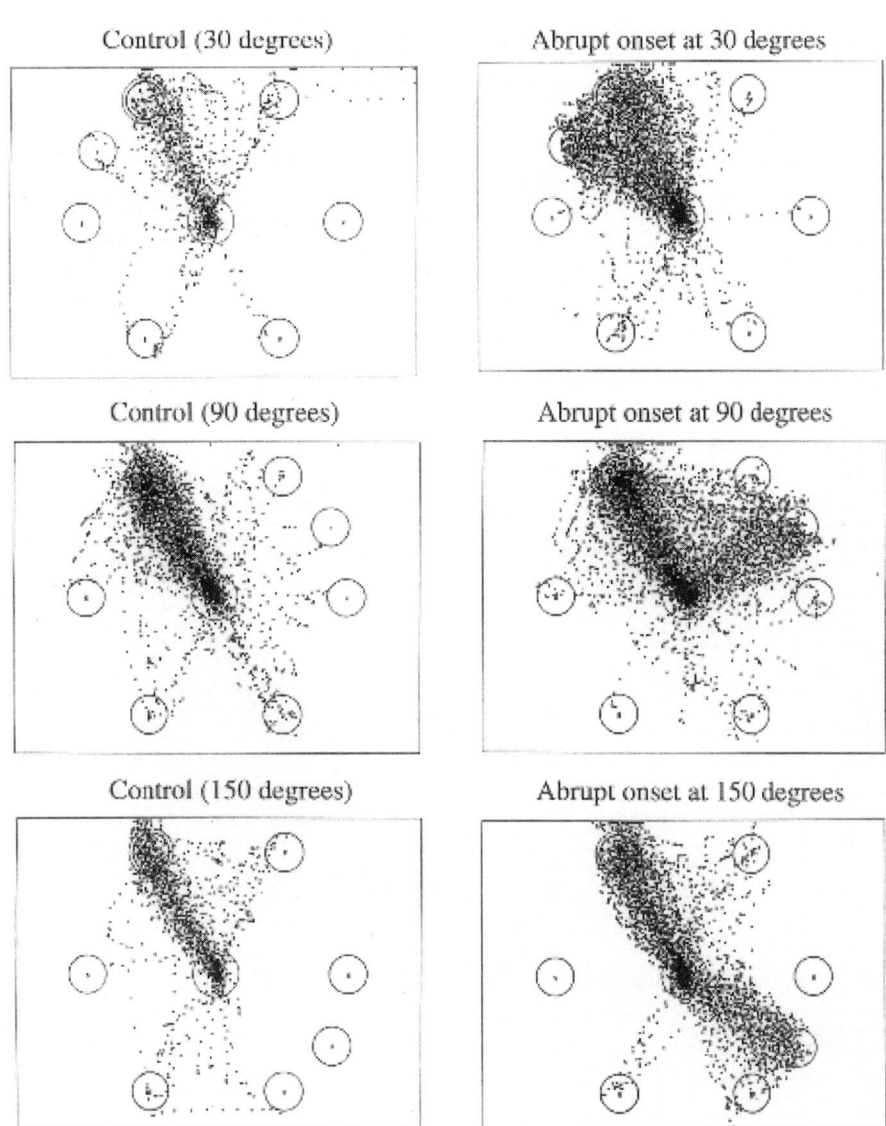

FIGURE 69.3 Oculomotor Capture. Data from Theeuwes et al. (1999). Eye movement behavior in the condition in which an abrupt onset was presented simultaneously with the target. The results are collapsed over all eight participants and normalized with respect to the position of target and onset. Sample points (every 4 ms) are taken from only the first saccade. Left panels: Eye movement behavior in the control condition in which no abrupt onset was presented. Right panels: Eye movement behavior in the condition in which an abrupt onset was presented; Either close to the target (top) somewhat away from the target (middle) and/or at the opposite side from the target (bottom). From Theeuwes et al. (1999).

References

Deubel, H., and Schneider, W. X. (1996). Saccade target selection and object recognition: Evidence for a common attentional mechanism. *Vision Research* **6**, 1827–1837.

Itti, L., and Koch, C. (2001). Computational modeling of visual attention. *Nature Reviews Neuroscience* **2**, 194–203.

Koch, C., and Ullman, S. (1985). Shifts in selective visual attention: towards the underlying neural circuitry. *Human Neurobiology* **4**, 219–227.

Klein, R. M. (2000). Inhibition of return. *Trends in Cognitive Science* **4**, 138–147.

Posner, M. I., and Cohen, Y. (1984). Components of visual orienting. In H. Bouma and D. Bouwhuis (Eds.), *Attention & Performance X*, pp. 531–556. Erlbaum, Hillsdale, NJ.

Sagi, D., and Julesz, B. (1985). Detection versus discrimination of visual orientation. *Perception* **14**, 619–628.

Simons, D. J. (2000). Attentional capture and inattentional blindness. *Trends in Cognitive Science* **4**, 147–155.

Theeuwes, J. (1991). Cross-dimensional perceptual selectivity. *Perception & Psychophysics* **50**, 184–193.

Theeuwes, J. (1992). Perceptual selectivity for color and form. *Perception & Psychophysics* **51**, 599–606.

Theeuwes, J. (1994). Stimulus-driven capture and attentional set: selective search for color and visual abrupt onsets. *Journal of Experimental Psychology: Human Perception and Performance* **20**, 799–806.

Theeuwes, J. (1996). Perceptual selectivity for color and form: On the nature of the interference effect. *In* A. F. Kramer, M. G. H. Coles, and G. D. Logan (Eds.), *Converging Operations in the Study of Visual Attention*, pp. 297–314. American Psychological Association, Washington DC.

Theeuwes, J., Kramer, A. F., Hahn, S., and Irwin, D. E. (1998). Our eyes do not always go where we want them to go: capture of the eyes by new objects. *Psychological Science* **9**, 379–385.

Theeuwes, J., Atchley, P., and Kramer, A. F. (2000). On the time course of top-down and bottom-up control of visual attention. *In* S. Monsell and J. Driver (Eds.), *Attention & Performance (Vol 18)*. MIT Press; Cambridge.

Theeuwes, J., and Godijn, R. (2002). Irrelevant singletons capture attention: evidence from inhibition of return. *Perception & Psychophysics* **64**, 764–770.

Treisman, A. M., and Gormican, S. (1988). Feature search in early vision: Evidence from search asymmetries. *Psychological Review* **95**, 15–48.

Yantis, S. (2000). Goal directed amd stimulus driven determinants of attentional control. *In* S. Monsell and J. Driver (Eds.), *Attention & Performance (Vol 18)*. MIT Press; Cambridge.

CHAPTER

70

Attentional Modulation of Apparent Stimulus Contrast

Julio C. Martinez-Trujillo and Stefan Treue

ABSTRACT

Voluntary attention modulates sensory processing in the visual system of primates by multiplying the response of neurons to attended/unattended stimuli. This effect resembles the modulation of neuronal responses caused by varying stimulus contrast. Additionally, attentional and contrast modulation of neuronal responses in monkey cortical area MT share a nonlinearity. Neuronal responses are more strongly modulated by attention and contrast changes for intermediate contrast stimuli than for low and high contrast stimuli. These similarities between attention and contrast suggests that they use similar or closely related mechanisms to modulate visual processing in the cortex.

I. EFFECTS OF ATTENTION AND STIMULUS CONTRAST ON NEURONAL RESPONSES

It has been demonstrated that directing attention toward a stimulus located inside the receptive field (RF) of a visual neuron multiplies the neuron's response to that stimulus (Treue and Martinez-Trujillo, 1999; McAdams and Maunsell, 1999). Additionally, directing attention to a particular stimulus feature increases the responses to sensory stimuli of neurons that are selective for that feature while it decreases the responses of neurons poorly selective for the same feature (Treue and Martinez-Trujillo, 1999). These two effects (space- and feature-based) can be explained by a feature-similarity gain model of attention in which the attentional modulation depends on the similarity between the attended stimulus properties (e.g., position, direction, orientation) and the sensory preferences (RF location, preferred feature (i.e., direction, orientation)) of the neuron (Treue and Martinez-Trujillo, 1999). The consequences of this attentional modulation will be an increase in the saliency of the attended stimulus representation and a corresponding suppression of the neuronal representation of unattended stimuli.

A similar relative enhancement or suppression of stimulus representations can be obtained when varying pure stimulus-dependent properties such as contrast. Moreover, such contrast changes evoke multiplicative changes in neuronal responses resembling those evoked by attention (Tolhurst, 1973; Dean, 1981; Albrecht and Hamilton, 1982; Sclar and Freeman, 1982; Treue and Martinez-Trujillo, 1999). This similarity between the effects of attention and the effects of varying stimulus contrast could simply reflect the similarity of two multiplicative mechanisms or it could indicate that the two processes are closely related and probably use similar mechanisms to modulate neuronal responses. In this latter case, one would expect that attentional and contrast modulation of neuronal responses would share the same properties.

II. MEASURING ATTENTIONAL MODULATION AS A FUNCTION OF STIMULUS CONTRAST

It has been demonstrated that changes in stimulus contrast are translated into changes in neuronal firing rate through a nonlinear (sigmoid) function (contrast response function, CRF) illustrated by the equation in Fig. 70.1 (Albrecht and Hamilton, 1982; Sclar et al., 1990). We hypothesized that if attentional and contrast modulations of cells' responses in area MT are interrelated (i.e., if the two signals are combined before being translated into effective firing rate), the effect of attention might be indistinguishable from a change in stimulus contrast. In this case the CRF would be shifted

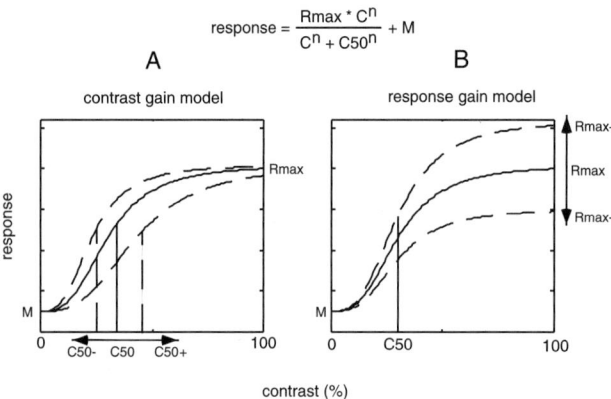

FIGURE 70.1 The two graphs show hypothetical contrast response functions (CRF) obtained by varying Rmax and C50 in the equation shown at the top (see methods for details). **A.** Attentional modulation of apparent stimulus contrast predicted by the contrast gain model (changes in the parameter C50); **B.** Attentional modulation of the cells' response predicted by the response gain model (changes in the parameter Rmax). The two-headed arrows indicate the axis along which the modulation occurs.

horizontally by attention and the largest response changes would occur along the central, steepest part of the CRF (contrast gain model, Fig. 70.1A). Such an effect can be achieved by changing the C50 value (contrast sensitivity) and has been reported in V4 neurons (Reynolds et al., 2000). If, on the other hand, contrast and attentional modulation are independent, the strength of attentional modulation of a given cell should reflect only the attentional state and should not depend on the stimulus contrast. This would cause a multiplication of the CRF by attention (response gain model, Fig. 70.1B) in a similar way as the multiplicative attentional modulation demonstrated for direction (Treue and Martinez-Trujillo, 1999) or orientation (McAdams and Maunsell, 1999) tuning functions.

A straightforward experimental design that can distinguish between these two possibilities consists of presenting one stimulus inside the RF of a neuron and a second identical stimulus outside. By changing the contrast of the two stimuli and recording the responses when the animal is attending either to one or the other stimulus one can measure attentional effects as a function of stimulus contrast. However, in such a design one confounding factor would be that changing the luminance of the attended stimulus (target) will create changes in task difficulty that might lead to changes in attentional effort and effects (Spitzer et al., 1988). We therefore exploited the finding that responses to unattended stimuli (distracters) are also modulated by attention (Moran and Desimone, 1985; Treue and Martinez-Trujillo, 1999) and introduced an additional stimulus into the RF that was always presented at full luminance and was the one the animal was instructed

to attend to. For symmetry, a similar paired stimulus arrangement was used outside the RF (see Fig. 70.2 for details). Each pair consisted of two moving random dot patterns (RDP), the potential target moving in the cells' anti-preferred/null direction (null pattern) and the distracter moving in the preferred direction (preferred pattern).

We varied the contrast of the distracter pattern moving in the cells' preferred direction and recorded the cells' responses in two attentional conditions: when the null pattern outside the RF was the target (attending outside), and when the null pattern inside the RF was the target (attending inside). The graph in Fig. 70.3A plots the average responses of one neuron (ordinate) as a function of the preferred pattern contrast (abscissa). The two data sets, attending inside (circles) and attending outside (squares), were fitted with a sigmoid function (see Fig. 70.1). The four parameters (Rmax, C50, n, and M) of the two CRFs are shown in the inserts. Attending inside the RF strongly increased C50 with only a negligible change in Rmax. This result agrees with the predictions of the contrast gain model (see Fig. 70.1A); that is, attentional effects are not a simple multiplication of a cell's response but vary nonlinearly as a function of stimulus contrast. One possible explanation that has been suggested for a similar nonlinearity found in V4 neurons (Reynolds et al., 2000) is that when the cell reaches its maximal firing rate, a further increase in response by attention cannot be achieved. This is clearly not applicable to our results because in our experiments the effect of attention is a suppression of the response. Additionally, in our sample the cells were capable of much higher firing rates if responses were not reduced by the presence of a null pattern in the RF. For the unit shown in Fig. 70.3A the response to the preferred direction RDP alone was ~55 spikes/second, more than twice the maximal firing rate for the 2-patterns configuration (~15 spikes/second).

We repeated the previous analysis in 34 of a sample of 63 MT units. The remaining 29 units were excluded because their C50 values in at least one of the attentional conditions were higher than the maximum contrast used during the experiments, which could cause an improper estimation of the CRF-parameters. For the 34 included cells there was no significant difference between the firing rate at the highest contrast and the fitted maximal firing rate. For all these neurons the goodness of fit of the model determined by the correlation coefficient between the data and the predicted values was larger than 0.8.

Next, we compared the differences between the normalized population responses across neurons in the two attentional conditions for different levels of

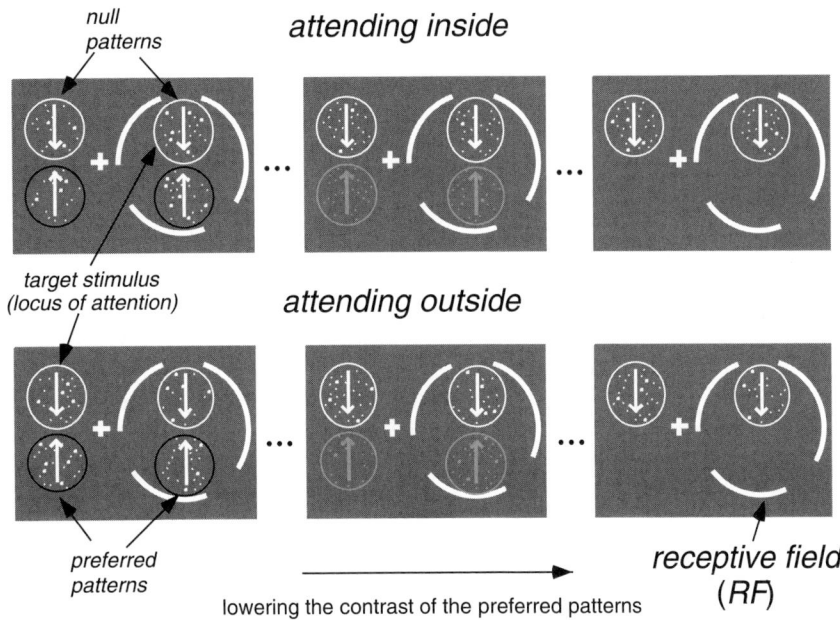

FIGURE 70.2 Experimental design. Two pairs of RDPs appeared on the screen, one positioned inside and the other outside the cells' RF (dashed circle). Each pair consisted of one preferred and one null pattern. In an attending inside condition (top row) the monkeys were attending to the null pattern inside the RF. In an attending outside condition (bottom row) the animals were attending to the null pattern located outside the RF. From left to right the panels illustrate decreasing luminance values of the preferred patterns leading to a decrease in responses.

stimulus contrast. Because the C50 value (contrast sensitivity) varied among neurons and therefore the steep portions of their CRFs was at different absolute contrast values, we normalized the abscissa using the formula (C50 − contrast)/(C50 + contrast). The effect of this procedure is to align the C50 values in all units (the central part of their CRFs) to the zero point (upper abscissa, Fig. 70.3B). The normalized contrast values can also be converted into percentage of the C50 value (lower abscissa, Fig. 70.3B). In such a plot, the contrast gain model predicts the largest differences in response between the two conditions at intermediate normalized contrast values. To test this hypothesis the normalized responses for each condition were grouped in bins and averaged across units (see Fig. 70.3B). Indeed, as predicted by the contrast gain model, the differences between the normalized responses in the two conditions (solid dark line in Fig. 70.3B) reached its highest significance in the middle portion of the graph.

III. A COMMON SUBSTRATE FOR ATTENTIONAL AND CONTRAST MODULATION OF RESPONSES

We have demonstrated that the response suppression to a distracter stimulus in MT neurons was stronger when that stimulus had intermediate, rather than low or high, contrast. This finding can be accounted for by an attentional influence that acts differentially on the gain of the various inputs converging on a given MT cell or even already on the firing rates of the neurons providing the input to MT (contrast gain model, see Fig. 70.1A). Since V1 provides a major direct sensory input to MT and attentional modulation in V1 neurons seems to be weaker than in later areas of the visual cortex, the attentional effects observed in MT might result from the modulation of input gain. This effect would be similar to a change in the contrast of the stimulus, raising the interesting issue of whether and how the contrast and attention can be disentangled perceptually. This recently has been addressed in a study by Carrasco et al. (2004), who showed that directing attention toward a stimulus increases its apparent contrast.

Another important implication of our findings is that the attentional suppression of unattended (distracter) stimuli allows the visual system to achieve a response modulation even when target (attended) stimuli have high-contrast and attention cannot further increases the neuronal responses to them (Reynolds et al., 2000). By reducing the response in neurons carrying a representation of distracter stimuli,

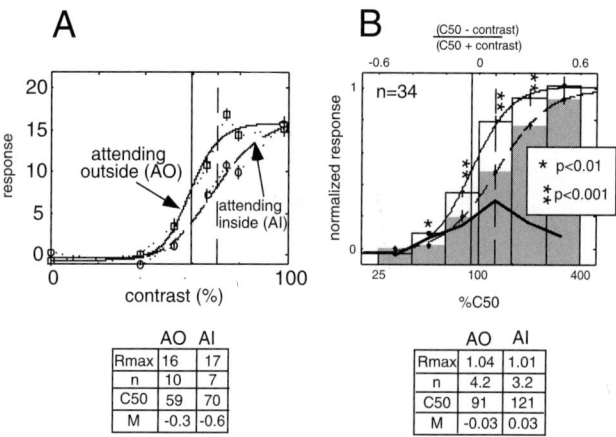

FIGURE 70.3 **A.** Average responses of one MT neuron (ordinate) to different contrast levels (abscissa) in the attending outside (AO, squares) and attending inside (AI, circles) conditions. The solid line represents the fit through the data points for the attending outside and the dashed line for the attending inside conditions. The vertical lines indicate the C50 value for each curve and the tables show the values of the four parameters (Rmax, C50, n, and M) of the CRFs. The error bars indicate the standard errors. **B.** Average normalized responses after aligning the contrast response functions in all units to their respective C50 values in the attending outside condition. The upper abscissa represents the values of the index (C50 − contrast)/(C50 + contrast) and the lower abscissa the same values converted to percentage of the C50 units. The ordinate represents the normalized response. The error bars represent the standard errors and the stars indicate responses that differ significantly from each other in the two conditions. The bars indicate the magnitude of the absolute differences in response for each bin and the gray solid line represents the values of the index (response attending inside − response attending outside)/(response attending inside + response attending outside).

the visual system is able to achieve an enhanced saliency, even for high contrast targets.

Other studies have found a similar contrast dependence of the attentional modulation of single cell responses in area V4 of the *temporal* pathway (Reynolds et al., 2000). These authors measured the attentional *enhancement* of neuronal responses in this area as a function of the contrast of a single stimulus inside the RF when it was the target or the distracter, respectively. In agreement with our results, the *enhancement* of responses to the target was stronger for intermediate than for high- or low-contrast stimuli.

This shared nonlinearity between the CRF and the magnitude of attentional modulation across different areas of the dorsal and ventral visual pathways indicates a tight link between attentional mechanisms and the mechanisms responsible for contrast encoding. This tight link does not seem to exist for the encoding mechanisms of other stimulus properties. For example, MT neurons are also monotonically modulated by the coherence in random dot patterns (Britten and Newsome, 1998). Because this modulation is a linear function of coherence, the nonlinear attentional effects we observed are not functionally equivalent to a change in stimulus coherence. Thus the mechanism responsible for the encoding of motion coherence (or any other stimulus property that is encoded in a quasi-linear way) does not seem to be directly modulated by attention.

When viewing the results of the present study in light of evidences of multiplicative attentional modulation of direction tuning curves it becomes apparent that attention cannot be simply thought of as a mechanism that creates changes in neural responses. For example, the same intermediate response of an MT neuron to a high contrast stimulus moving in a direction close to the preferred direction will be poorly modulated by attention but will be strongly modulated if it is evoked by a low contrast stimulus moving in the cell's preferred direction.

In summary, the current findings establish new constraints to mechanistic models of attention and suggest that attention can be seen as a mechanism aimed at modulating stimulus saliency while leaving other perceived stimulus properties relatively unchanged.

References

Albrecht, D. G., and Hamilton, D. B. (1982). Striate cortex of monkey and cat: contrast response function. *J. Neurophysiol.* **48**, 217–237.

Britten, K. H., and Newsome, W. T. (1998). Tuning bandwidths for near-threshold stimuli in area MT. *J. Neurophysiol.* **80**, 762–770.

Dean, A. F. (1981). The variability of discharge of simple cells in the cat striate cortex. *Exp. Brain Res.* **44**, 437–440.

Carrasco, M., Ling, S., and Read, S. (2004). Attention alters appearance. *Nature Neurosci.* **7**, 308–313.

McAdams, C., and Maunsell, J. H. R. (1999). Effects of attention on orientation-tuning functions of single neurons in macaque cortical area V4. *J. Neurosci.* **19**, 431–441.

Moran, J., and Desimone, R. (1985). Selective attention gates visual processing in the extrastriate cortex. *Science* **229**, 782–784.

Reynolds, J. H., Pasternak, T., and Desimone, R. (2000). Attention Increases Sensitivity of V4 Neurons. *Neuron* **26**, 703–714.

Sclar, G., and Freeman, R. D. (1982). Orientation selectivity in the cat's striate cortex is invariant with stimulus contrast. *Exp. Brain Res.* **46**, 457–461.

Sclar, G., Maunsell, J. H., and Lennie, P. (1990). Coding of image contrast in central visual pathways of the macaque monkey. *Vision Res.* **30**, 1–10.

Spitzer, H., Desimone, R., and Moran, J. (1988). Increased attention enhances both behavioral and neuronal performance. *Science* **240**, 338–340.

Tolhurst, D. J. (1973). Separate channels for the analysis of the shape and the movement of a moving visual stimulus. *J Physiol* **231**, 385–402. London.

Treue, S., and Martínez-Trujillo, J. C. (1999). Feature-based attention influences motion processing gain in macaque visual cortex. *Nature* **399**, 575–579.

CHAPTER 71

Attentional Suppression Early in the Macaque Visual System

Orban, G.A., Pauwels, K., Van Hulle, M.M., and Vanduffel, W.

ABSTRACT

We review the results of a double label 2-deoxyglucose study in the awake monkey demonstrating that attention can suppress activity at early levels in the primate visual system, in regions surrounding the representation of the attended stimulus. These findings are in agreement with human imaging results and were modeled in a dynamic simulation, highlighting the role of the reticular thalamic nucleus in these suppressive effects.

I. INTRODUCTION

This chapter is devoted to the effects of spatial attention at very early levels of the visual system. A series of publications including both single cell studies in the monkey (e.g., Reynolds et al., 2000; Treue and Maunsell, 1999) and human imaging studies (e.g., Tootell et al., 1998; Kastner et al., 1998) have generated the view that attention effects in the visual system 1) are mainly positive effects enhancing visual responses (but see Brefczynski and DeYoe, 1999); 2) occur relatively late in the hierarchy, at the level of V4 or beyond (but see O'Connor et al., 2002); and 3) have their source in parieto-frontal networks (Corbetta et al., 1998; Wardak et al., 2002). In contrast, the study (Vanduffel et al., 2000) reviewed in the present chapter provides evidence for suppressive attention-dependent modulation of activity at very early levels of the visual system: the dorsal Lateral Geniculate Nucleus (LGN) and striate cortex (or V1). This study also suggests that a subcortical structure, the reticular thalamic nucleus (RTN), might play a role in these attentional effects.

The Vanduffel study also stands out in the use of the relatively uncommon double label 2-deoxyglucose (2-DG) technique (Livingstone and Hubel, 1981; Geesaman et al., 1997), with which the metabolic activity evoked during two behavioral conditions can be compared at a very high spatial resolution (50 micron or better) throughout the brain. The double label 2-DG lacks the temporal resolution of optical recording, but is applicable to all parts of the monkey visual system. The technique has a better spatial resolution than monkey fMRI, but lacks the versatility of fMRI since only two conditions can be compared. That is the principal reason that we, and others, developed monkey fMRI (Logothetis et al., 1999; Vanduffel et al., 2001), which enables the investigator to sample, albeit indirectly, brain activity over a large number of conditions in the same monkey. The high resolution of the 2-DG technique was crucial, however, for the purpose of the study reviewed.

In single cell studies or functional imaging, the dominant paradigm consists of looking for effects of attention in those neurons, or the population of neurons, that process the stimuli to which attention is directed. Here, we observed a modulation of metabolic activity in the neurons representing regions in visual space surrounding the attended stimulus. Although monkeys fixated nearly identical displays containing a central grating, they either used or did not use the grating to make their behavioral response (an eye movement to a left or right target). Thus, attention was directed either *toward* or *away* from the central grating. Tritiated 2-DG labeled neurons that were activated during the attention to the grating condition, whereas ^{14}C-labeled 2-DG marked neurons that were activated in the attention away from the grating condition (the respective order was randomized between animals). When the monkey used the grating to guide its behavior (i.e., was paying attention to the grating), we observed a ring of suppressed metabolic activity in the parts of LGN and V1 surrounding the retinotopic representation of the grating.

When the monkey uses the grating, he pays attention to this region of visual space because he has to process the grating orientation. Thus, both spatial and featural attention are likely to be engaged in this condition. In the other condition, only the position of the target matters and thus, only spatial attention (directed to other parts of the display, away from the grating) comes into play.

In the discussion of the experimental findings, we will compare suppression of metabolic activity with decreased activity in the fMRI. We will also introduce a simple dynamic network, showing that the known connections between geniculate relay cells, cortical layers 4 and 6 and reticular neurons can generate the suppressive ring when a stimulus is attended, exactly as observed in our experiments.

II. A RING OF METABOLIC SUPPRESSION

The main data obtained in these experiments consist of differences in 2-DG concentration, reflecting differences in metabolic activity evoked during the two behavioral conditions. Figure 71.1 shows the concentration of radioactive 2-DG sampled over trajectories covering different parts and different layers of striate cortex in one of the four monkeys participating in the experiment. This figure illustrates the basic finding of the study: differences in 2-DG concentration were observed between the two attention conditions in certain layers and in certain parts of V1. Further analysis revealed that 1) these differences occurred mainly in the magnocellular input layer 4Cα; and 2) they were restricted to an annular region surrounding the representation of the grating. The laminar specificity of the effect allowed us to disambiguate the signs of these differences. Since only two conditions were compared, the reduced metabolic activity in the annular region surrounding the representation of the grating could have represented either a reduced activation in the attention to the grating condition or alternatively an increased activation in the attention-away condition. We compared the ratio of activity in layer 4Cα and a control layer (layer 4B) for each condition at different parts of V1 (inside the ring, over the ring, outside the ring). This procedure enabled us to establish that in the attention-to-the-grating condition, metabolic activity in the ring was lower in layer 4Cα than in layer 4B, whereas the ratio of these activities was similar in all parts of V1 in the other condition (see Fig. 11 of Vanduffel et al., 2000). Thus, our main finding was an attention-dependent suppression of metabolic activity surrounding the representation of the grating, rather than an enhancement of activity in the attention-away condition.

Figure 71.2 shows the distribution of the differential 2-DG uptake of the two conditions in layer 4Cα for the four different animals as a function of eccentricity, along with the radius of the grating used in each animal. This figure indicates that the ring of suppression changed in diameter with that of the grating, confirming that the suppression surrounded the stimulus representation. Note, however, that the strength of the suppression varies little with stimulus diameter.

Remarkably, similar observations were made in the LGN, at least in the magnocellular layers. Again, the suppression varied with the grating diameter (see Fig. 13 of Vanduffel et al., 2000), but in the LGN, the depth of this suppression increased with the diameter. In addition, another subcortical change was observed in the visual thalamus: metabolic activity of the reticular thalamic nucleus increased in the attention-to-the-grating condition relative to the attention-away condition.

III. DISCUSSION AND COMPUTATIONAL MODELING

The results reviewed here clearly indicate that suppressive changes occur in those parts of V1 and LGN that surround the representation of an attended stimulus. This finding is remarkable because of the level in the system at which it occurs, its retinotopic location, and its sign. Human fMRI studies have also reported suppressive effects in human V1 (e.g., Tootell et al., 1998; Smith et al., 2000) and even in human LGN (O'Connor et al., 2002). The effects occur in regions outside the representation of the attended stimulus (typically the fovea when attending a peripheral stimulus), rather than in the representation of the attended stimulus itself. The belief that attention modulates only the representation of the attended stimulus can be so entrenched that analysis is sometimes restricted to this part of cortex (e.g., Gandhi et al., 1999). Only a single study (Smith et al., 2000) has provided hints of a ring of suppression surrounding the representation of the attended stimulus. Human fMRI has also indicated (modestly) increased activation at the level of the stimulus representation, even in V1 (Tootell et al., 1998; Brefczynski and De Yoe, 1999; Gandhi et al., 1999; O'Connor et al., 2002). Such effects were weak in our data and rarely reached significance. There are obvious differences between the two types of experiments:

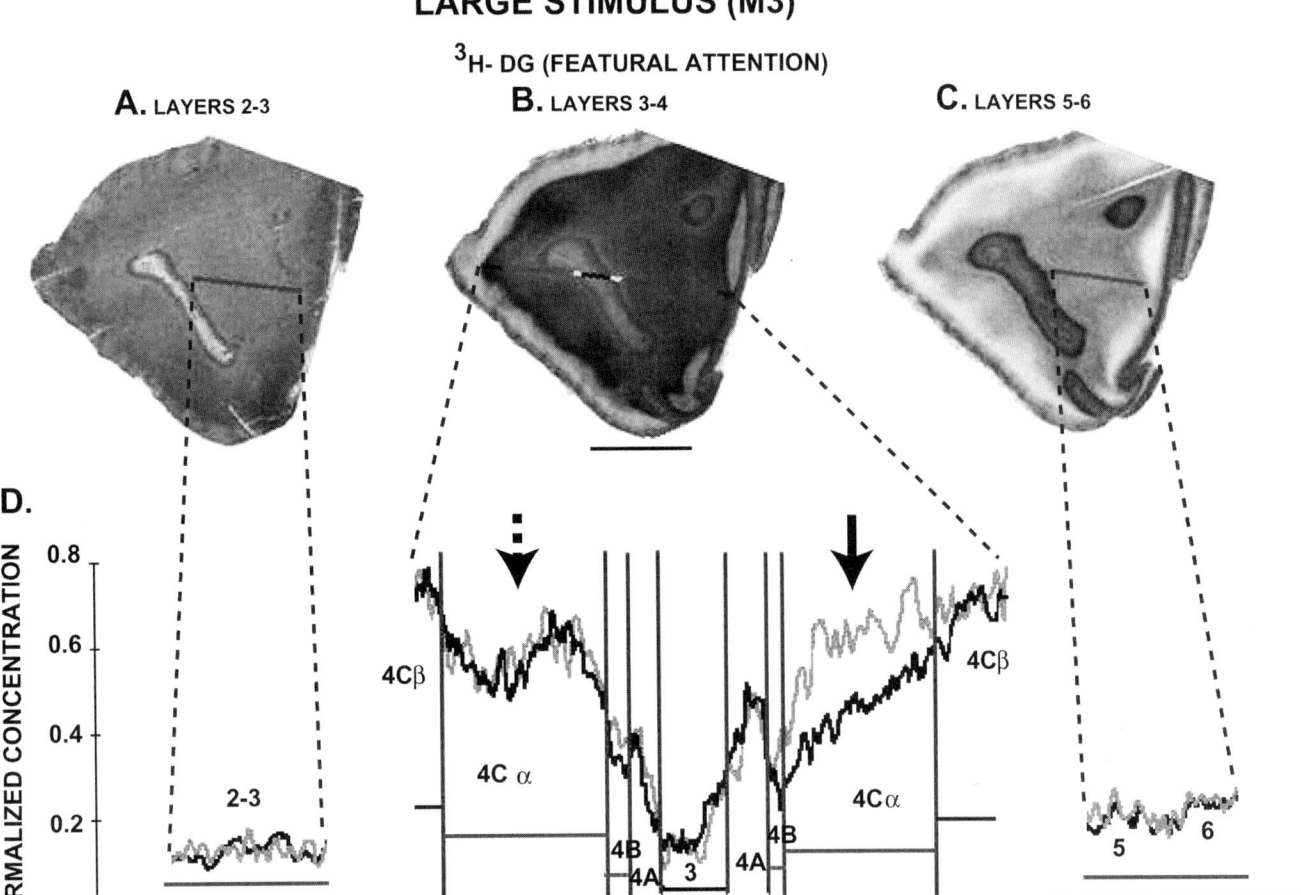

FIGURE 71.1 Plots of normalized [^{14}C]DG and [^{3}H]DG concentrations as a function of eccentricity in different layers of flattened area V1. **A–C.** The ^{3}H signals (related to the featural-attention task, i.e., the attention-to-the-grating condition) from single V1 sections through layers 2–3, 3–4, and 5–6, respectively. No orientation columns or visually driven enhanced activity can be observed. In the three sections, the fovea is represented toward the left, with more peripheral visual field representations toward the right. The upper visual field is represented in the lower portion of the section. **D.** Plots of normalized [^{3}H]DG and [^{14}C]DG concentrations as measured along the lines indicated in **A–C** (along the representation of the horizontal meridian from foveal to more peripheral visual field representations). Note the suppressed [^{3}H]DG concentration in more peripheral (solid black arrow in **D**), but not foveal (dotted black arrow in **D**) visual field representations of layer 4Cα.

different species, differences between metabolic and hemodynamic measurements, and different control conditions. To underscore the importance of the latter factor, it is worthwhile to consider some of the evidence for enhanced geniculate responses to attended stimuli (O'Connor et al., 2002). The authors compared two conditions: one in which the subjects attend a central stimulus (to count letters) and one in which the subjects attend a peripheral checkerboard stimulus (to detect its dimming). They compared responses to the checkerboard in the two conditions and reported that the activity evoked by the checkerboard was larger when the subjects attend to the checkerboard. Since only two conditions were compared, an alternative interpretation of their data would be that there is a smaller response to the checkerboard when the central stimulus is attended. The suppressive effects they described would then be an enhancement of this suppressive effect when the attention to the central stimulus is further loaded.

The difference between hemodynamic measurements (fMRI) and metabolic measurements should not

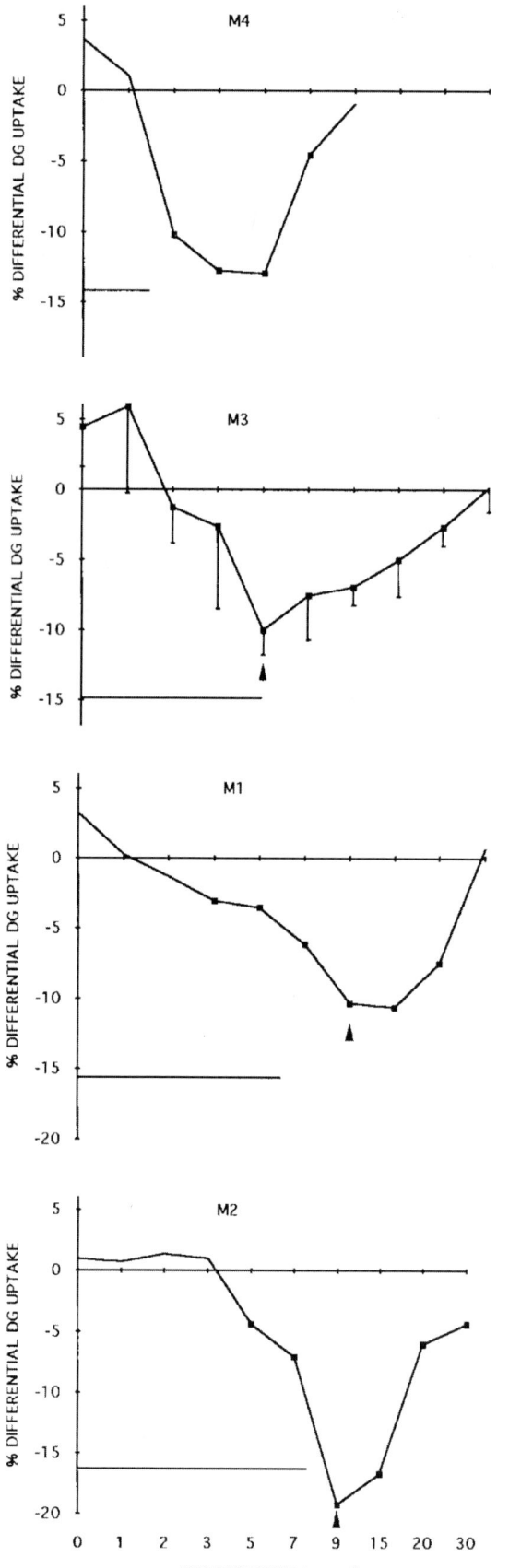

be underestimated. One reason why suppressive effects are less frequently reported in the fMRI studies, is that they could arise from purely vascular changes. This so-called plumbing hypothesis states that an increase in blood flow in an active brain region reduces blood flow in other less active parts of the brain. Such a hemodynamic effect cannot account for the suppressive effects reported here, supporting the view of those who claim that negative 'BOLD' reflects reduced neuronal activity. The metabolic measurements of the present study also clarify another point of contention in the interpretation of fMRI data; that is, the contribution of inhibitory neurons to the fMRI signal. Our results suggest that inhibitory neurons contribute to the metabolic brain activity (and hence hemodynamic responses) as well as excitatory neurons. Indeed, structures such as the RTN, which include only inhibitory neurons projecting massively to a target structure (here the LGN), show changes in differential metabolic activity opposite in sign to those of the target structure. Activity in the LGN decreased with attention to the grating, whereas activity in the RTN increased. Similar overall reversals in the sign of differential metabolic activity were observed in the basal ganglia (Vanduffel, unpublished PhD thesis) where two inhibitory projections are known to occur in sequence: the sign of the differential 2-DG uptake changed twice when going from the ventral putamen over the ventral pallidum to the medio-dorsal thalamic nucleus.

The results reviewed here also point to the involvement of the reticular thalamic nucleus in the generation of attentional modulation at the early levels of the visual system, in agreement with earlier suggestions (Crick, 1984; Montero, 1999). In order to provide further support for this view we modeled the experimental results using four building blocks: the relay cells of the LGN, the RTN, and layers 4Cα and 6 of V1. Each block consists of a chain of 300 neurons, which

FIGURE 71.2 Normalized differential DG uptake (using the spatial attention task as baseline) in layer 4Cα of V1 as a function of eccentricity is plotted for all four subjects. Percentage differential DG uptake = [(the normalized DG concentration related to the featural attention task) − (the normalized DG uptake related to the spatial attention task)] / (the normalized DG uptake related to the spatial attention task) × 100. The measurements along the horizontal and vertical meridians are averaged and thus, each data point represents the average of 90 isotope concentration measurements. The standard deviations are shown for monkey M3. Squares indicate the eccentricities for which the difference in isotope concentration reached a P-value of <0.05; two-tailed t-test. The radius of the gratings that were presented in each experiment is indicated by the lines at the bottom of each panel. Note that the shift of lower featural-attention DG uptake from parafoveal to larger eccentricities is correlated with the size of the stimulus.

summate weighted inputs linearly and compute their output using a sigmoidal transfer function. All neurons, except those of the RTN, are excitatory and no assumptions about the time courses of the synapses were included. The model (see Fig. 71.3) bears some resemblance to earlier attempts of Bazhenov et al. (1998) and Montero (1999), but differs from these by including more anatomical data (albeit from lower animals such as cats and rats) and in the dynamic nature of the model. The connections assumed between the four blocks are indicated in Fig. 71.3. Connectivity between any set of two blocks was constant. Noteworthy are the direct inhibition of relay cells by the reticular axons, in agreement with Wang et al. (2001), as well as the larger spread of the projections to and from the RTN compared to those to and from the relay cells, in agreement with the anatomical results of Bourassa and Deschênes (1995) and of Yen et al. (1985). For simplicity other inputs to the RTN have been omitted. The attention signal is assumed to be applied to layer 6 of V1, in agreement with a large body of anatomical results indicating that feedback connections end outside layer 4. One other distinguishing feature of the model is the nature of the attention gating signal, which can be uniform rather than spatially restricted. The main mechanism of the attentional modulation is the diffusion of stimulus-driven relay cell activity to RTN, which in turn injects inhibition into regions of the LGN surrounding the stimulus representation. Critical to the model is the large spread of the RTN connections. The fixed-point states, obtained after 10 iterations, reproduce the experimental findings nicely (see Fig. 71.3). The diameter of the suppressive ring increases with stimulus diameter and the depth of suppression depends on stimulus size much more in LGN than in V1. One apparent discrepancy is the absence in the 2-DG study of an attentional modulation in layer 6. This is in all likelihood due to the low level of spontaneous activity in layer 6. In regions surrounding the grating representation, no stimulus was present and therefore, a modulation could be observed only in regions in which spontaneous activity is high enough, such as the LGN and layer 4C of V1. Thus, this model, which is supported by our experimental data, opens new avenues in our thinking about the nature and source of the signals that

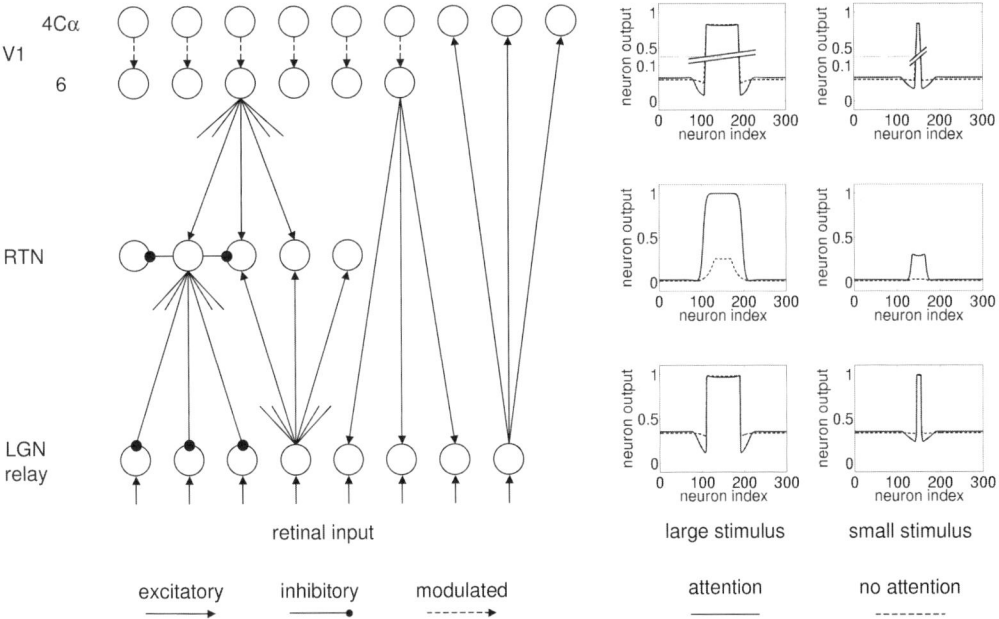

FIGURE 71.3 Model for attentional modulation in the thalamocortical network. Left panel shows synaptic connections between LGN relay neurons, RTN, and V1 layers 4Cα and 6 neurons. For each layer, a chain of 300 model neurons with sigmoidal output functions is used. Connectivity to and from RTN is broader (indicated by larger numbers of connections). All connections are excitatory except those coming from RTN; the dashed lines indicate the uniform attentional input in layer 6. Right panel shows the resulting neural output as a function of the neuron's position in the layer in the case of attention (solid lines) and no attention (dashed lines), and for two stimulus sizes (left and right columns). Upper row corresponds to layer 4Cα, middle row to RTN, and bottom row to LGN relay neurons. The thalamocortical model forms a recurrent network with fixed-point iteration dynamics. The plots in the right panel represent the fixed-point states, which are reached after 10 iterations.

cause the attentional modulation in the primate visual system.

IV. CONCLUSION

Using double label 2-deoxyglucose, Vanduffel et al. (2000) observed a ring of metabolic suppression surrounding the representation of the attended stimulus. The involvement of the reticular thalamic nucleus in the generation of this suppression is supported by a dynamic neuronal model.

Acknowledgments

WV is a fellow of the FWO. This work was supported by FWO and GOA 2000/11.

References

Bazhenov, M., Timofeev, I., Steriade, M., and Sejnowski, T. J. (1998). Computational models of thalamocortical augmenting responses. *J. Neurosci.* **18**, 6444–6465.

Bourassa, J., and Deschênes, M. (1995). Corticothalamic projections from the primary visual cortex in rats: a single fiber study using biocytin as anterograde tracer. *Neuroscience* **253**, 253–263.

Brefczynski, J. A., and De Yoe, E. A. (1999). A physiological correlate of the 'spotlight' of visual attention. *Nat. Neurosci.* **2**, 370–374.

Corbetta, M., Akbudak, E., Conturo, T. E., Snyder, A. Z, Ollinger, J. M., Drury, H. A., Linenweber, M. R., Petersen, S. E., Raichle, M. E., Van Essen, D. C., and Shulman, G. L. (1998). A common network of functional areas for attention and eye movements. *Neuron* **21**, 761–773.

Crick, F. (1984). Function of the thalamic reticular complex: the searchlight hypothesis. *Proc. Nat. Acad. Sci. USA* **81**, 4586–4590.

Gandhi, S. P., Heeger, D. J., and Boynton, G. M. (1999). Spatial attention affects brain activity in human primary visual cortex. *Proc. Nat. Acad. Sci. USA* **96**, 3314–3319.

Geesaman, B. J., Born, R. T., Andersen, R. A., and Tootell, R. B. (1997). Maps of complex motion selectivity in the superior temporal cortex of the alert macaque monkey: a double label 2-deoxyglucose study. *Cerebr. Cort.* **7**, 749–757.

Kastner, S., De Weerd, P., Desimone, R., and Ungerleider, L. G. (1998). Mechanisms of directed attention in the human extrastriate cortex as revealed by functional MRI. *Science* **282**, 108–111.

Livingstone, M. S., and Hubel, D. H. (1981). Effects of sleep and arousal on the processing of visual information in the cat. *Nature* **291**, 554–561.

Logothetis, N. K., Guggenberger, H., Peled, S., and Pauls, J. (1999). Functional imaging of the monkey brain. *Nat. Neurosci.* **2**, 555–562.

Montero, V. M. (1999). Amblyopia decreases activation of the corticogeniculate pathway and visual thalamic reticularis in attentive rats: a 'focal attention' hypothesis. *Neuroscience* **91**, 805–817.

O'Connor, D. H., Fukui, M. M., Pinsk, M. A., and Kastner, S. (2002). Attention modulates responses in the human lateral geniculate nucleus. *Nat. Neurosci.* **5**, 1203–1209.

Reynolds, J. H., Pasternak, T., and Desimone, R. (2000). Attention increases sensitivity of V4 neurons. *Neuron* **26**, 703–714.

Smith, A. T., Singh, K. D., and Greenlee, M. W. (2000). Attentional suppression of activity in the human visual cortex. *Neuroreport* **2**, 271–277.

Tootell, R. B. H., Hadjikani, N., Hall, K. E., Marett, S., Vanduffel, W., Vaughan, J. T., and Dale, A. M. (1998). The retinotopy of visual spatial attention. *Neuron* **21**, 1409–1422.

Treue, S., and Maunsell, J. H. R. (1999). Effects of attention on the processing of motion in macaque middle temporal and medial superior temporal visual cortical areas. *J. Neurosci.* **19**, 7591–7602.

Vanduffel, W., Tootell, R. B. H., and Orban, G. A. (2000). Attention-dependent suppression of metabolic activity in the early stages of the macaque visual system. *Cerebr. Cort.* **10**, 109–126.

Vanduffel, W., Fize, D., Mandeville, J. B., Nelissen, K., Van Hecke, P., Rosen, B. R., Tootell, R. B., and Orban, G. A. (2001). Visual motion processing investigated using contrast agent enhanced fMRI in awake behaving monkeys. *Neuron* **32**, 565–577.

Wang, S., Bickford, M. E., Van Horn, S. C., Erisir, A., Godwin, D. W., and Sherman, S. M. (2001). Synaptic targets of thalamic reticular nucleus terminals in the visual thalamus of the cat. *J. Comp. Neurol.* **440**, 321–341.

Wardak, C., Olivier, E., and Duhamel, J. R. (2002). Saccadic target selection deficits after lateral intraprietal area inactivation in monkeys. *J. Neurosci.* **22**, 9877–9884.

Yen, C. T., Conley, M., Hendry, S. H. C., and Jones, E. G. (1985). The morphology of physiologically identified gabaergic neurons in the somatic sensory part of the thalamic reticular nucleus in the cat. *J. Neurosci.* **5**, 254–268.

CHAPTER 72

Attentional Modulation in the Human Lateral Geniculate Nucleus and Pulvinar

Sabine Kastner, Keith A. Schneider, and Daniel H. O'Connor

ABSTRACT

Attentional mechanisms are important for selecting relevant and filtering out irrelevant information from cluttered visual scenes. Selective attention has been shown to affect neural activity in extrastriate and even in striate visual cortex. Here, we review evidence from functional magnetic resonance imaging studies (fMRI) demonstrating that attentional response modulation is not confined to cortical processing, but can occur as early as at the thalamic level. Neural activity in the human lateral geniculate nucleus (LGN) was modulated by attention in multiple ways. The LGN, traditionally viewed as the gateway to visual cortex, may play an important role as a gatekeeper in controlling attentional response gain. In the human pulvinar, a dorso-medial region was identified that responded to attended but not to unattended visual stimuli. Taken together, these studies demonstrate that fMRI can be used effectively to study attentional mechanisms of thalamo-cortical circuits in the human brain.

I. INTRODUCTION

One classical debate in the cognitive psychology of attention is that of early versus late selection. The early selection account (e.g., Broadbent, 1958) proposed that the neural representations of attended and unattended stimuli are affected differently at early sensory processing stages. As a result, the processing of attended stimuli will be facilitated and that of unattended stimuli will be suppressed. The late selection account (e.g., Duncan, 1980) suggested that both types of stimuli are more or less fully processed at sensory levels and that the rejection of unattended stimuli occurs at a stage beyond sensory coding (e.g., in memory systems). Given the massive feedforward and feedback interconnectivity of the brain, it should be stressed that the terms early and late cannot properly refer to the sequence of temporal activation, but rather a sequence of increasing complexity, with lower levels of the cortical hierarchy representing simple sensory features and higher levels representing more complex objects. With this in mind, one still-pertinent question that emerges from this classical debate is at which stage of representation within the visual pathway selective attention first affects neural processing. At the neural level, attentional response modulation was originally demonstrated in extrastriate but not in striate cortex. Recent evidence has shown that neural activity in striate cortex can also be affected, depending on certain task-related factors, such as the attentional demands or the need to integrate contextual information. Here, we will review evidence that attentional response modulation is not confined to cortical processing stages but occurs already at the thalamic level.

II. ATTENTIONAL MODULATION IN THE HUMAN LGN

A. Anatomy of the LGN

The LGN is the thalamic stage in the retinocortical projection and has traditionally been viewed as a simple relay between the retina and the visual cortex (Sherman and Guillery, 2001). The LGN consists of six eye-specific layers, four of which receive inputs from parvocellular retinal ganglion cells, and two of which receive magnocellular inputs. Each layer is organized into a precise retinotopic map. In addition to these retinal afferents, the LGN receives input from other sources including the primary visual cortex (V1), thalamic reticular nucleus (TRN), superior colliculus (SC), and brainstem. The functional role of these feedback inputs to the LGN is not well understood. Given its compact nature and retinotopic organization, the LGN would be an ideal stage in the processing hierarchy to efficiently modulate the sensory processing of information from specific and perhaps large portions of the visual field. Here, we review studies that used functional magnetic resonance imaging (fMRI) to investigate attentional response modulation in the human LGN (O'Connor et al., 2002).

At the cortical level, converging evidence from single-cell recording studies in monkeys and functional brain mapping studies in humans shows that selective attention modulates visually evoked activity in multiple ways. First, neural responses to attended visual stimuli are enhanced relative to the same stimuli when ignored (attentional enhancement) (Moran and Desimone, 1985). Second, neural responses to ignored stimuli are attenuated depending on the load of attentional resources engaged elsewhere (attentional suppression) (Rees et al., 1997). And third, directing attention to a location in the absence of visual stimulation and in anticipation of stimulus onset increases neural baseline activity (attention-related baseline increases) (Kastner et al., 1999). These effects of selective attention were investigated in the LGN in a series of three experiments.

In all experiments, flickering checkerboard stimuli were used in order to maximally activate the LGN. Checkerboard stimuli of high or low contrast were presented in alternation to the left or right visual hemifield while subjects maintained fixation, activating the right or left LGN, respectively (see Fig. 72.1). Activity also was observed in areas V1, V2, V3/VP, V4, TEO, V3A, and MT/MST of the visual cortex, as determined on the basis of retinotopic mapping (see the section II.E).

B. Attentional Response Enhancement in the LGN

To investigate attentional response enhancement in the LGN (Experiment 1), checkerboard stimuli of high- or low-contrast were presented to the left or right hemifield while subjects directed attention to the stimulus (attended condition) or away from the stimulus (unattended condition). In the unattended condition,

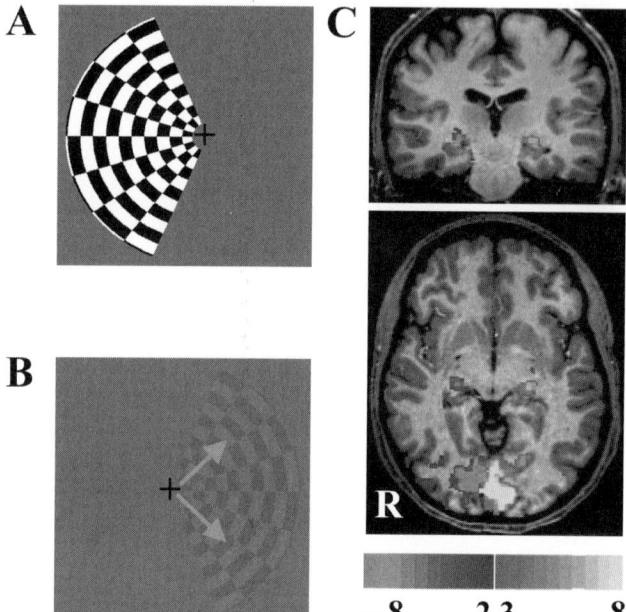

FIGURE 72.1 Visual stimuli and LGN activation. **A, B**. High- or low-contrast checkerboard stimuli were presented to the left or right of a central fixation (+) point. In Experiments 1 and 3, subjects covertly directed attention to a checkerboard arc (blue arrows) and detected randomly occurring luminance changes along that arc. The detection of luminance changes in low-contrast and high-contrast checkerboard stimuli was not matched for task difficulty. **C**. Axial slice showing activations of the right (green-blue) and left (yellow-orange) LGN and visual cortex evoked by the checkerboard stimuli. Activations in the thalamus were restricted to the LGN; no activations within the pulvinar were obtained. Scale indicates z-score values of activations in colored regions. R = right hemisphere. (Adapted from O'Connor et al., 2002.) See color plate.

attention was directed away from the stimulus by having subjects count letters at fixation. The letter counting task ensured proper fixation and effectively prevented subjects from covertly attending to the checkerboard stimuli. In the attended condition, subjects were instructed to covertly direct attention to the checkerboard stimulus and to detect luminance changes that occurred randomly in time at an eccentricity of 10 degrees of visual angle (see Fig. 72.1B).

In our statistical model, stimulation of the left visual hemifield was contrasted with stimulation of the right visual hemifield. Thereby, the analysis was restricted to voxels activated by the peripheral checkerboard stimuli, excluding foveal stimulus representations. Relative to the unattended condition, the mean fMRI signals evoked by the high-contrast stimulus increased significantly in the attended condition (see Fig. 72.2A). Similar attentional response enhancement was found with activity evoked by the low-

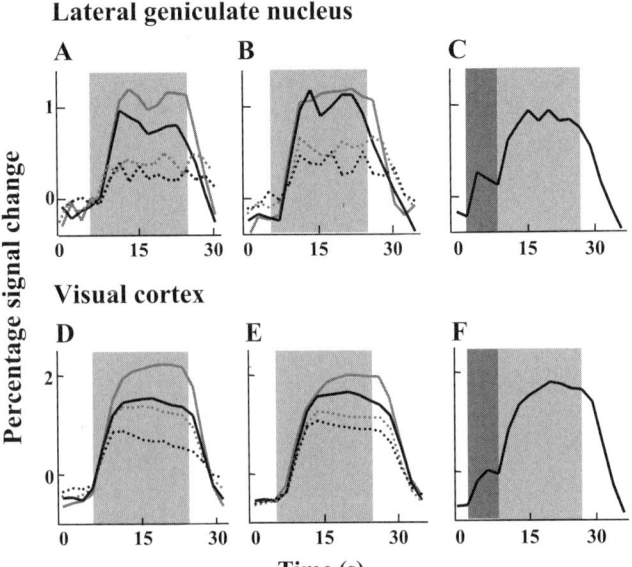

FIGURE 72.2 Time series of fMRI signals in the LGN and in visual cortex. Group analysis ($n = 4$). Data from the LGN and visual cortex were combined across left and right hemispheres. Activity in visual cortex was pooled across areas V1, V2, V3/VP, V4, TEO, V3A, and MT/MST. **A, D.** Attentional enhancement. During directed attention to the stimuli (gray lines), responses to both the high-contrast stimulus (100%, solid lines) and low-contrast stimulus (5%, dashed lines) were enhanced relative to an unattended condition (black lines). **B, E.** Attentional suppression. During an attentionally demanding difficult fixation task (black lines), responses evoked by both the high-contrast stimulus (100%, solid lines) and low-contrast stimulus (10%, dashed lines) were attenuated relative to an easy attention task at fixation (gray lines). **C, F.** Baseline increases. Baseline activity was elevated during directed attention to the periphery of the visual hemifield in expectation of the stimulus onset (darker gray shaded area). The lighter gray shaded area indicates the periods of checkerboard presentation. (Adapted from O'Connor et al., 2002.)

contrast stimulus (see Fig. 72.2A). These results suggest that attention facilitates visual processing in the LGN by enhancing neural responses to an attended stimulus relative to those evoked by the same stimulus when ignored.

Because this attention experiment was not designed to test for the spatial specificity of attentional modulation, it is possible that the attention effects found in the LGN reflected nonspecific attentional states such as arousal, rather than mechanisms of selective attention. General arousal affects the visual system in topographically nonspecific ways, whereas selective attention has been shown to be spatially specific. To test for the spatial specificity of attentional enhancement, a control experiment was performed in which high-contrast checkerboard stimuli were presented simultaneously to both hemifields. Subjects were instructed to direct attention either to the left or right checkerboard and to detect luminance changes as above, while ignoring the checkerboard stimulus in the respective contralateral hemifield. In the LGN, mean fMRI signals evoked by the checkerboard stimuli were significantly larger when attended than when ignored. The size of the attentional response enhancement was similar compared to that obtained in Experiment 1. These findings indicate that attentional response enhancement in the LGN was spatially selective and rules out the possibility that the modulation was due to unspecific attention effects such as arousal.

C. Attentional Response Suppression in the LGN

To investigate attentional-load dependent suppression in the LGN (Experiment 2), high- and low-contrast checkerboard stimuli were presented to the left or right hemifield while subjects performed either an easy attention task or a difficult attention task at fixation and ignored the peripheral checkerboard stimuli. During the easy attention task, subjects counted infrequent, brief color changes of the fixation cross. During the difficult attention task, subjects counted letters at fixation as in Experiment 1. Behavioral performance was $99\pm1\%$ (mean $\pm$ SEM) correct on average in the easy attention task and $54\pm7\%$ in the difficult attention task ($t(3) = 7.98$, $p < 0.01$), demonstrating the differences in attentional load.

Relative to the easy task condition, mean fMRI signals evoked by the high-contrast and by the low-contrast stimuli decreased significantly in the difficult task condition (see Fig. 72.2B). These fMRI signals reflect only activity evoked by the peripheral checkerboard stimuli when processed under conditions of different attentional load and are not confounded with activity evoked by the foveal stimuli. This finding suggests that neural activity evoked by ignored stimuli in the LGN is attenuated as a function of the load of attentional resources engaged elsewhere. According to accounts of attentional load (Rees et al., 1997), the degree to which ignored stimuli are processed is determined by the amount of spare attentional capacity. In our design, the low load task presumably left a greater amount of spare attentional capacity than the high load task, which may account for the observed attenuation of responses evoked by the ignored stimuli during the high load condition. The attentional-load dependent suppression of LGN activity is consistent with the effects of attentional load upon visual processing in area MT. Rees et al. (1997) found that fMRI responses in area MT evoked by an irrelevant motion stimulus presented peripherally were reduced during a high-load linguistic task compared to a low-load task at fixation.

D. Attention-Related Baseline Increases in the LGN

Selective attention not only modulates neural responses to visual stimuli, but also affects neural activity in the absence of visual stimulation. Directing attention to a peripheral target location in anticipation of a stimulus has been shown to elevate (spontaneous) baseline activity in visual cortex (Kastner et al., 1999). Such baseline increases may indicate a bias of neural responses in favor of an attended location. To investigate attention-related baseline increases in the LGN (Experiment 3), subjects were cued to covertly direct attention to the periphery of the left or right visual hemifield and to expect the onset of the stimulus. The expectation period, during which subjects were attending to the periphery without receiving visual input, was followed by attended presentations of a high-contrast checkerboard. During the attended presentations, subjects counted the occurrence of luminance changes as in Experiment 1. During the expectation period, fMRI signals increased by 0.3% relative to the preceding blank period in which subjects were fixating but not directing attention to the periphery. This elevation of baseline activity was followed by a further response increase evoked by the visual stimuli (see Fig. 72.2C). This finding suggests that neural activity in the LGN can be affected by attentional signals even in the absence of visual stimulation. Interestingly, Chen et al. (1998) found that fMRI activity was increased in both V1 and the LGN during a visual imagery task, further supporting the idea that neural activity in the LGN can be driven by pure top-down signals.

E. Comparison of Attention Effects in the LGN and in Visual Cortex

Similar effects to those observed in the LGN were also found throughout visual cortex. Visual areas V1, V2, V3, V4, TEO, V3A, and MT were identified based on standard retinotopic mapping for each participant. The attention effects found at the thalamic and at the cortical level were compared by normalizing the mean fMRI signal evoked in the LGN and in each activated cortical area. The resulting index values are a quantitative measure of the magnitude of attentional effects in each area and for each experiment (see Fig. 72.3). For all indices, larger index values indicate larger effects of attention. This analysis revealed two important results (see Fig. 72.3A–C). First, and in accordance with

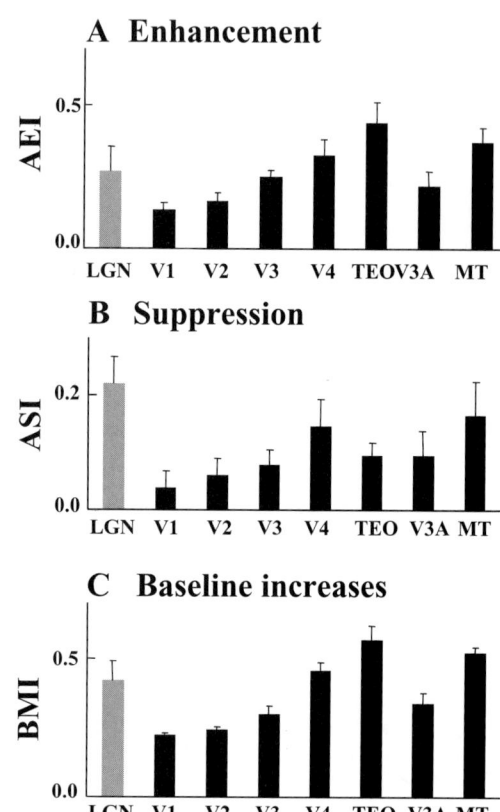

FIGURE 72.3 Attentional response modulation in the LGN and in visual cortical areas V1, V2, V3/VP, V4, TEO, V3A, and MT/MST. Attentional effects were quantified and normalized by defining several indices: **A.** attentional enhancement index (AEI, Experiment 1); **B.** attentional suppression index (ASI, Experiment 2); **C.** baseline modulation index (BMI, Experiment 3). For all indices, larger values indicate larger effects of attention. Index values were computed for each subject based on averaged signals obtained in the different attention conditions and are presented as averaged index values from four subjects. In visual cortex, attentional effects increased from early to later processing stages. Attentional effects in the LGN were larger than in V1. Vertical bars indicate S.E.M. across subjects. (Adapted from O'Connor et al., 2002.)

previous findings (as reviewed in Kastner and Ungerleider, 2000), the attentional effects of enhancement, suppression, and baseline elevations increased from early to more advanced processing levels along both the ventral and dorsal pathways of visual cortex (see Fig. 72.3A–C). Second, all three attentional effects tended to be stronger in the LGN than in striate cortex. This finding suggests that attentional response modulation in the LGN may be influenced not only by corticothalamic feedback from striate cortex, but also by sources such as the brainstem or the TRN, especially because hemodynamic responses appear to reflect the synaptic input to an area. Other possibilities that may explain the differences in magnitude of the modulation between the LGN and V1 include regional disparities underlying the blood oxygenation level dependent signal or nonlinearities in thalamo-cortical signal transmission.

F. The LGN: An Early Gatekeeper?

The studies reviewed thus far indicate that selective attention modulates neural activity in the LGN by enhancing neural responses to attended stimuli, by attenuating those to ignored stimuli, and by increasing baseline activity in the absence of visual stimulation. These findings challenge the classical notion that attentional effects are confined to cortical processing. Further, they suggest the need to revise the traditional view of the LGN as a mere gateway to the visual cortex. In fact, due to its afferent input, the LGN may be in an ideal strategic position to serve as an early gatekeeper in attentional gain control. In addition to corticothalamic feedback projections from V1, the LGN receives inputs from the SC, which is part of a distributed network of areas controlling eye movements, and the TRN. For several reasons, the TRN has long been implicated in theoretical accounts of selective attention (Crick, 1984). First, all feedforward projections from the thalamus to the cortex as well as their reverse projections pass through the TRN. Second, the TRN receives not only inputs from the LGN and V1, but also from several extrastriate areas and the pulvinar. Thereby, it may serve as a node where several cortical areas and thalamic nuclei of the visual system can interact to modulate thalamocortical transmission through inhibitory connections to LGN neurons (Sherman and Guillery, 2001). And third, the TRN contains topographically organized representations of the visual field and can thereby modulate thalamocortical or corticothalamic transmission in spatially specific ways. Even though much remains to be learned about the complex thalamic circuitry that may subserve attentional gain control in the LGN, it appears that attentional top-down signals modulate the processing of visual information in the LGN, the first stage beyond the retina.

III. ATTENTIONAL MODULATION IN THE HUMAN PULVINAR

Several lines of evidence indicate that the pulvinar is part of a distributed network subserving visuo-spatial attention (Kastner and Ungerleider, 2000). First, when the pulvinar is inactivated by muscimol, monkeys show impairments in filtering out irrelevant information from distracter stimuli (Desimone et al., 1990). A similar role of the pulvinar in filtering unwanted information has also been suggested by neuroimaging studies in humans (LaBerge and Buchsbaum, 1990). And second, patients with lesions in the pulvinar exhibit visuo-spatial hemineglect, an impairment in directing attention to the contralateral hemifield (Karnath et al., 2002).

In the macaque monkey, the pulvinar contains in its inferior, lateral, and medial subdivisions at least four visual areas that can be distinguished on the basis of their visuotopic organization and connectivity with cortical visual areas. Two of these areas are located in the inferior and lateral part of the pulvinar. These areas contain clearly organized retinotopic maps and are connected with cortical areas V1, V2, V4, and MT. A third area, described on the basis of its connections with area MT, is located in the medial portion of the inferior subdivision and does not appear to have a well-defined retinotopic map. A fourth area, known as Pdm, has been identified dorsal to these areas and also does not show much visuotopic organization. Importantly, neural responses in Pdm are related to visuo-spatial attention. It has been shown in particular that neural responses in Pdm were enhanced when an animal attended to a visual stimulus compared to when the same stimulus was ignored (Petersen et al., 1985). Here, we review a series of experiments that were aimed at identifying a homologue for area Pdm in the human pulvinar (Kastner et al., 2004).

Flickering checkerboard stimuli were presented simultaneously to both hemifields interleaved with blank periods or unilaterally in alternation to the right or left hemifield while subjects maintained fixation. In one set of studies, subjects were instructed by an arrow presented at fixation to direct attention either to the left or right checkerboard and to detect luminance changes that occurred at random times at an eccentricity of 10 degrees of visual angle. In another set of studies, subjects were instructed to passively view the stimuli without performing any specific task. During bilateral attended presentations of the checkerboard stimuli, two regions within the pulvinar, observed to be located superior and medial to the LGN activations (see Fig. 72.1), were consistently activated across subjects. The region in the right pulvinar is depicted for an individual subject in Fig. 72.4A. Mean Talairach coordinates across all subjects were x = 17, y = –24, z = 12 for the activated region in the right pulvinar and –12, –24, 8 for the activated region in the left pulvinar. Given the individual variability of the locations of activated regions within the pulvinar, it is not clear whether the regions in the left and right pulvinar represent corresponding visual areas. Importantly, our attention-related activity in dorso-medial regions of the human pulvinar was located in the same region of

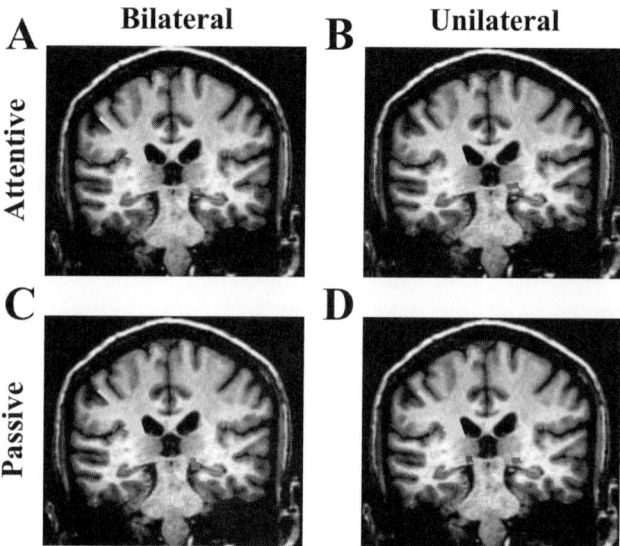

FIGURE 72.4 Functional activation of the pulvinar under different conditions. Results are shown from one subject tested in four different experiments. **A.** Bilateral attentive. The arrow indicates functional activation of the right pulvinar. The subject directed attention to a checkerboard stimulus presented simultaneously to the right and left visual hemifield. **B.** Unilateral attentive. No functional activation was found in the pulvinar when the stimuli were presented unilaterally and in alternation to the right and left hemifield while subjects directed attention to the checkerboards. **C.** Bilateral passive. No functional activation was obtained in the pulvinar when the stimuli were presented bilaterally and passively viewed by the subjects while maintaining fixation. **D.** Unilateral passive. No functional activation was found in the pulvinar in Experiment 1, in which checkerboard stimuli were presented in alternation to the right or left hemifield. Other conventions as in Fig. 72.1. (Adapted from Kastner et al., 2004.) See color plate.

Talairach space as one of the subcortical sites where lesions are consistently associated with visual neglect (Karnath et al., 2002). When subjects directed attention to checkerboard stimuli presented unilaterally in alternation to the right and left hemifield, no activations in the pulvinar were found (see Fig. 72.4B). This finding suggests that visually responsive neurons in the activated region of the pulvinar had large receptive fields (RFs) extending over the entire visual field. To test the possibility that the visually evoked pulvinar activity was strongly modulated by selective attention, the same experiments were performed while subjects did not direct attention to the checkerboard stimuli, but rather passively viewed the stimuli while maintaining fixation. No activation in the pulvinar was found under these conditions, as illustrated in Fig. 72.4C for the bilateral presentations and in Fig. 72.4D for the unilateral presentations.

Several of our findings support the idea that the activated regions in the left and right dorso-medial pulvinar may be homologous to area Pdm of the macaque pulvinar. First, the human pulvinar areas were strongly modulated by visuo-spatial attention. Such attentional modulation has been described for Pdm, but not for other visual areas of the macaque pulvinar (Petersen et al., 1985). Second, the human pulvinar areas appeared to have a coarse visuotopic organization with large RFs extending over the entire visual field. Neurons with large RFs of greater than 30 degrees of visual angle extending across the vertical meridian have been found in macaque Pdm, but not in other pulvinar areas (Petersen et al., 1985). And third, it is unlikely that the human pulvinar activations occurred in retinotopically organized pulvinar areas because they were not evoked by alternating hemifield stimuli. It is not clear why retinotopically organized areas of the human pulvinar were not activated in our studies. However, given the small size of the activated volumes within the pulvinar, this may reflect a lack of sensitivity and spatial resolution of the fMRI methods used here. Taken together, our findings support the notion of a functional role of the pulvinar in visuo-spatial attention.

IV. CONCLUSION

It is difficult to obtain fMRI signals from subcortical nuclei, owing to their small sizes and deep locations. In this chapter, we have reviewed studies demonstrating modulation of visually evoked activity by selective attention in the human LGN and pulvinar using experimental designs optimized to activate these neural structures. The results of these studies challenge the notion that attentional top-down effects in sensory processing areas are confined to the cortical level and suggest an important role of the thalamus in attentional gain control. Future imaging studies at higher field strength using high-resolution techniques may further advance our understanding of complex thalamo-cortical circuits in the human brain.

References

Broadbent, D. E. (1958). *Perception and Communication.* Pergamon Press, London.

Chen, W., Kato, T., Zhu, X. H., Ogawa, S., Tank, D. W., and Ugurbil, K. (1998). Human primary visual cortex and lateral geniculate nucleus activation during visual imagery. *NeuroReport* **9**, 3669–3674.

Crick, F. (1984). Function of the thalamic reticular complex: the searchlight hypothesis. *Proc. Natl. Acad. Sci. USA* **81**, 4586–4590.

Desimone, R., Wessinger, M., Thomas, L., and Schneider, W. (1990). Attentional control of visual perception: cortical and subcortical mechanisms. *Cold Spring Harb. Sym.* **55**, 963–971.

Duncan, J. (1980). The locus of interference in the perception of simultaneous stimuli. *Psych. Rev.* **87**, 272–300.

Karnath, H-O., Himmelbach, M., and Rorden, C. (2002). The subcortical anatomy of human spatial neglect: putamen, caudate nucleus and pulvinar. *Brain* **125**, 350–360.

Kastner, S., and Ungerleider, L. G. (2000). Mechanisms of visual attention in the human cortex. *Ann. Rev. Neurosci.* **23**, 315–341.

Kastner, S., O'Connor, D. H., Fukui, M. M., Fehd, H. M., Herwig, U., and Pinsk, M. A. (2004). Functional imaging of the human lateral geniculate nucleus and pulvinar. *J. Neurophysiol.* **91**, 438–448.

Kastner, S., Pinsk, M. A., De Weerd, P., Desimone, R., and Ungerleider, L. G. (1999). Increased activity in human visual cortex during directed attention in the absence of visual stimulation. *Neuron* **22**, 751–761.

LaBerge, D., and Buchsbaum, M. S. (1990). Positron emission tomographic measurements of pulvinar activity during an attention task. *J. Neurosci.* **10**, 613–619.

Moran, J., and Desimone, R. (1985). Selective attention gates visual processing in the extrastriate cortex. *Science* **229**, 782–784.

O'Connor, D. H., Fukui, M. M., Pinsk, M. A., and Kastner, S. (2002). Attention modulates responses in the human lateral geniculate nucleus. *Nature Neurosci.* **5**, 1203–1209.

Petersen, S. E., Robinson, D. L., and Keys, W. (1985). Pulvinar nuclei of the behaving rhesus monkey: visual responses and their modulation. *J. Neurophysiol.* **54**, 867–886.

Rees, G., Frith, C. D., and Lavie, N. (1997). Modulating irrelevant motion perception by varying attentional load in an unrelated task. *Science* **278**, 1616–1619.

Sherman, S. M., and Guillery, R. W. (2001). *Exploring the thalamus.* Academic Press, San Diego.

CHAPTER 73

Transient Covert Attention Increases Contrast Sensitivity and Spatial Resolution: Support for Signal Enhancement

Marisa Carrasco

ABSTRACT

Three studies provide experimental evidence indicating that transient attention can improve early visual processes—contrast sensitivity and spatial resolution—via signal enhancement. To assess the effects of transient attention we compared performance with a peripheral cue, which appeared adjacent to the target location, to that with either a central-neutral cue, which appeared either in the center of the display, or a distributed-neutral cue, which was designed to spread observers' attention across the possible target locations. In all three studies we evaluated whether the magnitude of the cueing effect varied as a function of location uncertainty; this was manipulated by varying the stimulus contrast, the performance level, and the presence of a local post-mask, and was assessed by performance in a localization task. Given that we used suprathreshold stimuli, excluded all sources of added external noise—distracters, global masks, local masks—and that we ruled out location uncertainty as the mediator of the attentional benefit, we conclude that transient attention can increase contrast sensitivity and spatial resolution via signal enhancement.

Visual attention allows us to select a certain aspect of a visual scene and grant it priority in processing. Spatial covert attention is enhanced processing of visual information at a cued location without eye movement. Covert attention can be allocated to a given location voluntarily, according to goals (sustained attention) or involuntarily, in a reflexive manner, to a stimulus that appears suddenly in the visual field (transient attention). Cognitive and psychophysical studies have shown that we can differentially engage these systems by using different spatial cues: A cue in the center of the visual field (endogenous cue) directs attention in a conceptually-driven fashion in ~300 ms and engages sustained attention, whereas a peripheral cue presented in a location near the relevant location (exogenous) draws attention in a stimulus-driven automatic manner in ~100 ms and engages transient attention. Whereas the shifts of attention by sustained cues appear to be under conscious control, it seems that it is hard or practically impossible for observers to ignore transient cues.

My lab has been particularly interested in characterizing the effects of transient attention on early visual processes and we have found that attention has the capacity to increase both contrast sensitivity (Cameron, Tai, and Carrasco, 2002; Carrasco, Penpeci-Talgar, and Eckstein, 2000; Carrasco, Talgar, and Cameron, 2001; including apparent or perceived contrast, Carrasco, Ling, and Read, 2004), and spatial resolution (Carrasco, Williams, and Yeshurun, 2002; Talgar and Carrasco, 2000; Yeshurun and Carrasco, 1998, 1999, 2000).

Prominent hypotheses that have been proposed to explain attentional effects include external noise reduction and signal enhancement. The external noise reduction hypothesis maintains that attention diminishes the impact of stimuli that are outside its focus. Noise-limited models incorporate external noise resulting from distracters and masks, as well as internal noise arising from such sources as spatial and temporal uncertainty of targets and distracters. According to these models, performance decreases as the number of distracters and spatial uncertainty increase, because the noise they introduce can be confused with the target signal (Baldassi and Burr, 2000; Dosher and Lu, 2000; Morgan et al., 1998; Pashler, 1998). Several studies have attributed attentional facilitation to reduction of external noise, either because a suprathreshold target could be confused with suprathreshold distracters or because a near-threshold target presented alone could be confused with empty locations.

The signal enhancement hypothesis proposes that attention directly improves the quality of the stimulus representation, by enhancing either contrast or spatial resolution. We have conducted a series of studies to evaluate whether signal enhancement occurs above and beyond external noise reduction. An attentional benefit can be attributed with certainty to signal enhancement only when we eliminate all the variables that, according to the external noise reduction model, might be responsible for the attentional effects. Presenting a suprathreshold target alone, without distracters or local or multiple masks, and eliminating spatial uncertainty, has allowed us to conclude that transient attention can increase contrast sensitivity (Cameron et al., 2002; Carrasco et al., 2000, 2001) and spatial resolution (Carrasco, Williams, and Yeshurun, 2002; Yeshurun and Carrasco, 1999) via signal enhancement.

I. TRANSIENT ATTENTION INCREASES SENSITIVITY ACROSS THE CONTRAST SENSITIVITY FUNCTION

Using a modified staircase procedure (QUEST) we compared the stimulus contrast necessary for observers to perform an orientation discrimination task at a given performance level when the target location was preceded by a peripheral cue appearing adjacent to the target location, and when it is preceded by a neutral cue appearing at fixation, which indicates that the target is equally likely to occur at any of the eight isoeccentric locations. Because the peripheral precue always indicated target location and appeared equally often adjacent to a stimulus of either orientation, it did not increase the likelihood of one of the orientations and observers could not rely on its presence to respond correctly (Carrasco et al., 2000, 2001, 2002). We assessed the effect of transient attention across a wide range of spatial frequencies and found that it increases sensitivity across the contrast sensitivity function (see Fig. 73.1A).

We found that external noise reduction signal detection (SDT) model could account for the cueing benefit in an easy discrimination task of vertical vs. horizontal Gabor patches. However, such a model could not account for this benefit when location uncertainty was reduced, either by increasing overall performance level, increasing stimulus contrast to enable fine discriminations of slightly tilted suprathreshold stimuli, or presenting a local post-mask. An SDT model that incorporates intrinsic uncertainty (the observers' inability to perfectly use information about the elements' spatial or temporal positions, sizes, or spatial frequencies) revealed that the cueing effect could not be explained by the mere reduction of location uncertainty. Given that the attentional benefits occurred under conditions that exclude all variables predicted by the external noise reduction model, the results support the signal enhancement model of attention (Carrasco, Penpeci-Talgar, and Eckstein, 2000).

We also investigated whether transient attention would differentially affect performance fields (shape depicted by percent correct performance at particular isoeccentric locations in the visual field) for orientation discrimination, detection, and localization tasks, while manipulating a number of visual factors. Although attention improved performance—less contrast is needed for the precued target to attain the same performance level as in the neutral condition—it did not affect the overall shape of the performance fields. The two patterns observed—a horizontal-vertical anisotropy (HVA) and a vertical meridian asymmetry (VMA)—became more pronounced as spatial frequency and eccentricity increased. We concluded that discriminability performance fields are determined by visual constraints (see Fig. 73.1B).

II. TRANSIENT ATTENTION INCREASES SENSITIVITY ACROSS THE CONTRAST PSYCHOMETRIC FUNCTION

We examined the effect of transient attention across a range of performance levels, from sub- to suprathreshold, when the target was presented alone at one of eight isoeccentric locations. Consistent with a contrast gain mechanism, we found that transient attention decreased the threshold of the psychometric function for contrast sensitivity in an orientation discrimination task even when the target was presented alone (see Fig. 73.2A). Given that we used suprathreshold stimuli, excluded all sources of added external

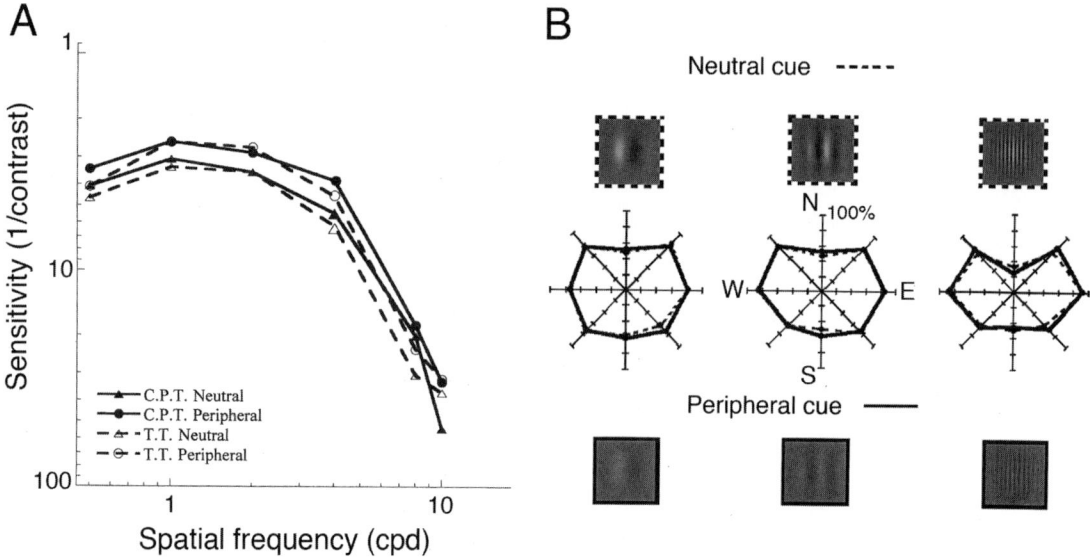

FIGURE 73.1 **A.** Data for two individual observers illustrating that for a target of constant contrast, precueing the target location enhances sensitivity across the contrast sensitivity function (Carrasco et al., 2000, Figure 3); **B.** the stimulus contrast necessary to attain the same performance level for a range of spatial frequencies is lower when the target location is precued by a peripheral cue (bottom squares) than by a neutral cue (top squares). The contrast differences depicted in the Gabor patches are based on data reported by Carrasco et al., 2000). In addition, this figure illustrates that contrast sensitivity is higher for the locations at the horizontal than vertical meridian, and lowest at the north location, and that the effect of the cue on performance is similar at these isoeccentric locations. Note that although the peripheral cue lowered the contrast threshold, the shape of the performance field is the same for both cueing conditions (Carrasco, Talgar, and Cameron, 2001).

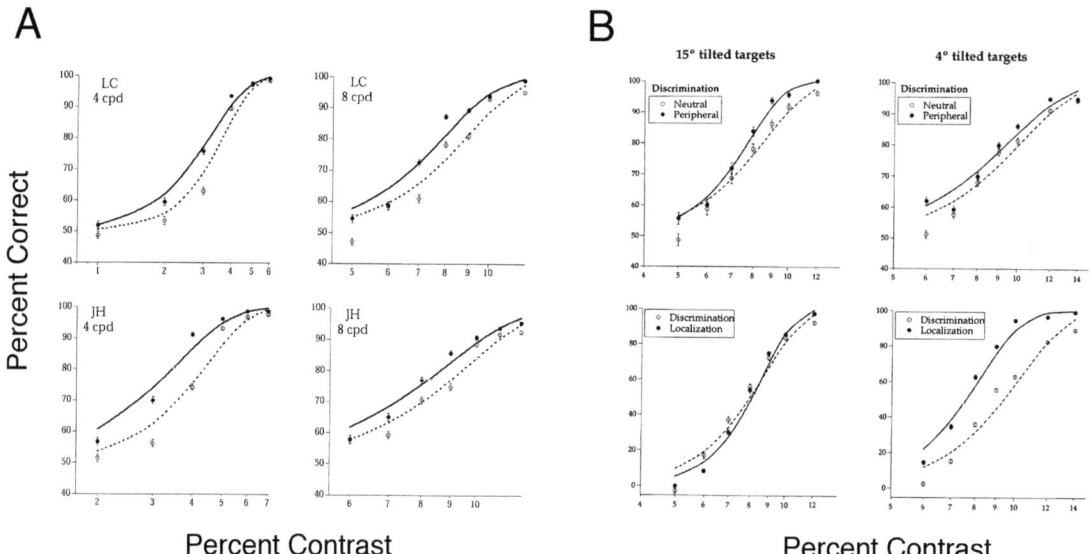

FIGURE 73.2 **A.** Psychometric functions (percent correct as a function of target contrast) for two of the spatial frequencies used (4 and 8 cpd), for two observers. Neutral precued condition is represented with open symbols and dotted lines; peripheral precue condition is represented with filled symbols and solid lines. Attention shifts the psychometric function to the left, and in some instances, makes the slope shallower (Cameron, Tai, and Carrasco, 2002; Figure 3). **B.** The peripheral cue increases contrast sensitivity throughout the psychometric function of contrast sensitivity to the same extent in an orientation discrimination task (top panels) for stimuli that differ in spatial uncertainty, 4° vs. 15° tilted (bottom panels), as assessed by localization performance (Cameron, Tai, and Carrasco, 2002; Figure 10).

noise—distracters, local and global masks—and showed experimentally that spatial uncertainty cannot explain this decrease in threshold, the observed attentional benefit is consistent with a signal enhancement mechanism. Once again discriminability showed the HVA and the VMA and the cue improved discriminability to a similar extent at all isoeccentric locations (Cameron, Tai, and Carrasco, 2002).

To investigate the role of spatial uncertainty in the precue effect, we conducted two control experiments. First, we made the discrimination task harder by decreasing the tilt of the targets from 15° to 4°. Observers required higher stimulus contrasts to perform this discrimination task, and this in turn diminished spatial uncertainty. Even though the target contrast was higher, an attentional effect of similar magnitude was observed. In addition, to assess directly the ease with which observers can localize the stimulus, we also performed a localization task. When the target was tilted 15°, discrimination and localization performance were tightly coupled. However, when the targets were tilted 4°, performance on the localization task was much better than performance on the discrimination task. Notwithstanding the superior localization performance on the 4° discrimination task, the attentional effect was comparable for both orientation conditions. Importantly, at contrasts that yielded perfect localization, there was still an attentional effect in the discrimination tasks. These data indicate that spatial uncertainty cannot fully explain the precue effects observed in this study.

III. TRANSIENT ATTENTION INCREASES APPARENT CONTRAST

Does attention alter appearance? Whether attention can actually affect the perceived intensity of a stimulus has been a matter of debate dating back to the founding fathers of experimental psychology and psychophysics. Whereas Helmholtz and James believed that attention intensifies a sensory impression, Fechner argued that attention is unable to alter sensory impressions. Although recent studies characterizing the effects of attention on early visual processes suggest effects on appearance, for instance the finding that transient attention increases contrast sensitivity (Cameron et al., 2002; Carrasco et al., 2000), which is consistent with a contrast gain mechanism, very little direct empirical evidence has addressed this issue (Carrasco, Ling, and Read, 2004).

To directly address this issue we developed a novel paradigm that allowed us to objectively assess observers' subjective experience while circumventing methodological limitations of previous studies. Observers were briefly presented with either a peripheral or neutral cue, followed by two Gabor patches (tilted 45° to the left or right) at 4° eccentricity to the left and right of fixation. The contrast of one of the Gabors was presented at a fixed contrast (Standard), whereas the other varied in contrast randomly from a range of values around the Standard (Test patch). We asked the observers: "What is the orientation of the grating that looks higher in contrast?" The orientation discrimination task served as a 'cover story' task, which de-emphasized the fact that we were interested in their subjective experience. We showed that when a Gabor was peripherally cued, the point of subjective equality (PSE) was shifted—the apparent contrast of the stimulus for which transient attention had been drawn to was higher than when attention was not drawn there. That is to say, when observers attend to a stimulus, they perceive it to be of much higher contrast than when they perceive the same stimulus without attention.

To rule out any alternative explanations for the effect, such as observers being biased toward the cue, control experiments were conducted. For instance, given the ephemeral nature of transient attention (it peaks at ~100ms and drops off rapidly afterward), we extended the SOA to 250ms. When transient attention was extinguished, there was no difference between the peripheral and neutral conditions.

This study provides evidence for a contrast gain model (Reynolds, Pasternak, and Desimone, 2000) in which attention allows for greater neuronal sensitivity, suggesting that attention changes the strength of a stimulus by enhancing its "effective contrast" or salience. It is as if attention boosts the actual stimulus contrast (Carrasco, Ling, and Read, 2004).

IV. TRANSIENT ATTENTION IMPROVES ACUITY

We investigated whether covert attention can enhance spatial resolution via signal enhancement in a visual acuity task. We used a suprathreshold target (Landolt-square), which appeared at one of four possible eccentricities along the vertical or horizontal meridian and asked observers to indicate which side of the Landolt-square had a gap (Carrasco, Williams, and Yeshurun, 2000). A transient cue improved observers' accuracy and speed even though we eliminated all sources of added external noise from the display; that is, distracters, global masks, and local masks (see Fig. 73.3).

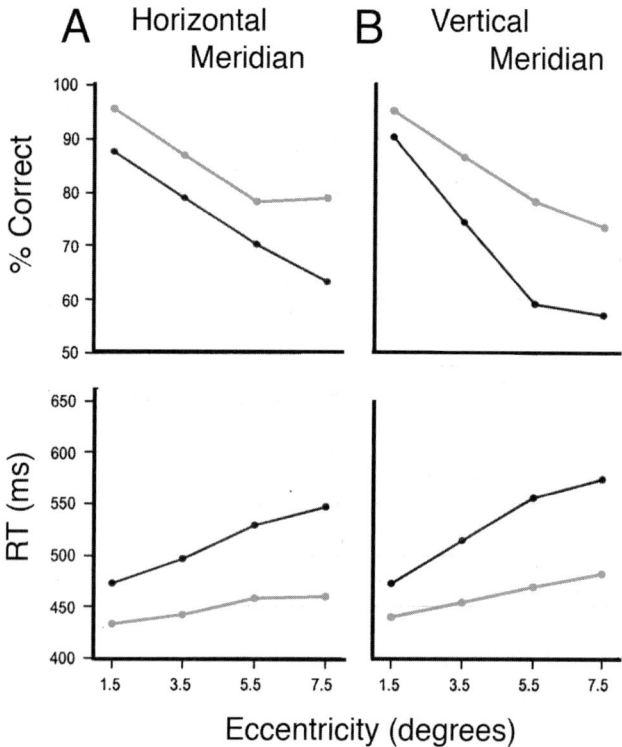

FIGURE 73.3 Percentage of trials correctly discriminated (top row) and the associated mean response times (bottom row) for spread-neutral and peripheral trials, as a function of eccentricity, when the target appeared along the **A** horizontal and **B** vertical meridian. (Carrasco, Williams, and Yeshurun, 2002, Figure 9).

To assess the effect of transient attention in any given task, it is necessary to compare performance when the target follows a peripheral cue and a neutral cue. Several authors have used a central-neutral cue that indicates the target onset but conveys no information regarding the target location (Cameron et al., 2002; Carrasco et al., 2000, 2002; Dosher and Lu, 2000; Yeshurun and Carrasco, 1998, 1999). Some authors have suggested that this central-neutral cue may reduce the extent of the attentional spread (Pashler, 1998). To rule this out, we used a neutral cue designed to spread observers' attention across the display, which consisted of four copies of the peripheral cue, simultaneously presented at the centers of each of the four quadrants. The performance difference between the peripheral cue and the spread-neutral cue was at least as pronounced as the difference between the peripheral cue and the central-neutral cue.

These results indicate that attention enhanced the signal for the following reasons; (a) all sources of added external noise were eliminated; (b) the target was suprathreshold, to diminish the location uncertainty associated with stimuli at contrast threshold; (c) the extent of the attentional benefit was the same regardless of the presence of a local post-mask, which reduces the target location uncertainty (Smith, 2000); (d) even though the peripheral cue would reduce the effective set size from the number of possible target locations 16 to 1 by excluding the empty locations, it could not affect performance substantially, because according to SDT models (Verghese, 2001) the high-contrast target presented alone would result in a high target-distracter discriminability, and the number of empty locations would have a negligible effect on performance.

Given that performance in the Landolt-square task indexes resolution, this benefit indicates that the enhanced stimulus representation can bring about finer spatial resolution (Lee et al., 1999; Morgan et al., 1998; Yeshurun and Carrasco, 1998, 2000). Consistent with previous studies, the decrement in performance with eccentricity was more pronounced along the vertical than the horizontal meridian, the magnitude of the cueing effect increased with eccentricity (Yeshurun and Carrasco, 1999), but the magnitude of this effect was similar at different isoeccentric locations (Cameron et al., 2002; Carrasco et al., 2001). The idea that attention enhances resolution has inspired a recent neural network model that implements the role that visual attention plays in object recognition (Deco and Zihl, 2001) and has also been captured by a computational model proposing that the interactions among visual spatial filters result in both increased gain and sharpened tuning (Lee, Itti, Koch, and Braun, 1999).

V. CONCLUSION

By eliminating all sources of added noise, the psychophysical studies reported here have shown that transient attention can increase contrast sensitivity (Cameron et al., 2002; Carrasco et al., 2000, 2001) and spatial resolution (Carrasco et al., 2002) via signal enhancement. Performance varied across isoeccentric locations, with superior performance at the horizontal than at the vertical meridian, and at the lower than at the upper region of the vertical meridian. (For a discussion of possible neurophysiological correlates to these visual asymmetries see Cameron et al., 2002; Carrasco et al., 2001, 2002). However, the extent of the attentional effect on discriminability was similar at isoeccentric locations.

We have shown that the attentional effect goes beyond the reduction of location uncertainty. For instance, although location uncertainty produces a greater degradation at low than at high performance levels, the magnitude of the attentional benefit is similar regardless of the likelihood of observers confusing the target with blank locations. Attention increases sensitivity throughout the psychometric function of contrast sensitivity to the same extent for stimuli that differ in spatial uncertainty (Cameron et al., 2002) and even when localization performance indicates that there is no uncertainty with regard to the target location (Carrasco et al., 2000). In addition, the presence of a local post-mask, which reduces location uncertainty, does not affect the magnitude of the attentional benefit (Carrasco et al., 2000, 2002). Likewise, with brief displays (100 ms), other authors have found that cueing the target location improves performance more than predicted by a signal-detection model of spatial uncertainty (Morgan et al., 1998). Finally, a spatial uncertainty model does not account for the effect of the near absence of attention on visual thresholds (Lee et al., 1999). Together, these studies indicate that even though spatial uncertainty reduction can play a role in performance, it cannot be the sole source of the attentional effect reported here.

It is reasonable to assume that attentional modulation may reflect a combination of mechanisms, such as reduction of external noise, reduction of spatial uncertainty, and signal enhancement (Carrasco et al., 2000, 2002; Dosher and Lu, 2000). These hypotheses find support in a growing body of physiological studies that have shown the instantiation of attention at the level of sensory representation. Single-cell recordings have demonstrated that directing attention toward the stimulus can alter responses of V1 neurons and result in stronger and more selective responses in both V4 and MT/MST neurons (Reynolds & Chelazzi, 2004), and fMRI studies have shown attentional modulation in striate and extrastriate visual cortex (Ress, Backus, and Heeger, 2000). As remarkable as the human visual and cognitive systems may be, inevitably we are still limited by both bandwidth and processing power. Visual attention is crucial in optimizing the systems' limited resources.

References

Baldassi, S., and Burr, D. C. (2000). Feature-based integration of orientation signals in visual search. *Vision Research* **40**, 1293–1300.

Cameron, E. L., Tai, J. C., and Carrasco, M. (2002). Covert attention affects the psychometric function of contrast sensitivity. *Vision Research* **42**, 949–967.

Carrasco, M., Ling, S., and Read, S. (2004). Attention alters appearance. *Nature Neuroscience* **7**, 308–313.

Carrasco, M., Penpeci-Talgar, C., and Eckstein, M. (2000). Spatial attention increases contrast sensitivity across the CSF: Support for signal enhancement. *Vision Research* **40**, 1203–1215.

Carrasco, M., Penpeci-Talgar, C., and Cameron, E. L. (2001). Characterizing visual performance fields: Effects of transient covert attention, spatial frequency, eccentricity, task and set size. *Spatial Vision* **15**, 61–75.

Carrasco, M., Williams, P. E., and Yeshurun, Y. (2002). Covert attention increases spatial resolution with or without masks: Support for signal enhancement. *Journal of Vision* **2(6)**, 467–479.

Deco, G., and Zihl, J. (2001). A neurodynamical model of visual attention: Feedback enhancement of spatial resolution in a hierarchical system. *Journal of Computational Neuroscience* **10**, 231–253.

Dosher, B. A., and Lu, Z.-L. (2000). Mechanisms of perceptual attention in precuing of location. *Vision Research* **40**, 1269–1292.

Lee, D. K., Itti, L., Koch, C., and Braun, J. (1999). Attention activates winner-take-all competition among visual filters. *Nature Neuroscience* **2**, 375–381.

Lu, Z., and Dosher, B. A. (2000). Spatial attention: Different mechanisms for central and peripheral temporal precues? *Journal of Experimental Psychology: Human Perception and Performance* **26(5)**, 1534–1548.

Morgan, M. J., Ward, R. M., and Castet, E. (1998). Visual search for a tilted target: Tests of spatial uncertainty models. *The Quarterly Journal of Experimental Psychology* **51A**, 347–370.

Reynolds, J. H., and Chelazzi, L. (2004). Attentional modulation of visual processing. *Annual Review of Neuroscience* **27**, 611–647.

Ress, D., Backus, B. T., and Heeger, D. J. (2000). Activity in primary visual cortex predicts performance in a visual detection task. *Nature Neuroscience* **3**, 940–945.

Verghese, P. (2001). Visual search and attention: A signal detection theory approach. *Neuron* **31**, 523–535.

Yeshurun, Y., and Carrasco, M. (1998). Attention improves or impairs visual performance by enhancing spatial resolution. *Nature* **396**, 72–75.

Yeshurun, Y., and Carrasco, M. (2000). The locus of attentional effects in texture segmentation. *Nature Neuroscience* **3(6)**, 622–627.

CHAPTER 74

External Noise Distinguishes Mechanisms of Attention

Zhong-Lin Lu and Barbara Anne Dosher

ABSTRACT

Attention to objects or features of objects affects performance on perceptual tasks such as detection, recognition, or identification. The speed or accuracy of task performance can be significantly influenced by changes in the state of attention, especially in complex task environments. The state of attention may be manipulated by specific environmental cues, or by decisions about allocation of attention induced by task demands. The challenge is to understand and predict these attention-mediated changes in task performance, and to identify the mechanisms by which attention is operating. A theoretical and empirical framework has been developed to directly assess the mechanisms of attention by systematically manipulating the amount and/or characteristics of the external noise added to the signal stimuli and measuring modulations of perceptual discriminability (signal and noise levels) in the cognitive processes. Three classes of attention mechanisms—stimulus enhancement, external noise exclusion, and internal noise reduction, each with its signature performance pattern, can be distinguished. Empirically, the two mechanisms of stimulus enhancement and external noise exclusion are shown to occur in different circumstances. A task-taxonomy of attention mechanisms is obtained by partitioning experiments in the literature according to several factors. The framework has also been extended to the assessment of spatial and temporal characteristics of the attention window, the coordination of multimodal auditory-visual cues in the performance of visual tasks, and the action of multitask load in performance.

I. INTRODUCTION

Attention to objects or features of objects affects performance on perceptual tasks such as detection, recognition, or identification. The speed or accuracy of task performance can be significantly influenced by changes in the state of attention, especially in complex task environments. The question is: How does attention improve human performance? A number of functional metaphors of attention have been proposed (see LaBerge, 1995 for a review): an early sensory filter, orienting in space, a moving spotlight, a spatio-temporal gate to memory and the glue conjoining multiple features, distribution of mental resources, a control process of short-term memory. These metaphoric models of attention make strong suggestions about how attention operates, and in certain cases even admit quantitative applications. An alternative approach is to develop a formal perceptual decision structure and test models of attention effects by studying modulations of perceptual discriminability (signal and noise levels) in the cognitive processes (Dosher and Lu, 2000b; Lu and Dosher, 1998).

In the domain of signal processing, there are essentially three ways to improve the signal-to-noise ratio: amplification, improved filtering, and modified gain control. In physiology, attention has been shown to increase cellular response sensitivity (Reynolds, Pasternak, and Desimone, 2000), sharpen signal (e.g., orientation/spatial frequency) selectivity (Haenny, Maunsell, and Schiller, 1988), and to exclude unwanted information through competitive interaction, resulting in apparent shrinking of neuronal receptive fields (Desimone and Duncan, 1995). At the behavioral level, spatial attention has been postulated to reduce decision uncertainty (Palmer, Ames, and Lindsey, 1993), enhance the attended stimulus (Posner, Nissen, and Ogden, 1978) and/or change contrast sensitivity (Carrasco, Penpeci-Talgar, and Eckstein, 2000), exclude external noise or distractors (Shiu and Pashler, 1994), and reduce contrast-gain control (Lee, Itti, Koch, and Braun, 1999). Motivated by neurophysiology and signal processing considerations, behavioral research based on external noise manipulations and the Perceptual Template Model (Lu and Dosher, 1998) distinguishes three mechanisms underlying performance improvements in perceptual tasks: stimulus enhancement, external noise exclusion, and multiplicative noise reduction. This review focuses on mechanisms of attention using paradigms that eliminate structural decision uncertainty; that is, observers are explicitly informed of the target location in all experimental conditions. Quantitative studies of the role of decision uncertainty, where the location of the target is not known, in paradigms involving tasks such as the classical Posner paradigm and visual search, can be found in Palmer et al. (1993).

II. THE PERCEPTUAL TEMPLATE MODEL (PTM) APPROACH

Observer models have been powerful tools in several areas of vision science. In the PTM approach, effects of attention are measured as a joint function of observer attention state and the amount and/or characteristics of the external noise added to the signal stimuli. Mechanisms of attention are identified as changes of observer characteristics through the perceptual template model (Dosher and Lu, 2000b; Lu and Dosher, 1998).

A. Perceptual Template Model of the Observer

Limited by various sources of noise such as intrinsic stimulus variability, receptor sampling errors, randomness of neural responses, and loss of information during neural transmission, perceptual processes exhibit various inefficiencies. Such inefficiencies at an overall system level can be modeled by characterizing perceptual processes as "perfect," noise-free computations with separate, equivalent internal noise—the amount of random internal noise necessary to produce the degree of inefficiency exhibited by the perceptual system. The PTM (Lu and Dosher, 1999) is an extension of many similar models of the human observer in the literature (Pelli, 1980).

The perceptual template model (see Fig. 74.1A) consists of five components: (1) A perceptual template with certain tuning characteristics (e.g., a spatial frequency filter $F(f)$ with a center frequency and a bandwidth such that a range of frequencies adjacent to the center frequency pass through with smaller gains). The template is normalized such that it passes the noise with gain 1.0 and the signal stimulus with gain β. (2) A nonlinear transducer function with the form $output = \| input \|^{\gamma}$. (3) A multiplicative internal noise that is Gaussian distributed with mean 0 and a standard deviation that is proportional (with a coefficient of N_m) to the total energy in the input stimulus. Multiplicative noise is a natural way of characterizing tasks in which, for example, perceived sensory variability, or perceived differences, are proportional to signal strength (Weber-law situations). (4) An independent additive internal noise that is Gaussian distributed with mean 0 and a fixed standard deviation N_a. The existence of an absolute sensory threshold for every perceptual process suggests that the perceptual system is limited by an additive noise whose amplitude doesn't depend on the amount of input. (5) A decision process that operates on the noisy internal representation of the stimulus. Depending on the task, the decision could reflect either detection or discrimi-

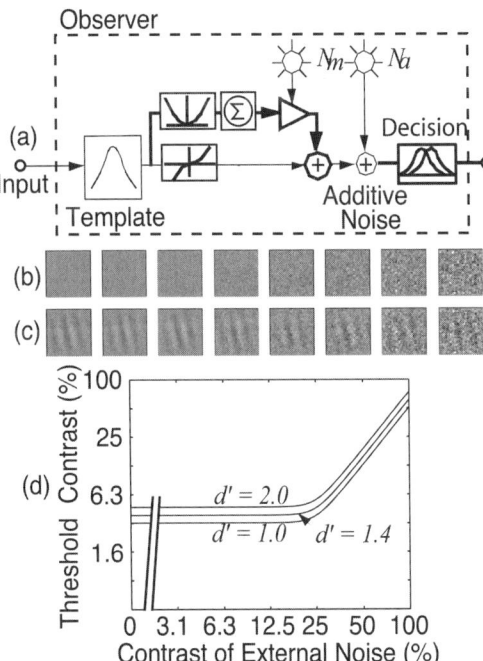

FIGURE 74.1 a. A noisy perceptual template model. b. Samples of eight levels of external noise. c. A Gabor signal embedded in the external noises shown in b. d. Simulated threshold versus external noise contrast (TVC) functions for a perceptual template model at three criterion performance levels ($d' = 1.0, 1.4, 2.0$).

nation, and could take the form of either N-alternative forced choice or yes/no, possibly with confidence ratings.

The observer can be characterized by systematically manipulating the amount of external noise added to the signal stimulus (see Fig. 74.1B,C) and observing how threshold—signal stimulus energy required for an observer to maintain a given performance level—depends on the amount of external noise (the threshold versus contrast, or TVC, function; Fig. 74.1D). In a typical application, the model parameters, N_a, N_m, β, and γ, are unknown quantities that can be estimated from TVC data such as that in Fig. 74.1D by nonlinear estimation techniques, or, alternatively, simple equations can be derived that allow us to compute estimates of several of the parameters from certain relations in the data. Two or three measured threshold levels are required (Lu and Dosher, 1999). Although this characterization does not distinguish between various sources for the inefficiency, it does allow us to quantify the overall efficiency of the perceptual system, and to compare the efficiency of the perceptual system in different perceptual tasks.

B. The External Noise Plus Attention Paradigm and Signatures of Attention Mechanisms

The theoretical performance signatures of attention mechanisms can be derived by studying the possible ways attention can affect various components of the PTM and generate model TVC functions for proposed attention mechanisms: stimulus enhancement, external noise exclusion, and internal multiplicative noise reduction (see Fig. 74.2).

1. Stimulus Enhancement

Enhanced performance due to attention in clear (noise-free) conditions corresponds to claims of perceptual enhancement (Posner et al., 1978). In the

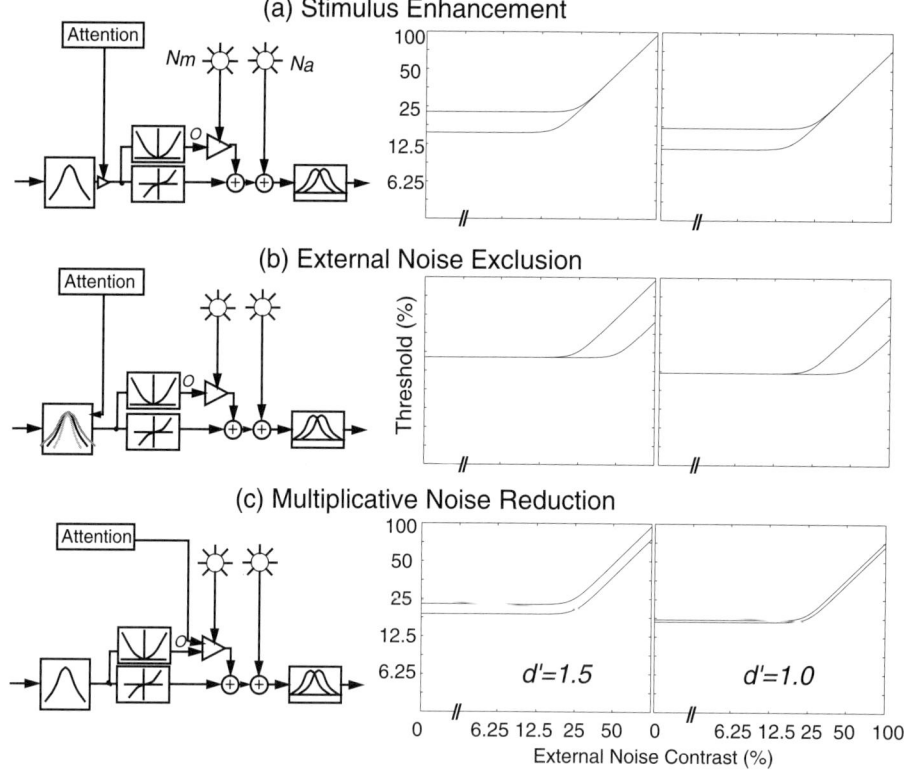

FIGURE 74.2 Signature performance patterns for three mechanisms of attention within the framework of a perceptual template model (PTM) at two performance criterion levels ($d' = 1.5$ and 1.0). **a.** Stimulus enhancement. It improves performance only in zero or low external noise. **b.** External noise exclusion. It modulates performance only at high levels of external noise. **c.** Internal multiplicative noise reduction. It affects performance at all levels of external noise, but increasingly so as external noise increases. In both **a** and **b**, the magnitude of attention effects does not depend on the performance criterion. In **c**, however, the magnitude of the attention effects depends critically on the performance criterion.

context of the PTM, stimulus enhancement is mathematically equivalent to internal additive noise reduction. The behavioral signature for this mechanism is performance improvement (reduced thresholds or lower curves) in the region of low or zero external noise (see Fig. 74.2A). This is because stimulus enhancement affects both the signal and the external noise in the input stimulus in the same way.

2. External Noise Exclusion

One key way in which attention improves performance is by focusing perceptual analysis on the appropriate time, spatial region, and/or content characteristics of the signal stimulus. This focusing serves to eliminate external noise from further processing, and is related to claims by Shiu and Pashler (1994) and others. The behavioral signature for this mechanism is performance improvements in the region of high external noise (see Fig. 74.2B), where there is external noise to exclude.

3. Internal Noise Reduction

Another possible mechanism of attention would involve the reduction of internal noise. The reduction of additive noise is formally equivalent to the enhancement of stimulus (see earlier). Multiplicative noise increases with increasing contrast in the stimulus display. Reduction of multiplicative internal noise produces a signature of improvements in both high and low levels of external noise, with slightly larger effects in high external noise (see Fig. 74.2C). To date, we have not empirically observed a case of multiplicative noise reduction by attention.

4. Distinguish Mechanism Mixtures

A direct comparison of the experimental data and the signature patterns of the PTM model may be sufficient to identify the underlying mechanism of attention in certain situations. In other situations, mixtures of more than one mechanism may underlie improvements in performance associated with attention in a task. In particular, a mixture of stimulus enhancement (low noise effects) and external noise exclusion (high noise effects) must be discriminated from multiplicative internal noise suppression (effects in low and high noise). This can be accomplished via measuring threshold versus external noise contrast at multiple criterion performance levels. A higher level of threshold performance, for example a d' of 1.5 instead of 1.0, requires higher contrast signals to achieve. For conditions differing in stimulus enhancement (see Fig. 74.2A), or in external noise exclusion (see Fig. 74.2B), the magnitude of the attention effect is the same (on the log contrast axis) at both the higher, more stringent threshold and the lower, less stringent threshold performance level. However, threshold contrast differences between two conditions at lower and higher criterion threshold values depend strongly upon criterion performance level in the conditions that differ in internal multiplicative noise reduction (see Fig. 74.2C). Thus, measuring TVC functions at two or more criterion performance levels resolves the individual contribution of each mechanism in a mixture situation (Dosher and Lu, 1999; Lu and Dosher, 1999).

III. EMPIRICAL RESULTS AND TAXONOMY OF MECHANISMS OF SPATIAL ATTENTION

Pure cases of stimulus enhancement (Lu and Dosher, 1998; Lu, Liu, and Dosher, 2000) and external noise exclusion (Dosher and Lu, 2000a; Dosher and Lu, 2000b; Lu and Dosher, 2000; Lu, Lesmes, and Dosher, 2002) have been documented in different situations as the mechanism of location-cued visual attention. Pure external noise exclusion (see Fig. 74.2B) occurs when attention improves performance only in high noise, yet has little or no effect in conditions with low or no external noise (Dosher and Lu, 2000a; Dosher and Lu, 2000b; Lu and Dosher, 2000; Lu et al., 2002). Stimulus enhancement (see Fig. 74.2A) is demonstrated by improvements due to attention in the absence of or presence of low external noise, but cannot improve performance when it is limited by external noise (Lu and Dosher, 1998; Lu et al., 2000).

Dosher and Lu (2000b) demonstrated that valid precueing one of four (or eight) widely separated spatial locations dramatically improved Gabor orientation identification relative to invalid precues in high external noise conditions. Four Gabor stimuli appeared on an annulus, one in each quadrant (see Fig. 74.3A). A 62.5% valid central arrow precue (150ms prior to the signal) indicated the likely report location. A simultaneous report cue indicated the to-be-reported target location. Full psychometric functions were measured for each external noise and attention condition. Threshold versus external noise contrast (TVC) functions at 62.5% correct is shown for the average observer in Fig. 74.3B. The primary mechanism of attention in this central cueing paradigm was external noise exclusion, implying a retuning of the perceptual template.

Two distinct attention systems have been postulated (Posner, 1980). Central cues, which point at the target location, are said to activate the *endogenous* attention system. Peripheral cues, which occur near the target location, are said to activate the *exogenous* attention

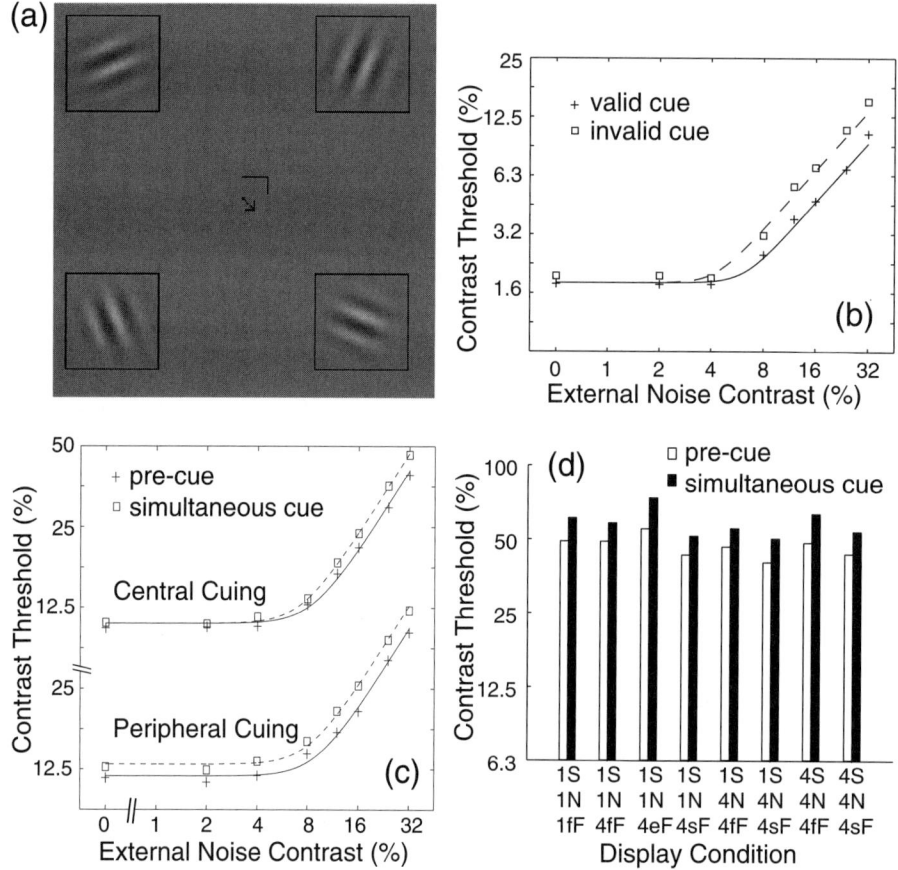

FIGURE 74.3 a. Sample layout of an invalidly cued trial. The precue (arrow) cues attention to one of four stimulus locations, and the caret cues the location to be reported (Dosher and Lu, 2000). b. Thresholds at 62.5% correct as a function of external noise level for the average observer in the valid and invalid conditions. Cue validity has a large effect at high levels of external noise (Dosher and Lu, 2000). c. Average TVC functions at $d' = 1.24$ across all the observers in central and peripheral cuing (Lu and Dosher, 2000). In both b and c, smooth curves are best fitting PTM predictions. d. Average contrast thresholds at 62.5% correct for eight display conditions in high external noise (Lu, Lesmes, and Dosher, 2002). Denoted by $jSkNltF$, in each display condition, signal stimuli, external noise, and frames in $t = stationary$, $flash$, or $elaborated$ style occurred in $j = 1$ or 4, $k = 1$ or 4 and $k \geq j$, and $l = 1$ or 4 and $l \geq k$ locations.

system. The two attention systems are thought to differ in both qualitative and quantitative ways. In a new study, Lu and Dosher (2000) found that thresholds in high external noise but not zero or low external noise are improved in central precuing, whereas thresholds in both high and low external noise are improved in peripheral precuing. The results suggest that the endogenous and exogenous attention systems invoke different mechanisms of attention: external noise exclusion for the endogenous system; external noise exclusion plus stimulus enhancement for the exogenous system (see Fig. 74.3C).

Task difficulty effects associated with increasing the number of orientation templates from two to four were essentially entirely accounted for by the statistical properties of the decision task (Dosher and Lu, 2000a).

The magnitude of external noise exclusion increases with display size—external noise exclusion played little role in two-location displays; but it drastically reduced contrast thresholds in eight-location displays. Lu, Lesmes, and Dosher (2002) compared effects of precuing in a wide variety of conditions. In the absence of external noise, precuing produced only marginal performance improvements in a small random subset of display conditions; in the presence of high external noise, precuing improved task performance by the same amount in all the display conditions; and the magnitude of spatial attention effects, as gauged by contrast threshold reduction, is nearly constant across all the display conditions (see Fig. 74.3D). Furthermore, there was little or no evidence for "cross-talk" between the nontarget regions and the response in either the precued or the simultaneously cued conditions. The results suggest that spatial attention excludes external noise in the target region.

To summarize, external noise exclusion has been shown to be a major mechanism of spatial attention in complex multiple-location displays with either central or peripheral location cues, whereas stimulus enhancement was shown to be associated primarily with peripheral cuing of location. The magnitude of attention effects depend on the energy of external noise (masks) added to the stimuli, the number of potential target locations, and the type of (peripheral versus central) cuing. These factors, together with decision uncertainty analysis, provide an effective framework to recast and reorganize the existing literature on spatial attention (Dosher and Lu, 2000a).

IV. OTHER APPLICATIONS AND EXTENSIONS

The PTM approach has recently been applied to identify the mechanisms of performance improvements in perceptual learning, object-based attention, and cross-modal cuing of spatial attention. The PTM model and the external noise paradigm have also been further extended to characterize the tuning characteristics of the perceptual template in spatial frequency, space, and time under attended and unattended states. These and other studies will further clarify the functional nature of external noise exclusion, and provide important links to results from neurophysiology.

Acknowledgment

This research was supported by US Air Force Office of Scientific Research, Visual Information Processing Program.

References

Carrasco, M., Penpeci-Talgar, C., and Eckstein, M. (2000). Spatial covert attention increases contrast sensitivity across the CSF: Support for signal enhancement. *Vision Research* **40**, 1203–1215.
Desimone, R., and Duncan, J. (1995). Neural mechanisms of selective visual attention. *Annual Review of Neuroscience* **18**, 193–222.
Dosher, B. A., and Lu, Z.-L. (1999). Mechanisms of perceptual learning. *Vision Research* **39**, 3197–3221.
Dosher, B. A., and Lu, Z.-L. (2000a). Mechanisms of perceptual attention in precuing of location. *Vision Research* **40**, 1269–1292.
Dosher, B. A., and Lu, Z.-L. (2000b). Noise exclusion in spatial attention. *Psychological Science* **11**, 139–146.
Haenny, P. E., Maunsell, J. H. R., and Schiller, P. H. (1988). State dependent activity in monkey visual cortex. *Experimental Brain Research* **69**, 245–259.
LaBerge, D. (1995). Attentional processing: The brain's art of mindfulness. *Perspectives in cognitive neuroscience* (pp. x, 262). Harvard University Press, Cambridge, MA, US.
Lee, D. K., Itti, L., Koch, C., and Braun, J. (1999). Attention activates winner-take-all competition among visual filters. *Nature Neuroscience* **2**, 375–381.
Lu, Z.-L., and Dosher, B. A. (1998). External noise distinguishes attention mechanisms. *Vision Research* **38**, 1183–1198.
Lu, Z.-L., and Dosher, B. A. (1999). Characterizing human perceptual inefficiencies with equivalent internal noise. *Journal of the Optical Society of America A Special Issue: Noise in imaging systems and human vision* **16**, 764–778.
Lu, Z.-L., and Dosher, B. A. (2000). Spatial attention: Different mechanisms for central and peripheral temporal precues? *Journal of Experimental Psychology: Human Perception and Performance* **26**, 1534–1548.
Lu, Z.-L., Lesmes, L. A., and Dosher, B. A. (2002). Spatial attention excludes external noise at the target location. *Journal of Vision* **2**, 312–323.
Lu, Z.-L., Liu, C. Q., and Dosher, B. A. (2000). Attention mechanisms for multi-location first- and second-order motion perception. *Vision Research* **40**, 173–186.
Palmer, J., Ames, C. T., and Lindsey, D. T. (1993). Measuring the effect of attention on simple visual search. *Journal of Experimental Psychology: Human Perception & Performance* **19**, 108–130.
Pelli, D. G. (1980). Effects of visual noise. (Ph. D. dissertation) University of Cambridge, Cambridge, England.
Posner, M. I., Nissen, M. J., and Ogden, W. C. (1978). Attended and unattended processing modes: The role for spatial location. In "Modes of perceiving and processing information" (N. H. L. Pick and I. J. Saltzman Eds.), (pp. 137–157). Erlbaum, Hillsdale, NJ.
Reynolds, J. H., Pasternak, T., and Desimone, R. (2000). Attention increases sensivity of V4 neurons. *Neuron* **26**, 703–714.
Shiu, L., and Pashler, H. (1994). Negligible effect of spatial precuing on identification of single digits. *Journal of Experimental Psychology: Human Perception and Performance* **20**, 1037–1054.

CHAPTER

75

Attentional Modulation and Changes in Effective Connectivity

Christian Büchel

ABSTRACT

The nonlinear nature of integration among cortical brain areas renders the effective connectivity between them inherently dynamic and context-sensitive. A compelling example is attentional modulation of responses in functionally specialized sensory areas. The aim of this chapter is to demonstrate that parietal regions may mediate selective attention to motion by modulating the effective connectivity from early visual cortex to the motion-sensitive area V5/hMT. Using functional magnetic resonance imaging, and an analysis of effective connectivity based on regression models and nonlinear system identification, we show that backward modulatory influences from the posterior parietal cortex are sufficient to account for a significant component of attentional modulation of V5/hMT responses to inputs from V1/V2. By explicitly modeling interactions among inputs to V5/hMT, we are able to make inferences about the influences of V1/V2 inputs and their concomitant activity-dependent modulation by parietal afferents. The latter effects embody dynamic changes in effective connectivity that may underlie attentional mechanisms and provide empirical evidence that attentional effects may be mediated by backward connections, of a modulatory sort, in humans.

I. INTRODUCTION

In systems neuroscience, the effect of attention usually is assessed on the basis of activity changes in one or more brain regions, while experimentally controlling the attentional set. Activity can either be spiking activity in single cell recordings (Treue and Maunsell, 1996) or an indirect hemodynamic measure of neuronal activity like the blood oxygenation level dependent contrast (BOLD) in functional magnetic resonance imaging (fMRI) (O'Craven and Savoy, 1995). Despite the great success of this approach, it usually neglects the possibility that attention modulates the interaction of (i.e., the effective connectivity between) areas rather than the activity in a single region per se. Thus attention can be seen as a modulation or change in effective connectivity as expressed in a quotation by LaBerge (La Berge, 1995): "The expression of attention in a brain area appears to be described effectively as an enhancement of activity in the attended set of pathways relative to the unattended set of pathways." Enhancing a pathway can either be accomplished by adding a certain bias to this pathway (tonic modulation) or by multiplying the intrinsic activity by a certain gain factor (phasic modulation). The former mechanism is additive, that is, linear, whereas the latter mechanism is nonlinear and resembles an interaction of intrinsic activity with a control mechanism (i.e., attention) and is thus a flexible mechanism for attentional modulation.

In this chapter we exemplify these ideas by using the framework of attention to visual motion. In particular, we examine the modulatory effect of higher areas (posterior parietal cortex; PPC) on the effective connectivity among lower visual areas in humans.

A. V5/hMT Function

Area V5/hMT (humans) or MT (nonhuman primates) is specialized for motion processing (Tootell et al., 1995; Zeki et al., 1991). It receives parallel inputs from the lateral geniculate via early visual cortex (V1/V2), and extrageniculate pathways involving the pulvinar (ffytche et al., 1995). Neuroimaging (Corbetta et al., 1991; O'Craven and Savoy, 1995) and unit-electrode recordings (Treue and Maunsell, 1996) show that V5/hMT responses can be modulated by attention. A likely source of this modulation is the posterior parietal cortex (PPC) (Assad and Maunsell, 1995; Mountcastle et al., 1981). The PPC is part of a distributed system, subserving visual attention, that includes the frontal eye fields, cingulate cortex, prefrontal cortex, thalamic (pulvinar), and other regions. These areas have been implicated on the basis of electrophysiological studies (Goldberg and Segraves, 1987) and lesion studies (Mesulam, 1981).

Anatomically V5/MT is reciprocally connected to V2 and the PPC (Ungerleider and Desimone, 1986). The specific hypothesis we wanted to test was that PPC exerts a (backward) modulatory influence over the forwards driving connections from V1/V2 to V5/hMT. To do this, we employed various nonlinear models of effective connectivity that are able to describe this modulatory effect.

First, we identify regions that show differential activations in relation to attentional set using a standard categorical analysis. In the second stage, changes in effective connectivity to these areas are assessed using a simple regression model implementing a psycho-physiological and physio-physiological interaction. Finally, we show how nonlinear system identification can explain these attention-dependent changes in effective connectivity by the modulatory influence of parietal areas.

B. Dataset

The fMRI experiment was performed on a 2T MR system equipped with a head volume coil. fMRI images were obtained every 3.2 seconds with echo-planar imaging (32 slices in each volume). The subject was scanned during four different conditions: fixation, attention, no attention, and stationary. Each condition lasted 32 seconds giving 10 volumes per condition. We acquired a total of 360 images. During all conditions the subjects looked at a fixation point in the middle of a screen. In this section we are interested in only the two conditions with visual motion (attention and no attention), where 250 small white dots moved radially from the fixation point, in random directions, toward the border of the screen, at a constant speed of 4.7° per second. The difference between attention and no attention lays in the explicit command given to the subject shortly before the condition: Just look indicated no attention and Detect changes, the attention condition. Both visual motion conditions were interleaved with fixation. No response was required. Before scanning subjects were trained in the psychophysical lab to detect subtle changes in speed of the dots. In this prescanning experiment we reduced the amount of change until subjects performed at chance level (i.e., indicated four to five changes in speed at random interval during a 20s block).

Regions of interest were defined by a categorical analysis comparing attention and no attention. As predicted, given a stimulus consisting of radially moving dots, we detected activation of the posterior thalamus, early visual cortex (V1/V2), motion sensitive area V5/hMT, and the posterior parietal complex (PPC). For the subsequent analysis of effective connectivity, we defined regions of interest (ROI) with a diameter of 8 mm, centered around the most significant voxel as revealed by the categorical comparison. A single time-series, representative of this region, was defined by the first eigenvector of all the voxels in the ROI (Büchel and Friston, 1997).

II. EFFECTIVE CONNECTIVITY ANALYSES

A. Regression Analyses

A simple regression analysis revealed that activity in V1/V2 and V5/hMT were highly dependent. High activity in V1/V2 was predictive of high activity in V5/hMT and vice versa. Given our research question that attention modulates the connection between V1/V2 and V5/hMT, we had to take this nonlinear effect into account. If attention exerts a positive modulatory influence on the path between V1/V2 and V5/hMT, the influence of V1/V2 on V5/hMT should depend on the attentional state. This can be tested, by splitting the observations into two sets, one containing observations during the attention condition and the other containing the observation during the no-attention condition. It is now possible to perform separate regressions of V5/hMT on V1/V2 using each set. If the hypothesis of a positive modulation is true, the slope of the regression of V5/hMT on V1/V2 should be steeper under attention. This is shown in Fig. 75.1a. In essence this model embodies the idea that a psychological variable can modulate the contribution of one area onto another and was thus called a psychophysiological interaction (Friston et al., 1997).

Given the hypothesis that attentional modulation stems from PPC, we can now finesse our model and substitute the psychological factor, with a physiological factor, namely PPC activity. This converts the psycho-physiological interaction model to a physio-physiological model (Friston et al., 1997). Instead of using a psychological factor (i.e., attention) we now

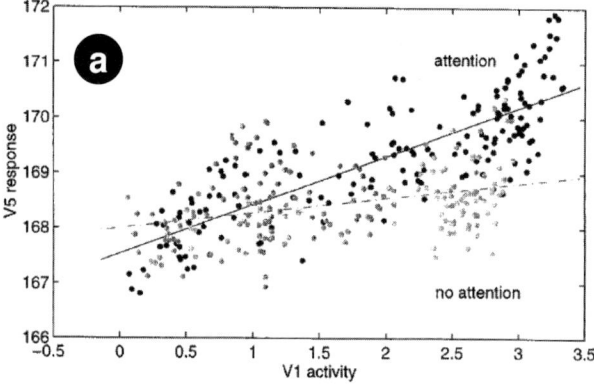

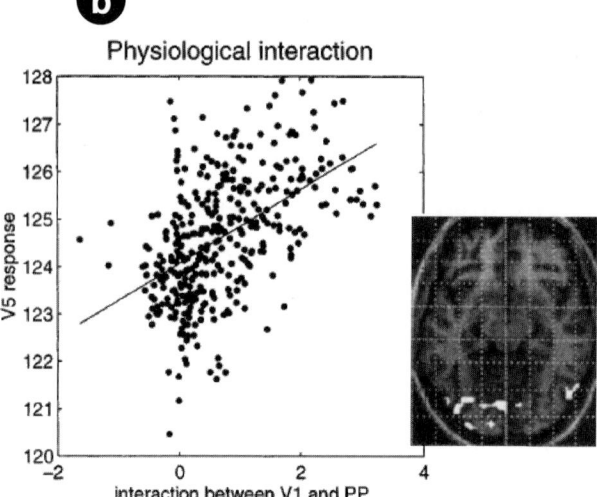

FIGURE 75.1 Changes in effective connectivity as assessed with regression analyses. **a.** V5/hMT activity is plotted as a function of V1/V2 for the attention and the no attention condition. The steeper slope of the regression curve under the attention condition indicates a stronger contribution of V1/V2 to V5/hMT under attention. **b.** A similar type of interaction is demonstrated; however, here the data is not grouped by experimental condition but by the activity in the parietal cortex. In this case, it can be seen that the parietal cortex exerts a positive modulatory influence on the connection between V1/V2 and V5/hMT.

assess the influence of PPC on the connection between V1/V2 and V5/hMT using a physio-physiological interaction. In this case a multiple linear regression model is set up with the following explanatory variables (i.e., regressors): (i) The time-series of V1/V2; (ii) the time-series of PPC activity; (iii) the interaction term between (i) and (ii), which is the element by element product of the two time-series after they have been normalized. If the hypothesis of a modulation of the V1 → V5 effective connection is true, the interaction term (iii) in the previous equation should explain a significant amount of variance in V5/hMT. Fig. 75.1b shows the regression of V5/hMT activity on activity in V1/V2 multiplied by the activity level in PPC; that is, the physio-physiological interaction term.

B. Volterra Series

A system of regression equation can be useful to describe very simple linear input-output systems. In more complex systems, this linearity assumption does not hold and one has to turn to nonlinear systems. The simplest form of nonlinearity (i.e., an interaction term) was integrated into the regression equation in the last section. This can be extended to describe more complex nonlinear systems. The model we are dealing with in this section is based on general nonlinear system identification. We assume that the activity in one region can be explained by a nonlinear convolution or filtering of the dynamics in regions that contribute inputs. The implicit nonlinear mixing of inputs over time allows source regions to cause responses in a target region that may be more enduring than the input or, indeed, delayed. Furthermore, the nonlinearities allow for inputs to interact and sensitize the target to other inputs, again with a different time course. In theory, this convolution model can model any driving or modulatory effect that distant regions exert over a target and characterizes them in terms of Volterra kernels. Because the kernels are high-order they embody interactions among inputs. The influence of one region a on another b therefore can be divided into two components: (i) the driving influence of a on b, irrespective of the activities elsewhere; and (ii) an activity-dependent component that represents an interaction with inputs from the remaining regions. These two aspects of effective connectivity correspond to terms that: (i) involve only activity in area a; and (ii) the remaining terms that model an interaction between activity in a and other regions. By using second order approximations, the kernels can be estimated with ordinary least squares which, in turn, facilitates the use of standard F tests for statistical testing. Connections (driving or modulatory) are assessed by testing the null hypothesis that the associated kernel is zero (i.e., the associated regressor in the general linear model does not explain a significant amount of additional variance). This can be repeated at every voxel in some prespecified target region (e.g., V5/hMT) to produce a statistical parametric map that reflects the significance of the connection. When making inferences about modulatory effects one is testing the null hypothesis that the interactions among inputs, causing a response in an area, are negligible. Having established that an effect is significant, the influence of remote regions (either driving or modulatory) can be characterized by using simulated inputs and the estimated kernels.

In contradistinction to SEM and regression approaches, this Volterra formulation does not necessarily rely on constrained anatomical models. It is furthermore possible by selecting the relevant regions (and their respective time-series) to infer the architecture of the system on that basis. The emphasis is not on estimating the parameters of an assumed connectivity architecture but on making statistical inferences about the integration of multiple inputs to a single area that elaborates a response.

The regions identified by the categorical comparison comprised the thalamus, V1/V2, V5/hMT, and the PPC. Significant contributions were found between V1/V2 and V5/hMT and between the thalamus and V1/V2 (see Fig. 75.2a). Most importantly, an interaction was detected between the PPC and the connection of V1/V2 to V5/hMT. We were able to show that this modulatory effect of PPC on V1/V2 inputs to V5/hMT was significant and replicated this finding four times in independent studies of further subjects. Figure 75.2b shows a characterization of this modulatory effect in terms of the increase in V5/hMT responses, to a simulated V1/V2 input, when PPC activity is zero (broken lines) and when it is high (solid lines), for the first three subjects. The broken lines in Fig. 75.2b represent the driving effect. In this example, the simulated input from V1/V2, was a 500-ms square wave convolved with a hemodynamic response function and scaled to unit height. The solid curves represent the same response when PPC activity is unity. It is evident that V1/V2 has a driving effect on V5/hMT and that PPC increases the responsiveness of V5/hMT to these inputs (compare the solid and dashed lines in Fig. 75.2b). Quantitatively, there is an increase of about 30% in the response to V1/V2 inputs for a unit increase in PPC activity. By virtue of the normalization applied to the time-series, the (dimensionless) activities are expressed in terms of the standard deviation of each region.

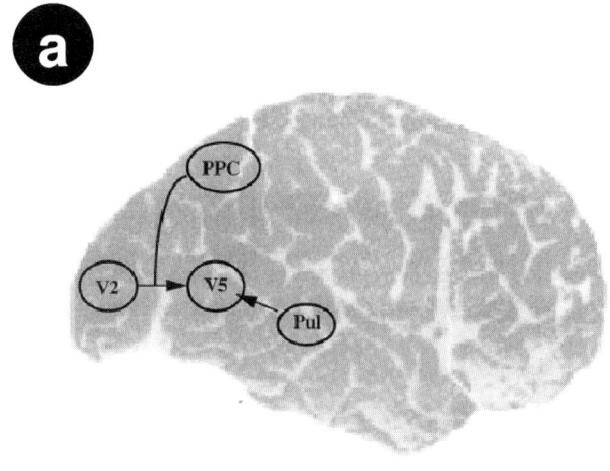

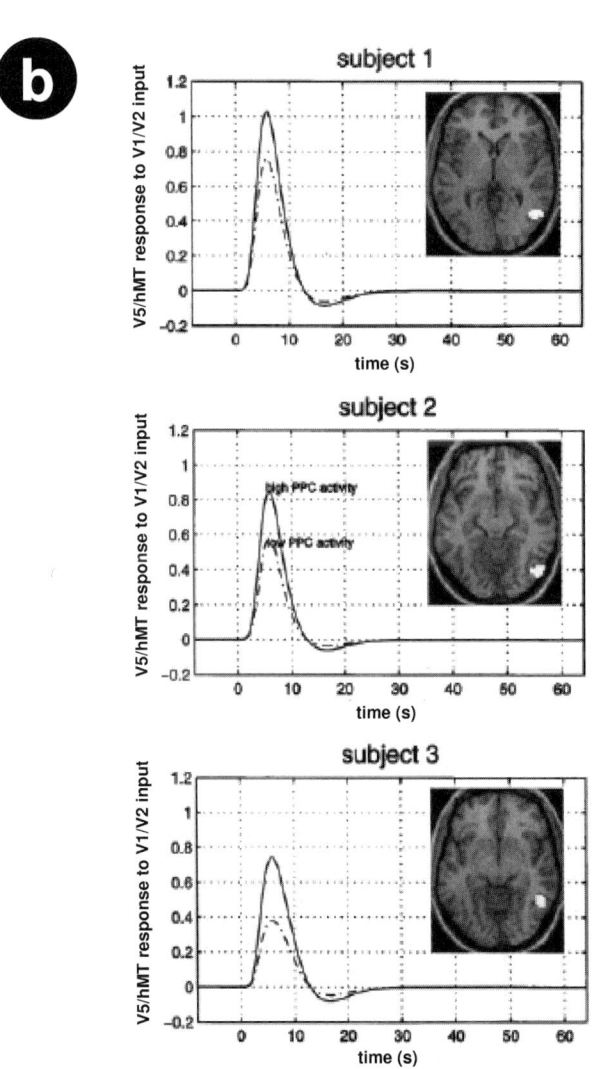

FIGURE 75.2 Volterra series analysis. **a.** All connections that exceeded the significance level of $p < 0.05$ are shown. Note that all connections are bidirectional (e.g., from V1/V2 to V5/hMT and vice versa). This analysis revealed that the data are consistent with the notion that PPC modulates the contribution of V1/V2 to V5/hMT. **b.** The behavior of the system is exemplified with artificial inputs into V1/V2 and PPC. The graph shows the resulting response in V5/hMT for three subjects. Dashed lines indicate the response to V1/V2 input when the modulator (PPC) is silent, that is, its activity is zero. The solid lines indicate the response in V5/hMT when the PPC is active. Note the increase of up to 30% depending on the state of the PPC.

III. DISCUSSION

Using different techniques for the estimation of effective connectivity from neuroimaging time-series, we have shown a modulatory influence of the PPC on the connection between V1/V2 and V5/hMT. It should be noted that the conclusions reached here are constrained inherently by the hypothesis tested. Alternative architectures could have been considered that may have yielded similar results. These analyses should not be construed as an exploratory characterization of network interactions, but represent a test of a specific hypothesis. The most that can be concluded is that the activity in PPC is sufficient to explain a significant component of attentional modulation of V5/hMT responses. It cannot be concluded that other afferents do not play a role or that PPC modulation is necessary for these effects. A second limitation is imposed by the spatial and temporal resolution of neuroimaging data. This precludes inferences about laminar-specific interactions or anything more refined than large-scale population dynamics. However, the wealth of data implicating PPC in an attentional role (Assad and Maunsell, 1995; Cavada and Goldman Rakic, 1989; Mountcastle et al., 1981) renders the current analysis congruent with a large body of convergent evidence.

In the first part of the chapter we have shown how a simple extension of regression models can be used to capture interactions between PPC and the connection between V1/V2 and V5/hMT. This enabled hypothesis testing about second-order interactions among inputs to an area, or indeed interactions between experimental factors and activity in a modulating source (i.e., psychophysiological interactions). However, these models assume that the activity in one region causes a response in another instantaneously. To allow for (*i*) temporal precedence of influences among cortical regions and (*ii*) the endogenous variability in the delay of the hemodynamic response, the Volterra formulation was developed. The Volterra approach explicitly models interactions over time and has been established in the analysis of fMRI time-series from single voxels (Friston et al., 1998). As in the regression analysis we were able to demonstrate an interaction of PPC activity with the contribution from V1/V2 to V5/hMT.

Interestingly, we were able to show that the same principles apply at subsequent stages. For example, backward connections from prefrontal areas may modulate V5/hMT inputs to the PPC (Büchel and Friston, 1997). This leads to the notion of a serially coupled hierarchy, where each stage is driven by lower stages but, at the same time, modulates these driving areas. This has important implications for conceptual and mathematical models of information processing because the dynamics at any level are some nonlinear function of activity in both lower and higher levels. This precludes serial transformations of the sensory input as assumed in many information theoretic accounts of early visual processing but is much more consistent with a "generative" model approach.

IV. EFFECTIVE CONNECTIVITY VERSUS CATEGORICAL COMPARISONS

One obvious advantage of the assessment of effective connectivity is that it allows one to test hypotheses about the integration of cortical areas. For example, in the presence of modulation, the categorical comparison between attention and no attention might reveal prestriate and parietal activations. However, the only statement possible is that these areas show higher cortical activity during the attention condition as opposed to the no attention condition. The analysis of effective connectivity revealed two additional results. First, attention affects the pathway from V1/V2 to V5/hMT. Second, the introduction of nonlinear interaction terms allowed us to test a hypothesis about how these modulations are mediated. The latter analysis suggested that the posterior parietal cortex exerts a modulatory influence on the connection between area V1/V2 and V5/hMT.

The measurements used in all examples in this chapter were hemodynamic in nature. This limits an interpretation at the level of neuronal interactions. However, the analogy between the form of the nonlinear interactions described earlier and voltage-dependent (i.e., modulatory) connections is a strong one. It is possible that the modulatory impact of PPC on V5 is mediated by predominantly voltage-dependent connections. We know of no direct electrophysiological evidence to suggest that extrinsic backward PPC to V5 connections are voltage-dependent; however, our results are consistent with this. An alternative explanation for modulatory effects, which does not necessarily involve voltage-dependent connections, can be found in the work of (Aertsen and Preissl, 1991). These authors show that effective connectivity varies strongly with, or is modulated by, background neuronal activity. The mechanism relates to the efficacy of subthreshold EPSPs in establishing dynamic interactions. This efficacy is a function of post-synaptic depolarisation, which in turn depends on the tonic background of activity.

SUMMARY

This chapter has reviewed the basic concepts of effective connectivity in neuroimaging of attention. We have introduced several methods to assess effective connectivity; that is, multiple linear regression, psychophysiological interactions and Volterra-series analysis. We have presented evidence for the hypothesis that changes in attentional set are associated with augmented responses of V5/hMT to driving inputs from V1/V2 and that the activity of posterior parietal cortex is sufficient to account for this modulation. This does not preclude the role of other modulatory effects (e.g., those mediated by cortico-thalamic loops), nor does it imply that parietal modulation is necessary. However, these results clearly show that parietal influences are sufficient to explain a significant component of attentional modulation in V5/hMT.

References

Aertsen, A., and Preissl, H. (1991). *Dynamics of activity and connectivity in physiological neuronal networks*. VCH Publishers, Inc., New York.

Assad, J. A., and Maunsell, J. H. (1995). Neuronal correlates of inferred motion in primate posterior parietal cortex. *Nature* **373**, 518–521.

Büchel, C., and Friston, K. J. (1997). Modulation of connectivity in visual pathways by attention: Cortical interactions evaluated with structural equation modeling and fMRI. *Cerebral Cortex* **7**, 768–778.

Cavada, C., and Goldman Rakic, P. S. (1989). Posterior parietal cortex in rhesus monkey: I. Parcellation of areas based on distinctive limbic and sensory corticocortical connections. *J. Comp. Neurol.* **287**, 393–421.

Corbetta, M., Miezin, F. M., Shulman, G. L., and Petersen, S. E. (1991). Selective attention modulates extrastriate visual regions in humans during visual feature discrimination and recognition. *Ciba. Found Symp.* **163**, 165–175.

ffytche, D. H., Guy, C. N., and Zeki, S. (1995). The parallel visual motion inputs into areas V1 and V5 of human cerebral cortex. *Brain* **118**, 1375–1394.

Friston, K. J., Büchel, C., Fink, G. R., Morris, J., Rolls, E., and Dolan, R. J. (1997). Psychophysiological and modulatory interactions in neuroimaging. *Neuroimage* **6**, 218–229.

Friston, K. J., Josephs, O., Rees, G., and Turner, R. (1998). Nonlinear event-related responses in fMRI. *Magnetic Resonance in Medicine* **39**, 41–52.

Goldberg, M. E., and Segraves, M. A. (1987). Visuospatial and motor attention in the monkey. *Neuropsychologia* **25**, 107–118.

La Berge, D. (1995). *Attentional processing*. Harvard University Press, Cambridge, MA.

Mesulam, M. M. (1981). A cortical network for directed attention and unilateral neglect. *Ann. Neurol.* **10**, 309–325.

Mountcastle, V. B., Andersen, R. A., and Motter, B. C. (1981). The influence of attentive fixation upon the excitability of the light-sensitive neurons of the posterior parietal cortex. *J. Neurosci.* **1**, 1218–1225.

O'Craven, K. M., and Savoy, R. L. (1995). Voluntary attention can modulate fMRI activity in human MT/MST. *Invest. Ophthalmol. Vis. Sci. (suppl.)* **36**, S856.

Tootell, R. B., Reppas, J. B., Kwong, K. K., Malach, R., Born, R. T., Brady, T. J., et al. (1995). *J. Neurosci.* **15**, 3215–3230.

Treue, S., and Maunsell, H. R. (1996). Attentional modulation of visual motion processing in cortical areas MT and MST. *Nature* **382**, 539–541.

Ungerleider, L. G., and Desimone, R. (1986). Cortical connections of visual area MT in the macaque. *J. Comp. Neurol.* **248**, 190–222.

Zeki, S., Watson, J. D., Lueck, C. J., Friston, K. J., Kennard, C., and Frackowiak, R. S. (1991). A direct demonstration of functional specialization in human visual cortex. *J. Neurosci.* **11**, 641–649.

CHAPTER 76

Attentional Modulation of Surround Inhibition

Barbara Zenger-Landolt

ABSTRACT

Contrast detection of an isolated target is an example of a visual task that does not require attention. However, when other task-irrelevant stimuli (a *surround*) are presented in the display, even a simple contrast-detection task can require attention, depending on the specific surround configuration.

In a series of psychophysical experiments, observers had to detect a Gabor target, which was surrounded by six Gabor patches (the surround). Attention was manipulated by means of a double-task paradigm: in one condition observers had to perform concurrently a central letter-discrimination task, and the contrast-detection task was then only poorly attended, whereas attention was fully available in the other condition. Under poorly attended conditions targets could be detected only when the target was salient and visually popped out due to a high contrast (above the surround contrast) or a distinct orientation (more than 10–15° different from the surround). Detection of nonsalient targets required attention.

I. INTRODUCTION

Converging evidence in the psychophysical literature demonstrates that simple contrast-detection and contrast-discrimination tasks require very little—if any—attention. In a study by Palmer (1994), observers had to find a bright target among dimmer targets, and thresholds were measured. As is often the case in visual search studies, performance decreased when observers had to pay attention to more display elements (Treisman and Gelade, 1980); however, the threshold increase with increasing set size was rather small, and could be attributed solely to the fact that additional distracters contribute noise at the decision stage. Similarly, Foley and Schwarz (1998) carried out visual search tasks where observers had to find a high-contrast Gabor patch among Gabor patches of lower contrast. Again, no attentional limitations were found, suggesting that contrast-detection and discrimination do not require attention (see Chapter 17 for a review on the use of visual search to study attention).

Similar results were obtained when attention was manipulated using a dual-task paradigm instead of a visual search task (Lee, Itti, Koch, and Braun, 1999). In these experiments, observers had to do a contrast detection or discrimination task on a peripheral Gabor patch. In addition, there was a central letter discrimination task that observers either had to perform concurrently with the peripheral task (which was then poorly attended), or that they could ignore (allowing them to fully attend the peripheral task). The results show again that the effects of attention on contrast detection and discrimination are relatively small (see Chapter 70 for attentional effects on the physiology of contrast processing).

Here, we report a series of psychophysical experiments (Zenger, Braun, and Koch, 2000) showing that even a simple contrast detection task may require attention when the target is not the most salient element in the display, but is surrounded by other salient elements (a surround). (For further tasks that do improve with attention see Chapter 73.) These experiments were motivated by suggestions that attention is relevant mostly in displays that contain several competing elements (Desimone and Duncan, 1995), and that attention acts by biasing this competition in favor of the attended element (see Chapter 50).

It is already well known that contextual elements can affect performance in simple contrast-detection tasks. For example, placing two collinear flankers next to the target patch reduces the target's detection threshold considerably (Polat and Sagi, 1993), a finding that has been attributed to facilitatory interactions in primary visual cortex. Other contextual elements such as oblique gratings presented at the target location generally deteriorate performance in contrast detection and discrimination tasks (Foley, 1994) and are believed to reflect local inhibitory interactions in primary visual cortex.

Much less is known about how these context effects are modulated by attention. Recent work by Freeman, Sagi, and Driver (2001) shows that the facilitatory flanker effects, in fact, do require that the observer attends the flankers (see Chapter 79). Physiological work suggests that, in the periphery, flanker effects become stronger in the absence of attention (Ito and Gilbert, 1999). In contrast to these studies, the experiments described here were concerned primarily with attentional modulation of inhibitory surround effects (Zenger et al., 2000). The results suggest that attention reduces the contribution of the surround to the target's gain control. Furthermore, the results point to the striking relevance of target salience (see Chapter 38 for more on the concept of salience): salient targets that pop out visually from the surround do not benefit from the availability of attention, whereas low-saliency targets can benefit enormously, with thresholds being reduced in some configurations by a factor of 4 or more.

II. METHODS

The experiments had a dual-task design with (1) an early-vision task that required observers to detect peripheral targets embedded in a surround; and (2) an attention-demanding letter-discrimination task, to allow for manipulation of the observer's attention toward or away from the early-vision task.

A. Early-Vision Task

Stimuli consisted of localized grating patches (Gabor patches). The target patch was presented in one of the four quadrants at 4° eccentricity. Target orientation varied in different conditions between 0° and 30°. Each of the four possible target locations was surrounded by six vertical Gabor patches (the surround) arranged in a hexagon around the target location (see Fig. 76.1). The distance between target and surround patches (and between neighboring surround patches) was always 1° of visual angle, corresponding to 4 times the carrier spatial frequency of the Gabor patches (which was 4 cyc/deg). The 24 surround patches all had the same contrast, which varied between 0% and 80% in different conditions. The spatial phase of the surround patches was randomized.

During data collection, observers first fixated a fixation cross and then initiated the trial by pressing a space bar. Four circular cues appeared for 180 ms to indicate the four possible target locations. After a blank interval of 500 ± 100 ms, the stimulus was presented for 83 ms and then replaced by a blank stimulus (with fixation cross). We did not use any backward masking. Observers had to determine which of the four quadrants contained the target (spatial 4AFC). Thresholds were estimated with a standard staircase procedure.

In Experiment 1, target detection thresholds were measured as a function of surround contrast, which was varied between 0% and 80%. The target was either vertical like the surround (Experiment 1A) or was tilted by 30° (Experiment 1B). In Experiment 2, surround contrast was fixed at 40%, and target orientation was varied between 0° and 30°.

B. Attentional Manipulation

To investigate the effects of attention, the observers were asked to perform each condition twice, with the early-vision task either fully attended or poorly attended.

In the *poorly attended* condition observers had to perform concurrently to the early-vision task a letter-discrimination task: five letters (L and T's) were presented in the center (see Fig. 76.1), and observers had to determine whether all letters were the same, or whether one was different from the others (binary yes/no task). The L's and T's were then masked with F's (SOA period varied between 164 and 236 ms for different observers). This task efficiently engages attention (Lee et al., 1999). Observers were instructed to give priority to the central letter-discrimination task, which also was the task they responded to first. In the *fully attended* condition observers were instructed to perform only the peripheral early-vision task. Note that visual stimulation was identical in the fully and poorly attended conditions and that the only difference between these two conditions lies in the instruction.

To ensure that observers perform optimally in the letter-discrimination task, *control* conditions with only the letter-discrimination task (and no early-vision task) were included as well. Blocks in the dual-task condition in which the central-task performance dropped clearly below the control condition performance were discarded from the analysis and the observer was required to repeat this condition immediately. Only a

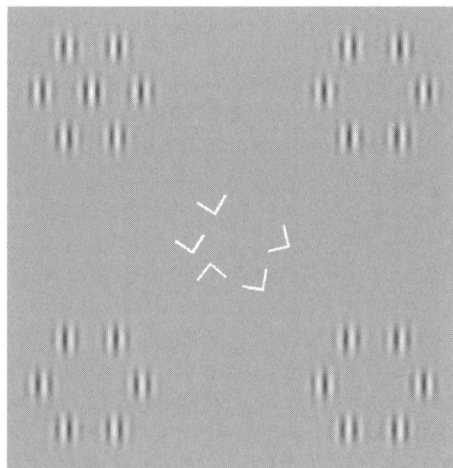 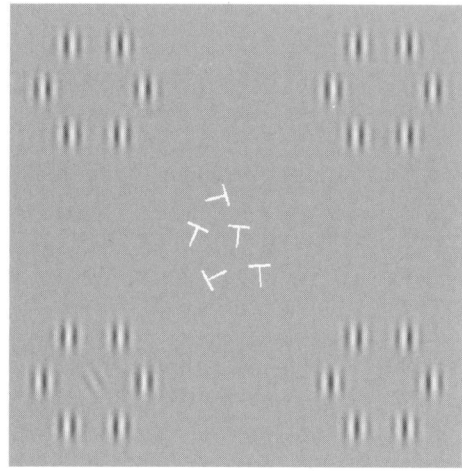

FIGURE 76.1 Example stimuli. Each display quandrant contains a hexagon consisting of six vertical Gabor patches (the surround). Only one of the four hexagons contains a central Gabor target, which is either vertical (Experiment 1A) or tilted by 30° (Experiment 1B). The display also comprises five rotated letters (L's and T's). To withdraw attention from the periphery, observers have to perform the central task; that is, they have to decide whether all letters are the same or whether one is different from the others.

few blocks had to be discarded (less than 5% of all dual-task blocks). For the remaining blocks we find that central-task performance levels in the control condition (on average 88.7% correct) is comparable to central-task performance in the dual-task situation (on average 88.2% correct), suggesting that in the dual-task situation attention was well engaged in the center.

C. Apparatus and Observers

Experiments were controlled by an O_2 Silicon Graphics workstation, and stimuli were displayed on a 19" raster monitor, which was gamma-corrected. Viewing distance was 125 cm. Five observers participated in the experiments. Observers first practiced for several sessions, since naïve observers may show a dual-task interference even with tasks thought to be preattentive (Joseph, Chun, and Nakayama, 1997). Further details concerning methods are described in Zenger et al. (2000).

III. RESULTS

The results of Experiment 1 are presented in Fig. 76.2. Contrast detection thresholds are plotted as a function of surround contrast. Attentional effects are present when there is a difference between the fully attended condition (solid lines) and the poorly attended condition (dashed lines). In Experiment 1A (top panel of Fig. 76.2), where observers had to detect a vertical target embedded in a vertical surround, attention clearly makes a difference, unless the surround contrast is zero (no surround). Attentional effects increase with increasing surround contrast and can become rather prominent, with threshold changes of a factor of 4 and more.

One important observation is that, in the poorly attended condition, targets are detected only when their contrast is considerably above the surround contrast, a finding that was very consistent across all five observers. In other words, a collinear target is invisible to the observer in the poorly attended condition, unless its contrast exceeds the surround contrast (causing contrast pop-out).

The results of Experiment 1B are shown in the lower panel of Fig. 76.2. This experiment was very similar to Experiment 1A, except that the target was tilted by 30° rather than being vertical like the surround. This small manipulation led to a dramatically different pattern of results: first, performance is largely independent of surround contrast; second, attentional effects were much smaller (though still significant). The introspective reason for the difference between Experiments 1A and B is that in the latter case the target pops out, even at low contrast, due to its orientation difference and is thus always easy to see (whether attended or not).

Experiment 2 was carried out to document the transition between the pattern of results found in Experiments 1A and 1B. The surround contrast was kept constant at 40% and target angles were varied between 0° tilt and 30° tilt. Results are shown in Fig. 76.3, with the detection thresholds plotted against the tilt angle. As we would have expected based on the previous

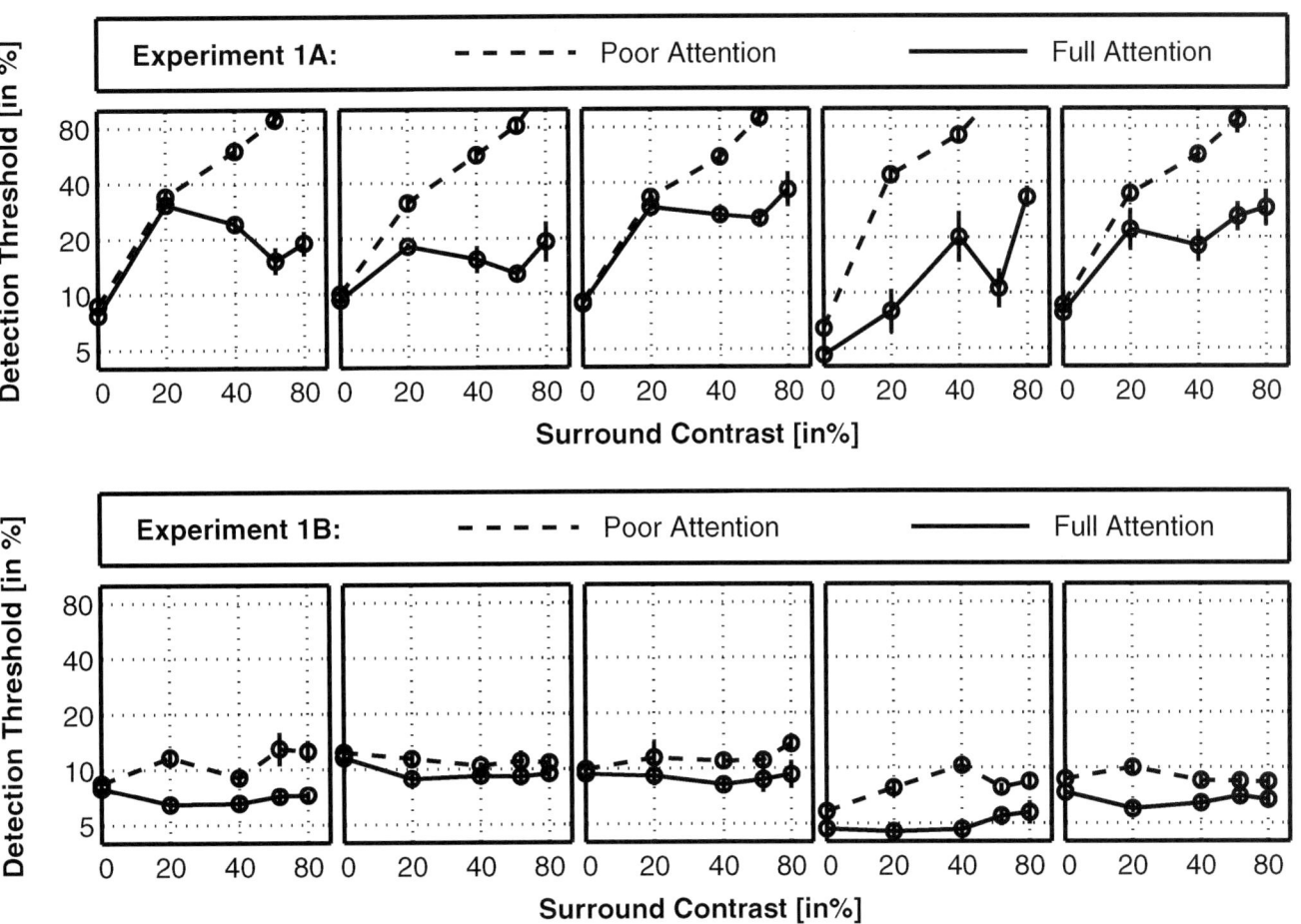

FIGURE 76.2 Contrast detection thresholds for the five different observers in Experiment 1. The error bars denote the standard error of the mean across sessions. Top: Contrast detection thresholds of vertical targets embedded in a vertical surround (Experiment 1A). In the poorly attended condition, detection is possible only when the target contrast is above the surround contrast, leading to a contrast popout. Availability of attention reduces thresholds considerably (in some cases more than fourfold). Bottom: Contrast detection thresholds of targets tilted by 30° (Experiment 1B). Detection thresholds are largely independent of surround contrast, and attentional effects are rather weak.

experiments, attentional effects are large when target orientation is equal to the surround orientation, but are small for tilt angles of 30°. The transition occurs typically at a tilt angle of around 10°–15° varying slightly between observers. Note that very small changes in the stimulus (increasing the tilt angle from 10° to 15°) can cause surprisingly large threshold differences (in particular for observers LG, LK, and VS).

IV. DISCUSSION

The experiments reported here show a strong dichotomy with regard to context effects and their attentional modulation. If the target is salient, either because its orientation differs sufficiently from the surround or because its contrast exceeds the contrast of the surround elements, detection of the target is possible even in the poorly attended condition. On the other hand, when the target does not pop out, that is, when the target is nonsalient, target detection is impossible in the poorly attended condition. In this condition, contrast detection thresholds can be improved by a factor of 4 or more by making attention available. In other words, only low-saliency targets benefit from the availability of attention, but not high-saliency targets.

These results remind us of the enormous literature on popout and visual search, which states that target detection is preattentive when the target differs from the distractors in a basic feature (such as orientation or color) but that target detection requires attention if there is no such feature difference (Treisman and Gelade, 1980; see Chapter 17). (See Chapter 93 for a

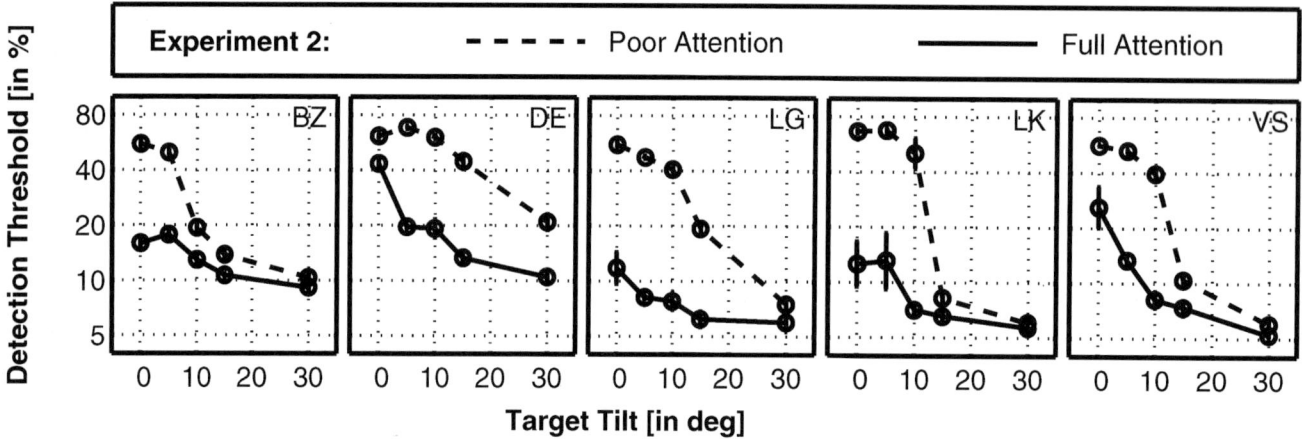

FIGURE 76.3 Contrast detection thresholds as a function of tilt angle (surround contrast is fixed at 40%) for five different observers. The error bars denote the standard error of the mean across sessions. The transition between the patterns of results observed in Experiments 1A and 1B is often rather sharp, and occurs at a tilt angle around 10° to 15°.

model on how salience may be computed in V1.) The difference between our study and classical pop-out studies is subtle, but critical: in a classical pop-out experiment observers do not have any information about where the target is, therefore, they cannot make judgments about the target unless it pops out (simply because they don't know where the target is). In our study, on the other hand, there is much less positional uncertainty. Observers knew where to look for the target: inside the hexagon! Surprisingly, this knowledge does not seem to help the observers; they apparently cannot render judgments about nonattended, nonsalient elements. In other words, the detection of nonsalient elements requires attention even when the possible target locations are known. This result could not have been predicted based on classical pop-out studies.

In a recent study very relevant to the experiments reported here, observers had to do a contrast-discrimination task on a peripheral Gabor target flanked with two Gabor patches (Zenger-Landolt and Koch, 2001). Flankers were found to strongly interfere with the contrast discrimination performance only when they shared the target orientation and when the contrast of the target was below the flanker contrast. Suppression was small or absent as soon as the target contrast exceeded the flanker contrast, or when the flanker orientation was orthogonal to the target orientation. A simple model quantitatively accounted for the data by assuming that flankers contribute to the target's gain control; that is, the neurons in the brain responding to the flankers divide the response to the units responding to the target. However, this gain control is effective only in a restricted contrast range, namely, when the target contrast is below the surround contrast. Once the target contrast exceeds the flanker contrast, the inhibition provided by the flankers is subtractive instead of divisive, and thus does not affect the target gain. Furthermore, gain control is weak when the target orientation is different from the flanker orientation.

The framework of this model suggests a rather simple and straightforward role for attention: to reduce the efficacy of gain control. For collinear targets, absence of attention would render the surround's gain control on the target very strong and basically annihilate the target response, explaining why detection is impossible under these conditions. Detection becomes possible only once the target exceeds the surround contrast, and gain control becomes ineffective (see upper panel of Fig. 76.2). For tilted targets, gain control by the surround is weak to begin with (even in the absence of attention), but is decreased once attention becomes available, explaining the small but significant performance improvement (see lower panel of Fig. 76.2).

Our results are consistent with the notion that attentional effects are observable mostly when several elements are present in the display, and that attention biases the competition among the different elements in favor of the attended element (Desimone and Duncan, 1995; see Chapter 50). In our experiments, it would appear that salient elements win the competition anyhow, and do not require any attentional boost, but nonsalient elements, that is, elements that are inhibited by surrounding elements, can benefit from the attentional bias that reduces the inhibitory interactions that make detection difficult.

Where in the brain does this competition occur? Reynolds, Chelazzi, and Desimone (1999) have observed competitive interactions between different elements in the display in cells in macaque area V2 and V4. Furthermore, these competitive interactions were altered by attention in a way that made the neurons respond more as if only the attended item was present in the display, but not the unattended distracter. Other studies show that in quadrants where the monkey has a lesion in area V4 or TEO, the monkey cannot detect a nonsalient target (Schiller and Lee, 1991). The animals cannot make precise judgments about the orientation of a target in the presence of distracters located presumably within the same receptive field (De Weerd, Peralta, Desimone, and Ungerleider, 1999); however, the monkey can solve all these tasks in the normal quadrant. Given the similarities between the effects of these lesions and the effects of withdrawing attention in our study, it is attractive to attribute the attentional effects observed here to a processing stage approximately related to V4. However, it is clear that future work will be required to establish a clear link between the psychophysical effects observed here and the mechanisms in the human brain from which they arise.

NOTES

Section II.B, the last two paragraphs of Section III, and Figures 76.1, 76.2, and 76.3 are reprinted in slightly adapted form from Zenger et al. (2000), with permission from Elsevier.

References

De Weerd, P., Peralta, M. R., 3rd, Desimone, R., and Ungerleider, L. G. (1999). Loss of attentional stimulus selection after extrastriate cortical lesions in macaques. *Nat. Neurosci.* **2**, 753–758.

Desimone, R., and Duncan, J. (1995). Neural mechanisms of selective visual attention. *Annu. Rev. Neurosci.* **18**, 193–222.

Foley, J. M. (1994). Human luminance pattern-vision mechanisms: masking experiments require a new model. *J. Opt. Soc. Am. A* **11**, 1710–1719.

Foley, J. M., and Schwarz, W. (1998). Spatial attention: The effect of position uncertainty and number of distractor patterns on the threshold versus contrast function for contrast discrimination. *J. Opt. Soc. Am. A* **15**, 1036–1047.

Freeman, E., Sagi, D., and Driver, J. (2001). Lateral interactions between targets and flankers in low-level vision depend on attention to the flankers. *Nat. Neurosci.* **4**, 1032–1036.

Ito, M., and Gilbert, C. D. (1999). Attention modulates contextual influences in the primary visual cortex of alert monkeys. *Neuron* **22**, 593–604.

Joseph, J. S., Chun, M. M., and Nakayama, K. (1997). Attentional requirements in a "preattentive" feature search task. *Nature* **387**, 805–807.

Lee, D. K., Itti, L., Koch, C., and Braun, J. (1999). Attention activates winner-take-all competition among visual filters. *Nat. Neurosci.* **2**, 375–381.

Palmer, J. (1994). Set-size effects in visual search: the effect of attention is independent of the stimulus for simple tasks. *Vision Res.* **34**, 1703–1721.

Polat, U., and Sagi, D. (1993). Lateral interactions between spatial channels: suppression and facilitation revealed by lateral masking experiments. *Vision Res.* **33**, 993–999.

Reynolds, J. H., Chelazzi, L., and Desimone, R. (1999). Competitive mechanisms subserve attention in macaque areas V2 and V4. *J. Neurosci.* **19**, 1736–1753.

Schiller, P. H., and Lee, K. (1991). The role of the primate extrastriate area V4 in vision. *Science* **251**, 1251–1253.

Treisman, A. M., and Gelade, G. (1980). A feature-integration theory of attention. *Cognit. Psychol.* **12**, 97–136.

Zenger, B., Braun, J., and Koch, C. (2000). Attentional effects on contrast detection in the presence of surround masks. *Vision Res.* **40**, 3717–3724.

Zenger-Landolt, B., and Koch, C. (2001). Flanker effects in peripheral contrast discrimination—psychophysics and modeling. *Vision Res.* **41**, 3663–3675.

CHAPTER 77

Attentional Processes in Texture Perception

Charles Chubb

ABSTRACT

An everyday texture judgment takes as input the image of a surface and produces a decision statistic gauging the surface's suitability for a particular purpose (e.g., whether to trust a rock with one's boot while hiking). It is proposed that (1) human vision performs such judgments in real time through the use of a modest number of neural arrays, each of which measures the level at each point in the visual field of a distinct *visual substance*; and (2) human vision can only measure decision statistics produced by spatially averaging linear combinations of these visual substances. Research is reviewed whose purpose is to catalogue the substances of human vision. One case study is considered: the visual substance *blackshot*, so called because blackshot level reflects the density in a texture of only those texture elements of Weber contrast very near −1.0. Support for proposal (2) is derived from an experiment investigating attentional control of texture orientation judgments.

I. TEXTURE JUDGMENTS IN DAILY LIFE

Many daily decisions require quick visual judgments about different sorts of surfaces and materials, At a glance you discriminate cotton from nylon from satin from wool. Without touching it, you sense how a fabric will feel against your skin; you see the dampness of a washcloth and the softness of a towel. You also see that a certain wall will get you dirty if you lean against it, and that a certain post may give you a splinter if you grab it.

These and many similar visual judgments are made so effortlessly that you barely notice them. In each case, however, a computation is executed that takes as input a visually complex field or surface and rapidly produces as output a value that enables a decision. As you hike, for example, you assess with each step whether a given rock face will hold your boot without slipping. How do you make such a judgment?

The following general picture seems unavoidable: The image R of the rock face is fed as input to a visual process G that produces a decision statistic G(R) reflecting the "grippiness" of the rock face. A high value of G(R) leads you to trust the rock face with your boot, whereas a low value leads you to avoid it or to trust it with heightened caution.

Crucially, G(R) must be computed in real time between one step and the next, How might a visual statistic such as G(R) be computed at a glance? Retinal processing suggests an answer, Each of the L, M, and S (Long-, Medium-, and Short-wavelength sensitive) cone arrays can be viewed as a spatially parallel processor. Consider, for example, the L cone array. A given L cone measures a specific property (which might be called the "long wavelength level") of the light impinging on the point in the retina where that L cone is situated. Moreover, all across the central region of the retina are stationed L cones, each of which continuously measures the long-wavelength level of the light impinging on its point of the retina. Thus the pattern of activations generated by the array of L cones composes a neural image (Robson, 1980) that provides a dynamic map of the long wavelength level of the light playing across the central visual field.

We propose that an analogous, spatially parallel architecture is used to make texture judgments; specifically, that human vision embodies some number N of retinotopically organized neural arrays, $A_1, A_2, \ldots, A_N$, each of which yields a time-varying neural image (analogous to those produced by the L, M, and S cone arrays). For k = 1, 2, ..., N, we think of the neural image generated by A_k as giving the instantaneous spatial map of a particular visual substance. This follows the proposal of Adelson and Bergen (1991) that the purpose of the initial stages of visual processing is to measure "the amounts of various kinds of visual 'substances' present in the image."

The neuron in array A_k corresponding to location (x,y) is assumed to monitor a local region of the retina centered at (x,y), continuously measuring in that region the changing level of the visual substance sensed by array A_k. For example, in standard back-pocket models (Chubb and Landy, 1991) of preattentive texture segregation (so called because researchers routinely pull such models from their back pockets to explain new instances of texture segregation), the visual substance sensed by a given array A_k is typically something like "local texture energy at some orientation θ_k and spatial frequency ω_k." In this case, a given neuron in A_k should be selectively activated by texture components of orientation θ_k and spatial frequency ω_k falling within its monitored region.

It should be noted, however, that the specific visual substances hypothesized by previous models have been largely intuitive and *ad hoc*. Only recently have methods been developed to psychophysically isolate and characterize human visual substances.

The substance-sensing arrays A_k offer speed; however, they require substantial neural hardware. Each array comprises many neurons, each of which is dedicated exclusively to gauging the level of a particular visual substance within a small region of the retina. These considerations suggest that the number N of elementary substances sensed by human vision is likely to be small. Thus, one might reasonably hope to compile a complete catalogue of human visual substances. Such a catalogue would aim to answer the following questions:

1. What is the number N of substance-sensing arrays in human vision?
2. For k = 1, 2, ..., N, what substance is sensed by array A_k?

The answer to the second question requires a precise description of the computation achieved by an arbitrary neuron of A_k, for k = 1, 2, ..., N. The next section describes the progress made in addressing questions 1 and 2.

II. CASE STUDY: THE ISOLATION AND ANALYSIS OF BLACKSHOT

As a testing ground for developing the psychophysical methods needed to isolate and functionally analyze visual substances, we have taken the space of textures that can be generated by tiling the stimulus field with small squares (called texels) of different gray-levels (Chubb and Landy, 1991; Chubb, Econopouly, and Landy, 1994; Chuhb, Landy, and Econopouly, 2004). We use 17 gray-levels, referred to here as gray-levels 0, 1, ..., 16, corresponding to Weber contrasts increasing in equal steps from –1 to 1.

A given texture is determined by the proportions p(k), k = 0, 1, ..., 16, of different gray-levels that it comprises. (Of course, for finite-sized texture patches, there exist only finitely many probability distributions p such that the proportions of different gray-levels in a patch conform exactly to p. However, if patches are large, arbitrary distributions can be well-approximated.) One generates a patch of texture with distribution p by assigning proportions of different gray-levels given by p and then randomly scrambling the locations of the texels in the patch. Accordingly, we call such textures *scrambles*. One additional constraint: we consider only scrambles with distributions p for which $p(k) \leq 2/17$ for all gray-levels k. This rules out degenerate textures dominated by only a few gray-levels.

We make the following basic assumption: Any two scrambles are discriminable at a glance to human vision if and only if they produce significantly different levels of at least one visual substance—that is, if and only if the two scrambles produce significantly different average levels of activation in at least one substance-sensing array A_k.

Figure 77.1 shows examples of different scrambles. Corresponding bar graphs give the probability distributions of the scrambles.

The scrambles in Figs. 77.1a and 77.1b differ in mean gray-level but are equal in variance, whereas those in Figs. 77.1c and 77.1d differ in variance but are equal in mean. Pilot studies reveal that if two scrambles differ moderately in mean gray-level, it is not possible to adjust their relative variances to null the evident difference in brightness between them. Nor is it possible to null a moderate difference in variance by adjusting the relative means of two scrambles. This implies that at least two visual substances enable discriminations between scrambles differing in gray-level mean and variance. Pilot studies suggest first that one of these substances, which might be called texture brightness, reflects primarily the mean gray-level of the scramble—that is, the varying levels of this substance produced by different scrambles follow closely the varying means of those different scrambles. Pilot studies also suggest that a second substance, which we might call texture energy, reflects primarily the

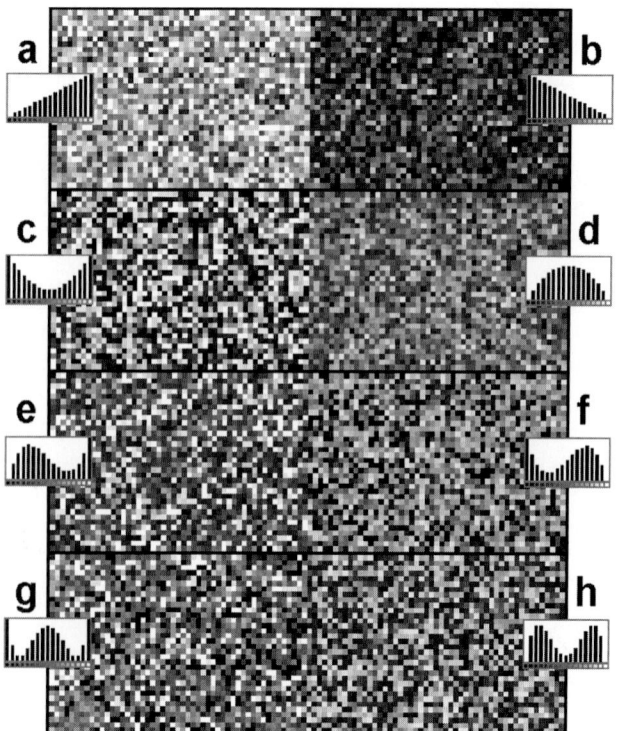

FIGURE 77.1 *Examples of scrambles.* Bar graphs give gray-level histograms. Scrambles (a) and (b) differ in mean; (c) and (d). differ in variance. All of (e), (f), (g), and (h) have mean and variance equal to the mean and variance of the uniformly distributed scramble (with equal proportions of all 17 gray-levels).

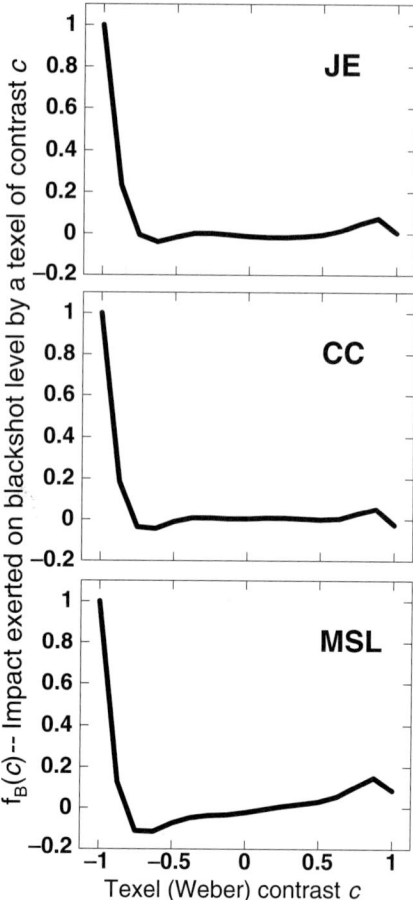

FIGURE 77.2 *Blackshot impact functions.* 7th order polynomial approximations for three observers. Impact functions are defined only up to arbitrary additive and multiplicative constants.

variance of the scramble. Detailed functional analysis of these two substances remains to be carried out.

There is, however, a third human visual substance enabling discriminations between scrambles about which more is known. It is this third substance (almost exclusively) that generates the perceptual differences between the scrambles in Figs. 77.1e,f,g, and h. These four scrambles are equated in gray-level mean and variance. Specifically, all are in the space Γ of scrambles whose mean and variance are equal to the mean and variance of the uniformly distributed scramble (i.e., the scramble with equal proportions of all 17 gray-levels).

Chubb, Econopouly, and Landy (1994) described the psychophysical method of *histogram contrast analysis* (for methodological details see Chubb, 1999; Chubb et al. 1994; Chubb et al., 2004), and used this technique to show that discriminations between scrambles in Γ were driven by variations in only one visual substance. Let B be the neural array used to sense this third substance. The results of Chubb et al. (1994) combined with additional, more recent measurements (Chubb, Landy, and Econopouly, 2004), provide estimates for three observers of the function f_B giving the impact exerted on neurons in B by texels of different gray-levels (see Fig. 77.2). (Note, however: these estimates are defined only up to arbitrary additive and multiplicative constants, which have been chosen to clarify the nature of the substance sensed by B.)

Note that f_B takes the value 1.0 for texels of (Weber) contrast −1.0, is perhaps slightly elevated for texels of contrast −0.875, but remains near 0.0 for texels with contrasts from −0.75 to 1.0. Within a scramble, then, any texel of contrast −1.0 boosts by one unit the level of the substance sensed by B, whereas any texel with contrast greater than −1.0 leaves the level of this substance largely unchanged. Thus, the level of this substance produced by a scramble (i.e., the average value of f_B applied to the scramble's texels) reflects with remarkable precision the proportion of texels in the scramble that have Weber contrast −1.0. In summary: this substance is influenced only by the very blackest components spattered (often relatively sparsely) throughout a predominantly bright, heterogenous scene. To suggest this pattern of sensitivity, we call this substance *blackshot*.

All the experiments done to this point have used small, square texels (11.2 min. in width). Thus, nothing is known about the degree to which blackshot-level may or may not be influenced by texture elements of different shapes. It is even possible that human vision may embody several blackshot-based substances tuned to texture elements of different shapes; for example, distinct substances influenced selectively by very black, elongated, texture elements of different orientations.

The general method used to isolate and functionally characterize blackshot is called *histogram contrast analysis* (Chubb, 1999; Chubb et al., 1994; Chubb and Landy, 1991; Chubb et al., 2004; Chubb and Nam, 2001; Chubb and Talevich, 2002; Nam and Chubb, 2001). This method can also be used to obtain functions analogous to f_B for texture brightness and texture variance (see Fig. 72.2).

Finally, it should be noted that histogram contrast analysis can be used to analyze textures made of arbitrary elements (not just small squares of different gray-levels). For example, texture elements might differ in color, shape, spatial frequency, orientation, binocular disparity, local motion, and so on. In all cases, however, one aims to discover (1) the number of distinct visual substances enabling discrimination between textures made of those elements, and (2) a function (analogous to f_B) for each visual substance showing how different types of texture elements influence the level of that substance produced by a texture.

III. ATTENTIONAL CONTROL OF TEXTURE JUDGMENTS

What is the role of attention in controlling the texture judgments common in daily life? Previous texture processing models have little to say about this for the following reason: the aim of all prior models was to account for *preattentive* texture segregation—the spontaneous emergence of boundaries due to texture differences. As suggested by the modifier preattentive, the computations yielding such boundaries were assumed to be strictly data-driven and hence bottom-up. The possible role of top-down control in texture processing has not been considered.

It is impossible, however, to ignore attentional influences in everyday texture judgments. One decision statistic is required to pick a dry towel from a rack instead of a damp one, whereas a different statistic is required to choose a juicy peach instead of a mealy one. But to compute different decision statistics, the observer must make different uses of the substance maps. Such situation-dependent adaptability requires top-down control.

How is such flexibility achieved? The simplest possibility is the following: To judge whether a surface is suitable for a given purpose, it might be that the observer must select one of the substance-sensing arrays $A_1, A_2, \ldots, A_N$ to do the job. Then the chosen array A_k is applied to the image Q of the surface. The observer uses the surface if (and only if) the average level across Q of the substance sensed by A_k falls within some criterion range. Under this scenario, human vision can compute only N statistics—the average levels of the N visual substances.

However, the wide variety of subtle judgments people seem able to make suggests that observers have greater flexibility in synthesizing decision statistics. Our working hypothesis is that observers can compute various linear mixtures of visual substances. Under this story, to assess the suitability of a surface for a given purpose, the observer selects weights $w_1, w_2, \ldots, w_N$ and computes the average of the substance mixture $w_1A_1 + w_2A_2 + \ldots + w_NA_N$ across the image of the surface. Undoubtedly, the weights $w_1, w_2, \ldots, w_N$ are bound by certain constraints. For example, it seems likely that $|w_1| + |w_2| + \ldots + |w_N| \leq 1$.

However, the mechanisms underlying the flexibility people exhibit in making texture-based judgments might be significantly more complicated than this. One possibility is that people may be able to combine visual substances nonlinearly in fashioning decision statistics. A more drastic possibility is that people can attentionally adjust the very computations performed by the neural arrays $A_1, A_2, \ldots, A_N$ themselves.

To our knowledge, only one study has directly investigated the role of attention in controlling texture judgments. Chubb and Talevich (2002) examined the flexibility of observers in synthesizing decision statistics for textures composed of elongated elements, half oriented slightly right of vertical, and the others slightly left of vertical. The oppositely oriented texture elements were intermixed on a mean gray background to generate ambiguously oriented textures.

In different (separately blocked) conditions, observers made different judgments about these textures (with feedback). In one condition, observers judged which orientation (up-right vs. up-left) had more total energy. In another condition, observers judged which orientation had greater bright-side energy (i.e., observers tried to judge which orientation had more energy, ignoring texture elements of negative contrast polarity). In an analogous condition, observers judged which orientation had greater dark-side energy.

Histogram contrast analysis was used to determine, in each condition, the impact exerted on judgments by texture elements of different contrasts. When judging which orientation had greater total energy, observers were equally sensitive to positive and negative contrast polarity information. However, when judging which orientation had greater bright-side energy, observers were influenced very little by texture elements of negative polarity, and vice versa. In addition, observers performed all three tasks very efficiently suggesting that (1) they have separate access to two visual substances that might be called positive-polarity orientation strength and negative-polarity orientation strength, and (2) they can synthesize decision statistics that linearly mix these two substances.

By contrast, when asked to make judgments that could not be performed using a linear mixture of these two substances, observers' performance efficiency plummeted, suggesting that they can synthesize *only* decision statistics that linearly combine positive-polarity and negative-polarity orientation strength.

IV. CHALLENGES

We have argued that everyday texture judgments require the extraction of some modest number of neural maps, each reflecting the distribution throughout the visual field of a particular visual substance. Our understanding of vision would be greatly deepened if we knew exactly how many substances there are, and what each map reflects. The functional characterization of blackshot illustrates how these questions can be addressed using histogram contrast analysis.

A complete catalogue of visual substances must be the cornerstone of any model of surface perception. Given such a catalogue, methods akin to those of Chubb and Talevich (2002) could be used to determine how people combine information from different substance maps in making visual judgments about surfaces.

References

Adelson, E. H., and Bergen J. R. (1991). The plenoptic function and the elements of early vision. *In* Michael S. Landy and J. Anthony Movshon. *Computational models of visual processing.* MIT Press, Cambridge, MA, US. 3–20.

Beck, J. (1966). Perceptual grouping produced by changes in orientation and shape. *Science* **154**, 538–540.

Chubb, C. (1999). Texture-based methods for analyzing elementary visual substances. *J. Math, Psych.* **43**, 539–567.

Chubb, C., Econopouly, J., and Landy, M. S. (1994). Histogram contrast analysis and the visual segregation of IID textures. *Journal of the Optical Society of America A* **11**, 2350–2374.

Chubb, C., and Landy, M. S. (1991). Orthogonal distribution analysis: A new approach to the study of texture perception. *In* Michael S. Landy and J. Anthony Movshon, *Computational models of visual processing.* MIT Press, Cambridge, MA, US. 291–301.

Chubb, C., Landy, M. S., and Econopouly, J. (2004). A visual mechanism tuned to black. *Vision Res.* **44**, 3223–3232.

Chubb, C., and Nam, J-H. (2000). The variance of high contrast texture is sensed using negative half-wave rectification. *Vision Res.* **40**, 1695–1709.

Chubb, C., and Talevich, J. (2002) Attentional control of texture orientation judgments. *Vision Research* **42**, 311–330.

Julesz, B. (1962). Visual pattern discrimination, *IRE Transactions of Information Theory* **IT-8**, 84–92.

Julesz, B. (1975). Experiments in the visual perception of texture. *Scientific American* April, 34–43.

Nam, J-H., and Chubb, C. (2000). Texture luminance judgments are approximately veridical. *Vision Res.* **40**, 1677–1694.

Robson, J. G. (1980). Neural images: The physiological basis of spatial vision. *In* C. S. Harris, *Visual Coding and Adaptability*, 177–214. Erlbaum, Hillsdale, NJ.

CHAPTER

78

Mechanisms of Perceptual Learning

Barbara Anne Dosher and Zhong-Lin Lu

ABSTRACT

Perceptual learning—improvement in the performance of a perceptual task as a function of practice or training—is a widely observed phenomenon that may have important practical and theoretical consequences. Perceptual learning may reflect plasticity in different levels of perceptual analysis, including changes in early visual, auditory, or somatosensory cortices, as well as higher-order changes in the weighting of information in task performance. Perceptual learning, as distinct from cognitive learning or strategy selection, often exhibits significant specificity to the trained stimuli or tasks, and is assessed by transfer (or, conversely, generalization) tests. At a behavioral level, the effects of perceptual learning on an observer's performance are characterized by external noise tests within the framework of noisy ideal observer models. In visual perceptual tasks, behavioral analysis, combined with evidence from neuroscience, supports perceptual learning at several levels that has the function of improving two separable mechanisms: tuning of the task relevant perceptual template (external noise exclusion) and enhancing the stimulus (reducing absolute threshold). These two mechanisms of improvement are separable in certain circumstances, but often coexist, albeit with decoupled magnitude. Many improvements due to perceptual learning reflect retuning through reweighting of unchanged early sensory representations.

I. PERCEPTUAL LEARNING

Large improvements in performance on even the simplest perceptual tasks as a result of practice or training have been observed in adult humans in virtually every sensory modality. Improvement in performance as a result of practice can have a profound impact on the speed or accuracy of performance in perceptual tasks. Perceptual learning has been documented in a wide variety of tasks, but the mechanism(s) by which performance is improved have been more difficult to identify. Transfer of perceptual learning to modified forms of the same task or to different related tasks has been the primary tool for discovering what is learned and inferring the physiological basis of that learning (Fahle and Poggio, 2002; Fine and Jacobs, 2002). Recently, however, transfer methods have been augmented by the use of observer models and external noise tests that identify more precisely the consequences and mechanisms of perceptual learning (Dosher and Lu, 1998; Dosher and Lu, 1999). The focus in this review is on visual perceptual learning.

A. Perceptual Learning in Visual Tasks

Since Gibson's (1969) influential review, learning effects in adults have been reported for the detection or discrimination of visual gratings, stimulus orientation judgments, motion direction discrimination, texture discrimination, time to perceive random dot stereograms, stereoacuity, hyperacuity, and vernier tasks (see Ahissar and Hochstein, 1998; Dosher and Lu, 1999 for reviews). Improvements in performance are claimed to reflect perceptual learning, as opposed to cognitive learning, strategy selection, motor learning, or automatization of stimulus-response relations (Schneider and Shiffrin, 1997) whenever the performance is shown to be specific to either a retinal location or to a basic stimulus dimension such as orientation or spatial frequency (Karni and Sagi, 1991).

B. Specificity of Perceptual Learning

Specificity to a retinal location, or to a stimulus feature such as orientation does provide a strong argument for the perceptual nature of learning. However, it is more difficult to unambiguously identify the mechanism of perceptual learning. It is more difficult still to infer the functional locus of that learning (Dosher and Lu, 1998; Mollon and Danilova, 1996), although strong claims have been made about plasticity in early visual areas (Karni and Sagi, 1991). Perceptual learning has been evaluated for specificity (or conversely, transfer) to retinal position, eye of origin, orientation or spatial frequency, and retinal size. Specificity of learning to retinal position is often observed in texture discrimination, phase discrimination, orientation discrimination, and visual search (see Fahle and Poggio, 2002). Position specificity is generally evaluated by visual quadrant (Dosher and Lu, 1999; Karni and Sagi, 1991); finer location specificity of one degree or less has also been reported in several tasks. Similarly, orientation discrimination and motion direction discrimination are at least partially specific to orientation, although sometimes transfer occurs for mirror reversed orientations or homologous locations across the midline. Whether task specificity or generalization obtains in a task has recently been postulated to depend on task difficulty (Ahissar and Hochstein, 1997; Liu and Vaina, 1998).

Such location specificity is associated by some (Karni and Sagi, 1991) with early locus in the visual system, perhaps V1, with small receptive fields (but see Mollon and Danilova, 1996 for a critique). Orientation specificity is also often associated with early visual areas (V1, V2). Specificity to the trained eye has been seen in a very few cases, and this also specifies early visual cortex as the relevant region of perceptual coding. Specificity of perceptual learning to a retinal location or orientation does implicate representations early in the visual system, but the conclusion that such specificity implicates plasticity or perceptual retuning in those early visual areas is by no means obligatory (Dosher and Lu, 1998; Dosher and Lu, 1999; Mollon and Danilova, 1996). Instead, it may be the connections between the early visual system representations and decision processes at higher levels that embody system plasticity. This issue is relevant in the consideration of the physiological evidence.

C. Task Compatibility

Protocols for learning two or more interrelated perceptual tasks may also provide constraints on the inferred level of perceptual learning. If perceptual learning has retuned the neurons representing certain orientations in V1, for example, then transfer to a nearby but different orientation should be difficult, and retesting of the original orientation after new training on a second orientation should show alteration in performance. However, two apparently conflicting sets of targets can sometimes be learned either successively or simultaneously without significant interference. This implies that, although the relevant stimulus features may be coded early in visual system, learning may consist of changes in connectivity from the output of those early areas to an interaction structure or to a decision structure (Dosher and Lu, 1998; Dosher and Lu, 1999).

II. MECHANISMS OF PERCEPTUAL LEARNING

A converging methodology for the investigation of perceptual learning characterizes the limitations in performance of the observer and then identifies the aspects of performance that have improved with practice or training. Observer models and external noise tests are useful for this purpose. They allow the characterization of the dependence of perceptual performance in zero noise (absolute threshold) and in relation to limits on performance from high external noise in the stimulus. Improvements in performance can then be classified as stimulus enhancement improvements in zero or low noise or as external noise exclusion when the system is retuned to exclude limiting external noise.

A. Observer Models

Perceptual performance is limited by such factors as intrinsic stimulus variability, receptor sampling errors, randomness of neural responses, and loss of information during neural transmission. At an overall system level, these inefficiencies can be quantified in terms of a noisy ideal observer limited by *equivalent* internal noise—random internal noise necessary to produce the degree of inefficiency exhibited by the perceptual system (e.g., Lu and Dosher, 1999; Pelli, 1981). The amount of *equivalent* internal noise is estimated by systematically manipulating the amount of *external* noise (like TV snow or auditory white noise) added to the signal stimulus and observing how threshold—signal stimulus energy required for an observer to maintain a given performance level—depends on the amount of external noise (see Lu and Dosher, 1999 for a review). These methods characterize the overall limitations of the perceptual system, and allow comparisons of the efficiency of the perceptual system in different perceptual tasks. In fact, specification of internal noise has become a requirement of any computational model of human perception (Sperling, 1989).

The noisy perceptual template model (PTM) (Lu and Dosher, 1999) includes (1) a perceptual template with signal stimulus gain β; (2) a power function nonlinearity γ, (3) a multiplicative internal noise N_{mult} that is proportional to the energy in the stimulus; (4) an additive internal noise N_{add}, that accounts for absolute threshold; and finally, (5) a decision process that operates on the noisy internal representation of the stimulus. The amount of noise in the external stimulus is N_{ext}.

This model leads to a fundamental signal to noise equation for the observer system with three possible mechanisms of learning:

$$d' = \frac{(\beta c)^{\gamma}}{\sqrt{A_f^{2\gamma} N_{ext}^{2\gamma} + A_m^2 N_{mult}^2 \left((\beta c)^{2\gamma} + A_f^{2\gamma} N_{ext}^{2\gamma}\right) + A_a^2 N_{add}^2}}$$

The three learning improvement factors are A_f ($0 \leq A_f \leq 1$), which reduces external noise N_{ext}, A_a ($0 \leq A_a \leq 1$), which reduces internal additive noise N_a, and A_m ($0 \leq A_m \leq 1$), which reduces internal multiplicative noise N_m. This equation is rewritten to give the contrast threshold for a criterion d':

$$c_\tau = \frac{1}{\beta}\left[\frac{(1+A_m^2 N_{mult}^2)(A_f N_{ext})^{2\gamma} + (A_2^2 N_{add}^2)}{1/d'^2 - A_m^2 N_m^2}\right]^{\frac{1}{2\gamma}}$$

The PTM observer model, its relation to earlier linear observer models, and tests for mechanism mixtures (Lu and Dosher, 1999), are described in more detail in chapter 79, which develops the mechanisms of attention within the same framework.

B. Mechanism Signatures

There are three distinct mechanisms of perceptual learning: *stimulus enhancement*, *external noise exclusion*, and changes in *gain control*. These mechanisms are analogous to three aspects of signal processing: amplification, filtering, and gain control modification. Each of the three mechanisms has a key signature in perceptual task performance.

Learning through external noise exclusion is seen as improvement (A_f) in the ability to filter out external noise added to the stimulus, resulting in reduced thresholds in high external noise. It reflects learning to focus on the appropriate time, spatial region, and/or feature content of the signal stimulus. Learning through stimulus enhancement is seen as improvement (A_a) in the amplification of the stimulus,

resulting in reduced thresholds in zero or low external noise (e.g., absolute threshold). Learning through multiplicative noise reduction or through a change in nonlinear transducer are learning-induced changes (A_m) in gain control, and would be seen in all external noise conditions.

Changes in external noise exclusion and in stimulus enhancement both show effects that are the same (on a log scale of performance) regardless of the criterion performance level used to define threshold, c_τ. Changes in gain control, and specifically in multiplicative internal noise, lead to effects that appear larger (on a log scale) at higher criterion performance levels (Dosher and Lu, 1999). To date, changes in gain control due to perceptual learning have not been observed, but both isolated cases of external noise exclusion and mixtures of external noise exclusion and stimulus enhancement have been observed in perceptual learning.

C. Observed Mechanisms

Perceptual learning may improve external noise exclusion (filtering), stimulus enhancement (amplification), or both. When both forms of improvement occur, one mechanism of learning may be more effective than the other. Consider several examples.

Figure 78.1 shows an example (Dosher and Lu, 1998; Dosher and Lu, 1999) in which both mechanisms of learning are prominent, although the magnitudes are partially decoupled. Performance in an orientation discrimination task in visual periphery was dramatically improved with practice. Figure 78.1a shows the spatial layout of the stimulus, which used a dual-task format: the character 5/S was discriminated in a rapid stream of characters at fovea, whereas the orientation (top tilted ±12 deg from vertical) of a small sine wave patch (Gabor) was discriminated in the periphery. Contrast thresholds were measured using two different adaptive staircases (see Figs. 78.1b and c) yielding 70.3% and 70.7% accuracies. Performance was measured in a range of external noise conditions (see Fig. 78.1d).

The threshold signal contrasts depend upon criterion accuracy, external noise level, and practice (see Fig. 78.1e). Higher criterion accuracy demands higher contrast thresholds. In the high-noise region where external noise is the limiting factor in performance, contrast thresholds increase with increasing external noise. Practice reduces contrast thresholds by a downward vertical shift (in the log) with practice in both zero and in high external noise (albeit with slightly different magnitudes). We also observed a shift relationship between thresholds at the two performance criteria. These strong shift properties in the log

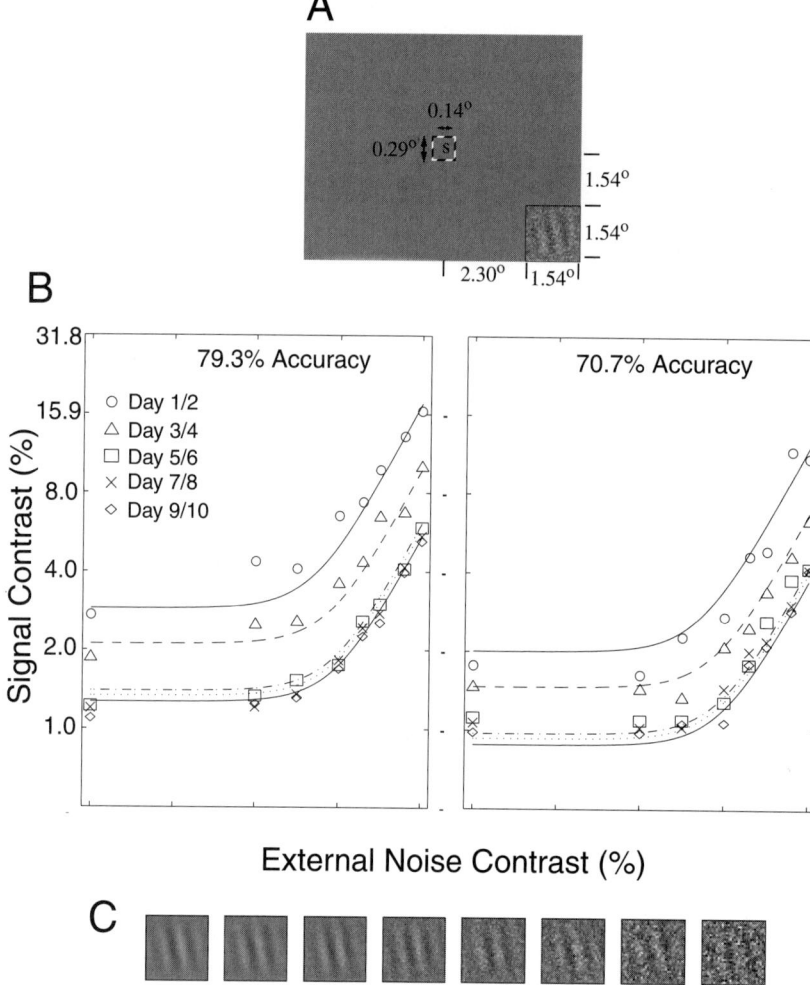

FIGURE 78.1 A perceptual learning task using the external noise paradigm. **A.** Spatial layout of the task, including the peripheral orientation discrimination Gabor stimulus, and a central letter stimulus for a secondary task. **B.** Contrast threshold (Gabor signal contrast corresponding to the criterion accuracy) as a function of the external noise in the stimulus. Threshold is a systematic function of criterion, external noise, and practice (data from Dosher and Lu, 1998). **C.** Examples of a signal of constant contrast embedded in increasing amounts of external noise.

contrast threshold as a function of criterion are a special characteristic of the perceptual template model. These results were subsequently replicated in face and in texture identification tasks (Gold, Bennett, and Sekuler, 1999).

Other studies using the external noise paradigm and framework have clearly documented the independence of these two mechanisms of learning in high and low external noise. For example, in certain conditions, such as training orientation discrimination in fovea, external noise exclusion is isolated as a separate mechanism. Figure 78.2 illustrates the results of practice in an orientation discrimination task at the fovea. Oriented sinewave patches (Gabors) of 45° ± 8° were discriminated (Lu and Dosher, 2004). This result is important because it isolates learning to exclude high external noise as distinct from enhancement of the stimulus. This rules out explanations of perceptual learning as an improvement in calculation efficiency.

D. Attention in Perceptual Learning

Many behavioral tests of perceptual learning in vision require the observer to maintain fixation while carrying out the critical perceptual task(s) in the periphery. Sometimes (Dosher and Lu, 1998; Karni and Sagi, 1991), fixation is ensured by the presence of a task at fovea. This requires the division of perceptual

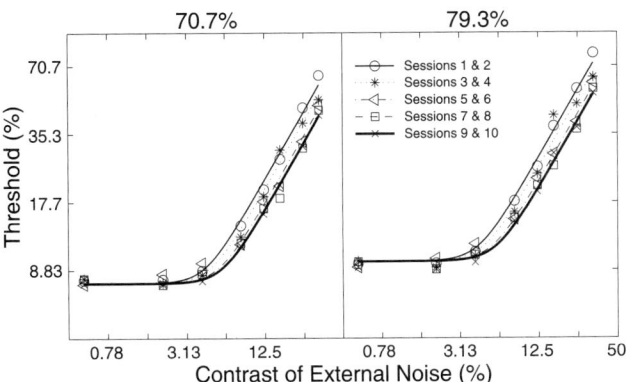

FIGURE 78.2 Perceptual learning in fovea that isolates external noise exclusion, reflecting learning about when, where, and what to look for in the target stimulus. These data (Lu and Dosher, 2004) measured the threshold signal contrast to discriminate the orientation (45° ± 8°) of a sine wave patch at fovea at two different performance criteria.

resources across the task at fixation and the task in the periphery, and/or division between overt direction of the eye and attention. The importance of learning to attend to the perceptual task, especially when this involves sharing resources between fovea and periphery, deserves further systematic investigation. Even in cases where fixation and perception are not divided, learning has been reported to be restricted to the relevant or attended feature of a perceptual stimulus (Ahissar and Hochstein, 1998), whereas irrelevant variations that later become the aspect to be judged may benefit little if at all from exposure. Alternatively, some researchers have claimed that perceptual sensitivity is improved by practice even when the relevant stimuli are subliminal adjuncts to the primary task, although learning of unattended stimuli appears to require correlation with an attended task (Seitz and Watanabe, 2003).

III. RETUNING VERSUS REWEIGHTING

Visual perceptual learning may reflect plasticity in early visual areas that alters the basic sensory coding of the stimulus (Karni and Sagi, 1991) or it may reflect plasticity of the connections between the sensory coding of the stimulus and decision units that may reside at higher cortical levels (Dosher and Lu, 1998; Mollon and Danilova, 1996). If retinal or stimulus feature specificity is demonstrated, then this implicates the selection of early visual representations in the connections to decision, but it does not require that early visual representations be altered by training. Plasticity

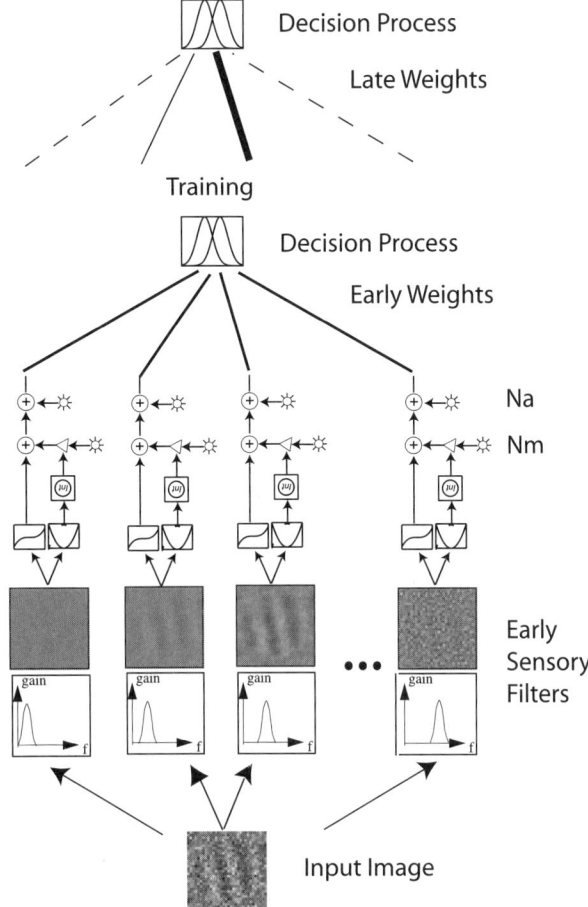

FIGURE 78.3 Schematic diagram of the perceptual system and perceptual learning via reweighting. The visual input is represented in early sensory filters, and this information is then fed to decision units. Perceptual learning that is specific to an attribute of early sensory filters (e.g., location, orientation, scale) may reflect retuning of the sensory representation (not shown), or a constant sensory representation with changes in the feed-forward weights to decision. After Dosher and Lu (1998).

based on reweighting has the additional advantage that early visual representations are left unchanged, so that perceptual learning of one task need not impact on another task and is hence compatible with observed abilities for compatible training of multiple perceptual tasks.

A schematic of reweighting due to practice is shown in Fig. 78.3. Early sensory filters specified for different spatial frequencies (and orientations) improve their connections to decision without changing the properties of the filters themselves. Perceptual learning may consist in learning optimal weights for inputs from early sensory coding. Network models are beginning to be developed to account for perceptual learning in specific tasks (Fahle and Poggio, 2002).

IV. PHYSIOLOGICAL CORRELATES OF LEARNING

Alteration of early sensory representations through training or exposure has been widely reported in auditory (Bakin and Weinberger, 1990) and somatosensory cortex (Recanzone, Merzenich, Jenkins, Grajski, and Dinse, 1992). In these cases, repeated experience with certain stimuli causes either narrowed tuning of the early cortical cells, or recruitment of additional cells into responding to the trained stimuli.

In the case of visual perceptual learning, however, evidence for substantial alterations in early visual cortices with training or practice is lacking. In three separate studies (Crist, Li, and Gilbert, 2001; Ghose, Yang, and Maunsell, 2002; Schoups, Vogels, Qian, and Orban, 2001) in awake behaving monkeys, behavioural performance has improved significantly, these improvements are to some degree specific, yet there is little (Schoups et al., 2001) or no (Crist et al., 2001; Ghose et al., 2002) evidence for significant changes in tuning of cells in V1 or V2. These results suggest that, minimally, perceptual learning must involve plasticity at many levels, certainly at a range of levels above V1 and V2. On the other hand, location and orientation specificity of learning identifies these early visual areas as the locus of the relevant sensory representations that are read-out to higher levels.

A combined analysis and correlation of physiological evidence, sophisticated behavioral analysis—including observer models and external noise tests—and explicit (network) models of the learning process in future developments will allow a full specification of perceptual learning at the neural and the behavioural levels.

V. IMPACT OF PERCEPTUAL LEARNING

The mechanisms of perceptual learning have wide ranging implications for the nature of plasticity in adult systems. The existence of perceptual learning also has possible practical implications for many technical areas of assisted perception and recovery from injury or damage. For example, perceptual learning processes can be very important following the introduction of hearing aids or cochlear implants. Technological environments with assisted perception displays may require training and perceptual learning to support optimized performance. How to optimize perceptual learning remains an important question for many practical applications.

References

Ahissar, M., and Hochstein, S. (1997). Task difficulty and the specificity of perceptual learning. *Nature* **387**, 401–406.

Ahissar, M., and Hochstein, S. (1998). Perceptual learning. In V. Walsh and J. Kulikowski, Eds., *Perceptual constancy: Why things look as they do.* (pp. 455–498). Cambridge University Press, Cambridge.

Bakin, J. S., and Weinberger, N. M. (1990). Classical-Conditioning Induces Cs-Specific Receptive-Field Plasticity in the Auditory-Cortex of the Guinea-Pig. *Brain Research* **536**, 271–286.

Crist, R. E., Li, W., and Gilbert, C. D. (2001). Learning to see: Experience and attention in primary visual cortex. *Nature Neuroscience* **4**, 519–525.

Dosher, B., and Lu, Z.-L. (1998). Perceptual learning reflects external noise filtering and internal noise reduction through channel reweighting. *Proceedings of the National Academy of Sciences of the United States of America* **95**, 13988–13993.

Dosher, B., and Lu, Z.-L. (1999). Mechanisms of perceptual learning. *Vision Research* **39**, 3197–3221.

Fahle, M., and Poggio, T. (2002). *Perceptual learning.* MIT Press, Cambridge.

Fine, I., and Jacobs, R. A. (2002). Comparing perceptual learning across tasks: A review. *Journal of Vision* **2**, 190–203.

Ghose, G. M., Yang, T., and Maunsell, J. H. R. (2002). Physiological correlates of perceptual learning in monkey V1 and V2. *Journal of Neurophysiology* **87**, 1867–1888.

Gibson, E. (1969). *Principles of perceptual learning.* Appleton-Century-Crofts, New York.

Gold, J., Bennett, P. J., and Sekuler, A. B. (1999). Signal but not noise changes with perceptual learning. *Nature* **402**, 176–178.

Karni, A., and Sagi, D. (1991). Where Practice Makes Perfect in Texture-Discrimination—Evidence for Primary Visual-Cortex Plasticity. *Proceedings of the National Academy of Sciences of the United States of America* **88**, 4966–4970.

Liu, Z., and Vaina, L. M. (1998). Simultaneous learning of motion discrimination in two directions. *Cognitive Brain Research* **6**, 347–349.

Lu, Z.-L., and Dosher, B. (2004). Perceptual learning retunes the perceptual template. *Journal of Vision* **4**, 44–56.

Lu, Z.-L., and Dosher, B. A. (1999). Characterizing human perceptual inefficiencies with equivalent internal noise. *Journal of the Optical Society of America A Special Issue: Noise in imaging systems and human vision* **16**, 764–778.

Mollon, J. D., and Danilova, M. V. (1996). Three remarks on perceptual learning. *Spatial Vision* **10**, 51–58.

Pelli, D. (1981). Effects of visual noise. PhD. dissertation. University of Cambridge, Cambridge, England.

Recanzone, G. H., Merzenich, M. M., Jenkins, W. M., Grajski, K. A., and Dinse, H. R. (1992). Topographic reorganization of the hand representation in cortical area 3b of owl monkeys trained in a frequency-discrimination task. *Journal of Neurophysiology* **67**, 1031–1056.

Schneider, W., and Shiffrin, R. M. (1997). Controlled and automatic human information processing: I. Detection, search, and attention. *Psychological Review* **84**, 1–66.

Schoups, A., Vogels, R., Qian, N., and Orban, G. (2001). Practising orientation identification improves orientation coding in V1 neurons. *Nature* **412**, 549–553.

Seitz, A. R., and Watanabe, T. (2003). Is subliminal learning really passive? *Nature* **422**, 36.

Sperling, G. (1989). Three stages and two systems of visual processing. *Spatial Vision* **4**, 183–207.

CHAPTER 79

Lateral Interactions between Targets and Flankers Require Attention

Elliot Freeman

ABSTRACT

What we see from moment to moment depends on the current sensory context, and also on the behavioural context of the task in which we are engaged. However, little is understood about how these different factors interact. Integration of visual context has often been studied using the phenomenon of lateral interactions, in which perceptual sensitivity and also neural activity in early visual cortex to a central oriented target is enhanced in the presence of collinear flanking patches (lined up to form a virtual contour). Our recent psychophysical studies suggest that task-directed attention to the flankers can directly modulate lateral interactions, affecting integration of the whole flanker-target configuration, rather than just the perceptibility of the individual parts. This attentional modulation is highly specific to collinear configurations and also dependent on tasks involving discrimination of global rather than local stimulus properties. Our results suggest that attention plays a critical role in the selective integration of task-relevant groups, directly modulating early visual mechanisms sensitive to collinear structures.

I. STIMULUS AND TASK-RELATED CONTEXT EFFECTS IN EARLY VISION

In everyday activity, the visual system routinely faces the problem of disentangling contours relating to objects of interest from those caused by reflections, shadows, or occlusions by other irrelevant objects. What we ultimately experience depends not only on stimulus-driven grouping processes, sensitive to the geometrical relationships between different parts of a scene, but also on task-driven attentional factors sensitive to which aspects of the scene are currently relevant. Grouping has traditionally been seen as a separate stage preceding grouping (Driver, Davis, Russell, Turatto, and Freeman, 2001). However, recent data suggest that grouping and selective processes can both operate at similarly early stages in visual processing (see Chapters 107 and 47).

Much has been learned about stimulus-determined context-integration, from the phenomenon of lateral interactions (Polat, 1999). For example, perceptibility of an oriented target element, and neural activity in early visual cortex (e.g., V1), can both be enhanced in the context of flanking elements (positioned outside the target cell's classical receptive field), when all are lined up to form a virtual contour (Kapadia, Ito, Gilbert, and Westheimer, 1995). Evidence suggests that this collinearity effect reflects spreading of excitatory impulses via horizontal connections, which tend to connect V1 neurones along lines parallel to their orientation preference (Callaway, 1998).

We devised a method to test whether this phenomenon of lateral interactions might be modulated as a function of changing attentional demands (Freeman, Sagi, and Driver, 2001). Our results suggest that the mechanisms by which local contextual components are integrated into global groups can indeed be selectively modulated by attention, depending on the relevance of specific perceptual groups to the task at hand. Given the neural underpinnings of lateral interactions, any such modulation should shed light on the mechanisms by which grouping and selective processes interface with each other.

II. BASIC DUAL-TASK DUAL-AXIS PROCEDURE

Our basic method used a novel dual-axis stimulus, with four suprathreshold flankers surrounding a central near-threshold target (see Fig. 79.1a). The orientation of the central target was collinear with the pair of flankers along just one of the two axes (though with small misalignments, see later), but was orthogonal to the orientation of each of the flankers in the pair along the other axis. The primary task was to indicate in which of the two successive intervals the near-threshold central target was presented. To direct covert attention to one or other flanker-pair, we imposed a secondary Vernier discrimination task to be performed concurrently with the central target detection task. Subjects were instructed to judge the relative position of the flankers on just one prespecified axis, indicating in which of two successive intervals they were displaced in a prespecified direction. The two flankers on the other axis were to be ignored. Two conditions may thus be compared: attending to the collinear flankers versus attending to the orthogonal flankers, while the stimuli and attention to the central target are kept constant.

This approach differs from most others, where the intensity or spatial distribution of attention is manipulated over the whole stimulus display, for example by varying the difficulty of a secondary task or by directing attention to one or several spatial positions (e.g., Gilbert, Ito, Kapadia, and Westheimer, 2000; see also Chapters 76 and 74). Such manipulations might alter local processing of display elements in a number of ways, affecting spatial uncertainty, spatial resolution, or internal noise levels within restricted regions (see Chapter 73). By controlling these factors it becomes possible, in principle, to measure specifically how attention influences lateral interactions between flankers and target, while eliminating the possible confounding effect of local attentionally driven changes in the processing of the target.

This method was successful in demonstrating that lateral interactions depend on attention to the flankers, in the absence of any changes in the stimulus (Freeman et al., 2001). The graph in Fig. 79.2a (filled symbols) shows target contrast thresholds averaged across 16 subjects (all naïve except for EF). Central target detection was facilitated (i.e., target contrast thresholds were reduced) when the collinear flankers were attended, but not when the orthogonal flankers were attended instead, despite the continued presence of ignored collinear flankers. This result indicates that attention can modulate the lateral interactions phenomenon. On their own, however, these data would be consistent with a range of different accounts for how this attentional modulation might be achieved. Our new results help to narrow down these possibilities, characterizing the functional role played by attention, its constraints, and the likely mechanisms of modulation.

III. STRENGTH AND SPECIFICITY OF ATTENTIONAL MODULATION

A. Collinearity vs. Local Feature Similarity

The defining property of classic lateral-interaction phenomena is their sensitivity to *collinear* configurations in particular. If our attentional effect specifically modulates the underlying mechanisms of lateral interactions, it should likewise be configuration-specific, found only when stimulus elements have strictly collinear arrangements. An alternative account of the attentional effects, which would not be configuration-specific, can be derived from current influential accounts in terms of feature-based attention (e.g., Saenz, Buracas, and Boynton, 2002; Chapter 49). On such accounts, attention to a given feature at one location automatically enhances perception of all similar features throughout the visual field. This might explain our initial results, given that in collinear configurations the central target has the same local orientation as each of the collinear flankers. The central target might therefore in principle have been facilitated by attending just the local orientation of the relevant flankers, rather than by a global collinear arrangement *per se*.

A test of this feature-specific account was recently reported in an independent study by Saenz and Boynton (2002). This study used stimuli in which target and flankers on one axis were arranged parallel to each other (e.g., | | |), so that they had the same local orientations but no global collinearity. In apparent support for the feature-specific hypothesis, target detection improved slightly when this axis was attended, compared to attention to the other orthogonal axis. However, some past studies have reported target facilitation even with this parallel target-flanker configuration, though in the absence of attentional manipulations and with only one pair of flankers present (see Polat, 1999). We have tested the feature-specific account using a configuration with which facilitation has never previously been found (Freeman, Driver and Sagi, 2004). We disrupted collinearity by rotating all patches by 45 degrees, so that local orientation similarity was preserved on one axis (see Fig. 79.1b, right) while the other axis remained orthogonal

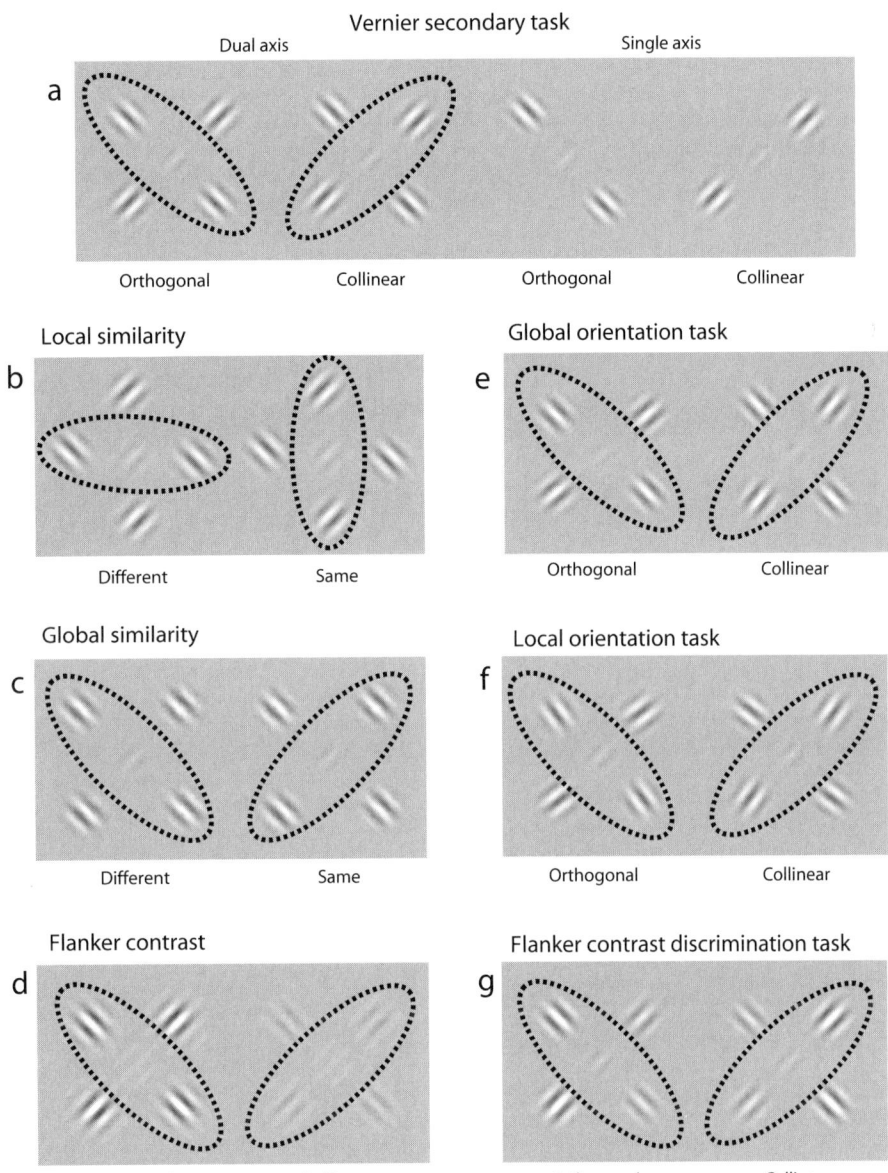

FIGURE 79.1 Sample stimuli, composed of a central low-contrast target Gabor patch (Gaussian envelope distribution and carrier wavelength both equal to 0.15 degrees), surrounded by one (single-axis) or two (dual-axis) pairs of flankers (flanker-target separation approx. four wavelengths). Dotted ellipses illustrate the two attentional conditions. Observers perform a task on one flanker-pair while ignoring the other, at the same time as detecting the central target (2 × 2-interval forced-choice, with 80 ms display intervals and 500 ms ISI). **A.** Stimuli for the Vernier task, with the two leftmost images depicting dual-axis stimuli in attend-orthogonal and attend-collinear conditions; the rightmost two images show single-axis orthogonal and collinear stimuli. **B.** Local similarity manipulation: target has either the same (right) or 90 degrees different (left) orientation to the flankers attended for the Vernier task. **C.** Global similarity: target has either the same or different orientation as the virtual contour connecting the task-relevant flankers. **D.** Flanker contrast manipulation: same as in **A** but with high or low contrast flankers. **E.** Global orientation task: observers discriminate between two rotations of a specified flanker-axis. **F.** Local orientation task: observers discriminate rotation of flankers around their own centre. **G.** Flanker-contrast discrimination task: observers decide which flanker of a relevant pair has the higher contrast.

to the target orientation (see Fig. 79.1b, left). We found that the attentional effect was completely eliminated for five out of six subjects (subject mean shown in Fig. 79.2b), compared to our standard stimuli (Figs. 79.1a and 79.2a) containing collinear elements. Feature-specific attention therefore cannot account for all of the attentional modulation effects found using the present lateral interactions paradigm.

B. Collinearity vs. Global Feature Similarity

We tested the specificity of the attentional modulation for collinearity further by manipulating similarity at the global level. Note that in a collinear configuration the central target shares not only the local orientation of the flankers, but also the global orientation of the virtual contour along which the target and collinear flankers are all aligned. A generalized feature-specific account might argue that attending to the collinear flankers improves detection of targets aligned with this global orientation. We tested this by rotating the flankers on one axis by 90 degrees, so that the target still had the same global orientation as the virtual contour drawn through this axis, but without having local orientation similarity (see Fig. 79.1c, right, and control stimuli on the left). This disruption of collinearity again resulted in a complete elimination of

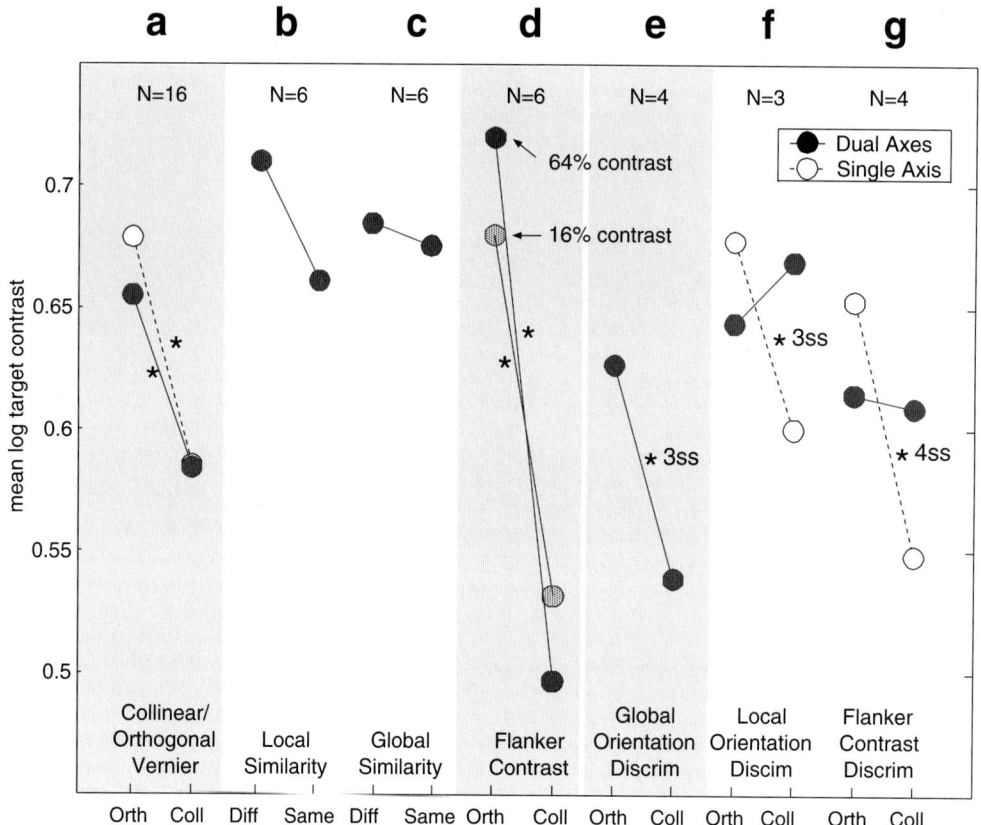

FIGURE 79.2 Target contrast thresholds averaged across observers, in log units, estimated using method of constant stimuli (except for **D**; see text). Refer to Figs. 79.1a–g for example dual-axis stimuli corresponding respectively to conditions labelled A–G here. Shaded regions indicate conditions where a significant attentional effect was found for dual-axis stimuli. **A.** Standard attend-collinear versus attend-orthogonal condition. All 16 observers experienced this plus one or more of the other conditions, but with the exception of EF were naïve to the purpose of the studies. **B, C.** Attentional effects were abolished following disruption of target-flanker collinearity by manipulating stimulus geometry at global or local levels. **D.** Similar stimuli and task to **A**, showing that the significant attentional effect is not diminished by increased flanker contrast. Differences in threshold-estimation method might account for the increased apparent effect-size for just this condition; see main text. Asterisks indicate significant difference between conditions ($p < .01$) in group analyses. **E–G.** Dual-axis attentional effect depends critically on the specific task performed on the flankers, while the single-axis stimulus effect remained unaffected. Here, statistical reliability is assessed by individual: numbers by asterisks indicate how many observers in each sample showed significant differences between conditions ($p < .01$).

attentional modulation effects (means for six subjects in Fig. 79.2c).

Neither local nor global feature similarity can explain away these critical results. We therefore may conclude that attention can specifically modulate sensitivity to collinear configurations.

C. Strength of Modulation

Although attention may modulate sensitivity to collinearity, this does not necessarily imply that attention directly modulates the integrative lateral interactions mechanisms themselves that underlie this sensitivity. Attention might instead just affect how information relating to collinear structures is subsequently used after the initial analysis of collinearity is complete. For example, one possibility is that the target reciprocally influences the contrast response to collinear flankers (Solomon, Watson, and Morgan, 1999). This might provide an extra cue in the flankers for detecting the target, which might be detectable only when the flankers are attended. Another possibility is that the secondary Vernier task is easier with collinear flankers, leading to a redistribution of attentional resources in favour of the primary target detection task (though note there was no evidence of consistent configuration-dependent variations in Vernier accuracy in any of our studies). Both of these accounts predict attentional modulation that is specifically dependent on collinearity. However, they assume no direct attentional interaction with the mechanisms underlying flanker-target facilitation. Such indirect attentional effects therefore should ride on top of stimulus-driven effects of collinearity. If this were the whole explanation of the observed modulation, it should always be possible to measure some residual facilitation from collinear flankers even when they are unattended, because the mechanisms underlying this facilitation would still be functioning as normal.

Evidence from two sources indicates that there can be no such residual facilitation. First, in our original studies, we compared dual-axis stimuli (filled symbols in Fig. 79.2a) with single-axis stimuli (see Fig. 79.1a, right) in which only one collinear or one orthogonal pair of flankers was displayed (open symbols). The physical presence or absence of task-irrelevant collinear flankers made no difference to target detection. The effect of ignoring collinear flankers (i.e., reduction of their impact on the target) was therefore equivalent in size to physically removing them altogether. Second, in the previous experiment manipulating global congruity (Freeman et al., 2004), we could compare the influence of *unattended* flankers that are either collinear or noncollinear with the target (compare leftmost points of Figs. 79.2a and 79.2c and corresponding stimuli left of Figs. 79.1a and 79.1c). This comparison showed there was no significant residual target facilitation from ignored collinear flankers, indicating that attentional exclusion of their influence is complete. This strong modulation appears consistent with attention directly modulating the processing of collinear configurations, rather than merely filtering the output of this processing at some later stage.

IV. MECHANISM OF ATTENTIONAL MODULATION

Our psychophysical evidence for attentional modulation of lateral interactions appears consistent with recent influential single-cell studies, which showed not only that the response of neurons in primary visual cortex to an oriented target falling inside the classical receptive field (CRF) can be influenced by a collinear flanker outside the CRF, but also that this influence can be modulated by manipulations of spatial attention to the target (Gilbert et al., 2000). This neurophysiological result was taken to imply that attention can modulate flanker-target interactions by weighting the horizontal lateral connections between cells responding to target and flankers.

Such a *connection-weighting* account is not the only possible explanation of these attentional modulation effects, however. A mere shift in the distribution of spatial attention may be sufficient to modulate perceptual sensitivity to a stimulus, by amplifying the response of cells whose CRF includes the attended stimulus (Treue, 2001). Attended elements would thus have a greater effective contrast. Unlike the Gilbert et al. (2000) study, our experiments by design kept spatial attention to the target constant, ensuring that any observed attentional effects most likely reflect modulation of flanker effects on the target. However, our attentional manipulation did involve directing spatial attention toward or away from the collinear flankers, which on the previous account might have modulated the local contrast response to unattended versus attended flankers. Thus, the response to unattended flankers might have been suppressed to such an extent that they no longer provided any facilitating input to the central target. This *flanker-modulation* account would imply that attention does not change the lateral-interaction mechanisms themselves, but just controls the input to them.

We devised a psychophysical method for distinguishing between these connection-weighting and flanker-modulation accounts (Freeman, Driver, Sagi, and Zhaoping, 2003). We derived divergent predictions from a known nonlinearity relating flanker contrast to psychophysical target detection threshold. Facilitation of central target detection by collinear flankers has been shown to increase to a maximum as flanker contrast crosses contrast threshold; thereafter, further flanker contrast increments produce no further improvements in target visibility (Polat, 1999). If the flanker-modulation account of our attentional effects were true, then ignored collinear flankers would have to be suppressed to their contrast threshold before their facilitation of the target is appreciably reduced. This account therefore predicts elimination of the attentional effect when physical flanker contrast is too high to be suppressed to threshold levels by being ignored (see Fig. 79.1d for examples of stimuli with high and low flanker contrast). Figure 79.2d plots mean thresholds for six subjects, showing significant attentional effects at low flanker contrasts (16%), but with no significant reduction in the attentional effect at higher flanker contrast and (64%). Increasing flanker contrast further to 80% did nothing to reduce the attentional effect. Note that for this study only, thresholds were estimated from d' measured at fixed target contrast across a range of flanker contrasts, whereas all other studies employed the method of constant stimuli. This methodological difference might explain the increased effect-size apparent in the graph.

To explain this result, the flanker modulation hypothesis would have to assume that even at these high contrasts, flankers can still be suppressed to below their detection threshold when ignored. Such strong modulation seems highly implausible in the light of existing data on attention and contrast (Carrasco et al., 2000; Lee, Itti, Koch, and Braun, 1999; Zenger, Braun, and Koch, 2000). On the other hand, the connection-weighting account described earlier predicts exactly the independence of the attentional effect from suprathreshold changes in flanker contrast that we observed. According to this account, attention should amplify or attenuate the impact of flanker on the target by the same amount regardless of the actual contrast of the flankers. Our results thus allow us to conclude that attention modulates integration of the whole flanker-target configuration, rather than just the processing of the individual parts, thus providing the first clear psychophysical evidence that attention can directly influence contextual integration in early vision.

V. FUNCTIONAL ROLE OF LATERAL INTERACTIONS AND ATTENTIONAL MODULATION THEREOF

A. Task Specificity vs. Object-based Attention

Lateral-interaction mechanisms commonly are assumed to perform the function of grouping local elements into a global pattern. If attention directly modulates lateral interactions, then this modulation might be strongest whenever observers are required to judge the properties of the global pattern formed by the flankers, compared to when only local properties of the individual flankers are task-relevant. This task-specific function of attention would contrast with the less specific but influential notion of object-based attention (Driver et al., 2001). On such object-based accounts, all the visual properties of an attended object are extracted and integrated into global descriptions, whether these properties are individually task-relevant or not.

We have found that our attentional effect successfully generalized from the original Vernier flanker task to a *global orientation* discrimination task. In this task, flankers on each axis are positioned along a virtual contour that can have two global orientations, switching unpredictably between display intervals. Flankers therefore rotate around a single central origin, and observers decide in which interval they have a pre-specified rotation (see Fig. 79.1e). Mean thresholds are plotted in Fig. 79.2e, with significant attentional effects for three out of four observers tested. However, the attentional effect was completely abolished (i.e., no difference between attend-collinear and attend-orthogonal conditions; see means for three observers in Fig. 79.2f) given a task requiring purely *local orientation* discrimination of the relevant flankers, where each flanker rotates around its own centre (see Fig. 79.1f). A similar null effect of attention was observed for discrimination of relative flanker contrast (see Fig. 79.1g and filled symbols in Fig. 79.2g, for four observers) and also flanker colour (Freeman and Driver, in press), despite care being taken to ensure that both flankers had to be simultaneously attended. This task-dependency runs counter to the predictions of object-based attention models, in which attention to collinear flankers should be sufficient to encode all their local and global properties. Instead, it implies that the global properties of the whole target-flanker configuration may be encoded (via lateral interactions) only when these properties are specifically task-relevant. Furthermore, these results confirm the popular but hitherto untested assumption that lateral interactions play a critical role in perceptual grouping, challenging accounts in which this phenomenon is merely a by-product of local perceptual processing, resulting, for example, from signal averaging within spatial analysers (e.g., Solomon et al., 1999).

B. Dependence of Task-Specificity on Rival Flanker Axes

Interestingly, the preceding task-specific results were found only when two pairs of flankers are present, with one pair attended for the flanker task, and one pair ignored. The nature of the flanker task had no effect on target facilitation when only a *single* pair of flankers was present. Significant differences in threshold were found between single-axis collinear versus orthogonal stimuli effects for all observers tested (see open symbols in Fig. 79.2f and 79.2g). This reveals an important limitation of what top-down modulation by the imposed task can achieve: it cannot arbitrarily switch on or off lateral interactions whenever collinear stimuli are present. Rather, attentional modulation arises only when observers have to select between multiple sets of flankers. This result concurs with physiological data showing that weak attentional modulation of cell responses to an isolated stimulus can be greatly enhanced when other irrelevant stimuli are simultaneously presented, as if selective attention is biasing an active competition between them (Desimone and Duncan, 1995; see also Chapter 50). In the present case, attentional modulation may reflect top-down bias of the competition between alternative perceptual groupings (formed by each of the two flanking axes), in favor of one that is relevant.

VI. PHYSIOLOGICAL CORRELATES OF LATERAL INTERACTIONS AND ATTENTIONAL MODULATION

Given the known neurobiology of lateral interactions, the most likely mechanism for attentional modulation would involve direct modulation of the integrative horizontal connections in early visual cortex. Evidence from two new EEG studies supports such an early anatomical locus. First, we found early event-related potentials (ERPs) over the occipital midline (180–230 ms) are correlated with target facilitation from collinear versus orthogonal flankers in single-axis displays (Khoe, Freeman, Woldorff, and Mangun, 2004). Second we found a later component (180–230 ms) that is dependent on attention to collinear (versus orthogonal) context in dual-axis displays (Khoe, Freeman, Woldorff, and Mangun in preparation). The relatively late timing of these attention-related ERP effects is consistent with a role for top-down feedback in modulating the stimulus-driven response to collinear configurations.

VII. CONCLUSIONS

Our findings to date reveal both the power and the limitations of attention in modulating processing of collinear structures, which taken altogether allow us to be more precise about how and for what purpose task-driven attentional factors might interface with mechanisms of context integration in early vision.

First, the strong and specific modulation of collinear configurations, and the independence of attentional effects from flanker contrast, are all consistent with attention directly modulating the integration of target and flankers into a contour, rather than just affecting the visibility of the individual stimulus elements. Second, although it seems that task-directed attention might have the power to exert direct control over integrative circuitry in early vision, the scope of its influence is strongly constrained by the precise sensory and behavioral context in which the stimulus is encountered. The modulation we observed does not depend simply on where the subject is attending or to what particular features, but is highly specific to stimuli containing collinear elements, and then again only when performing a global task that arguably requires perception of these elements as a continuous contour with supervening global properties. A further important criterion may be the presence of competing stimuli or stimulus groupings to select between; in our case, two intersecting virtual contours.

These constraints hint at the likely functional role played by this top-down control of lateral interactions, in *selectively integrating* local image elements with their behaviorally relevant context. Thus, given a typical scene containing multiple overlapping contours, the task of selecting one object (e.g., looking at one's reflection in the window while ignoring the background) may be achieved by adjusting the neural circuitry so that only the relevant contour segments are integrated with each other, and other competing contours are excluded from the perceived group.

Acknowledgments

This research was supported by a project grant to EF from the BBSRC (UK) grant number 31/S13736. Thanks to Jon Driver for helpful criticism.

References

Callaway, E. M. (1998). Local circuits in primary visual cortex of the macaque monkey. *Annual Review of Neuroscience* **21**, 47–74.

Desimone, R., and Duncan, J. (1995). Neural mechanisms of selective visual attention. *Annual Review of Neuroscience* **18**, 193–222.

Driver, J., Davis, G., Russell, C., Turatto, M., and Freeman, E. (2001). Segmentation, attention and phenomenal visual objects. *Cognition* **80**, 61–95.

Freeman, E., Sagi, D., and Driver, J. (2001). Lateral interactions between targets and flankers in low-level vision depend on attention to the flankers. *Nature Neuroscience* **4**, 1032–1036.

Freeman, E., Driver, J., Sagi, D., and Zhaoping, L. (2003) Top-down modulation of lateral interactions in early vision. Does attention affect integration of the whole or just perception of the parts? *Curr Biol.* **13**, 985–989.

Freeman, E., Sagi, D., and Driver, J. (2004) Configuration-specific attentional modulation of flanker-target lateral interactions. *Perception* **33**, 181–194.

Freeman, E., Driver, J. (In press). Task-dependent modulation of target-flanker lateral interactions: effects of manipulating attention to spatial and non-spatial flanker attributes. *Perception and Psychophysics*.

Gilbert, C., Ito, M., Kapadia, M., and Westheimer, G. (2000). Interactions between attention, context and learning in primary visual cortex. *Vision Research* **40**, 1217–1226.

Kapadia, M. K., Ito, M., Gilbert, C. D., and Westheimer, G. (1995). Improvement in visual sensitivity by changes in local context: Parallel studies in human observers and in V1 of alert monkeys. *Neuron* **15**, 843–856.

Khoe, W., Freeman, E., Woldorff, M. G., and Mangun, G. R. (2004). Probing lateral interactions in early visual areas: an event-related potential study of collinear and orthogonal stimulus configurations. *Vision Research* **44(14)**, 1659–1673.

Khoe, W., Freeman, E., Woldorff, M. G., and Mangun, G. R. (In prep). Longer Latency Effects of Attention on Integrative Processes in Early Visual Areas.

Lee, D. K., Itti, L., Koch, C., and Braun, J. (1999). Attention activates winner-takes-all competition among visual filters. *Nature Neuroscience* **2**, 375–381.

Polat, U. (1999). Functional architecture of long-range perceptual interactions. *Spatial Vision* **12**, 143–162.

Saenz, M., Buracas, G. T., and Boynton, G. M. (2002). Global effects of feature-based attention in human visual cortex. *Nature Neuroscience* **5**, 631–632.

Saenz, M., and Boynton, G. M. (2002). Feature-based attention modulates lateral interactions in contrast detection. Society for Neuroscience, 32nd annual meeting.

Solomon, J. A., Watson, A. B., and Morgan, M. J. (1999). Transducer model produces facilitation from opposite-sign flanks. *Vision Research* **39**, 987–992.

Treue, S. (2001). Neural correlates of attention in primate visual cortex. *Trends in Neurosciences* **24**, 295–300.

Zenger, B., Braun, J., and Koch, C. (2000). Attentional effects on contrast detection in the presence of surround masks. *Vision Research* **40**, 3717–3724.

CHAPTER
80

Attention and Changes in Neural Selectivity

Scott O. Murray

ABSTRACT

Attention increases our perceptual awareness of the details of a stimulus, but the neurophysiological basis for this improvement is yet to be understood. One possibility is that attention increases the specificity of, or "sharpens", the population of neurons that encode a stimulus. We tested this idea by using adaptation of the fMRI signal to infer the amount of overlap in the neural populations representing changes in object rotation. These measurements were made both with and without attention directed to the objects. It was assumed that if attention sharpens the population response there would be less overlap in the neural populations representing two different object views, resulting in less adaptation and a greater measured signal. We observed a pattern of signal changes consistent with this prediction. The implications of these findings are discussed as well as the possible single-neuron mechanisms that could lead to greater population selectivity.

I. INTRODUCTION

It is well established that attention can increase the efficiency of information processing. For example, our ability to perceive the details of a visual stimulus can be significantly improved with attention (see Chapter 73). One way to characterize the perceptual improvements associated with attention is that the differences between two stimuli are more likely to be perceived with attention than without attention. How might these perceptual improvements be accomplished at the neural level? One strategy might be to maximize the uniqueness, or specificity, of the population of neurons encoding a stimulus by enhancing the contribution of those neurons that code for specific features of the stimulus and reducing the contribution of all other neurons in a population. In a recent study (Murray and Wojciulik, 2004) we examined this idea by measuring selectivity for object rotations under different attention conditions in the human lateral occipital complex (LOC), a cortical area previously shown to be important for processing object shape (Malach et al., 1995).

In contrast to previous studies that have addressed changes in neural selectivity at the single neuron level (e.g., Spitzer, Desimone, and Moran, 1988; Haenny and Schiller, 1988; McAdams and Maunsell, 1999; Treue and Martinez Trujillo, 1999), our study examined changes in *population selectivity*—that is, changes in the specific population of neurons that are active in response to a given stimulus. To illustrate the point of population selectivity, consider a simple example—first without attention—where the population of neurons that respond to a given stimulus is measured. If a second, slightly different stimulus is presented, we assume that it will activate a different but overlapping population of neurons. If we consider the exact same stimuli, now with attention, what we envision is that attention "sharpens" the population response resulting in significantly less overlap in the neural populations representing the two stimuli. In other words, the perceptual improvements associated with attention reflect more specificity in the population response. It is this increase in specificity that we consider to be an increase in *population selectivity*.

One technique that is sensitive to the amount of overlap in neural populations is fMRI adaptation (Grill-Spector and Malach, 2001; Kourtzi and Kanwisher, 2001). Though there are a variety of ways in which the technique can be used, the logic rests on the observation that there is an overall reduction in neural activity when a stimulus is repeated (Miller, Li, and Desimone, 1991). In an event-related design, such as in our experiment, two stimuli are presented in close temporal proximity and the event-related response to the combined image pair is measured in a specified region-of-interest. There is typically a decreased response on trials when the two stimuli are identical versus when the two stimuli are different. The proposed explanation for this differential response is that the amount of adaptation is dependent on the overlap in the neural populations representing the two stimuli.

II. EXPERIMENTAL FINDINGS

Using the adaptation logic we made our measurements in the LOC in response to changes in object orientation. On each trial, an image of a novel three-dimensional shape was presented for 400 ms; then, after a 200 ms interstimulus interval, a second object image was presented that either could be: (1) identical, (2) rotated 15 degrees, or (3) rotated 45 degrees (see Fig. 80.1). An important point is that previous neurophysiology (Logothetis and Pauls, 1995) and behavioral (Bülthoff and Edelman, 1992) results have suggested that changes in viewpoint are represented by different populations of neurons. We capitalized on this previous work because it gave us an easily parameterized dimension; we could assume that as we progressively changed the amount of rotational difference between the first and second objects that we would activate progressively different populations of neurons.

Importantly, the objects could be either unattended or attended. In the unattended condition, subjects attended to the fixation dot and reported if there was a change in color of the dot during the trial. The task was designed to be moderately difficult (approximately 85 percent correct) and subjects reported having little awareness of either the shape or the orientation of the objects while performing the fixation task. In the attended condition subjects reported whether the second object was rotated to the right, left, or not rotated at all with respect to the first object.

The design of the experiment made specific predictions with regard to the hypothesis that attention increases population selectivity. To make these predictions explicit consider the diagram in Fig. 80.2. In both attention conditions, the 0-degree (i.e., identical) condition will activate the same neural populations and, because of neural adaptation, we expected to measure the smallest signal. As we progressively increased the amount of rotation difference between the first and second images, we expected to activate progressively different populations of neurons and therefore measure less adaptation—that is, our measured signal should progressively increase as a function of the amount of rotation difference between the first and second presentation. Importantly, if attention increases population selectivity, we expected more specificity

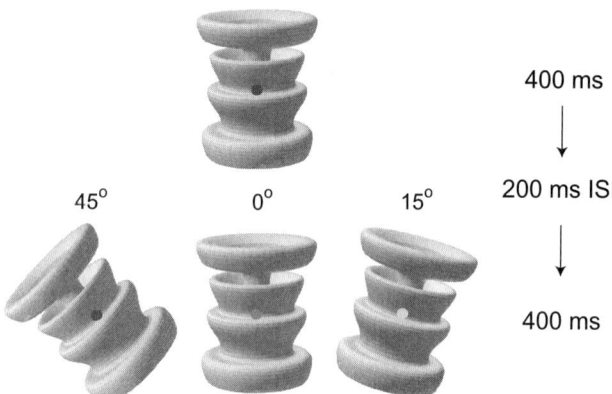

FIGURE 80.1 A schematic of the event-related design. Each trial contained a pair of object images, each shown for 400 ms and separated by a 200 ms interstimulus interval. The second stimulus was in one of three orientations with respect to the first stimulus—rotated 0, ±15, or ±45 degrees. Subjects either attended to the fixation dot and performed a same-different color-matching task (objects unattended) or attended to the shape and reported if it had been rotated to the left, right, or not rotated (objects attended).

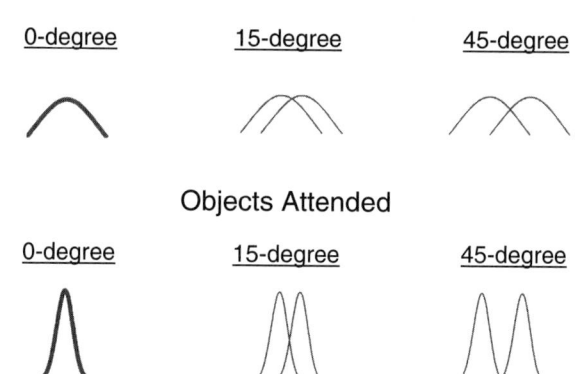

FIGURE 80.2 The expected pattern of overlap in the neural populations representing the two objects. The distributions represent activity across a population of neurons (not necessarily spatially contiguous neurons). In the 0-degree condition (when the two objects are identical) each presentation of the image will activate the same population of neurons and, because of neural adaptation, we expected to measure the smallest signal for both attention conditions. If attention increases population selectivity we expected significantly less overlap in the neural populations representing the two stimuli in the 15 and 45-degree conditions, resulting in a greater rotation-dependent response in the attended versus the unattended condition.

and consequently less overlap in the populations representing the two stimuli. In other words, we expected to measure a greater rotation-dependent response in the attended versus the unattended condition.

The event-related averages for nine subjects are shown in Fig. 80.3. We took as our measured response the peak occurring five seconds after stimulus presentation. These peaks were then normalized to remove scaling differences between the attended and unattended conditions and are shown in Fig. 80.3 (right). There are two important points to note. The first is that we found a significant progressive increase in signal as a function of the rotation difference between the objects, indicating (as expected) that different neural populations represent changes in object viewpoint. The second, and more important point, is that the rotation-dependent signal was significantly greater for the attended versus the unattended condition. The adaptation functions represent the degree to which a new population of neurons is activated as function of rotation. The significant increase—or sharpening—of the attention condition is consistent with the idea that attention is increasing population selectivity.

In a follow-up experiment we examined the generality of the effect. We were observing a significant increase in the rotation-dependent effect with attention, but we were using a rotation task to direct subjects' attention toward the objects. Because the objects were the same shape on each trial it was possible that they were attending only to the symmetry axis of the objects and not to the shape of the objects. In the second experiment, we used a very similar design but, in addition to rotation differences between the first and second object, there could also be a change in shape. Subjects were told to ignore any rotation differences and respond whether the second object was the same or different shape. We observed a rotation-dependent response that was nearly equivalent to what was observed in the first experiment suggesting that whenever attention is directed toward the objects—independent of the specific task—there is an increase in population selectivity.

III. IMPLICATIONS

It does appear that attention increases population selectivity by reducing the overlap in the neural populations representing two stimuli. What might this look like at the level of neural activity? What we envision is that attention selectively increases activity in those neurons that code for features of an attended object and reduces activity in all other neurons. This idea is consistent with our results if one considers the pattern of adaptation that would occur if one noisy (i.e., nonspecific) response pattern was followed by a second noisy response pattern. There is necessarily going to be a large amount of overlap in the neural populations, leading to more adaptation and a reduced signal. With the sharpened population response with attention, however, the increased specificity of the neural responses will lead to less overlap, therefore less adaptation, and a greater measured signal—similar to what we observed with our data.

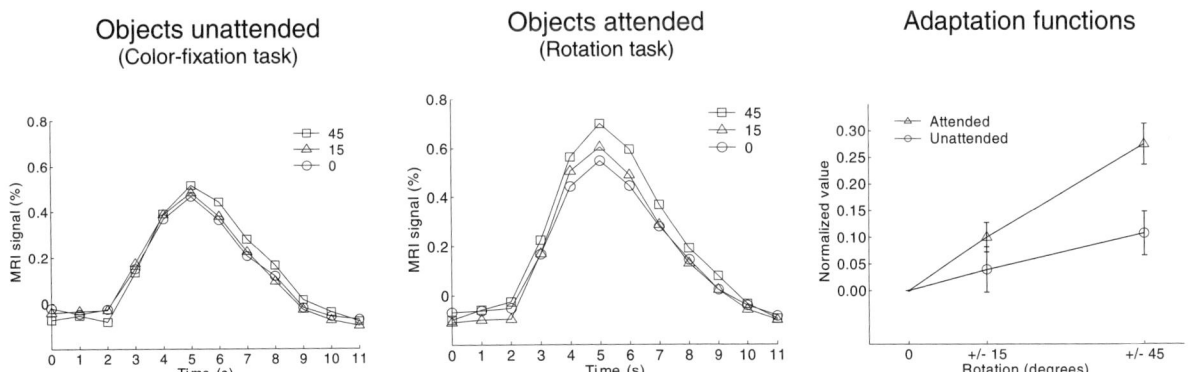

FIGURE 80.3 Average time course of percent signal change from a fixation baseline for the three different rotation conditions are shown for the unattended (left) and attended (middle) conditions. Trials start at time = 0 seconds. During both conditions there is a rotation dependent effect with the 45-degree rotation condition larger than the 15-degree larger than the 0-degree conditions. However, the rotation-dependent effect is increased significantly for the attended condition. Adaptation functions (right) were calculated by normalizing the peak responses to the peak in the 0-degree condition. The functions represent the extent to which a new population of neurons was activated as a function of the angular relationship between the pairs of objects. The function is significantly higher for the attended as compared to the unattended condition, indicating greater population selectivity.

An important consideration is the potential implications our results have for response properties of individual neurons. Considerable research has addressed how attention may affect responses at the single neuron level. In general, these studies have shown that attention may reduce the size of receptive fields (Moran and Desimone, 1985) as well as increase neural firing rate in restricted cortical areas when attention is directed to particular locations or features (e.g., Hillyard, Vogel, and Luck, 1998). However, the extent to which attention affects neural tuning for specific visual features, such as orientation, is not clear. Although earlier studies appeared to demonstrate narrower tuning functions with attention (Spitzer, Desimone, and Moran, 1988; Haenny and Schiller, 1988), more recent investigations have not confirmed this finding, showing only a multiplicative gain of the sensory response (McAdams and Maunsell, 1999; Treue and Martinez Trujillo, 1999). In addition, these recent studies have pointed to potential methodological problems in the earlier work in how tuning curve width was defined. There are, however, other recent studies that have suggested a need for narrower tuning functions to account for certain behavioral effects (Lee et al., 1999) and there are also very recent neurophysiological studies that have indeed shown significant increases in selectivity (Wegener et al., Society for Neuroscience, 2003) and narrowing of tuning curves with attention (Sharma, Lyon, and Sur, Society for Neuroscience, 2003) at the individual neuron level.

Can the results from our study weigh in on this issue of whether attention narrows individual neuron tuning curves? This is an important question because it speaks to the possible mechanisms that could give rise to increased population selectivity. Although it is certainly the case that narrower tuning functions could lead to the greater population selectivity that we observed, we measured a population-level response and the reverse conclusion is not necessarily implied. That is, we cannot make definitive statements about tuning changes of individual neurons. In fact, any mechanism that differentially affects gain could account for our findings.

To illustrate this point, consider a recent neurophysiological study that did not show any change in the width of tuning curves as a function of attention (Treue and Martinez Trujillo, 1999). This study measured motion direction tuning in neurons in area MT in the macaque monkey and demonstrated that both feature and spatial manipulations of attention affected the gain but not the tuning width of individual neurons. However, an important result from this study is that there were demonstrated increases *and* decreases in the gain of individual neurons. When attention was directed toward the stimulus used to measure the neuron's tuning curve there were significant increases in gain relative to the baseline sensory response of the neuron. In addition, when attention was directed away from the stimulus used to characterize the neuron's tuning curve, there were significant decreases in gain relative to the sensory response. The authors used these findings to posit a "feature similarity gain model," "in which the up- or downregulation of the gain of a sensory neuron reflects the similarity of the features of the currently behaviourally relevant target and the sensory selectivity of the neuron...". Applying this principle of stimulus-specific up and down-regulations of gain to a population of neurons would result in a sharpened population response—similar to what we measured in our study—and does not require a change in the tuning width of individual neurons.

There are other potential mechanisms occurring at the level of individual neurons that could account for our finding. For example, one monkey neurophysiology study examined orientation tuning of individual V4 neurons with and without attention (McAdams and Maunsell, 1999). The authors emphasized that the predominant effect of attention was a change in gain in the neural response with no change in tuning width. However, a considerable number of neurons (16%) had no identifiable tuning and low overall response without attention, yet showed normal tuning functions with attention. If a similar phenomenon occurs in the human LOC, it could account for our findings. The existence of neurons with such a response pattern would result in both an increase in overall activity and an increase in population selectivity, both of which we observed in our data.

IV. CONCLUSION

Though it is still an open question whether (and, probably more importantly, under what circumstances) attention changes the tuning width of individual neurons, it does appear that attention does sharpen the population-level response. We expect that, ultimately, there will be a variety of mechanisms occurring at the individual neuron level giving rise to greater population selectivity. Whatever specific mechanisms are employed, our finding reveals an important computational strategy of attention. Restricting activity to those neurons that code for specific features of a stimulus would have clear perceptual benefits. As well, this strategy may be particularly useful for a distributed parts-based population code thought to be used for representing complex objects. Future research, at both the single neuron and population level, will help determine the extent to which increases in population selectivity are used as a mechanism to improve the efficiency of information processing.

References

Bülthoff, H. H., and Edelman, S. Y. (1992). Psychophysical support for a two-dimensional view interpolation theory of object recognition. *Proc. Natl. Acad. Sci. U.S.A.* **89**, 60–64.

Grill-Spector, K., and Malach, R. (2001). fMR-adaptation: a tool for studying the functional properties of human cortical neurons. *Acta Psychol (Amst)* **103**, 293–321.

Haenny, P. E., and Schiller, P. H. (1988). State dependent activity in monkey visual cortex. I. Single cell activity in V1 and V4 on visual tasks. *Exp. Brain Res.* **69**, 225–244.

Hillyard, S. A., Vogel, E. K., and Luck, S. J. (1998). Sensory gain control (amplification) as a mechanism of selective attention: electrophysiological and neuroimaging evidence. *Phil. Trans. R. Soc. Lond. B.* **353**, 1257–1270.

Kourtzi, Z., and Kanwisher, N. (2001). Representation of perceived object shape by the human lateral occipital complex. *Science* **293**, 1506–1509.

Lee, D. K., Itti, L., Koch, C., Braun, J. (1999). Attention activates winner-take-all competition among visual filters. *Nature Neuroscience* **2**, 375–381.

Logothetis, N. K., and Pauls, J. (1995). Psychophysical and physiological evidence for viewer-centered object representations in the primate. *Cereb. Cortex* **3**, 270–288.

Malach, R. et al. (1995). Object-related activity revealed by functional magnetic resonance imaging in human occipital cortex. *Proc. Natl. Acad. Sci. U.S.A.* **92**, 8135–8139.

McAdams, C. J., and Maunsell, J. H. R. (1999). Effects of attention on orientation-tuning functions of single neurons in macaque cortical area V4. *J. Neurosci.* **19**, 431–441.

Miller, E. K., Li, L., and Desimone, R. (1991). A neural mechanism for working and recognition memory in inferior temporal cortex. *Science* **254**, 1377–1379.

Moran, J., and Desimone, R. (1985). Selective attention gates visual processing in the extrastriate cortex. *Science* **229**, 782–784.

Murray, S. O., and Wojciulik, E. (2004). Attention increases neural selectivity in the human lateral occipital complex. *Nature Neuroscience* **7**, 70–74.

Spitzer, H., Desimone, R., and Moran, J. (1988). Increased attention enhances both behavioral and neuronal performance. *Science* **240**, 338–340.

Treue, S., and Martinez Trujillo, J. C. (1999). Feature-based attention influences motion processing gain in macaque visual cortex. *Nature* **399**, 575–579.

CHAPTER

81

Attentional Effects on Motion Processing

Amy A. Rezec and Karen R. Dobkins

ABSTRACT

In the last decade, much has been learned about the influence of attention on visual motion processing. Many studies have shown that directing spatial attention to a motion stimulus or attending to a particular feature of a motion stimulus significantly alters the processing of that stimulus. In this review, we begin by discussing results from human psychophysical experiments investigating the effects of attention on motion perception. This is followed by a summary of single-unit neurophysiological experiments in monkeys and functional magnetic resonance imaging (fMRI) studies in humans investigating the effects of attention on neural motion responses. In the last section, we discuss the results of experiments that have attempted to establish links between psychophysical and neural effects of attention.

I. INTRODUCTION

The ability to detect and discriminate visual motion has played a critical role in evolutionary survival; for example, by allowing for the detection of prey/predators and by providing information essential in guiding self-locomotion. Over the last several decades, a wealth of studies has elucidated the mechanisms underlying different aspects of motion processing, ranging from low-level to high-level motion phenomena. In the last several years, researchers have begun to ask how attention influences motion processing. We review this literature by summarizing results from psychophysical studies (in humans and monkeys), neurophysiological studies (single-unit recordings in monkeys and fMRI in humans), and the efforts that have been made to establish links between the two.

II. PSYCHOPHYSICAL STUDIES

Psychophysical studies of attentional influences on motion processing have focused on the effects of *selective spatial* attention, *divided spatial* attention, and *feature-based* attention.

A. Selective Spatial Attention

One clear example of the effects of spatial attention on perceptual motion processing comes from a study by Chaudhuri (1990), demonstrating that diverting attention away from a motion stimulus (by requiring subjects to perform a difficult letter task at the center of gaze) significantly diminishes the motion aftereffect resulting from that stimulus. This finding suggests that attention serves to enhance the sensory processing of motion signals at a relatively early level in visual processing. More recently, Dobkins and Bosworth (2001) used a spatial precueing paradigm to investigate the effects of spatial attention on coherent motion thresholds. In these studies, visual performance was compared under conditions in which a spatial precue directed subjects' attention to the location of a to-be-presented motion stimulus versus when no precue was provided. Here, a small but significant benefit of precueing was found for short stimulus durations (<150 ms). In line with the previously mentioned effect of attention on the motion after-effect, this result suggests that directing attention to a motion stimulus may enhance the sensory processing of that stimulus (but see Lu, Liu, and Dosher, 2000, for results suggesting minimal effects of attention on *contrast sensitivity* for moving stimuli).

These studies all employed stimuli defined by luminance motion. More recently, the effects of spatial attention have been investigated for chromatic (red/green, isoluminant) motion (Thiele, Rezec, and Dobkins, 2002). Results from these experiments show that removing spatial attention from a moving stimulus (by requiring subjects to perform a difficult letter task at the center of gaze) affects luminance and chromatic motion processing equally. Although these findings suggest no greater effect of *spatial* attention on chromatic motion, other studies have argued that *feature-based* attention plays a special role in mediating chromatic versus luminance motion processing (see Thiele, Rezec, and Dobkins, 2002 for discussion).

B. Divided Spatial Attention

In contrast to *selective* spatial attention studies, *divided* attention studies address how dividing attention across multiple stimuli in a display affects stimulus processing. Dobkins and Bosworth (2001) investigated divided attention effects on motion processing by employing a "set-size effects" paradigm (see Fig. 81.1A). In this study, subjects' coherent motion thresholds obtained for a "target" motion stimulus presented among three confusable noise distracters (set-size = 4) were compared to thresholds obtained for a target motion stimulus presented alone (set-size = 1). Although the addition of distracters was found to impair motion performance significantly, these set-size effects could be accounted for by a simple model based on signal detection theory, which assumes *unlimited* attentional capacity such that the quality of sensory processing does not decline as the number of items in the visual scene increases (see Fig. 81.1B). Rather, this model demonstrates that visual performance worsens with increasing set-size because the presence of distracters increases the probability of an error occurring at the *decision level* (see Palmer, 1994, for similar effects of divided attention in other domains of vision, e.g., orientation and color). In sum, these results suggest that attention can be divided across multiple moving stimuli simultaneously without any limitations on sensory processing, *per se*.

Interestingly, results from a recent psychophysical study have suggested that attention to multiple

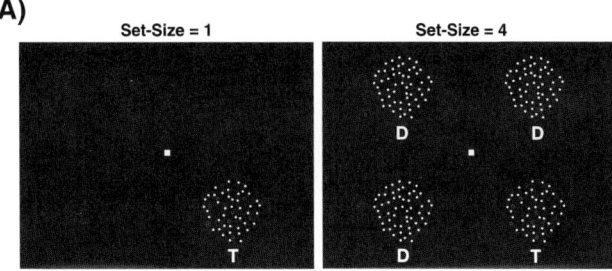

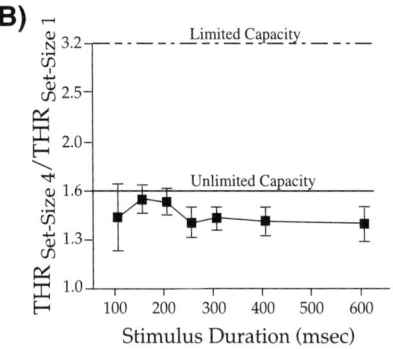

FIGURE 81.1 Effects of Divided Attention on Motion Processing. A. Set-Size Stimuli and Design (Dobkins and Bosworth, 2001). A proportion of dots (i.e., "signal" dots, shown here as *open* circles) move in a coherent direction ("leftward" or "rightward") while the others (i.e., "noise" dots, shown here as *filled* circles) move in a random fashion (note, all dots were filled in the actual display). The signal proportion is varied across trials to obtain a *coherent motion threshold* (i.e., the percentage of signal dots required to yield 75% correct direction discrimination). Thresholds are obtained for set-size = 1, where a "target" motion stimulus (T) is presented alone (*left panel*) and set-size = 4, where the motion stimulus is presented among three "distracters" (D) containing no motion signal (*right panel*). The motion target appears randomly in one of four regions of visual space (superior left, superior right, inferior left, inferior right). **B. Set-Size Data.** Group mean threshold ratios (THR set-size1/THR set-size4) are plotted as a function of stimulus duration. Error bars denote standard errors of the means across subjects (n = 9). Threshold ratios are greater than 1.0, indicating better performance in the set-size 1 condition. Across all durations, threshold ratios fall extremely close to the predictions for an Unlimited Capacity model (solid line), yet far from the predictions for a Limited Capacity model, which assumes that the quality of sensory processing declines as the number of items in the display increases (dashed line).

motion stimuli may not be *evenly* divided across space; there appears to be a greater weighting of attention in the inferior, as compared to the superior, visual field (Rezec and Dobkins, 2004).

C. Feature-based Attention

In addition to the effects of spatial attention, other studies have investigated the effects of *feature-based* attention on motion processing. One way this has been addressed has been to present two moving stimuli superimposed (which controls for the effects of spatial attention), and require subjects to attend to one or the other. For example, Lankheet and VerStraten (1995) asked subjects to attend to one of two oppositely moving (and superimposed) dot fields, and found that the motion after-effect produced by a subsequently presented dynamic motion stimulus was in the direction opposite to the attended motion field (and see Alais and Blake, 1999 for a similar finding). This result demonstrates that simply shifting attention to a particular feature (*direction of motion*, in this experiment) within a given region of space can enhance the processing of that feature. More evidence for feature-based attention effects on motion processing comes from a recent study measuring speed discrimination (Saenz, Buracas, and Boynton, 2003). In this study, subjects were required to make concurrent speed discrimination judgments for two different motion stimuli presented at spatially separate locations. Performance on this task was significantly better when the two stimuli moved in the *same* versus *different* directions. These findings suggest that attending to a particular feature—direction of motion—facilitates the processing of other stimuli in the visual field containing the same feature (and see the section "Feature-Based Attention," later, for a neural correlate of these perceptual findings).

III. NEUROPHYSIOLOGICAL STUDIES

Several studies have investigated the *neural* effects of attention on motion processing, both in macaque monkeys and humans, with the main focus on motion areas MT and MST. Complementing the psychophysical studies described previously, these neural studies address the effects of *selective spatial* attention and *feature-based* attention, which are addressed in turn, next.

A. Selective Spatial Attention

1. Attentional Modulation of Neural Responses

The effects of spatial attention on neural motion processing have been addressed in several single-unit neurophysiological studies in macaque monkeys, which compare neural responses to a moving stimulus when spatial attention is directed to, versus away, from that stimulus. In these experiments, two spatially segregated moving stimuli are presented; one that the monkey is instructed (via a precue) to respond to (and thus, by definition, *attend* to) and one that the monkey ignores. Using this paradigm, Seidemann and Newsome (1999) found that when both moving stimuli fell within a neuron's receptive field (one in the neuron's preferred, and one in the null, direction), attending to the preferred motion stimulus increased the responses of MT/MST neurons by approximately 9%. The magnitude of the attention effect was roughly the same when one stimulus was placed inside, and the other outside, the receptive field. Using a similar paradigm, Treue and Maunsell (1996, 1999) found much larger attentional effects. When one moving stimulus was inside the neuron's receptive field, attention was found to increase responses by 20% in MT and 40% in MST. When both moving stimuli were within a neuron's receptive field (one in the preferred, and one in the null, direction), attention increased responses by 70% in MT and 100% in MST. The discrepancy in magnitude of attentional effects between the Treue and Maunsell and Seidemann and Newsome studies may be due to stimulus or task differences between the two. For example, the task used in the Treue and Maunsell studies may have involved the additional recruitment of *feature*-based attention, thereby resulting in greater overall attentional effects (see Treue and Maunsell, 1999 for discussion).

These effects of selective spatial attention revealed in single-unit studies are supported by results from fMRI experiments conducted in human subjects. For example, Rees et al. (1997) demonstrated that responses in area MT+ to a peripherally presented motion stimulus varied depending on how much attention was diverted away from that stimulus.

In a more recent study, Cook and Maunsell (2002) investigated the effects of spatial attention by employing a valid/invalid precue paradigm while recording from neurons in areas MT and VIP. On each trial, two spatially segregated noise stimuli appeared, only one of which (the target) would subsequently contain a coherent motion signal (and only one of which was presented in the neuron's receptive field). Monkeys were trained to detect, and respond to, the onset of coherent motion in the target. As a means of manipulating the amount of spatial attention placed on the motion target, the location of the target was validly cued 80% of the time. Thus, for the 20% invalidly cued trials, the monkey's attention to the motion target was expected to be considerably less. In contrast to the aforementioned neurophysiological studies, the advantage of the valid/invalid precue approach is that it allows behavioral performance to be measured for both well- and *poorly*-attended stimuli, and thus allows for direct comparisons between psychophysical and neural effects of attention. In area MT, neural responses were approximately 18% greater in the well-attended (i.e., valid precue) than in the poorly-attended (i.e., invalid precue) condition, demonstrating neural effects of attention that are on the order of those previously observed (Treue and Maunsell, 1996, 1999; Seidemann and Newsome, 1999). In addition to reporting the increase in neural response due to attention, Cook and Maunsell also measured the effective change in motion signal strength due to attention, by determining the change in signal strength (measured in percent coherent motion) of a well-attended stimulus required to yield the same neural response as a poorly-attended stimulus. Here, they found that for MT neurons, attention increased the effective signal strength by roughly 3%. In VIP, attention increased the effective signal strength by a larger amount, approximately 14%. In order to determine which brain area mediated the psychophysical effects, these values in MT and VIP neurons were compared to simultaneously obtained psychophysical data (see the section, "Neural and Psychophysical Data," later).

2. Attentional Effects on Direction Selectivity and Contrast Coding

In addition to studies measuring the overall effects of spatial attention on neural responses, others have investigated whether selective spatial attention acts to sharpen the direction selectivity of MT neurons and whether attention serves to alter the effective luminance contrast of a moving stimulus (see Chapter 70). With respect to *directional tuning*, it has been shown that spatial attention increases responses by a constant factor across the entire direction tuning curve (i.e., attention acts as a response gain), thereby producing no change in the sharpness of tuning (as measured by the bandwidth of the tuning curve) (see Chapter 49). These findings indicate that attention influences MT responses to a moving stimulus independently of the direction encoding of that stimulus; that is, that attention and direction do not interact. The pattern of results is quite different for the encoding of stimulus *contrast*, however. Martinez-Trujillo and Treue (2002) addressed this issue by measuring the effect of spatial attention for different contrasts of a moving stimulus. If attention and contrast act independently on neural responses, attention should increase responses by a constant multiplicative factor across stimulus contrasts (i.e., the response gain model of attention). Alternatively, if attention and contrast interact, the relationship between attention and contrast should not be con-stant. Specifically, attention might serve to shift the contrast response function leftward along the horizontal axis, which would result in the largest effects of attention occurring for stimuli of intermediate contrasts. In this scenario, attention can be described as altering the effective contrast of the stimulus (referred to as the contrast gain model of attention). The results from this MT study support the contrast gain model, indicating that bottom-up changes in contrast are, in essence, interchangeable with top-down influences of attention.

B. Feature-based Attention

The neural effects of feature-based attention on motion processing have been addressed in a single-unit study by Treue and Martinez-Trujillo (1999). Similar to the *spatial* attention experiments described earlier, these experiments presented two moving stimuli—one inside, and one outside—the receptive field of an MT neuron, with the directions of these two stimuli being either the same or opposite from one another. The relevant results are for the situation in which the monkey was required to perform a motion task on (and thus attend to) the stimulus outside the receptive field, such that responses to an ignored motion stimulus in the receptive field were recorded. Here, the results revealed larger responses to the ignored motion stimulus when the direction of that stimulus was the *same* as that of the stimulus the monkey attended to outside the receptive field (average increase = 13%). This finding thus reflects a feature-based attentional mechanism for motion that can exert effects of roughly the same magnitude as shifting spatial-based attention. Results from human fMRI studies have also demonstrated neural effects of feature-based attention. Saenz, Buracas, and Boynton (2002) obtained fMRI responses while subjects performed a speed discrimination task on one of two oppositely moving (and superimposed) dot fields in a given region of space. Simultaneously, in a separate region of space, an ignored moving stimulus was presented, moving in either the same or opposite direction as the attended stimulus. In line with the single-unit data of Treue and Martinez-Trujillo, and mirroring the results from their psychophysical paradigm (described earlier), fMRI responses in MT+ were found to be larger when the ignored motion stimulus moved in the same direction as the attended motion stimulus. Evidence for feature-based attention has also been provided in fMRI studies by O'Craven et al. (1997). In these experiments, responses in area MT+ to the same stimulus were modulated solely by having subjects shift their attention between a moving set of dots and a stationary set of dots that were superimposed.

IV. THE RELATIONSHIP BETWEEN NEURAL AND PSYCHOPHYSICAL EFFECTS OF ATTENTION

Although the effects of attention on both motion perception and neural responses in MT/MST have been amply demonstrated, few studies have attempted to establish direct links between neural and psychophysical effects. The most useful approach has been to quantify and compare neural and psychophysical effects within the same monkey subject or to compare neural effects in monkeys with psychophysical effects in humans, which are addressed in turn, next.

A. Neural and Psychophysical Data Obtained from the Same Monkey Subjects

To date, a single study has compared neural and psychophysical effects of spatial attention in the same monkey subject. The dearth of studies of this sort can be attributed, in part, to the fact that attention studies require a comparison between responses to attended versus *ignored* stimuli. Although obtaining neural responses to ignored stimuli is easily achieved, the same is not true for perceptual responses; asking a subject to behaviorally respond to an ignored stimulus will likely make that stimulus no longer ignored. For human psychophysical studies, this is less of an issue since experimenters can explain the subtleties of "ignoring yet responding" to a stimulus, however, this is very hard to realize in monkey subjects. Cook and Maunsell (2002) circumvented this problem by employing a valid/invalid precueing paradigm, which allowed them to obtain perceptual responses to both a well- and poorly-attended moving stimulus (see previous description of paradigm). Using this method, they compared the neural effects of spatial attention (in areas MT and VIP) with psychophysical effects obtained from the same monkey subject. For both neural and psychophysical data, attending to a motion stimulus was found to be equivalent to increasing the motion signal strength of that stimulus by a given amount. However, the magnitude of the psychophysically derived change in effective signal strength (approximately 6%) matched *neither* the neurally derived estimate in area MT or in area VIP; psychophysical effects were *larger* than neural effects in MT (3%) and *smaller* than those in VIP (14%). This result led the authors to speculate that area MST (which is situated between areas MT and VIP) may be the brain area whose responses correlate with (and thus underlie) the psychophysical effect of attention.

These results and conclusions should be viewed with some caution, however, since the estimated magnitude of attention effect is likely to be influenced by the particular analysis parameters employed (e.g., linear versus nonlinear functions fit to the data sets, and/or which response epoch was chosen for the measurements; see Cook and Maunsell, 2002, for discussion). Small errors in deriving these estimates of attention effect could result in fallacious matches/mismatches between neural and psychophysical effects.

B. Psychophysical Data from Humans Compared to Neural Data from Monkeys

In a recent human psychophysical study, Rezec, Krekelberg, and Dobkins (2004) quantified the effects of spatial attention on motion processing as a function of luminance contrast, with the goal of comparing their psychophysical data with the neural results of Martinez-Trujillo and Treue (2002) in area MT (described earlier). Like the psychophysical design employed by Chaudhuri (1990) (see earlier), this study compared the duration of the motion after-effect produced by a moving grating stimulus that the subject was allowed to attend to versus one that the subject ignored (by requiring subjects to perform a difficult letter task at the center of gaze during the presentation of the motion stimulus).

As expected, diverting attention away from the motion stimulus significantly diminished the duration of the motion after-effect produced by that stimulus,

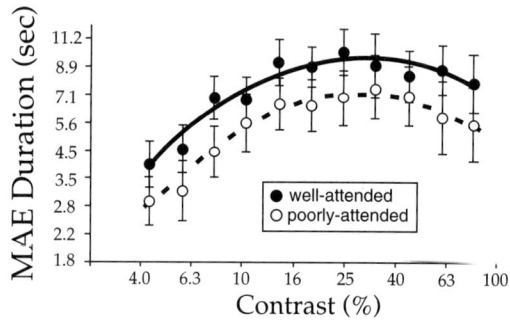

FIGURE 81.2 Effects of Attention as a Function of Luminance Contrast: Psychophysical Data. Mean MAE durations (second) are plotted for well-attended (*filled circles*) and poorly-attended (*open circles*) moving grating stimuli as a function of stimulus contrast (Rezec, Krekelberg, and Dobkins, 2004). Error bars denote standard errors of the means across subjects (n = 6). The effect of attention obtained by dividing MAE duration for well-attended stimuli by that for poorly-attended stimuli was 1.4-fold and did not vary significantly as a function of contrast (p = 0.23). A model incorporating the effects of *adaptation* to motion, in addition to the effects of attention to motion, suggests that the data are consistent with the contrast gain model of attention (see text for details).

indicating that attention serves to alter the sensory processing of motion signals at a relatively early level in visual processing. And, the effect of attention was found to be roughly *constant* (about 1.4-fold) across a wide range of stimulus contrasts (4–80%), which included contrasts where the duration effect had saturated (see Fig. 81.2). At first glance, these psychophysical results might appear consistent with the response gain model. However, the authors instead showed that a *contrast gain model* of attention—one that incorporated the effects of *adaptation* to motion, in addition to the effects of attention to motion—could easily account for their psychophysical findings. Specifically, the model assumes that adapting to a moving stimulus shifts the contrast response function *rightward* (as has been shown for area MT neurons, Kohn and Movshon, 2002), and that this shift is greater under attend versus ignore conditions. These psychophysical findings are therefore in line with the contrast gain effects of attention observed in area MT.

V. CONCLUSION

The psychophysical and neural studies described provide evidence about how and which areas of the brain may mediate the effects of visual attention revealed perceptually. However, since attentional effects are observed within several different cortical areas along the motion pathway (e.g., MT, MST, and VIP), it is perhaps naïve to believe that a single brain area underlies the perceptual effects. Rather, perceptual effects may be mediated by joint activation in several brain areas and/or the contribution of different areas may be weighted depending on the particular task at hand. Future experiments will continue to elucidate the relationship between neural and psychophysical effects of attention on motion, and other aspects of visual, processing.

References

Alais, D., and Blake, R. (1999). Neural strength of visual attention gauged by motion adaptation. *Nat. Neurosci.* **2**, 1015–1018.

Chaudhuri, A. (1990). Modulation of the motion aftereffect by selective attention. *Nature* **344**, 60–62.

Cook, E. P., and Maunsell, J. H. (2002). Attentional modulation of behavioral performance and neuronal responses in middle temporal and ventral intraparietal areas of macaque monkey. *J. Neurosci.* **22**, 1994–2004.

Dobkins, K. R., and Bosworth, R. G. (2001). Effects of set-size and selective spatial attention on motion processing. *Vision Res.* **41**, 1501–1517.

Lankheet, M. J., and Verstraten, F. A. (1995). Attentional modulation of adaptation to two-component transparent motion. *Vision Res.* **35**, 1401–1412.

Lu, Z. L., Liu, C. Q., and Dosher, B. A. (2000). Attention mechanisms for multi-location first- and second-order motion perception. *Vision Res.* **40**, 173–186.

Kohn, A., and Movshon, J. A. (2003). Neuronal adaptation to visual motion in area MT of macaque. *Neuron* **39**, 681–691.

Martinez-Trujillo, J., and Treue, S. (2002). Attentional modulation strength in cortical area MT depends on stimulus contrast. *Neuron* **35**, 365–370.

O'Craven, K. M., Rosen, B. R., Kwong, K. K., Treisman, A., and Savoy, R. L. (1997). Voluntary attention modulates fMRI activity in human MT-MST. *Neuron* **18**, 591–598.

Palmer, J. (1994). Set-size effects in visual search: the effect of attention is independent of the stimulus for simple tasks. *Vision Res.* **34**, 1703–1721.

Rees, G., Frith, C. D., and Lavie, N. (1997). Modulating irrelevant motion perception by varying attentional load in an unrelated task. *Science* **278**, 1616–1619.

Rezec, A. A., Krekelberg, B., and Dobkins, K. R. (2004). Attention Enhances Adaptability: Evidence from Motion Adaptation Experiments. *Vision Research* **44**, 3035–3044.

Rezec, A. A., and Dobkins, K. R. (2004). Attentional Weighting: A Possible Account of Visual Field Asymmetries in Visual Search? *Spatial Vision*, **17**, 269–293.

Saenz, M., Buracas, G. T., and Boynton, G. M. (2002). Global effects of feature-based attention in human visual cortex. *Nat. Neurosci.* **5**, 631–632.

Saenz, M., Buracas, G. T., and Boynton, G. M. (2003). Global feature-based attention for motion and color. *Vision Res.* **43**, 629–637.

Seidemann, E., and Newsome, W. T. (1999). Effect of spatial attention on the responses of area MT neurons. *J. Neurophysiol.* **81**, 1783–1794.

Thiele, A., Rezec, A., and Dobkins, K. R. (2002). Chromatic input to motion processing in the absence of attention. *Vision Res.* **42**, 1395–1401.

Treue, S., and Martinez-Trujillo, J. C. (1999). Feature-based attention influences motion processing gain in macaque visual cortex. *Nature* **399**, 575–579.

Treue, S., and Maunsell, J. H. (1996). Attentional modulation of visual motion processing in cortical areas MT and MST. *Nature* **382**, 539–541.

Treue, S., and Maunsell, J. H. (1999). Effects of attention on the processing of motion in macaque middle temporal and medial superior temporal visual cortical areas. *J. Neurosci.* **19**, 7591–7602.

CHAPTER 82

ERP Studies of Selective Attention to Nonspatial Features

Alice Mado Proverbio and Alberto Zani

ABSTRACT

This paper concentrates on electrophysiological data concerning selective attention to nonspatial attributes (spatial frequency, color, shape, orientation, etc.), and the way these attributes are combined into a unified percept, so that it becomes identified as an object.

Feature-based and object-based mechanisms of the brain as investigated with ERPs are analyzed. An overview is provided of studies reporting the differential activation of two cortical subsystems of the visual brain, the so-called dorsal, or "Where," and ventral, or "What" systems, in conditions in which stimulus attributes must be selectively attended to separately and/or conjointly.

Efforts are made to demonstrate the task-related relative segregation and complex interactions of the aforementioned systems during the separate or conjoint processing of stimulus attributes.

I. NONSPATIAL SELECTIVE ATTENTION

For a long time it was believed that the primary projection areas of brain cortex acted as simple analyzers of input features and were not directly involved in the so-called *top-down* selection mechanisms; that is mechanisms based on higher level cognitive strategies. Only recently has this conception been challenged as a result of new findings provided by hemodynamic bioimaging and neurophysiological and electromagnetic techniques. These techniques are able, on the one hand, to determine the functional activation of the cortical and subcortical areas, and, on the other, to reveal the early timing of the attentional influences on the processing stages.

Event-related potential (ERP) studies on the selection of single nonspatial stimulus attributes (such as color, orientation, texture, shape, or spatial frequency) have indicated that the timing of attention modulation for such processing starts as early as 80–100 ms poststimulus, and this is manifested in the form of a modulation of P/N80 or P120 components (see Chapter 85), continuing with a prominent increase in negativity at N1 and N2, called selection negativity (SN), and a large P300 response to targets.

For example, the selection of checkerboard patterns based on their check-size produces an increase in amplitude of the P1 and N115 early responses recorded at electrodes O1 and O2 corresponding to mesial occipital areas (Zani and Proverbio, 1995). Likewise, selecting gratings on the basis of their spatial frequency (Zani and Proverbio, 1997) and orientation (Karayanidis and Michie, 1997; Proverbio, Esposito, and Zani, 2002), or selecting alphanumeric characters on the basis of their shape produces an increase in the evoked response at the sensory level. Color selection is also reported to occur at a very early latency, often even earlier than the selection of other nonspatial features such as size, shape, or orientation. It has also been reported that color selection specifically activates the ventral stream, unlike other stimulus attributes such as motion (Anllo-Vento and Hillyard, 1996).

II. DORSAL AND VENTRAL STREAMS

Several neuroimaging and neurometabolic studies have shown the existence of separate pathways for object recognition and spatial localization. These pathways are named the ventral and the dorsal streams of the visual system (Arrington et al., 2000; Fink et al., 1997; Haxby et al., 1991; Olson, 2001; Ungerleider and Mishkin, 1982): one projects to the inferior temporal cortex (*What system*), and the other to the parietal cortex (*Where system*).

A. Properties

Evidence has accumulated to indicate that the dorsal stream handles information on spatial position and motion of environmental stimuli, as it possesses collicular afferents (including ipsilateral ones), whereas the ventral system analyses physical features such as orientation, color, spatial frequency, and texture (Ungerleider and Mishkin, 1982). Although the former mostly receives afferent fibers from large magnocellular gangliar cells, the latter receives afferents from small parvocellular cells. There is evidence that these two systems are related to scotopic and peripheral vision, as opposed to photopic and foveal vision, to the vision of low as opposed to high spatial frequencies and, more generally, to visual attention mechanisms based on space rather than on the object (Fink et al., 1997).

Hemodynamic functional anatomical studies have clearly shown that visual attention modulates the activity of both systems (O'Craven et al., 1999; Arrington et al., 2000; Wang et al., 1999).

B. ERP Studies

This modulation of activity has also been observed by measuring changes in amplitude, latency, and scalp topography of event-related potentials (ERPs) of the brain to visual stimuli as a function of task relevance and attention condition (e.g., Annlo-Vento and Hillyard, 1996; Martin-Loeches et al., 1999; Zani and Proverbio, 1995; Wang et al., 1999). Attention mechanisms based on spatial location and mostly involving the dorsal stream are extensively reviewed in Chapter 84.

1. Visual Evoked Potentials

Modern studies of visual evoked potentials have provided robust evidence of a topographical and functional dissociation between visual areas devoted to the analysis of low versus high spatial frequency content of luminance-modulated gratings. In detail, there is greater activity of lateral occipital areas during processing of low spatial frequencies and greater activity of mesial occipital areas during processing of high spatial frequencies (Proverbio et al., 1996), as shown in the voltage and scalp current density maps (SCD) of Fig. 82.1. Dipole-source analysis performed on ERP data has indicated secondary and primary visual cortices as possible generators of the surface electrical activity recorded at lateral occipital and mesial occipital scalp sites, respectively (Kenemans et al., 2000; see also Chapter 84). These data indicate the possible starting point of the dorsal and ventral streams of visual processing in the visual cortex, investigated by means of the electrophysiological techniques.

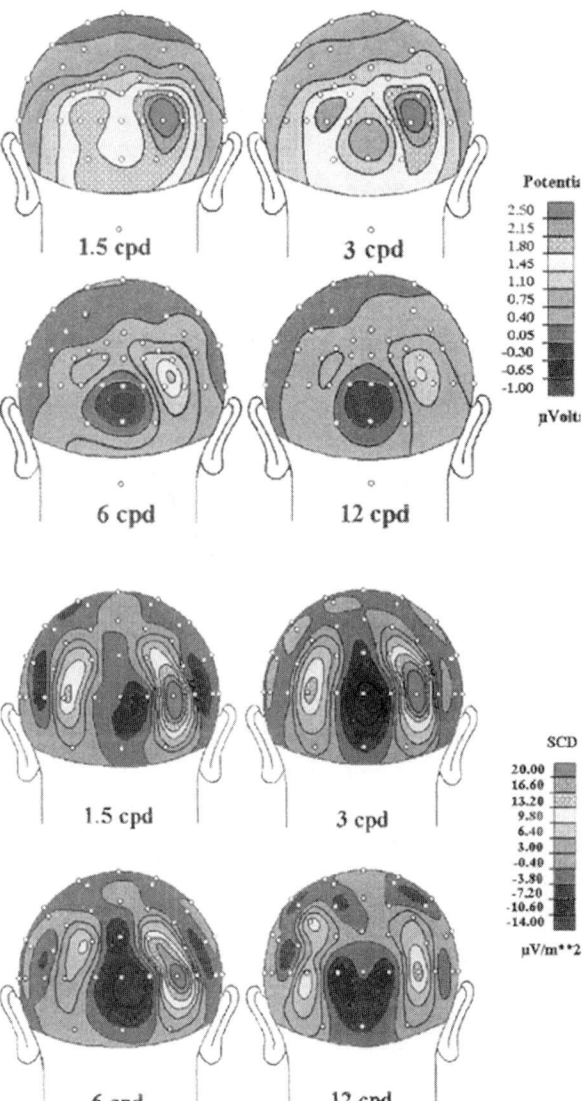

FIGURE 82.1 Grand-average voltage (top) and Scalp Current Density (SCD) maps (bottom) of brain activity recorded in response to four different luminance-modulated gratings going from a low (1.5 cycles per degree, cpd) up to a high (12 cpd) spatial frequency, presented in the central visual field during passive viewing. Note the topographic dissociation between the pattern of response to high vs. low frequency gratings, reflecting the equivalent distinction between the dorsal/ventral type of activation. Reprinted from *Brain Topography* 9; A. M. Proverbio, A. Zani, and C. Avella; Differential activation of multiple current sources of foveal VEPs as a function of spatial frequency, pp. 59–69. Copyright (1996) with permission from Kluwer Academic/Plenum Publishers and authors. (See color plate)

2. ERPs Studies on Selective Attention

Space-based and Frequency-based Attentional Selection The two mechanisms of object- and space-based selective attention normally work in close interaction. Yet, since they are partly functionally segregated, and probably based on the activation of nonoverlapping visual neural areas, to some degree it is possible to investigate these areas separately in order to unveil their neurofunctional activation. Recently we carried out a series of experiments (see Chapter 85) in which we were able to observe the two functional mechanisms by inducing different types of attentional set in the viewers. This was done by adopting the same set of visual stimuli in different tasks—lateralized isoluminant gratings of two spatial frequencies, namely 1 and 7 cpd—and modifying the experimental instructions. In different sessions the same individuals were instructed to pay attention and respond to different stimulus properties (frequency or location) while brain-evoked responses were recorded with a 32-channel montage. In the first paradigm, participants were requested to pay selective attention to a spatial frequency, and to respond to the target frequency whatever its spatial location, whereas in the second they had to attend and respond to all the gratings solely on the basis of their spatial location, thus ignoring the frequency. The results revealed a very early attentional selection effect in both cases, although with completely different morphology and topographical activation (see Fig. 82.2a,b).

Indeed, location selection affected the P1 component of the ERP at lateral occipital scalp sites. This effect, commonly reported in the literature on spatial attention (e.g., see Mangun, 2002, for an exhaustive review of studies reporting this effect), was followed by a large enhancement in positivity called P300. Conversely, the selection of spatial frequency produced a very early P/N80 followed by a considerable negative deflection (N1/N2 complex, strongly modulated by *selection negativity*) and a somewhat delayed P300. At the same time, the reaction times in the spatial selection task were about 100 ms faster than those obtained in the frequency selection task. These data confirm the partial functional independence of the two visual feature selection systems, both based on a very early sensory filter (early selection) although dependent on two anatomically and functionally separate neural streams. Observation of the topographic distribution of the attentional effect of ERP differences—obtained by subtracting the response to the nontargets from that to the same stimuli when they were targets—for the selection of spatial frequency identified a filter that was strongly linked to visual processing, first of area 17 (at the level of P80) and then of areas 18 and 19 (at the level of *selection negativity*), whereas the selection of the spatial position was apparently based on the functionality of the dorsal visual pathway.

The 3D maps presented in Fig. 82.3 show the scalp distribution of attention-related activity in the range of selection negativity, probably reflecting the selective modulation of the ventral stream. Indeed, this activity was recorded only in the frequency-relevant task, and was absent during the location-relevant task.

Object-based Attentional Selection and Selection Negativity Much evidence has been provided that attention can be directed to a conjunction of features (object-based attention) and that object representation actually is encoded during the earliest stages of sensory processing (still improperly called preattentive stages) as shown, for example, by Valdes-Sosa et al. (1998), who found that the P1 component of ERPs (100 ms) was affected by object-based attention even if the stimulus fell at an unattended space location. Again, O'Craven et al. (1999) used an fMRI study to demonstrate that attending to one attribute of an object (such as the motion of a moving face) enhanced the neural representation not only of that attribute but also of another attribute of the same object (for example, the face), thus providing physiological evidence that whole objects are selected even when only one visual attribute is relevant. Many advances have been made in the comprehension of the neural systems implicated in object processing (Olson, 2001), and there is currently agreement on the view that object knowledge is represented in the inferior temporal cortex (ventral stream).

In a further work on feature-conjunction selection (Proverbio and Zani, 2002) we investigated the mechanisms of the combined selection of frequency and spatial location using the same gratings adopted in the previous attend-location and attend-frequency paradigms and recording ERPs with a 32-channel montage. The task consisted in selectively attending to both frequency and location dimensions of the gratings. The results indicated that selection of the frequency occurred much earlier than had previously been believed, that is, within 60–80 ms post-stimulus (at this regard see also Chapter 85). Scalp current density maps (SCD) of attention effects showed a lateral occipital activation for location selection (independent of frequency relevance) and a mesial occipital activation for frequency selection (independent of location relevance), probably suggesting a dissociation between secondary and primary visual areas. These data support the hypothesis that the two attentional mechanisms may act by differentially modulating the

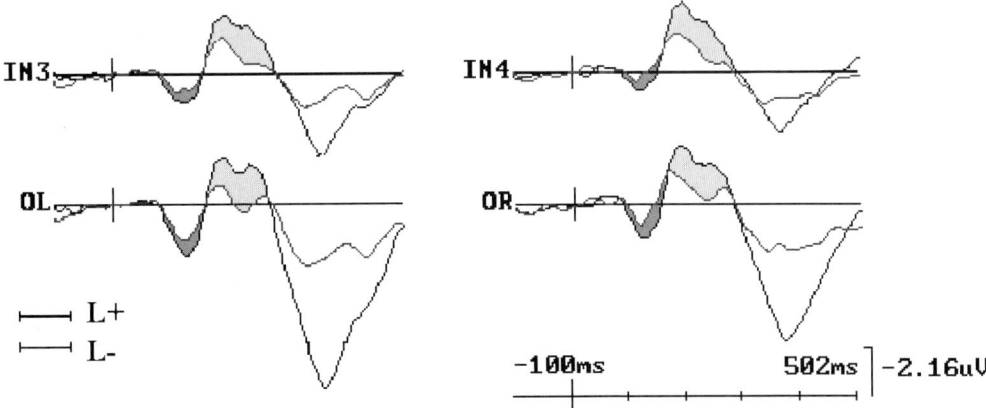

FIGURE 82.2 Grand-average event-related potentials ERPs recorded in response to luminance-modulated gratings of different spatial frequency presented at different space locations, displayed as a function of stimulus relevance. In **A** only the spatial frequency of the stimulus was task-relevant and, respectively, target (F+) or nontarget (F−). In **B** only stimulus location was relevant: target (L+) or nontarget (L−). In **C** both frequency and location had to be selectively attended to in a conjoined attention task. Note that ERP waveforms shown here as a function of location-relevance (target: L+, or nontarget: L−) are grand-averages across responses to stimuli independent of frequency relevance.

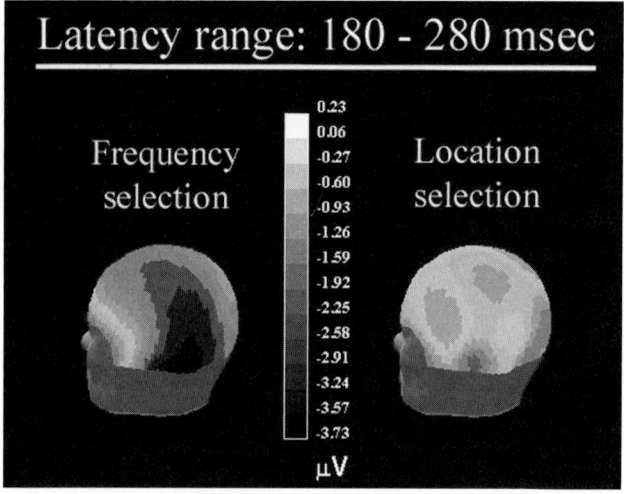

FIGURE 82.3 Realistic head three-dimensional voltage maps of brain activity recorded during the attend-frequency (left) and attend-location (right) tasks in the latency range of selection negativity (i.e., 180–280 ms). Maps were computed on the difference waves obtained from the target—nontarget subtraction to show the effect of selective attention on ERPs. Note how space-selection did not produce any attention-related selection negativity, and how the map for frequency selection pointed at a focus centered over the baso-occipito-temporal area consistent with a ventral activation.

activity of the dorsal and ventral streams for spatial and nonspatial selective attention, respectively.

Interestingly, frequency selection depended on location relevance, in that ERPs were larger to stimuli relevant in both features (L+F+) rather than only in location (L+F−), whereas frequency relevance *per se* was not sufficient to enhance the ERP response when the location was irrelevant (L−F+). More importantly, stimuli relevant only in location elicited a large attentional selection negativity, similar to that elicited in the attention to frequency paradigm described previously, which was instead absolutely absent in the attend-location paradigm (see Fig. 82.2c). Three-dimensional maps of voltage and scalp current density (SCD) for selective negativity provided evidence of a strong similarity in the distribution of this response across the conjoined-selection paradigm (also when spatial frequency was not relevant (L+F−), although still attentionally evaluated by the viewer) and the independent spatial frequency-selection paradigm for frequency-relevant gratings (F+). This result suggests that the selection mechanisms of the two features operate in parallel right from the earliest stages of analysis (see also Chapter 67), and that feature selection, rather than being temporally preceded by space selection, depends on it, in that is centered on precise coordinates of the attended receptive location. On the other hand, location relevance, in a paradigm in which frequency and location must be conjointly attended to, is affected by stimulus spatial frequency even if this is not task-relevant (L+F−), since it activates sensory attentional mechanisms (reflected by selection negativity) usually involved in spatial frequency but not location selection.

So far we have seen how it is possible to use laboratory investigations to study the way in which the visual system processes and attentionally selects one or more visual features (spatial frequency, depth, stereopsis, color, orientation, texture, luminance) of the surrounding environment, separately. Of course, in actual fact, we perceive a unitary environment and not a separate series of objects or individual attributes. (We do not address binding issues here; see Chapters 24 and 47). This perception of the unitary nature derives from the interaction between the "Where" and "What" systems which, although partially anatomically and functionally distinct, operate in parallel and in very close coordination. Clear evidence of this interdependence comes from neuroimaging and neuropsychological literature. For example, the clinical neuropsychological study by Friedman-Hill et al. (1995) indicated that patients with focused bilateral lesions of the parietal cortex are unable to combine correctly color and shape of stimuli presented in the two visual hemifields. This suggests that the integrity of the "Where" system is essential for the correct recognition of objects. On the other hand, a large body of very recent neuropsychological, neurophysiological, and behavioral data indicates that the spatial and nonspatial intentional mechanisms are probably not separated at all. Overall, these data support the view that object-directed or feature-directed selection, rather than being preceded by a space selection, is centered on line on precise coordinates of the attended space (i.e., *object-centered space receptive fields*; see Olson, 2001). In our view, our electrophysiological data on conjoined selection of frequency and space reported previously are in line with this view derived from other research lines in cognitive neuroscience.

Object Selection: Color and Shape Processing In a conjoined selective attention study we recently studied the mechanism subserving the processing of color and shape of familiar objects, in order to provide ERP indices of attention effects during ventral stream activation (Proverbio et al., 2004). On the whole, compared to color selection, shape selection produced slower response times, and larger and later N2, Frontal Selection Positivity (FSP), and P300 components, thus further supporting the viewpoint that color dimension selection is faster and easier than shape selection. Unprecedented in the literature, however, was the result that color-related selection negativity (SN) was significantly affected by shape content, but not vice versa. When color was the attended feature (e.g., yellow) the SN response was significantly greater if the stimulus shape, which was the irrelevant dimension, was canonically color-related (e.g., a chick) rather than unrelated (e.g., an elephant). This greater SN response was particularly marked over the left hemisphere. The opposite pattern did not occur in the attend-shape task, in which the stimulus color content did not affect target shape selection.

The data suggest a strong interaction between the two mechanisms subserving the selection of shape and color, and indeed provide evidence that multiple features of an object are processed automatically as a perceptual whole and not independently of each other. The fact that, in our study, shape content affected color processing even when shape was the irrelevant dimension and despite color, as an independent feature, being processed faster than shape, strongly suggests that object processing does not depend on the processing output of independent visual features (in this case shape and color), but that it is carried out in parallel right from the earliest stages of processing within the occipito-temporal pathway, as supported by the findings of recent neuroimaging literature. We propose that the left occipito-temporal scalp area, where the selection negativity was shown to be maximally affected by stimulus prototypicity (e.g., a green artichoke as compared with a pink one), reflects the underlying activity of structures belonging to the ventral stream, which are involved in the brain's representation of object-color knowledge.

III. CONCLUSION

The view that emerges from the literature indicates that the dorsal and ventral streams of the visual system, although partially anatomically segregated, may be activated in parallel and in an independent or conjoined mode depending on the attentional demands and task requirements. In addition, the perception of multidimensional objects is accomplished through an active binding process of spatial and nonspatial features. This mechanism would not be based on hierarchically organized independent processes, but rather on the horizontal processing of visual cells that takes place at very early stages of analysis as advanced by models (see Chapter 24) assuming that object attributes may be initially conjoined in a single representation, while being separately analyzed in parallel, dimension by dimension.

References

Anllo-Vento, L., and Hillyard, S. A. (1996). Selective attention to the color and direction of moving stimuli: electrophysiological correlates of hierarchical feature selection. *Percept. Psychophys* **58**, 191–206.

Arrington, C. M., Carr, T. H., Mayer, A. R., and Rao, S. M. (2000). Neural mechanisms of visual attention: object-based selection of a region in space. *J. Cogn. Neurosci.* **12**, 106–117.

Fink, G. R., Dolan, R. J., Halligan, P. W., Marshall, J. C., et al. (1997). Space-based and object-based visual attention: shared and specific neural domains. *Brain* **120**, 2013–2028.

Friedman-Hill, S. R., Robertson, L. C., and Treisman, A. (1995). Parietal contributions to visual feature binding: evidence from a patient with bilateral lesions. *Science* **269**, 853–855.

Haxby, J. V., Grady, C. L., Horwitz, B., Ungerleider, L. G., et al. (1991). Dissociation of object and spatial visual processing pathways in human extrastriate cortex. *Proc. Natl. Acad. Sci. USA* **88**, 1621–1625.

Karayanidis, F., and Michie, P. T. (1997). Evidence of visual processing negativity with attention to orientation and color in central space. *Electroencephalogr. Clin. Neurophysiol.* **103**, 282–297.

Kenemans, J. L., Baas, J. M., Mangun, G. R., Lijffijt, M., et al. (2000). On the processing of spatial frequencies as revealed by evoked-potential source modeling. *Clin. Neurophysiol.* **111**, 1113–1123.

Mangun, G. R. (2002). Neural mechanisms of attention. *In* "The Cognitive Electrophysiology of Mind and Brain," (A. Zani and A. M. Proverbio, Eds.), pp. 247–258. Academic Press, San Diego.

O'Craven, K. M., Downing, P. E., and Kanwisher, N. (1999). fMRI evidence for objects as the units of attentional selection. *Nature* **401**, 584–587.

Olson, C. R. (2001). Object-based vision and attention in primates. *Curr. Opin. Neurobiol.* **11**, 171–179.

Proverbio, A. M., and Zani, A. (2002). Visual selective attention to object features. *In* "The cognitive electrophysiology of mind and brain," (A. Zani and A. M. Proverbio, Eds.), pp. 275–308. Academic Press/Elsevier, San Diego.

Proverbio, A. M., Burco, F., Del Zotto, M., and Zani, A. (2004). Blue piglets? Electrophysiological evidence for the primacy of shape over color in object recognition. *Cogn. Brain Res.* **18**, 288–300.

Proverbio, A. M., Esposito, P., and Zani, A. (2002). Early involvement of temporal area in attentional selection of grating orientation: an ERP study. *Cogn. Brain Res.* **13**, 139–151.

Proverbio, A. M., Zani A., and Avella, C. (1996). Differential activation of multiple current sources of foveal VEPs as a function of spatial frequency. *Brain Top* **9**, 59–69.

Ungerleider, L. G., and Mishkin, M. (1982). Two cortical visual systems. *In* "Analysis of visual behavior (D. J. Ingle, M. A. Goodale, and R. J. W. Mansfield, Eds.), pp. 549–586. The MIT Press, Cambridge, Mass.

Valdes-Sosa, M., Bobes, M. A., Rodriguez, V., and Pinilla, T. (1998). Switching attention without shifting the spotlight: object-based attentional modulation of brain potentials. *J. Cogn. Neurosci.* **10**, 137–151.

Wang, J., Zhou, T., Qiu, M., Du, A., et al. (1999). Relationship between ventral stream for object vision and dorsal stream for spatial vision: an fMRI + ERP study. *Hum. Brain Mapp.* **8**, 170–181.

Zani, A., and Proverbio, A. M. (1995). ERP signs of early selective attention effects to check size. *Electroencephalogr. Clin. Neurophysiol.* **95**, 277–292.

Zani, A., and Proverbio, A. M. (1997). Attention modulation of short latency ERPs by selective attention to conjunction of spatial frequency and location. *J. Psychophysiol.* **11**, 21–32.

Effects of Attention on Figure-Ground Responses in the Primary Visual Cortex during Working Memory

Hans Supèr

ABSTRACT

Recently neural activity related to visual working memory has been described in the primary visual cortex. These responses are observed during a delayed figure-ground response task where neural activity that segregates figure from ground continues after the removal of the figure. Attention influences the maintenance of the figure-ground signal during the memory period. Two effects on neural activity are found: an enhancement of activity at the memorized location and a strong global suppression of activity. Therefore, these results indicate that attention has a dual effect on neural responses in the primary visual cortex during working memory.

I. ATTENTION AND WORKING MEMORY

The visual system is confronted with a huge amount of sensory information from the surrounding visual environment. The visual system, however, has a limited capacity to process the incoming signals and thus a selection of relevant information is required. Several mechanisms contribute to such a selection; one of them is attention. Attention typically is categorized as bottom-up or top-down effects on visual processing. High-contrast stimuli prevail lower-contrast stimuli for visual processing, which reflects a bottom-up process, and top-down influences bias the competition in favor of attended locations or objects within the visual scene compared to ignored locations or objects.

A further selection is made when visual information needs to be stored in memory. Working memory is the selective maintenance of relevant information and is under control of attention (Awh and Jonides, 2001). Only attended objects enter working memory whereas unattended or ignored stimuli will not be stored into memory. However, at what point in the stream of visual processing does attention start to exert its influence, and to what extent is the visual information processed in the absence of attention? This chapter will discuss the affects of attention on the neural responses in the primary visual cortex that represent figure-ground perception and in particular, working memory.

II. FIGURE-GROUND SEGREGATION

Neurons in the primary visual cortex (V1 or striate cortex) receive a small portion of the visual scene through their receptive fields (RFs), which are retinotopically arranged. Only when a stimulus falls within the neuron's RF will the neuron respond. Therefore, visual processing starts with a mosaic of sensory information. In order for the subject to perceive the visual scene as a whole, the piecewise visual information needs to be appropriately combined into coherent objects. Neural responses that are related to such type of perceptual processing are also present in the primary visual cortex (Paradiso, 2002). For example, during figure-ground segmentation the figure is segregated from the background by an enhanced firing rate of V1 neurons when the figure falls on the RF of a neuron compared to when background covers the RF. This figure-ground activity corresponds to the complete figure; that is, border and surface information (Lamme, 1995), and is observed in the late (>100 ms.) modulated activity of a V1 neuron and not in the early part of the stimulus evoked response.

So we perceive the visual environment as a compilation of visual objects. However, at a certain moment we consciously perceive only a small part of the visual scene (Rensink, 2002). This selection is believed to be under the control of attention, which selects information to enter visual awareness. At a neural level, activity that represents the perception of a stimulus is found in the primary visual cortex. In a figure-ground detection task where animals have to report the presence or absence of a figure, the late modulated activity that signals figure-ground segregation is absent when the animals fail to report the stimulus (Supèr et al., 2001a). Also, the strength of the figure-ground signal correlates with the saliency of the figure. Do these findings mean that attention controls figure-ground segregation?

Visual filling-in of an image is preattentive (Mattingley et al., 1997) and recent evidence shows that figure-ground segmentation occurs in the absence of focal attention toward the stimulus (Scholte et al., 2001). This suggests that attention selects visual information after the occurrence of figure-ground activity. However, focal attention may not be required for figure-ground segmentation, attention influences figure-ground segregation. Scene segmentation requires perceptual learning, even for pop-out stimuli, which is observed in V1 (Lee et al., 2002), and attention enhances visual detection in the primary visual cortex (Ress et al., 2000). Recent results show that during moments of inattention figure-ground segregation in the primary visual cortex does not occur (Supèr et al., 2003) suggesting a preparatory role for attention to set the visual cortex ready for processing incoming visual information up to perceptual level. Thus (after learning), figure-ground segregation may occur in the absence of focal attention toward the stimulus, but global or sustained attention seems to be necessary for the occurrence of figure-ground activity in the primary visual cortex and for reporting the stimulus; that is, reaching awareness.

III. WORKING MEMORY IN THE PRIMARY VISUAL CORTEX

Once visual objects are perceived it is sometimes necessary to maintain the sensory information for a short period of time into memory, for example, when an action still needs to be executed after the disappearance of the stimulus. Working memory is a system for the maintenance of behaviorally relevant information and it stores complete objects rather than individual elements of an object (Luck and Vogel, 1997). Working memory has a limited capacity to store visual information, which is under the control of attention where only attended objects enter the domain of working memory.

The neural basis of visual working memory is found in parietal and frontal cortex. Neurons in these areas start to enhance their firing rate when a relevant stimulus that falls within their RF disappears from the visual scene (Chafee et al., 1998). These enhanced responses are maintained during the period the location or attributes of the stimulus is held in memory. However, it is not clear whether the sustained activity represents a kind of attentional control or whether it represents the storage of the actual sensory information (Supèr, 2002). Since the areas of parietal and frontal cortex are believed to be the source of top-down attention and control the way cortical information is processed and executed, it may be argued that the sustained responses in these areas reflect a top-down attentional control for maintaining the perceptual information during a brief period and thus do not carry the visual information. The fact these enhanced responses in these areas occur generally at the moment that the stimulus disappears from the visual scene supports this notion.

The neural representation of a stimulus in the primary visual cortex may indicate that this area provides the working memory system with the sensory information that needs to be stored. Evidence for this comes from the observations of sustained responses in the primary visual cortex during a figure-ground delayed response task (Supèr et al., 2001b). In this task animals fixated a central dot on the monitor while a textured figure within a background was presented briefly. Following removal of the figure a homogenous texture was displayed and the animals maintained fixation. After some delay (up to two seconds) the animals were instructed to report the location of the previously presented figure by making an eye movement toward that location. During such a task late modulated activity that segregated the figure from background continued after the disappearance of the figure (see Fig. 83.1). Thus the continuation of the signal that segregates the figure from ground, that is, the neural activity that represents the neural correlate of the perception of the figure, operates as a memory signal once the figure is removed from the visual scene.

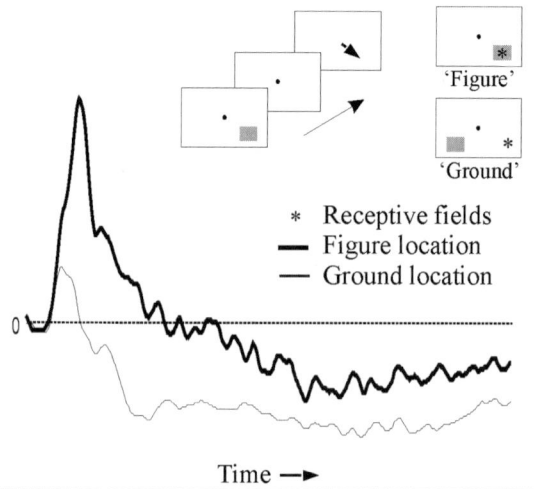

FIGURE 83.1 Neural responses during a delayed figure-ground detection task. Animals fixated at a central dot while a motion defined figure appeared briefly (~30 ms). After removal of the figure the animal maintained fixation until cued to saccade toward the location of the previously presented figure. The figure was positioned such that it covered the receptive fields of the recorded neurons ("figure" condition) or such that background covered the receptive fields ("ground" condition). Neural responses show a sustained response difference between figure and ground trials until a saccade is made. Note that during the delay period neural responses to both figure and ground conditions become suppressed.

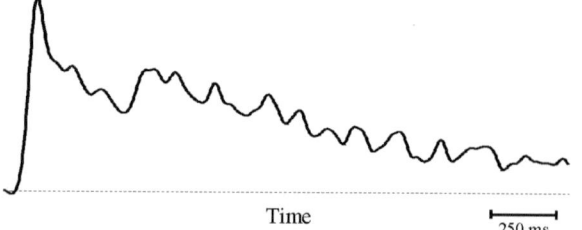

FIGURE 83.2 Example of figure-ground activity (=figure response-ground response) during a delayed figure-ground response task. The figure-ground signal is strongest at the beginning of figure-ground segregation and declines gradually during the delay period.

IV. ATTENTION CONTROLS WORKING MEMORY

The effects of attention, either bottom-up or top-down, on neural responses are observed in many cortical areas of the visual system; even at the earliest stage in the primary visual cortex, which shows that attention can act at all levels of visual processing (Treue, 2001). The role of attention is not restricted to biasing the selection of the visual information to be processed, it also is involved in the maintenance of the attended information for a short time period (Schmidt et al., 2002).

In a delayed figure-ground response task the bottom-up signal of attention is absent since the RF stimuli are identical for the figure and ground location, therefore top-down attention likely controls the continuation of the figure-ground activity. However, the sustained figure-ground responses decline over time so that for longer delay periods the figure-ground signal becomes weaker (see Fig. 83.2). This decay over time may indicate that the maintenance of the figure-ground response is not an active process and therefore does not reflect a selective gating mechanism by attention. It appears that the strength of the perceptual signal, that is, the amount of figure-ground signal, determines the strength of the memory signal (Sheth, 2003). This is consistent with the notion that figure-ground segregation can occur in the absence of focal attention toward the figure. To test whether attentional control is required for the continuation of the figure-ground signal a pop-out stimulus was presented in the vicinity of the figure location during the delay period in the figure-ground delayed response task. The sudden onset of the pop-out stimulus automatically captures attention (Schmidt et al., 2002). If the figure-ground responses were not affected by such a salient stimulus it would indicate that the sustained figure-ground response during the delay period is not controlled by any attentional mechanism. The results however show that when the pop-out stimulus is presented it has a huge effect on sustained neural responses in the primary visual cortex but not necessarily on the figure-ground signal. After the presentation of the pop-out stimulus neural activity is strongly suppressed (not shown). This effect of the pop-out stimulus shows that neurons in the primary visual cortex remain sensitive for changes in the visual environment. Moreover, since the pop-out stimulus is presented outside the RFs of the recorded neurons this result provide neural support for the finding that contextual information influences memory (Awh and Jonides, 2001).

The presentation of the pop-out stimulus during the delay period influences the continuation of the figure-ground signal depending on the demands of the task (see Fig. 83.3). When the pop-out stimulus is a distracter and is thus irrelevant for the animal (but it will nevertheless capture attention) it suppresses the figure and ground responses equally so that the figure-ground signal is maintained. This signifies that attention affects neural responses during memory but it does not affect the continuation of the memory signal,

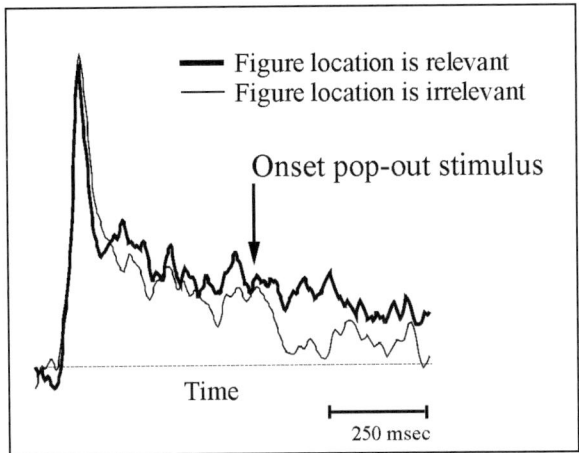

FIGURE 83.3 Figure-ground activity and shift of attention during delay period. During the delay period a pop-out stimulus was presented. This pop-out stimulus captures attention and could be either relevant or irrelevant to the animal. The pop-out stimulus resulted in a global decrease of neural responses in the figure and ground condition (not shown) Only when the pop-out stimulus was relevant the figure-ground signal disappeared (thin trace); otherwise, the figure-ground signal continued (thick trace).

that is, figure-ground activity. However, when the pop-out stimulus is the target and therefore the figure location becomes irrelevant the figure-ground signal disappears after the appearance of the pop-out stimulus. Thus in this case the pop-out stimulus differentially affects figure and ground responses. The effects of the pop-out stimulus on the figure-ground responses occur rapidly (<50 ms. after stimulus presentation) indicating a bottom-up effect. The differential influences on the figure and ground responses, however, require top-down control of attention. In conclusion, the continuation of the figure-ground signal in the primary visual cortex is an active process and presumably under the control of top-down attention from higher visual areas, which would entail recurrent interactions through feedback projections.

V. EFFECTS OF ATTENTION ON THE NEURAL RESPONSES DURING DELAY PERIOD

The general effects of attention on the processing of sensory information are twofold. First, directing attention toward the stimulus enhances the neural processing at that particular location. Second, neural responses to a stimulus are suppressed when the focus of attention is away from the stimulus. Recently, feature-based attentional modulation that reaches far beyond the confines of the spatial RF has been reported (Treue and Martinez Trujillo, 1999). The activity of MT neurons is larger when the animal is attending to a preferred direction stimulus versus an anti-preferred one, even when the attended stimulus is far from the RF. Attending to a particular feature, such as a direction of motion, thus seems to enhance the responsiveness of all neurons that prefer this stimulus feature, not just of those whose RF includes the attended stimulus. Attention is thus linked to the location and properties of the RF. So, directing attention results in a push-pull mechanism where neural responses to attended locations/elements become enhanced and unattended locations/elements suppressed. These effects on the neural responses can act globally as well as locally.

How does attention affect neural responses during working memory in the primary visual cortex? In the higher visual areas neurons show enhanced firing rates for the period the stimulus is remembered. Studies investigating neural responses in lower visual areas found suppression of activity during a working memory task where many neurons show reduced activity (Nakamura and Colby, 2000). Increased neural activity is considered as a neural marker for memorized objects/locations. The strong global suppression of neural responses may have resulted in the failure in imaging studies to notice the involvement of early visual areas in working memory. Also the reduced activity during the delay period can be considered as an argument that the lower areas do not participate in the retention of visual information and are thus not part of the memory organization.

In the primary visual cortex both enhancement and suppression of neural activity are observed during the memory period in a delayed figure-ground response task. The enhancement of neural responses is specific for the neurons that have their RFs on the location of the object that is maintained in memory. That is, neurons that encode the figure position have stronger responses compared to the responses when background covers the RFs, and therefore the enhanced responses are limited to the RF location. Interestingly, this differential activity does not depend on the context of the stimulus in the RF, like during figure-ground segregations, since the RF stimuli during the delay period are identical for both figure and ground conditions. Besides this spatial specific enhancement of neural responses a strong global suppression of neural activity occurs during the delay period. This suppressive is observed when the figure was presented onto or outside the RFs of the recorded neurons. This means that the strong suppressive effects on neural activity are not spatially confined but instead act globally, affecting most likely the whole visual area.

The suppressive influence on the neural activity of V1 neurons is related to the continuation of the figure-ground activity where the inhibitory effect accompanies the maintenance of the figure location into memory. When an incorrect behavioral response is given, that is, when the animal failed to remember the figure location, the figure-ground signal starts to diminish gradually so that no signal is left when a behavioral report has to be made. The fading of the memory signal is paralleled by the disappearance of the global inhibitory influences on the neural responses. Thus, the enhanced responses that signal the memorized location in the primary visual cortex is accompanied by a global suppression of neural responses. Therefore to maintain visual information into memory attention results in interplay of facilitatory and inhibitory influences that take place during the period the stimulus is stored into memory.

Thus top-down attention has a dual effect on the neural responses in the primary visual cortex during working memory. On the one hand a strong suppression of activity is evoked that affects presumably the entire visual cortex, and on the other hand a spatial specific enhanced responses of the neurons that signal the relevant location. Both type of effects are required, and most likely closely related, in order to store visual information into working memory.

VI. CONCLUSION

Neurons in the primary visual cortex signal the figure perception in a figure-ground detection task. This signal continues after the removal of the figure and represents a neural correlate of working memory. During the delay period the memory signal is maintained through a dual effect of attention on the neural activity in the primary visual cortex. On the one hand attention results in the enhancement of activity at the relevant location, and on the other hand attention causes an overall suppression of neural activity of the primary visual cortex. These differential responses may combine different neural mechanisms where top-down attention represents a push-pull mechanism.

References

Awh, E., and Jonides, J. (2001). Overlapping mechanisms of attention and spatial working memory. *Trends in Cognitive Sciences* **5**, 119–126.

Lamme, V. A. F. (1995). The neurophysiology of figure-ground segregation in primary visual cortex. *Journal of Neuroscience* **15**, 1605–1615.

Lee, T. S., Yang, C. F., Romero, R. D., and Mumford, D. (2002). Neural activity in early visual cortex reflects behavioral experience and higher-order perceptual saliency. *Nature Neuroscience* **5**, 589–597.

Mattingley, J. B., Davis, G., and Driver, J. (1997). Preattentive filling-in of visual surfaces in parietal extinction. *Science* **275**, 671–674.

Nakamura, K., and Colby, C. L. (2000). Visual, saccade-related, and cognitive activation of single neurons in monkey extrastriate area V3A. *Journal of Neurophysiology* **84**, 677–692.

Paradiso, M. A. (2002). Perceptual and neuronal correspondence in primary visual cortex. *Current Opinion Neurobiology* **12**, 155–161.

Rensink, R. A. (2002). Change detection. *Annual Reviews Psychology* **53**, 245–277.

Ress, D., Backus, B. T., and Heeger, D. J. (2000). Activity in primary visual cortex predicts performance in a visual detection task. *Nature Neuroscience* **3**, 940–945.

Schmidt, B. K., Vogel, E. K., Woodman, G. F., and Luck, S. J. (2002). Voluntary and automatic attentional control of visual working memory. *Perception & Psychophysics* **64**, 754–763.

Scholte, H. S., Spekrijse, H., and Lamme, V. A. F. (2001). Neural correlates of global scene segmentation are present during inattentional blindness. *Journal of Vision* **1**, 346a.

Sheth, B. R., and Shimojo, S. (2003). Signal strength determines the nature of the relationship between perception and working memory. *Journal of Cognitive Neuroscience* **15**, 173–184.

Supèr, H., Spekreijse, H., and Lamme, V. A. F. (2001a). Two sensory modes of signal processing in monkey primary visual cortex (V1). *Nature Neuroscience* **4**, 304–310.

Supèr, H., Spekreijse, H., and Lamme, V. A. F. (2001b). Neuronal correlate of working memory in the monkey primary visual cortex. *Science* **293**, 120–124.

Supèr, H. (2002). Cognitive functions in the primary visual cortex; from perception to memory. *Review in the Neurosciences* **13**, 287–298.

Supèr, H., Van der Togt, C., Spekreijse, H., and Lamme, V. A. F. (2003). Cortical state in primary visual cortex predicts figure-ground perception. *Journal of Neuroscience* **23**, 3407–3414.

Treue, S. (2001). Neural correlates of attention in primate visual cortex. *Trends in Neurosciences* **24**, 295–300.

Treue, S., and Martinez Trujillo, J. C. (1999). Feature-based attention influences motion processing gain in macaque visual cortex. *Nature* **399**, 575–579.

> # CHAPTER
84

Electrophysiological and Neuroimaging Approaches to the Study of Visual Attention

Antígona Martínez and Steven A. Hillyard

I. INTRODUCTION—ERPS

To characterize the neural operations that underlie cognitive systems such as attention it is essential to analyze the temporal sequences of information processing within the brain regions involved. Blood-flow imaging methods such as positron emission tomography (PET) and functional magnetic resonance imaging (fMRI) have proven very useful in their ability to provide fine-grained spatial information about the anatomical areas and neural networks that participate in specific cognitive processes. However, due to the intrinsically sluggish nature of hemodynamic responses, these methods do not provide adequate information regarding the time course of sensory and cognitive operations in the brain.

The temporal structure of neural activation patterns can be studied electrophysiologically in humans by means of noninvasive recordings of event related potentials (ERPs). ERPs are scalp-recorded voltage fluctuations within the ongoing electroencephalogram (EEG) that arise from the post-synaptic activity of large populations of similarly oriented cortical neurons firing synchronously during the processing of information (Regan, 1989). The firing patterns of these cortical neurons can be triggered by an external stimulus and/or elicited in conjunction with specific sensory, motor, or cognitive processes. Over the past 30 years, ERPs and their magnetic field counterparts have been used extensively to study the brain mechanisms underlying perceptual and cognitive processes including selective attention (Rugg and Coles, 1995).

Because ERP amplitudes are typically very small (on the order of microvolts) in relation to the EEG fluctuations within which they are embedded, data processing techniques are required in order to extract the ERP signal from the background EEG noise (Regan, 1989). Among the analytic techniques routinely used to isolate ERP signals, simple averaging of the time-domain waveform is by far the most common, but other approaches, such as Fourier analysis, wavelet decomposition, and principal or independent component analysis are also employed. Averaged ERP waveforms consist of a temporal sequence of positive and negative voltage deflections (the ERP components) occurring at specific latencies following stimulus onset. ERP components may be characterized by 1) their post-stimulus latencies, indicating the timing of the underlying neural activity; 2) their topographical distributions over the scalp; and 3) by their amplitudes in microvolts, which is a reflection of the net summation of synaptic activity of the participating cortical neurons (Regan, 1989). In some cases, ERP components may also be defined as the waveform features associated with a particular cognitive operation or arising from a particular anatomical generator.

II. ANALYZING THE NEURAL SOURCE(S) OF ERP COMPONENTS

In order to make inferences about the neural activity patterns engaged during a given cognitive task, the surface-recorded ERP changes recorded on the scalp during task performance must be mapped to specific intracranial generator sources. One simple approach is to visualize the topographical distribution of voltage and scalp current density (SCD) over the surface of the head at different time points. Although this method is useful in identifying the general regions of the brain that are active, it provides limited information as to the location of the neural generators that are involved.

Calculating intracranial generator locations based on surface-recorded ERPs involves the so-called inverse problem, which is in principle unsolvable in that multiple generator configurations can give rise to any given pattern of surface potentials. However, under appropriate conditions, good approximations of source locations can be achieved through the use of computer algorithms for dipole modeling that take into account physical models of the head in addition to constraints such as generator orientation with respect to the brain (Scherg, 1990). Furthermore, the validity of estimated dipolar sources can be improved significantly by integrating dipole modeling with information obtained from anatomical and functional MRI (Mangun, Hinrichs et al., 2001).

III. ERP MEASURES OF SPATIAL ATTENTION

The cortical mechanisms of visual-spatial attention in humans have been investigated using ERPs in numerous studies (reviewed in Hillyard and Anllo-Vento 1998). A general finding is that stimuli falling within the focus of attention evoke markedly larger amplitudes for several of the short-latency visual ERP components. The earliest of these attention-sensitive components are labeled P1 (post-stimulus onset latency of 80–120 ms) and N1 (140–190 ms), respectively. The amplitude enhancement of the P1 and N1 components elicited by attended stimuli (relative to unattended stimuli) is initially largest over occipital scalp sites contralateral to the visual field of the eliciting stimulus and occurs without producing a change in their latencies, wave shapes, or scalp distributions (see Fig. 84.1). Attention-related modulations of the P1 and N1 components have been reported in a variety of spatial attention paradigms and stimulus configurations, and have been considered a unique electrophysiological signature of spatially-focused attentional selection in humans (Hillyard and Anllo-Vento, 1998).

The finding that only the amplitudes of the early sensory-evoked components are affected by focused attention favors the view that spatial attention operates as a gain control mechanism that amplifies the neural response that is evoked automatically by a particular stimulus (Hillyard and Anllo-Vento, 1998). According to this hypothesis, focused spatial attention influences the responsiveness of the same neurons that process the physical sensory characteristics of the

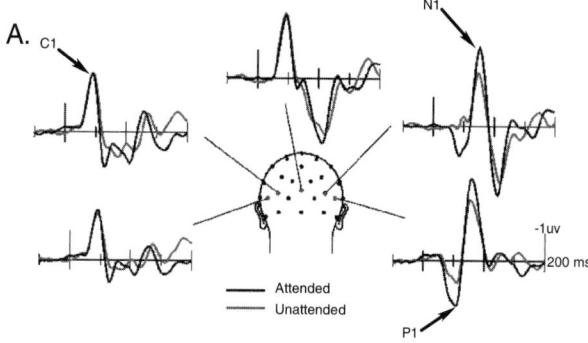

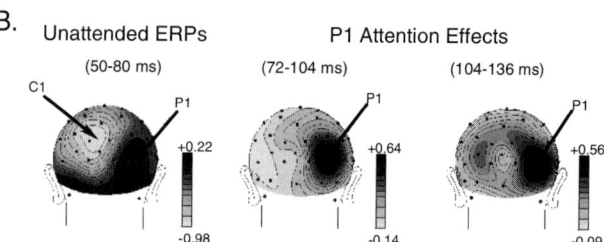

FIGURE 84.1 A. Grand-averaged ERP waveforms from selected scalp sites in response to standard (nontarget) stimuli in the left visual field (LVF) in the study by Martínez et al. (1999). ERPs shown were recorded from electrodes at occipitotemporal (T01/T02), temporal (T5/T6) and midline (IPz) sites. B. Voltage topographies of early ERP components derived from grand average waveforms. The mean voltage over the time window 50–80 ms for ERPs elicited by unattended stimuli in the LVF are shown on the left. The maps of the P1 attention effects over early (middle) and late (right) time windows were derived from the difference waves formed by subtracting the ERPs elicited by standard stimuli when unattended from the ERPs to the same stimuli when attended. Amplitude scales are in microvolts.

eliciting stimulus rather than recruiting a different population of neurons. The relative amplification of sensory inputs from attended spatial locations can presumably improve the signal-to-noise ratios of neurons that code the attended target stimuli.

IV. SOURCE MODELING OF ATTENTION-RELATED ERP COMPONENTS

To relate scalp-recorded ERP attention effects with specific brain structures, several studies have combined electrophysiological recordings and inverse dipole modeling techniques with structural and functional neuroimaging (Clark and Hillyard, 1996; Martínez, Anllo-Vento et al., 1999; Mangun, Hinrichs et al., 2001; DiRusso, Martínez et al., 2003). These studies have found that the earliest sensory-evoked ERP component, termed C1, which has an onset latency of 50–60 ms, does not change significantly with spatial attention (see Fig. 84.1). The C1 has maximal amplitude over midline occipital scalp sites and remains invariant as attention is shifted to and from the location of the evoking stimulus. Dipole modeling of the C1 component suggests that it is generated by a source in primary visual cortex (area V1) and reflects the initial volley of sensory input into visual cortex. This is consistent with other characteristics of the C1, including its topography, short latency, and polarity-inversion characteristics, all pointing to a neural generator in striate cortex (DiRusso, Martínez et al., 2003). Thus, the amplitude modulation of the P1 (beginning at 70–80 ms) appears to represent the earliest effect of spatial attention on visual processing.

To investigate the time course of enhanced neural activity in different visual areas during spatial attention, Martínez et al. (1999) combined ERP recordings and dipole source modeling with fMRI. Identical stimulus and task conditions were utilized in separate ERP-recording and fMRI sessions. Stimuli were delivered in randomized sequences to either the left (LVF) or right visual field (RVF), while subjects covertly attended to one field at a time and detected the occurrence of infrequent target stimuli in the attended visual field. Functional MRI revealed attention-related modulations of activity in several visual cortical areas including area V1 and in the majority of subjects in areas V2, V3, VP, and V4v (see Fig. 84.2). Additionally, most subjects had significant activation in a portion of the middle occipital gyrus, anterior to area V3A, in the posterior fusiform gyrus anterior to area V4v, and in the contralateral posterior parietal cortex. These fMRI activations indicated that spatial attention was associated with enhanced neural activity in circumscribed zones of multiple visual-cortical areas of the hemisphere contralateral to the attended visual field.

Electrophysiological data obtained from the same group of participants revealed enhanced amplitudes of the P1 and N1 components over occipital scalp areas contralateral to the side of the eliciting stimulus (see Fig. 84.1). The grand-average topographical data was used to estimate the neural generators of the attention-insensitive C1 component and the P1 attention effect using dipole modeling (Scherg, 1990). A single dipole in each hemisphere within the calcarine fissure accounted for the C1 component's voltage topography. In contrast, the early phase of the P1 attention effect was localized close to the fMRI activation foci in area V3 and the middle occipital gyrus, and the late phase of the P1 attention effect corresponded with the fMRI activations in area V4 and the posterior fusiform gyrus (see Fig. 84.2). Similarly placed dipoles in dorsal and

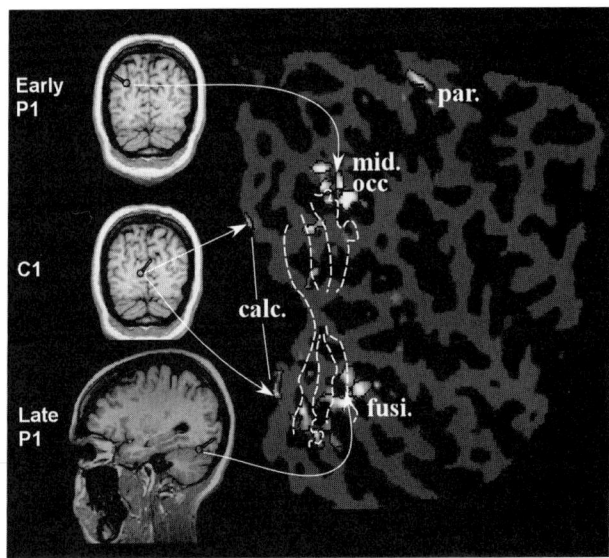

FIGURE 84.2 fMRI activations from a single subject during attention to stimuli in the left visual field are shown projected onto the flattened cortical representation of the contralateral (right) hemisphere (sulcal cortex, dark gray; gyral cortex, light gray). Dashed white lines demarcate the boundaries of each retinotopic visual area obtained by calculating the visual field sign after mapping eccentricity and polar angle. Significant clusters of attention-related activations (outlined in black) are shown in the calcarine fissure (calc.), posterior fusiform gyrus (fusi.), middle occipital gyrus (mid. occ.), and posterior parietal cortex (par.). Anatomical coronal and sagittal MRI images on left indicate the locations of fitted dipoles from the grand-average ERP model in response to the same stimuli. Data from Martínez et al. (1999).

ventral extrastriate areas also accounted for the posterior N1 component in the 160–200 ms latency range (DiRusso, Martínez et al., 2003).

The lack of attention-related changes in the amplitude of the short-latency C1 component stands in sharp contrast to the robust attentional modulations observed in striate cortex as shown by fMRI. In particular, it remains to be explained why it is that the C1 component having its putative origin in striate cortex is unaffected by attention, whereas fMRI indicates the presence of robust attentional modulations in the very same striate areas. This question was explored by further dipole modeling analyses, which revealed a long-latency attention effect at 150–250 ms that was localized to the same striate source that accounted for the early C1 component (Martínez, DiRusso et al., 2001). In a separate study that also combined fMRI with ERP recordings, DiRusso et al. (2003) confirmed a delayed (150–225 ms) attention effect arising from the dipolar source that accounted for the attention-invariant C1 component; furthermore, this attention effect, like the C1, inverted in polarity for upper versus lower field stimuli, consistent with a neural generator in area V1. Similar delayed attention effects localized to striate cortex have also been reported in studies using recordings of magnetic event-related fields (Noesselt, Hillyard et al., 2002).

This delayed enhancement of V1 activity is consistent with recent single-unit recording studies in monkeys that have reported modulations of V1 neuronal activity beginning later than 100ms post-stimulus, well beyond the peak of the initial sensory-evoked response, which was found to remain unchanged as a function of the animal's attentional state (Lamme, Super et al., 2000). Similarly delayed responses in lower visual areas V1 and V2 were found to occur subsequent to attention-related enhancements of neural activity in area V4 (reviewed in Schroeder, Mehta et al., 2001) and may represent modulatory reentrant signals from higher extrastriate areas.

An alternative hypothesis that may account in part for the mismatch between the hemodynamic and electrophysiological effects in striate cortex is that the V1 activity observed with fMRI represents a top-down bias signal, which produces sustained attention-related increases in striate activity without necessarily modulating the initial stimulus-evoked response. Tonic bias effects such as these have been observed only in extrastriate neurons (reviewed in Martínez, DiRusso et al., 2001), but it is conceivable that the same or similar mechanisms might exist in striate cortex. If present, such tonic changes in neuronal activity may not be detectable using stimulus time-locked ERP recording techniques.

In sum, the data reviewed here provide evidence that the primary visual cortex does play a role in spatial attention, but this area does not serve as the locus of initial sensory gain control where attended visual inputs are first selectively enhanced. This essential function of attentional amplification (Posner and Dehaene, 1994), which improves the perceptibility of stimuli at attended locations, seems to be initially carried out in retinotopically organized extrastriate visual areas including V3-V4. These amplified signals may then be conveyed back to area V1 by reentrant feedback projections, which may function to improve the figure/ground segregation and salience at attended stimuli (Lamme, Super et al., 2000). Ultimately, these enhanced signals are routed to higher visual areas including those of the occipito-temporal ventral stream to gain preferential access to limited-capacity stages of feature analysis and pattern recognition. These pathways, as well as the pathways implicated in top-down control of attention, are shown in Fig. 84.3.

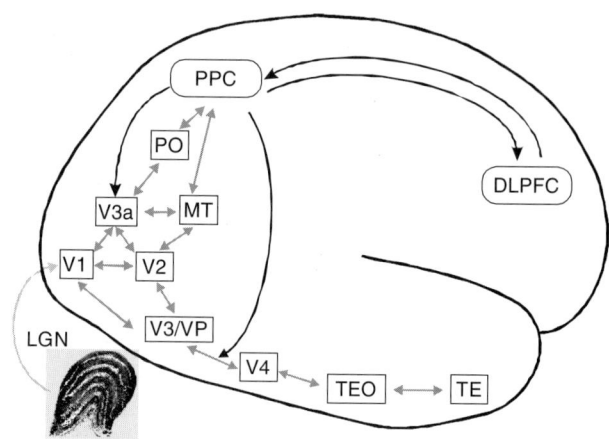

FIGURE 84.3 Proposed pathways mediating spatial attention based on ERP, MEG, and fMRI evidence. A top-down attentional control network consisting of interconnected dorsolateral prefrontal (DLPFC) and posterior parietal (PPC) cortical areas is proposed to modulate incoming visual information via projections to both dorsal and ventral extrastriate visual pathways. Evidence summarized in this chapter suggests that attended inputs are first enhanced dorsally in the region of area V3/middle occipital gyrus and ventrally in the vicinity of area V4/fusiform gyrus in the time range 80–130 msec after stimulus onset. These enhanced signals are then fed forward to higher visual areas and back to lower areas including V1.

V. ERP STUDIES OF ATTENTION TO NONSPATIAL FEATURES

In contrast to spatial attention, the selection of stimuli based on nonspatial attributes such as color or shape is not associated with amplitude enhancements of the sensory-evoked P1 and N1 components (reviewed in Hillyard and Anllo-Vento, 1998). Instead, feature-specific attention effects typically are characterized by a broad negative deflection over posterior electrode sites, the selection negativity (SN), which begins 140–180ms after stimulus onset and persists for 200ms or more. A smaller selection positivity (SP) occurring in the same interval as the SN but with a topographical distribution over anterior scalp regions, is also often observed. Selection negativities and positivities varying slightly in onset latency and scalp distribution have been reported for a number of different nonspatial features such as orientation, color, spatial frequency, direction of motion and shape (Hillyard and Anllo-Vento, 1998; Martínez, DiRusso et al., 2001; Baas, Kenemans et al., 2002).

The SN and the SP can be observed in the ERP difference waves formed by subtracting ERP waveforms elicited by stimuli having the attended feature dimension from waveforms elicited by the same stimuli when unattended. As visualized in these difference waves, the onset of the SN and SP (the point at which the difference potentials differs from zero) provides a good measure of the timing of when the attended feature begins to be selectively processed. Additionally, inverse source localization of the SN and SP can provide information about the specific neural generator(s) that participate in the selection of different stimulus features.

In a recent study (Martínez, DiRusso et al., 2001), we investigated the timing and neuroanatomical sources of the SN and associated ERPs during attention to spatial frequency. Subjects were presented with a stream of randomly intermixed gabor patches consisting of either high or low spatial frequencies. In separate conditions, the task was to selectively focus attention on one of the two spatial frequencies (high or low) and detect the occurrence of occasional targets having a slightly different spatial frequency.

ERP components associated with selective attention to spatial frequency were identified in difference waves formed by subtracting the ERPs elicited by one spatial frequency stimulus when it was not attended from the ERPs elicited by the same stimulus when it was attended. The earliest ERP signs of selection for spatial frequency began at approximately 100 ms and were characterized initially by a negative deflection in the attentional difference wave (ND120) for high spatial frequencies and a positive-going (PD130) component for low spatial frequencies. These short-latency effects were localized by dipole modeling to dorsal medial and dorso-lateral portions of extrastriate cortex, respectively (see Fig. 84.4).

The early attention effects ND120 and PD130 began about 40 ms after the onset latencies of the exogenous C1 and P1 components, and hence, did not appear to reflect amplitude modulations of these short-latency stimulus-evoked ERPs. Moreover, the topographical distributions and modeled dipolar sources of the early spatial frequency attention effects differed from those of the evoked P1 and C1 waves, with the C1 being localized in or near striate cortex as in previous studies.. These temporal and topographical differences suggest that the PD130 and ND120 reflect stimulus selection processes that do not involve a simple modulation of the initial sensory evoked response. Thus, in contrast with some reports (Zani and Proverbio, Chapter 85) but in agreement with others (Baas, Kenemans et al., 2002), no evidence of early modulation of sensory-evoked activity in the striate cortex was obtained for stimulus selection based on spatial frequency.

The most prominent component in the difference waves was a selection negativity with a bilateral dipolar source in ventral extrastriate cortex elicited by both high and low frequency stimuli when they were attended (see Fig. 84.4). This SN resembled that elicited by stimuli selected on the basis of other nonspatial stimulus features and may index neuronal activity within ventral extrastriate areas specialized for the processing of different visual features including spatial frequency information. Consistent with current models of hemispheric specialization for spatial frequency processing, the high and low frequency SNs were asymmetrically distributed, with higher amplitudes over the left and right hemispheres, respectively.

VI. CONCLUSION

It is evident that spatial frequency selection is characterized by a different pattern of ERP modulation than is selection for location, indicating that these two processes are mediated by qualitatively different cortical mechanisms. Consistent with this observation is the hypothesis that the integration of object features into coherent objects is contingent upon the prior selection of spatial location (Hillyard and Anllo-Vento, 1998). In this view, spatial attention is "special" and has primacy over the selection of other, nonspatial stimulus features.

The data reviewed here suggests that spatial attention plays a unique role in selecting information from the environment and involves an early gain control over perceptual processing. The weight of the evidence thus far indicates that extrastriate cortical areas are the initial loci of attentional selection and that selection of elementary nonspatial features is hierarchically tied to the initial segmentation of the visual world along spatially defined dimensions. Recent advances in neuroimaging techniques make possible the integration of physiological measures that register spatial and temporal aspects of neuronal activity underlying perception and attention. Such combined data will be important for testing alternative conceptualizations of selective attention that have emerged from decades of behavioral studies and will provide an essential link between human and animal studies of attention.

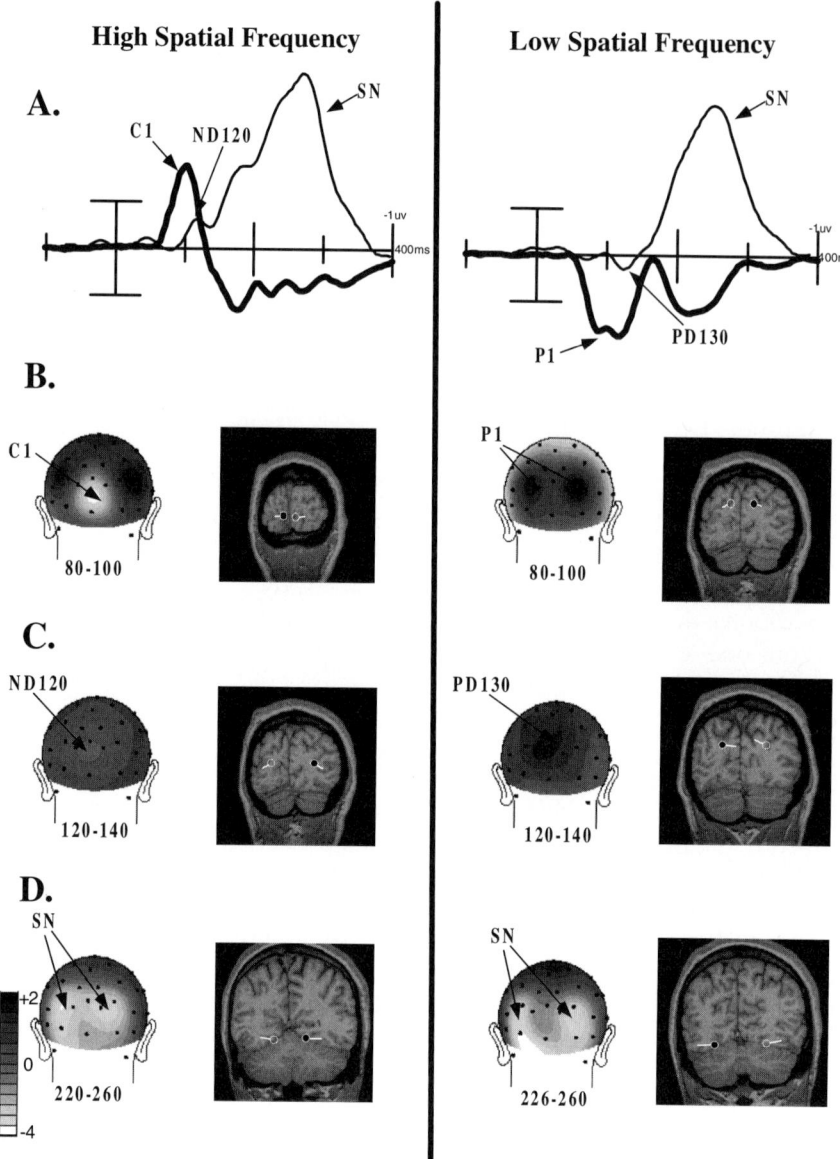

FIGURE 84.4 A. Grand-averaged ERPs elicited by high and low spatial frequency gratings (thick tracings) and the corresponding attentional difference wave (thin tracing), recorded from a representative electrode site over the posterior scalp. B. Topographical distributions of early ERPs in the 80–100 ms latency range to unattended gratings are shown to the left of coronal sections showing positions of fitted dipoles accounting for the unattended C1 (high frequency) and P1 (low frequency) components. C. Topographies of early ND120 and PD130 attention effects are shown to the left of corresponding dipole models in the 120–140 ms time range. D. Voltage topographies of selection negativity (SN) in the high and low spatial frequency attentional difference waves at 220–260 ms (left) and localization of source dipoles on corresponding coronal sections (right). Voltage scale is in microvolts. Data from Martínez et al. (2001).

References

Baas, J. M., Kenemans, J. L., et al. (2002). Selective attention to spatial frequency: an ERP and source localization analysis. *Clinical Neurophysiology* **113**, 1840–1854.

Clark, V. P., and Hillyard, S. A. (1996). Spatial selective attention affects early extrastriate but not striate components of the visual evoked potential. *Journal of Cognitive Neuroscience* **8**, 387–402.

DiRusso, F., Martínez, A., et al. (2003). Source analysis of event-related cortical activity during visuo-spatial attention. *Cerebral Cortex* **13**, 486–499.

Hillyard, S. A., and Anllo-Vento, L. (1998). Event-related brain potentials in the study of visual selective attention. *Proceedings of the National Academy of Sciences* **95**, 781–787.

Lamme, V. A., Super, H., et al. (2000). The role of primary visual cortex (V1) in visual awareness. *Vision Research* **40**, 1507–1521.

Mangun, G. R., Hinrichs, H., et al. (2001). Integrating electrophysiological and neuroimaging of spatial selective attention to simple isolated visual stimuli. *Vision Research* **41**, 1423–1435.

Martínez, A., Anllo-Vento, L., et al. (1999). Involvement of striate and extrastriate visual cortical areas in spatial attention. *Nature Neuroscience* **2**, 364–369.

Martínez, A., DiRusso, F., et al. (2001). Electrophysiological analysis of cortical mechanisms of selective attention to high and low spatial frequencies. *Clinical Neurophysiology* **112**, 1980–1998.

Martínez, A., DiRusso, F., et al. (2001). Putting spatial attention on the map: timing and localization of stimulus selection processes in striate and extrastriate visual areas. *Vision Research* **41**, 1437–1457.

Noesselt, T., Hillyard, S. A., et al. (2002). Delayed striate cortical activation during spatial attention. *Neuron* **35**, 575–587.

Posner, M. I., and Dehaene, S. (1994). Attentional networks. *Trends in Neurosciences* **17**, 75–79.

Regan, D. (1989). *Human brain electrophysiology: Evoked potentials and evoked magnetic fields in science and medicine.* Elsevier, New York.

Rugg, M. D., and Coles, M. G. H. Eds. (1995). *Electrophysiology of Mind-Event-Related Potentials and Cognition.* Oxford University Press, Oxford.

Scherg, M. (1990). Fundamentals of dipole source analysis. In "Auditory Evoked Magnetic Fields and Electric Potentials" (F. Grandori, M. Hoke, and G. L. Roman). Basel, Karger, 40–69.

Schroeder, C. E., Mehta, A. D., et al. (2001). Determinants and mechanisms of attentional modulation of neural processing. *Frontiers in Bioscience* **6**, D672–684.

CHAPTER 85

The Timing of Attentional Modulation of Visual Processing as Indexed by ERPs

Alberto Zani and Alice Mado Proverbio

ABSTRACT

The most debated matter in electrophysiological investigations of neural mechanisms governing attentional selection of visual information is the timing of attentional modulation of processing in the visual pathways. Indeed, because of their high temporal resolution (in the order of milliseconds), but their scarce spatial resolution in localizing neural processing, event-related potentials (ERPs) of the brain have contributed predominantly to determining the time at which a particular attentional manipulation affects visual processing, and to providing evidence for the debate on early versus late selection. After a review of our own ERP studies and *in vivo* cellular, ERPs/ERFs, and fMRI studies of other authors, we advance the opinion that although modulation of spatial attention processing might start after relatively longer latencies and in extrastriate areas, feature-directed attention affects striate cortical, if not subcortical, regions at a very early time point.

I. INTRODUCTION

Scalp-recorded event-related potentials (ERPs) and event-related fields (ERFs) of the brain are voltage and magnetic field fluctuations generated by changes in membrane potentials of large cell assemblies in the underlying brain tissue. These fluctuations can be recorded using external sensors, which, unlike for ERFs, for ERPs recording are called electrodes. Usually, they are fixed on elastic caps that can accommodate a minimum of 16, and a maximum of 256, electrodes that are placed all over the human scalp following standard coordinates more or less guaranteeing their positioning in certain cortical areas. The electrodes are indicated by one or more letters, denoting the area(s) in which they fall (e.g., O for occipital, T for temporal, etc.), and a number, depending, among other factors, on the distance of the site from the medial sagittal line.

What is recorded as a unitary continuous waveform at scalp sites over the various brain lobes actually consists of the sum of multiple positive and negative phasic deflections or potentials whose polarity is indicated by the letters P and N, respectively, followed by increasing numbers denoting the temporal progression of their appearance in time (e.g., P1, N1, P2, etc.). Earlier potentials are much more strongly affected by physical features of sensory events than the later components, which reflect higher order cognitive or behavioral responses.

Besides polarity (negative or positive), ERP/ERF components are characterized by manyfold attributes: amplitude, latency, scalp topography, their underlying cerebral source(s), and their specific relationship with task variables. Manipulation of cognitive tasks may affect both amplitude and latency of a component besides its topographic scalp distribution. ERPs/ERFs have been recorded during the same types of tasks as those that have been used in attentional behavioral studies. A strong claim was made, and later confirmed, that, because ERPs/ERFs provide a continuous measure of post-stimulus processing, the latency of these components reflects the temporal sequence of processing and offers more specific information on the timing and duration of processing stages than do behavioral responses, such as reaction times, which merely reflect the total output of the processing stages.

ERPs/ERFs have been, and are still being, used extensively to study visual selective processing by comparing the ERPs to attended and ignored space locations or stimuli. When relevant and irrelevant stimuli are presented in a random sequence, the difference in the corresponding ERPs reflects a difference in the stimulus set for the relevant versus irrelevant stimulus attribute. The point in time, or stage, of the longstanding distinction between the early sensory (i.e., evoked) and late endogenous (i.e., invoked) components of the ERPs can be determined when selective attention intervenes to modulate information processing in the visual pathways. The high temporal resolution (in the order of milliseconds) of ERPs makes these an excellent technique for determining the time at which a particular attentional manipulation affects visual processing, and for providing evidence for the debate on early versus late selection.

II. ERPS AND THE TIMING OF SPACE- AND FEATURE-DIRECTED VISUAL ATTENTION

Countless ERP studies on visual spatial attention tasks have shown modulation of the major early ERP/ERF responses measured at scalp sites over the occipital lobes and termed P1 (onset at 70–100 msec) and N1 (onset at 150–200 ms.) Current dipole mapping demonstrates that these components originate in extrastriate visual cortex (see, for instance, Clark and Hillyard, 1996; Martinez et al., 2001; or Mangun, 2002 for a recent review). It is also true, however, that an ERF study examining the effects of attending a conjunction of location and spatial frequency provided some evidence of a strong effect of attention at about 150 ms, in this case localized to striate cortex. Based on this finding, the conclusions advanced by the authors were that this was a sign of feedback into striate cortex, determining a reactivation of this area as a function of attention (Aine et al., 1995). In addition, although Oakley and Eason (1990) reported an intriguing, yet unreplicated, very early negative attention effect within the 40–70 ms range, which these authors considered as evidence of precortical gating, there are consistent findings of a lack of any influence of spatial attention on the so-called visual C1 response (latency range 50–110 ms post-stimulus) (Martinez et al., 2001; Mangun, 2002). This C1 response, which precedes the P1, appears as a negative or positive wave in function of upper or lower hemiretinal stimulation, respectively, and is consistent with the crossed retinotopic organization of visual pathways and the calcarine fissure. Indeed, several dipole localization studies have found that this component arises entirely (Clark and Hillyard, 1996; Martinez et al., 2001; Mangun, 2002), or just initially (Foxe and Simpson, 2002), from striate cortex. Based on the aforementioned ERP findings the undisputed view is shared among ERPers that spatial selection does affect visual processing starting beyond the lowest striate levels and the earliest sensory-evoked recordable cortical activity.

Conversely, just how early in time nonspatial selection (e.g., color, motion, spatial frequency, etc.) can be is still much debated, because the effects reported have not always been replicated (at this regard see Proverbio et al., 2002, for a review of some major reasons for this; see also Chapters 82 and 84). A first nonspatial attention study by our group (Zani and Proverbio, 1995) presented a set of six checkerboards producing foveal stimulation at spatial frequencies ranging between 6 and 0.5 cycles per degree (cpd) of visual angle. Volunteers selectively paid heed to one spatial frequency, which was varied from time to time in different runs, and ignored all the others. ERPs were measured from O1 and O2, mesial-occipital, and OL and OR, lateral-occipital, homologous electrode sites. Our findings indicated that attention to check size was able to modulate ERPs as early as 90 msec poststimulus. Indeed, when the check-sizes were attended to rather than ignored, a larger positivity was recorded within 70–110 ms (P90) at the lateral-occipital electrodes and a larger negativity within an 80–130 msec latency range (N115) at the mesial-occipital electrodes. Interestingly, these effects somehow also were found for the smallest check-size differences within the bandwidth. Though of tiny magnitude, these relatively early latency effects proved to be highly robust. Since these effects concerned the size-specific channels of the visual cortex, and since some invasive single cell studies in monkeys had suggested that even neurons near the beginning of the visual-processing pathways can tell whether an animal is paying attention to a particular visual stimulus (e.g., Motter, 1998, for a review of these early and later studies), we proposed that our findings indicated (a) an attentional modulation of nonspatial visual processing, and (b) that attention influences neuronal activity in V1, or striate cortex, as well as extrastriate cortex.

In a further investigation (Zani and Proverbio, 1997), in which attention to a conjunction of location (L) and spatial frequency (F) was investigated, seven volunteers were shown a series of four sinusoidal gratings (0.75, 1.5, 3, and 6 cpd) randomly presented in the four quadrants of the visual field. In different runs, the volunteers had to pay heed and respond to a conjunction of the two features (e.g., 0.75 cpd in the URQ), and to ignore all the other frequencies and quadrants. Thus, in separate conditions one and the same stimulus could be relevant in both features, relevant in location–irrelevant in frequency, irrelevant in location–relevant in frequency, and irrelevant in both features. Though the C1 response hinted at some attention effects, we did not venture to make systematic analyses of these effects. Since most previous ERP studies of visual attention had concerned the P1 peak, we measured this wave within the latency range of 60–140 ms post-stimulus as a function of experimental conditions. As shown in Fig. 85.1, the conjoined processing of both relevant location and frequency elicited a larger P1 amplitude than processing of either single relevant feature, and neither relevant feature (for features binding issues see Chapter 24; for the multiplicative nature of the effects of space- and feature-directed conjoined attention, see Chapter 49; see also the "Discussion" section of Zani and Proverbio, 1997). Very interestingly, processing of a frequency irrelevant grating within the attended location elicited a significantly larger P1 than processing neither relevant feature. This held true independently of the visual quadrant. Our findings of feature-selection dependent increases in the P1 response at the attended location indicated that, unlike what was thought before, relatively early changes in extrastriate activation, and in turn in perceptual sensitivity, may also result from a selection of nonspatial features. This conclusion is supported by findings of object-based P1 enhancements by Weber et al. (1997).

To confirm our previous results and to compare spatial and nonspatial selection further, we later set up a between-subjects attention paradigm (Zani et al., 1999) in which 1 and 7 cpd spatial frequency gratings were presented in the upper quadrants of the visual field while subjects were instructed to pay heed and respond to the location independently of the spatial frequency, or vice versa. As can be seen in Fig. 85.2, attention to frequency clearly manifested as early as within the latency range of C1 deflection (i.e., 60–80 ms post-stimulus), whereas, consistent with evidence in the literature, attention to location effects started at the P1 level, beyond the sensory-evoked afferents to the primary cortex, reflected by the C1. In the light of the dipole localization evidence reported in the literature, these feature-selection-based C1 findings demonstrated that the striate cortex was activated by nonspatial attention and that this activation was modulated at very early time points.

In seeking confirmation of our findings of this very early nonspatial attention effect at the C1 level, we recently reran the conjoined selection task of location and spatial frequency (Zani and Proverbio, 1997). This time 21 volunteers were tested, and instead of simply measuring the amplitude of the P1 peak, we measured the mean amplitude of the early time portion of ERP waveforms in four 20 ms-long latency windows starting as early as 60 ms post-stimulus (60–80, 80–100, 100–120, and 120–140 ms). Figure 85.3 reports the grand-average ERPs for the four attentional conditions with an expanded time scale to highlight these earliest latency responses. As clearly visible, the C1 wave changed in polarity as a function of stimulation of upper or lower quadrants, whereas the longer latency P1, though of smaller amplitude and shorter latency in the lower quadrants than in the upper ones, consistently appeared as a positive response independently of stimulus location. Most interestingly, early latency visual processing appeared to be modulated by frequency relevance, *per se* or in interaction with location, starting as early as in the 60–80 ms time window and further continuing throughout the following P1-related ones. Overall, this

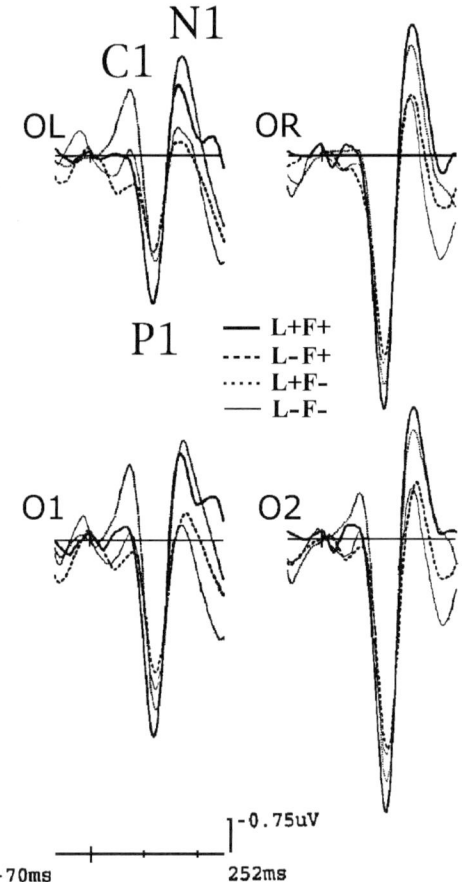

FIGURE 85.1 Grand average ERPs from mesial (O1 and O2) and lateral occipital (OL and OR) homologous sites in response to 0.75 cpd black-and-white sinusoidal gratings presented in the upper left quadrant (ULQ) as a function of feature relevance during a conjoined selection task of spatial location (L) and frequency (F): L+F+: both features relevant; L−F+: location irrelevant, frequency relevant; L+F−: location relevant, frequency irrelevant; L−F−: both features irrelevant. Note that ERP waveforms are drawn with an expanded time scale to highlight the early timing of visual processing. Reproduced and modified with permission from the *Journal of Psychophysiology*, Vol. 11 (1) 1997, pp. 21–37.

Attention to spatial location

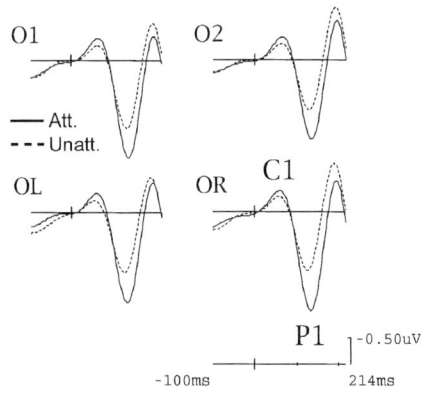

Attention to spatial frequency

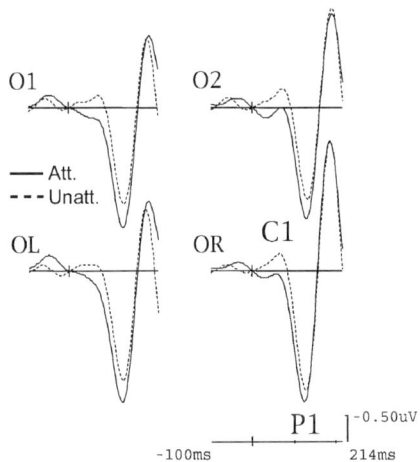

FIGURE 85.2 Top: Grand average ERPs from mesial and lateral occipital sites to 7 cpd when falling at an attended (Att.) or unattended (Unatt.) location. Bottom: Grand average ERPs to 6 cpd when attended and unattended independently of the spatial location. Note that while in the former condition attentional modulation started at P1 latency, in the latter case attentional modulation started as early as at C1 latency.

modulation consisted in a larger positivity for frequency relevant than frequency irrelevant gratings. Conversely, location relevance effects started in the later 80–100 ms window and continued steadily throughout the following three time windows (Zani and Proverbio, in preparation).

All in all, these findings indicate that the selection of a stimulus feature, such as spatial frequency, at an attended location is carried out through the modulation of visual pathways at the earliest post-stimulus time points (60–80 ms). This seems to hold even if, as recently advanced by Foxe and Simpson (2002), only

0.7 cpd grating over the horizontal meridian

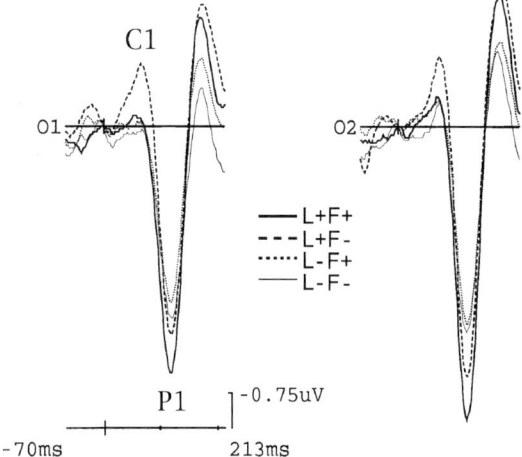

0.7 cpd grating below the horizontal meridian

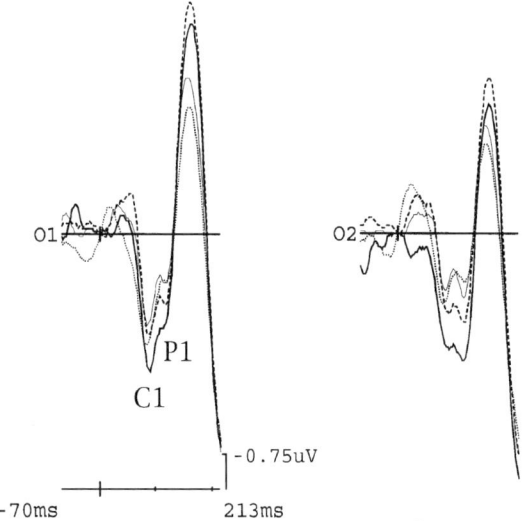

FIGURE 85.3 ERPs to 0.7 cpd recorded at mesial occipital sites while paying attention to a conjunction of spatial frequency (F) and location (L) and grand averaged across the left and right quadrants of the visual field falling, respectively, above and below the horizontal meridian as a function of attention condition: relevant (+), irrelevant (−). Note that although the C1 wave changed in polarity across the horizontal meridian, overall the attention effects manifested as an increase in positivity, as they also did at the longer P1 latency.

the initial portion—about the first 10–15 ms—of the C1 wave (C1e: mean onset latency of activity over occipital cortex at 56 ms) represents a response predominated by V1 activity, and the remaining portion of it is actually derived from multiple visual generators, thus making it possible that, besides P1 and N1, C1 too probably reflects relatively later processing than the initial volley of sensory afferents to the primary visual cortex.

Interestingly, hemodynamic evidence collected with fMRI indicates that the striate occipital cortex too is affected by visual spatial attention (see Kanwisher and Wojciulik, 2000, for a review of these data; see also Chapters 61, 71, and 51). It must be said that the lack of high-temporal resolution of fMRI measurements leaves it open to doubt whether the modulation represents activation of this occipital area at very early points in time. In this regard, it has been proposed that the attentional modulation of striate activity found with fMRI, but not with ERPs in a study using both techniques, might represent a delayed feedback modulation of the initial striate response from higher visual extrastriate or parietal cortices (Martinez et al., 2001).

In support of this hypothesis there are studies based on single cell recordings indicating that neuronal attentional activation in V1 starts as late as 139 ms post-stimulus in a simple tracing, and as late as 159 ms in a search and tracing composite task (Roelfsema et al., 2003). Again, there are MEG data localizing a relatively longer latency component in the 130–160 ms range to the calcarine fissure, interpreted as a sign of feedback reactivation of this region, during conjoined attention to location and spatial frequency (Aine et al., 1995).

Notwithstanding this electrophysiologically timed evidence, unprecedented fMRI findings of a modulation of the thalamic lateral geniculate nuclei during visuo-spatial selective processing have also been recently reported (for a review of fMRI evidence in favor of such earliest gating mechanisms see Chapters 72 and 71).

III. CONCLUSION

The findings reported here open up several intriguing scenarios. The first is that both striate cortex and, before that, the earlier thalamic levels of the visual pathways are modulated by spatial attention, but that this intracranial modulation is harder to pick up by ERPs/ERFs perhaps because it is too small or because it represents a prestimulus baseline increase in neural activity, as tentatively suggested by Kanwisher and Wojciulik (2000). However, in the light of the earliest small, though robust, ERP findings obtained in our spatial frequency selection tasks as well as in our conjoined features selection task, this caveat concerning ERP methodology does not seem to hold (for a more extensive review of ERP early effects for feature-directed visual attention see Proverbio et al., 2002; see also Chapter 82).

The second scenario is that the fMRI signs of modulation of early levels in the visual streams by spatial attention are actually a reflection of a delayed feed back from higher brain areas. Unlike previously ever thought, this delayed feedback is directed to both the striate cortex and subcortical districts of the visual streams. Thus, the fMRI and ERP findings truly reflect different levels of visual processing and attending a spatial location does not modulate the activity of early levels of the visual streams early in time, as suggested by the longer latency effects (70–140 msec) found by ERPs/ERFs. This second scenario seems to be supported by the lack of any earlier effects, with the exception of those recorded in Oakley and Eason's study (1990), than the P1 latency in the countless ERP studies on spatial attention, including our own study (Zani et al., 1999).

If this is true, however, a further scenario might well be advanced on the basis of the different results we obtained from the tasks requiring attention to location or frequency and to location and frequency. Indeed, it is possible that while attending a spatial location the processing activity might be modulated at a longer latency, as reflected by P1 effects, because this task does not impose such a heavy processing load as to require the activation of the lowest levels of visual pathways at the earliest time points. Conversely, while selecting an object (or an object feature such as spatial frequency) from among several similar ones presented, discrimination or selection of the relevant object features must be carried out in conjunction with other stimulus dimensions, creating a significant work load that may require modulation of activity of the lowest levels, too, of the visual routes at the earliest time, as possibly reflected by the C1 effects recorded at the scalp. This view seems consistent with a theory of visual attention suggesting that the stage of selection depends on the processing load, occurring at later levels with low loads and at early levels with high loads (Lavie, 1995). Consistent with this theory, there are also findings based on cellular recordings in alert monkeys showing that the degree of attentional modulation of V1 increases with the number of items present in the visual field (Motter, 1998).

References

Aine, C. J., Supek, S., and George, J. S. (1995). Temporal dynamics of visual-evoked neuromagnetic sources: effects of stimulus parameters and selective attention. *Int. J. Neurosci.* **80**, 79–104.

Clark, V. P., and Hillyard, S. A. (1996). Spatial selective attention affects early extrastriate but not striate components of the visual evoked potentials. *J. Cogn. Neurosci.* **8**, 387–402.

Foxe, J. J., and Simpson, G. V. (2002). Flow of activation from V1 to frontal cortex in humans. A framework for defining "early" visual processing. *Exp. Brain Res.* **142**, 139–150.

Kanwisher, N., and Wojciulik, E. (2000). Visual attention: insights from brain imaging. *Nature Rev. Neurosci.* **1**, 91–100.

Lavie, N. (1995). Perceptual load as a necessary condition for selective attention. *J. Exp. Psychol.: Hum. Percept Perform* **21**, 451–468.

Mangun, G. R. (2002). Neural mechanisms of attention. In "The Cognitive Electrophysiology of Mind and Brain," (A. Zani and A. M. Proverbio, Eds.), pp. 247–258. Academic Press, San Diego.

Martinez, A., Di Russo, F., Anllo-Vento, L., Sereno, M. I., Buxton, R. B., and Hillyard, S. A. (2001). Putting spatial attention on the map: timing and localization of stimulus selection processes in striate and extrastriate visual areas. *Vision Res.* **41**, 1437–1457.

Motter, B. C. (1998). Neurophysiology of visual attention. In "The Attentive Brain," (R. Parasuraman, Ed.), pp. 51–69. The MIT Press, Cambridge, Mass., London, England.

Oakley, M. T., and Eason, R. G. (1990). Subcortical gating in the human visual system during spatial selective attention. *Int. J. Psychophysiol.* **9**, 105–120.

Proverbio, A. M., Esposito, P., and Zani, A. (2002). Early involvement of temporal area in attentional selection of grating orientation: an ERP study. *Cogn. Brain Res.* **13**, 139–151.

Roelfsema, P. R., Khayat, P. S., and Spekreijse, H. (2003). Subtask sequencing in the primary visual cortex. *PNAS* **100**, 5467–5472.

Weber, T. A., Kramer, A. F., and Miller, G. A. (1997). Selective processing of superimposed objects: an electrophysiological analysis of object-based attentional selection. *Biol. Psychol.* **45**, 159–182.

Zani, A., and Proverbio, A. M. (1995). ERP signs of early selective attention effects to check size. *Electroencephalogr. Clin. Neurophysiol* **95**, 277–292.

Zani, A., and Proverbio, A. M. (1997). Attention modulation of short latency ERPs by selective attention to conjunction of spatial frequency and location. *J. Psychophysiol.* **11**, 21–32.

Zani, A., Avella, C., Lilli, S., and Proverbio, A. M. (1999). Scalp current density (SCD) mapping of cerebral activity during object and space selection in humans. *Biomedizin. Tech.* **44**, 162–165.

CHAPTER 86

Selective Visual Attention Modulates Oscillatory Neuronal Synchronization

Pascal Fries and Robert Desimone

ABSTRACT

When we attend to one object in a natural scene with many items, the attended item is fully processed in the ventral processing pathway of visual cortex, whereas distracting items are filtered from the receptive fields of neurons along this same pathway. It is therefore likely that, at any given level of the pathway, neuronal responses to attended stimuli have a greater impact on downstream neurons, compared to unattended stimuli. We tested whether this processing advantage for attended items might be achieved by a high-frequency synchronization of the activity of neurons whose receptive field contains the attended stimulus. Neurons in cortical area V4 were recorded while macaque monkeys attended to behaviorally relevant stimuli and ignored distracters. Neurons activated by the attended stimulus showed increased gamma frequency (35–90 Hz) synchronization but reduced low frequency (<17 Hz) synchronization, compared to neurons at nearby V4 sites activated by distracters. We propose that the modulation of oscillatory neuronal synchronization is a general mechanism for rapidly modulating the downstream impact of neurons carrying behaviorally relevant information.

I. ATTENTIONAL MECHANISMS MODULATE NEURONAL IMPACT

We have a remarkable capacity to recognize objects in spite of variations in their location or appearance. This capacity is probably related to the response characteristics of neurons in high-level visual areas that show selectivity for complex stimulus features but are fairly invariant to object size, retinal position, and so on. As one consequence of this invariance, neurons in inferior temporal cortex commonly have receptive fields (RFs) that span much of the central visual field. In a typical visual scene, such an extended RF will contain multiple distinct objects. In the absence of selective visual attention, firing rates under those conditions are a weighted average of the responses to the individual stimuli and thus may be hard to interpret by downstream neurons (Reynolds, Chelazzi, and Desimone, 1999). Attentional mechanisms therefore are needed to bias access to higher visual cortical areas in favor of input from the attended stimulus.

One way to accomplish this bias is to enhance the postsynaptic impact of neurons in early visual areas that are activated by the attended stimulus (Reynolds et al., 1999). Neurons can enhance their impact on postsynaptic targets in a number of ways, the two most obvious ones being an increase in firing rate and an increase in precise synchronization among their action potentials. Single cell recordings in monkey visual cortex demonstrate that firing rates often are not affected by attention when a single stimulus occupies the RF (Luck, Chelazzi, Hillyard, and Desimone, 1997), unless the stimulus is of low contrast, that is, within the contrast-response range of the cell (Reynolds, Pasternak, and Desimone, 2000). However, it is still possible that attention modulates spike synchronization, which has been shown in *in vivo* intracellular recordings to be an effective modulator of postsynaptic responses (Azouz and Gray, 2003). For those reasons, we set out to test whether selective visual attention modulates oscillatory synchronization among neurons in area V4 of two awake macaque monkeys (Fries, Reynolds, Rorie, and Desimone, 2001).

II. SELECTIVE ATTENTION MODULATES OSCILLATORY NEURONAL SYNCHRONIZATION

We recorded both spikes from small clusters of neurons (multiunit activity) and local field potentials (LFP) simultaneously from multiple V4 sites with overlapping receptive fields (RFs). The monkeys were trained to keep fixation on a central spot. On separate trials, the monkeys' attention was directed either into the RF of the recorded neurons or to an equally eccentric location in another quadrant of the visual field. The cue directing attention was either a short (0.75°) line next to the fixation spot, pointing to the location of the upcoming target, or alternatively, the fixation spot color, red cueing the upper, green the lower upcoming stimulus. The delay between cue and stimulus onset was 1500–2000 ms and the cue remained throughout the trial. In a subset of recordings, we used a blocked trial design without any explicit cue and obtained essentially identical results.

The stimuli were two patches of moving square wave gratings that were presented at equal eccentricity, one inside and one outside the RFs (see Fig. 86.1C). After a random interval of 500–5000 ms, the white stripes of the cued (target) stimulus changed to isoluminant yellow. The monkey was rewarded if it maintained fixation throughout the trial and released a bar within 650 ms of the color change. On half the trials, the same color change occurred for the uncued (distracter) stimulus. Responses to distracter changes resulted in a timeout without reward. Performance was 83–87% correct and we recorded 100 to 300 correct trials per attention condition. The recorded neurons were exclusively activated by the stimulus inside the RF and we compared neuronal activity between the two attention conditions for correctly performed trials. We refer to the condition with attention into the RF as "with attention," always implicitly comparing to identical sensory conditions but with attention outside the RF.

Figure 86.1 shows one example pair of recording sites. The response histograms (see Fig. 86.1D) demonstrate stimulus-evoked responses but attention does not have any clear effect on the firing rates, either before stimulus onset or during the sustained stimulus driven response. Next, we tested whether attention had any effect on local neuronal synchronization. To this end, we calculated spike-triggered averages (STAs) of the LFP (Fries, Roelfsema, Engel, König, and Singer, 1997; Fries, Schröder, Roelfsema, Singer, and Engel, 2002). Spikes and LFPs were taken from different electrodes, separated by at least 650 μm. Whenever the spike electrode recorded a spike, the LFP trace

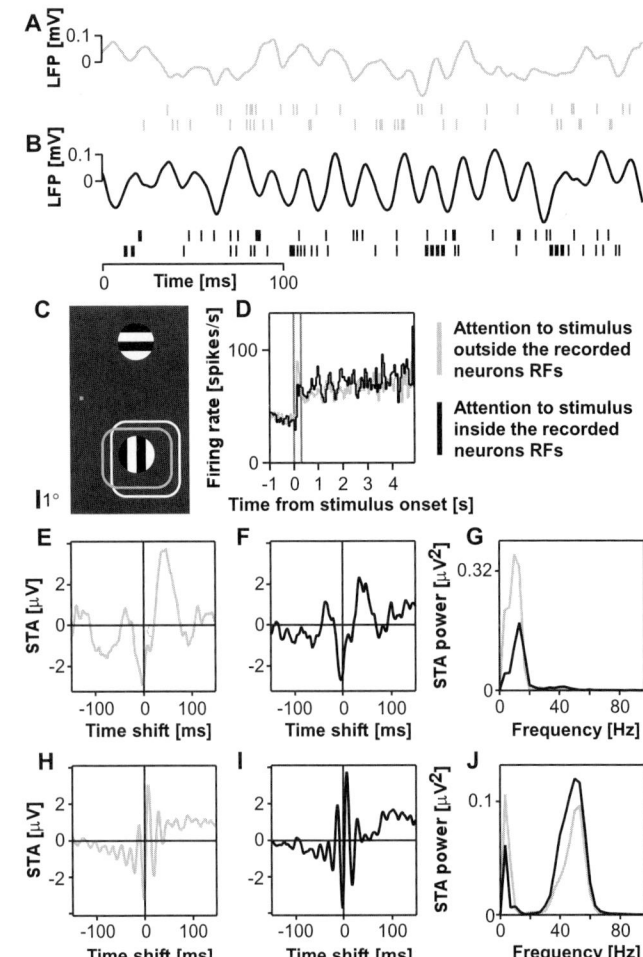

FIGURE 86.1 Attentional modulation of oscillatory synchronization between spikes and LFP from two separate electrodes. Raw stimulus-driven LFP and multiunit activity with attention outside the RF (A) and into the RF (B). C. RFs (not visible to monkey; white: spike recording site, grey: LFP recording site), fixation point and grating stimuli are to scale. The RFs for both recording sites were determined from the multiunit activity and included only one of the two stimuli. In separate trials, this stimulus was either attended or ignored. Data are from 300 correct trials per attention condition. D. Firing rate histograms. Vertical lines indicate stimulus onset and 300 ms after stimulus onset. Delay period was the 1 s interval before stimulus onset, and stimulus period was from 300 ms after stimulus onset until one of the stimuli changed its color. Delay period STAs for attention outside the RF (E) and into the RF (F) and respective power spectra (G). Stimulus period STAs for attention outside the RF (H) and into the RF (I) and respective power spectra (J).

from the other electrode recorded from 128 ms before to 128 ms after that spike was added to the STA and later, this sum was divided by the total number of spikes. If neuronal activity at the two sites were independent of each other, the STA would turn out flat. However, the STAs typically showed a pronounced negative central peak and a clear oscillatory modulation. Since LFP negativities generally reflect depolarizations of pre- or postsynaptic neuronal elements, this STA modulation demonstrates oscillatory synchronization between neuronal activities at the two recordings sites. This oscillatory synchronization was present both during the delay (see Fig. 86.1E, F) and the stimulus period (see Fig. 86.1H, I). To further quantify the strength of this oscillatory synchronization, we calculated the power spectra of the STAs, giving the strength of oscillatory modulation as a function of frequency. During the delay, the power spectra of the STAs (see Fig. 86.1G) were dominated by frequencies below 17 Hz. Attention reduced this low frequency synchronization. During the stimulus-evoked response, the power spectra of the STAs typically showed two distinct bands (see Fig. 86.1J). One band was below 10 Hz, and the other extended from 35 to 60 Hz, which corresponds to the classical gamma band typically observed in visual cortex upon appropriate stimulation. Attention led to a reduction in <10 Hz synchronization, and, conversely, to an increase in gamma frequency synchronization.

To test whether these changes in synchronization were precisely localized within V4, we made additional recordings with the stimulus outside the RF very close to the RF border (data not shown). Even with closely-spaced stimuli, we found the same attentional modulation of synchronization as with the second stimulus far away. In addition to these changes in synchronization, firing rates to the RF stimulus were also moderately suppressed when attention was directed to the surround stimulus, consistent with previous studies of competitive interactions between stimuli in V4 RFs with medium-contrast stimuli. Large firing rate changes with attention occurred only with a competing stimulus very near to the RF border.

The presented examples were typical for the entire set of recordings from two hemispheres of two monkeys. For the statistics, we determined for each pair of recording sites the spike-field coherence (SFC) (Fries et al., 1997; Fries et al., 2002). The SFC is computed by taking the power spectrum of the STA and dividing it by the average power spectrum of all the LFP pieces that had been averaged to obtain the STA. Since the SFC is normalized for both spike rate and LFP power it is immune to changes in those parameters. The SFC measures phase-synchronization between spikes and LFP oscillations as a function of frequency. It ranges from 0 for a complete lack of synchronization, to 1 for perfect phase synchronization. We analysed the SFC for the frequency bands that were obvious from the SFC spectra and pooled data for the stimulus configurations in which the distracters were near and far to the RF.

For the delay period (see Fig. 86.2A, B), low frequency SFC was reduced by a median of 51% with attention (160 decreases, 23 increases, $p < 10^{-6}$) (Worden, Foxe, Wang, and Simpson, 2000). The delay period STAs and the corresponding SFCs did not show clear gamma-frequency modulations (see Fig. 86.1E–G). However, statistically, the gamma band SFC (35 to 60 Hz) increased by a median 10% with attention (106 increases, 77 decreases, $p < 0.02$). Delay firing rates were increased by a median 5% with attention, which was not significant (35 increases, 26 decreases, $p = 0.13$). During the stimulus period (see Fig. 86.2C, D), low frequency SFC was reduced by a median of 23% with attention (142 decreases, 65 increases, $p < 10^{-6}$), and gamma frequency SFC increased by a median of 19% (167 increases, 40 decreases, $p < 10^{-6}$). Firing rates were enhanced by a median of 16% with attention (68 increases, one decrease, $p < 10^{-6}$).

This analysis of the sustained response excluded the first 300 ms after stimulus onset to avoid response onset transients. We separately analysed the poststimulus time course of firing rates and LFP. For the recording site shown as an example in Fig. 86.3B, attention did not affect the mean firing rate until about 420 ms after stimulus onset, consistent with other recent

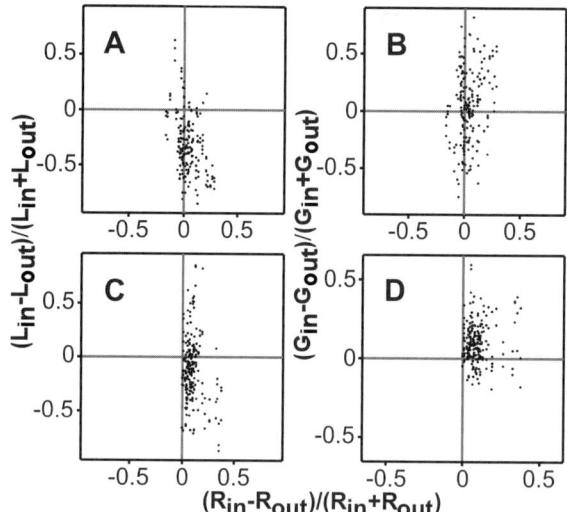

FIGURE 86.2 Population measures of attentional effects on the SFC. Scatter plots compare attentional effects on low and gamma frequency SFC and on firing rates. Each dot represents one pair of recording sites. X- and Y-axis values are attentional indices defined as AI(P) = [P(in) − P(out)] / [P(in) + P(out)], with P being one of the three parameters under study, low frequency synchronization (L), gamma frequency synchronization (G), and firing rates (R). P(in) is the value of the parameter with attention directed into the RF, P(out) with attention directed outside the RF. A, B. Activity from the 1 s delay period before stimulus onset. C, D. Activity from the stimulus period.

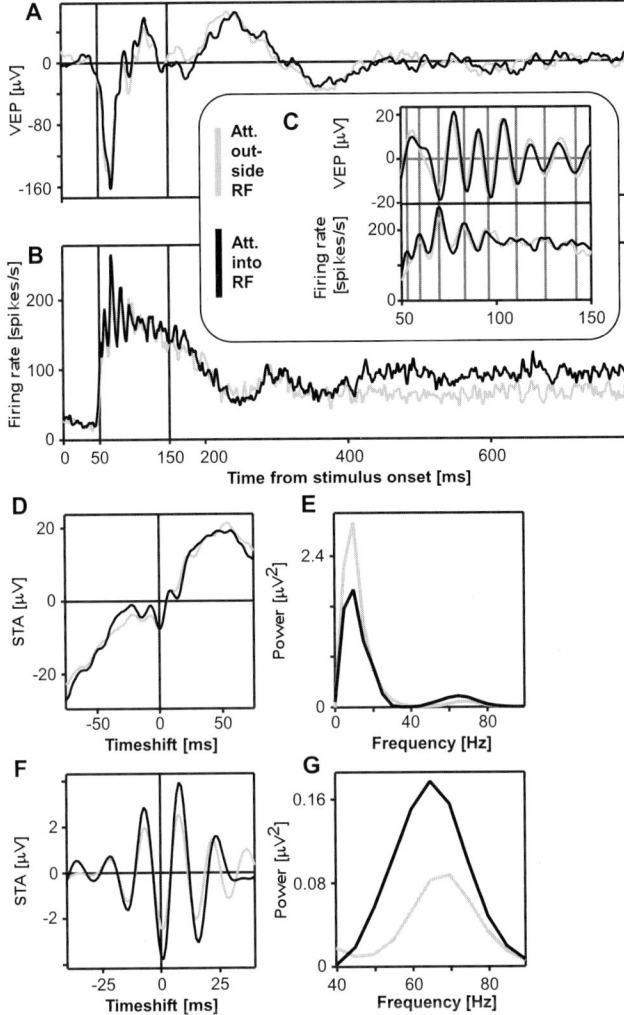

frequency peak at around 65 Hz that was enhanced by attention (see Fig. 86.3F, G). Both the visual evoked potential (VEP) (see Fig. 86.3A) and the spike histogram (see Fig. 86.3B) contained strong stimulus locked gamma frequency oscillations in the first 100 ms of the response (see Fig. 86.3C). Thus, this very early gamma frequency synchronization was at least partially locked to stimulus onset, whereas oscillatory synchronization during the later, sustained visual response was not stimulus locked.

Similar observations were made across the population. Attention did not modulate mean firing rates in the period from 50–150 ms after stimulus onset (median decrease 0.5%; 32 decreases, 29 increases, p = 0.35) and significant sustained attentional effects on mean firing rate did not begin until about 450 ms. By contrast, low frequency SFC in the 50–150 ms period was reduced by a median 8% (108 decreases, 75 increases, p < 0.01) with attention, and gamma frequency (40–90 Hz) synchronization was enhanced by a median 16% during this period (114 increases, 69 decreases, p < 0.0005). Thus, attentional effects on SFC preceded firing rate changes. VEPs showed low frequency power in the 50–150 ms period reduced by a median 12% (45 decreases, 19 increases, p < 0.001) with attention, and gamma frequency power was increased by a median 19% (48 decreases, 16 increases, p < 0.00005). Spike histograms showed a median 15% increase in gamma frequency power with attention (38 increases, 23 decreases, p < 0.05) but only a weak tendency for reduced low frequency power (–2%, 33 decreases, 28 increases, p = 0.26).

FIGURE 86.3 Attention effects in early response. Data are from 300 correct trials per attention condition. VEPs (**A**) and spike histograms (**B**) from two separate electrodes as a function of time after stimulus onset. Vertical lines indicate time period for which STAs (**D, F**) were calculated. The modulation of firing rate by attention starts only about 420 ms after stimulus onset. From 50–150 ms after stimulus onset, there are stimulus-locked gamma frequency oscillations in firing rate synchronized with LFP fluctuations. Gamma frequency oscillations are shown in detail (**C**) with the LFP filtered (40–90 Hz) and vertical lines indicating peaks of the rhythmic population activity. **D.** STAs for 50–150 ms after stimulus onset and (**E**) the respective power spectra. **F.** The STA from (**D**), filtered (40–90 Hz) and (**G**) the respective part of the power spectrum.

III. CHANGES IN NEURONAL SYNCHRONIZATION MAY BE A GENERAL MECHANISM TO MODULATE NEURONAL IMPACT

In summary, attention increased gamma frequency and reduced low frequency synchronization among those V4 neurons that contained the focus of attention in their receptive field. Qualitatively similar effects were observed during the delay period before stimulus onset, for the first few hundred milliseconds after response onset and for the period of sustained stimulus driven neuronal activation. During the first two of those periods, the delay period and the early poststimulus period, firing rates were not significantly affected by attention. Earlier studies have described gamma frequency synchronization also in the absence of focused attention and even under anesthesia (Eckhorn, Bauer, Jordan, Brosch, Kruse, Munk, and Reitboeck, 1988; Gray, König, Engel, and Singer, 1989;

studies using a single high contrast stimulus in the RF (Reynolds et al., 2000). By contrast, synchronization was modulated by attention very early in the response. STAs for the 100 ms period after response onset (starting 50 ms poststimulus) contained large low frequency modulations with superimposed gamma frequency modulations (see Fig. 86.3D). The low frequency (10 Hz) synchronization was reduced by attention (see Fig. 86.3E). Conversely, there was a smaller gamma

Maldonado, Friedman-Hill, and Gray, 2000). Our findings are in keeping with those reports, because also the ignored distracter stimulus induced substantial gamma frequency synchronization among the recorded neurons. Earlier studies also showed that gamma-frequency synchronization in the visual cortex can be enhanced by brainstem stimulation (Munk, Roelfsema, König, Engel, and Singer, 1996). However, the mechanisms that mediate the effects of selective visual attention presented here are not yet clear. A plausible mechanism will have to explain the high spatial selectivity of the effects that was on the order of the RF sizes in V4. In some experiments (data not shown) we also found circumstantial evidence that feature (color) selective attention modulated oscillatory neuronal synchronization in V4 in a similar way as the spatially selective attention that we describe here. Thus, if both effects are mediated by the same mechanism, this mechanism has to show high spatial selectivity as well as selectivity for other features like color. Future studies will have to address this issue.

The observed changes in synchronization are likely to enhance the impact of the affected neurons on postsynaptic targets (Salinas and Sejnowski, 2000; Engel, Fries, and Singer, 2001; Salinas and Sejnowski, 2001; Salinas and Sejnowski, 2002). It has recently been shown with intracellular *in vivo* recordings that the firing rate of a neuron is mainly determined by the gamma-frequency components in its synaptic input while low-frequency components are actively compensated for (Azouz et al., 2003). Thus, in order to endow a group of neurons representing the attended stimulus with enhanced impact, the most effective mechanism is probably an increased gamma-frequency synchronization among those neurons. This is exactly what we found in macaque V4.

Although the interpretation of attentional effects on gamma frequency synchronization is straightforward, we can only speculate in interpreting the effects of attention on low-frequency synchronization. We hypothesize that the observed *de*-synchronization in the low frequency range also enhances postsynaptic efficacy of the affected spikes. Low frequency synchronization causes spikes to co-occur within 50–100 ms. However, there are a number of biophysical mechanisms, such as spike frequency adaptation, that reduce the impact of spikes that occur late in a train of input spikes and those mechanisms typically have time constants between 20 and 50 ms (Ahmed, Anderson, Douglas, Martin, and Whitteridge, 1998). Although spikes coincident at gamma frequency remain unaffected, spikes that co-occur within 50–100 ms will mutually reduce each other's impact on the postsynaptic target. Thus, an impact enhancement requires low frequency synchronization to be reduced, in contrast to the required enhancement of gamma frequency synchronization.

A common finding in the field of attention is that the attentional effects on firing rate are much larger when there are two or more stimuli within the receptive field of the studied neuron as compared to when there is only one stimulus (Moran and Desimone, 1985; Treue and Maunsell, 1996). It turns out that when there are two stimuli within the RF, directing attention to one of them will drive the firing rate of the neuron toward the firing rate that is found when this attended stimulus is shown in isolation (Reynolds et al., 1999). When attention is shifted from a stimulus evoking strong responses to a stimulus evoking weak responses, then this shift of attention within the RF will cause the firing rates to be reduced. Thus, attention does not simply have an enhancing or suppressive effect, but causes the neuronal responses to be dominated by the attended stimulus while the influence of ignored stimuli is diminished. Reynolds et al. (1999) have proposed a mathematical model that explains those competitive interactions within receptive fields quantitatively. This model incorporates two layers. In the upper layer, neurons have larger RFs that contain two or more stimuli. In the lower layers, neurons have smaller RFs that contain only one of the two stimuli. Two stimuli contained in two different low-level RFs will fall within one high-level RF. When attention is switched from one stimulus to the other, the model predicts little or no effect of attention on firing rates in the lower level (depending on stimulus contrast), but at the same time, attention should cause neurons at the higher level to respond as if there were only the attended stimulus in the RF. This is achieved solely by enhancing the impact of those lower-level neurons that contain the attended stimulus in their RF. We hypothesize that the changes in synchronization described here are the neuronal substrate for this enhancement in impact of the lower-level neurons. Future experiments will test whether firing rate changes at the higher level can be directly correlated to synchronization changes at the lower level.

An increased impact of a neuronal population on its postsynaptic targets is equivalent to an increase in effective synaptic gain. Previous studies have proposed an increase in synaptic gain to explain a wide variety of behavioral influences on neuronal firing rates (Andersen, Essick, and Siegel, 1985; Olshausen, Anderson, and Van Essen, 1993). Increasing the effective synaptic gain by modulating synchronization at precise locations in the cortex therefore might be a fundamental neuronal mechanism for amplifying signals.

References

Ahmed, B., Anderson, J. C., Douglas, R. J., Martin, K. A., and Whitteridge, D. (1998). Estimates of the net excitatory currents evoked by visual stimulation of identified neurons in cat visual cortex. *Cereb. Cortex* **8**, 462–476.

Andersen, R. A., Essick, G. K., and Siegel, R. M. (1985). Encoding of spatial location by posterior parietal neurons. *Science* **230**, 456–458.

Azouz, R., and Gray, C. M. (2003). Adaptive coincidence detection and dynamic gain control in visual cortical neurons in vivo. *Neuron* **37**, 513–523.

Eckhorn, R., Bauer, R., Jordan, W., Brosch, M., Kruse, W., Munk, M., and Reitboeck, H. J. (1988). Coherent oscillations: a mechanism of feature linking in the visual cortex? Multiple electrode and correlation analyses in the cat. *Biol. Cybern.* **60**, 121–130.

Engel, A. K., Fries, P., and Singer, W. (2001). Dynamic predictions: oscillations and synchrony in top-down processing. *Nat. Rev. Neurosci.* **2**, 704–716.

Fries, P., Reynolds, J. H., Rorie, A. E., and Desimone, R. (2001). Modulation of oscillatory neuronal synchronization by selective visual attention. *Science* **291**, 1560–1563.

Fries, P., Roelfsema, P. R., Engel, A. K., König, P., and Singer, W. (1997). Synchronization of oscillatory responses in visual cortex correlates with perception in interocular rivalry. *Proc. Natl. Acad. Sci. U.S.A* **94**, 12699–12704.

Fries, P., Schröder, J. H., Roelfsema, P. R., Singer, W., and Engel, A. K. (2002). Oscillatory neuronal synchronization in primary visual cortex as a correlate of stimulus selection. *J. Neurosci.* **22**, 3739–3754.

Gray, C. M., König, P., Engel, A. K., and Singer, W. (1989). Oscillatory responses in cat visual cortex exhibit inter-columnar synchronization which reflects global stimulus properties. *Nature* **338**, 334–337.

Luck, S. J., Chelazzi, L., Hillyard, S. A., and Desimone, R. (1997). Neural mechanisms of spatial selective attention in areas V1, V2, and V4 of macaque visual cortex. *J. Neurophysiol.* **77**, 24–42.

Maldonado, P. E., Friedman-Hill, S., and Gray, C. M. (2000). Dynamics of striate cortical activity in the alert macaque: II. Fast time scale synchronization. *Cereb. Cortex* **10**, 1117–1131.

Moran, J., and Desimone, R. (1985). Selective attention gates visual processing in the extrastriate cortex. *Science* **229**, 782–784.

Munk, M. H., Roelfsema, P. R., König, P., Engel, A. K., and Singer, W. (1996). Role of reticular activation in the modulation of intracortical synchronization. *Science* **272**, 271–274.

Olshausen, B. A., Anderson, C. H., and Van Essen, D. C. (1993). A neurobiological model of visual attention and invariant pattern recognition based on dynamic routing of information. *J. Neurosci.* **13**, 4700–4719.

Reynolds, J. H., Chelazzi, L., and Desimone, R. (1999). Competitive mechanisms subserve attention in macaque areas V2 and V4. *J. Neurosci.* **19**, 1736–1753.

Reynolds, J. H., Pasternak, T., and Desimone, R. (2000). Attention increases sensitivity of V4 neurons. *Neuron* **26**, 703–714.

Salinas, E., and Sejnowski, T. J. (2000). Impact of correlated synaptic input on output firing rate and variability in simple neuronal models. *J. Neurosci.* **20**, 6193–6209.

Salinas, E., and Sejnowski, T. J. (2001). Correlated neuronal activity and the flow of neural information. *Nat. Rev. Neurosci.* **2**, 539–550.

Salinas, E., and Sejnowski, T. J. (2002). Integrate-and-fire neurons driven by correlated stochastic input. *Neural Comput.* **14**, 2111–2155.

Treue, S., and Maunsell, J. H. (1996). Attentional modulation of visual motion processing in cortical areas MT and MST. *Nature* **382**, 539–541.

Worden, M. S., Foxe, J. J., Wang, N., and Simpson, G. V. (2000). Anticipatory biasing of visuospatial attention indexed by retinotopically specific alpha-band electroencephalography increases over occipital cortex. *J. Neurosci.* **20**, RC63.

# CHAPTER 87

Putative Role of Oscillations and Synchrony in Cortical Signal Processing and Attention

Wolf Singer

ABSTRACT

Evidence is presented that precise synchronization of neuronal discharges is used as a mechanism complementary to the modulation of discharge rates in order to raise the saliency of neuronal responses. It is proposed that this mechanism serves a variety of functions such as perceptual grouping, attention dependent response selection, and the definition of relations in short- and long-term memory. Spike sequences can become synchronized either by temporally structured stimuli (stimulus locking) and/or by internal synchronizing mechanisms. The latter process is often associated with an oscillatory patterning of responses in various frequency ranges. In case of attention dependent response selection the oscillatory patterning of neuronal activity precedes the appearance of the expected stimuli. It is proposed that these self-generated dynamic states are matched with incoming sensory signals and that these interactions permit a rapid translation of rate coded sensory information into distributed sparse temporal codes in which precise synchrony serves as a signature of relatedness both in the context of signal processing and Hebbian learning.

I. INTRODUCTION

In cognition, astronomical numbers of relations among the features of perceptual objects have to be analyzed and represented. Similar combinatorial problems arise when the execution of adapted movements requires the coordination of muscles in ever-changing constellations. Therefore, mechanisms are required that permit rapid and flexible definition of relations between the responses of large numbers of distributed neurons.

A common and well-documented strategy for the binding of distributed responses is the implementation of conjunction-specific neurons. Representing relations by conjunction units is rapid because it can be realized by convergence in feed-forward architectures and it is reliable because the responses of a particular cell always signal the same set of relations (labeled line coding). However, this coding strategy, if not complemented by additional binding mechanisms, meets with a number of problems. In cognition, exceedingly large numbers of conjunction units would be required for the exhaustive analysis and representation of the manifold intra- and cross-modal feature constellations of real world objects. Also, it is hard to see how novel objects and hence entirely new relations among features can be recognized and represented as this would require prewired units for novel conjunctions or at least rapid reconfiguration of input connections to previously uncommitted cells. Finally, unresolved problems arise with the representation of the nested relations that are constitutive of virtually all cognitive objects, be they visual scenes, melodies, or sentences (Singer, 1999). Likewise, it is difficult to see how the distributed activation patterns of motor neurons can be orchestrated by the responses of a few conjunction-specific command neurons.

A complementary coding strategy is needed, therefore, that permits a more flexible and dynamic definition of relations than can be achieved with hard-wired conjunction units.

II. TAGGING RESPONSES AS RELATED

The mechanism that tags distributed responses as related must assure that these responses are processed and evaluated together at subsequent processing stages and are not confounded with other, unrelated responses. One option to achieve this goal is to raise selectively the saliency of the responses that should be related and processed jointly. There are essentially three ways to increase the relative impact of selected responses. First, nonselected responses can be inhibited and thereby excluded from further processing. Second, the discharge frequency of the selected responses can be enhanced, and third, the selected cells can be made to discharge in precise temporal synchrony. Single cell studies have provided evidence for the implementation of the first two mechanisms. However, these strategies are associated with constraints that reduce the system's capacity for parallel processing. If differences between the discharge rates of neurons are used to identify and group related responses, the speed is limited at which different sets of relations can be defined successively. The rate limiting factor is the duration of the interval over which EPSPs have to be integrated until EPSP trains of different frequency produce significantly different postsynaptic responses. Estimates based on reaction times, evoked potentials, and latencies of single cell responses suggest that it takes maximally a few tens of milliseconds per processing stage to perform the computations necessary for the analysis and recognition of patterns of average complexity (Thorpe et al., 1996). Given the relatively low discharge rates of cortical neurons, with interspike intervals ranging typically between 20 to many hundreds of milliseconds, this implies that grouping operations must be accomplished on the basis of a small number of spikes per neuron. This precludes evaluation of rate changes of individual neurons and necessitates population coding. Rate fluctuations become detectable only within the required short intervals if they occur simultaneously in a sufficiently large number of neurons and if downstream neurons average across these population responses. In this case minor but well-synchronized increases in the discharges of individual neurons can sum up to highly significant increases of cross-channel spike density. Such increases in the spike density of population responses can be brought about by two, in principle, independent mechanisms. They can be generated by insertion of additional well-timed spikes, which is equivalent with a coherent increase of the discharge rate of individual cells. Cross-channel spike density can also be transiently enhanced by coordination of spike timing; that is, by synchronization. In this case spike density of the population response is increased within short time intervals by advancing and delaying discharges of individual neurons so that more spikes than predicted by average discharge rates coincide within a given interval. Such coordination of spike timing can occur in the absence of measurable changes in the discharge rate of individual neurons. Evidence suggests that both strategies are used by the nervous system to jointly raise the saliency of responses and to thereby tag them as related. In bottom-up feed-forward processing, coherent increases in cross-channel spike density usually are generated by stimulus-locked increases in discharge frequency. If an object suddenly appears or if it moves, the relatedness of features belonging to this object is signaled by the synchrony of the stimulus locked responses to the respective features. However, there are numerous conditions where such coordination or binding of responses by stimulus locking is not possible. This is the case when stimuli lack temporal structure or when responses to temporally offset stimuli need to be bound and tagged as related. Finally, there are all those cases where features are selected and bound together by top-down, attention dependent processes. Here, the increase in cross-channel spike density that is required for the definition of relations needs to be coordinated by internal mechanisms that operate independently of stimulus timing. Again, these mechanisms could act by inducing precisely timed joint rate increases in the selected channels or they could coordinate spike timing by synchronization. Numerous studies in trained animals provide evidence that attention dependent selection of stimulus features is associated with a moderate increase in discharge rate (Engel et al., 2001). However, as most of these studies relied on the analysis of averaged single cell responses, they could not assess the temporal coherence of these rate increases across channels nor could they detect rate independent increases in discharge synchrony. It is only recently that multielectrode recordings have been performed in order to search for cross-channel coordination of spike density and to distinguish between stimulus locked and internally induced covariation of discharge rates and spike timing. These studies revealed the existence of synchronizing mechanisms that can transiently augment cross-channel spike densities above the level predicted by average discharge rate. They indicate further that this grouping of discharges occurs frequently on the basis of an oscillatory patterning of the discharge sequences (Singer, 1999; Engel et al., 2001).

III. RATE CODES VERSUS TEMPORAL CODES

Because the distinctions between rate and temporal codes are controversial they warrant a brief comment. According to classical conventions, discharge rates of individual neurons are measured over time spans long enough to count at least 3 to 4 spikes; that is, 100 to several hundred ms depending on discharge rate. However, if what matters is the spike density across numerous input channels, rate fluctuations can be evaluated with much higher temporal resolution—provided that they are highly coherent across channels. Neurons can discriminate between synchronous and temporally dispersed inputs with a resolution in the millisecond range (Usrey and Reid, 1999). Thus, they are sensitive to fluctuations of cross-channel spike densities at time scales at least one order of magnitude shorter than the average interspike intervals in the converging inputs. However, once the relevant integration windows become considerably shorter than the average interspike interval of the afferent spike trains it begins to matter whether the spikes in the input connections are distributed randomly or clustered in time. The postsynaptic cell's response, to a large extent, will depend on the precise timing of the spikes in the converging inputs. Hence, timing relations among spikes can be used to encode information and this is what the term "temporal code" alludes to. Because synchronization of discharges raises with high temporal resolution and selectivity, the saliency of only those response segments that contain precisely synchronized events, this code is particularly well suited to define relations on short time scales a prerequisite for dynamic response grouping. Per definition, this code is relational and can therefore be analyzed only by studying the temporal relations among the discharges of simultaneously recorded neurons. And here a distinction needs to be made between measurements of cross channel spike densities and actual search for coincident firing in identified subsets of channels. By simply counting events across channels, as is commonly done for the calculation of population vectors, the potentially important information is lost which cells contributed at which time to the measured coincidences.

Using a temporal code for the definition of relations has several advantages. One is that it permits specification of relations independently of firing rate. The discharge rates of neurons depend on numerous variables such as stimulus energy or the match between stimulus and receptive field properties and it is not obvious how these modulations of response amplitude can be distinguished from those signaling the relatedness of responses. Not all strong responses are necessarily related. Because synchrony can be modulated independently of rates by temporal regrouping of discharges it can be used as a tag of relatedness that is orthogonal to rate fluctuations. Response amplitudes could thus be reserved to signal the presence and saliency of features and synchronicity could be used in parallel to signal whether these features are related.

Another advantage of selecting responses by synchronization is that the timing of synchronized input events is preserved with high precision in the output activity of cells because synchronized input is transmitted with minimal latency jitter (Diesmann et al., 1999). This, in turn, could be exploited to preserve the signature of relatedness across processing stages, which is of great importance in parallel, distributed processing. Finally, synchronization enhances processing speed also by accelerating synaptic transmission *per se* because synchronized EPSPs trigger action potentials with minimal delay.

IV. SYNCHRONY AS A CODE FOR THE DEFINITION OF RELATIONS

Following the discoveries that 1) cortical neurons often engage in synchronous oscillatory activity that is not stimulus locked but caused by internal interactions (Gray and Singer, 1989); 2) neurons distributed both within and across cortical areas can synchronize their discharges with a precision in the millisecond range; and 3) synchronization probability reflects common Gestalt-criteria of perceptual grouping; it had been proposed that the cerebral cortex imposes a temporal micro-structure on otherwise sustained responses and uses this temporal patterning to express through synchronization the degree of relatedness of the responses (Singer, 1999; Engel et al., 2001).

If it were the case that not only stimulus-locked response synchronization but also internally generated synchronization serves to tag responses as related, internally generated synchrony needs to meet several criteria. First, its precision should be in the millisecond range to match the temporal windows for effective spatial summation. Second, synchrony must be established and dissolved at a rate fast enough to be compatible with known processing speed. Third, synchronized activity must be more effective than nonsynchronized activity in driving cells in target structures. Fourth, the occurrence and structure of synchronization patterns should be closely related with perceptual or motor processes. Fifth, disruption or artificial enhancement of synchrony should lead to changes in cognitive or motor functions.

The postulate that the precision of internally generated synchronization should match the windows for effective synaptic summation is supported by cross-correlation data. The widths of the correlation peaks at half height are typically in the range of less than 10 ms, in particular, when the global EEG is in a desynchronized state. In this case internally generated synchronization is often associated with an oscillatory patterning of the respective responses in the high β- or the γ-frequency range (from 20 to 60 Hz). This limits the duration of the episodes of joint firing to the peaks of the oscillations and causes a sharp temporal demarcation of synchronous events because the oscillation troughs tend to be free of spikes.

The second postulate that internally generated synchronization must be established and dissolved very rapidly (within maximally a few tens of milliseconds) has received theoretical and experimental support only recently. Simulations with spiking neurons revealed that networks of appropriately coupled units can indeed undergo very rapid transitions from uncorrelated to synchronized states and vice versa (Singer, 1999).

Experimental support is also available for the third postulate that synchronized activity has a stronger impact in target structures than temporally dispersed firing. Simultaneous recordings from coupled neuron triplets along thalamo-cortical and intracortical pathways in the visual system by Reid and colleagues (Usrey and Reid, 1999) have revealed that EPSPs synchronized within intervals below 5 ms are much more effective than EPSPs dispersed over longer intervals. Similar conclusions are suggested by simulation studies, *in vitro* experiments, and *in vivo* recordings. Multielectrode recordings from several sites of the cat visual cortex and retinotopically corresponding loci in the superior colliculus indicated that the impact that a particular group of cortical cells has on target cells in the colliculus increases substantially whenever the cortical cells synchronize their discharges with other cortical cell groups projecting to the same site in the tectum. Enhanced saliency of synchronized responses can also be inferred from the tight correlation between the strength of neuronal response synchronization and perception that was observed in experiments on binocular rivalry in cats and human subjects (Fries et al., 2002). Finally, there is evidence from multisite microstimulation in the superior colliculus of awake cats that slight changes (<5 ms) in the synchronicity of the applied pulse trains lead to significantly different saccadic eye movements (Brecht et al., 2004).

V. SYNCHRONIZATION AND FEATURE BINDING

The hypothesis that internal synchronization of discharges serves to group responses for joint processing predicts that synchronization probability should reflect some of the basic Gestalt-criteria according to which the visual system groups related features during scene segmentation. In agreement with this prediction a series of studies provided evidence that stimulus configurations that comply with criteria such as continuity, proximity, similarity in the orientation domain, colinearity, and common fate evoke synchronized responses with higher probability than configurations that are devoid of groupable features (Singer, 1999).

A particularly close correlation between neuronal synchrony and perceptual grouping has recently been observed in experiments with plaid stimuli (see Fig. 87.1A). Here is a case where dynamic changes in perceptual grouping result in very different percepts. In the case where component motion is perceived, responses evoked by the two superimposed gratings must be segregated from each other and responses evoked by the same grating must be grouped and processed jointly; in the case of pattern motion, responses to all contours must be bound together to represent a single surface. If this grouping of responses is initiated by selective synchronization, two predictions must hold (see Fig. 87.1B): First, neurons that prefer the direction of motion of one of the two gratings and have colinearly aligned receptive fields should always synchronize their responses because they respond always to contours that belong to the same surface. Second, neurons that are tuned to the respective motion directions of the two gratings should synchronize their responses in case of pattern motion because they then respond to contours of the same surface, but they should not synchronize in case of component motion because their responses are then evoked by contours belonging to different surfaces.

Cross-correlation analysis of responses from cell pairs distributed either within or across areas 18 and the posterior medio-lateral suprasylvian sulcus (PMLS) of the cat visual cortex confirmed these predictions (Castelo-Branco et al., 2000) (see Fig. 87.1C). Dynamic changes in synchronization, thus, could serve to encode in a context dependent way the rela-

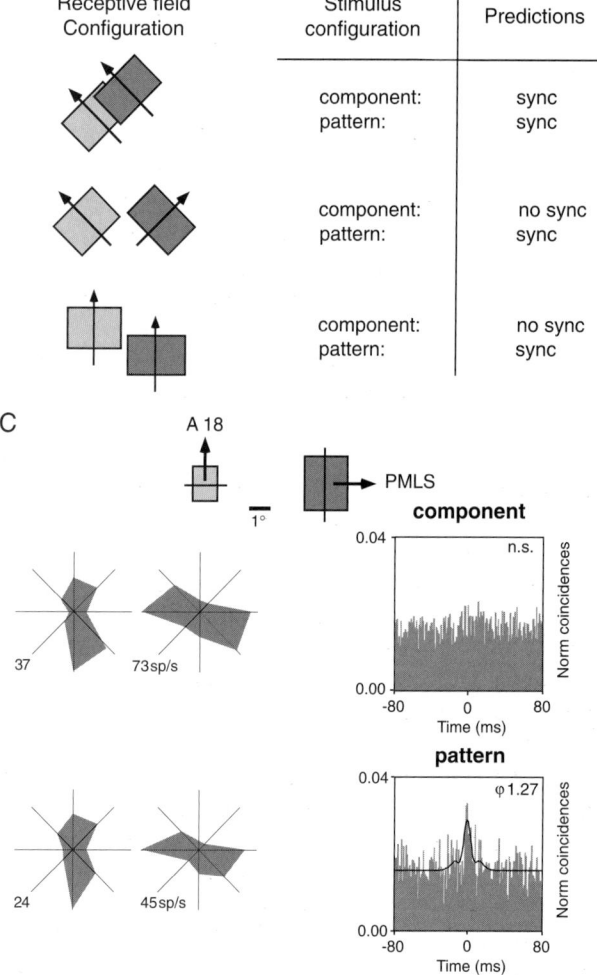

FIGURE 87.1 **A.** Two superimposed gratings that differ in orientation and drift in different directions are perceived either as two independently moving gratings (component motion) or as a single pattern drifting in the intermediate direction (pattern motion), depending on whether the luminance conditions at the intersections are compatible with transparency. **B.** Predictions on the synchronization behavior of neurons as a function of their receptive field configuration (left) and stimulation conditions (right). **C.** Changes in synchronization behavior of two neurons recorded simultaneously from areas 18 and PMLS that were activated with a plaid stimulus under component (upper graph) and pattern motion conditions (lower graph). The two neurons preferred gratings with orthogonal orientation (see receptive field configuration, top, and tuning curves obtained with component and pattern, respectively) and synchronized their responses only when activated with the pattern stimulus (compare cross-correlograms on the right). (Adapted from Castelo-Branco et al., 2000.)

tions among the simultaneous responses to spatially superimposed contours and thereby bias their association with distinct surfaces.

VI. THE ROLE OF RESPONSE SYNCHRONIZATION IN RESPONSE SELECTION AND ATTENTION

The notion that spike synchronization and oscillatory patterning of responses in the beta- and gamma-frequency range are related with arousal and attention is suggested by the evidence that both phenomena are particularly well expressed when the brain is in an activated state; that is, when the EEG is desynchronized and exhibits low power in the frequency range below 10 Hz. Such EEG-patterns are characteristic for the aroused, attentive as well as for the dreaming brain and thus for states in which sensory representations can be activated. In the same direction point the numerous observations both in animals and human subjects that synchronous oscillations in the γ-frequency range and their synchronization become more prominent during states of focused attention or when subjects are engaged in cognitive tasks that put strong demands on feature binding or short term memory functions (Tallon-Baudry and Bertrand, 1999; Engel et al., 2001; Waltz et al., 2002).

Multielectrode recordings from awake cats and monkeys, trained to perform discrimination tasks, indicate that attentional mechanisms enhance neuronal synchrony in anticipation of the expected task. One observes an increase in oscillatory activity in the beta- and gamma-frequency range that is associated with increased coherence between the spontaneous discharges of cells and the oscillations of the local field potential. These attentional effects appear to be selective and confined to cortical areas and sites that need to be engaged for the execution of the anticipated tasks (Roelfsema et al., 1997) (see Fig. 87.2). This priming could have the effect that responses to attended stimuli synchronize faster and with greater precision than responses to nonattended stimuli. Anticipatory synchronization, thus, could be one of the mechanisms by which expected responses are made more salient and selected for further processing. Recent evidence suggests in addition that internally generated, not stimulus-driven, gamma-oscillations exhibit distinct patterns of coherence that reflect the functional architecture of intracortical association fibers. These oscillations have a strong influence on the latencies of responses to stimuli and cause through coordinated latency shifting a rapid and feature-specific synchro-

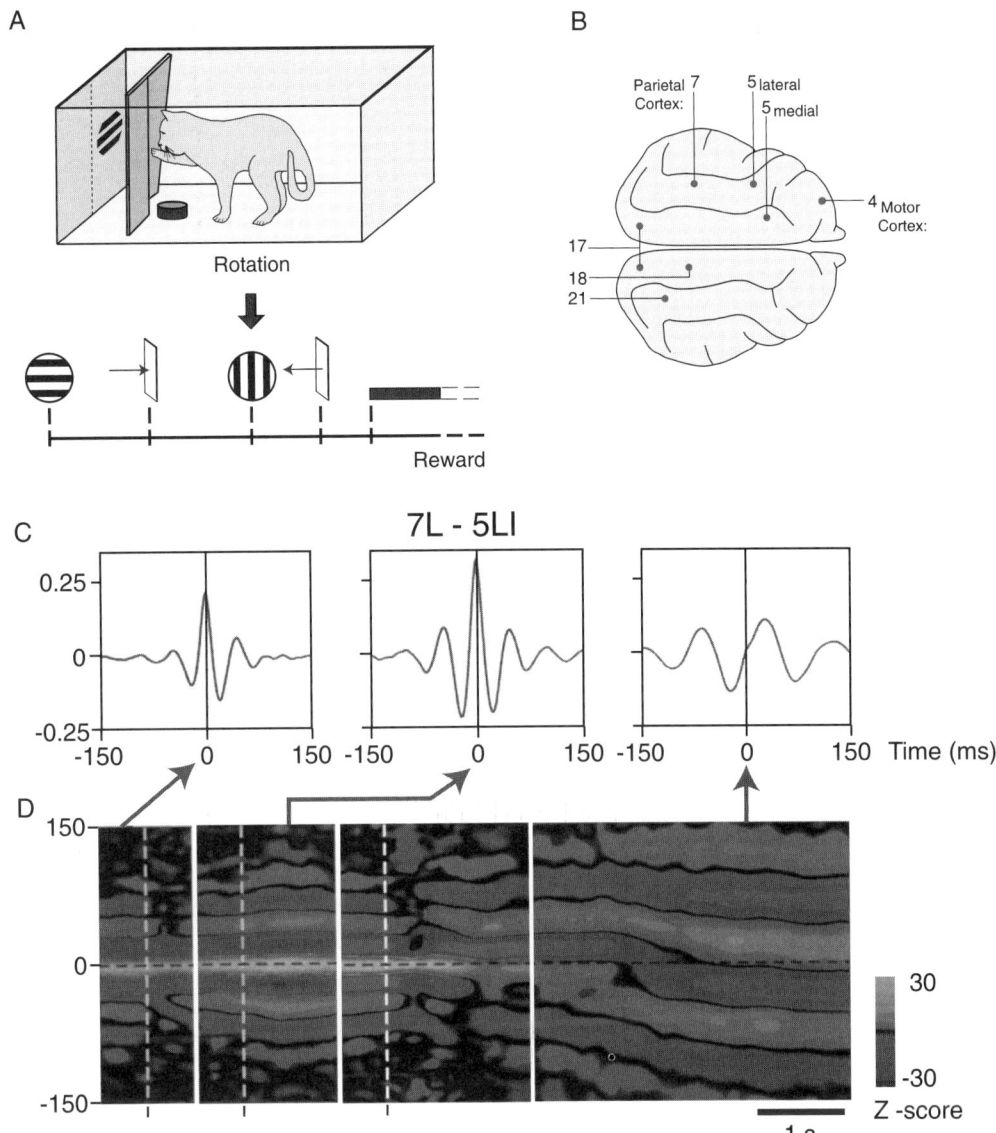

FIGURE 87.2 Expectation-related synchrony in visual cortex in awake cats during a visuomotor coordination task (Roelfsema et al., 1997). **A.** Cats had to respond to a visual stimulus by pressing and holding a transparent door as they waited for a rotation of the stimulus. **B.** Local field potentials (LFPs) were recorded from visual, parietal, and motor cortex. **C.** Cross-correlograms between responses recorded from electrodes 7 and 5 during various phases of the trial. Left: prior to target presentation; Middle: after target presentation; Right: during consumption of reward. **D.** The contour plot shows how the temporal correlation between parietal areas 5 and 7 evolves over time. Abscissa: time relative to stimulus appearance. Zero on the ordinate indicates synchrony between the recording sites. Red–yellow, positive correlation; Green–blue, negative correlation. The high-amplitude zero peak shows that synchrony occurs before the appearance of the stimulus (**C**, left), presumably reflecting expectancy, and increases during attentive visuomotor processing (**C**, middle). Synchrony breaks down completely after the cat has responded to rotation of the pattern by releasing the door (**C**, right). (Adapted from Engel et al., 2001) (See color plate.)

nization of the very first spikes of the responses to patterns (Fries et al., 2001). It has been proposed, therefore, that the knowledge stored in the functional architecture of the neocortex is permanently translated into dynamic states that permit very rapid conversion of rate coded sensory input into temporal codes. Signal processing in the cortex, thus, would consist essentially of a matching operation in which afferent sensory activity is compared with expectancies that are expressed by specific dynamic states. This anticipatory nature of cortical computations is probably one reason why processing of sensory signals is so fast (Engel et al., 2001).

A close correlation between response synchronization and stimulus selection has also been found in experiments on binocular rivalry that were performed in strabismic animals (Fries et al., 2002). Due to experience-dependent modifications of processing circuitry perception in nonamblyopic strabismic subjects always alternates between the two eyes. We have exploited this phenomenon of rivalry to investigate how neuronal responses that are selected and perceived differ from those that are suppressed and excluded from supporting perception (see Fig. 87.3). These experiments revealed that the rate responses of neurons in areas 17 and 18 were neither enhanced nor attenuated when they did or did not support perception. A close and highly significant correlation existed, however, between changes in the strength of response synchronization and the outcome of rivalry. Cells mediating responses that were perceived and used for the control of eye movements increased the synchronicity of their discharges, whereas the reverse was true for cells mediating responses that were excluded from perception. The synchronized responses exhibited an oscillatory modulation in the gamma-frequency range and differential changes of synchrony were especially pronounced in the spike-field coherence, which is a particularly sensitive measure of population synchrony. Thus, in early visual areas selection of responses for further processing appears to be achieved by modulating the degree of synchronization among large populations of neurons rather than by modifying the amplitude of responses. It is only at later processing stages that the poorly synchronized responses to the suppressed stimuli eventually fail to elicit suprathreshold responses and that cells respond only to the selected stimulus (Leopold and Logothetis, 1996).

Support for a role of synchronization in attention dependent top-down selection of responses is provided by multielectrode recordings from the somato-sensory (Steinmetz et al., 2000) and the visual cortex (Fries et al., 2001a) of monkeys performing discrimination tasks. When monkeys that are engaged in the discrimination of tactile stimuli focus their attention to the task, synchronicity among neuronal responses in somato-sensory cortex shows a significant increase (Steinmetz et al., 2000). Similar observations have been made in visual cortex area V4 during a visual discrimination task (Fries et al., 2001a). Here, additional evidence was obtained, using the sensitive synchronicity measure of spike-field coherence, that synchrony among the discharges of neurons increased in

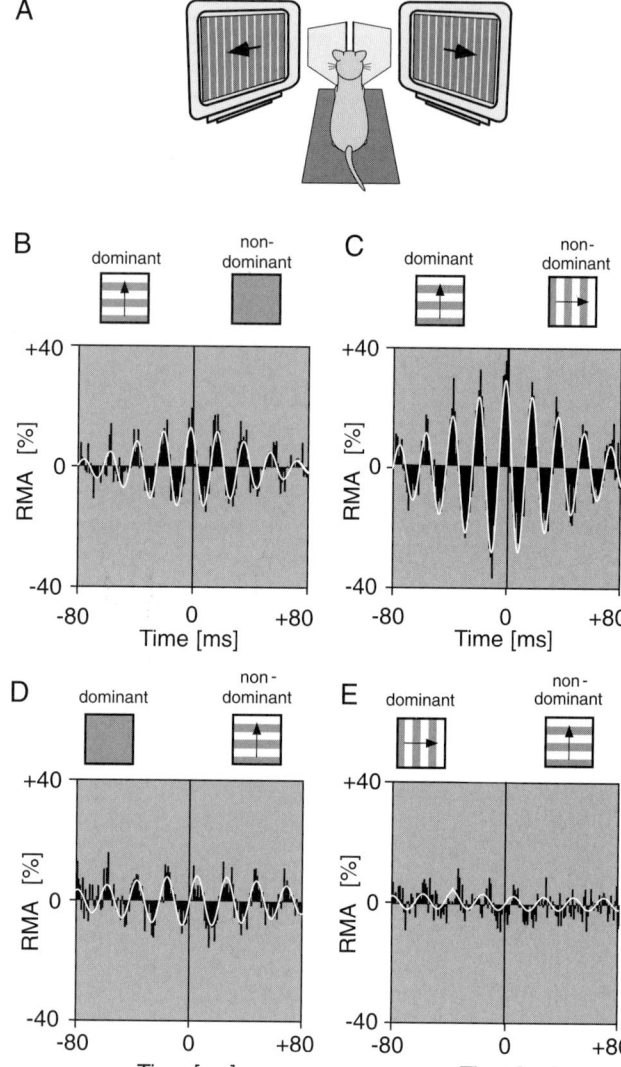

FIGURE 87.3 Neuronal synchronization under conditions of binocular rivalry. **A**. Using two mirrors, different patterns were presented to the two eyes of strabismic cats. Panels (**B–E**) show normalized cross-correlograms for two pairs of recording sites activated by the eye that won (**B, C**) and lost (**D, E**) in interocular competition, respectively. Insets above the correlograms indicate stimulation conditions. Under monocular stimulation **B**, cells driven by the winning eye show a significant correlation which is enhanced after introduction of the rivalrous stimulus to the other eye, **C**. The reverse is the case for cells driven by the loosing eye (compare conditions **D** and **E**). The white continuous line superimposed on the correlograms represents a damped cosine function fitted to the data. RMA, relative modulation amplitude of the center peak in the correlogram, computed as the ration of peak amplitude over offset of correlogram modulation. This measure reflects the strength of synchrony. (Modified from Fries et al., 2002)

a retinotopically selective way already in anticipation of the task and before the stimuli were presented.

Recent evidence indicates that synchronous oscillatory activity in the beta- and gamma-frequency range also plays an important role in short-term memory functions. In parietal and prefrontal areas, oscillatory activity in this frequency range increases during the retention period (Tallon-Baudry and Bertrand, 1999). This activity correlates with memory load and its amplitude and persistence predicts performance (Waltz et al., 2002). At present, it is difficult to decide to which extent these synchronization phenomena are related to attentional processes or to memory related binding functions because of the close relation between attention and working memory.

Taken together these results suggest that response synchronization and hence temporal coding is used as a mechanism complementary to rate modulation in order to select and group responses in the context of both attention independent, bottom-up and attention-dependent top-down selection and binding operations. In both cases precise spike synchrony across processing channels appears to be used to jointly raise the saliency of brief response segments and to tag these response episodes as related. However, the mechanisms by which precise synchrony is produced may be quite different. In bottom-up processing some of the synchronicity is caused by time locking of discharges to temporally modulated stimuli (external timing). Most of synchrony observed in cortical recording is, however, caused by internal adjustments of spike timing and/or induction of transient, highly coherent rate increases. Some coincident firing is imported from the retina where independent synchronizing mechanisms are at action. These provide temporal codes for stimulus energy, motion, and continuity (Singer, 1999). However, the bulk of synchronous firing is due to intracortical synchronizing mechanisms, whereby most of the long distance synchrony appears to be achieved by the generation of oscillatory responses in the beta- and gamma-frequency range that can become phase locked. This intracortical synchronization can arise from local circuit interactions within circumscribed cortical regions in response to sensory activation, but it can also be imposed by attentional mechanisms in the absence of sensory stimuli. To date it is unclear to which extent these attention-dependent synchronization phenomena are mediated directly via cortico-cortical feedback connections or indirectly via subcortical loops involving the so-called nonspecific thalamic nuclei and/or the striatum and claustrum.

References

Brecht, M., Singer, W., and Engel, A. K. (2004). Amplitude and direction of saccadic eye movements depend on the synchronicity of collicular population activity. *J. Neurophysiol.* **92**, 424–432.

Castelo-Branco, M., Goebel, R., Neuenschwander, S., and Singer, W. (2000). Neural synchrony correlates with surface segregation rules. *Nature* **405**, 685–689.

Diesmann, M., Gewaltig, M.-O., and Aertsen, A. (1999). Stable propagation of synchronous spiking in cortical neural networks. *Nature* **402**, 529–533.

Engel, A. K., Fries, P., and Singer, W. (2001). Dynamic predictions: oscillations and synchrony in top-down processing. *Nature Rev. Neurosci.* **2**, 704–716.

Fries, P., Neuenschwander, S., Engel, A. K., Goebel, R., and Singer, W. (2001). Rapid feature selective neuronal synchronization through correlated latency shifting. *Nature Neurosci.* **4**, 194–200.

Fries, P., Reynolds, J. H., Rorie, A. E., and Desimone, R. (2001a). Modulation of oscillatory neuronal synchronization by selective visual attention. *Science* **291**, 1560–1563.

Fries, P., Schröder, J.-H., Roelfsema, P. R., Singer, W., and Engel, A. K. (2002). Oscillatory neuronal synchronization in primary visual cortex as a correlate of perceptual stimulus selection. *J. Neurosci.* **22**, 3739–3754.

Gray, C. M., and Singer, W. (1989). Stimulus-specific neuronal oscillations in orientation columns of cat visual cortex. *Proc. Natl. Acad. Sci. USA* **86**, 1698–1702.

Leopold, D. A., and Logothetis, N. K. (1996). Activity changes in early visual cortex reflect monkeys' percepts during binocular rivalry. *Nature* **379**, 549–553.

Roelfsema, P. R., Engel, A. K., König, P., and Singer, W. (1997). Visuomotor integration is associated with zero time-lag synchronization among cortical areas. *Nature* **385**, 157–161.

Singer, W. (1999). Neuronal synchrony: a versatile code for the definition of relations? *Neuron* **24**, 49–65.

Steinmetz, P. N., Roy, A., Fitzgerald, P. J., Hsiao, S. S., Johnson, K. O., and Niebur, E. (2000). Attention modulates synchronized neuronal firing in primate somatosensory cortex. *Nature* **404**, 187–190.

Tallon-Baudry, C., and Bertrand, O. (1999). Oscillatory gamma activity in humans and its role in object representation. *Trends Cognitive Sci.* **3**, 151–162.

Thorpe, S., Fize, D., and Marlot, C. (1996). Speed of processing in the human visual system. *Nature* **38**, 520–522.

Usrey, W. M., and Reid, R. C. (1999). Synchronous activity in the visual system. *Annu. Rev. Physiol.* **61**, 435–456.

Waltz, J. A., Linden, D. E. J., Bittner, R., Muckli, L., Goebel, R., Singer, W., and Munk, M. H. J. (2002). Working memory load-dependent changes in the power of evoked and induced oscillations in the EEG. *Soc. Neurosci. Abstr.* **2002**, 416.2.

CHAPTER

88

Attention to Tactile Stimuli Increases Neural Synchrony in Somatosensory Cortex

P. N. Steinmetz, S. S. Hsiao, K. O. Johnson and E. Niebur

ABSTRACT

Perception of tactile stimuli requires that the neural activity evoked by stimuli that are relevant for behavior are efficiently selected. Attention influences the firing rates of neurons in the brainstem, thalamus, and cortex. At the level of secondary somatosensory cortex (SII), the firing rates of about 90% of all neurons are modulated by selective attention. In addition to modulating the firing rates of neurons, selective attention also affects the temporal synchrony of firing between neurons. In tasks in which macaques switched their attention toward and away from tactile and visual stimuli, more than 40% of the neurons in SII cortex changed the degree of synchronous firing between neurons when attending to the tactile stimuli. The effects cannot be accounted for by simple changes in firing rate. Computational models show that changes in synchrony between neurons representing a stimulus is a potential mechanism of selective attention.

I. THE NEED FOR SOMATOSENSORY SELECTION

As in all of the sensory systems, the role of attention in the somatosensory system is to select which stimuli get processed by the neural mechanisms leading to perception. Among the senses, the need for selection is perhaps most acute in the somatosensory system because of the large number of different kinds of receptors arrayed over the body surface. At a particular time the sensory information relevant for behavior may come from thermoreceptors, pain receptors, proprioceptors (joint and limb position), or tactile receptors and the relevant site may come from anywhere on the body including the head, a finger, a single limb, or multiple limbs. For a review of psychophysical and neurophysiological studies of tactile attention, see Hsiao and Vega-Bermudez (2002).

Psychophysical studies of attention to tactile stimuli show that attention functions like a cognitive spotlight that can be rapidly focused and moved to different locations on the body surface. Stimuli within the focus of attention are perceived more rapidly and accurately than stimuli outside the focus of attention. The spotlight analogy can be appreciated by sitting still and switching one's attention between different locations on the body. For example, focus exclusively and intently on the tongue for a period and then on the right index finger. Notice how sensations from the attended body parts appear and disappear. Recent psychophysical studies have shown that the focus of attention may have a minimal size that is modality dependent: tactile tasks based on the information provided by one receptor type have a minimal focus of attention that includes the entire hand, whereas tactile tasks based on the information provided by another receptor type have a minimal focus of attention that includes only a single finger pad.

The neural mechanisms of selective attention are not well understood. However, neurophysiological studies in nonhuman primates and imaging studies in humans have shown that attention affects the neural activity in all of the cortical and subcortical areas related to tactile sensation. Sensations from all body parts begin with the activation of multiple receptor types. These receptors provide information about skin temperature, pain, itch, body position and movement, and touch. The sensations of touch are based on inputs from four types of mechanoreceptive nerve fibers that innervate the skin. The different fiber types have different, specialized endings that provide them with specific sensitivities to different features of tactile stimuli (for a review see Johnson, 2001). Information from all of these fiber types is relayed through the dorsal column nuclei of the brainstem and the ventrobasal complex of the thalamus to primary sensory cortex (SI). This information is then distributed in ventral pathways to neurons in the parietal ventral, secondary somatosensory (SII) cortex and insula and in dorsal pathways to neurons in areas 5 and 7 of the parietal cortex as illustrated in Fig. 88.1 (Hsiao et al., 2003). It is thought that the division between these ventral and dorsal streams is analogous to the division between ventral (what) and dorsal (where) pathways in vision (Friedman et al., 1986).

Neural activity throughout these pathways (except for primary nerves) is affected by the focus and state of attention. The effects are small in the subcortical pathways, intermediate in SI cortex, and large in SII cortex. In SII cortex changes in the focus of attention cause substantial changes (increases and decreases) in neuronal firing rates in more than 90% of neurons.

In this chapter, we discuss neurophysiological studies of SII neurons in animals switching their attention between tactile and visual stimuli. These studies show that changes in the focus of attention modify the amount of synchronous firing between pairs of neurons as well as the firing rates of individual neurons. Computational studies have shown that changing firing synchrony is a potentially powerful way to select relevant stimuli and these studies demonstrate that attention could function with such a mechanism (Niebur et al., 1993; Niebur and Koch, 1994). The idea is that higher cognitive areas select which stimuli are to be processed by modifying the degree of synchronous firing between those neurons that represent the relevant stimuli and desynchronize the firing between neurons that do not. The importance of synchrony is that synchronous action potentials produce larger EPSPs (excitatory postsynaptic potentials) than do asynchronous action potentials and have a larger potential for producing an action potential.

II. CANDIDATE NEURAL MECHANISMS OF SOMATOSENSORY SELECTION

Two theoretical studies have proposed that a mix of rate and temporal coding may play a role in attentional

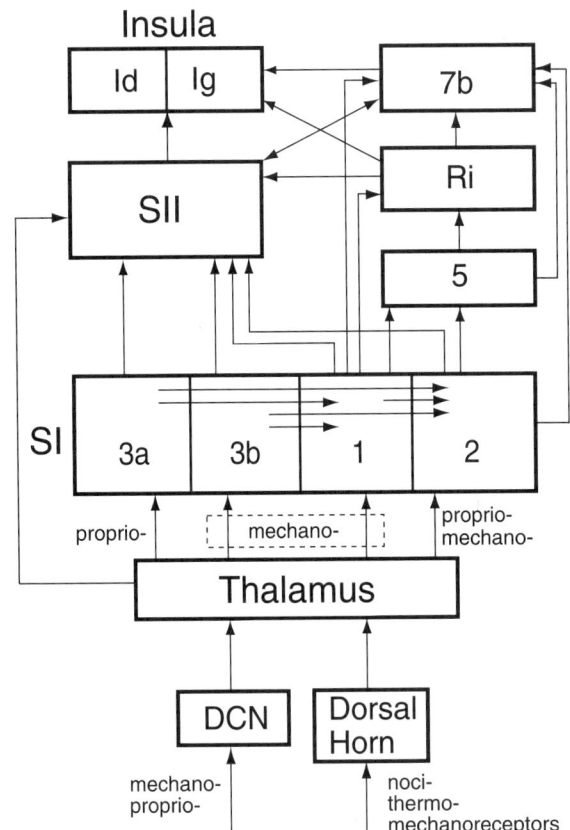

FIGURE 88.1 Schematic diagram of the somatosensory system showing only the feed-forward projections. Projections to frontal and motor cortical areas are not shown. DCN—dorsal column nuclei located in the base of the brainstem where the spinal cord meets the brainstem; relays tactile and propriospinal (position and movement) information. Dorsal horn—dorsal arm of the gray matter of the spinal cord, which relays information from the small myelinated and unmyelinated peripheral neurons. Projections above the thalamus show connections related only to the information from the DCN. The projections of information about pain, temperature, and other information from the small fibers are different. Areas 3a, 3b, 1, and 2 within SI cortex are arrayed across cortex in a mediolateral strip behind motor cortex. SII and insular cortex lie within the lateral fissure below SI cortex. Areas 5, Ri, and 7b lie posterior to SI cortex. SII cortex, the object of the study discussed in this chapter, is critical for object recognition. From Hsiao, Johnson, and Yoshioka, "Neural mechanisms of tactile perception," in *Comprehensive Handbook of Psychology*, Volume 3: Biological Psychology (Gallagher and Nelson, eds); ©2003, Wiley, New York. This material is used by permission of John Wiley & Sons, Inc.

selection. The models show that the temporal component could be based on either periodic firing (Niebur et al., 1993) or on aperiodic synchrony between neurons representing the selected stimuli (Niebur and Koch, 1994). Experimental evidence supporting the first mechanism in the visual system is provided by Fries (Fries et al., 2001 and Chapter 86). Experimental evidence supporting the second mechanism was recently reviewed by Niebur (Niebur et al., 2002; Niebur, 2002) (see Bichot et al., 2003, for more recent results) and is discussed briefly in this chapter.

Periodic Synchrony

In primate somatosensory cortex, periodic correlations of neural activity have been observed in several experiments in which the awareness of stimuli was changing. Ahissar and Vaadia (1990) reported oscillations in SII cortex during performance of a simple auditory detection task that were suppressed by somatosensory stimulation. Similarly, Lebedev and Nelson (1995) found shared oscillations in pairs of SI neurons while a monkey performed wrist movements cued by vibration; these oscillations were present while waiting for the cue and were suppressed by the vibration and movement. Murthy and Fetz (1996) reported oscillatory correlations between pairs of somatosensory and motor neurons that were greatest during exploratory arm movements; the activity of these pairs was phase-locked to local field potential oscillations.

Although none of these experiments was specifically designed to test mechanisms of selective attention, the results show that periodic oscillations are present in somatosensory cortex when the subject has a heightened awareness of its tactile environment, and that the oscillations are decreased or eliminated once the tactile cues are no longer pertinent to a task.

A. Nonperiodic Synchrony

In nonperiodic synchrony there is no evident periodicity in the synchronous firing between neurons. The effects of changes in the focus of attention on nonperiodic synchrony were studied first in SII cortex in monkeys trained to switch attention between tactile and visual discrimination tasks (Steinmetz et al., 2000). In that study, embossed letters were scanned across a restrained finger. In the tactile task, the animal had to detect a particular letter shape to obtain a reward. In the visual task it had to detect when an illuminated square dimmed. During the visual task, the drum continued to rotate and present letters on the fingerpad, but they were irrelevant for obtaining a reward.

During the experiment, the activity of up to 7 neurons in SII cortex was recorded simultaneously. The joint activity of pairs of neurons was examined using shift-predictor corrected cross correlograms. A substantial percentage (66%) of such pairs showed a

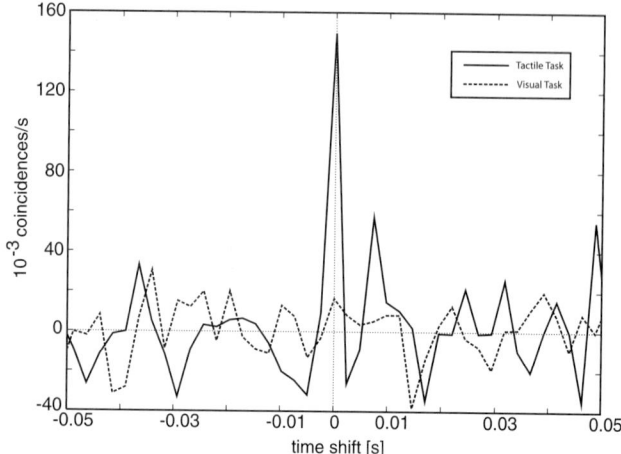

FIGURE 88.2 Shift-predictor corrected cross-correlogram (SCCC) for a pair of neurons in SII. Solid line: SCCC while the monkey discriminated embossed letters presented to its finger tip. Dashed line: SCCC while the monkey detected dimming of white squares presented on a visual display.

statistically significant peak in their cross correlogram as illustrated in Fig. 88.2. More interestingly, of the pairs with a significant peak, 17% changed the degree of synchronous firing when the monkey shifted its focus of attention between the tactile and visual tasks. This change was in addition to changes in synchrony expected of changes in firing rate, as observed previously (Hsiao et al., 1993).

III. POPULATION ACTIVITY

Assuming random sampling of neurons, the total percentage of neurons involved in synchronous firing is greater than the percentage of pairs of neurons since two neurons are involved in each pair (though the exact percentage depends on the relative timing between three or more neurons). For example, if 40% of neurons in a population are synchronized and change their synchrony together, only 16% of neuron pairs sampled at random will come from this subset and, therefore, display the change in synchrony. Thus a substantial majority of neurons in SII are involved in synchronous firing and more than 40% of these neurons change their synchrony when shifting attention between tasks. These results are potentially significant since previous computational studies (Niebur and Koch, 1994) showed that a change of this magnitude is capable of modulating the activity of neurons in higher cortical areas in an attentionally dependent manner.

IV. SUMMARY

The somatosensory system, like the auditory and visual systems, must filter a vast amount of sensory information to obtain that which is relevant for any particular behavioral objective. This filtering, or attentional selection, may be implemented by synchronizing the firing of pairs of neurons, thereby labeling stimuli that are relevant for the animal's behavior. The available evidence suggests that the somatosensory cortex accomplishes this filtering and attentional selection through changes in synchrony as well as changes in the firing rates of neurons in somatosensory cortex.

References

Ahissar, E., and Vaadia, E. (1990). Oscillatory activity of single units in a somatosensory cortex of an awake monkey and their possible role in texture analysis. *PNAS* **87**, 8935–8939.

Bichot, N. P., Rossi, A. F., and Desimone, R. (2003). Cue, feature and saccade related modulation of neuronal synchronization in Macaque area V4 during free-viewing visual search. *Society for Neuroscience Abstracts*, 551.8.

Friedman, D. P., Murray, E. A., O'Neill, J. B., and Mishkin, M. (1986). Cortical connections of the somatosensory fields of the lateral sulcus of macaques: Evidence for a corticolimbic pathway for touch. *J Comp Neurol* **252**, 323–347.

Fries, P., Reynolds, J. H., Rorie, A. E., and Desimone, R. (2001). Modulation of oscillatory neuronal synchronization by selective visual attention. *Science* **291**, 1560–1563.

Hsiao, S. S., O'Shaughnessy, D. M., and Johnson, K. O. (1993). Effects of selective attention on spatial form processing in monkey primary and secondary somatosensory cortex. *Journal of Neurophysiology* **70**, 444–447.

Hsiao, S. S., Johnson, K. O., and Yoshioka, T. (2003). Processing of tactile information in the primate brain. In "Handbook of Psychology (3): Biological Psychology" (Gallagher and Nelson, Eds.), pp. 211–236. Wiley, New York.

Hsiao, S. S., and Vega-Bermudez, F. (2002). Attention in the somatosensory system. In "The Somatosensory System: Deciphering the Brain's Own Body Image" (Nelson, Ed.), pp. 197–217. CRC Press, Boca Raton.

Johnson, K. O. (2001). The roles and functions of cutaneous mechanoreceptors. *Curr Opin Neurobiol* **11**, 455–461.

Lebedev, M. A., and Nelson, R. J. (1995). Rhythmically firing (20–50 Hz) neurons in monkey primary somatosensory cortex: activity patterns during initiation of vibratory-cued hand movements. *Journal of Computational Neuroscience* **2**, 313–334.

Murthy, V. N., and Fetz, E. E. (1996). Synchronization of neurons during local field potential oscillations in sensorimotor cortex of awake monkeys. *J. Neurophysiology* **76**, 3968–3982.

Niebur, E. (2002). Electrophysiological correlates of synchronous neural activity and attention: a short review. *Biosystems* **67**, 157–166.

Niebur, E., Hsiao, S. S., and Johnson, K. O. (2002). Synchrony: a neuronal mechanism for attentional selection? *Current Opinion in Neurobiology* **12**, 190–194.

Niebur, E., and Koch, C. (1994). A model for the neuronal implementation of selective visual attention based on temporal correlation among neurons. *Journal of Computational Neuroscience* **1**, 141–158.

Niebur, E., Koch, C., and Rosin, C. (1993). An oscillation-based model for the neuronal basis of attention. *Vision Research* **33**, 2789–2802.

Steinmetz, P. N., Roy, A., Fitzgerald, P. J., Hsiao, S. S., Johnson, K. O., and Niebur, E. (2000). Attention modulates synchronized neuronal firing in primate somatosensory cortex. *Nature* **404**, 187–190.

CHAPTER 89

Crossmodal Attention in Event Perception

Katsumi Watanabe and Shinsuke Shimojo

ABSTRACT

Recent studies have revealed how attention operates across modalities and affects the detection of, the orientation to, and the perceptual localization of, stimuli. In contrast, attentional modulation of crossmodal *event perception* is relatively unexplored. We have investigated the involvement of attentional process in crossmodal event perception using an ambiguous motion display. Two identical visual targets moving across each other can be perceived either to bounce off or to stream through each other. A brief sound at the moment the targets coincide was shown to bias perception toward bouncing. Our investigations revealed that auditory, tactile, and visual transients all bias perception toward bouncing when coincident with the visual crossing. This implies the possibility that automatic shifts of crossmodal attention from the moving stimuli to the sensory transient mediate the disambiguation of the streaming/bouncing motion display. Further experiments directly manipulating endogenous attention supported this attention account. The streaming percept arises when attentional resource is available around the moment of the visual coincidence and the bouncing percept results from the lack of attentional resource. Thus, dynamic allocation of crossmodal attentional resource can mediate crossmodal event perception.

I. CONVENTIONAL CROSSMODAL PHENOMENA AND CROSSMODAL SPATIAL ATTENTION

Since different senses are attuned to the different aspects of the environment, crossmodal interactions in theory can reduce perceptual ambiguity that may result from relying on a single sensory modality. The behavioral and perceptual outcome of crossmodal interaction are exemplified by the detection facilitation of bisensory stimuli (Stein and Meredith, 1993), localization bias under conflicting crossmodal information (ventriloquist effect; Howard and Templeton, 1966), and categorization bias of complex auditory stimuli with additional visual inputs (McGurk effect; McGurk and MacDonald, 1976).

Recently, crossmodal interaction has gained renewed interest with fresh perspectives of attention. Duncan et al. (1997) have shown that a task in one modality does not interfere with another task in another modality (under some conditions at least) and suggested that attentional resources are not shared by different modalities. On the other hand, Driver, Spence, and colleagues have been rigorously exploring how automatic shifts of spatial attention affect the processing of stimuli in different modalities (Driver and Spence, 1998). These studies suggest that the way attention affects crossmodal interaction varies very much depending on stimuli, context, and task.

II. ATTENTIONAL EFFECTS ON EVENT PERCEPTION

Compared with crossmodal attentional modulation on detection and localization, attentional effects on *crossmodal event perception* are relatively unexplored (Calvert et al., 1998). The paramount reason for this seems to be the lack of adequate psychophysical paradigm on crossmodal events. Although the McGurk effect is a robust phenomenon, the effect inherently depends on the complex temporal structure of auditory (speech) and visual (facial movement) stimuli.

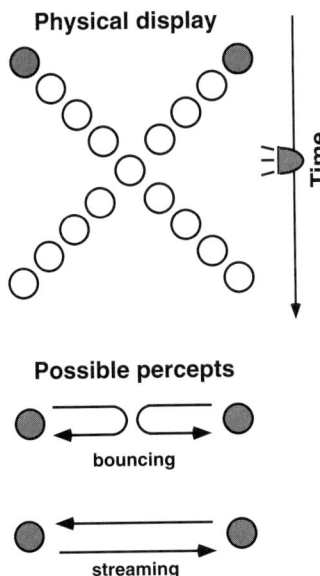

FIGURE 89.1 Sekuler's ambiguous motion display (Sekuler et al., 1997). Because of correspondence ambiguity, there are two possible and mutually exclusive interpretations: streaming and bouncing. Although the streaming percept usually dominates, a sound presented at the moment of coincidence biases perception toward bouncing.

With this regard, Sekuler et al. (1997) devised a simple yet interesting paradigm to show a compelling effect of sound on visual motion perception. In a two-dimensional display, two identical visual targets moving across each other can be perceived either to bounce off or to stream through each other (see Fig. 89.1). Despite this ambiguity in the visual stimulus, observers normally show a strong bias toward seeing the streaming percept. However, a brief sound at the moment the targets coincide biases perception toward bouncing. This suggests that crossmodal interaction is involved in solving the streaming/bouncing visual motion ambiguity. The success of Sekuler's display requires (1) that the visual stimulus is *ambiguous*; (2) that the possible interpretations are *mutually exclusive yet effortless to perceive*; (3) that it uses a *simple* auditory stimulus, and opens the possibility of investigations on the unexplored topic of auditory influence on visual event perception. Our series of investigation revealed that crossmodal attention plays a central role in disambiguation of the motion event, as summarized in the following.

A. Auditory-Context Dependency of the Bounce-Inducing Effect

The first clue to the involvement of attention in the disambiguation process comes from its dependency on auditory context and saliency (Watanabe and Shimojo, 2001). The bounce-inducing effect by a simultaneous sound can be attenuated (from 80% to 30–50%, approximately) when the sound is embedded in a sequence of other identical sounds and therefore its perceptual saliency is reduced. Subsequent experiments show that the bounce-inducing effect can be recovered if the simultaneous sound pops out from the embedding sound sequence (due to pitch or intensity difference). These results suggest that the saliency of auditory transients, not just the local, physical changes of stimulus energy, is critical for the bounce-inducing effect. Furthermore, the bounce-inducing effect turned out to be a suprathreshold effect; the magnitude of the effect closely correlates with the detection of a salient auditory stimulus (Watanabe, 2001). These opening results raise several critical questions. If it is indeed saliency that matters, then would a transient in a modality other than auditory also enhance bouncing? And, more relevantly to the theme of this book, what is the relationship between attentional processes and salient sensory transients that induce the bouncing percept?

B. Bounce-Inducing Effect by Auditory, Tactile, and Visual Transients

The second clue to the involvement of crossmodal attention derives from bounce-inducing effects by sensory transients in modalities other than auditory. An intuitive account for the bounce-inducing effect may be that the simultaneous sound increases the probability of seeing a bouncing event because people have experienced such synchronized events so many times due to physical bouncing in the real world. In other words, associative memory developed via repeated experience enables the brain to perform an inverse inference (audiovisual association account). A strict interpretation of this account would predict that the effect should be restricted only to a transient stimulus in audition, but not in other modalities such as vision or touch, because not many collision events are accompanied with a flash or a touch in the real world. On the contrary, we found that auditory, tactile, and visual transients around the moment of the visual coincidence all bias visual perception toward bouncing (Watanabe, 2001; Fig. 89.2). This result clearly contradicts the simple association account based on auditory-visual coupling of physical bouncing events. Thus, the bounce-inducing effect is a function of *supermodal* saliency and synchrony in general, which requires a common crossmodal mechanism that mediates the disambiguation of the streaming/bouncing motion display. An obvious candidate is crossmodal attention.

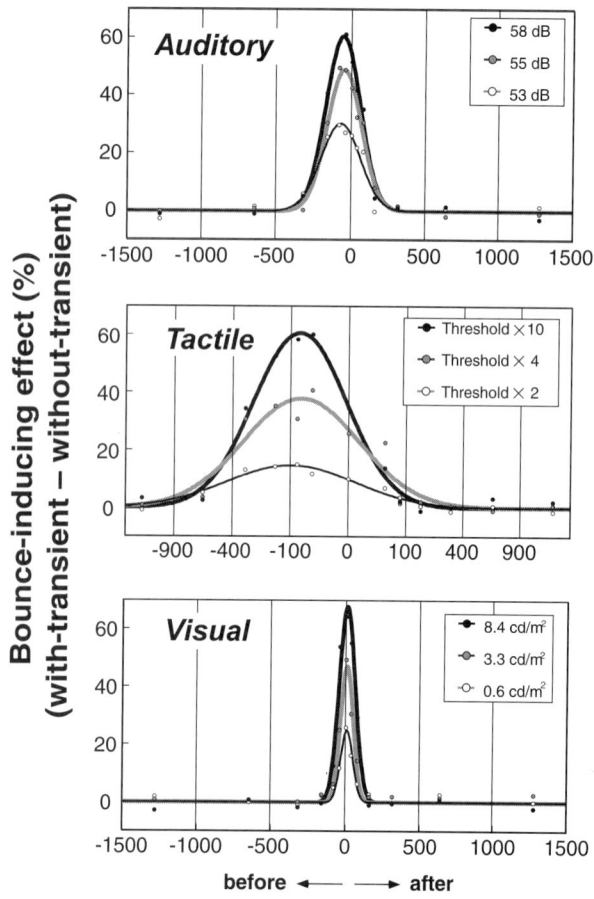

FIGURE 89.2 Dependency of the bounce-inducing effect on intensity (saliency) of various sensory transients. The results were approximated with a Gaussian function without any parameter constraint. For the results with tactile transients, in order to correct the skewed distributions of the results, a square-root transformation, with the 0-ms asynchrony being the origin, was applied on the time axis. Note that there are temporal asymmetries in the case of auditory and tactile transients, but not with the visual transients. Adapted from Watanabe (2001).

C. Attention Account Based on Automatic Capture of Crossmodal Attention

Motivated by the results of supermodal saliency dependency, we propose the attention hypothesis for the bounce-inducing effect. Generally, a sudden sensory event automatically attracts attention (about attentional capture, see Chapters 69 and 12). The attention hypothesis states that the streaming percept requires attentional resource. One possibility is that attention to motion directly enhances the directional bias based on the past stimulation of local motion detectors, thereby producing a tendency of motion as continuing to occur in the same direction as in the past (or "motion inertia"; Anstis and Ramachandran, 1986).

Auditory, tactile, and visual transients automatically capture crossmodal attention (Driver and Spence, 1998). Attentional capture reduces attentional resource available to motion processing, which in turn leads to the enhancement of the bouncing percept. The magnitude of attentional capture depends on the saliency of the stimulus. This fits very well with the saliency dependency of the bounce-inducing effect. In addition, most motion perceptions are modulated by attention (see Chapter 81). Since the Sekuler's display involves ambiguous object motion and salient sensory transient, it is highly possible that automatic attentional capture plays a role in the disambiguation of the ambiguous motion event.

D. Enhancement of Bouncing Percept without Sensory Transients

William James distinguished between two forms of attention: active (*endogenous*) versus passive (*exogenous*) attention. In the view of the attention account, the bounce-inducing effect by salient transients occurs because of *exogenous* distraction of attention from visual motion stimuli. Effectiveness of *endogenous* distraction of attention on motion ambiguity solving, then, is clearly worth examining as the ultimate test of the attention account (Watanabe and Shimojo, 1998).

Endogenous components of attention can be investigated by the measurement of interference between concurrent tasks. In the basic concurrent task paradigm, observers are asked to perform two tasks (Task A and Task B) at the same time, but primarily concentrating on one of the tasks (e.g., Task B). Task B is usually an attention-demanding task. In a separate session of the experiment observers are asked to perform only Task A, but the stimuli from Task B are also presented. Performance of Task A with the demand of Task B is compared to performance without it. Since the stimulus configuration is identical in both cases, if the demand of Task B has influence on the performance of Task A, Task A may be concluded to require attentional resources. Applying this basic logic, we investigated whether an endogenous shift of visual attention would alter the relative frequency of streaming and bouncing percepts by asking observers an attention-demanding task (detecting a brief, small gap presented on the bull's-eye fixation stimulus) around the time of the visual coincidence. Since observers did not know when exactly the brief gap would happen but knew that there would be only a single presentation of the gap for each trial, attention would be concentrated on the fixation stimulus until the task

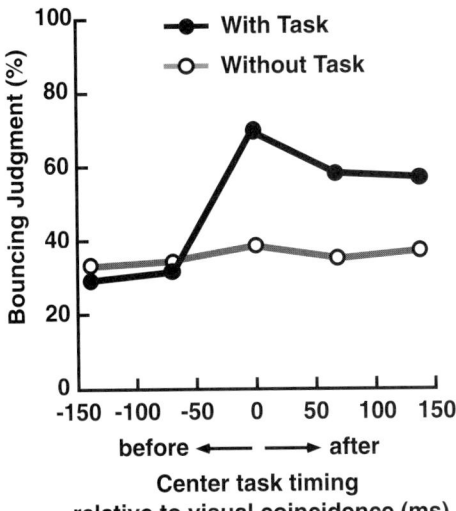

FIGURE 89.3 Effects of the central task timing (attention availability) on the relative frequency of the bouncing percept. The gray line indicates the condition without the center task and the black line indicates the condition with the center task. Adapted from Watanabe and Shimojo (1998).

occurred. Therefore, if the central task demand was imposed *at* or *after* the visual coincidence in the periphery, attention would have to be focused on the fixation stimulus at the moment of the visual coincidence. The attention account predicts that the bouncing percept would increase in such a condition because of less attention being available for processing motion perception. In contrast, if the stimulus for the central task occurred *before* the streaming/bouncing event, attention would be finished with the central task and already available for motion processing at the time of the visual coincidence. Thus, there would be no effect of the central task. Results exactly followed this predicted pattern (see Fig. 89.3); when attention has to be concentrated on the central task at the moment of the visual coincidence (*at* and *after*), the bouncing percept becomes dominant (the right-shoulder in the black curve of Fig. 89.3).

E. Involvement of Dynamic Attention Allocation in Perceptual Disambiguation

Using the concurrent task paradigm, we demonstrated that endogenous distraction of attention at the visual coincidence biases visual perception toward bouncing, despite the fact that there is no transient at the time of the visual coincidence. The results suggest (1) that availability of attentional resource to visual motion contributes to the bias toward the perception of streaming whereas distractions of attention increased the frequency of the bouncing percept, and (2) that a dynamic attentional process is involved in perceptually disambiguating the streaming/bouncing motion display. Also, when more attentional resources are available, attentional resources automatically are used for processing motion perception. This automaticity can be inferred because the observers were instructed not to intentionally (attentionally) track one of the targets, and motion is believed to attract attention automatically.

The attention account for the bounce-inducing effect is in accord with the idea that attention is a limited resource. The proposed automaticity of attentional allocation (i.e., that default is for moving stimuli, then attentional-grabbing factors take it away from them) resembles Lavie's theory of attention (Lavie, 1995); people cannot simply ignore a particular object or aspect of a stimulus; in order to ignore a stimulus we have to attend to another task. A functional magnetic resonance experiment has recently provided supporting evidence (Rees et al., 1999). Observers performed an attention-demanding visual task while a task-irrelevant random-dot field moved in the same screen. Activation of area MT (visual motion area in humans) inversely correlated with the task load, but MT activation did not correlate with the observer's intention to ignore the random-dot field. Attentional resource available to visual motion depends on how much the distracting task uses up the attentional resource.

F. Temporal Asymmetry of the Crossmodal Interaction Window

To appreciate the dynamic allocation of attention within and across modalities, it is worth noting that the bounce-inducing effect increases as the saliency of the sensory transient increases but the temporal window for the bounce-inducing effect is fixed in time (see Fig. 89.2). And more importantly, with the transient sound and vibration, peculiar shifts of the temporal window center were observed. The peaks were shifted such that the transients presented before the visual coincidence produced a stronger bounce-inducing effect than those presented after the visual coincidence. The shifts were about 60 ms in the auditory-induced effect and about 80 ms in the tactile-induced effect, respectively. Such a temporal asymmetry is virtually nonexistent in the bounce-inducing effect by visual transients.

Our attention account is consistent with this temporal asymmetry of the crossmodal interaction window. The visual coincidence must occur slightly after the auditory and tactile transients in order to achieve the maximal bounce-inducing effect. This may be due to the known time cost of attentional switching between two modalities (modality-shifting effect; LaBerge, 1973). In the streaming/bouncing motion display, attention is automatically directed to the moving stimulus in the visual modality. An auditory or tactile transient attracts attention, but it takes time to shift attention between modalities. Hence, in order to produce the maximal attentional capture at the moment of the visual coincidence, a nonvisual transient has to occur before the visual coincidence in order to capture maximum attentional resources on time. The absence of the temporal asymmetry in the bounce-inducing effect by visual transients may be due to the absence of the modality-shifting delay, and the much faster attentional switching within the visual modality.

III. DEVELOPMENTAL ASSAY UTILIZING THE CROSSMODAL DISPLAY

A developmental study of the bounce-inducing effect in our laboratory has recently demonstrated that at six months of age the bounce-inducing effect by sound becomes functional (Scheier et al., 2003). By employing the classical habituation-dishabituation technique, we found that the six- and eight-month-old, but not the four-month-old, infants responded to the ambiguous motion display with a sound synchronized with the visual coincidence and the identical visual display with a sound no longer synchronized (2.3 s before or after the visual coincidence) as categorically different. Importantly, the older infants treated the ambiguous motion display with a synchronized sound (habituation display) and the physically bouncing, unambiguous display (test display) as equivalent by not exhibiting dishabituation. Whereas earlier onset ages are reported for sensitivity to audio-visual spatiotemporal relations (e.g., Lewkowicz, 2000), what unique in our study are the perceptual ambiguity and, possibly, the involvement attention in solving it. Thus, the developmental emergence of the crossmodal effect seem to reflect developmental changes in the attention-related mechanism (see Chapters 14 and 34). The relatively late onset of this particular aspect of crossmodal interaction may be interpreted in that crossmodal associative learning first utilize and tune up crossmodal (spatiotemporal) coincidences, and then attention mechanisms interact in resolving perceptual ambiguities in crossmodal event perception.

IV. CROSSMODAL EVENT PERCEPTION BY DYNAMIC ATTENTIONAL ALLOCATION

Our results emphasize that investigation on crossmodal event perception requires scrutiny on the precise roles of attentional processes as well as the intuitive associative learning interpretation. At the first glance, the attention account may sound counterintuitive, given the intuitiveness of the audiovisual association account. However, it may be worth emphasizing that these explanations are not mutually exclusive as both may reflect the likelihood of the combination of a physical collision and a transient event in the natural environment. The results from our experiments imply only that associative learning based on experience is not an exclusive way to implement a perceptual event coupling. As long as an algorithm can compute and represent joint probabilities given sensory events within and between modalities, any mechanism would suffice. It still serves as a specialized collision detector, yet implemented in a more general-purpose attention mechanism, presumably because it is a cost-efficient solution with regard to biological resources and evolution. Evolution and development of attention mechanism may have been due to biological necessity of selection and integration across, as well as within, modalities.

References

Anstis, S. M., and Ramachandran, V. S. (1986). Entrained path deflection in apparent motion. *Vision Res.* **26**, 1731–1739.

Calvert, G. A., Brammer, M. J., and Iversen, S. D. (1998). Crossmodal identification. *Trend. Cog. Sci.* **2**, 247–253.

Driver, J., and Spence, C. (1998). Attention and the crossmodal construction of space. *Trend. Cog. Sci.* **2**, 254–262.

Duncan, J., Martens, S., and Ward, R. (1997). Restricted attentional resource within but not between sensory modalities. *Nature* **387**, 808–810.

Howard, I. P., and Templeton, W. B. (1966). *Human Spatial Orientation*. Wiley, London.

LaBerge, D. (1973). Identification of two components of the time to switch attention: A test of a serial and a parallel model of attention. In S. Kornblum (Ed.), *Attention and Performance IV* (pp. 71–85). Academic Press, New York.

Lavie, N. (1995). Perceptual load as a necessary condition for selective attention. *J. Exp. Psychol. Hum. Percept. Perform.* **21**, 451–468.

Lewkowicz, D. J. (2000). The development of intersensory temporal perception: An epigenetic systems/limitations view. *Psychological Bulletin* **126**, 281–308.

McGurk, H., and MacDonald, J. W. (1976). Hearing lips and seeing voices. *Nature* **264**, 746–748.

Rees, G., Frith, C. D., and Lavie, N. (1999). Modulating irrelevant motion perception by varying attentional load in an unrelated task. *Science* **278**, 1616–1619.

Scheier, C., Lewkowicz, D. J., and Shimojo, S. (2003). Sound induces perceptual reorganization of an ambiguous motion display in human infants. *Developmental Sci.* **6**, 233–241.

Sekuler, R., Sekuler, A. B., and Lau, R. (1997). Sound alters visual motion perception. *Nature* **385**, 308.

Stein, B. E., and Meredith, M. A. (1993). *The Merging of the Senses*. MIT Press, Cambridge, MA.

Watanabe, K. (2001). *Crossmodal interaction in humans*. Ph.D thesis. Computation and Neural Systems, California Institute of Technology, Pasadena.

Watanabe, K., and Shimojo, S. (1998). Attentional modulation in perception of visual motion events. *Perception* **27**, 1041–1054.

Watanabe, K., and Shimojo, S. (2001). When sound affects vision: effects of auditory grouping on visual motion perception. *Psychological Sci.* **12**, 109–116.

… SECTION IV

SYSTEMS

CHAPTER

90

The FeatureGate Model of Visual Selection

Kyle R. Cave, Min-Shik Kim, Narcisse P. Bichot and Kenith V. Sobel

ABSTRACT

The FeatureGate model is designed to account for the results from a number of studies in visual attention, including parallel feature searches and serial conjunction searches, variations in search slope with variations in feature contrast and individual subject differences, attentional gradients triggered by cueing, feature-driven spatial selection, split attention, inhibition of distractor locations, and flanking inhibition. The model is implemented in a neural network consisting of a hierarchy of spatial maps. Attentional gates control the flow of information from each level of the hierarchy to the next. The gates are jointly controlled by a bottom-up system favoring locations with unique features and a top-down system favoring locations with features designated as target features. The gating of each location depends on the features present there, hence the name FeatureGate.

I. INTRODUCTION

The FeatureGate model of visual selection is motivated by principles of feature-driven selection and distractor inhibition (Kim and Cave, 1995; Shih and Sperling, 1996; Cave and Zimmerman, 1997; Cepeda et al., 1998; Bichot et al., 1999), along with other basic assumptions about attention supported by earlier experiments using visual search (Treisman and Gelade, 1980; Wolfe et al., 1989), spatial cuing (Downing and Pinker, 1985), and distractor interference (Eriksen and Eriksen, 1974). In developing the model, we began with the assumption that if high-level mechanisms receive information about multiple objects simultaneously, the information from each object can interfere with processing and identification of the other objects. Interference could occur within mechanisms for complex visual processing, response selection, or both. Most visual tasks require intervention from an attentional system that controls the flow of information from different objects into these high-level mechanisms to prevent interference.

The model is implemented as a simple neural network, illustrated in Fig. 90.1. The input is encoded as a pattern of activation across a two-dimensional spatial arrangement of neural units. For each location there is a set of units, with each unit registering the presence of a primitive visual feature (color, orientation, etc.) at that location. Connections from all of the input units across different locations converge to feed a single set of output units that represent the presence of visual features and objects but not their locations. This location-independent output is thus available for higher-level visual processing. Attentional gates control the flow of information across several units at a lower level to a single unit in the next-higher level. Because the features present across several locations govern the extent to which the associated gates are opened, we named the model FeatureGate.

II. SELECTING LOCATIONS IN FEATUREGATE

In FeatureGate, all the features at a single location share the same fate: They are all facilitated or inhibited to the same extent. All the features from just one location (color, orientation, etc.) pass through an open attentional gate to the output while at the same time closed gates prevent the features at other locations from appearing in the output. The gates' settings result

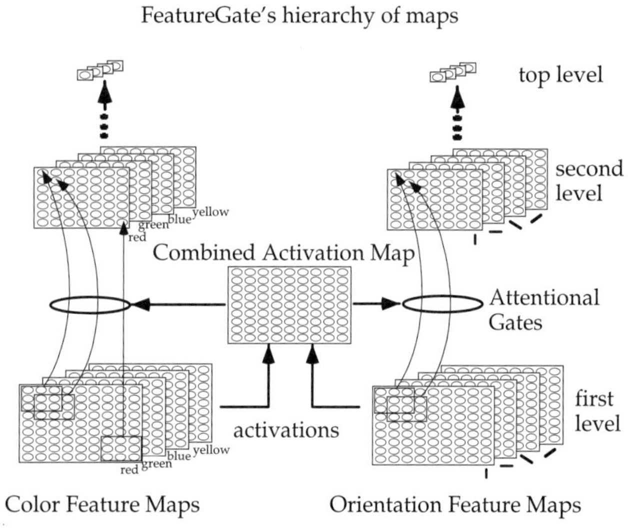

FeatureGate's hierarchy of maps

FIGURE 90.1 General architecture of the model.

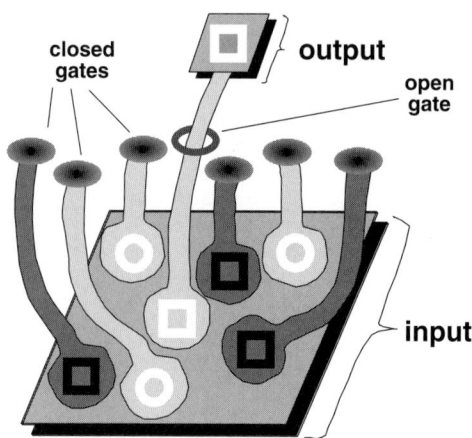

FIGURE 90.2 As distractors are inhibited, only features from the target location are output by the model.

from a competition to select the one location that is likeliest to contain important information. Figure 90.2 illustrates how the inhibition of information from each of the distractor locations produces an output containing only the features of the selected target location.

The importance of each location for visual processing is determined by two distinct factors. In some circumstances, the viewer is seeking a particular visual object for which some or all of the distinguishing visual features are known (as in a search for the person wearing red stripes). But oftentimes, a location with distinct features is very informative and deserving of attention, even though the object at that location may not have been expected. In order to meet both of these attentional goals, the competition between different locations for attention is jointly governed by two subsystems, which we have named the top-down system and the bottom-up system.

The top-down system favors locations with known target features by closing gates at locations with non-target features to inhibit processing of information from those locations. The more the features at a location differ from the target features, the more its gate will be closed and the more that location will be inhibited, as seen in Kim and Cave's (1995) experiment. The top-down system is engaged only when a target with known features is being sought, and the variation in attentional strength in Cave and Zimmerman's (1997) study suggests that top-down inhibition can be turned up or down according to the needs of the task.

The bottom-up system compares the features at each location to those in neighboring locations and holds gates open at those locations with features that differ from the features at surrounding locations. It is driven solely by stimulus properties, independently of any task-related goals. This system ensures that a location with a featural singleton, such as a red object in a field of green objects, will not have its gate closed. The more the features at a given location differ from the surrounding features, the stronger the influence from the bottom-up system to keep its gate open. In some circumstances the bottom-up system will hold the gate open at a singleton distractor's location, even when the subject intends to ignore it (Pashler, 1988; Theeuwes, 1992).

The top-down and bottom-up influences take the form of a set of activations assigned to each location. First the top-down and bottom-up systems each calculate a separate activation for each location. Within each subsystem, these activations are determined separately for each feature dimension (color, orientation, etc.). The top-down activations for each dimension combine to form an overall top-down activation for each location. The bottom-up system sums activations from across several locations as well, and then the top-down and bottom-up activations are summed (just as in the original Guided Search model in Cave and Wolfe, 1990) to form an overall activation for that location. Finally, the gate at the location with the highest activation in the local group is thrown open while gates at other locations are partially closed. Thus, the top-down and bottom-up systems work together to control the attentional gates so that only the features at the location with the highest activation are strongly represented in the output and are made available for high-level processing.

III. HIERARCHICAL STRUCTURE

To work quickly and efficiently within a neural network, FeatureGate should be implemented with a minimum of long connections from one end of a spatial map to another. Thus, whenever possible, FeatureGate uses local operations that compare features and activations near each other in the visual field. Rather than compare the activation at one location to that of every other location in the input, the visual field is split up into small regions, or neighborhoods. The competition previously described is carried out separately within each region, with the gate for the location within the region with the highest activation being fully open and the gates for locations with other stimuli being partially closed. As a result, the winning location from each of these regions is combined with all the other winning locations to form a new spatially organized representation at the next level. The locations in each region of this new level then compete, with the winning locations forming a third set of spatially organized representations, and the same procedure continues. Thus, the FeatureGate network, unlike the original Guided Search model, consists of a hierarchy of spatial representations, or maps, as shown in Fig. 90.3. The maps at each level have fewer units than those in the level below; consequently, each higher-level unit receives information from a larger area of the visual field than any unit at a lower level. At the upper levels of the hierarchy, each location in the maps actually represents a single location selected from a broad region of the visual field. At the top level is just a single location. The features at this level are the output of the model and represent the features from the single location with the highest activation, which allowed it to win a competition at every level in the hierarchy.

This hierarchy of spatial maps corresponds roughly to the organization of visual areas in cortex. The earliest cortical area, V1, is very large and is made up of cells with small receptive fields. Going up the hierarchy of brain areas, through V2, V4, and into the inferior temporal lobe, the receptive fields become increasingly larger, just as they do for the units in the higher levels of FeatureGate. The cells in the upper levels also respond to more complex stimulus properties. Currently, FeatureGate includes no specific claims about what complex properties are represented by these high level cells, but presumably they encode combinations of the basic features represented at lower levels, as in Mozer's (1991) model.

In this hierarchical architecture, a target will compete with nearby distractors in the lower levels, and with more distant distractors in the upper levels. Given that in Cave and Zimmerman's (1997) experiment the nearby distractor locations were inhibited more strongly than those farther from the target, we can infer that in that task the distractor inhibition (or at least the inhibition generated by the top-down system, which is more important in Cave and Zimmerman's letter search) is stronger in the lower levels. Perhaps the simple shape features that differentiate the letters are represented most explicitly at that level, making it more important to prevent distractor interference there.

IV. INHIBITION OF RETURN AND SERIAL SEARCH

In standard feature and conjunction searches, FeatureGate behaves much like the Guided Search model. In the course of the competition among different stimulus locations, information from all but one of the stimulus locations will be inhibited somewhere in the hierarchy. Through this inhibition, the model selects a single location to be represented in the output. As in Guided Search, the bottom-up and top-down systems work to select the target location, but in some circumstances the target will be inhibited and distractors will be selected instead.

A distractor will rarely be selected over a target when the target differs from distractors by a single salient feature (e.g., red among green). In these feature

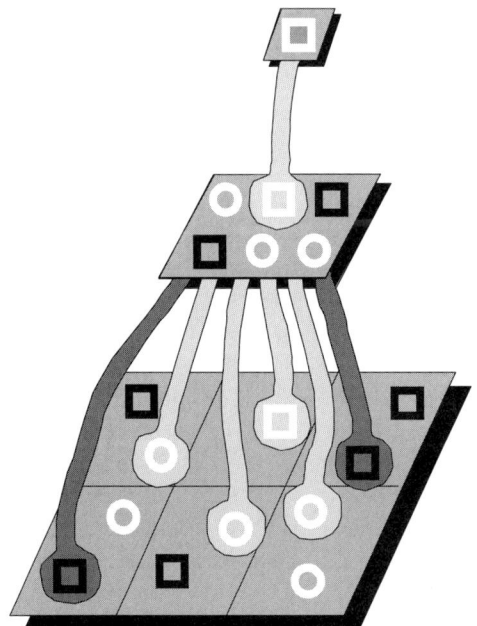

FIGURE 90.3 Hierarchical representations produced by the model.

searches, FeatureGate's bottom-up system allocates a strong dose of activation to the target's location but not to the distractor locations. Thus information from all the distractor locations will be blocked, whereas the target location is not blocked and its features are represented in the output with little interference. As a result, feature searches are easy for FeatureGate (just as they are for human subjects; Treisman and Gelade, 1980) because the target location is usually selected immediately.

However, when the target is defined by a conjunction of features (e.g., red horizontal among red vertical and green horizontal distractors), the selected location often does not contain the desired target. The same sort of selection errors occur in the Guided Search model in conjunction searches. In Guided Search, these errors are caused by random noise added to the activation values. The same random noise can be added to the activations of the units within FeatureGate, but under some conditions FeatureGate already makes errors in selecting conjunction targets without the noise. The bottom-up system activates locations with features differing from other features in the display and, because the bottom-up feature comparisons at the bottom levels are done within local regions, a distractor of one type can be highly activated if it happens to be near distractors of another type. Thus, if a red vertical distractor is positioned close to a number of green horizontal distractors, the red vertical distractor can be selected instead of the red horizontal target.

Once the distractor location has been selected, processed, and determined to be a nontarget, that location can be inhibited via inhibition of return (Posner and Cohen, 1984). This inhibition reduces the activation at the winning location, without affecting any of the other activations in the network. After the winning location is inhibited, a new cycle of competitions is initiated among the different locations. Because winners from previous competitions are now inhibited, they take no part in any competitions that they won in previous cycles, and some other location will now win and be represented in the output. This repeated selecting and inhibition of return represents serial search. In the search for a red horizontal target, FeatureGate may select a series of red vertical and green horizontal distractors before selecting the target. In general, the more distractors there are present in a conjunction search display, the more distractors will be selected before the target. Thus, in FeatureGate the time to find conjunction targets generally rises with the number of distractors, just as it does in Guided Search (Cave and Wolfe, 1990) and in human subjects (Treisman and Gelade, 1990; Wolfe et al., 1989).

Subjects in Wolfe and colleagues' (1989) conjunction searches exhibited a wide range of performance. For some participants, the search latency hardly increased with display size, whereas for others response time rose sharply with display size. Guided Search explained these differences by assuming that subjects with steep search slopes had more noise in their attentional activations. In FeatureGate, strong bottom-up activations assume the role of noise in Guided Search. Subjects with steep search functions are assumed to have strong bottom-up systems relative to those subjects with comparatively shallow search slopes.

V. ACCOUNTING FOR FEATURE-DRIVEN SELECTION AND DISTRACTOR INHIBITION

Kim and Cave (1995) showed that in a conjunction search for a target defined by both color and shape, locations with the target color and shape received the most attention, locations with the nontarget color and shape received the least attention, and the locations with one target feature and one nontarget feature received an intermediate amount of attention. This type of feature-driven selection is consistent with the predictions of Guided Search, in which each target feature present at a location adds to that location's activation. FeatureGate produces the same general pattern of feature-driven selection as Guided Search, although FeatureGate goes beyond the original Guided Search model to show how feature-driven selection can be implemented in a neural network.

In FeatureGate, the activation for each location decreases if it contains a distractor feature while a neighboring location contains a target feature along the same dimension. Thus, during a search for a red horizontal target, a location with green will suffer a reduction in activation if a red stimulus is nearby. Because the activations are first calculated separately for color and orientation before being summed, a location with red and horizontal will have the highest total activation, a location with green and vertical will have the lowest, and locations with red and vertical or green and horizontal will usually have intermediate activations. These activations control the attentional gates, so gates for locations with green and vertical will generally be closed the most and locations with either green or vertical (but not both) will be closed an intermediate amount. Once the gates are set to control the processing of the search array, they will also affect the processing of an attentional probe that appears at one of the locations. Thus, if a probe appears at a location with one nontarget feature and a gate that is somewhat closed, information about the probe will be partially blocked in its flow up the hierarchy and it will take longer for it to trigger a response than a probe at a location with both target features and a fully open gate. If the probe appears at a location with two nontarget features, its information flow is blocked even more and it will take even longer to trigger a response. In other words, the FeatureGate model is generally consistent with the probe response times from Kim and Cave (1995).

The closing of gates at distractor locations also explains the slow responses to spatial probes at distractor locations in experiments by Cepeda and colleagues (1998). These results suggest that the inhibition is applied very sparingly so as not to inhibit locations that are not currently interfering with processing of the target. In this experiment, subjects searched for a single red target among green distractors. Because the target was a color singleton and because subjects knew the target color in advance, both the bottom-up and top-down systems produced higher activation at the red location than at the green locations. Only locations with green were inhibited. Areas that form part of a uniform background were not as strongly inhibited. Because of this lack of inhibition, if an unexpected stimulus suddenly appeared at one of these blank locations, it could be detected relatively quickly. Consistent with these data, FeatureGate does not inhibit locations that form part of a uniform background.

If there is a single target in a display full of distractors, information from distractors near the target will converge with target information at a low level of the hierarchy, whereas information from distractors farther away will converge with the target information at a higher level. Cave and Zimmerman (1997) found that distractor locations near the target were inhibited more than those far from the target, leading to the conclusion that inhibition is stronger in lower levels of the hierarchy. This result may reflect a general pattern of attentional inhibition or may, instead, indicate that the spatial configurations that distinguish the letters used in Cave and Zimmerman's search are encoded in the lower levels and require more inhibition at those levels to prevent interference.

When searching for a single target, FeatureGate employs a winner-take-all competition within each region so that only a single location passes its features up to the next level. For some other visual tasks, however, it would be useful to combine features from different locations in the top level of the hierarchy. In their experiment demonstrating attention at two locations simultaneously, Bichot and colleagues (1999) designed such a task. Subjects viewed an array of squares and circles and had to determine whether two target objects in the array had the same shape. This comparison can be performed by representing features from both target locations simultaneously at the top level. If features belonging to squares and circles are both active, then the two target shapes clearly do not match and a "no" response is triggered. For this task, the winner-take-all competition in FeatureGate is relaxed so that every location with the target color is allowed to send its features to the top. Thus, FeatureGate can split its attention across two locations, just as Bichot and colleagues' subjects did.

The hierarchy of spatial maps in FeatureGate shares some properties with the structure of the model proposed by Itti and Koch (2000), although they have limited their model to bottom-up control of attention and thus it cannot search for a specified target among similar distractors in the way that FeatureGate does. Other models (Bundesen, 1990; Rao et al., 2002) share other properties with FeatureGate, but focus only on the top-down control of attention, without explaining bottom-up detection of unique features. In addition to its integration of bottom-up and top-down systems, FeatureGate differs from all of these models in that the competition among different objects and locations is distributed across different levels of the hierarchy. This structure allows the competition at the lower levels to be broken down into many local competitions that can be conducted independently and allows the individual units to exhibit responses similar to those found by Moran and Desimone (1985). For a fuller consideration of the experimental evidence behind the model and comparisons with other models, see Cave (1999).

References

Bichot, N. P., Cave, K. R., and Pashler, H. (1999). Visual selection mediated by location: Feature-based selection of noncontiguous locations. *Perception Psychophys.* **61**, 403–423.

Bundesen, C. (1990). A theory of visual attention. *Psychol. Rev.* **97**, 523–547.

Cave, K. R. (1999). The FeatureGate model of visual selection. *Psychol. Res.* **62**, 182–194.

Cave, K. R., and Wolfe, J. M. (1990). Modeling the role of parallel processing in visual search. *Cogn. Psychol.* **22**, 225–271.

Cave, K. R., and Zimmerman, J. M. (1997). Flexibility in spatial attention before and after practice. *Psychol. Sci.* **8**, 399–403.

Cepeda, N. J., Cave, K. R., Bichot, N. P., and Kim, M.-S. (1998). Spatial selection via feature-driven inhibition of distractor locations. *Perception Psychophys.* **60**, 727–746.

Downing, C. J., and Pinker, S. (1985). The spatial structure of visual attention. *In* "Attention and Performance XI: Mechanisms of Attention" (M. I. Posner and O. S. M. Mann, eds.), pp. 171–187. Erlbaum, Hillsdale, NJ.

Eriksen, B. A., and Eriksen, C. W. (1974). Effects of noise letters upon the identification of a target letter in a nonsearch task. *Perception Psychophys.* **16**, 143–149.

Itti, L., and Koch, C. (2000). A saliency-based search mechanism for overt and covert shifts of visual attention. *Vis. Res.* **40**, 1489–1506.

Kim, M.-S., and Cave, K. R. (1995). Spatial attention in visual search for features and feature conjunctions. *Psychol. Sci.* **6**, 376–380.

Moran, J., and Desimone, R. (1985). Selective attention gates visual processing in the extrastriate cortex. *Science* **229**, 782–784.

Mozer, M. C. (1991). "The Perception of Multiple Objects." MIT Press, Cambridge, MA.

Pashler, H. (1988). Cross-dimensional interaction and texture segregation. *Perception Psychophys.* **43**, 307–318.

Posner, M. I., and Cohen, Y. (1984). Components of visual orienting. *In* "Attention and Performance X" (H. Bouma and D. Bowhuis, eds.), pp. 531–556. Erlbaum, Hillsdale, NJ.

Rao, R. P. N., Zelinsky, G. J., Hayhoe, M. M., and Ballard, D. H. (2002). Eye movements in iconic visual search. *Vis. Search* **42**, 1447–1463.

Shih, S.-I., and Sperling, G. (1996). Is there feature-based attentional selection in visual search? *J. Exp. Psychol. Hum. Perception Performance* **22**, 758–779.

Theeuwes, J. (1992). Perceptual selectivity for color and form. *Perception Psychophys.* **51**, 599–606.

Treisman, A. M., and Gelade G. (1980). A feature integration theory of attention. *Cogn. Psychol.* **12**, 97–136.

Wolfe, J. M., Cave, K. R., and Franzel, S. L., (1989). Guided search: An alternative to the feature integration model for visual search. *J. Exp. Psychol. Hum. Perception Performance* **15**, 419–433.

CHAPTER

91

Probabilistic Models of Attention Based on Iconic Representations and Predictive Coding

Rajesh P. N. Rao and Dana H. Ballard

ABSTRACT

We describe two models of attention that use probabilistic principles to compute task-relevant variables. In the first model, objects and visual scenes are represented iconically using spatial filters at multiple scales. A maximum likelihood–based approach is used to compute the location of a target in a given scene. The eye movements generated by such a strategy are shown to be similar to human eye movement patterns elicited during visual search in naturalistic scenes. The second model is based on the statistical concept of predictive coding. It assumes that top-down feedback from higher cortical areas conveys predictions of expected activity at lower levels while the errors in prediction are conveyed through feedforward connections. The model explains how multiple objects in a scene can be recognized sequentially without an explicit spotlight of attention. An extension of the model provides an interpretation of object-based versus spatial attention in terms of interactions between "what and "where" networks in the visual pathway.

I. INTRODUCTION

Animals receive a vast amount of sensory information in their interactions with the natural world. The brain's limited processing resources permits only a fraction of this information to be processed at any given moment in time. Furthermore, this information is typically noisy and the animal's knowledge of its world is almost always incomplete. The fundamental challenge in such an environment is to be able to select and process only those portions of the sensory inputs that are relevant to the particular task at hand and to the animal's continued survival. Attention is nature's answer to this challenge.

Attention is often classified as being either overt or covert. Overt visual attention typically involves making eye movements to shift gaze to interesting or task-relevant parts of a scene. Covert attention, on the other hand, involves the ability to preferentially process an object or location in a visual scene without shifting gaze. A useful metaphor for understanding attention has been the notion of a spotlight or search light that can he focused on specific portions of a visual scene (see Desimone and Duncan, 1995; Newsome, 1996, for reviews). Numerous models have been proposed for simulating an attentional spotlight, two prominent examples being saliency maps and hierarchical routing circuits (Hinton, 1981; Koch and Ullman, 1985; Olshausen et al., 1993; Tsotsos et al., 1995; Niebur and Koch, 1996; Itti and Koch, 2000).

In this article, we describe models of attention formulated at two different levels of abstraction. The first model explains overt attention during visual search in terms of saliency maps and iconic representations; the second model, which is formulated closer to the neural implementation level, provides an interpretation of covert shifts of attention without the use of an explicit spotlight. The two models share a common foundation in that both acknowledge the noisy and uncertain nature of the environment by using probabilistic principles for achieving their goals.

II. PROBABILISTIC CONTROL OF ATTENTION USING ICONIC REPRESENTATIONS

Human vision relies extensively on the ability to make saccadic eye movements to orient the high-acuity foveal region of the eye over targets of interest in a visual scene. Many studies have shown that this overt form of attention is controlled by the ongoing cognitive demands of the task at hand (see Rao et al., 2002, for references). A key problem in most visual tasks is saccadic targeting: How are points of interest selected as targets for eye movements?

The targeting problem can be better understood within the context of a task such as visual search in which a subject executes eye movements to find a memorized target in the current visual scene. Three important computational problems need to be solved: (1) the target object and the visual scene need to be represented using an efficient visual code, (2) the contents of the visual scene need to compared with the memorized target object to find potential matches, and (3) an eye movement needs to be executed to the location deemed most likely to contain the target object. We discuss below a model proposed in (Rao et al., 1996, 2002) that addresses these three problems using iconic representations, saliency maps, and maximum likelihood estimation, respectively.

A. Iconic Representation of Objects

The naive method of representing objects as grey-level images is clearly impractical given the high dimensionality of such a representation and the lack of invariance to transformations and view changes. A more efficient alternative is to encode objects iconically using a set of basis functions or spatial filters (Lades et al., 1993; Rao and Ballard, 1995; Itti and Koch, 2000). Such a representation approximates the transformations imposed by the receptive fields of neurons in the primary visual cortex. The model proposed in (Rao and Ballard, 1995; Rao et al., 1996, 2002) uses a set of oriented derivatives of Gaussians (Fig. 91.1A, top panel):

$$G_i^{\theta_j}, \ i = 1, 2, 3, \theta_j = 0, \cdots, m\pi/(i+1), \ m = 1, \cdots, i \quad (1)$$

where i denotes the order of the derivative and θ_j refers to the preferred orientation of the filter. The response of an image patch I centered at (x_0, y_0) to a particular basis filter $G_i^{\theta_j}$ can be obtained by convolving the image patch with the filter:

$$r_{i,j}(x_0, y_0) = \int\int G_i^{\theta_j}(x_0 - x, y_0 - y) I(x, y) dx dy \quad (2)$$

The iconic representation for the local image patch centered at (x_0, y_0) is formed by combining into a high-dimensional vector the responses from all these basis filters at different scales:

$$\mathbf{r}(x_0, y_0) = [r_{i,j,s}(x_0, y_0)] \quad (3)$$

where i denotes the order of the filter, j denotes the orientation, and $s = s_{min}, \ldots, s_{max}$ denotes the scale of the filter. For computational efficiency, a Gaussian pyramid representation of the image was used to generate multiscale responses from a set of basis filters at a fixed scale. As an example, Fig. 91.1A shows the filter-based responses at a given location in a cluttered scene for a set of five filters at five spatial scales. It can he shown that the filter response vector at an image location provides an almost unique representation of the local image region surrounding that location when compared with response vectors from other locations or images (Rao and Ballard, 1995).

B. Targeting Eye Movements in Visual Search

We now summarize the model proposed in Rao et al. (1996, 2002) for characterizing human eye movements in visual search. Suppose that objects of interest are represented by a set of memorized filter response vectors $\mathbf{r}_s^m$, where m denotes a particular target object in memory and s denotes the scale of the filters. Given a new input image and a target object T, the model computes a saliency map $S(x, y)$ that stores, at each image location (x, y), the similarity between the response vector for that location and the memorized target response vector $\mathbf{r}_s^T$. Furthermore, the model assumes that the computation of the saliency map proceeds in a coarse-to-fine fashion: Responses from larger spatial scale filters are compared before the smaller scale responses. Finally, the most likely location of the target object is chosen probabilistically according to the Boltzmann distribution computed from the similarity values in the saliency map (for those familiar with the Boltzmann distribution, the energy function is assumed to be given by the saliency map outputs). The entire targeting process can be summarized as follows:

1. Set the initial scale of analysis k to the largest scale, k = max. Set $S(x, y) = 0$ for all (x, y).
2. Compute the current saliency map across all locations (x, y) based on filter responses from the current scale k up to the maximum scale:

$$S(x, y) = \sum_{s=k}^{max} \|\mathbf{r}_s(x, y) - \mathbf{r}_s^m\|^2 \quad (4)$$

I. INTRODUCTION

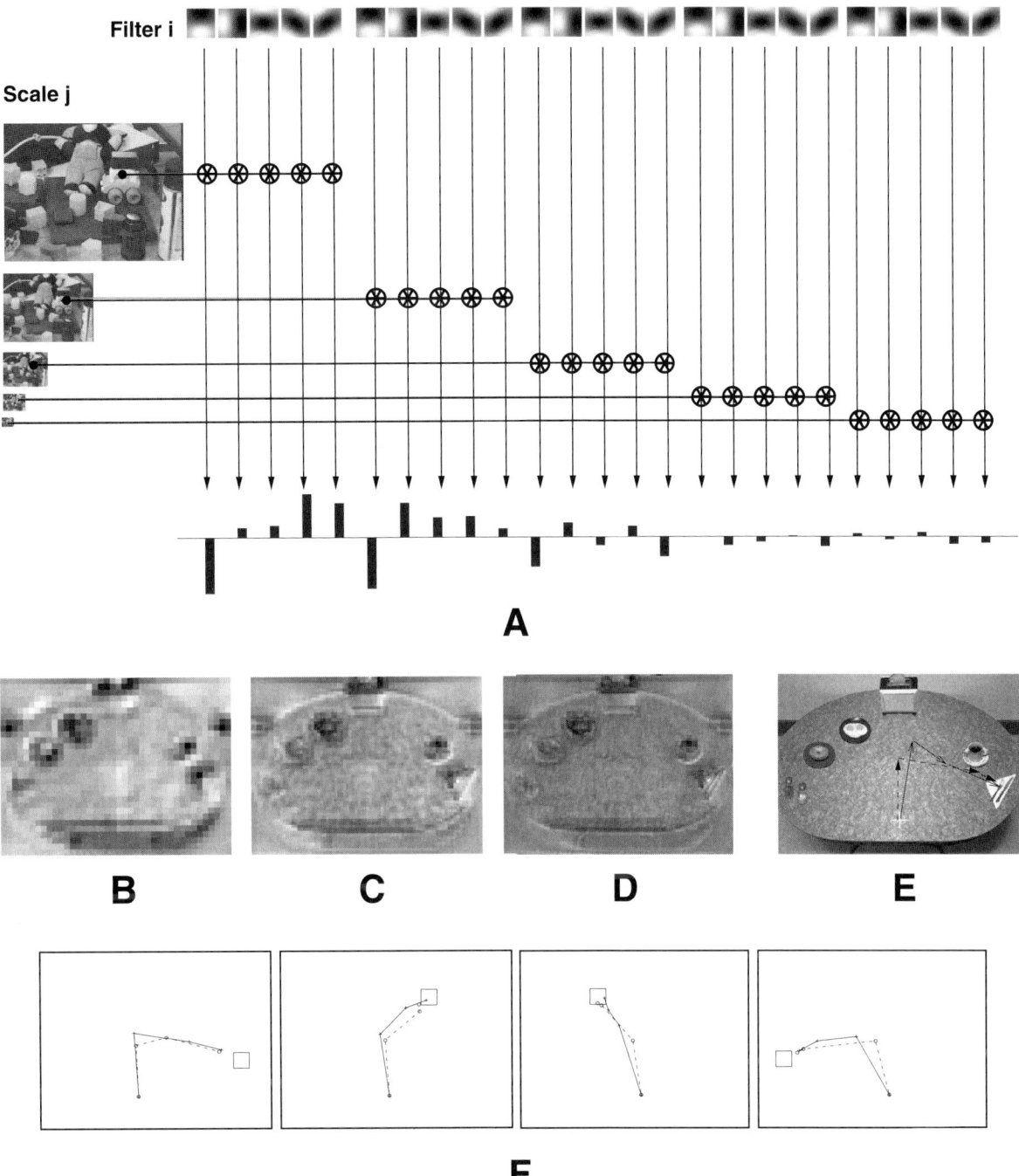

FIGURE 91.1 Iconic representations and saliency maps for overt attention. (A) Iconic representation for characterizing local image patches. A set of five oriented filters (repeated five times) is shown at the top. A cluttered image is shown on the left at five different spatial scales (Gaussian pyramid representation). The iconic representation for a given image location is the vector of 25 spatial filter responses obtained at that location. This representation is depicted at the bottom as a histogram. Positive responses are represented as upward bars and negative responses as downward bars. (B), (C), and (D) show the saliency map $S(x, y)$ after the inclusion of the largest, intermediate, and smallest scale filter responses respectively (only three scales were used in these simulations). The brightest points are the closest matches to the target object (in this case, a fork and knife on a napkin). (E) shows three successive eye movements as determined from maximum likelihood weighted averaging of the saliency maps in (B), (C), and (D) respectively. For comparison, saccades from a human subject are depicted as dashed lines with arrows. (F) Comparison of model and human eye movements to four different target locations averaged over subjects and target objects. The square box denotes a 1-deg region centered around each target. The dashed line segments correspond to human data while the solid line segments correspond to model data.

$S(x, y)$ is the square of the Euclidean distance between the filter response vector $\mathbf{r}_s$ for image location (x, y) and the memorized target response vector $\mathbf{r}_s^m$, summed over the scales $s = k, \ldots,$ max.

3. The location for the next eye movement is given by a weighted average determined from the following maximum likelihood scheme (cf. Nowlan, 1990):

$$(\hat{x}, \hat{y}) = \sum_{(x,y)} (x, y) \cdot \frac{e^{-S(x,y)/\lambda(k)}}{\sum_{(x,y)} e^{-S(x,y)/\lambda(k)}} \quad (5)$$

where $\lambda(k)$ is a temperature parameter that is decreased with k. Decreasing $\lambda(k)$ allows the search to evolve from an initial state where all target locations compete equally for an eye movement to a final state where only a few most likely target locations remain.

4. Repeat steps (2) and (3) for $k = $ max $- 1,$ max $- 2,$ ... until either the target object has been foveated or the number of scales has been exhausted. In the former case, a recognition process signals the successful termination of the search. In the latter case, the object is deemed to be absent in the given image.

C. Comparison with Human Eye Movement Patterns

The model just described was tested in a series of eye-tracking experiments involving human subjects performing visual search in naturalistic scenes (Rao et al., 1996, 2002; Zelinsky et al., 1997). Figure 91.1B–D shows the saliency maps for one such scene (a dining table scene) after including one, two, and three different spatial scales in the iconic representation. Figure 1E shows the sequence of fixations generated by the model for this image, together with those recorded from a human subject. The target (composed of the fork and the knife) was the same in both cases. As can be seen, the locations predicted by the model for successive eye movements are similar to those seen in the fixation pattern that the human subject generated for this image.

A detailed comparison of the model to human data can be found in Rao et al. (2002). We briefly summarize time results here. The comparison was based on 480 search trials pooled over four subjects. An average path to each of six possible target locations was computed by averaging the fixations over subjects and search scenes. The model data were averaged over the different targets for each location. A comparison of the average paths generated by the model amid by human subjects is shown in Fig. 1F. The box in each subfigure represents a 1-deg region centered on each target location. As is evident, there is good agreement between the model and human data for each location. The number of errors made by the model was found to be close to the number of errors made by human subjects (Rao et al., 2002). The average standard deviation for the subjects, averaged over all fixations, was 1.5 deg, whereas the deviation between model and the average subject fixations was 0.7 deg, indicating that the model's behavior is within the profile expected of an individual subject.

These results suggest that human eye-movement patterns during visual search can be understood in terms of a maximum likelihood procedure for computing the most likely location of a target in a coarse-to-fine manner. This model can be viewed as a systems-level model of overt attention in that it is formulated in terms of higher-level abstractions such as saliency maps. In the next section, we describe a probabilistic model of attention that attempts to bridge the gap between systems-level modeling and neural modeling.

III. PREDICTIVE CODING MODEL OF ATTENTION

The model in the previous section focused on overt attentional control using spatial filter-based representations. The choice of the filters themselves was arbitrary. We may be inclined to ask whether there exist methods to learn appropriate representations of objects and natural scenes directly from their images. Thus can be accomplished through the probabilistic notion of generative models of images. We show in this section that various forms of attention may be regarded as emergent properties of predictive coding networks that use generative models for representing images. Such networks also suggest functional roles for feedback and feedforward connections in the dorsal and ventral visual pathways in the mammalian brain.

A. Generative Models and Predictive Coding

Assume that an image, denoted by a vector $\mathbf{I}$ of n pixels, can be represented as a linear combination of a set of k basis vectors $U_1, U_2, \ldots, U_k$:

$$\mathbf{I} = \sum_{j=1}^{k} U_j r_j + \mathbf{n} \quad (6)$$

$$= U\mathbf{r} + \mathbf{n} \quad (7)$$

where $\mathbf{n}$ is a zero-mean Gaussian white noise process, U is the $n \times k$ matrix whose columns consist of the basis vectors U_j, and $\mathbf{r}$ is the $k \times 1$ vector consisting of coefficients r_j. In a neurobiological setting, the values in the

ith row of U can be regarded as synaptic strength of the ith model neuron whereas the coefficients r_j denote the presynaptic activities received by these neurons.

Our goal is to estimate the coefficients **r** for any given image and, on a longer time scale, τ learn appropriate basis vectors in U directly from the input image stream. Consider the following squared-error optimization function for minimization:

$$E = (\mathbf{I} - U\mathbf{r})^T S (\mathbf{I} - U\mathbf{r}) \qquad (8)$$

where the superscript T denotes vector (or matrix) transpose and S is a diagonal weighting matrix. Given that **n** is Gaussian, it can be shown that minimizing E is equivalent to maximizing the log likelihood of the observed data **I** with respect to model parameters U amid **r** (see, for example, Rao and Ballard, 1997; Rao, 1999). For the purposes of modeling attention, it is useful to choose the diagonal entries in the matrix S as:

$$S^{i,i} = \min\left\{1, c/(\mathbf{I}^i - U^i\mathbf{r})^2\right\}$$

Here, $\mathbf{I}^i$ is the ith pixel of **I**, U^i is the ith row of U, and c is a threshold parameter. Note that S effectively clips the ith summand in E to a constant saturation value c whenever the squared error $(\mathbf{I}^i - U^i\mathbf{r})^2$ exceeds c. Thus, statistical outliers (i.e., image regions containing distracting irrelevant, or unknown objects) are prevented from influencing the optimization process due to the large errors that they produce (see later discussion for an example).

We can minimize E with respect to **r** and U using gradient descent to obtain the following differential equations:

$$\dot{\mathbf{r}} = -\frac{k_1}{2}\frac{\partial E}{\partial \mathbf{r}} = k_1 U^T G(t)(\mathbf{I} - U\mathbf{r}) \qquad (9)$$

$$\dot{U} = -\frac{c_1}{2}\frac{\partial E}{\partial U} = c_1 G(t)(\mathbf{I} - U\mathbf{r})\mathbf{r}^T \qquad (10)$$

where $\dot{\mathbf{r}}$ and $\dot{U}$ denote the temporal derivatives of **r** and U respectively, and k_1 and c_1 are positive the constants that determine the rate of descent toward a minimum of E. For a given static image, U is typically kept fixed until **r** converges to a stable value; this value of **r** is then used to update U as specified in Eq. (10). The matrix $G(t)$ in the equations is an $n \times n$ diagonal matrix whose diagonal entries at time t are given by:

$$G^{i,i}(t) = \begin{cases} 0 & \text{if } (\mathbf{I}^i(t) - U^i\mathbf{r}(t))^2 > c(t) \\ 1 & \text{otherwise} \end{cases} \qquad (11)$$

G can be regarded as the sensory residual gain or gating matrix. It determines the gain on the various components of the incoming sensory residual error $(\mathbf{I} - U\mathbf{r})$. By effectively excluding any high residual errors, G allows the model to ignore the corresponding outliers (occluding objects or clutter) in the input **I**, thereby enabling it to robustly estimate **r**. In fact, Eq. (9) can be interpreted as implementing an approximate form of the robust Kalman filter (Rao, 1998).

Figure 91.2A depicts a recurrent network that implements Eq. (9). The network can be regarded as a predictive coding circuit wherein feedback connections carry predictions ($U\mathbf{r}$) of lower level inputs (**I**) while feedforward connections carry filtered error signals $((\mathbf{I}(x) - U\mathbf{r}))$. Predictive coding has previously been used to model visual cortical response properties such as contextual and nonclassical receptive field effects (see Rao and Ballard, 1999).

B. Visual Attention without a Spotlight

We now illustrate how the predictive coding model previously discussed can be used to model attentional shifts. We assume a training phase in which objects are shown to the recognition system without occlusions or background clutter (e.g., Fig. 91.2B). These objects are learned by alternating between Eqs. (9) and (10) during repeated exposures to the objects, until the basis matrix U stabilizes.

Now consider the case where a familiar object from the training database occurs with another occluding object or background clutter in an input image (Fig. 91.2C, leftmost image). When the object vector **r** is calculated using Eq. (9); the model predicts only the familiar object, causing relatively large errors in the areas of the image that do not match the predictions. These regions of the image are treated as outliers, and the gating matrix G prevents these regions front influencing the estimation of **r**. The system is thus able to focus attention on a familiar object despite occlusions and background clutter as shown in Fig. 91.2C.

More interestingly, the outliers (shown in white) produce a crude segmentation of the occluder and background clutter, which can subsequently be used to focus attention on previously ignored objects and recover their identity. In particular, an outlier mask **m** can be defined by taking the complement of the diagonal of G (i.e., $\mathbf{m}^i = 1 - G^{i,i}$). By replacing the diagonal of G with **m** in Eq. (9) and repeating the estimation process, the network can attend to the image region(s) that were previously ignored as outliers. As shown in Fig. 91.2D, the network first recognizes the dominant object, typically the object occupying a larger area of the input image or possessing regions with higher contrast. The outlier mask **m** is subsequently used for switching attention and extracting the identity of the second object (Fig. 91.2D, lower arrow and rightmost image).

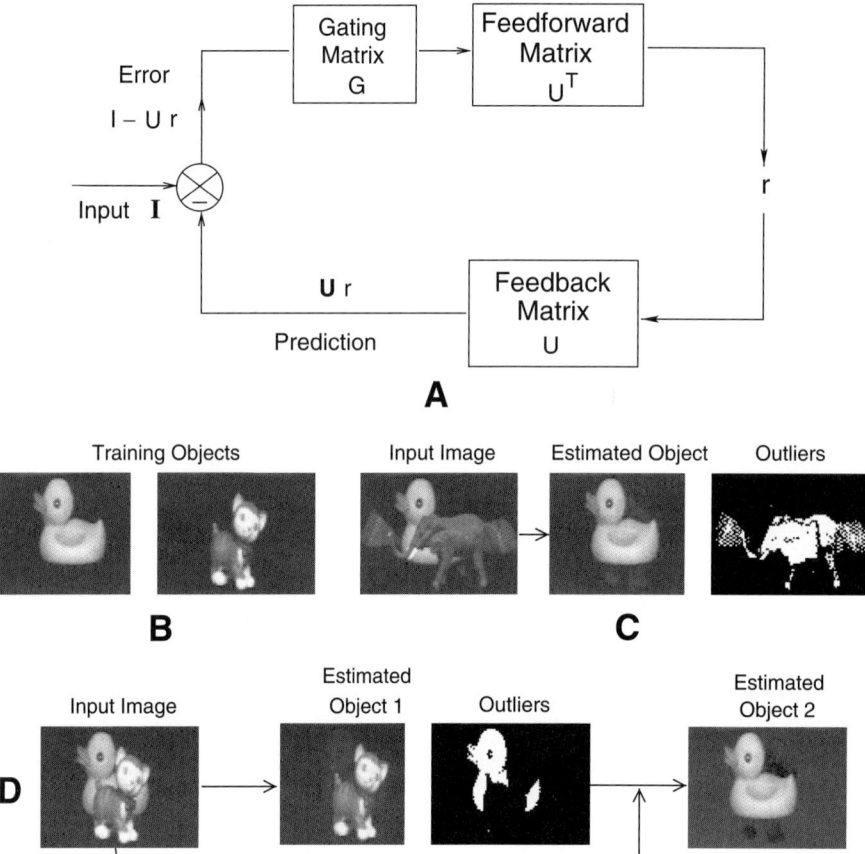

FIGURE 91.2 Attention in a predictive coding model of visual processing. (A) An implementation of the predictive coding model (Eq. 9) in the form of a recurrent neural network. The matrices U and U^T are represented by the synaptic weights of linear feedback and feedforward neurons respectively. The gating matrix G is implemented by a set of threshold nonlinear neurons with binary outputs. (B) Example images used to train a predictive coding network. (C) Given a cluttered image, the network treats occlusions and background objects as outliers (white regions in the third image, depicting the diagonal of the gating matrix G). This allows the network to "attend to" and recognize a training object (duck) despite clutter, as indicated by the relatively accurate final reconstructed image ($U\mathbf{r}$) shown in the middle. (D) In the more interesting case of the training objects occluding one another, the network converges to one of the objects (the dominant one in the image—in this case, the object in the foreground). Having recognized one object, the second object is attended to and recognized by taking the complement of the outliers (diagonal of G) and repeating the estimation process (third and fourth images).

C. Object-Based versus Spatial Attention

The predictive coding model previously discussed can be extended to account for transformations of objects in an image using a generative model based on the Taylor series expansion of a new image $\mathbf{I}(\mathbf{x})$ in terms of a canonical image $\mathbf{I}$:

$$\mathbf{I}(\mathbf{x}) = \mathbf{I} + \frac{\partial \mathbf{I}}{\partial \mathbf{I}}\mathbf{x} + \mathbf{n} \quad (12)$$

$$= U\mathbf{r} + DI_d\mathbf{x} + \mathbf{n} \quad (13)$$

where D is a matrix of differential operators to be learned from data and I_d is a diagonal matrix containing the appropriate number of copies of the image $\mathbf{I} = U\mathbf{r}$ along the diagonal (Rao and Ballard, 1998).

The generative model in Eq. (13) can be used to derive equations similar to Eqs. (9) and (10) for estimating both $\mathbf{r}$ and $\mathbf{x}$ and for learning U and D (see Rao and Ballard, 1998, for details). Fig. 91.3A shows an implementation of this model using two parallel but cooperating networks, one estimating object identity $\mathbf{r}$ ("what") and the other estimating object transformation $\mathbf{x}$ ("where"). This functional dichotomy between object recognition and transformation estimation is reminiscent of the well-known division of labor between the ventral and dorsal streams in the primate visual cortex (Felleman and Van Essen, 1991).

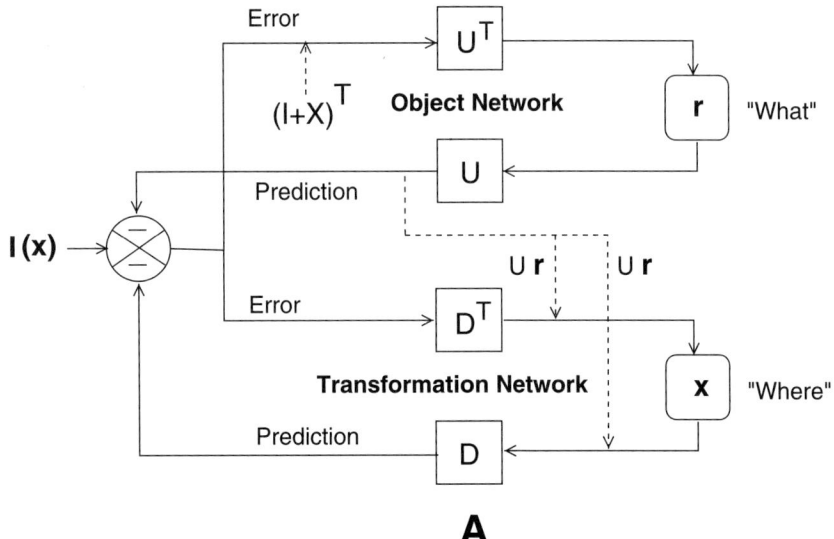

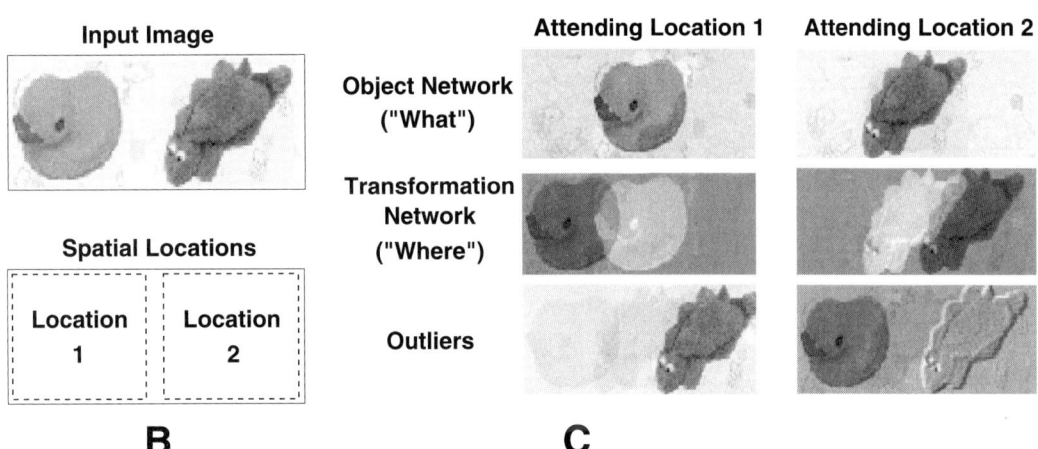

FIGURE 91.3 "What-Where" networks and spatial attention. (A) A pair of predictive coding networks. The one at the top computes the identity of objects through object features **r** ("what") and the one at the bottom computes the transformation **x** ("where") (see Rao and Ballard 1998, for more details). The gating matrix G is not shown. Fixing x in the "where" network causes the "what" network to converge to the identity of the object in that spatial location (a form of spatial attention), whereas fixing **r** in the "what" network causes the "where" network to converge to the object's most likely spatial location (a form of object-based attention). (B) A test image containing two training objects in locations 1 and 2. The network was trained on images containing only one object in the center of an image. (C) Fixing the spatial position vector x to location 1 causes the object network to converge to the duck object (left panels "Attending Location 1"). The pixels containing the dinosaur object are treated as outliers. On the other hand, fixing x to location 2 focuses attention on the object in location 2, that is, the dinosaur object (right panels "Attending Location 2").

Given an input image, the pair of networks in Fig. 91.3A simultaneously estimate an object and its transformation by jointly optimizing the generative model in Eq. (13). Therefore, fixing a particular spatial location x in the transformation network should cause the object network to converge to the identity **r** of the object in that spatial location (a form of *spatial attention*). On the other hand, fixing an object's identity in the object network should cause the transformation network to converge to its most likely spatial location in the image (a form of object-based attention).

Figures 91.3B and C illustrate an example of spatial attention using an input image containing two training objects simultaneously, one in location 1 and the other in location 2. The networks were trained on images containing only one object in the center of an image. As shown in Fig. 91.3C, when the transformation vector **x** is set to location 1 (left panels, "Attending Location 1"), the object vector **r** converges to the canonical representation of the object in location 1; the image predicted by the object network is that of object 1 in its central (canonical) position. Setting **x** to location 2 causes the object network to converge to object 2 (right panels, "Attending Location 2"). In both cases, pixels containing the second object are treated as outliers (shown here in grayscale rather than in binary form). Thus, spatial attention emerges as a consequence of a top-down signal (for example, from short-term or working memory) that constrains the activity of the transformation network to be a memorized value. Likewise, object-based attention emerges in the network as a consequence of constraining the activity of the object network (not shown). These results suggest an interpretation of spatial and object-based attention in terms of specific constraints being placed on activities in the dorsal or ventral visual pathway by memory-related neurons in prefrontal cortex and other areas implicated in working memory.

IV. SUMMARY AND CONCLUSION

In this article, we have reviewed two models of attention, both based on probabilistic principles but formulated at two different levels of abstraction. The first model relies on iconic representations and the concept of saliency maps to predict eye movements during visual search in naturalistic scenes. Saliency maps have played an important role in models of attention, especially those focusing on extracting interesting locations in a scene based on bottom-up sensory information (Koch and Ullman, 1985; Niebur and Koch, 1996; Itti and Koch 2000). The model discussed in this article combines both bottom-up scene representations and a top-down target representation to generate a saliency map. The model further assumes that the saliency map is computed in a coarse-to-fine manner such that larger scale filter responses are compared first. Motivation for coarse-to-fine computation of saliency maps comes from several studies that show that lower spatial frequencies influence visual perception earlier than higher spatial frequencies (e.g., Navon, 1977; Schyns and Oliva, 1994). For a given saliency map, the model computes the most likely target location as the weighted average of all locations, the weight being determined by the location's saliency. This procedure is motivated by previous work in probabilistic reasoning and learning based on the Boltzmann and related distributions (Hinton and Sejnowski, 1986; Nowlan, 1990). The saliency map and the weighted averaging scheme in the model may have correlates in the posterior parietal cortex and the superior colliculus, respectively (Desimone and Duncan, 1995; McIlwain, 1991). The model explains experimental results showing that humans make successive eye-movements to the center-of-gravity of clusters of objects before landing on a most likely object location (Zelinsky et al., 1997). The model assumes that the oculomotor system is ready to move before all the scales can be matched, and thus the eyes move to the current best target position, thereby increasing the chances of an early match. These results suggest that the human visual system uses a probabilistic method based on maximum likelihood (or more generally, maximum *a posteriori*) estimation to shift gaze to points of interest in a natural scene.

The second model is based on the probabilistic notion of generative models. By hypothesizing a mathematical model for how images are synthesized using a set of basis functions, a network can be derived for learning these basis functions and estimating their coefficients. This network implements a form of predictive coding in which top-down feedback is used to predict a lower-level signal (e.g., an image) while the feedforward signals convey the prediction errors. The network computes an optimal set of coefficients that serve to represent the contents of an image in a compact and efficient manner. Such predictive coding networks have proved useful in modeling visual cortical response properties (Rao and Ballard, 1999). We have discussed how attention emerges in such networks as a consequence of selective filtering of predictive error signals. This allows the network to focus attention on a single object or switch attention to another object without using an explicit spotlight of attention.

The predictive coding model can be extended to account for transformations of objects in images, resulting in "what-where" networks that can simultaneously recognize an object and estimate its pose. We have discussed how object-based attention and spatial attention are emergent properties of such networks, caused by placing constraints on the "what" or "where" network, respectively. The model thus provides a unifying explanation for well-known spatial attention results as well as more recent results on object-based attention showing that subjects can reliably track an object superimposed with a distractor object occupying the same spatial location (Kanwisher and Wojciulik, 2000).

The functional dichotomy in the predictive coding model between the "what" and the "where" networks resembles the well-known division of labor between cortical networks in the ventral and dorsal visual pathway. This observation leads to several potentially testable predictions. Substantial connections exist between the dorsal amid ventral visual pathways (Felleman and Van Essen, 1991), but their function is unknown. These connections form an integral part of the joint optimization process in the model (see Fig. 91.3A; Rao and Ballard, 1998). The model predicts that damage to dorsal areas should produce noticeable effects in object-based attentional tasks, whereas damage to ventral areas should produce significant deficits in spatial attention tasks. This is interesting, given that dorsal and ventral areas are traditionally associated only with spatial and object-related perception, respectively. Similarly, damage to frontal cortical areas, the presumed source of top-down constraints on the "what" and "where" networks, should adversely affect both spatial as well as object-based attention tasks.

The predictive coding model emphasizes the global nature of attention. The different forms of visual attention are interpreted as emerging from the constrained optimization of a generative model thought to be encoded jointly within the dorsal and ventral visual cortical pathways.

References

Desimone, R., and Duncan, J. (1995). Neural mechanisms of selective visual attention. *Annu. Rev. of Neurosci.* **18**, 193–222.

Felleman, D. J., and Van Essen, D. C. (1991). Distributed hierarchical processing in the primate cerebral cortex. *Cerebral Cortex* **1**, 1–47.

Hinton, G. (1981). A parallel computation that assigns canonical object-based frames of reference. *In* "7th International Joint Conference on Artificial Intelligence", Vancouver BC, Canada. pp. 683–685.

Hinton, G., and Sejnowski, T. (1986). Learning and relearning in Boltzmann machines. *In* "Parallel Distributed Processing", (D. Rumelhart and J. McClelland, eds.), vol. 1, pp. 282–317. MIT Press, Cambridge, MA.

Itti, L., and Koch, C. (2000). A saliency-based search mechanism for overt and covert shifts of visual attention. *Vis. Res.* **40**, 1489–1506.

Kanwisher, N., and Wojciulik, E. (2000). Visual attention: Insights from brain imaging. *Nat. Rev. Neurosci.* **1**, 91–100.

Koch, C., and Ullman, S. (1985). Shifts in selective visual attention: Toward the underlying neural circuitry. *Hum. Neurobiol.* **4**, 219–227.

Lades, M., Vorbruggen, J. C., Buhmann, J., Lange, J., von der Malsburg, C., Wuntz, R. P., and Konen, W. (1993). Distortion invariant object recognition in the dynamic link architecture. *IEEE Trans. Comp.* **42**, 300–311.

Mcllwain, J. T. (1991). Distributed spatial coding in the superior colliculus: A review. *Vis. Neurosci.* **6**, 3–13.

Navon, D. (1977). Forest before trees: The precedence of global features in visual perception. *Cogn. Psychol.* **9**, 353–383.

Newsonne, W. T. (1996). Spotlights, highlights and visual awareness. *Curr. Biol.* **6**, 357–360.

Niebur, E., and Koch, C. (1996). Control of selective visual attention: Modeling the "where" pathway. *In* "Advances in Neural Information Processing Systems 8" (D. Touretzky, M. Mozer, and M. Hasselmo, eds.), pp. 802–808. MIT Press, Cambridge, MA.

Nowlan, S. (1990). Maximum likelihood competitive learning. *In* "Advances in Neural Information Processing Systems 2" (D. Touretzky, ed.), pp. 574–582. Morgan Kaufmann, San Mateo, CA.

Olshausen, B. A., Van Essen, D. C., and Anderson, C. H. (1993). A neurobiological model of visual attention and invariant pattern recognition based on dynamic routing of information. *J. Neurosci.* **13**, 4700–4719.

Rao, R. P. N. (1998). Correlates of attention in a model of dynamic visual recognition. *In* "Advances in Neural Information Processing Systems 10" (M. I. Jordan, M. J. Kearns, and S. A. Solla, eds.), pp. 80–86. MIT Press, Cambridge, MA.

Rao, R. P. N. (1999). An optimal estimation approach to visual perception and learning. *Vis. Res.* **39**, 1963–1989.

Rao, R. P. N., and Ballard, D. H. (1995). An active vision architecture based on iconic representations. *Artificial Intelligence* **78**, 461–505.

Rae, R. P. N., and Ballard, D. H. (1997). Dynamic model of visual recognition predicts neural response properties in the visual cortex. *Neu. Comp.* **9**, 721–763.

Rao, R. P. N., and Ballard, D. H. (1998). Development of localized oriented receptive fields by learning a translation-invariant code for natural images. *Network: Comp. Neural Syst.* **9**, 219–234.

Rao, R. P. N., and Ballard, D. H. (1999). Predictive coding in the visual cortex: A functional interpretation of some extra-classical receptive field effects. *Nat. Neurosci.* **2**, 79–87.

Rao, R. P. N., Zelinsky, G. J., Hayhoe, M. M., and Ballard, D. H. (1996). Modeling saccadic targeting in visual search. *In* "Advances in Neural Information Processing Systems 8" (D. Touretzky, M. Mozer, and M. Hasselmo, eds.), pp. 830–836. MIT Press, Cambridge, MA.

Rao, R. P. N., Zelinsky, G. J., Hayhoe, M. M., and Ballard, D. H. (2002). Eye movements in iconic visual search. *Vis. Res.* **42**, 1447–1463.

Schyns, P. G., and Oliva, A. (1994). From blobs to boundary edges: Evidence for time- and spatial-scale-dependent scene recognition. *Psychol. Sci.* **5**, 195–200.

Tsotsos, J., Culhane, S., Wai, W., Lai, Y., Davis, N., and Nuflo, F. (1995). Modeling visual attention via selective tuning. *Artificial Intelligence* **78**, 507–545.

Zelinsky, G. J., Rao, R. P. N., Hayhoe, M. M., and Ballard, D. H. (1997). Eye movements reveal the spatio-temporal dynamics of visual search. *Psychol. Sci.* **8**, 448–453.

CHAPTER 92

The Selective Tuning Model for Visual Attention

John K. Tsotsos

ABSTRACT

The Selective Tuning Model (STM) is a proposal for the explanation at the computational and behavioral levels of visual attention in humans and primates. Key characteristics of the model include a top-down coarse-to-fine winner-take-all selection process, a unique winner-takes-all formulation with provable convergence properties and region rather than point selection, a task-relevant inhibitory bias mechanism, selective inhibition in both spatial and feature dimensions for the elimination of signal interference that leads to a suppressive surround for attended items, a task-specific executive controller, and an extensive set of predictions, many of which have now been supported by experiment. In this short space it is not possible to do justice to the many other models; in terms of comparison nor citations. As result, only a short bibliography is provided.

I. INTRODUCTION

A number of models have been put forward that attempt to explain attentive behavior in humans and primates, and many are described elsewhere in this volume. This article presents one such model, the Selective Tuning Model (STM), which embeds attentive functionality within a complete vision system. Explanations of attentive performance arise at the computational as well as behavioral and neurobiological levels, and predictions that have been experimentally confirmed have accompanied those explanations. The model foundations were presented in Tsotsos (1988, 1993A, 1993B), predating most current models, and focused on theoretical arguments on the nature of attention, the capacity of neural processes, and algorithms and strategies for overcoming capacity limitations. As a result, the model features several unique characteristics that distinguish it from other theories.

We show here from first principles which qualitative form visual processing must take and define a theory and an accompanying computer simulation that can perform at least qualitatively in the same manner as human vision performs. It features a theoretical foundation of provable properties based in the theory of computational complexity (see Chapter 1). The first principles arise because vision is formulated as a search problem (given a specific input, what is the subset of neurons that best represent the content of the image?), and complexity theory is concerned with the cost of achieving solutions to such problems. This foundation suggests a specific biologically plausible architecture as well as its processing stages (for a more detailed account see Tsotsos et al. 1995).

II. THE SELECTIVE TUNING MODEL

A. The Model

The visual processing architecture is pyramidal in structure with units within this network receiving both feedforward and feedback connections. When a stimulus is first applied to the input layer of the pyramid, it activates in a feedforward manner all of the units within the pyramid to which it is connected; the result is a diverging cone of activity within the processing pyramid. It is assumed that response strength of units in the network is a measure of goodness-of-match of stimulus to model and of relative importance of the contents of the corresponding receptive field in the scene.

Selection relies on a hierarchy of winner-takes-all (WTA) processes. WTA is a parallel algorithm for finding the maximum value in a set. First, a WTA process operates across the entire visual field at the top layer; It computes the global winner, the units with largest response (see Sect. IIC for details). The fact that the first competition is a global one is critical to the method because otherwise no proof could be provided of its convergence properties. The WTA can accept guidance for areas or stimulus qualities to favor if that guidance is available but operates independently otherwise. The search process then proceeds to the lower levels by activating a hierarchy of WTA processes. The global winner activates a WTA that operates only over its direct inputs. This localizes the largest response units within the top-level winning receptive field. Next, all the connections of the visual pyramid that do not contribute to the winner are pruned (inhibited). The top layer is not inhibited by this mechanism. However, with this inhibition the input to the higher level unit changes, and thus its output changes. This refinement of unit responses is an important consequence because the goal of attention is to reduce or eliminate signal interference. The final output of the attended units at the top layer will be the same as if the attended stimulus appeared on a blank field. This strategy of finding the winners within successively smaller receptive fields layer by layer in the pyramid and then pruning away irrelevant connections is applied recursively through the pyramid. The end result is that from a globally strongest response; the cause of that largest response is localized in the sensory field at the earliest levels. The paths remaining may be considered the pass zone of the attended stimulus, whereas the pruned paths form the inhibitory zone of an attentional beam. The WTA does not violate biological connectivity or time constraints.

An executive controller is responsible for implementing the following sequence of operations:

1. Acquire target as appropriate for the task; store in working memory.
2. Apply top-down biases, inhibiting units that compute task irrelevant quantities.
3. See the stimulus, activating feature pyramids in feedforward manner.
4. Activate top-down WTA process at top layers of feature pyramids.
5. Implement a layer-by-layer top-down search through the hierarchical WTA based on the winners in the top layer.
6. After completion of step 5, permit time for refined stimulus computation to complete a second feedforward pass. Note that this feedforward refinement does not begin with the completion of the lowermost WTA process; rather, it occurs simultaneously with completing WTA processes (step 5) as they proceed downward in the hierarchy. On completion of the lowermost WTA, some additional time is required for the completion of the feedforward refinement.
7. Extract the output of the top layers and place it in working memory for task verification.
8. Inhibit pass-zone connections to permit next most salient item to be processed.

This multipass process may not seem to reflect the reality of biological processes, which seem very fast. However, it is not claimed that all these steps are needed for all tasks. Several different levels of tasks may be distinguished, defined as:

detection   Is the target present in the stimulus, yes or no?
localization   Detection plus accurate location.
recognition   Localization plus accurate description of stimulus.
understanding recognition plus role of stimulus in the context of the scene.

The executive controller is responsible for the choice of task. If detection is the task, then the winner after step 4, if it matches the target, will suffice and the remaining steps are not needed. Thus, detection in this framework requires only a single feedforward pass. If a localization task is required, then all steps up to 7 are required because (as argued in Sect. IIB) the top-down WTA is needed to isolate the stimulus and remove the signal interference from nearby stimuli. This clearly takes a longer time to accomplish. If recognition is the task, then all the steps, and perhaps several iterations of the procedure, are needed in order to provide a complete description. The understanding task has similar requirements, although this is not within the scope of the model at this point.

B. Top-Down Selection

STM features a top-down selection mechanism based on a coarse-to-fine WTA hierarchy. Why is a purely feedforward strategy not sufficient? There seems to be no disagreement on the need for top-down mechanisms if task or domain knowledge is considered, although few nontrivial schemes seem to exist. Both biological evidence and complexity arguments suggest the visual architecture consists of a multilayer hierarchy with pyramidal abstraction. One task of selective attention is to find the value, location, and extent of the most salient image subset.

Considering a purely feedforward scheme there are three cases. In a pyramid with:

1. fixed-size receptive fields with no overlap, it is able to find the largest single input with local WTA computations for each receptive field, but location is lost and extent cannot be considered.
2. fixed-size receptive fields with overlap, it suffers from the spreading winners problem; location is lost and extent is ambiguous.
3. all possible receptive-field sizes in each layer, it becomes intractable due to combinatorics.

Case 1 might be useful for certain computer vision detection tasks only, and it cannot be considered as a reasonable proposal for biological vision. Case 3 is not plausible. Only case 2 reflects a biologically realistic architecture. Only a top-down strategy can successfully determine the location and extent of a selected stimulus in such a constrained architecture as used in STM.

C. Winner-Takes-All and Saliency

The WTA scheme within STM is defined as an iterative process that can be realized in a biologically plausible manner insofar as time to convergence and connectivity requirements are concerned. The basis for its distinguishing characteristic comes from the fact that it implicitly creates a partitioning of the set of unit responses into bins of width determined by a task-specific parameter, θ. The partitioning arises because inhibition between units is not based on the value of a single unit but rather on the difference between pairs of unit values. Further, this WTA process is not restricted to converging to single points, as in all other formulations; rather, contiguous regions are found as winners.

First, competition depends linearly on the difference between unit strengths in the following way. Unit A will inhibit unit B in the competition if the response of A, denoted by $\rho(A)$, satisfies $|\rho(A) - \rho(B)| > \theta$. Otherwise A will not inhibit B. The overall impact of the competition on unit B is the weighted sum of all inhibitory effects, each of whose magnitude is determined by $|\rho(A) - \rho(B)|$. It has been shown (Tsotsos et al. 1995) that this WTA is guaranteed to converge, has well-defined properties with respect to finding strongest items, and has well-defined convergence characteristics. The time to convergence, in contrast to any other iterative or relaxation-based method is specified by a simple relationship involving θ and the maximum possible value Z across all unit responses. The reasons for this is that because the partitioning procedure uses differences of values, the smallest units will be inhibited by all other units, whereas the largest valued units will not be inhibited by any other units. As a result, the small units are reduced to zero very quickly, whereas the time for the second largest units to be eliminated depends only on the values of those units and the largest units. A two-unit network is easy to characterize as a result. The time to convergence is given by

$$\log_2\left(\frac{A-\theta}{A-B}\right)$$

for A being the largest value and B the second largest value. This is also quite consistent with behavioral evidence; the closer in response strength the two units are, the longer it takes to distinguish them.

Second, the competition depends linearly on the topographical distance between units, (i.e., the features they represent). The larger the distance, the greater the inhibition. This strategy will find the largest, most spatially contiguous subset within the winning bin. A spatially large and contiguous region will inhibit a contiguous region of similar response strengths but of smaller spatial extent. At the top layer, this is a global competition; at lower layers, there is competition only within receptive fields. The implementation currently utilizes this for the units in the winning bin for efficiency reasons only. In practice, the weights depend strongly on the types of computations the units represent. There may also be a task-specific component included in the weights. Finally, a rectifier is needed for the whole operation to ensure no unit values go below zero. The iterative update continues until there is only one bin of positive response values remaining and all other bins contain units whose values have fallen below θ.

The key question is: How is the root of the WTA process determined? Let F be the set of feature maps at the top layer of each feature pyramid, F^i, $I = 1$ to n. Values at each x,y location within map i are represented by $M^i_{x,y}$. The root of the WTA computation is set by a competition at the top layers of the pyramid, depending on network configuration (task biases can weight each computation). The winning value is W, and this is determined as follows.

1. If there is only a single active feature pyramid f,
$$W = \max_{x,y} M^f_{x,y} \qquad (1)$$

2. If F contains more than one feature map, representing mutually exclusive features, then
$$W = \max_{x,y}\left(\max_{i \in F} M^i_{x,y}\right) \qquad (2)$$

3. If F contains more than one feature map representing features that can coexist at each point, then

there are several WTA processes, all rooted at the same location but operating through different feature pyramids:

$$W = \max_{x,y}\left(\sum_{i \in F} M^i_{x,y}\right) \quad (3)$$

4. If F contains subsets representing features that are mutually exclusive (the set A, as in case 2) as well as complementary (the set B, as in case 3), the winning locations are determined by the sum of the strongest response among set B (following method 3) plus the strongest response within set A (using method 2). Thus, a combination of the strategies is used and there are more than one WTA process, all rooted at the same location but operating through different feature pyramids:

$$W = \max_{x,y}\left|\sum_{b \in B} M^b_{x,y} + \max_{a \in A}(M^a_{x,y})\right| \quad (4)$$

As a result, there is no single saliency map in this model as there is in all other models. Indeed, there is no single WTA process necessarily. Saliency is a dynamic and task-specific determination and one that may differ even between processing layers as required. Further, this does not imply that a feature map must exist for any possible combination of features. Features are encoded separately in a predefined set of maps and the relationships of competition or cooperation among them provide the potential for combinations. The four types of competitions then select which combinations are to be further explored.

D. An Example

Figure 92.1 shows a hypothetical visual processing pyramid. There are four layers, each unit connected to seven units in the layer above it and seven units in the layer below it. The input layer (bottom layer) is numbered 1, and the output layer (top layer) is numbered 4. The examples that follow are intended to illustrate the structure and time course of the application of attentional selection in the model. The first example shows the structure that results if a single stimulus is placed in the visual field. The second example shows the time course of attentional selection if two stimuli are placed in the visual field. Only the feedforward connections are shown; the feedback connections are analogous.

In the second example, an instruction to attend to one stimulus rather than the other is required. How can this be accomplished? In STM, the executive can accept a sequence of stimuli. If the task is such that the executive knows that the first image will contain a location cue, then it can be processed in the following manner. Process the cue image in the same manner as described previously, determine the winning location and the feature pyramid where it is found (there is only one stimulus so it will be the winner) in the top layers of the feature pyramids, and store that location. Then, accept the test image, and process it, skipping the WTA stage (step 4), and use the location and feature pyramids determined from the previous image. Then proceed normally.

The second example illustrates two important aspects of the model. First, it shows that the retinotopic distance between the two stimuli in the input layer is important. If the nonattended stimulus were one unit closer, none of its signal would reach the output layer after the application of the attentional beam. On the other hand, if it were one unit farther away, the conflict region would be smaller. The degree of attentional modulation is dependent on stimulus separation; the extent of the attentional inhibition due to the inhibit zone of the attentional beam is limited. This has been shown experimentally (see Cutzu and Tsotsos 2003).

Second, the example shows the modulatory effect over time. Assume that the two stimuli are different so that no unit in the hierarchy is selective to both; in other words, if they are both within a single receptive field the neuron's response will be nonoptimal or ambiguous. Suppose one records from one of the units in layer 3 during the entire process. What will the response look like over time? On stimulus onset, the response will rise from zero to some level; then, as the beam is applied, it will change. The changes are most apparent in this hypothetical example in layers 2 and 3. Some units will become inhibited as the beam passes through the layer from top to bottom. Others will become enhanced because the second stimulus that would otherwise be inhibitory in their receptive fields will be eliminated. As a result, both enhancement and suppression can be observed but only in specific portions of each visual processing layer. This spatiotemporal structure of attentional modulation (illustrated more abstractly in Figure 92.2) represents a strong prediction of the model. (See Tsotsos et al. 2001 for an elaboration on this example.)

III. COMPUTATIONAL AND BIOLOGICAL IMPLICATIONS

A. The Simulation

The model has been implemented and tested in several labs for various computer vision and robotics tasks. Current model structure is shown in Fig. 92.3. The executive controller and working memory, the motion pathway (V1, MT, MST, MST), the peripheral

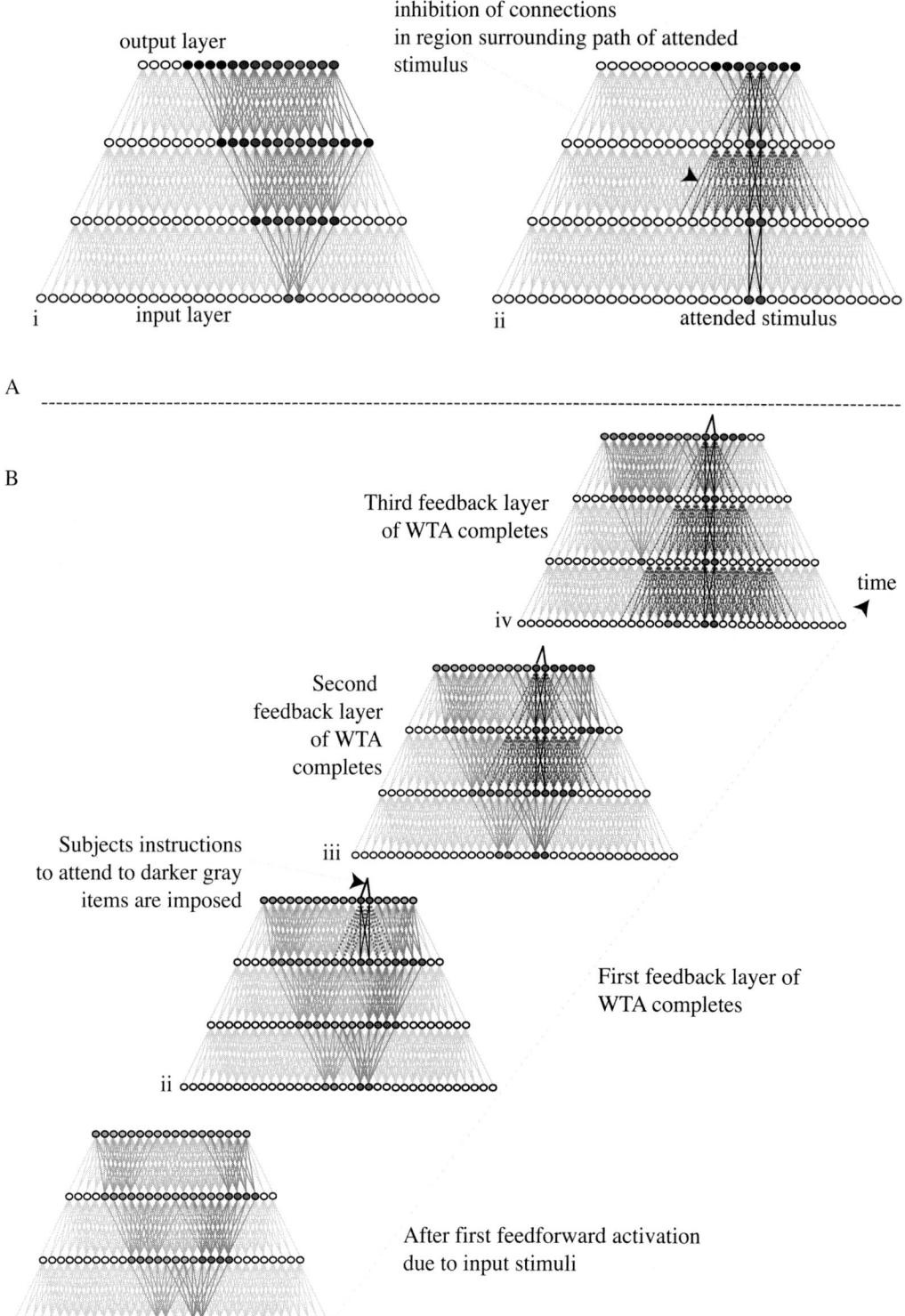

FIGURE 92.1 (A). (i) A hypothetical visual processing pyramid showing the portion of the pyramid activated due to the initial feedforward stimulation of a single input stimulus. (ii) The final configuration of the attentional beam reacting to a single input stimulus. (B). (i) The visual processing pyramid at the point where the activation due to two separate stimuli in the input layer has just reached the output layer. No attentional effects are yet in evidence. (ii) Attention is focussed at the location of the output layer corresponding to the location of the selected input and the first level of inhibition due to the attentional beam completes. The feedforward flow of the black connections is inhibited. (iii) The second level of attentional inhibition. Several units in layer 3 now receive no input and thus do not provide signals to the output layer. (iv) The third and final level of inhibition due to the attentional beam.

IV. CONCLUSION

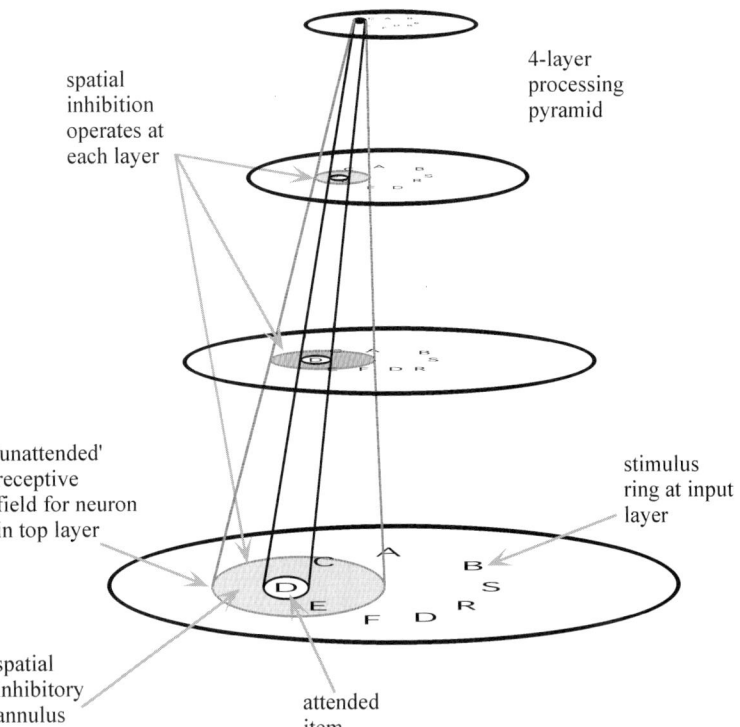

FIGURE 92.2 Convergence of neural connections that causes signal interference. This also is a good illustration of why attentional effects should be seen anywhere there is a many-to-one neural convergence because it shows the nature of signal interference that is always present within a receptive field except if a stimulus is on a blank field.

target area PO, the gaze WTA, and gaze controller have all been implemented and examples of performance can be seen in Culhane and Tsotsos (1992) and Tsotsos et al. (1995, 2002). Work is currently underway for the object pathway (V1, V2, V4, IT) and for binocular stimuli, as well as extensions of the executive controller and recognition layers.

B. Biological and Behavioral Predictions

In Tsotsos (1990), the first description of the overall structure of the model appeared along with most of the basic predictions. These included:

- Suppression around attended items in space as well as in the feature dimension.
- Top-down process; attentional guidance and control are integrated into the visual processing hierarchy rather than being centralized in some external brain structure, implying that the latency of attentional modulations *decreases* from lower to higher visual areas.
- Preattentive and attentive visual processing occur in the same neural substrate, which contrasts with the traditional view that these are wholly independent mechanisms.
- Attentional modulation appears wherever there is many-to-one, feedforward neural convergence.
- Topographic distance between attended items and distractors affects the amount of attentional modulation.

Significance evidence has accrued in the intervening years supporting these predictions. The supporting evidence is recounted in Tsotsos et al. (2001).

IV. CONCLUSION

The STM was derived in a first-principles fashion. The model not only displays performance compatible with experimental observations, but also does so in a self-contained manner. That is, input to the model is a set of real, digitized images and not preprocessed data. Several examples of these ideas using the implemented model with real images obtained from a robot head have appeared. The predictive power of the model seems broad. STM incorporates a number of

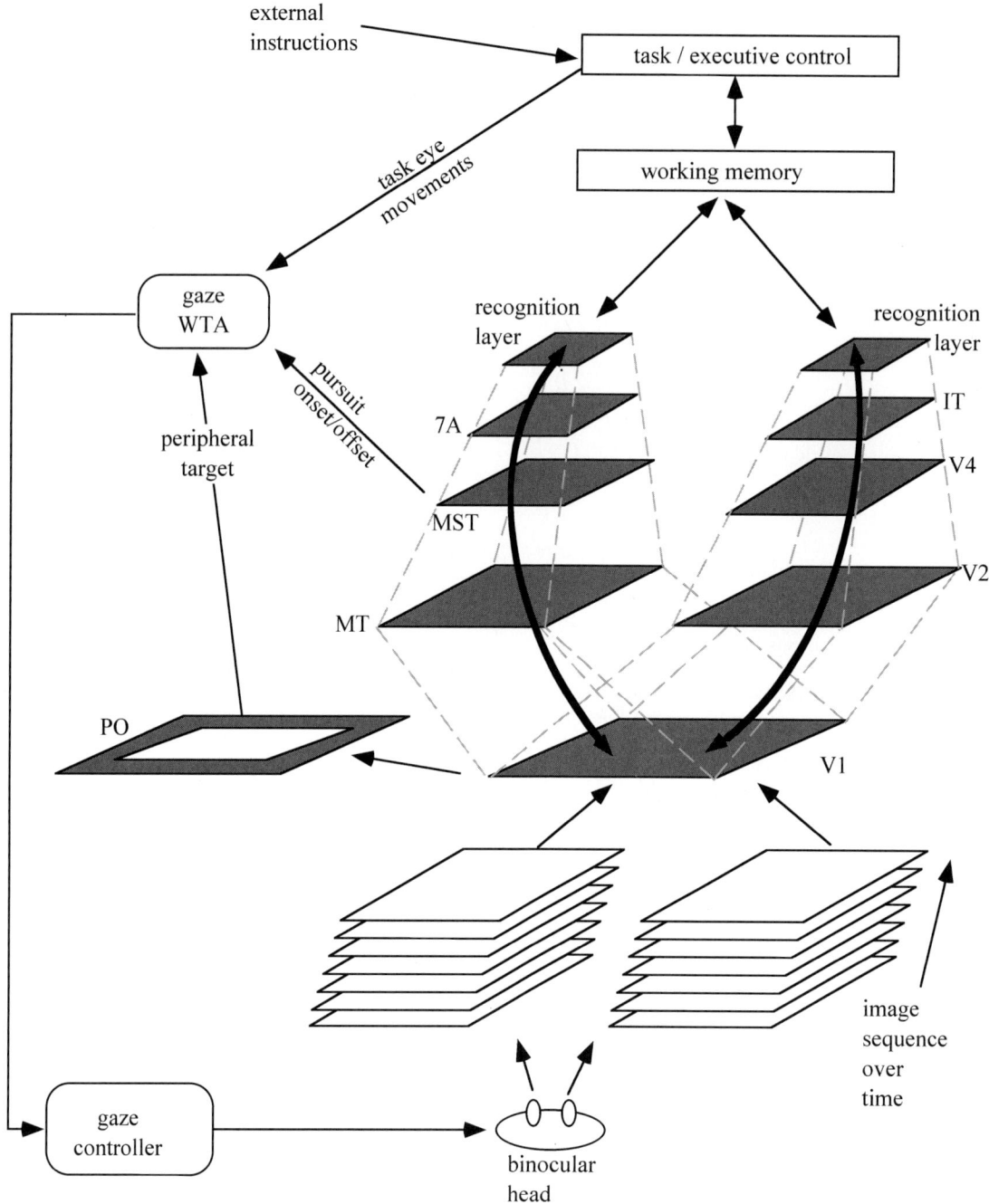

FIGURE 92.3 The overall structure of the vision system on which STM operates.

concepts that distinguish it from all other models, namely, a top-down, coarse-to-fine, WTA selection process, a unique WTA formulation with provable convergence properties and region rather than point selection, a task-relevant inhibitory bias mechanism, selective inhibition in both spatial and feature dimensions for the elimination of signal interference that leads to a suppressive surround for attended items, and a task-specific executive controller. These distinguishing characteristics also appear to provide most of its power and biological plausibility.

References

Culhane, S., and Tsotsos, J. K. (1992). An attentional prototype for early vision. In "Second European Conference on Computer Vision, Lecture Notes in Computer Science 588" (I. G. Sandini, Ed.), pp. 551–560. Springer Verlag, Heidelberg, Germany.

Cutzu, F., and Tsotsos, J. K. (2003). The selective tuning model of visual attention: Testing the predictions arising from the inhibitory surround mechanism, *Vis. Res.* **43**, 205–219.

Tsotsos, J. K. (1988). How does human vision beat the computational complexity of visual perception? *In* "Computational Processes in Human Vision: An Interdisciplinary Perspective" (Z. Pylyshyn, Ed.), pp. 286–338. Ablex Press, Norwood, NJ.

Tsotsos, J. K. (1990). A complexity level analysis of vision. *Behav. Brain Sci.* **13**, 423–455.

Tsotsos, J. K. (1993A). An inhibitory beam for attentional selection. *In* "Spatial Vision in Humans and Robots" (L. Harris and M. Jenkin, Eds.), pp. 313–331, Cambridge University Press. New York, NY.

Tsotsos, J. K. (1993B). The role of computational complexity in understanding perception. *In* "Foundations of Perceptual Theory" (S. Masin, Ed.), pp. 261–296, North-Holland Press, Amsterdam.

Tsotsos, J. K. (1995). Towards a Computational Model of Visual Attention. *In* "Early Vision and Beyond" (T. Papathomas, C, Chubb, A. Gorea, and E. Kowler, Eds.), pp. 207–218. MIT Press/Bradford Books, Cambridge, MA.

Tsotsos, J. K. (1999). Triangles, pyramids, connections and attentive inhibition, *PSYCHE* **5**.

Tsotsos, J. K., Culhane, S., and Cutzu, F. (2001). From theoretical foundations to a hierarchical circuit for selective attention. *In* "Visual Attention and Cortical Cicrcuits" (J. Braun, C. Koch, and J. Davis, Eds.), pp. 285–306. MIT Press, Cambridge, MA.

Tsotsos, J. K., Culhane, S., Wai, W., Lai, Y., Davis, N., and Nuflo, F. (1995). Modeling visual attention via selective tuning. *Artificial Intelligence* **8**, 507–547.

Tsotsos, J. K., Pomplun, M., Liu, Y., Martinez-Trujillo, J., and Simine, E. (2002). Attending to motion: localzing and labelling simple motion patterns in image sequences. *In* "Biologically Motivated Computer Vision. Second International Workshop, Lecture Notes in Computer Science 2525" (H. H. Bülthoff, S.-W. Lee, T. Poggio, and C. Wallraven, Eds.), pp. 439–452. Springer-Verlag, Heidelberg, Germany.

CHAPTER 93

The Primary Visual Cortex Creates a Bottom-up Saliency Map

Li Zhaoping

ABSTRACT

It has been proposed that the primary visual cortex (V1) creates a saliency map using autonomous intracortical mechanisms. This saliency of a visual location describes the location's ability to attract attention without top-down factors. It increases monotonously with the firing rate of the most active V1 cell responding to that location. Given the prevalent feature selectivities of V1 cells (many tuned to more than one feature dimension), no separate feature maps, or any subsequent combinations of them, are needed to create a saliency map. This proposal has been demonstrated in a biologically based V1 model. By relating the saliencies of the visual search targets or object (texture) boundaries to the eases of the visual search or segmentation tasks, the model accounted for behavioral data such as how task difficulties can be influenced by image features and their spatial configurations. This proposal links physiology with psychophysics, thereby making testable predictions, some of which are subsequently confirmed experimentally.

I. THE BOTTOM-UP SALIENCY MAP REGARDLESS OF VISUAL FEATURES SIGNALLED BY FEATURE-SELECTIVE CELLS IN V1

A saliency map aids the selection of visual inputs for further processing given limited computational resources. To better understand the selection, we separate bottom-tip from top-down mechanisms (see Chapters 17 and 90 for alternative approaches) and consider a saliency map of the visual field constructed by bottom-up mechanisms only, such that a location with a higher scalar value in this map is more likely to attract attention and be further processed. The primary visual cortex receives many top-down inputs from higher visual areas. Hence, a bottom-up saliency map in V1 is an idealization when the top-down influences are ineffective (Chapter 69), such as very shortly after visual presentation (Chapter 39) and without specific top-down knowledge or when the animal is under anesthesia. Furthermore, the saliency value is regardless of the visual features such as color and orientation (Treisman and Gelade, 1980) so that, the saliency of a red dot, for example, can he compared with that of a vertical moving bar (see Chapter 38). This property may have led to a common belief, as implicitly or explicitly expressed in previous works on saliency maps (Koch and Ullman, 1985; Itti et al., 1998), that saliency must be signaled by cells untuned to features, such as cells in parietal cortex (Gottlieb et al., 1998; see Chapter 94) and that the saliency map must be outside V1 whose cells are feature tuned. However, just as the purchasing power of UK pound sterling is the same regardless of the holder's nationality or gender, the firing rate of V1 cells could be a universal currency for saliency with or without simultaneously decoding the input features from them. Finally, using V1's output for saliency signal (to direct eye movements perhaps by sending outputs to the superior colliculus) does not preclude V1 from sending its outputs to other visual areas and contributing to other visual computations such as object recognition.

The response of a V1 cell to inputs within its classical receptive field (CRF) can be influenced by contextual inputs near but outside the CRF, due to the long but finite range intracortical interactions linking nearby cells (Knierim and van Essen, 1992; Kapadia et al., 1995). Hence, the saliency of a location is determined both by the input strength (or contrast) at that location *and* by its context, as expected (see Chapters 37 and 38). Furthermore, any visual location can evoke responses from many V1 cells whose CRFs overlap. For instance, a small vertical red bar may excite cells tuned to vertical orientation, or cells tuned to red but untuned to orientation, or cells whose optimal orientation is 5 deg from vertical and whose tuning width is 15 deg. The proposed saliency of a location is determined by the firing rate of the most responsive cell to it, regardless of the cell's optimal feature value. (This way, no or minimal computation is needed to decide how cells contribute to signaling the saliency of a location.) Hence, the saliency of the red vertical bar are likely signaled by a cell tuned to vertical, a cell tuned to red, or a cell tuned to both, depending on the context, but less likely to be signaled by a cell tuned to 10 deg from vertical. Given the population firing rates from all responding cells (regardless of their optimal features) to the whole image, the saliency of a location may be phenomenologically measured by the z score, $z \equiv (S - \bar{S})/\sigma$, where S is the (highest) evoked response to that visual location, $\bar{S}$ is the mean response to the image, and σ is the standard deviation in the population response (Li, 2002).

II. DEMONSTRATING THE SALIENCY MAP BY A V1 MODEL

A biologically based V1 model (Li, 2002) is used to demonstrate and validate the saliency map. The model (Fig. 93.1) focuses on layers 2–3 of V1 where intracortical connections are prevalent. Each model pyramidal cell receives direct visual inputs within its CRF, monosynaptic excitation and disynaptic inhibition from local pyramidal cells tuned to similar orientations, and general orientation unspecific local surround suppression. The model produces the usual contextual influences observed physiologically. In particular, the response of a cell to an optimally oriented bar within its CRF is suppressed if the CRF is surrounded by contextual bars, with the strongest suppression from contextual bars oriented parallel to the central bar within the cell's CRF (termed iso-orientation suppression) and weakest suppression from contextual bars oriented orthogonally to the central bar (Knierim and van Essen, 1992). The cell's response can be enhanced when contextual bars align with the central bar to form a smooth contour—colinear facilitation (Kapadia et al., 1995).

When the model is presented with visual stimuli resembling those in visual search and texture segmen-

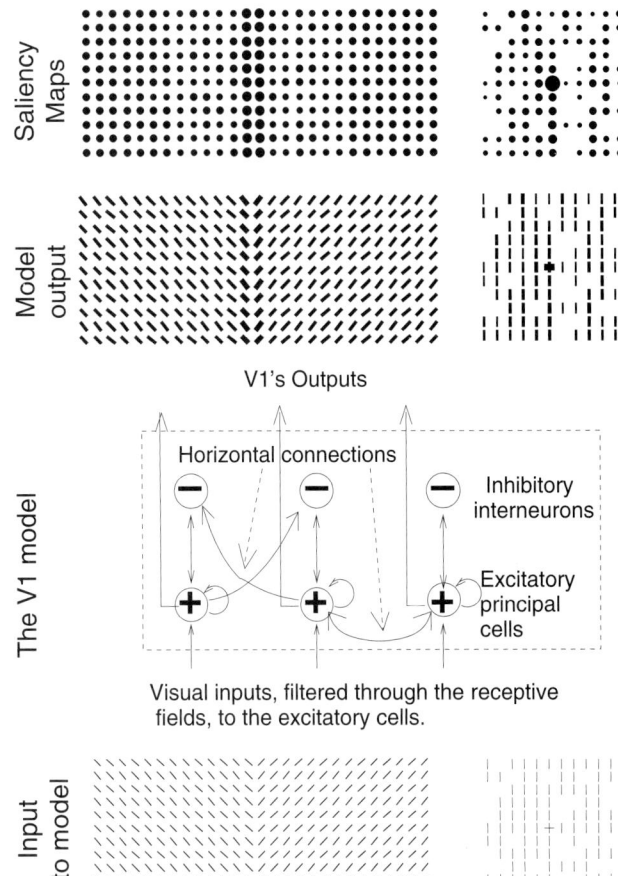

FIGURE 93.1 The V1 model and its function. Our model focuses on the part of V1 responsible for contextual influences: layers 2–3 pyramidal cells, interneurons, and horizontal intracortical connections. Pyramidal cells and interneurons interact with one another locally and reciprocally. A pyramidal cell can excite other pyramidal cells monosynaptically or inhibit them disynaptically by projecting to the relevant inhibitory interneurons. General and local normalization of activities are also included in the model. Shown also are two input images to the model and their output response maps. The input strengths are determined by the bar's contrast. Each input bar in each image has the same contrast. A principal (pyramidal) cell can only receive direct visual input from an input bar in its CRF. The output responses depend on both the input contrasts and the contextual stimuli of each bar due to contextual influences. Each input/output image plotted is only a small part of a large extended input/output image. In all figures in this paper, the thicknesses of the bars in each plot are plotted proportionally to their input/output strengths for visualization. At the top are saliency maps, where the size of the the circle at each location represents the firing rate of the most active cell responding to that visual location.

tation experiments, the strongest responses are located at or near the popout targets or texture boundaries. In Fig. 93.1 (right), the cross pops out among the bars because its horizontal bar, the only one that does not experience any iso-orientation suppression from other (vertical) bars in the image, evokes the highest response in the image. Hence, iso-feature suppression

(Li and Li, 1994) is the neural basis for the ease of feature search (Treisman and Gelade, 1980). Similarly, sufficient orientation contrast at the border between two textures of uniformly oriented bars can pop out because a border bar, having one-half as many iso-oriented contextual neighbors as those of bars away from the border, evoke relatively higher responses (Fig. 93.1, left). In Fig. 93.2A, the target vertical bar does not pop out among crosses because it is suppressed by other vertical bars in the neighboring crosses—hence, Treisman's phenomenological rule that a target lacking a feature present in the background does not pop out (Treisman and Gelade, 1980). Also, a unique conjunction of two orientations (bars) is difficult to search in a background that includes both orientations, because neither oriented bar escapes from the iso-orientation suppression experienced by all background bars (on average). Fig. 93.2A and the right image in Fig. 93.1 compose a trivial pair of search asymmetry, when the ease of search changes upon a target-distractor identity swap. Our V1 model also agrees (Li, 2002) with human vision in subtler or weaker examples of search asymmetry (used by Treisman and Gormican, 1988), such as searching for a circle among elipses and vice versa. These subtle asymmetries provided a severe test to our proposal and model because they cannot be explained simply by the mechanism of iso-orientation suppression. Colinear facilitation and general surround suppression also contribute in these examples. The relative ease of searching for a target in a more homogeneous background (with more similar distractors or more regular spatial configurations; Duncan and Hymphreys, 1989; Rubinstein and Sagi, 1990) is also understood in the model (Fig. 93.2C–E). Iso-feature suppression between the distractors is stronger when the distractors are the same or similar, thus reducing background responses to make the target relatively more conspicuous. Regular spatial locations of the distractors enable each distractor to have the same or similar contextual environment. This makes the background responses homogeneous or gives a small standard deviation σ in the population responses, thus giving a high z score, $z \propto 1/\sigma$, for the target (Li, 2002).

III. BASIC FEATURES AND CONJUNCTION SEARCHES EXPLAINED BY V1 SALIENCY MECHANISM

According to our model, a feature dimension is a basic dimension that enables popout when the follow-

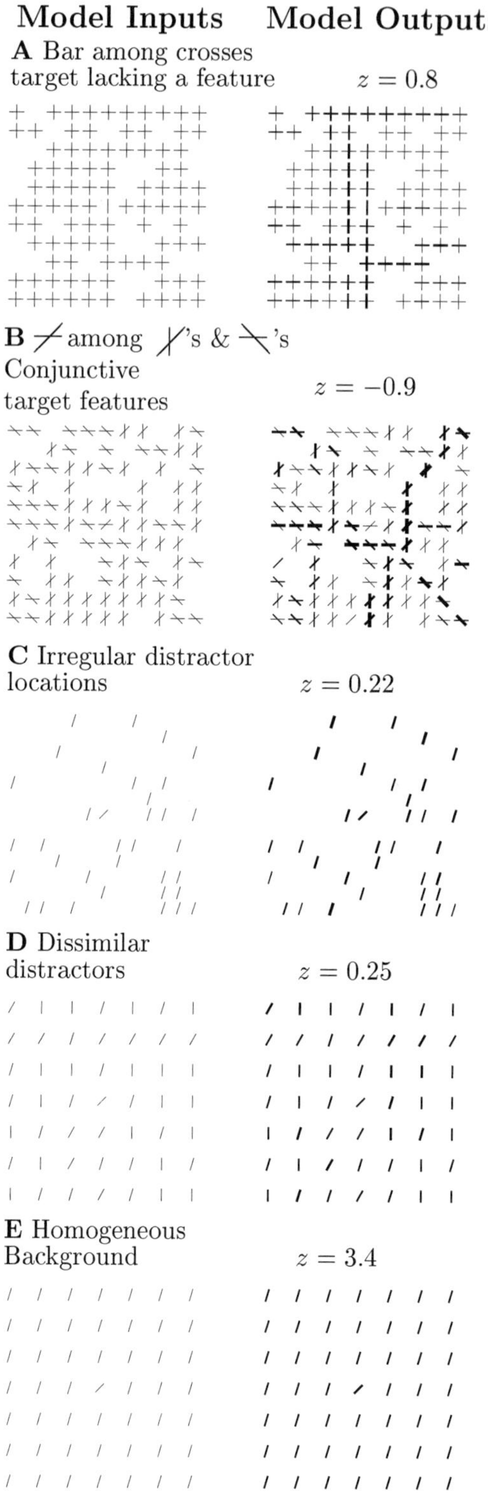

FIGURE 93.2 More examples of model performances. All search targets are at the center of the stimulus patterns. Left column: stimulus inputs; right column: model outputs. The z scores of the target, a measure of the target saliency, are indicated. A and the right example in Fig. 93.1 compose a trivial example pair of search asymmetry. C, D, and E illustrate how distractor dissimilarity or irregular distractor locations impair the search for a 45° oriented target bar.

ing two neural bases are present: (1) V1 cells should be tuned to this feature dimension (e.g., orientation) in order to signal it and (2) the intracortical connections should be tuned to that dimension so that they only link cells tuned to similar optimal feature values (e.g., orientation) to achieve the iso-feature (e.g., iso-orientation) suppression essential for popout. Thus, if conjunctions of features also meet these two criteria (i.e., both the cells and the intracortical connections are tuned to conjunctions of features) popout of a target defined by a conjunction of features, say a red vertical target among green vertical and red horizontal distractors, is feasible. Hence, a conjunction of two orientations (or more generally, two sufficiently different features within a single feature dimension) does not pop out because individual V1 cells are not simultaneously tuned to two sufficiently different orientations. Color-orientation conjunction search is typically difficult (Wolfe, 1998) because V1 cells are more likely to be either color or orientation tuned and less likely to be conjunctively tuned to both (Livingstone and Hubel, 1984; Li, 2002). That a conjunction of motion and orientation (or stereo and motion; Wolfe, 1998) can pop out is consistent with V1 physiology that cells are conjunctively tuned to motion and orientation (or stereo and motion). Furthermore, these psychophysical observations predict that the intracortical connections must be conjunctively tuned to optimal features of the linked cells in both feature dimensions—motion direction and orientation (Li, 2002). With input of a unique conjunction target among many distractor conjunctions, the conjunction cell tuned to the target has the highest response by escaping iso-conjunction suppression that is present on cells responding to distractors.

IV. SALIENCY AND INTERACTIONS AMONG FEATURE DIMENSIONS

The visual search for a 45 deg target bar among 135 deg distractor bars is easy (even for infants; Chapter 34); so is the segmentation between two textures of uniformly oriented bars at 45 deg and 135 deg. Snowden (1998) observed that, although these tasks depend only on the orientation feature of the bars, the texture segmentation became difficult when each stimulus bar was randomly assigned a color from two choices (say, red and green), whereas the orientation search task remained easy under the same color randomization (Fig. 93.3 see color plate in book). (Nothdurft 1997 also observed interference of luminance variations on orientation-based texture segmentation.)

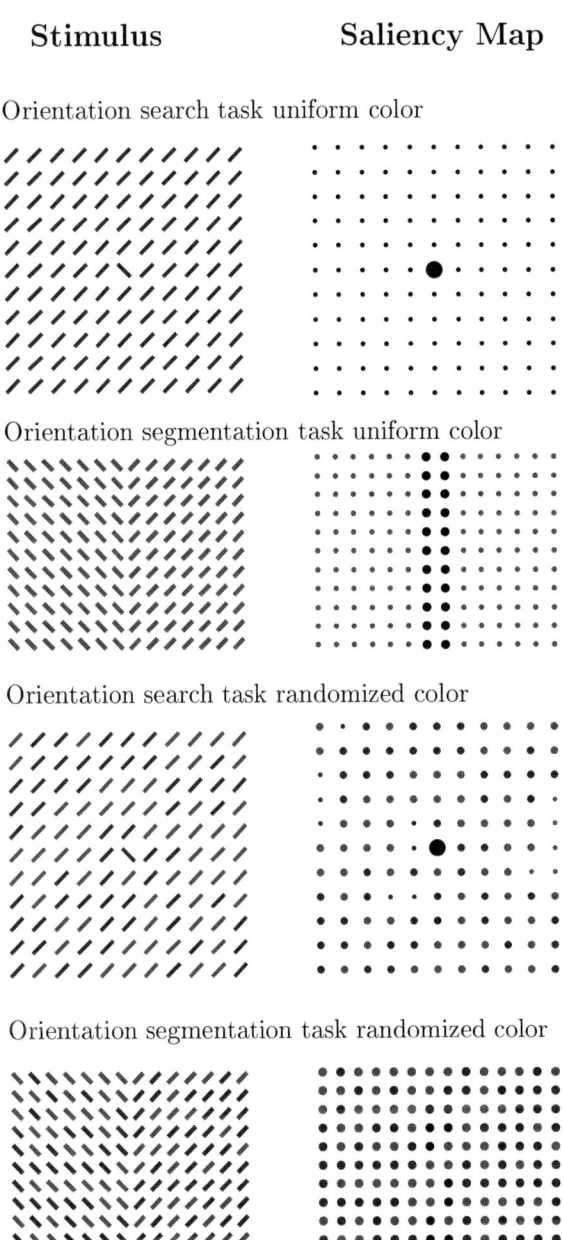

FIGURE 93.3 Color inference in orientation feature-based tasks. The two random colors of the stimulus bars are visualized in the grey scale images as black and white bars. Each bar evokes responses from color-tuned cells *and* from orientation-tuned cells. Randomizing the colors of the stimulus bars increases the response levels (and variations) in the color tuned cells, submerging the responses from the orientation tuned cells to the texture border but not that to the search target. Hence, the texture segmentation task, but not the search task, is impaired. Note that in uniform color stimuli, the saliency of the target is much higher than that of the texture border against their respective background (see color plate).

Such color interference is comprehensible in our framework by noting the changes in saliencies of the target bar or the texture border under the color randomization. Let each colored bar evoke responses from both orientation-tuned cells insensitive to color and color-tuned cells insensitive to orientation (omitting for simplicity, without changing our conclusion, the minority of cells tuned conjunctively to color and orientation). We consider stimuli such that each colored bar, when presented alone, evokes comparable responses in the corresponding color- and orientation-tuned cells, and assume that the iso-orientation and iso-color suppressions have comparable strength. With uniform color stimuli, both the color- and orientation-tuned cells responding to a background bar (i.e., away from the target or texture border) experience iso-feature (iso-color or iso-orientation) suppression and give suppressed responses of similar levels. Meanwhile, the orientation-tuned cells responding to the target or the texture border bars are relatively more active because they experience no or weaker iso-orientation suppression, respectively, making the target or the border pop out. Note, however, the orientation-tuned cell responding to the target bar, the only one with no contextual neighbors of the same orientation, is much more active than those responding to the texture border bars (Fig. 93.3). When the bar colors are randomized, the number of iso-color, contextual, neighbors of any bar is halved on average, making the color-tuned cell less suppressed. Furthermore, their iso-color suppression is of similar magnitude as the iso-orientation suppression experienced by the orientation-tuned cells responding to the texture border because each border bar also has one-half of its contextual neighbors of the same feature (orientation). Thus, the response to the border (from the orientation tuned cells) is submerged by the background responses from the color-tuned cells, making the border less conspicuous. Meanwhile, in the single-target search stimulus, the orientation-tuned cell responding to the target is still the most active against the background of the more active color-tuned cells because it is the only excited cell not experiencing any iso-feature suppression (Fig. 93.3). Hence, popout is not impaired. Therefore, the essential reasons for color interference in these orientation feature based tasks are (1) object saliency rather than subject scrutiny plays a larger role in such tasks and (2) saliency is regardless of the feature dimension(s) of cells signaling it—hence, the activity of a color-tuned cell signaling saliency of one bar is compared with the activity of an orientation-tuned cell signaling saliency of another bar to see which bar is more salient. Note that activities of the color-tuned cell and the orientation-tuned cell responding to the same location (bar) are not summed up linearly or nonlinearly to signal the saliency of this location (bar) (See Chapter 38 and Nothdurft 1997 for a different perspective.) Otherwise, the border should be more salient than predicted because the border highlight from the orientation activities superposed on a planar background of color activities, albeit enhanced, still leads to the relative border highlight. From our analysis, we can predict that drawing color randomly from more color choices should make the tasks even more difficult by further reducing the iso-color suppresion on the color-tuned cells. This prediction was recently confirmed (Zhaoping and Snowden, 2003).

V. DISCUSSION

V1 is the largest visual area in the brain. The cells' CRFs are smaller and under weaker attentional influences than those in higher visual areas. Hence, the saliency map should be roughly stable and have a satisfactory spatial resolution. Neurons beyond V1 were found to have their activities correlate with saliencies (Gottlieb, 1998). It is desirable to find out whether these signals are relayed from lower visual areas or whether there is a hierarchy of saliency maps from different visual areas (Chapter 45). By guiding top-down visual attention, the bottom-up saliency map should have its effects visible in tasks with significant top-down influences. Furthermore, a well-understood bottom-up saliency map should definitely help to elucidate the mechanisms of attention.

References

Duncan, J., and Humphreys, G. (1989). Visual search and stimulus similarity. *Psychol. Rev.* **96**, 1–26.

Gottlieb, J. P., et al. (1998). The representation of visual salience in monkey parietal cortex. *Nature* **391**, 481–484.

Itti, L., et al. (1998). A model of saliency-based visual attention for rapid scene analysis. *IEEE Trans. Patt. Anal. Mach. Intell.* **20**, 1254–1259.

Kapadia, M. K., Ito, M., Gilbert, C. D., and Westheimer G. (1995). Improvement in visual sensitivity by changes in local context: Parallel studies in human observers and in V1 of alert monkeys. *Neuron* **15**, 843–856.

Knierim, J. J., and van Essen, D. C. (1992). Neuronal responses to static texture patterns in area V1 of the alert macaque monkeys. *J. Neurophysiol.* **67**, 961–980.

Koch, C., and Ullman, S. (1985). Shifts in selective visual attention: Towards the underlynig neural circuitry. *Hum. Neurobiol.* **4**, 219–227.

Li, Z. (2002). A saliency map in primary visual cortex *Trends in Cog. Sci.* **6**, 9–16.

Li, C. Y., and Li, W. (1994). Extensive integration field beyond the classical receptive field of cat's striate cortical neurons—classification and tuning properties. *Visi. Res.* **34**, 2337–2355.

Livingstone M. S., and Hubel, D. H. (1984). Anatomy and physiology of a color system in the primate visual cortex. *J. Neurosci.* **4**, 309–356.

Nothdurft, H-C. (1997). Different approaches to the coding of visual segmentation. *In* Harris L. and Jenkins M. "Computational and Psychophysical Mechanisms of Visual Coding" (L. Harris and M. Jenkins, eds.), pp. 20–43. Cambridge University Press, New York.

Rubenstein, B., and Sagi, D. (1990). Spatial variability as a limiting factor in texture discrimination tasks: Implications for performance asymmetries. *J. Opt. Soc. Am. A* **9**, 1632–1643.

Snowden, R. J. (1998). Texture segregation and visual search: A comnparison of the effects of random variations along irrelevant dimensions. *J. Exp. Psychol.: Hum. Perception Performance* **24**, 1354–1367.

Treisman, A., and Gelade, G. (1980). A feature integration theory of attention. *Cogn. Psychol.* **12**, 97–136.

Treisman, A., and Gormican, S. (1988). Feature analysis in early vision: Evidence for search asymmetries. *Psychol. Rev.* **95**, 15–48.

Wolfe, J. M. (1998). Visual search, a review. *In* "Attention" (H. Pashler, ed.), pp. 13–74. Psychology Press.

Zhaoping, L., and Snowden R. J. (2003). A psychophysical test of the saliency map in V1. Paper presented to the Vision Science Society annual meeting, Sarasota, Florida, May 2003.

CHAPTER 94

Models of Bottom-up Attention and Saliency

Laurent Itti

ABSTRACT

Visually conspicuous, or so-called salient, stimuli often have the capability of attracting focal visual attention toward their locations. Several computational architectures subserving this bottom-up, stimulus-driven, spatiotemporal deployment of attention are reviewed in this chapter. The resulting computational models have applications not only to the prediction of visual search psychophysics, but also, in the domain of machine vision, to the rapid selection of regions of interest in complex, cluttered visual environments. We describe an unusal such application, to the objective evaluation of advertising designs.

I. INTRODUCTION

One of the most important functions of selective visual attention is to rapidly direct our gaze toward objects of interest in our visual environment. From an evolutionary standpoint, this rapid orienting capability is critical in allowing living systems to quickly become aware of possible prey, mates, or predators in their cluttered visual world. It has become clear that attention guides where to look next based on both bottom-up (image-based) and top-down (task-dependent) cues (James, 1890/1981). As such, attention implements an information processing bottleneck, allowing only a small part of the incoming sensory information to reach short-term memory and visual awareness (see Chapter 107). That is, instead of attempting to fully process the massive sensory input in parallel, nature has devised a serial strategy to achieve near real-time performance despite limited computational capacity: Attention allows us to break down the problem of scene understanding into rapid series of computationally less demanding, localized visual analysis problems.

Developing computational models of how attention is deployed onto complex visual scenes has been a longstanding challenge for fundamental neuroscience, with additional motivation provided by numerous potential applications in artificial vision, for tasks including surveillance, automatic target detection, navigational aids, and robotics control. Here we focus on biologically plausible computational modeling of bottom-up guidance of attention toward salient image locations, while Chapter 33 casts these models within broader frameworks that combine bottom-up and top-down attention control signals.

II. PREATTENTIVE FEATURES AND SALIENCY MAP

Development of computational models of attention started with the feature integration theory of Treisman and Gelade (1980), which proposed that only simple visual features are computed in a massively parallel manner over the entire visual field. Attention is then necessary to bind those early features into a united object representation, and the selected bound representation is the only part of the visual world that passes though the attentional bottleneck. This idea was further specified into a neurally plausible computational architecture by Koch and Ullman (1985) and later by Itti et al. (1998) (Fig. 94.1). These models are centered around a "saliency map," that is, an explicit two-dimensional topographic map that encodes for stimulus conspicuity, or salience, at every location in the visual scene. The saliency map receives

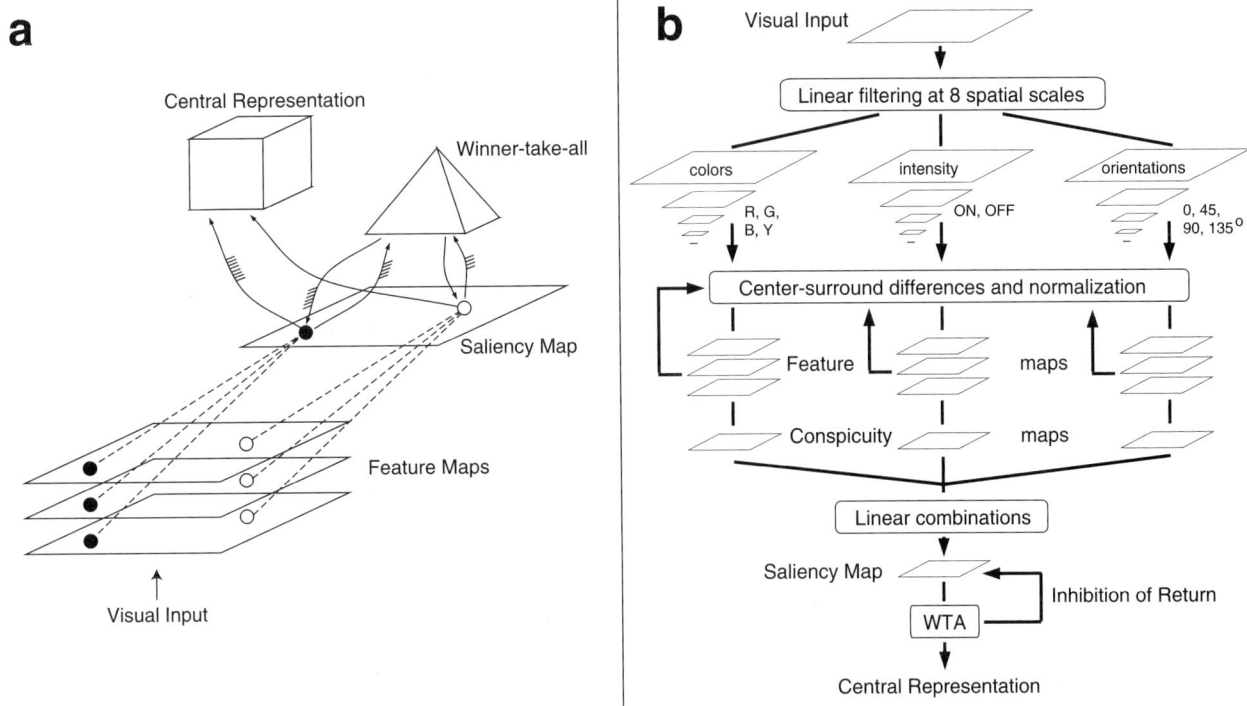

FIGURE 94.1 (a) Koch and Ullman's (1985) proposal for a computational model of bottom-up attention. Early visual features are computed, in a massively parallel manner, in a set of preattentive feature maps receiving retinal input. Activity from all feature maps is combined at each location, giving rise to responses in the topographic saliency map. A winner-take-all network detects the most salient location and directs attention toward it, such that only features from this location are routed toward further analysis and central representation. (b) Extended architecture of such a scheme proposed by Itti et al. (1998). It builds directly on the architecture proposed in (a), but provides a complete implementation of all processing stages. Visual features are computed using linear filtering at eight spatial scales, followed by center–surround differences among six pairs of scales, which compute local spatial contrast in each feature dimension, for a total of 42 maps. An iterative lateral inhibition scheme instantiates nonclassic surround competition for salience within each feature map. After competition, feature maps are combined into a single "conspicuity map" for each feature type. The seven conspicuity maps then are summed into the unique topographic saliency map. The saliency map is implemented as a two-dimensional lattice of artificial leaky integrator neurons. The winner-take-all (WTA) network, also implemented using leaky integrators, detects the most salient location and directs attention toward it. An inhibition-of-return mechanism transiently suppresses this location in the saliency map, such that attention is autonomously directed to the next most salient image location.

inputs from early visual processing, and provides an efficient control strategy by which the focus of attention simply scans the saliency map in order of decreasing saliency.

Several important lessons have been learned from these models and the empirical studies that helped better specify them. First, different features contribute with different strengths to perceptual salience (Nothdurft, 2000), and this relative feature weighting can be influenced in a task-dependent manner, top-down (Wolfe, 1994) and through training. Second, at a given visual location, there is little evidence for strong interactions across different visual modalities, such as color and orientation (Nothdurft, 2000). This is not too surprising from a computational standpoint, as one would otherwise expect these interactions to also be trainable and modulable top-down, resulting in the ability to learn to efficiently detect conjunctive targets, which we lack (Fig. 94.2). Third and most importantly, what appears to matter in guiding bottom-up attention is feature contrast, not local absolute feature strength. Indeed, not only are most early visual neurons tuned to some type of local spatial contrast (such as center–surround or oriented edges), but neuronal responses are also strongly modulated by context, in a manner that extends far beyond the range of the

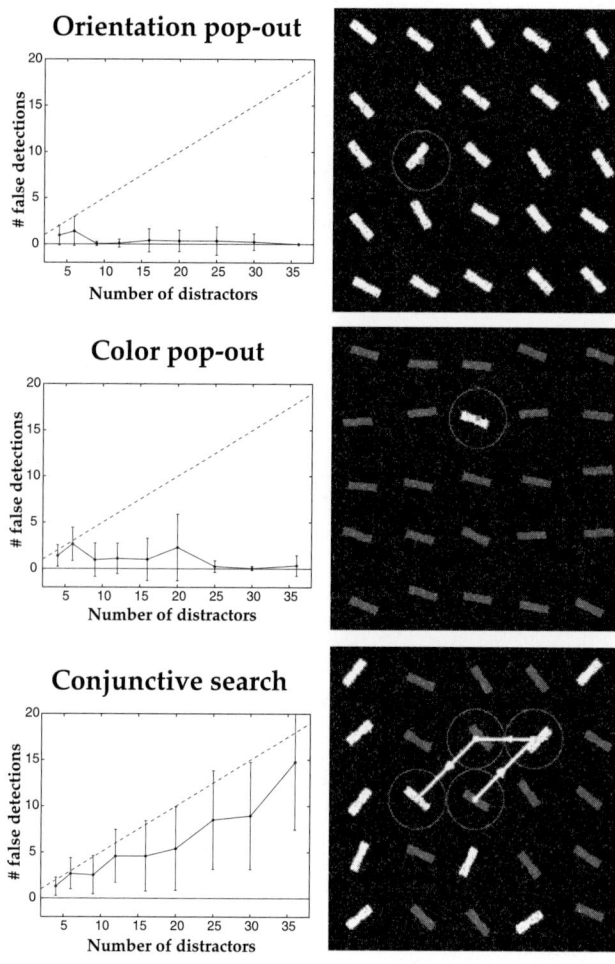

FIGURE 94.2 Model performance on noisy versions of popout and conjunctive tasks of the type pioneered by Treisman and Gelade (1980). Dark and light bars in the figure represent isoluminant red- and green-colored bars in the original displays, with strong speckle noise added. Dashed lines show chance value, based on the size of the simulated visual field and the size of the candidate recognition area (corresponds to the performance of an observer who would scan, on average, half of the distracters prior to finding the target). Solid lines (mean ± S.D.) show performance of the model. The typical search slopes of human subjects in feature search and conjunction search, respectively, are successfully reproduced by the model: search is easy and slopes are at in the popout cases (when the target differs from the distracters in one feature dimension, e.g., color), while search is more difficult and search time increases with the number of items in the display when the target differs from the distracters by a conjunction of two features (e.g., here the target is the only bar that is both light and tilted counterclockwise from vertical). Each stimulus was drawn inside a 64 × 64-pixel box, and the radius of the focus of attention was fixed to 32 pixels. For a fixed number of stimuli, we tested 20 randomly generated images in each task; the saliency map and winner-take-all were initialized to zero (corresponding to a uniformly black visual input) prior to each trial.

classical receptive field (Allman et al., 1985). In a first approximation, the computational consequences of nonclassical surround modulation are twofold: First, a broad inhibitory effect is observed when a neuron is excited with its preferred stimulus but that stimulus extends beyond the neuron's classic receptive field, compared with when the stimulus is restricted to the classic receptive field and the surrounding visual space either is empty or contains nonpreferred stimuli (Sillito et al., 1995). Second, long-range excitatory connections in V1 appear to enhance responses of orientation-selective neurons when stimuli extend to form a contour (Gilbert et al., 2000). These interactions are thought to play a critical role in perceptual grouping. The net result is that activity in early cortical areas is surprisingly sparse when monkeys are free-viewing natural scenes, compared with the vigorous responses that can be elicited by small laboratory stimuli presented in isolation.

III. IMPLEMENTED ARCHITECTURES

Many successful models for the bottom-up control of attention are architectured around a saliency map. What differentiates the models, then, is the strategy employed to prune the incoming sensory input and extract salience.

One strategy proposed by Wolfe (1994) has been to rely on top-down knowledge to emphasize those features that may distinguish a target item being searched for from surrounding clutter: if you are searching for a red object, emphasize the contribution of color to your saliency map, while toning down the influence of orientation. The FeatureGate model of Cave et al. provides a full neural network implementation of a system that combines this idea with bottom-up mechanisms (see Chapter 90), so that a target may attract attention based on combined bottom-up salience and top-down expectations. In a similar vein, Rao et al. (see Chapter 91) have proposed that saliency could be computed from the Euclidean distance between a target feature vector (i.e., a target signature, in terms of respective amounts of various colors, orientations, etc., present in the target) and feature vectors extracted at every location in the visual input. Extending these ideas in a Bayesian framework, Torralba has proposed that a coarse global analysis of the entire scene's gist may provide contextual guidance cues (see Chapter 96): for example, when searching for cars, rapidly realizing that you are looking at a city scene and having coarsely understood its layout may help you more rapidly focus onto street surfaces, where cars are likely to be.

Studying richer interactions between low-level vision and a saliency map, Tsotsos and colleagues (see Chapter 92) implemented attentional selection using a combination of a feedforward bottom-up feature extraction hierarchy and a feedback selective tuning of these feature extraction mechanisms. In this model, the target for attention focusing is selected at the top level of the processing hierarchy (the equivalent of a saliency map), based on feedforward activation and on possible additional top-down biasing for certain locations or features. That location is then propagated back through the feature extraction hierarchy, through the activation of a cascade of winner-take-all networks embedded within the bottom-up processing pyramid. Spatial competition for salience is thus refined at each level of processing, as the feedforward paths not contributing to the winning location are pruned (resulting in the feedback propagation of an "inhibitory beam" around the selected target).

In view of the affluence of models based on a saliency map, it is important to note that postulating centralized control based on such a map is not the only computational alternative for the bottom-up guidance of attention. For example, Zhaoping (see Chapter 93) proposed that salience may not necessarily be computed in a separate brain area from the low-level visual features, and that instead it may be expressed as a modulation onto feature responses observed in V1. Similarly, Desimone and Duncan (1995) argued that salience is not explicitly represented by specific neurons, but instead is implicitly coded in a distributed modulatory manner across the various feature maps. Attentional selection is then performed based on top-down weighting of the bottom-up feature maps that are relevant to a target of interest. Several models have successfully applied this strategy to synthetic stimuli (see Chapter 98).

Although originally a theoretical construct supported by sparse experimental evidence, the idea of a unique, centralized saliency map appears today to be challenged by the multiplicity of candidate neural correlates recently unraveled, including areas in the lateral intraparietal sulcus of the posterior parietal cortex, the frontal eye fields, the inferior and lateral subdivisions of the pulvinar, and the superior colliculus (Kustov and Robinson, 1996; Gottlieb et al., 1998). One possible explanation for this multiplicity could be that some of the neurons in all those areas indeed are concerned with the explicit computation of salience, but are found at different stages along the sensorimotor processing stream. For example, other functions have also been assigned to the posterior parietal cortex, such as that of mapping retinotopic to head-centered coordinate systems and of memorizing targets for eye or arm movements (Andersen et al., 1990; Dominey and Arbib, 1992). More detailed experimental studies are thus needed to tell apart possible subtle differences in the functions and representations found in those brain areas. Most probably, the main difference between those brain regions is the balance between their roles in perception and action (see Chapter 22).

IV. ATTENTION AND RECOGNITION

So far, we have reviewed computational modeling and supporting experimental evidence for a basic architecture concerned with the bottom-up control of attention: Early visual features are computed in a set of topographic feature maps; spatial competition for salience prunes the feature responses to preserve only a sparse representation of the few most conspicuous locations; all feature maps are then combined into a unique scalar saliency map; and, finally, the saliency map is scanned by the focus of attention through the interplay between winner-take-all and inhibition-of-return. Although such simple computational architecture may accurately describe how attention is deployed within the first few hundreds of milliseconds after the presentation of a new scene, it is obvious that a more complete model of attentional control must include top-down, volitional biasing influences as well. The computational challenge, then, lies in the integration of bottom-up and top-down cues, such as to provide coherent control signals for the focus of attention, and in the interplay between attentional orienting and scene or object recognition.

A very interesting example of integration in a model was recently provided by Schill et al. (see Chapter 109). Their model aims at performing scene (or object) recognition, using attention (or eye movements) to focus on those parts of the scene being analyzed that are most informative in disambiguating its identity. To this end, a hierarchical knowledge tree is trained into the model, in which leaves represent identified objects, intermediary nodes represent more general object classes, and links between nodes contain sensorimotor information used for discrimination between possible objects (i.e., bottom-up feature response to be expected for particular points in the object and eye movements targeted at those points). During the iterative recognition of an object, the system programs its next fixation toward the location the will maximally increase information about the object being recognized, in that it will best allow the model to discriminate between the various current candidate object classes.

Rybak et al. (see Chapter 108) proposed a related model, in which scanpaths (containing motor control directives stored in a "where" memory and locally expected bottom-up features stored in a "what" memory) are learned for each scene or object to be recognized. When presented with a new image, the model starts by selecting candidate scanpaths based on matching bottom-up features in the image to those stored in the "what" memory. For each candidate scanpath, the model deploys attention according to the directives in the "where" memory, and compares the local contents of the "what" memory at each fixation to the local image features. This model has demonstrated strong ability to recognize complex gray-scale scenes and faces, in a translation-, rotation- and scale-independent manner.

A more extreme view at the basis of the models just mentioned is the "scanpath theory" of Stark (Noton and Stark, 1971), in which the control of eye movements is almost exclusively under top-down control. The theory proposes that what we see is only remotely related to the patterns of activation in our retinas, as suggested by our permanent illusion of crisp perception over our entire visual environment, although only the central 2° of our foveal vision provides such crisp sampling of the visual world. Rather, the scanpath theory argues that a cognitive model of what we expect to see is the basis for our percept; the sequence of eye movements that we make to analyze a scene, then, is mostly controlled top-down by our cognitive model of that scene. This theory has had a number of successful applications to robotics control, in which an internal model of a robot's working environment was used to restrict the analysis of incoming video sequences to a small number of circumscribed regions important for a given task (see Chapter 48).

One important challenge for combined models of attention and recognition consists of finding suitable neuronal correlates. Despite a biological inspiration in their architectures, the models just reviewed in this section, indeed, do not relate in much detail to biological correlates of object recognition. Although a number of biologically plausible models have been proposed for object recognition in the ventral "what" stream (see Chapter 46), their integration into biological models concerned mostly with attentional control in the dorsal "where" stream remains an open issue. This integration will, in particular, have to account for the increasing experimental support for an object-based rather than purely spatial focus of attention (see Chapter 66).

V. APPLICATIONS

Attention is a desirable mechanism not only for biological organisms, but also for efficient allocation of resources in artificial systems (see Chapter 95). Several chapters in this volume indeed explore how an attentional component may serve to build better sensing machines (see Chapters 102–106).

Here we relate a rather anecdotical but unusual application of attention models to the quantitative evaluation of advertising designs. In particular, we explore the problem of selecting among several candidate images for a magazine's cover. In creative and advertising design, current evaluation and selection criteria for candidate artwork creations rely mostly on focus groups, where groups of expert and/or naive human subjects compare the qualities and drawbacks of several competing design proposals. A question that arises, then, is whether a more objective metric could be developed to rank competing proposals. One such metric, particularly popular in the 1980s, has been to use eye-tracking devices to monitor eye movements of human subjects viewing the candidate designs. However, the high cost and difficulty of implementation of such evaluation have limited its applicability, and few advertising agencies appear to have developed eye-tracking setups in-house.

Computational models of attention, to the extent to which they may indeed relatively well predict the visual attractiveness of different locations in an artwork piece, may provide a simpler and more cost-effective solution to the problem of developing rapid, objective, and unbiased evaluation criteria. The assumption is that locations marked as highly salient by the model ought to attract the gaze of a majority of potential customers. A better design, then, is one where information that matters to the advertiser is conveyed at these locations.

Post-sales market analysis in the domain of magazine sales has revealed that what sells a given magazine issue is not the quality or beauty of the artwork present on the cover, but the catchy text messages that tease casual viewers about the contents of the magazine (source: Peter Walker of McCann-Erickson, a large multinational advertising agency). This allows one to set a fairly simple quantitative criterion for the evaluation of various cover proposals: whichever candidate design has the highest average model-predicted salience over the important text messages wins. An example of such application is shown in Fig. 94.3. While in one design most of the text messages are reliably marked as salient by the model, in the other approximately the lower half of the text is ignored, because of its low contrast against a textured background.

This simple application certainly has limitations, in particular because a purely bottom-up model with no object recognition or text reading capability was here used to evaluate the covers, but helps in opening the horizon of potential applications for attention models. In a similar vein, such models could also be used to assist merchandising experts in optimally arranging

FIGURE 94.3 Example application to the quantitative evaluation of advertising designs. Each row shows, from left to right, the first 10, 20, and 30 attention shifts predicted by our model. For the cover shown on the top row, good coverage of all text messages is predicted after 30 shifts, indicating that the text is likely to attract attention and gaze. Under the assumption that these text messages are what sells the magazine, the top cover hence is evaluated as a good design by the model. In contrast, after 30 attention shifts on the cover of the bottom row, large regions of text have remained unvisited by the model. This is not too surprising, as the ignored portions of text are masked by the strongly textured background and are hard to read. Nevertheless, both of the covers shown here have actually been published in two different issues of the British magazine *Good Housekeeping*. Hence, current quality control and design selection processes, based mostly on subjective evaluations by focus groups, could benefit from the addition of an objective measure like average saliency of text regions in the images. Photographs courtesy of Peter Walker, McCann-Erickson.

their products on supermarket shelves, or to assist product placement experts in optimizing the exposure of sponsor products in films or televised sports.

VI. SUMMARY AND CONCLUSION

We have reviewed recent work on biologically plausible computational models of attention, with a particular emphasis on bottom-up control of attentional deployment. Although much progress has been made in the past few decades, the field of attention modeling still is very young and appears full of promise. In particular, one of the grand challenges for the coming years will be in further developing system-level models where attention interacts with object recognition, gist analysis, symbolic knowledge and top-down task demands during complex goal-driven scene understanding (see Chapter 33). Yet, already today, attention modeling enjoys many exciting applications which also deserve to be further explored, from automatic target detection to advertising design.

References

Allman, J., Miezin, F., and McGuinness, E. (1985). Stimulus specific responses from beyond the classical receptive field: neurophysiological mechanisms for local-global comparisons in visual neurons. *Annu. Rev. Neurosci.* **8**, 407–430.

Andersen, R. A., Bracewell, R. M., Barash, S., Gnadt, J. W., and Fogassi, L. (1990). Eye position effects on visual, memory, and saccade-related activity in areas LIP and 7a of macaque. *J. Neurosci.* **10**, 1176–1196.

Desimone, R., and Duncan, J. (1995). Neural mechanisms of selective visual attention. *Annu. Rev. Neurosci.* **18**, 193–222.

Dominey, P. F., and Arbib, M. A. (1992). A cortico-subcortical model for generation of spatially accurate sequential saccades. *Cereb. Cortex.* **2**, 153–175.

Gilbert, C., Ito, M., Kapadia, M., and Westheimer, G. (2000). Interactions between attention, context and learning in primary visual cortex. *Vision Res.* **40**, 1217–1226.

Gottlieb, J. P., Kusunoki, M., and Goldberg, M. E. (1998). The representation of visual salience in monkey parietal cortex. *Nature* **391**, 481–484.

Itti, L., Koch, C., and Niebur, E. (1998). A model of saliency-based visual attention for rapid scene analysis. *IEEE Trans. Pattern Anal. Machine Intell.* **20**, 1254–1259.

James, W. (1890/1981). "The Principles of Psychology." Harvard University Press, Cambridge, MA.

Koch, C., and Ullman, S. (1985). Shifts in selective visual attention: towards the underlying neural circuitry. *Hum. Neurobiol.* **4**, 219–227.

Kustov, A. A., and Robinson, D. L. (1996). Shared neural control of attentional shifts and eye movements. *Nature* **384**, 74–77.

Nothdurft, H. (2000). Salience from feature contrast: variations with texture density. *Vision Res.* **40**, 3181–3200.

Noton, D., and Stark, L. (1971). Scanpaths in eye movements during pattern perception. *Science* **171**, 308–311.

Sillito, A. M., Grieve, K. L., Jones, H. E., Cudeiro, J., and Davis, J. (1995). Visual cortical mechanisms detecting focal orientation discontinuities. *Nature* **378**, 492–496.

Treisman, A. M., and Gelade, G. (1980). A feature-integration theory of attention. *Cogn. Psychol.* **12**, 97–136.

Wolfe, J. M. (1994). Visual search in continuous, naturalistic stimuli. *Vision Res.* **34**, 1187–1195.

CHAPTER 95

Saliency in Computer Vision

Gérard Medioni and Philippos Mordohai

ABSTRACT

The goal of computer vision is to develop algorithms for image understanding for computers and not necessarily to emulate the human vision system in biologically plausible ways. Nevertheless, research in computer vision has understandably looked to the human visual system for inspiration and intuition. One key aspect of human perception is saliency, the property of certain arrangements conspicuously standing out from a cluttered background. Over the past several years, we have developed a computational framework to detect salient perceptual structures in 2D, 3D, or N-D data sets, even under severe noise corruption. In the framework, data tokens are represented by tensors and the saliency of each token is computed based on information propagated among neighboring tokens via tensor voting. The Tensor Voting Framework enables us to cast computer vision problems as perceptual organization ones whose solution is the most salient perceptual structures.

I. INTRODUCTION

As pointed out by Max Wertheimer in a classic article (Wertheimer, 1923), the perception of wholes comprising many parts is not a property of the parts themselves, but a property of the arrangement of these parts. Gestalt psychologists investigated the principles that guide the arrangements and divisions of stimuli. These include proximity, similarity, good continuation, uniform destiny, closure, and simplicity. They operate at the pre-attentive bottom-up stages of perception. Psychological experiments documented the contribution of the gestalt principles to a universal perception of most stimuli by human observers. In many cases, they reinforce one another, but, in other cases, one factor dominates and produces a certain arrangement instead of another competing possibility. For instance, for a given set of stimuli, similarity may prevail over proximity to produce a more salient organization. The conflicts that often occur between gestalt principles are obstacles to the development of artificial systems for perceptual organization. Most researchers limit the number of principles they consider and develop ways to adjust their relative weights to achieve the desired organization under different circumstances.

Grossberg and Mingolla (1985) and Grossberg and Todorovic (1988) have developed the Boundary Contour System and the Feature Contour System that can group fragmented and even illusory edges to form closed boundaries and regions by feature cooperation. Shashua and Ullman (1988) detect globally salient curves in cluttered environments using locally connected networks. The measure of saliency they selected increases with respect to the length and smoothness of a curve and attenuates with gaps and curvature. Heitger and von der Heydt (1993) group elementary curves into contours via convolution with a set of orientation selective kernels whose responses decay with distance and difference in orientation. Junctions, corners, line ends, and illusory contours can also be explicitly detected.

The motivation behind the Tensor Voting Framework was the development of a general computational framework for computer vision problems. They are addressed within a gestalt framework, where the primitives are grouped according to principles to give rise to salient structures. This is compatible with the most generally applicable constraint in vision, that the "matter is cohesive," proposed by Marr (1982). We chose a local, model-free approach over a global, model-based alternative. With the latter, we cannot discriminate between local model misfits and noise. Furthermore, a hierarchy of more abstract descriptions can be derived from the low-level local one, if that is desired, whereas the opposite is not always feasible. The framework is noniterative, very robust to noise, has no initialization requirements, and has only one critical parameter, the scale of voting.

II. THE TENSOR VOTING FRAMEWORK

In this section we briefly review the tensor voting framework that was originally presented in Medioni et al. (2000). The two main aspects of the framework are the representation by second-order symmetric tensors and the information propagation mechanism by tensor voting. It is described in 2D here, but it is suitable for perceptual organization in any dimension. The novelty of our approach is that there is no objective function that is explicitly defined and optimized according to global criteria. Instead, tensor voting is performed locally and the saliency of perceptual structures is estimated as a function of the support tokens receive from their neighbors. Tokens with compatible orientations that can form salient structures reinforce one another. A key property of the framework that sets it apart from all of the methods developed to date is that we define saliency as a tensor, as opposed to a scalar. The tensor conveys considerably richer information, such as the type of the perceptual structure and its most likely orientation.

The representation of a token consists of a symmetric second-order tensor that encodes perceptual saliency. The tensor essentially indicates the saliency of each type of perceptual structure the token belongs to and its preferred orientation. Tensors were first used as a signal processing tool for computer vision applications by Granlund and Knutsson (1995). In 2D, a symmetric second-order tensor can be viewed as an ellipse, or a 2×2 matrix, that can be decomposed as in the following equation:

$$T = (\lambda_1 - \lambda_2)\hat{e}_1\hat{e}_1^T + \lambda_2(\hat{e}_1\hat{e}_1^T + \hat{e}_2\hat{e}_2^T) \quad (1)$$

where λ_i are the eigenvalues in descending order and $\hat{e}_i$ are the corresponding eigenvectors. The first term corresponds to an elongated stick tensor that encodes certainty of orientation normal to $\hat{e}_1$ with saliency $\lambda_1 - \lambda_2$, and the second term corresponds to an isotropic ball tensor that encodes uncertainty of orientation with saliency λ_2.

The inputs consist of tokens, which can be points or points with an associated orientation (curvels). The former are encoded as ball tensors and the latter as stick tensors. We propose to combine the information contained by the arrangement of these tokens by tensor voting, a method of information propagation in which tokens convey their orientation preferences to their neighbors in the form of votes, which are also tensors that are cast from token to token. Each vote has the orientation the receiver would have if the voter and receiver were part of the same smooth perceptual structure, which we choose to be the arc of the osculating circle at the voter that goes through the receiver. The saliency decay function we have chosen is the following (see also Fig. 95.1):

$$\mathrm{DF}(s, \kappa, \sigma) = e^{-\left(\frac{s^2 + c\kappa^2}{\sigma^2}\right)} \quad (2)$$

where s is the arc length OP, κ is the curvature, c is a constant, and σ is the scale of voting, which determines the effective neighborhood size. According to the gestalt principle of proximity, the strength of the votes attenuates with distance. The strength of the vote also decreases with increased curvature of the hypothesized structure, making straight continuations preferable to curved ones following the principles of smooth continuation and simplicity. Votes are cast by the stick and ball component of each tensor using the precomputed voting fields shown in Fig. 95.1.

Each token casts votes to its neighbors and collects votes from them. Votes are accumulated by tensor addition and result in a generic tensor. An analysis of the results is performed by decomposing the tensors according to Eq. (1) and generating stick and ball saliency maps. Junctions can be detected as distinct local maxima of ball saliency, region inliers are characterized by high ball saliency, and curve inliers are

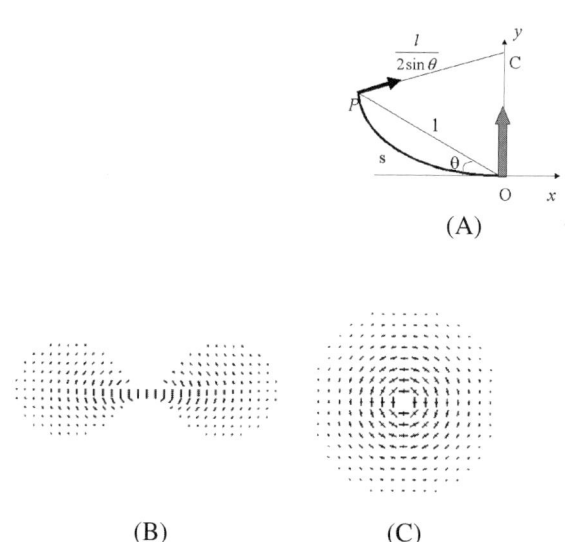

FIGURE 95.1 (A) Vote generation from a unit stick voter and (B) the stick and (C) ball fields.

II. THE TENSOR VOTING FRAMEWORK

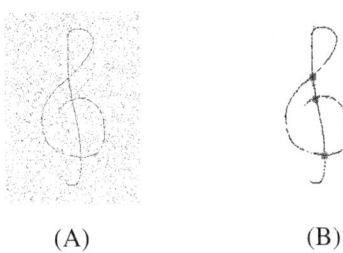

FIGURE 95.2 (A) Input, (B) Curves and junctions. Curves and junctions from a noisy point set. Junctions have been enlarged and marked as squares.

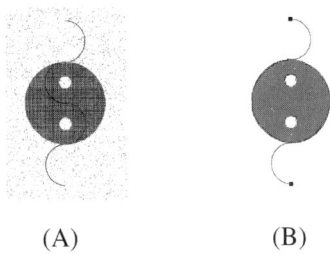

FIGURE 95.3 (A) Input, (B) Regions and curves. Regions and curves from a noisy point set. Region boundaries and curve end points are marked in black.

characterized by high curve saliency. Results are illustrated in Fig. 95.2. We also perform first order voting (Tong et al., 2001) to detect curve end points and region boundaries. Figure 95.3. demonstrates the simultaneous detection of a region and a curve, as well as their terminations, from a set of unoriented points.

Note that virtually the same results can be obtained using a wide range of voting scales.

We have briefly presented a generic framework that provides a novel definition of and implementation of saliency. It can be applied to curves, regions, and also problems in higher dimensions. It is efficient, is robust to noise, and has only one critical parameter.

References

Granlund, G., and Knutsson, H. (1995). "Signal Processing for Computer Vision." Kluwer, Dordrecht.

Grossberg, S., and Mingolla, E. (1985). Neural dynamics of form perception: Boundary completion, illusory figures, and neon color spreading. *Psychol. Rev.* **92**, 173–211.

Grossberg, S., and Todorovic, D. (1988). Neural dynamics of 1-d and 2-d brightness perception: A unified model of classical and recent phenomena. *Perception Psychophys.* **43**, 723–742.

Heitger, F., and von der Heydt, R. (1993). A computational model of neural contour processing: Figure-ground segregation and illusory contours. *In* Proceedings of the International Conference on Computer Vision (H. Nagel, T. Huang, Y. Shirai, Eds.), IEEE, pp. 32–40, Berlin.

Marr, D. (1982). "Vision." Freeman Press.

Medioni, G., Lee, M., and Tang, C. (2000). "A Computational Framework for Segmentation and Grouping." Elsevier, Amsterdam.

Shashua, A., and Ullman, S. (1988). Structural saliency: The detection of globally salient structures using a locally connected network. *In* Proceedings of the International Conference on Computer Vision (R. Bajcsy, S. Ullman, Eds.), IEEE, pp. 321–327, Tampa.

Tong, W., Tang, C., and Medioni, G. (2001). First order tensor voting and application to 3-d scale analysis. *In* Proceedings of the International Conference on Computer Vision and Pattern Recognition (P. Flynn, Ed.), IEEE, pp. I:175–182, Kauai, Hawaii.

Wertheimer, M. (1923). Routledge and Kegan Paul translated by W. Ellis, Vol. **4**, 301–350, London.

CHAPTER 96

Contextual Influences on Saliency

Antonio Torralba

ABSTRACT

This article describes a model for including scene/context priors in attention guidance. In the proposed scheme, visual context information can be available early in the visual processing chain, in order to modulate the saliency of image regions and to provide an efficient shortcut for object detection and recognition. The scene is represented by means of a low-dimensional global description obtained from low-level features. The global scene features are then used to predict the probability of presence of the target object in the scene, and its location and scale, before exploring the image.

I. INTRODUCTION

What is the role of contextual information in object recognition and detection tasks? What is the influence of the scene on determining the way that attention is deployed when trying to solve a task? How is the saliency of different image regions enhanced or reduced as a function of high-level scene information?

A number of studies have shown the importance of scene factors in object search and recognition. Studies by Biederman et al. (1982) and Palmer (1975) highlight the effect of contextual information in the processing time for object recognition. Rensink et al. (1997) have shown that changes in real-world scenes are noticed most quickly for objects or regions of interest, thus suggesting a preferential deployment of attention to these parts of a scene. Henderson and Hollingworth (1999) have reported results suggesting that the choice of these regions is governed not merely by their low-level saliency but also by scene semantics. Chun and Jiang (1998; Chapter 40) showed that visual search is facilitated when a correlation exists across different trials between the contextual configuration of the display and the target location. In a similar vein, several studies support the idea that scene semantics can be available early in the chain of information processing (Schyns and Oliva, 1994; Thorpe et al., 1996) and suggest that scene recognition may not require object recognition as a first step (Schyns and Oliva, 1994; Oliva and Torralba, 2001; Chapter 41).

Here a scheme is described in which visual context information can be available early in the visual processing chain in order to modulate the saliency of image regions and to provide an efficient shortcut for object detection and recognition. Context consists in a global description of the scene obtained from low-level features. In the proposed scheme, contextual information is used to predict the presence and absence of the target before scanning the image and to select the image regions that are relevant for the task.

II. THE SCENE CONTEXT

In Fig. 96.1A, observers describe the scenes as (left) a pedestrian in the street, (center) a car in the street, and (right) some food on a table. However, in the three images, the blob is identical (the pedestrian blob is the same shape as the car except for a 90-degs rotation). When object intrinsic information is reduced so much that an object cannot be identified based on local information, the object recognition system is not invariant to changes in pose, orientation, and location and context plays a mayor role in recognition.

In saliency models of attention (see Chapter 94), the context of the target object is considered as a collection of distractors. Figure 96.1B (left), shows a display with a salient target, in which context (distractors) does not affect target processing (Treisman and Gelade, 1980). In the central image, a person is embedded in the back-

FIGURE 96.1 (A) When object intrinsic information is reduced, then context plays a major role in recognition. Now, the object recognition system is not invariant to pose, orientation, location, and background. (B) Effects of context in masking and providing priors for finding the target.

ground. The person is masked by the context and is difficult to find. In these two examples, context is noninformative and its only effect on the search is the ability of the background to mask the target. But context can also provide information about the presence of the target. In Fig. 96.1B (right) the context, instead of masking the person, provides priors about what are the expected locations and scales in which we can find the target. In the canyon scene, a person could be almost in any location. However, in the street scene, the environment imposes strong constraints on the typical locations in which people is expected to be. This use of context is the one we are interested in modeling here.

III. THE REPRESENTATION OF SCENES

We can define the context of a particular object in terms of other previously recognized objects within the scene. There, the context representation is object-centered and requires object recognition as a first step.

The context representation described here does not require parsing the image to build a representation of the scene. As suggested in (Oliva and Torralba, 2001), it is possible to build a description of the scene that bypasses object identities, in which the scene is represented as a single entity. The representation proposed is based on identifying a number of properties that are related to the scene and that do not refer to individual objects. Our goal here is to use such a scheme for including context information in object representations and to demonstrate its role in facilitating object detection (Torralba, 2003a, 2003b).

As illustrated in Fig. 96.2, the analysis of the image is performed using two parallel pathways: one local (e.g., objects) and one global (e.g., scenes). Here we describe the features that can be used in both pathways.

A. Local Features

Most models of attention and object recognition rely on the definition of sets of local features. In the local

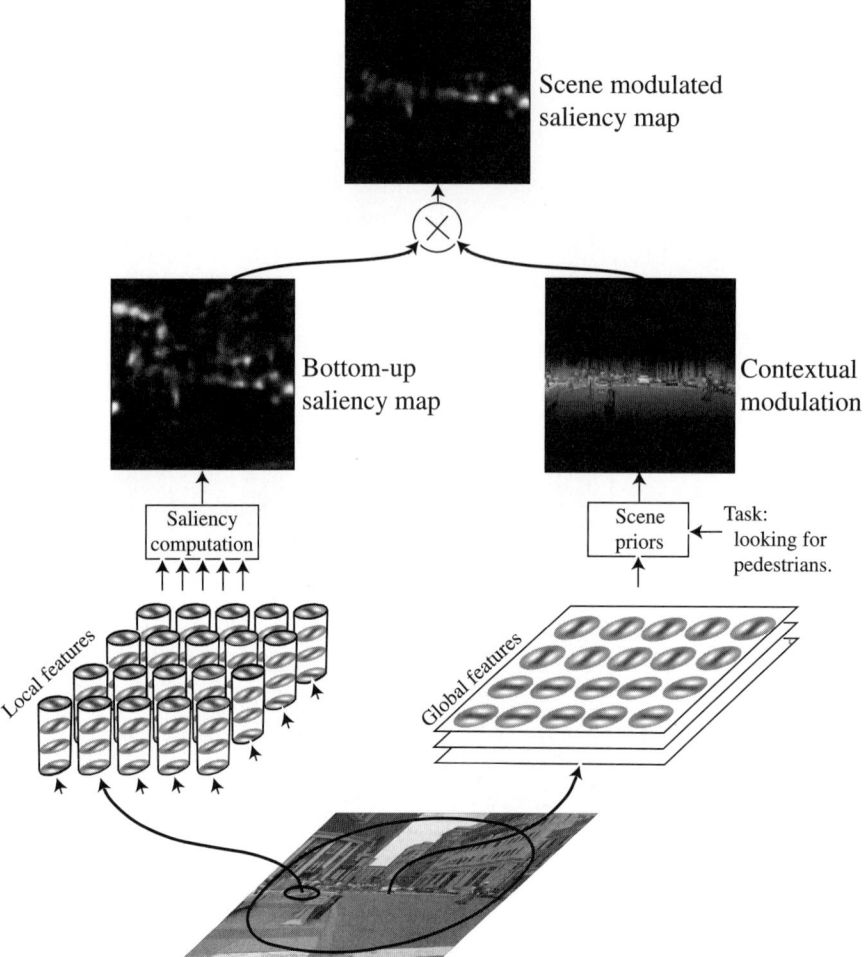

FIGURE 96.2 Contextual and local pathways for pre-attentive object search. This scheme incorporates contextual information for modulating image saliency. The scheme consists in two parallel pathways: the first one processes local image information; the second one encodes globally the pattern of activation of the feature maps. When looking for a person in the image, the saliency map, which is task independent, selects image regions that are salient in terms of local orientations and spatial frequencies. However, the contextual priming (task dependent) drives attention to the image regions that can contain the target object (sidewalks for pedestrian). The final attentional map, obtained as the product of both maps, selects the salient locations in side the image region relevant for the task.

pathway, each location is represented by a vector of features that describes local image properties. It could be a collection of templates (e.g., object detection) or a vector composed by the output of wavelets at different orientations and scales (e.g., saliency models of attention).

For instance, in Fig. 96.2, each local feature vector is a jet of filter responses: $v_1(x) = \{g_1(x), g_2(x), \ldots, g_N(x)\}$. Following the structure of the receptive fields of simple and complex cells in V1, the features $g_k(x)$ used here are obtained as: $g_k(x) = |\Sigma_{x'} I(x') h_k(x' - x)|^2$, where $I(x)$ is the input image and $h_k(x)$ is a Gabor-like wavelet tuned in orientation and scale.

B. Global Features

In the global pathway, the entire image is represented by a unique set of features that summarizes the appearance of the scene without encoding specific objects or regions. In the example in Fig. 96.2, the global feature shown responds to a combination of the output of oriented filters at different image locations (Oliva and Torralba, 2001): $\mathbf{v}_c = \{\Sigma_x \Sigma_k g_k(x) \phi_m(x, k); m = 1, M\}$, where $\phi_m(x, k)$ is a set of weights that specify how to combine the outputs of the local features $g_k(x)$ to build a global feature v_c. M is the total number of global features.

In the toy example in Fig. 96.2, the global feature responds strongly to images with horizontal structures in the bottom half of the image and vertical structures in the upper half of the image (this organization corresponds to the typical structure of a street scene). The global image representation is built by a collection of such kind of features.

In the next section, we describe how both local and global features can be combined to introduce contextual factors in attention.

IV. MODEL FOR SCENE PRIORS AND THE MODULATION OF SALIENCY

Here we describe a Bayesian framework (e.g., Kersten and Yuille, 2003) for object search that integrates saliency, object appearance, and scene priors in order to guide attention (Torralba 2003a, 2003b). In a statistical framework, when looking for a target (o represents the object category), at each image location (x) and scale of analysis (σ) a probability of containing the target is assigned: $p(o, x, \sigma, \alpha | \mathbf{v}_l, \mathbf{v}_c)$. α is a vector whose parameters describe the appearance of the target (e.g., point of view). The probability is conditional on the local and global image features. The object probability function can be decomposed applying Bayes rule as:

$$p(O|\mathbf{v}_l, \mathbf{v}_c) = \frac{1}{p(\mathbf{v}_l|\mathbf{v}_c)} p(\mathbf{v}_l|O, \mathbf{v}_c) p(O|\mathbf{v}_c) \quad (1)$$

For simplicity of notation we have grouped all the variables that describe the appearance of the object in the image as: $O = \{o, x, \sigma, \alpha\}$. The three factors in Eq. (1) provide a simplified framework for representing three levels of attention guidance (Torralba, 2003b).

A. Saliency

The normalization factor, $1/p(\mathbf{v}_l|\mathbf{v}_c)$, does not depend on the target or task constraints and therefore is a bottom-up factor. It provides a measure of how unlikely it is to find a set of local measurements $\mathbf{v}_l$ within the context $\mathbf{v}_c$. We can define local saliency as $S(x) = 1/p(v_l(x)|\mathbf{v}_c)$. This probabilistic definition of saliency fits more naturally with object detection and recognition formulations.

This formulation follows the hypothesis that frequent image features are more likely to belong to the background, whereas rare image features are more likely to be diagnostic features for the detection of (interesting) objects. Note that the term $S(x)$ does not incorporate any information about the appearance of the target. We approximate $S(x)$ by fitting a Gaussian to the distribution of local features in the image.

B. Target-Driven Control of Attention

The second factor, $p(\mathbf{v}_l | O, \mathbf{v}_c)$, gives the likelihood of the local measurements $\mathbf{v}_l$ when the object O is present in a particular context. This factor represents the top-down knowledge of the target appearance and how it contributes to the search (Rao et al., 1996). Regions of the image with features unlikely to belong to the target object are vetoed and regions with attended features are enhanced (see Chapter 17). Note that when the object properties O fully constraint the object appearance, it is possible to approximate $p(v_l | O, \mathbf{v}_c) \simeq p(v_l | O)$. This approximation allows the dissociation of the contribution of local image features and global (contextual) image features.

C. Scene Priors

The third factor, the PDF $p(O | \mathbf{v}_c)$, provides context-based priors on object class, location, scale, and appearance. This term is of capital importance for ensuring reliable inferences in situations in which the local image measurements $\mathbf{v}_l$ produce ambiguous interpretations. This factor does not depend on local measurements or target models.

Using the definition of an object in a scene, $O = \{o, x, \sigma, \alpha\}$, contextual influences become more evident if we apply Bayes rule successively in order to split the PDF $p(O | \mathbf{v}_c)$ into several factors that model different kinds of scene priors for object search:

$$p(O|\mathbf{v}_c) = p(\alpha|x, \mathbf{v}_c, o) p(\sigma|x, \mathbf{v}_c, o) p(x|\mathbf{v}_c, o) P(o|\mathbf{v}_C) \quad (2)$$

According to this decomposition of the PDF, the contextual modulation of target saliency is a function of four factors:

- *Object-class priming*: $P(o | \mathbf{v}_c)$ provides the probability of presence of the object class o in the scene. If $P(o | \mathbf{v}_c)$ is very small, then object search need not be initiated.
- *Contextual control of focus of attention*: $p(x | o, \mathbf{v}_c)$. This PDF gives the most likely locations for the presence of object o given context information.
- *Contextual selection of scale*: $p(\sigma | x, o, \mathbf{v}_c)$. This gives the likely size of the object o in the context $\mathbf{v}_c$. When looking for an object, the expected size of the target determines the scanning resolution that needs to be used when exploring the image.
- *Contextual selection of target appearance*: $p(\alpha | x, o, \mathbf{v}_c)$. This gives the expected shapes (point of views, aspect ratio) of the object.

Most popular computational models of object recognition focus on modeling the probability function $p(O | \mathbf{v}_l)$ without taking into account contextual priors.

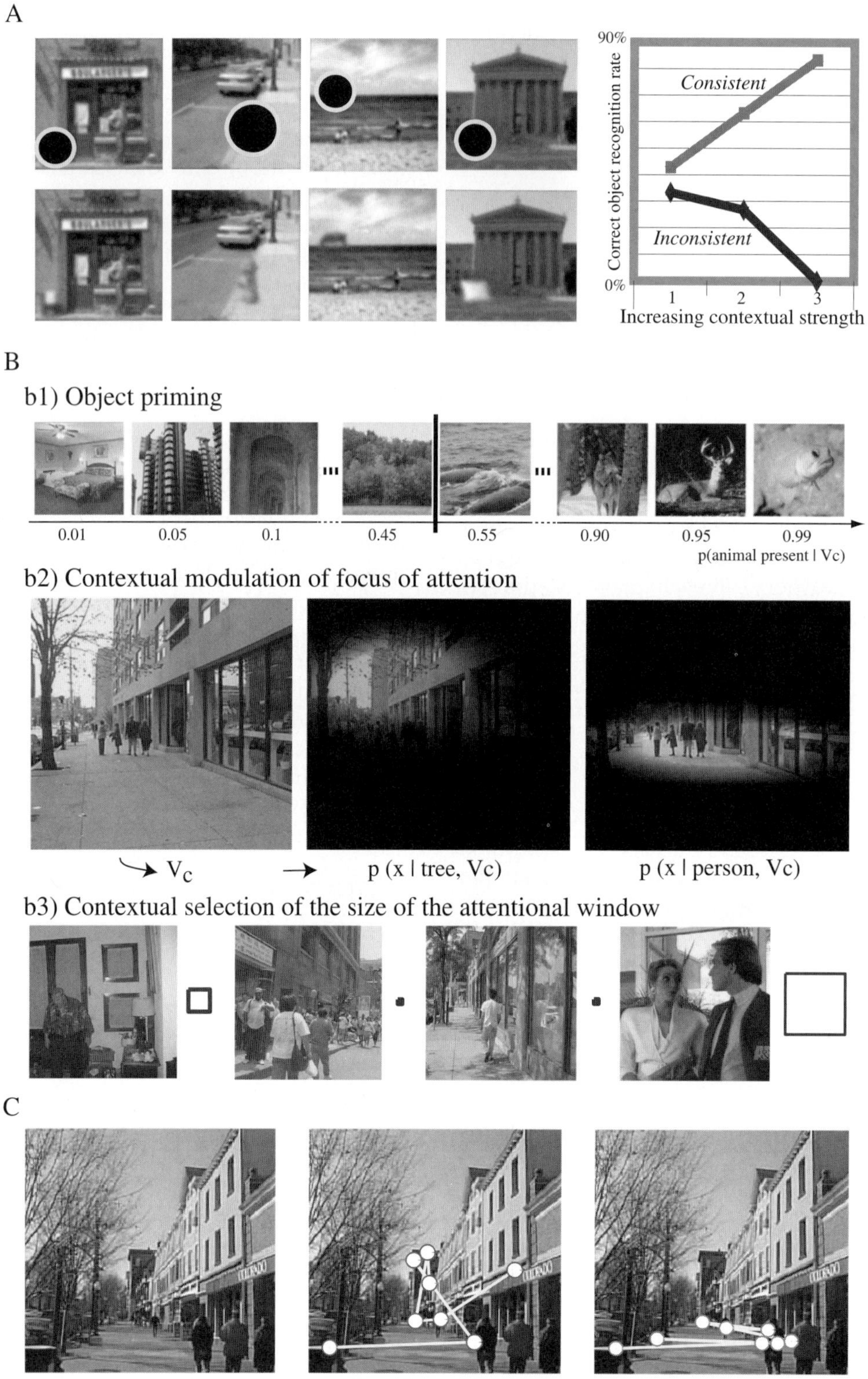

FIGURE 96.3 (A) Role of contextual priors on object recognition by subjects. (B) Scene priors obtained from global image features. (C) Examples of salient locations obtained from saliency alone (center) and combining both context and saliency (right). Including scene priors provides better predictions for the location of the target.

V. RESULTS

Figure 96.3A shows the effect that contextual priors $p(O \mid \mathbf{v}_c)$ have on subject performances for recognition. First, subjects are asked to guess the identity of the objects behind the masks (Fig. 96.3A, top). That experiment allows us to evaluate the distribution of objects that subjects are considering for each scene: $P(objects \mid x, \sigma, scene)$. Then, we can sort the scenes according to the strength of the priors (by measuring the entropy of the distributions). In a second experiment, we show how the strength of these priors affect recognition. We ask subjects to recognize blurred objects when they are placed in consistent and inconsistent backgrounds. The results (Fig. 96.3A, right) show that observer's performance on a recognition task is correlated with the strength of the priors (Bar, 2003; Chapter 25).

Figure 96.3B summarizes the results of the contextual model. The role of the contextual priors in modulating attention is to provide information about past search experience in similar environments and the strategies that were successful in finding the target. In this model, we assume that the contextual features $\mathbf{v}_c$ already carries all the information needed to identify the scene and that the scene is identified at a glance, without requiring eye movements. The eye movements are only required in order to analyze in detail regions of the image that are relevant for a task (i.e., to find somebody). The contextual priors $p(O \mid \mathbf{v}_c)$ contain the information about how the scene features $\mathbf{v}_c$ were related to the target properties O (image location, scale, and pose) during the past experience. The system is trained by our first providing the system a collection of images in which the target has been already located. The PDF is learned using a mixture of gaussians and the EM algorithm (Torralba, 2003a). Once the system has learned the relationship between scenes and objects, it can predict the expected locations for several objects in new scenes (Fig. 96.3B, right).

Figure 96.3B provides examples of the results of global contextual priming for:

(b1) Predicting the presence or absence of objects. Here we show the results for solving the task of animal present/absent using only scene priors, before scanning the image (Torralba and Oliva, 2003). The system has an 80% correct prediction rate on this task.

(b2) Focus of attention (e.g., expected locations of people and trees). Contextual priors for location reduce the area of the image that needs to be explored when looking for the target.

(b3) Scale selection (e.g., expected size of face in the image).

Finally, Fig. 96.3C compares the salient points and the region of interest predicted by a model using only bottom-up saliency maps (center) and when combining saliency and the scene priors for location (right). When including scene priors, the candidate locations are only within the image region that has a high probability of containing the target. Experiments show that including scene priors provides better predictions of human eye movements than saliency alone (Oliva et al., 2003).

VI. CONCLUSION

The model proposed includes scene priors for object search early in the visual processing chain. Therefore, the scene priors constitute an effective shortcut for object detection as it provides priors for the object's presence/absence before scanning the image.

From an algorithmic point of view, contextual control of the focus of attention is important because it avoids expending computational resources in spatial locations with a low probability of containing the target based on prior experience. It also provides criteria for rejecting possible false detections or salient features that fall outside the primed region.

References

Bar, M. (2003). A cortical mechanism for triggering top-down facilitation in visual object recognition. *J. Cogn. Neurosci.* **15**, 600–609.

Biederman, I., Mezzanotte, R. J., and Rabinowitz, J. C. (1982). Scene perception: Detecting and judging objects undergoing relational violations. *Cogn. Psychol.* **14**, 143–177.

Chun, M. M., and Jiang, Y. (1998). Contextual cueing: Implicit learning and memory of visual context guides spatial attention. *Cogn. Psychol.* **36**, 28–71.

Henderson, J. M., and Hollingworth, A. (1999). High level scene perception. *Annu. Rev. Psychol.* **50**, 243–271.

Itti, L., Koch, C., and Niebur, E. (1998). A model of saliency-based visual attention for rapid scene analysis. *IEEE Trans. Pattern Analysis and Machine Vis.* **20**, 1254.

Kersten, D., and Yuille, A. (2003). Bayesian models of object perception. *Curr. Opin. Neurobiol.* **13**, 150–158.

Oliva, A., and Torralba, A. (2001). Modeling the shape of the scene: A holistic representation of the spatial envelope. *Int. J. Comp. Vis.* **42**, 145–175.

Oliva, A., Torralba, A., Castelhano, M. S., and Henderson, J. M. (2003). Top-down control of visual attention in object detection. *IEEE Proceedings of the International Conference on Image Processing*, Barcelona (Spain). **1**, 519–526.

Palmer, S. E. (1975). The effects of contextual scenes on the identification of objects. *Memory and Cogn.* **3**, 519–526.

Rao, R. P. N., Zelinsky, G. J., Hayhoe, M. M., and Ballard, D. H. (1996). Modeling saccadic targeting in visual search. "NIPS'95." (D. Touretzky, M. Mozer and M. Hasselmo, Eds.), Advances in neural information processing systems. **8**, 830–836. MIT press, Cambridge, MA, USA.

Rensink, R. A., O'Regan, J. K., and Clark, J. J. (1997). To see or not to see: The need for attention to perceive changes in scenes. *Psychol. Sci.* **8**, 368–373.

Schyns, P. G., and Oliva, A. (1994). From blobs to boundary edges: evidence for time and spatial scale dependent scene recognition. *Psychol. Sci.* **5**, 195–200.

Thorpe, S., Fize, D., and Marlot, C. (1996). Speed of processing in the human visual system. *Nature* **381**, 520–522.

Torralba, A. (2003a). Contextual priming for object detection. *Int. J. Comp. Vis.* **53**, 169–191.

Torralba, A. (2003b). Modeling global scene factors in attention. *J. Opt. Soc. Am. A.* **20**, 1407–1418.

Torralba, A., and Oliva, A. (2003). Statistics of natural image categories. *Network: Computation Neural Syst.* **14**, 391–412.

Treisman, A., and Gelade, G. (1980). A feature integration theory of attention. *Cogn. Psychol.* **12**, 97–136.

Wolfe, J. M. (1994). Guided search 2.0: A revised model of visual search. *Psychonomic Bull. Rev.* **1**, 202–228.

CHAPTER 97

A Neurodynamical Model of Visual Attention

Gustavo Deco, Edmund T. Rolls and Josef Zihl

ABSTRACT

We describe a computational neuroscience approach to the role of attention in visual object perception. Neurodynamics, based on the mean field and integrate-and-fire methods, is used to provide a quantitative formulation for the dynamical evolution of single neurons, neural networks, and coupled modules of networks. We explicitly model the feedforward (bottom-up) and feedback (top-down) interactions between posterior brain areas (V1; V2; V4, inferior temporal cortex, (IT); and parietal cortex, (PP) and anterior (prefrontal cortex, (PFC)) brain areas. We assume that top-down processes operate using biased competition to implement attentional effects in earlier cortical processing areas. The computational model shows how attentional effects important in, for example, visual search may be implemented in the brain, shows that some serial attentional effects can be accounted for by different difficulties of constraint satisfaction in a purely parallel dynamical system, and makes predictions at the single-neuron, fMRI, and brain-lesion levels of investigation.

I. INTRODUCTION

This chapter describes a neural computation approach to attention to link together much empirical evidence from neurophysiology, the biophysics of single neurons, neuroimaging, psychophysics, and neuropsychology and to produce an understanding of how vision and attention are implemented by processing in the brain. The empirical evidence that is brought to bear is largely from nonhuman primates and from humans because of the considerable similarity of their visual systems, and the overall aims are to understand how visual perception is implemented in the human brain and the disorders that arise after brain damage.

II. VISUAL ATTENTIONAL MECHANISMS

The dominant neurobiological hypothesis to account for attentional selection is that attention serves to enhance the responses of neurons representing stimuli at a single relevant location in the visual field. This enhancement model is related to the metaphor for focal attention in terms of a spotlight (Treisman, 1988). According to this assumption, the concept of attention is based on explicit serial mechanisms. The feature integration theory of visual selective attention (Treisman, 1988) explains numerous psychophysical experiments on visual search and offers an interpretation of the binding problem. In the feature integration theory, the first, preattentive, process runs in parallel across the complete visual field extracting single primitive features without integrating them. The second, attentive, stage corresponds to the serial specialized integration of information from a limited part of the field at any one time. Evidence for these two stages of attentional visual processing comes from psychophysical experiments using visual search tasks in which subjects examine a display containing randomly positioned items in order to detect an *a priori* defined target.

The biased competition model (Duncan, 1996) postulates an alternative mechanism for selective attention. According to this model, the enhancing effect of attention on neuronal responses is understood in the context of competition among all the stimuli in the visual field. The biased competition hypothesis states that the multiple stimuli in the visual field activate populations of neurons that engage in competitive mechanisms. Attending to a stimulus at a particular location or with a particular feature biases this competition in favor of neurons that respond to the location, or the features, of the attended stimulus. This attentional effect is produced by generating signals within areas outside the visual cortex that are then fed back to extrastriate visual cortical areas, where they bias the competition such that, when multiple stimuli appear in the visual field, the cells representing the attended stimulus win, thereby suppressing cells representing distracting stimuli (Duncan, 1996). According to this line of work, attention appears as an emergent property of competitive interactions that work in parallel across the visual field. Neurophysiological experiments are consistent with this hypotheses in showing that attention serves to modulate the suppressive interaction between two or more stimuli within the receptive field (Reynolds et al., 1999; Chelazzi, 1999, Reynolds and Desimone, 1999). Further evidence comes from functional magnetic resonance imaging (fMRI) in humans (Kastner et al., 1999), which indicates that when multiple stimuli are present simultaneously in the visual field their cortical representations within the object recognition pathway interact in a competitive, suppressive fashion; this is not the case when the stimuli are presented sequentially. It was also observed that directing attention to one of the stimuli counteracts the suppressive influence of nearby stimuli.

III. A UNIFYING NEURODYNAMICAL COMPUTATIONAL MODEL

In order to provide a quantitative understanding of these attentional phenomena and their relation to working memory, we have developed a computational model that uses a short-term memory module (in the prefrontal cortex) to hold in memory the target of attention and to provide biased competition of earlier visual cortical processing stages in the ventral and dorsal visual streams. Interactions between these streams take place in early visual cortical areas. The whole system is implemented in a model with mean field-based or integrate-and-fire dynamics, so that the time courses of the attentional processes in what is a parallel distributed processing system can be understood. A fuller description of the model is provided by Rolls and Deco (2002).

The overall systems-level representation of the model is shown in Fig. 97.1. The system is essentially composed of six modules (V1, V2–V4, IT, PP, v46, and d46), structured such that they resemble the two known main visual paths of the mammalian visual cortex: the "what" and "where" paths (Deco and Zihl, 1998, 2001; Rolls and Deco, 2002). These six modules represent the minimum number of components to be taken into account within this complex system in order to describe the desired visual attention mechanism. Information from the retino–geniculo–striate pathway enters the visual cortex through area V1 in the occipital lobe and proceeds into two processing streams. The occipital-temporal stream leads ventrally through V2–V4 and IT (inferotemporal cortex), and the occipito-parietal stream leads dorsally into PP (posterior parietal complex). In this model system, we choose to model with our ventral stream one particular aspect of inferior temporal cortex function—translation-invariant object recognition. We choose to model with our dorsal stream one particular aspect of parietal cortex function—encoding of visual space in retinotopic coordinates. This is obviously a great simplification of the complex hierarchical architecture and functions of the primate visual system, but the model is sufficient to test and demonstrate our fundamental proposals about how the object and spatial streams interact to implement visual attentional processes. We assume that top-down feedback connections from prefrontal cortex area 46 (modules d46 and v46) provide the external top-down bias that specifies the processing conditions of earlier modules. Concretely, the feedback connections from area v46 to the IT module specify the target object in a visual search task; and the feedback connections from area d46 to the PP module generate the bias for a targeted spatial location in an object recognition task. In this version of the dynamical model the activity A_i of a given cortical area is described with a mean field approximation (Usher and Niebur, 1996; Rolls and Deco, 2002). For a given pool of neurons, the dynamics of the activity $A_i(t)$ is mathematically expressed by:

$$\tau \frac{dA_i(t)}{dt} = -A_i(t) + aF(A_i(t)) - bF(A^I(t)) \quad (1)$$
$$+ I_0 + I_i^E(t) + I_i^A(t) + \nu$$

where $F(x)$ is the response or activation function (transforming current into discharge rate), $A^I(t)$ is the activity of the common inhibitory pool, I_0 is a diffuse spontaneous background input, $I_i^E(t)$ is the external sensory input to the cells in pool i, ν is Gaussian noise, and $I_i^A(t)$ represents an external input coming from higher modules, called attentional top-down bias. The dynamic evolution of activity in the cortical area level can be simulated in the framework of the present

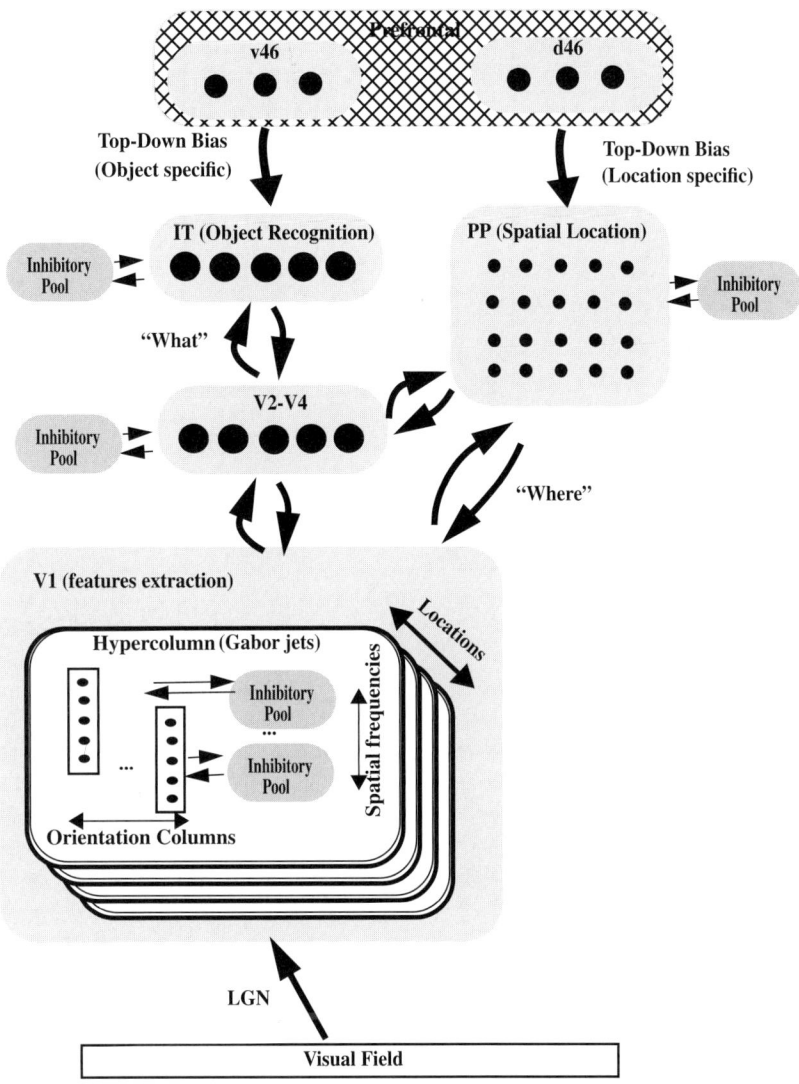

FIGURE 97.1 The systems-level architecture of a model of the cortical mechanisms of visual attention. The system is essentially composed of six modules structured so that they resemble the two known main visual pathways of the primate visual cortex. IT, inferior temporal visual cortex; PP, posterior parietal visual cortex; LGN, lateral geniculate nucleus.

model by integrating the neuronal population (or pool) activity in a given area over space and time (Rolls and Deco, 2002).

IV. EXPERIMENTAL VERIFICATION

A. Simulation of Single-Cell Experiments

With the neurodynamical model presented in Sec. III, Deco and Lee (2002) and Corchs and Deco (2002) were able (see Fig. 97.2) to produce results that were qualitatively similar to those Reynolds et al. (1999) obtained during investigations of the effects of attention on single neurons recorded in V2. The competitive interactions in the absence of attention are due to the intramodular competitive dynamics. The modulatory biasing corrections in the attended conditions are caused by the intermodular interactions between the V1 and PP pools and between the PP pools and prefrontal top-down modulation.

B. FMRI Experiments

The neurodynamical model of Sec. III can be used to simulate and understand neuroimaging data

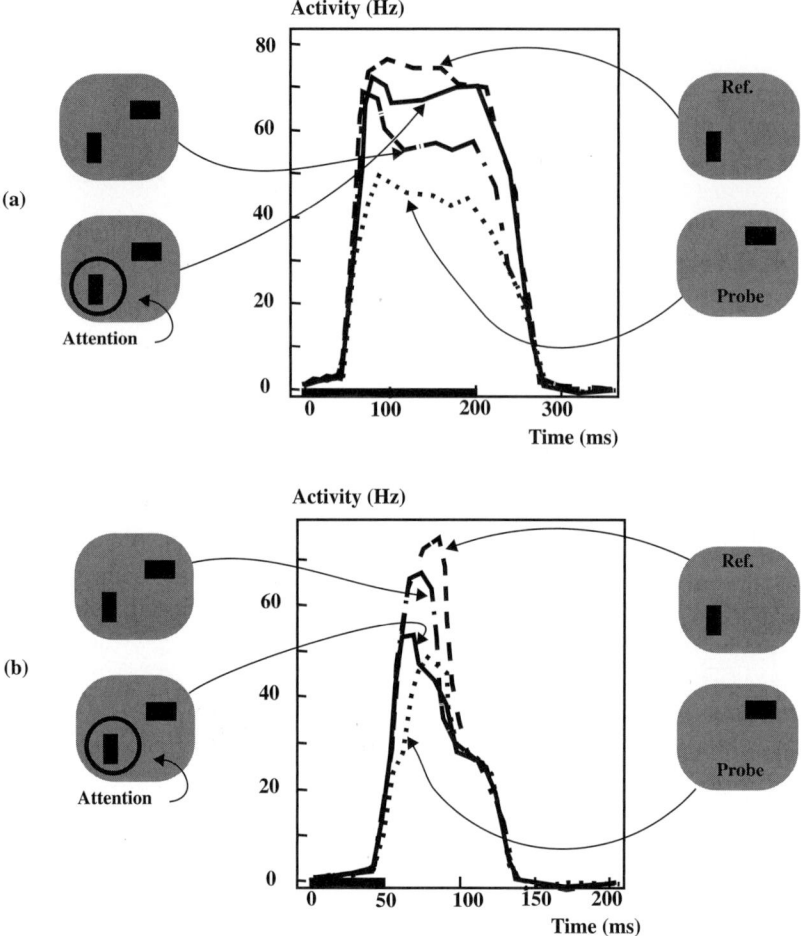

FIGURE 97.2 Simulation of the experiment of Reynolds et al. (1999). (A) The stimulus was presented for 200 ms. When an optimal reference stimulus was presented alone, the cell's response (dashed line) was much stronger than its response (dotted line) when a suboptimal probe stimulus was presented alone in the receptive field. Simultaneous presentation of both the reference and the probe stimuli produced an intermediate response (dashed-dotted line), indicating that the probe was producing competitive suppression of the response to the reference stimulus. However, when spatial attention was directed toward to the reference stimulus, the suppression due to the probe was largely eliminated and the neuronal response returned to the level at which the reference was presented alone (continuous line). (B) The same manipulation as in (A) except that the stimulus was presented for only 50 ms.

obtained with, for example, fMRI by integrating the activity across all the neuronal populations or pools in a given brain area over space and time. The integration over space yields an average activity of a brain area at a given time. Integration over time is performed to simulate the temporal resolution of fMRI experiments. Corchs and Deco (2002) simulated fMRI signals from V4 under the experimental conditions defined in an fMRI investigation of attention by Kastner et al. (1999). Figure 97.3 shows the results of the simulations. The simulations showed that the fMRI signals were smaller in magnitude during simultaneous (SIM) than during sequential (SEQ) presentations in unattended conditions because of the mutual suppression induced by the four competitively interacting stimuli. On the other hand, the average fMRI signals with attention increased more strongly for simultaneously presented stimuli than did the ones for sequentially presented

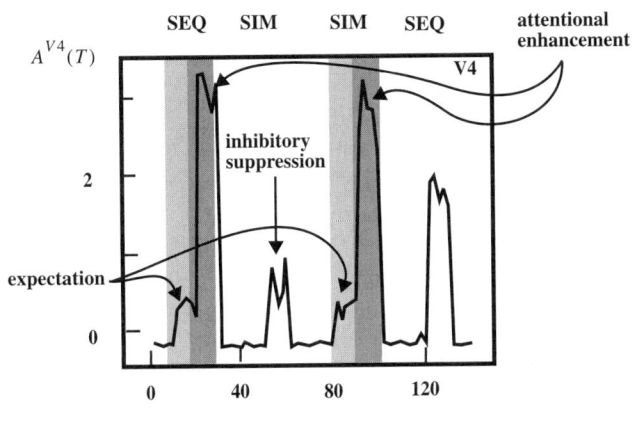

FIGURE 97.3 Computer simulations of fMRI signals in module V4 based on the experiments of Kastner et al. (1999). Light shading indicates the expectation periods, dark shading the attended presentations, and blocks without shading correspond to the unattended condition. SEQ, sequential presentation; SIM, simultaneous presentation.

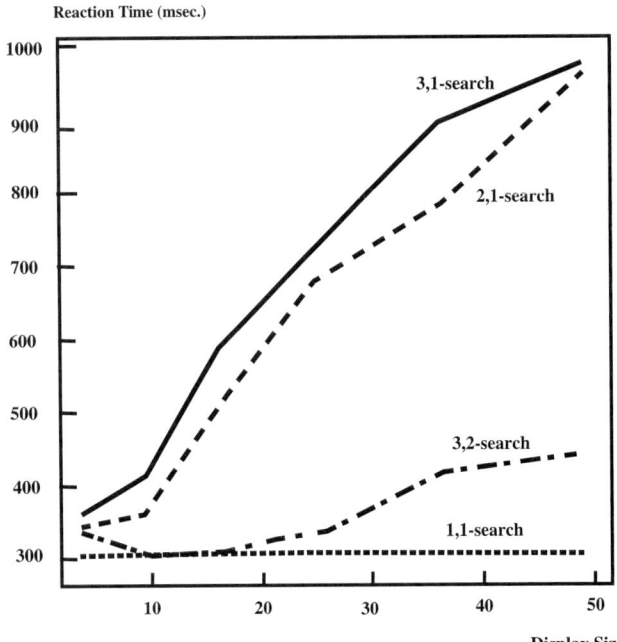

FIGUER 97.4 Search times for feature and conjunction searches obtained using the extended computational cortical model. 1,1-search is a feature search; 3,1-, 2,1-, and 3,2-search are conjunction searches.

stimuli. Thus, the suppressive interactions were partially cancelled out by attention. The results demonstrate that the model can capture and provide an account of what is found in attention experiments when the measurements are made at the fMRI level. Details of the study are provided by Rolls and Deco (2002).

C. Simulation of Psychophysical Experiments: Parallel Search versus Serial Search

The model is also able to capture and account for psychophysical studies on visual search. In search tasks in which search time is measured as a function of the number of items in the display, it is found that the time needed to find a target is independent of the number of distractors in a feature search condition in which the target differs from the distractors in one single feature (e.g., color). In contrast, in a conjunction search task the target is defined by a conjunction of features, and each distractor shares at least one of those features with the target. The search time then increases linearly with the number of items in the display, implying a serial process. Deco and Zihl (2001) and Deco and Lee (2002) showed that these properties of feature and conjunction search are displayed by the attentional architecture described in Sec. III, as shown in Fig. 97.4. The items were defined by three feature dimensions (e.g., size, orientation, and color), each having two values (e.g., size: big/small, orientation: horizontal/vertical, color: white/black). Figure 97.4 shows examples for each kind of search.[1] For each display size, the experiment was repeated 100 times, each time with different randomly generated targets at random positions and randomly generated distractors. The mean value T of the 100 simulated search times is plotted as a function of the display size S. The slopes for all simulations are consistent with existing experimental results (Quinlan and Humphreys, 1987).

The implication of these computational results is that, although the network performs search across the visual field in parallel, there are differences in the latencies of the neural responses in the different conditions related to how easily the dynamical system can perform the constraint satisfaction for the different conditions. In particular, effects that mimic serial focal attention can be demonstrated in these completely parallel system. The model demonstrates that neither explicit serial focal search, nor saliency maps needs to be assumed and that parallel search and serial search

---

[1] Using the terminology of Quinlan and Humphreys (1987), feature search corresponds to a 1,1-search; a standard conjunction search corresponds to a 2,1-search; and a triple conjunction search can be a 3,1 or a 3,2-search, depending on whether the target differs from all distractor groups by one or two features, respectively.

might not be two very different and independent stages, as previously thought (Treisman, 1988). In the computational model described in Sec. III, the two stages of processing (pre-attentive and attentive) involve the same mechanism and feature integration is accomplished dynamically by the interaction between the ventral IT module and the early V1 module. Feature integration is an emergent phenomenon due to interactive activation between the cortical areas rather than a separate stage of visual processing or involving a separate visual area.

D. Simulation of Neuropsychological Impairments in Attention

By artificially damaging the connections in the model, it is possible to account for a number of attentional deficits occurring in humans after brain damage. For example, with graded damage increasing in severity toward the right of the topologically organized parietal module, the model reproduces some of the symptoms of the left hemineglect of space that occurs after right parietal cortex damage in humans (Rolls and Deco, 2002). If this lesion is combined with local lateral inhibition (rather than global inhibition) within a processing module, then it is even possible to account for object-based neglect, in which the left half only of each of a series of objects arranged across the visual field is not seen (Deco and Rolls, 2003b).

V. CONCLUSION

The concept of biased competition has been used to account for object attention (Usher and Niehur, 1996) and spatial attention in the ventral stream (Reynolds et al., 1999). The neurodynamical model reviewed in this chapter advances these ideas by bringing in the dorsal stream and the early visual areas to coordinate the organization of attention in a unified system (Deco and Zihl, 2001; Rolls and Deco, 2002; Chapter 98). The two modes of attention emerge depending simply on whether a top-down bias is introduced to either the dorsal stream PP module or the ventral stream IT module. The spatial attention effect and competition interaction effect observed in the experiments of Moran and Desimone (1985) and Reynolds et al. (1999) can be accounted for by this model. The model also shows and explains the dynamical competition and attention modulation effects found in attention experiments at the level of gross brain area activation as measured with fMRI (Kastner et al., 1999; Corchs and Deco, 2002). In the context of visual search, the model shows that Treisman's feature integration (Treisman, 1988) can be implemented as an emergent phenomenon arising from the interaction between early visual cortical areas and the various extrastriate areas in the ventral and dorsal visual streams (Deco and Zihl, 2001). The system works across the visual field in parallel, but, due to the different dynamic latencies, resembles the two apparent different modes of visual attention: serial focal search and parallel search.

In summary, computational neuroscience provides a mathematical framework for studying the mechanisms involved in brain function, such as visual attentional mechanisms. The neurodynamical model analyzed here is based on evidence from neurophysiological and psychological findings. The simulations obtained with this theoretical model successfully reproduce the results of neurophysiological and fMRI experiments on spatial attention, as well as studies on serial and parallel search, and neuropsychological phenomena such as object-based visual neglect that can follow brain damage.

Acknowledgments

Support for this work was contributed by ICREA and by the Medical Research Council.

References

Chelazzi, L. (1999). Serial attention mechanisms in visual search: A critical look at the evidence. *Psychol. Res.* **62**, 195–219.

Corchs, S., and Deco, G. (2002). Large-scale neural model for visual attention: Integration of experimental single-cell and fMRI data. *Cerebral Cortex* **12**, 339–348.

Deco, G., and Lee, T. S. (2002). An unified model of spatial and object attention based on inter-cortical biased competition. *Neurocomputing* **44–46**, 769–774.

Deco, G., and Rolls, E. T. (2003a). Attention and working memory: A dynamical model of neuronal activity in the prefrontal cortex. *Eur. J. Neurosci.* **18**, 2374–2390.

Deco, G., and Rolls, E. T. (2003b). Object-based visual neglect: A computational hypothesis. *Eur. J. Neurosci.* **16**, 1994–2000.

Deco, G., and Zihl, J. (1998). A neural model of binding and selective attention for visual search. *In* "Proceedings of the 5th Neural Computation and Psychology Workshop," pp. 262–271. Springer Verlag, London.

Deco, G., and Zihl, J. (2001). Top-down selective visual attention: A neurodynamical approach. *Vis. Cogn.* **8**, 118–139.

Duncan, J. (1996). Cooperating brain systems in selective perception and action. *In* "Attention and Performance XVI" (T. Inui and J. L. McClelland, eds.), pp. 549–578. MIT Press, Cambridge, MA.

Kastner, S., Pinsk, M., De Weerd, P., Desimone, R., and Ungerleider, L. (1999). Increased activity in human visual cortex during directed attention in the absence of visual stimulation. *Neuron* **22**, 751–761.

Moran, J., and Desimone, R. (1985). Selective attention gates visual processing in the extrastriate cortex. *Science* **229**, 782–784.

Quinlan, P. T., and Humphreys, G. W. (1987). Visual search for targets defined by combination of color, shape and size: An examination of the task constraints on feature and conjunction searches. *Perception Psychophys.*, **41**, 455–472.

Reynolds, J., Chelazzi, L., and Desimone, R. (1999). Competitive mechanisms sub-serve attention in macaque areas V2 and V4. *J. Neurosci.* **19**, 1736–1753.

Reynolds, J., and Desimone, R. (1999). The role of neural mechanisms of attention in solving the binding problem. *Neuron* **24**, 19–29.

Rolls, E. T., and Deco, G. (2002). "Computational Neuroscience of Vision." Oxford University Press, Oxford.

Treisman, A. (1988). Features and objects: The fourteenth Barlett memorial lecture. *Q. J. Exp. Psychol.*, **40A**, 201–237.

Usher, M., and Niebur, E. (1996). Modeling the temporal dynamics of IT neurons in visual search: A mechanism for top-down selective attention. *J. Cogn. Neurosci.* **8**, 311–327.

CHAPTER 98

How the Detection of Objects in Natural Scenes Constrains Attention in Time

Fred H. Hamker

ABSTRACT

If we want to explain attention, we ultimately have to explore how we perceive natural scenes, the environment primates typically encounter. The task of detecting an object in a natural scene constrains the involvement of attention differently than in artificial scenes. I suggest that attention emerges through time on the systems level based on three general principles, and I demonstrate their feasibility in a computational model. In this model, attention itself is not a prerequisite for object recognition but feedback constrains feedforward processing and improves target discrimination. As a result, a state evolves that allows the linking of areas involved in planning to early areas responsible for scene analysis.

I. INTRODUCTION

Attention has been investigated in numerous tasks using displays with isolated items. In those experiments and related models, the spotlight model of attention has been shown to explain a number of findings. In natural scenes, however, a subject faces the additional problem of object detection and segmentation. A pure spatially based form of attention (e.g., a spotlight of attention) hardly improves the discrimination of the object of interest against the background. I suggest the following three principles that allow the detection of objects in natural scenes: (1) High-level processing and object detection are possible without spatial attention; that is, spatial attention does not gate processing; (2) a form of feature-based attention enhances features of interest in parallel and thus enhances an object of interest against the background; and (3) focal processing needs time to develop and emerges through reentrant processing from occulomotor areas.

These three principles are routed in one general rule of perception: Convergent zones in one brain area provide an expectation about feature or space. When an expectation is sent to other areas, it enhances the gain, given that a match with the input of this area occurs. This concept of a population-based inference is related to Bayesian inference, but avoids the computation of probabilities. Competition cleans up the population activity in higher stages from all unimportant stimuli so that a full recognition can take place. As a result, attention emerges on the network level. There is no area in the brain that is solely devoted to computing attention.

II. THE MODEL

In order to explain attention as a distributed, competitive resource I have suggested that attention emerges through interactions (Hamker, 1999). The presented computational model consisted of an area with large receptive fields inferior temporal cortex (IT) that is responsible for a largely location-invariant scene description, an area with small receptive fields (V1–V4) that encodes the features of objects within a small to intermediate spatial scale and an area of spatial processing (PP, FEF, SC) that encodes information about the location of objects. Similar as in the Selective Tuning Model (Chapter 92) top-down connections within the ventral pathway play an important role (Chapters 25, 83, and 107), but here a selection in space is localized outside the ventral stream. Such a tri-modular architecture has been recently shown to account for various attention effects based on competition (Hamker, 2000; Chapter 97).

As an extension to this general model, I here describe a new population-based computational approach that aims at modeling specific areas of the brain, including their temporal dynamics (Fig. 98.1). This model has been developed to be consistent with a range of electrophysiological findings. The model V4 area (Hamker, 2004a) has been demonstrated to quantitatively account for receptive field competition in V4 (Reynolds et al., 1999) and for multiplicative effects on the tuning curve (Chapter 49). A slightly simplified version of the proposed systems model (Hamker, 2003) has been shown to match the time course of IT and V4 activity in visual search (Chelazzi et al., 1993, 1998). The model is also consistent with findings in the FEF (Schall, 2002) and psychophysical data (Hamker, 2004b), and it predicts that the FEF provides a spatially organized reentry signal to extrastriate visual areas (Hamker, 2001).

Evidence for the latter prediction has been given by stimulating the FEF, which influenced stimulus-related activity in V4 (Moore and Armstrong, 2003). By fitting the model data with the experimental data gained by Chelazzi et al. (1998), I identified movement cells in the FEF as a possible convergent zone, which provides the spatial reentry signal (Hamker, 2003). Their timing and selectivity corresponds with the observed target discrimination in the ventral stream. In such a movement plan model, activity of the movement cells is required to produce a reentry signal. A potential problem could arise in explaining covert attention. During fixation, movement neurons might be inhibited by fixation cells and thus are presumably inactive, whereas visual neurons are not inhibited and therefore can provide both a reentry signal that modulates visual processes in extrastriate cortex and the target selection signal to the movement neurons. However, no experiment has clearly ruled out that the movement cells are inactive during covert attention. It is possible that fixation cell activity is reduced, which in turn allows movement cells to be active but below the level that elicits an eye movement. Others have proposed a visual selection model (Chapter 22). A potential problem of the visual selection model is its low signal-to-noise ratio. Although the visual cells show a target selection, distractor activity is initially almost equally strong. If these activities are directly fed back, spatial attention would be initially distributed to all stimuli. In addition, the target selection in visual cells appears very early as compared to the late occurrence of spatial attention in some psychophysical experiments. At present, it is not possible to rule out either model describing how the FEF might be involved in attention.

The model (Fig. 98.1) consists of the following components. Consistent with the idea of stimulus-driven salience (Chapter 39), a saliency module extracts features from the natural scene and weights their initial conspicuity by computing center-surround differences in parallel. Please note, such center-surround differences have been used by Itti and Koch (2000) to compute a saliency map. However, a saliency map that selects a location on a very fast time scale, which would be necessary to explain a popout perception on basis of a spatial selection, has not been found in the brain. Thus, according to principle 1, I suggest an alternative to an external saliency map: Stimulus-driven saliency emphasizes, but does not select, unique features in parallel within the ventral stream. For simplicity, the model computes the center-surround

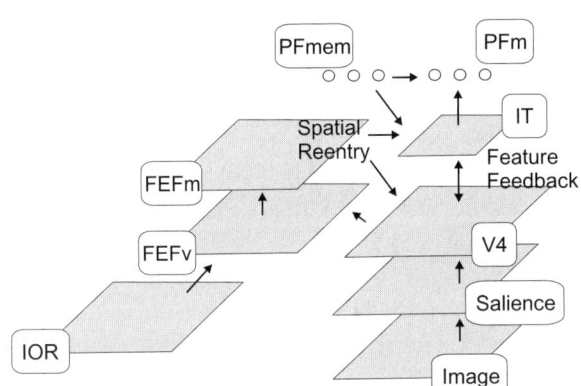

FIGURE 98.1 Model for top-down guided detection of objects. First, information about the content and its low-level stimulus-driven salience is extracted. This information is sent further upward to V4 and to IT cells that are broadly tuned to location. The target template is encoded in prefrontal memory cells (PFmem). Prefrontal match cells (PFm) indicate by comparison of PFmem with IT whether the target is actively encoded in IT. Feedback from PFmem to IT increases the strength of all features in IT matching the expected features. Feedback from IT to V4 sends the information about the target downward to cells with a higher spatial tuning. Frontal eye-field visuomovement cells (FEFv) combine the feature information across all dimensions and indicate salient or relevant locations in the scene. A competition among frontal eye-field movement cells (FEFm) determines the expected location of a target. Even during this competition, the movement cells provide a reentry signal to V4 and IT, which enhances the gain for all features at locations where the receptive field overlaps with the movement field. The inhibition of return (IOR) map memorizes recently visited locations and inhibits the FEFv cells.

differences prior to V4 (see also Chapters 45 and 93). The model is consistent with the idea that stimulus-driven conspicuity can be determined at higher levels as well (Hochstein and Ahissar, 2002). I combine the feature value with its corresponding conspicuity into a population code (Hamker and Worcester, 2002), which is then continuously modified. At each location x, I construct a space, whose axes are defined by the represented features and by one additional conspicuity axis. The encoded feature is then defined by the subset of active cells within a set of neurons i sampling the feature space. The present version computes five parallel channels: intensity, orientation, red-green opponency, blue-yellow opponency, and spatial frequency.

Each V4 layer receives the obtained features, weighted by the initial conspicuity value, as input. Feature-specific feedback from IT cells and spatial reentry from the frontal eye-field movement cells both control the gain of the bottom-up input. V4 cells compete in representing their encoded stimuli.

The populations from different locations in V4 project to IT, but only within the same channel. I simulate a map containing nine populations (sets of i neurons) with overlapping receptive fields. For simplicity, the complexity of features is not increased from V4 to IT. Thus, the model IT populations represent the same feature space as model V4 populations. The receptive field size, however, increases in the model so that several populations in V4 converge onto one population in IT. IT receives feature-specific feedback from the prefrontal memory and location-specific feedback from the frontal eye-field movement cells, which again control the gain. Principle 2 is implemented in the model by feedback from prefrontal memory to IT and further back to V4.

The FEFv neurons receive convergent afferents from V4 and IT and add up the activity across all channels. The information from the target template, in addition, enhances the locations that result in a match between target and encoded feature at all locations simultaneously. This allows the biasing of specific locations by the joint probability that the searched features are encoded at a certain location. The firing rate of FEFv cells represent the saliency and task relevance of location (Chapter 21), pooled over different channels, whereas the conspicuity of each feature is encoded in V4 and IT.

The effect of the FEFv cells on the FEFm cells is a feedforward excitation and surround inhibition. Thus, by increasing their activity slowly over time FEFm cells determine the expected location of the target. According to principle 3, the FEFm activity provides a delayed reentry signal to extrastriate areas.

There is currently no clear indication where cells that ensure an inhibition of return are located (chapter 16). We regard each location x as inspected, dependent on the selection of an eye movement or when a match in the PFm cells is lost. In this case, the inhibition of return (IOR) cells are charged at the location of the strongest FEFm cell for a period of time. This causes a suppression of the recently attended location in the FEFv map. IOR cells slowly decay.

I now show how the FEF and IT might contribute to the detection of an object in natural scenes.

III. RESULTS

I first demonstrate how the model operates in a free-viewing task, which is only driven by the stimulus saliency (Fig. 98.2). The overt scanning behavior is

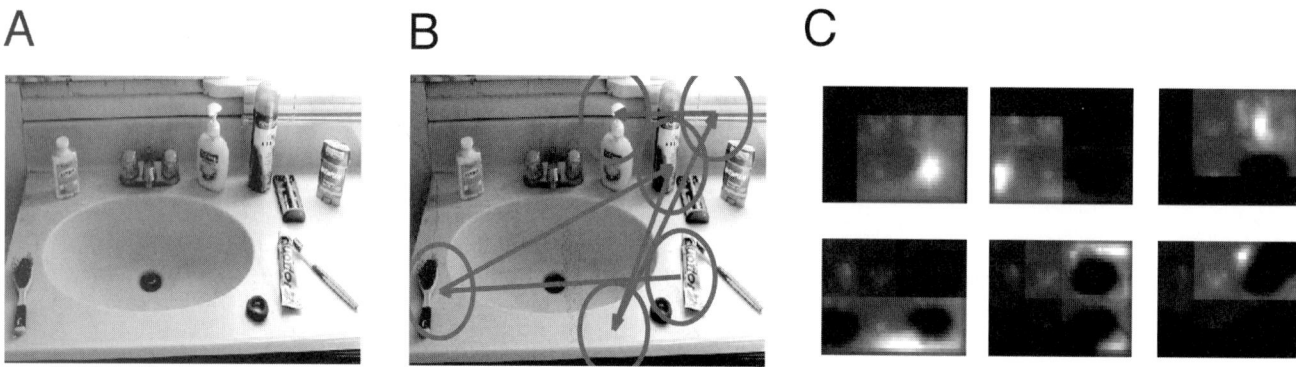

FIGURE 98.2 Results of a free-viewing task. (A) Natural scene. (B) Scanpath. The scan starts on the toothpaste and visits the hairbrush, the shaving cream, two salient edges, and then the soap. (C) Activity of FEFv cells prior to the next scan. By definition, they represent locations, which are actively processed in the V4 and IT map and, thus, represent possible target locations. An IOR map inhibits FEFv cells at locations that were recently visited (causing the black holes in the activity landscape).

similar to pure feedforward approaches (Chapter 94). The major difference is that the saliency is actively constructed within the network prior to each shift (Fig. 98.2C). I now demonstrate how the behavior of the model changes when it searches for a specific object in the scene. To mimic the activation of a search template, I present the model objects from which it generates very simple templates (Fig. 98.3A). This template, which is hold in PFmem cells, guides perception by changing the sensitivity of IT cells due to a feature-specific feedback. When presenting the search scene, initially IT cells reflect conspicuous features, but over time those features that match the target template get further enhanced (Fig. 98.3B). Thus, the features of the object of interest are enhanced prior to any spatial focus of attention. The frontal eye-field visual cells encode salient locations. At approximately 85–90ms, all areas that contain objects are processed in parallel. Spatial reentry then enhances all features at the selected location at approximately 110ms after scene onset. As a result, the initial top-down guided information is altered to process all the features of the target object. For example, the very red color of the asprin bottle is only encoded in IT after the emergence of the

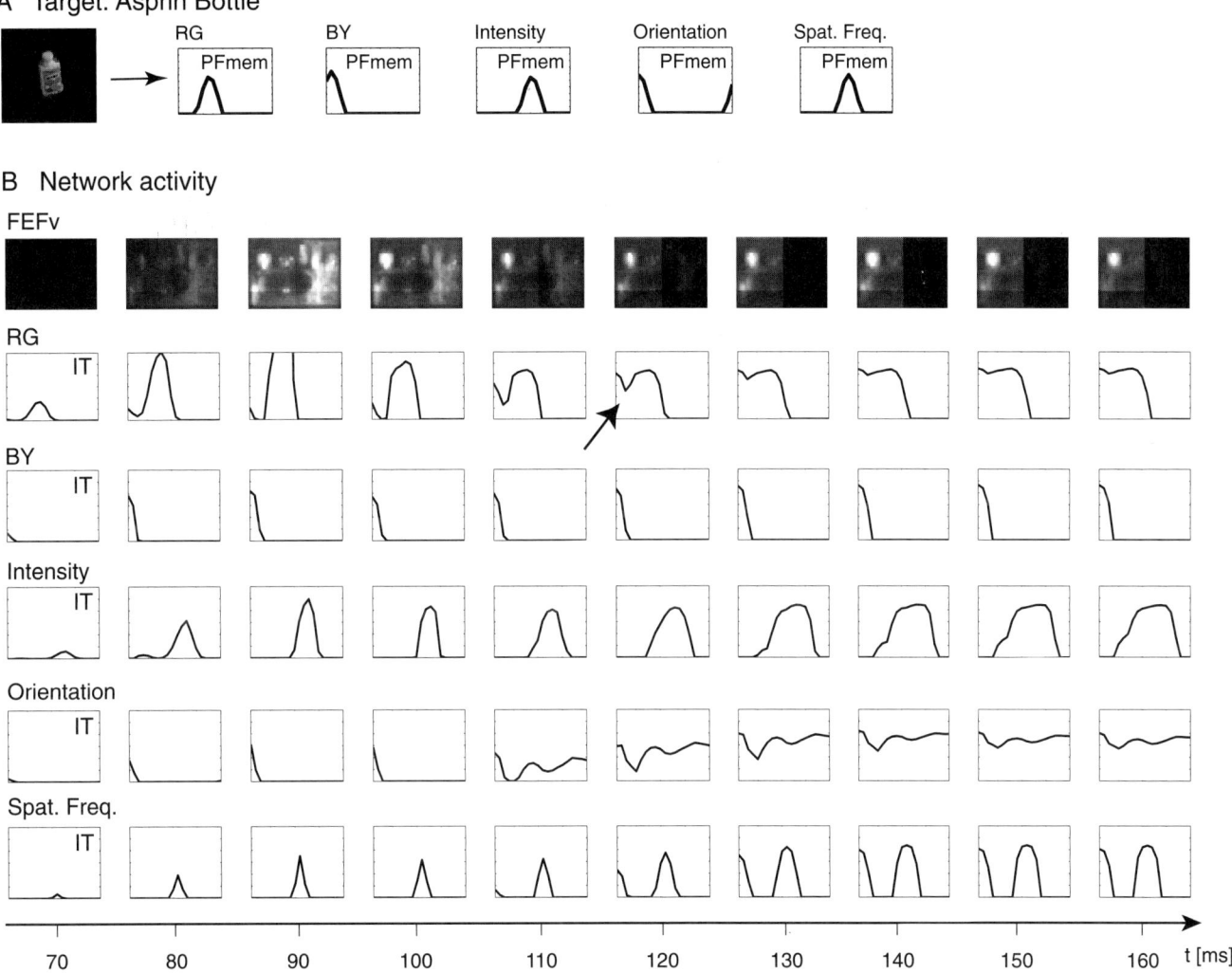

FIGURE 98.3 (A) After the presentation of a target object, PFmem cells memorize in each channel the most conspicous feature. (B) The temporal process of a goal-directed object detection task in a natural scene. The frontal eye-field visual cells indicate preferred processing, which is not identical with a spatial focus of attention. At first they reflect salient locations, whereas later they discriminate target from distractor locations. The activity of IT cell populations with a receptive field covering the target initially show activity that is inferred by the search template. Later activity is dominated by the emerging spatial focus and reflects other features of the object that were not searched for. The arrow indicates the enhancement of the cells encoding the red color due to spatial reentry.

spatial reentry signal because those features were not among of the expected features (arrow in Fig. 98.3B).

The model does not always search in parallel. If the target does not sufficiently discriminate from the background, reentrant processing can be misguided and the model automatically switches into a serial search mode.

IV. DISCUSSION

I have presented an approach to model perception in natural scenes based on three general principles of computation in the brain. (1) I postulate that high-level vision does not require spatial selection. This is consistent with the finding that some scenes allow the parallel detection of categories, such as animals, in the near absence of spatial attention (Li et al., 2002). (2) Because the spatial resolution of high-level vision is poor, the function of massive feedback projections is to provide cells in early location-specific areas information about the feature of interest. (3) Such enhanced activity in the "what" pathway is picked up by maps in the "where" pathway, which locates the object for action preparation. Reentrant activity, for example, from the FEF (Moore and Armstrong, 2003), then enhances all features of the object in order to allow a more detailed analysis. The model predicts that object identification begins before the eyes actually fixate on the object.

My model suggests that the brain uses reentrant processing to constrain processing in some areas by decisions in convergent areas. A match of the bottom-up input with the expectation increases the gain, which can alter the interpretation of a visual scene by tuning the population response. As a result, suppressive and facilitatory effects occur, commonly referred to as attention. My model is consistent with the idea of Biased Competition (Desimone and Duncan, 1995; Chapter 50), but population-based inference extends the idea of a mere competition toward the high-level guidance of low-level processing. The described mechanisms implement a dynamic filter that allows the connection of planning processes with the physical world and presumably the elaboration of the content of awareness (Chapter 29). Such an approach unifies recognition and attention as interdependent aspects of one network.

References

Chelazzi, L., Duncan, J., Miller, E. K., and Desimone, R. (1998). Responses of neurons in inferior temporal cortex during memory-guided visual search. *J. Neurophysiol.* **80**, 2918–2940.

Chelazzi, L., Miller, E. K., Duncan, J., and Desimone, R. (1993). A neural basis for visual search in inferior temporal cortex. *Nature* **363**, 345–347.

Desimone. R., and Duncan, J. (1995). Neural mechanisms of selective attention. *Annu. Rev. Neurosci.* **18**, 193–222.

Hamker, F. H. (1999). The role of feedback connections in task-driven visual search. *In* "Connectionist Models in Cognitive Neuroscience" (D. Heinke et al., Eds.) pp. 252–261. Springer Verlag, London.

Hamker, F. H. (2000). Distributed competition in directed attention. *In* "Proceedings in Artificial Intelligence" (G. Baratoff and H. Neumann, Eds.), Vol. 9, pp. 39–44. Akademische Verlagsgesellschaft, Berlin.

Hamker, F. H. (2001). Attention as a result of distributed competition. *Soc. Neurosci. Abstr.* **27**, 348.10.

Hamker, F. H. (2003). The reentry hypothesis: linking eye movements to visual perception. *J. Vis.* **11**, 808–816.

Hamker, F. H. (2004a). Predictions of a model of spatial attention using sum- and max-pooling functions. *Neurocomputing* **56C**, 329–343.

Hamker, F. H. (2004b). A dynamic model of how feature cues guide spatial attention. *Vis. Res.* **44**, 501–521.

Hamker, F. H., and Worcester, J. (2002). Object detection in natural scenes by feedback. *In* "Biologically Motivated Computer Vision" (H. H. Bülthoff et al., Eds.), pp. 398–407. Springer Verlag, New York.

Hochstein S, and Ahissar M. (2002). View from the top: hierarchies and reverse hierarchies in the visual system. *Neuron* **36**, 791–780.

Itti, L., and Koch, C. (2000). A saliency-based search mechanism for overt and covert shifts of visual attention. *Vis. Res.* **40**, 1489–1506.

Li, F.-F., VanRullen, R., Koch, C., and Perona, P. (2002). Rapid natural scene categorization in the near absence of attention. *Proc. Natl. Acad. Sci. U.S.A.* **99**, 9596–9601.

Moore, T., and Armstrong, K. M. (2003). Selective gating of visual signals by microstimulation of frontal cortex. *Nature* **421**, 370–373.

Reynolds, J.H., Chelazzi, L., and Desimone, R. (1999). Competetive mechanism subserve attention in macaque areas V2 and V4. *J. Neurosci.* **19**, 1736–1753.

Schall, J. D. (2002). The neural selection and control of saccades by the frontal eye field. *Phil. Trans. R. Soc. Lond. B* **357**, 1073–1082.

CHAPTER 99

Memory-Driven Visual Attention: An Emergent Behavior of Map-Seeking Circuits

David W. Arathorn

ABSTRACT

The term "attention," like the term "recognition," seems to convey a specific function until one tries to define a mechanism to perform it. By generalizing the attentional behavior which is an emergent property of map-seeking circuits, one obtains a computationally explicit definition of top-down driven attention as well as a neurally plausible mechanism to implement a broad range of the psychophysical behaviors to which that label is attached. Top-down or memory-driven attentional behavior, in this definition, consists of directly locating and segmenting the subjects within the visual field having a contextually valid mapping to a model contained in a primed memory within or across sensory modalities. Cross-modal priming of visual attention is demonstrated to be an inherent behavior of a more general visual mechanism as characterized by a map-seeking circuit.

I. INTRODUCTION

To begin the discussion, a simple distinction is made between bottom-up- and top-down-driven attentional processes. Bottom-up-driven attention relies on some simple visual property of a target within the visual field that distinguishes it from all, or most, of the other potential targets in the field. Top-down-driven attention relies on the ability of the visual system to segregate a target in the visual field that corresponds in some way to a representation already captured. As a convention here, we assume that representation resides in a neuronal structure that can be thought of as memory—hence, memory-driven attention.

The visual properties that can be exploited for bottom-up driven attention are very general, such as high contrast or motion, and their utility depends on their uniqueness in the current visual field. It is presumed that the early visual system implements a variety of filters sensitive to these properties to produce afferent signals that represent their presence and hence their assumed uniqueness in the scene directs the attention of the visual system to the location of the target possessing those properties. However, natural visual tasks often require discerning important targets that are embedded in a visually complex scene and that have no distinguishing low-level properties. Attention to these targets must be driven by a memory representation of the form or shape of the entirety (or at least large parts) of the target. Particularly in natural environments, it is the normal case that a target presents a different view to the observer than that which may reside in memory, so the visual mechanism must be able to establish the correspondence of the two under a wide range of visual transformations. In addition, in many circumstances it must be assumed that a number of different potential target types deserve attention, so the mechanism must be able to establish the correspondence to a number of memory candidates concurrently. Furthermore, those memory candidates may not be exclusively visual representations.

Memory-driven attention, therefore, consists of locating and segmenting the subjects within the visual field having a contextually valid mapping to the shape and/or other attributes of a representation contained in a primed memory representation within or across sensory modalities. Two classes of solution have been proposed to this problem: one class based on identifying the transformation and the other class based on invariant encodings. The second class has been thwarted by the lack of a method to construct robust invariants. Further, it does not consider the situation in which the memory contains a representation that is not visual but, for example, kinesthetic. The solution discussed here, the map-seeking circuit (Arathorn, 2002), belongs to the first class of solutions. It avoids the problem of finding anchor points by using the entire pattern as the basis of correspondence between targets and memory representations related by a composition of transformations. It is also capable of establishing a correspondence between a target and one of a number of memories concurrently.

II. MAP-SEEKING CIRCUITS

The map-seeking circuit, which is proposed as a generic cortical module, exploits the bidirectional flow of information through multiple layers of transformation. The transformations along the two pathways are inverse and reciprocal to one another. Multiple layers of transformation allow a wide range of aggregate transformations to be composed from a limited repertoire in each layer. To avoid combinatorial explosion in time or neuronal resource, the circuit exploits a mathematical ordering property of superpositions that permits all the transformations in each layer to be applied concurrently to produce a signal superposition from which the correct sequence of transformations is selected by a process of convergence. The convergence is fast enough to be a plausible mechanism for neuronal circuits.

A multilayer circuit is shown in Fig. 99.1. (Note: "Layer" in this context is not intended to imply anatomical equivalence to cortical layers.) As can be seen, mapping layers constitute the repeatable modules from which multilayer circuits are constructed. The circuit, whether in neuronal or algorithmic form, is constructed from a few basic operations—map, match, attenuate, superposition, compete, scaling, and nonlinearity—that as in the case of neuronal circuits, may be inherent in the implementation.

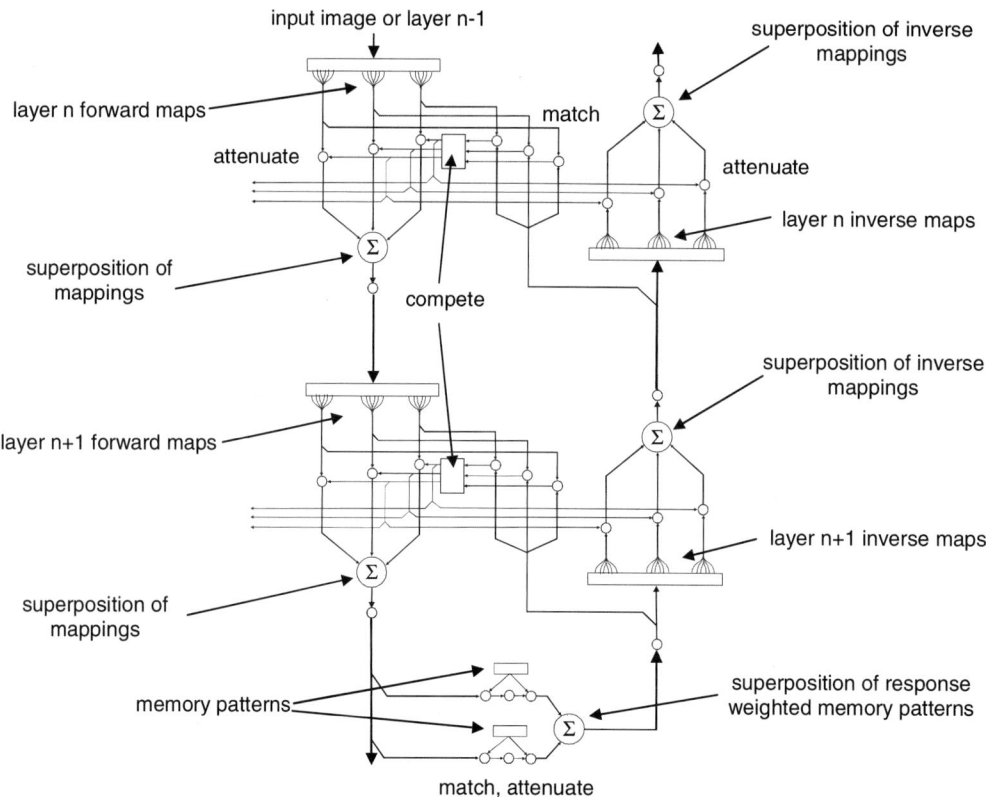

FIGURE 99.1 Map-seeking circuit with two mapping layers and memory layer. Basic operations indicated.

These general labels, as opposed to specific mathematical expressions, are used here because function of the circuit is not sensitive to the particular characteristics of these operations: Neuronal implementations with quite different closures from algorithmic implementations nevertheless behave almost identically. The superposition of all the active mappings in each layer provides the input to the next layer in both directions, and a superposition of all the active memories feeds the backward path. A matching operation occurs in the memory layer and in each mapping layer and produces a coefficient that controls the magnitude of each component added to the corresponding superposition. In each layer, every mapping in the forward path has a corresponding inverse mapping in the backward path, and each pair is controlled by the same coefficient. The convergence of the circuit occurs as a result of the competition between the mappings in each layer, gradually pushing all but one of the coefficients toward zero and leaving as a result one mapping contributing to the superposition in each layer. At the same time, competition between memories leaves one active memory contributing to the superposition driving the backward path. Thus, the fixed point leaves both the memory and transformations fully determined or, in the failure case, no mappings and no memories contributing (i.e., nonrecognition state).

Therefore, the inherent behavior of the map-seeking circuit is exactly what was defined earlier to be memory-driven attention: It finds a target in the visual field that corresponds through some set of composed transformations to one of a number of patterns stored in memory. The specific mappings selected convey important information independent of identity: the translation mapping locates the target in the visual field, the rotation and scaling mappings indicate approximate perspective. The parameters of 3D mappings indicate pose or orientation relative to the viewer.

III. ATTENTION SHIFTS

To complete the attentional behavior it is necessary to add some circuitry to shift attention within the field once a single target is recognized. This is achieved simply by subtracting the inverse mapping of the recognized memory pattern from the input field and allowing the circuit to recognize targets in whatever is left. This continues until there is no pattern in the input field that corresponds to the memories patterns driving the attentional process. Examples of this as performed both by neuronal and algorithmic circuitry can be seen in Arathorn (2002).

IV. CROSS-MODAL ATTENTION: FINDING 2D TARGETS USING 3D MEMORIES

A practical map-seeking circuit for recognizing 2D targets from a set of 2D memory patterns typically comprises three mapping layers: a translation mapping layer, a rotation mapping layer, and a scaling layer (although different combinations are possible). A circuit for recognizing 2D target images memories of another modality involves adding one more layer that maps from the 2D projection to the new representation. Examples of such recognition tasks are common: interpreting limb pose and visually recognizing a shape from a kinesthetic memory. In the example shown in Fig. 99.2, a 3D model of a pig is stored in memory, encoded as surface normals in three space. In neuronal terms, this can be accomplished by encoding each normal in a small nonlinear region of a space-filling dendrite. In this encoding, there is no visually recognizable set of edges in the 3D model to compare with the edge-filtered input image of the pig among distractors. Instead, the 2D image from the third layer, which has undergone translation, rotation, and scaling, is mapped by multiple 3D projections onto the forward path to the 3D model memory layer. The presence of an edge in the 2D image is transformed into an encoding of the line of view vector for a range of viewing angles. Each local region of the model responds independently to the superposition of the view vector encodings, the response being proportional to perpendicularity of the view vector and the local normal. The local responses of the memory layer are inversely mapped back into a superposition of 2D projections in the fourth layer. The matching between forward and backward signals takes place in the 2D domain, but yields the 3D projection of the model that best matches the 2D target view. The behavior of the circuit is thus to discover not just the translation, rotation in plane, and 2D scaling but also the correct 3D projection that matches the input image to the 3D model, as seen in Fig. 99.2E. The geometric parameters of the discovered mappings are:

L1: Y = 112, X = 111
L2: Rot = −13.000
L3: Scale = 1.275, ScaleYcor = 1.150, ScaleXcor = 1.200
L4: HorizAngle = −55.000, VerticalAngle = 10.000

As mathematical as this process sounds, it is entirely implementable in neuronally plausible, low-precision operations.

In this task, it is assumed that pigs, not cars or cats, deserve attention, and hence the pig memory is primed. As can be seen in Fig. 99.2, the circuit locates

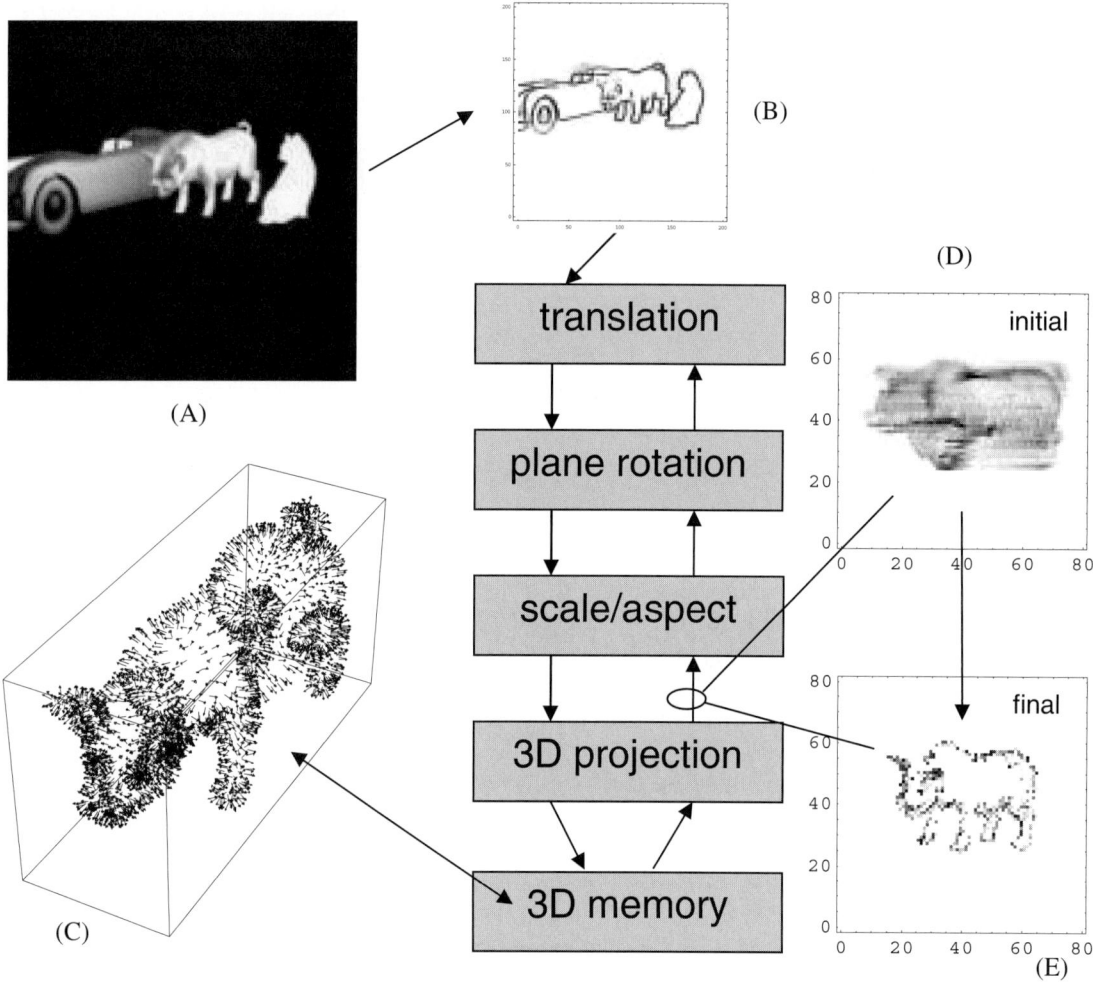

FIGURE 99.2 Locating 2D target using a 3D surface normal model. (A) Input raw image imput, (B) edge-filtered input image, (C) rendering of 3D surface normal memory pattern, (D) layer 4 backward superposition—initial iteration, and (E) layer 4 backward superposition—final iteration. The parameters of the winning mapping in each of four layers on final iteration provide target location (in plane rotation, scale, and aspect ratio) and 3D projection. Models courtesy 3DCafe.com

and identifies the pig among the other targets in the input image and determines its 3D orientation relative to the viewer. Many examples of recognizing targets under transformation and amid clutter from primed 2D memory representations can be seen in Arathorn (2002).

V. BIOLOGICAL EVIDENCE

What is the biological evidence for the existence of memory-driven attention? There is a substantial literature of evidence for top-down contribution to attention (see references in Chapter 25). A compelling demonstration is provided by Sheinberg and Logothetis (2001). Particular temporal cortical neurons in a macaque respond to highly specific, trained target images when the animal is searching for those targets embedded in a natural scene. There is no aspect of the trained patterns that is distinguishable by bottom-up processing alone. The cells exhibit a partial response when a saccade brings the target within a few degrees of centered, and this partial response immediately precedes the final saccade to target. This indicates (1) that the response precipitates attentional behavior when the target is within a sufficiently high-resolution radius of visual field and (2) that an accurate displacement signal is obtained during this penultimate saccade. The latter corresponds exactly to the information provided by a translational mapping as suggested here.

There is both neurophysiological and psychophysical evidence, such as Lee and Nguyen (2001), Supèr et al. (2001), and Hupé et al. (1998) for a cortical architecture consistent with map-seeking circuits, but not conclusive evidence, as discussed in Arathorn (2002). However, more recently new research has provided increasingly specific support. In particular, the retinotopic reciprocity of the fan-outs of feedforward and feedback pathways between V1 and a number of other visual cortical areas has been established by Angelucci et al. (2002). Murray et al. (2002) demonstrate a decrease in cortical activity on convergence to a recognition state, as seen in the neuronal map-seeking circuit models.

VI. RELATIONSHIP TO BOTTOM-UP ATTENTIONAL MECHANISMS

The inherent top-down attentional behavior of the mechanism proposed here does not preclude the presence of or need for bottom-up attentional mechanisms in a complete visual system. The geometries of human vision in the periphery compel the use of feature-based bottom-up attentional mechanisms to direct foveal vision onto target candidates for any but the nearest objects. A target at a distance of several hundred times its diameter is projected onto the human fovea at the lower limit of resolution capable of supporting practical recognition (Johnson, 1958). Even for closer targets projected at greater eccentricity, retinal resolution is too low to support shape-based matching. Although the map-seeking mechanism is capable of complex scene segmentation within the retinotopic area of high resolution, it still requires a cooperative bottom-up attentional mechanism to steer that high-resolution region with reasonable probability into the vicinity of, although not necessarily directly onto, interesting targets. The combination of mechanisms produces the search-and-leap behavior reported by Sheinberg and Logothetis (2001). The extensive list of feature types that appear to drive human bottom-up attention (Chapter 17) provides a wide repertoire for task-dependent search, but the length of the list raises the question of whether they are all hard-wired or whether at least some of them are the product of top-down-influenced configuration or tuning of a few filter primitives in the early visual cortices. Sigman and Gilbert (2000) provide some experimental evidence supporting this possibility. From a functional point of view, however, once configured for the attentional task at hand by top-down inputs, the operation is identical to a hard-wired bottom-up pre-attentive mechanism.

VII. CONCLUSION

Generalizing the attentional behavior inherent in map-seeking circuits leads to a computationally explicit definition: Memory-driven, or top-down, attention is the locating and segmenting of subjects within the visual field having a contextually valid mapping to one or more of the attributes of a model contained in a primed memory within or across sensory modalities. The fact that such behavior is inherent and is a mechanism capable of a broad range of visual task suggests that top-down attention may not be the function of a distinct system but is instead an emergent property of the general visual mechanism.

References

Angelucci, A., Levitt, J. B., Walton, E. J. S., Hupé, J-M., Bullier, J., and Lund, J. S. (2002). Circuits for local and global signal integration in primary visual cortex. *J. Neurosci.* **22**, 8633–8646.

Arathorn, D. W. (2002). "Map-Seeking Circuits in Visual Cognition." Stanford University Press, Palo Alto, CA.

Hupé, J. M., James, A. C., Payne, B. R., Lomber, S. G., Girard, P., and Bullier, J. (1998). Cortical feedback improves discrimination between figure and background by V1, V2, and V3 neurons, *Nature* **394**, 784–787.

Johnson, J. (1958). Analysis of image forming systems. In Image Intensifier Symposium, Fort Belvior, Virginia, October 6–7, pp. 249–273.

Lee, T. S., and Nguyen, M. (2001). Dynamics of subjective contour formation in the early visual cortex. *Proc. Nat. Acad. Sci. U.S.A.* **98**, 1907–1911.

Murray, S., Kersten, D., Olshausen, B., Schrater, P., and Woods, D. (2002). Shape perception reduces activity in human primary visual cortex, *Proc. Nat. Acad. Sci. U.S.A.* **99**, 15164–15169.

Sheinberg, D. L., and Logothetis, N. K. (2001). Noticing familiar objects in real world scenes: The role of temporal cortical neurons in natural vision. *J. Neurosci.* **21**, 1340–1350.

Sigman, M., and Gilbert, C. D. (2000). Learning to find a shape. *Nat. Neurosci.* **3**, 264–269.

Supèr, H., Spekreijse, H., and Lamme, V. A. F. (2001). Two distinct modes of sensory processing observed in monkey visual cortex (V1). *Nat. Neurosci.* **4**, 304–310.

Models courtesy 3DCafe.com

CHAPTER 100

The Role of Short-Term Memory in Visual Attention

Gustavo Deco and Edmund T. Rolls

ABSTRACT

We describe a computational neuroscience investigation of how a short-term memory in the prefrontal cortex could hold a rule in memory, and use this to provide the attentional bias in a biased competition model of how mapping from stimuli to responses can he switched by the context that is held in the short-term memory. The sensory stimuli are mapped through intermediate neurons responding to combinations of the sensory stimuli and the mapping required for motor responses, and the bias operates at the intermediate neuron level to switch the mapping. The mapping network consists of hierarchically organized sensory—intermediate—motor attractor pools of neurons. Each of the different global attractors of the whole mapping network corresponding to the different cue-response—context situations are composed of different and distributed single-pool attractors, and emerge from the cooperation and competition between the interacting single pools. The model is implemented using integrate-and-fire neurons to produce realistic spiking dynamics. The spiking in the model is similar to that of neurons recorded in the primate prefrontal cortex showing task-specific firing activity during working memory delay periods. The model thus implements a theory of rapid task switching using an attention-like mechanism, including a short-term memory in the prefrontal cortex.

I. INTRODUCTION

Complex context-dependent behavior can arise from the integration of sensory, memory, and motor processing in cognitive processes implemented in the prefrontal cortex (PFC). The aim of the research described here is to model, and therefore help to understand, the underlying mechanisms that implement the short-term-memory-related activity observed in neurons in the primate PFC in the context-dependent stimulus-response (associative) and delayed spatial response tasks investigated by (Asaad et al., 2000). The model builds on the integrate-and-fire attractor network treatment of Brunel and Wang (2001) by introducing a hierarchically organized set of different attractor network pools in the dorsolateral prefrontal cortex (PFC). The hierarchical structure is organized within the general framework of the biased competition model of attention (Duncan, 1996; Chelazzi, 1999; Rolls and Deco, 2002). The general approach was developed further by Rolls and Deco (2002), who introduced a neurodynamical theoretical framework for biased competition in the context of visual attention.

II. ATTENTION AND WORKING MEMORY

We explicitly model some of the working memory functions of the prefrontal cortex that are the source of the biased competition for the posterior perceptual areas in the parietal and temporal cortex (Chapter 97). Moreover, we use an approach to the dynamics in which the each neuron in the network is modeled at the integrate-and-fire level, so that we can produce spiking from the neurons in the network that can be directly compared with recordings from single neurons.

Working memory refers to an active system for maintaining and manipulating information in mind, held during a short period, usually of seconds. The prefrontal cortex is involved in at least some types of working memory, and neuronal activity in it continues during short-term memory periods (Fuster, 2000), Asaad et al. (2000) investigated the functions of the prefrontal cortex in working memory by analyzing neuronal activity when a monkey performs two different working memory tasks using the same stimuli and responses. In a conditional object-response (associative) task with a delay, the monkey was shown one of two stimuli and, after a delay, had to make either a right-ward or leftward oculomotor saccade response, depending on which stimulus was shown. In another experiment, recordings were made both during the object-response task and during a delayed spatial response task, in which the same stimuli were used, but the rule required was different, namely to respond toward the location where the stimulus had been shown (Asaad et al., 2000). The main motivation for such studies was the fact that for real-world behavior, the mapping between a stimulus and a response is typically more complicated than a one-to-one mapping. The same stimulus can lead to different behaviors depending on the situation, or the same behavior may be elicited by different cueing stimuli. In the performance of these tasks, prefrontal cortex neurons were found that respond in the delay period to the stimulus object, the stimulus position (sensory pools), combinations of the response and the stimulus object or position (intermediate pools), and the response required, left or right (premotor pools).

A. A Model of the Prefrontal Cortex

The model is designed to help understand the underlying mechanisms that implement the working-memory-related activity observed, in neurons in the primate PFC in the context-dependent stimulus-response (associative) and delayed spatial response tasks investigated by Asaad et al. (2000). We build on the integrate-and-fire attractor network treatment of Brunel and Wang (2001) and introduce a prefrontal cortex model with a hierarchically organized set of different attractor network pools. The hierarchical structure is organized within the general framework of the biased competition model of attention (Duncan, 1996; Chelazzi, 1999; Rolls and Deco, 2002). The operation of the neurons in the network is described in Sec. IIB.

Figure 100.1 shows schematically the synaptic structure assumed in the PFC network. The network is composed of N_E (excitatory) pyramidal cells and N_I inhibitory interneurons. In our simulations, we use $N_E = 1600$ and $N_I = 400$. The neurons are fully connected. There are different populations or pools of neurons in the PFC network. Each pool of excitatory cells contains fN_E neurons, where f, the fraction of the neurons in any one pool, was set at 0.05. There are four excitatory pools: sensory, task- or rule-specific, premotor, and nonselective. These correspond to the types of neuron recorded in the PFC during the performance of this task (Asaad et al., 2000). The sensory pools encode information about objects or spatial location. The premotor pools encode the motor response (in our case the leftward or rightward oculomotor saccade). The intermediate pools are task-specific and perform the mapping between the sensory stimuli and the required motor response. The intermediate pools respond to combinations of the sensory stimuli and the response required (e.g., to object 1 requiring a left oculomotor saccade). The intermediate pools receive an external biasing input that reflects the current rule (e.g., on this trial, when object 1 is shown make the left response after the delay period). The remaining excitatory neurons do not have specific sensory, response, or biasing inputs and are in a nonselective pool. All the inhibitory neurons are clustered into a common inhibitory pool, so that there is global competition throughout the network.

We assume that the synaptic coupling strengths between any two neurons in the network are established by Hebbian learning. As a consequence of this, neurons within a specific excitatory pool are mutually coupled with a strong weight $w_s = 2.1$. Neurons in the inhibitory pool are mutually connected with an intermediate weight $w = 1$. They are also connected with all excitatory neurons within the same intermediate weight $w = 1$. The connection strength between two neurons in two different specific excitatory pools is weak and given by $w_w = 1 - 2f(w_s - 1)/(1 - 2f) = 0.8778$ unless otherwise specified. (This functional form for w_w was chosen so that the spontaneous level of neuronal firing is largely unaffected by synaptic modifications.) Neurons in a specific excitatory pool are connected to neurons in the nonselective pool with a feedforward synaptic weight $w = 1$ and a feedback synaptic connection of w_w. The connections between the different pools are set up to achieve the required mapping from the sensory input pools through the intermediate pools to the premotor pools, assuming Hebbian learning based on the activity of individual pools while the different tasks are being performed. The forward connections (input to intermediate to output pools) are $w_s = 2.1$. The corresponding feedback synaptic connections are slightly weaker ($w_f = 1.7$ for the feedback synapses between rule-specific and sensory pools and w_w for the feedback synapses between the premotor and rule-specific pools).

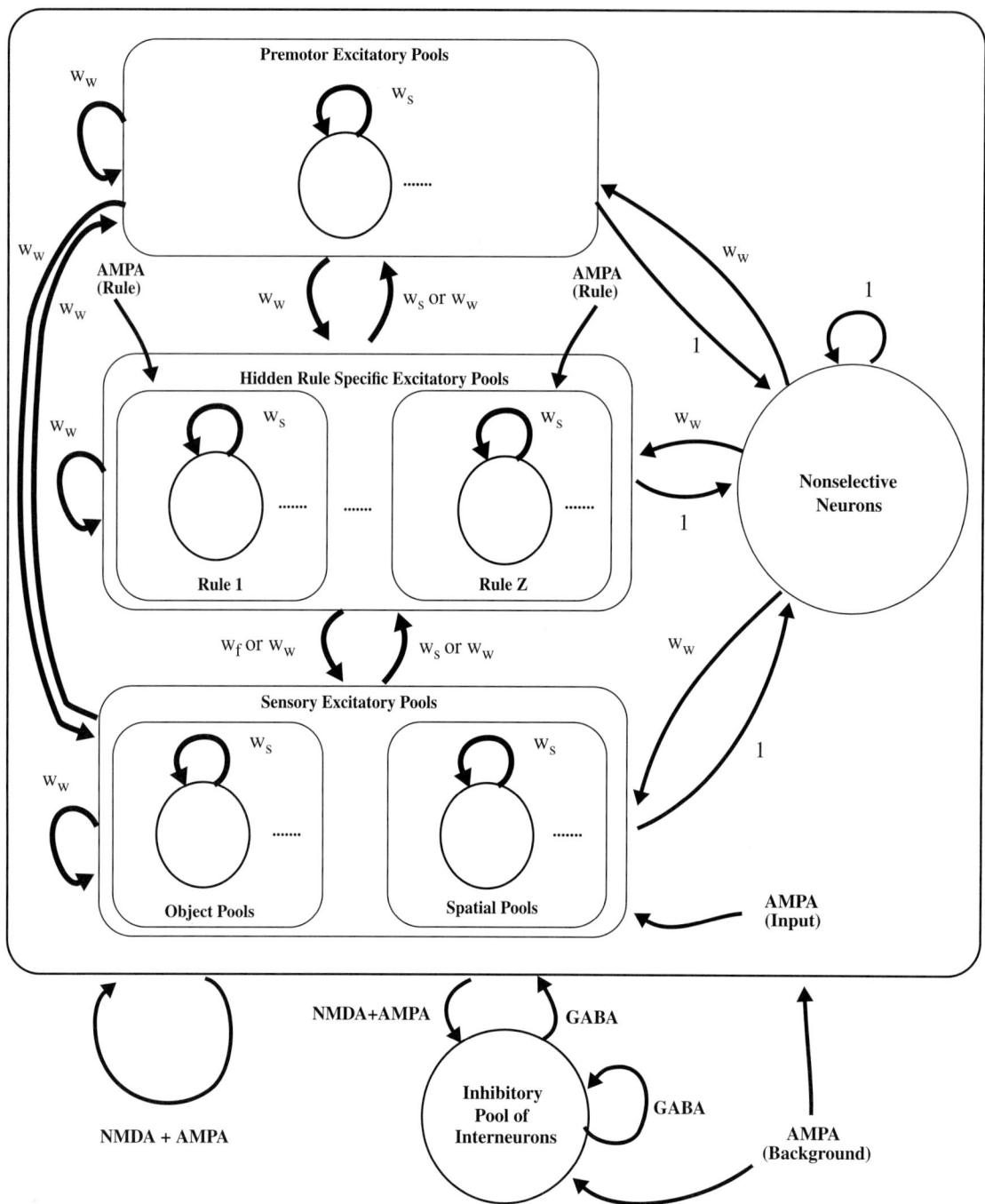

FIGURE 100.1 Prefrontal cortical module.

Each neuron (pyramidal cells and interneurons) receives $N_{ext} = 800$ excitatory AMPA synaptic connections from outside the network. These connections provide three types of external interactions: (1) a background noise due to the spontaneous firing activity of neurons outside the network, (2) a sensory related input, and (3) a rule or context-related bias input that specifies the task. The external inputs are given by a Poisson train of spikes. In order to model the background spontaneous activity of neurons in the network (Brunel and Wang, 2001), we assume that Poisson spikes arrive at each external synapse with a rate of $v_{ext} = 3\,\text{Hz}$, consistent with the spontaneous activity observed in the cerebral cortex. The sensory input is encoded by increasing the external input Poisson rate v_{ext} to $v_{ext} + \lambda_{input}$ to the neurons in the appropriate specific sensory pools. We used λ_{input} = 100 Hz. Finally, the biasing specification of the context (i.e., which rule is active) is modeled by assuming that each neuron in each of the pools in the group of intermediate pools associated with the active task receives external Poisson spikes with an increased rate from v_{ext} to $v_{ext} + \lambda_{rule}$ throughout the trial. We use λ_{rule} = 120 Hz. This external top-down rule-specific input probably comes from the external prefrontal neurons that directly encode abstract rules (Wallis et al., 2001), which in turn are influenced by the reward system (in the orbitofrontal cortex and amygdala) to enable the correct rule to be selected during for example reversal. During the last 100 ms of the response period, the external rate to all neurons is increased by a factor of 1.5 in order to take into account the increase in afferent inputs due to behavioral responses and reward signals (Brunel and Wang, 2001).

We performed numerical simulations of the experiment of Asaad et al. (2000) by means of a PFC architecture that includes two premotor pools of response neurons, one corresponding with leftward saccade responses (L) and the other corresponding to rightward saccade responses (R); four sensory pools (two object-selective neuronal pools, one corresponding to object O1 and the other corresponding to object O2; and two pools with selectivity for the spatial location of the stimulus, one corresponding to location S1 and the other corresponding to the location S2); and four intermediate neuronal pools, one for each of the four possible stimulus-response combinations. The intermediate pools are considered as being in two groups, one for the object-response associative task and the other for the delayed spatial response task. Figure 100.2 plots the temporal evolution of the averaged population activity for three neural pools, namely the premotor pool L, the intermediate spatial pool S1-L, and the intermediate associative pool O1-L. Cue, response, and selective context-specific associative activity is explicity maintained during the short-term memory delay period by the recurrent connections. The left panel corresponds to the delayed spatial response condition and the right panel to the conditional object-response associative task condition. Each graph shows two curves corresponding to the two possible response directions (the blue upper trace corresponds to L and the red to R). Figure 100.2A shows activity in the premotor pool L, which was response direction-selective in both tasks. B shows activity in the intermediate spatial pool S1-L, which was response-direction-selective (to the L, blue curve) in only the delayed spatial response task. C shows activity in the intermediate associative pool O1-L, which was direction-selective in only the conditional object-place associative task. All the three types of neurons found experimentally by Asaad et al. (2000) can be identified with pools in our prefrontal network. (See color plate.)

Figure 100.3 plots the rastergrams of randomly selected neurons for each neuronal population (or pool) in the network. The spatio-temporal spiking activity shows that during the short-term memory delay period only the relevant sensory cue, associated future oculomotor response, and intermediate neurons maintain persistent activity and build up a stable global attractor in the network. The underlying biased competition mechanisms are very explicit in this experiment. Note that neurons in pools for the irrelevant input sensory dimension (location for the object-response associative task, and object for the delayed spatial response task) are inhibited during the sensory cue period and are not sustained during the short-term memory delay period. Only the relevant single-pool attractors, given the rule context, that are suitable for the cue-response mapping survive the competition and are persistently maintained with high-firing activity during the short-term memory delay period. This suppression effect has been recently observed by recording the activity of prefrontal neurons in monkeys carrying out a focused attention task (Everling et al., 2002). In their spatial cueing task, they observed strong filtering of the PFC response to unattended targets. These attentional modulation effects are well-known in posterior areas of the visual system (Chelazzi, 1999). Our previous (Rolls and Deco, 2002) and present computational simulations suggest that, in the PFC, filtering of ignored inputs may reach a level commensurate with time strong global effects of selective attention in human behavior and that this selection in the PFC is the basis of the attentional modulation found in more posterior sensory cortical areas, implemented through backprojections from the prefrontal cortex to the more posterior cortical areas.

B. The Integrate-and-Fire Neurons in the Model

This section provides an overview of the level of implementation of the model of attention described here, to enable direct comparisons to be made with the spiking activity of neurons in the brain.

The basic circuit of an integrate-and-fire model consists of the cell membrane capacitance C_m, in parallel with the cell membrane resistance R_m driven by a synaptic current (excitatory or inhibitory postsynaptic potential, EPSP or IPSP). If the voltage across the capacitor reaches a threshold θ, the circuit is shunted and a δ-pulse (spike) is generated and transmitted to

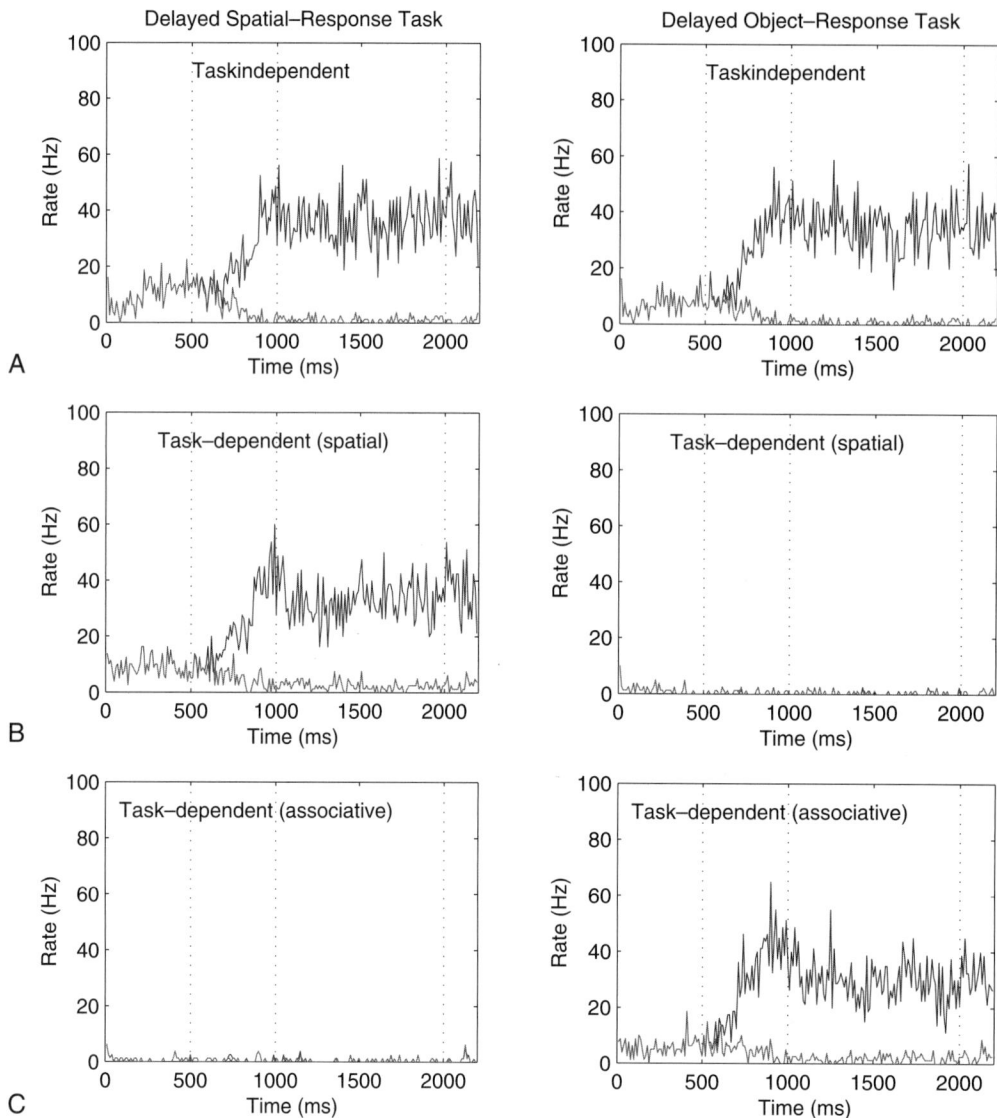

FIGURE 100.2 Simulation results corresponding to the experimental paradigm of Asaad et al. (2000). The temporal evolution of the averaged population activity for three neural pools during the execution of the object-response (associative) and delayed spatial response task. Cue, response, and selective context-specific associative activity is explicitly maintained during the short-term-memory-related delay period in the period 1000–2000 ms. The stimulus was shown in the period 500–1000 ms. The left panel corresponds to the delayed spatial response task condition and the right panel to the object-response (associative) task condition. Each picture plots two curves corresponding to the two possible responses (dark corresponds to L and light to R). (A) the premotor pool L, which was response direction-selective in both tasks. (B) the intermediate spatial pool S1-L, which was direction-selective (to the L, dark curve) in only the delayed spatial response task. (C) the intermediate associative pool O1-L, which was direction-selective in only the object-response (associative) task. (See color plate)

other neurons. The incoming presynaptic δ-pulse from other neurons is basically low-pass filtered first by the synaptic and membrane time constants before it is used as an EPSP or IPSP in the one-compartment neuronal model. We use biologically realistic parameters. We assume for both kinds of neuron a resting potential $V_L = -70$ mV, a firing threshold $\theta = -50$ mV, and a reset potential $V_{reset} = -55$ mV. The membrane capacitance C_m is 0.5 nF for the pyramidal neurons and 0.2 nF for the interneurons. The membrane leak conductance g_m is 25 nS for pyramidal cells and 20 nS for interneurons. The refractory period τ_{ref} is 2 ms for pyramidal cells and 1 ms for interneurons. Consequently, the membrane time constant $\tau_m = C_m/g_m$ is 20 ms for pyram-

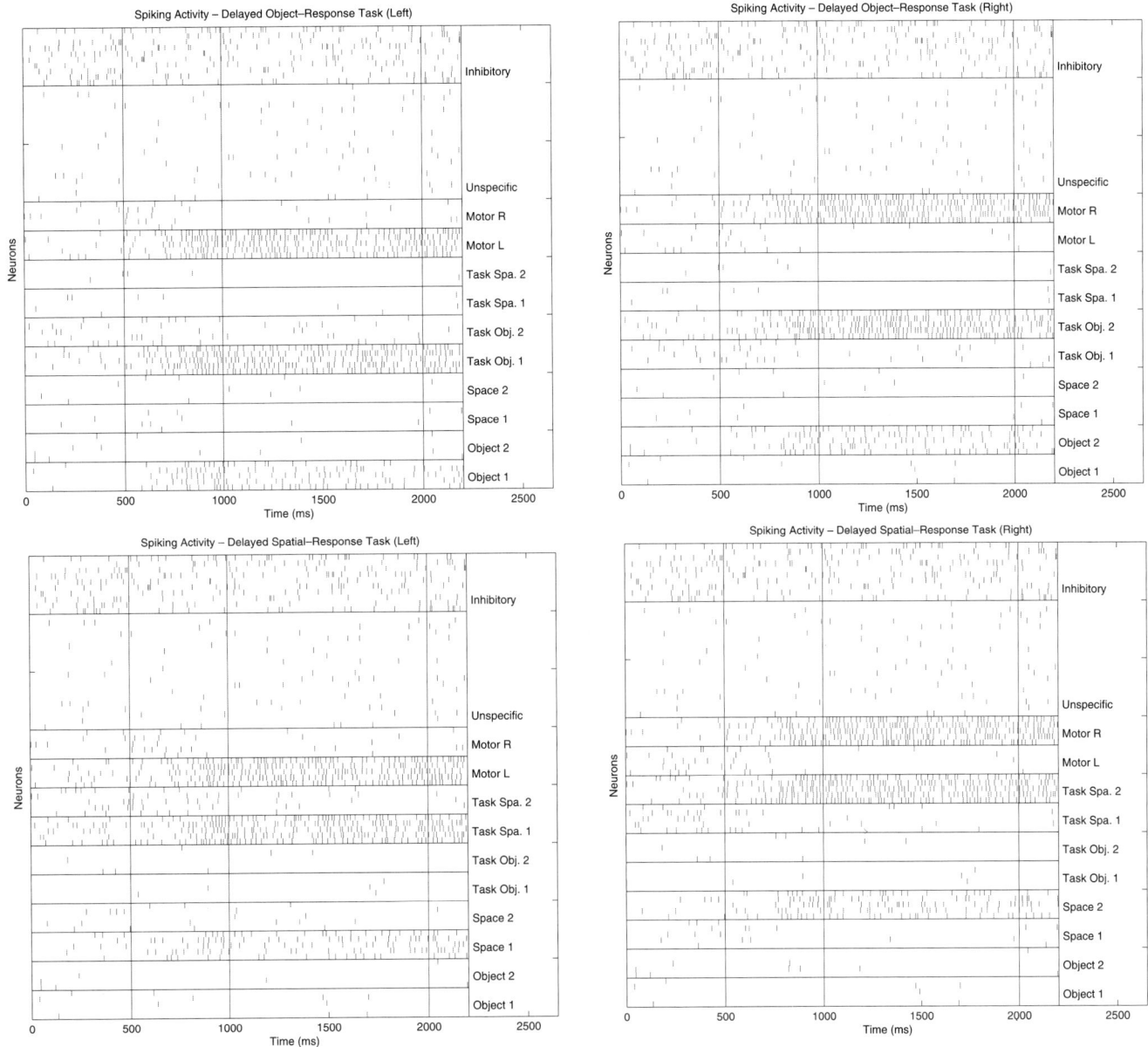

FIGURE 100.3 Rastergrams of randomly selected neurons for each pool in the PFC network (five for each sensory, intermediate, and premotor pool, 20 for the nonselective (unspecific) excitatory pool, and 10 for the inhibitory pool) and for all task conditions after the experimental paradigm of Asaad et al. (2000). The spatiotemporal spiking activity shows that during the short-term-memory delay period only the sensory cue (e.g., object 1 and space 1) associated future oculomotor response (e.g., Motor R), and intermediate (e.g., Task Obj 1) neurons maintain persistent activity and build up the stable global attractor of the network. The underlying biased competition mechanisms are explicit (see Deco amid Rolls, 2003).

idal cells and 10 ms for interneurons. More specifically, the subthreshold membrane potential V(t) of each neuron evolves according to time following equation:

$$C_m \frac{dV(t)}{dt} = -g_m(V(t) - V_L) - I_{syn}(t) \quad (1)$$

where $I_{syn}(t)$ is the total synaptic current flow into the cell.

The synaptic current flows into the cells are mediated by three distinct families of receptors. The total synaptic current is given by the sum of glutamatergic excitatory components (NMDA and AMPA) and inhibitory components (GABA, I_G). We consider that external excitatory contributions are produced through AMPA receptors (I_{Ae}) and that the excitatory recurrent synapses are produced through AMPA and

NMDA receptors (I_A and I_N). The total synaptic current is therefore given by:

$$I_{syn}(t) = I_{Ae}(t) + I_A(t) + I_N(t) + I_G(t) \quad (2)$$

where

$$I_{Ae}(t) = g_{Ae}(V(t) - V_E) \sum_{j=1}^{N_e} s_j^{Ae}(t) \quad (3)$$

$$I_A(t) = g_A(V(t) - V_E) \sum_{j=1}^{N_E} w_j s_j^A(t) \quad (4)$$

$$I_N(t) = \frac{g_N(V(t) - V_E)}{\left(1 + C_{Mg^{2+}} + e^{\left(\frac{-0.062V(t)}{3.57}\right)}\right)} \sum_{j=1}^{N_E} w_j s_j^N(t) \quad (5)$$

$$I_G(t) = g_G(V(t) - V_I) \sum_{j=1}^{N_I} s_j^G(t) \quad (6)$$

In the preceding equations, $V_E = 0\,\text{mV}$ and $V_I = -70\,\text{mV}$. The fractions of open channels s are given by:

$$\frac{ds_j^{Ae}(t)}{dt} = -\frac{s_j^{Ae}(t)}{\tau_A} + \sum_k \delta(t - t_j^k) \quad (7)$$

$$\frac{ds_j^A(t)}{dt} = -\frac{s_j^A(t)}{\tau_A} + \sum_k \delta(t - t_j^k) \quad (8)$$

$$\frac{ds_j^N(t)}{dt} = -\frac{s_j^N(t)}{\tau_{N,d}} + \alpha x_j(t)(1 - s_j^N(t)) \quad (9)$$

$$\frac{dx_j(t)}{dt} = -\frac{x_j(t)}{\tau_{N,r}} + \sum_k \delta(t - t_j^k) \quad (10)$$

$$\frac{ds_j^G(t)}{dt} = -\frac{s_j^G(t)}{\tau_G} + \sum_k \delta(t - t_j^k) \quad (11)$$

where the sums over k represent a sum over spikes emitted by presynaptic neuron j at time t_j^k. The value of $\alpha = 0.5\,\text{ms}^{-1}$.

The values of the conductances (in nanosiemens) for pyramidal neurons were $g_{Ae} = 2.08$, $g_A = 0.052$, $g_N = 0.164$, and $g_G = 1.13$ and for interneurons were g_{Ae} 1.62, $g_A = 0.0405$, $g_N = 0.129$, and $g_G = 0.87$. We consider that the NMDA currents have a voltage dependence that is controlled by the extracellular magnesium concentration, $[Mg^{2+}] = 1\,\text{mM}$. We neglect the rise time of both AMPA and GABA synaptic currents because they are typically extremely short (<1 ms). The rise time for NMDA synapses is $\tau_{N,r} = 2\,\text{ms}$. The decay time for AMPA synapses is τ_A 2 ms, for NMDA synapses $\tau_{N,d} = 100\,\text{ms}$, and for GABA synapses $\tau_G = 10\,\text{ms}$.

III. CONCLUSION

We have reviewed a detailed theoretical neurodynamical analysis of the spiking and synaptic mechanisms underlying behavior that requires complex context-dependent mapping between sensory stimuli and actions (Deco and Rolls, 2003). Our neurodynamical architecture of the PFC unifies attentionally biased competitive mechanisms and recurrent excitatory mechanisms that support short-term memory-related neuronal activity. Overall, the network has the architecture of a single-attractor network with multiple activated populations or pools of neurons. These different pools engage in competitive interactions, are organized with some hierarchy imposed by the asymmetrically strong forward and backward connections, and receive biasing inputs to influence the relative activity of the different pools, thus implementing attention-based or rule-based mapping from sensory inputs to motor outputs. Even more, the integrate—and fire implementation of the network enables us to make explicit predictions of the effect of neuromodulation by manipulation of the dopamine level on the conditional object-response and delayed spatial-response tasks. A decrease in NMDA-related conductances produced by an increase in D2 receptor activation or a decrease in Dl receptor activation weakens and shortens the persistent neuronal activity in short-term memory periods. In addition, we predict that the same pharmacological manipulations produce more response errors as a consequence of the more similar level of neuronal firing in the competing neuronal pools.

In summary, computational neuroscience provides a mathematical framework for studying time mechanisms involved in brain function, such as visual attentional mechanisms. The neurodynamical model analyzed here is based on evidence from neurophysiological and psychological findings. The simulations obtained with this theoretical model successfully reproduce the results of neurophysiological and fMRI experiments on spatial attention, as well as studies on serial and parallel search and on neuropsychological phenomena such as object-based visual neglect that can follow brain damage. The simulations also make predictions, for example about the effects of pharmacological agents on short-term memory and attention.

Acknowledgments

Support for this research was contributed by ICREA and the Medical Research Council.

References

Asaad, W., Rainer, G., and Miller, E. (2000). Task-specific neural activity in the primate prefrontal cortex. *J. Neurophysiol.* **84**, 451–459.

Brunel, N., and Wang, X. (2001). Effects of neuromodulation in a cortical networks model of object working memory dominated by recurrent inhibition. *J. Computational Neurosci.* **11**, 63–85.

Chelazzi, L. (1999). Serial attention mechanisms in visual search: A critical look at the evidence. *Psychol. Res.* **62**, 195–219.

Deco, G., and Rolls, E. T. (2003). Attention and working memory: A dynamical model of neuronal activity in the prefrontal cortex. *Eur. J. Neurosci.* **18**, 2374–2390.

Duncan, J. (1996). Cooperating brain systems in selective perception and action. *In* "Attention and Performance XVI" (T. Inui and J. L. McClelland, Eds.), MIT Press, Cambridge, MA.

Everling, S., Tinsley, C., Gaffan, D., and Duncan, J. (2002). Filtering of neural signals by focussed attention in the monkey prefrontal cortex. *Nat. Neurosci.* **5**, 671–676.

Fuster, J. (2000). Executive frontal functions. *Exp. Brain. Res.* **133**, 66–70.

Rolls, E. T., and Deco, G. (2002). "Computational Neuroscience of Vision." Oxford University Press, Oxford.

Wallis, J., Anderson, K., and Miller, E. (2001). Single neurons in prefrontal cortex code abstract rules. *Nature* **411**, 953–956.

CHAPTER
101

Scene Segmentation through Synchronization

Günther Palm and Andreas Knoblauch

ABSTRACT

Even when creating a biologically realistic model for an apparently very simple cognitive task such as seeking a certain object in the visual field, we are confronted with severe problems concerning the binding of distributed representations. In this work, we present simulation results from a model of two reciprocally coupled visual cortical areas. One area is a peripheral visual area where local object features are represented; the other is a more central visual area where whole objects are recognized. In our model, correct binding is achieved by the simultaneous switching of the activation state of corresponding neuron groups. We relate our simulations to neurophysiological findings concerning attention and biased competition and demonstrate how these findings can be explained very naturally by assuming different kinds of bindings between neuron groups in different areas as produced by our model. Although the binding is fluctuating in the absence of attention, it becomes static by the attentional bias.

I. INTRODUCTION

The problem of scene segmentation of complex visual images containing many objects close to one another and sometimes partially overlapping is still one of the hardest problems in computer vision. One idea to solve this problem originates from experimental observations in the visual cortex: Visual feature detectors of features belonging to the same object synchronize their temporal activity (Singer and Gray, 1995). The question is: How this synchronization can be achieved. This is shown in our model.

We have modeled the interaction of two visual cortical areas in order to get a better understanding of a number of experimentally observed neural response patterns in primary, or low-level, visual cortical areas, which may have to do with the perceptual interpretation of visual scenes, in particular also with local competition of different interpretations and with attention biasing this competition. The guiding idea behind these simulations is that perception is not a one-way feedforward process from the sense organs to some higher brain centers that perform the semantic categorization of the input; it rather is a feedback matching process in which high-level categorical expectations are matched with the sensory input, the matching being realized by a kind of resonance in the mutual activation of neural patterns in the primary and the corresponding higher cortical area. This idea is probably quite plausible in itself; it is also motivated by the cell assembly theory of brain function that was initiated by Donald Hebb (1949) and taken up by many brain scientists (e.g., Palm, 1982). Similar ideas have also been formulated under different names (Marr, 1970; Grossberg, 1980; Abeles, 1982).

From the experimental neurophysiological point of view, the primary visual cortices in cats and monkeys seemed to be well understood in rather fine detail due to many experiments in the 1960s and 1970s. The Nobel Prize had been awarded to Hubel and Wiesel for their achievements in this understanding. Meanwhile, an increasing number of experiments have shown that this picture is not so simple. For example, most neurons even in V1 and V2 of the macaque monkey do react to stimuli outside their classical receptive fields; they can even react to illusory contours (von der Heydt et al., 1984). More generally, there appears to be a substantial influence of high-level processes, such as gestalt principles, perceptual switching, or attention in the appropriate experimental conditions. The feedback from higher cortical areas to the lower areas appears to be the most natural explanation for these effects.

A more theoretical motivation for our simulations was to study the superposition problem. Can we activate several assemblies at the same time (i.e., their superposition) and still tell them apart? Strictly speaking, we cannot, and there should always be only one assembly active at the same time (Palm, 1982), but do we not need this when we perceive several objects in a visual scene at the same time? The answer to this puzzle may lie in a closer look at the time scales involved. It could be that within 0.5 or 1s we see several things in an image, but at a time scale of milliseconds, or perhaps even up to 100 ms, we activate only one assembly at a time. This is what actually happens in our simulations.

A version of this solution to the binding problem was called the temporal correlation hypothesis (TCH) by von der Malsburg (1986). The TCH was experimentally supported by the observation of synchronized high-frequency gamma oscillations (40–60 Hz) in the visual cortex, apparently reflecting global stimulus properties (Eckhorn et al., 1988; Singer and Gray, 1995). Although some methodological doubts have been raised with respect to the ubiquity and significance of the oscillations (e.g., Ghose and Freeman, 1992), the basic findings could be reproduced and more evidence for this oscillatory activity was found in EEG or local field potential recordings and also in single-unit auto- or cross-correlograms (see also Chapter 87). Although the general idea of interpreting coincidence in assemblies on the time scale of milliseconds or spikes has been pursued by many if not most researchers elaborating the Hebbian assembly concept in the temporal domain (Abeles, 1982; von der Malsburg, 1986; Palm, 1990; Ritz et al., 1994; Chapter 87), the special use of fine timing made in the context of the visual binding problem, also related to the idea of phase coding (activating different assemblies at different phases of a single underlying oscillation), has been subject to considerable controversies (Shadlen and Movshon, 1999; Knoblauch and Palm, 2002a, 2002b).

The main purpose of this paper is to show that the original Hebbian ideas and the corresponding models based on associative memory almost automatically and naturally solve the binding problem by temporal correlation when they are implemented in biologically realistic networks of spiking neurons.

II. MODELING

For simulations, we used spiking neurons similar to common integrate-and-fire models, including refractoriness and habituation. The network consists of three areas—retinal (R), primary visual contical (P), and central associative (C)—each composed of several neuron populations.

In R, input patterns corresponding to stimulus objects in the visual field are represented in a 100×100 bitmap. The two cortical areas P and C are both modeled with one excitatory and two inhibitory neuron populations, where only one of them receives extra-areal input. Only the excitatory connections are specific according to Hebbian learning. This architecture was motivated by the requirement of an efficient threshold control, especially for C, and is also biologically plausible.

Each neuron population in area P has size 100×100. Connections from R and inside P are modeled corresponding to the subsystem of orientation selective columns in the primary visual cortex. The internal excitatory connections in P couple specifically neurons that have similar orientation preferences and in addition, are near-neighbored or collinearly aligned. This is as expected from Hebbian correlation learning during the presence of stimuli rich of contours and results in a patchy representation of stimulus objects. At a certain location in P, only the neurons with orientation preferences best matching the pattern in R get strongly activated while neighboring cells are suppressed by recurrent inhibition.

Area C (size 40×40) is modeled as a fully connected auto-associative network, according to Hebbian learning of topographic random representations of stimulus objects. Areas P and C are reciprocally connected by a topographic heteroassociation, according to Hebbian learning of corresponding stimulus representations in P and C.

We use simple shapes as test stimuli (e.g., a triangle, a rectangle, or an ellipse). To investigate more global model properties we simulated also a larger variant of the model in which also area C is organized topographically. Details of the neuron and network model are described in Knoblauch and Palm (2002a, 2002b).

III. SIMULATION RESULTS

A. Segmentation Trough Synchronization

TCH suggests synchronous fast (gamma) oscillations to solve the binding problem. In this section, we test our model using a superposition of three test stimuli; a triangle, an ellipse, and a rectangle. Here the representation of each stimulus in the primary visual area P is a certain set of orientation-selective patches and the binding problem coordinates the activity of patches that belong to one representation in area P as well as corresponding representations in areas P and C.

Figure 101.1 shows simultaneous recordings of spike activity from neurons in areas P and C. Multiunit spike activity is depicted separately for the different (sub)assemblies representing the three stimuli (triangle, ellipse, and rectangle). The recordings of one assembly in C (Fig. 101.1A) show periods of fast oscillatory activity lasting for a few hundred milliseconds alternating with longer periods of essential silence. Comparing the three recordings from C, we observe that only one assembly is strongly activated at a time.

Multiunit activity from different patches belonging to the three assemblies in area P reveal similar results (Fig. 101.1B) to area C. The recordings show periods of fast and precise (little phase jitter) oscillatory activity (fast state) lasting again for a few hundred milliseconds, alternating with periods of relatively slow and unordered activity (slow state). When looking at different patches of one assembly we observe that activity is synchronized only during the fast state but asynchronous during the slow state. Moreover, corresponding assemblies in P and C are simultaneously either in the fast or in the slow state.

Although on the level of mean firing rates, the three assemblies appear to be activated at the same time, correlation analysis of spike records reveals synchronization (central peaks) on a fine time scale (a few milliseconds) *within* each assembly, whereas activity from different assemblies remains uncorrelated (flat correlograms). This is consistent with findings of specifically peaked and flat correlograms in neurophysiological experiments investigating the representation of multiple objects (Singer and Gray, 1995), whereas common phase-coding models (Ritz et al., 1994) would instead predict shifted peaks in the correlations of different assemblies.

In addition to narrow peaks on the fine time scale, we observed in the correlograms broader peaks similar to the C(astle)-peaks found in neurophysiological recordings (Nelson et al., 1992), indicating a second type of nonoscillatory synchronization on a larger time scale. The widths of the broad peaks depend on the duration of the fast state of oscillatory activity, and the switching between the slow and fast state is obviously due to habituation of activated neurons. In further simulations (Knoblauch and Palm, 2002b), we manipulated the duration of the fast state by varying the habituation parameters of the neurons. For strong habituation we obtained durations down to 20–50 ms

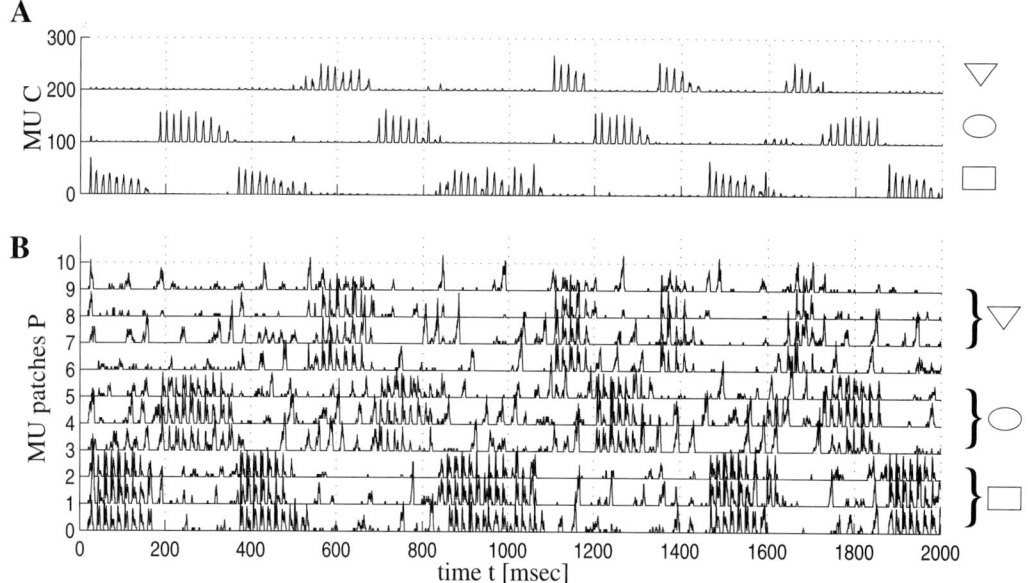

FIGURE 101.1 Analysis of multiunit (MU) spike activity in the model of two reciprocally connected areas when stimulating with a superposition of three stimuli (triangle, ellipse, and rectangle). (A) MU spike activity in central visual area C for the triangle (top row), ellipse (middle row), and rectangle assembly (bottom row). (B) MU spike records from 10 different activated locations in primary visual area P representing the triangle (rows 6–9), the ellipse (rows 3–5), and the rectangle (rows 0–2). (Modified from Knoblauch and Palm 2002a.)

where the enhanced period soften degenerated to a single retrieval cycle. For weak habituation the duration of the fast state periods increased up to seconds, and could result in the sustained activation of a single assembly without switching between different assemblies.

The correlograms found in neurophysiological experiments were either flat or showed central peaks, whereas oscillatory modulations or central troughs were observed rarely. Further simulations showed that these results can also be reproduced by our model if many assemblies are activated at the same time. A possible conclusion could be that the real brain is actually in a many-activated-assemblies regime even without stimulation, where ongoing activity wanders through all the local assemblies (Tsodyks et al., 1999). Indeed, the behavior of our model without stimulation but with enhanced noise level is very similar to the case of many simultaneously presented stimuli. In this view, a stimulus causes the corresponding assembly, spending more time in the fast state than the other assemblies. This is in line with recent findings of stimulus-dependent two-state fluctuations of membrane potentials (Anderson et al., 2000) and could also explain the neurophysiological results described in the next section.

B. Attention and Biased Competition

The results from the previous sections suggest an involvement of attentional processes in scene segmentation. The prevalence of one of the three stimulus objects was switched on a time scale of tens to hundreds of milliseconds (fast and slow state). This may be interpreted as self-generated attention switching serially from one object to the next (Treisman, 1998). For the following simulations, we modeled a top-down attentional bias explicitly as additional tonic excitation to the neurons of one selected assembly in area C. In order to create a scenario as in experiments by Reynolds and Desimone (1999; see also Chapters 50 and 86) we stored the two shapes shown in Fig. 101.2 in the memory. In this scenario, two stimuli in the receptive field of a single neuron were used, a preferred and a poor stimulus. When only the preferred stimulus is presented the neuron exhibits a strong response, whereas if only the poor stimulus is used the response is weak. Interestingly, for the superposition of the two stimuli, the response lies between the responses for single stimuli, indicating competition between the two stimuli. If attention is directed to one of the stimuli, the neuron responds as if only the attended stimulus were present. Thus, the effect of attention is similar to a filter that eliminates or weakens unattended stimuli. Another neurophysiological result reproduced by our model is that the described effect scales with the receptive field size; that is, the effect was stronger in higher visual areas (area C, or, e.g., V4 or IT) and weaker in lower visual areas (area P, e.g., V1 or V2).

Our model explains the competition and the filter property by two binding modes corresponding to the two activation states (fast and slow states). For one stimulus, the binding of the representations in the two areas is static, whereas for two competing stimuli,

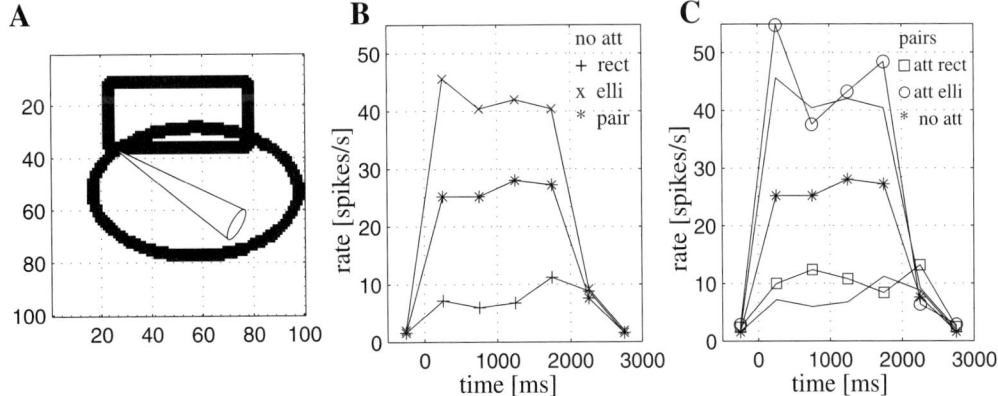

FIGURE 101.2 Effects of attention on single unit (SU) responses (PSTH) when stimulating with one or two objects. (A) Stimulus configuration (rectangle and ellipse). (B) SU response (PSTH) in the absence of attention when stimulating with the rectangle (+), the ellipse (×), and both the rectangle and ellipse (*). (C) SU response when stimulating with both the rectangle and ellipse. Results without attention (*) corresponds to A. If attention is directed to the rectangle (□) or ellipse (○), the neuron responds as if the nonattended stimuus were absent. PSTHs (bin size 500 ms) were computed from five trials of length 3s (where stimulus was active from 0–2s). (Modified from Knoblauch and Palm, 2002b.)

binding fluctuates. With the bias of attention, binding becomes static again.

C. Integration on Larger Space and Time Scales

Our model should be interpreted as the bidirectional connection of two small patches of cortex. In the real-visual system, many such structures are arranged in parallel. To account for more global interactions we simulated a model variant with larger areas (area P, size 30×500; area C, size 15×250), where area C is modeled as a topographic associative memory (Knoblauch and Palm, 2002b) with only local internal connections. As stimuli we used three long bars that extended over the whole modeled areas and were represented by three learned assemblies in area C. Fig. 101.3 shows results from a simulation of the model stimulated with all three bars simultaneously. At time $t = 4$ sec attention was directed for three seconds to the upper bar, while attention was absent in the rest of the simulation.

When attention is directed to one of the three bars, the whole corresponding assembly enters the fast state, similar to when only the attended bar is present in the visual field. In addition, the globally synchronized oscillations of the smaller model (Fig. 101.1) mutate to waves of activity moving quickly along the bar. At any given time, the oscillation phase is a continuous function of the location along the bar representation (Fig. 101.3B) reflecting the global shape of the stimulus. However, fast activity in P and C is synchronized only locally, corresponding to the feedback range of intra-areal and inter-areal connections, and correlation analysis reveals that the central peaks of (long-term) correlograms decay with distance (Knoblauch and Palm, 2002b). Moreover, short-term correlograms (time window 50 ms) show that the central peaks of the long-term correlograms result mainly from averaging over phase shifts that increase with distance from the recording sites. A similar effect can also be observed for the broad peaks.

Without attention, the representations of the three bars compete in the higher area C. At a fixed location the situation is very similar to the local model (Fig. 101.1). The assembly corresponding to the most salient stimulus wins and enters the fast state. After some time (e.g., some hundred milliseconds), the assembly habituates and returns to the slow state, so that another assembly can enter the fast state. However, for the global model it turns out that the assembly switching is a rather local property. At different locations, different assemblies can be in the fast state at the same time.

Correlation analysis reveals that synchronization on the larger time scale extends further (Knoblauch and Palm, 2002b) than synchronization on the finer time scale. Actually it turns out that the synchronization range depends mainly on the duration of the fast state, and is therefore independent of the extension of the underlying lateral synaptic connections. It could therefore be manipulated, e.g., by modulatory synaptic input or attention. This long-range synchronization on a larger scale of space and time might be an important property of the visual system to integrate local features of possibly new objects in the visual field. Although synaptic connectivity is local, global synchronization can be achieved as soon as neighboring local stimulus (sub)representations (e.g., in a topographical area) are compatible with each other (cf., [BINDING CONTOUR SEGMENTS INTO SPATIALLY EXTENDED OBJECTS]).

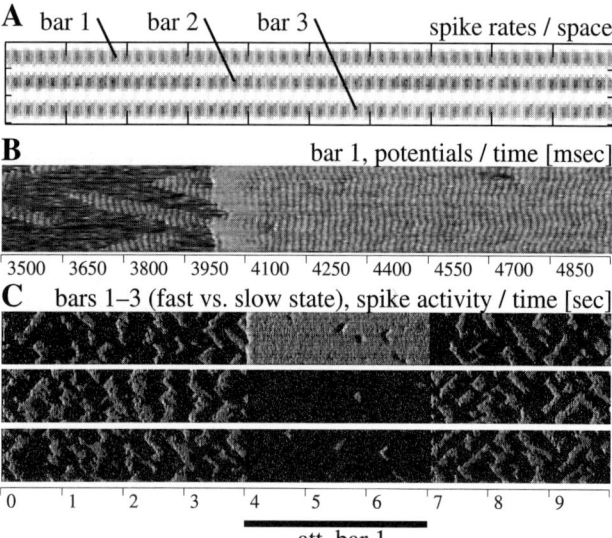

FIGURE 101.3 Activity in a larger model variant (size 30×500) when stimulating with three elongated bars simultaneously Between 4s and 7s after stimulus onset, attention is directed to the (upper) bar 1. (A) Spike rates in area P. Fifty activated patches (black) reflect each of the bar stimuli. (B) Plot of the membrane potentials versus time for the central neurons of the activated patches in area P for bar 1 (plot contains one line for each neuron). (C) Spike activity of three bar representations. Upper plot is similar to (B) but larger time scale. Middle and lower plots correspond to bars 2 and 3. White patches indicate strong activation of neurons (fast state). Without attention, the fast state is split up among the three bar representations due to competition (slow waves, fluctuating binding). Attention selects one of the bars (static binding).

IV. CONCLUSIONS

We have investigated the synchronization of distributed neural assemblies in a model for scene segmentation and binding comprising two interacting cortical areas. Our results are compatible with several lines of experimentally observed phenomena, such as synchronization on a fast time scale (Eckhorn et al., 1988; Singer and Gray, 1995; Chapter 87), synchronization on a slow time scale (Nelson et al., 1992), ongoing activity as measured by optical recording (Tsodyks et al., 1999), two-state fluctuations of membrane potentials (Anderson et al., 2000), and attention and biased competition (Reynolds and Desimone, 1999; Chapters 50 and 86).

Acknowledgments

This research was partially supported by the Volkswagen foundation (VW-I/77 355) and the MirrorBot project of the European Union.

References

Abeles, M. (1982). "Local Cortical Circuits". Springer, Berlin.

Anderson, J., Lampl, I., Reichova, I., Carandini, M., and Ferster, D. (2000). Stimulus dependence of two-state fluctuations of membrane potential in cat visual cortex. *Nat. Neurosci.* **3**, 617–621.

Eckhorn, R., Bauer, R., Jordan, W., Brosch, M., Kruse, W., Munk, M., and Reitboeck, H. (1988). Coherent oscillations: A mechanism of feature linking in the visual cortex? Multiple electrode and correlation analyses in the cat. *Biol. Cybernetics* **60**, 121–130.

Ghose, G. M., and Freeman, R. D. (1992). Oscillatory discharge in the visual system: Does it have a functional role? *J. Neurophysiol.* **68**, 1558–1574.

Grossberg, S. (1980). How does a brain build a cognitive code? *Psychol. Rev.* **1**, 1–51.

Hebb, S. (1949). "The Organization of Behavior. A Neuropsychological Theory." Wiley, New York.

Knoblauch, A., and Palm, G. (2002a). Scene segmentation by spike synchronization in reciprocally connected visual areas, I. Local effects of cortical feedback. *Biol. Cybernetics* **87**, 151–167.

Knoblauch, A., and Palm, G. (2002b). Scene segmentation by spike synchronization in reciprocally connected visual areas, II. Global assemblies and synchronization on larger space and time scales. *Biol. Cybernetics* **87**, 168–184.

Marr, D. (1970). A theory for cerebral neocortex. *Proc. R. Soc. Lond. B* **176**, 161–234.

Nelson, J., Salin, P., Munk, M., Arzi, M., and Bullier, J. (1992). Spatial and temporal coherence in corticocortical connections: A cross-correlation study in areas 17 and 18 in the cat. *Vis. Neurosci.* **9**, 21–38.

Palm, G. (1982): "Neural Assemblies. An Alternative Approach to Artificial Intelligence." Springer, Berlin.

Palm, G. (1990). Cell assemblies as a guideline for brain research. *Concepts Neurosci.* **1**, 133–148.

Reynolds, J., and Desimone, R. (1999). The role of neural mechanisms of attention in solving the binding problem. *Neuron* **24**, 19–29.

Ritz, R., Gerstner, W., Fuentes, U., and van Hemmen, J. (1994). A biologically motivated and analytically soluble model of collective oscillations in the cortex, II. Applications to binding and pattern segmentation. *Biol. Cybern.* **71**, 349–358.

Singer, W., and Gray, C. (1995). Visual feature integration and the temporal correlation hypothesis. *Annu. Rev. Neurosci.* **18**, 555–586.

Shadlen, M. N., and Newsome, W. T. (1999). Sychrony unbound: A critical evaluation of the temporal binding hypothesis. *Neuron* **24**, 67–77.

Treisman, A. (1998). Feature binding, attention and object perception. *Phil. Trans. R. Soc. London B* **353**, 1295–1306.

Tsodyks, M., Kenet, T., Grinvald, A., and Arieli, A. (1999). Linking spontaneous activity of single cortical neurons and the underlying functional architecture. *Science* **286**, 1943–1946.

von der Heydt, R., Peterhans, E., and Baumgartner, G. (1984). Illusory contours and cortical neuron responses. *Science* **224**, 1260–1262.

von der Malsburg, C. (1986). Am I thinking assemblies? *In* "Brain Theory" (G. Palm and A. Aertsen, eds.), pp. 161–176. Springer-Verlag, Berlin/Heidelberg.

Willshaw, D., Buneman, O., and Longuet-Higgins, H. (1969). Non-holographic associative memory. *Nature* **222**, 960–962.

CHAPTER 102

Attentive Wide-Field Sensing for Visual Telepresence and Surveillance

James H. Elder, Fadi Dornaika, Bob Hou and Ronen Goldstein

ABSTRACT

Physical and computational constraints limit the spatial resolution and field of view (FOV) achievable in any single sensor. In the human eye, a compromise has evolved in which resolution is high near the optical axis, falling off with eccentricity. The success of this design depends on visual mechanisms for rapidly detecting interesting events in the periphery and on motor mechanisms for accurately redirecting the fovea to peripheral targets for more detailed analysis.

Prototype artificial visual systems based conceptually on these principles are being developed in our laboratory. We partition visual sensing into two components. A pre-attentive component provides a large, fixed FOV at low resolution, allowing detection of events of interest over an entire visual environment. An attentive component provides a much smaller, shiftable FOV at high resolution, designed to recognize and interpret events detected by the pre-attentive system. These prototype sensors serve as useful testbeds for computational models of visual attention and may lead to new telepresence and surveillance technologies.

I. INTRODUCTION

Typical machine vision systems employ a single sensor with relatively small field of view (FOV). This can be sufficient for narrow applications (e.g., assembly-line inspection), where the location of the object of interest is strongly constrained. However, machine vision research is increasingly concerned with more humanlike visual tasks (e.g., surveillance of a large, open environment), and this has led to increasing interest in wide FOV machine vision sensing, particularly panoramic sensing (e.g., Oh and Hall, 1987; Yagi and Kawato, 1990; Ishiguro et al., 1992; Nayar, 1997).

The advantages of a wide FOV for surveillance and teleconferencing applications are clear; however, these advantages come at the expense of resolution. Switching from the 14 deg FOV of a typical lens to the 360 deg FOV of a panoramic camera results in a 26-fold reduction in linear resolution. For a standard 640 × 480 pixel camera, horizontal resolution is reduced to roughly 0.5 deg/pixel, a factor of 60 below human foveal resolution.

The human visual system has evolved a bipartite solution to the FOV/resolution trade-off. The FOV of the human eye is roughly 160 × 175 deg—nearly hemispheric. Central vision is served by roughly 5 million photoreceptive cones that provide high resolution, chromatic sensation over a 5-deg FOV, while roughly 100 million rods provide relatively low-resolution achromatic vision over the remainder of the visual field (Wandell, 1995). The effective resolution is extended by fast gaze-shifting mechanisms and a memory system that allows a form of integration over multiple fixations (Irwin and Gordon, 1998).

Variations on this architecture are found in other species. Many insects have panoramic visual systems; the springing spider, for example, has four eyes that capture movement over the entire viewing sphere and two small-FOV high-resolution eyes used in predation and mating (Moller et al., 2001).

Exploration of such heterogeneous visual architectures for computer vision is just beginning. Early attempts have focused on the design and fabrication of space-variant (foveated) sensor chips (Spiegel et al.,

1989; Ferrari et al., 1995; Pardo et al., 1997; Wodnicki et al., 1997). However, because the density of photoreceptive elements on these sensors is no greater than for regular sensors, they do not provide a resolution advantage over traditional chips and hence do not address the fundamental resolution/FOV trade-off.

The problem is more squarely addressed by mosaicing systems that compose mosaics from individual overlapping high-resolution images obtained by a single camera rotated about its optical center (Szeliski, 1994; Kumar et al., 1995). Such systems are useful for recording high-resolution still-life panoramas, but are of limited use for dynamic scenes, because the instantaneous FOV is typically small. An alternative is to compose the mosaic from images simultaneously recorded by many identical cameras with overlapping fields of view. A disadvantage of this approach is the multiplicity of hardware and independent data channels that must be integrated and maintained. For example, a standard 25-mm lens provides a FOV of roughly 14×10 deg. Allowing for 25% overlap between adjacent images to support accurate mosaicing, achieving this resolution over a hemispheric FOV would require on the order of 260 cameras!

Consideration of biological visual systems inspires a more practical solution to the FOV/resolution dilemma for machine vision. Instead of employing dozens or hundreds of identical sensors, it may be feasible to employ a small number of very different sensors with complementary properties. For example, in a human-inspired machine vision system, one sensor with low resolution but large FOV may play the role of the peripheral visual system, whereas a second sensor, with high-resolution but small FOV, serves as the fovea.

In the human these two subsystems share a single sensor substrate (the retina) and common optics; their machine vision analogs are more easily realized as distinct physical components. Because technological constraints limit the total number of photoreceptive elements on a reasonably priced chip, achieving fovealike resolution in a machine sensor depends on telephoto optics that limit the FOV. Obtaining the large FOV required by the peripheral component requires a much shorter focal length, hence separate optics.

This physical split, although necessary, introduces significant complications. The displacement of the optical centers of the two subsystems introduces parallax, making accurate correspondence of information in the two sensors nontrivial. This correspondence problem is further complicated by the disparity in resolution between the sensors. Interestingly, this problem is exactly that faced by the human visual system in attempting to accurately land a saccade on a peripheral target (Chapter 20). The efficacy of the human saccadic system thus provides some evidence that the proposed approach may be feasible.

As for the human eye, an artificial fovea is only useful if it can be rapidly directed to important visual events in the scene. This obviously requires a fast, accurate, and reliable 2D rotational (e.g., pan and tilt) motion platform. But it also requires a set of fast visual mechanisms capable of identifying and localizing these important visual events at relatively low resolution.

In this article, we outline recent research in our laboratory to design, build, and evaluate attentive wide-field sensors based on these principles. These prototypes, presently at an early stage of development in terms of their attention capabilities, are fully functioning, real-time active vision systems and, as such, can serve as interesting testbeds for theories of attention such as those proposed in this volume.

Attentive wide-field sensing may also be useful in a number of application areas. In telepresence applications, the attentive sensor can be directed toward events or activities of interest to a remote human observer, whereas low-resolution data from the pre-attentive sensor continues to provide the observer with a sense of context or situational awareness (Geisler and Perry, 1998). Recent work with saccade-contingent displays (Loschky and McConkie, 1999) has shown that video data viewed in the periphery of the human visual system can be substantially subsampled with negligible subjective or objective impact. Although our prototype sensors are not eye-slaved, this prior work suggests that attention-contingent sampling for human-in-the-loop video is feasible and potentially useful.

Attentive wide-field sensing may also be useful in autonomous surveillance applications. Events detected in the pre-attentive sensor may generate saccade commands to allow more detailed inspection and verification at the higher resolution of the attentive sensor.

II. PHYSICAL DESIGN

Figure 102.1 shows two wide-field attentive vision systems designed and built in our laboratory. Both prototypes employ two sensors. The pre-attentive sensors have large, fixed FOVs and play the role of the peripheral visual system. The attentive sensors have small FOVs, but are mounted on pan and tilt platforms, allowing gaze to be rapidly redirected. In both systems, the attentive sensor is based on a 25-mm lens with a FOV of roughly 13 deg, providing an angular

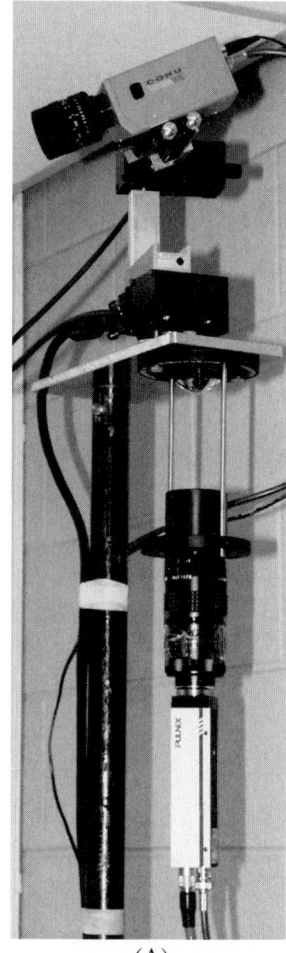

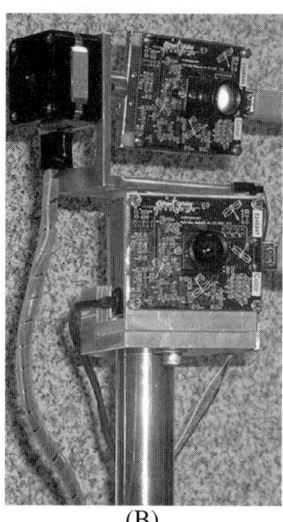

FIGURE 102.1 Two prototype attentive wide-field sensors. (A) Sensor 1. Attentive camera is mounted on a gimbaled pan and tilt platform at the top of the sensor. A Pre-attentive camera is mounted on the bottom, imaging a panoramic view reflected from a parabolic mirror. Apparatus shown measures roughly 75 cm in height. (B) Sensor 2. Two identical digital board cameras are mounted in close proximity on a vertical axis, with the attentive sensor on top. Apparatus shown measures roughly 20 cm in height.

resolution of 1.2 arcmin, roughly half the resolution of the human visual system. An example attentive image is shown in Fig. 102.2C.

The principal difference between the two systems lies in the pre-attentive sensor. In Sensor 1 (Fig. 102.1A), the pre-attentive sensor is based on a parabolic catadioptric video sensor (Nayar, 1997) purchased from Cyclovision Technologies (now RemoteReality™). This sensor provides a panoramic FOV (Fig. 102.2A), permitting surveillance from the center of a visual environment. In Sensor 2 (Fig. 102.1B), the pre-attentive sensor is based on a more conventional wide-angle lens with 2.1-mm focal length, providing a FOV of 130 deg (Fig. 102.1B), somewhat less than the human eye. This FOV is well-suited to surveillance of a visual environment from a corner location. The second system is also designed to be lighter, faster, cheaper, and smaller.

The raw video frames from the pre-attentive sensors are geometrically distorted (Fig. 102.1A–B). These distortions are caused by the curvature of the parabolic mirror in Sensor 1 (Nayar, 1997) and by radial and tangential error of the preattentive lens in Sensor 2. In both systems, these distortions are removed online in software prior to further processing.

In both sensors, the motion platforms have been designed so that the axes of rotation intersect approximately at the optical center of the attentive visual system, thus minimizing motion parallax. However, the physical separation of the pre-attentive and attentive sensors creates significant parallax between the two video streams. In Sensor 1, the optical baseline is 22 cm. In sensor 2, our efforts at compressing the package resulted in a baseline of only 7.5 cm, much closer to the average human interocular separation of 6 cm. This reduction is important in limiting the search region required for online, dynamic fusion of the pre-attentive and attentive visual streams.

III. FUSION

In telepresence systems, continuous data from remote sensors are displayed for a human observer. In autonomous surveillance systems, these data are interpreted directly by computer vision algorithms. In either case, incorporating wide-field attentive sensing raises the problem of data fusion. For example, in a remote surveillance system monitored by human security specialists, pre-attentive and attentive visual data could be displayed on separate monitors. But this would require observers to shift their gaze back and forth between the displays, mentally integrating the two disparate video streams. Accurate registration of

FIGURE 102.2 Wide-field attentive sensing imagery. (A) Raw, uncorrected pre-attentive image from Sensor 1. (B) Raw, uncorrected pre-attentive image from Sensor 2. (C) Example attentive image from Sensor 1. (D) Corresponding image from Sensor 2. Note the gross difference in resolution. (E) Fused imagery based on precalibrated homography. (F) Improved registration based on direct dynamic fusion algorithm. Circular vignetting has been used to smooth the transition between resolutions.

the visual data from the two streams raises the possibility of integrating the visual data into a single seamless window, lowering the workload for the human observers.

When an event of interest is detected in the pre-attentive stream, either by a human observer or by an automatic algorithm, the attentive sensor must be rapidly and accurately redirected to the corresponding location. Further automatic interpretation of a pre-attentively detected event may also depend on an estimate of the location of the event within the attentive FOV based on pre-attentive data. These computations depend on the accurate registration of pre-attentive and attentive visual streams.

Accurate registration in turn depends on computing correspondences between the two visual streams. In a typical computer vision stereo sensor, the intrinsic parameters of the cameras are identical and their extrinsic parameters are fixed and can be estimated in advance. This greatly simplifies correspondence, facilitating feature comparison and restricting the search space to fixed epipolar lines.

Computing correspondence for wide-field attentive sensing is more complicated, because the attentive sensor is not fixed and the intrinsic parameters of the two sensors are very different. On the other hand, the system is not completely unconstrained: the pre-attentive sensor is fixed, and the attentive sensor has only two rotational degrees of freedom. These constraints, coupled with nonuniform prior distributions on the distance of surfaces from the sensor, simplify the fusion problem considerably.

Most critically, visual scenes are coherent: The distance of visible surface points projecting to neighboring sensor pixels are highly correlated. To exploit this property, we model the scene as piecewise planar and approximate the correspondence between attentive and pre-attentive coordinate frames using a table of 2D projective mappings (homographies), indexed by the pan and tilt coordinates of the attentive sensor.

A. Calibration

For a static scene, eight-parameter homographies can be estimated using a manual technique and a simple calibration rig. Pre-attentive/attentive image pairs of the scene are captured at regular pan and tilt intervals of the attentive sensor. Twelve- to sixteen-point pairs are manually localized in each image pair, and the corresponding least-squares homographies are estimated using standard techniques. The result is a table of homographies, indexed by pan and tilt coordinates of the attentive sensor. In operation, given the arbitrary pan and tilt coordinates, the corresponding homography can be estimated from the table using bilinear interpolation.

For each image pair, we also store the projection of the attentive image center into pre-attentive coordinates. This allows the construction of a second table, indexed by pre-attentive coordinates, which provides the pan and tilt coordinates required to center the attentive sensor at a specific pre-attentive location. Bilinear interpolation is used to generate a pan and tilt command given an arbitrary saccadic target in pre-attentive coordinates, generated either by human or machine attention algorithms.

This fixed system of coordinate transforms between the two sensors will be accurate only if the viewing distance is large or if dynamic variations in depth are small relative to viewing distance. Because neither condition holds in general, we calibrate the system for intermediate distances and then use a dynamic fusion algorithm to obtain more precise registration during operation.

B. Dynamic Fusion

Automatic image registration is a well-studied problem, but the problem of registering pre-attentive and attentive images presents new challenges. The main difficulty is the extreme difference in resolution between the sensors (from 10:1 to 16:1 linear resolution ratio in our prototypes). Figure 102.1C–D shows an example of the same subject viewed in attentive and pre-attentive streams. The problem is complicated further by resolution inhomogeneities of the pre-attentive sensor (more pronounced for Sensor 1 than Sensor 2).

We address the problem in a three-stage approach. In the first stage, we use the pan and tilt coordinates of the attentive sensor to index the calibration table and determine the precomputed homography relating the two sensor images. Figure 102.2E shows an example of fusion based on precalibration. Because the subject is out of the local plane of calibration, there is significant registration error (note, for example, the mislocation of the shoulder of the subject on the right). Nevertheless, the translation and scaling parameters of the precalibrated homography are good enough to seed a search region within which the true homography should lie, within predefined confidence limits.

In the second stage, we compute a coarse registration within this search region using a parametric template-matching technique on a multiresolution representation of the attentive image that accommodates the inhomogeneity of the pre-attentive image. This provides an estimate of the translation and scale factors between the two streams.

In the third stage, this coarse registration is used to bootstrap a refinement process in which a full 2D projective mapping is computed. We have studied two refinement methods. The first recovers point matches between either high-gradient pixels or interest points and then uses a robust estimation procedure (RANSAC; Fischler and Bolles, 1981) to estimate the complete 2D projective transformation. The second method directly estimates the geometric and photometric transforms between the images by minimizing intensity discrepancies. In empirical evaluations, we have found the second direct method to be superior in both accuracy and reliability. An example of the resulting fusion is shown in Fig. 102.2F. Note that visual features in the attentive and pre-attentive streams join smoothly at the boundary of the attentive image. Circular vignetting has been used to further smooth the transition between the streams. All coordinate transforms and blending are performed using standard PC graphics hardware using OpenGL, allowing the systems to operate at roughly 17 frames per second (fps).

IV. SACCADIC CONTROL

In telepresence mode, the gaze of the attentive sensor can be controlled by a human observer monitoring the fused display. When an interesting event is noted in the pre-attentive field, the observer uses a point-and-click mouse interface on the pre-attentive field to drive the attentive sensor to the event location.

In automatic surveillance mode, saccades are determined by automatic attention algorithms. In human vision, one of the most powerful exogenous attention cues is visual change (Hillstrom and Yantis, 1994). A fundamental issue in change detection is how to select the spatial scale of analysis. In our case, the purpose of detection is to drive the attentive sensor to the point of interest to resolve the change. Thus, it is natural to match the scale of analysis to the FOV of the attentive sensor in pre-attentive coordinates. In this way, saccades will resolve the greatest amount of motion energy.

In our change-detection algorithm, successive pre-attentive RGB image pairs are differenced, rectified, and summed to form a primitive motion map. This map is convolved with a separable kernel that approximates the FOV of the attentive sensor in panoramic coordinates and thresholded to prevent generation of saccades due to sensor noise and vibration. The location of the maximum of the resulting change map determines the next fixation.

V. TRACKING AND SMOOTH PURSUIT

Human tracking data can be useful in evaluating the relative importance of simultaneous events of interest detected in the pre-attentive sensor. In a security application, for example, these data may be used to distinguish normal horizontal walking motion, from the motion of someone running or scaling a wall.

Tracking data are also important for accurate interception of events by the attentive sensor. If the velocity of the target is not taken into account in saccadic planning, the attentive sensor will perpetually lag the target. Tracking data from both the pre-attentive and attentive streams can also be used for smooth pursuit, that is, to control attentive gaze to remain locked to the moving target.

Automatic visual tracking of human activity has been the subject of intensive research in recent years (Aggarwal and Cai, 1997). Traditional methods use conventional cameras and are based on extracting and/or matching features such as occluding contours, skin color, and head regions.

For wide-field attentive-sensing applications, we are particularly interested in techniques that can be applied to low-resolution imagery, allowing the tracking of activity in the pre-attentive stream. At such resolutions, tracking techniques based on detailed modeling of the human body are unlikely to work because individual body parts may not be resolved.

In our prototypes, we have been studying tracking algorithms based on relatively low-level features such as intensity, intensity gradients, and color. We have found that combining these multiple features leads to a tracking system that is much more robust to occlusions, scale changes, and nonrigid motions than systems based on a single feature. An example of human body tracking based on this multifeature approach is shown in Fig. 102.3A. The tracking algorithm operates in real time (17 fps).

VI. MEMORY

What information the human visual system retains over a sequence of fixations is a subject of significant debate (e.g., Rensink et al., 1997; Chapter 13). There is no question, however, that humans have some forms of visual memory (iconic, short-term, long-term) (Chapter 33).

In any wide-field attentive sensing system, there is a trade-off between spatial and temporal resolution. The pre-attentive sensor provides good temporal resolution (17 fps in our prototypes) but poor spatial resolution. The attentive sensor provides good spatial

FIGURE 102.3 (A) Pre-attentive tracking of human motion. Rectangle indicates estimated body position and size. (B) Trading off spatial and temporal resolution using a primitive form of trans-saccadic memory.

resolution but poor temporal resolution, in that it may visit a particular portion of the visual scene relatively infrequently. By titrating the information from the two stream in different ways over space and time, we can adjust this trade-off and potentially adapt it to the dynamics of a particular visual environment. For example, in a completely static scene, a high-resolution visual image can be built up over time using a sequence of attentive fixations tiling the viewing sphere. In a more dynamic scene, changing portions of the mosaic may be updated using more immediate low-resolution data from the pre-attentive sensor.

We have implemented a primitive version of this idea. In our attentive wide-field sensor prototypes, the display duration of attentive images from previous fixations is determined by a memory parameter. At one extreme, previous attentive data are immediately replaced by more recent low-resolution data from the pre-attentive sensor. At the other extreme, a sequence of fixations builds up a persistent high-resolution mosaic. In intermediate modes, attentive data from previous fixations gradually fade into more recent low-resolution pre-attentive data (Fig. 102.3B). In this way, resolution in space can be traded off against resolution in time, depending on the nature of the application and the nature of the visual scene.

VII. FUTURE DIRECTIONS

It seems likely that the human attention and saccadic systems have co-evolved with the systematic fall-off in photoreceptor density, resolution, and sensitivity as a function of eccentricity. Attentive wide-field sensing platforms, which can mimic both the high resolution of the human fovea and the wide FOV of the human peripheral visual system, thus provide an interesting testbed on which to evaluate models of human attention.

It remains to be seen how effective and immersive an experience an attentive wide-field sensor can deliver in a telepresence application. Given the speed of eye movements and the intolerance of the visual system to lag, eye-slaved systems are impractical in many telepresence applications. We wish to investigate the degree to which intelligent attention algorithms and system memory can be used to provide an effective visual experience in situations where lags are significant and bandwidth is limited.

In the near-term, applications are likely in more loosely coupled telepresence environments where the attentive sensor is not directly slaved to the eye. In a remote learning application, for example, an attentive sensor at the lecture site might track an instructor, keeping her face and gestures in clear view to the remote students. A raised hand might be detected by a pre-attentive sensor at the classroom site, directing an attentive sensor to provide the instructor with high-resolution video of a student who wishes to ask a question.

Surveillance applications are also feasible in the nearer term. Fairly simple face detection and tracking algorithms in the pre-attentive stream may be sufficient to guide the attentive sensor to deliver high-resolution imagery of human activities. Even without effective face recognition or activity interpretation algorithms, such technologies could be useful in improving the quality of archival databases and reducing the workload of human security specialists.

Acknowledgements

This work was supported by the Ontario Centre for Research in Earth and Space Technology and by the Canadian Institute for Robotics and Intelligent Systems.

References

Aggarwal, J., and Cai, Q. (1997). Human motion analysis: a review. *In* "Nonrigid and Articulated Motion Workshop Proceedings," pp. 90–102. IEEE, San Juan, Puerto Rico.

Ferrari, F., Nielsen, J., Questa, P., and Sandini, G. (1995). Space variant imaging. *Sensor Rev.* **15**, 17–20.

Fischler, M. A., and Bolles, R. C. (1981). Random sample consensus: A paradigm for model fitting with applications to image analysis and automated cartography. *Commun. ACM* **24**, 381–395.

Geisler, W. S., and Perry, J. S. (1998). A real-time foveated multiresolution system for low-bandwidth video communication. *In* "Human Vision and Electronic Imaging" (B. Rogowitz and T. Pappas, eds.), pp. 294–305. SPIE, San Jose, CA.

Hillstrom, A. P., and Yantis, S. (1994). Visual motion and attentional capture. *Perception Psychophys.* **55**, 399–411.

Irwin, D. E., and Gordon, R. D. (1998). Eye movements, attention and trans-saccadic memory. *Vis. Cogn.* **5**, 127–155.

Ishiguro, H., Yamamoto, M., and Tsuji, S. (1992). Omni-directional stereo. *IEEE Trans. Pattern Analysis Machine Intell.* **14**, 257–262.

Kumar, R., Anandan, P., Irani, M., Bergen, J., and Hanna, K. (1995). Representation of scenes from collections of images. *In* "Proceedings of the IEEE Workshop on Representation of Visual Scenes," pp. 10–17. IEEE, Los Alamitos, CA.

Loschky, L. and McConkie, C. W. (2002). Investigating spatial vision and dynamic attention selection using a gaze-contingent multiresolutional display. *Journal of Experimental Applied* **8(2)**, 99–117.

Moller, R., Lambrinos, D., Roggendorf, T., Pfeifer, R., and Wehner, R. (2001). Insect strategies of visual homing in mobile robots. *In* "Biorobotics—Methods and Applications" (B. Webb and T. Consi, eds.), AAAI Press/MIT Press, pp. 37–66, Cambridge, MA, USA.

Nayar, S. (1997). Catadioptric omnidirectional camera. *In* "Proceedings of the IEEE Conference on Computer Vision Pattern Recognition," pp. 482–488.

Oh, S. J., and Hall, E. L. (1987). Guidance of a mobile robot using an omnidirectional vision navigation system. *Proc. Soc. Photo-Opti. Instrum. Eng.* **852**, 288–300.

Pardo, F., Dierickx, B., and Scheffer, D. (1997). CMOS foveated image sensor: Signal scaling and small geometry effects. *IEEE Trans. Elec. Dev.* **44**, 1731–1737.

Rensink, R. A., O'Regan, J. K., and Clark, J. J. (1997). To see or not to see: The need for attention to perceive changes in scenes. *Psychol. Sci.* **8**, 368–373.

Spiegel, J. van der, Kreider, G., Claeys, C., Debusschere, I., Sandini, G., Dario, P., Fantini, F., Belluti, P., and Soncini, G. (1989). A foveated retina-like sensor using CCD technology. *In* "Analog VLSI Implementation of Neural Systems" (C. Mead and M. Ismail, eds.), pp. 294–305. Kluwer, Boston.

Szeliski, R. (1994). "Image Mosaicing for Tele-reality Applications." Tech. Rep. No. CRL 94/2. Cambridge Research Laboratory, Cambridge, MA.

Wandell, B. (1995). "Foundations of Vision." Sinauer, Sunderland, MA.

Wodnicki, R., Roberts, C. W., and Levine, M. (1997). Design and evaluation of a log-polar image sensor fabricated using a standard 1.2μm ASIC CMOS process. *IEEE J. Solid-State Circuits* **32**, 1274–1277.

Yagi, Y., and Kawato, S. (1990). Panoramic scene analysis with conic projection. *In* "Proceedings of the International Conference on Robots and Systems." *IEEE* **1**, 181–187.

CHAPTER

103

Neuromorphic Selective Attention Systems

Giacomo Indiveri

ABSTRACT

Selective attention mechanisms allow sensory systems with limited processing capacity to function in real time, independently of the size of the stimulus input space. They can be particularly useful in artificial vision systems, where the amount of information provided by the sensors typically exceeds the system's processing capacity. We describe a very large-scale integration (VLSI) neuromorphic device and a spike-based communication infrastructure for constructing multichip vision systems based on models of selective attention. These hardware models reproduce many of the behaviors of biological selective attention systems. We demonstrate the feasibility of this approach by showing experimental results from a hardware two-chip vision system that performs active tracking and selective scanning tasks.

I. INTRODUCTION

Selective attention is a mechanism used to sequentially select and process salient subregions of the input space while suppressing inputs arriving from nonsalient regions. By processing small amounts of sensory information in a serial fashion rather than attempting to process all the sensory data in parallel, this mechanism overcomes the problem of flooding limited processing capacity systems with sensory inputs. It is a mechanism found in many biological systems and can be a useful engineering tool for developing artificial systems that need to process real-time sensory data. We present a custom, mixed-mode analog-digital, very large-scale integration (VLSI) device useful for constructing artificial neuromorphic, multichip selective attention systems, able to operate in real time on real-world stimuli. These hardware systems are neuromorphic in the sense that they contain biologically plausible computational elements arranged and interconnected using the same organizing principles found in biological neural systems.

One of the key computational operations performed by selective attention mechanisms is the winner-takes-all (WTA) operation. This highly nonlinear operation is used to select the strongest input, corresponding to the most salient feature, and to suppress all other inputs. In many models of selective attention, a WTA network selects the most salient input, locks on to it (attends to it) for a set time interval, deselects it, and subsequently suppresses it, through an inhibition of return mechanism (Itti et al., 1998; see Chapter 94), allowing successive inputs of decreasing strength to be selected. The VLSI device we present implements this WTA mechanism, including the inhibition of return dynamics, and uses an asynchronous communication protocol useful for constructing hierarchical multichip selective attention systems.

II. MULTICHIP SELECTIVE ATTENTION MODELS

Several single-chip systems for implementing visual selective attention mechanisms in real time have been proposed (Morris et al., 1998; Brajovic and Kanade, 1998; Indiveri, 1999). These systems contain photosensing elements and processing elements in the same focal plane and typically apply a competitive selection process to visual stimuli. Unlike these systems, multichip systems decouple the sensing stage from the selective attention/competition stage. Therefore input signals need not come only from visual sensors, but could represent a wide variety of sensory stimuli obtained from a variety of sources. In multichip selective attention systems, the visual sensors used for generating the saliency map are relatively high-resolution silicon retinas and do not have the small-fill factors that single-chip 2D attention systems are burdened with. Furthermore the signals encoding the bottom-up generated saliency map sent to a selective attention chip can be merged with top-down modulatory signals (e.g., from associative memory modules) to bias the competition process (see Chapters 99 and 100).

In these types of multichip systems, analog signals are transmitted between chips using an asynchronous communication protocol based on the address-event representation (AER; see Fig. 103.1) In this representation, analog signals are converted into streams of stereotyped nonclocked digital pulses (spikes) and encoded using pulse-frequency modulation (spike rates). When a spiking element on a VLSI device generates a pulse, its address is encoded and placed on a digital bus using asynchronous logic (see Fig. 103.1). In this asynchronous representation, time represents itself and analog signals are encoded by the interspike intervals between the addresses of their sending nodes.

The device that allows us to construct hierarchical selective attention systems, receiving input from AER sensors and providing output to higher processing modules, is the selective attention chip.

III. THE SELECTIVE ATTENTION CHIP

The selective attention chip we present here was implemented using a standard VLSI (CMOS) technology of 1.6 μm and contains an array of 8 × 8 cells of the type shown in Fig. 103.2. Because processing occurs fully in parallel, in principal arrays of arbitrary size could be implemented. In practice, the maximum array size is limited by the maximum area of silicon available.

Each cell of the 2D array comprises an excitatory silicon synapse, an inhibitory silicon synapse, a current-mode hysteretic WTA circuit (Indiveri, 2001), an output integrate and fire (spiking) neuron, and two position-to-voltage (P2V) circuits that convert the spatial position of the winner into two continuous-time analog voltages. A detailed description of all of these circuits, together with quantitative analysis and a description of their response properties is described in Indiveri (2000).

The excitatory synapse of each WTA cell receives off-chip address events and integrates them into an excitatory current I_{ex}. The cell in the WTA network that wins the competition supplies a current to the P2V circuits and to its output neuron. The neuron's spikes used to transmit the pixel's address off chip are also integrated by the WTA cell's inhibitory synapse into an inhibitory current I_{ior} (to implement the inhibition of return mechanism). As the integrated inhibitory current I_{ior} increases, the cell's net input current (I_{ex}—I_{ior}) decreases. Eventually the net input current to the

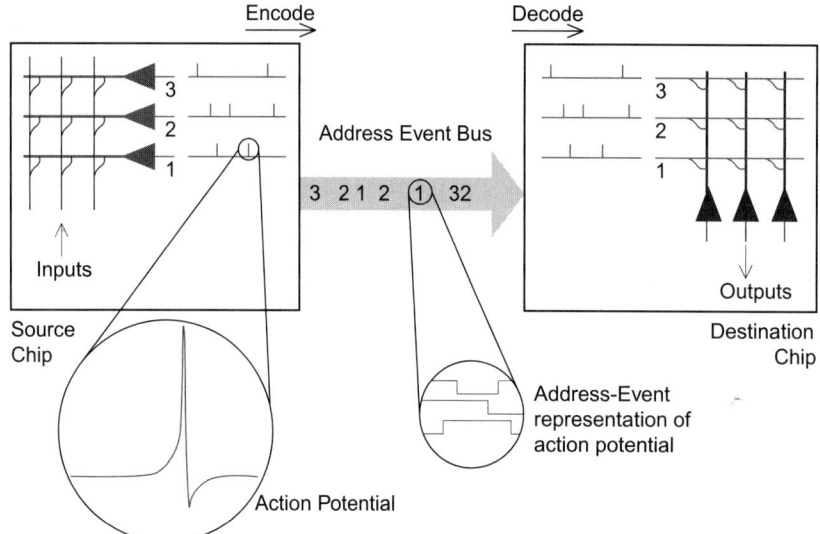

FIGURE 103.1 Schematic diagram of an AER chip-to-chip communication example. As soon as a sending node on the source chip generates an event, its address is written on the Address-Event Bus. The destination chip decodes the address-events as they arrive and routes them to the corresponding receiving nodes.

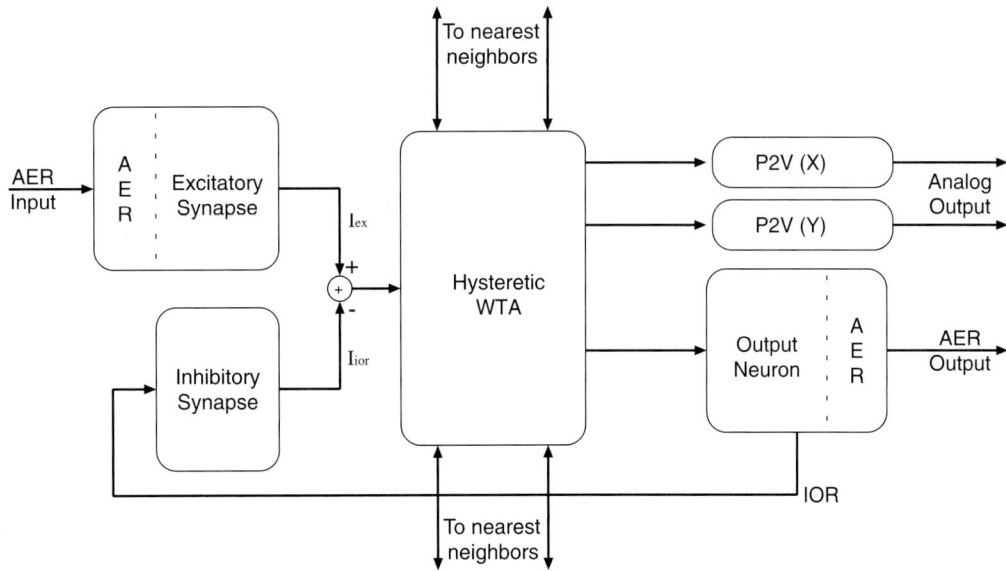

FIGURE 103.2 Block diagram of a basic cell of the 8 × 8 selective attention architecture.

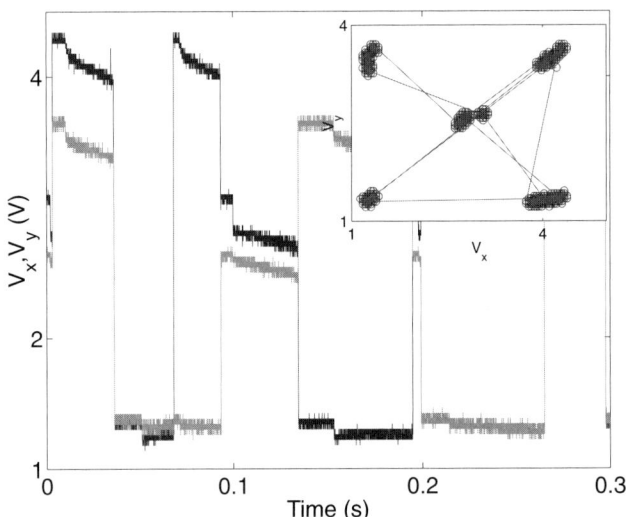

FIGURE 103.3 Output of the P2V x circuit and y circuit of the selective attention chip in response to a test stimulus exciting four corners of the input array at a rate of 30 Hz and a central cell at a rate of 50 Hz. The inset shows the x output trace versus the y output trace. (See color plate)

synapses, on the input stimuli, and on the frequency of the output neuron, the WTA network will switch the selection of the winner between the strongest input and the next-strongest, or between the strongest and more inputs of successively decreasing strength, generating focus of attention scan paths (see also Chapter 48). Figure 103.3 shows an example of a focus of attention scanpath measured from the selective attention chip.

IV. A TWO-CHIP ACTIVE VISION SYSTEM

We constructed a two-chip selective attention system by interfacing an AER silicon retina that responds to contrast transients (Kramer, 2002) to the selective attention chip (see Fig. 103.4). The AER retina was mounted on a pan and tilt unit and was controlled (via a workstation) by the selective attention chip's output. The address-events produced by the silicon retina report the position of high-contrast moving objects. The selective attention chip receives these events as its inputs and selects the locations with highest contrast moving objects, serially cycling through them. A workstation reads the AER output of the selective attention chip uses and the results to drive the motorized pan and tilt unit, centering the selected locations on the sensor's imaging array. The system therefore makes quick saccadic sensor movements following the focus of attention scanpaths. An example of the behavior of such a system, in response to a subject waving his fingers in front of the system is

winning cells decreases below the value of the net input current exciting a different cell, and the WTA network switches state, selecting the other cell as the winner. When the old winning cell is deselected, its corresponding output neuron stops firing and its inhibitory synapse recovers, decreasing the inhibitory current I_{ior} back to zero. Depending on the time constants and strength of the excitatory and inhibitory

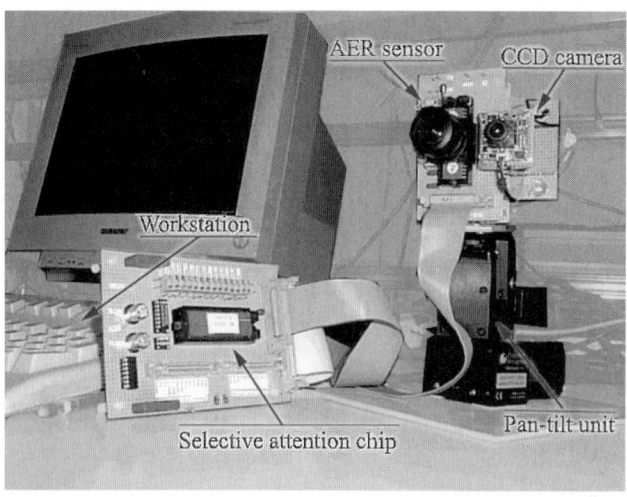

FIGURE 103.4 Active vision system setup. The AER retina mounted on a pan and tilt unit transmits its output to the selective attention chip. The latter sends the results of its computations to a host computer that uses this data to drive the pan and tilt unit's motors. A CCD camera is mounted next to the retina to display the retina's field of view.

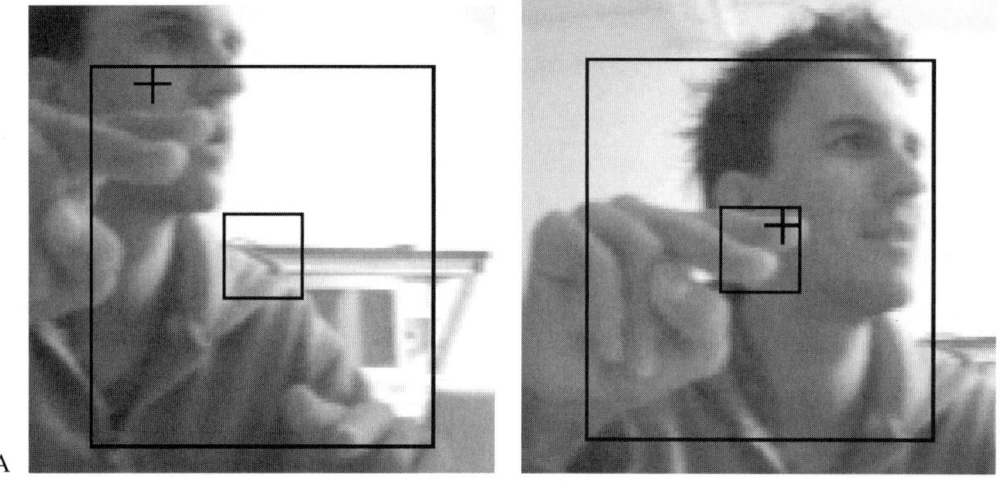

FIGURE 103.5 (A) Silicon retina view captured by a standard CCD camera mounted next to the retina, just before the pan and tilt unit movement; the location selected by the selective attention chip is indicated by the cross in the top left corner. (B) Silicon retina view just after the movement; the selected target is centered with the retina's view.

shown in Fig. 103.5. The details of the system and of its behavior are described in (Indiveri et al., 2001).

V. CONCLUSION

Hardware multichip selective attention systems that process real-world stimuli in real time can be built using neuromorphic VLSI devices containing silicon neurons, silicon synapses, and dynamic competitive networks (such as WTA circuits). We have presented a VLSI device of this type and described a two-chip selective attention system able to select and attend to high-contrast moving objects. This simple example shows how the circuits developed can be reliably used in massively paralle VLSI networks of spiking neurons and how it is possible to build AER multichip systems able to reliably receive, transmit, and process signals represented as spikes.

Combining the flexibility offered by the digital AER communication infrastructure with the tunable properties of the network's competition mechanism, it is possible to construct more complex hierarchical systems (see Chapter 94). These hardware systems can be useful real-time tools for scientific investigation, useful engineering tools for commercial applications, or both, in the true spirit of neuromorphic engineering.

References

Brajovic, V., and Kanade, T. (1998). Computational sensor for visual tracking with attention. *IEEE J. Solid State Circuits* **33**, 1199–1207.

Indiveri, G. (1999). Neuromorphic analog VLSI sensor for visual tracking: Circuits and application examples. *IEEE Trans. Circuits Syst. II* **46**, 1337–1347.

Indiveri, G. (2000). Modeling selective attention using a neuromorphic analog VLSI device. *Neural Computation* **12**, 2857–2880.

Indiveri, G. (2001). A current-mode hysteretic winner-take-all network, with excitatory and inhibitory coupling. *Analog Integrated Circuits Signal Proc.* **28**, 279–291.

Indiveri, G., Mürer, R., and Kramer, J. (2001). Active vision using an analog VLSI model of selective attention. *IEEE Trans. Circuits Syst. II* **48**, 492–500.

Itti, L., Koch, C., Niebur, E. and (1998). A model of saliency-based visual attention for rapid scene analysis. *IEEE Trans. Pattern Analysis Machine Intell.* **20**, 1254–1259.

Kramer, J. (2002). An integrated optical transient sensor. *IEEE Trans. Circuits Systems II* **49**, 612–628.

Morris, T. G., Horiuchi, T. K., and DeWeerth, S. P. (1998). Object-based selection within an analog VLSI visual attention system. *IEEE Trans. Circuits Syst. II* **45**, 1564–1572.

CHAPTER

104

The Role of Visual Attention in the Control of Locomotion

M. Anthony Lewis

ABSTRACT

A primary factor in the evolution of vision was undoubted the need for perceptual support of gait modulation and decision making related to locomotion. The ability to attend and respond to objects meaningful in a locomotory sense is particularly important as animal species increase in size and agility. In this work, we show how a generic architecture can very quickly learn to attend to objects in the environment that are meaningful to an agent, in this case a robot. These objects are important because they affect the robot's primary purpose in life—to walk. The robot responds to meaningful objects, not just novel ones. Thus, under this paradigm attention is not driven by bottom-up processing but by the need to support a behavior.

I. INTRODUCTION

Why should one object in the environment be more deserving of attention than any other? The objective of this article is to consider the purpose of vision in guiding legged locomotion. From this perspective, objects are interesting if they influence locomotory goals.

The perceptual needs of animals increase along two dimensions: size and speed. Slow-moving animals may integrate proprioceptive information to a much greater degree than quick moving animals. Fast, large animals must make use of exteroceptive sensors to anticipate and recover from collisions with the environment. Due to their larger size and speed, these animals are slower to respond and may injure themselves much more readily while running at high speeds. How do fast-moving, heavy animals negotiate the environment agilely? This brings us to the issue of vision. Vision is an ideal exteroceptive sensor. It allows the animal to anticipate collisions with the environment and to accommodate potentially destabilizing objects. Clearly, these potentially destabilizing objects must be attended to for the safety of the animal.

What do we know about vision and locomotion? Visual information is not necessary to walk on firm, flat surfaces. We can walk on such a surface in the dark or with our eyes closed. However, a firm, flat surface is an atypical case in nature. Vision helps with balance and is a preferred modality over other senses such as vestibular and proprioceptive. Vision assists in navigation and route planning. Significantly, vision assists in adjusting stride to avoid and surmount obstacles. These different uses of vision are apparently carried out in parallel. Information enters a single sensory portal, the eyes, is processed and influences the common resource of the motor system; the purpose of movement can variously be described as the results of the harmonious relationship of these locomotory objectives. An excellent review of the role of vision during locomotion can be found in a review by Patla (1997).

Although the derivation of percepts related to balance and navigation is demonstrated by relatively simple insect systems, we might conjecture that these functions are not as difficult as stepping over and avoiding obstacles. One key problem in using vision to adjust gait over rough terrain is that the perceptual system must know which aspects of the object are relevant to locomotion. There must be a system that links together raw sensory data with meaning in terms of motor action. We describe a model of a system that does this here. This model learns to make a link between perception and action based on experience. The system begins its existence with a *tabula rasa* (blank slate), not knowing what is meaningful in the environment and what is not. Through encounters and mistakes, it builds a rudimentary perceptual apparatus that can detect potentially destabilizing features of the environment (i.e. obstacles). In this way, meaning is given to one aspect of the environment over another. An interesting feature of this model is that different perceptual modes can be used (i.e., stereopsis, optic flow) within the same framework.

II. MODEL

Much of this model has previously been described (Lewis and Simó, 1999, 2001; Lewis, 2002). This embodied model uses a robotic device that walks, collects data, and a effects the environment. The robot device used is shown in Fig. 104.1. This robot is a tabletop robot, about 20 cm tall.

FIGURE 104.1 Two-legged robot used to implement and test our model.

A. Processing Overview

A motor system generates commands to the robot as shown in Fig. 104.2. The resulting movement creates a changing environment for the robot. This environment is viewed by a single camera or stereo pair of cameras. The system uses traveling fixation; that is, the cameras remain at a fixed angle.

The camera information is processed to produce a coarse optic flow field, reflecting the movement of the image under the robot or a set of vectors giving a measure of disparity at points in the image in the travel path of the robot. Simultaneously, a prediction is made of the visual cues that the robot should be sensing at a particular portion of its stride. For example, when the robot is in the dual-support phase (two legs on the ground), the points on a surface in front of the robot will have a certain tangential velocity, detected by the optic flow vector, and a certain height to the eye, determined by the stereo disparity measure.

A comparison is made between the expected and the actual visual cue. The difference between the expected and actual cue is called novelty. Novelty can be enhanced or ameliorated by a gain factor. The gain is varied so as to keep the time-averaged appearance of novelty at a stable rate. Any resulting novelty is then used to drive a change in stride length or other aspect of the motor structure. The mapping between the novelty and motor action is accomplished via a sensorimotor map.

Learning and adaptation occurs in several places in the model. First, the prediction of visual cues involves

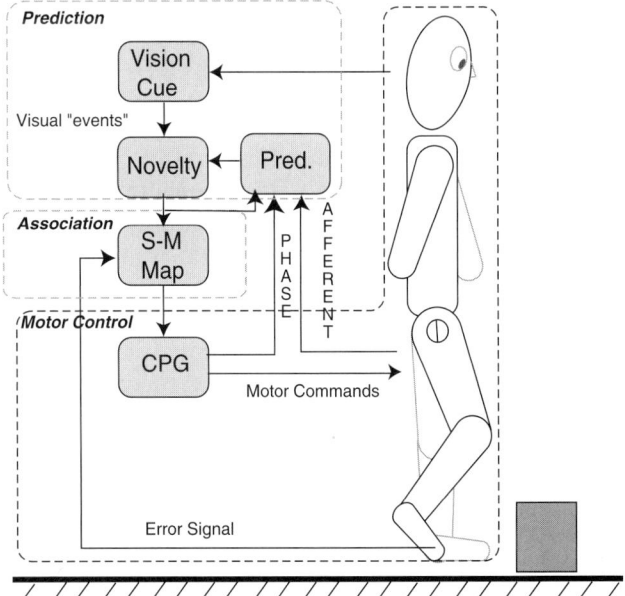

FIGURE 104.2 Schematic diagram of the model's architecture.

learning a mapping between the phase of the robots stride, roughly corresponding to the robots configuration and the visual cues it should expect. Second, after novelty is detected, the gain factor is adjusted up or down, depending on the rate of occurrence of novelty. If the system is unable to reliably predict the surface in front of it, it sees novelty all the time and the gain factor is lowered, and the apposite occurs if the surface is predictable vice-versa. The effect of this is to make the system less sensitive to time when prediction is less reliable and to make the system more sensitive to novelty when prediction can be made more reliably.

Finally, learning takes place in the sensorimotor map. This map uses a measure of robot destabilization as an error signal. If the robot hits an object, it triggers a tactile switch. This tactile switch triggering indicates an error. This error is used to modify the sensorimotor map in such as way that the robot will be more likely not to have a tactile trigger at some point in the future.

B. Motor Systems

At the heart of the motor system is a central pattern generator or CPG. The CPG, in this case, is a system of distributed limit-cycle oscillators. By appropriate coupling between these oscillators, various phase relationships between the oscillators can be maintained and altered at will. The output of the CPG is passed through an output function that maps phase (i.e. θ ranging from 0 to 0.1) to a motor output that represents a force generated around a joint. Each oscillator can generate either a negative or positive torque. To achieve the full range of control, oscillators are grouped into agonist/antagonist pairs to control the joint of a limb.

The CPG can be thought of as an element that controls coordination of joints and can filter sensory-derived actions. To assist in the second role, the phase or output of the CPG is transmitted up to a cerebellum like structure in our model. This cerebellum receives input from both the CPG system and visual input. In this model, the CPG activates various regions of the cerebellum.

Interestingly, in humans during demanding tasks, eye movement is linked to the locomotor cycle. Hollands and colleagues found tight coupling between eye movement and leg movement (Hollands et al., 1995; Hollands and Marple-Horvat, 2001). In Hollands and Marple-Horvat (2001), subjects were made to walk over randomly distributed stepping stones. The investigators found that subjects made saccades to points where the foot would land. As the foot approached the target stone, the eyes would saccade to the next stone. This was accomplished with surprising regularity, implying synchronization between the spinal CPG and eye movement.

C. Prediction Structure

The predictive structure consists of three modules: a visual cue, a prediction of this visual cue, and a module that extracts novelty. We describe each of these in term.

1. Prediction. The "Pred" box in Fig. 104.2 receives phase information as well as afferent information from the robot, including information encoding joint position and so forth. The prediction circuit is divided into discrete slices. Learning occurs here in the prediction model to create a prediction of sensory events based on proprioceptive and phase information from the CPG.

2. Visual Cues. Visual cues enter from two cameras mounted on the robot. In the case of our experiments, we have used both optic flow information (Lewis, 2002) and stereopsis information (Lewis and Simó, 2001).

3. Novelty. This information is then to predict what the robot should see at a particular instant. To achieve this, predicted information is compared between predicted and actual. The novelty structure incorporates adaptation. The system adapts to the novelty in the following way. We assume that under most circumstances novelty should occur no more frequently than a given base-line rate. In our experiments, this rate was chosen as 5%. Based on this, the system adapts its gain to novel stimuli. By doing so, the system learns to reject points in the gait where sensor information is hard to predict (i.e., everything seems novel) and will become more sensitive to regions where there is low novelty. This reminds the author of the "pins and needles" feeling when a limb falls asleep. The limb becomes hypersensitive. This can happen in this model as well if sensory input is denied or is too predictable.

D. Sensorimotor Action

The sensorimotor (S-M) map is a map that will weight novelty outputs and to creates a signal that can modulate the CPG. In our example, this is a 2D map. One dimension is the phase of the gait; thus, different motor commands can be generated at different phases of the gait. The second dimension is retinal distance from the foot to the horizon; this allows variation in command depending on the distance to the object.

A tactile trigger on the front of the foot causes a change in the motor map. A tactile trigger event can occur in two ways. First, if the foot is lifting, the foot may collide with an object. Second, if the foot is descending and hits an object prematurely, this indicates an error as well.

The sensorimotor map must be modified. To do this, we have to identify the cells that could have been active a second or so before the collision and modify their contribution to controlling the CPG. In this model, we use an eligibility trace (Sutton and Barto, 1998). The eligibility trace we use is a second-order eligibility trace. This means that the dynamics of the eligibility trace are described by a second-order nonlinear differential equation. To give a gist of how this works, if a novel event is identified by the prediction system, this novelty is used to drive the second-order differential equation associated with that particular cell. The differential equation is arranged so that a pulse input causes an output that peaks at some time later, in this model between 300 and 1000 ms. The time of the peak can be adjusted by varying parameters in the equation describing the eligibility trace.

An output of the eligibility trace indicates that a particular cell is eligible for modification. The greater the output of the eligibility trace, the greater the eligibility. Therefore, when the eligibility trace reaches its peak, the associated cell is maximally eligible for plastic modification.

The direction of the synaptic weight modification is critical. The appearance of novelty can trigger either a lengthening or a shortening of the stride. If a trigger occurrs while the foot was on its way up, it might indicate that the stride should be shortened in the future. This will cause the footfall to occur further behind the obstacle. Although this mapping is ad hoc, it seems clear that a more general-purpose method might be used, but with a consequent increase in learning time.

E. Simulation

The basic framework of this model has remained stable from the early simulation studies (Lewis and Simó, 1999). In those studies, we found that a robot (or human) learned to step over obstacles after very few trials—11 mistakes in 100 trials, for example. We found that the footfall variance decreased as the subject approached the object. Footfall variance measures how regularly the foot placement is one step, two steps, and three steps before an obstacle. This accords quantitatively with human data in the long jump (Lee et al., 1982).

F. In Real Robots

We have used the same system in real robots (Lewis and Simó, 2001; Lewis, 2002). In this case, we found that the robot learned to step over obstacles with very few trials (approximately 10–20 examples). Examining the novelty field, we found that the robot could identify very fine features of the environment such as very small obstacles. We found that this system worked with both stereoscopic as well as optic flow cues with no modification to the basic architecture.

III. CONCLUSION

We have demonstrated how the motor needs of a system can lead to the assignment of meaning (via the sensorimotor map) to novel events. These novel events, in turn, are detected by comparing expected versus actual stimuli.

Notice that information processing occurs bottom up. There is no top-down direction of attention. However, we found the need for a gait or sensorimotor map to transform novelty to action. This map also gates or limits the behavioral expression of novel sensory stimuli, allowing the motor action to be a response to meaningful stimuli and suppressing stimuli with no value in maintaining the goal of stable locomotion.

References

Hollands, M. A., and Marple-Horvat, D. E. (2001). Coordination of eye and leg movements during visually guided stepping. *J. Motor Behav.* **33**, 205–216.

Hollands, M. A., Marple-Hovat, D. E. et al. (1995). Human eye movements during visually guided stepping. *J. Motor Behav.* **27**, 155–163.

Lee, D. N., Lishman, J. R. et al. (1982). Regulation of gait in long jumping. *J. Expe. Psychol.: Hum. Perception Performance* **8**, 448–459.

Lewis, M. A. (2002). *Detecting surface features during locomotion using optic flow*. In "Proceedings of the 2002 IEEE International Conference on Robotics and Automation (ICRA)," May 11–15, Washington, D.C., USA, pp. 305–310.

Lewis, M. A., and Simó, L. S. (1999). Elegant stepping: A model of visually triggered gait adaptation. *Connect. Sci.* **11**(3–4): 331–344.

Lewis, M. A., and Simó, L. S. (2001). Certain principles of biomorphic robots. *Autonomous Robots* **11**, 221–226.

Patla, A. E. (1997). Understanding the roles of vision in the control of human locomotion. *Gait Posture* **5**, 54–69.

Sutton, R. S., and Barto, A. G. (1998). "Reinforcement Learning: An Introduction". Adaptive Computation and Machine Learning. MIT Press, Cambridge, MA.

CHAPTER 105

Attention Architectures for Machine Vision and Mobile Robots

Lucas Paletta, Erich Rome and Hilary Buxton

ABSTRACT

Computer vision systems that are applied for image understanding in real-world environments require the capability to focus operations on task relevant events in an ongoing input stream of visual information. Attentive systems must indirectly provide solutions to characteristic challenges in real-world processing, such as the complexity in input imagery and uncertainty in the acquired information. We address successful methodologies on saliency and feature selection, describe attentive systems with respect to object and scene recognition, and review saccadic interpretation under decision processes. In robotic systems, we understand attention embedded in the context of optimizing sensorimotor behavior and multisensor-based active perception. We present an overview on system architectures that play a crucial role in attentive robots, with emphasis on multimodal information fusion and humanoid robots.

I. INTRODUCTION

Vision systems with the task of operating in real-world environments must tackle challenges that arise from the specific conditions and the uncertainty in the visual information of the captured images. Selective attention is necessary to focus on information that is relevant to a current task and mandatory to make a choice for the processing of appropriate data sources in response to a specific system state in space and time. In machine vision, the development of enabling technologies such as video surveillance systems, miniaturized mobile sensors, and ambient intelligence systems involves the real-time analysis of enormous quantities of data. Knowledge has to be applied concerning what needs to be attended to and when and what to do in a meaningful sequence, in correspondence with visual feedback. Methods on attention and control are mandatory to render computer vision systems more robust. In mobile robots, the embodied actuators may affect perception in an even greater sense, introducing mobility and thereby deciding the observer's viewpoint, linking physical presence with specificity in its sensoric experience. The following sections review attention architectures in machine vision and robotic systems, with an emphasis on research results presented at the International Workshop on Attention and Performance in Computer Vision 2003 (Paletta et al., 2003).

II. ATTENTIVE COMPUTER VISION SYSTEMS

Attention architectures in machine vision are introduced from the view of bottom-up processing in terms of saliency operators. In analogy to top-down paths in human perception (Braun et al., 2001), task-dependent modulation of feature extraction is a relevant issue, in particular regarding the task of object search in real-world scenes. Contextual modulation of processing enables us to take advantage of context cues in order to focus processing on most promising, such as salient image information. In addition, saccadic integration operates on the process chain of contextually related local interpretations of a global object or scene information. Finally, in dynamic vision, the extraction of visual motion plays an outstanding role in providing essential cues and attracting attention.

A. Saliency from Feature Selection

Saliency of a local image area must be defined on the basis of the specific visual information and, accordingly, on the basis of an appropriate feature detector. Attentive processing in computer vision initially used saliency-based models, in which the strength of the response of feature detectors determined candidate locations by matching (Clark and Ferrier, 1988). In a model of purely bottom-up information processing, the Culhane-Tsotsos feature detector (Culhane and Tsotsos, 1992) builds a hierarchy of representations relying on the assumption that input cell values directly reflect how salient a specific location is. A saliency operator based on information measures with respect to spatial locations and scales of objects in an image is provided by Jagersand (1995). It results from the expected information gain from Kullback contrasts between successive resolution lengths.

More elaborated models of attentive stimulus-driven search were proposed by Itti et al. (1998), combining first multiscale image features into a single topographical saliency map. Competition among neurons in this map give rise to a single winning location that corresponds to the next attended target (Sec. IIB). The underlying model is based on the Feature Integration Theory from Treisman and Gelade (1980). A referring implementation that is invariant to similarity transformation is described by Lionelle and Draper (2003). The saliency of local image regions has more recently become relevant in object recognition and wide-base line stereo (Fraundorfer and Bischof, 2003), taking into account the three closely interrelated aspects of saliency, scale, and content. The detector is translation, rotation, and scale invariant.

In current machine attention, bottom-up selection plays an important role in providing early cues in a multistage competetive scheme of attention processing (Navalpakkam and Itti, 2002). Backer and Mertsching (2003) introduced a cascaded computation by selecting a small number of discrete items in a preattentive phase analyzing symmetry, eccentricity, color contrast, and depth, and then applied smiattentive processes of tracking and information accumulation until a single cue of interest could be more efficiently selected.

B. Object and Scene Recognition

Top-down processing has been emphasized in computer vision tasks related to object search. The main focus of research is on how to integrate bottom-up and top-down information to attain an efficient decision on where to focus attention to within the input image. A critical task in computer vision is the recognition of entities of interest in the visual appearance (i.e., object recognition). In real-world environments, object recognition must cope with a high number of degrees of freedom (pose, scale, illumination, etc.), and uncertainty in the visual information plays a major role. Attention supports object search by constraining the dimension of the search space by limiting the visual information to a restricted set of hypotheses bounded to regions.

One central thesis is that attention acts to optimize the search procedure inherent in a solution to vision (Tsotsos et al., 1995). The accordingly proposed selective tuning model emphasized the role of top-down processing in attention mechanisms. Analogously, task-based attention (de Laar et al., 1997) operates in top-down information paths, adjusting a processing layer of intermediate features in a goal-driven manner.

An influential model in attention modeling is to feed bottom-up information into a competitive processing layer, where feature responses are competing according to a winner-takes-all (WTA) strategy for priority in interpretation (Itti et al., 1998; Navalpakkam and Itti, 2002). Ramstrøm and Christensen (2002) proposed a distributed control layer inspired by market principles in which bottom-up feature responses are competing according to a game theoretical model—the competetive equilibrium—with top-down information.

Visual attention in terms of an interactive process has been considered by Lee et al. (2003), in which attentional behavior should be biased with respect to particular objects, spatial locations, and time. A spiking neural network, for example, can bias selective behavior in such a way that it speeds up (facilitates) or slows down (interferes) with the processing of a given visual stimulus, gradually relating "where" and "when" access of information in the network. The network can then manipulate the amount of bottom-up and top-down influence on a search task to investigate the dynamic and modulatory aspects of selective attention.

A face detection task demonstrates the new functionality (Fig. 105.1), allocating the focus of attention to possible target locations. Skin color, facial features, and ellipse-like shape determine a bottom-up map that is correlated with cued features. The net input $\text{net}_j(t)$ of a spiking neural network node j at time t is,

$$\text{net}_j(t) = \alpha \sum_{i=1}^{n} w_{ij}(t) x_i^B + \beta \sum_{i=1,r=1}^{n,m} u_{irj}(t) (x_i^B)^2 x_r^T \quad (1)$$

where B and T stand for the bottom-up and top-down inputs, n and m are the dimensions of the bottom-up and top-down inputs, and w and u are the bottom-up

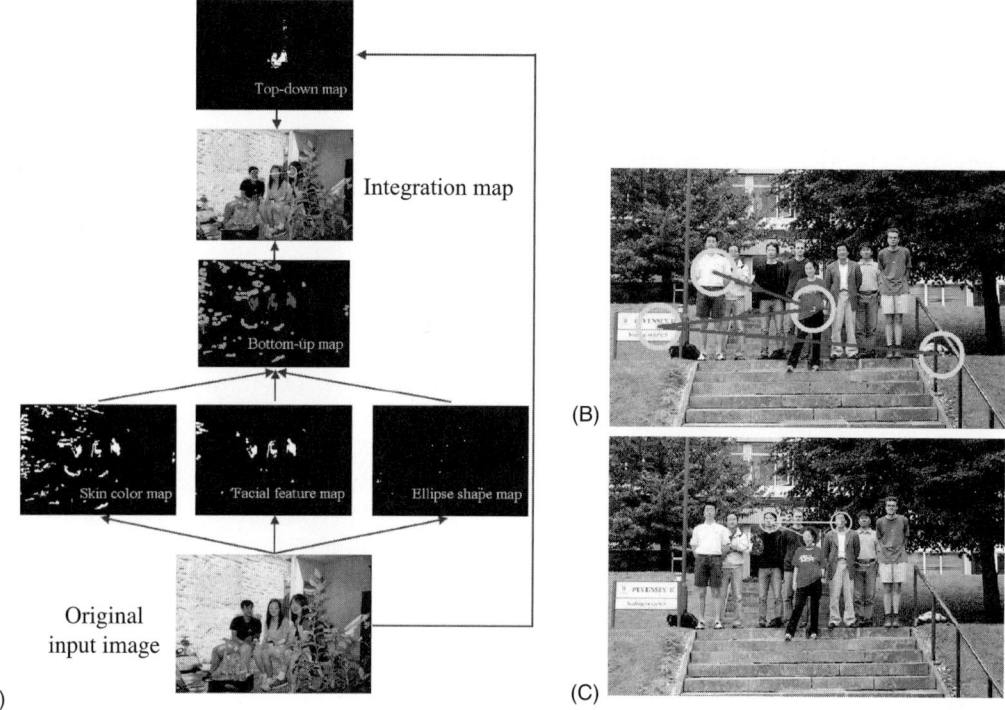

FIGURE 105.1 Example of interactive processing with natural images containing faces (Lee et al., 2003). The model (A) allocates focus of attention to possible target locations. The more task-relevant the target location with respect to the cue, the more likely the location is selected early in the attentional trajectory. Results from bottom-up WTA attention (B) are contrasted to scanpaths from graded interactive top-down attention processing (C).

and multiplicative weights, respectively. The $(x_i^B)^2$ term reinforces correlation between the bottom-up and top-down information stream, depending on whether they are consistent or not. The impact of the top-down information is illustrated by the different behaviors in Fig. 105.1B and C, showing the refined selection of targets not only due to generic visual features (B) but also from the maximization of information between extracted and object-cued features (C).

These computational models of selective attention in the visual search task support the spotlighting of a sequence of regions of interest. Applications are in the detection of objects of interest and scene recognition that is simply based on the occurrence of objects within. Any geometrical or probabilistic relation between serially focused information is not investigated here. However, context analysis would be mandatory for any more extended analysis of the scene or part-based object recognition.

C. Contextual Cueing and Attention Strategies

In computer vision, we face the challenging task of detecting objects of interest in outdoor environments. Changing illumination, different weather conditions, and noise in the imaging process are the most important issues that require a truly robust detection system. Recent work on real-time video interpretation therefore considers attentional mechanisms (Navalpakkam and Itti, 2002) and cascaded systems (Viola and Jones, 2001) to coarsely analyze the complete video frame in a first step, reject irrelevant hypotheses, and iteratively apply increasingly complex classifiers with appropriate level of detail (Sec. IIA). Attention from context priming (Torralba and Sinha, 2001; Ogris and Paletta, 2003) makes sense out of globally defined environmental features to set priors on object-related observable variables to obtain spatial pointers to regions of interest and therefore significantly improves the quality of service in real-time interpretation. Cascaded processing serves as fundamental methodology in sequential saccadic interpretation, giving rise to improved analysis from any updating evidence.

1. Context Priming

Investigations on the binding between scene recognition and object localization made in experimental psychology have produced clear evidence that highly local features play an important role into facilitating detection from predictive schemes (Hollingworth and Henderson, 2002). In particular, the visual system infers knowledge about stimuli occurring in certain locations, leading to expectancies regarding the most probable target in the different locations (location-specific target expectancies). Related work on scene recognition concerns the contextual mapping from objects to objects (Rimey, 1993) and from global scene features to object hypotheses (Torralba and Sinha, 2001) with respect to a static environment.

The extraction of scene landmarks has recently been applied for priming tasks in attentive object detection (Ogris and Paletta, 2003). Landmarks can be detected more robustly and rapidly than arbitrary objects. The use of landmark configurations inherently includes local spatial context and thus becomes more locally discriminative and predictive than using global features (Fig. 105.2A). Landmark configurations are then mapped to visual object events in local proximity. Defining the mapping in terms of a probabilistic estimation of direction β,

$$p(\mathbf{x}_i | l_j, o_k, \beta(l_j, o_k)) \propto N_{l_i}(\mu_\beta, \sigma_\beta) \quad (2)$$

where image pixel $\mathbf{x}_i$ provides confidence p for being on a line to landmark l_i, and $\lambda_{i,t} \in l_i$ is an associated landmark appearance sampled at time t. This confidence is exponentially decreasing—according to a Gaussian $N(\cdot)$—with increasing angle β from the reference line μ_β (from landmark l_i to object o_k). The approach integrates, then, those landmarks l_j that have been consecutively visited in an observation sequence and been selected as estimators for the next object location, by

$$p(\mathbf{x}_i | l_1, \cdots | l_j, \cdots, o_k, \beta(l_j, o_k)) = \prod_{j=1}^{J} p(\mathbf{x}_i | l_j, o_k, \beta(l_j, o_k)) \quad (3)$$

Figure 105.2B illustrates the result of a recursive estimation of attention from predictions of individual landmarks. Experimental results on frames of a video scene demonstrate an average prediction error of ≈3° in the attention direction and a hit rate of ≈94% (Ogris and Paletta, 2003).

2. Saccadic Information Integration

The framework of active and purposive vision laid the conceptual basis to understand computer vision tasks from the perspective of acquiring and optimizing sequential decisions in goal-driven tasks. In this context, Bandera et al. (1996) introduced reinforcement learning to improve the performance of foveal visual attention for the task of model-based object recognition. Recent attempts to model Markovian decision processes for automatic gaze control in face and scene perception (Henderson et al., 2001) and attentive viewpoint control (Paletta and Pinz, 2000) demonstrate the potential for automatic saccade and viewpoint control for the interpretation of objects and scenes.

III. ATTENTION IN ROBOTIC SYSTEMS

The understanding of how to design intelligent robotic systems has seen a paradigm shift during the last decade. Robot architectures are no longer control driven by symbolic artificial intelligence (sense-plan-act) and do not rely on universal perceptual reconstructions of the environment by purposive sensor-driven control (sense-act) and interpretation schemes. Instead, with reference to situatedness, embodiment, and context-relatedness of task performance, visual perception has been relocated as a functional basis for behavior-based control (Arkin, 1998). Attention has been playing an increasingly important role in providing solutions to the control of a growing stream of sensory input. Mobile robots are often 'thrown' into a complex environment where they have to apply their knowledge and find out about what

(a)   (b)

FIGURE 105.2 Spatial attention from local context (Ogris and Paletta, 2003). (A) Landmark configurations are discriminated locally to provide (B) attentive pointers to nearby object locations.

needs to be attended to and when and what to do in correspondence with visual feedback.

A. Autonomous Multisensor Robots

Common tasks in the control of autonomous mobile robots are collision avoidance, navigation, and tracking and manipulation of objects (Coelho et al., 2001). In order to execute these tasks correctly, the robot needs to detect objects and free space in its environment quickly and reliably, emphasizing the role of machine attention to find points of interest.

Multisensor data provide new challenges on selective attention to autonomous and mobile robot systems. In particular, humanoid robots are involved in a multitude of sensory inputs, such as from image feature, motion, and audio signal streams (Vijayakumar et al., 2001). Vijayakumar et al. (2001) applied a WTA network for saliency computation according to Itti et al.'s (1998) model on a 30 degrees of freedom (DOF) humanoid with pan and tilt peripheral and foveal vision, demonstrating motion and person detection in video-rate attention control.

Objects usually have range discontinuities at their borders that can help to detect them. Models comprising depth as a feature typically use stereo vision to compute it, which is computationally expensive, and only a fraction of the image pixels contribute to the computed 3D point clouds. Frintrop et al. (2003) describe attention from a multimodal 3D laser scanner. The attention system has a similar structure as the Neuromorphic Vision Toolkit (Itti et al., 1998), but uses only intensity and orientation as features. Depth information is intensity coded, and the system is capable of simultaneously processing both remission and depth value images, generating a single saliency map for both. Figure 105.3 illustrates how focus of attention from reflectance and range images contributes to classification and detection using fusion of the individual sensor maps.

Cognitive aspects in robot attention were implemented by Dickinson et al. (1997). Here, an active object recognition strategy combined the use of an attention mechanism for viewpoint selection to disambiguate recovered object features. The attention mechanism consisted of a probabilistic search through a hierarchy of predicted feature observations. Paletta and Rome (2000) describe a robotic system that uses task-specific knowledge in order to direct attention to certain regions of interest. Object detection is performed by an appearance-based classification of the regions of interest (ROIs). Because the domain objects (sewer pipe inlets) are often very similar, the system learns an information-fusion approach to disambiguate objects. The classification is performed iteratively with images taken from different viewpoints,

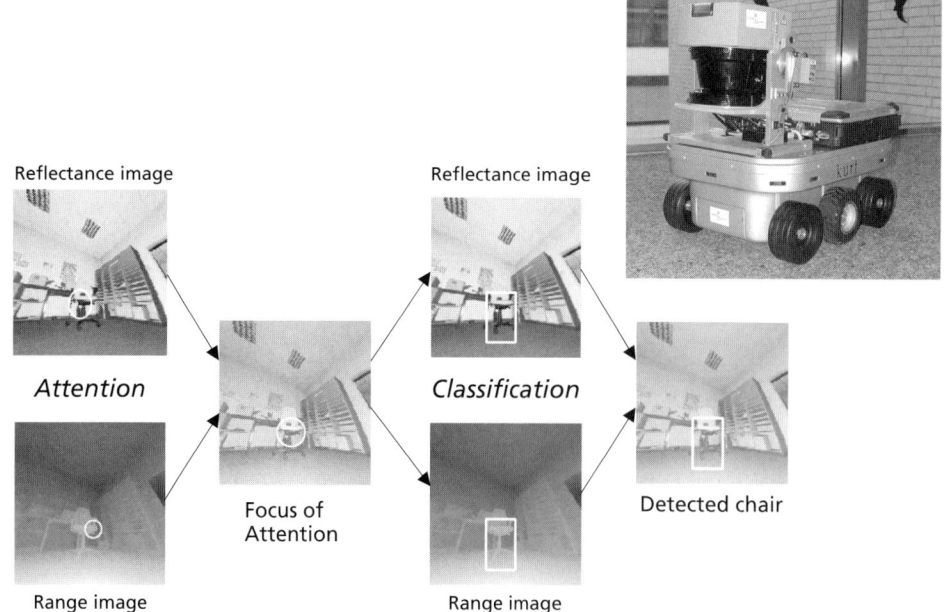

FIGURE 105.3 The custom 3D range finder mounted on top of the mobile robot KURT2 (upper right). Office scene imaged with the 3D scanner in range- and remission-value modes. The system, described in (Frintrop et al., 2003, 2004), performs attention-based ROI selection and successive classification within the ROI.

until the confidence is sufficient for making a decision. The robot KISMET (Breazeal and Scasselatti, 1999) integrated perception, attention, drives, emotions, behavior arbitration, and expressive acts in order to interact socially with humans. The attention system was based on Wolfe's model (Wolfe, 2000) and integrated perceptions with habituation effects and influences from the robot's motivational and behavioral state to create a context-dependent attention activation map. Finally, joint attention is highly relevant in the supervised learning stages of human development, specifically for the capability to interact with the environment. Nagai et al. (2003) demonstrated the finding of and attending to a salient object in the robot's view and a sensorimotor coordination when the visual attention succeeds. Based on these mechanisms, the robot learns the sensorimotor coordination when the robot can watch the salient object by shifting its gaze direction from the caregiver's face to the object.

B. Visuomotor Attention

Oculomotor control is not just important for spatial but also for sequential attention. Because eye movements must be executed in a sequential manner, it is crucial to focus visual attention at the right time on the right targets so that subsequent information processes, in particular motor planning and execution, receive relevant information sufficiently fast to update ongoing processes. From this viewpoint, oculomotor control may, for example, be a crucial constraint on how movement of other body parts are planned. Miyashita et al. (1996) showed that anticipatory saccades in sequential procedural learning in monkeys are tightly coupled to the limb-motor system. Similend, Shibata et al. (2001) developed a biomimetic gaze stabilization that is mused for attentional mechanisms in a humanoid robot (Vijayakumar et al., 2001).

Actually, robot visuomotor attention is in its early stages. Coelho et al. (2001) demonstrated the importance of learning of visual features that the observer has to attend to for more efficient grasping. They constructed constellations of visual features to predict relative hand-object postures that lead reliably to haptic utility.

IV. CONCLUSION

Attention is an ongoing research topic in computer vision that gains increasingly in importance under the guidance of cognitive vision system research. The classic application fields, such as video analysis, surveillance, and robotics, and not the only ones to depend on attention methodologies. Emerging technologies, such as mobile vision, wearable computing, and service robotics, are by nature limited in computational resources but should operate in everyday's complex contexts and work environments. These upcoming challenges to both computational and semantic complexity provide the motivation to reinforce research on attentional mechanisms in machine vision and mobile robot perception.

References

Arkin, R. C. (1998). "Behavior-Based Robotics." MIT Press, Cambridge, MA.

Backer, G., and Mertsching, B. (2003). Two selection stages provide efficient object-based attentional control for dynamic vision. *In* "Proceedings of the Workshop on Attention and Performance in Computer Vision," pp. 9–16. Joanneum Research, Graz, Austria.

Bandera, C., Vice, F. J., Bravo, J. M., Harmon, M. E., and Baird, L. C. (1996). Residual Q-learning applied to visual attention. *In* "Proceedings of the International Conference on Machine Learning," (Lorenza Saitta Ed.) pp. 20–27. Morgan Kaufmann Publishers, San Francisco, CA: Bari, Italy.

Braun, J., Koch, C., Lee, D. K., and Itti, L. (2001). Perceptual consequences of multilevel selection. *In* "Visual Attention and Cortical Circuits" (J. Braun, C. Koch, and J. L. Davis, Eds.), pp. 215–242. MIT Press, Cambridge, MA.

Breazeal, S., and Scasselatti, B. (1999). A context-dependent attention system for a social robot. *In* "Proceedings of the International Joint Conference on Artificial Intelligence," (Thomas Dean, Ed.) Vol. 2, pp. 1146–1153. Morgan Kaufmann, San Francisco, CA: Stockholm, Sweden.

Clark, J., and Ferrier, N. (1988). Modal control of an attentive vision system. *In* "Proceedings of the International Conference on Computer Vision", *IEEE*, pp. 514–523. Tampa, Florida.

Coelho, J., Piater, J., and Grupen, R. (2001). Visual perceptual categories for reaching and grasping with a humanoid robot. *Robotics Autonomous Syst.* **37**, 195–219. Santa Margherita Ligure, Italy, Springer.

Culhane, S., and Tsotsos, J. (1992). An attentional prototype for early vision. *In* "Proceedings of the European Conference on Computer Vision," pp. 551–560. Sauta Maighenta Ligure, Italy, Springer.

de Laar, P. V., Heskes, T., and Gielen, C. (1997). Task-dependent learning of attention. *Neural Networks* **10**, 981–992.

Dickinson, S. J., Christensen, H. I., Tsotsos, J. K., and Olofsson, G. (1997). Active object recognition integrating attention and viewpoint control. *Comput. Vis. Image Understanding*, **67**, 239–260.

Fraundorfer, F., and Bischof, H. (2003). Utilizing saliency operators for image matching. *In* "Proceedings of the Workshop on Attention and Performance in Computer Vision", pp. 17–24. Joanneum Research, Graz, Austria.

Frintrop, S., Nüchter, A., and Surmann, H. (2004). Visual attention for object recognition in spatial 3D data. *In* "Proceedings of the 2nd Workshop on Attention and Performance in Computational Vision," pp. 75–82. Joanneum Research, Graz, Austria.

Frintrop, S., Rome, E., Nüchter, A., and Surmann, H. (2003). An attentive, multi-modal laser "eye". *In* "Proceedings of the International Conference on Computer Vision Systems," pp. 202–211. Joanneum Research, Graz, Austria.

Henderson, J. M., Falk, R., Minut, S., Dyer, F. C., and Mahadevan, S. (2001). Gaze control for face learning and recognition by humans and machines. *In* "From Fragments to Objects: Segmentation and Grouping in Vision" (T. Shipley and P. Kellman, Eds.), pp. 463–481. Elsevier Science, Oxford.

Hollingworth, A., and Henderson, J. (2002). Accurate visual memory for previously attended objects in natural scenes. *J. Expe. Psychol. Hum. Perception Performance* **28**, 113–136.

Itti, L., Koch, C., and Niebur, E. (1998). A model of saliency-based visual attention for rapid scene analysis. *IEEE Trans. Pattern Analysis Machine Intell.* **20**, 1254–1259.

Jagersand, M. (1995). Saliency maps and attention selection scale and spatial coordinates: An information theoretic approach. *In* "Proceedings of the International Conference on Computer Vision," IEEE, pp. 195–202. Boston, MA.

Lee, K., Buxton, H., and Feng, J. (2003). Selective attention for cue-guided search using a spiking neural network. *In* "Proceedings of the Workshop on Attention and Performance in Computer Vision," pp. 55–63. Graz, Austria.

Lionelle, A., and Draper, B. (2003). Evaluation of selective attention under similarity transforms. In "Proceedings of the Workshop on Attention and Performance in Computer Vision," pp. 31–38. Graz, Austria.

Miyashita, K., Kato, R., Miyauchi, S., and Hikosaka, O. (1996). Anticipatory saccades in sequential procedural learning monkeys. *J. Neurophysiol.* **76**, 1361–1365.

Nagai, Y., Hosoda, K., Morita, A., and Asada, M. (2003). Joint attention emerges through bootstrap learning. *In* "Proceedings of the International Conference on Intelligent Robotic Systems," IEEE, pp. 168–173. Las Vegas, Nevada.

Navalpakkam, V., and Itti, L. (2002). A goal oriented attention guidance model. *In* "Proceedings of the International Workshop on Biologically Motivated Computer Vision," Heinrila H. Bulthoff, Seong-Whau Lee, Tomah A. Poggio, Christian Wallranean (Eds.) pp. 453–461. Tübingen, Germany, Springer.

Ogris, G., and Paletta, L. (2003). Contextual cueing for spatial attention in object detection from video. *In* "Proceedings of the Workshop on Attention and Performance in Computer Vision," pp. 64–72. Joanneum Research, Graz, Austria.

Paletta, L., Humphreys, G., and Fisher, R. (Eds.) (2003). "Proceedings of the International Workshop on Attention and Performance in Computer Vision." Joanneum Research, Graz, Austria.

Paletta, L., and Pinz, A. (2000). Active object recognition by view integration and reinforcement learning. *Robotics Autonomous Syst.* **31**, 71–86.

Paletta, L., and Rome, F. (2000). Learning fusion strategies for visual object detection. *In* "Proceedings of the International Conference on Intelligent Robots and Systems," IEEE, pp. 1446–1452. Takamatsu, Japan.

Ramstrøm, O., and Christensen, H. (2002). Visual attention using game theory. *In* "Proceedings of the International Workshop on Biologically Motivated Computer Vision," Heinrila H. Bulthoff, Seong-Whau Lee, Tomah A. Poggio, Clairliau Wallranean (Eds.) pp. 462–471.

Rimey, R. D. (1993). Control of selective perception using bayes nets and decision theory. Technical Report TR468. Computer Science Department, University of Rochester.

Shibata, T., Vijayakumar, S., Conradt, J., and Schaal, S. (2001). Biomimetic oculomotor control. *Adaptive Behav.* **9**, 189–207.

Torralba, A., and Sinha, P. (2001). Statistical context priming for object detection. *In* "Proceedings of the International Conference on Computer Vision," pp. 763–770. Vancouver, Canada. IEEE.

Treisman, A., and Gelade, G. (1980). A feature integration theory of attention. *Cogn. Psychol.* **12**, 97–136.

Tsotsos, J. K., Culhane, S. M., Wai, W, Y. K., Lai, Y., Davis, N., and Nuflo, F. (1995). Modelling visual attention via selective tuning. *Artificial Intell.* **78**, 507–545.

Vijayakumar, S., Conradt, J., Shibata, T., and Schaal, S. (2001). Overt visual attention for a humanoid robot. *In* "Proceedings of the International Conference on Intelligent Robots and Systems," Maui, HI. Vol. 4, pp. 2332–2337.

Viola, P., and Jones, M. (2001). Rapid object detection using a boosted cascade of simple features. *In* "Proceedings of the Conference on Computer Vision and Pattern Recognition," Kauai, HI. pp. 511–518. IEEE.

Wolfe, J. M. (2000). Visual attention. *In* "Seeing" (K. K. DeValois, Ed.), pp. 335–386. Academic Press, San Diego, CA.

CHAPTER

106

Attention for Computer Graphics Rendering

Hector Yee and Sumanta Pattanaik

ABSTRACT

We present a method to accelerate global illumination computation in prerendered animations by taking advantage of visual saliency and the limitations of the human visual system. A visual importance map, constructed from a saliency and visual sensitivity model, is used to accelerate rendering. Results indicate an order of magnitude improvement in computational speed.

I. EFFICIENT REALISTIC RENDERING OF SYNTHETIC IMAGES

Accurate lighting computation is one of the key elements to realism in rendered images. In the past 2 decades, many physically based global illumination algorithms have been developed for accurately computing the light. These algorithms simulate the propagation of light in a three-dimensional environment and compute the distribution of light to desired accuracy. Unfortunately, the computation times of these algorithms are high. For a reasonably complex scene with dynamic objects, physically accurate global illumination computation for each image frame can take many hours. Thus, one of the main research goals in the field of rendering is to develop algorithms for faster and more realistic rendering. In this chapter, we describe one such algorithm (Yee et al., 2001), in which we improve the speed of the rendering computation by allocating computational effort to various parts of a scene in proportion with the visual saliency of the areas of the scene.

The basic idea behind this work is that only a small fraction of a viewed scene is visually important. Thus, we save significantly on time by terminating the computation early in areas of the image that are not visually important.

II. VISUAL ATTENTION–BASED RENDERING ALGORITHM

The first step in our algorithm is to derive the saliency of the areas of the scene visible in the current view. To be able to do this, we first obtain a quick estimate of the final image by hardware rendering a preview image of the scene. The preview image contains important visual cues such as the color of objects, the textures of the objects, and approximate contrasts. We apply a model of visual attention to the preview image to obtain a saliency map (see Chapter 94). The next step of our algorithm is to apply a model of human visual system to the preview image to obtain visual sensitivity map for the viewed scene (Ramasubramanian et al., 1999). The model of the human visual system takes into account the loss of contrast sensitivity in the presence of high spatial frequencies and motion. We modulate the saliency map with the visual sensitivity map to obtain a map of visual importance. We use this map of visual importance to guide our global illumination computation. We allocate less computational effort to areas of less importance and more to areas of higher importance. Figure 106.1A shows an image computed using our efficient global illumination algorithm. Figure 106.1B is the importance map for the image, generated using visual saliency and the model of the human visual system. The brighter pixels correspond to low visual importance and received less computational effort. Darker pixels correspond to high visual importance

FIGURE 106.1 (A) Image drawn by Global Illumination. (B) Visual importance map derived from visual salience and contrast sensitivity. (Bright areas are less important.)

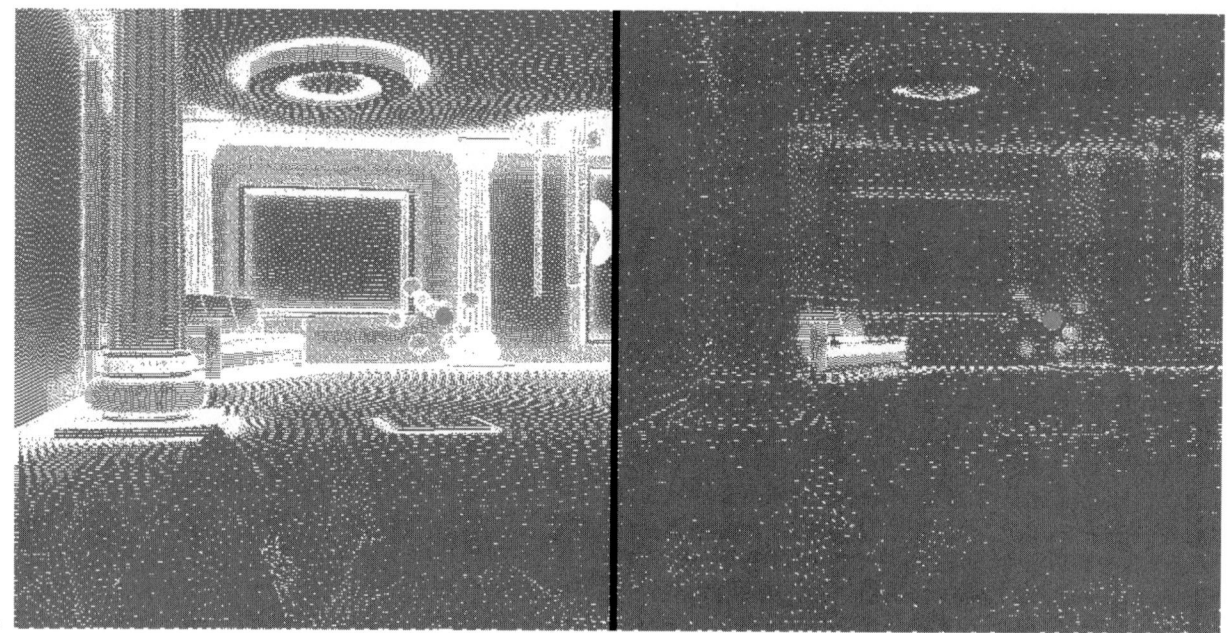

FIGURE 106.2 Sampling difference between (A) the reference and (B) the importance-guided solutions.

and hence received more computational effort. For example, the sofa near the center received more computer time. The rational behind this is that the sofa stands out in the corner of the scene as salient and it has very little high frequency content to mask the error; hence, correspondingly more computation effort is needed to ensure that the result looks accurate to the visual system.

By using a model of visual attention to guide the global illumination algorithm, we found that we can speed up calculation by an order of magnitude. Figure 106.2 shows the distribution of the computational effort. A bright dot in the image corresponds to a unit amount of effort spent by our global illumination algorithm. The density of the white dots is proportional to the computational effort spent by the algorithm. In

Fig. 105.2B we have the effort distribution by our algorithm. For comparison, in Fig. 105.2A we have the effort distribution for a reference solution generated by switching off the attention-based acceleration. As can be seen, our algorithm used far less processing power to compute the lighting with equal visual accuracy. Furthermore, computation was concentrated in visually important regions such as the sofa, the flower petals, and the bouncing spheres.

III. RELATED WORK IN THE FIELD

We have come across Haber et al.'s (2001) work that makes attempts similar to our work by using visual attention to speed up high-quality rendering computation. It uses the saliency map to enhance interactive graphics by performing ray tracing in areas that are deemed to be visually important. This allows them to simulate computationally expensive effects such as glossy surfaces at interactive rates.

References

Haber, J., Myszkowski, K., Yamauchi, H., and Seidel, H.-P. Perceptually Guided Corrective Splatting. (2001). In EuroGraphics (Manchester, UK, September 4–7), pp. 142–152.

Ramasubramanian, M., Pattanaik, S. N., and Greenberg, D. P. (1999). A perceptually based physical error metric for realistic image synthesis. *In* "Proceedings of SIGGRAPH'99," Alyn Rockwood, Ed. pp. 73–82. Addison Wesley Longman, Los Angeles.

Yee, H., Pattanaik, S. N., and Greenberg, D. P. (2001). Spatio-temporal sensitivity and visual attention for efficient rendering of dynamic environments. *ACM Trans.* Graphics **20**, 39–65.

CHAPTER 107

Linking Attention to Learning, Expectation, Competition, and Consciousness

Stephen Grossberg

ABSTRACT

The concept of attention has been used in many senses, often without clarifying how or why attention works as it does. Attention, like consciousness, is often described in a disembodied way. The present article summarizes neural models and supportive data about how attention is linked to processes of learning, expectation, competition, and consciousness. A key theme is that attention modulates cortical self-organization and stability. The perceptual and cognitive neocortex is organized into six main cell layers, with characteristic sublamina. Attention is part of a unified design of bottom-up, horizontal, and top-down interactions among identified cells in laminar cortical circuits. Neural models clarify how attention may be allocated during processes of visual perception, learning, and search; auditory streaming and speech perception; movement target selection during sensory-motor control; mental imagery and fantasy; and hallucinations during mental disorders, among other processes.

I. INTRODUCTION

Attention is a behavioral concept, but one whose properties arise from brain mechanisms. To fully understand how attention works, we need to mechanistically link brain mechanisms to the attentive behavioral functions that they control. Building brain-behavior links for processes of attention is particularly challenging because attention is typically a modulatory process that can sensitize, or prime, an observer to expect an object to occur at a given location or with particular stimulus properties (Posner, 1980; Duncan, 1984). Were attention, by itself, able to routinely activate fully formed perceptual representations, then we could not tell the difference between external reality and internal fantasy or hallucination. Thus, to fully understand attention, we need to explain the brain processes that attention is modulating. A rapidly growing number of models can now quantitatively simulate the neurophysiologically recorded dynamics of identified nerve cells in known anatomies *and* the behaviors that they control, and these models naturally include attentional processes.

This article emphasizes models and data about how attention is realized within the laminar circuits of neocortex. Neural system models have also clarified how attention may be allocated during many different tasks.

II. LINKING ATTENTION TO LEARNING, EXPECTATION, COMPETITION, SYNCHRONIZATION, AND CONSCIOUSNESS

Neural models of perception and cognition have predicted that top-down attention is a key mechanism for solving the stability-plasticity dilemma (Grossberg, 1980, 1999b), which concerns the fact that brains can rapidly learn enormous amounts of information throughout life without just as rapidly forgetting what they already know. How do attentive processes within neocortex help to stabilize cortical learning and memory through time so that they are not catastrophically overwritten by the new stimuli with which they are continually bombarded?

Adaptive Resonance Theory (ART) proposes to explain how attention helps to solves the stability-plasticity dilemma by modeling how bottom-up signals activate top-down expectations whose signals are matched against bottom-up data. Both the bottom-up and top-down pathways contain adaptive weights, or long-term memory traces, that may be modified by experience (Fig. 107.1A). The learned top-down expectations focus attention on information that matches them (Fig. 107.1B). They select, synchronize, and amplify the activities of cells within the attentional focus while suppressing the activities of irrelevant cells that which could otherwise be incorporated into previously learned memories and thereby destabilize them. The cell activities that survive such top-down attentional focusing rapidly reactivate bottom-up pathways (Fig. 107.1A), thereby generating a feedback resonance between bottom-up and top-down signal exchanges. Such resonances rapidly bind distributed information at multiple levels of brain processing into context-sensitive representations of objects and events. These resonances are proposed to support slower processes of learning, hence the name adaptive resonance. ART also predicts that "All conscious states are resonant states." Thus, ART links attention to processes of learning, expectation, competition, synchronization, and consciousness.

Since these predictions were made in the 1970s, many experimental and modeling studies have provided support for them. Some relevant experiments are summarized here. Other chapters in this volume provide additional supportive evidence (e.g., Chapters 25, 44, 49, 50, 66, 81, and 88).

Mathematical analyses have proved how easily learning can lead to catastrophic forgetting in response to a changing world and how top-down attention can stabilize learning if it satisfies four properties (Carpenter and Grossberg, 1991), which together are called the ART Matching Rule:

Bottom-up automatic activation. A cell, or cell population, can become active enough to generate output signals if it receives a large enough bottom-up input, other things being equal. Such an input can drive the cell to suprathreshold levels of activation.

Top-down priming. A cell cannot fire if it receives only a large top-down expectation input. Such a top-down signal can modulate, prime, or sensitize the cell and thereby prepare it to react more quickly and vigorously to subsequent bottom-up inputs that approximately match the top-down expectation. A top-down signal can also shift the baseline firing rate of the cell. It cannot, however, generate large behaviorally significant output signals from the cell.

Match. A cell can fire if it receives large convergent bottom-up and top-down inputs. Such a matching process can generate enhanced and synchronized cell activation as resonance takes hold.

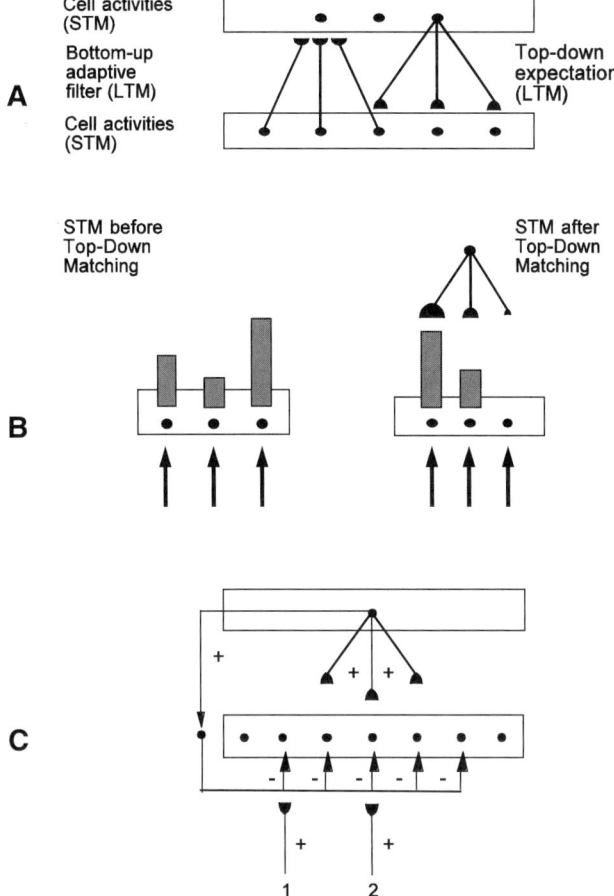

FIGURE 107.1 (A) Patterns of activation, or short-term memory (STM), on a lower processing level send bottom-up signals to a higher processing level. These signals are multiplied by adaptive weights, or learned long-term memory (LTM) traces, which influence the activation of the cells at the higher processing level. These latter cells, in turn, activate top-town expectation signals that are also multiplied by learned LTM traces. These top-down expectations are matched against the STM pattern that is active at the lower level. (B) This matching processes, confirms, amplifies, and synchronizes STM activations that are supported by large LTM traces in an active top-down expectation and suppresses STM activations that do not get top-down support, thereby focusing attention on the representations encoded by the selected cells. The size of the hemidisks at the end of the top-down pathways represents the strength of the learned LTM trace that is stored in that pathway. (C) The ART Matching Rule may be realized by a top-down modulatory on-center off-surround network. (Reprinted with permission from Grossberg, 1999a).

Mismatch. Cell activity is suppressed, even if the cell receives a large bottom-up input, if it also receives only a small, or zero, top-down expectation input.

The simplest mathematically possible circuit (Carpenter and Grossberg, 1991), a top-down modulatory on-center off-surround network (Fig. 107.1), has successfully been used to simulate a variety of behavioral and brain data (Grossberg, 1999b). In such a circuit, when only bottom-up signals are active, all cells can fire that receive large enough inputs. When only top-down attention is active, cells in the off-surround that receive inhibition but no excitation can get strongly inhibited, whereas cells in the on-center that receive a combination of excitation and inhibition can get at most subliminally activated due to an approximate balance between excitation and inhibition. When bottom-up and top-down inputs match (pathway 2 in Fig. 107.1C), the two excitatory sources of excitation (bottom-up and top-down) that converge at the cell can overwhelm the one inhibitory source; it is a case of two against one that can lead to synchronous firing. When bottom-up and top-down inputs mismatch (pathway 1 in Fig. 107.1C), the top-down inhibition can neutralize the bottom-up excitation; it is a case of one against one.

III. ATTENTION IS MODULATORY

The ART Matching Rule predicts that top-down attention accomplishes modulatory priming and matching by using competitive mechanisms such as the top-down modulatory on-center off-surround network in Fig. 107.1C. Data compatible with this prediction were gradually reported over the years, with an acceleration of experiments during the past 5 years. For example, Zeki and Shipp (1988, 316) wrote that "backward connections seem not to excite cells in lower areas, but instead influence the way they respond to stimuli." Likewise, the data of Sillito et al. (1994, 479–482) on attentional feedback from V1 to LGN led them to conclude that "the cortico-thalamic input is only strong enough to exert an effect on those dLGN cells that are additionally polarized by their retinal input. . . . the feedback circuit searches for correlations that support the 'hypothesis' represented by a particular pattern of cortical activity." Their experiments demonstrated all of the properties of the ART Matching Rule—they found in addition that "cortically induced correlation of relay cell activity produces coherent firing in those groups of relay cells with receptive-field alignments appropriate to signal the particular orientation of the moving contour to the cortex. . . . this increases the gain of the input for feature-linked events detected by the cortex." In other words, top-down signaling, by itself, cannot fully activate LGN cells; it needs matched bottom-up retinal inputs to do so, and those LGN cells whose bottom-up signals support cortical activity may get synchronized and amplified by this feedback. In addition, anatomical studies have shown that the top-down V1 to LGN pathway realizes a top-down on-center off-surround network (Dubin and Cleland, 1977; Weber et al., 1989; see Fig. 107.2D).

IV. LAMINAR ORGANIZATION OF BOTTOM-UP, HORIZONTAL, AND TOP-DOWN CONNECTIONS

How are top-down attentional circuits realized within the brain, in particular within perceptual and cognitive neocortex? All the sensory and cognitive neocortex is organized into six main layers of cells. A recent family of LAMINART models (Fig. 107.2) proposes a detailed answer to this question for the interblob stream of visual cortex and, by extension, to other neocortical areas by characterizing how bottom-up, top-down, and horizontal interactions are organized within cortical layers to generate percepts of visual form. In particular, LAMINART shows how these interactions help visual cortex to realize: (1) stable development and learning of circuit connections and weights in response to a changing environment, (2) coherent grouping or binding of distributed information into boundary representations of objects and events without a loss of analog sensitivity—the property of analog coherence, and (3) attentional focusing on important object representations at the expense of less important representations (Grossberg, 1999a; Grossberg, Mingolla, et al., 1997; Grossberg and Raizada, 2000; Grossberg and Seitz, 2003; Grossberg and Williamson, 2001; Raizada and Grossberg, 2001). Three important implications of this result are as follows.

First, biological vision systems are not merely bottom-up filtering devices, as Hubel and Wiesel proposed in their classical analysis of early vision. Rather, even early stages of visual cortex join together bottom-up filtering, horizontal grouping, and top-down attention. Perceptual grouping, the process that binds spatially distributed and incomplete information into 3D object representations, starts at an early cortical stage; see Fig. 107.2C. These grouping interactions are often cited as the basis of nonclassical receptive fields that are sensitive to the context in which individual features are found (Bosking et al., 1997; Grosof et al.,

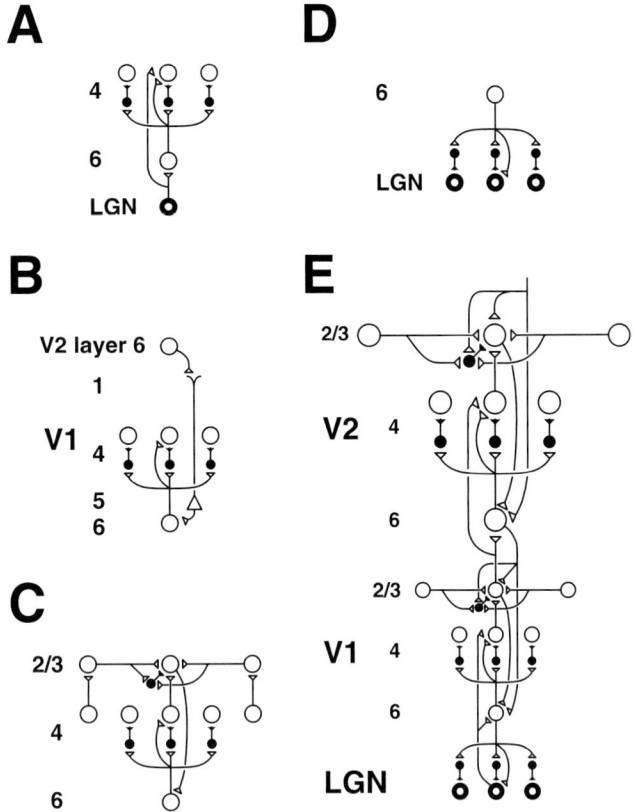

FIGURE 107.2 How known cortical connections join layer 6 → 4 and layer 2/3 circuits to form a laminar V1 and V2 model. Inhibitory interneurons are shown as solid black. (A) The LGN provides bottom-up activation to layer 4 via two routes. First, it makes a strong connection directly into layer 4. Second, LGN axons send collaterals into layer 6 and thereby also activate layer 4 via the 6 → 4 on-center off-surround path. The combined effect of the bottom-up LGN pathways is to stimulate layer 4 via an on-center off-surround, which provides divisive contrast normalization (Grossberg, 1980; Heeger, 1992) of layer 4 cell responses. (B) Folded feedback carries attentional signals from higher cortex into layer 4 of *V1* via the modulatory 6 → 4 path. Corticocortical feedback axons tend preferentially to originate in layer 6 of the higher area and to terminate in layer 1 of the lower cortex, where they can excite the apical dendrites of layer 5 pyramidal cells whose axons send collaterals into layer 6. The triangle represents such a layer 5 pyramidal cell. Several other routes through which feedback can pass into V1 layer 6 exist. Having arrived in layer 6, the feedback is then folded back up into the feedforward stream by passing through the 6 → 4 on-center off-surround path. (C) Connecting the 6 → 4 on-center off-surround to the layer 2/3 grouping circuit. Like-oriented layer 4 simple cells with opposite contrast polarities compete (not shown) before generating half-wave-rectified outputs that converge onto layer 2/3 complex cells in the column above them. Just like attentional signals from higher cortex, as shown in (B), groupings that form within layer 2/3 also send activation into the folded feedback path to enhance their own positions in layer 4 beneath them via the 6 → 4 on-center and to suppress input to other groupings via the 6 → 4 off-surround. Direct layer 2/3 → 6 connections in macaque V1 exist, as well as indirect routes via layer 5. (D) Top-down corticogeniculate feedback from V1 layer 6 to LGN also has an on-center off-surround anatomy, similar to the 6 → 4 path. The on-center feedback selectively enhances LGN cells that are consistent with the activation that they cause, and the off-surround contributes to length-sensitive (end-stopped) responses that facilitate grouping perpendicular to line ends. (E) The entire V1/V2 circuit. V2 repeats the laminar pattern of V1 circuitry but at a larger spatial scale. In particular, the horizontal layer 2/3 connections have a longer range in V2, allowing above-threshold perceptual groupings between more widely spaced inducing stimuli to form. V1 layer 2/3 projects up to V2 layers 6 and 4, just as LGN projects to layers 6 and 4 of V1. Higher cortical areas send feedback into V2, which ultimately reaches layer 6, just as V2 feedback acts on layer 6 of V1 (Sandell and Schiller, 1982). Feedback paths from higher cortical areas straight into V1 (not shown) can complement and enhance feedback from V2 into V1. Top-down attention can also modulate layer 2/3 pyramidal cells directly by activating both the pyramidal cells and inhibitory interneurons in that layer. The inhibition tends to balance the excitation, leading to a modulatory effect. These top-down attentional pathways tend to synapse in layer 1, as shown in (B). Their synapses on apical dendrites in layer 1 are not shown, for simplicity. (Reprinted with permission from Raizada and Grossberg 2001, where supportive data references are cited).

1993; Kapadia et al., 1995; Knierim and van Essen, 1992; Peterhans and von der Heydt, 1989; Polat et al., 1998; Sheth et al., 1996; von der Heydt et al., 1984; Sillito et al., 1995).

Second, even early visual processing is modulated by system goals via top-down expectations and attention (Motter, 1993; Roelfsema et al., 1998; Sillito et al., 1994; Somers et al., 1999; Watanabe et al., 1998). In particular, Fig. 107.2B illustrates how layer 6 of a higher cortical area can modulate layer 4 of a lower cortical area via a top-down on-center off-surround circuit. It can do so, for example, by activating apical dendrites in layer 1 of layer 5 cells, which activate layer 6 cells. Layer 6 cells, in turn, can activate layer 4 through a modulatory on-center off-surround network. Such a circuit exemplifies folded feedback (Grossberg, 1999a); namely, top-down signals are folded back into the feedforward flow of visual information processing. The 6-to-4 network is thus predicted to be an interface, called the preattentive-attentive interface, where data-driven bottom-up pre-attentive processing and task-directed top-down attentive processing are fused together via a shared decision circuit.

This layer 6-to-4 modulatory decision circuit realizes at least three functional roles in the model: contrast normalization of bottom-up inputs from earlier processing levels (Fig. 107.2A), selection of winning groupings that start to form in layer 2/3 via horizontal connections while preserving their analog coherence in response to intracortical feedback (C), and attentional priming in response to intercortical feedback from a higher cortical level (B). In particular, attention shares the same decision circuit as pre-attentive filtering and grouping, which is how attention can do its work. Attention can also directly modulate layer 2/3 groupings by activating the dendrites in layer 1 of excitatory and inhibitory cells in layer 2/3 (Lund and Wu, 1997; Rockland, and Virga, 1989). A balance between excitation and inhibition has been predicted to be a basic design principle in perceptual grouping (Grossberg, 1999a; Grossberg and Raizada, 2000). By activating both excitatory and inhibitory cells in layer 2/3, inhibitory interneurons that synapse on excitatory cells may balance their activation, thereby enabling attention to directly modulate the responses of grouping cells in layer 2/3.

Third, mechanisms governing property (1) in the infant lead to properties (2) and (3) in the adult. Thus, mechanisms that enable the cortex to learn in a stable way *define* key properties of adult visual information processing. This last result shows that learning and information processing need to be codesigned for either to work well in a novel environment.

V. ATTENTION, COMPETITION, AND MATCHING

Both ART and LAMINART predict that attention from higher cortical areas, such as area V2, acts on cells in area V1 via a top-down modulatory on-center off-surround network. Experiments of Hupé et al. (1997, 1031) support this prediction by showing that "feedback connections from area V2 modulate but do not create center-surround interactions in V1 neurons." More generally, the prediction that top-down attention has an on-center off-surround characteristic has received a considerable amount of psychological and neurobiological empirical confirmation in the visual system (Bullier et al., 1996; Caputo and Guerra, 1998; Downing, 1988; Mounts, 2000; Reynolds et al., 1999; Smith et al., 2000; Somers et al., 1999; Sillito et al., 1994; Steinman et al., 1995; Vanduffell et al., 2000). In particular, the claim that bottom-up sensory activity is enhanced when matched by top-down on-center signals is in accord with an extensive neurophysiological literature showing the facilitatory effect of attentional feedback (Luck et al., 1997; Roelfsema et al., 1998; Sillito et al., 1994), but not with models in which matches with top-down feedback cause suppression (Mumford, 1992; Rao and Ballard, 1999). ART predicts that on-center off-surround attentional feedback should exist in all sensory and cognitive systems that are capable of stable online learning. In particular, feedback from auditory cortex to the medial geniculate nucleus (MGN) and the inferior colliculus (IC) also has an on-center off-surround form (Zhang et al., 1997), as does feedback in the rodent barrel system (Temereanca and Simons, 2001).

Top-down attention through competitive matching has also been used to explain data about 3D figure-ground separation (Kelly and Grossberg, 2000), visual object learning and recognition (Bradski and Grossberg, 1995; Carpenter and Ross, 1995; Grossberg, 1999b), visual search (Grossberg et al., 1994), visual motion perception (Chey et al., 1997; Grossberg, Mingolla, and Viswanathan, 2001), auditory streaming (Grossberg, 1999b), speech perception and word recognition (Grossberg et al., 1997; Grossberg and Myers, 2000; Grossberg and Stone, 1986), selection of eye movement targets (Grossberg, Roberts, et al., 1997), and imagery, fantasy, and hallucinations (Grossberg, 2000).

The ART prediction that attention is mediated through competitive mechanisms has recently been restated in terms of the concept of biased competition (Desimone, 1998; see Chapter 50), in which attention biases the competitive influences within the network.

Figure 107.3 summarizes data in Reynolds et al. (1999) and a simulation of these data from Grossberg and Raizada (2000) that illustrate the on-center off-surround character of attention in macaque V2.

VI. OBJECT-BASED ATTENTION VIA THE PRE-ATTENTIVE-ATTENTIVE INTERFACE

When images that contain unambiguous groupings are processed, the laminar circuit in Fig. 107.2E can react quickly with a fast feedforward sweep of activation through layers 4-to-2/3 in one cortical area then to 4-to-2/3 in a higher cortical area, and so on (Thorpe et al., 1996). When ambiguous and complex scenes are being processed, competitive interactions in layers 4 and 2/3 are predicted to attenuate amplitude and processing rate of cell activation in layer 2/3. Intracortical feedback from layer 2/3-to-6 to 4-to-2/3 enables stronger groupings in layer 2/3 to be contrast-enhanced while they quickly inhibit weaker groupings and then to fire vigorously to higher cortical levels.

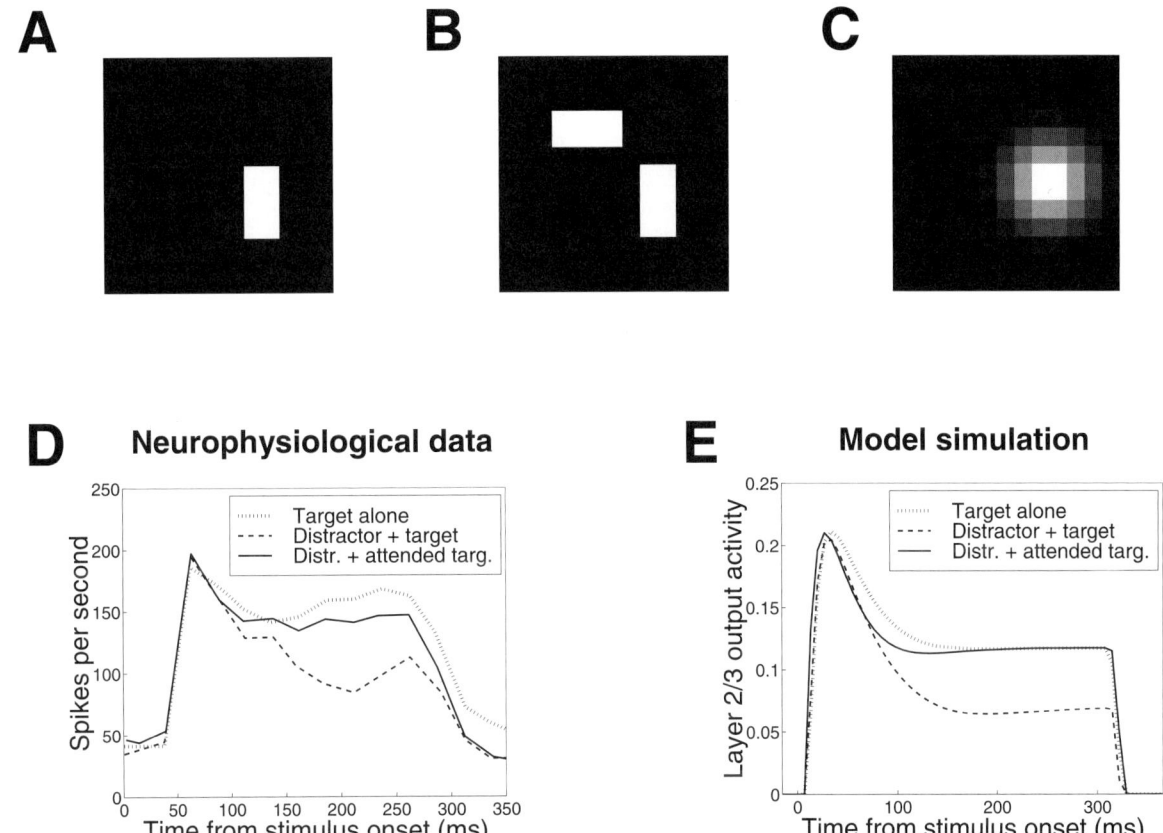

FIGURE 107.3 The effect of attention on competition between visual stimuli. A target stimulus, presented on its own (A), elicits strong neural activity at the recorded cell. When a second, distractor stimulus is presented nearby (B), it competes against the target and activity is reduced. Directing spatial attention to the location of the target stimulus (C), protects the target from this competition and restores neural activity to the levels elicited by the target on its own. The stimuli shown here, based on those used in the neurophysiological experiments in Reynolds et al. (1999), were presented to the model neural network. Spatial attention (C), was implemented as a Gaussian of activity fed back into layer 6. (D) Neurophysiological data from macaque V2 that illustrate the recorded activity patterns just described: strong responses to an isolated target (dotted line), weaker responses when a competing distractor is placed nearby (dashed line), and restored levels of activity when the target is attended (solid line). (Adapted with permission from Reynolds et al., 1999, Fig. 107.5.) (E) Model simulation of the Reynolds et al. (1999) data. The time courses illustrated show the activity of a vertically oriented cell stimulated by the target bar. If only the horizontal distractor bar were presented on its own, this cell would respond very weakly. If both target and distractor were presented, but with the horizontal distractor attended, the cell would respond but more weakly than the illustrated case in which the distractor and target are presented together, with neither attended. (Reprinted with permission from Grossberg and Raizada, 2000.)

Because the cortex uses the same circuits to select groupings (Fig. 107.2C), and to prime attention (B), attention can selectively focus on an entire object by flowing along the perceptual groupings that define the object boundary (Roelfsema et al., 1998; see Fig. 107.4A, B). In particular, when attention causes an excitatory modulatory bias at some cells in layer 4, groupings that form in layer 2/3 can be enhanced by this modulation via their positive feedback loops from 2/3-to-6-to-4-to-2/3. The direct modulation of layer 2/3 by attention can also enhance these groupings. Figure 107.4C summarizes a LAMINART simulation of the Roelfsema et al. (1998) data. LAMINART also simulates the spread of attention along an illusory contour (Raizada and Grossberg, 2001), consistent with experimental data in Moore et al. (1998), thereby illustrating how the cortex can attend incomplete object data.

The ability of attention to selectively light up entire object representations has an obviously important survival value. It is thus of interest that the intracortical and intercortical feedback circuits that control this property have been shown in modeling studies to help stabilize infant development and adult perceptual learning within multiple cortical areas, including cortical areas V1 and V2 (Carpenter and Grossberg, 1991;

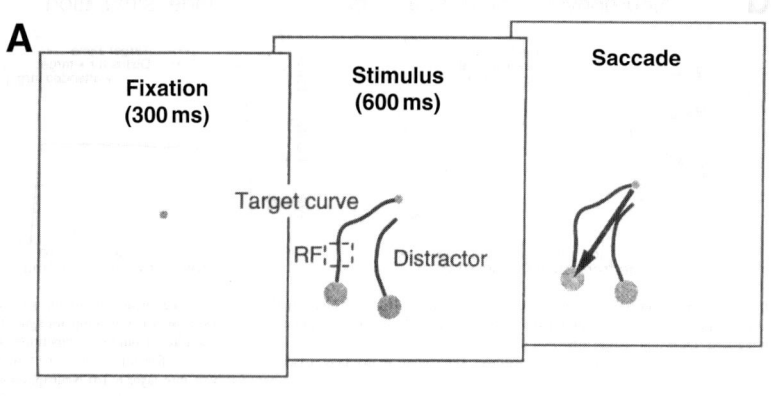

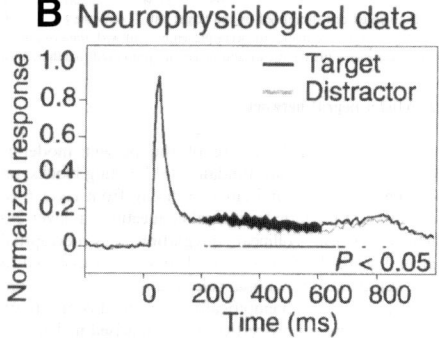 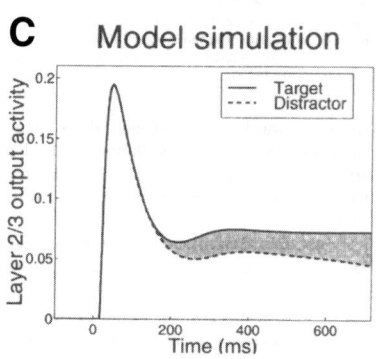

FIGURE 107.4 Spread of visual attention along an object boundary grouping, from an experiment by Roelfsema et al. (1998). (A) The experimental paradigm. Macaque monkeys performed a curve-tracing task, during which physiological recordings were made in V1. A fixation spot was presented for 300 ms, followed by a target curve and a distractor curve presented simultaneously. The target was connected at one end to the fixation point. While maintaining fixation, the monkeys had to trace the target curve and then, after 600 ms, make a saccade to its end point. (B) Neurophysiological data showing attentional enhancement of the firing of a neuron when its receptive field (RF) lay on the target curve, as opposed to the distractor. Note that the enhancement occurs approximately 200 ms after the initial burst of activity. Further studies have indicated that the enhancement starts later in distal curve segments, far from the fixation point, than it does in proximal segments, closer to fixation (Pieter Roelfsema, pers. comm.). This suggests that attentional signals propagate along the length of the target curve. (A and B adapted with permission from Roelfsema et al., 1998.) (C) Model simulation of the Roelfsema et al. data. (Reprinted with permission from Grossberg and Raizada, 2000.)

Grossberg, 1999a, 1999b; Grossberg and Williamson, 2001).

VII. THE LINK BETWEEN ATTENTION AND LEARNING

The ART proposal that attention helps to stabilize cortical development and learning, thereby preventing catastrophic forgetting, suggests that top-down attentional mechanisms should be present in *every* cortical area where these processes occur. The ART solution to the stability-plasticity problem is to allow neural representations to be modified only by those incoming stimuli with which they form a sufficiently close match. If the match is close enough, then resonance and learning occurs. Precisely because the match is sufficiently close, such learning fine-tunes the memories of existing representations. In this way, outliers cannot cause a radical overwriting of an already learned representation.

ART also proposes how, at higher levels of perceptual and cognitive processing, including inferotemporal and prefrontal cortex, a learning individual can flexibly vary the criterion of how good a match is needed between bottom-up and top-down information in order for presently active recognition categories and their top-down expectations to be refined through learning (Carpenter and Grossberg, 1987; Grossberg, 1999b). This can be achieved by a process called vigilance control, which can alter the criterion of how good a match is needed for resonance to occur. When coarse matches are allowed, a top-down expectation, say from prefrontal to inferotemporal cortex, can learn a prototype that is capable of focusing attention on general and abstract information. When only fine matches are allowed, learned prototypes are more specific and concrete and can focus attention even on individual exemplars, such as particular views of particular faces (Desimone and Ungerleider, 1989; Gochin et al., 1991; Harries and Perrett, 1991).

If the active top-down prototype does not match well enough with the bottom-up input, then its neural activity is extinguished and hence unable to cause plastic changes. Suppression of an active representation enables a memory search to ensue, whereby a different representation can become active instead through bottom-up signaling. This representation, in turn, reads out top-down signals that either gives rise to a match, thereby allowing learning, or a nonmatch, causing the search process to repeat until either a match is found or the incoming stimulus causes a totally new representation to be formed. ART proposes how such a memory search can be mediated by corticohippocampal interactions (Carpenter and Grossberg, 1991, 1993; Grossberg and Merrill, 1996).

This summary of ART-based recognition learning shows how the focus of object attention may become either abstract or concrete, depending on the task-constraints that are imposed. Compatible data from recordings in inferotemporal cortex were reported by Spitzer et al. (1988), who exposed monkeys to easy and difficult discriminations, and showed "in the difficult condition the animals adopted a stricter internal criterion for discriminating matching from nonmatching stimuli.... the animals' internal representations of the stimuli were better separated, independent of the criterion used to discriminate them.... increased effort appears to cause enhancement of the responses and sharpened selectivity for attended stimuli" (339–340).

Other experiments have also supported the predicted link between attention and learning. Psychophysically, the role of attention in controlling adult plasticity and perceptual learning was demonstrated by Ahissar and Hochstein (1993). Gao and Suga (1998) reported physiological evidence that acoustic stimuli caused plastic changes in the *IC* of bats only when the IC received top-down feedback from auditory cortex. These authors also reported that plasticity is enhanced when the auditory stimuli were made behaviorally relevant, consistent with the ART proposal that top-down feedback allows attended, and thus relevant, stimuli to be learned while suppressing unattended irrelevant ones. Evidence that cortical feedback also controls thalamic plasticity in the somatosensory system has been found by Krupa et al. (1999) and by Parker and Dostrovsky (1999). These findings are reviewed by Kaas (1999).

Studies of *inter*cortical attention-activated feedback and *intra*cortical grouping-activated feedback have also shown that either type of feedback can rapidly synchronize the firing patterns of higher and lower cortical areas (Grossberg and Grunewald, 1997; Grossberg and Somers, 1991; Yazdanbakhsh and Grossberg, 2004). ART puts this result into perspective by suggesting that resonance may lead to synchronization, which may facilitate cortical learning by enhancing the probability that "cells that fire together wire together." Engel et al. (2001) review data about top-down cortical feedback and synchrony (see also Chapter 87). The cortex also includes circuits that enable development and learning to self-stabilize without top-down *inter*cortical attention by using the *intra*cortical pathway from layer 2/3-to-6 to 4-to-2/3 (see Fig. 107.2C) to act as a selection circuit that inhibits outliers before they can cause catastrophic forgetting; see Raizada and Grossberg (2003) for further discussion.

VIII. DIVIDED, OBJECT VERSUS SPATIAL, AND HIERARCHICAL ATTENTION

Although bottom-up inputs that arrive in the off-surround of an active top-down attentional prime may be suppressed, inputs outside the off-surround may not be suppressed. This is already clear in some conditions of the Reynolds et al. (1999) experiment that is summarized in Fig. 107.3. In fact, many studies have shown that attention may be simultaneously divided among several targets (e.g., Pylyshyn and Storm, 1988; Yantis, 1992). In addition, both object and spatial attention may influence visual perception (Duncan, 1984; Posner, 1980). The distinction between object and spatial attention reflects the organization of visual cortex into parallel "what" and "where" processing streams. Many cognitive neuroscience experiments support the hypotheses of Ungerleider and Mishkin (1982) and Goodale and Milner (1992) that inferotemporal cortex and its cortical projections learn to categorize and recognize what objects are in the world, whereas parietal cortex and its cortical projections learn to determine where they are in space and how to act with respect to them. Because the "what" stream strives to generate invariant object representations that are independent of spatial coordinates, whereas the "where" stream generates representations of object location, these streams must interact to control actions aimed at recognized objects. Indeed, both object and spatial attention are needed to search for visual targets amid distractors. Grossberg et al. (1994) quantitatively fit a large human psychophysical database about visual search with a Spatial Object Search (SOS) model that proposes the way that 3D boundary groupings and surface representations interact with object attention and spatial attention to find targets amid distractors. In this analysis, object and spatial attention must be sensitive to perceptual groupings as well as to surface properties such as all occurrences of a color on a prescribed depth plane (Grossberg, 1994).

The present article focuses on the microarchitecture of attention but is consistent with, and clarifies, how attention may be globally organized across many brain regions acting together. In particular, laminar cortical circuits (Fig. 107.2E) clarify how attention can leap between brain regions via their layers 6 and thereby modulate cells in multiple cortical areas without firing them.

Acknowledgment

Supported in part by the Air Force Office of Scientific Research (AFOSR F49620-01-1-0397) the National Science Foundation (SBE-0354378) and the Office of Naval Research (ONR N00014-01-1-0624).

References

Ahissar, M., and Hochstein, S. (1993). Attentional control of early perceptual learning. *Proc. Natl. Acad. Sci. U.S.A.* **90**, 5718–5722.

Bosking, W. H., Zhang, Y., Schofield, B., and Fitzpatrick, D. (1997). Orientation selectivity and the arrangement of horizontal connections in the tree shrew striate cortex. *J. Neurosci.* **17**, 2112–2127.

Bradski, G., and Grossberg, S. (1995). Fast-learning VIEWNET architectures for recognizing three-dimensional objects from multiple two-dimensional views. *Neural Net.* **8**, 1053–1080.

Bullier, J., Hupé, J. M., James, A., and Girard, P. (1996). Functional interactions between areas V1 and V2 in the monkey. *J. Physiol.* **90**, 217–220.

Caputo, G., and Guerra, S. (1998). Attentional selection by distractor suppression. *Vis. Res.* **38**, 669–689.

Carpenter, G. A., and Grossberg, S. (1987). A massively parallel architecture for a self-organizing neural pattern recognition machine. *Comput. Vis. Graphics Image Processing* **37**, 54–115.

Carpenter, G. A., and Grossberg, S. (eds.) (1991). "Pattern Recognition by Self-Organizing Neural Networks." MIT Press, Cambridge, MA.

Carpenter, G. A., and Grossberg, S. (1993). Normal and amnesic learning, recognition, and memory by a model of cortico-hippocampal interactions. *Trends Neurosci.* **16**, 131–137.

Carpenter, G. A., and Ross, W. D. (1995). ART-EMAP: A neural network architecture for object recognition by evidence accumulation. *IEEE Trans. Neural Net.* **6**, 805–818.

Chey, J., Grossberg, S., and Mingolla, E. (1997). Neural dynamics of motion grouping: From aperture ambiguity to object speed and direction. *J. Opt. Soc. Am. A1* **4**, 2570–2594.

Desimone, R. (1998). Visual attention mediated by biased competition in extrastriate visual cortex. *Phil. Trans. Roy. Soc. London* **353**, 1245–1255.

Desimone, R., and Ungerleider, L. G. (1989). Neural mechanisms of visual processing in monkeys. In "Handbook of Neuropsychology" (F. Boller and J. Grafman, eds.), Vol. 2, pp. 267–299. Elsevier, Amsterdam.

Downing, C. J. (1988). Expectancy and visual-spatial attention: Effects on perceptual quality. *J. Exp. Psych. Hum. Perception Performance* **14**, 188–202.

Dubin, M. W., and Cleland, B. G. (1977). Organization of visual inputs to interneurons of lateral geniculate nucleus of the cat. *J. Neurophysiol.* **40**, 410–427.

Duncan, J. (1984). Selective attention and the organization of visual information. *J. Exp. Psych. Gen.* **113**, 501–517.

Engel, A. K., Fries, P., and Singer, W. (2001). Dynamic predictions: Oscillations and synchrony in top-down processing. *Nat. Rev. Neurosci.* **2**, 704–716.

Gao, E., and Suga, N. (1998). Experience-dependent corticofugal adjustment of midbrain frequency map in bat auditory system. *Proc. Natl. Acad. Sci. U.S.A.* **95**, 12663–12670.

Gochin, P. M., Miller, E. K., Gross, C. G., and Gerstein, G. L. (1991). Functional interactions among neurons in inferior temporal cortex of the awake macaque. *Exp. Brain Res.* **84**, 505–516.

Goodale, M. A., and Milner, D. (1992). Separate visual pathways for perception and action. *Trends Neurosci.* **15**, 10–25.

Grosof, D. H., Shapley, R. M., and Hawken, M. J. (1993). Macaque V1 neurons can signal "illusory" contours. *Nature* **365**, 550–552.

Grossberg, S. (1980). How does a brain build a cognitive code? *Psych. Rev.* **87**, 1–51.

Grossberg, S. (1994). 3-D vision and figure-ground separation by visual cortex. *Perception Psychophys.* **55**, 48–120.

Grossberg, S. (1999a). How does the cerebral cortex work? Learning, attention, and grouping by the laminar circuits of visual cortex. *Spat. Vis.* **12**, 163–187.

Grossberg, S. (1999b). The link between brain learning, attention, and consciousness. *Consci. Cogn.* **8**, 1–44.

Grossberg, S. (2000). How hallucinations may arise from brain mechanisms of learning, attention, and volition. *J. Intl. Neuropsych. Soc.* **6**, 583–592.

Grossberg, S., Boardman, I., and Cohen, M. A. (1997). Neural dynamics of variable-rate speech categorization. *J. Exp. Psych. Hum. Perception Performance* **23**, 481–503.

Grossberg, S., and Grunewald, A. (1997). Cortical synchronization and perceptual framing. *J. Cogn. Neurosci.* **9**, 117–132.

Grossberg, S., and Merrill, J. W. L. (1996). The hippocampus and cerebellum in adaptively timed learning, recognition, and movement. *J. Cogn. Neurosci.* **8**, 257–277.

Grossberg, S., Mingolla, E., and Ross, W. D. (1994). A neural theory of attentive visual search: Interactions of boundary, surface, spatial, and object representations. *Psych. Rev.* **101**, 470–489.

Grossberg, S., Mingolla, E., and Ross, W. D. (1997). Visual brain and visual perception: How does the cortex do perceptual grouping? *Trends Neurosci.* **20**, 106–111.

Grossberg, S., Mingolla, E., and Viswanathan, L. (2001). Neural dynamics of motion integration and segmentation within and across apertures. *Vis. Res.* **41**, 2521–2553.

Grossberg, S., and Myers, C. W. (2000). The resonant dynamics of speech perception: Interword integration and duration-dependent backward effects. *Psych. Rev.* **4**, 735–767.

Grossberg, S., and Raizada, R. D. (2000). Contrast-sensitive perceptual grouping and object-based attention in the laminar circuits of primary visual cortex. *Vis. Res.* **40**, 1413–1432.

Grossberg, S., Roberts, K., Aguilar, M., and Bullock, D. (1997). A neural model of multimodal adaptive saccadic eye movement control by superior colliculus. *J. Neurosci.* **17**, 9706–9725.

Grossberg, S., and Seitz, A. (2003). Laminar development of receptive fields, maps, and columns in visual cortex: The coordinating role of the subplate. *Cereb. Cortex.* **13**, 852–863.

Grossberg, S., and Somers, D. (1991). Synchronized oscillations during cooperative feature linking in a cortical model of visual perception. *Neural Net.* **4**, 453–466.

Grossberg, S., and Stone, G. O. (1986). Neural dynamics of word recognition and recall: Attentional priming, learning, and resonance. *Psych. Rev.* **93**, 46–74.

Grossberg, S., and Williamson, J. R. (2001). A neural model of how horizontal and interlaminar connections of visual cortex develop into adult circuits that carry out perceptual groupings and learning. *Cereb. Cortex* **11**, 37–58.

Harries, M. H., and Perrett, D. I. (1991). Visual processing of faces in temporal cortex: Physiological evidence for a modular organization and possible anatomical correlates. *J. Cogn. Neurosci.* **3**, 9–24.

Heeger, D. J. (1992). Normalization of cell responses in cat striate cortex. *Vis. Neurosci.* **9**, 181–197.

Hupé, J. M., James, A. C., Girard, D. C., and Bullier, J. (1997). Feedback connections from V2 modulate intrinsic connectivity within V1. *Soc. Neurosci. Abstracts* 406.15, 1031.

Kaas, J. H. (1999). Is most of neural plasticity in the thalamus cortical? *Proc. Natl. Acad. Sci. U.S.A.* **96**, 7622–7623.

Kapadia, M. K., Ito, M., Gilbert, C. D., and Westheimer, G. (1995). Improvement in visual sensitivity by changes in local context: parallel studies in human observers and in V1 of alert monkeys. *Neuron* **15**, 843–856.

Kelly, F., and Grossberg, S. (2000). Neural dynamics of 3-D surface perception: Figure-ground separation and lightness perception. *Perception Psychophys.* **62**, 1596–1618.

Knierim, J. J., and Van Essen, D. C. (1992). Neuronal responses to static texture patterns in area V1 of the alert macaque monkey. *J. Neurophysiol.* **67**, 961–980.

Krupa, D. J., Ghazanfar, A. A., and Nicolelis, M. A. (1999). Immediate thalamic sensory plasticity depends on corticothalamic feedback. *Proc. Natl. Acad. Sci. U.S.A.* **96**, 8200–8205.

Luck, S. J., Chelazzi, L., Hillyard, S. A., and Desimone, R. (1997). Neural mechanisms of spatial selective attention in areas V1, V2, and V4 of macaque visual cortex. *J. Neurophysiol.* **77**, 24–42.

Lund, J. S., and Wu, C. Q. (1997). Local circuit neurons of macaque monkey striate cortex: IV. Neurons of laminae 1–3A. *J. Comp. Neurol.* **384**, 109–126.

Moore, C. M., Yantis, S., and Vaughan, B. (1998). Object-based visual selection: Evidence from perceptual completion. *Psych. Sci.* **9**, 104–110.

Motter, B. C. (1993). Focal attention produces spatially selective processing in visual cortical areas V1, V2 and V4 in the presence of competing stimuli. *J. Neurophysiol.* **70**, 909–919.

Mounts, J. R. W. (2000). Evidence for suppressive mechanisms in attentional selection: Feature singletons produce inhibitory surrounds. *Perception Psychophys.* **62**, 969–983.

Mumford, D. (1992). On the computational architecture of the neocortex, II. The role of corticocortical loops. *Biol. Cybernet.* **66**, 241–251.

Parker, J. L., and Dostrovsky, J. O. (1999). Cortical involvement in the induction, but not expression, of thalamic plasticity. *J. Neurosci.* **19**, 8623–8629.

Peterhans, E., and von der Heydt, R. (1989). Mechanisms of contour perception in monkey visual cortex, II. Contours bridging gaps. *J. Neurosci.* **9**, 1749–1763.

Polat, U., Mizobe, K., Pettet, M. W., Kasamatsu, T., and Norcia, A. M. (1998). Collinear stimuli regulate visual responses depending on cell's contrast threshold. *Nature* **391**, 580–584.

Posner, M. I. (1980). Orienting of attention. *Q. J. Exp. Psychol.* **32**, 2–25.

Pylyshyn, Z. W., and Storm, R. W. (1988). Tracking multiple independent targets: Evidence for a parallel tracking mechanism. *Spat. Vis.* **3**, 179–197.

Raizada, R., and Grossberg, S. (2001). Context-sensitive bindings by the laminar circuits of V1 and V2: A unified model of perceptual grouping, attention, and orientation contrast. *Vis. Cogn.* **8**, 341–466.

Raizada, R., and Grossberg, S. (2003). Towards a theory of the laminar architecture of cerebral cortex: Computational clues from the visual system. *Cereb. Cortex* **13**, 100–113.

Rao, R. P. N., and Ballard, D. H. (1999). Predictive coding in the visual cortex: A functional interpretation of some extra-classical receptive field effects. *Nat. Neurosci.* **2**, 79–87.

Reynolds, J., Chelazzi, L., and Desimone, R. (1999). Competitive mechanisms subserve attention in macaque areas V2 and V4. *J. Neurosci.* **19**, 1736–1753.

Rockland, K. S., and Virga, A. (1989). Terminal arbors of individual "feedback" axons projecting from area V2 to V1 in the macaque monkey: A study using immunohistochemistry of anterogradely transported phaseolus vulgaris-leucoagglutinin. *J. Comp. Neurol.* **285**, 54–72.

Roelfsema, P. R., Lamme, V. A. F., and Spekreijse, H. (1998). Object-based attention in the primary visual cortex of the macaque monkey. *Nature* **395**, 376–381.

Sandell, J. H., and Schiller, P. H. (1982). Effect of cooling area 18 on striate cortex cells in the squirrel monkey. *J. Neurophysiol.* **48**, 38–48.

Sheth, B. R., Sharma, J., Rao, S. C., and Sur, M. (1996). Orientation maps of subjective contours in visual cortex. *Science* **274**, 2110–2115.

Sillito, A. M., Grieve, K. L., Jones, H. E., Cudeiro, J., and Davis, J. (1995). Visual cortical mechanisms detecting focal orientation discontinuities. *Nature* **378**, 492–496.

Sillito, A. M., Jones, H. E., Gerstein, G. L., and West, D. C. (1994). Feature-linked synchronization of thalamic relay cell firing induced by feedback from the visual cortex. *Nature* **369**, 479–482.

Smith, A. T., Singh, K. D., and Greenlee, M. W. (2000). Attentional suppression of activity in the human visual cortex. *Neurorepetition* **11**, 271–277.

Somers, D. C., Dale, A. M., Seiffert, A. E., and Tootell, R. B. (1999). Functional MRI reveals spatially specific attentional modulation in human primary visual cortex. *Proc. Natl. Acad. Sci. U.S.A.* **96**, 1663–1668.

Spitzer, H., Desimone, R., and Moran, J. (1988). Increased attention enhances both behavioral and neuronal performance. *Science* **240**, 338–340.

Steinman, B. A., Steinman, S. B., and Lehmkuhle, S. (1995). Visual attention mechanisms show a canter-surround organization. *Vis. Res.* **35**, 1859–1869.

Temereanca, S., and Simons, D. J. (2001). Topographic specificity in the functional effects of corticofugal feedback in the whisker/barrel system. *Soc. Neurosci. Abstracts* 393.6.

Thorpe, S., Fize, D., and Marlot, C. (1996). Speed of processing in the human visual system. *Nature* **381**, 520–522.

Ungerleider, L. G., and Mishkin, M. (1982). Two cortical visual systems: Separation of appearance and location of objects. *In* "Analysis of Visual Behavior" (D. L. Ingle, M. A. Goodale, and R. J. W. Mansfield, eds.), pp. 549–586. MIT Press, Cambridge, MA.

Vanduffel, W., Tootell, R. B., and Orban, G. A. (2000). Attention-dependent suppression of meta-bolic activity in the early stages of the macaque visual system. *Cereb. Cortex* **10**, 109–126.

von der Heydt, R., Peterhans, E., and Baumgartner, G. (1984). Illusory contours and cortical neuron responses. *Science* **224**, 1260–1262.

Watanabe, T., Sasaki, Y., Nielsen, M., Takino, R., and Miyakawa, S. (1998). Attention-regulated activity in human primary visual cortex. *J. Neurophysiol.* **79**, 2218–2221.

Weber, A. J., Kalil, R. E., and Behan, M. (1989). Synaptic connections between coricogeniculate axons and interneurons in the dorsal lateral geniculate nucleus of the cat. *J. Comp. Neurol.* **289**, 156–164.

Yantis, S. (1992). Multielement visual tracking: Attention and perceptual organization. *Cogn. Psychol.* **24**, 295–340.

Yazdanbakhsh, A., and Grossberg, S. (2004). Fast synchronization of perceptual grouping in laminar visual cortical circuits. *Neural Networks* **17**, 707–718.

Zeki, S., and Shipp, S. (1988). The functional logic of cortical connections. *Nature* **335**, 311–317.

Zhang, Y., Suga, N., and Yan, J. (1997). Corticofugal modulation of frequency processing in bat auditory system. *Nature* **387**, 900–903.

Attention-Guided Recognition Based on "What" and "Where" Representations:
A Behavioral Model

Ilya A. Rybak, Valentina I. Gusakova, Alexander V. Golovan, Lubov N. Podladchikova and Natalia A. Shevtsova

ABSTRACT

We describe the model of attention-guided visual perception and recognition previously published in *Vision Research* (Rybak et al., 1998). The model contains (1) a low-level subsystem that performs a fovealike transformation and detection of primary features (edges) and (2) a high-level subsystem that includes separated "what" (sensory memory) and "where" (motor memory) subsystems. In the model, image recognition occurs under top-down control of visual attention during the execution of a behavioral recognition program formed during the primary viewing of the image. The recognition program contains both programmed movements of an attention window (stored in the motor memory) and predicted image fragments (stored in the sensory memory) for each consecutive fixation. The model shows the ability to recognize complex images (e.g., faces) invariantly with respect to shift, rotation, and scale.

I. INTRODUCTION

During visual perception and recognition, human eyes move and successively fixate at the most informative parts of the viewed image or scene. Because the density of photoreceptors in the retina decreases from in the central area (fovea) to the periphery, the resolution of image representation in the visual cortex is the highest for the part of the image projected onto the fovea and decreases rapidly with the distance from the fovea projection. The major function of visual attention is to actuate and control eye movement and hence to perform a problem-oriented selection and processing of information from the visible world (Burt, 1988; Neisser, 1967; Noton and Stark, 1971; Posner and Presti, 1987; Treisman and Gelade, 1980; Yarbus, 1967). Despite a relatively low resolution of image representation in the retinal periphery, it provides important information that is used by the attention mechanisms for selecting the next eye position and directing the consecutive eye movements. This leads to the view that visual perception and recognition are actually behavioral processes that probably cannot be completely understood in the limited frames of neural computations without taking into account the behavioral and cognitive aspects of perception and the role of attention in visual perception and recognition.

A. A Behavioral Paradigm in Visual Perception and the Role of Visual Attention

From the behavioral point of view, an internal representation (model) of new circumstances is formed in the brain during conscious observation and active examination. The active examination is aimed toward the finding and memorizing of functional relationships between the applied actions and the resulting changes in sensory information. An external object becomes known and may be recognized when the system is able to subconsciously manipulate the object and predict the object's reactions to the applied actions. According to this paradigm, the internal object representation contains chains of alternating traces in motor and sensory memories. Each of these chains reflects an alternating sequence of elementary motor actions and sensory (proprioceptive and exteroceptive) signals that are expected to arrive in response to each action. The brain uses these chains as behavioral programs in subconscious behavioral recognition when the object is (or is assumed to be) known. This behavioral recognition has two stages: conscious selection of the appropriate behavioral program and the subconscious execution of the program. Matching the expected (predicted) sensory signals to the actual sensory signals, arriving after each motor action, is the essential procedure in the program execution.

This behavioral paradigm has been formulated and developed in the context of visual perception and recognition in the series of conceptually significant works (Didday and Arbib, 1975; Noton and Stark, 1971; Kosslyn et al., 1990; Yarbus, 1967). Using Yarbus's approach, Noton and Stark (1971) compared the individual scanpaths of human eye movements in two phases: during memorizing and during subsequent recognition of the same image. They found these scanpaths to be topologically similar and suggested that each object is memorized and stored in memory as an alternating sequence of object features and eye movements required to reach the next feature. The results of Noton and Stark (1971) and Didday and Arbib (1975) prompted the consideration of eye movement scanpaths as behavioral programs for recognition. The process of recognition was supposed to consist of an alternating sequence of eye movements (recalled from the motor memory and directed by attention) and verifications of the expected image fragments (recalled from the sensory memory).

Ungerleider and Mishkin (1982), Mishkin et al. (1983), and Kosslyn et al. (1990) presented neuroanatomical and psychological data complementary to this behavioral concept. It was found that the higher levels of the visual system contain two major pathways for visual processing, called the "where" and "what" pathways. The "where" pathway leads dorsally to the parietal cortex and is involved in processing and representing spatial information (spatial locations and relationships). The "what" pathway leads ventrally to the inferior temporal cortex and deals with processing and representing object features (Kosslyn et al., 1990; Mishkin et al., 1983; Ungerleider and Mishkin, 1982). The behavioral concept joined with this neuroanatomical theory provides (1) the explicit functional coupling between the low-level vision (foveal structure of the retinocortical projection, orientation selectivity in the visual cortex, etc.) and the high-level brain structures involved in visual perception and recognition and (2) the clear functional role of visual attention in the coupling the low- and high-levels of the visual system.

B. Image Features, Invariant Representation, and Frames of Reference

Beginning with the classic Hubel and Wiesel (1962) work, neurophysiological studies have demonstrated that neurons in the primary visual cortex can detect elementary image features such as local orientations of line segments or edges. Therefore, most theories of vision assume that the visual system detects simple features (e.g., local line segment or edge orientation but not spatial relations between them) at the preattentive stage and uses some attention mechanisms of a serial type to bind the simple features into more complex shape features (Neisser, 1967; Treisman and Gelade, 1980).

One key issue of visual recognition is the mechanism used for invariant image representation. Marr (1982), Palmer (1983), Hinton and Lang (1985), and others assumed that the visual system uses an object-based frame of reference attached to the center of the object. However, the object-based reference paradigm has several significant disadvantages. First, this paradigm presumes that the object is isolated and does not have a complex background. Second, if a part of an object is missing or occluded, or an additional part is present, the center of the object may be shifted, that makes it difficult to recognize the object. In addition, this paradigm presumes that the object is small and simple enough to be recognized during one gaze fixation. As a result, previous models of recognition that used the object-based frame of reference could demonstrate invariant recognition of only simple objects (letters, binary objects without background, etc.) to which such a frame of reference is easily attached.

In the model reviewed here (Rybak et al., 1998), a spatial pattern of edges is extracted from the retinal image at each fixation. These edges are considered as elementary or first-order features of the image. Image representation at the fixation point is based on the assumption that the first-order features (edges) extracted from the retinal periphery perform two distinct functions. One function of the peripheral edges is to provide potential targets for the next gaze fixation (which was used in most active vision models; see, for example, Burt, 1988). The other function of peripheral edges, as suggested in the reviewed model, is to provide a context (context features) for the basic feature (edge) in the center of the fovea.

In the reviewed model, the relative orientations of context edges and their relative angular locations with respect to the basic edge at each point of fixation are used as second-order invariant features of the image. Thus, instead of the object-based frame of reference, the model uses a feature-based frame of reference attached to the basic edge at the fixation point. Because both the retinal images at the fixation points and the sequential shifts of the fixation point are represented in this invariant form, the entire image is invariantly represented in the memory of the system. Moreover, the feature-based frame of reference coupled with the multiplicity of fixation points along the scanpath may allow the system to recognize an image from its part (from a fraction of the scanpath belonging to this part) when the image is partly perturbed or the object of recognition in the image is occluded.

II. A MODEL OF ATTENTION-GUIDED RECOGNITION BASED ON "WHAT" AND "WHERE" REPRESENTATIONS

A. Model Description

A functional diagram of the model is shown in Fig. 108.1. The attention window (AW) performs a primary transformation of the image into a retinal image at the fixation point (see Figs. 108.2 and 108.3B). This transformation provides a decrease in resolution of the retinal image from the center to the periphery of the AW similar to that in the cortical projection of the retinal image. The retinal image in the AW is used as an input to the module for primary feature detection, which performs a function similar to the primary visual cortex. This module contains a set of neurons with orientationally selective receptive fields (ORFs) tuned to different orientations of the local edge. Neurons with the ORF, centered at the same point but with different orientation tuning, interact competi-

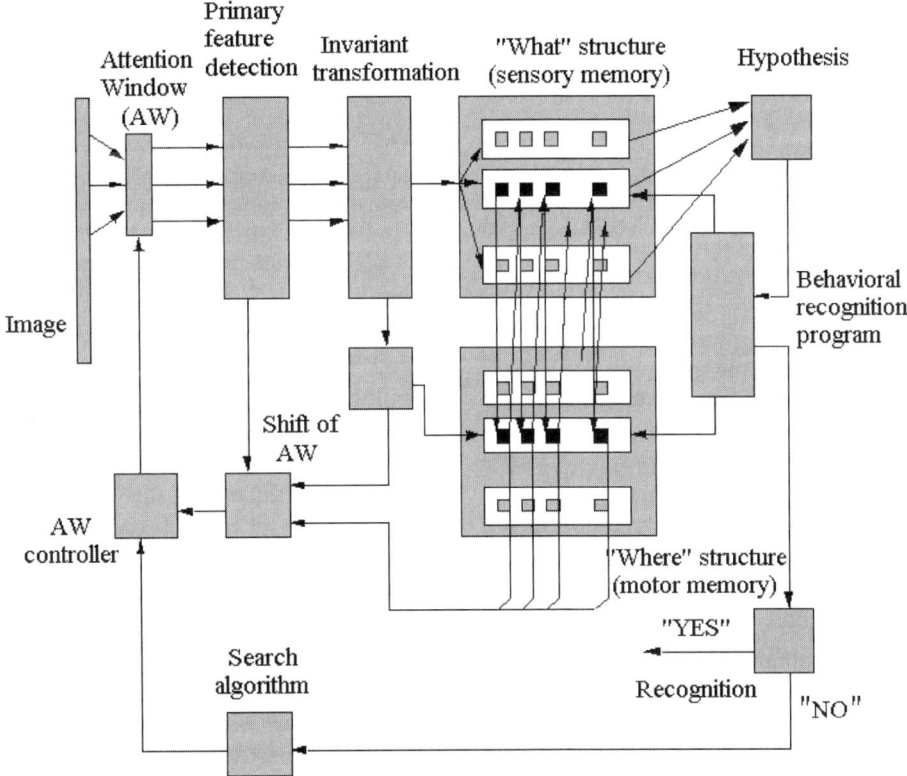

FIGURE 108.1 Schematic of the model.

(a) MEMORIZING

(b) RECOGNITION

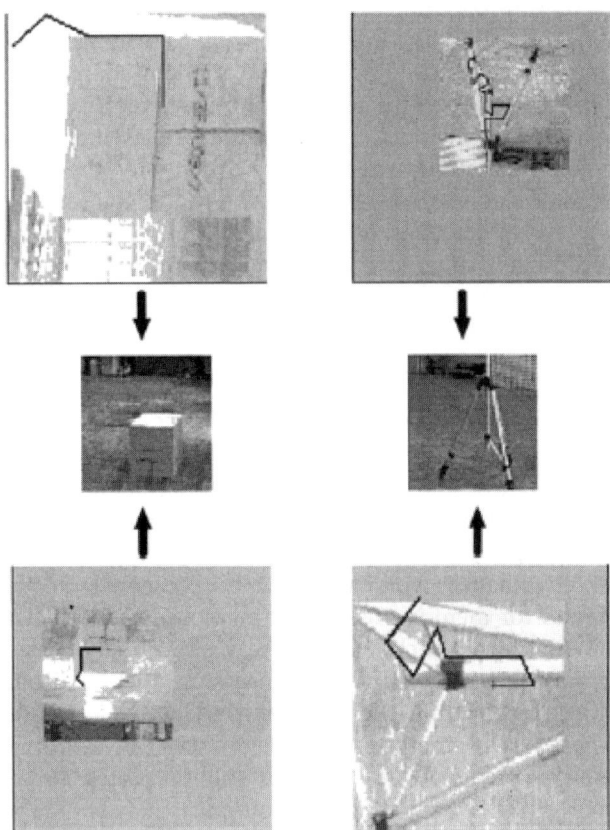

FIGURE 108.4 Examples of recognition of test images invariantly with respect to shift, rotation, and scale (a) The test images with the scanpaths in the memorizing mode. (b) The images presented for recognition (with the scanpaths during the recognition mode). These images were obtained from the test images by shifting, rotation, and scaling. The results of recognition are shown by arrays.

ioral program associated with the accepted hypothesis about the object. The scanpaths of viewing during the recognition mode were topologically similar to the scanpaths of viewing during the memorization of the same images (see Figs. 108.4b and 108.5b).

III. DISCUSSION: INVARIANT IMAGE REPRESENTATION AND RECOGNITION

The exciting ability of natural visual systems of invariant recognition has attracted the attention of scientists for more than 100 years. For several decades, the property of invariant recognition has been one of the major objectives and test criteria in different scientific areas from artificial neural networks to computer and robot vision. However, from a behavioral point of view, the ability of invariant recognition is not an ultimate goal but rather a tool that helps the system to plan and execute actions. This problem-oriented (task-driven) behavior requires a problem-oriented perception and recognition. According to Didday and Arbib (1975), the goal is "not to build a little internal copy of the visual scene, but rather to assist making the decision of what action (if any) the organism should next initiate." From this point of view, the absolute invariant recognition; in which exactly the same internal representation is achieved for an object irrespective of its size, location, orientation in space, and so forth, is useless. In that case, the organism will learn that the object is present in the scene, but will not know how to manipulate it. For example, the system will know that a cup is present, but will not know how to take it (where the handle is) and whether it is possible to pour some tea into it (how the cup is oriented in space). Thus, recognition should be considered as a process (behavior) during which the system either actively manipulates a noninvariant object representation in memory (by transforming it to match the external image view) or manipulates the external image using active eye and head movements. The resultant manipulations, used to fit the model to the object, give the system additional information about object location, orientation, size, and so forth.

On the other hand, a lack of invariant representations makes the task of recognition practically unresolved. How would the system know which one of the majority of noninvariant models stored in memory to take for manipulations in order to match the object and how long to manipulate the selected model before making the decision to take another model? The natural visual system evidently solves this dilemma by way of some "smart" combination of these two opposite approaches.

A possible (and hopefully plausible) way for such a combination may be based on the invariant representation of object's elements in the vicinity of fixation points (within the AW) and on the use of object manipulations (eye movements and shifts of the AW) in order to represent spatial relationships. The latter idea was used in the reviewed model. The model holds

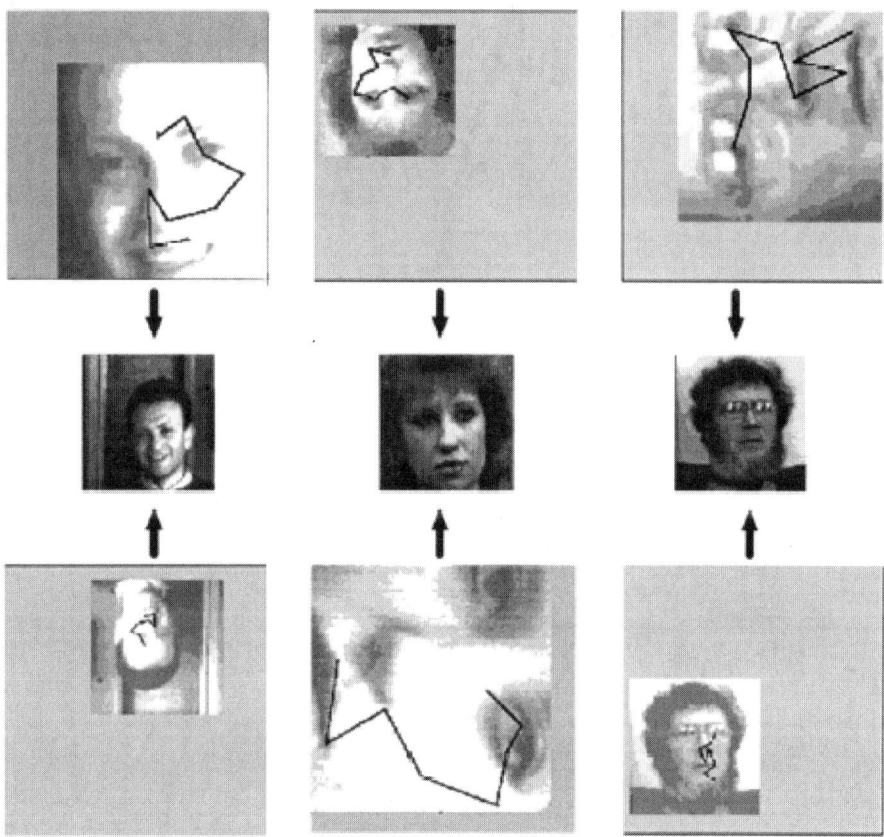

FIGURE 108.5 Examples of recognition of faces invariantly with respect to shift, rotation, and scale. (a) The test images with scanpaths in memorizing mode. (b) The images presented for recognition (with the scanpaths during the recognition mode).

invariantly represented image fragments in the sensory memory. Each fragment is associated with a certain object and with a certain action that the system should execute according to the behavioral recognition program that is associated with the object. The initial identification of a known fragment (invariantly represented in memory) gives a start of the behavioral recognition process, which is executed under the top-down control of attention. The recognition process provides information about object location and orientation in space. The system architecture used in the model, coupled with the behavioral algorithms of image memorizing and recognition and with the feature-based reference frame, allows the system both to recognize objects invariantly with respect to their position and orientation in space and to manipulate objects in space using absolute parameters of the basic feature at the fixation point and relative spatial relationships recalled from the motor memory.

In the reviewed model, the algorithm for the invariant representation of the retinal image within the AW is based on the encoding of relative (with respect to the basic edge at the fixation point) orientations and angular positions of the detected edges. In other words, each basic edge at the fixation point is considered in the context of a set of other edges in the retinal image. With the decrease in resolution toward the retinal periphery, a more detailed and precise representation of image partition in the vicinity of the fixation point is considered in the context of a coarser generalized representation of a lager part of the image (or the entire image).

In order to memorize a particular object in the image or scene containing several objects and/or a complex background, the model should select only the points of fixation that belong to the same object. The current version of our model does not do this in general. In order to make this certain, in the memorizing mode the model should deal with images containing single objects with a uniform background. Then, in the recognition mode, the model should be able to recognize these objects in multiobject scenes with complex backgrounds. In contrast, the natural visual system uses special mechanisms that provide object separation independent of or even before object recognition (stereopsis and binocular depth perception; analysis of occlusions during head and body movements, color and texture analysis, etc.). Additional mechanisms that separate the objects in the image from one another and from the background should be incorporated in the model to allow memorizing objects in complex multiobject images. These mechanisms will prevent a selection of fixation points outside the object of interest.

In conclusion, the reviewed model provides important insights into the role of behavioral aspects for invariant pattern recognition. The basic algorithmic ideas of the model and approach used may be applied to computer and robot vision systems aimed toward invariant image recognition.

References

Burt, P. J. (1988). Smart sensing within a pyramid vision machine. *Proc. IEEE* **76**, 1006–1015.

Didday, R. L., and Arbib, M. A. (1975). Eye movements and visual perception: A two visual system model. *Int. J. Man-Machine Studi.* **7**, 547–569.

Hinton, G. E., and Lang, K. J. (1985). Shape recognition and illusory conjunctions. *In* "Proceedings of the 9th International Joint Conference on Artificial Intelligence" (Aravind K. Joshi Ed.), pp. 252–259. Morgan Kaufmann, Los Angeles, CA.

Hubel, D. H., and Wiesel, T. N. (1962). Receptive fields, binocular integration and functional architecture in the cat's visual cortex. *J. Physiol.* **160**, 106–154.

Kosslyn, S. M., Flynn, R. A., Amsterdam J. B., and Wang, G. (1990). Components of high-level vision: A cognitive neuroscience analysis and account of neurological syndromes. *Cognition* **34**, 203–277.

Marr, D. (1982). "Vision." W. H. Freeman, New York.

Mishkin, M., Ungerleider, L. G., and Macko, K. A. (1983). Object vision and spatial vision: Two cortical pathways. *Trends Neurosci.* **6**, 414–417.

Neisser, V. (1967). "Cognitive Psychology." Appleton, New York.

Noton, D., and Stark, L. (1971). Scanpaths in eye movements during pattern recognition. *Science* **171**, 72–75.

Palmer, S. E. (1983). The psychology of perceptual organization: A transformational approach. *In* "Human and Machine Vision" (J. Beck, B. Hope, and A. Rosenfeld, Eds.), pp. 545–567. New York: Academic Press.

Posner, M. I., and Presti, D. E. (1987). Selective attention and cognitive control. *Trends Neurosci.* **10**, 13–17.

Rybak, I. A., Gusakova, V. I., Golovan, A. V., Podladchikova, L. N., and Shevtsova, N. A. (1998). A model of attention-guided visual perception and recognition. *Vis. Res.* **38**, 2387–2400.

Treisman, A. M., and Gelade, G. (1980). A feature integration theory of attention. *Cogr. Psychol.* **12**, 97–136.

Ungerleider, L. G., and Mishkin, M. (1982). Two cortical visual systems. *In* "Analysis of Visual Behavior" (D. J. Ingle, M. A. Goodale, and R. J. W. Mansfield, Eds.), pp. 549–586. MIT Press, Cambridge, MA.

Yarbus, A. L. (1967). "Eye Movements and Vision." Plenum, New York.

CHAPTER 109

A Model of Attention and Recognition by Information Maximization

Kerstin Schill

ABSTRACT

The perception of an image by a human observer is a highly selective procedure in which only a small subset of image locations is selected and processed by the precise and efficient neural machinery of foveal vision. To understand the principles behind this selection of the informative regions of images, we have developed a hybrid system, which consists of a combination of a top-down knowledge-based reasoning system with a low-level preprocessing by linear and nonlinear neural operators (Schill, 1997; Schill et al., 2001). In the analysis of a scene, the system generates a fixation sequence by calculating in each step which feature has to be selected next to reach a maximum of information gain. The information gain calculation is based on the Dempster-Shafer belief theory, which is also used to establish a hypothesis about the scene.

I. INTRODUCTION

Biological vision systems have developed an efficient design in which the pattern recognition capabilities are concentrated in a small region of the visual field, the central fovea, whereas the periphery has only limited optical resolution and processing power. Hence, with a static eye, we can only see a small spot of the environment with satisfactory quality, but this spot can be rapidly moved with fast saccadic eye movements of up to 700 deg/s toward all the relevant regions of a scene. This selection process is determined by bottom-up processes on the input scene as well as by top-down processes determined by the memory, internal states, and current tasks (Yarbus, 1967).

The modeling of the strategies for the control of eye-movements and for related attentional processes has recently attracted increasing interest, but in our view the detailed analysis of the interactions between bottom-up and top-down processes has not been a central issue in most of these systems.

A powerful system that relies on a pure bottom-up approach has been suggested by Itti and Koch (2001). This system includes a number of low-level features such as local contrast, color, and texture and can already achieve a considerable correlation with the eye movements of human observers; but it does not include a top-down strategy or reasoning component. A system for planning saccadic eye movements in the context of a search task that considers both processing directions to a certain degree has been developed by Rao et al. (1997); however, the task of the top-down component is restricted to determining a target description matching the image.

Classic attention models address the top-down part, but only in a limited fashion with respect to reasoning or exploration strategies. Olshausen et al. (1993) developed a system for visual attention in the context of scale- and position-invariant pattern recognition. The top-down aspect is realized as an associative memory that influences the selection process of image regions provided by the control neurons and aids in the completion of the patterns. The system of Tsotsos et al. (1995) is based on the work of Koch and Ullman (1985) and models the influence of task-specific aspects on attention processes. However, expectations about the object or scenes and the influence of such expectations on bottom-up processes are not considered.

The system of Egner (1997) considers the computation of an expectation about the current image by computing a most probable hypothesis about the current scene. This hypothesis influences a zooming mechanism that determines a detailed analysis of a subregion. This system is not intended to handle the evolution and the influence of multiple competing hypotheses. The calculation of a most probable hypothesis is also their basic strategy in the approach of Chernyak and Stark (2001). Segments represent scenes, characterized by sensory data such as brightness or hue value and their spatial arrangement. Positional information about a segment is treated in a similar way as other features, so the "where" and "what" pathways are available and both aspects can be used in the representation. A "where" and "what" system is also modeled in the high-level stage of the system by Rybak (Rybak et al., 1998; Chapter 108). The search mode in this system scans an image until a feature is found that is also present in one of the stored images. This principle is based on the feature ring thesis of Noton and Stark (1971) and assumes a deterministic relation between features and objects that makes it difficult to deal with imperfect or ambiguous data in a flexible fashion.

A general characteristic feature of these systems is the restricted interaction between bottom-up and top-down influences. In particular, the top-down influence of higher-level cognitive processes and the associated reasoning strategies have not been a central issue.

In order to integrate bottom-up and top-down processing into an architecture that is able to cope with the typical incompleteness, ambiguity, and inconsistency in sensory input data, we have developed a hybrid system (Schill, 1997; Schill et al., 2001). It consists of a knowledge-based inference component that uses the Dempster-Shafer (D-S) theory (Shafer, 1976) for uncertain reasoning. This top-down reasoning component interacts with a neural-network preprocessing stage, which reflects the computations performed in the early stages of the visual system (Zetzsche and Krieger, 2001). This architecture is augmented by a control strategy that analyzes a scene by determining a sequence of saccadic fixations promising maximum information gain.

II. LINEAR AND NONLINEAR PREPROCESSING

The aim of the preprocessing stage is to identify highly relevant and informative candidate locations within the scene, which can be the goal of saccadic fixations, and to provide information about the actual fixated local pattern that can be used to improve the ongoing analysis of the scene by the knowledge-based system. Thus, one question is: What kind of candidate features should potentially deserve further inspection (see also Chapter 94)? In experiments on eye movements, human subjects showed a clear bias to fixate image regions with frequency components of multiple orientations (e.g., image regions with curved edges or occlusion patterns); (see Zetzsche and Krieger, 2001; Chapter 37). These results fit well with the concept of intrinsic dimensionality (Zetzsche and Krieger, 2001), according to which the least redundant information in local image configurations is given by intrinsically two-dimensional (i2D) signals. The extraction of these highly informative features can be provided by nonlinear i2D selective operators (Fig. 109.1), as developed by Zetzsche and colleagues (see Zetzsche and Krieger, 2001; Chapter 37).

The preprocessing stage of our system consists of a wavelet-like image decomposition by size- and orientation-specific filters with a spatial resolution that is maximal in the center and decreases in the periphery. This subsystem is used for two purposes: (1) the computation of a multidimensional feature vector,

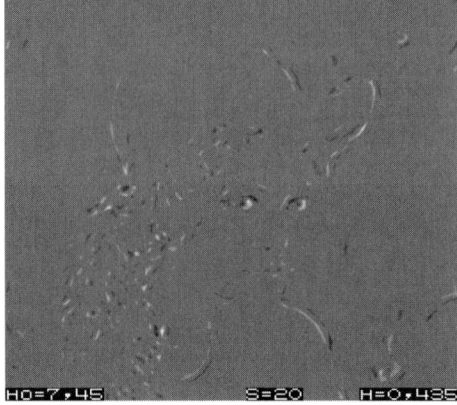

FIGURE 109.1 Potential fixation candidates (salient features) as derived from nonlinear i2D selective neural operators.

which provides information about the current foveal input to the knowledge-based system, and (2) the computation of potential fixation points, which is currently achieved by a nonlinear 2D specific combination of peripheral filter outputs.

The operation of our system thus combines two effects. Nonlinear i2D selective operators are applied to the whole peripheral image in order to extract possible fixation candidates. Out of these locations, the top-down strategy then selects the actual fixations made by the system. At the actual fixated locations, the outputs of linear orientation selective filters are combined into a feature vector that is a characteristic of the local image region.

III. SENSORIMOTOR FEATURES AND HIGH-LEVEL SCENE CONCEPTS

One view in computer vision is that scenes or objects are represented purely by afferent sensory features. In contrast to this, the representation in our system is based on both sensoric data and motor actions. The basic sensorimotor features are triples of the form: {current sensory feature, eye movement, target sensory feature}, where the sensory features are found by the linear and nonlinear preprocessing and the motor component is represented by the relative change of gaze. These sensorimotor features provide the link between the bottom-up preprossesing and the top-down knowledge-based component that includes a knowledge-base constituted by concepts of scenes organized in a hierarchical manner. The uncertainty with which a sensorimotor feature supports a set of scenes in the hierarchy are learned by the system. It is expressed by a belief measure as defined in the D-S theory (Shafer, 1976).

The D-S theory can be considered an extension of probability theory, which for example, enables the distinction between nonsupporting evidence and the lack of knowledge. An important concept is the frame of discernment θ, which is the set of all possible singelton hypotheses in the considered domain, in our system the set of all possible scenes considered. A basic function is the basic probability assignment (bpa) $m: 2^\theta \to [0, 1]$ (Shafer 1976). It obeys the following three axioms:

$$m: 2^\theta \to [0, 1] \quad (1)$$

$$m(\varnothing) = 0 \quad (2)$$

$$\sum_{A \subseteq \theta} m(A) = 1 \quad (3)$$

The heart of the theory is the Dempster Rule of Combination, which combines two basic belief assignments:

$$\forall C \in 2^\theta, C \neq \varnothing$$

$$A_i, B_j \in 2^\theta : m_{12}(C) = \frac{\sum_{A_i \cap B_j = C} m_1(A_i) * m_2(B_j)}{K} \quad (4)$$

with the normalization constant

$$K = 1 - \sum_{A_i \cap B_j = \varnothing} m_1(A_i) * m_2(B_j) \quad (5)$$

The Dempster Rule of Combination is the basis for combining a number of belief measures for scene hypotheses as induced by a number of feature vectors being selected by eye movements. Because our system is based on a hierarchical hypothesis space T, we use an approximation of this rule for treelike knowledge structures (Gordon and Shortliffe, 1985) with which a corresponding m_T value for each set of scenes in the tree can be calculated.

How can we achieve such a basic probability assignment m with which a sensorimotor feature triple supports a hypotheses of a scene (or a set of scenes)? This is learned by the system during the analysis of a scene (see V). After this analysis of a scene, the long-term memory of the system (i.e., information over all analyzed scenes) is updated using the content of the short-term memory (i.e., all feature vectors generated on this scene). The long-term memory consists of a matrix $M = \{x_{ij}\}$, which relates sensorimotor feature vectors f_i and possible hyptheses H_j by the relative frequency of a feature vector f_i selected in a scene H_j of the hierarchy:

$$\forall (H_j \in T) : r_i(H_i) = \frac{x_{ij}}{\sum x_{ik}} \quad (6)$$

We have shown that $r: T \to [0, 1]$ corresponds to a basic belief assignment (bpa) $m: 2^\theta \to [0, 1]$ of the D-S theory and can be used to express the belief that a certain feature triple {current foveal feature, eye movement, target feature} yields for a hypothesis of a scene. The strategy with which the system generates a sequence of fixations on a scene is based on the principle of information gain.

IV. THE INFERENCE STRATEGY IBIG

The basic principle of our strategy IBIG (inference by information gain) is to determine those data that, when collected next, will yield the largest information gain with respect to the actual belief distribution in the hypotheses' space (Schill, 1997). Applied to scene analysis, the inference strategy determines the region of interest, where the expected target feature promises the maximum information gain with respect to all activated scene hypothesis (Schill et al., 2001). The information gain is calculated by the difference between the current belief m_T of a scene hypothesis H_j in the hierarchy and the potential belief $\hat{m}_T$ this hypothesis could reach:

$$l = |m_T - \hat{m}_T| \qquad (7)$$

The current belief m_T is the combination of all bpas corresponding to yet-selected feature vectors inducing belief for a hypothesis of a set of scenes (or a singleton scene) in the hierarchy T. Accordingly the potential belief $\hat{m}_T$ is calculated by combining the bpas corresponding to yet-uncollected data. Thus, it is the belief that the scene hypothesis H_j could potentially be achieved in the subsequent step by selecting a sensorimotor feature.

Because the system is working with a hierarchical knowledge structure, the information gain calculation has to consider all possible intersections between the sets of hypothesis H_i and H_j in the hierarchy, as well as both types of potential beliefs (supporting and non-supporting). A set of equations has been derived corresponding to the possible cases (Schill, 1997). The information gain is calculated with respect to all activated scene hypotheses, thus all hypotheses with current belief. The hypothesis H_j whose yet-uncollected features promise the largest information gain is the one which determines the feature selection process in the next step. After a new sensorimotor feature has been selected, the new current belief distribution is recalculated and the information gain calculation is applied again. Thus, the system calculates the eye movement whose potential sensory target will provide the largest information gain with respect to the current belief distribution in the knowledge-base and the current sensory input (see Fig. 109.2).

Simulations of the information gain strategy show that it acts in parallel on the represented knowledge while it organizes itself depending on the evidence distribution as induced by the available data. The reasoning behavior is determined by the permanently changing relationship between the knowledge structure and the growing corpus of actual available data, which leads to efficient strategies in terms of required system steps and input data (for details, see Schill, 1997).

V. SYSTEM BEHAVIOR

The following stages describe one circle of the system behavior. Assume a region of interest in the

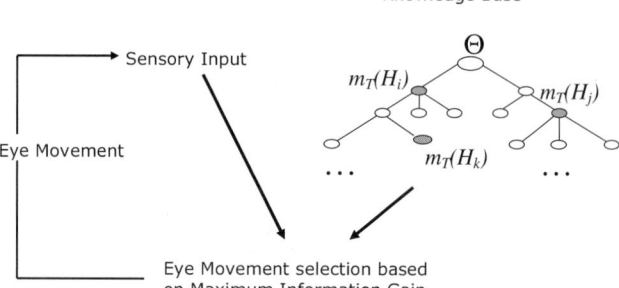

FIGURE 109.2 Based on the bottom-up processed current sensory input and the top-down activation of scenes in the knowledge-base, the system computes the eye movement whose target feature promises the maximum information gain. (Adapted from Schill, 2001)

scene has been determined. The fovea is then directed toward this location by a saccadic movement. At the new foveal position, the local feature vector is computed on the basis of the orientation-selective filter outputs of the preprocessing stage. IBIG then computes, depending on the already available feature triples and the new incoming information, the updated current belief distribution with respect to the long-term memory. Then the eye movement (i.e., the next potential target feature and the associated relative change of gaze) that promises the maximum information gain with respect to this current belief situation is determined. For this, only those eye movements are considered whose stored foveal feature vector is consistent with the actual feature vector at the given foveal position and whose relative target position is within the spatial range of one of the candidate locations, as provided by the nonlinear preprocessing. Then the saccade is performed on this new location, the local information is extracted, and the system starts to calculate a new eye movement based on the new current sensory input (i.e., the updated belief distribution in the scene hierarchy). This cycle is repeated until a certain belief threshold for a hypothesis is reached and the system suggests this hypothesis (i.e., the label of a scene) as the result of its analysis. Via supervised learning, the system gets feedback about the correct label of the current scene. This label, together with the content of the short-term memory, is transferred to the long-term memory in order to generate an updated knowledge-base. Thus, the analysis of a scene and the learning is performed with the same principle, namely information gain. After a certain classification threshold is reached, the supervised learning mode can be switched off. Figure 109.3 shows a screenshot of the system during the analysis of a polygon pattern.

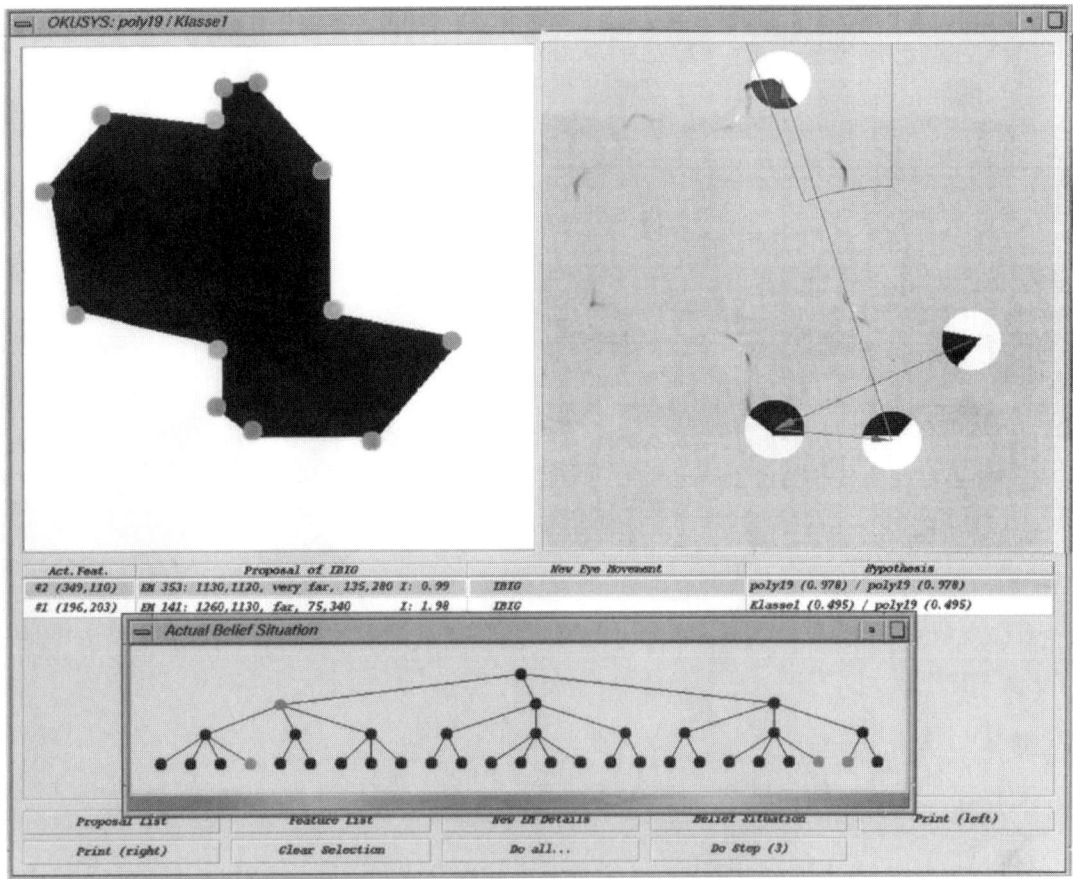

FIGURE 109.3 Screen dump of the system analyzing polygons. On the left side, the sequence of fixation on i2D selective features is shown.

VI. DISCUSSION

In order to understand more about the principles of the selection of informative regions in saccadic scene analysis, we have suggested some first steps toward a computational model of these processes. Because this selection process is often dependent on both bottom-up processes and higher cognitive components such as internal states, memory, and tasks, a hybrid system consisting of a preprocessing stage linked to a knowledge-based reasoning component has been developed.

The top-down and bottom-up processes rely on a common principle (information gain) and are integrated into an architecture that is able to cope with the typical incompleteness and uncertainty in sensory input data. The support that a sensorimotor feature promises for a scene is learned by the system and is expressed in terms of the D-S-theory. The bottom-up process identifies intrinsically 2D-dimensional signals as possible fixation candidates and reflects the computations performed in the early stages of the visual system. The reasoning strategy works in a parallel and adaptive way by calculating the sensorimotor feature promising the maximal information gain with respect to the activated scene hypothesis and the current input available.

Our further developments will be guided by an ongoing evaluation with empirical investigations in which the systems behavior will be compared with the recorded eye movements of human subjects. Although the correspondence between human behavior and certain bottom-up gained candidate features can already be shown, the main task of these experiments will be to investigate the interaction between the bottom-up processes and the top-down selection processes. The impressive efficiency of human scene analysis is the inspiring and challenging goal of this modeling.

References

Chernyak, D. A., and Stark, L. W. (2001). Top-down guided eye movements? *IEEE Trans. Syst. Man Cyberneti.* **31** (4), 514–522.

Egner, S. (1997). Zur Rolle der Aufmerksamkeit für die Objekterkennung: Modellierung, Simulation, Empirie. PhD diss. University of Hamburg.

Gordon, J., and Shortliffe, E. H. (1985). A method for managing evidential reasoning in a hierarchical hypothesis space. *Artificial Intell.* **26**, 323–357.

Itti, L., and Koch, C. (2001). Feature combination strategies for saliency-based visual attention systems. *J. Electron. Imaging* **10**, 152–160.

Koch, C., and Ullman, S. (1985). Shifts in selective visual attention: towards the underlying neural circuitry. *Hum. Neurobiol.* **4**, 219–227.

Noton, D., and Stark, L. (1971). Scanpaths in saccadic eye movements while reviewing and recognizing pattens. *Vis. Ras.* **11**, 929–942.

Olshausen, B., Anderson, C. H., and van Essen, D. (1993). A neurobiological model of visual attention and invariant pattern recognition based on dynamic routing of information. *J. Neurosc.* **13**, 4700–4719.

Rao, R. P., Zelinsky, G. J., Hayoe, M. M., and Ballard, D. H. (1997). "Eye movements in visual cognition: A computational study" Technical Report, 97.1. Computer Science Department, University of Rochester.

Rybak, I., Gusakova, V., Golovan, A., Podladchikova, L., and Shevtsova, N. (1998). A model of attention-guided visual perception and recognition. *Vis. Res.* **38**, 2387–2400.

Shafer, G. (1976). "A Mathematical Theory of Evidence." Princeton University Press, Princeton, NJ.

Schill, K. (1997). Decision support systems with adaptive reasoning strategies. *In* "Foundations of Computer Scienes: Theory, Cognition, Application" (C. Freksa, M. Jantzen, and R. Valk, eds.), pp. 417–427. Springer, Berlin.

Schill, K., Umkehrer, E., Beinlich, S., Krieger, G., and Zetzsche, C. (2001). Scene analysis with saccadic eye movements: Top-down and bottom-up modeling. *J. Electron. Imaging* **10**, 152–160.

Tsotsos, J. K., Culhane, S. M., Wai, W. Y. K., Lai, Y., Davis, N., and Nuflo, F. (1995). Modelling visual attention via selective tuning. *Artificial Intell.* **78**, 507–545.

Yarbus, A. L. (1967). "Eye Movements and Vision." Plenum Press, New York.

Zetzsche, C., and Krieger, G. (2001). Nonlinear mechanisms and higher-order statistics in biological vision and electronic image processing: Review and perspective. *J. Electron. Imaging* **10**, 56–99.

Index

NUMBERS

2.5D sketch stage, 172
2-DG technique, 429
2D memory patterns, 607
2D shape recognition, 274
3D attention, 112
3D model memory layer, 607
3D scenes, 110
3D space and attention allocation, 109–13
　depth, 110–12
　　and aging, 111–12
　　cues for, 112
　　objects and surfaces, 111
　overview, 109
　research on, 109–10
　viewer-centered versus object-centered representations, 112
5CSRT (five-choice serial reaction time task), 58, 59, 60

A

AB (attentional blink) bottleneck, 384–85
absolute criteria, 153, 155
acallosal patients, 364
access awareness, 172, 173
accurate registration, 628
acetylcholine (ACh), 58–59, 216
active attention, 540
active vision system, two-chip, 635–37
activity interpretation algorithms, 631
acuity and transient attention, 445–47
Adaptive Resonance Theory (ART), 653, 654, 659
addiction, attention, 395
additional singleton paradigm, 418–19
address-event representation, 634
adrenergic drugs, 60
adult visual search and pop-out, 207–8
AER (address event representation) silicon retina, 635
afferents, 283
AFmaps, 376
aging, depth and attention control, 111–12
AGM (attention-guidance map), 201
AIT (anterior part of the inferotemporal cortex), 266
algorithmic circuitry, 607
alternative perceptual groupings, 483
ambiguous motion perceptions, 540
AMI (attentional modulation index), 301
amnesic search, 265
AMPA receptors, 615
amplitude modulations, 511
amygdala, 161, 162–63, 164
analog sensitivity, 654
anaphora and structure of discourse, 325–27
angle detector, 27–28
angular gyrus, 409
annular region, 430
annulus-disk configuration, 377
ANOVA, 373, 408
anterior callosal sections, 365
anterior cingulate, 385, 409
anterior electrodes, 341
anterior part of the inferotemporal cortex (AIT), 274
anticipatory synchronization, 530
anti-extinction, 270
antisaccade pop-out visual searches, 120
aperiodic synchrony, 535
apical dendrites, 655
apparent contrast, 445
architecture, *see also* machine vision architecture; putative functional architecture
　circuit, of dynamic routing, 13–14
　constraints on in model of human vision, 6
　implemented, of bottom-up attention and saliency models, 579–80
　triadic, 79
aROIs (algorithmic regions of interest), 299
arousal level and noradrenergic system, 51–54
ART (Adaptive Resonance Theory), 653, 654, 659
artificial fovea, 625
artificial vision systems, 633
ART-like neural network, 20
assisted perception displays, 475
associative memory, 539
astronomical numbers, 526
asymmetry
　hemispheric, 31–32
　temporal, of cross-modal interaction window, 541–42
asynchronous communication protocol, 633
asynchronous cortical activity, 365
attend-collinear conditions, 482
attend-frequency paradigms, 498
attend-location paradigms, 498, 500
attend-null direction, 302
attend-orthogonal conditions, 482
attend-preferred direction, 302
attentional blink (AB), neural basis of, 383–88
　AB bottleneck, 384–85
　effects of real and virtual brain lesions, 386–88
　neural fate of T2, 385–86
　overview, 383
attentional bottleneck, 576
attentional capture, 542
attentional complexes
　contents of, 78–79
　independence of, 78
attentional components in visual searches, 395–97
attentional control, 38, 39, 67, 343, 358, 362, 372, 396
attentional difference wave, 511
attentional disengagement, 59
attentional field map (AFmap), 372, 375
attentional gain control, 439
attentional links, 652–62
　to competition, 652–57
　divided, object versus spatial, and hierarchical, 660–62

attentional links (*continued*)
  laminar organization of bottom-up, horizontal, and top-down connections, 654–56
  to learning, 652–54, 659
  modulation of, 654
  object-based, via pre-attentive-attentive interface, 657–59
  overview, 652
attentional load and reflexive attentional effects, 222
attentional manipulation, 461–62
attentional mechanisms, 609
attentional modulation, 302, 375, 405, 425, 433, 435–41, 565; *see also* effective connectivity and attentional modulation; lateral interactions; motion processing; surround inhibition; timing of attentional modulation of visual processing
  in lateral geniculate nucleus, 435–39
    anatomy of, 435–36
    baseline increases in, 438
    comparison to visual cortex, 438–39
    overview, 435
    response enhancement in, 436–37
    response suppression in, 437–38
  overview, 435
  in pulvinar, 439–41
  of stimulus contrast, 425–28
    common substrate for, 427–28
    effects of on neuronal responses, 425
    overview, 425
attentional modulation index (AMI), 301
attentional movements, 396
attentional network, 393
attentional probe, 550
attentional response gain, 435
attentional selection, 18, 43, 420–21
attentional shift, 58
attentional suppression, in Macaque visual system, 429–34
  computational modeling, 430–34
  overview, 429–30
  ring of metabolic suppression, 430
attentional top-down feedback, 161
attentional visual processing, 593
attentional weight, 416–17
attention–awareness model, 18–23
attention-away condition, 430
attention-based acceleration, 651
attention capture, 69–75
  automatic cross-modal, 540
  cross-modal, and cortical visual processing, 223–25
  explicit, and inattentional blindness, 72–74
  by feature singletons, 418–24
    attentional selection and inhibition of return, 420–21
    and eye movements, 422–24
    overview, 418
    spatial attention, 420, 421–22

  implicit, 69–72, 74
  overview, 69
attention-dependent modulation, 429
attention-dependent suppression, 430
attention-guidance map (AGM), 201
attention-guided visual perception, 663
attention mechanisms, 448
attention modulation, 596
attention-related activation, 371–74
attention-related baseline, 438
attention-related foci, 374
attention-related mechanism, 542
attention-related modulation, 375, 508
attention scan paths, 635
attention shifts, 298
attention switch, 343
attention system, 647
attention-to-the-grating condition, 430
attention vector-sum (AVS) model, 334
attention window (AW), 665
attentive computer vision systems, 642–45
  contextual cueing and attention strategies, 644–45
    overview, 644
    priming, 644–45
    saccadic information integration, 645
  object and scene recognition, 643–44
  overview, 642
  saliency from feature selection, 643
attentive coordinate frames, 628
attentive mechanisms, 211–12
attentive sensors, 625
attentive viewpoint control, 645
attentive vision, 290
attraction, criterion, *see* criterion attraction and unique internal representation
attractor network pools, 611
audio-visual spatio-temporal relations, 542
auditory-context dependency of bounce-inducing effect, 539
auditory cortex, 656, 659
auditory-induced effect, 541
auditory perceptual organization, 317–23
  electrophysiological measures, 318–21
    manipulation of attention, 318–21
    overview, 318
  hierarchical decomposition model, 321
  neuropsychological approach, 321–23
  overview, 317
  streaming, 317–21
    indirect effects of on competing tasks, 319
    outside focus of attention, 318
auditory streaming, 317, 318
auditory transients and bounce-inducing effect, 539–40
auditory-visual coupling, 539
automatic algorithm, 628
automatic amygdala activation, 162–63
automatic gaze control, 645
automatic image registration, 628
automatic processing, 412
automatic responses, 342
automatic stimulus categorization, 385

automatic surveillance mode, 629
automatic visual tracking, 629
autonomous control of information flow, 14–15
autonomous multisensor robots, 646–47
AVS (attention vector-sum) model, 334, 335
awareness, 19, 22, 165–66, 171
  degree to which unexpected objects intrude on, 73
  difference from attention, 167
  model of attention, 18–23
  and visual attention, 167–74
    overview, 167
    phenomenal, 172–74
    processing and memory, 167–69
    recurrent processing, 169–70
    selection of conscious experiences, 167
AW (attention window), 665

B

backward-looking center, 326
backward masking paradigms, 161
backward path, 607
backwards blocking, 216
Balint's syndrome, 136–38, 269–70, 346
ball saliency, 584
ball tensors, 584
basal ganglia, 386, 391
baseline activity in visual cortex, 308–10
baseline increases in lateral geniculate nucleus, 438
baseline sensory response, 488
baseline shift, 308, 354
base response, 311, 312
basic visual dimension, 418
Bayesian framework, 24, 589
behavior, system, 674–75
behavioral algorithms, 669
behavioral aspects of visual extinction and hemispatial neglect, 351–52
behavioral concept, 664
behavioral definition of visual saliency, 273–74
behavioral evidence for cross-modal spatial attention, 187–89
behavioral paradigm in visual perception and visual attention, 663–64
behavioral predictions of selective tuning model for visual attention, 567–69
behavioral programs, 664, 667
behavioral properties of inhibition of return, 96–97
behavioral recognition process, 669
behavioral results of fMRI of object-based attention, 403–5
behavioral studies of temporal orientation, 257–59
behavior-based control, 645
behavior-brain correlations and spatial attention, 30–31
behavioral context, 477
belief distribution, 674
beta-frequency range, 530

between-subjects attention paradigm, 516
biased competition model, 137, 161, 303, 380, 410, 593, 594, 604, 621
biasing competition in human visual cortex, 305–10
　limited processing capacity and, 305
　neural basis for among multiple stimuli, 305–7
　overview, 305
　top-down, 307–10
　　filtration of unwanted information, 307–8
　　increased baseline activity, 308–10
　　overview, 307
biasing signals for emotional stimuli, 163–65
bilateral dipolar source, 511
bilateral posterior cortical damage, 346
Bilinear interpolation, 628
binding, 135–39
　feature, and oscillation and synchrony, 529–30
　implicit and explicit, 137–39
　　of features and objects, 137–38
　　of objects to location, 138–39
　overview, 135
　problems with, 79, 135–36
　visual feature, neuropsychology of, 269–71
binocular disparity, 112
biological evidence of memory-driven visual attention, 607–8
biological implications of selective tuning model for visual attention, see selective tuning model for visual attention
biological models, 580
biological vision systems, 625, 654, 671
biophysical mechanisms, 524
bisensory stimuli, 538
bispectra, 231
blackshot, 466, 467–69
black trace, 390–91
blindness, inattentional, 72–74, 161; see also change blindness
blink, attentional, see attentional blink
blood-flow imaging methods, 507
blood oxygenation level dependent (BOLD), 36, 366, 454
BOLAR (bank of linear analyzer responses) vector, 397–99
Boltzmann distribution, 554–56, 560
bottlenecks in visual attention, 77–78
bottom-up attention, 608–9
bottom-up attentional mechanisms, 387, 401, 418, 421, 493, 502, 504, 533, 548, 551, 672
bottom-up automatic activation, 653
bottom-up connections, laminar organization of, 654–56
bottom-up control in split-brain patients, 361–62
bottom-up-driven attention, 605
bottom-up pathways, 653
bottom-up pre-attentive processing, 656

bottom-up preprocessing, 673
bottom-up processing pyramid, 579
bottom-up saliency maps, 591, 634
bottom-up selection, 240
bounce-inducing effect, 539
　auditory-context dependency of, 539
　by auditory, tactile, and visual transients, 539–40
　enhancement of without sensory transients, 540–41
bouncing motion display, 538
Boundary Contour System, 583
bounded visual search, 4
brain, see also split-brain patients
　brain–behavior correlations and spatial attention, 30–31
　brain-evoked responses, 498
　functional and neural locus of redundancy gain in normal, 365–67
　imaging, 185, 364, 365
　lesions to, effects of real and virtual, 386–88
　regions of, functional differences between different, 37
bright-side energy, 469
buffer model, 268
B-Y coding, 397

C

calibration and wide-field sensing, 628
callosotomies, 359
canonical region, 385
canonical representation, 560
capacity
　limits for memory, 268
　limits for spatial discrimination, 8–10
　　feature detection, 9–10
　　feature integration, 9
　　feature segregation, 8–9
　　overview, 8
　limits for visual attention, 77–78
　selective, of attention, 347–48
capacity-limited attentional processes, 387, 562
capacity-limited attention-demanding stage, 385–86
capacity-limited processing stages, 383–84
capacity-unlimited processing stages, 383–84
capture, attention, see attention capture
categorical comparisons, 458–59
CDF (cumulative distribution function), 364
cellular response sensitivity, 448
center frequency, 449
centering theory, 326
central fixation bias, 241
central grating, 429
central-neutral cue, 442, 446
central pattern generator (CPG), 640
central sounds, 347
central spatial cueing, 410
central stimulus, 84–85, 431

central-task performance levels, 462
central visual area, 618
centroid computation, 397–98
cerebral blood flow, decreased, 353
cerebral hemispheres, 358, 362
CFQ (Cognitive Failures Questionnaire), 337
change blindness, 76–81, 161
　nature of visual attention, 77–79
　　and attentional complexes, 78–79
　　binding problems, 79
　　capacity limits and bottlenecks, 77–78
　　scene perception, 79–81
　overview, 76–77
change detection, 77–78, 629
channels, attentional enhancement of selected, 340–41
checkerboard stimuli, 355, 436
chip, selective attention, 634–35
cholinergic attentional effects, 50
cholinergic modulation of memory/learning processes, 54
cholinergic system, 54–56, 350
　direct effects of, 54–56
　modulation of, 58–60
　　effects of systemic agents, 59–60
　　lesions of nucleus basalis, 59
　　memory and learning, 54
　　overview, 58–59
　overview, 54
chromatic motion processing, 491
cingulate sulcus, 409
circuit architecture of dynamic routing, 13–14
classical gamma band, 522
classical receptive field (CRF), 481, 570, 577, 579
classic attention models, 671
clonidine, 51
CNV (contingent negative variation), 259
co-activation model, 363
coarse-before-fine sequence, 252
coarse optic flow field, 639
codes, rate and temporal, 528
Cognitive Failures Questionnaire (CFQ), 337
cognitive neocortex, 652, 654
cognitive neuroscience, 93–95, 346
cognitive processing, 448, 507, 659
cognitive research and temporal orientation, 261–63
cognitive systems, 352, 442, 507
coherence field, 201
coherent motion signal, 492
co-hydroxyolopamine (6-OHDA), 60
collicular afferents, 497
collicular neurons, 365
collinear configurations, 478, 480, 481
collinear flankers, 478
collinearity, 481
　and global feature similarity, 480–81
　and local feature similarity, 478–80
collinear targets, 462, 464

color, role of in express recognition of scene gist, 253
color-nondiagnostic scenes, 254
color-orientation conjunction search, 573
color-tuned cells, 573–74
combinatorial explosion, 606
compatibility, task, 472
competition
 biased, 621
 links to attention, 652–54
complementary coding strategy, 526
complementary hybrid stimuli, 252
complex context-dependent behavior, 610
complexity theory, 562
complex multiple-location displays, 453
complex nonlinear systems, 456
computational capacity, 576
computational foundations for attentive processes, 3–7
 complexity of vision, 4–7
  constraints on models, 6–7
  intractable problems, 5
  objections to analysis of, 5
  overview, 4
 overview, 3
 theoretical background, 3–4
computational implications of selective tuning model for visual attention, *see* selective tuning model for visual attention
computational model, 582, 600
computational model of spatial language apprehension, 331–33
computation of surprise, 25–26
computer graphics rendering of attention, 649–51
computer vision, *see* attentive computer vision systems; saliency
conceptual nature of gist of scene, 251–52
conditional object-response, 611
conditioned stimuli (CS), 213
conditioning, 213–18
 and learning, 214–16
  overview, 213–14
 and prediction, 216–17
conjoined-selection paradigm, 500
conjunction searches, 120, 211
conjunction-specific neurons, 526
conjunction units, 526
conjunction visual search, 121
connected image elements and contour grouping, 289–90
connection-weighting account, 481–82
conscious experiences, selection of, 167
consciousness, 18, 652–54
conscious processing, 352
conscious scene perception, 385
conscious target report, 386
conspicuous stimuli and visual selection, 118–20
conspicuous stimuli and visual selection, popout visual search, 118–20
 effect of singleton distracter during, 118–20

feature expectancy during, 118
 priming during, 118
 pro- and antisaccade, 120
context-dependent stimulus-response tasks, 611
context priming, 644–45
context representation, 587
contextual bars, 571
contextual cueing, 246, 247, 579; *see also* machine vision architecture
contextual guidance of visual attention, 246–50
 cueing task, 246–48
  object, 248
  overview, 246–47
  spatial, 247–48
  temporal, 248
 overview, 246
 statistical learning, 248–49
contextual influences on saliency, 586–92
 overview, 586
 scenes, 586–87
  priors, model for, and modulation, 589–90
  representation of, 587–89
contextual mapping, 645
contextual model, 591
contextual modulation of salience of feature contrast
 in area V1, 235–36
 in perception of, 236–39
  by distance, 236
  orientation contrast versus onset or offset, 236
  overview, 236
contextual receptive field, 557
contingent negative variation (CNV), 259
continuous data, 26
continuous performance-type (CPT), 57
contour grouping, 288–95
 of connected image elements, 289–90
 neurophysiology of, 290–92
 overview, 288–89
 psychology of, 292–95
contralateral hemifield, 358, 359–60, 437, 439
contralateral posterior parietal cortex, 509
contralateral target location, 372
contralateral ventral occipital areas, 407
contralesional information, 351
contralesional left/right neglect, 31
contralesional visual stimuli, 352, 355
contrast
 apparent, 445
 coding, 493
 contrast-dependent response modulation, 45–47
 contrast-dependent response modulations, 46
 contrast-detection tasks, 460
 contrast-discrimination tasks, 460, 464
 contrast gain model, 426, 427, 445, 493, 495
 contrast matching procedure, 156

 contrast psychometric function, 443–45
 contrast sensitivity function, 443
 contrast signals, 412, 415
  effective, and spatial attention, 47–49
  effect of elevating, 47
  identifying areas of overlap among categories of, 66
  orientation versus onset or offset, 236
  response threshold, 42–43
  sensitivity function, 443
 contrast response function (CRF), 45, 379, 425–26, 495
control signals, 12
convergence characteristics, 564
convergence properties, 568
convergent zones, 600
convolution model, 456
coordinate frames, 393
coordinate system, 666
corpus callosum, 358–59, 363, 364, 365
correlates, physiological, of lateral interactions and attentional modulation, 483
correlation, measure of, 241
correlation analysis, 622
correlograms, 620
cortex, *see* object recognition in cortex
cortical activation, 372
cortical architecture, 608
cortical areas, 496, 656, 657
cortical bases of spatial attention development, 85–88
cortical development, 659
cortical enhancement, 375
cortical feedback, 659
cortical hemispheres, 381
cortical layers, 430, 660
cortical learning, 652
cortical levels, 438
cortical localization, 366
cortical mechanisms, 508
cortical neurons, 507, 527
cortical parvocellular system, 365
cortical processing, 366, 439
cortical projection, 665
cortical regions, 458
cortical self-organization, 652
cortical visual areas, 439
cortical visual processing, 223–25
cortical voxel, 372–73
cortico-cortical feedback connections, 533
cortico-subcortical interactions, 363, 365
corticothalamic feedback, 438, 439
corticothalamic transmission, 439
cost-efficient solution, 542
covert attention and saccadic eye movements, 114–16
 and overt attention, 114–15
 overview, 114
 and use of salience map, 115
covert orienting, 58, 61, 82–88, 553
 during central stimulus attention, 84–85
 cortical bases of spatial attention development, 85–88

overview, 82
to peripheral stimuli, 82–84
CPG (central pattern generator), 640
CPT (continuous performance-type), 57
CRF (classical receptive field), 481, 570, 577, 579
CRF (contrast response function), 45, 379, 425–26, 495
CRF-parameters, 426
criterion attraction, 152
criterion attraction and unique internal representation, 155–57
criterion performance level, 451, 473
critical perceptual task, 474
cross-channel spike density, 527
cross-correlation analysis, 529
cross-dimension search, 413, 414, 416
cross-modal attention, in event perception, 538–43
  automatic capture of, 540
  bounce-inducing effect, 539–41
  developmental assay utilizing display, 542
  dynamic attentional allocation, 541–43
  overview, 538
  spatial, 538
  temporal asymmetry of interaction window, 541–42
cross-modal attentional capture, 223–25
cross-modal cuing, 453
cross-modal interaction, 157, 541
cross-modal spatial attention, 187–96
  behavioral evidence for effects, 187–89
  ERP evidence for sensory effects of, 189–91
  neuroimaging evidence for modulation of sensory cortex by, 191–96
  overview, 187
crowding phenomena, 170
CTOA (cue-target onset asynchrony), 389
"cued" trials, 361
cueing, *see also* machine vision architecture
  contextual, 246–48
    object, 248
    overview, 246–47
    spatial, 247–48
    temporal, 248
  cues for depth and attention control, 112
  cuing studies, 65
  nonspatial, 283–87
  spatial, 283
cue-response mapping, 613
cue-target onset asynchrony (CTOA), 389
cue validity effect, 131
cue versus active (CvsA) category, 65–66
cue versus baseline (CvsB) category, 65
cue versus passive (CvsP) category, 65
Culhane-Tsotsos feature detector, 643
cumulative distribution function (CDF), 364
cumulative distribution functions, 267
CvsA (cue versus active) category, 65–66
CvsB (cue versus baseline) category, 65
CvsP (cue versus passive) category, 65

D

data processing techniques, 507
decidability, 4
decision behaviors, 152–59
  criterion attraction and unique internal representation, 155–57
  historical background, 152–53
  overview, 152
  signal detection theory, 153–54
decision level, 491
decision/response criterion, 158
decision statistic G(R), 466
declarative memory, 21
decomposition model, hierarchical, 321
defection, feature, 9–10
delayed spatial response tasks, 611, 613, 616
Dempster-Shafer (D-S) theory, 671, 672, 673
depth and attention control, 110–12
  and aging, 111–12
  cues for, 112
  objects and surfaces, 111
de-synchronization, 524
detection of objects in natural scenes, 600–604
  model of, 600–602
  overview, 600
deviance detection, 343
dexmedotomidine, 53
diagonal matrix, 558
digital AER (address event representation) communication infrastructure, 637
dimensional feature contrast signals, 415
dimensionality, intrinsic, 228–30
dimensionally redundant targets, 415
dimensional weighting effects, 414
dimension-based attention, *see* pop-out search
dimension-based coding, 417
dimension-based processing, 413
dimension-specific costs, 414
dimension weighting account, 414, 416
dipole mapping, 515
dipole modeling, 508
dipole-source analysis, 497
dipole-source modeling, 509
directed attention, 308
directional motor deficit, 346
directional tuning, 426, 493
direction-selective neurons, 300
direction selectivity, 493
Dirichlet model, 26
disambiguation, perceptual, 541
disambiguation process, 539
discourse, *see* structure of discourse
discourse focus, 326
discrete data, 26
discrimination, 153
disengagement of location targets, 222
dissociation of target selection from saccade production
  in saccade choice, 126–27
  in time, 125–26
distance and perception of salient feature contrast, 236

distractor inhibition in FeatureGate model, 550–52
distractors, 359, 372, 550
distractors, rejected, *see* visual search
distributed-neutral cue, 442
disynaptic inhibition, 571
divided attention, 363–67, 660–62
  overview, 363
  redundancy gain, 363–64
    functional and neural locus of in normal brain, 365–67
    paradoxical interhemispheric, in split brain, 364–65
divided spatial attention, 490
DNAB lesions, 60
domain objects, 646
dopamine, 60–61
dopamine D1-receptors, 350
dopamine D2-receptors, 350
dopaminergic system, 51, 350
dorsal and ventral streams, 496–501
  ERP studies of, 497–501
    overview, 497
    visual evoked potentials, 497
  overview, 496–97
dorsal network, 30
dorsal noradrenergic bundle, lesions of, 60
dorsal pathways, 393, 428, 438, 498, 535, 560–61
dorsal premotor cortex, 363, 366
dorsal premotor regions, 366
dorsal streams, 44, 351, 497, 501, 558, 596
dorsolateral prefrontal cortex, 610
dorso-medial regions, 435
dorsoventral patches, 374
downstream neurons, 520
drugs, adrenergic, 60
dual-axis stimuli, 481
dual-task performance in split-brain patients, 360–61
dual-task situation, 462
dynamical competition, 596
dynamical system, 595
dynamic attentional allocation, 541, 542–43
dynamic attentional control signals, 37–41
  feature-based, 38–39
  location-based, 37–38
  object-based, 39–41
  overview, 37
dynamic attentional process, 541
dynamic fusion and wide-field sensing, 628–29
dynamic motion stimulus, 492
dynamic neuronal model, 434
dynamic routing, 12–14

E

early sensory processing, 222
early vision
  stimulus and task-related context effects in, 477
  and surround inhibition, 461
early-vision task, 461

early visual cortex, *see also* nonsensory signals in early visual cortex
   spatially-specific attentional modulation in, 377–78
eccentricity experiment, 375
edge-filtered input image, 607
EEG (electroencephalogram), 318, 483, 507, 530
effective connectivity and attentional modulation, 454–59
   analyses of, 455–57
      overview, 455
      regression, 455–56
      Volterra series, 456–57
   versus categorical comparisons, 458–59
   dataset, 455
   overview, 454–55
   V5/hMT function, 454–55
effective contrast and spatial attention, 47–49
efMRI (event-related functional magnetic resonance imaging), 259
electrical scalp recordings, 365, 366
electrical stimulation, 391
electrodes, 514
electroencephalogram (EEG), 318, 483, 507, 530
electrophysiological measures of auditory perceptual organization, 318–21
   manipulation of attention, 318–21
   overview, 318
electrophysiological studies of visual attention, *see* visual attention
electrophysiology, 343, 345, 364, 391, 407, 455, 497, 500, 514
electrophysiology of reflexive attention, 219–25
   cross-modal attentional capture and cortical visual processing, 223–25
   and ISI effects
      long, 222–23
      short, 220–22
   overview, 219–20
elementary motor actions, 664
emotional perception, 160; *see also* visual attention
empirical evaluations, 629
empirical framework, 448
endogenous attention, 168, 451–52
endogenous distraction, 540
endogenous saliency, 277
enhanced firing rate, 294
enhancement model, 593
enhancement of response in lateral geniculate nucleus, 436–37
entry, prior, *see* prior entry
environmental cues, 448
environmental representations, spatial processing of, 146–51
   hierarchical model of, 146–47
   and nested environments, 148–50
   overview, 146
environmental spatial reference frames, 393
environmental stimuli, 497

EPSPs (excitatory postsynaptic potentials), 535
equidistant peripheral streams, 380
ERFs (event-related fields), 514
ERPs, *see* event-related potentials
event-onset asynchrony, 408
event perception, *see* cross-modal attention
event-related fields (ERFs), 514
event-related functional magnetic resonance imaging (efMRI), 259
event-related potentials (ERPs), 85–87, 93–94, 99, 163, 164, 187, 189–91, 219, 259, 261, 327, 339, 348, 384, 405, 483, 496, 497, 507, 511, 514; *see also* dorsal and ventral streams; prefrontal damage and ERP measures; visual attention
   timing of space and feature-directed visual attention, 515–19
event-related timecourses, 403
excitatory connections, 579
excitatory modulation, tonic, 340
excitatory neurons, 432
excitatory pool, 611
excitatory postsynaptic potentials (EPSPs), 535
excitatory synapse, 634
excitotoxic lesions, 59
exogenous cross-modal spatial cuing, 189, 190
exogenous saccade, 418
expectation, links to attention, 652–54
explicit attention capture, 72–74
explicit binding, 137–39
explicit report, 270
external noise, 448–53, 474
   exclusion, 451, 453, 473
   overview, 448–49
   perceptual template model, 449–51
      attention paradigm and mechanisms, 450–51
      of observer, 449–50
      overview, 449
   taxonomy of mechanisms of spatial attention, 451–53
external prefrontal neurons, 613
external top-down rule-specific input, 613
extinction, 156
extinction, visual, *see* visual extinction and hemispatial neglect
extracellular magnesium concentration, 616
extrageniculate pathways, 454
extrastriate activation, 516
extrastriate areas, 439
extrastriate cortex, 365, 511, 515, 518, 594, 601
extrastriate neural activity, 340
eye fixation, prediction of, 296–99
   overview, 296
   saliency predicts informativeness, 299
   scanpath and attention, 296–99

eye-gaze orienting, 361
eye movements, 422, 532, 554, 591
   and attention capture by feature singletons, 422–24
   and iconic representations
      comparison with patterns of, 556
      targeting of in visual search, 554–56
   in natural scenes, 240–42
   relationship between stimulus features and, 241
   saccadic, *see* covert attention and saccadic eye movement; saccadic eye movements
   and spatial attention, 183
   during visual search, neurocomputational model of, 397–98, 397–400
eye position, influence on attention orienting, 184
eye-slaved systems, 631
eye-tracking devices, 582

F

face recognition, 631, 667–68
face-selective regions, 353
false alarms, 314
featural attention, 286
feature-based attention, 38, 302–3, 490, 491–92, 493, 496, 505
   attentional control, 38–39
   psychophysical studies of motion processing, 491–92
feature-based targeting of spatial attention, 407–11
feature binding
   neuropsychology of, 269–71
   and oscillation and synchrony, 529–30
feature-conjunction selection, 498
Feature Contour System, 583
feature contrast, *see* saliency
feature-cueing trials, 409
feature defection, 9–10, 274–75
feature dimensions, interactions among, 573–74
feature-directed visual attention, 515–19
feature-driven selection, 550
feature expectancy in popout visual searches, 118
FeatureGate model of visual selection, 547–52
   feature-driven selection and distractor inhibition, 550–52
   hierarchical structure, 549
   inhibition of return and serial search, 549–50
   overview, 547
   selecting locations in, 547–48
feature integration, 9
Feature Integration Theory (FIT), 137, 269, 401, 413, 593, 643
feature search, 120, 208–10
feature segregation, 8–9, 208–9
feature selection, saliency from, 643

feature similarity gain model, 300–304, 425
  MT area
    effects of feature-based attention in, 302–3
    effects of spatial attention in, 300–302
    overview, 300
feature singletons, *see* attention capture
feature-specific feedback, 602
feature-specific inter-trial effects, 414
feature-specific synchronization, 530–31
feedback, 286, 288, 351, 505, 562, 565, 593, 611, 618
feedforward, 170, 279, 282–83, 286, 526, 553, 560, 562, 565, 593, 611, 656
feedforward sweep (FFS), 169–70, 172, 288
FEFm cells, 602
FEFs, *see* frontal eye fields (FEFs)
FEFv cells, 602
FFS (feedforward sweep), 169–70, 172, 288
field of view (FOV), 624
figure-ground responses, 502–6
  overview, 502
    segregation, 502–3
    and working memory, 502–5
figure/ground segregation, 510
filter responses, 588
filtration of unwanted information, 307–8
fine-grained spatial information, 507
fire (spiking) neuron, 634
Fish Film program, 324
FIT (Feature Integration Theory), 137, 269, 401, 413, 593, 643
five-choice serial reaction time task (5CSRT), 58, 59, 60
fixated region statistics, 227–28
  analysis of spatial variance, 227
  higher-order, 228
  overview, 227
  second-order, 227–28
fixation, 116, 122, 125, 203, 666
fixed-volume "pool", 360
flanker axes, rival, and task specificity, 483
flanker-modulation account, 481–82
flankers, 464
flanker-target, 481, 482
flat hazard functions, 265
flexible windows of spatial attention, 377–78
flickering checkerboard stimuli, 439
flicker paradigm, 77, 331
fMRI, *see* functional magnetic resonance imaging (fMRI)
focal lesions, 345
folded feedback, 656
foraging facilitator, inhibition of return as, 97–98
formal perceptual decision structure, 448
Fourier analysis, 507
fovea, 114, 296, 377–78, 474, 609, 625, 631, 674
fovea-like transformation, 663
foveal stimuli, 259, 436, 437, 515
FOV (field of view), 624
frames of reference, 664–65

free recall, 265
free will, 168
frequency-relevant gratings, 500
frequency selection, 498
frontal cortex, 386, 387, 503
frontal eye fields (FEFs), 30, 117, 124–25, 391, 602, 603; *see also* pop-out search
  role in saccade production, 124–25
  role in visual processing, 125
  and target selection, 125
  and visual selection, 117–23, 124–25
    based on knowledge, 120–21
    and conspicuous stimuli, 118–20
    control of overt orienting, 121–22
    overview, 117
    role in saccade production, 124–25
    role in visual processing, 125
    and target selection, 125
fronto-parietal cortex, 385, 386
fronto-parietal locus, 387
fronto-parietal regions, 383
fronto-parietal response, 342
functional brain mapping studies, 436
functional connectivity, 193, 353
functional dichotomy, 560
functional heterogeneity, 31
functional imaging, 31, 129, 347, 349
functional locus of redundancy gain in normal brain, 365–67
functional magnetic resonance imaging (fMRI), 35, 65, 187, 306, 353, 365, 366, 372, 376, 384, 407, 429, 430, 436, 454, 490, 507, 594
  experiments in neurodynamical model of visual attention, 595–97
  of object-based attention, 402–5
    behavioral results, 403
    event-related, 403–5
    overview, 402–3
functional neuroimaging, 508
fundamental neuronal mechanism, 524
fusiform cortex, 353
Fusiform Face Area (FFA), 402, 403
fusion of wide-field sensing, *see* wide-field sensing for visual telepresence and surveillance

G

GABA synaptic currents, 616
Gabor-like wavelet, 588
Gabor orientation identification, 451
Gabor patches, 443, 445, 460, 461, 464, 511
Gabor stimuli, 451
gain control mechanisms, 473, 508
gain factor, 639–40
galvanic skin response (GSR), 343
gamma band, 365
gamma frequency, 520, 522, 523, 524, 530
gamma-oscillations, 530
gating, 79, 274, 276, 504, 557
Gaussian filter responses, 397
Gaussian pyramid, 554
Gaussian white noise process, 556

gaze-contingent displays, 116
gaze direction, 194
generative model, 558, 561
generic vision, 3–4, 644
geniculate relay cells, 430
geometric parameters, 607
geons, 255, 256
gestalt principles, 352, 583
gestures, 20
gist of scene, 251–56
  holistic representation of, 255–56
  nature of, 251–54
    conceptual, 251–52
    overview, 251
    perceptual, 252–54
  overview, 251
global contextual priming, 591
global features
  in scene representation, 588–89
  similarity of and collinearity, 480–81
  variation, 234–35
global groups, 477
global illumination algorithms, 649, 650
global image representation, 589
global masks, 445
global meaning, 246
global pathway, 588
global precedence, 141
global stimulus properties, 477, 619
glue conjoining multiple features, 448
glutamatergic excitatory components, 615
goal-directed behavior, 339
goal-driven attention, 387
graphics, computer, *see* computer graphics rendering of attention
grating diameter, 430
gray-scale images, 667
grouping, contour, *see* contour grouping
GSR (galvanic skin response), 343
guidance, 103
guided search, 199, 413, 548, 549–50

H

habituation and surprise, 26–27
habituation-dishabituation technique, 542
hand-eye interaction effect, 175
hazard function, 258
head-centered coordinate systems, 579
heart rate deceleration condition, 84–85
hemiretinal stimulation, 515
hemispatial neglect, 29–33, 345, 351, 386; *see also* nonspatially lateralized mechanisms in hemispatial neglect; visual extinction and hemispatial neglect
hemispherectomy, 365
hemispheric asymmetry, 31–32, 359
hemispheric control, 358
hemispheric independence, 362
hemodynamic bioimaging, 496
hemodynamic effect, 432
hemodynamic response, 438, 457, 458
heterogeneous disorder, 345

heterogeneous visual architectures, 624
hierarchical attention, 660–62
hierarchical decomposition model, 321
hierarchical routing circuits, 553
hierarchical structure in FeatureGate model, 547–48
high-contrast stimuli, 428
high curve saliency, 585
higher-level cognitive processes, 672
higher-order fixated region statistics, 228
higher-order spectra (bispectra), 231
higher visual areas, 621
high-frequency gamma oscillations, 619
high-frequency synchronization, 520
high-gradient pixels, 629
high-level categorization/identification, 387
high-level scene concepts, 673
high-level subsystem, 663
high-load linguistic task, 437
high-resolution foveal vision, 296
high-resolution mosaic, 631
high-saliency targets, 463
high spatial frequencies, 141
histogram contrast analysis, 468, 469, 470
holistic representation of gist of scene, 255–56
homographies, 628
homologous electrode sites, 515
horizontal connections, 288
horizontal connections, laminar organization of, 654–56
horizontal-vertical anisotropy (HVA), 443
Human Regions-of-Interest (hROIs), 296–97, 298
human vision, models of
  computational, 3–4
  constraints on, 6–7
    architecture, 6
    information routing, 6
    overview, 6
hybrids, 252
hypothetical attention movement, 396

I

i2D-selective operators, 230
i2D signals, 230
IBIG (inference by information gain), 673, 674
iconic memory, 172
iconic representations, 554–56
  and eye movements
    comparison with patterns of, 556
    targeting of in visual search, 554–56
  of objects, 554
  overview, 554
identified nerve cells, 652
IIN (ipsilateral invalid negativity), 221, 222
illusory conjunctions, 136, 269
illusory contours, 352
image decomposition, 672
image elements, connected, and contour grouping, 289–90
image features, 589, 664–65
image memorizing, 669
imaging studies, 429
impairments in attention, neuropsychological, 598–99
implemented architectures of bottom-up attention and saliency models, 579–80
implicit attention capture, 69–72, 74
implicit binding, 137–39
implicit residual processing, 352–53
inattentional blindness, 72–74, 161
increasing slope pattern, 331
independence of attentional complexes, 78
independent spatial frequency-selection paradigm, 500
indexing reference and located objects, 333
index values, 438
infant covert orienting, see covert orienting
inference strategy IBIG, 673–74
inferior colliculus (IC), 656
inferior frontal cortex, 386, 387
inferior lateral frontal, 385
inferior parietal cortex, 386
inferior parietal lobules, 31, 345, 410
inferior posterior parietal cortex, 352
inferotemporal cortex (IT), 274, 294, 659
information and surprise, 24–25
information flow, see visual attention
information integration, saccadic, 645
information maximization, see recognition
information routing, 6
informativeness predictions by saliency, 299
inhibition, distractor, in FeatureGate model, 550–52
inhibition of return (IOR), 96–100, 219, 222–23, 421, 602
  and attentional selection, 420–21
  behavioral properties of, 96–97
  in FeatureGate model of visual selection, 549–50
  as foraging facilitator, 97–98
  model task for exploration of, 96
  neural implementation of, 98–100
  overview, 96
inhibition of selective attention, 339–40
inhibitory components, 615
inhibitory inputs, 47
inhibitory neurons, 376
inhibit zone, 565
input sensory dimension, 613
input signals, 634
integrate-and-fire methods, 593
integrate-and-fire neurons in prefrontal cortex, 613–17
integration, 9, 294, 593, 622–23
integrative lateral interactions mechanisms, 481
intention, 168, 240
interactions, lateral, see lateral interactions
interaction skeleton, 289
inter-areal connections, 622
intercortical attention-activated feedback, 659
intercortical feedback circuits, 658
interference effect, 185
interhemispheric redundancy gain in split brain, 364–65
interhemispheric transfer time, 365
intermanual reaction time, 366
intermediate associative pool, 613
intermediate-level processing, 666
intermediate neuron level, 610
intermediate pools, 611
intermediate region, 381
internal additive noise, 473
internal cognitive model, 297
internal noise reduction, 451
internetwork relationships and spatial attention, 32
interstimulus intervals (ISI), 486
  long effects of, 222–23
  short effects of, 220–22
    attentional load and reflexive attentional effects, 222
    disengagement and reorientation, 222
    overview, 220–22
inter-trial effects, 414, 415, 416
inter-trial memory and redundancy gains, 416–17
intra-areal connections, 622
intracortical connections, 573
intracortical feedback, 656, 657
intracortical pathways, 529, 659
intracortical synchronizing mechanisms, 533
intracranial generator, 507
intractable problems, 5
intradimensional response devices, 417
intranetwork specializations and spatial attention, 31
intra-parietal cortex, 385–86, 387
intraparietal sulcus (IPS), 35, 63, 67, 349, 385, 409
intrinsic dimensionality, 228–30
intuition, 152
invariant image representation, 664–65, 668–70
invariant recognition, 668
involuntary attention shift, 341–44
  and deviance detection, 343
  and novelty detection, 343–44
  overview, 341–43
IOR, see inhibition of return (IOR)
ipsilateral activation, 372
ipsilateral extrastriate cortex, 340
ipsilateral hemisphere, 372
ipsilateral invalid negativity (IIN), 221, 222
ipsilateral visual hemifield, 360
IPS (intraparietal sulcus), 35, 63, 67, 349, 385, 409
irrelevant color distractor singleton, 419
irrelevant featural singleton, 419
irrelevant singleton task, 419
irrelevant sound effect (ISE), 320
irrelevant tones, 340
ISI, see interstimulus intervals (ISI)
iso-eccentricity bands, 378

isoeccentric locations, 443, 446
iso-feature suppression, 572, 574
isolation and texture perception, 467–69
iso-orientation suppression, 571
isotropic ball tensor, 584
iterative method, 564
iterative recognition, 580

J

joint optimization process, 561
just noticeable difference (JND), 90–91

K

Kalman filter, 214–15, 557
kinesthetic memory, 607
knowledge-based system, 673
knowledge-based visual selection, 120–21

L

label spreading process, 289
laminar circuits, 657, 660
laminar organization, 654–56
Landolt-square, 445, 446
language, spatial, see spatial language and attention
language and attention, 324–29
    overview, 324
    semantics, 327–29
        overview, 327
        reading, 328–29
        word association, 327–28
    structure of discourse, 324–27
        assignment of syntactic roles, 324–25
        overview, 324
        reference and anaphora, 325–27
latency, 83, 84, 94
lateral connections, 288
lateral frontal cortex, 385, 386
lateral geniculate nucleus (LGN), 429, 430, 432, 435–39, 654
    anatomy of, 435–36
    baseline increases in, 438
    comparison to visual cortex, 438–39
    overview, 435
    response enhancement in, 436–37
    response suppression in, 437–38
lateral interactions, 477–84
    attentional modulation, 478–84
        collinearity, 478–81
        overview, 478
        physiological correlates of, 483–84
        task specificity, 482–83
    overview, 477
    stimulus and task-related context effects in early vision, 477
lateral intraparietal area, 391
lateral intraparietal sulcus, 579
lateralized flashes, 364
lateralized mechanisms, see nonspatially lateralized mechanisms in hemispatial neglect

lateral occipital cortex (LOC), 280, 485
lateral-occipital electrodes, 515
lateral prefrontal (LPFC), 327, 340
lateral visual hemifields, 359, 360
learning, see also perceptual learning
    and attention in conditioning, 214–16
    links to attention, 652–54, 659
    statistical, 248–49
learning modulation by cholinergic system, 54
LED (light-emitting diode) cues, 177
left cerebral hemisphere, 359
left frontal operculum, 409
left hemineglect, 596
left hemisphere lesions, 340
left occipito-temporal scalp area, 501
left parietal cortex, 353
left posterior brain, 327
left-right spatial gradient, 349–50
left-sided stimuli, 407
left visual field (LVF), 509
left visual hemifield, 366, 436, 438
leftward saccade responses, 613
lesions, 181, 339–40, 407
    of cholinergic nucleus basalis, 59
    of dorsal noradrenergic bundle, 60
lesions, brain, effects of real and virtual, 386–88
letter-discrimination task, 461
LFPs (local field potentials), 521
LGN, see lateral geniculate nucleus (LGN)
light-emitting diode (LED) cues, 177
limb-motor system, 647
limit-cycle oscillators, 640
limited-capacity (focal) attention, 412
limited-capacity memory, 268
limited-capacity resources, 416
limited processing capacity, 305
linear input-output systems, 456
linear neural network, 20
linear neural operators, 671
linear preprocessing, 672–73
line segments, 419
local contextual components, 477
local features
    contrast, 234–35, 419
    in scene representation, 587–88
    similarity of and collinearity, 478–80
local field potentials (LFPs), 521
local image patch, 554
local masks, 445
local-mismatch detection, 420
local neuronal synchronization, 521
local orientation discrimination, 482
local post-mask, 442, 446
local stimulus properties, 477
located objects, indexing, 333
location-based attentional control, 35–36, 37–38
location-based targeting of spatial attention, 407–11
location-cued visual attention, 451
location frequency, 516
location probability effects, 249

location-relevant task, 498
locations, selection of in FeatureGate model, 547–48
location selection, 498
location-specific feedback, 602
locomotion control, see visual attention
locus coeruleus (LC), 51
logic/planning system, 20, 22
long-term memory, 77, 168, 201, 653, 673
long-wavelength level, 466
low-contrast stimuli, 428
low-dimensional global description, 586
lower visual areas, 621
low-frequency components, 524
low frequency modulations, 523
low frequency synchronization, 524
low-level subsystem, 663
low-resolution imagery, 629
low-resolution peripheral vision, 296
low-saliency targets, 461, 463
low spatial frequencies, 141
luminance contrast, 273, 274, 494
luminance-modulated gratings, 497
luminance motion, 491

M

macaque pulvinar, 440
Macaque visual system, see attentional suppression
machine learning, 19–21
machine vision architecture, 642–48
    attentive computer vision systems, 642–45
        contextual cueing and attention strategies, 644–45
        object and scene recognition, 643–44
        overview, 642
        saliency from feature selection, 643
    overview, 642
    robotic systems, 645–48
        autonomous multisensor, 646–47
        overview, 645–46
        visuomotor attention, 647–48
machine vision systems, 624
magnetic event-related fields, 510
magnocellular gangliar cells, 497
magnocellular layers, 430
magnocellular pathway, 141
manipulation, attentional, 461–62
manual neural networks, 393
mapping layer, 606
mapping network, 610
map-seeking circuits, 605, 606–7, 608
Markovian decision processes, 645
Markov models, 26
matching, links to attention, 656–57
matching operation, 606
maximal independence, 239
MAX operation, 280, 283
MAX pooling function, 280–81
McCollough effect, 138
mean attentional effect, 373
mechanisms governing property, 656

mechanisms of perceptual learning, *see* perceptual learning
mechanoreceptive nerve fibers, 535
medial frontal area of brain, 327–28
medial geniculate nucleus (MGN), 656
medial occipital cortex, 375
median-case analysis, 5
medio-dorsal thalamic nucleus, 432
MEG data localizing, 518
membrane capacitance, 614
membrane leak, 614
membrane potentials, 622
memorizing mode, 666, 667, 669
memory, 168; *see also* visual search
   of awareness and visual attention, 167–69
   iconic memory, 172
   inter-trial, and redundancy gains, 416–17
   long-term memory, 77, 168, 201, 653, 673
   modulation by cholinergic system, 54
   short-term memory, 384, 673
   and wide-field sensing, 629–31
   working memory, 168, 172, 201, 360
memory-driven attention, 605, 607
memory-driven search, 265
memory-driven visual attention, 605–9
   attention shifts, 607
   biological evidence, 607–8
   and bottom-up attentional mechanisms, 608–9
   cross-modal attention, 607
   map-seeking circuits, 606–7
   overview, 605
memoryless process, 265
memory-related neurons, 560
memory signal, 504
memory trace, 343
meridian effect, 182
mesial occipital activation, 498
mesial occipital areas, 496
mesial occipital electrodes, 515
meta-analysis of top-down attentional control neural systems, 65–66
metabolic brain activity, 432
metabolic measurements, 431
metabolic suppression, ring of, 430
metric relations, 331
MGN (medial geniculate nucleus), 656
midazolam, 53
middle occipital gyrus, 509
middle temporal area (MT), 300, 375
   effects of feature-based attention in, 302–3
   effects of spatial attention in, 300–302
mid-eccentricity, 381
midline occipital scalp sites, 509
minimal independence, 239
mismatch negativity (MMN), 318, 343
mobile conjugate reinforcement paradigm, 209
modality-shifting delay, 542
modality-specific sensory processing, 190
model-based object recognition, 645
model pyramidal cell, 571

modulation, spatially-specific attentional, *see* spatially-specific attentional modulation
modulation of attention, 654
modulation of memory and learning by cholinergic system, 54
modulation of neuronal impact, *see* selective visual attention
modulation of saliency, *see* contextual influences on saliency
modulation of sensory cortex, 187–89
monosynaptic excitation, 571
Monte Carlo methods, 27
mosaicing systems, 625
motion after-effect, 491, 494
motion coherence, 428
motion parallax, 112, 626
motion pathway, 495, 565–67
motion processing, 454, 490–95, 540
   neural and psychophysical effects of attention, 494–95
      overview, 494
      studies of monkeys, 494–95
   neurophysiological studies, 492–93
      feature-based attention, 493
      overview, 492
      selective spatial attention, 492–93
   overview, 490
   psychophysical studies, 490–92
      feature-based attention, 491–92
      and spatial attention, 490–91
motion-sensitive area, 454
motion stimulus, 490
motion vs. change, 76
motoric stage, 366
motor neurons, 536
motor response, 363
motor selection process, 124
motor systems, 640
movement neurons, 125
MT area, *see* middle temporal area (MT)
multibottleneck view, 387
multichip selective attention models, 633–34
multicomponent deficit, 383
multidimensional objects, 501
multielectrode recordings, 527, 529, 530
multilayer circuit, 606
multimodal auditory-visual cues, 448
multimodal stimuli, 365
multipass process, 563
multiple internal representations, 154
multiple linear regression, 456, 459
multiple response units, 417
multiple stimuli, 363
multiple-target visual search, 266–67
multiple visual-cortical areas, 509
multiplicative gain model, 379
multiplicative internal noise, 449
multiplicative mechanisms, 425
multiplicative model, 379
multiplicative modulation, 301
multiplicative noise, 451, 473
multisensor-based active perception, 642

multisensor data, 646
multisensor robots, autonomous, 646–47
multisensory integration, 189
multisensory prior entry research, 91
multisensory spatial correspondence, 193
multistage competitive scheme, 643
muscarinic receptor antagonists, 59

N

natural scene statistics, 226–32; *see also* stimulus-driven guidance in natural scenes
   detectors of salient visual features, 228–30
      concept of intrinsic dimensionality, 228–30
      i2D-selective operators, 230
      overview, 228
   of fixated regions, 227–28
      analysis of spatial variance, 227
      higher-order, 228
      overview, 227
      second-order, 227–28
   overview, 226–27
navigation in nested environments, 148–49
NBM (nucleus basalis magnocellularis), 59
negative polarity, 470
negative stimuli, 163
neglect, hemispatial, *see also* visual extinction and hemispatial neglect
neglect symptoms, 352
neocortex, 531
nervous system, 527
nested environments, 147
   navigation in, 148–49
   spatial updating in, 148–49
network configuration, 564
network for spatial attention, 30
neural and psychophysical effects of attention, 494–95
   overview, 494
   studies of monkeys, 494–95
neural arrays, 469
neural correlates, 167, 506
   of visual attention, 311–13
   of visual perception, 313–16
neural development, theory of, 211–12
neural implementation of inhibition of return, 98–100
neural locus of redundancy gain in normal brain, 365–67
neural mechanisms, 514, 534
neural mechanisms of somatosensory selection, *see* tactile stimuli
neural modeling, 556, 652
neural-network preprocessing stage, 672
neural population, 380
neural representation, 389, 503
neural responses, attentional modulation of, 492–93
neural salience, 392
neural selectivity, 485–89
neural sources of ERP components, 507–8

neural substrates and mechanisms of
   temporal orientation, 259–60
neural systems of top-down attentional
   control, 63–68
   method of analysis of, 65–66
      cuing studies, 65
      meta-analysis, 65–66
   overview, 63–65
   results of studies of, 66
neural tuning, 488
neuroanatomical overlap, 407
neuroanatomical theory, 664
neurobiological hypothesis, 593
neurobiological setting, 556
neurobiology of salience of feature
   contrast, 235
neurochemical studies, 58–61
   cholinergic modulation, 58–60
      effects of systemic agents, 59–60
      lesions of nucleus basalis, 59
      overview, 58–59
   dopamine, 60–61
   noradrenaline, 60
      adrenergic drugs, 60
      lesions of dorsal noradrenergic bundle,
         60
      overview, 58
neurocomputational model of eye
   movements during visual searches,
   397–400
neurodynamical model, 595, 596
neurodynamical model of visual attention,
   593–99
   experiments, 595–99
      FMRI, 595–97
      overview, 595
      simulation of neuropsychological
         impairments in attention, 598–99
      simulation of psychophysical, 597–98
      simulation of single-cell, 595
   overview, 593
   unification of, 594–95
neurodynamical theoretical framework, 610
neurofunctional activation, 498
neuroimagery, see also visual attention
   and modulation of sensory cortex,
      187–89
   of spatial topography, 372–75
      overview, 372
      retinotopic coordinates, 373–75
      for single target, 372–73
   studies, 386, 410, 496–97
   techniques, 511
   of visual extinction and hemispatial
      neglect, 353–57
neurometabolic studies, 496–97
neuromorphic selective attention systems,
   633–37
   chips, 634–35
   multichip models, 633–34
   overview, 633
   two-chip active vision system, 635–37
Neuromorphic Vision Toolkit, 646
neuromorphic VLSI devices, 636

neuronal circuitry, 14, 607
neuronal firing rate, 425
neuronal implementations, 606
neuronal interactions, 458
neuronal map-seeking circuit models, 608
neuronal population, 594, 613
neuronal receptive fields, 448
neuronal recording studies, 348
neuronal responses, 425, 427–28, 529
neuronal substrate, 524
neuronal synchronization, see selective
   visual attention
neuronal synchrony, 529, 530
neuron encoding, 485
neuron population, 619
neurons, integrate-and-fire, in prefrontal
   cortex, 613–17
neuron's receptive field, 42
neuropharmacology of attention, 57–62
   neurochemical studies, 58–61
      cholinergic modulation, 58–60
      dopamine, 60–61
      noradrenaline, 60
   overview, 57
   selective spatial attention and attentional
      shift, 58
   sustained attention, 57–58
neurophysiology, 371–76
   of contour grouping, 290–92
   mechanisms, 371
   motion processing studies, 492–93
      feature-based attention, 493
      overview, 492
      selective spatial attention, 492–93
   neuroimaging spatial topography, 372–75
      overview, 372
      retinotopic coordinates, 373–75
      for single target, 372–73
   overview, 371
   of reflexive orienting of spatial attention,
      389–94
   signals, 390
   studies, 488, 490, 492, 534, 664
   visual spatial selectiveness, 371–72,
      375–76
neuropsychology
   auditory perceptual organization, 321–23
   impairments in attention, 598–99
   of visual feature binding, 269–71
neuroscience, cognitive, and study of prior
   entry, 93–95
neutral cues, 258, 414, 446
"neutral" trials, 361
nicotine, as cognitive enhancer, 59
nicotinic modulation, 54–55
NMDA receptors, 616
noise, see external noise; internal noise
   reduction
noise tests, 471
nonamblyopic strabismic subjects, 532
nonattentional pathways, 200
nonclassical receptive field, 557
nonclocked digital pulses, 634
nonlateralized components, 346

nonlateralized functions, 349
nonlateralized selective attention, 347
nonlinear approach, 231
nonlinear i2D selective operators, 672–73
nonlinear interaction, 458
nonlinear neural operators, 671
nonlinear preprocessing, 672–73
nonlinear (sigmoid) function, 425
nonlinear system identification, 454
nonlinear transducer function, 449
non-oculomotor attention, 185–86
nonoverlapping visual neural areas, 498
nonperiodic synchrony, 536
nonsalient elements, 464
nonselected responses, 527
nonselective pool, 611
nonsensory signals in early visual cortex,
   311–16
   neural correlates
      of visual attention, 311–13
      of visual perception, 313–16
   overview, 311
nonspatial attentional orienting, 67–68
nonspatial cueing, 283–87
nonspatial features, ERP studies of, 510–13
nonspatially lateralized mechanisms in
   hemispatial neglect, 345–50
   detection of salience, 348–49
   overview, 345
   selective attention capacity, 347–48
   spatially lateralized deficits, 345–46,
      349–50
   sustained attention, 348
   trans-saccadic spatial working memory,
      349
nonspatial selective attention, 496, 500
nontarget distractors, 359
noradrenaline, 51, 60
noradrenergic a2 receptor, 51
noradrenergic attentional effects, 50
noradrenergic system, 51–54, 350
   and arousal level, 51–54
   effects of on attention, 51
   overview, 51
norepinephrine, 216
normalization factor, 589
normative statistical models, 214
novel dual-axis stimulus, 478
novel stimuli, 343
novelty detection, 343–44
"novelty seeking", 343
nucleus basalis, lesions of cholinergic, 59
nucleus basalis magnocellularis (NBM),
   59

O

object-based attention, 15, 39–40, 130–34,
   376, 401–6, 482, 498, 560, 660–62
   fMRI of, 402–5
      behavioral results, 403
      event-related, 403–5
      overview, 402–3
   overview, 130, 401

object-based attention (*continued*)
  relationship to space-based attention, 133–34
  versus task specificity, 482–83
  via pre-attentive-attentive interface, 657–59
object-based attentional control, 39–41
object-based mechanisms, 496
object-based mode, 130
object-based reference paradigm, 664
object boundary, 658
object-centered processing, 15
object-centered representations of three-dimensional attention, 112
object-class priming, 589
object contextual cueing, 248
object-cued features, 644
object detection, 600
object file theory of trans-saccadic memory, 200
object intrinsic information, 586
object pathway, 567
object recognition in cortex, 279–87
  experimental results, 279–80
  overview, 279
  Standard Model, 280–83
    limitations of feedforward approach, 282–83
    overview, 280–82
  top-down attentional and task-dependent modulations, 283–87
    nonspatial cueing, 283–87
    overview, 283
    spatial cueing, 283
object-response associative task, 613
objects
  depth and attention control, 111
  detection of in natural scenes, 600–604
    model of, 600–602
    overview, 600
  iconic representation of, 554
  indexing of reference and located, 333
  recognition of, 643–44, 667–68
  representations in prefrontal cortex, 142–43
object (texture) boundaries, 570
object-tuned cells, 280
observed mechanisms of perceptual learning, 473–74
observer intention and knowledge in scene perception, 80
observer models and perceptual learning, 472–73
occipital cortex, 54–55, 377, 407, 409
occipital lobe, 381, 594
occipital pole, 374
occipital visual cortex, 371, 375
occipito-parietal cortex, 372
occipito-temporal cortex, 387
occipito-temporal pathway, 501
occipito-temporal ventral stream, 510
occlusion, 112
oculomotor attention, 185
oculomotor capture, 418, 422

oculomotor control, 647
oculomotor domain, 267
oculomotor measure, 396
oculomotor network, 391
oculomotor neural networks, 393
oculomotor saccade response, 611
oculomotor signaling, 183–85
oculomotor system, 177, 183
"oddball" paradigm, 348, 349
offset contrast, 236
one-shot paradigm, 77
onset contrast, 236
optic flow cues, 641
optimal feature values, 571, 572
orbital PFC, 142
orientation, *see also* covert orienting; overt orienting control
  sense of, 150
  spatial, 358–59
orientationally selective receptive fields (ORFs), 665
orientation contrast, 236
orientation discrimination task, 473
orientation search task, 573
orientation-selective neurons, 579
orientation-selective patches, 620
orientation tuning functions, 426
orienting, reflexive, of spatial attention, 389–94
orthogonal cuing methodology, 92
orthogonal flankers, 478
orthogonal stimuli, 483
oscillations and synchrony, 526–33
  and definition of relations, 528–29
  and feature binding, 529–30
  overview, 526
  rate codes versus temporal codes, 528
  response selection and attention, 530–33
oscillatory cycle, 640
oscillatory modulation, 521–22, 532
oscillatory neuronal synchronization, *see* selective visual attention
oscillatory responses, 533
oscillatory synchronization, 520, 522
oscillatory systems, 365
outlier mask, 557
output integrate, 634
overall-saliency units, 415, 416
overt attention, 114–15
overt orienting control, 121–22
overt search behavior, 396
overt shifts, 185
overt visual attention, 553

P

P1 validity effect, 85, 87
P3a component, 343
panoramic sensing, 624
paradoxical interhemispheric redundancy gain in split brain, 364–65
parahippocampal complex (PHC), 274
parahippocampal place area (PPA), 385, 402, 403

parallel algorithm, 563
parallel-coactive accounts, 415
parallel distributed processing system, 594
parallel dynamical system, 593
parallel feature searches, 547
parallel processing, 101
parallel-race accounts, 415
parallel search, 120, 210–11, 595
parallel target-flanker configuration, 478
parametric template-matching technique, 628
parietal activations, 458
parietal cortex, 54–55, 372, 386, 387, 497, 503, 518, 570, 610
parietal lobes, 136–37, 391
parietal lobules, 31
parietal mechanisms of attentional control, 35–41
  cuing studies of, 35–37
  dynamic signals for, 37–41
    feature-based, 38–39
    location-based, 37–38
    object-based, 39–41
  overview, 35
parietal module, 596
parietal syndrome, 30
parieto-frontal networks, 385, 429
parieto-premotor networks, 366
parts-based population code, 488
parvocellular cells, 436, 497
passive attention, 540
passive viewing, 379
pathological biases, 355
Pdm area, 439
perception, 311; *see also* cross-modal attention; scenes, perception of; texture perception; top-down facilitation of perception
  behavioral paradigm in visual, and visual attention, 663–64
  neural correlates of visual, 313–16
perceptual analysis, 471
perceptual coding, 472
perceptual disambiguation, 541
perceptual extinction, 351
perceptual grouping, 483, 529, 654
perceptual learning, 471–76
  mechanisms of, 472–75
    attention in, 474–75
    observed, 473–74
    observer models, 472–73
    overview, 472
    signatures of, 473
  overview, 471
  physiological correlates of, 476
  retuning versus reweighting, 475–76
  specificity of, 472
  task compatibility, 472
  in visual tasks, 471
perceptual motion processing, 490
perceptual nature of gist of scene, 252–54
perceptual neocortex, 652, 654
perceptual organization, auditory, *see* auditory perceptual organization

perceptual priming (robust effect), 118
perceptual processing, 449, 659
perceptual saliency, 577, 584
perceptual sensitivity, 475, 477
perceptual template model (PTM) and external noise, 449–51
   attention paradigm and mechanisms, 450–51
      internal noise reduction, 451
      overview, 450
      stimulus enhancement, 450–51
   of observer, 449–50
   overview, 449
periodic synchrony, 536
peripheral checkerboard stimuli, 431, 437
peripheral cue, 442, 446
peripheral spatial cueing, 408–9, 410
peripheral stimuli, 82–84, 408
peripheral vision, 497
peripheral visual area, 618
peripheral visual system, 625, 631
PET (positron emission tomography), 353, 507
PFC network, 611
PFC (prefrontal cortex), 141, 142, 143, 274, 280, 361, 560, 593, 610, 611
PFm cells, 602
phase-locked interhemispheric oscillations, 365
phase-synchronization, 522
phasic activation, target-related, 341
PHC (parahippocampal complex), 274
phenomenal awareness, 172–74
phenomenological experiences, 393
phenomenology of salience of feature contrast, 233–34
phonological system, 328
photopic vision, 497
photosensing elements, 633
physical design of wide-field sensing, 625–26
physiological studies, 447
physio-physiological interaction, 455, 456
pipe stimuli, 111
PIT (posterior part of the inferotemporal cortex), 274
place code, 31
point of subjective equality (PSE), 445
point of subjective simultaneity (PSS), 90, 91, 445
Poisson spikes, 612–13
polar angle dimension, 375
polarity information, 470
pop-out search, 118–20, 125, 127, 208, 209, 210, 234–35, 503, 504–5, 571
   dimension-based attention in, 412–17
      feature contrast signals, 415–16
      inter-trial memory and redundancy gains, 416–17
      overview, 412
      target definition, 412–14
      top-down modulable processing, 414–15

   effect of singleton distracter during, 118–20
   feature expectancy during, 118
   priming during, 118
   pro- and antisaccade, 120
population-based inference, 600
population coding, 175, 276, 527, 602
population-level response, 488
population response, 487
population selectivity, 485, 486, 488
positron emission tomography (PET), 353, 507
Posner paradigm, 182, 449
possessing regions, 557
posterior callosal sections, 365
posterior cortex lesions, 341
posterior fusiform gyrus, 509
posterior medio-lateral suprasylvian sulcus (PMLS), 529
posterior parietal cortex (PPC), 63, 348, 454–55, 459, 560, 579
posterior part of the inferotemporal cortex (PIT), 274
posterior thalamus, 455
post-sales market analysis, 582
post-stimulus latencies, 507
post-stimulus processing, 514
post-stimulus time points, 517
postsynaptic cell response, 528
postsynaptic neuronal elements, 522
postsynaptic responses, 527
postsynaptic targets, 520, 524
power spectra, 522
PPA (parahippocampal place area), 385, 402, 403
PPC modulation, 458
PPC (posterior parietal cortex), 63, 348, 454–55, 459, 560, 579
PP (posterior parietal complex), 594, 595
pre-attentive-attentive interface, 657–59
pre-attentive binding, 138
pre-attentive coordinate frames, 628
preattentive dimension, 101
preattentive features, 101–4, 576–79
pre-attentive field, 629
preattentive guidance, 103
preattentive mechanisms, 207–8, 211–12, 421
preattentive parallel encoding, 420
preattentive parallel level, 418
preattentive parallel search, 419
preattentive parallel stage, 418
preattentive segmentation, 422
preattentive sensors, 625–26
preattentive texture segregation, 467, 469
precomputed voting fields, 584
precortical gating, 515
pre-cues, 9, 10
prediction
   and attention in conditioning, 214–16
   of human eye fixations, 296–99
      overview, 296
      by saliency of informativeness, 299
      scanpath and attention, 296–99

prediction structure of visual attention and locomotion control model, 640
predictive coding, 557, 558, 560, 561
predictive power, 567
prefrontal cortex, see also working memory
   and origin of top-down perception facilitation, 142–43
prefrontal cortex (PFC), 141, 142, 143, 274, 280, 361, 560, 593, 610, 611
prefrontal damage and ERP measures, 339–44
   involuntary attention shift, 341–44
      and deviance detection, 343
      and novelty detection, 343–44
      overview, 341–43
   overview, 339
   selective attention, 339–41
      facilitation, 340–41
      inhibition, 339–40
      overview, 339
prefrontal memory, 602
premotor pools, 611, 613
premotor regions, 409
premotor theory of attention, 117, 181–86
   brain imaging studies and oculomotor areas, 185
   non-oculomotor, 185–86
   overview, 181–82
preprocessing, 378, 672–73, 675
prestriate activations, 458
primary cortex, 516
primary motor cortex, 366
primary sensory cortex, 535
primary visual cortex, 353, 375, 377, 378, 381, 461, 481, 502, 554, 664
   bottom-up saliency map, 570–75
      interactions among feature dimensions, 573–74
      overview, 570
      in V1 area, 571–72
   working memory and figure-ground responses in, 503–4
prime attention, 658
priming during popout visual searches, 118
primitive motion map, 629
prior entry, 89–95
   cognitive neuroscience approaches to study of, 93–95
   early research on, 89–91
   overview, 89
   response bias confound, 91–93
   spatial confound in multisensory research, 91
priors, scene, see contextual influences on saliency
probabilistic method, 560
probabilistic models of attention, 553–61
   overview, 553
   predictive coding, 556–61
      generative models and, 556–57
      object-based versus spatial attention, 558–61
      overview, 556

probabilistic models of attention (*continued*)
  visual attention without spotlight, 557–58
  using iconic representations, 554–56
    of objects, 554
    overview, 554
probability function, 589
probability theory, 264
problem-oriented perception, 668
problem-oriented recognition, 668
problem-oriented selection, 663
processing, *see also* motion processing; spatial processing
  of awareness and visual attention, 167–69
  cortical visual, 223–25
  implicit residual, 352–53
  limited capacity for, 305
  overview of model of visual attention and locomotion control, 639–40
  recurrent, 169–70
  top-down modulable dimensional, 414–15
  visual, 125
processing elements, 633
processing pyramid, 562
projective mappings, 628
prosaccade popout visual searches, 120
prototype artificial visual systems, 624
prototype sensors, 625
proximity effect, 9
PSS (point of subjective simultaneity), 90, 91, 445
psychological factor, 455
psychology of contour grouping, 292–95
psychometric function, contrast, 443–45
psychometric functions, 443, 447, 451
psychopharmacology, 51
psychopharmacology of human attention, 50–56; *see also* neural and psychophysical effects of attention
  cholinergic system, 54–56
    direct effects of, 54–56
    modulation of memory and learning, 54
  noradrenergic system, 51–54
    and arousal level, 51–54
    effects of on attention, 51
    overview, 50–51
psychophysical experiments, 460, 597–98
psychophysical methods, 467
psychophysical paradigm, 493, 538
psychophysical studies, 371, 442, 477, 490, 534
psychophysical studies of motion processing, 490–92
  feature-based attention, 491–92
  overview, 490
  and spatial attention
    divided, 491
    selective, 490–91
psychophysical target detection, 482
psychophysics, 152
psycho-physiological interactions, 455, 459

PTM, *see* perceptual template model (PTM) and external noise
pulse-frequency modulation, 634
pulvinar, 53, 439
pulvinar, attentional modulation in, 439–41
pure tones, 317
pursuit, smooth, and wide-field sensing, 629
push-pull mechanism, 505
putative functional architecture, 197–203
  overview, 197–201
  phases of, 201–3
putative neural systems, 362
pyramidal neurons, 614, 616

Q

quantitative evaluation, 581
QUEST, 443

R

race model, 363, 364–65
race model inequality (RMI), 415, 416
random dot patterns (RDP), 300, 426
randomized visual search, 266
random search model, 265, 266
rapid serial visual presentation (RSVP), 252, 272, 377, 378, 380–81, 383
"rapid switching" hypothesis, 381
rapid visual categorization paradigms, 272
rate coded sensory information, 526
rate codes, 528
raw sensory data, 638
raw video frames, 626
RDP (random dot patterns), 300, 426
reaches and saccade trajectories, 175–80
reaction times and overt orienting control, 121
reaction times (RTs), 58, 101, 257, 363, 413–14, 415
reading and semantics, 328–29
real-time sensory data, 633
real-time video interpretation, 644
real-visual system, 622
receptive field properties, 528
receptive fields (RFs), 300, 382, 425, 492, 502, 520, 562, 563, 594, 600, 621
receptive location, 500
reciprocal inhibitory interconnections, 666
recognition, 580
  attention-guided, 663–70
    behavioral paradigm in visual perception and visual attention, 663–64
    invariant image representation and, 668–70
    model of, 665–68
    overview, 663–65
  by information maximization, model of, 671–76
    inference strategy IBIG, 673–74
    linear and nonlinear preprocessing, 672–73

overview, 671–72
sensorimotor features and high-level scene concepts, 673
system behavior, 674–75
  of objects in cortex, *see* object recognition in cortex
recognition layers, 567
recognition mode, 667, 668, 669
recognition process, 556
recognition program, 663
recognition system, 557
recurrent connections, 288, 290–92
recurrent processing (RP), 169–70
redundancy gain, 363–64, 416–17
  functional and neural locus of in normal brain, 365–67
  paradoxical interhemispheric, in split brain, 364–65
Redundant Target Effect (RTE), 363–64
redundant targets, 366
reference and structure of discourse, 325–27
reference frame, spatial, selection of, 334
reference frames, 332
reference objects, indexing, 333
reflective attention, *see* electrophysiology of reflexive attention
reflex arch, 168
reflexive attention, 219, 220, 222
reflexive orienting, 361, 389–94
region inliers, 584
region of interest (ROI), 11, 15, 455
regression analyses, 455–56
reinforcement learning, 134
rejected distractors, *see* visual search
relaxation-based method, 564
relevance, 28, 73
reorientation of location targets, 222
representations, *see* environmental representations; iconic representations
residual processing, implicit, 352–53
residual unconscious processing, 353
response bias confound, 91–93
response enhancement in lateral geniculate nucleus, 436–37
response generation and overt orienting control, 121–22
response histograms, 521
response inhibition test, 337
response modulation, contrast-dependent, 45–47
response pattern, 487
response selection and attention, 530–33
response suppression, 44, 437–38
reticular axons, 433
reticular neurons, 430
reticular thalamic nucleus (RTN), 429, 432, 433, 434
retina, 391
retinal images, 665, 667, 669
retinocortical projection, 435
retino–geniculo–striate pathway, 594
retinotopic accuracy, 375
retinotopic coordinates, 373–75

retinotopic cortex, 405
retinotopic cortical activation, 377
retinotopic locations, 371, 373, 374, 375, 430
retinotopic mapping, 306, 351, 436, 438, 439
retinotopic metric, 375
retinotopic organization, 372
retuning, 475–76
reweighting, 475–76
RFs (receptive fields), 300, 382, 425, 492, 502, 520, 562, 563, 594, 600, 621
R-G coding, 397
right angular gyrus, 409, 410
right cerebral hemisphere, 359
right frontal lobe, 348
right-handed subjects, 408
right-hemisphere, 345, 348, 353
right inferior parietal lobule, 407, 409, 410
right inferior parietal region, 409
right middle frontal gyrus, 409
right occipital cortex, 372
right parietal cortex, 596
right-sided stimuli, 407
right visual field (RVF), 509
right visual hemifield, 436, 438
rightward saccade responses, 613
ring of metabolic suppression, 430
rival flanker axes and task specificity, 483
RMI (race model inequality), 415, 416
robotic systems, 645–48
  autonomous multisensor, 646–47
  overview, 645–46
  visuomotor attention, 647–48
robots in visual attention and locomotion control model, 641
robot visuomotor attention, 647
rodent barrel system, 656
ROImaps, 373
rotation-dependent effect, 487
rotation mapping layer, 607
routing, 6; see also dynamic routing
RP (recurrent processing), 169–70
RSVP (rapid serial visual presentation), 252, 272, 377, 378, 380–81, 383
RTE (Redundant Target Effect), 363–64
RTN (reticular thalamic nucleus), 429, 432, 433, 434
RTs (reaction times), 58, 101, 257, 363, 413–14, 415
RT x set size function, 266
rule-based mapping, 616
RVF (right visual field), 509

S

saccade-contingent displays, 625
saccade deviations, 177
saccade initiation routines, 397
saccade landing position, 397
saccades, 114, 115–16, 122, 127, 391
saccadic eye movements, 671; see also covert attention and saccadic eye movement
  control of and wide-field sensing, 629
  frontal eye field's role in production of, 124–25
  information integration, 645
  reaches and trajectory of, 175–80
  and target selection, 125–28
  visual selection without, 127–28
saccadic fixations, 672
saccadic interpretation, 644
saccadic movement, 674
saccadic reaction times, 389–90, 391
saccadic scene analysis, 675
saccadic sensor movements, 635
saccadic targeting, 554
salience maps, 115
saliency, 168; see also contextual influences on saliency; contextual modulation of salience of feature contrast; primary visual cortex; visual saliency
  in computer vision, 583–85
    overview, 583
    tensor voting framework, 584–85
  detection of, 348–49
  of feature contrast, 233–39
    neurobiology of, 235
    overview, 233
    phenomenology of, 233–34
    popout from, 234–35
    role of in vision, 239
  from feature selection, 643
  and prediction of informativeness, 299
  in selective tuning model for visual attention, 564–65
  of stimulus, 242–45
  of visual features, 228–30
saliency-based models, 643
saliency decay function, 584
saliency imbalance, 156
saliency maps, 395, 397–98, 399, 413, 415, 420–21, 553, 554, 560, 565, 576–77, 634, 649
saliency signals, 412, 413, 415, 416
salient elements, 464
salient object, 418
salient perceptual structures, 583
salient singletons, 418
salient stimuli, 51, 348, 361
salient targets, 461
same-color stimuli, 407
sampling in visual search, 264–65
  Amnesic Model, 265
  overview, 264
  Standard Model, 264–67
saporin lesions, 59
SART, see Sustained Attention to Response Test (SART)
scalp current density (SCD), 497, 498, 500, 507–8
scalp-recorded event-related potentials, 514
scalp-recorded signatures, 339
scalp-recorded voltage fluctuations, 507
scanpath and prediction of eye fixation, 296–99
scanpaths, 200, 268, 296, 580, 667
scenes, 197–203; see also contextual influences on saliency; gist of scene
  basic components of, 197–201
  high-level concepts of, 673
  overview, 197
  perception of, 79–80
    observer intention and knowledge, 80
    overview, 79
    triadic architecture, 79
    vision outside focus of attention, 80–81
  putative functional architecture, 197–203
    overview, 197–201
    phases of, 201–3
  recognition of in attentive computer vision systems, 643–44
scene statistics, see natural scene statistics
segmentation of through synchronization, 618–23
  models, 619
  overview, 618–19
  simulation results, 619–23
scopolamine, 59–60
scotopic vision, 497
scrambles, 467
SDT models, 446
search, serial, in FeatureGate model, 549–50
search, visual, see visual search
search asymmetries, 102
search mode, 667
secondary auditory cortex, 343
secondary somatosensory (SII) cortex, 534, 535, 536
second-order differential equation, 641
second-order fixated region statistics, 227–28
second-order interactions, 458
second-order symmetric tensors, 584
segmentation, scene, see scenes
segregation, 8–9, 208–10, 502–3
selection, see also frontal eye fields (FEFs); target selection; visual selection
  of conscious experiences, 167
  somatosensory, see tactile stimuli
  and spatial attention, 42–44
selection initiation routines, 397
selection negativity (SN), 496, 501, 510
selection positivity (SP), 510
selective attention, 156, 158, 213, 305, 339–41, 407, 496–501, 633, 635; see also neuromorphic selective attention systems
  attention–awareness model, 18–23
  capacity, 347–48
  dorsal and ventral streams, 496–501
    ERP studies of, 497–501
    overview, 496–97
  facilitation of, 340–41
    enhancement of selected channel, 340–41
    overview, 340
    target-related phasic activation, 341
    tonic excitatory modulation, 340
  inhibition, 339–40

selective attention (*continued*)
  nonspatial, 496
  overview, 18, 339, 496
  and top-down attention control, 106
selective bias, 107
selective coupling, 422
selective spatial attention, *see* spatial attention
selective tuning model for visual attention, 562–69
  computational and biological implications, 565–69
    biological and behavioral predictions, 567–69
    overview, 565
    simulation, 565–67
  overview, 562
  top-down selection, 563–64
  winner-takes-all and saliency, 564–65
Selective Tuning Model (STM), 562, 601, 643
selective visual attention, and oscillatory neuronal synchronization, 520–25
  modulation of neuronal impact, 520, 523–25
  overview, 520
self-generated attention, 621
self-generated dynamic states, 526
semantics, 327–29
  overview, 327
  reading, 328–29
  semantic categories, 147
  semantic contrast, 273
  semantic priming, 352, 384
  semantic processing, 326
  word association, 327–28
seminal primate electrophysiology studies, 378–79
sense organs, 618
sensitivity, *see* transient attention
sensitivity, distinction from decision criterion, 153
sensorimotor action, 640–41
sensorimotor behavior, 642
sensorimotor coordination, 647
sensorimotor features, 673, 674, 675
sensorimotor map, 640, 641
sensorimotor neurons, 392
sensorimotor (or parietofrontal) dichotomies, 30
sensorimotor (S-M) map, 640
sensory attentional mechanisms, 500
sensory coding, 435, 475
sensory context, 477
sensory cortex, 187–89
sensory effects of cross-modal spatial attention, 189–91
sensory enhancement mechanisms, 313
sensory-evoked recordable cortical activity, 515
sensory field, 563
sensory filters, 448, 475, 498
sensory gain control, 510
sensory impression, 445

sensory information, 503, 505
sensory input, 612
sensory interference, 157
sensory memory, 666, 669
sensory neuron, 488
sensory pools, 611
sensory portal, 638
sensory preferences, 425
sensory processing, 167, 435, 498
sensory response, 488
sensory stimuli, 391, 610
sensory-to-motor axis, 391–92
sensory transients and bounce-inducing effect, 540–41
sequential bayesian learning, 26–27
serial conjunction searches, 547
serial focal attention, 595
serial mechanisms, 593
serial search, 549–50, 595, 604
SFC (spike-field coherence), 522, 532
SFS (superior frontal sulcus), 65, 67
Shannon's entropy, 28
shift-related activity, 40
shifts of attention, 58, 607
short-latency effects, 511
short-term memory, 384, 413, 448, 532, 576, 594, 610–17, 673
  overview, 610
  working memory, 610–17
    model of prefrontal cortex, 611–17
    overview, 610–11
SI cortex, 535
sigmoidal transfer function, 433
sigmoid function, 425, 426
signal detection theory, 153–54
signal detection theory (SDT), 8, 9, 152–53, 313, 443, 447
signal enhancement, 442, 443
signal integration, 416
signal processing, 449
signal stimulus, 450, 451
signal superposition, 606
signal-to-noise ratio, 344, 380, 448
signature characteristics of visual attention, 312
signatures of perceptual learning mechanism, 473
SII (secondary somatosensory) cortex, 534, 535, 536
silicon synapse, 634
simulation, *see also* scenes
  of neuropsychological impairments in attention, 598–99
  of psychophysical experiments, 597–98
  in selective tuning model for visual attention, 565–67
  of single-cell experiments, 595
  in visual attention and locomotion control model, 641
sinewave patches, 474
single-attractor network, 616
single-axis stimuli, 481
single-cell experiments, 595
single-cell recordings, 180, 447

single-cell responses, 527
single-cell studies, 481
single-chip systems, 633
single-neuron mechanisms, 485
single-pool attractors, 610
single-target search stimulus, 574
singleton distracters in popout visual searches, 118–20
singletons, *see* attention capture
single-unit recording studies, 510
sit and wait strategies, 266
smooth pursuit and wide-field sensing, 629
SN (selection negativity), 496, 501, 510
somatosensory cortex, 532, 536, 537
somatosensory selection, *see* tactile stimuli
somatosensory system, 534, 537, 659
SOS (Spatial Object Search) model, 660
source modeling of attention-related ERP components, 508–10
space-based attention, 130–34
space-based attentional processing, 133
space-based mode, 130
space-directed visual attention, 515–19
space-variant (foveated) sensor chips, 624
spatial attention, 29–34, 181, 182, 300–302, 402, 407, 430, 490, 492, 493, 560, 604, 660–62
  brain–behavior correlations, 30–31
  capture of by feature singletons, 420–22
  contrast-dependent response modulation, 45–47
  cortical bases of development, 85–88
  cross-modal, 187–96, 538
    behavioral evidence for effects, 187–89
    ERP evidence for sensory effects of, 189–91
    neuroimagery and modulation of sensory cortex by, 191–96
    overview, 187
  and effective contrast, 47–49
  ERP measures of, 508
  and eye movements, 183, 185
  hemispatial neglect, 29–33
  hemispheric asymmetry, 31–32
  internetwork relationships, 32
  intranetwork specializations, 31
  large-scale network for, 32
  location or feature-based targeting of, 407–11
  network for, 30
  neurophysiological studies of motion processing, 492–93
  overview, 29
  psychophysical studies of motion processing, 490–91
  reflexive orienting of, 389–94
  selective, and attentional shift, 58
  taxonomy of mechanisms of, 451–53
  using 3D scenes to study, 110
  and visual cortical circuits, 42–49
    facilitation and selection, 42–44
    overview, 42
spatial-based attention, 493

spatial confound in multisensory prior entry research, 91
spatial contrast, 577
spatial cueing, 82, 247–48, 283, 361, 407, 409, 410, 547
spatial discrimination, capacity limits for, 8–10
   feature defection, 9–10
   feature integration, 9
   feature segregation, 8–9
   overview, 8
spatial filters, 397, 554, 556
spatial focused attention, 420
spatial frequencies, 141, 472, 475, 497, 511, 516–17
spatial frequency filter F(f), 449
spatial frequency gratings, 516
spatial indexing, 332, 333
spatial interference paradigm, 385
spatial language and attention, 330–36
   computational model of apprehension, 331–33
   construction of template, 334–36
   indexing reference and located objects, 333
   overview, 330
   selection of reference frame, 334
spatial localization, 366
spatial location, 498, 518, 559, 643
spatially contiguous subset, 564
spatially-focused attentional selection, 508
spatially lateralized deficits, 345–46, 349–50
spatially lateralized functions, 349
spatially lateralized mechanisms, 346–47
spatially lateralized models, 346
spatially-specific attentional modulation, 377–82
   flexible windows of attention, 380–82
   mechanisms of selection, 378–80
   overview, 377
   in VI and other early visual cortical areas, 377–78
spatially uniform sampling, 13
spatial maps, 467, 549
spatial neglect, 407
Spatial Object Search (SOS) model, 660
spatial orientation, 260–61, 358–59
spatial pointers, 644
spatial precueing paradigm, 490
spatial probes, 551
spatial processing, of environmental representations, 146–51
   hierarchical model of, 146–47
   and nested environments, 148–50
spatial reentry signal, 601
spatial representations, 549
spatial resolution, 514, 629–31
spatial routing of information, 276
spatial scales, 554, 556
spatial selection, 371–72, 375–76, 407, 498, 604
spatial selective attention, 500
spatial templates, 332
spatial uncertainty model, 447

spatial updating process, 148
spatial variance, 227
spatial visual attention, 109
spatio-temporal data, 396
spatio-temporal dynamics, 396, 399
species-specific behavior, 324
specificity, see also task specificity
   of perceptual learning, 472
SPECT techniques, 353
spike-field coherence (SFC), 522, 532
spike frequency adaptation, 524
spike histograms, 523
spike sequences, 526
spike timing and visual saliency, 276–77
split-brain patients, 358–62
   overview, 358
   paradoxical interhemispheric redundancy gain in, 364–65
   resources and dual-task performance, 360–61
   spatial orientation, 358–59
   top-down versus bottom-up control, 361–62
   visual search, 359–60
spread-neutral cue, 446
SP (selection positivity), 510
square-root law, 9
stability-plasticity dilemma, 652, 659
Standard Model of visual processing in cortex, 280–83
   limitations of feedforward approach, 282–83
   overview, 280–82
Standard Model of visual search sampling, 264–67
   empirical tests of, 265–67
      cumulative distribution functions, 267
      multiple-target search, 266–67
      overview, 265
      randomized search, 266
standard signal detection theory (SDT), 154
Standard (Test patch), 445
statistical learning, 248–49
statistical parametric map, 456
statistic gauging, 466
statistics, scene, see natural scene statistics
stereoscopic flow cues, 641
stimuli, tactile, see tactile stimuli
stimuli processing, 160–61
stimulus attention, central, and infant covert orienting, 84–85
stimulus contrast, attentional modulation of, 425–28
   common substrate for, 427–28
   effects of on neuronal responses, 425
   overview, 425
stimulus-dependent properties, 425
stimulus-determined context-integration, 477
stimulus dimension, 472
stimulus drive levels, 379
stimulus-driven activations, 374
stimulus-driven grouping processes, 477

stimulus-driven guidance in natural scenes, 240–45
   eye movements, 240–42
   overview, 240
   salience of, 242–45
stimulus-driven relay cell, 433
stimulus-driven saliency, 601
stimulus enhancement, 448, 450–51, 473
stimulus objects, 619
stimulus onset asynchronies (SOAs), 82, 90, 318
stimulus-press-stimulus-press mode, 337
stimulus-related context effects in early vision, 477
stimulus segments, 373
stimulus selection processes, 511
stimulus spatial frequency, 500
STM (Selective Tuning Model), 562, 601, 643
stop signals, 366
strabismic animals, 532
strategies, attention, see machine vision architecture
streaming, auditory, 319–21
   indirect effects of on competing tasks, 319
   outside focus of attention, 318
striate cortex, 429, 430, 435, 438, 509, 510, 511, 515, 518
striate occipital cortex, 518
striatum, 533
stroke patients, 156
Stroop effect, 270
structure of discourse, 324–27
   assignment of syntactic roles, 324–25
   overview, 324
   reference and anaphora, 325–27
subcortical areas, 496, 518
subcortical connections, 361
subcortical inputs, 362
subcortical regions, 345
subcortical sites, 440
subcortical structures, 361, 365, 429
substance-sensing arrays, 467, 469
substrate for attentional modulation of stimulus contrast, 427–28
subthreshold activation, 352
subthreshold contrast, 43
subthreshold membrane potential, 615
superior colliculus (SC), 85, 365, 389, 391, 436, 529, 560
superior frontal sulcus (SFS), 65, 67
superior parietal lobe (SPL), 347
superior parietal lobules, 31, 410
superior temporal sulcus, 409
supermodal saliency, 539
super-ordinate saliency representation, 412
suppression, attentional, see attentional suppression
suppression, ring of metabolic, 430
suppression of response in lateral geniculate nucleus, 437–38
suprathreshold distracters, 443
suprathreshold effect, 539

suprathreshold stimuli, 443
suprathreshold target, 443
surfaces, depth and attention controls, 111
surprise and attentional mechanisms, 24–28
 computation of, 25–26
  continuous data, 26
  discrete data and Dirichlet model, 26
 and habituation, 26–27
 overview, 24–25
surround contrast, 464
surround inhibition, attentional modulation of, 460–65
 attentional manipulation, 461–62
 early-vision task, 461
 overview, 460–61
surveillance, see wide-field sensing for visual telepresence and surveillance
surveillance systems, 626
sustained attention, 57–58, 348
Sustained Attention to Response Test (SART), 337–38
 clinical validity of, 337
 ecological validity of, 337
 neural basis of performance of, 338
 overview, 337
 psychometric properties, 338
 validity of, 337–38
symbolic artificial intelligence, 645
symmetric divergence, 25
synaptic connections, 383
synaptic coupling, 611
synaptic current, 615, 616
synaptic gain, 524
synaptic strength, 557
synaptic weight modification, 641
synchronization, see also oscillations and synchrony; scenes; selective visual attention
 links to attention, 652–54
 periodic and nonperiodic, 536
synchronizing mechanisms, 527
synchronous fast (gamma) oscillations, 619
synchronous firing, 536, 654
synchronous oscillatory activity, 528, 530, 532
synesthesia, 270
syntactic processing, 326
syntactic roles, 324–25
system architecture, 669
systematic search model, 265
systematic visual search, 268
system behavior, 674–75
systemic cholinergic agents, 59–60
systems-level modeling, 556, 594

T

T2 area, 385–86
tabula rasa (blank slate), 639
tactile-induced effect, 541
tactile stimuli, 534–37
 overview, 534
 population activity, 536–37
 somatosensory selection
  need for, 534–35
  neural mechanisms of, 535–36
tactile transients and bounce-inducing effect, 539–40
tactile trigger, 640–41
tagging mechanism, 421
Talairach space, 440
tangential velocity, 639
target contrast, 464, 478
target-defining dimensions, 414, 416, 417
target definition in pop-out search, 412–14
target detection thresholds, 461
target-distracter discriminability, 385
target-distracter interference, 387
target-driven control of attention, 589
target-flanker configuration, 482
targeting eye movements in visual search, 554–56
target motion stimulus, 491
target processing, 586
target-related bias, 399
target-related phasic activation, 341
target-related response, 390–91
target selection, 125–28
task compatibility, 472
task-dependent modulation, 642
task-directed attention, 483
task-irrelevant stimuli, 460
task priorities, maintenance of, 106–7
task-related context effects in early vision, 477
task-related goals, 548
task-related relative segregation, 496
task-relevance map (TRM), 200
task-relevant inhibitory bias mechanism, 568
task-relevant variables, 553
task-specific component, 564
task-specific determination, 565
task-specific executive controller, 562, 568
task specificity
 dependence of on rival flanker axes, 483
 versus object-based attention, 482–83
Taylor series expansion, 558
TCH (temporal correlation hypothesis), 619
TD cognitive-spatial model, 297–98
telephoto optics, 625
telepresence, visual, see wide-field sensing for visual telepresence and surveillance
telepresence mode, 629
telepresence systems, 626
temperature parameter, 556
template, spatial, 334–36
template-enhancing mechanisms, 313
template match operation, 280–81
temporal asymmetry of cross-modal interaction window, 541–42
temporal coding, 526, 528, 533, 535
temporal contextual cueing, 248
temporal correlation hypothesis (TCH), 619
temporal cortex, 497, 610
temporal cortical neurons, 607
temporal domain, 619
temporal dynamics, 601
temporal expectancies, 258–59, 261
temporal orientation, 257–63
 behavioral studies of, 257–59
 comparison of spatial orientation and, 260–61
 implications of for cognitive research, 261–63
 neural substrates and mechanisms of, 259–60
 overview, 257
temporal pathway, 428
temporal phase mapped design, 375
temporal resolution, 629–31
temporo-parietal junction, 387
temporo-parietal lesions, 343
tensor voting framework, 584–85
texels, 467–68
texture-based judgments, 469
texture boundaries, 571
texture energy, 467–68
texture identification tasks, 474
texture-like line arrays, 234
texture perception, 466–70
 isolation and analysis of blackshot, 467–69
 overview, 466
texture segmentation, 102, 207
thalamic lateral geniculate nuclei, 518
thalamic levels, 435, 518
thalamic nuclei, 439, 533
thalamic plasticity, 659
thalamic reticular nucleus (TRN), 436, 439
thalamo-cortical circuits, 435
thalamo-cortical pathways, 529
thalamo-cortical signal transmission, 439
thalamocortical transmission, 439
thalamus, 391, 457
theoretical performance signatures, 450
three-dimensional space and attention allocation, 109–13
 depth, 110–12
  and aging, 111–12
  cues for, 112
  objects and surfaces, 111
 overview, 109
 research on, 109–10
 viewer-centered versus object-centered representations, 112
three-dimensional model memory layer, 607
three-dimentional scenes, 110
threshold level, 315, 380
threshold signal contrasts, 473
tilted targets, 464
time-varying neural image, 466
timing of attentional modulation of visual processing, 514–19
 ERPs and timing of space and feature-directed visual attention, 515–19
 overview, 514–15
TMS (transcranial magnetic stimulation), 136, 387
tonic bias effects, 510

tonic deficit, 350
tonic excitatory modulation, 340
top-down attention, 105–8, 283–87; see also neural systems of top-down attentional control
   control of selective attention, 106
   and maintenance of task priorities, 106–7
   nonspatial cueing, 283–87
   overview, 105, 283
   spatial cueing, 283
top-down attentional mechanisms, 387, 401, 415, 418, 469, 483, 493, 496, 502, 504, 506, 510, 532, 547, 548, 551, 672
top-down attentive processing, 656
top-down bias in visual cortex, 307–10
   filtration of unwanted information, 307–8
   increased baseline activity, 308–10
   overview, 307
top-down connections, laminar organization of, 654–56
top-down control in split-brain patients, 361–62
top-down control signals, 311
top-down cortical feedback, 659
top-down-driven attention, 605
top-down facilitation of perception, 140–45
   cortical origin of, 142–43
   overview, 140–41
top-down feature biasing, 199
top-down knowledge-based component, 673
top-down manner, 101
top-down modulable dimensional processing, 414–15
top-down modulating signals, 634
top-down priming, 653
top-down selection, 240, 563–64
topical classes, 227
topographical distributions, 511
topographical saliency map, 643
topographic associative memory, 622
topographic distribution, 498
topographic heteroassociation, 619
topographic scalp distribution, 514
topography, spatial, see neuroimagery
total energy, 469
Tower of Hanoi problem, 20
tracking and wide-field sensing, 629
tractability, 4
trajectory of saccades, see saccadic eye movements
transcranial magnetic stimulation (TMS), 136, 387
transient attention, 442–47
   and contrast psychometric function, 443–45
   and contrast sensitivity function, 443
   improvement of acuity, 445–47
   and increase in apparent contrast, 445
   overview, 442–43
transients, sensory, and bounce-inducing effect, 540–41
trans-saccadic memory, 200, 349
traveling fixation, 639

triadic architecture, 79, 197
trimodular architecture, 601
TRM (task-relevance map), 200
TRN (thalamic reticular nucleus), 436, 439
tuning, selective, model for visual attention, see selective tuning model for visual attention
Turing machine, 4
TVC functions, 450
two-chip active vision system, 635–37
two-chip selective attention system, 635, 636
two-dimensional topographic map, 420, 576
two-dimensional memory patterns, 607
two-dimensional shape recognition, 274
two-point correlation function, 241

U

unattended condition, 163
unbounded visual search, 4, 6
uncertainty theory, 158
unconditioned stimulus, 213
unconscious processing, 352
"uncued" trials, 361
unilateral contralateral stimuli, 354
unilateral lesions, 348–49
unilateral neglect, 321
unimodal stimuli, 261, 365
unimodal studies, 189
unique criterion, 155, 157
unitary internal representation, 152, 155–57

V

V1 area, 377–78
   and bottom-up saliency map in primary visual cortex, 571–72
   basic features and conjunction searches, 572–73
   feature-selective cells in, 570–71
   contextual modulation, 235–36
V5/hMT function, 454–55
valence-dependent responses, 164
valid cues, 258
valid/invalid precueing paradigm, 494
variance, spatial, 227
vector code, 31
ventral extrastriate areas, 511
ventral extrastriate cortex, 511
ventral network, 30
ventral occipital cortex, 375, 410
ventral occipitotemporal activity, 410
ventral occipitotemporal areas, 409
ventral occipitotemporal pathways, 355
ventral pallidum, 432
ventral pathways, 351, 438, 520, 535
ventral putamen, 432
ventral streams, 44, 496–501, 558, 596, 601
   ERP studies of, 497–501
   overview, 497
   visual evoked potentials, 497
   overview, 496–97

ventral visual information, 117
ventral visual pathways, 274–76, 353, 355, 428, 560–61
ventrobasal complex, 535
ventrolateral PFC, 142
VEPs (visual-evoked potentials), 259, 497
vertical meridian, 446
vertical meridian asymmetry (VMA), 443
very large-scale integration (VLSI), 633
viewer-centered representations of three-dimensional attention, 112
viewer-to-object relationships, 148
view-tuned cells, 280
virtual contour, 477
virtual representation, 79
vision, see also attentive computer vision systems; saliency
   complexity of computational, 4–7
      constraints on models, 6–7
      intractable problems, 5
      objections to analysis of, 5
      overview, 4
   early
      stimulus and task-related context effects in, 477
      task and surround inhibition, 461
   models of human, 3–4
   outside focus of attention, 80–81
   role of salience of feature contrast in, 239
   system, two-chip active, 635–37
visual acuity task, 445
visual attention, see also memory-driven visual attention; neurodynamical model of visual attention; selective tuning model for visual attention; selective visual attention; short-term memory
   and awareness, 167–74
      overview, 167
      phenomenal, 172–74
      processing, 167–70
      selection of conscious experiences, 167
   behavioral paradigm in visual perception and, 663–64
   electrophysiological and neuroimaging studies of, 507–13
   and emotional perception, 160–66
      and awareness, 165–66
      biasing signals, 163–65
      overview, 160
      stimuli processing, 160–61
      test of automatic amygdala activation, 162–63
   and information flow, 11–17
      autonomous control, 14–15
      dynamic routing, 12–14
      neurobiological substrates and mechanisms, 15–16
      overview, 11–12
   and locomotion control, 638–41
      model of, 639–41
      overview, 638–39
   nature of, 77–79

visual attention (*continued*)
   and attentional complexes, 78–79
   binding problems, 79
   capacity limits and bottlenecks, 77–78
   overview, 77
   scene perception, 79–80
  neural correlates of, 311–13
  rendering algorithm for computer graphics, 649–51
  and spatial selectiveness, 371–72, 375–76
  timing of space and feature-directed, 515–19
visual attention mechanism, 594
visual awareness, 576
visual callosal fibers, 365
visual cognition, 377
visual coincidence, 541
visual conical processing, 377
visual context, 477, 586
visual cortex, 38, 353, 372, 427, 435, 455, 472, 532, 594, 654, 660; *see also* biasing competition in human visual cortex; nonsensory signals in early visual cortex; primary visual cortex
  comparison to lateral geniculate nucleus, 438–39
  primary, working memory and figure-ground responses in, 503–4
visual cortical areas, 520
visual cortical circuits and spatial attention, 42–49
  contrast-dependent response modulation, 45–47
  and effective contrast, 47–49
  facilitation and selection, 42–44
  overview, 42
visual cortical neurons, 379
visual cortical responses, 355
visual crossing, 538
visual dimensions, 413
visual-evoked potentials (VEPs), 259, 497
visual extinction and hemispatial neglect, 351–57
  behavioral and anatomical aspects, 351–52
  functional neuroimaging, 353–57
  implicit residual processing, 352–53
  overview, 351
visual feature binding, neuropsychology of, 269–71
visual feature detectors, 618
visual feature search, 407
visual field, 470, 570
visual generators, 517
visual importance map, 649
visual judgments, 466
visual modalities, 577
visual motion perception, 539
visual neurons, 577
visual pathways, 274–76, 514, 515, 518
visual perception, neural correlates of, 313–16
visual perceptual learning, 475–76

visual presentation, 570
visual processing, 125, 425
  cortical, and cross-modal attentional capture, 223–25
  timing of attentional modulation of, 515–19
visual psychophysics, 377
visual quadrant, 516
visual reaction time paradigm, 343
visual-related neurons, 122
visual saliency, 272–78
  behavioral definition of, 273–74
  overview, 272–73
  spike timing and, 276–77
  in ventral visual pathway, 274–76
visual search, 115, 359–60, 395–400; *see also* Standard Model of visual search sampling
  attention addiction, 395
  attentional components in, 395–97
  guidance, 101–4
    overview, 101
    and preattentive features, 101–4
  memory and rejected distractors, 264–68
    costs of systematic search, 268
    limited-capacity, 268
    in oculomotor domain, 267
    overview, 264
    sampling, 264–65
  neurocomputational model of eye movements during, 397–400
  overview, 395
  and popout in infancy, 207–12
    versus adult mechanisms, 207–8
    conjunction search, 211
    feature search and segregation, 208–10
    overview, 207
    parallel search in toddlers and, 210–11
    preattentive and attentive mechanisms, 211–12
  psychophysics, 576
  targeting of eye movements in, 554–56
  targets, 570
visual search task, 419
visual selection, 124–29; *see also* FeatureGate model of visual selection; frontal eye fields (FEFs)
  overview, 124
  with saccade, 125–28
  without saccade, 127–28
visual selective attention, 593
visual selective processing, 514
visual sensitivity map, 649
visual spatial filters, 446
visual substance, 466
visual system, 401, 504, 605
visual system, Macaque, *see* attentional suppression
visual tasks and perceptual learning, 471
visual telepresence, *see* wide-field sensing for visual telepresence and surveillance
visual transformations, 605

visual transients and bounce-inducing effect, 539–40
Visual Translator (VITRA), 200–201
visual word form system, 328
visual working memory (VWM), 15
visuomotor attention, 647–48
visuomotor neurons, 389
visuomotor systems, properties of, 175
visuo-spatial attention, 361, 372, 383, 385, 439, 440
visuo-spatial hemineglect, 439
visuotopic organization, 439
VITRA (Visual Translator), 15, 200–201
VLSI (very large-scale integration), 633
VMA (vertical meridian asymmetry), 443
volitional orienting conditions, 361
voltage deflections, 507
voltage-dependent connections, 458
Volterra formulation, 457, 458
Volterra kernels, 456
Volterra series of analyses, 456–57, 459
voluntary attention, 219
voluntary covert orienting, 63
voluntary orienting paradigm, 64
voting framework, tensor, 584–85
voxels, 375, 436, 455

W

Weber contrast, 468
"what" pathway, 604, 660
"where" pathway, 604, 660
white snow paradox, 25
wide-field sensing for visual telepresence and surveillance, 624–32
  fusion, 626–29
    calibration, 628
    dynamic, 628–29
    overview, 626–28
  memory, 629–31
  overview, 624–25
  physical design, 625–26
  saccadic control, 629
  tracking and smooth pursuit, 629
wide-field sensors, 625
willed intention, 107
windows of spatial attention, 377–78
winner-takes-all in selective tuning model for visual attention, 564–65
winner-take-all (WTA), 562, 563, 568, 633, 635, 643
within-dimension search, 413, 414
word association, 327–28
working memory, 168, 172, 201, 360, 610–17
  and figure-ground responses, 502–5
  model of prefrontal cortex, 611–13
  overview, 610–11
worst-case analysis, 5

Z

zombie system, 18, 21